P9-AFZ-280

Biotechnology of Biopolymers

From Synthesis to Patents

Edited by A. Steinbüchel and Y. Doi

Biotechnology of Biopolymers. From Synthesis to Patents. Edited by A. Steinbüchel and Y. Doi

ISBN: 3-527-31110-6

Related Titles

A. Steinbüchel

Biopolymers. 10 Volumes + Index

2003

ISBN 3-527-30290-5

Single "Biopolymers" Volumes:

A. Steinbüchel, M. Hofrichter

Vol. 1: Lignin, Humic Substances and Coal

2001

ISBN 3-527-30220-4

T. Koyama, A. Steinbüchel

Vol. 2: Polyisoprenoids

2001

ISBN 3-527-30221-2

Y. Doi, A. Steinbüchel

Vol. 3a: Polyesters I – Biological Systems and Biotechnological Production

2002

ISBN 3-527-30224-7

Y. Doi, A. Steinbüchel

Vol. 3b: Polyesters II – Properties and Chemical Synthesis

2002

ISBN 3-527-30219-0

A. Steinbüchel, Y. Doi

Vol. 4: Polyesters III – Applications and Commercial Products

2002

ISBN 3-527-30225-5

A. Steinbüchel, E. J. Vandamme, S. De Baets

Vol. 5: Polysaccharides I – Polysaccharides from Prokaryotes

2002

ISBN 3-527-30226-3

A. Steinbüchel, S. De Baets, E. J. Vandamme

Vol. 6: Polysaccharides II – Polysaccharides from Eukaryotes

2002

ISBN 3-527-30227-1

A. Steinbüchel, S. R. Fahnestock

Vol. 7: Polyamides and Complex Proteinaceous Materials I

2003

ISBN 3-527-30222-0

S. R. Fahnestock, A. Steinbüchel

Vol. 8: Polyamides and Complex Proteinaceous Materials II

2003

ISBN 3-527-30223-9

A. Steinbüchel, S. Matsumura

Vol. 9: Miscellaneous Biopolymers and Biodegradation of Synthetic Polymers

2003

ISBN 3-527-30228-X

A. Steinbüchel

Vol. 10: General Aspects and Special Applications

2003

ISBN 3-527-30229-8

A. Steinbüchel

Cumulative Index

2003

ISBN 3-527-30230-1

H. F. Mark

Encyclopedia of Polymer Science and Technology

Part 1: Volumes 1–4, **2003**, ISBN 0-471-28824-1
Part 2: Volumes 5–8, **2003**, ISBN 0-471-28781-4
Part 3: Volumes 9–12, **2004**, ISBN 0-471-28780-6

Biotechnology of Biopolymers

From Synthesis to Patents

Edited by A. Steinbüchel and Y. Doi

Volume 1
Lignin, Coal, Polyisoprenoids, Polyesters and Polysaccharides

WILEY-VCH Verlag GmbH & Co. KGaA

Biotechnology of Biopolymers. From Synthesis to Patents. Edited by A. Steinbüchel and Y. Doi

ISBN: 3-527-31110-6

Editors:

Prof. Dr. Alexander Steinbüchel
Institut für Mikrobiologie
Westfälische Wilhelms-Universität
Corrensstrasse 3
48149 Münster
Germany

Prof. Dr. Yoshiharu Doi
Polymer Chemistry Laboratory
RIKEN Insitute
2-1 Hirosawa, Wako-shi
Saitama 351-0198
Japan

Cover Illustration:
The dextran molecule is a fine example for a biotechnologically produced biopolymer. Scaling up biotechnical production from the detail to the plant goes hand in hand with economic success.
(Photograph with kind permission of Bio-engineering AG, Wald, Switzerland)

Library of Congress Card No: applied for

British Library Cataloguing-in-Publication Data: A catalogue record for this book is available from the British Library

Bibliographic information published by Die Deutsche Bibliothek
Die Deutsche Bibliothek lists this publication in the Deutsche Nationalbibliografie; detailed bibliographic data is available in the Internet at <http://dnb.ddb.de>.

Printed in the Federal Republic of Germany
Printed on acid-free paper.

Composition, Printing, and Bookbinding:
Konrad Triltsch
Print und digitale Medien GmbH,
Ochsenfurt-Hohestadt

ISBN 3-527-31110-6

Preface

Biopolymers and their derivatives are diverse, abundant, and important for life, exhibit fascinating properties and are of increasing importance for various applications. Living matter is able to synthesize an overwhelming variety of polymers, which can be divided into eight major classes according to their chemical structure: (i) nucleic acids, such as ribonucleic acids and deoxyribonucleic acids, (ii) polyamides, such as proteins and poly(amino acids), (iii) polysaccharides, such as cellulose, starch and xanthan, (iv) organic polyoxoesters, such as poly(hydroxyalkanoic acids), poly(malic acid) and cutin, (v) polythioesters, which were reported only recently, (vi) inorganic polyesters with polyphosphate as the only example, (vii) polyisoprenoides, such as natural rubber or Gutta Percha and (viii) polyphenols, such as lignin or humic acids. Biopolymers may occur in any organism, and contribute in most of them by far to the major fraction of the cellular dry matter. Biopolymers fulfill a wide range of quite different essential or beneficial functions for these organisms, including conservation and expression of genetic information, catalysis of reactions, storage of carbon, nitrogen, phosphorus and other nutrients and of energy, defense and protection against attacks by other cells or hazardous environmental or intrinsic factors, sensors of biotic and abiotic factors, communication with the environment and other organisms, mediators of the adhesion to surfaces of other organisms or of non-living matter, and many more. In addition, many of them are structural components of cells, tissues and entire organisms.

To fulfill all these different functions, biopolymers need to exhibit rather diverse properties. They must very specifically interact with a large variety of different substances, components and materials, and often they need to have extraordinary high affinities to them. Finally, they must have a high degree of strength. Some of these properties are utilized directly or indirectly for various applications. This and the possibility to produce them from renewable resources, as living matter mostly does, make biopolymers interesting candidates for industry.

Basic and applied research has already contributed much knowledge about the enzyme systems catalyzing the biosynthesis, degradation and modification of biopolymers, as well as about the properties of biopolymers. This has also resulted in an increased interest in biopolymers for various applications in industry, medicine, pharmacy, agriculture, electronics and in many other areas. However, taking into account the developments over the last two decades and a review of the literature shows that our knowledge is still scarce. The genes for the biosynthesis pathways of many biopolymers are still not available or were

Biotechnology of Biopolymers. From Synthesis to Patents. Edited by A. Steinbüchel and Y. Doi

ISBN: 3-527-31110-6

identified only recently, many new biopolymers have been described only recently, and the biological, chemical, physical and material properties have only been investigated for a minor fraction of the biopolymers. Often promising biopolymers are not available in sufficient amounts for this purpose. Nevertheless, polymer chemists, engineers and material scientists in academia and industry have discovered biopolymers as chemicals and materials for many new applications, or are considering biopolymers as model compounds to design novel synthetic polymers.

The complexity and relevance of biopolymers initiated the publication of the ten-volume handbook "Biopolymers", which comprehensively reviews and compiles information regarding (i) occurrence, synthesis, isolation and production, (ii) properties and applications, (iii) biodegradation and modification not only of natural but also of synthetic polymers, and (iv) the relevant analysis methods to reveal the structures and properties. This book series was published between 2001 and 2003. We are very grateful to the publisher Wiley-VCH for recognizing the demand for such a book series and to undertake this large, new project. Publication of the "Biopolymers" was very successful, and the entire book series has even recently been translated into Chinese. Special thanks are due to Karin Dembowsky and Dr. Andreas Sendtko and many others at Wiley-VCH for their initiatives, constant efforts, helpful suggestions, constructive criticism and wonderful ideas.

From this book series we selected 39 chapters, which deal with biotechnologically relevant aspects of biopolymers, and which are now published in this seperate, new book. By doing this, interested readers from academia and industry will gain access to a very comprehensive book covering the most important applied aspects of biopolymers and their production. We are very grateful to the authors of the selected chapters for allowing the contents of their "Biopolymers" contributions to be included in this new title.

Münster and Saitama
September 2004

Alexander Steinbüchel
Yoshiharu Doi

Contents

Volume 1

Biotechnology of Biopolymers. From Synthesis to Patents. Edited by A. Steinbüchel and Y. Doi

ISBN: 3-527-31110-6

Volume 2

I
Lignin and Coal

Biotechnology of Biopolymers. From Synthesis to Patents. Edited by A. Steinbüchel and Y. Doi

ISBN: 3-527-31110-6

1
Synthesis of Lignin in Transgenic and Mutant Plants

Dr. Jeffrey F. D. Dean
Daniel B. Warnell School of Forest Resources, University of Georgia, Athens, GA 30602-2152, USA; Tel: +01-7065421710; Fax: +01-7065428356; E-mail: jeffdean@uga.edu

Biotechnology of Biopolymers. From Synthesis to Patents. Edited by A. Steinbüchel and Y. Doi

ISBN: 3-527-31110-6

4-CL	4-coumarate, CoA ligase
AldOMT	5-hydroxyconiferyl aldehyde *O*-methyltransferase
bm	*b*rown *m*idrib
C3H	4-coumarate 3-hydroxylase
C4H	cinnamate 4-hydroxylase
CAD	cinnamyl alcohol dehydrogenase
CAGT	5′-diphosphoglucose:coniferyl alcohol glucosyltransferase
CAld5H	coniferyl aldehyde 5-hydroxylase
CBG	coniferin-β-glucosidase
CCoA3H	4-coumaryl:CoA hydroxylase
CCoAOMT	caffeoyl-CoA 3-*O*-methyltransferase
CCR	cinnamoyl-CoA reductase
CoA	coenzyme A
COMT	caffeate *O*-methyltransferase
DAHP	3-deoxy-D-*arabino*-heptulosonate 7-phosphate
F5H	ferulate 5-hydroxylase
fah	*f*erulic *a*cid *h*ydroxylase
IAA	indole acetic acid
LAC	laccase
MYB	transcription factor related to avian myeloblastosis transforming gene
NMR	nuclear magnetic resonance
PAL	phenylalanine ammonia lyase
rol	*Agrobacterium tumefaciens* R_i-plasmid localized *ro*ot *l*ocus oncogenes
SAH	S-adenosylhomocysteine
SAM	S-adenosylmethionine

1 Introduction

These are exciting and contentious times for investigators working to understand the structure and biosynthesis of lignin. Technological advances are enabling us to discern with high resolution under nearly in situ conditions previously unappreciated structures and linkages within native lignin polymers (Dean, 1997; Argyropoulos, 1999). Meanwhile, mutants blocked at specific points in the lignin biosynthetic pathway are being identified in natural populations, or are being created in the laboratory through genetic engineering (MacKay et al., 1997; Provan et al., 1997; Baucher et al., 1998a). At the confluence of these advances there again rages a debate as to the very nature of what does and does not constitute lignin, and recent findings suggest that the possibilities for manipulating lignin characteristics in vivo are far greater than could have been imagined even a few years ago (Lewis, 1999; Ralph et al., 1999a). Lignin researchers appear cursed to live in interesting times, indeed.

In the most general of terms, lignin is a chemically recalcitrant polymer of phenylpropanoid units linked together in a complex and irregular pattern which varies from species to species, tissue to tissue, and cell to cell. Vascular plants use lignin to line their conductive tissues as a barrier to water loss; thus, lignin was instrumental in the spread of these plants throughout the terrestrial landscape. Plants have subsequently harnessed lignin to bind cells together, rigidify their lamellate cell walls into microcomposite structures of remarkable strength, and provide a physical barrier against invading microorganisms. Such chemical characteristics as hydrophobicity and stable, irregular cross-links make lignin an ideal material to limit water loss and stifle pathogen invasion. At the same time, these characteristics have only begrudgingly yielded to chemical analyses and structural dissection, and rapid advancement in the area of structural analysis has really only begun to accelerate in the past decade.

Lignin is most abundant in wood, where it comprises 20–30% of the total dry weight, and constitutes the principal barrier to the production of pulp and paper. Historically, efforts to manufacture paper more economically have provided the principal impetus for studies of lignin structure and biosynthesis. Consequently, the accumulated information about lignin produced in arborescent plants tends to skew our perception of what lignin is and does. Thus, Freudenberg (1968) defined lignin as the heteropolymer resulting from the dehydrogenation of a mixture of three *p*-hydroxycinnamyl alcohols – *p*-coumaryl, coniferyl, and sinapyl alcohols – best exemplified by spruce milled-wood lignin. This definition was drawn specifically to delineate a perceived difference between the polymer isolated from trees and other higher-order vascular plants and the aromatic polymers that can be isolated from various bryophytes and pteridophytes.

However, studies of lignin mutants in both herbaceous and woody plants have more recently provided us with a greater appreciation of the true diversity of function to which plants have adapted this polymer. Consequently, Freudenberg's definition of what constitutes lignin seems increasingly restrictive. No doubt further compositional and structural surprises await those who will extend the latest techniques for micro-scale and subcellular lignin analysis (multi-dimensional NMR, pyrolysis-mass spectrometry, UV microspectrometry, etc.) to lower plants in order to address lingering questions pertinent to the evolution of lignin. Specifically, what is the true chemical nature of the polyphenolic materials identified in mosses and algae (Siegel, 1969; Miksche and Yasuda, 1978; Delwiche et al., 1989), and is it possible that these compounds and their biosynthetic pathways share a common ancestor with angiosperm and gymnosperm lignin? These are questions that need to be addressed at the level of individual cells (e.g., hydroid cells in hornworts), and the answers have great potential to extend further our ideas as to how lignin structure and composition might be modified through genetic manipulation of existing metabolic pathways in plants. Given the observations of Ralph and co-workers (1999a) that plants can incorporate a much wider variety of phenolic precursors into lignin than would be anticipated from Freudenberg's definition, it is tempting to speculate that lignin composition and structure could be changed even more drastically by drawing on our expanding knowledge of developmental processes that parallel lignification in other organisms, e.g., sclerotization of insect cuticle (Andersen et al., 1996) and melanization of fungal rhizomorphs (Butler and Day, 1998). The introduction of new metabolic pathways from these organisms into woody plants could lead to materials having unique and valuable properties.

1.1 Potential Applications

1.1.1 Pulp and Paper

Based on value of shipments, the pulp and paper industry ranks eighth among all U.S. manufacturing industries, and pulp, paper, and paperboard mills account for about 12% of total manufacturing energy use in the U.S. (Nilsson et al., 1995). Although the industry produces more than half of all the energy it consumes, energy still constitutes about 17% of total costs to the industry (this number does not account for the value of co-generated energy from burning wastes and residues). Since the production of pulp from wood is almost entirely a matter of disrupting the lignin matrix between and within fibers, it would seem that modifications to lignin would have the potential for providing this industry with substantial energy savings. However, with respect to lignin modification, it is important to note that there are two very different pulping processes – mechanical and chemical (e.g. Kraft) – and the energy demands for these two processes are radically different.

In the case of mechanical pulping, physical grinding is used to disrupt the interfiber lignin matrix and separate the wood fibers, but little or no lignin is removed from the pulp. Mechanical pulp is used primarily for the production of newsprint and other low-grade papers, and its manufacture consumes enormous amounts of energy. For example, a medium-sized mechanical pulp mill having the capacity to produce sufficient newsprint for just five or six metropolitan newspapers can consume enough energy to power a residential suburb of 250,000 people. Although mechanical pulp only represents about 20% of the pulp produced in the U.S., it is by far the most energy-consumptive product manufactured in this sector of the economy. Studies have shown that introduction of ionizable groups in lignin, such as occurs during sulfite treatment of wood chips prior to chemimechanical pulping, reduces energy consumption and improves the efficiency of the chip refining process (Htun and Salmen, 1996). This suggests that genetic manipulations resulting in trees which incorporate phenolic acids into their lignin, similar to the situation noted for a loblolly pine mutant lacking cinnamyl alcohol dehydrogenase activity (MacKay et al., 1997), could be used to improve trees for specific use in mechanical pulping. Depending upon the economics, it might also be advantageous to reduce the total amount of lignin in trees destined for mechanical pulping. However, it must be recognized when proposing such modifications that mechanical pulping is a high-volume, high-yield process with low profit margins; thus, it might be difficult to justify the cost of genetically engineering trees specifically for mechanical pulping.

In contrast to mechanical pulp, more than 80% of all paper is produced from chemical or Kraft pulp, and the Kraft process, overall, generates more energy than it consumes. In the Kraft process, lignin extracted at high temperature under highly alkaline conditions (pH >11.0) is burned so as to recover the caustic chemicals, while at the same time generating enough electrical power to run the process, as well as excess power that can be sold to local utilities. Thus, trees modified specifically to suit the needs of the Kraft pulp industry should contain at least as much lignin as normal trees, but that lignin should be more easily and thoroughly extracted under the initial pulping conditions. In the

Kraft pulping process, as much as 90% of the starting lignin is removed during the initial cook. The remaining 10% *residual* lignin imparts a brown color to the pulp, necessitating bleaching treatments if the pulp is destined for the production of white paper. Some of the residual lignin is covalently attached to fibers as a result of chemical reactions occurring during the pulping process, but there is evidence that some covalent linkages are formed during lignin biosynthesis (Helm, 2000). As pulp bleaching processes are expensive and usually have a negative environmental impact, genetic manipulations to reduce the naturally occurring covalent cross-links between lignin and the other fiber polymers would be of significant commercial interest. As Kraft pulps generally command higher prices than mechanical pulps, the economics for modifying lignin by genetic engineering might be easier to justify in trees destined for Kraft pulping.

1.1.2
Solid Wood Products

The properties of wood most relevant to structural use are strength, dimensional stability, and resistance to decay (Whetten and Sederoff, 1991). Although each of these characteristics is impacted by the lignin composition and content of wood, there are relatively few studies which directly relate lignin and structural wood qualities in ways that would suggest what changes to lignin might be desirable for solid wood products. However, the need for such studies is growing as solid wood industries come to rely increasingly on trees from intensively managed plantations, as well as second-growth, rather than old-growth, forests. The juvenile wood from these sources is generally considered inferior for solid wood products, in part because it has a lower lignin content and lower strength, coupled with higher microfibril angles (Kennedy, 1995). The latter is particularly problematic for solid wood products, as high microfibril angle causes directional shrinkage which is manifested as warping and twisting in structural lumber. Studies, such as those by Tjeerdsma et al. (1998) showing how heat treatment of lumber can bring about autocondensation of lignin and thereby improve dimensional stability, may suggest what types of chemical alterations would be desirable in lignin. However, even with such information it will still be necessary to identify and introduce into the trees genes capable of bringing about the correct modifications to lignin precursors.

1.1.3
Textile Fibers

Industrial interest in bast fibers, such as those produced from flax, kenaf, ramie, hemp, or jute, has increased in recent years (Smeder and Liljedahl, 1996). Lignin has both positive and negative effects on the quality of bast fibers used in the production of textiles and cordage. Individual fiber cells are stiffened by the lignin within in their secondary walls, while bundles of these fiber cells, which constitute the material most often used in woven materials, are strengthened by the inter-fiber lignin matrix (see for example, Angelini et al., 2000). Lignin also contributes to the coloration of these fibers, necessitating bleaching processes to produce white cloth (Amin et al., 1998). Although techniques are available for the genetic engineering of flax (Dong and McHughen, 1993) and ramie (Dusi et al., 1993), there is as yet insufficient information relating lignin localization, composition and content to fiber characteristics to allow for predictions as to how lignin biosynthesis might best be modified to improve these fibers. On the other hand, it may be useful to consider the commercial potential for naturally red-colored fibers that could be developed by manipulating the expression of certain enzymes involved in monolignol biosynthesis (Higuchi et al., 1994; Tsai et al., 1998).

1.1.4
Biomass Conversion

Studies have shown that lignin content is inversely correlated with the conversion of glucans to ethanol in combined saccharification/fermentation processes (Vinzant et al., 1997). A similar relationship appears to limit the fodder value of forages grazed by ruminants (Buxton and Redfearn, 1997; Hatfield et al., 1999). Although there have not been any reports of improved fuel production from plants that have been manipulated with respect to lignin, lines of alfalfa having modified lignin composition are being tested for improved digestibility by livestock (Baucher et al., 1999).

1.1.5
Novel Products

The aromatic nuclei of the lignin polymer have long been eyed as a potential renewable source of valuable chemical feedstocks (Faix, 1992). Recent examples of lignin-derived products having potential commercial value include alkylated lignins that have useful properties as dispersing and emulsifying agents (Kosikova et al., 2000), lignins used as thermoplastic copolymers (Li and Sarkanen, 2000), and liquid aromatic hydrocarbons obtained by catalytic cracking of lignin (Thring et al., 2000). Modifications to the lignin biosynthetic pathway that result in the incorporation of different precursors could potentially expand the range of useful products to be derived from lignin.

1.2
Challenges

Efforts to modify the lignin biosynthetic pathway through genetic engineering have so far been pursued in a relatively aggressive fashion, not only because of the potential commercial value to such industries as pulp and paper manufacturing, but also because of the common perception that this metabolic pathway is relatively well understood. In fact, as highlighted by the unusual and unexpected lignin composition in a mutant of loblolly pine (Ralph et al., 1997), our understanding of the metabolic networks underlying lignin biosynthesis is as yet insufficient to predict accurately any changes in the lignin polymer resulting from specific genetic changes. Given that we understand even less about the ways in which lignin composition and quantity contribute to the bulk properties of wood (i.e., wood quality), it will require much more testing before we can predictably generate genotypes from which to produce superior products. Along the way there will also be questions of environmental fitness that will need to be answered. For example, how far can lignin content be reduced before damage by pathogens, insects and wind becomes too great a problem? Perhaps the biggest challenge will be to develop vectors with promoters and terminators that will limit transgene expression to specific lignifying tissues so as not to disrupt growth and development in other parts of the plant (Boudet, 2000).

2
Historical Outline

2.1
Chemistry

Against the current debate concerning a chemical definition for lignin (Lewis, 1999; Sederoff et al., 1999), it seems appropriate to consider for a moment the origin and definition of the term, lignin. Derived from the Latin for wood (*lignum*), the first use of the word, lignin in the English language is ascribed to Imison (1822), who was describing what was then considered an elemental and nondivisible material remaining after wood fibers were first boiled in water and then in alcohol. It was de Candolle (1832) who first described wood fibers as a composite of cellulose in a matrix of *ligneuse* (Fr.). Using various combinations of sulfurous acid and ammonia to fractionate woody fibers, Payen (1838) resolved wood into a material isomeric with starch (cellulose) and a carbon-rich encrusting material (lignin). However, it was actually Schulze (1856) who first applied the term lignin to these carbon-rich encrusting materials, noting that they permeated cellulose fibers to varying degrees. Lignin thus came to be defined as the insoluble polymeric material remaining when the polysaccharides and extractives are removed from wood, and its constituent residues (*p*-coumaryl, guaiacyl and syringyl) were those that were readily detectable in wood.

This definition works reasonably well from a chemist's point of view as it provides a convenient basis on which to discriminate between lignin and the wide array of polyphenolic materials in plants. However, knowing the potential promiscuity of coupling reactions that can occur via the free radical mechanism used for monolignol polymerization, it should not surprise a chemist that in the richly aromatic world of plant tissues, lignin polymers might contain constituents other than the three dictated by definition. There was never a question as to whether the 5-hydroxyguaiacyl units found in brown midrib monocots were to be considered part of the lignin, although direct evidence of this has only recently been published (Kim et al., 2000), and now evidence shows that such residues are common constituents of the lignins from a variety of plant species (Suzuki et al., 1997). It seems likely that new analytical techniques (e.g., Ralph and Lu, 1998) and more careful surveys of the phenylpropanoid polymers from a wider range of plant species will provide further challenges to the strict chemical definition of lignin.

2.2 Biology

Several excellent review articles discussing variations in lignin content and composition, both natural and engineered are available (Campbell and Sederoff, 1996; Monties 1998). Strategies for manipulating the lignin biosynthetic pathway in transgenic plants for selected end uses have also been presented in great detail (Dixon et al., 1996; Grima-Pettenati and Goffner, 1999; Boudet, 2000). One area of research that bears close watching for those interested in applying some of these strategies to modifying in lignin biosynthesis is metabolic channeling of phenylpropanoids through multienzyme complexes (Rasmussen and Dixon, 1999; Winkel-Shirley, 1999). Such complexes have been anticipated for many years (Stafford, 1981), and given some of the unexpected changes that have resulted from initial efforts to change lignin in transgenic plants, it seems likely that such complexes may have a profound impact on the success or failure of specific types of modifications.

3 Lignin Mutants

3.1 Brown Midrib Monocots

Brown midrib (*bm*) mutants of maize, easily identified by the reddish-brown color of their central leaf vein (Figure 1), have been known for more than 75 years, and similar mutants have been found in sorghum, sudangrass, and pearl millet (Barriere and Argillier, 1993). These mutants have generated sig-

Fig. 1 Stem and midvein of a brown midrib mutant (left) and a normal (right) maize plant. This particular mutant is CAD-deficient (*bm1*). (Photograph courtesy of Cold Spring Harbor Laboratory Press.)

nificant agronomic interest because their tissues are more easily digested by ruminants, providing better nutrition for livestock (Cherney et al., 1991). However, the known varieties of these plants are not widely grown as they often suffer from slow growth, increased susceptibility to pests, and an increased tendency to lodge, all of which lead to decreased yields.

Kuc and Nelson (1964) were the first to demonstrate that brown midrib mutations, specifically the maize the *bm1* mutation, resulted in the production of abnormal lignin. There are four classes of maize *bm* mutants, each affecting a different component of the lignin biosynthetic pathway (Lechtenberg et al., 1972). *Bm1* mutants are affected in the expression of coniferyl alcohol dehydrogenase (CAD) activity (Halpin et al., 1998), while *bm3* mutants have altered caffeate *O*-methyltransferase (COMT) activity (Vignols et al., 1995). Specific enzyme activities have not yet been associated with the other two classes of brown midrib mutants in maize. The reduction of CAD activity in maize *bm1* mutants leads to large increases in the hydroxycinnamylaldehyde content of the lignin (Provan et al., 1997), and such aldehydes have been implicated in formation of the red chromophore responsible, in part, for the defining coloration of the mutants (Higuchi et al., 1994). In *bm3* mutants, the lignin also contains increased amounts of 5-hydroxyguaiacyl units, but in addition has a greatly reduced ratio of syringyl to guaiacyl units, as well as decreases in *p*-coumaric acid esters and overall lignin content (Chabbert et al., 1994). With regard to this last point, it is important to note that lignin quantification techniques are notoriously sensitive to compositional changes in the lignin, and thus should be interpreted with caution (Dean, 1997).

3.2 Loblolly Pine

A null allele for the loblolly pine CAD gene was identified in a genetic mapping experiment using haploid megagametophyte tissues from a heterozygous parent (MacKay et al., 1997). Mutants homozygous for this allele, obtained by selfing the heterozygous parent, were easily distinguished by the red-brown color of their xylem tissue. Analysis of the lignin from a 12-year-old mutant identified in a breeding population showed that it contained high levels of hydroxycinnamaldehydes, similar to the *bm1* mutants of maize (Ralph et al., 1997). The lignin was also distinguished by the incorporation of other subunits not previously noted in lignin (Figure 2), including dihydroconiferyl alcohol, vanillin (4-hydroxy-3-methoxybenzaldehyde) (Ralph et al., 1999a; Sederoff et al., 1999), and arylpropane-1,3-diols (Ralph et al., 1999b).

The overall lignin content of CAD-deficient pines was not significantly different than in

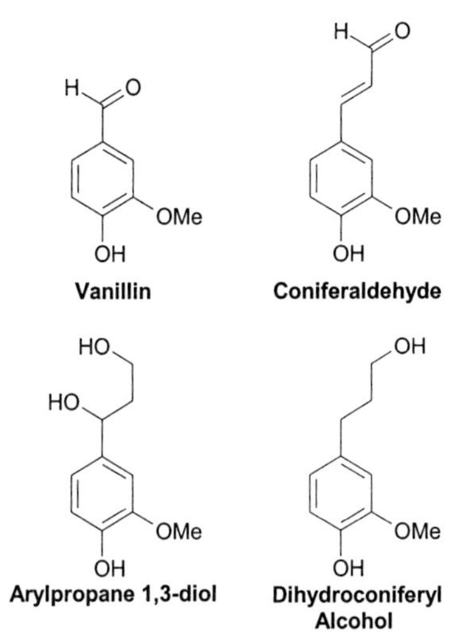

Fig. 2 A selection of unusual lignin constituents identified in CAD-deficient loblolly pine

wild-type pines. Wood chips from the CAD-deficient trees were more easily delignified in a soda pulping process, although not under Kraft pulping conditions (MacKay et al., 1999). Soda pulping does have some advantages over the Kraft process in that it does not produce large amounts of the volatile organic sulfur compounds that lead to the noxious odors associated with Kraft mills. However, it remains to be seen whether CAD-deficient conifers will generate significant commercial interest.

3.3 *Arabidopsis*

As a model system for plant development, *Arabidopsis* has proven a powerful genetic tool with which to dissect complex processes, including lignification (Anderson and Roberts, 1998). A mutant (*fah1*) lacking leaf fluorescence characteristic of sinapyl-glucosides was identified, and radiotracer experiments indicated that the mutation blocked the phenylpropanoid biosynthetic pathway between ferulate and 5-hydroxyferulate (Chapple et al., 1992). This genetic block resulted in the mutant producing a lignin containing no syringyl residues. Subsequent gene tagging experiments led to the cloning of a novel cytochrome P450-dependent monooxygenase, the elusive ferulate 5-hydroxylase that had previously proven impossible to identify through biochemical means (Meyer et al., 1996). In a surprise to most researchers in the field, recent work has demonstrated that the in vivo substrate for this enzyme is, in fact, coniferyl aldehyde, rather than ferulate (Humphreys et al., 1999; Osakabe et al., 1999; Li et al., 2000). These findings, in conjunction with studies of genetically engineered plants, have led to significant revisions in our perception of the final steps in monolignol biosynthesis (Figure 3).

A variety of *Arabidopsis* mutants altered with respect to lignin localization have also been identified (for example, Dawson et al., 1999; Boudet, 2000 and references therein; Liljegren et al., 2000; Zhong et al., 2000). Most of these mutations appear to affect developmental controls that govern a variety of metabolic and structural pathways in addition to lignin biosynthesis. Thus, in most cases it remains to be determined whether the mutations influence lignin deposition directly or indirectly.

4 Lignin Transgenics

4.1 The Shikimate Pathway and Phenylalanine Ammonia Lyase

In response to a variety of developmental signals and environmental conditions, carbon flux through the shikimate pathway is regulated in a co-ordinated fashion with lignin biosynthesis (Weaver and Herrmann, 1997). Because there are numerous branch points leading from this pathway to impor-

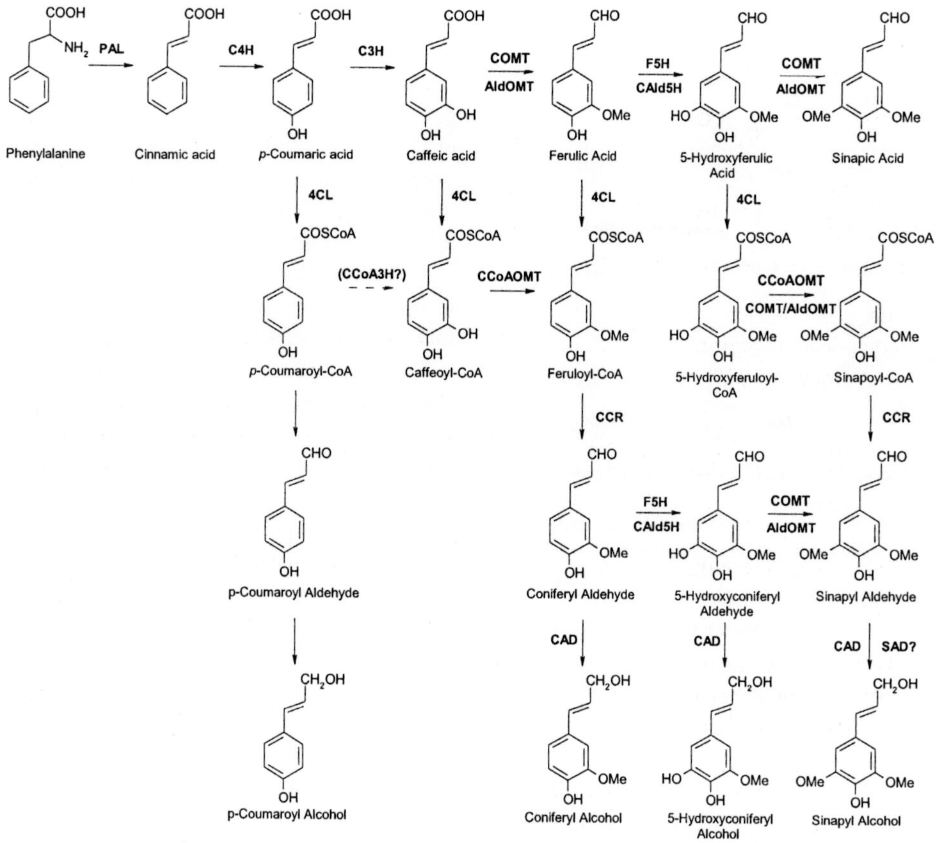

Fig. 3 Biosynthetic pathway for the monolignol precursors of lignin. (Adapted from Meyermans et al., 2000.)

tant metabolites, the pathway is unlikely to provide many useful targets for those wishing to alter lignin. However, transgenic potato plants in which an antisense construct was used to down-regulate the first enzyme of the shikimate pathway, 3-deoxy-D-*arabino*-heptulosonate-7-phosphate (DAHP) synthase, contained less lignin, but also had aberrant stem length and girth (Jones et al., 1995). Over-expression in transgenic potato of a tryptophan decarboxylase gene produced an artificial metabolic sink for shikimate pathway products, resulting in a phenotype similar to that produced by down-regulation of DAHP synthase (Yao et al., 1995) (Figure 4).

Phenylalanine ammonia lyase (PAL) is the gateway through which products of the shikimate pathway enter phenylpropanoid metabolism. When PAL activity in stems of transgenic tobacco plants was reduced more than three-fold by the expression of an antisense gene, lignin levels were significantly reduced (Bate et al., 1994). Although the plants were more susceptible to disease, apparently due to reduced levels of phytoalexins (Maher et al., 1994), cell wall digestibility was significantly enhanced (Sewalt et al., 1997a). Lignin from PAL-suppressed tobacco had an increased syringyl/guaiacyl ratio, suggesting that there might be divergent biosynthetic pathways for coniferyl and

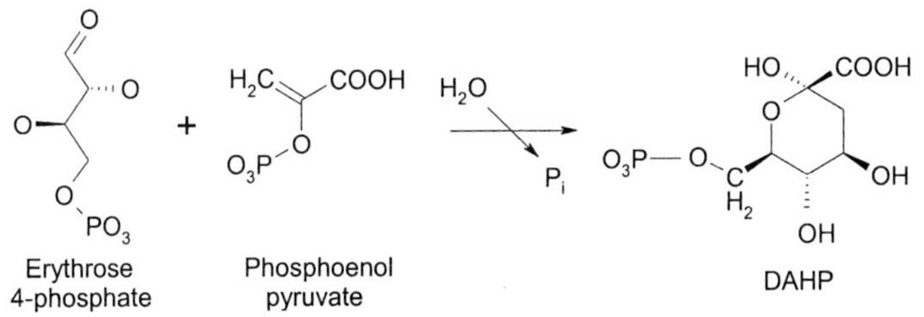

Fig. 4 Reaction scheme for 3-deoxy-D-*arabino*-heptulosonate 7-phosphate (DAHP) synthase

sinapyl alcohols beginning even at the level of PAL (Sewalt et al., 1997b) (Figure 5).

4.2 Cinnamate 4-Hydroxylase and 4-Coumarate-3-Hydroxylase

In contrast to the results obtained when PAL activity was reduced in tobacco stems, reduction of cinnamate 4-hydroxylase (C4H) activity in transgenic tobacco, using either antisense expression or sense co-suppression, reduced the syringyl/guaiacyl ratio (Sewalt et al., 1997b). However, reduction of C4H activity also reduced the total lignin content of stems, just as did reduction of PAL activity (Blount et al., 2000) (Figure 6).

Similar to the situation with ferulate 5-hydroxylase (F5H) prior to identification of the *Arabidopsis fah1* mutant (Chapple et al., 1992), the enzyme and gene responsible for the second hydroxylation step in the lignin pathway have not yet been definitively identified. The results of Wang et al. (1997), who purified an enzyme responsible for a 4-coumaroyl:CoA hydroxylase (CCoA3H) activity only to find that it was likely a polyphenoloxidase, underscore the difficulties inherent in working with these enzymes. A mutagenesis approach, screening for *Arabidopsis* mutants lacking caffeic acid derivatives, may again be necessary to positively identify the correct gene. Note that two of the patents listed in Table 1 claim to cover so-called 4-coumarate-3-hydroxylase (C3H) genes, yet there is insufficient evidence available to be certain that the sequences described actually encode the correct enzymes.

4.3 4-Coumarate:CoA Ligase

Antisense expression was used to reduce the 4-coumarate:CoA ligase (4CL) activity in transgenic tobacco, resulting in stem xylem tissues that were brown in color and contained reduced levels of lignin (Kajita et al.,

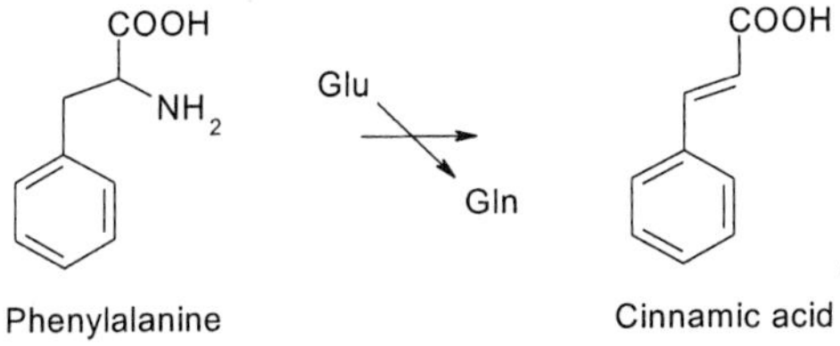

Fig. 5 Reaction scheme for phenylalanine ammonia lyase (PAL)

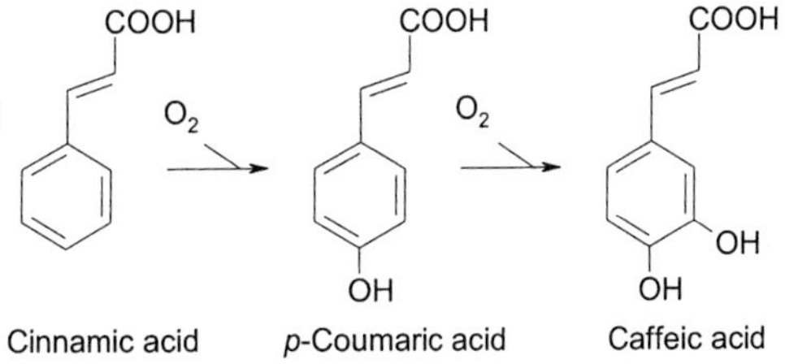

Fig. 6 Sequential hydroxylation reactions catalyzed by cinnamate 4-hydroxylase (C4H) and 4-coumarate 3-hydroxylase (C3H), respectively

Tab. 1 Patents issued for the use of lignin biosynthetic genes to manipulate lignin in transgenic plants (as of June 1, 2000). This information was retrieved via the IBM Intellectual Property Network (http://www.patents.ibm.com/)

Gene(s)	*Applicant*	*Patent number*	*Date issued-filed (day/month/year)*
Pathways			
PAL CAD	Genesis R&D Corp.	US5850020	15/12/98–11/09/96
C4H CGT	Fletcher Challenge	US5952486	14/09/99–21/11/97
C3H CBG	Forests	WO9811205A3	20/08/98–10/09/97
4CL POX		WO9811205A2	19/03/98–10/09/97
OMT PHL		EP929682A2	21/07/99–10/09/97
CCR LAC			
C4H CCR	Pioneer Hi-Bred	WO9910498A2	04/03/99–24/08/98
C3H CAD	International		
4CL LAC			
F5H			
COMT			
CCoAOMT			
4CL	International Paper	WO9931243A1	24/06/99–16/12/98
bi-OMT			
P450-1			
P450-2			
Single genes			
4CL	Michigan Technological University	WO9924561A2	20/05/99–12/11/98
F5H	Purdue University	US5981837	09/11/99–18/06/98
		WO9723599A2	03/07/97–19/12/96
		WO9803535A1	29/01/98–18/07/97
OMT	ICI Plc	US5959178	28/09/99–27/10/94
	Zeneca Limited	WO9305160A1	18/03/93–09/09/92
		EP603250A1	29/06/94–09/09/92
OMT	Michigan Technological University	US5886243	23/03/99–18/09/96
CCR	Zeneca Limited	US5866791	02/02/99–13/03/96
		WO9839454A1	11/09/98–25/02/98
CCR	CNRS	WO1996FR0001544	10/04/97–03/10/96
		US6015943	18/01/00–24/02/97
CAD	Zeneca Limited	US5451514	19/09/95–28/12/93
		US5633439	27/05/97–28/02/95
CAD	N.C. State University	US5824842	20/10/98–26/07/96
CBG	University of British Columbia	US5973228	26/10/99–23/07/98
LAC	CNRS	WO1997FR0000948	04/12/97–30/05/97

1996). The syringyl/guaiacyl ratio of the lignin was reduced in these mutants, and the reduced lignin content appeared to be related to increased numbers of collapsed vessel cells in the xylem tissues (Kajita et al., 1997a). In some plants transformed with sense constructs of 4CL, co-suppression was not uniform throughout the xylem, but showed a sectored pattern (Kajita et al., 1997b) (Figure 7).

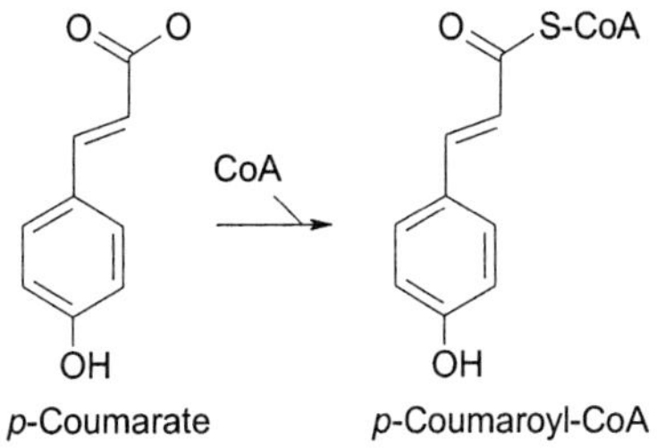

p-Coumarate → *p*-Coumaroyl-CoA (CoA)

Fig. 7 Reaction scheme for 4-coumarate:CoA ligase (4CL)

In contrast to the situation with tobacco, antisense repression of 4CL activity in *Arabidopsis* depressed the guaiacyl content of the plants, but not the syringyl content, leading to increased syringyl/guaiacyl ratios (Lee et al., 1997). Overall lignin content was reduced in the antisense 4CL *Arabidopsis* plants. The most striking results obtained to date for antisense reduction of 4CL gene expression were reported by Hu et al. (1999) in transgenic aspen. Aspen has at least two different 4CL genes: one is expressed in the leaf epidermis and stem, apparently involved in the synthesis of flavonoids; and the second is expressed exclusively in lignifying xylem tissue. In trees expressing an antisense construct of the second gene, lignin levels were reduced by as much as 45% with no apparent alteration of lignin composition. Unexpectedly, these reductions in lignin content were accompanied by enhanced stem, leaf and root growth. Not surprisingly, these results have garnered a great deal of interest from the pulp and paper industry.

4.4 *O*-Methyltransferases

Heterologous expression of an antisense COMT gene from alfalfa in transgenic tobacco was reported to result in significant reductions in stem lignin content, but no apparent changes in lignin composition (Ni et al., 1994). In contrast, heterologous expression of an aspen COMT in tobacco reduced the syringyl content of the lignin in plants having depressed levels of COMT activity (Dwivedi et al., 1994). Atanassova et al. (1995) transformed tobacco with both sense and antisense constructs prepared using a tobacco COMT gene. In their study, neither antisense inhibition nor sense suppression led to a reduction in lignin content, and reductions in COMT activity of up to 56% from wild-type had no effect on lignin composition. However, in lines retaining $\leq$12% of wild-type levels of COMT activity, the syringyl content of the lignin was reduced by up to 90% and significant amounts of 5-hydroxyguaiacyl residues were recovered. No color changes were seen in the xylem tissues of even the most severely affected plants. Vailhe et al. (1996b) also reported reduced syringyl content in the lignin from transgenic tobacco plants in which the COMT activity was reduced using either sense co-suppression or antisense repression. Although the lignin content of these plants was reported to be unchanged, their dry matter degradability (ruminant digestibility) was reportedly increased. Sewalt et al. (1997a) also reported increased digestibility of tobacco in which COMT activity was reduced by antisense gene expression. However, they also reported an overall decrease in lignin content accompanied by an increase in the syringyl/guaiacyl ratio (Figure 8).

With respect to woody species, down-regulation of COMT activity in poplar by up to 90%, while not affecting overall lignin content, did decrease the syringyl/guaiacyl ratio significantly (Van Doorsselaere et al., 1995). In lines having the greatest reduction in COMT activity, the lignin also contained 5-hydroxyguaiacyl residues, and the xylem was described as being pale rose in color. Further studies with these trees indicated that the increased proportion of guaiacyl residues

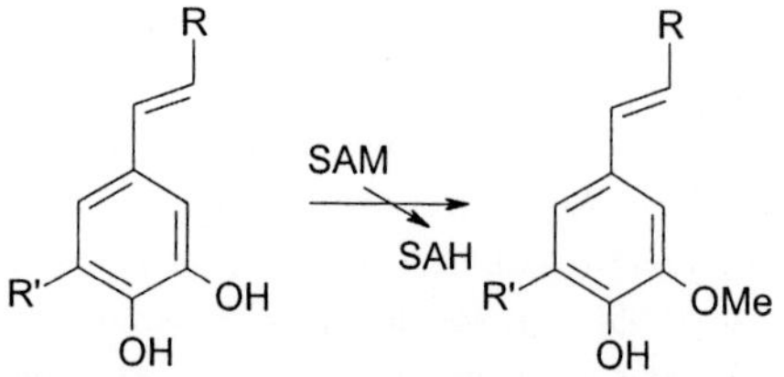

Fig. 8 Reaction scheme for caffeate O-methyltransferase/5-hydroxyconiferyl aldehyde O-methyltransferase (COMT/AldOMT), where R represents either the acid or aldehyde structure and R' represents either H (caffeoyl) or OMe (5-hydroxyferuloyl)

resulted in an increased content of condensed linkages which made the wood less amenable to Kraft pulping (Baucher et al., 1998b; Lapierre et al., 1999). In contrast to the results in poplar, sense suppression of COMT expression in transgenic aspen yielded trees having reddish-brown wood, similar to what is seen in the maize *bm3* mutants (Tsai et al., 1998). The coloration was correlated with an increased content of coniferyl aldehyde (5-hydroxyguaiacyl residues) and a decreased syringyl/guaiacyl ratio in the lignin. In some lines the coloration was mottled, suggesting that sense suppression did not occur with equal efficiency throughout the xylem tissue. This observation bore a strong resemblance to what was seen when sense suppression was used in tobacco to alter 4CL activity (Kajita et al., 1997b). Interestingly, antisense inhibition of the aspen COMT activity by up to 60% did not lead to changes in wood color or lignin content (Boerjan et al., 1997). However, lignin composition was altered by the incorporation of 5-hydroxyconiferyl alcohol into the polymer (Ralph et al., 2000).

Recently, Li et al. (2000) noted that at least some of the *O*-methyltransferases which are active on the free acid cinnamates generally appear more active when the cognate aldehydes are provided as substrates. Focusing on the conversion of 5-hydroxyferuloyl aldehyde to sinapyl aldehyde, these researchers have redesignated this enzyme AldOMT, and have suggested that working in conjunction with CAld5H (see below), this activity is a regulating factor for sinapyl alcohol biosynthesis in angiosperms.

A second family of *O*-methyltransferases, the caffeoyl-coenzyme A 3-*O*-methyltransferases (CCoAOMT), have also been shown to be involved in lignification (Ye and Varner, 1995; Ye, 1997). Antisense repression of CCoAOMT activity in tobacco led to a decrease in lignin content, and altered lignin composition by increasing the proportion of syringyl residues (Zhong et al., 1998). In the same study, reduction of COMT activity did not affect lignin content, but did result in lignin having a higher guaiacyl content. Simultaneous reduction of both *O*-methyltransferases in tobacco led to a reduction in lignin content beyond that obtained by blocking CCoAOMT activity alone. Antisense inhibition of CCoAOMT activity in poplar led to a modest decrease in lignin content, a slight increase in the syringyl to guaiacyl ration, and a dramatic increase in the levels phenolate glucosides (Meyermans et al., 2000) (Figure 9).

As a cautionary note for those attempting to draw general conclusions from the work to date as to how lignin might be altered by manipulating *O*-methyltransferases, it is important to appreciate the gene family complexity and likely diversity of function for these enzymes in plants (Maury et al.,

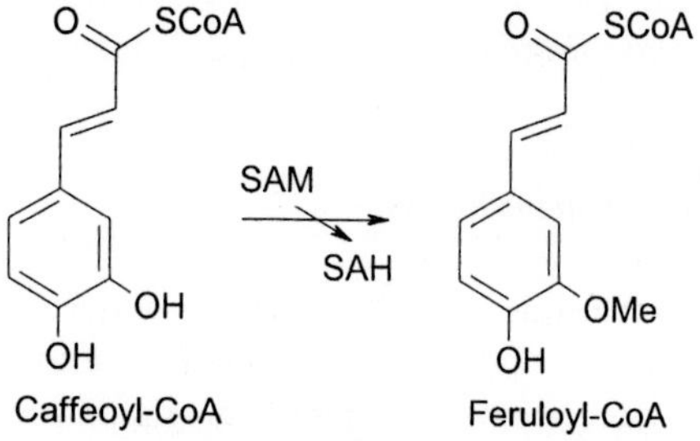

Fig. 9 Reaction scheme for caffeoyl-CoA 3-*O*-methyltransferase (CCoAOMT)

1999). It will likely require many more careful studies using homologous sense and antisense expression of specific gene family members to gain a full appreciation of how these genes and their products influence lignin content and composition.

4.5 Cinnamoyl-CoA Reductase

Tobacco plants transformed with an antisense cinnamoyl-CoA reductase (CCR) gene had reduced lignin content, increased syringyl/guaiacyl ratio, and xylem that was orange-brown in color (Piquemal et al., 1998) (Figure 10).

NMR analysis verified that these plants had a substantially reduced guaiacyl content, but at the same time contained significant amounts of tyramine ferulate, which appeared to provide a sink for the feruloyl-CoA units that accumulated due to the metabolic block at CCR (Ralph et al., 1998). Similar changes have evidently been seen in *Arabidopsis* plants transformed with an antisense CCR gene (as noted in Boudet, 2000).

4.6 Ferulate 5-Hydroxylase

When the *Arabidopsis* F5H gene was overexpressed using the 35S cauliflower mosaic virus promoter in transgenic lines derived from the *fah1 Arabidopsis* mutant, the syringyl content of the lignin was restored to near wild-type levels (Meyer et al., 1998). However, when expression was driven using the *Arabidopsis* C4H promoter, the lignin from transgenic lines was composed almost entirely of syringyl units. NMR analysis verified this latter observation (Figure 11), and placed this lignin as the most syringyl-rich ever reported (Marita et al., 1999). These results underscore the need for increased efforts to use tissue-specific, rather than constitutive promoters to obtain the desired lignin phenotypes.

Note that subsequent to the initial reports, this enzyme was shown to be more active when coniferyl aldehyde was supplied as a substrate (Humphreys et al., 1999; Osakabe et al., 1999). Consequently, the latter group has elected to refer to refer to the enzyme as coniferyl aldehyde 5-hydroxylase (CAld5H), rather than F5H.

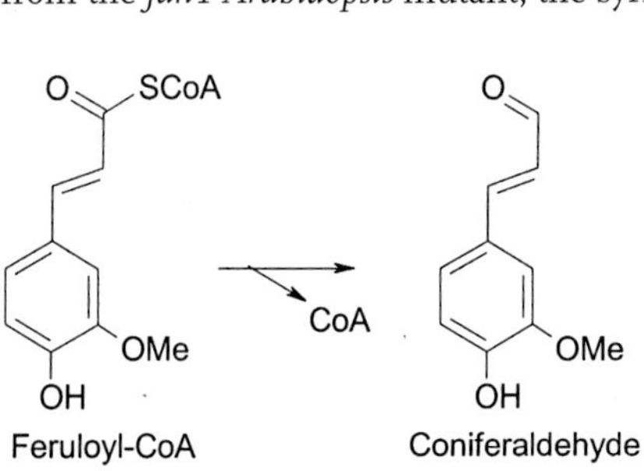

O SCoA OMe OH Feruloyl-CoA → CoA O OMe OH Coniferaldehyde

Fig. 10 Reaction scheme for cinnamoyl-CoA reductase (CCR)

4.7 Cinnamyl Alcohol Dehydrogenase

Antisense repression of cinnamyl alcohol dehydrogenase (CAD) in tobacco did not decrease lignin content, but did lead to increased incorporation of cinnamaldehydes into the lignin (Halpin et al., 1994). Consequently, the xylem in these transgenic plants was red-brown in color, similar to that in maize *bm1* mutants. The lignin from CAD-deficient tobacco was more easily extracted under mild alkaline conditions, similar to the recent observations for CAD-deficient loblolly pine (MacKay et al., 1999). A similar effect was obtained in tobacco using heterologous expression of an antisense CAD gene from *Aralia cordata* (Hibino et al., 1995). Spectroscopic analyses suggested that the lignin in CAD-deficient tobacco plants was less condensed, i.e., it contained fewer cross-links (Stewart et al., 1997), and this change appeared to be largely responsible for increasing

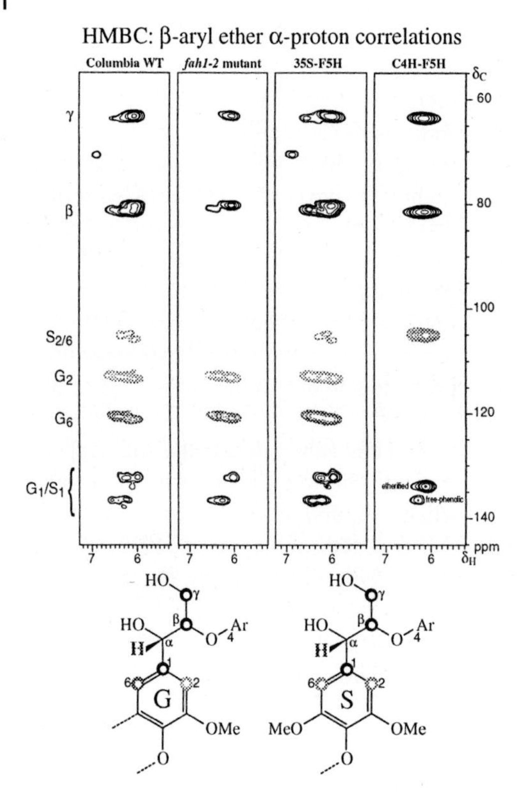

Fig. 11 Portions of two-dimensional NMR spectra showing α-proton correlations in β-aryl ether units of lignin in wild-type and transgenic *Arabidopsis*. Loss of syringyl residues in the *fah1* mutant is shown by the loss of the $S_{2/6}$ signal (second column), while hyper-accumulation of syringyl residues and loss of guaiacyl residues in the F5H over-expressing transgenic line can be seen in the $S_{2/6}$, and G_2 and G_6 signals (fourth column), respectively. (Figure from Marita et al. (1999), courtesy of J. Ralph and C.C.S. Chapple.)

the digestibility of tissues from these plants (Vailhe et al., 1996a, 1998). Baucher et al. (1999) have shown that down-regulation of alfalfa CAD leads to similar changes in lignin composition and increases in digestibility. The levels of cinnamaldehydes and benzaldehydes in lignin from CAD-deficient tobacco were shown to increase in direct proportion to reductions in CAD activity, compensating directly for reductions in the content of guaiacyl and syringyl residues (Ralph et al., 1998; Yahiaoui et al., 1998). The changed properties of the lignin affected the longitudinal tensile modulus of the xylem tissue in CAD-deficient tobacco, reducing it from 2.8 GPa to 1.9 Gpa, suggesting that tobacco xylem tissue cell walls are more sensitive to changes in the properties of the matrix than can be predicted using current cell wall mechanical models (Hepworth and Vincent, 1998) (Figure 12).

Antisense and co-suppression strategies were both used to reduce the CAD activity in transgenic poplar (Baucher et al., 1996). In trees having a 70% reduction in CAD activity, lignin levels, as well as syringyl and guaiacyl

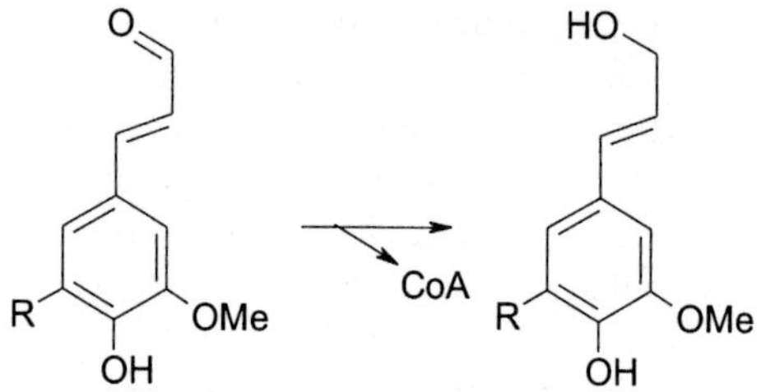

Fig. 12 Reaction scheme for cinnamyl alcohol dehydrogenase (CAD), where R represents either H (coniferaldehyde) or OMe (sinapylaldehyde)

levels, were unchanged. However, the xylem tissue was colored red, and was found to be more readily extractable under mild alkaline conditions. The wood from these trees was shown to contain increased levels of syringaldehyde and diarylpropane residues, as well as free phenolic groups, and as a consequence was significantly easier to pulp under Kraft pulping conditions (Lapierre et al., 1999, 2000).

4.8
Coniferin β-Glucosidase

A gene for coniferin β-glucosidase (CBG) has been cloned from lodgepole pine, but there have not yet been any reports describing the use of this or homologous genes in transgenic plants (Dharmawardhana et al., 1999). No gene has yet been identified for the uridine 5′-diphosphoglucose:coniferyl alcohol glucosyltransferase (CAGT) activity described by Savidge and Forster (1998).

4.9
Peroxidases and Other Oxidases

For efforts to decrease total lignin content in transgenic plants, the extracellular oxidative polymerization reactions have been an attractive target due to the decreased likelihood that downstream products might have critical physiological functions. So far, however, it has proven much easier to increase than decrease lignin levels via this route. Lagrimini (1991) was the first to show that over-expression of an anionic tobacco peroxidase in transgenic tobacco led to increased deposition of lignin-like polymers when the plants were wounded. However, antisense expression of the same gene had no effect on vascular lignin levels, suggesting that the gene was not directly involved in vascular lignin biosynthesis (Lagrimini et al., 1997). Over-expression of a tomato peroxidase in transformed tomato yielded similar results (El Mansouri et al., 1999). On the other hand, expression of a cucumber peroxidase in transgenic potato did not change levels of tissue phenolics, nor did it increase disease resistance (Ray et al., 1998). Transgenic tobacco plants engineered to increase endogenous indole acetic acid (IAA) levels were found to contain increased lignin (Sitbon et al., 1992). Evidence suggests that overproduction of IAA led to increased ethylene production, which induced peroxidase expression, and this led to the increased lignin content (Sitbon et al., 1999). In another approach, a fungal glucose oxidase expressed in transgenic potato led to increased H_2O_2 production and consequent increases in stem and root lignin content (Wu et al., 1997)

Blue copper oxidases of the laccase (LAC) family have been over-expressed in transgenic yellow-poplar plants where they led to increased deposition of polyphenolic materials and severe stunting of regenerated plantlets (Dean et al., 1998; LaFayette et al., 1999). However, transgenic trees expressing antisense laccase genes have shown no obvious phenotypes (Dean et al., 1998; Boudet, 2000). Antisense inhibition of one laccase gene in *Arabidopsis* resulted in plants that grew very poorly, but did not have any obvious alterations in lignin (Halpin et al., 1999). Recent evidence from the author's laboratory, in fact, suggests that the enzymes may not directly involved in monolignol polymerization (J.F.D. Dean, unpublished results).

A further oxidase, so-called coniferyl alcohol oxidase, has not yet been cloned or tested in transgenic plants, although a partial N-terminal sequence does exist for the enzyme from spruce (Udagamarandeniya and Savidge, 1995).

4.10 Transcription Factors and Other Targets

Substantial evidence exists to suggest that lignin biosynthetic genes are co-ordinately regulated at the level of transcription; thus, transcription factors may have great potential for altering expression of multiple pathway genes simultaneously. Over-expression of an antirrhinum MYB-related transcription factor led to a decrease in lignin content in transgenic tobacco, although the plants also displayed undesirable modifications to their growth and development (Tamagnone et al., 1998). The *rol* genes found in T-DNA from *Agrobacterium rhizogenes* can substantially alter levels of various phytohormones when they are expressed in plant tissues. *RolC* expression can increase levels of cytokinins, and in transformed aspen, *rolC* caused atypical latewood formation with reduced lignin content and discolored wood (Grunwald et al., 2000).

5 Outlook and Perspectives

As the commercial enterprise most directly affected with lignin, the pulp and paper industry has made the most substantial investment in attempts at genetic modification of lignin biosynthesis (Merkle and Dean, 2000). Yet, as pointed out by Mullin and Bertrand (1998), the economics underlying these investments can be difficult to justify as the returns are based on materials harvested many years in the future, by which time chemists and engineers may have already solved the problems that are in focus today. However, if we are to be successful in shifting the basis for the manufacturing sector of the world's economy toward renewable resources, we must learn how to manipulate the metabolic pathways in plants with much greater skill. The lignin biosynthetic pathway provides an excellent target on which tomorrow's metabolic engineers may learn and hone their skills.

6 Patents

As shown in Table 1, broad scope patents covering many of the known lignin biosynthetic pathway genes have been issued. One set, issued to Genesis Research and Development Corp. and Fletcher Challenge Forests, has, as a preferred embodiment, the use of these genes in woody plants, such as radiata pine and eucalyptus. A second set, issued to Pioneer Hi-Bred, focuses on the use of these genes in agricultural crops. A third multigene patent, issued to International Paper, focuses on the genes necessary to introduce syringyl residues into the lignin produced in coniferous trees. Other patents cover many of the individual genes associated with the pathway. Obviously, there appear to be some overlapping claims, but to date no legal challenges have been made or settled.

7 References

Amin, M. N., Begum, M., Shahjahan, M., Rahman, M. S., Baksh, T. (1998) Bleaching process for increasing whiteness and lightfastness of kenaf and kenaf-jute blended fabrics, *Cell. Chem. Technol.* **32**, 497–504.

Andersen, S. O., Peter, M. G., Roepstorff, P. (1996) Cuticular sclerotization in insects, *Comp. Biochem. Physiol. B* **113**, 689–705.

Anderson, M., Roberts, J. (1998) *Arabidopsis*. Boca Raton, FL: CRC Press.

Angelini, L. G., Lazzeri, A., Levita, G., Fontanelli, D., Bozzi, C. (2000) Ramie (*Boehmeria nivea* (L.) Gaud.) and Spanish Broom (*Spartium junceum* L.) fibres for composite materials: agronomical aspects, morphology and mechanical properties, *Ind. Crop. Prod.* **11**, 145–161.

Argyropoulos, D. S. (1999) *Advances in Lignocellulosics Characterization*. Atlanta: Tappi Press.

Atanassova, R., Favet, N., Martz, F., Chabbert, B., Tollier, M. T., Monties, B., Fritig, B., Legrand, M. (1995) Altered lignin composition in transgenic tobacco expressing *O*-methyltransferase sequences in sense and antisense orientation, *Plant J.* **8**, 465–477.

Barriere, Y., Argillier, O. (1993) Brown-midrib genes of maize – A review, *Agronomie* **13**, 865–876.

Bate, N. J., Orr, J., Ni, W. T., Meromi, A., Nadler-hassar, T., Doerner, P. W., Dixon, R. A., Lamb, C. J., Elkind, Y. (1994) Quantitative relationship between phenylalanine ammonia-lyase levels and phenylpropanoid accumulation in transgenic tobacco identifies a rate-determining step in natural product synthesis, *Proc. Natl. Acad. Sci. USA* **91**, 7608–7612.

Baucher, M., Chabbert, B., Pilate, G., Van Doorsselaere, J., Tollier, M. T., Petit-Conil, M., Cornu, D., Monties, B., Van Montagu, M., Inze, D., Jouanin, L., Boerjan, W. (1996) Red xylem and higher lignin extractability by down-regulating a cinnamyl alcohol dehydrogenase in poplar, *Plant Physiol.* **112**, 1479–1490.

Baucher, M., Monties, B., Van Montagu, M., Boerjan, W. (1998a) Biosynthesis and genetic engineering of lignin, *Crit. Rev. Plant Sci.* **17**, 125–197.

Baucher, M., Christensen, J. H., Meyermans, H., Chen, C. Y., Van Doorsselaere, J., Leple, J. C., Pilate, G., Petit-Conil, M., Jouanin, L., Chabbert, B., Monties, B., Van Montagu, M., Boerjan, W. (1998b) Applications of molecular genetics for biosynthesis of novel lignins, *Polym. Degrad. Stabil.* **59**, 47–52.

Baucher, M., Bernard-Vailhe, M. A., Chabbert, B., Besle, J. M., Opsomer, C., Van Montagu, M., Botterman, J. (1999) Down-regulation of cinnamyl alcohol dehydrogenase in transgenic alfalfa (*Medicago sativa* L.) and the effect on lignin composition and digestibility, *Plant Mol. Biol.* **39**, 437–447.

Blount, J. W., Korth, K. L., Masoud, S. A., Rasmussen, S., Lamb, C., Dixon, R. A. (2000) Altering expression of cinnamic acid 4-hydroxylase in transgenic plants provides evidence for a feedback loop at the entry point into the phenylpropanoid pathway, *Plant Physiol.* **122**, 107–116.

Boerjan, W., Baucher, M., Chabbert, B., Petit-Conil, M., Leple, J. C., Pilate, G., Cornu, D., Monties, B., Van Montagu, M., Van Doorsselaere, J., Tsai, C.-J., Podila, G. K., Joshi, C. P., Chiang, V. L. (1997) Genetic modification of lignin biosynthesis in quaking aspen and poplar, in: *Micropropagation, Genetic Engineering, and Molecular Biology of Populus* (Klopfenstein, N. B., Chun, Y. W., Kim, M. S., Ahuja, M. R., Eds.), pp. 193–205. Fort Collins, CO: U.S. Department of Agriculture, Forest Service, Rocky Mountain Forest and Range Experiment Station.

Boudet, A. M. (2000) Lignins and lignification: Selected issues, *Plant Physiol. Biochem.* **38**, 81–96.

Butler, M. J., Day, A. W. (1998) Fungal melanins: a review, *Can. J. Microbiol.* **44**, 1115–1136.

Buxton, D. R., Redfearn, D. D. (1997) Plant limitations to fiber digestion and utilization, *J. Nutr.* **127**, S814–S818.

Campbell, M. M., Sederoff, R. R. (1996) Variation in lignin content and composition – Mechanism of control and implications for the genetic improvement of plants, *Plant Physiol.* **110**, 3–13.

Chabbert, B., Tollier, M. T., Monties, B., Barriere, Y., Argillier, O. (1994) Biological variability in lignification of maize – expression of the brown midrib *bm3* mutation in three maize cultivars, *J. Sci. Food Agric.* **64**, 349–355.

Chapple, C. C. S., Vogt, T., Ellis, B. E., Somerville, C. R. (1992) An *Arabidopsis* mutant defective in the general phenylpropanoid pathway, *Plant Cell* **4**, 1413–1424.

Cherney, J. H., Cherney, D. J. R., Akin, D. E., Axtell, J. D. (1991) Potential of brown-midrib, low-lignin mutants for improving forage quality, *Adv. Agron.* **46**, 157–198.

Dawson, J., Sozen, E., Vizir, I., Van Waeyenberge, S., Wilson, Z. A., Mulligan, B. J. (1999) Characterization and genetic mapping of a mutation (*ms35*) which prevents anther dehiscence in *Arabidopsis thaliana* by affecting secondary wall thickening in the endothecium, *New Phytol.* **144**, 213–222.

Dean, J. F. D. (1997) Lignin analysis, in: *Plant Biochemistry/Molecular Biology Laboratory Manual* (Dashek, W. V., Ed.), pp. 199–215. Boca Raton, FL: CRC Press, Inc.

Dean, J. F. D., LaFayette, P. R., Rugh, C., Tristram, A. H., Hoopes, J. T., Merkle, S. A., Eriksson, K-E. L. (1998) Laccases associated with lignifying tissues, *ACS Symp. Ser.* **697**, 96–108.

De Candolle, A.-P. (1832) *Physiologie Végétale, Pt. II*, p. 165. Paris, Béchet Jeune.

Delwiche, C. F., Graham, L. E., Thompson, N. (1989) Lignin-like compounds and sporopollenin in *Coleochaete*, an algal model for land plant ancestry, *Science* **245**, 399–401.

Dharmawardhana, D. P., Ellis, B. E., Carlson, J. E. (1999) cDNA cloning and heterologous expression of coniferin beta-glucosidase, *Plant Mol. Biol.* **40**, 365–372.

Dixon, R. A., Lamb, C. J., Masoud, S., Sewalt, V. J. H., Paiva, N. L. (1996) Metabolic engineering: prospects for crop improvement through the genetic manipulation of phenylpropanoid biosynthesis and defense responses – A review, *Gene* **179**, 61–71.

Dong, J. Z., McHughen, A. (1993) Transgenic flax plants from *Agrobacterium*-mediated transformation – incidence of chimeric regenerants and inheritance of transgenic plants, *Plant Sci.* **91**, 139–148.

Dusi, D. M. A., Dubald, M., Dealmeida, E. R. P., Caldas, L. S., Gander, E. S. (1993) Transgenic plants of ramie (*Boehmeria nivea* Gaud) obtained by *Agrobacterium*-mediated transformation., *Plant Cell Rep.* **12**, 625–628.

Dwivedi, U. N., Campbell, W. H., Yu, J., Datla, R. S. S., Bugos, R. C., Chiang, V. L., Podila, G. K. (1994) Modification of lignin biosynthesis in transgenic *Nicotiana* through expression of an antisense *O*-methyltransferase gene from *Populus*, *Plant Mol. Biol.* **26**, 61–71.

El Mansouri, I., Mercado, J. A., Santiago-Domenech, N., Pliego-Alfaro, F., Valpuesta, V., Quesada, M. A. (1999) Biochemical and phenotypical characterization of transgenic tomato plants overexpressing a basic peroxidase, *Physiol. Plant.* **106**, 355–362.

Faix, O. (1992) New aspects of lignin utilization in large amounts, *Papier* **46**, 733–740.

Freudenberg, K. (1968) *Constitution and Biosynthesis of Lignin* (Freudenberg, K., Neish, A. C., Eds.), pp. 114–115. New York: Springer-Verlag.

Grima-Pettenati, J., Goffner, D. (1999) Lignin genetic engineering revisited, *Plant Sci.* **145**, 51–65.

Grunwald, C., Deutsch, F., Eckstein, D., Fladung, M. (2000) Wood formation in *rolC* transgenic aspen trees, *Trees-Struct. Funct.* **14**, 297–304.

Halpin, C., Knight, M. E., Foxon, G. A., Campbell, M. M., Boudet, A. M., Boon, J. J., Chabbert, B., Tollier, M. T., Schuch, W. (1994) Manipulation of lignin quality by down-regulation of cinnamyl alcohol-dehydrogenase, *Plant J.* **6**, 339–350.

Halpin, C., Holt, K., Chojecki, J., Oliver, D., Chabbert, B., Monties, B., Edwards, K., Barakate, A., Foxon, G. A. (1998) Brown-midrib maize (*bm1*) – a mutation affecting the cinnamyl alcohol dehydrogenase gene, *Plant J.* **14**, 545–553.

Halpin, C., Barakate, A., Abbott, J., El Amrani A. (1999) *Arabidopsis* laccases, *Forest Biotechnology 99, July 11–16, 1999*, University of Oxford, England (poster 29).

Hatfield, R. D., Ralph, J., Grabber, J. H. (1999) Cell wall structural foundations: molecular basis for improving forage digestibilities, *Crop Sci.* **39**, 27–37.

Helm, R. F. (2000) Lignin-polysaccharide interactions in woody plants, *ACS Symp. Ser.* **742**, 161–171.

Hepworth, D. G., Vincent, J. F. V. (1998) The mechanical properties of xylem tissue from tobacco plants (*Nicotiana tabacum* 'Samsun'), *Ann. Bot.* **81**,751–759.

Hibino, T., Takabe, K., Kawazu, T., Shibata, D., Higuchi, T. (1995) Increase of cinnamaldehyde groups in lignin of transgenic tobacco plants carrying an antisense gene for cinnamyl alcohol-dehydrogenase, *Biosci. Biotechnol. Biochem.* **59**, 929–931.

Higuchi, T., Ito, T., Umezawa, T., Hibino, T., Shibata, D. (1994) Red-brown color of lignified tissues of transgenic plants with antisense CAD gene – wine-red lignin from coniferyl aldehyde, *J. Biotechnol.* **37**, 151–158.

Htun, M., Salmen, L. (1996) The importance of understanding the physical and chemical properties of wood to achieve energy efficiency in mechanical pulping, *Woch. Papierfabrik.* **124**, 232–235.

Hu, W. J., Harding, S. A., Lung, J., Popko, J. L., Ralph, J., Stokke, D. D., Tsai, C. J., Chiang, V. L. (1999) Repression of lignin biosynthesis promotes cellulose accumulation and growth in transgenic trees, *Nature Biotechnol.* **17**, 808–812.

Humphreys, J. M., Hemm, M. R., Chapple, C. (1999) New routes for lignin biosynthesis defined by biochemical characterization of recombinant ferulate 5-hydroxylase, a multifunctional cytochrome P450-dependent monooxygenase, *Proc. Natl. Acad. Sci. USA*, **96**, 10045–10050.

Imison, J. (1822) *Elements of Science and Art, Vol. II.*, p. 131. London: A. & R. Spottiswoode.

Jones, J. D., Henstrand, J. M., Handa, A. K., Herrmann, K. M., Weller, S. C. (1995) Impaired wound induction of 3-deoxy-D-*arabino*-heptulosonate-7-phosphate (DAHP) synthase and altered stem development in transgenic potato plants expressing a DAHP synthase antisense construct, *Plant Physiol.* **108**, 1413–1421.

Kajita, S., Katayama, Y., Omori, S. (1996) Alterations in the biosynthesis of lignin in transgenic plants with chimeric genes for 4-coumarate:coenzyme A ligase, *Plant Cell Physiol.* **37**, 957–965.

Kajita, S., Hishiyama, S., Tomimura, Y., Katayama, Y., Omori, S. (1997a) Structural characterization of modified lignin in transgenic tobacco plants in which the activity of 4-coumarate:coenzyme A ligase is depressed, *Plant Physiol.* **114**, 871–879.

Kajita, S., Mashino, Y., Nishikubo, N., Katayama, Y., Omori, S. (1997b) Immunological characterization of transgenic tobacco plants with a chimeric gene for 4-coumarate:CoA ligase that have altered lignin in their xylem tissue, *Plant Sci.* **128**, 109–118.

Kennedy, R. W. (1995) Coniferous wood quality in the future – concerns and strategies, *Wood Sci. Technol.* **29**, 321–338.

Kim, H., Ralph, J., Yahiaoui, N., Pean, M., Boudet, A.-M. (2000) Cross-coupling of hydroxycinnamyl aldehydes into lignins, *Org. Lett.* **2**, 2197–2200.

Kosikova, B., Duris, M., Demianova, V. (2000) Conversion of lignin biopolymer into surface-active derivatives, *Eur. Polym. J.* **36**, 1209–1212.

Kuc, J., Nelson, O. (1964) The abnormal lignins produced by the *brown-midrib* mutants of maize. I. The *brown-midrib-1* mutant, *Arch. Biochem. Biophys.* **105**, 103–113.

LaFayette, P. R., Eriksson, K-E. L., Dean, J. F. D. (1999) Characterization and heterologous expression of laccase cDNAs from xylem tissues of yellow-poplar (*Liriodendron tulipifera*), *Plant Mol. Biol.* **40**, 23–35.

Lagrimini, L. M. (1991) Wound-induced deposition of polyphenols in transgenic plants overexpressing peroxidase, *Plant Physiol.* **96**, 577–583.

Lagrimini, L. M., Gingas, V., Finger, F., Rothstein, S., Liu, T. T. Y. (1997) Characterization of antisense transformed plants deficient in the tobacco anionic peroxidase, *Plant Physiol.* **114**, 1187–1196.

Lapierre, C., Pollet, B., Petit-Conil, M., Toval, G., Romero, J., Pilate, G., Leple, J. C., Boerjan, W., Ferret, V., De Nadai, V., Jouanin, L. (1999) Structural alterations of lignins in transgenic poplars with depressed cinnamyl alcohol dehydrogenase or caffeic acid *O*-methyltransferase activity have an opposite impact on the efficiency of industrial kraft pulping, *Plant Physiol.* **119**, 153–163.

Lapierre, C., Pollet, B., Petit-Conil, M., Pilate, G., Leple, C., Boerjan, W., Jouanin, L. (2000) Genetic engineering of poplar lignins: Impact of lignin alteration on kraft pulping performances, *ACS Symp. Ser.* **742**, 145–160.

Lechtenberg, V. L., Muller, L. D., Bauman, L. F., Rhykerd, C. L., Barnes, R. F. (1972) Laboratory and in vitro evaluation of inbred and F_2 populations of brown midrib mutants of *Zea mays* L, *Agron. J.* **64**, 657–660.

Lee, D., Meyer, K., Chapple, C., Douglas, C. J. (1997) Antisense suppression of 4-coumarate:coenzyme A ligase activity in *Arabidopsis* leads to altered lignin subunit composition, *Plant Cell* **9**, 1985–1998.

Lewis, N. G. (1999) A 20th century roller coaster ride: a short account of lignification, *Curr. Opin. Plant Biol.* **2**, 153–162.

Li, Y., Sarkanen, S. (2000) Thermoplastics with very high lignin contents, *ACS Symp. Ser.* **742**, 351–366.

Li, L. G., Popko, J. L., Umezawa, T., Chiang, V. L. (2000) 5-Hydroxyconiferyl aldehyde modulates enzymatic methylation for syringyl monolignol formation, a new view of monolignol biosynthesis in angiosperms, *J. Biol. Chem.* **275**, 6537–6545.

Liljegren, S. J., Ditta, G. S., Eshed, H. Y., Savidge, B., Bowman, J. L., Yanofsky, M. F. (2000) SHATTERPROOF MADS-box genes control seed dispersal in *Arabidopsis*, *Nature* **404**, 766–770.

MacKay, J. J., O'Malley, D. M., Presnell, T., Booker, F. L., Campbell, M. M., Whetten, R. W., Sederoff, R. R. (1997) Inheritance, gene expression, and lignin characterization in a mutant pine deficient in cinnamyl alcohol dehydrogenase, *Proc. Natl. Acad. Sci. USA* **94**, 8255–8260.

MacKay, J., Presnell, T., Jameel, H., Taneda, H., O'Malley, D., Sederoff, R. (1999) Modified lignin and delignification with a CAD-deficient loblolly pine, *Holzforschung* **53**, 403–410.

Maher, E. A., Bate, N. J., Ni, W. T., Elkind, Y., Dixon, R. A., Lamb, C. J. (1994) Increased disease susceptibility of transgenic tobacco plants with suppressed levels of preformed phenylpropanoid products, *Proc. Natl. Acad. Sci. USA* **91**, 7802–7806.

Marita, J. M., Ralph, J., Hatfield, R. D., Chapple, C. (1999) NMR characterization of lignins in *Arabidopsis* altered in the activity of ferulate 5-hydroxylase, *Proc. Natl. Acad. Sci. USA* **96**, 12328–12332.

Maury, S., Geoffroy, P., Legrand, M. (1999) Tobacco O-methyltransferases involved in phenylpropanoid metabolism. The different caffeoyl-coenzyme A/5-hydroxyferuloyl-coenzyme A 3/5-O-methyltransferase and caffeic acid/5-hydroxyferulic acid 3/5-O-methyltransferase classes have distinct substrate specificities and expression patterns, *Plant Physiol.* **121**, 215–223.

Merkle, S. A., Dean, J. F. D. (2000) Forest tree biotechnology, *Curr. Opin. Biotechnol.* **11**, 298–302.

Meyer, K., Cusumano, J. C., Somerville, C., Chapple, C. C. S. (1996) Ferulate-5-hydroxylase from *Arabidopsis thaliana* defines a new family of cytochrome P450-dependent monooxygenases, *Proc. Natl. Acad. Sci. USA* **93**, 6869–6874.

Meyer, K., Shirley, A. M., Cusumano, J. C., Bell-Lelong, D. A., Chapple, C. (1998) Lignin monomer composition is determined by the expression of a cytochrome P450-dependent monooxygenase in *Arabidopsis*, *Proc. Natl. Acad. Sci. USA* **95**, 6619–6623.

Meyermans, H., Morreel, K., Lapierre, C., Pollet, B., De Bruyn, A., Busson, R., Herdewijn, P., Devreese, B., Van Beeumen, J., Marita, J. M., Ralph, J., Chen, C., Burggraeve, B., Van Montague, M., Messens, E., Boerjan, W. (2000) Modifications in lignin and accumulation of phenolic glucosides in poplar xylem upon down-regulation of caffeoyl-coenzyme A *O*-methyltransferase, an enzyme involved in lignin biosynthesis, *J. Biol. Chem.* (in press).

Miksche, G. E., Yasuda, S. (1978) Lignin of 'giant' mosses and some related species, *Phytochemistry* **17**, 503–504.

Monties, B. (1998) Novel structures and properties of lignins in relation to their natural and induced variability in ecotypes, mutants and transgenic plants, *Polym. Degrad. Stabil.* **59**, 53–64.

Mullin, T. J., Bertrand, S. (1998) Environmental release of transgenic trees in Canada – potential benefits and assessment of biosafety, *Forest. Chron.* **74**, 203–219.

Ni, W. T., Paiva, N. L., Dixon, R. A. (1994) Reduced lignin in transgenic plants containing a caffeic acid O-methyltransferase antisense gene, *Trans. Res.* **3**, 120–126.

Nilsson, L. J., Larson, E. D., Gilbreath, K., Gupta, A. (1995) Energy Efficiency and the Pulp and Paper Industry, *ACEEE Summer Study on Energy Efficiency in Industry: 'Partnership, Productivity, and Environment'*, August 1–4, Grand Island, NY (http://www.aceee.org/pubs/ie962.htm)

Osakabe, K., Tsao, C. C., Li, L. G., Popko, J. L., Umezawa, T., Carraway, D. T., Smeltzer, R. H., Joshi, C. P., Chiang, V. L. (1999) Coniferyl aldehyde 5-hydroxylation and methylation direct syringyl lignin biosynthesis in angiosperms, *Proc. Natl. Acad. Sci. USA* **96**, 8955–8960.

Payen, A. (1838) Study of the composition of the natural tissue of plants and of lignin, *Comptes Rendues* **7**, 1052.

Piquemal, J., Lapierre, C., Myton, K., O'Connell, A., Schuch, W., Grima-Pettenati, J., Boudet, A. M. (1998) Down-regulation of cinnamoyl-CoA reductase induces significant changes of lignin profiles in transgenic tobacco plants, *Plant J.* **13**, 71–83.

Provan, G. J., Scobbie, L., Chesson, A. (1997) Characterisation of lignin from CAD and OMT deficient Bm mutants of maize, *J. Sci. Food Agric.* **73**, 133–142.

Ralph, J., Lu, F. C. (1998) The DFRC method for lignin analysis. 6. A simple modification for identifying natural acetates on lignins, *J. Agric. Food Chem.* **46**, 4616–4619.

Ralph, J., MacKay, J. J., Hatfield, R. D., O'Malley, D. M., Whetten, R. W., Sederoff, R. R. (1997) Abnormal lignin in a loblolly pine mutant, *Science* **277**, 235–239.
Ralph, J., Hatfield, R. D., Piquemal, J., Yahiaoui, N., Pean, M., Lapierre, C., Boudet, A. M. (1998) NMR characterization of altered lignins extracted from tobacco plants down-regulated for lignification enzymes cinnamyl-alcohol dehydrogenase and cinnamoyl-CoA reductase, *Proc. Natl. Acad. Sci. USA* **95**, 12803–12808.
Ralph, J., Hatfield, R. D., Marita, J. M., Lu, F., Peng, J., Kim, H., Grabber, J. H., MacKay, J. J, O'Malley, D. M., Sederoff, R. R., Chapple, C., Chiang, V., Boudet, A. M. (1999a) Lignin structure in lignin-biosynthetic-pathway mutants and transgenics; new opportunities for engineering lignin?, *10th International Symposium on Wood and Pulping Chemistry*, June 7–10, Yokohama, Japan (*http://www.dfrc.ars.usda.gov/FullTextPDFs/ISWPC99-Mutants.pdf*)
Ralph, J., Kim, H., Peng, J., Lu, F. (1999b) Arylpropane-1,3-diols in lignins from normal and CAD-deficient pines, *Org. Lett.* **1**, 323–326
Ralph, J., Lapierre, C., Lu, F., Marita, J. M., Pilate, G., Van Doorsselaere, J., Boerjan, W., Jouanin, L. (2000) NMR evidence for benzodioxane structures resulting from incorporation of 5-hydroxyconiferyl alcohol into lignins of O-methyl-transferase-deficient poplars, *J. Agric. Food Chem.* (in press).
Rasmussen, S., Dixon, R. A. (1999) Transgene-mediated and elicitor-induced perturbation of metabolic channeling at the entry point into the phenylpropanoid pathway, *Plant Cell* **11**, 1537–1551.
Ray, H., Douches, D. S., Hammerschmidt, R. (1998) Transformation of potato with cucumber peroxidase: expression and disease response, *Physiol. Mol. Plant Pathol.* **53**, 93–103.
Savidge, R. A., Forster, H. (1998) Seasonal activity of uridine 5′-diphosphoglucose: coniferyl alcohol glucosyltransferase in relation to cambial growth and dormancy in conifers, *Can. J. Bot.* **76**, 486–493.
Schulze, F. (1856) Zur kenntnis des lignins und seines vorkommens im planzenkörper. Festschrift zur 400sten jubelfeier der Universität Greifswald.
Sederoff, R. R., MacKay, J. J., Ralph, J., Hatfield, R. D. (1999) Unexpected variation in lignin, *Curr. Opin. Plant Biol.* **2**, 145–152.
Sewalt, V. J. H., Ni, W. T., Jung, H. G., Dixon, R. A. (1997a) Lignin impact on fiber degradation: Increased enzymatic digestibility of genetically engineered tobacco (*Nicotiana tabacum*) stems reduced in lignin content, *J. Agric. Food Chem.* **45**, 1977–1983.
Sewalt, V. J. H., Ni, W. T., Blount, J. W., Jung, H. G., Masoud, S. A., Howles, P. A., Lamb, C., Dixon, R. A. (1997b) Reduced lignin content and altered lignin composition in transgenic tobacco down-regulated in expression of L-phenylalanine ammonia-lyase or cinnamate 4-hydroxylase, *Plant Physiol.* **115**, 41–50.
Siegel, S. M. (1969) Evidence for the presence of lignin in moss gametophytes, *Am. J. Bot.* **56**, 175–179.
Sitbon, F., Hennion, S., Sundberg, B., Little, C. H. A., Olsson, O., Sandberg, G. (1992) Transgenic tobacco plants coexpressing the *Agrobacterium tumefaciens iaaM* and *iaaH* genes display altered growth and indole acetic acid metabolism, *Plant Physiol.* **99**, 1062–1069.
Sitbon, F., Hennion, S., Little, C. H. A., Sundberg, B. (1999) Enhanced ethylene production and peroxidase activity in IAA-overproducing transgenic tobacco plants is associated with increased lignin content and altered lignin composition, *Plant Sci.* **141**, 165–173.
Smeder, B., Liljedahl, S. (1996) Market oriented identification of important properties in developing flax fibres for technical uses, *Ind. Crop Prod.* **5**, 149–162.
Stafford, H. A. (1981) Compartmentation in natural product biosynthesis by multienzyme complexes, in: *The Biochemistry of Plants, Vol. 7* (Conn, E. E., Ed.), pp. 117–137. New York: Academic Press.
Stewart, D., Yahiaoui, N., McDougall, G. J., Myton, K., Marque, C., Boudet, A. M., Haigh, J. (1997) Fourier-transform infrared and Raman spectroscopic evidence for the incorporation of cinnamaldehydes into the lignin of transgenic tobacco (*Nicotiana tabacum* L) plants with reduced expression of cinnamyl alcohol dehydrogenase, *Planta* **201**, 311–318.
Suzuki, S., Lam, T. B. T., Iiyama, K. (1997) 5-Hydroxyguaiacyl nuclei as aromatic constituents of native lignin, *Phytochemistry* **46**, 695–700.
Tamagnone, L., Merida, A., Parr, A., Mackay, S., Culianez-Macia, F. A., Roberts, K., Martin, C. (1998) The AmMYB308 and AmMYB330 transcription factors from *Antirrhinum* regulate phenylpropanoid and lignin biosynthesis in transgenic tobacco, *Plant Cell* **10**, 135–154.
Thring, R. W., Katikaneni, S. P. R., Bakhshi, N. N. (2000) The production of gasoline range hydrocarbons from Alcell (R) lignin using HZSM-5 catalyst, *Fuel Proc. Technol.* **62**, 17–30.

Tjeerdsma, B. F., Boonstra, M., Pizzi, A., Tekely, P., Militz, H. (1998) Characterisation of thermally modified wood: molecular reasons for wood performance improvement, *Holz Als Roh-Und Werkstoff* **56**, 149–153.
Tsai, C. J., Popko, J. L., Mielke, M. R., Hu, W. J., Podila, G. K., Chiang, V. L. (1998) Suppression of O-methyltransferase gene by homologous sense transgene in quaking aspen causes red-brown wood phenotypes, *Plant Physiol.* **117**, 101–112.
Udagamarandeniya, P. V., Savidge, R. A. (1995) Coniferyl alcohol oxidase – a catechol oxidase, *Trees-Struct. Funct.* **10**, 102–107.
Vailhe, M. A. B., Cornu, A., Robert, D., Maillot, M. P., Besle, J. M. (1996a) Cell wall degradability of transgenic tobacco stems in relation to their chemical extraction and lignin quality, *J. Agric. Food Chem.* **44**, 1164–1169.
Vailhe, M. A. B., Migne, C., Cornu, A., Maillot, M. P., Grenet, E., Besle, J. M., Atanassova, R., Martz, F., Legrand, M. (1996b) Effect of modification of the *O*-methyltransferase activity on cell wall composition, ultrastructure and degradability of transgenic tobacco, *J. Sci. Food Agric.* **72**, 385–391.
Vailhe, M. A. B., Besle, J. M., Maillot, M. P., Cornu, A., Halpin, C., Knight, M. (1998) Effect of down-regulation of cinnamyl alcohol dehydrogenase on cell wall composition and on degradability of tobacco stems, *J. Sci. Food Agric.* **76**, 505–514.
Van Doorsselaere, J., Baucher, M., Chognot, E., Chabbert, B., Tollier, M. T., Petit-Conil, M., Leple, J. C., Pilate, G., Cornu, D., Monties, B., Van-Montagu, M., Inze, D., Boerjan, W., Jouanin, L. (1995) A novel lignin in poplar trees with a reduced caffeic acid 5-hydroxyferulic acid *O*-methyltransferase activity, *Plant J.* **8**, 855–864.
Vignols, F., Rigau, J., Torres, M. A., Capellades, M., Puigdomenech, P. (1995) The brown midrib3 (*bm3*) mutation in maize occurs in the gene encoding caffeic acid O-methyltransferase, *Plant Cell* **7**, 407–416.
Vinzant, T. B., Ehrman, C. I., Adney, W. S., Thomas, S. R., Himmel, M. E. (1997) Simultaneous saccharification and fermentation of pretreated hardwoods – Effect of native lignin content, *Appl. Biochem. Biotechnol.* **62**, 99–104.
Wang, Z. X., Li, S. M., Loscher, R., Heide, L. (1997) 4-Coumaroyl coenzyme A 3-hydroxylase activity from cell cultures of *Lithospermum erythrorhizon* and its relationship to polyphenol oxidase, *Arch. Biochem. Biophys.* **347**, 249–255.
Weaver, L. M., Herrmann, K. M. (1997) Dynamics of the shikimate pathway in plants, *Trends Plant Sci.* **2**, 346–351.
Whetten, R., Sederoff, R. (1991) Genetic engineering of wood, *Forest Ecol. Manag.* **43**, 301–316.
Winkel-Shirley, B. (1999) Evidence for enzyme complexes in the phenylpropanoid and flavonoid pathways, *Physiol. Plant.* **107**, 142–149.
Wu, G. S., Shortt, B. J., Lawrence, E. B., Leon, J., Fitzsimmons, K. C., Levine, E. B., Raskin, I., Shah, D. M. (1997) Activation of host defense mechanisms by elevated production of H_2O_2 in transgenic plants, *Plant Physiol.* **115**, 427–435.
Yahiaoui, N., Marque, C., Myton, K. E., Negrel, J., Boudet, A. M. (1998) Impact of different levels of cinnamyl alcohol dehydrogenase down-regulation on lignins of transgenic tobacco plants, *Planta* **204**, 8–15.
Yao, K. N., Deluca, V., Brisson, N. (1995) Creation of a metabolic sink for tryptophan alters the phenylpropanoid pathway and the susceptibility of potato to *Phytophthora infestans*, *Plant Cell* **7**, 1787–1799.
Ye, Z. H. (1997) Association of caffeoyl coenzyme A 3-*O*-methyltransferase expression with lignifying tissues in several dicot plants, *Plant Physiol.* **115**, 1341–1350.
Ye, Z. H., Varner, J. E. (1995) Differential expression of 2-*O*-methyltransferases in lignin biosynthesis in *Zinnia elegans*, *Plant Physiol.* **108**, 459–467.
Zhong, R. Q., Morrison, W. H., Negrel, J., Ye, Z. H. (1998) Dual methylation pathways in lignin biosynthesis, *Plant Cell*, **10**, 2033–2045.
Zhong, R. Q., Ripperger, A., Ye, Z. H. (2000) Ectopic deposition of lignin in the pith of stems of two *Arabidopsis* mutants, *Plant Physiol.* **123**, 59–69.

2
Biotechnological Applications of Lignin-Degrading Fungi (White-Rot Fungi)

Prof. Dr. Gary M. Scott[1], Dr. Masood Akhtar[2]

[1] Faculty of Paper Science and Engineering, State University of New York, College of Environmental Science and Forestry, One Forestry Drive, Syracuse, NY 13210, USA, Phone: + 1-3154706523, Fax: + 13154704745, Email: gscott@esf.edu

[2] Biopulping International, Inc., P.O. Box 5463, Madison, WI 53705, USA, Phone: + 1-6082319484, Fax: + 1-6082319543, Email: makhtar@facstaff.wisc.edu

BOD biological oxygen demand
COD chemical oxygen demand

Biotechnology of Biopolymers. From Synthesis to Patents. Edited by A. Steinbüchel and Y. Doi

ISBN: 3-527-31110-6

CSF	canadian standard freeness
FPL	Forest Products Laboratory
MyCoR	mycelial color removal
RBC	rotating biological contactor
Tappi	Technical Assocation of the Pulp and Paper Industry
TMP	thermomechanical pulp
USDA	United States Department of Agriculture
VOC	volatile organic compound
vvm	volume/volume/min

1
Introduction

A primary user of the wood resources in the world is the pulp and paper industry. This industry is a growing portion of the world's economy, especially in developing nations. However, the industry is also very capital intensive, which tends to slow the incorporation of new technologies into the mills. In addition, the paper industry is under constant environmental pressure in regards to their air and water. To meet these increasing production and environmental demands, the industry often looks towards incremental improvements in their existing technology. However, it may be necessary to look to more revolutionary technologies to meet the needs of the industry.

Currently, the pulp and paper industry uses two main pulping technologies: mechanical and chemical. Mechanical pulping processes are electrical energy intensive, resulting in paper with poor brightness stability and poor strength. However, they have the advantage of being high yield and producing pulp with some desirable optical properties. On the other hand, chemical pulping results in stronger paper as a result of the chemical dissolution of the lignin from the wood. The primary chemical process, accounting for approximately 75% of the pulp production worldwide, is the kraft process, which uses a mixture of sodium hydroxide and sodium sulfide to remove the lignin. Because of this dissolution of the lignin, yields can easily be under 50% for this process and the pulp requires extensive bleaching. In addition, the use of sulfur in the pulping process causes some environmental concerns. Another chemical pulping process, the sulfite process, also uses sulfur compounds as the pulping chemicals. While sulfite pulp generally requires less bleaching chemicals than kraft, the process also has some environmental concerns in relation to the disposal of the spent chemicals.

Both of the chemical processes tend not to be very specific in their removal of lignin from the wood. A significant amount of cellulose degradation also takes place. Many microorganisms are involved in the decomposition of wood. However, the organisms that most effectively biodegrade lignin are basidiomycete fungi. Since wood is a complex biological material, a variety of mechanisms are used in the degradation of the structure. One class of wood-decay fungi are the lignin-degrading species called white-rot fungi, because they typically turn wood white as they decay it. The bright white color is the remaining undegraded cellulose after the lignin has been removed. There is potential for these fungi to be used in the pulping, bleaching, and effluent-treatment processes to reduce the environmental impact of the industry.

This chapter briefly reviews the history of the use of these fungi in various processes related to the pulp and paper industry. The remainder of the chapter discusses the current state of the research in several important areas using the white-rot fungi.

Biopulping, the treatment of wood chips prior to pulping, is probably the most studied area. Other areas studied include biobleaching, the use of these fungi or their enzymes to brighten pulp, and effluent treatment. The chapter concludes with an outlook of the future of this environmentally-benign technology for the pulp and paper industry and the engineering, technological, and biological challenges that need to be addressed to bring this technology to the industry.

2 Historical Outline

Lignin-degrading fungi have been used for several applications within the paper industry. Biopulping is probably the most studied application in which the fungi are used as a pretreatment to the pulping of wood or as a pulping method itself. These fungi have also been used for the treatment of pulp for both pulping and bleaching applications with some success. Other applications of these fungi and the enzymes that they produce include bleaching, fiber property restoration, pitch reduction, and effluent treatment. The history of the past work of these applications are detailed below.

2.1 Biopulping

The use of white-rot fungi for the biological delignification of wood was first seriously considered by Lawson and Still (1957) at the West Virginia Pulp and Paper Company research laboratory (now Westvaco Corporation). Subsequent work showed that paper strength properties increased with the extent of natural degradation of pine by white-rot fungi (Kawase, 1962; Reis and Libby, 1960). In 1972, Kirk and Kringstad showed that pretreatment of aspen wood chips with *Rigidoporus ulmarius* reduced the electrical energy requirements during refiner mechanical pulping and produced stronger paper than did control chips (Forest Products Laboratory Internal Report). Related work was done at a Swedish research laboratory (STFI) in Stockholm, and the first published report on biopulping demonstrated that fungal treatment could result in significant energy savings for mechanical pulping (Ander and Eriksson, 1975; Eriksson et al., 1976). Considerable efforts by the Swedish group were directed toward developing cellulase-less mutants of selected white-rot fungi to improve the selectivity of lignin degradation and, thus, the specficity of biopulping (Johnsrud and Eriksson, 1985; Eriksson et al., 1983; Erikkson, 1990). Samuelsson and coworkers (1980) applied the white-rot fungus *Phlebia radiata* and its cellulase-less mutant to chips and pulp. Eriksson and Vallander (1982) treated wood chips (with or without added glucose) with white-rot fungi, in most cases a cellulase-less mutant of *Phanerochaete chrysosporium*. The energy required for defibration and refining decreased somewhat with increasing treatment time.

Bar-Lev and coworkers (1982) showed that treatment of coarse thermomechanical pulp with the white-rot fungus *P. chrysosporium* prior to secondary refining reduced the energy requirement by 25 to 30% and increased paper strength properties. Similarly, Pilon and coworkers (1982) reported an increase in paper strength properties when refiner mechanical pulp was treated with several fungi. Subsequently, Pellinen et al. (1989) subjected chemithermomechanical pulp to the white-rot fungus *P. chrysosporium* for two weeks and reported improved paper strength properties and lower energy requirements in a PFI mill. Many other researchers have also reported significant increases in paper strength properties or energy savings in refining with fungal treatments of wood

(Akamatsu et al., 1984; Nishibe et al. 1988, 1992; Setliff et al. 1990; Kashino et al., 1991, 1993; Patel et al., 1994).

Pearce and coworkers (1995) screened 204 isolates of wild-type wood decay fungi for biomechanical pulping of eucalyptus chips. Some of these fungi saved 40 to 50% electrical energy in refining and resulted in greater brightness of unbleached pulp as compared with that of pulps from untreated control chips. Some of the strains were found to be effective on unsterilized wood chips. In another study, Schmidt and coworkers (1996) reported a novel method for growing selected white-rot fungi on unsterilized wood chips. Chips in compressed bales inoculated with *P. chrysosporium* showed reduced energy requirements for mechanical pulping. Other details on biopulping research have been described in review articles and the literature cited therein (Akhtar et al., 1997a; Eriksson and Kirk, 1985; Kirk et al., 1992, 1993, 1994; Messner and Srebotnik, 1994). Tables 1 and 2 summarize the various species of fungus used in biopulping research for mechanical pulping and chemical pulping, respectively. Many white-rot fungi also reduce the pitch and extractive content in wood. These are summarized in Table 3.

2.2 Biobleaching

Lignin is the primary component of wood that imparts the light brown color to wood. Cellulose, in its pure form, is a bright white compound. White-rot fungi are so named because, at advanced stages of decay, the wood is nearly pure cellulose and thus results in the remaining wood being white. This effect gave early researchers the idea that this class of fungi would be effective at bleaching.

As with biopulping, early work concentrated in the white-rot fungus, *P. chrysosporium*, which still is the most studied wood-decay fungus. Under optimized conditions, this fungus greatly delignified the pulp, thus reducing the amount of bleach chemicals necessary in subsequent process steps.

Researcher found other fungi that were also effective as a pretreatment to the bleaching of pulp. For example, *Trametes (Coriolus) versicolor* significantly decreased the lignin content (as measured by kappa number, a measure of the relative lignin content of pulp) and brightened the pulp as determined by several researchers (Paice et al., 1989; Reid et al., 1990; Reid and Paice, 1994b; Ho et al., 1990; Reid et al., 1990). Unfortunately, in many cases, the fungal attack is not selective to just the lignin component of wood. In many cases, the cellulose is also depolymerized by the enzymes secreted by the fungi. Extensive screening programs have been done by several researchers to find effective fungi for bleaching. In several cases, other fungi have been identified as effective for bleaching. A screening of 2068 samples based on phenol oxidase activity isolated strain IZU-154 as a lignin degrader (Nishida et al., 1988). Another screening of 1212 samples discovered that *P. sordida* reduces the bleaching chemicals needed in subsequent bleaching steps (Hirai et al., 1994). A final extensive screening of 1758 cultures isolated strain SKB-1152 which brightened pulps considerably (Iimori et al., 1994). Table 4 summarizes the various species of white-rot fungi used for biobleaching.

2.3 Effluent Treatment

White-rot fungi were recognized early as being useful for the treatment of bleaching effluents. The earliest work for this application was done by Fukuzumi and others (1977). In their work, the fungi were grown in shaking flasks containing nutrients and the spent liquor. Decolorization of the spent

Tab. 1 Lignin-degrading fungi applied to mechanical pulping

Fungus species	***Reference***
Antrodiella sp. RK1	Patel et al (1994)
Ceriporiopsis subvermispora	Setliff et al. (1990); Akhtar et al. (1997a), Akhtar et al. (1998)
Coprinus cinereus	Nishibe et al. (1988)
Trametes (Coriolus) hirsuta	Akamatsu et al. (1984)
Dichomitus squalens	Akhtar et al. (1998)
Hyphodontia setulosa	Akhtar et al. (1998)
Isolate IZU-154	Kashino et al. (1991)
Perenniporia medulla-panis	Akhtar et al. (1998)
Phanerochaete chrysosporium	Schmidt et al. (1996); Kirk et al. (1993); Eriksson and Vallander (1982); Johnsrud et al. (1987); Bar-Lev et al. (1982); Abuhasan et al. (1988); Nishibe et al. (1988); Setliff et al. (1990)
Phellinus pini	Akhtar et al. (1998)
Phlebia brevispora	Akhtar et al. (1998)
Phlebia radiata	Samuelsson et al. (1980)
Phlebia subserialis	Akhtar et al. (1998)
Phlebia tremellosa	Akhtar et al. (1998)
Rigidoporus ulmarius	Kirk and Kringstad (1972)
Trametes coccinea	Akamatsu et al. (1984)
Trametes sanguinea	Akamatsu et al. (1984)
Trametes versicolor	Pilon et al. (1982)

Tab. 2 Lignin-degrading fungi applied to chemical pulping

Fungus species	***Reference***
Ceriporiopsis subvermispora	Messner et al. (1998); Messner et al. (1992); Messner and Srebotnik (1994); Scott et al. (1995a,b); Bosshard and Holzkunde (1984)
Trametes (Coriolus) versicolor	Wolfaardt et al. (1996)
Dichomitus squalens	Messner et al. (1992)
Phanerochaete chrysosporium	Messner et al. (1992); Oriaran et al. (1989, 1990, 1991)
Phellinus pini	Adamski et al. (1987)
Phlebia brevispora	Messner et al. (1992)
Phlebia tremellosa	Messner et al. (1992)
Pycnoporus sanguineus	Wolfaardt et al. (1996)

liquor was the measurement made. Many other researchers showed results for the decolorization of effluent in the subsequent years with several screening studies being done and many different fungi. Table 5 summarizes the fungi used for bleach effluent decolorization. As with many of the other applications, *P. chrysosporium* is probably the most studied fungus for this application.

3 Biopulping

The pulp and paper industry utilizes mechanical, chemical, or a combination of the two pulping methods to produce pulp of desired characteristics. Mechanical pulping involves the use of mechanical force to separate the wood fibers. The pulps produced

Tab. 3 Fungus species used for pitch reduction

Fungus species	*Reference*
Trichoderma lignorum	Nilsson and Asserson (1966)
Lecythophora sp.	Chen et al. (1995b); Gao (1996)
Gliocladium roseum	Nilsson and Asserson (1966)
Gliocladium deliquescens	Nilsson and Asserson (1966)
Gliocladium viride	Nilsson and Asserson (1966)
Rhizopus arrhizus	Nilsson and Asserson (1966)
Ophiostoma piliferum	Breuil et al. (1998)
Ceriporiopsis subvermispora	Breuil et al. (1998); Fischer et al (1994)
Trametes versicolor	Breuil et al. (1998)
Phanerochaete chrysosporium	Breuil et al. (1998)
Pleurotus ostreatus	Breuil et al. (1998)
Trichaptum biforme	Breuil et al. (1998)
Schizophyllum commune	Breuil et al. (1998)
Phlebia tremellosa	Breuil et al. (1998)
Ophiostoma paceae	Chen et al. (1995b); Gao (1996)
Ophiostoma ainoae	Chen et al. (1995b); Gao (1996)

Tab. 4 Lignin-degrading fungi for bleaching

Fungus species	*Reference*
Ganoderma sp.	Hirai et al. (1994)
Isolate IZU-154	Nishida et al. (1988); Fujita et al. (1991)
Phanerochaete sordida	Kondo et al. (1994b); Harai et al. (1994)
Isolate SKB-1152	Iimori et al. (1994)
Trametes versicolor	Paice et al. (1989); Reid et al. (1990); Reid and Paice (1994b); Addleman et al. (1995)

are high yield (90–95%) and produce paper with high bulk, good opacity, and excellent printability. However, these processes are energy intensive and produce paper with lower strength and high color reversion as compared to the chemical processes. Chemical pulping involves the use of chemicals to degrade and dissolve lignin from the wood, releasing the cellulose fibers. Chemcial pulping processes yield pulps with high strength; however, these processes are low yield (40–50%), expensive, and capital intensive. The predominant chemical process worldwide is the kraft process, using sodium hydroxide and sodium sulfide as the active chemicals. The sulfite process, which is more common in Europe, uses sodium, magnesium, or calcium sulfite and sulfurous acid as the active chemicals. Many papers are blends of these two types of pulps to produce sheets with the proper balance of strength and optical properties.

Biopulping appears to have the potential to overcome some of the problems associated with conventional mechanical and chemical pulping methods. Biopulping is an environmentally-friendly technology that substantially increases mill throughput or reduces electrical energy consumptions at the same throughput for mechanical pulping. By producing stronger pulp with longer fibers and increased fibrillation, biomechanical pulping reduces the amount of kraft pulp required to achieve a given paper strength.

Tab. 5 Lignin-degrading fungi applied to bleaching effluents

Fungus species	Reference
Ganoderma lacidum	Wang et al. (1992)
Lentinus edodes	Lee et al. (1994); Esposito et al. (1991)
Phanerochaete chrysosporium	Livernoche et al. (1983); Eaton et al. (1980, 1982); Mittar et al. (1992); Sundman et al. (1981); Joyce and Pellinen (1990); Messner et al. (1990); Kang et al. (1996)
Phlebia brevispora	Eaton et al. (1982)
Phlebia subserialis	Eaton et al. (1982)
Pleurotus ostreatus	Livernoche et al. (1983)
Polyporus sanguineous	Lambais (1988)
Trametes (Polyporus) versicolor	Livernoche et al. (1983);Lambais (1988)
Poria cinerascens	Eaton et al. (1982)
Ramaria sp.	Galeno and Agosin (1990)
Tinctoporia sp.	Fukuzumi et al. (1977)

3.1 Mechanical Pulping

Fungal pretreatment for mechanical pulping has certain benefits including reduced refining energy, stronger fibers, and removal of pitch. A comprehensive evaluation of biomechanical pulping began in 1987 at the USDA Forest Service, Forest Products Laboratory (FPL) under the Biopulping Consortium. The Consortium involved the FPL, the Universities of Wisconsin and Minnesota, and up to 22 pulp, paper, and related companies. The goal of the Consortium was to evaluate the technical feasibility of fungal pretreatment for mechanical pulping. From this respect, the Consortium was a success. The rest of this subsection describes the research of the Consortium and the engineering work done after the end of the consortium.

Biopulping is a simple concept, but harnessing lignin-degrading fungi in a commercially attractive process is not simple. Many biological variables were examined and optimized; major ones included fungus species and strains, inoculum form and amount, wood species, chip pretreatments, inoculum form and size, wood species, chip pretreatments, incubation time, ventilation, and nutrients. Each variable was examined using an assay procedure of treating chips, making pulp in the refiner, measuring the refining energy, and determining the resulting pulp and paper properties. Three biological variables were found to be of paramount importance for successful implementation of biopulping: fungus selection, chip surface decontamination, and inoculum.

Laboratory procedures. The laboratory treatments employed simple and inexpensive aerated static-bed bioreactors (Leatham, 1983). Chips (1500 g dry weight basis) were introduced into each 21-liter reactor and sterilized by steaming. Chip moisture content was adjusted to 55 to 60% on a wet weight basis, the optimum range for the growth of the fungus. The chips were then inoculated with the fungus and incubated at the appropriate temperature (27 °C for *Ceriporiopsis subvermispora*) for two weeks while being

Tab. 6 Results of lignin and wood sugar analysis of pine wood decayed for 12 weeks by various white-rot fungi (Akhtar et al., 1998).

			Loss (%)		
Fungus	***Biomass***	***Lignin***	***Glucose (Glucan)***	***Xylose (Xylan)***	***Mannose (Mannan)***
C. subvermispora	25	50	3	48	13
P. chrysosporium	20	31	4	44	0
P. brevispora	27	49	11	60	16
P. pini	27	45	8	55	40

Composition of sound *Pinus strobus*: 33% lignin, 41% glucose, 8% xylose, and 11% mannose.

aerated continuously with humidified air at a rate of 0.0227 vvm (Akhtar et al., 1992a,b, 1993; Kirk et al., 1993; Myers et al., 1988).

After harvest, the chips were fiberized by multiple passes through a Sprout–Waldron Model 105-A 300 mm-diameter single rotating disk atmospheric refiner and the energy consumption measured. Pulps were refined to just above and below 100 ml of Canadian Standard Freeness (CSF), which is an industry standard measurement of the drainage rate of pulp (Tappi, 1999). Handsheets, a standard laboratory papermaking protocol, were made with both control and fungus-treated pulp and tested for physical and optical properties using standard methods (Tappi, 1999). Energy values and physical properties were regressed to 100 ml CSF for comparison (Akhtar et al., 1992a,b, Akhtar 1994; Myers et al., 1988).

Fungus selection. Selection of the fungus species concentrated on the lignin-degrading white-rot fungi. Since lignin is essentially the "glue" that holds the fibers together, these fungi should reduce the electrical energy requirements significantly in the mechanical pulping processes. Many white-rot fungi can remove significant amounts of lignin while leaving the cellulose virtually untouched as can be seen in Table 6. As the table shows, *C. subvermispora* is very selective towards lignin and hemicellulose, only degrading 3% of the cellulose (glucose) over a twelve-week period. Of course, twelve-week incubations are not practical from an industrial standpoint and commercial applications of these fungi to delignify to this extent will not be realized. However, measurements of the energy savings by some of these fungi indicate that the delignification achieved during shorter incubation times is sufficient for industrial applications as shown in Table 7. For example, the biomass loss during a two-week incubation with *C. subvermispora* is only 2–5%. Many of these fungi sufficiently weaken the structure of the wood to cause significant energy savings in mechanical pulping (Akhtar et al., 1998).

There are other aspects that also need to be considered when selecting a fungus for biomechanical pulping besides the energy savings. The quality of the resulting pulp is another such factor. Table 7 also shows the resulting tear strength improvement due to the fungal pretreatment. As can be seen, most of these fungi improve the tear strength of the paper by 20 to 50% for a two-week treatment. The other strength properties also show a similar trend to that seen for tear strength. Another factor is the robustness of the fungus to both the species of wood and the competition from other microorganisms on the wood. Being able to outcompete the other organisms improves the chances of survival and growth of the biopulping fungus when it is applied to the wood. *C. subvermispora* has

Tab. 7 Energy savings and tear strength improvement for biomechanical pulping of various species of wood by various white-rot fungi (Akhtar et al., 1998).

Fungus	*Wood*	*Incubation Time (weeks)*	*Energy Savings (%)*	*Tear Index Improvement (%)*
C. subvermispora	Aspen	2	38	48
C. subvermispora	Aspen	4	40	131
C. subvermispora	Loblolly pine	2	34	46
C. subvermispora	Loblolly pine	4	42	67
C. subvermispora	Mixed softwoods	2	30	29
C. subvermispora	Mixed softwoods	2	37	48
C. subvermispora	Spruce	1	14	28
C. subvermispora	Spruce	2	24	52
D. squalens	Loblolly pine	2	18	41
H. setulosa	Aspen	2	36	16
P. brevispora	Aspen	2	38	19
P. brevispora	Loblolly pine	2	16	13
P. medulla-panis	Loblolly pine	2	19	34
P. subserialis	Aspen	2	40	0
P. subserialis	Loblolly pine	2	33	44
P. tremellosa	Aspen	2	27	24
P. tremellosa	Loblolly pine	2	21	–

been found to be very effective on a variety of both hardwood and softwood species. This fungus is also a fairly selective lignin degrader and fairly aggressive in terms of competition (Akhtar et al., 1998).

Phlebia subserialis has also been identified that gives results comparable to those obtained with *C. subvermispora*, while addressing some of its shortcomings. Of importance is that *P. subserialis* has a broader temperature range than does *C. subvermispora*. For example, Table 8 gives the energy savings as a function of the incubation temperature for these two fungi. The effective temperature range for *P. subserialis* extends several degrees higher than for *C. subvermispora*. In addition, *P. subserialis* produces less aerial hyphae which also has engineering ramifications as well as producing chips that are not as soft and compressible. These advantages over *C. subvermispora* should reduce the cost of biotreatment significantly during scale-up through a decreased need for temperature control and greater ease of pile ventilation (Scott et al., 1998).

Chip surface decontamination. In early studies, sterilization of the wood chips prior to fungal inoculum was carried out by autoclaving the wood chips. This step was necessary because wood chip surfaces are normally contaminated with cells and spores of many other fungi and bacteria which can hamper biopulping. While sterilization is possible on a laboratory scale, it is increasingly difficult at larger scales and impossible at the commercial scale.

Chemical decontamination is one possible solution, but in most cases they are expensive and also reduce the growth of the desired microorganism. There are, however, some relatively inexpensive chemicals that not only decontaminate wood chips but also stimulate the growth of the biopulping fungus. Of the various chemical tested, sodium bisulfite, sodium meta-bisulfite, and sodium hydrosulfite were found to be effective (Leatham, 1983; Akhtar et al., 1995).

Although bisulfite is effective, it was subsequently found that a simple steaming is

Tab. 8 Energy savings as a function of the incubation temperature for C. *subvermispora* and *P. subserialis* (Akhtar et al., 1998).

Fungus	*Energy Savings over Control (%)*				
	22 °C	*27 °C*	*32 °C*	*35 °C*	*39 °C*
C. subvermispora	18	32	33	0	0
P. subserialis	17	33	34	18	12

equally so. Although this is not a sterilization of the wood chips as in autoclaving, the decontamination of the surfaces of the wood chips is sufficient to allow the biopulping fungus to establish itself and outcompete the other organisms. However, the length of the steaming necessary to allow good growth of the fungus is also a function of the exposure of the chips` surfaces to the atmospheric steam. Figure 1 shows this relationship. In the laboratory bioreactors, 10 minutes of steaming is needed; however, the chips are quiescent and undisturbed during the steaming. At the other extreme, when the chips are fluidized by the steam and there is excellent exposure of the surface of the chips to the steam, the decontamination time is greatly reduced to 15 seconds. Intermediate to those two extremes is the screw conveyor which impart some tumbling action to the chips. With this method, which is the most industrially feasible method, about 1 minute of steaming is required.

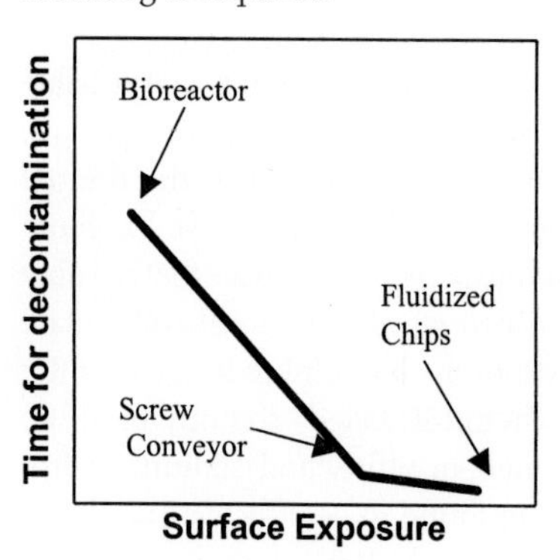

Fig. 1 Relationship between amount of surface exposure and time needed for decontamination when using atmospheric steam.

Inoculum. In any microbiological process, the inoculum is of key importance. As with the other variables, a number of inoculum variables are involved including amount and physical form. In early work, the inoculum took the form of precolonized wood chips using *P. chrysosporium*. With this type of inoculum, levels from 2.5 to 20% (dry weight basis) were effective for mechanical pulping. Addition of nitrogen as a nutrient had a beneficial effect in terms of energy savings but caused greater weight loss pulping (Kirk et al., 1993).

With *C. subvermispora*, a liquid inoculum was used which consisted of macerated mycelial mats from liquid surface cultures. In the initial work, 3 kg of fungus was used per ton of wood chips (dry weight basis). At this level, the energy savings and property improvements reported above could be achieved (Fischer et al., 1994). However, this level of inoculum was too high to be practical. A remarkable reduction was eventually achieved by adding corn steep liquor to the inoculum suspension.

Corn steep liquor is a condensed fermented corn extract that is produced in the corn wet-milling process when the dry corn is steeped in warm sulfurous acid solution. During the process, the grain solubles are released and undergo a mild lactic acid fermentation by naturally occuring bacteria. Corn steep liquor is currently used as a feed supplement for ruminants, a nutrient source for poultry, and a nutrient in industrial fermentation processes. In addition, corn steep liquor is relatively inexpensive and available throughout the United States and abroad. The miscellaneous fractions of corn steep liquor contain metal ions, amino acids, vitamins, and other compounds in small quantities. When 0.5% corn steep liquor is added, the amount of inoculum can be reduced from 3 kg ton^{-1} to 5 g ton^{-1}, and in some cases even lower. The component or components of corn steep liquor responsible for the beneficial effect are not known (Akhtar et al., 1997b, Akhtar, 1997).

Engineering and scale-up. Taking the biopulping process from the laboratory to the commercial scale required several engineering and scale-up steps. Most of the challenges involved taking a successful laboratory procedure and redesigning it to be practical on a larger scale. These challenges occur in two main areas: (1) preparing and inoculating the chips and (2) maintaining the proper growth conditions for the fungus during incubation.

On a laboratory scale, each step in the process is done in a batchwise fashion. On a larger, commercial scale the steps involved must be done on a continuous basis. Figure 2 shows the biopulping process and how it fits into an existing mill's wood yard operation. After the normal harvesting, chipping, and screening operations, the chips are steamed as previously discussed to decontaminate the chip surfaces. After steaming, the chips are usually near 100 °C, at least at the surface. Thus, the chips must be cooled sufficiently to allow the inoculum to be added. Complete cooling is not needed before the inoculum is added; however, the chips must be within the temperature growth range of the fungus within a relatively short period of time after adding the inoculum. Hence, cooling can take place in two stages: before inoculation and after the chips are placed into storage that has a ventilation system for further cooling.

Once the chips are in the pile, the proper conditions for the growth of the fungus must be maintained. In the laboratory, this was done by placing the bioreactors into a fixed temperature incubator for the time needed. On the larger scale, a more active approach is needed. The biopulping fungus has an optimum growth temperature range. Furthermore, the fungal metabolism is exothermic, and the fungus is not self-regulating in terms of temperature. For example, when biopulping was performed in a relatively small 1 ton pile without forced ventilation, the pile center reached about 42 °C with 48 hours after inoculation as a result of metabolic heat generated by the fungus. This is 10 °C higher than the optimum growth temperature of *C. subvermispora*. Similar heat buildup is seen in commercial chip piles simply because of the natural microorganisms present in the wood. Forced air is very effective for controlling both the moisture and the temperature throughout the pile. This required an understanding of the air flow through the chip bed, the heat generation of the fungus, the changes in the chip structure because of the fungus, and the nutrient and oxygen requirements of the fungus (Scott et al., 1998).

A treatment system was built based on two screw conveyors that transported the chips and acted as treatment chambers as shown in

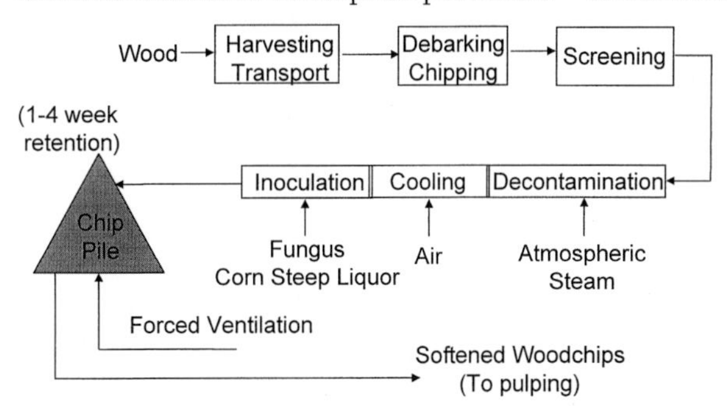

Fig. 2 Overview of proposed biopulping process showing how the biotreatment process fits into an existing mill's wood handling system.

Figure 3. Steam was injected into the first screw conveyor, which heated and decontaminated the wood chip surfaces. A surge bin was located between the two screw conveyors to act as a buffer. From the bottom of the surge bin, a second screw conveyor removed the chips, which were subsequently cooled with filtered air blown into the second conveyor. In the second of this conveyor, the inoculum suspension was applied and mixed thoroughly with the chips through the tumbling action in the screw conveyor. From the second screw conveyor, the chips fell into a pile for the two-week incubation.

Equipment of this design was used to treat 50 tons of spruce chips (dry weight basis) with *C. subvermispora* at FPL at a throughput of 2 tons per hour. During the subsequent two-week period, the chip pile was ventilated with conditioned air to maintain the proper growth temperature and chip moisture throughout the pile. Figure 4 is a photo of one of the 50 ton trials with the equipment beside it. The chips from this trial, along with control chips, were refined through a thermomechanical pulp (TMP) mill producing lightweight coated paper. The fungal pretreatment saved 33% electrical energy and improved paper strength properties significantly compared to the control (Akhtar et al., 2000).

Often, mills blend the TMP with groundwood and kraft pulp to produce paper with the

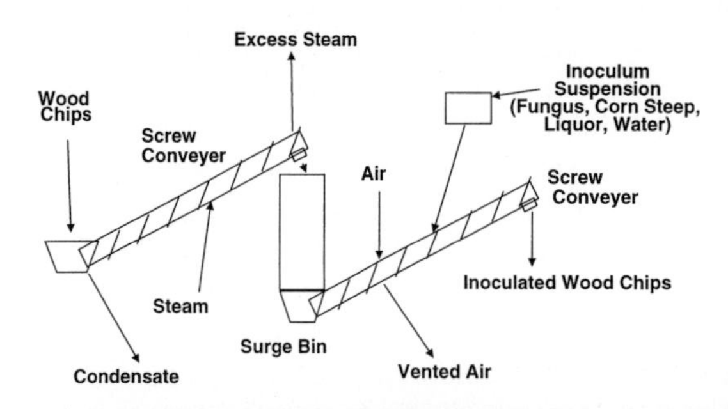

Fig. 3 Overview of a continuous treatment process for decontaminating, cooling, and inoculating wood chips. The system is based on two screw conveyors with a surge bin between them.

Fig. 4 Overview of the large 50 ton trial held at the Forest Products Laboratory. To the left of the pile are the ventilation units, and to the right is the treatment apparatus.

desired characteristics. Researchers performed such blending studies at the laboratory scale and confirmed our findings on a pilot paper machine. Strength properties improved even when blending the TMP and groundwood with 40 to 50% softwood kraft. Figure 5 shows the tensile index for control and treated pulps as a function of the amount of kraft pulp. The same tensile strength as control with 50% kraft can be obtained with 10% less kraft when biotreated TMP is used. Similar results were obtained with respect to the tear index. Thus, biotreated pulp allows the substitution of the biotreated TMP for more expensive kraft pulp. As has been the case throughout biopulping research, a darkening of the chips occurs, resulting in a loss of brightness, but bleaching can regain most of this brightness (Scott et al., 1998). The economics of such a process also look quite favorable (Scott and Swaney, 1998).

3.2 Chemical Pulping

As with mechanical pulping, a pretreatment of chips with white-rot fungi can also be effective for chemical pulping.

Sulfite pulping. Studies at FPL in Madison demonstrated that fungal pretreatment has a significant effect on both calcium- and sodium-based sulfite pulping (Scott et al., 1995a,b). This pretreatment involved two strains of *C. subvermispora* (CZ-3 and SS-3). The focus of these studies was on the yield, kappa number, liquor composition, pulp bleachability, and effluent analysis. In the case of sodium bisulfite pulping, both the kappa number and the yield were reduced at the same cooking time and chemical charge as you may see in Table 9. Both strains of the fungus performed quite similarly. For calcium based pulping, one strain of the fungus (CZ-3) reduced the kappa number by a greater amount after pulping while maintaining the yield. Again, the cooking times and chemical charges were the same. Additional studies showed that shorter cooking times could be used to reach the same kappa number as the control, thus potentially increasing throughput and reducing energy

Tab. 9 Yield and kappa number for the various based sulfite cooking methods using *C. subvermispora* (Akhtar et al., 1998).

Parameters	*Control*	*Treatment (CZ-3)*	*Treatment (SS-3)*
	Calcium acid sulfite		
Total cook time (hr)	9.50	9.50	9.50
Pulp yield (%)	47.63	47.70	47.80
Kappa number	26.75	13.72	21.05
	Calcium acid sulfite		
Sodium bisulfite			
Total cook time (hr)	5.25	5.25	5.25
Pulp yield (%)	49.93	48.16	48.20
Kappa number	31.27	22.84	22.90
	Calcium acid sulfite		
Magnesium sulfite			
Total cook time (hr)	5.25	4.00	–
Pulp yield (%)	n/a	n/a	–
Kappa number	24.8	25.4	–

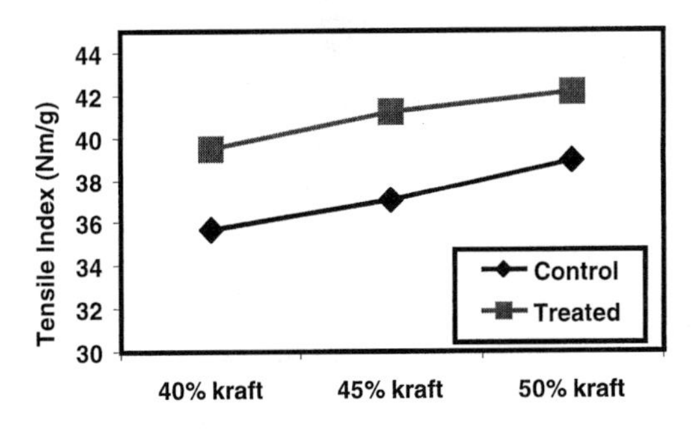

Fig. 5 Tensile index of TMP blended with different levels of kraft pulp. Note that the same tensile strength can be achieved with less kraft pulp in the furnish when bio-treated TMP is used.

consumption. For the calcium-based pulping, liquor titrations indicated that the same amount of cooking liquor was consumed for both the treated and untreated cases. Subsequent bleaching experiments with both the treated and untreated pulps indicated that although there was an initial 5 point brightness loss due to the fungal pretreatment, this was easily regained in the subsequent bleaching steps.

Messner and coworkers (1998) also looked at magnesium-based sulfite pulping. Very similar results, are seen for this base compared to the others as shown in Table 9. Essentially the same kappa number is reached with a significantly shorter cooking time. With the magnesium-based process, only a slight brightness loss was seen (about 1.5 points) which was easily regained with bleaching. At the same cooking time, the kappa number was reduced by about 50%, similar to the results seen for calcium acid pulping using the same strain of the fungus. Interestingly enough, when this pulp was subsequently bleached, no additional brightness gain was achieved, even with the reduction in kappa number. As with mechanical pulping, the fungal pretreatment produces chromophores which tend to darken the resulting pulp.

Kraft pulping. Over the past decade, several researchers have studied the use of fungal pretreatments for kraft pulping. A study using the white-rot fungi *Phellinus pini* and *Streum hirsutum* showed that the treatment reduced the subsequent refining energy of the pulp (Adamski et al., 1987). A more extensive study of *P. chrysosporium* on red oak and aspen wood was carried out by Oriaran and coworkers (1990). After a 30 day fungal treatment, higher yields were obtained when pulped to similar kappa numbers. Furthermore, the pulp responded better to refining and showed increased tensile and burst indices. A study of many South African fungi on pine wood was shown to reduce the kappa number by 17%. However, the yield also decreased and active alkali consumption increased (Messner et al., 1998).

The effect of *T. (Polystictus) versicolor* and *Pleurotus ostreatus* on radiata pine has also been studied (Molina et al., 1996, 1997). Both of these fungi made kraft pulping easier, resulting in a reduction in the amount of energy need for pulping ranging from 11 to 14%. In the case of *T. (P.) versicolor*, a decrease in the mechanical properties of the pulp was observed. Kraft pulping using *S. hirsutum*, *Pycnoporus sanguineus*, and *T. (C.) versicolor* has been investigated by Wolfaardt and coworkers (1996) as being promising fungi for pretreatment. Chen and Schmidt (1995) used the well-studied fungus, *P. chrysosporium*, and found increases in burst and tear with some decrease in brightness.

Early work by Hunt (1976, 1978a,b) looked at the effect of using naturally decayed wood in the kraft pulping process. In his work, he found that the presence of rot had a significant effect on the resulting strength properties of the pulp. However, this was found to be dependent on the particular type of fungus that caused the rot. In most cases, the less specific in regards to the degradation of carbohydrates and lignin, the more detrimental the effect of the fungus. A more recent study of beetle-killed spruce in the Kenai Peninsula of Alaska indicated that the degree of decay had a slight effect on the refining energy and strength properties (Scott et al., 2000). The presence of sap rot was found to be an important indicator of pulping efficiency and resultant pulp quality.

A study done in collaboration with Pramod Bajpai and Partima Bajpai of Thapar Research Group, India, on bio-kraft pulping of eucalyptus chips demonstrated some of the potential benefits of fungal pretreatment. The fungal pretreatment reduced the cooking time from 90 to 30 minutes without affecting the quality of the final product as shown in Table 10. The treated chips had similar strength properties to the untreated chips at the much shorter cooking time and the same chemical charge (17% active alkali and 23% sulfidity). Control chips cooked for only 30 minutes at temperature were only partially cooked, resulting in a very high shive content (Akhtar et al., 1999).

4 Biobleaching

The primary lignin-degrading enzymes produced by white-rot fungi are laccase, manganese peroxidase, and lignin peroxidase, which are the so-called phenol-oxidizing enzymes. Of the three, manganese peroxidase is the enzyme most often associated with bleaching activity. Table 11 shows the effect of enzyme activity on the bleaching effect of two of these enzymes for bleaching of hardwood kraft pulp. In the work of Addleman et al. (1995), the enzymes were produced by mutants of *T. versicolor.* It was also noted in this work that mutants of this fungus that did not produce these two enzymes did not bleach the kraft pulp. The work of Hirai et al. (1994) used a species of *Ganoderma* that produced a variant of manganese peroxidase that adhered strongly to the pulp surface and was not easily detected using standard assay techniques. It must be remembered that in each of these cases both manganese peroxidase and lignin peroxidase were produced, and thus it is impossible to separate the effects of the two enzymes (Reid, 1998).

The bleaching effect using white-rot fungi is through the removal of the residual lignin from the pulp. Using *T. versicolor,* the kappa number is reduced from about 15 to 9 in three days and further reduced to 5 in fourteen days for O_2-delignified hardwood (Reid and Paice, 1994a,b). Since the kappa number is directly proportional to the fraction of lignin in the pulp, about two-thirds of the residual lignin is removed in a fourteen-day treatment. Softwoods see a similar dramatic decrease in the lignin content from an initial value of 26 to 22 in three days and to 12 in fourteen days. The brightness is inversely related to the kappa number: a hardwood with a kappa number of 15 has a brightness of about 35. This increases to over 55 when the kappa number is reduced to below 10 (Reid and Paice, 1994a,b). Further work by Tsuchikawa et al. (1995) shows a similar inverse relationship between the

Tab. 10 Bio-kraft pulping of eucalyptus chips with *C. subvermispora* in a two-week treatment (Akhtar et al., 1999).

Parameters	*Control*	*Treatment*
Ramp time (min)	90	90
Time at temperature (min)	30	30
Pulp yield (%)	46	46
Brightness (%)	88.6	90.5
Burst Index (kNg^{-1})	4.6	4.8
Tear Index (mNm^2g^{-1})	7.8	8.0
Tensile Index (Nmg^{-1})	68.9	70.5

Tab. 11 Bleaching effect of white-rot fungi on hardwood pulp as a function of the enzyme activity taken from correlations based on the referenced data (Reid, 1998).

Enzyme	*Brightness Increase Enzyme Activity* 10^1	10^2	10^3	*Reference*
Manganese Peroxidase	10.5	23.0	–	Hirai et al. (1994)
Manganese Peroxidase	1.5	7.5	15.0	Addleman et al. (1995)
Laccase	10.5	25.5	–	Addleman et al. (1995)

kappa number and brightness with treatment with *P. sordida*. In their work, the brightness of hardwood pulps increases from 35 to 60 while the kappa number decreases from 14 to 6 with a ten-day treatment.

The major drawback of using fungal treatments on pulp for bleaching is the length of time needed for the reaction to occur. As with biopulping, treatment times on the order of 5 to 15 days are needed to achieve a significant degree of delignification. In a commercial setting, the storage of pulp for this length of time could be prohibitive. The direct use of enzymes, as discussed in a following section, could address this concern.

5 Effluent Treatment

In the production of paper, a great deal of water is used. Before this water can be returned to the receiving waters, the residual chemicals and process byproducts must be removed from the effluent or rendered harmless. Bleaching effluents are often of greater concern since they can contain chlorine (used for bleaching) as well as the residual lignin that is removed from the pulp, giving the effluent a dark brown color. Furthermore, it is difficult to reuse this water elsewhere in the mill due to corrosion problems, and thus, it must be treated and released. Typically, this effluent is treated by biological oxidation, which unfortunately, does not significantly decrease the color of the effluent. Effective treatment is expensive and difficult.

Since the color of the effluent is directly related to the amount of chlorolignin compounds present, it can be used as a measure of the efficacy of effluent treatment methods. In comparison, conventional methods do very little to reduce the effluent color. Initial work indicated that *P. chrysosporium* reduced the color of bleach plant effluent by 60% over a five-day period (Eaton et al., 1980) and 70% in a seven-day period (Mittar et al., 1992). Reductions in the BOD and COD were 40% and 50%, respectively. Since the color is reduced more than the amount of residual material in the effluent as measured by BOD and COD, there must be a destruction of chromophores while they are still part of the residual lignin molecules (Sundman et al., 1981; Kondo, 1998).

Two laboratory-scale processes have been created based on these concepts. The MyCoR RBC process developed by Joyce and coworkers uses a rotating biological contactor reactor with the fungus *P. chrysosporium*. The RBC unit is designed to be installed between the primary treatment and secondary clarification stages of a conventional treatment plant. For lignin degradation, however, the system required the addition of nutrients and especially nitrogen. Furthermore, a high oxygen concentration was required to achieve economical levels of decolorization (Kondo, 1998). Using air instead of oxygen reduces the cost of operation but also lowers the decolorization rate (Prouty, 1990).

The trickling filter bioreactor of the MYCOPOR process is a second implementation using lignin-degrading fungi for effluent treatment (Messner et al., 1990). In this implementation, *P. chrysosporium* is immobilized on polyurethane foam which is placed in a tubular reactor. Air and the effluent are introduced into the reactor in a counterflow process. Twelve hours of cycling of the effluent in such a reactor reduced the color by 60%. As with the MyCoR process, this system also required supplementation of the fungus with both oxygen and nutrients. With both of these processes, the major barrier to implementation is the need and expense of the nutrients and the oxygen supplementation that is needed (Kondo, 1998).

Table 12 summarizes some of the different white-rot fungi that have been used for

Tab. 12 Color reduction by white-rot fungi used for the treatment of bleach plant effluent.

Fungus	*Treatment Time*	*Color Reduction (%)*	*BOD reduction (%)*	*Reference*
Isolate IZU-154	5 da	66		Lee et al. (1993a,b)
Isolate KS-62	10 da	80		Lee et al. (1994)
Isolate KS-62	7 da	70		Lee et al. (1994)
L. edodes	5 da	73		Esposito et al. (1991)
P. chrysosporium (MYCOPOR)	12 hr	60	–	Messner et al. (1990)
P. chrysosporium	5 da	60	40	Eaton et al. (1982)
P. chrysosporium	7 da	70	50	Mittar et al. (1992)
T. versicolor	17 h	45		Royer et al. (1983)
T. versicolor	3 da	80	–	Livernoche et al. (1981, 1983)
T. versicolor	3 da	88	–	Mehna (1995)
T. versicolor	6 da	60	–	Livernoche et al. (1983)
T. versicolor	6 da	78		Yin et al. (1989a)

effluent decolorization. More information regarding these fungi can be found in a recent publication (Kondo, 1998).

6 Application of Ligninolytic Enzymes

The major drawback of using fungi in industrial applications is the time required for the fungi to grow and to effect the changes in the wood and pulp. As discussed in the previous section, biopulping is a two-week process, bleaching with fungus takes 5 to 15 days, and effluent treatment takes 1 to 10 days. For biopulping, the early phase of the process is dedicated to building up the fungal biomass; very little biopulping and enzyme secretion is done. It is only after several days that the enzyme production is sufficient to modify and remove the lignin in the wood and delignification proceeds. If enzymes could be produced and isolated, the growth phase can be eliminated and the treatment time reduced. The use of lignin-degrading enzymes for bleaching and effluent treatment are discussed below.

6.1 Bleaching

Treatment of wood pulp with the extracted enzymes rather than the fungal culture can greatly speed up the initial reaction and brightening of the pulp. The work of Paice and coworkers (1993) treated wood pulp with *T. versicolor* and the enzymes extracted from the culture filtrate. The enzymes rapidly increased the brightness by about 5 points in one day. During the same time period, the pulp treated with the fungus showed no brightness increase. However, after 5 days, the enzymes had brightened the pulp by only 6 points while the fungus increased the brightness by over 20 points. The results of this experiment show that a rapid initial increase in brightness can be achieved using the lignin-degrading enzymes in a manner of hours after application. It is thought that manganese peroxidase is the primary enzyme responsible for the brightening effect (Kondo et al., 1994a; Harazono et al., 1996). Repeated applications of managanese peroxidase can increase the pulp brightness to 75, which is acceptable for high grade paper (Kondo et al., 1994a).

6.2 Effluent Treatment

As with bleaching, the enzymes are actually the driving force in the decolorization of the effluent. These enzymes can be immobilized on a variety of supports including glass beads, Celite, and activated agarose. Table 13 summarizes some of the work done with using enzymes for decolorization of effluents. As we saw with the bleaching using lignin-degrading enzymes, the treatment times are somewhat reduced when compared with using the fungus, but the efficacy was also not as great. Some of these enzymes also tended to reduce the resulting COD of the effluent through the degradation of the residual lignin compounds found in the effluent. This is still a relatively new area of research that needs greater investigation (Kondo, 1998).

7 Outlook and Perspectives

Many of the areas discussed above need further research into the basic understanding of using these wood-decay fungi in industrial applications. However, the use of lignin-degrading fungi as a pretreatment for mechanical pulping is the closest of these technologies to commercial application.

Several issues need to be considered in making the final scale-up to the industrial levels, which can range from 200 to 2000 tons (dry) or more of chips being processed on a daily basis. The larger scale with a two-week treatment time would require the routine storage of 28,000 tons of wood for a 2000 ton per day plant. Although some mills do store and manage inventories in these ranges, others may need to make changes in their wood yard operations to take advantage of this technology. Chip rotation has to be controlled with a first-in, first-out policy to maintain a consistent furnish to the pulp mill as is usually the case.

Another concern is the variation in the fungal treatment in different parts of the piles. As temperatures in the pile vary, so does the efficacy of the biopulping process (Scott et al., 1998). Near the edges of the piles, contamination with other microorganisms may increase competition and reduce the biopulping efficacy. In larger piles, where the surface-to-volume ratio is quite low, the outer chips represent only a small fraction of the pile. Furthermore, untreated chips in large industrial piles often heat to more than 50 °C because of indigenous microbial growth, leading to variation of the chip quality throughout the pile, with the hotter center of the pile being more affected by this growth. Furthermore, some indigenous organisms also degrade the cellulose in the wood, leading to pulp quality reductions and varia-

Tab. 13 Color reduction by immobilized lignin-degrading enzymes used in the treatment of bleach plant effluent.

Enzyme	*Treatment Time (h)*	*Color Reduction (%)*	*Reference*
Horseradish peroxidase	48	37	Dezotti et al. (1995)
Horseradish peroxidase	72	38	Davis and Burns (1990)
Horseradish peroxidase	72	61	Ferrer et al. (1991)
Laccase	80	40	Davis and Burns (1990)
Lignin peroxidase	48	12	Denzotti et al. (1995)
Lignin peroxidase	72	46	Ferrer et al. (1991)

tions (Parham, 1983). With biopulping, this suite of naturally-occurring organisms is replaced with a single lignin-specific fungus that is grown under controlled conditions. The single organism, together with the better control of chip-pile conditions, should lead to a number of quality improvements including a reduction in the pitch content of the wood chips by *C. subvermispora*.

On an industrial scale, suitable equipment is available for this technology. For example, chip steaming and decontamination could be easily accomplished in a presteaming vessel similar to that used for Kamyr digesters (Smook, 1982) or in a vertical, pressurized steaming bin. Cooling and inoculation will likely take place at atmospheric pressure. Air conveying will naturally cool the chips during transport, thus requiring the inoculation to be done at the end of the conveying system and before being incubated. Mills using other conveying methods, such as belts or screw conveyers, may require the addition of some type of ventilation. In our pilot-scale work, the cooling and inoculation of the chips were done through ventilation in a screw conveyer. Pile ventilation strategies are given in Scott et al. (1998).

Currently, it is estimated that losses of approximately 1% per month of wood occur in outside chip storage systems (Parham, 1983). This loss is mainly due to the blowing of fines, respiration of the living wood cells, and the activity of microorganism. The blowing of fines and sawdust as well as microorganism growth can also cause environmental difficulties in the vicinity of the chip piles. Thus, indoor storage should also be considered as an option for incorporating a biopulping operation into a mill. Enclosing the chip storage operation will significantly reduce blowing dust and other environmental concerns. Furthermore, better control of the environment for the growth of the fungus would be maintained throughout the year. Enclosing the chip storage would also allow the recovery of the heat produced by the fungus for use in conditioning the incoming air. The geometry of the enclosed storage would also tend to reduce the blower costs. These factors could result in substantial energy savings, especially during the winter months in northern climates. Mills that do not have sufficient space for the biopulping operation at their mill site could consider setting up a satellite chipping and treatment operation. In this way, chips could be treated prior to shipment to the pulp mill at a site that is convenient to the harvesting operations.

Mill-scale refining of fungus-treated chips gave results similar to those obtained using the laboratory-scale bioreactors. With this information, a complete process flowsheet has been established for the commercial operation of the process. Based on the electrical energy savings and the strength improvements, the process economics looks very attractive. Several independent economic evaluations of biopulping have now been completed by both university and industry economists and engineers and are in agreement. Based on 33% energy savings and a 5% reduction in kraft pulp in the final product, a savings of about $ 5 million each year can be realized. The additional benefits of increased throughput, and reduced pitch content and environmental impact improve the economic picture for this technology even further (Scott and Swaney, 1998)

The biopulping process has now passed through the science and the engineering evaluation phases, and is in the realm of business. One chemical company has already agreed to produce and supply fungal inoculum on a commercial scale. No adverse effects of lignin-degrading fungi on humans have been reported in the literature. These fungi are natural wood decayers. However, a biopulping operation would entail producing substantial amounts of fungus in a pile on a routine basis. For that reason, the best biopulping fungus, *C. subvermisprora*, was tested by professionals for adverse effects on humans. It was concluded that the fungus is safe for use on a commercial scale. One of the paper companies has hired a professional company to look into the effects of the technology on the environment. The company is currently analyzing volatile organic compounds (VOC's) given off during biopulping and comparing them with those from a standard chip pile storage. An engineering and construction company has also agreed to provide engineering and construction services and performance guarantees on the equipment needed for full-scale implementation of the technology.

8 Patents

Patents in this area of research have been issued for both the microorganisms that are used in the process, as well as for the process itself. In many cases, specific aspects of the process are patented. The key patents for biopulping are held by Blanchette et al. (1991), which covers the use of *C. subvermispora* for the treatment of wood chips, and Akhtar (1997, 1998), which covers the use of the nutrient supplement corn steep liquor during the treatment. The corn steep liquor patents are particularly important as they allowed the amount of inoculum required to be reduced to commercially-viable levels. The other biopulping-related patents cover various applications or enhancements to the process. For example, the series of patents by Blanchette et al. (1995, 1996, 1998) cover the use of white-rot fungi for the removal of pitch from wood chips.

Wastewater treatment is of primary concern for industries and especially for the pulp and paper industry. The early work in using fungi for the treatment and decolorization of wastewater are given in the patents of Blair and Davis (1981) and Chang et al. (1985, 1987). These patents claim that the white-rot fungi are especially effective in treating the effluent from bleaching processes in the paper industry. Finally, for several applications, researchers found that the enzymes produced by the white-rot fungi are effective for bleaching and other applications. The patent of Call (1995) covers the production of these enzymes from white-rot fungi.

The patent list below includes many of the key patents in this field in addition to the ones discussed above.

9
Patent List

Akhtar, M., Attridge, M.C., Koning Jr., J.W., Kirk, T.K. (1995) Method of pulping wood chips with a fungus using sulfite salt-treated wood chips (U.S. patent no. 5,460,697).

Akhtar, M. (1997). Method of enhancing biopulping efficacy (U.S. patent no. 5,620,564).

Akhtar, M. (1998). Method of enhancing biopulping efficacy (U.S. patent no. 5,750,005).

Akhtar, M., Scott, G.M., Lentz, M. (1996) Biopulping with high temperature fungus *Phlebia subserialis* (pending).

Akhtar M., Lentz, M.J., Lightfoot, E.N., Scott, G.M., Swaney, R., Kirk, T.K. (1997) Method and apparatus for commercial scale biopulping (pending).

Akhtar M., Scott, G.M., Ahmed, A., Lentz, M.J., Horn, E.G. (1997) Biopulping industrial wood waste (pending).

Akhtar M., Lentz, M.J., Lightfoot, E.N., Scott, G.M., Swaney, R., Kirk, T.K. (1998) Method and apparatus for commercial scale biopulping (pending).

Back; S., Lazorisak, N.W., Smeltzer, N.L., Schmitt, J.F. (1995) Production of soft paper products from old newspaper (U.S. patent no. 6,027,610).

Bajpai, P., Bajpai, K.P., Akhtar, M. (1998) Eucalyptus bio-kraft pulping process (pending).

Blair, J.E., Davis, L.T. (1981) Process for decolorizing pulp and paper mill wastewater and microorganism capable of same (U.S. patent no. 4,266,035).

Blanchette, R.A., Brush, T.S., Farrell, R.L., Krisa, K.A., Mishra, C. (1995) Process for treatment of waste stream using *Scytinostroma galactinum* fungus ATCC 20966 (U.S. patent no. 5,545,544).

Blanchette, R.A., Brush, T.S., Farrell, R.L., Krisa, K.A., Mishra, C. (1995) White-rot fungus and uses thereof (U.S. patent no. 5,427,945).

Blanchette, R.A., Brush, T.S., Farrell, R.L., Krisa, K.A., Mishra, C. (1996) White-rot fungus and uses thereof (U.S. patent no. 5,554,535).

Blanchette, R.A., Burnes, T.A., Farrell, R.L., Iverson, S. (1998) Pitch degradation with wood colonizing bacteria (U.S. patent no. 5,766,926).

Blanchette, R.A., Farrell, R.L., Behrendt, C.J. (1995) Biological control for wood products and debarking (U.S. patent no. 5,518,921).

Blanchette, R.A., Farrell, R.L., Iverson, S. (1995) Pitch degradation with white rot fungus (U.S. patent no. 5,472,874).

Blanchette, R.A., Farrell, R.L., Iverson, S. (1995) Pitch degradation with white rot fungi (U.S. patent no. 5,476,790).

Blanchette, R.A., Iverson, S., Behrendt, C.J. (1998) Pitch and lignin degradation with white rot fungi (U.S. patent no. 5,705,383).

Blanchette, R.A., Leatham, G.F., Attridge, M.C., Akhtar, M., Myers, G.C. (1991) Biomechanical pulping with *C. subvermispora* (U.S. Patent no. 5,055,159).

Buswell, J. A., Odier, E. (1989) Microorganisms of the *Phanerochaete chrysosporium* strain and their use (U.S. patent no. 4,889,807).

Call, H.-P. (1995) Process for the production of ligniolytic enzymes by means of white rot fungi (U.S. patent no. 5,403,723).

Chang, H.-M., Joyce, T.W., Kirk, T.K., Huynh, V.-B. (1985) Process of degrading chloro-organics by white-rot fungi (U.S. patent no. 4,554,075).

Chang, H.-M., Joyce, T.W., Kirk, T.K. (1987) Process of treating effluent from a pulp or papermaking operation (U.S. patent no. 4,655,926).

Eriksson, K.-E., Ander, P., Henningsson, B., Nilsson, T., Goodell, B. (1976) Method for producing cellulose pulp (U.S. patent no. 3,962,033).

Farrell, R.L., Kirk, T.K., Tien, M. (1987) Novel enzymes which catalyze the degradation and modification of lignin (U.S. patent no. 4,687,741).

Nghiem, N.P. (1995) Biological process to remove color from paper mill wastewater (U.S. patent no. 5,407,577).
Nishida, T., Kashino, Y., Mimura, A., Takahara, Y., Sakai, K. (1992) Method for producing pulp by treatment using a microorganism, and its related enzymes (U.S. patent no. 5,081,027).
Nishida, T., Kashino, Y., Mimura, A., Takahara, Y., Sakai, K. (1992) Enzymes employed for producing pulps (U.S. patent no. 5,149,648).
Nishida, T., Takahara, Y., Sakai, K., Kondo, R., Lee, S.-H. (1995) Method for treating liquid waste after pulp bleaching can be achieved (U.S. patent no. 5,431,820).
Shen, H.-P., Mou, D.-G., Lim, K.-L., Feng, P., Chen, C.-H. (1992) Microbial decolorization of wastewater (U.S. patent no. 5,091,089).
Zimmerman, W.C., Farrell, R.L. (1997) Fungi for pitch reduction and their preparation (U.S. patent no. 5,607,855).

10
References

Akamatsu, I., Yoshihara, K., Kamishima, H., Fujii, T. (1984) Influence of white-rot fungi on poplar chips and thermo-mechanical pulping of fungi-treated chips, *Mokuzai Gakkaishi* **30**, 697.

Adamski, Z., Gawecki T., Zielinski, M.H. (1987) Combined biological and chemical delignification of wood, in: *Funkcne Integrovane Obhospodarovanie Lesoy a Komplexne Vyuztie Dreva* (Medzinarodna Vedecka Konferencia). Poznan, Poland: Acad. Agric.

Archibald, F.S., Addleman K., Dumonceaux T., Paice M.G., Bourbonnais, R. (1995) Production and characterization of *Trametes versicolor* mutants unable to bleach hardwood kraft pulp, *Appl. Environ. Microbiol.* **61**, 3687.

Akhtar, M., Attridge M.C., Blanchette R.A., Myers G.C., Wall M.B., Sykes M.S., Koning, J.W. Jr., Burgess R.R., Wegner T.H., Kirk, T.K. (1992a) The white-rot fungus *Ceriporiopsis subvermispora* saves electrical energy and improves strength properties during biomechanical pulping of wood, in: *Biotechnology in the Pulp and Paper Industry* (Kuwahara, M., Shimada M., Ed.), pp. 3–8. Kyoto, Japan: UNI Publishers Co., Ltd.

Akhtar, M., Attridge, M.C., Myers, G.C., Kirk, T.K., Blanchette, R.A. (1992b) Biomechanical pulping of loblolly pine with strains of the white-rot fungus *Ceriporiopsis subvermispora, Tappi J.* **75**, 105.

Akhtar, M., Attridge, M.C., Myers, G.C., Blanchette, R.A. (1993) Biomechanical pulping of loblolly pine chips with selected white-rot fungi, *Holzforschung* **47**, 36.

Akhtar, M. (1994) Biomechanical pulping of aspen wood chips with three strains of *Ceriporiopsis subvermispora, Holzforschung* **48**, 199.

Akhtar, M., Blanchette R.A., Burnes, T.A. (1995) Using Simons stain to predict energy savings during biomechanical pulping, *Wood Fiber Sci.* **27**, 258.

Akhtar, M., Blanchette, R.A., Kirk, T.K. (1997a) Microbial delignification and biomechanical pulping, in: *Adv. in Biochem. Eng./Biotechnol.*, Vol. 57, pp. 159–195. Heidelberg: Springer-Verlag.

Akhtar, M., Lentz, M.J., Blanchette, R.A., Kirk, T.K. (1997b) Corn steep liquor lowers the amount of inoculum for biopulping, *Tappi J.* **80**, 161.

Akhtar, M., Blanchette, R.A., Myers, G.C., Kirk, T.K. (1998) An overview of biomechanical pulping research, in: *Environmentally Friendly Technologies for the Pulp and Paper Industry* (Young, R.A., Akhtar, M., Ed.), pp. 309–340 . New York: John Wiley & Sons.

Akhtar, M., Horn, E.G., Lentz, M.J., Scott, G.M., Sykes, M.S., Myers, G.C., Kirk, T.K. (1999) Towards commercialization of biopulping, *Paperage* **115**(2), 17.

Akhtar, M. Scott, G.M., Swaney, R.E., Shipley, D.F. (2000) Biomechanical pulping: a mill-scale evaluation, *Resources, Conservation and Recycling* **28**, 241.

Ander, P., Eriksson, K.-E. (1975) Mekanisk massa fran forrotad flis-en inledande undersokning *Svensk Papperstidning* **18**, 641.

Archibald, F., Paice, M.G., Jurasek, L. (1990) Decolorization of kraft bleachery effluent chromophores by *Coriolus (Trametes) versicolor, Enzyme Microbiol. Technol.* **12**(11), 846.

Bar-Lev, S.S., Kirk, T.K., Chang, H.-M. (1982) Fungal treatment can reduce energy requirements for secondary refining of TMP, *Tappi J.* **65**, 111.

Bergbauer, M., Eggert, C. (1992) Differences in the persistence of various bleachery effluent lignins against attack by white-rot fungi, *Biotechnol. Lett.* **14**, 869.

Bosshard, H.H., Holzkunde, M.H. (1984) *Aspekte der Holzbearbeitung und Holzverwertung*, Basel: Birkhäuser Verlag.

Breuil, C., Iverson, S., Gao, Y. (1998) Fungal treatment of wood chips to remove extractives, in: *Environmentally Friendly Technologies for the Pulp and Paper Industry* (Young, R.A., Akhtar, M., Ed.), pp. 541–565. New York: John Wiley & Sons.

Chen, Y., Schmidt, E.L. (1995) Improving aspen kraft pulp by a novel low-technology fungal pretreatment, *Wood Fiber Sci.* **27**, 198.

Davis, S., Burns, R.G. (1990) Decolorization of phenolic effluents by soluble and immobilized phenol oxidases, *Appl. Microbiol. Biotechnol.* **32**, 721.

Dezotti, M., Innocentini-Mei, L.H., Duran, N. (1995) Silica immobilized enzyme catalyzed removal of chlorolignins from eucalyptus kraft effluent, *J. Biotechnol.* **43**, 161.

Eaton, D.C., Chang, H.-M., Kirk, T.K. (1980) Fungal decolorization of kraft bleach effluents, *Tappi* **63**(10), 103.

Eaton, D.C., Chang, H.-M., Joyce, T.W., Jeffries, T.W., Kirk, T.K. (1982) Method obtains fungal reduction of the color of extraction-stage kraft bleaach effluents, *Tappi* **65**(6), 89.

Eriksson, K.-E. (1990) Biotechnology in the pulp and paper industry, *Wood Sci. Technol.* **24**, 79.

Eriksson, K.-E., Kirk, T.K. (1985) Biobleaching and treatment of kraft bleaching effluents with white-rot fungi, in: *Comprehensive Biotechology: The Principles, Applications and Regulations of Biotechnology in Industry, Agriculture and Medicine* (Murray, M.-Y., Ed.), pp. 271–294. New York: Pergamon Press.

Eriksson, K.-E., Vallander, L. (1982) Properties of pulps from thermomechanical pulping of chips pretreated with fungi, *Svensk Papperstid.* **85**, R33.

Eriksson, K.-E., Gunewald, A., Vallander, L. (1980) Studies of growth condition in wood for three white-rot fungi and their cellulaseless mutants, *Biotechnol. Bioeng.* **22**, 363.

Eriksson, K.-E., Johnsrud, S.C., Vallander, L. (1983) Degradation of lignin and lignin model compounds by various mutants of the white-rot fungus *Sporotrichun pulverulentum, Arch. Microbiol.* **135**, 161.

Esposito, E., Canhos, P., Duran, N. (1991) Screening of lignin-degrading fungi for removal of color from kraft mill wastewater with no additional extra carbon-source, *Biotechnol. Lett.* **13**, 571.

Ferrer, I., Dezotti, M., Duran, N. (1991) Decolorization of Kraft effluent by free and immobilized lignin peroxidases and horseradish peroxidase, *Biotechnol. Lett.* **13**, 577.

Fischer, K., Akhtar, M., Blanchette, R.A., Burnes, T.A., Messner, M., Kirk, T.K. (1994) Reduction of resin content in wood chips during experimental biological pulping processes, *Holzforschung* **48**, 285.

Fujita, K., Kondo, R., Sakai, K., Kashino, Y., Nishida, T., Takahara, Y. (1991) Biobleaching of kraft pulp using white-rot fungus IZU-154, *Tappi J.* **74** (11), 123.

Fukuzumi, K., Nishida, A., Aoshima, K., Minami, K. (1977) Decolourization of kraft waste liquor with white-rot fungi. I. Screening of the fungi and culturing conditions for decolourization of kraft waste liquor, *Mokuzai Gakkaishi* **23**, 290.

Galeno, G.D., Agosin, E.T. (1990) Screening of white-rot fungi for efficient decolourization of bleach pulp effluents, *Biotechnol. Lett.* **12**(11), 869.

Gao, Y. (1996) Biodegradation of lipids by wood sapstaining *Ophiostoma* spp, *Ph.D. thesis*, Department of Wood Science, Faculty of Forestry, University of British Columbia, Canada.

Harazono, K., Kondo, R., Sakai, K. (1996) Bleaching of hardwood kraft pulp with manganese peroxidase from *Phanerochaete sordida* YK-624 without addition of MnSO sub(4), *Appl. Environ. Microbiol.* **62**, 913.

Hirai, H., Kondo, R., Sakai, K. (1994) Screening of lignin-degrading fungi and their ligninolytic enzyme activities during biological bleaching of kraft pulp, *Mokuzai Gakkaishi* **40**, 980.

Ho, C., Jurasek, L., Paice, M.G. (1990) The effect of inoculum on bleaching of hardwood kraft pulp with *Coriolus versicolor, J. Pulp Paper Sci.* **16**, J78.

Hunt, K. (1976) Kraft pulp of beta-irradiated Douglas fir chips, *Tappi* **59**(10), 89.

Hunt, K. (1978a). Kraft pulping of decayed wood: western red cedar and alpine fir, *Pulp Paper Can.* **79**(7), 30.

Hunt, K. (1978b). Pulping western hemlock decayed by white-rot fungi, *Pulp Paper Can.* **79**(6), 75.

Iimori, T., Kaneko, R., Yoshikawa, H., Machida, M., Yoshioka, H., Murakami, K. (1994) Screening of pulp-bleaching fungi and bleaching activity isolated fungus SKB-1152, *Mokuzai Gakkaishi* **40**, 73.

Johnsrud, S.C., Eriksson, K.-E. (1985) Cross-breeding of selected and mutated homokaryotic strains of Phanerochaete K3: New cellulase deficient strains with increased ability to degrade lignin, *Appl. Microbiol. Biotechnol.* **21**, 320.

Joyce, T.W., Pellinen, J. (1990) White-rot fungi for treatment of pulp and paper industry wastewater, *Tappi Environmental Conf.*, pp. 1–13. Tappi, Atlanta.

Kang, C.H., On, H.K., Won, C.H., Srebotnik, E., Messner, K. (1996) Studies on the treatment of paper mill wastewater by *Phanerochaete chryso-*

sporium, in: *Biotechnology in the pulp and paper industry* (Srebotnik, E., Messner, K., Ed.), pp.263–266. Vienna: Facultas-Universitätsverlag.

Kashino, Y., Nishida, T., Takahara, Y. (1991) Biomechanical pulping by the hyperligninolytic fungus IZU-154, *Proc. 6th Int'l Symp. on Wood and Pulping Chem.* Vol. 2, pp. 291–294.

Kashino, Y., Nishida, T., Takahara, Y., Fujita, K., Kondo, R., Sakai, K. (1993) Biomechanical pulping using white-rot fungus IZU-154, *Tappi J.* **76**(12), 167.

Kawase, K. (1962). Chemical components of wood decayed under natural conditions and their properties, *J. Fac. Agri. Hokkaido Univ.* **52**, 186.

Kirk, T.K., Burgess, R.R., Koning, J.W. Jr. (1992) Use of fungi in pulping wood: an overview of biopulping research, in: *Frontiers in industrial mycology. Proceedings of Industrial Mycology Symposium*, Routledge, (Leatham, G.F., Ed.), Chap. 7, pp. 99 ff. New York: Chapman, & Hall.

Kirk, T.K., Koning, J.W. Jr., Burgess, R.R. et al. (1993). Biopulping: A glimpse of the future, *Res. Rep.FPL-RP-523*, Madison, WI.

Kirk, T.K., Akhtar, M., Blanchette, R.A. (1994) Biopulping: seven years of consortia research, *Proc. 1994 TAPPI Biological Sciences Symposium*, Tappi, Atlanta, pp. 57–66.

Kondo, R.(1998). Waste treatment of kraft effluent by white-rot fungi, in: *Environmentally Friendly Technologies for the Pulp and Paper Industry* (Young, R.A., Akhtar, M., Ed.), pp. 515–539. New York: John Wiley & Sons.

Kondo, R., Harazono, K., Sakai, K. (1994a). Bleaching of hardwood kraft pulp with manganese peroxidase secreted from *Phanerochaete sordida* YK-624, *Appl. Environ. Microbiol.* **60**, 4359.

Kondo, R., Kurashiki, K., K., Sakai, K. (1994b) *In vitro* bleaching of hardwood kraft pulp by extracellular enzymes excreted from white rot fungi in a cultivation system using a membrane filter, *Appl. Environ. Microbiol.* **60**, 921.

Lambais, M.R. (1988) Biodecolorization of the effluent of a bleached pulp plant, *Rev. Microbiol.* Sao Paulo **19**, 425.

Lawson, L.R. Jr., Still, C.N. (1957) The biological decomposition of lignin-literature survey, *Tappi* **40**, 56A.

Leatham, G.F. (1983). A chemically defined medium for the fruiting of *Lentinus edodes*, *Mycologia* **75**, 905.

Lee, S.-H., Kondo, R., Sakai, K., Nishida, T., Takahara, Y. (1993a) Treatment of kraft bleaching effluents by lignin-degrading fungi I. Decolorization of kraft bleaching effluents by the lignin-degrading fungus IZU-154, *Mokuzai Gakkaishi* **39**, 470.

Lee, S.-H., Kondo, R., Sakai, K., Nishida, T., Takahara, Y. (1993b) Treatment of kraft bleaching effluents by lignin-degrading fungi II. Detection of nucleic acid constituents in the effluent treated with the lignin-degrading fungus IZU-154, *Mokuzai Gakkaishi* **39**, 1089.

Lee, S.-H., Kondo, R., Sakai, K. (1994). Treatment of kraft bleaching effluents by lignin-degrading fungi III. Treatment by newly found fungus KS-62 without additional nutrients, *Mokuzai Gakkaishi* **40**(6), 612.

Livernoche, D., Jurasek, L., Desrochers, M., Veliky, I.A. (1981) Decolorization of a kraft mill effluent with fungal mycelium immobilized in calcium alginate gel, *Biotechnol. Lett.* **3**, 701.

Livernoche, D., Jurasek, L., Desrochers, M., Dorica, J., Veliky, I.A. (1983) Removal of color from kraft mill wastewaters with cultures of white-rot fungi and with immobilized mycelium of *Coriolus versicolor*, *Biotechnol. and Bioengg.* **25**, 2055.

Mehna, A., Bajpai, P., Bajpai, P.K. (1995). Studies on decolorization of effluent from a small pulp mill utilizing agriresidues with *Trametes versicolor*, *Enzyme Microb. Technol.* **17**, 18.

Messner, K., Srebotnik, E. (1994) Biopulping: An overview of developments in an environmentally safe paper-making technology, *FEMS Microbiology Reviews* **13**, 351.

Messner, K., Ertler, G., Jaklin-Farcher, S., Boskovsky, P., Regensberger, U., Blaha, A. (1990). Treatment of bleach plant effluents by the Mycopor system, in: *Biotechnology in Pulp and Paper Manufacture* (Kirk, T.K., Chang, H.-M., Ed.), pp. 245–251. Stoneham: Butterworth Heinemann.

Messner, K., Masek, S., Srebotnik, E., Techt, G. (1992) Fungal Pretreatment of Wood Chips for Chemical Pulping, Proc. 5th Int'l Conf. On Biotech, in: *Biotechnology in the Pulp and Paper Industry*, (Kuwahara, M., Shimada, M., Ed.), pp. 9–13. Tokyo: UNI Publishers Co., Ltd.

Messner, K., Koller, K., Wall, M.B., Akhtar, M., Scott, G.M. (1998). Fungal pretreatment of wood chips for chemical pulping, in: *Environmentally Friendly Technologies for the Pulp and Paper Industry* (Young, R.A., Akhtar, M., Ed.), pp. 385–419. New York: John Wiley & Sons.

Mittar, D., Khanna, P.K., Marwaha, S.S., Kennedy, J.F. (1992).Biobleaching of pulp and paper mill effluents by *Phanerochaete chrysosporium*, *.J. Chem. Tech. Biotechnol.* **53**, 81.

Molina, J.G., Donosi, J.E., Gomez, A.D. (1996). Biopulping for kraft pulp of *Pinus radiata*, in: 50th

Appita Annual General Conference (Appita), pp. 57–63. Carlton, Australia.

Molina, J.G., Colonelli, P., Moena, R.G. (1997) Refining biokraft pulp of *Pinus radiata*, in: 51st *Appita Annual General Conference* (Appita), pp. 199–206. Carlton, Australia.

Myers, G.C., Leatham, G.F., Wegner, T.H., Blanchette, R.A. (1988) Fungal pretreatment of aspen chips improves strength of refiner mechanical pulp, *Tappi J.* **73**, 105.

Nishibe, F., Okubo, K., Ishikawa, H. (1988) Screening of white-rot fungi and characteristics of biodegraded TMP, *Japan Tappi* **42**, 383.

Nishida, T., Kashino, Y., Mimura, A., Takahara, Y. (1988) Lignin biodegradation by wood-rotting fungi I. Screening of lignin-degrading fungi, *Mokuzai Gakkaishi* **34**, 530.

Oriaran, T.P., Labosky Jr., P., Royse, D.J. (1989) Lignin degradation capabilities of *Pleurotus estreatus, Lentinula edodes,* and *Phanerochaete chrysosporium, Wood Fiber Sci.* **21**(2), 183.

Oriaran, T.P., Labosky Jr., P., Blankenhorn, P.R. (1990) Kraft pulping and papermaking properties of *Phanerochaete chrysosporium*-degraded aspen, *Tappi* **73**, 147.

Oriaran, T.P., Labosky Jr., P., Blankenhorn, P.R. (1991) Direct biological bleaching of hardwood kraft pulp with the fungus *Coriolus versicolor, Wood Fiber Sci.* **23**, 316.

Paice, M.G., Jurasek, L., Ho, C., Bourbonnais, R., Archibald, F. (1989) Direct biological bleaching of hardwood pulp with the fungus *Coriolous versicolor, Tappi J.* **72**(5), 217.

Paice, M.G., Reid, I.D., Bourbonnais, R., Archibald, F.S., Jurasek, L. (1993) Manganese peroxidase, produced by *Trametes versicolor* during pulp bleaching, demethylates and delignifies kraft pulp, *Appl. Environ. Microbiol.* **59**, 260.

Parham, R.A. (1983) Pulp and paper manufacture, Vol. 1, *Joint Textbook Committee of the Paper Industry*: Atlanta, GA.

Patel, R.N., Thakker, G.D., Rao, K.K. (1994) Potential use of a white-rot fungus *Antrodiella* sp. RK1 for biopulping, *J. Biotechnol.* **36**, 19.

Pearce, M.H., Dunlop, R.W., Falk, C.J., Norman, K. (1995) Screening lignin degrading fungi for biomechanical pulping of eucalypt wood chips, in: *Proc. 49th Appita Annual General Conf.*, Australia, pp. 347–351.

Pellinen, J., Abuhasan, M.J., Joyce, T.W., Chang, H.-M. (1988) Biological delignification of pulp by *Phanerochaete chrysosporium, J. Biotechnol.* **10**, 161.

Pilon, L., Barbe, C., Desrochers, M., Juraske, L., Neumann, P.J. (1982) Fungal treatment of mechanical pulps: Its effect on paper properties, *Biotechnol. Bioeng.* **24**, 2063.

Prouty, A. (1990) Bench-scale development and evaluation of a fungal bioreactor for color removal from bleach effluents, *Appl. Microbiol. Biotechnol.* **32**, 490.

Reid, I.D. (1998) Bleaching kraft pulps with white-rot fungi, in: *Environmentally Friendly Technologies for the Pulp and Paper Industry* (Young, R.A., Akhtar, M., Ed.), pp. 505–514. New York: John Wiley & Sons.

Reid, I.D., Paice, M.G. (1994a) Biological bleaching of kraft pulps by white-rot fungi and their enzymes, *FEMS Microbiology Reviews* **13**, 369.

Reid, I.D., Paice, M.G. (1994b) Effect of residual lignin type and amount on bleaching of kraft pulp by *Trametes versicolor, Appl. Environ. Microbiol.* **60**, 1395.

Reid, I.D., Paice, M.G., Ho, C., Jurasek, L (1990) Biological bleaching of softwood kraft pulp with the fungus *Trametes (Coriolus) versicolor, Tappi* **73**(8), 149.

Reis, C.J.,.Libby, C.E (1960) An experimental study of the effect of *Fonnes pini* (Thure) Lloyd on the pulping qualities of pond pine *Pinus serotina* (Michx) cooked by the sulfate process, *Tappi J.* **43**, 489.

Royer, G., Livernoche, D., Desrochers, M., Jurasek, L., Rouleau, D., Mayer, R.C. (1983) Decolorization of kraft mill effluent: kinietics of a continuous process using immobilized *Coriolus versicolor, Biotechnol. Lett.* **5**, 321.

Samuelsson, L., Mjober, P.J., Harler, N., Vallander, L., Eriksson, K.-E. (1980) Influence of fungal treatment on the strength versus energy relationship in mechanical pulping, *Svensk Papperstidning* **8**, 221.

Schmidt, E.L., Olsen, K.K., Akhtar, M. (1996) Compression of nosterile green wood chips as an aid to fungal pretreatment (biopulping), in: *Proc. 211th American Chemical Society National Meeting,* New Orleans.

Scott, G.M., Akhtar, M., Lentz, M.J. et al. (1995a) Fungal pretreatment of wood chips for sulfite pulping, in: *Proc. 1995 Tappi Pulping Conf.*, Tappi, Atlanta.

Scott, G.M., Akhtar, M., Lentz, M.J., Sykes, M., Abubakr, S. (1995b) Environmental aspects of biosulfite pulping, in: *Proc. 1995 International Environmental Conf.*, Tappi, Atlanta.

Scott, G.M., Akhtar, M., Lentz, M.J., Swaney, R.E. (1998) Engineering, scale-up, and economic aspects of fungal pretreatment for wood chips, in: *Environmentally Friendly Technologies for the Pulp and Paper Industry* (Young, R.A., Akhtar, M., Ed.), pp. 341–383. New York: John Wiley & Sons.

Scott, G.M., Akhtar, M., Lentz, M.J., Kirk, T.K., Swaney, R.E. (1998a) New technology for papermaking: commercializing biopulping *Tappi J*. **81** 220.

Scott, G.M., Akhtar, M., Lentz, M.J., Horn, E., Swaney, R.E., Kirk, T.K. (1998b) An overview of biopulping research: discovery and engineering *Korean Tappi* **30**(4), 18.

Scott, G:M., Swaney, R.E. (1998) New technology for papermaking: biopulping economics *Tappi J*. **81**, 153–157.

Scott, G.M., Bormett, D.W., Ross Sutherland, N., Abubakr, S. (2000). Bettle-killed spruce utilization in the Kenai Peninsula *Tappi J*. **83**(6), 52.

Setliff, E.C., Eriksson, K.-E., Marton, R., Granzow, S.G. (1990) Biomechanical pulping with white-rot fungi *Tappi J*. **73**, 141.

Smook ,G.A. (1982) *Handbook for Pulp and Paper Technologists*. Atlanta: Tappi Press.

Sundman, G., Kirk, T.K. (1981) Fungal decolorization of kraft bleach plant effluent *Tappi* **64**(9), 145.

Tappi (1999), *Official Test Methods*. Atlanta: Tappi Press.

Wang, S.-H., Ferguson, J.F., McCarthy, J.L. (1992) The decolorization and dechlorination of kraft bleach plant effluent solutes by use of three fungi: *Ganoderma lacidum, Coriolus versicolor*, and *Hericium erinaceu, Holzforschung* **46**(3), 219.

Wolfaardt, J.F., Bosman, J.L., Jacobs, A., Male, J.R, Rabie, C.J. (1996) Bio-kraft pulping of softwood, in: *Proc. Of the 6th Int'l Conf. On Biotechnology in the Pulp and Paper Industry* (Srebotnik, E., Messner, K., Ed.), pp. 211–216. Vienna: Facultas-Universitätsverlag.

Yin, C.-F., Joyce, T.W., Chang, H.-M. (1989a) Role of glucose in fungal decolorization of wood pulp bleaching effluents, *J. Biotechnol.* **10**, 77.

3
Biotechnological Conversion of Coals into Upgraded Products

Dr. Horst Meyrahn[1]
Prof. Dr. Alexander Steinbüchel[2]
[1] RWE Rheinbraun AG, Department of Environmental Protection, Stüttgenweg 2, 50935 Köln, Germany; Tel: +49-221-48020542, Fax: +49-221-48023650, E-mail: horst.meyrahn@rwerheinbraun.com
[2] Institut für Mikrobiologie, Westfälische Wilhelms-Universität Münster, Corrensstraße 3, 48149 Münster, Germany; Tel: +49-251-8339821; Fax: +49-251-8338388; E-mail: steinbu@uni-muenster.de

Biotechnology of Biopolymers. From Synthesis to Patents. Edited by A. Steinbüchel and Y. Doi

ISBN: 3-527-31110-6

ABTS	2,2′-Azina-bis(3-ethylthiazoline-6-sulfonate)
'BoA'	Braunkohlekraftwerk mit optimierter Anlagentechnik (German abbreviation for lignite-based power plant with optimized plant engineering)
EU	European Union
3HB	3-Hydroxybutyrate
3HD	3-Hydroxydecanoate
3HDD	3-Hydroxydodecanoate
3HHx	3-Hydroxyhexanoate
3HO	3-Hydroxyoctanoate
3HV	3-Hydroxyvalerate
PHA	Polyhydroxyalkanoate
TOC	Total organic carbon

1 Introduction

Coal has for many years been the basis for mankind's technological and industrial development. During this time, coal has been the main energy source in industrial processes. Coal is distributed world-wide, and in terms of available amounts is most likely the greatest non-renewable energy resource. Nevertheless, for almost 100 years great efforts have been made to transform coal into upgraded and more easily handled products, by the application of various physico-chemical methods. In addition to physico-chemical methods, biological processes have been considered as an alternative to improve coal, as well as other fossil fuels (Fakoussa, 1992; Cámara et al., 1997).

During recent years, biotechnology has rapidly succeeded in constantly gaining new applications, in particular in the chemical and pharmaceutical industry and in the environmental sector. Examples of the latter are waste water treatment and soil remediation, where use is made of the capability of microorganisms to convert organic matters into liquid substances, and finally gaseous substances such as carbon dioxide and methane (Ehrler et al., 1987; Schacht et al., 1991). Only a few years ago, a large number of substances were considered unable to be attacked by means of biology. Extensive screening, however, resulted in the discovery of a constantly increasing number of novel microorganisms, which are capable of converting even complex and large molecules that otherwise are difficult to degrade. These more complex substances also include coal. The difficulties, however, rise with an increasing degree of carbonization, i.e. it is much easier biologically to convert lignite into gaseous and liquid products than, for example, hard coal.

This chapter describes two aspects of the biotechnological conversion of coals into upgraded products. One aspect refers to the conversion of lignite into liquid products for the energy and chemical sectors. Another aspect is the production of biopolymers for manufacturing biodegradable plastic materials. Both aspects have in common that in the first step of treatment the coal must be partially degraded and converted into soluble products. These products represent an aqueous solution of a mixture of heterogeneous chemical compounds. In the case of fuel production, the necessary upgrading steps are de-watering and de-ashing in order to produce a high calorific liquid fuel. For the production of bioplastics, the solubilized products are used by bacteria as a chemical feedstock, and are converted into well-defined chemical compounds such as polyhydroxyalkanoic acids (PHA).

2 Historical Outline

The first investigations into the action of fungi and bacteria on coal were carried out as early as 1927 at the Max-Planck Institute of Coal Chemistry in Mülheim (Fischer and Fuchs, 1927). The methods applied at that time did not, however, succeed in clarifying the question of whether, and to what extent, degradation of the macromolecular coal structure occurs. For this reason, coal had for many years been considered inaccessible to the action of microorganisms. However, in 1981, the studies of Fakoussa – who showed for the first time that both fungi and bacteria can utilize coal as the sole carbon source – and later of Cohen and Gabrielle (1982) – who described the microbial production of water-soluble products from coal – demonstrated that coal is indeed a substrate for certain microorganisms. These findings encouraged several laboratories during the mid-1980s to intensify their research on the biotechnological conversion of coal. All studies carried out so far are still at a basic stage of research.

The major task in coal microbiology has been the screening and identification of suitable microorganisms, which live for example in coal seams, overburden dumps, or abandoned mine galleries. Various fungi (mycelium-forming fungi, yeasts) and bacteria (e.g., *Pseudomonas* sp., actinomycetes) have been found to be capable of solubilizing or depolymerizing coal (Rehm, 1992). These include many organisms that attack lignin – the plant polymer that provided the strength and rigidity to wood. This is quite conclusive, as during the coalification process, coal was formed, inter alia, mostly from lignin.

3 Biotechnological Conversion of Lignite to Liquid Products for the Energy and Chemical Sectors

One of the key factors for every industrial activity is energy. Thus, a secure supply of appropriately priced energy is a major and indispensable element of economic activity. The European Union (EU) is already the world's largest importer of oil, natural gas, and coal. According to estimates of the European Commission, by 2020, its dependence on Third-World countries for its energy supply could increase further, to about two-thirds overall, to about three-quarters for gas, and almost completely for oil. High and increasing dependency of the EU on imports for its energy supplies is risky, and depends on geopolitical trends which are difficult to predict. Hard coal and lignite accordingly are – and will continue to be – important pillars of a balanced energy mix. One reason for this is that their use for electricity generation is absolutely essential if continuity of production and moderation in prices are to be guaranteed.

In Germany, the energy supply for power generation has been characterized by a cost-cutting and risk-minimizing mix for 40 years. In 1999, hard coal (26%), lignite (24%) and nuclear energy (31%) made almost equal contributions to power generation, while natural gas (10%), mineral oil (1%), water and others (8%) were used to a lesser extent. With its 39% share in total primary energy production, lignite contributes some 24% to the country's power generation. Lignite's particular significance to Germany's energy sector is due to the fact that it is the only domestic energy source which is available at competitive conditions in the long term.

In order to increase the competitiveness of lignite, one aim in biotechnological lignite upgrading is the production of liquid products suitable for power production. As an alternative to direct combustion and thermochemical conversion, attempts have been made to convert lignite into liquid fuels by means of bacteria and fungi. Due to the 'mild' reaction conditions (atmospheric pressure, room temperature) prevailing during microbiological conversion processes, favorably priced techniques seem to be feasible. Compared with the solid coal fuel, conversion into liquid and gaseous coal products would permit a simpler handling and provide technically easy processing options that include corresponding cost advantages (Reich-Walber et al., 1997). An extended range of upgraded coal products is also expected to open new applications for lignite.

3.1 Chemical versus Biotechnological Conversion

Chemical/thermal processes for lignite liquefaction and gasification had been developed at an industrial scale by the early 1980s (Teggers, 1987). In this context, coal gasification and coal hydrogenation were aimed at producing chemical products or fuels, respectively. These chemical/thermal processes are characterized by the application of high temperatures and pressures, which correspondingly require expensive plants and lead to high operating costs. This is one important reason why lignite-based liquid and gaseous products were so far lacking economic efficiency as compared with competitive products derived from crude oil and natural gas.

In contrast to this, biological conversions take place under 'mild' ambient conditions with respect to pressure and temperature. Because of this, relatively simple and favorably priced technologies are to be expected for biological conversions of lignite. However, this will also result in lower conversion rates per unit time as compared with that of already established lignite upgrading processes. An essential prerequisite for a biotechnological alternative to the chemical/thermal processes is the availability of microorganisms that are capable of converting lignite into suitable liquid products at an industrial scale.

3.2 Chemical Lignite Conversion in Aqueous Solution

Conversion of solid lignite into a solubilized coal product (coal solubilization = formation of black liquids from solid coal particles) can be performed purely on a chemical basis in an alkaline environment. This is due to the fact that the degree of carbonization of lignite is lower than that of hard coal, and lignite contains a considerable percentage of humic acids (30–70%, wt/wt). Humic acids – in terms of water solubility – are characterized by a high content of carboxyl groups (-COOH) which tend to form salts under alkaline conditions:

$$[\text{Humic acid}]\text{-COOH} + \text{MeOH} \rightarrow [\text{Humic acid}]\text{-COOMe} + H_2O$$

This process converts coal humic acids into water-soluble substances. For an industrial large-scale process, however, the alkaline solubilization of lignite is economically not feasible, because the reactions occur only on a 'molecule by molecule basis', and such a process would therefore require very large amounts of alkaline substances. In addition, the lignite skeleton (coal matrix), in having a higher degree of carbonization and also a higher molecular weight than humic acids, is not solubilized by alkali. Therefore, the degree of coal conversion via alkaline solubilization is limited.

Cohen et al. (1990) have reported that the chelating agent ammonium oxalate is the major coal-solubilizing agent of the fungus *Trametes versicolor.* Chelators remove complex-forming metal ions from the coal texture causing – at least in part – disintegration of the coal structure. However, Fakoussa (1994) has found that chelators solubilize only highly oxidized coals, such as leonardite. For other coals, which are oxidized to a lesser extent and have a low ash content (e.g., German lignites), it was found that the application of chelators alone is insufficient to solubilize lignite to a significant extent (Cohen, 1995). Fakoussa (1994) suggested that conversions, which could alter the carbon skeleton of certain coals, are probably catalyzed by oxidative enzymes expressed in particular by fungi. Therefore, extensive screening tests have been carried out to find suitable 'coal microorganisms' which can be used for the solubilization and depolymerization of coal.

3.3 Enzymatically Induced, Chemical Coal Conversion

As an intermediate result of a research project on microbial lignite conversion (Meyrahn et al., 1996), 160 L of solubilized lignite (in aqueous solution) were produced by employing an extract obtained from the cells of a bacterial consortium isolated from soils (working designation BRIU4; individual strains were not further specified) (Gaddy, 1996). Chemical analysis and investigations into de-watering and de-ashing of the solubilized coal were also carried out.

3.3.1 Solubilization Procedure

The lignite used for the solubilization experiment was a mixture of run-of-mine coals (solid) from the Garzweiler and Hambach open-cast mines (Rheinbraun AG, Germany). The particles were ground to sizes of <1 mm (37%, wt/wt), <4 mm (42%) and <10 mm (21%). For the results of chemical analyses, see Table 1. Coal solubilization was performed using a mixed bacterial culture (BRIU4; Gaddy, 1996), in a technical process that could be scaled up. In order to achieve high degrees of solubilization ($\geq 50\%$), the blended coal had to be pretreated with 1% (vol/vol) H_2O_2; without H_2O_2 pretreatment, solubilization was only about 20%. Within the scope of the research program, 160 L of solubilized coal were produced in three process steps (Figure 1). Steps I, II, and III show the solubilization of the non-pretreated blended coal. A two-stage filtration (first stage at 20 μm, second stage at 5 μm) in step II was used to separate the non-converted residual coal from the solubilized coal. After filtration, the solublilized coal contained about 2% (wt/wt) coal (moisture free). In step III, the liquid was concentrated by a factor of 3.5 by applying a light vacuum in order to reduce the volume for the transport from the USA to Germany. The products obtained from step III were the solubilized coal (sample A) and the condensate. Solubilized coal from blended coal pretreated with 1% H_2O_2 (sample B) was produced analogously (steps Ia to IIIa).

3.3.2 Chemical Analyses of Solubilized Coal

Proximate and ultimate analyses of the solubilized coal were performed using the samples A and B. The ultimate analysis was carried out with the product of the precipitation of samples A and B with HCl (pH 2). The precipitated solid material containing the solubilized coal was separated from water by

Tab. 1 Proximate and ultimate analysis of liquefied and blended coal.

Analysis		***Liquefied Coal Sample A***	***Liquefied Coal Sample B***	***Blended Coal***
Proximate				
Water	% (wt/wt)	92.3	93.4	56.4
Ash (450 °C)	% (wt/wt)	8.93	6.38	7.16
pH–value		7.35	7.40	–
total organic carbon (TOC)	mg/Liter	25300	29600	–
Ultimate (dry)		#	#	
Carbon	% (wt/wt)	66.8	62.2	64.7
Hydrogen	% (wt/wt)	5.44	5.14	4.65
Nitrogen	% (wt/wt)	1.55	2.49	0.80
Sulphur	% (wt/wt)	0.80	0.78	0.80
Oxygen	% (wt/wt)	23.5	26.6	21.9
Ash (450 °C)	% (wt/wt)	1.87	2.80	–

Separation of solids after HCl precipitation.

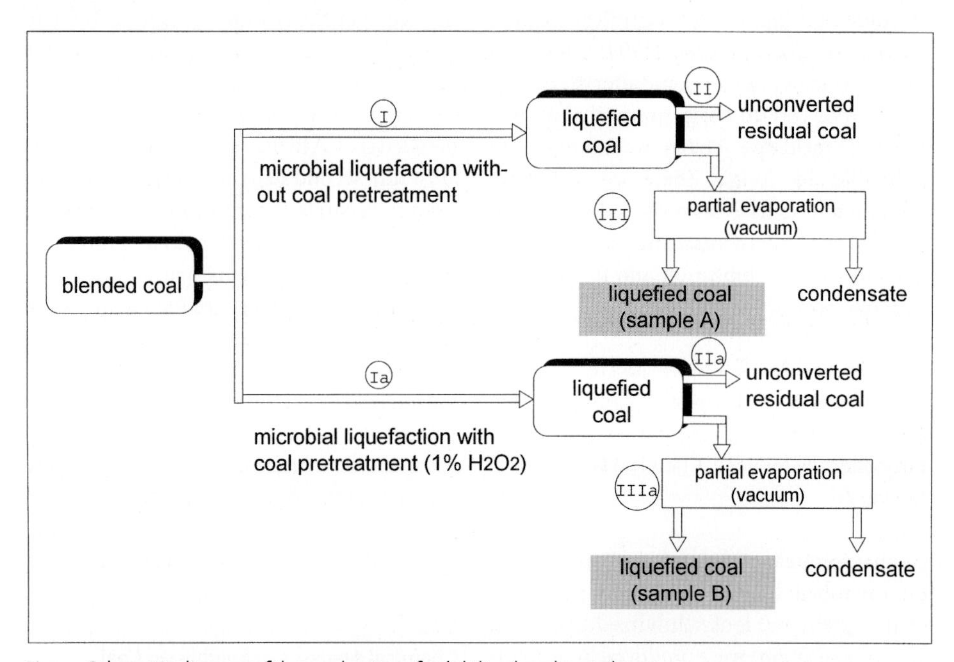

Fig. 1 Schematic diagram of the production of solubilized coal samples.

30 min centrifugation at 3000 rpm, and was subsequently cleaned three times with acidified water (HCl, pH 3.5). In order to check for a possible change of the ash content resulting from microbial treatment and acid precipitation, the ash content of the precipitate was also analyzed.

The results of the proximate and ultimate analyses are given in Table 1. The solubilized coal is characterized by a high water content

(93% wt). In terms of water-free substance (dry), solubilized coal sample A had an ash content of 8.93% (wt/wt), whereas that of sample B was only 6.38% (wt/wt). The blended coal, by comparison, had an ash content (dry) of 7.16% (wt/wt). The solubilized coal exhibited a significantly higher nitrogen content than the blended coal (sample A: 1.55%; sample B: 2.49%; blended coal: 0.8%, wt/wt). The content of hydrogen of the liquefied coal was slightly higher than that of the blended coal (sample A: 5.44%; sample B: 5.14%, blended coal: 4.65%, wt/wt). The differences in the carbon contents of liquefied and blended coal were minimal ($\pm 2\%$, wt/wt), and the sulfur contents remained unchanged. The oxygen content of liquefied coal from sample A (23.5%, wt/wt) remained unchanged within the experimental error ($\pm 2\%$, wt/wt), whereas in case of sample B, the oxygen content was significantly higher (26.6%, wt/wt) than that of the blended coal (21.9%, wt/wt). The pH value of the liquefied coal was about pH 7.4. In the condensate, a maximum of 44 mg total organic carbon (TOC) per liter was measured. Compared with the TOC contents of sample A (25,300 mg L^{-1}) and sample B (29,600 mg L^{-1}), only a very small and thus negligible portion of organic carbon ($< 0.2\%$) was removed by applying the vacuum to samples A and B.

3.3.3
Preparation Tests with Solubilized Coal

Because of the high water content (93%), the solubilized coal must be considered as a preproduct from which water must be removed before use as a fuel. The maximum water content of the fuel should not exceed 20%. Calculated on the basis of 1000 kg solubilized, 916 kg of water have to be separated, in order to produce 84 kg of a fuel with a water content of 14 kg (= 20%, wt/wt). With respect to the implementation of the separation of the solubilized coal from water at likewise low costs, an additional attempt was to achieve a higher hydrophobicity of the coal material by enzymatic decarboxylation within in the bioreactor. This would converting the solubilized coal, which consisted mainly of hydrophilic coal-derived humic acids, into a hydrophobic product via the following reaction, according to Fakoussa et al. (1999):

$$[\text{Humic acids}]\text{-COO}^- + [\text{H}^+] \rightarrow \text{enzymatic reaction} \rightarrow [\text{Humic acids}]\text{-H} + CO_2$$

The hydrophobic product should be more easily separable from the water by phase separation. In view of the economics of such a process, the enzyme (decarboxylase) must be highly unspecific, stable, and extracellular. However, at present only intracellular decarboxylases are available, and these are difficult to handle. Therefore, with the presently available technique, an enzyme-mediated increase in the hydrophobicity of solubilized coal is not feasible on an industrial scale. Nonetheless, studies are being continued to find more suitable decarboxylases, which could be used in future technical processes (Fakoussa et al., 1999).

Because enzymatic decarboxylation was not available as a method, other more favorably priced water treatment processes (which are already well established on an industrial scale) were investigated for the separation of the solubilized coal phase from the aqueous culture filtrate. Since it was not known whether the solubilized coal product occurred in a liquid, was dissolved, or was in a colloidal state in the aqueous medium, the following treatment processes were investigated at the laboratory scale (Meyrahn et al., 1998):

1) Precipitation/flocculation
2) Demulsification

3) Membrane filtration
4) Extraction
5) Electrocoagulation.

The initial product for the individual approaches was solubilized coal, which had been produced from the non-pretreated run-of-mine coal (sample A).

It was found that with precipitation/flocculation, demulsification, membrane filtration, extraction and electrocoagulation, an efficient drainage and reduction of the ash content of the liquefied coal was not feasible. The phase separation of the solubilized coal product from the aqueous medium cannot simply be implemented by applying industrial water treatment processes. With regard to the separation processes, the main problem is that the liquefied coal product is too hydrophilic and soluble in water, and therefore very difficult to drain. This indicates that mainly polar humic acids were extracted from the coal during fermentation, probably by ammonia produced by the bacteria. In the presence of weak acids (e.g., acetic acid, carbonic acid) it was possible to reduce this hydrophilicity and to precipitate the liquefied coal product as a separate solid phase. This required the addition of large quantities of chemicals, which can be implemented at a laboratory scale but is not economically feasible at an industrial scale. For example, for a large-scale lignite throughput of 4,000 t/h for power generation, 96,000 t of liquefied lignite must be treated per day. To adjust the pH value to 4.0 using acetic acid would require a daily consumption of about 2000 t of acid.

The results of the chemical analyses provide indications about the reaction mechanism of the bacterial lignite solubilization. It was assumed previously that above all, oxidative enzymes are responsible for microbial lignite solubilization by introducing additional functional groups containing oxygen into coal (Fakoussa, 1992). It was therefore surprising that during the solubilization of the non-pretreated coal, no significant amounts of oxygen were incorporated into the coal molecule. This phenomenon can be explained by the fact that no oxidative processes occurred during the bacterial solubilization reactions. Oxygen incorporation into the coal was only detected if the coal was pretreated with H_2O_2. Because of the high costs, the use of large amounts of H_2O_2 (1% of the coal) is not suited for a coal treatment at an industrial scale. Due to the increased nitrogen content of the solubilized coal, it is assumed that urea, which is added to the medium as a nitrogen source for the bacteria, plays an important role in the solubilization process performed by the bacterial culture BRIU4. From detailed studies it was further concluded that the mechanism of coal solubilization with the isolate BRIU4 results only indirectly from an enzymatic reaction that converts urea in the medium to ammonia, which subsequently promotes the removal of humic acids from the coal due to its alkalinity. Such a reaction could, for example, be catalyzed by a urease, which is widespread among bacteria. All in all, it is unlikely therefore that the coal was directly attacked by bacterial enzymes in this process. Similar results were found for certain filamentous fungi (molds; Hofrichter et al., 1997): The 'typical' coal solubilization depends mainly on the nitrogen content of the medium (resulting in higher pH values), as well as on the degree of oxidation of coal, which can be enhanced, e.g., by pretreatment with H_2O_2. For this process, extracellular enzymes such as oxidases and peroxidases seem to play only a minor role – if at all.

With regard to the quality of the product for energetic uses, it was not possible to process the humic acid-like, solubilized coal product from the bacterial culture into a liquid fuel. The solubilized coal product was not upgradable as intended. The water content of the best product, which was derived via electrocoagulation, was still 88% (wt/wt), and therefore showed no improvement when compared with the run-of-mine coal used as a starting material and which had a water content of only 56.4% (wt/wt). The final conclusion is, that coal liquefaction using the bacterial isolate BRIU4 did not yield the desired material; rather, this required the application of alternative strategies for the bioconversion of coal in order to optimize the product.

3.4 Enzymatic Conversion of Coal into Liquids

Based on the results described above, it was necessary to intensify basic studies for the bioconversion of lignite into liquid products, with an emphasis on the use of ligninolytic enzymes – especially laccases and peroxidases. The objective was to cleave the coal molecules into fragments which are smaller than those obtained during the alkaline extraction of humic acids from the coal. This must involve the cleavage of C–C and/or C–O bonds.

The enzyme manganese peroxidase isolated from fungi demonstrated a general suitability for this depolymerization for humic acids derived from coal (Ziegenhagen et al., 1997). However, the application of manganese peroxidase for the conversion of coal at an industrial scale is impossible for cost reasons, since the required addition of stoichiometric amounts of hydrogen peroxide (H_2O_2) is too expensive. The cost for H_2O_2 alone would amount to ~$20 per ton of coal converted.

Laccases, by contrast, have the advantage that atmospheric oxygen is used as the oxidizing agent. There is evidence that the degradation of humic acids is related to the laccase activity of the fungus *Trametes versicolor* (Fakoussa et al., 1997). During the attack on the heterogeneous coal molecules, the enzyme employed must react highly unspecifically. The main problem is that coal, being a high-molecular weight substrate, cannot enter the catalytic center of the enzyme; alternately, the enzyme cannot penetrate the macromolecular network of the coal. The problem could be solved by using so-called low-molecular mass mediators that are able to penetrate the complex structure of coal. From studies on the enzymatic removal of lignin from cellulose, it is known that the basidiomycete *Pycnoporus cinnabarinus* obtained from decaying pine wood is able to degrade lignin by using laccase and the fungal metabolite 3-hydroxyanthranilate as a mediator (Eggert et al., 1996a). This mechanism might also be applicable to the bioconversion of coal, since lignite is formed through the geochemical transformation of wood and still possesses lignin-like chemical structures. A schematic diagram of the laccase-catalyzed coal conversion reaction is shown in Figure 2. The depolymerization and decolorization of lignite humic acids in liquid cultures was recently demonstrated for several basidiomycetes by Temp et al. (1999). During this reaction, fulvic acid-like compounds of lower molecular mass were formed. Since the growth rates of white-rot fungi are too low for large-scale application of microbial fuel conversions, the use of enzyme preparations is mandatory. The knowledge of key enzymes and their cofactors involved in the conversion of lignite is required to optimize in-vitro systems for technical applications. Methods for the isolation of large amounts of laccase from fungal cultures have been developed, and the enzyme was found to be highly stable (Eggert et al., 1996b). Therefore, it should be possible to perform in-vitro coal depolymerization tests by using isolated laccase. The main advantage of such a process for the coal bioconversion at an industrial scale would be that the cost-intensive adjust-

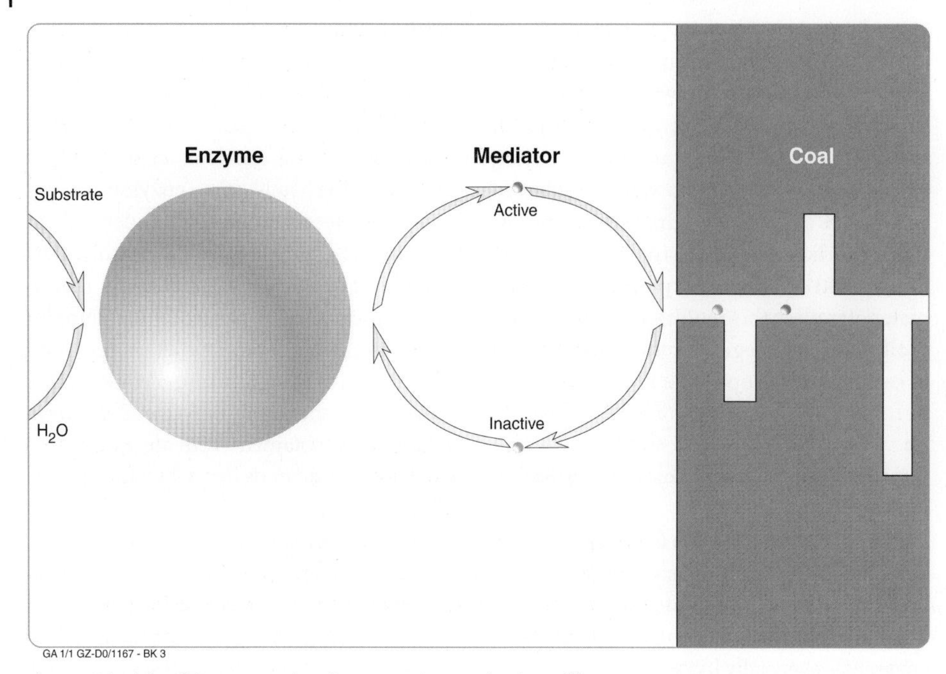

Fig. 2 Principle of the enzyme/mediator reaction mechanism of laccases.

ment of optimal growth conditions for the fungus is not required. A study was therefore performed to determine whether in-vitro conversion/degradation of lignite by laccases in combination with mediators is able to produce a product suitable for the production of a lignite-based liquid fuel.

Extracellular laccase from *Pycnoporus cinnabarinus* (ATCC 200478) was produced in a bioreactor (Eggert and Temp, 1999). After purification of the enzyme, in-vitro lignite conversion was performed after the addition of a buffer solution (for pH adjustment) and of redox mediators. In addition to the enzyme of *P. cinnabarinus*, another fungus-derived laccase (NOVO laccase SP 807) from NOVO Nordisk, Denmark, was also used. The experiments showed, that the addition of redox mediators is necessary to obtained a lignite conversion higher than 2%. The depolymerization of the lignite matrix or of lignite particles in one step was increased up to 8% using redox mediators such as ABTS (2,2′-azina-bis[3-ethylthiazoline-6-sulfonate], which was one of the best redox-mediators enabling laccase to convert lignite. The maximum lignite conversion rate achieved was 20% by using subsequent enzymatic steps and introducing fresh laccase and mediator (Götz et al., 1999).

3.5 Use of Biotechnologically Solubilized Lignite for Electricity Generation

A technical application of biotechnologically solubilized lignite at a larger scale might be the production of electricity. The technical concept was developed with the aim of assessing the economic efficiency of power generation based on biotechnologically solubilized lignite (Reich-Walber et al., 1995, 1997). Starting with conventional lignite extraction on the basis of the open-cast mining technology, this concept included, inter alia, coal solubilization in bioreactors located near the deposit, transport of the liquefied coal with reduced contents of moisture, ash and sulfur via pipelines, and power generation in the power plant. Compared with the state-of-the-art, viz. 'BoA' (German abbreviation for lignite-based power plant with optimized plant engineering), and despite the additionally occurring costs due to coal bioconversion, considerable financial incentives would arise due to the reductions of costs for coal transport and power plant technologies. Therefore, the activities in respect of microbiological lignite upgrading have been intensified with the aim of scaling-up laboratory results to a technically assessable scale (Meyrahn et al., 1996). The most efficient biotechnological process developed in the research project was established at the laboratory scale using the enzymatic laccase mediator system described in Section 3.4. The highest rate of lignite conversion into a liquid product was 20% (wt/wt). This conversion rate was only reached when chemical mediators such as ABTS (2,2'-azino-bis[3-ethylthiazoline-6-sulfonate]) were used. The costs for the mediators amounted to $22,000 per ton of coal. Under these circumstances, the industrial use of the biotechnological lignite conversion for the preparation of liquid fuels is economically not feasible. The present bottlenecks identified from an industrial viewpoint are: (1) the rate of coal conversion is too low (20%, wt/wt); (2) the costs of chemicals/mediators) are too high ($35 to $22,000/t coal); and (3) efficient drainage of liquid coal products is not possible.

3.6 Use of Biotechnologically Liquefied Lignite as Input Material in the Refinery Sector

Biotechnologically upgraded lignite could extend the range of products, and open up new lignite applications. One new application might be the use of solubilized lignite as input material in the refinery sector to replace oil. Based on the results described above, the use of liquid lignite products is so far not possible because: (1) the contents of water (88%, wt/wt) and oxygen (26%, wt/wt, on dry basis) are too high; (2) the product is not an oil-like liquid; and (3) the liquefaction procedure is, at present, too expensive.

4 Use of Biotechnologically Liquefied Lignite as Chemical Feedstock for the Production of Bioplastics

Polyhydroxyalkanoic acids (PHA) are a complex class of naturally occurring polyesters, which consist of various hydroxyalkanoic acids linked by oxo ester linkages formed between the hydroxyl of one constituent and the carboxyl group of another constituent (Doi, 1990; Steinbüchel and Valentin, 1995). They are synthesized by a wide range of bacteria, which deposit PHAs as cytoplasmic inclusions for storage of energy and carbon (Anderson and Dawes, 1990; Steinbüchel et al., 1995). PHAs are thermoplastic and/or elastomeric, exhibit a high degree of polymerization, consist of only the *R* stereoisomers of hydroxyalkanoic acids if the carbon atom with the hydroxyl group is chiral, and are generally biodegradable (Müller and Seebach, 1993; Jendrossek et al., 1996). In addition, many of the PHAs can be produced from renewable resources (Steinbüchel and Füchtenbusch, 1998; Madison and Huisman, 1999). Therefore, the industry has an increasing interest in PHAs, to use them not only for the manufacture of biodegradable packaging materials but also for various medical and pharmaceutical applications, and as a source for the synthesis of enantiomeric pure chemicals (Hocking and Marchessault, 1994).

PHAs are synthesized from many different carbon sources. In general, the carbon source must be converted into a coenzyme A thioester of a hydroxyalkanoic acid by the metabolism of the PHA-accumulating bacterium. The hydroxyacyl moieties of these thioesters are then polymerized to the polyester by PHA synthases, which are the key enzymes of PHA biosynthesis. The biochemistry and molecular biology of PHA biosynthesis and of PHA synthases have been thoroughly investigated during the past few years. These studies have revealed much knowledge on the physiology as well as on the enzymology, and well-characterized PHA biosynthesis genes are now available from many different bacteria (Rehm and Steinbüchel, 1999).

4.1 Reasons to Investigate the Conversion of Coal into PHAs

There are several advantages of processes resulting in the conversion of lignite to an intracellular polymeric product. Normally, the complex structure, heterogeneous composition, and other features of coal do not allow its use as a feedstock for fermentation processes that mostly result in the formation of extracellular products. In contrast to glucose or fatty acids, which are widely used by the industry as carbon sources for microbial fermentations, the structures of lignite and other coals are rather heterogeneous, and could result in the formation of several undesired by-products from the various constituents. In addition, the downstream processing, i. e., the isolation of extracellular products from the medium must be rather tedious because the product has to be separated from many components, including the residual components of the lignite.

On the other hand, lignite and other coals are cheap and abundantly available feedstocks. In addition, PHAs offer the advantage that the product occurs intracellularly and thus can initially be easily separated from the medium components and the remaining coal liquefaction products simply by harvesting the cells. Subsequently, the PHAs can be released from the cells by established processes within the industry. Furthermore, the product has a uniform chemical structure, allowing the conversion of a highly complex substrate into a somewhat homogeneous product consisting of repeating units of enantiomerically pure hydroxyalkanoic acids. These arguments would also apply to some other intracellular products, such as triacylglycerols, which are accumulated to a large extent by many Gram-positive bacteria belonging to the genus *Rhodococcus* and other actinomycetes (Alvarez et al., 1996; Wältermann et al., 2000), or polysaccharides. However, so far the biotechnological production of triacylglycerols or polysaccharides from coals has not been investigated.

4.2 Screening of Laboratory Strains which Grow on Lignite Depolymerization Products and Convert them into PHAs

Until recently, no PHA-accumulating bacteria were known that could use lignite or other coals directly as a carbon source. However, a wide range of bacteria have now been screened for their capability to use the chemically heterogeneous, low-rank coal solubilization and depolymerization products obtained from lignite treated with fungi (*Trichoderma atroviride, Clitocybula dusenii*) or with chemicals, as sole carbon source for growth and to accumulate PHAs (Füchtenbusch and Steinbüchel, 1999).

Pseudomonas oleovorans turned out to be the most promising candidate. The growth of this bacterium depended strongly on the concentration of coal degradation products isolated from *T. atroviride* in the medium, and no inhibitory effect of these degradation products on the growth of *P. oleovorans* was observed at concentrations up to 2.5% (w/v). The maximum rate of conversion of the fed depolymerization products into bacterial dry mass was approximately 30% during submersed cultivation (Füchtenbusch and Steinbüchel 1999). In addition, *P. putida, Rhodococcus opacus, R. ruber, Nocardia opaca* and *N. corallina* used these solubilized products also for growth, though to a less extent. *Burkholderia cepacia* used only depolymerization products obtained by chemical treatment, whereas *R. erythropolis* and *R. fascians* did not use either lignite depolymerization product.

The accumulation of PHAs from coal bioconversion products was studied in detail in *P. oleovorans* and *R. ruber.* In mineral salts medium and under starvation for ammonium, the wild-type of *P. oleovorans* accumulated PHAs up to 8% of the cell dry weight. The accumulated PHAs represented a copolyester consisting of 3-hydroxydecanoate (3HD) as main constituent, plus 3-hydroxyhexanoate (3HHx) and 3-hydroxydodecanoate (3HDD) as minor constituents. A recombinant strain of this *P. oleovorans* harboring the PHA biosynthesis operon of *Ralstonia eutropha* accumulated PHAs up to 6% of the cell dry weight. The accumulated PHAs consisted most probably of a blend of two different polyesters. One was poly(3-hydroxybutyrate) homopolyester, contributing to approximately 9%, and the other was a copolyester consisting of 3HHx and 3-hydroxyoctanoate (3HO) as main constituent plus 3HD and 3HDD as minor constituents (Füchtenbusch and Steinbüchel, 1999). *R. ruber* accumulated under the same conditions a copolyester consisting of 3-hydroxybutyrate (3HB) plus 3-hydroxyvalerate (3HV); however, the polyester contributed to only approximately 2–3% of the cell dry weight (Füchtenbusch and Steinbüchel, 1999).

The main bottlenecks in the synthesis and production of PHAs from depolymerization products are the scaling-up of substrate production and the low conversion rate and low yield of PHAs obtained (Klein et al., 1999). This is mainly due to the current need to carry out the conversion in two separate steps, with the solubilization of lignite by *T. atroviride* in the first step, and the conversion of the solubilized products into PHAs in the second step by a suitable bacterium. In addition, the isolation of the solubilized lignite from the hyphae of the fungus or from the medium is tedious. Therefore, it would be desirable to have microorganisms available that could convert lignite directly into PHA in a one-step process. One possibility would be to establish PHA biosynthesis in the lignite-solubilizing fungus by transfer of the genes for PHA biosynthesis, e.g., from *R. eutropha* or *P. oleovorans*, but this has not yet been carried out. Another possibility would to identify a bacterium that is also capable of solubilizing and depolymerizing coal. If such a bacterium were not able to synthesize PHAs, it would not be difficult to establish PHA biosynthesis in this bacterium by using genetic methods. Recently, such a bacterium was successfully identified (Füchtenbusch et al., 2001). In a mineral salts medium, and using untreated lignite as sole carbon source, various Gram-positive bacteria (e.g., *Mycobacterium fortuitum, Micromonospora aurantiaca* and three isolates belonging to the genus *Gordonia*, most probably to the species *polyisoprenivorans*) were able to grow. Analysis of the remaining carbon source revealed significant changes in the structure of lignite, indicating that growth had occurred at the expense of the carbon provided by lignite (H. Schmiers, B. Füchtenbusch and A. Steinbüchel, unpublished results). Interestingly, all these bacteria are capable of utilizing natural rubber as well as some synthetic rubbers as sole carbon sources for growth, and were previously isolated as such. For this reason, these bacteria are currently also under investigation in our laboratory (Linos et al., 1999, 2000; Berekaa et al., 2000). Since these bacteria are unable to synthesize PHAs, genetic transfer systems must be developed that allow the transfer and expression of PHA biosynthesis genes from other bacteria.

5
Outlook and Perspectives

The investigations carried out during the past decade on the use of biotechnologically solubilized lignite for power production or as input material for the refinery sector have revealed new and scientifically interesting information. However, most of these studies did not provide a breakthrough for applications at an industrial scale; neither did they provide a perspective on how these problems might be resolved in the near future. As yet, the coal conversion rates obtained have been too low, and the enzymes involved have turned out to be difficult to handle and/or to require expensive chemicals in order to function. Therefore, the biotechnological conversion of lignite into upgraded products cannot be performed at competitive costs at present. The conversion of the structurally heterogeneous coal or lignite into uniform intracellular products by microorganisms appears promising, and may provide an interesting alternative biotechnological use of these fossil carbon sources. On the basis of the current data it must be concluded that the biotechnological conversion of lignite cannot be developed as a commercial method within the short and medium term. It appears that further research is required in this respect, and that the variety of microorganisms and their metabolic potential must be more broadly utilized.

6
Patents

Fakoussa, R.M. (1999) Behandlung von Braunkohlebestandteilen zum Zwecke der Veredlung. German patent application 199 45 975.4.

Füchtenbusch, B., Linos, A., Steinbüchel, A. (1999) Produktion von Braunkohle-Verflüssigungsprodukten durch kautschukabbauende Mikroorganismen. German patent application 10009696.4.

Reich-Walber, M., Gaddy, J.L. (1997) Verfahren zur mikrobiellen Solubilisierung von festen fossilen kohlenstoffhaltigen Substanzen. German patent application 197 10 846 A1.

7
References

Alvarez, H. M., Mayer, F., Fabritius, D., Steinbüchel, A. (1996) Formation of intracytoplasmic lipid inclusions by *Rhodococcus opacus* strain PD630, *Arch. Microbiol.* **165**, 377–386.

Anderson, A. J., Dawes, E. A. (1990) Occurrence, metabolism, metabolic role, and industrial uses of bacterial polyhydroxyalkanoates, *Microbiol. Rev.* **54**, 450–472.

Berekaa, M. M., Linos, A., Reichelt, R., Keller, U., Steinbüchel, A. (2000) Effect of pretreatment of rubber material on its biodegradability by various rubber degrading bacteria, *FEMS Microbiol. Lett.* **184**, 199–206.

Cámara, Á., Laborda, F., Monistrol, I. F. (1997) Special Issue: 5th International Symposium on Biological Processing of Fossil Fuels, Madrid, *Fuel Process. Technol.* **52**, 1–284.

Cohen, M. S. (1995) Internal report for Rheinbraun AG (unpublished results).

Cohen, M. S., Gabrielle, P. D. (1982) Degradation of coal by the fungi *Polyporus versicolor* and *Poria monticola*, *Appl. Environ. Microbiol.* **44**, 23–27.

Cohen, M. S., Feldmann, K. A., Brown, C. S., Grey E. T. (1990) Isolation and identification of the coal-solubilizing agent produced by *Trametes versicolor*, *Appl. Environ. Microbiol.* **56**, 3285–3290.

Doi, Y. (1990) *Microbial Polyesters*. VCH Publishers, New York.

Eggert, C., Temp, U. (1999) Internal report for Rheinbraun AG (unpublished results).

Eggert, C., Temp, U., Dean, J. F. D., Eriksson, K. E. L. (1996a) A fungal metabolite mediates degradation of non-phenolic lignin structures and synthetic lignin by laccase, *FEBS Lett.* **391**, 144–148.

Eggert, C., Temp, U., Eriksson, K.E.L. (1996b) The ligninolytic system of the white rot fungus *Pycnoporus cinnabarinus*: purification and characterization of the laccase, *Appl. Environ. Microbiol.* **62**, 1151–1158.

Ehrler, P., Glöckler, R., Erken, M., Ritter, G. (1987) Unterstützung der aeroben biologischen Abwasserreinigung durch Braunkohlenkoks, *Korrespondenz Abwasser* **2**, 129.

Fakoussa, R.M. (1981) Coal as a Substrate for Microorganisms: Investigations of the microbial decomposition of untreated bituminous coals, Ph.D. Thesis, Friedrich-Wilhelms-Universität, Bonn.

Fakoussa, R. M. (1992) Mikroorganismen erschließen Kohle-Ressourcen, *BioEngineering* **8**, 21–28.

Fakoussa, R. M. (1994) The influence of different chelators on the solubilization/liquefaction of different pretreated and natural lignites, *Fuel Proc. Technol.* **40**, 183–189.

Fakoussa, R. M., Frost, P., Schwämmle, A. (1997) Investigations into the in vitro-liquefaction of brown coal, Proceedings of the 9th International Conference on Coal Science, September 7–12, 1997, Essen, Germany (Ziegler, A. et al., Eds), pp. 1591–1594. DGMK Tagungsberichte 9704.

Fakoussa, R. M., Lammerich, H. P., Götz, G. K. E., Tesch, S. (1999) The second step: increasing the hydrophobicity of coal-derived humic acids enzymatically, Proceedings of the 7th International Symposium on Biological Processing of Fossil Fuels, September 26–29, 1999, Madrid.

Fischer, F., Fuchs, W. (1927) Über das Wachstum von Pilzen auf Kohle, *Brennstoff-Chemie* **8**, 293.

Füchtenbusch, B., Steinbüchel, A. (1999) Biosynthesis of polyhydroxyalkanoates from low-rank coal liquefaction products by *Pseudomonas oleovorans* and *Rhodococcus ruber*, *Appl. Microbiol. Biotechnol.* **52**, 91–95.

Füchtenbusch, B., Schmiers, H., Steinbüchel, A. (2001) Production of humic and fulvic acids with the fungus *Trichoderma atroviride* in a stirred tank reactor and analysis of their chemical structure, *Appl. Microbiol. Biotechnol.*, submitted.

Gaddy, J. L. (1996) Biological conversion of rhenish brown coal, internal report for Rheinbraun AG (unpublished results).

Götz, G. K. E., Schwämmle, A., Temp, U., Eggert, C., Fakoussa, R. M. (1999) Mediator-assisted depolymerization of brown coal by redox enzymes: scope and limitations, Proceedings, 7th International Symposium on Biological Processing of Fossil Fuels, Madrid, September 26–29.

Hocking, P. J., Marchessault R. H. (1994) Biopolyesters. In: *Chemistry and Technology of Biodegradable Polymers* (Griffin, G. J. L., Ed.), pp. 48–96, Blackie Academic

Hofrichter, M., Bublitz, F., Fritsche, W. (1997) Fungal attack on coal II. Solubilization of low-rank coal by filamentous fungi, *Fuel Process. Technol.* **52**, 55–64.

Jendrossek, D., Schirmer A., Schlegel, H. G. (1996) Biodegradation of polyhydroxyalkanoic acids, *Appl. Microbiol. Biotechnol.* **46**, 451–463.

Klein, J., Catcheside, D. E. A., Fakoussa, R., Gazso, L., Fritsche, W., Höfer, M., Laborda, F., Margarit, I., Rehm, H.-J., Reich-Walber, M., Sand, W., Schacht, S., Schmiers, H., Setti, L., Steinbüchel, A. (1999) Biological processing of fossil fuels – Résumé of the Bioconversion Session of ICCS '97, *Appl. Microbiol. Biotechnol.* **52**, 2–15.

Linos, A., Steinbüchel, A, Spröer, C., Kroppenstedt, R. (1999) *Gordonia polyisoprenivorans* sp. *nov.*, a rubber-degrading actinomycete isolated from automobile tire, *Int. J. Syst. Bacteriol.* **49**, 1785–1791.

Linos, A., Berekaa, M. M., Keller, U., Reichelt, R., Schmitt, J., Flemming, H.-C., Kroppenstedt, R. M., Steinbüchel, A. (2000) Biodegradation of *cis*-1,4-polyisoprene rubbers by distinct actinomycetes: microbial strategies and detailed surface analysis, *Appl. Environ. Microbiol.* **66**, 1639–1645.

Madison, L. L., Huisman, G. W. (1999) Metabolic engineering of poly(3-hydroxyalkanoates): from DNA to plastic, *Microbiol. Mol. Biol. Rev.* **63**, 21–53.

Meyrahn, H.; Reich-Walber, M., Lenz, U. (1996) Braunkohle und Biotechnologie – Chancen für eine neue Generation der Kohleveredlung?, *Glückauf* **132**, 697–700.

Meyrahn, H., Reich-Walber, M., Felgener, G.W. (1998) Internal report for Rheinbraun AG (unpublished results).

Müller H. M., Seebach, D. (1993) Poly(hydroxyalkanoates): a fifth class of physiologically important organic biopolymers?, *Angew. Chem.* **32**, 477–502.

Rehm, B. H. A., Steinbüchel, A. (1999) Biochemical analysis of PHA synthases and other proteins required for PHA synthesis, *Int. J. Biol. Macromol.* **25**, 3–19.

Rehm, H.-J. (1992) Biochemie und Biotechnologie der Kohle und kohlestämmiger Verbindungen, *Erdöl und Kohle* **11**, 443.

Reich-Walber, M., Meyrahn, H., Lenz, U., Engelhard, J. (1995) Biotechnological conversion of lignite to liquid and gaseous products for the energy and chemical sectors, *Erdgas Erdöl Kohle* **10**, 423–426.

Reich-Walber, M., Meyrahn, H., Lenz, U. (1997) Rheinbraun's concept for power generation based on biotechnologically converted lignite, *Fuel Process. Technol.* **52**, 267–277.

Schacht, S., Pfeifer, F., van Afferden, M. (1991) Microbial strategies for the degradation of structurally persistent aromatic hydrocarbons, Proceedings, Third Symposium on Biotechnology of Coal and Coal-derived Substances, Essen, September 23–24, p. 171.

Steinbüchel, A., Füchtenbusch, B. (1998) Bacterial and other biological systems for polyester production, *Trends Biotechnol.* **16**, 419–427.

Steinbüchel, A., Valentin, H. E. (1995) Diversity of bacterial polyhydroxyalkanoic acids, *FEMS Microbiol. Lett.* **128**, 219–228.

Steinbüchel, A., Aerts, K., Babel, W., Föllner, C., Liebergesell, M., Madkour, M. H., Mayer, F., Pieper-Fürst, U., Pries, A., Valentin, H. E., Wieczorek, R. (1995) Considerations on the structure and biochemistry of bacterial polyhydroxyalkanoic acid inclusions, *Can. J. Microbiol.* **41** (Suppl. 1), 94–105.

Teggers, H. (1987) Forschung und Entwicklung für die Braunkohlenveredlung, *Braunkohle* **39**, 154.

Temp, U., Meyrahn, H., Eggert, C. (1999) Extracellular phenol oxidase patterns during depolymerization of low-rank coal by three basidiomycetes, *Biotechnol. Lett.* **21**, 281–287.

Wältermann, M., Luftmann, H., Baumeister, D., Kalscheuer, R., Steinbüchel, A. (2000) *Rhodococcus opacus* strain PD630 as a new source of high-value single-cell oil? Isolation and characterization of triacylglycerols and other storage lipids, *Microbiology* **146**, 1143–1149.

Ziegenhagen, D., Hofrichter, M., Fritsche, W. (1997) In-vitro depolymerisation of coal humic acids by manganese peroxidase of *Cliticybula dusenii* b11, Proceedings, 9th International Conference on Coal Science, September 7–12, 1997, Essen, Germany (Ziegler, A. et al., Eds), pp. 1631–1634. DGMK Tagungsberichte 9704.

II
Polyisoprenoids

Biotechnology of Biopolymers. From Synthesis to Patents. Edited by A. Steinbüchel and Y. Doi

ISBN: 3-527-31110-6

4 Biochemistry of Natural Rubber and Structure of Latex

Dr. Dhirayos Wititsuwannakul[1], Dr. Rapepun Wititsuwannakul[2]

[1] Department of Biochemistry, Faculty of Science, Mahidol University, Rama 6 Road, Bangkok 10400, Thailand; Tel: +66-022455195; Fax: +66-022480375; E-mail: scdwt@mahidol.ac.th

[2] Department of Biochemistry, Faculty of Science, Prince of Songkla University, Hat-Yai, Songkla 90110, Thailand; Tel: +66-074211030; Fax: +66-074446656; E-mail: wrapepun@ratree.psu.ac.th

Biotechnology of Biopolymers. From Synthesis to Patents. Edited by A. Steinbüchel and Y. Doi

ISBN: 3-527-31110-6

AOS active oxygen species
ATPase adenosine 5′-triphosphatase
BI bursting index
cDNA complementary DNA
DCPTA 2(3,4-dichlorophenoxy) triethylamine
DMADP dimethylallyl diphosphate
DOPA 3,4-dihydroxyphenylalanine
FDP farnesyl diphosphate
GDP geranyl diphosphate
GGDP geranylgeranyl diphosphate
GPC gel permeation chromatography
HMGCoA 3-hydroxy-3-methyl-glutaryl coenzyme A

HMGR	HMGCoA reductase
HMGS	HMGCoA synthase
IDP	isopentenyl diphosphate
kD	kilodalton
mRNA	messenger RNA
MVA	mevalonic acid
MWD	molecular weight distribution
NAD	nicotinamide adenine dinucleotide
NADH	nicotinamide adenine dinucleotide, reduced form
NADP	nicotinamide adenine dinucleotide phosphate
NADPH	nicotinamide adenine dinucleotide phosphate, reduced form
NRL	natural rubber latex
pI	isoelectric point
PIP-DP	polyisoprenyl diphosphate
PR proteins	pathogenesis-related proteins
REF	rubber elongation factor
SALB	South American leaf blight
SRPP	small rubber particle protein
TLC	thin layer chromatography
TSC	total solids content
UDP	uridine diphosphate

1 Introduction

Natural rubber and other polyisoprenoids obtained from plants are high-molecular weight hydrocarbon polymers consisting almost entirely of five-carbon isoprene units, C_5H_8. These polyisoprenoids are major components of the latex synthesized by specially differentiated cells of the plants and other living organisms. Detailed studies on structure of these polymers have shown that double bonds in the rubbers from *Hevea brasiliensis* (*Hevea* rubber tree) and *Parthenium argentatum* (Guayule) are in the *cis* configuration, *cis*-1,4-polyisoprene, and those from gutta and chicle are in the *trans* configuration, *trans*-1,4-polyisoprene. These natural polyisoprenes are synthesized by enzyme-catalyzed polymerization of isoprene units to various different degrees, resulting in a wide range of molecular weights that are dependent upon the sources from which they are derived. *Hevea* rubber is a typical high-molecular weight *cis*-polyisoprene with a very wide range of molecular weight distribution. Guayule rubber is also of high-molecular weight *cis*-polyisoprene, but having a narrow molecular weights distribution and physical properties slightly different from that of *Hevea* rubber. Gutta and chicle are *trans*-polyisoprenes with much lower molecular weights as compared to the other two rubbers (Tanaka, 1991). In general, the molecular weight of these different polyisoprenes can range from a few thousand up to several million Daltons.

Numerous plants belonging to several different families can form rubber latex. The rubber with high-molecular weight polyisoprenes are produced in the latex of about 300 genera of Angiosperms. The milky latex fluid will flow from these plants after a slight incision of the tissues. In *Parthenium argentatum* (Guayule), the latex is produced and stored in the parenchyma cells (Backhaus and Walsh, 1983). However, more commonly in most plants, the latex is produced and stored in the tubular structure known as laticifers. The most prominent and well studied among these plants is *Hevea brasiliensis.* The laticiferous system of the rubber tree has recently been described and extensively reviewed (de Fay et al., 1989).

The latex composition may vary to a large extent among the different plant species. The polymers in the suspension of which it is formed may contain varying proportions of rubber and of different compounds in addition. Of some 12,500 species of laticiferous plants, about 7000 are found to produce polyisoprenes. In most case the polyisoprene is mixed with resin, making it difficult to use when the content of the resin is high. A limited number of rubber-producing plants can be suitably utilized, and only a few species are cultivated and have economic importance. Among them, *H. brasiliensis* is proved to be the best rubber producer. A few hundred milliliters of latex can be obtained from each tree by simply incising the bark, the common practice of tapping the rubber trees.

Hevea brasiliensis is used commercially for the production of natural rubber that is used industrially for various finished products. High-yielding clones of *Hevea* have been selected and developed, leading to the high productivity of the cultivars in rubber plantations, especially in Southeast Asia. To a large extent, plant breeders have bred out the regulatory mechanism that controls the supply of photosynthetically derived carbon sources that are subsequently channeled into rubber formation. This has resulted in plants that produce far more rubber than the native clones of the *H. brasiliensis.* The discovery that latex production in *Hevea* can be stimulated by the plant hormone, ethylene, is clearly of advantage to rubber planters, and has led to a great deal of research into understanding the hormonal stimulation mechanism. Today, an ethylene generator (Ethephon) is commonly used to stimulate *Hevea* latex production. The effect of high rubber production on growth is quite considerable in rubber trees of high-yielding clones, as evidenced in the reduction in girth increment of the trees.

The latex, which contains the rubber particles, accumulates in specialized cells or vessels known as laticifers. In *Hevea,* the rubber is formed and stored in the rings of laticifers in the bark. Anatomoses between adjacent vessels (Figure 1) in the rings allow the latex from a large area of the cortex to drain upon tapping. The opening of the latex vessels by tapping cuts causes the latex to flow out due to the high turgor pressure inside. The latex flow will continue for a certain period of time and subsequently stop due to coagulation of the rubber particles and

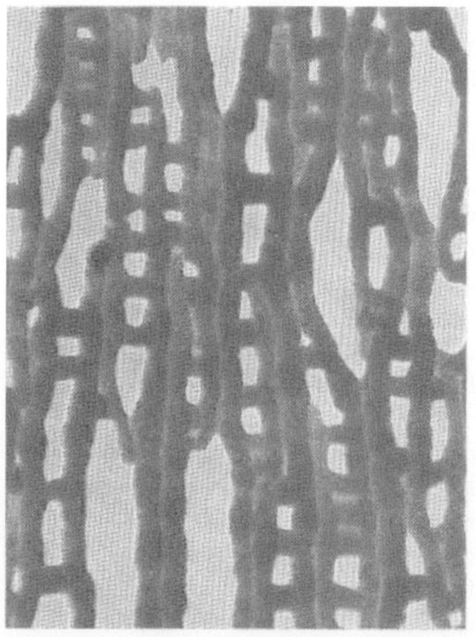

Fig. 1 The latex vessels of *Hevea brasiliensis* laticifers showing anastomoses between adjacent vessels which allow the latex from a large area of the cortex to drain upon tapping of the rubber tree. (Adapted from Zhao, X-Q. (1987) *J. Natl. Rubb. Res.* **2**, 94–98, with permission.).

formation of flocs that leads to plugging at the end of the vessels. The latex is specialized cytoplasm containing several different organelles in addition to rubber particles. These organelles include nuclei, mitochondria, fragments of the endoplasmic reticulum and ribosomes. In addition to these minor components, there are two other major specialized particles which are uniquely characteristic of *Hevea* latex, namely the lutoids and Frey-Wyssling particles.

Lutoids – a major component of latex – are osmotically active spheres which are 1 – 3 μm in diameter. They are gray in color, and surrounded by a single layer membrane. Biochemical analyses revealed that the lipids of lutoids membrane are rich in phosphatidic acids and saturated fatty acyl residues. The name lutoid (yellow) turned out to be a misnomer, because as originally isolated they were contaminated with the yellow Frey-Wyssling particles which contain β-carotene. Frey-Wyssling particles represent a minor component of latex as compared to lutoids. These Frey-Wyssling particles are surrounded by a double-layer membrane and contain many membrane or tubular structures (Figure 2) as well as β-carotene, which is responsible for the characteristic orange to yellow color layer on fractionation by ultracentrifugation of the fresh latex. However, the detailed functions of these particles in latex are still very little known at the present.

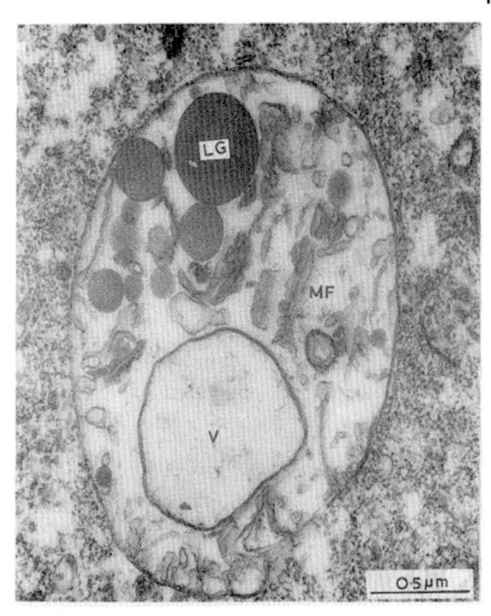

Fig. 2 Frey-Wyssling particles prepared from *Hevea* latex showing the bounded membrane with lipid globules (LG), vesicles (V) and membrane fragments (MF) inside the particles. (Adapted from Gomez, J. B., Hamzah, S. (1989) *J. Nat. Rubb. Res.* **4**, 75–85, with permission.).

This chapter will describe in detail the natural rubber and the latex of *Hevea brasiliensis*. Although the rubbers and latex of other plants and organisms have also been included, they are only covered briefly and less extensively as they will be covered in detail by other authors in this series. This chapter is aimed at providing only background information for comparison with *Hevea* latex. As most of the studies have been performed extensively with the *Hevea* latex over a long period, the focus will be on the plentiful information and details available on this material. Further emphasis will be placed on health problems related to the latex finished products, and to latex allergy caused by latex proteins. Reports of the natural rubber latex allergy have increased steadily since the first reported case in 1979. This is a problem of major health concern and deserves serious attention.

2
Historical Outline

The history of natural rubber and its industrial utilization can be described in two phases. The first phase is the early discovery of natural rubber and its uses in making simple waterproof products. This is followed by improvements in the properties of rubber to make it more suitable and versatile for better-quality products. The second phase is the transfer of the rubber tree from its original site and the establishment of the rubber plantations around the world, especially the predominant rubber plantation industry in Southeast Asia.

2.1 Discovery of Rubber and Improvement of Rubber Properties

Objects made from rubber were first noticed and recorded when Christopher Columbus and his entourage discovered America. The first botanical description of *Hevea* was made by Fresneau in 1747. Subsequently, the names of Hancock, Macintosh, Goodyear, Dunlop and several others have been associated with the history of rubber. Despite the long period since the first discovery of rubber usage by native Indians, rubber is still being regarded as being a strategic material attributed to our civilization. Industrial revolution doubtless provided a major input in contributing to the versatility of natural rubber with regard to the invention and manufacture of rubber products for our daily needs and well-being.

The historical perspective on rubber and rubber trees, as related to the development of the rubber industry, has recently been recounted (Truscott, 1995). Reference to the 'elastic matcrial' was first recorded in the 1500s, when Spanish explorers discovered native inhabitants of the New World using a milky substance to create bouncing toys and also to render their foot coverings and food-handling objects impermeable to water. The milky latex sap was collected from rubber trees, which oozed the rubber sap when their bark was slashed. Unfortunately, samples of these waterproof items did not fare well during the long return journey to Europe. The products degraded, and the resultant gummy substance was of little interest. Indeed, the material was virtually forgotten in recorded history until 1777, when Joseph Priestley pushed a piece of this elastic material across a lead pencil mark on a piece of paper and the mark was rubbed out – thus, the term 'rubber' was first coined. In 1823, Charles Macintosh dissolved lumps of rubber in benzene and poured rubber solution between two pieces of fabric. When the rubber solidified, the resulting material was waterproof and fairly flexible. This simple procedure thus marked the inauguration of the rainwear industry. Although of great use to people living in a temperate, wet climate, the rubberized material was seen to become stiff and brittle in the cold, and sticky and soft in the heat. Both these extremes of conditions led to material breakdown and were tolerated only because an alternative was not yet available.

In 1839, while attempting to create more thermally compatible rubber products, Charles Goodyear inadvertently dropped elemental sulfur into a pan of liquid rubber heating on a stove. The resultant product maintained its elasticity remarkably well when exposed to temperature extremes (Mernagh, 1986), and thus the process of 'vulcanization' was born. The introduction of a practical method of vulcanization, by both Thomas Hancock and Charles Goodyear, changed the nature of rubber and created a new demand for rubber with properties that were satisfactory for industrial use. The incorporation of sulfur, plus admixture of lead oxide and heat, produced a rubber which was proof against hot air and cold air temperatures, and also resistant to melting. Hence, the explosive growth of the rubber industry was started with vulcanized rubber (Brydson, 1978), and this has subsequently resulted in innumerable applications of rubber products, all of which have been supported by patient research not only into the properties of rubber but also improvements in existing manufacturing processes. Vulcanization was much improved by the use of organic accelerators which reduced both the time taken and the amount of sulfur required for the process.

The next hurdle was to move rubber and latex manufacturing into the industrial age. This feat was accomplished by John Dunlop in 1886 while developing a method to produce pneumatic tires. The birth of the tire industry intensified the demand for rubber, the progress of research, and launched the development in manufacturing of latex. Initially, latex was coagulated and processed into compressed blocks and large sheets to await solvent dissolution before pouring it into molds or making it into dipped products. Johnson introduced the method of preservation of liquid latex with ammonia in 1953. This method, together with development of a 50% reduction in the water content of bulk latex by centrifugation, has made feasible the commercial export of concentrated liquid latex (Allen and Jones, 1988).

2.2 History of Rubber Trees and Rubber Plantations

Initially, *Hevea* was not the only candidate for domestication among the many rubber-producing trees. Between 1870 and 1914 four main species underwent extensive trials, including *Ficus elastica* (Asia), *Castilla elastica* (Mexico and West Indies), Ceara rubber *Manihot glaziovii* (Tanganyika and Asia) and *Funtumia elastica* (Africa). None of these plants, however, was sufficiently productive to compete with *Hevea*. The era for rubber production from cultivated *Hevea* and the organized rubber plantation was thus begun.

Although more than 2000 species of plants produce latex, almost 99% of the latex used commercially is harvested from *H. brasiliensis* trees. In 1876, the British East India Company collected 70,000 *Hevea* seeds from their native habitat in the Amazon basin of South America. The seeds were transported to greenhouse facilities in England, where only 2500 germinated. The majority of seedlings were sent to Ceylon (now Sri Lanka), but 22 plants being were taken to Singapore where they formed the foundation for the plantations in Southeast Asia. In retrospect, this endeavor was fortunate because in the early 1900s, the fungus *Microcyclus ulei* virtually destroyed the indigenous-source wild rubber from the jungle trees of the Amazonian rain forest. The South American leaf blight (SALB) caused by this fungus has virtually wiped out commercial plantations there. The blight remain endemic today (de Camargo et al., 1976), and the region has ceased to be a major economic contributor to the world supply of rubber, with most rubber raw material now being produced in Southeast Asia. Thailand has become the number one rubber producer in the world since the early 1990s, though parts of West Africa, South China and Vietnam are increasingly contributing to the world rubber supply.

The organized production of natural rubber to meet industrial needs is now 150 years old. This was preceded by many centuries of localized rubber use in the tropics. Over these years the once simple activity of gathering wild rubber to make a few useful objects has developed into the extensive and complex plantation industry of today. Natural rubber production history (Baulkwill, 1989) can be described in four phases:

1) Pre-industrial but often sophisticated use of rubber from wild plants in tropical America.
2) The organized gathering and export of wild rubber in response to demand from nineteenth-century manufacturers.
3) The establishment of cultivated *Hevea* plantations as the era of the motor vehicle began, and the cultivated crop's domina-

tion of the market in the first half of the twentieth century.

4) The emergence of synthetic rubber in large quantities in the Second World War and its subsequent capture of two-thirds of the market.

An early rubber-based industry, namely the rubber clothing business, boomed and subsequently collapsed due to the unsatisfactory properties of raw rubber, and only the discovery of vulcanization was able to solve this problem. The most startling changes in the demand for rubber in industry came with the adaptation of Dunlop's pneumatic rubber tires for bicycles and motor-powered vehicles. In fact, after 1900 the automobile tire industry began to create a demand for the rubber raw material which 10 years later the wild-rubber industry would be unable to satisfy. This led to the new finding of raw material from wild to cultivated rubber. Early production of rubber was from the lower Amazon (Brazil's Para State), but the domestication of *Hevea brasiliensis* in the East (Asia) has been the most spectacular event in the rubber industry. During the space of only 40 years (1870–1914), a novel plantation industry of some 21 million acres was created to meet the new industrial demand. The successful transfer of *Hevea brasiliensis* to Asia, and the subsequent establishment of commercial rubber plantations there, was the result of many favorable circumstances, the ecological suitability for *Hevea* of several Asian countries in the tropical rain forest belt being the most important contributory factor.

3
Latex of *Hevea brasiliensis*

The *Hevea* latex collected by regular tapping consists of the cytoplasm expelled from the latex vessels, and is similar to the latex in situ. The cytoplasmic nature of tapped latex was firmly established by electron microscopy studies (Dickenson, 1969). Latex is the cytoplasm of an anastomosed cell system which is specialized in the synthesis of *cis*-polyisoprenes. The latex usually contains 25–50% dry matter, 90% of which is made up of rubber. Tapping severs a number of latex vessel rings, and the latex which flows out comprises the contents of vessels at different stages of development. All the organelles as occurring in the latex vessels can be found in the tapped latex. The major particles most common in latex are the rubber particles, the lutoids, and Frey-Wyssling particles which are less numerous than the other two. The composition of latex is about 30–40% rubber, 10–20% lutoids, and 2–3% other substances. The structure and composition of fresh latex has been elucidated by high-speed centrifugation (Moir, 1959). Generally, the latex can be fractionated into three distinct zones (Figure 3) by ultracentrifugation. The top fraction consists almost entirely of rubber, the middle fraction is the metabolic active aqueous phase of latex called C-serum, and the relatively heavy bottom fraction consists mainly of lutoids. The yellow, lipid-containing Frey-Wyssling complexes are normally found at the upper border of the sedimented bottom fraction.

3.1
Composition of *Hevea* Latex

Separation of the latex into three major zones is by ultracentrifugation, after which their chemical compositions can be determined. The top rubber fraction contains, in addition to the rubber hydrocarbon, the proteins and lipids associated with the rubber particles. The serum phase contains most of the soluble substances normally found in the cytosol of plant cells. The bottom fraction can be studied by repeated freezing and thawing of the

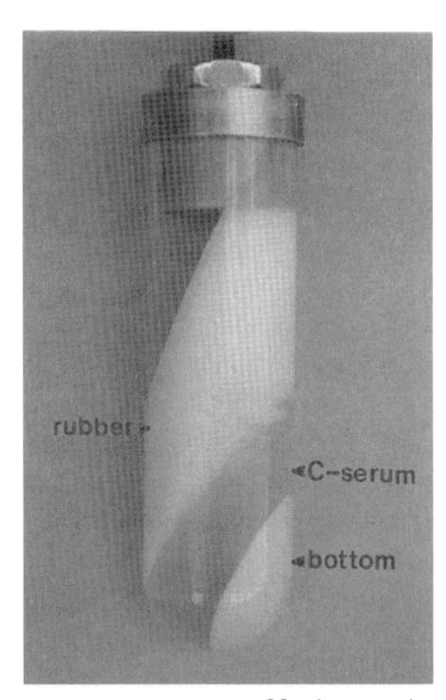

Fig. 3 Fractionation of fresh *Hevea* latex by ultracentrifugation into three major separated zones with the upper rubber layer, the aqueous phase C-serum and the bottom fraction containing lutoid particles.

lutoids. In this manner the membranes of the lutoids are ruptured and their liquid content, referred to as B-serum, can be analyzed. B-serum has been found to contain proteins and other nitrogen compounds as well as metal ions. It can thus be visualized that the latex is a cytoplasm system consisting of particles of rubber hydrocarbon dispersed in an aqueous serum phase. There are also numerous non-rubber particles called lutoids. The rubber particles are made up of rubber hydrocarbon surrounded by a protective membrane layer consisting of proteins and lipids. The total composition of fresh latex, apart from water, can be summarized as shown in Table 1. Besides the rubber hydrocarbon which is the major component of the latex, various other components (proteins, lipids, carbohydrates and inorganic substances) are also present which play important roles in the latex metabolism and functions.

The details on the compositions of the three major latex fractions (rubber particles, C-serum, and lutoids) and their functions will be extensively presented here. In general, the latex composition can be collectively categorized into two major components, the rubber and non-rubber constituents. The non-rubber constituents of latex will be briefly outlined first.

Tab. 1 Composition of fresh natural rubber latex

Component	*Content* (%, wt/vol)*
Rubber hydrocarbon	25–45
Protein	1–2
Carbohydrate	1–2
Lipids	0.9–1.7
Organic solutes	0.4–0.5
Inorganic substances	0.4–0.6

* The % content of the components varies according to clonal variations of the rubber clones.

3.2 Non-Rubber Constituents of *Hevea* Latex

As mentioned, in addition to the rubber, latex contains numerous non-rubber constituents (Figure 4), all of which are present and distributed in all three latex fractions. Proteins and lipids are found associated with the rubber particles, while C-serum contains substances normally found in the cytosol (carbohydrates, proteins, amino acids, inositols, enzymes and intermediates of various biochemical processes, including rubber biosynthesis). Lutoids contain specific substances unique to its functions. Details of the non-rubber constituents have been reviewed recently (Subramaniam, 1995). The following brief discussion will refer to proteins, carbohydrates, and lipids, as well as the inorganic substances.

3.2.1 Proteins

Apart from rubber hydrocarbon and water, proteins and carbohydrates are present in

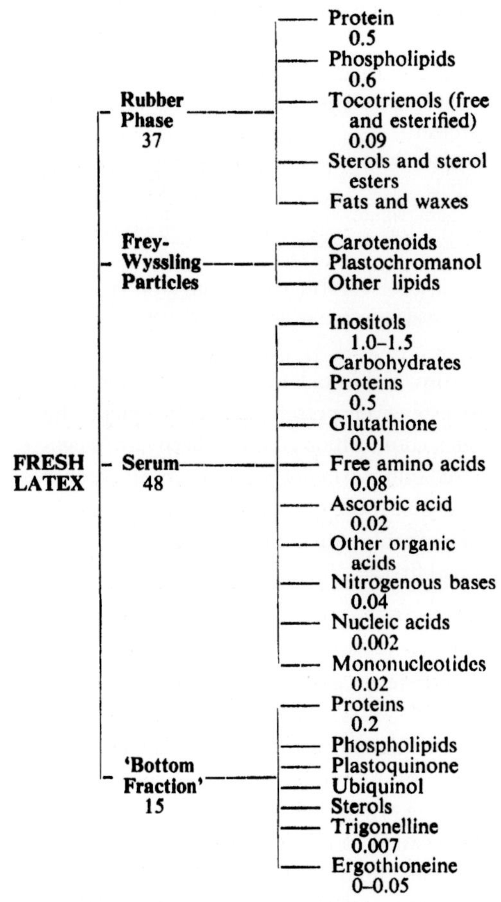

Fig. 4 Organic non-rubber consituents of fresh latex. Values positioned beneath the components indicate their approximate concentration in g per 100 g latex. (Adapted from Archer, B. L. et al. (1969) *J. Rubb. Res. Inst. Malaya* **21**, 560–569, with permission.).

highest proportion in latex. The proteins content in latex shows clonal variations and can range from 1% to more than 1.8% in different samples of latex. In a typical latex sample about 25–30% of the proteins are found in the rubber phase, with 45–50% in the serum phase and about 25% in the bottom fraction. The amount of proteins in the rubber phase is less variable than the total amount of proteins in different samples of latex. The serum proteins consists of around 19 anionic and five cationic proteins (Tata and Moir, 1964). The major protein is α-globulin with an isoelectric point (pI) of 4.8. There are seven anionic and six cationic proteins in the bottom fraction, the major proteins being hevein (>50%, pI 4) and hevamine (~30%, pI 9). The amino acid sequences of hevein (Walujono et al., 1975) and rubber elongation factor (REF), a 14 kDa protein in the rubber phase (Dennis et al., 1989), have been established. Recent report showed more than 200 different proteins are present in the latex (Alenius et al., 1994b), suggesting that the protein composition in the latex is highly complex.

3.2.2 Carbohydrates

Sucrose supply and utilization for latex production plays a very important role in the metabolism of latex, and has been reviewed by Tupy (1989). Of important note and quite unique to the latex is the presence of quebrachitol, an inositol derivative. Quebrachitol (1-methyl inositol) is the most abundant, and makes up about 75–95% of the total carbohydrates present in latex, being found mainly in the serum phase. The large amount and ubiquitous presence of this compound is a unique characteristic of the *Hevea* latex. The reason for its accumulation and its physiological function in latex is not known, but it has been postulated to serve an active role in rubber biosynthesis (Bealing, 1976). About six to seven other carbohydrates components are found in small amounts; these are mostly common sugars for various metabolic processes in the latex.

3.2.3 Lipids

Lipids in the latex play an important role in the stability of rubber particles. They are found not only associated with the rubber particles, but also throughout the latex fractions. The lipids content of latex also shows clonal variations (Ho et al., 1976). The neutral lipids constitute more than 59% of the total lipids, while the other components are mainly phospholipids and glycolipids. The presence of proteolipids have also been reported. The lipids occur mainly in the rubber particles and the bottom fraction. Several neutral lipids found are triglycerides (also mono- and diglycerides), free fatty acids and esters, sterols, and lipid-soluble vitamins (carotenoids and tocotrienols). Phosphatidylcholine is the major phospholipid, while phosphatidylethanolamine and phosphatidylinositol are found in smaller amounts. Phosphatidic acids are also reported as important components of the lutoid membrane (DuPont et al., 1976). Fatty acids in the latex occur mostly in the esterified form, the major fatty acids being C_{16}, C_{18} and C_{20} (palmitic, palmitoleic, stearic, oleic, linoleic, linolenic, and arachidonic acids). Also present is a rare furanoic acid (10,13-epoxy-11-methyloctadeca-10,12-dienoic acid) which is found mainly in the triglycerides fraction. Rubber latex is only the second known plant source of this furanoic acid (Hasma and Subramaniam, 1978). It is somewhat unusual to find this fatty acid in the latex, and its physiological function is not yet known. This situation is similar to that seen with quebrachitol, which is present in large quantities, though its function has yet to be determined.

3.2.4 Inorganic Substances

In addition to the major non-rubber constituents described above, several inorganic components are also present. Of these, potassium is the most abundant element in the latex, its concentration being of the order of a few thousand part per million (ppm). The next most common element (a few hundred ppm) is magnesium, this being mainly contained in the lutoids. Magnesium was found to reduce the mechanical stability of the latex (Philpott and Wesgarth, 1953). Magnesium and potassium also show clonal variations. Other elements occurring in much smaller concentrations are the more common sodium, calcium, iron, copper, manganese, and zinc. Also present is rubidium, though the function of this element in latex is not yet known.

3.3 Rubber Particles in *Hevea* Latex

Rubber hydrocarbon is the major component of *Hevea* latex, the rubber content varying from 25% to 45% as dry content of latex. The average molecular weight ranges from 200 to 600 kDa. The rubber molecules are found as particles in the latex (Figure 5), these consisting mainly of rubber (90%) in association with lipophilic molecules (mainly lipids and proteins), forming the film that encloses the rubber particles (Ho et al., 1976). This film carries negative charges and is responsible for the stability of rubber particles when suspended in aqueous serum. The size of the particles ranges from 5 nm to 3 μm, and they are spherical in shape. They also show plasticity as they have polygonal shapes in mature laticifers, where the particles are numerous. The size distribution, as determined by electron microscopy, showed a maximum distribution of 0.1 μm particles (Gomez and Moir, 1979), each of which may contain several hundred rubber molecules. Molecular weight analyses using gel permeation chromatography showed a bimodal distribution of rubber of low and high molecular weights, with average values of 100–200 kDa and 1000–2500 kDa, respectively (Subramaniam, 1976). The other main component of rubber particles is the enclosing membrane, which consists of lipids, proteins, and enzymes. These components contribute colloidal charge to the rubber

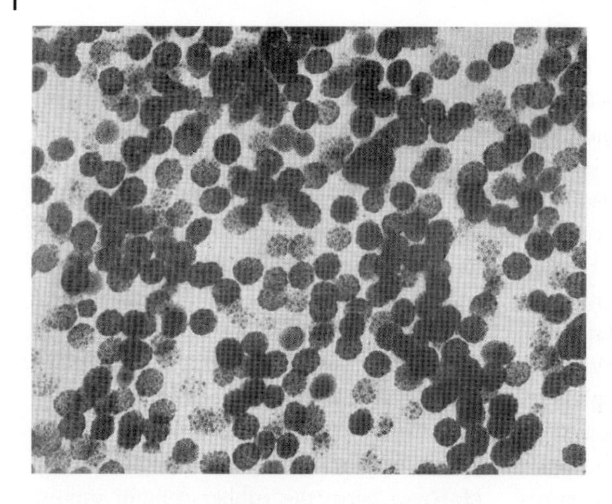

Fig. 5 Electron micrograph of rubber particles showing the spherical shape and uniform structure of the separated small rubber particles. (Adapted from Dickenson, P. B. (1969) *J. Rubb. Res. Inst. Malaya* **21**, 543–559, with permission.).

particle and their stability in the latex. Each component will be described in greater detail later in the chapter.

3.4 Rubber Particle Membrane

The rubber particles are commonly found in association with lipids, which is thought to be of membrane nature. Microscopically, the particles appear to have a uniform structure, with the rubber molecules enclosed by a thin film (Southorn, 1961). When examined by electron microscopy, the rubber particles appear homogeneous, and have a uniform internal structure, but are surrounded by a film that is more opaque than the polyisoprenes contained inside (Lau et al., 1986). Analyses of the nature of the film enclosing the rubber particles show the presence of phospholipids and proteins, together with neutral lipids similar to the membrane structure. The detailed description of rubber particles and other components has been reviewed (d'Auzac and Jacob, 1989), and the composition of the membrane of rubber particles separated and purified by ultracentrifugation analyzed. The membrane components comprise lipids, proteins, enzymes, and charges (see below).

3.4.1 Rubber Particle Lipids

Analyses of rubber particles purified by ultracentrifugation showed them to contain up to 3.2% total lipids, of which ~2.1% are neutral lipids expressed as rubber weight (Ho et al., 1976). Separation of neutral lipid showed it to be composed of at least 14 different substances. Triglycerides were the most abundant, accounting for almost 45% of the neutral lipids, while sterols, sterol esters and fatty acid esters constituted about 40%. Other neutral lipids present in trace amount were diglycerides, monoglycerides, and free fatty acids. In addition, tocotrienols and some phenolic substances were also found to be associated with rubber particles (Ho et al., 1976).

Phospholipids are important components of the rubber particles. Marked differences between clones in neutral lipids content of the rubber phase were noted. In contrast, the phospholipid content did not vary much among the different clones (Ho et al., 1976). Analyses of the phospholipids consistently showed three spots on thin-layer chromatography (TLC) separation; these were identified as a considerable quantity of phosphatidylcholine and smaller quantities of phosphatidylethanolamine and phosphatidylglycerol. Phosphatidic acid was found to be predominant in the membrane of lutoids, but was not detected on the rubber particles, even though the precaution was taken of inhibiting phospholipase D activity (Dupont et al., 1976). In addition, the presence of sphingolipids and glycolipids has also been reported. The stability of rubber particles suspension in latex is dependent on the negative charges film of proteins and phospholipids (Philpott and Wesgarth, 1953).

3.4.2
Rubber Particle Proteins

Proteins are found as indigenous components of the film enclosing the rubber particle. Together with lipids, these proteins form the membrane of particles which contribute to their stability. The pI of these proteins ranges from 3.0 to 5.0 which is characteristic for surface proteins. In an electric field, particles will move toward the anode, indicating that they have net negative charge on the surface (Verhaar, 1959). Anionic soaps do not affect the particles' colloidal stability, but cationic soaps cause flocculation, probably due to neutralization of the surface charge. These proteins can be considered as intrinsic or peripheral, depending on their binding and affinity with rubber particles. One of the most plentiful proteins in latex is α-globulin with pI of 4.5. It was found both in the cytosol and adsorbed onto the particles' surface, and might contribute to their colloidal stability in latex (Archer et al., 1963a). A protein group of hydrophobic nature was also found in rubber particles, and proteolipids have been isolated and characterized (Hasma, 1987). This protein was suggested to be a component of the polar lipid backbone that forms part of the membrane of rubber particles. The protein content of rubber particles was recently refined for a more accurate quantitative analysis (Yeang et al., 1995) in light of the concern over rubber protein allergens.

Other proteins with enzyme activity have also been described. One interesting and well-characterized enzyme is rubber transferase, which was first detected on the washed rubber particle surface (Audley and Archer, 1988). This enzyme is involved in the synthesis and formation of rubber molecules on the particles' surface, and was found to be distributed between the cytosol and rubber particles, similar to the case of α-globulin. It has been isolated and purified from latex C-serum, but was active only when adsorbed onto particles for the chain elongation process of rubber molecules (Light and Dennis, 1989). Another important protein found to be actively involved in rubber synthesis was a 14 kDa protein referred to as rubber elongation factor (REF). The amino acid sequence of REF has been determined, and the molecular cloning of the *REF* gene also carried out (Attanyaka et al., 1991). More recently, molecular cloning of a protein that is tightly bound onto the small rubber particles has been reported (Oh et al., 1999). The cloned cDNA encodes a 24 kDa protein which is tightly bound onto the rubber particles. This protein was suggested to be active in the synthesis of rubber, together with the REF. The 24 kDa protein has been reported previously as a potent latex allergen, and is always found together with REF, bound so tightly to particles that they cannot be removed even by extensive washing. In addition to these tightly bound and well-characterized proteins, some peripheral proteins have also been described, though their function is unknown. The implication of these particle-bound proteins causing allergic reactions towards rubber products, especially latex gloves, will be discussed later.

3.4.3
Rubber Particle Enzymes

Some of the rubber particle proteins with enzyme activity were described briefly in the preceding section. Washed rubber particles and the bound rubber synthesis enzymes have been studied and reviewed (Audley and Archer, 1988). The formation of rubber molecules, at least in terms of the elongation steps, occurs at the particles' surface (Archer et al., 1982), and rubber transferase is the enzyme responsible for this process. This enzyme was also found in the C-serum, and is most likely distributed between the two fractions (McMullen and McSweeney, 1966). It has been isolated from the C-serum and purified for enzyme characterization (Archer and Audley, 1967). The enzyme was without activity in the absence of washed rubber particles, remaining inactive while not adsorbed onto the particles, even when the latter have been purified by gel filtration and washed repeatedly. The reaction catalyzed by rubber transferase appears to be essentially chain extension or elongation of the pre-existing rubber molecule, although a role in the formation of new rubber molecules has also been suggested (Lynen, 1969). However, the mechanism by which new rubber particles are formed is still unknown at present.

To date, it has not been possible to demonstrate the polymerization of isopentenyl diphosphate (IDP) in vitro except on the pre-existing rubber particles. The reaction that occur in vivo must do so at some other sites as a prerequisite on initiation that preludes the formation of new rubber molecules. Some reports have suggested vehemently that membrane phospholipids may have a key role in enabling the combined functions of different transferase enzymes to operate in vivo for new rubber formation (Keenan and Allen, 1974; Baba and Allen, 1980). It appears that some phospholipids, membranes, or other amphipathic micelles are essential for the initiation and formation of new rubber molecules. The rubber particles' surface was also suggested as being the site of IDP isomerase (Lynen, 1969), the enzyme catalyzing the conversion of IDP into dimethylallyl diphosphate (DMADP). The IDP isomerase is essential for the formation of DMADP and the chain initiation of new rubber molecules. It has long been suggested to be present in latex, but only indirect evidence has been provided for its detection. More recently, the direct detection and characterization of this enzyme was reported (Koyama et al., 1996), the isomerase being located in the cytoplasm or C-serum of latex, and not with the rubber particles as was previously suspected. The enzyme was activated in the presence of reducing agents and detergents (Koyama et al., 1996). These findings might provide partial support for the possible initiation of rubber chain elongation at a site other than on the rubber particles.

3.4.4
Rubber Particle Charges

The colloidal stability of latex is attributed to the presence of surface charges on the rubber particles. The film or membrane surrounding the particles provide them with a negative charge, as shown by surface potential or zeta potential (Southorn and Yip, 1968). The particle membrane composition of proteins, phospholipids, and other substances has been determined as described above, and shown to be of a negatively charged nature (Ho et al., 1976). The reduction in phospholipid content of a cloned latex known for its instability has also been noted, and this observation was subsequently extended to show that the lipid content of rubber particles correlated positively with colloidal stability of the latex (Sherief and Sethuraj, 1978). The colloidal stability was reduced by magnesium released from the damaged or ruptured lutoids (Philpott and Wesgarth, 1953) as the surface charges were neutralized. The effect of inorganic cations was investigated in relation to flocculation of the rubber particles and to plugging of the latex vessels, with a negative effect on latex flow (Yip and Gomez, 1984).

3.5 C Serum of *Hevea* Latex

The C-serum fraction of the centrifuged latex (see Figure 3) is the aqueous phase of laticiferous cytoplasm, and this can be considered as the latex cytosol. This cytosol is not fundamentally different from the normal cytosol of plant cells. Analyses of the latex cytosol fraction obtained by ultracentrifugation show that it contains various different organelles and particles. All glycolytic enzymes (d'Auzac and Jacob, 1969) and other common cytosolic enzymes including those of the isoprenoid pathway (Suvachittanont and Wititsuwannakul, 1995; Koyama et al., 1996) have been detected, indicating that the cytosol is active in a number of metabolic processes.

The involvement of C-serum in rubber biosynthesis was also noted (Tangpakdee et al., 1997a), as was the importance of calcium binding protein (calmodulin) in controlling many different metabolic processes. Calmodulin was found to activate HMG-CoA reductase in the bottom fraction, and purification and characterization of calmodulin from C-serum (Wititsuwannakul et al., 1990b) showed it to have an important role in the regulation of latex metabolism. Moreover, a highly positive correlation between calmodulin level and latex yield was also demonstrated (Wititsuwannakul et al., 1990b). The composition of latex cytosol has been reviewed (d'Auzac and Jacob, 1989), and the presence of many high-molecular weight compounds, low-molecular weight organic solutes and mineral elements are also well documented (see below).

3.5.1 High-Molecular Weight Compounds in C Serum

The high-molecular weight compounds in C-serum (see Figure 4) are mainly proteins and specific nucleic acids. The distribution of proteins in whole latex is approximately 20% in the rubber phase, 20% in the bottom fraction, and 60% in the C-serum (Archer et al., 1963a). A large number of proteins are present, as revealed by gel electrophoresis (Tata and Edwin, 1970). One C-serum protein that is present at the highest level is α-globulin, the pI of this (4.55) being similar to that of α-globulin from other sources. α-Globulin has high binding affinity for adsorption onto rubber particles, and was suggested to be one of the particle proteins contributing to the particles' colloidal stability (Archer et al., 1963a). Several enzymes present in C-serum were reported, as described above; thus, those proteins with enzymatic activities are also the major component. These enzymes and their metabolic activities in latex will be discussed in details below.

Nucleic acids are another group of another high-molecular weight compounds present in the latex C-serum, while the presence of in-vitro active ribosomes in latex showed that latex cytosol also contains many nucleic acids (McMullen, 1962). These nucleic acids (soluble RNA and DNA) were found mainly in the serum fraction of centrifuged fresh latex (Tupy, 1969). Analyses of the nucleic acids show both clonal variation and treatment of rubber trees as expressed in the latex. Wound-induced accumulation of mRNA for hevein protein has also been reported (Broekaert et al., 1990), this being the defense response that results from tapping of the rubber trees. The latex cytosol was also found to show differential expression of laticifer-specific genes that have been extensively characterized (Kush et al., 1990). These genes were also found to be influenced by latex production and hormone treatments. Specific mRNAs separated as poly(A) RNA was shown to be induced and elevated by the effect of ethylene, with an ethylene-induced increase and accumulation of glutamine synthetase and mRNA level having been noted (Pujade-Renaud et al., 1994).

3.5.2 Low-Molecular Weight Compounds in C Serum

The metabolic importance of C-serum components is well noted (Jacob et al., 1989). Latex cytosol contain numerous groups of low-molecular weight organic solutes (see Figure 4) that are important in latex metabolic activities, either as precursors or metabolic intermediates. They comprise sugars, amino acids, nucleotides, and nitrogenous bases among many others. Sucrose is the main sugar in latex, together with smaller amounts of glucose, fructose and raffinose. These sugars are important for glycolysis and have been shown to be very active in latex (d'Auzac and Jacob, 1969; Tupy, 1973). In addition, inositols are also prominently present in latex cytosol. Quebrachitol (1-methyl inositol) is found in very high content, comprising >75% of the total carbohydrates in the latex. The presence of quebrachitol is closely correlated with the *cis*-polyisoprene level in the latex, and it was suggested that quebrachitol might undergo catabolism for rubber biosynthesis as an alternative to sucrose (Bealing, 1969, 1976). The pentose phosphate pathway was also found to be very active (Tupy and Resing, 1969) in supplying NADPH for synthesis of rubber in the latex. Channeling of sucrose for different metabolic processes in latex was supported from these findings (Bealing, 1976). C-serum contains all the common amino acids, but in different proportion (Yong and Singh, 1976). The concentration in cytosol (~30 mM) is similar to that of amino acids in lutoid B-serum. However, the distribution of neutral, acidic and basic amino acids in the cytosol is quite different from that in lutoid serum, the acidic amino acids being predominant in the C-serum and the basic amino acids in B-serum. The major amino acids in the cytosol are constituted mainly (up to 80%) by alanine, aspartic, and glutamic acid and its amide (Yong and Singh, 1976).

The latex cytosol contains all the known nucleotides for metabolic activities of latex. Adenine nucleotides attribute to an average energy charge of around 0.6 in the C-serum, while average values for ATP, ADP, and AMP were around 125, 133, and 55 nmol g^{-1} dry rubber, respectively (d'Auzac and Jacob, 1989). Concentrations of coenzymes in the latex (coenzyme A, NAD, and NADP) have been measured. Coenzyme A is indispensable in the biosynthesis of lipids and isoprenoids, while concentrations of cytosolic NAD and NADPH are 70 μM and 2 μM, respectively (Archer et al., 1969). The low content of NADPH in latex might serve as regulatory cofactor in latex. NADPH is a specific electron donor for the reduction of HMG-CoA to mevalonic acid (MVA) by the enzyme HMG-CoA reductase (HMGR), a possible rate-limiting step in the biosynthesis of rubber (Benedict, 1983). The important role of HMGR in *Hevea* rubber biosynthesis has been extensively investigated. A correlation between diurnal variations of HMGR activity and rubber content has been verified (Wititsuwannakul, 1986), as has the purification of *Hevea* latex HMGR (Wititsuwannakul et al., 1990a) and activation by cytosolic calmodulin (Wititsuwannakul et al., 1990b). Likewise, a correlation between levels of HMGR and rubber yield has also been documented (Wititsuwannakul et al., 1988).

In addition, the presence of UDP-glucosamine and UDP-galactosamine in latex have been reported (Archer et al., 1969). These nucleotides may involved in the glycosylation of glycoproteins in latex, though this was also implied for the lectin function of latex. The *Hevea* lectin was recently isolated, purified and characterized (Wititsuwannakul et al., 1998), and implicated in rubber coagulation and vessels plugging (Gidrol et al., 1994). In addition to these key molecules, the latex cytosol also contains several other small molecules, including many organic acids of which malic and citric acids comprise almost 90% of the total organic acids in latex. Several reducing agents are also present in latex cytosol, the major participants being glutathione, cysteine, and ascorbic acid (0.72, 0.44, and 1.9–3.9 mM, respectively) (d'Auzac and Jacob, 1989). These reducing compounds are essential for maintaining the redox potential of latex.

3.6 Lutoids of *Hevea* Latex

Lutoids are major membrane-bound particles that are sedimented in the bottom fraction (see Figure 3) of the centrifuged fresh latex (Moir, 1959). They are considered as polydispersed lysosomal vacuoles and have been extensively discussed (d'Auzac and Jacob, 1989). The lutoid content of latex is quite considerable, and has been assigned a number of important functions in the latex. Lutoids constitute ~20% by volume of fresh latex, whereas the rubber phase forms an average of ~30–40%. Lutoids are spherical in shape (Figure 6), and their diameter is larger than that of rubber particles. Lutoids are enclosed by single membrane, the lipids composition of which has been determined. They play an important role in the colloidal stability of latex due to the negatively charged membrane which is very rich in phosphatidic acids content (Dupont et al., 1976). Lutoids are considered to be similar to lysosomes, with a high acid hydrolase content (Jacob et al., 1976), and play a major role in the coagulation of rubber particles in colloidal suspensions of latex (Southorn and Edwin, 1968). The mechanism of this coagulant effect is due to the release of cations and proteins from ruptured lutoids (Pakianathan and Milford, 1973). The storage function of lutoids was shown in the accumulation of several proteins, enzymes and solutes with active membrane transport activity (Chrestin and Gidrol, 1986). Lutoids are an important latex component, both with regard to volume and the functions of their

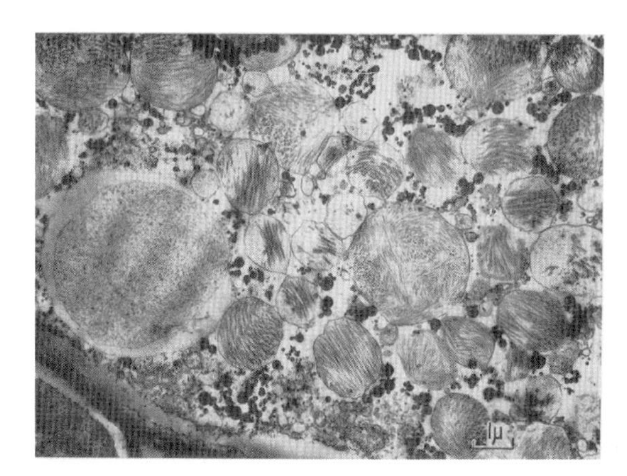

Fig. 6 Lutoid particles of the fresh *Hevea* latex showing the spherical shape of a larger diameter than the rubber particles, with full development of the enclosed proteinaceous microfibrils. The unit membrane surrounding the particles is contorted because of slight plasmolysis as the lutoids are quite osmosensitive. (Adapted from Dickenson, P. B. (1969) *J. Rubb. Res. Inst. Malaya* **21**, 543–559, with permission.).

chemicals and enzymes (d'Auzac and Jacob, 1989). Their composition can be considered as having two distinct components, namely the lutoids membrane and the internal contents, referred to as B-serum.

3.6.1
Lutoids Membrane

Cytological observation, together with biochemical and physiological research on lutoids membrane have provided data concerning the structure, composition, and role of these organelles in latex. Electron microscopy showed the micellar nature of the lutoids membrane structure (Gomez and Southorn, 1969), this being a highly osmosensitive, single-layer membrane of 8–10 nm thickness. Analysis of the membrane chemical composition showed the phospholipid content to constitute 37.5% of the weight of proteins. The lutoid membrane has been shown to be very rich in phosphatidic acids, these accounting for 82% of the total phospholipids fraction. The fatty acids composition (analyzed by methanolysis) of phospholipids showed a predominance of saturated fatty acids, mainly palmitic and stearic, with unsaturated fatty acids (oleic, linoleic acid) also present (Dupont et al., 1976). There were almost equivalent quantities of saturated and unsaturated fatty acids in the membrane. The exceptionally high phosphatidic acid content may explain the high negative surface charges of the lutoid (Southorn and Yip, 1968). The relative abundance of saturated fatty acids in lutoids stands out clearly from that in the membranes of other plant organelles. The relative rigidity and fragility of the lutoids membrane to osmotic shocks, and its low resistance to mechanical stress (Pakianathan et al., 1966), can be partly explained by the membrane's fatty acids composition. The analysis of rubber particles membrane carried out under the same conditions shows that the membrane is totally free of phosphatidic acids, but does contain mainly phosphatidylcholine and phosphatidylethanolamine (Ho et al., 1976). The other major lutoid membrane components are proteins, several of which are enzymes important in the lutoids' function. Thus, the lutoid membrane has been shown to play an essential and highly complex role in latex (d'Auzac and Jacob, 1989; Paardekooper, 1989).

3.6.2
Lutoids Membrane Proteins and Enzymes

Several proteins are present in the lutoid membrane, and many of these are active enzymes. One well-characterized membrane enzyme is ATPase (Moreau et al., 1975). The electron transport activity of lutoids has been linked to ATPase, which is activated by several anions, this in turn leading to an accumulation of anions within the lutoid compartment. ATPase also operates as a proton pump to maintain proton gradients between the lutoids and latex cytosol – a function which was demonstrated by Chrestin and Gidrol (1986) and extensively reviewed (Chrestin et al., 1986). Other membrane enzymes included NADH-cytochrome c reductase; this functions in an outward proton-pumping redox system that tends to reduce the concentration of protons in lutoids and hence acidify the cytosol (Moreau et al., 1975). NADH-quinone reductase (d'Auzac et al., 1986) has also been described as being responsible for the production of superoxide ions. The generation of free radicals and the consequent effect on lutoid integrity and latex colloid stability has been the subject of recent interest and speculation.

Recently, the enzyme HMG-CoA reductase has been purified from lutoids membrane by solubilization with mild detergent (Wititsuwannakul et al., 1990a). Characterization of the purified enzyme was carried out by determining its molecular structure and properties. The native enzyme was found to be a tetramer of four 44 kDa subunits – as found for other plant specimens – and membrane-bound (Bach, 1986). HMG-CoA was found to be activated by reducing agents for maximum activity. In C-serum, regulation of this enzyme by calmodulin was also studied in detail (Wititsuwannakul et al., 1990b), the results suggesting important interactions of C-serum contents in order to maintain optimal lutoids membrane activities. Rubber formation by the bottom fraction was also recently reported (Tangpakdee et al., 1997b), and this help to explain the participation of lutoids in the rubber biosynthesis process. A number of other different lutoid functions have been reviewed, and various lutoid membrane enzymes characterized, by d'Auzac and Jacob (1989), and therefore these aspects will be described here only briefly. Details of the structural proteins of the lutoid membrane are poorly understood, and still await further characterization.

3.6.3
B-Serum of Lutoids and Its Composition

The inside content of lutoids is called B-serum, and this comprises of several different components, including microfibrils, proteins and enzymes, as well as various small-molecule compounds. Microfibrils (Figure 7) are cluster of proteins with helical structure when seen under the electron microscope (Archer et al., 1963b; Dickenson, 1965). Characterization of their structure showed the presence of carbohydrate up to 4% and the presence of acidic proteins with a pI of 4 (Audley, 1965; Gomez, 1976). These fibril proteins may completely fill the space inside the lutoids, and it has been speculated that such proteins may function as a nitrogen reserve (Dickenson, 1969) that may be degraded by lutoid proteases. Microhelical, spring-shaped proteins have also been identified; these consist of basic proteins of 22 kDa and an acidic assembling protein of ~160 kDa in microhelical cluster formations (Gomez and Tata, 1977). Clonal variation of microhelical content was also observed. In addition to the clustered insoluble proteins, other soluble proteins have also been characterized, with electrophoretic analyses showing the presence of at least seven to eight different proteins. Among these, hevein was the most abundant, comprising 70% of the total proteins (Archer, 1960; Audley, 1966).

Hevein is quantitatively the most important anionic proteins, its special feature being that it is a very small protein (5 kDa). Hevein

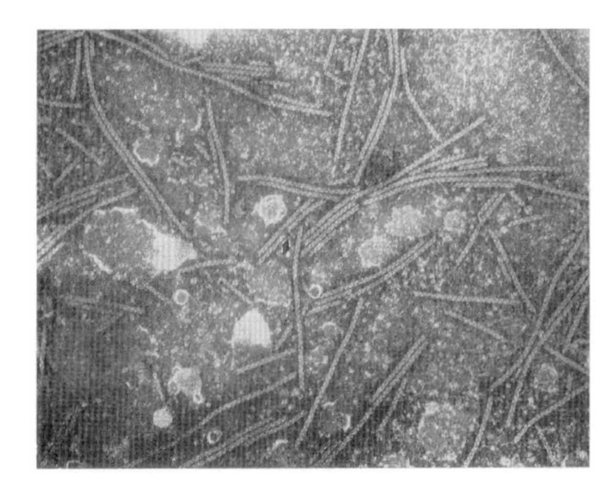

Fig. 7 The extracted microfibrils in B serum of the lutoids, showing the long proteinaceous helical microfibrils. The microfibrils are clusters of proteins with double-helical structure as observed by electron microscopy. (Adapted from Dickenson, P. B. (1969) *J. Rubb. Res. Inst. Malaya* **21**, 543–559, with permission.).

has been purified, crystallized, and characterized (Tata, 1976), and its amino acids composition and sequence determined (Walujono et al., 1975). Its high sulfur content is contributed by cystine alone. In light of its low molecular weight and high cysteine content, hevein has been considered possibly to function as a protease inhibitor similar to those detected in the vacuoles of potato and other *Solanaceae*. However, no protease inhibitor activity of hevein could be detected by a variety of tests (Walujono et al., 1975). The role of these fibril proteins and hevein, and the significance of their presence in lutoid, is not understood, although recently the involvement of hevein in the coagulation of rubber particles was suggested (Gidrol et al., 1994). Besides hevein, two other basic proteins that occur in high concentration in lutoids have been detected and characterized as hevamine A and hevamine B (Archer et al., 1969). The lutoids have considerable influence on the metabolism of latex and its regulation by means of the exchanges which take place between their contents and the latex C-serum. Several enzymes are present in B-serum that are important in metabolic processes and homeostasis of the latex.

3.6.4
B-Serum Enzymes of the Lutoids

Several typical vacuole enzymes can be detected in B-serum of the lutoids. Lysozyme, which is commonly found in egg white and hydrolyzes mucopolysaccharides and bacterial cell walls, was detected in the latex bottom fraction (Archer et al., 1969). It was located in B-serum and considered to be the same as the hevamines, which were shown to be abundant basic protein in lutoids (Tata et al., 1976). B-serum lysozyme has been purified and characterized, and the enzyme's kinetics investigated and partial amino acid composition and sequence determined (Tata et al., 1983). Lysozyme is generally implicated in the defense against antimicrobials, and several acid hydrolases (phosphodiesterase, acid phosphatase, ribonuclease, cathepsin, β-glucosidase and β-galactosidase) were reported to be present in B-serum (Archer et al., 1969). The presence of these enzymes shows similarity between lysosomes in animals, and lutoids. The role of the lutoids' content in stability and the flow of latex was demonstrated (Southorn and Edwin, 1968), and shown to cause rubber particles to coagulate, leading to the plugging of latex vessels and stopping of latex flow (see below). Acid phosphatase was found to be a lutoid-specific enzyme and to appear in the latex cytosol only after the lutoids had been damaged or ruptured. The ratio of free acid phosphatase and the total enzyme activities reflects the stability of lutoids; this ratio is referred to as the bursting index (BI), and influences the flow of latex. A high correlation between BI and vessel plugging, as measured by latex flow duration, has been documented and reported for seasonal and clonal variations (Yeang and Paranajothy, 1982).

Details of oxidation-reduction enzymes in B-serum have been reviewed (d'Auzac and Jacob, 1989). These enzymes include catalase, phenol oxidase and tyrosinase, while peroxidase enzymes such as monophenol, polyphenol and DOPA oxidases have been detected in B-serum as cationic proteins. These enzymes are always found together with their inhibitors, and it was suggested that compartmentalization in the lutoids might occur in order for these enzymes to be regulated. In addition, the enzymes of pathogenesis-related proteins such as chitinase and β-1,3-glucanase were also found to be accumulated in lutoids (Churngchow et al., 1995). These enzymes appeared to be induced as a wounding response following repeated tappings, and served as a defense against subsequent attack by pathogenic microbes. Normally, these enzymes are either absent or at very low level in healthy plants, and this can also be applied to the high level of lysozyme (Tata et al., 1983). Other B-serum enzymes have been investigated and their details reviewed (d'Auzac and Jacob, 1989).

3.7 Lutoids and Colloidal Stability of Latex

The contribution of lutoids to latex colloidal stability, which results from negative charges carried by the lutoid structural components, was referred to earlier. The more important of these structural components include rubber particles and lutoids, and integrity of the latter is recognized as being essential to maintain latex stability. Their content of acidic serum, divalent cations, and positively charged proteins might suggest that lutoids have a destabilizing role in latex, and this is certainly true while the lutoids remain undamaged or unruptured. However, the lutoids are highly osmosensitive, and consequently, after tapping, a certain amount of their content is released from the ruptured lutoids into latex cytosol. The role of lutoids as coagulants has been demonstrated by the formation of microflocs consisting of lutoids fragment and rubber particles (Southorn, 1969). The lutoids B-serum is effectively capable of provoking the microflocs formation of a dilute suspension of rubber (Southorn and Yip, 1968). It was shown that C-serum proteins are mainly anionic, while those of B-serum proteins are mainly cationic (Moir and Tata, 1960). It is clear that placing the B-serum proteins in contact with negatively charged particles of rubber results in neutralization of the surface charges, and destabilization of the latex colloidal solution (Southorn and Yip, 1968; Sherief and Sethuraj, 1978). These early studies form the basis for further analysis in rubber particle coagulation caused by purified hevein (Gidrol et al., 1994). The importance of lutoids with regard to latex flow duration, vessel plugging and the effect on rubber yield will be described in greater detail below.

3.7.1 Lutoids and Latex Vessels Plugging

Rubber yield can be ascribed to two parameters: latex flow duration; and latex synthesis or regeneration capacity. Lutoids are implicated to influence the time of latex flow and latex vessels plugging (Paardekooper, 1989), and thus integrity of the lutoids is essential for latex colloidal stability. Damage to, or rupture of, the lutoids causes coagulation of rubber particles, and this leads to stopping of latex flow. Several studies have provided evidence that the major cause of vessel plugging during latex flow is damage to the lutoids. This was caused by osmotic shock as a result of different turgor pressures between the inside and outside of vessels during flow after tapping. During the initial period of fast flow, the damaged lutoids are swept out of the vessels before they suffering irreversible damage. During the subsequent slower flow, the lutoids suffer greater damage within the vessels and aggregate with rubber particles to form flocs which accumulate near the cut ends, thus initiating the plugging process (Pakianathan, 1966; Pakianathan and Milford, 1973). In whole latex, breakage of lutoids by ultrasonic treatment resulted in the formation of flocs of rubber and a damaged lutoids fragment (Southorn and Edwin, 1968). Therefore, plugging within the vessels is caused by the release of B-serum from ruptured lutoids. Fresh latex always contains microflocs, the formation of which is confined to the area of damaged lutoids where the content of B-serum is momentarily high, and is limited by a stabilizing effect of the C-serum. The latex is envisaged as a dual colloidal system in which negatively charged particles, rubber particles and lutoids, are dispersed in the neutral C-serum containing anionic proteins (Southorn and Yip, 1968). The two antagonistic systems can only exist as long as they are separated by the intact lutoids membrane. Lutoids damage results in interaction between its cationic contents and anionic surfaces of the rubber particles, causing the formation of flocs.

It is observed that the enzyme acid phosphatase is released in the B-serum when lutoids are damaged. The ratio of free and total acid phosphatase – the bursting index – indicates the percentage of lutoids in a sample that had ruptured. The BI was inversely related to the osmolarity of latex, and the first fraction of latex collected after tapping was higher than subsequent fractions (Pakianathan et al., 1966). It was found that damage to the bottom fraction was greatest in the first flow fraction after tapping. A high correlation between plugging index, intensity of plugging, and the lutoid BI was found when studying seasonal variations in the flow pattern (Yeang and Paranjothy, 1982). A positive correlation between total cyclitols in C-serum and plugging index with increased lutoids damage was also noted (Low, 1978). Lutoids could be disrupted by the mechanical shearing forces to which the latex is subjected when flowing through the vessels under high-pressure gradients after tapping (Yip and Southorn, 1968). Difference between clones in the composition of protective film on rubber particles may be partly responsible for the flocs formation. Rubber particles are strongly protected by a complex film of protein and lipid materials (see above). Marked differences between clones in their neutral lipid content of the rubber phase in which phospholipids content did not differ much showed a negative correlation with the plugging index. Long-flowing or slow-plugging clones have a high neutral lipid content in the rubber phase, as much as five times that of the fast-plugging clones (Ho et al., 1976). A negative correlation between the phospholipid content of lutoid membrane and BI, leading to rapid plugging, was documented (Sherief and Sethuraj, 1978). These findings suggested that, in addition to the effect of lutoids behavior, latex vessel plugging is also influenced by the lipids of the rubber particles. Thus, it can be seen that the plugging of vessels and any subsequent effect on the latex yield is quite a complex process. Nonetheless, other factors in the process have also been speculated upon and are awaiting elucidation.

3.8 *Hevea* Latex Metabolism

Hevea latex is a unique system consisting of a specialized cytoplasm (Dickenson, 1965; Gomez and Moir, 1979), the organization and composition of which reflects the biological functions of *Hevea* specialized laticiferous tissue. It has long been shown that fresh latex can synthesize rubber in vitro from labeled precursors such as acetate or MVA (Park and Bonner, 1958; Kekwick et al., 1959). The enzymes' activity and their location with substrates and affectors make it possible to study metabolism and the biochemical function of latex as a 'rubber factory'. Latex as obtained by tapping is not so destructive compared with the general method for preparing cell cytoplasm, and this makes it very suitable to study latex metabolic functions. Latex metabolism has been studied in various different aspects, especially the rubber biosynthesis and other related metabolic processes (Jacob et al., 1989). Tapping of rubber trees for latex makes it necessary for laticiferous tissue to make up for the lost materials between successive tappings, and the regeneration of cell materials increases along with latex production. If regeneration is not sufficiently effective, this can be a limiting factor for latex production (Jacob et al., 1986). Intense metabolic activity and numerous enzymes are required to be involved in at least four important processes for latex regeneration:

- the catabolism which provide energy and reducing capacity for the anabolic processes;
- the activity of anabolic pathways for various syntheses including isoprenoids;
- the mechanisms associated with regulatory systems and homeostasis; and
- the supply of nutrients to the zones or subcellular components in which cell materials are regenerated.

The specificity of laticiferous tissue is so organized that its particular major function is directed to the production of rubber in the latex.

The formation of rubber in *Hevea* laticifers seems to be a very complex control system, and several questions regarding the regulation of rubber biosynthesis have been investigated. The continuing requirement of carbon, NADPH and the need of ATP for rubber biosynthesis must place a high degree of demands on the metabolic economy of the tissues. It has been calculated that the required rate of regeneration of rubber in alternate daily tapping is of the order of 1 μmol isoprene unit per mL latex per minute (Bealing, 1976). The capacity of latex to incorporate acetate and MVA has been found to fluctuate markedly with the season (Bealing, 1976). The control mechanism in the formation of rubber appears to be a complex and intricate process. The interesting feature of *Hevea* metabolism is not only that for rubber formation, but also that it can be stimulated to produce rubber and other many components of latex by repeated tapping. This replenishment is not called for in the untapped tree, and although a few terpenoids have been shown to suffer catabolism in plants, there is no evidence for the breakdown of rubber in vivo. The enhanced rubber yield by ethylene – a hormone associated with response to wounding in plants – has been extensively studied and reviewed (d'Auzac, 1989a,b). The activation of specific genes may shed some light on the mode of action of the tapping stimulus and ethylene activation, and this topic will be discussed in the following text.

3.8.1
Rubber and Isoprenoids Biosynthesis in *Hevea brasiliensis*

Hevea rubber biosynthesis and its control has been often reviewed (Archer and Audley, 1987; Kekwick, 1989; Kush, 1994; Tanaka et al., 1996), and so it is only briefly described here as related to the function of specialized laticifers, together with a discussion on the key regulatory enzymes. An interesting aspect of *Hevea* laticifers is the fine-tuning in terms of compartmentalization of function or the division of labor for rubber biosynthesis pathway. It was shown that the laticifers have a differential gene expression profile (Kush et al., 1990; Kush, 1994). The genes involved in rubber synthesis are highly expressed in the latex as compared to those in the leaves. The specialized differential expression serves a two-fold function. First, that the desired enzymes for rubber synthesis are expressed in the very tissues where formation is taking place. Localizing the rubber synthesis activity in laticifers allows other, different, metabolic processes in other tissues to operate at their optimum and to be well-balanced with the whole of the plant's functions. Second, specialized functions and well-coordinated divisions of labor thus appear to be well organized for specific channeling of precursors and metabolites for different metabolic pathways. The tissue and cell differentiations destined to perform certain functions to best fit the metabolic distribution can thus clearly be seen in *Hevea brasiliensis*. This is somewhat different for other rubber-producing plants such as guayule.

Hevea brasiliensis is unlike some plants, for example *Parthenium argentatum* (guayule), where rubber synthesis takes place in the cytosol of the parenchymal tissues along with other metabolic processes required for orderly parenchymal cell functions. The extraction of rubber from guayule plants is difficult because of the relatively low abundance of rubber particles in the cell, and the limits imposed by the cell volume (Backhaus and Walsh, 1983). The procedure of obtaining rubber from guayule is very destructive because of the subcellular localization of rubber. Guayule rubber can only be obtained by crushing the stems and extracted the rubber, together with all other cellular materials and impurities of all cell types. This is one of the prime reasons that the *Hevea* rubber tree is the only commercially viable source of natural rubber in the world. Even though some 2000 species of plants that produce rubber of varying types and quantities are known (Backhaus, 1985; Mahlberg, 1993), none is used or found comparable to *Hevea* with regard to the superior quality of the rubber produced. This is in contrast to the *Hevea* laticifers, which contain a high content of latex that gushes out after the latex vessels are opened, either by a small excision in the bark or by tapping (which is less destructive to the plant tissues). The flow of latex is due to the high turgor inside the laticifers compared with the outside. After latex has flown out of the cut vessels for some time, the flow will stop due to coagulation of rubber particles forming latex plugs either at the vessel ends or at the wounding site when tapped. A lectin-like small protein, hevein, which is localized in lutoids (Gidrol et al., 1994) has been shown to play an important role in the plugging of latex vessels. Lutoids have been shown to possess very fragile membranes that burst in response to tapping because of the difference in turgor pressure on the lutoid, and its properties. The tapped latex contains a vast number of intact organelles, which in turn makes it an excellent specimen for the study of differential and specialized metabolic functions in plants. As *Hevea* laticifers are an anastomosed system, the latex in essence represents the cytoplasm of a single cell type. The rubber yields are the results of two contributing factors, the latex flow properties and regenerating synthetic capacity.

3.8.2
Isoprenoids Biosynthesis

Besides rubber, *Hevea* laticifers also synthesize a number of diverse isoprenoids (Kush, 1994). Different plants have the capacity to synthesize certain isoprenoid compounds for specific functions, from simple isoprenoids to the more complicated versions such as natural rubber. The plants produce this wide range of isoprenoids in different amounts in specific organelles, and at different stages of growth and development. Since the diverse isoprenoid compounds are produced by a more or less conserved biosynthetic pathway (Randall et al., 1993), plants must execute a control mechanism to ensure that synthesis of the necessary isoprenoids occurs in the right place and at the right time. Such control is very likely mediated through regulatory enzymes, and attempts have been made to understand the regulatory mechanism of the isoprenoids and polyisoprenoids biosynthesis as well as their interrelationship (Kekwick, 1989; Mahlberg, 1993). The pathway used for the formation of isoprenoids in plants is similar to the sterol biosynthetic pathway that was elucidated in animals and yeast (Cornforth et al., 1966, 1972; Clausen et al., 1974; Taylor and Parks, 1978). The isoprenoid biosynthesis may be viewed as the pathway from acetyl-CoA via MVA and IDP to long-chain prenyl diphosphate. A large number of branch points can lead to a variety of diverse isoprenoids in plants (Kleinig, 1989; Randall et al., 1993).

Recently, the oligoprenoid and polyprenoid in *Hevea* latex were examined (Koyama et al., 1995; Tangpakdee et al., 1997a). Although the chain length of B-serum isoprenoids showed several components of C_{15}-C_{60} which were more or less of equal proportion, the C-serum isoprenoids were quite different. The major chain length in the C-serum isoprenoids was the C_{20}-GGDP as analyzed by autoradiography. Moreover, only a few isoprenoids were detected in the C-serum as compared with several in the B-serum. The C_{15}-farnesyl diphosphate (FDP) in C-serum was present in much less quantity than the C_{20}, though the difference between the two remains unclear. It can be assumed that the components in B-serum may be the intermediates preluding the formation of rubber, as was recently reported on rubber formation in the fresh bottom fraction of centrifuged latex (Tangpakdee et al., 1997b). The major C-serum of C_{20}-GGDP, and to a lesser extent C_{15}-FDP, might be the substrates for the prenylation of proteins, as the presence of prenylated proteins has been increasingly reported to occur in plant cells. However, the exact role of these isoprenoids in the two latex sera remains to be elucidated. More recently, the polyprenoids of the dolichols group and other group in *Hevea* latex were analyzed by two-dimensional TLC (Tateyama et al., 1999) It was found that the chain length of dolichols in *Hevea* ranges from C_{65} to C_{105}. The analysis of dolichols of the *Hevea* seeds, root, shoots, and leaves of different ages was also carried out, with comparison of the differences being examined to understand the changes associated with growth and development. The function of dolichols is commonly known to associate with the glycosylation process, and it is assumed that the presence of dolichols in *Hevea* is no exception. The role of glycoproteins has received much attention in the light of reports on the presence of lectin in *Hevea* latex (Wititsuwannakul, D. et al., 1997; Wititsuwannakul, R. et al., 1997), the glycoproteins having been found to play important roles in latex metabolism and colloidal stability.

The enzymes HMG-CoA synthase (HMGS) and HMG-CoA reductase (HMGR) have been implicated as essential regulatory enzymes in the biosynthesis of IDP (Brown and Goldstein, 1980; Bach, 1986; Goldstein and Brown, 1990). Similar roles have also been implicated in plants, although conclusive evidence for the regulatory role of HMGS is not yet available. Downstream of IDP in the formation of specific isoprenoids varies, depending on the end products and subcellular compartmentalization. The role of HMGR in the regulation of isoprenoids and rubber biosynthesis in *Hevea brasiliensis* has been well documented (Wititsuwannakul, 1986). Diurnal variations of HMGR levels and the dry rubber contents of latex were conclusively shown with high corresponding and positive correlations (Wititsuwannakul, 1986). The enzyme was located as membrane-bound and purified from the membrane of lutoids (Wititsuwannakul et al., 1990a). The purified enzyme was then analyzed and characterized, and found to be similar to the HMGR from other plant specimens. *Hevea* HMGR was activated by calmodulin, the calcium-binding protein involved in myriad regulatory processes, located in the latex C-serum (Wititsuwannakul et al., 1990b). A positive correlation between HMGR activity and calmodulin levels was demonstrated, corresponding with the levels of dry rubber content. Comparison between the high-yielding and low-yielding clones also showed clonal variations of calmodulin level in the same direction, and correspondingly. In the rubber tree, this compartmentalization is highly specialized for the syntheses of the rubber and isoprenoids, as demonstrated by the regulatory mechanism of lutoid HMGR by C-serum calmodulin in the control of rubber biosynthesis (Wititsuwannakul,

1986; Wititsuwannakul, 1990a,b) and the difference in isoprenoids distribution in B-serum and C-serum.

3.9 Factors Affecting Rubber and Latex Yields

Rubber yield is the result of two contributory factors, namely latex flow properties and regenerating synthetic capacity. It is generally observed that the latex yield after tapping depends first on the duration of latex flow. Subsequently, regeneration of the laticifer's content between the two tappings limits in turn the quantity of latex being collected. Therefore, flow and regeneration constitute the two most important limiting factors of latex production in *Hevea*. Put another way, we can examine this aspect as the synthesis capacity and the latex flow characters that influence the latex yield, and this has been shown to be a phenomenon of clonal variations.

3.9.1 Rubber Latex Regeneration Capacity

The utilization of sucrose as a carbon source, together with the enzymes of related metabolic pathways and the effect of pH changes, have been shown repeatedly to correlate with latex production. This subject has been extensively reviewed (Tupy, 1989), and will be outlined briefly to provide a cohesive view of the results referred to earlier in this chapter. Rubber biosynthesis, and the role of the key enzymes in various different pathways involved, has been outlined elsewhere. Here, we focus mainly on the regeneration of latex between successive tappings. Incision of the bark, causing the latex to exude from the latex vessels, results in a set of processes to be initiated and activated. Migration then occurs of reserves or their products from their production or accumulation zones to the demanding zones where regeneration of the rubber and latex constituents. These phenomena take place progressively, and the reconstitution of the latex to be collected at tapping requires a certain amount of time which depends on the amount being exuded. The essential function of laticiferous cells is the synthesis of *cis*-polyisoprene, which forms over 90% of cell contents. It is the metabolism of this highly specialized cellular environment and of the regeneration of latex between two tappings of *Hevea* which can be a limiting factor for the latex and rubber production.

Several apparent parameters were found to influence the latex regeneration metabolism. These controlling factors were commonly analyzed and compared for their effects in the tapped latex. The importance of sufficient availability of the sucrose in latex was extensively studied as precursor and having essential role for polyisoprenoids synthesis (Tupy, 1989). Sucrose metabolism and its utilization is controlled by invertase, and the levels of enzyme activity were positively correlated with latex production (Yeang et al., 1984; Low and Yeang, 1985). Some key enzymes affected by ions (Mg^{2+} and phosphate) and thiol (SH) groups for reducing conditions were documented in relation to the rubber biosynthesis. These included invertase, puruvate kinase and phosphoenol pyruvate (PEP) carboxylase. The activity levels of these enzymes influence the latex production capacity, as has been detailed and summarized earlier (Jacob et al., 1986; Tupy, 1989). The intracellular pH is the essential factor in isoprenoid metabolic regulation, while increased activities of invertase and PEP carboxylase were found under the alkaline pH of the laticifer contents, and this correlated with the latex and rubber yield (Yeang et al., 1986; Tupy, 1989).

Significant positive correlation between these parameters and latex production were documented. The relationships between the total solids content, reduced thiols content, pH of the latex, and the latex yields among different clones and the seasons have been analyzed and extensively characterized. Highly significant positive correlations were found between these three key parameters and the latex yield (Jacob et al., 1986). All correlations analyzed were of high statistical significance (P <0.001), as reported in extensive studies (Jacob et al., 1986), and hence these reflect the latex regeneration capacity of the laticifers which requires the coordination and interaction of several parameters in intricate manner. In addition, the induction and activation of protein synthesis was also noted for certain enzymes (Broekaert et al., 1990; Kush et al., 1990; Goyvaert et al., 1991; Kush, 1994; Pujade-Renaud et al., 1994) as the key factor on enzyme levels for latex regeneration. This is also the important key parameter in addition to the availability of sugars and the alkaline cytosol conditions of the laticifers and the latex cytosol (Tupy, 1989). Thus, it can be clearly seen that latex regeneration is a highly complex process that requires the participation of a myriad of parameters in a coordinated and synchronized manner for the orderly function of the laticifers and latex cytosol. Moreover, there are as yet some unidentified parameters to be added to the list as research progresses in this area.

3.9.2 Ethylene and Latex Yield

The mechanism of ethylene activation on rubber biosynthesis and enhanced latex yield has been the subject of many studies and has been recently reviewed (d'Auzac, 1989a,b; Kush, 1994). Ethylene has been implicated in mediating the signal for wounding response in the plant defense (Ryan, 1984), as well as in the stimulation of latex production (d'Auzac, 1989a,b). A protective role against various fungal and bacterial invasions has been proposed, this view on defense function having been developed in connection with the presence of high concentrations of antifungal and antibacterial substances in latex. The presence of lytic enzymes (chitinase, glucanase, lysozyme and others) has been demonstrated in the latex (John, 1993; Churngchow et al., 1995), and a protease inhibitor has also been found in latex, probably for defense purposes (Archer, 1983). The accumulation of these protective substances in latex is not constitutive, but rather an induction as a consequence of wounding responses to the tapping, for defense functions. Ethylene has been shown to play the key role in defense response (Koiwa et al., 1997), although its effect on the enhancement of latex yield remains poorly understood.

It is well documented that rubber trees, like any other plant, will respond to phytohormone treatment. The effect of ethylene on latex yield has been extensively investigated in *Hevea brasiliensis*, using a molecular biological approach. Although the biochemistry of *Hevea* polyisoprene synthesis is, in general, relatively well understood, it was only recently that studies on the molecular biology of the system were initiated (Broekaert et al., 1990; Kush et al., 1990). Cloning of various genes upstream of IDP was possible due to the availability of the heterologous probes (Dennis et al., 1989; Broekaert et al., 1990; Attanyaka et al., 1991; Chye et al., 1991) and to the highly conserved nature of this pathway. However, in the biosynthesis of natural rubber, molecular analysis downstream of IDP was more difficult because little was known or previously characterized. The cloning of prenyltransferase and REF (Light and Dennis, 1989; Attanyaka et al., 1991; Goyvaerts et al., 1991) that are involved in the final step of rubber biosynthesis are examples of studies of this type. Nonetheless, a number of questions remain unanswered about the process, some of which concern the exact mechanism of polyisoprene elongation of IDP when added in the *cis* configuration (Light et al., 1989; Cornish, 1993). In addition to these is the effect of hormone treatment on gene expression, as well as specific genes activation related to rubber biosynthesis and the increase in latex production stimulated by ethylene.

The stimulatory effect of ethylene on latex leads to various induced metabolic changes. It has been shown that a wide variety of substances and treatments can stimulate latex production. Each substance might have a specific role on the laticifers, and the end result is a combined, cumulative effect. It has been shown in vitro that the acid phosphatases released from lutoids hydrolyzed the key substances needed for the biosynthesis of rubber (Archer et al., 1963b). It would also be interesting to determine which molecules or enzymes are involved in the improvement of lutoids stability. It was noted that a direct correlation existed between the stability of lutoids and in vitro rubber biosynthesis, as well as a prolongation of latex flow (Archer et al., 1963b; Southorn and Edwin, 1968). Ethylene stimulates latex production, increasing duration of latex flow after tapping, by activating the metabolism involved in latex regeneration (Coupe and Chrestin, 1989). It is clear, therefore, that stimulation resulted in an increased stability of the lutoids. After tapping, regeneration in situ leads to the reconstitution of exuded latex before the next tapping. In quantitative terms, 100 mL of latex that is exuded is completely regenerated within ~60 h (Pujade-Renaud et al., 1994). This corresponds to a net synthesis of ~50 g of dry rubber and 1–2 g of protein during this period. Thus, highly intense rates of metabolic activity such as energy-generating catabolic pathways as well as anabolic processes are required. These include a large increase in glycolysis (Tupe, 1973), as well as increases in the adenylate pool and polysome and RNA contents (Amalon et al., 1992) – all of which are indicators of metabolic activation leading to increased latex production. The rise in the level of transcripts of several enzymes (chitinase, glutamine synthetase, and hevein) was shown to be increased by ethylene (Pujade-Renaud et al., 1994).

The response to ethylene can provide an ideal system to study the signal transduction in *Hevea*. An intriguing question is how the signal of ethylene becomes transduced. A number of speculative suggestions have been made, including the involvement of calmodulin or other calcium-binding proteins (Wititsuwannakul et al., 1990b). Another possibility is the role of protein prenylation. The rapidly expanding family of prenylated proteins include those that are involved in signal transduction (Cox and Der, 1992), as well as proteins involved with intracellular vesicular transport, cytoskeleton organization, and cell growth control and polarity (Cox and Der, 1992). It might be possible that a specific chaperone may need to be prenylated in order to act as a messenger, as has been documented in other plants (Zhu et al., 1993). It is therefore logical to consider that an important role for the prenylation in latex might exist. As elaborate metabolic machinery has been adapted for the biosynthesis of isoprenoids in specialized laticifers of *Hevea*, it is possible that the non-rubber isoprenoids may serve a regulatory function for metabolism. Intensive tapping or over-stimulation by ethylene leads to increased consumption of sucrose, leading in turn to an imbalance of carbon sources relationships. The exhaustion of carbohydrate reserves is one of principal reasons for physiological fatigue of stress and reduced capacity to hormone response. This results in the increased superoxide anion production, in turn causing free radical-based damage of various membranes, and especially those of the lutoids. Over-stimulation by ethylene can result in a phenomenon of coagulation of latex in situ called 'bark dryness', after which no more latex is obtained (Chrestin, 1989).

3.9.3
Latex Flow and Affecting Parameters

Latex flow has been outlined above briefly as being related to the roles of lutoids in latex. Latex regeneration and latex flow characters are interrelated, and the importance of flow with regard to latex yield or productivity is stressed by the fact that part of the effect on ethylene stimulation is prolongation of latex flow time (Abraham and Taylor, 1967; Abraham et al., 1968). Stimulation causes more intensive drainage from latex vessels due to the high turgor pressure inside the laticifers (Pakianathan et al., 1976). Ethylene stimulation was also found to increase sucrose utilization by activating the invertase enzyme and its level, leading to enhanced latex regeneration capacity (Low and Yeang, 1985). The enhanced capacity would certainly lead to the increased latex volume being stored in the laticifers, and the higher turgor pressure with the enhanced flow character (Pakianathan et al., 1976). Several mechanisms or causes have been suggested to explain the increased flow elongation as a consequence of enhanced latex regeneration capacity are (d'Auzac, 1989b), with a number of points identified as contributory factors:

1) An increase in osmotic and turgor pressures in the laticifers as a result of increased latex volume is due to an effect of plant growth regulators and ethylene stimulation.
2) The lowering of latex viscosity by a slight reduction in rubber concentration may explain the effect of ethylene on enhanced flow.
3) An increase in the elasticity of the laticifer's cell wall leads to a reduction in the collapse of vessels.
4) Modification of laticifer's permeability by ethylene will allow higher influx of water and solute, and hence better drainage of the latex vessels.
5) Plugging of the vessel ends is delayed by increased latex stability, hence leading to longer flow.
6) Ethylene stimulation increases the bark drainage area of the latex flow.
7) The increased lutoids stability by ethylene leads to a delay in coagulation of the rubber particles and hence a slowing effect on vessel plugging.

There is a limit to the enhancement of latex flow following ethylene stimulation. After a certain duration of flow time, the flow will stop as a result of the vessels becoming plugged; this is an attempt by the plant to contain the damage that might occur to laticifer metabolism due to exhaustion of its contents, and is analogous to the wounding response as a protective mechanism. Upon neutralization of their negative surface charges by various factors, rubber particles clump together to form microflocs. This leads to plugging of the vessels ends, and hence stops the latex flow. Floc formation caused by release of the contents of the lutoids that were damaged at the tapping cut, as well as during the flow, has been well documented (Southorn and Yip, 1968; Southorn, 1969, d'Auzac, 1989a). This was shown by the inverse relationship between the BI of the lutoids and the duration of latex flow (Yeang and Paranajothy, 1982; Yeang and David, 1984), as described above. Any delay in obstruction of the latex vessels due to slow plugging is associated with stability of luoids. Although the effects of ethylene stimulation on enhanced latex flow has been reviewed (Coupe and Chrestin, 1989), several other parameters may also have such a positive effect on latex flow and hence increase its yield. A high total solids content (TSC) in the latex can limit flow due to high viscosity, while increased water exchange in laticifers may cause a reduction in latex TSC, in turn lowering latex viscosity and promoting flow. Latex also contains thiols in the form of cysteine and glutathione which play an important role in the metabolism of laticifers, the thiols providing stability to the lutoids when they are damaged by free radicals generated in latex (Coupe and Chrestin, 1989). Thus, it is clear that latex flow is very much influenced by the lutoids, whether in terms of reduced latex flow time after lutoids rupture, or prolonged flow time by improved lutoid stability and integrity. Although much emphasis has been placed on ethylene however, several other factors are known to be involved in latex flow, notably the rubber particles and C-serum. In addition, other factors have yet to be identified that might provide a better understanding of this complex and complicate process.

4
Latex of Other Plants

Numerous plants can synthesize rubber in the latex, even though *Hevea* latex is the only suitable and viable source of natural rubber produced commercially. The formation of rubber in plants has been reviewed (Backhaus, 1985) and will be described only briefly here. It is commonly known that rubber latex is not only formed in the *Hevea brasiliensis*, but it is also synthesized and accumulated in over 2000 species of plants representing about 300 genera from seven families (Archer and Audley, 1973). The rubber-producing plant families are *Apocynaceae, Asteraceae, Asclepiadaceae, Euphorbiaceae, Loranthaceae, Moraceae,* and *Sapotaceae*. The latex and rubber composition varies among all these species. Rubber is composed of plant secondary metabolites that are produced in large quantities by certain plant cells, but has no known function to the plants. A metabolic function has been speculated upon, but specific details have yet to be indicated. The presence of rubber in various plants does not appear to deter herbivore feeding, insect attack or diseases caused by microbes. Rubber does not serve as an energy source, despite tremendous resources being allocated to its formation. There is no evidence that plants possess the enzymes capable of degrading rubber (Archer and Audley, 1973), and it appears that once rubber is formed, it remains in the cells. The enzyme responsible for the turnover of carbon in the rubber has yet to be identified and characterized. The ability of these plants to synthesize rubber is based on the presence of the key enzyme, rubber transferase, which causes *cis*-polymerization of isoprene units (IDP) into the long hydrocarbon chain of rubber. These are assembled and accumulated as rubber particles ranging in size from 5 nm to 3 μm. These particles are enclosed or surrounded by a lipophilic film. The presence of enzymes and the processes for rubber formation are well documented, and in some of these plants, for example *Parthenium argentatum* (guayule) and *Hevea brasiliensis* (rubber tree), detailed studies have been performed on the rubber synthesis process (Benedict, 1983). *P. argentatum* and *H. brasiliensis* represent plant species which accumulate large quantities of high-molecular weight rubber, albeit by different intercellular routes. Rubber synthesis takes place in parenchyma cells in guayule (Backhaus and Walsh, 1983), but in specialized latex vessel cells (laticifers) in *Hevea* (Archer et al., 1982). Other plants produce far less rubber than *Hevea* or guayule; they usually produce rubber of inferior quality and of a much lower molecular weight than *Hevea*, and this limits their commercial potential.

The molecular weights of rubber from various species were examined and compared. The average molecular weight of rubber molecules can vary considerably with the species. The reason for *Hevea* and guayule being unique among rubber-producing plants is that they produce rubber with a predominantly high molecular weight average of $3-7\times10^5$, which is critical for commercial utilization. Other rubber-producing plants produce inferior quality rubber with a molecular weight of $\sim5\times10^4$ or less. The molecular weight distribution (MWD) of rubber extracted from *Lactuca, Carissa, Candelilla,* and *Asclepias* compared with guayule, as demonstrated by gel-permeation chromatography (GPC), showed a value at least 100 times lower than that for guayule ($\leq10^4$ as compared with 10^6 for guayule). An interesting feature of molecular weight is that *Hevea* and guayule exhibit bimodal MWD, with a major peak at $\sim7\times10^5$ and a minor peak at $\sim0.5-1.0\times10^5$. This suggests either that polymerization in these two plants either undergoes a two-step process, or that two forms of rubber transferase may exist. In contrast, rubber in other plants uniformly shows a single peak of low molecular weight of $\sim10^4$ or less, indicating a lower degree of polymerization in these plants and different properties of the rubber transferase and activity in *Hevea* and guayule.

4.1 Latex of Guayule

Rubber biosynthesis has been extensively studied in *Hevea brasiliensis* (rubber tree) and *Parthenium argentatum* (guayule). Guayule rubber is a typical high-molecular weight *cis*-polyisoprene having a narrow MWD and physical properties that are slightly different from those of *Hevea* rubber. However, little is known of the structural differences between the rubbers from these two major sources. Recently, sunflower rubber was characterized and reported as a prominent rubber-producing member of higher plants (Tanaka, 1986). Low-molecular weight rubber from sunflower could serve as appropriate model for characterization of the chemical structure of *Hevea* and guayule rubbers. The findings showed that the fundamental structure of guayule rubber is similar to that of *Hevea* rubber, as deduced from the sunflower rubber structure elucidation (Tanaka, 1986). The guayule shrub synthesizes two types of secondary compounds – rubber and resin – in relatively large amounts. The rubber accumulates in parenchymal cells of the stem (Backhaus and Walsh, 1983) as opposed to the latex vessels in *Hevea*. Clearly, the amount of rubber that a guayule plant is capable of accumulating is limited by the total volume of its parenchymal tissues (Backhaus and Walsh, 1983). Cells which store rubber are found in the secondary cortex, phloem and xylem rays (Mehta, 1982; Mehta and Hanson, 1983). Treatment with a bioregulator, 2(3,4-dichlorophenoxy) triethylamine (DCPTA), was reported to increase the rubber content in guayule by two- to six-fold (Yokoyama et al., 1977), the increase of rubber in the phloem being more pronounced in winter samples in cells of the cortex.

An outstanding feature of rubber biosynthesis in guayule is that it is stimulated by low night-time temperatures (Bonner, 1967). Exposure of plants to 27°C during day-time, followed by 7°C at night for four months induced a three- to four-fold increase in rubber compared with control plants grown at a constant 27°C day and night. The low night temperature specifically induced the synthesis of *cis*-polyisoprene, with no effect on the rate of *trans*-isoprenoid resins formation (Goss et al., 1984). Exposure of guayule plants to low night-time temperatures stimulated the formation of rubber particles in cortical parenchymal cells. This suggested that the enzymatic potential for synthesizing rubber is greater in the stems from cold-exposed plants (Goss et al., 1984). The mechanism of the cold induction of guayule rubber is not known, but has been suggested as induced expression of genes coding for enzymes involved in rubber synthesis (Bonner, 1975). Regulation of guayule rubber content and biomass by water stress has also been noted (Bucks et al., 1984, 1985a, b). Several studies on guayule indicated that water stress will result in increased rubber content in the shrub (Miyamoto et al., 1984; Bucks, et al., 1985b). It is commonly known that rubber particles isolated from *Hevea* latex are enclosed with a film of phospholipids and proteins, as described above. It is clear that rubber particles contain protein factors necessary for the polymerization reactions in rubber formation. The same can also be observed for rubber particles of guayule (Backhaus and Walsh, 1983; Backhaus, 1985).

4.2
Possible Role or Functions of the Latex in Plants

Latex is produced in about 2000 plant species, and in varying degrees of quality and quantity. Rubber, which is the major and most important isoprenoid polymer in the latex, has no clear physiological function within the plants, and the specific and exact function of latex is the subject of much speculation. Details of the nature and possible role of latex rubber polyisoprene in plants have been presented only briefly (Archer, 1980), and solid evidence and strong support are still required to substantiate the roles as discussed. To date, several hypotheses and speculations have been proposed to explain the role or function played by rubber and latex in plants. The initial suggestion that the latex protected the plants from animal attack, but this has been discarded due to a lack of corroborative evidence. The proposition that rubber in latex acted as a reserve food supply or carbon storage reservoir was without strong support. This is not in accord with the fact that rubber in the latex, once it has been formed, is not further metabolized (Bonner and Galston, 1947). There is no evidence for the breakdown of rubber in vivo to substantiate this proposition. On the other hand, the suggestion that the formation of rubber in latex is a fermentative process analogous to the anaerobic production of ethanol has also been proposed (Ritter, 1954; Bealing, 1965). However, the absence of stoichiometric coupling of the oxidative and reductive phases of the biosynthesis processes has been extensively analyzed and discussed in detail (Archer and Audley, 1973), which precluded this explanation.

The main question needing to be addressed is why the plants, especially *H. brasiliensis*, expend so much energy and resources to synthesize natural rubber at all. This should lead to a search to determine whether latex has an important role in plant metabolism per se. The control mechanism in the formation of rubber appears to be a complex and intricate process. The interesting feature of *Hevea* metabolism is not only that for rubber formation, but also that it can be stimulated to produce rubber and other many components of latex by repeated tapping (Kush, 1994). This replenishment is not called for in the untapped tree, and although a few terpenoids have been shown to suffer catabolism in plants, there is no evidence for the breakdown of rubber in vivo. These are intriguing questions to most plant physiologists and biochemists, but we still do not have a clear answer to them. It was recently proposed that the latex may play a protective role against various fungal and bacterial attacks and the invasion of plant tissues by such pathogenic microbials (Kush, 1994). This view has been proposed because of the very high concentrations of antifungal and antibacterial substances such as chitinase, lysozyme and other pathogenesis-related proteins (PR proteins) being present in the latex (John, 1993). The latex of *H. brasiliensis*, which contains high levels of both chitinases and chitinases/lysozymes substantiates this suggestion (Martin, 1991). More recently, the role of *cis*-1,4-polyisoprene in the *Hevea* latex as having antioxidant function has been proposed (Tangpakdee et al., 1999). In latex, this compound is presumed to act as a free radicals acceptor or scavenger in laticifers to protect against damage caused by active oxygen species (AOS). However, the evidence to support this proposal remains superficial and is not sufficiently well documented to substantiate this claim.

Ethylene has been implicated in mediating signal for wounding response in plant defense (Ryan, 1984) and in the stimulation of latex production (d'Auzac, 1989b). A protective role for latex against various fungal and bacterial invasions has been proposed, this having been developed in connection with the presence of high concentrations of antimicrobial substances in the latex. Lytic enzymes (chitinase, glucanase, lysozyme and some other PR proteins) have been shown to be present in the latex (John, 1993; Churngchow et al., 1995), while the presence of a protease inhibitor in *Hevea* latex, probably for defense function, has also been reported (Archer, 1983). The accumulation of these protective substances in latex is not constitutive, but is induced as a consequence of the wounding response after tapping of the rubber trees. Ethylene has been shown to play a key role in defense response (Koiwa et al., 1997), although the effect of ethylene on enhanced latex yield is still not well understood. However, the role of latex as a result of induction by ethylene in defense function appears to be the most acceptable hypothesis at present. Thus, it seems likely that the end product of polyisoprene synthesis or rubber formation in latex has no known function per se, and the exact role of rubber and latex in plants remains an open question. In the case of laticiferous plants such as *H. brasiliensis*, it is possible that other processes occurring in laticifers may be of greater importance in terms of plant physiology than are polyisoprene synthesis and rubber formation.

5 Latex of Fungi

As mentioned earlier, the formation of polyisoprene rubber is confined to higher plants of dicotyledon angiosperms. Over 2000 species, representing about 300 genera of higher plants, are capable of the synthesis and accumulation of polyisoprene in the form of rubber latex. These natural polyisoprenes have been characterized as the *cis*- or *trans*-configuration by NMR analytical techniques. In addition, at last two fungal genera, *Lactarius* and *Peziza*, were also found capable of rubber biosynthesis (Stewart et al., 1955). Unknown species of fungal genera *Lactarius* and *Peziza* were reported to produce low-molecular weight polyisoprenes. Rubber extracted from four species of *Lactarius* was characterized by infrared analysis as *cis*-polyisoprene with an intrinsic viscosity of 0.43–1.02. Very few studies were carried out on the rubber from fungi, and hence the characterization of fungal rubber is little known. Here, discussion will be limited to the minimal information and reports available. Studies on the structure and biosynthetic mechanism of rubber from mushroom (Tanaka et al., 1990) have provided some details on the nature and characteristics of the rubber produced by the lower living organisms.

Fungal rubber was reviewed with regard to elucidating its structure and characterizing the initiation and terminal units of low molecular weight rubber. These simple studies with the fungal small molecule rubber may serve as a model for more advanced investigations into the rubber of higher plants, which has a much greater molecular weight. The most extensive studies were made with *Lactarius* rubber, using NMR analysis, to characterize the rubber biosynthetic process in fungi (Tanaka et al., 1990). A few species of *Hygrophorus*, *Russula*, as well as *Lactarius*, were found to produce rubber. Among different *Lactarius* species, *Lactarius subplinthogalus* was found to produce *cis*-polyisoprene rubber as high as 7% of the dry weight of sporocarps. The biosynthesis of *Hevea* natural rubber has been assumed to proceed from DMADP by the successive addition of IPP in the *cis* configuration. However, there was no direct evidence either to prove or verify the initiation step. Nor has any biochemical investigation been carried out on the termination mechanism. A ^{13}C-NMR method was used to analyze the structure of both terminal units, and the alignment of *cis* and *trans* isoprene units along the polymer chain. This has provided important information on *cis*- polyisoprene synthesized in the nonlaticiferous cells. The rubbers from goldenrod and sunflower leaves consist of a dimethylallyl group and about two or three isoprene units in the *trans* configuration, followed by a long sequence of isoprene units in the *cis* configuration, aligned in that order. This is the evidence that biosynthesis of rubber in these plants starts with *trans* GGDP or *trans* FDP as the initiating species. This initiation step is similar to that observed with other *cis*-polyisoprenes. Using ^{13}C-NMR analyses, the presence of terminal *trans* units were shown for *Hevea* rubber and other wild rubbers (Tanaka, 1989).

Rubbers from fungi were characterized both for the structure of terminal units and mechanism of rubber biosynthesis in the fungal latex. Comparison was then made with other rubber-producing species of higher plants, which contain mostly *cis*-polyisoprene as rubber in the latex. The rubbers from *Lactarius volemus* and *Lactarius chrysorrheus* (*Lactarius* rubber) were characterized (Tanaka et al., 1990) and found to be similar to *cis*-polyisoprenes in latex of other higher plants, as previously reported. The structure of terminal units and alignment of the isoprene units provide information on initiation and termination mechanisms of rubber formation in fungi. Polymerization of *Lactarius* rubber starts from *trans* FDP and proceeds by successive condensation of IDP to form isoprene units in the *cis* configuration. Termination is presumed to occur by direct esterification of polyisoprenyl diphosphate (PIP-DP), or after dephosphorylation followed by esterification with fatty acids. Chemical or biochemical modification of both terminal groups and the main chain also occur on the fungus rubber (Tanaka et al., 1990). Information on the termination mechanism may provide a clue as to the process of controlling the molecular weight of rubber. The study on low-molecular weight rubber from fungi may also serve as a model for elucidation of the molecular weight controlling mechanism in *Hevea* rubber, as well as in the rubber from other higher plants. It is expected that greater attention will be paid to studying fungal rubber as a possible novel source of natural products other than the rubber that might be present in the fungal latex. So far, information on the structure and composition of fungal latex has not been obtained in great quantity, except for the demonstration of the presence of rubber in fungal latex (Tanaka et al., 1990). The plentiful information from *Hevea* latex study would be very helpful to investigate the structure and composition of fungal latex, especially the nonrubber constituents. Whether the nonrubber constituents of fungal latex are similar or different from those of *Hevea* latex will be interesting, and this subject clearly requires further study.

6 Diseases Related to Natural Rubber Latex (Latex Allergy)

Natural rubber latex (NRL) from *Hevea brasiliensis* has been used for a very long time in the industrial production of items that are used on a daily basis and, being an inert material should pose no danger to health. However, the latex in which rubber forms a major component contains a myriad of substances that are foreign to our immune system. These non-rubber substances are also found tightly bound or associated with the rubber particles and hence will always be present with rubber in the finished products. As mentioned above, rubber particles are enclosed or surrounded by a protective film that is composed mainly of lipids and proteins. These lipids are similar to the common lipids in our body, and so would not be expected to be recognized as foreign substances by our immune system. However, the rubber particle proteins are quite different from proteins in our bodies, and hence will be recognized as 'foreign' by our immune system, and this is where health problems begin. Frequent exposure to these foreign proteins will cause sensitization in the individuals or users, and this in turn leads to an allergic response. Allergy to natural rubber has been identified as being caused by the proteins present in the rubber products, especially in the case of rubber gloves and other rubber products than are used frequently. Indeed, allergy to natural rubber has recently become a major health concern and will be discussed in detail below.

6.1 Latex Allergy

Reports of NRL immediate hypersensitivity have increased steadily since the first reported case in 1979 (Nutter, 1979; Granady and Slater, 1995), and NRL protein allergy has been the subject of recent review (Hamann, 1993). Specific immunoglobulin IgE and IgG4 antibodies to a series of NRL proteins have been detected (Alenius et al., 1991, 1992) in the sera of NRL-sensitized individuals. Different sources of NRL proteins, and purification methods coupled with nonstandardized pooled, sensitized sera have contributed to the detection of a wide range of antigens. Antigens in the regions of 14 kDa and 30 kDa seem the most reproducible. The initial presentation after cutaneous exposure is typically contact urticaria. Aerosolized protein bound to cornstarch powder may lead to rhinitis, conjunctivitis and bronchospasm. Mucosal and intraoperative parenteral exposure seem to lead to the greatest risk of massive histamine release and anaphylaxis. Although a standardized reagent is not available, the most sensitive diagnostic test remains the prick test. Treatment for the NRL protein-sensitized individual is NRL product avoidance. Non-NRL alternatives should be considered routine among risk group, such as spina bifida and atopic patients with hand eczema, in order to prevent unnecessary sensitization.

Each year, NRL-containing medical and consumer products are safely and efficaciously used on billions of occasions. The challenge remains to identify specifically the NRL allergens, and to reduce their levels below a threshold which either induces sensitivity or elicits a reaction. The increase in prevalence of latex hypersensitivity since 1979 remains an issue of debate, and it is most likely that the situation has arisen due to a complex combination of factors. The most significant contributor to the rise in prevalence has been the vast increase (by several hundred) in the number of (inexperienced) NRL glove manufacturers during the period 1986–7. This was due to public awareness of AIDS, coupled with an increased demand for protection against the condition. Allergen levels rapidly reached unprecedented levels and exceeded sensitizing thresholds. Vast numbers of people, who previously had been using NRL without problems for decades, were needlessly sensitized. Hypersensitivity reactions to rubber products are common. Most early references to latex as initiator of immune responses describe cases of contact dermatitis, a delayed hypersensitivity (type IV). This type of reaction, in cases involving latex exposure, is typically localized to the hands after the use of rubber gloves. The allergic lesions are localized to the areas of skin in direct contact with the rubber product. Between 1979 and 1988, numerous reports appeared in the literature of IgE-mediated (type I) immediate reaction to latex (Nutter, 1979; Dooms-Goossens, 1988), the studies pointing to the fact that the allergenic reactions were due to specific IgE antibodies directed to allergenic proteins in the *Hevea* latex (Wrangsjo et al., 1986).

6.2 Latex Protein Allergens

Characterization of NRL allergens has been recently reviewed (Kurup et al., 1992, 1995). A number of proteins have been detected in NRL and in latex finished products (Dalrymple and Audley, 1992; Makinen-Kiljunen et al., 1992). Several of these proteins have specific reactivity with patient serum antibodies. A large number of peptides has been reported in non-ammoniated *Hevea* latex. Although more than 240 different polypeptides can be separated by two-dimensional gel electrophoresis (2D-PAGE) of the latex serum, less than 25% of these peptides showed reactivity with IgE from patients with latex allergy (Alenius et al., 1994a). A protein with molecular weight of 58 kDa was isolated from latex gloves with the same monomers of 14.6 kDa, and was shown to have complete homology with REF (Czuppon et al., 1993a). REF was implicated as the major allergen in latex, and a 14-amino acid synthetic peptide representing the N-terminal domain of REF reacted with IgE antibodies in a majority of latex-sensitive patients (Czuppon et al., 1993b). Although REF was considered the major allergen in NRL, several other studies showed a large number of polypeptides associated with development of the latex allergy (Kurup et al., 1993a; Alenius et al., 1994a). The major polypeptide among these is hevein, which has chitin-binding properties. Significant variation in the number and quantity of proteins in various extracts of latex gloves has been reported (Chambeyron et al., 1992). SDS-PAGE demonstrated a wide range of peptides, from 5 kDa to 200 kDa in various extracts. IgE antibody in the sera of latex-allergic patients showed a wide range of reactivity in immunoblot studies. Predominant protein allergens showed molecular sizes ranging from 5 kDa to 100 kDa, and the 14 kDa peptide (REF) showed more frequent reactivity with patient sera (Yeang et al., 1996). Significant IgE binding to four latex proteins of 14, 20, 23, and 28 kDa was reported for sera from allergic patients (Yeang et al., 1996).

6.3 Proteins and Allergens in Latex Gloves

The question of whether the amount of eluting protein correlates with the allergenic reactions of a given NRL product is an issue for debate (Turjanmaa et al., 1995). It is generally agreed that both the protein and the allergen activity in latex gloves may vary considerably (Turjanmaa et al., 1988; Alenius et al., 1994b). The determination of total protein extractable from NRL products does not necessarily characterize NRL allergen activity. Besides, cross-reactivity of NRL and food allergens has been documented (Lavaud et al., 1992), with NRL-allergic patients showing IgE antibodies that bind to a number of banana and avocado proteins (Ahlroth et al., 1994). The characteristics of cross-reacting NRL and fruit allergens are as yet largely unknown. Allergens in the glove extract may be denatured or modified during the rubber aging and processing (Makinen-Kiljunen et al., 1992). Antigen cross-reaction with latex proteins is, therefore, relatively common.

Several studies have shown that there is cross-reactivity between latex and other allergens, these including a variety of fruits, vegetables, and grains. Such studies have shown variable levels of cross-reactivity between food allergens and gloves or latex (Kurup et al., 1993b). The cross-reactivity among these antigens may be explained in term of the process that is seen uniquely in plants. It was reported that the peptide sequence deduced from the mRNA accumulating in response to wounding or hormonal treatment shows homology to several chitin-binding proteins, such as the latex hevein (Broekaert et al., 1990). A chitin-binding domain has been suggested to be the building block of many proteins with diverse specificity (Chrispeels and Raikhel, 1991). Whether such a common molecule exists and can be implied in the cross-reactivity between these other allergens and latex is not known, or is not clear at present. It is possible (or likely) that a common molecule can be implicated in the cross-reactivity between these fruit or vegetable allergens and hevein, as the latter has been purified from latex and shown to have chitin-binding properties (Gidrol et al., 1994). Hevein was also found to be one of the major latex allergens. An association between latex hypersensitivity and fruit allergy would act as a precautionary measure, but this would not necessarily mean that all people allergic to fruit antigens would become allergic to latex, and vice versa.

6.4 Cross-Reactivity of Allergens

Cross-reactivity between latex and other plant source allergens has more recently been assessed in greater detail (Truscott, 1995). It was commonly observed that several proteins in NRL are similar to those found in other plant species. Hevamines, the cationic proteins in *Hevea* latex, are very similar in molecular weight, amino acid composition and cationic features to lysozymes found in fig (*Ficus*) and papaya (*Carica*) latex (Tata et al., 1976). The *Hevea* lutoid proteins, which increase during ethylene or ethylene oxide treatment as described above on ethylene stimulation of latex yield, are very similar to the lutoids found in latex of banana (*Musaceae*), avocado (*Lauraceae*), and several other fruits. These fruits have been reported to cross-react with *Hevea* latex in several clinical tests on allergenicity evaluations (Czuppens et al., 1992; Lavaud et al., 1992; Ross et al., 1992; DeCorres et al., 1993; Fisher, 1986, 1987, 1993). Interestingly, chestnuts are also found to be cross-reactive with allergenic proteins from *Hevea* (Anabarro et al., 1993; Lavaud et al., 1992). Chestnuts are often sterilized with ethylene oxide to prevent spoilage by microbial and insect infestation (Foegeding and Busta, 1991). It is conceivable that these somewhat similar lutoid proteins conjugate with the extremely reactive ethylene molecule, creating a mutually recognized hapten or carrier responsible for cross-sensitization or cross-reactivity. By taking these several observations to the next step, it is possible that ethylene becomes the common bridge linking 'transferred recognition' and subsequent cross-reactivity.

The nature of the similar (but not identical) proteins fractions in ethylene-treated fruits (avocado, banana, chestnuts) and the *Hevea* latex, which is exposed to ethylene released in the wounding response of tapping the rubber trees, might lead to the allergens' cross-reactivity. It then stands to reason that the allergenicity potential of latex is increased with further exposure to ethylene in the form of ethylene oxide, which conjugates with the non-haptenated latex proteins. The increased exposure to these hapten or carrier allergens as a whole increases the dose challenge that would be experienced from a single allergen, and this is a plausible contributing factor for reducing the amount of time required until a threshold sensitization level is reached. It may be that the sensitized healthcare worker would experience his/her first type I reaction at an earlier age when he/she would finally have had a sufficiently high exposure if the sensitization mechanism recognized each of these proteins as separate and unique allergens.

It will be interesting to note that this logical hypothesis can hold by examining other fruits and vegetables (potato, melon, carrot, radish and orange) treated with ethylene derivatives (Goren and Huberman, 1976; Christofferson and Laties, 1982; Apelbaum et al., 1984; Vreugdenhil et al., 1984). It is most probable that individuals will react to the raw vegetable rather than the cooked preparations. Receptors for ethylene must be present in these fruits and vegetables, as they are in tomato and avocado. Since they recognize the same hormone chemical switch (ethylene), the specific receptors must be very similar, potentiating the possibility of mutual immunological recognition.

6.5 Allergens and Antigens of Latex Products

The number and concentration of allergenic proteins remaining in the finished product are subjects currently undergoing serious and significant scrutiny. Processing variables that might conceivably contribute to this variability are manifold. Antigen isolation is difficult to harmonize because of the variability of antigen source, isolation techniques and the serum pools. Detection by immunological techniques usually involves incubation of the NRL-sensitive sera with various sources of antigens. These antigen sources include non-ammoniated, low-ammoniated, and high-ammoniated NRL. In addition, the antigen sources are either dry rubber latex, latex-finished products, or fresh rubber latex. These have potentially a distinct hydrolytic impact on the NRL antigen and contribute to the broad range of protein size showing serum antibodies with binding activity (Makinen-Kiljunen et al., 1992, 1993; Alenius et al., 1994a). Despite the variability of techniques used in currently published reports, proteins with molecular weight of ~14.5 kDa and 30 kDa appear with significant frequency to merit closer consideration (Yeang et al., 1996).

6.6 Remarks on *Hevea* Latex Usage and Allergy

Development of the medical glove as a natural rubber finished product has revolutionized the protection of both healthcare providers and patients. The manufacture of latex gloves involves the combination of a complex raw material, a variety of chemical additives, and multiparameter process to meet customer requirements. Dermal compatibility issues related to these chemicals have been known since the 1930s, when the first case of irritant and allergic contact dermatitis due to a variety of rubber products were first documented. With occupational exposure to hepatitis and HIV, there has been a re-emphasis on the importance of hand protection. As would be expected, the increased glove usage has resulted in an increase in the frequency of irritant and allergic contact dermatitis (delayed type IV). However, it is the occurrence and rising incidence of immediate type I hypersensitivity to allergenic proteins from NRL that have caused the greatest concern.

Research into this issue has been slow because sensitized individuals recognize different peptide allergens, and not all have routine (consistent) latex contact. This seems to indicate that the cause of latex allergy is varied and multifaceted. NRL allergy has proved to be an important allergy among both healthcare and nonhealthcare professionals. The issues to focus on at present include the development of more accurate in vivo and in vitro diagnostic methods, and the standardization of techniques for the allergenicity testing of various latex gloves and other NRL products. To reach these goals, the molecular structure of important NRL allergens should be identified. Future studies should elucidate whether previously characterized rubber proteins, such as hevein, hevamine and REF are significant allergens.

6.7 Latex Protein Allergens from cDNA Clones

Attempts have been made to isolate and purify NRL proteins through cDNA cloning (Broekaert et al., 1990; Attanyaka et al., 1991; Lee et al., 1991). The two major rubber latex proteins obtained by this method, hevein and REF, are being evaluated for their immunochemical characteristics. Hevein, the major latex lutoids protein, with molecular weight ~5 kDa, was shown to have functions involving rubber particle coagulation and microfloc formation (Gidrol et al., 1994), leading to latex vessel plugging. Hevein has been shown to be the major component of the lutoids and is present at the highest concentration, representing ~70% of the total lutoids soluble B-serum proteins (Audley, 1966).

Hevein is quantitatively the most important latex B-serum protein which has been purified and characterized in detail (Tata, 1976), its amino acid composition and sequence having long been determined (Walujono et al., 1975). Hevein is a chitin-binding latex protein that has been purified and crystallized (Archer, 1960; Audley, 1966), and has also been shown to accumulate as one of the highly induced proteins in the wounding response (Broekaert et al., 1990) of rubber trees following tapping for latex collection. The protein of hevein cDNA clones has been shown to have high reactivity to IgE antibody of patients with latex allergy (Broekaert et al., 1990; Makinen-Kiljunen et al., 1992). Hevein has been implicated as a common molecule in allergen cross-reactivity. The REF, a tightly bound rubber particle protein that is involved in the rubber biosynthesis process, has been characterized and its amino acid sequence determined (Dennis et al., 1989; Light and Dennis, 1989). The nucleotide sequence of REF has been reported for a 14.6 kDa protein (Attanyaka et al., 1991). This protein was shown to be the major allergen in latex, with very high reactivity to the serum IgE antibody of latex-allergic patients (Czuppon et al., 1993b) in several tests and allergenic studies.

Most recently, the cDNA clone for small rubber particle protein (SRPP) was reported in an extensive and detailed study (Oh et al., 1999). The isolation, characterization and functional analysis of the cDNA clone encoding a small rubber particle protein in *Hevea* latex was reported and assessed in great detail. It is generally found that more proteins of SRPP are tightly bound with greater affinity towards small rubber particles than larger particles. These proteins are still found to be tightly bound to rubber particles even after extensive washing of the particles, and centrifugation (Oh et al., 1999). The characterization and functional analysis of the cDNA clone encoding for this 24 kDa protein was elucidated and found to be involved in rubber biosynthesis, its function being similar to that of REF. The SRPP might work in conjunction, or in combination, with the REF which is also present on the surface of rubber particles, together with this 24 kDa protein.

The 24 kDa protein is one of the most tightly bound on the rubber particle surface, similar to REF (Dennis et al., 1989; Light and Dennis, 1989; Oh et al., 1999). Both the REF and 24 kDa protein have been reported to be major rubber latex allergens, with very strong allergenicity and high antigenicity (Makinen-Kiljunen et al., 1992; Yeang et al., 1996, 1998; Oh et al., 1999). These two proteins are the most frequently encountered major latex allergens and pose a serious threat to sensitized latex-allergic patients (Makinen-Kiljunen et al., 1992; Yeang et al., 1998). Developments in the molecular biology and cDNA cloning of these latex allergen proteins will clearly provide important and major input to the progress, and improve our understanding of this increasingly serious health problem. The outcome of these studies will help to ease our concern for effective protection and prevention from contamination or infection, and of allergies associated with NRL finished products encountered in daily life.

7
Outlook and Perspectives

Natural rubber, an important isoprenoid polymer with no known physiological function to the plant, is produced in about 2000 plant species, and in varying degrees of both quality and quantity. Rubber is the raw material of choice for heavy-duty tires and other industrial uses that require elasticity, flexibility and resilience. *Hevea brasiliensis* has, until now, been the only commercial source of natural rubber, mainly because of its abundance in the tree, its quality, and the ease of harvesting. However, the diminishing acreage of rubber plantations and life-threatening latex allergy to *Hevea* rubber, coupled with an increasing demand, have prompted research interest in the study of rubber biosynthesis and the development of alternative rubber sources. Despite this, no better alternative source of natural rubber has been obtained; neither can any alternative match the superior quality, economic advantage, and developed expertise in the rubber plantations of *Hevea brasiliensis*. The possibility of synthetic rubber has been raised, but no satisfactory synthetics can match the superior biological and physical properties of *Hevea* rubber, even though the economic consideration in terms of higher cost can be put aside. Natural rubber is an extraordinary product of polyisoprene synthesis, and *Hevea brasiliensis*, especially when compared with other plant, is the only commercially viable source. A minor exception is guayule rubber, though commercial production of this has yet to be developed. Rubber appears to be a complex of secondary metabolites for which the producing plant has no obvious use, yet is in itself of major commercial importance. For rubber planters, as well as for rubber plant breeders, the early molecular markers of latex production can be of vital use and have immense practical implication. This can only be achieved through an understanding of the biochemistry of rubber biosynthesis and the physiology of rubber trees during their production of latex and their capacity for regeneration upon tapping. The selection of superior rubber clones may be at the seedling stage, based on findings made in multidisciplinary research. However, improvements in rubber yield and rubber quality, as well as the response to ethylene stimulation, are priority criteria for the mutual interest of both the agricultural and industrial exploitation of rubber trees.

In *Hevea brasiliensis*, rubber biosynthesis takes place on the surface of rubber particles suspended in the latex, which is the cytoplasm of laticifers. The laticifers are specialized vessels located adjacent to the phloem of rubber tree. When severed during tapping, the high turgor pressure inside the laticifers expels latex containing 30–50% (w/w) *cis*-1,4-polyisoprene. The differential expression of several rubber biosynthesis-related genes in latex has been documented. REF, a protein (enzyme?) involved in rubber synthesis, is highly expressed in laticifers. These transcribed genes are actively translated into proteins, with some 200 distinct polypeptides being present in *Hevea* latex. Genes expressed in *Hevea* latex can be divided into three groups based on the proteins that they encode: (1) defense-related proteins such as hevein, chitinase, and β-1,3-glucanase; (2) rubber biosynthesis (RB)-related proteins such as REF, HMGR, and FDP synthase; and (3) latex allergen proteins such as hevein and REF, among several other known allergens (Hev b3, Hev b4, Hev b5, and Hev b7). The biological functions of the allergenic proteins are largely unknown, but continuing molecular biology developments and cDNA cloning in the study of these latex allergen proteins will clearly provide major input to progress and a better understanding of this increasingly serious health problem. In addition, benefit will be obtained with regard to effective protection and prevention against contamination or infection, and of allergies associated with the NRL finished products encountered during daily living. Moreover, in the long term it is possible to foresee the transformed rubber tree as a mini-industry in the production of biologically and pharmaceutically important molecules, along with natural rubber. To achieve this, the development of a transformation system for rubber tree is a requisite, in order to identify the laticifer-specific *cis* elements, probably in the promoter of genes which show laticifer specific expression. Laticifer-specific gene expression has been well documented, an example being the production of rubber particle proteins expressed exclusively in rubber plants. One might consider also the use of a model system such as dandelion (Russian rubber), a rubber-producing weed, having anastomosed laticifers like the rubber tree, for transformation and promoter analysis of latex and rubber formation. This situation is similar to that with *Arabidopsis* in our understanding of plant molecular biology. To comprehend the molecular and biological aspects of rubber biosynthesis, it is important to investigate the gene expression profile in latex. Among the genes most abundantly expressed in latex were found cDNA clones encoding major rubber particle proteins of 14.6 kDa and 24 kDa, both of which cause allergenic responses in sensitized patients. The amino acid sequence of the 24 kDa protein is highly homologous to that of REF, suggesting its potential involvement in rubber synthesis.

Because rubber formation takes place in laticifers, genes highly expressed in such tissues may code for enzymes involved in rubber biosynthesis. A number of genes that are highly expressed in latex (as compared with leaves) have been identified. The latex RNAs are highly enriched in transcripts encoding rubber biosynthesis-related enzymes (20- to 100-fold) as well as defense-related proteins (10- to 50-fold). These genes, along with defense genes, are among the most abundant transcripts. The latex has been suggested to play a protective role because of the high content of defense-related proteins that it contains. However, it remains to be answered why *Hevea* allocate excessive energy and resources to rubber biosynthesis and the formation of rubber particles that are dispersed in the latex as a colloidal suspension. Each of the rubber particles contains hundreds to thousands of rubber molecules within this enclosing interface. Analysis using GPC has revealed that small rubber particles contain rubber of higher molecular weight than do large particles. Rubber particles are not simply an inert ball of rubber; rather, analyses of the particles of four rubber-producing plants (including *Hevea brasiliensis*) show that the particle surface is a mosaic of proteins, conventional membrane lipids, and other components. Among many latex proteins, two main proteins remain associated with rubber particles after repeated washing, namely REF (14 kDa) with large particles, and the 24 kDa protein, with small particles. Rubber transferase (*cis*-prenyltransferase) activity has been reported for the rubber particles-bound proteins of three rubber-producing plants, *Hevea brasiliensis, Parthenium argentatum,* and *Ficus elastica.* Although at least two rubber particles proteins, REF and rubber transferase, have been suggested to be involved in rubber biosynthesis in *Hevea,* the detailed mechanism is still little known. Although the gene encoding for rubber transferase has not been cloned, a full-length cDNA encoding REF has been cloned. It has been established that REF plays a functional role in rubber polymerization; however, the actual role of REF and the nature of rubber transferase in rubber elongation has not been fully assessed.

The present major concern is the health problem related to latex allergy. Some latex proteins have been shown to be allergens which create a serious biomedical problem and deserve the due attention of molecular biologists and biochemists. In the rubber industry today, an enormous amount of protein in latex (~5 g L^{-1}) is wasted after the coagulation and separation of natural rubber. This rich source of nitrogen can be used as an important by-product after suitable treatment. Not only the proteins, but also numerous nonrubber constituents in the latex, are also overlooked and discarded. Indeed, close attention should be paid to the potential use of these precious chemicals, which not only contribute to the colloidal stability of latex but also to the maintenance of important metabolic functions in the cytoplasm. In addition to *cis*-polyisoprenoids, many nonrubber isoprenoids produced in the latex may offer commercial potential. The above outline forms only part of the vast interest in *Hevea* latex, and clearly some areas of investigation have been omitted here. Nonetheless, in our resource-scarce world, attention should be perhaps be paid in the short term to a more comprehensive utilization of this type of valuable industrial by-product. The discussion presented has attempted to highlight the importance of a highly specialized branch of isoprenoids biosynthesis that occurs naturally in rubber trees. The vast potential for improving the rubber industry, together with a potential for establishing spin-off industries, is clear, with outlooks and perspectives not confined to rubber but extending to the other, non-rubber, constituents of *Hevea* latex.

Two recent developments of interest in natural rubber research have been: (1) the reduction of protein in latex finished products to minimize allergic effects; and (2) vulcanization of rubber latex by radiation, rather than with sulfur. Both biological and physical approaches have been used to remove the proteins from rubber particles in latex before its use in manufactured products. Protease treatment of the latex to hydrolyze or digest the proteins away from the rubber particles has been attempted, but this appears ineffective as the proteins tend to localize in hydrophobic regions inaccessible to the proteases. Several enzymes have been used, but none has provided any satisfactory outcome, and problems concerning steric hindrance must be addressed if optimum protease treatment is to be realized. It appears paradoxical that exogenous protein (as proteases) be added to rid the latex of endogenous proteins; indeed, the suggestion has been made that the proteases might themselves act as additional allergens. A double centrifugation process to concentrate latex offers one physical means of reducing latex proteins, but whether this technique will remove tightly bound proteins from the rubber particles remains unclear. A second approach, which is 'environmentally friendly', is to use radiation for vulcanization, rather than sulfur. This would avoiding the production of sulfur-containing waste and be beneficial to the environment, though the performance of both vulcanized rubber types must be comparable.

8 Relevant Patents

Backhaus, R.A. and Pan, Z. (1997) Rubber particle protein gene from guayule. U.S. Patent. 5 633 433.

Beezhold, D.H. (1996) Methods to remove proteins from natural rubber latex. U.S. Patent. 5 563 241.

Cornish, K. (1998) Hypoallergic natural rubber products from parthenium argentatum (gray) and other non-*Hevea brasiliensis* species. U.S. Patent. 5 717 050.

Dove, J.S. (1997) Methods for reducing allergenicity of natural rubber latex articles and articles so produced. U.S. Patent. 5 691 446.

Dove, J.S. (1998) Methods for reducing allergenicity of the natural rubber latex articles. U.S. Patent. 5 741 885.

Ji, W. (1994) Methods for extracting polyisoprenes from plants. U.S. Patent. 5 321 111.

Lui, J.H. and Shreve, D.S. (1987) Rubber polymerase and methods for their production and use. U.S. Patent. 4 638 028.

Ong, C.O. (1998) Preservation and enhanced stabilization of latex. U.S. Patent. 5 840 790.

Raikhel, N. V. et al. (1999) cDNA encoding a polypeptide including a hevein sequence. U.S. Patent. 5 900 480.

Schloman, W. W. (1999) Reduced lipid natural rubber latex. U.S. Patent. 5 998 512.

Sikora, L.A. (1991) DNA fragment encoding a rubber polymerase and its use. U.S. Patent. 4 983 729.

Tanaka, Y. et al. (1999) Process for preparing deproteinized natural rubber latex molding and deproteinizing agent for natural rubber latex. U.S. Patent. 5 910 567.

Tanaka, Y. et al (1997) Means for mechanically stabilizing deproteinized natural rubber latex. U.S. Patent. 5 610 212.

Trautman, J.C. (1998) Method of neutralizing protein allergens in natural rubber latex product formed thereby. U.S. Patent. 5 777 004.

Umland, H. and Petri, C. (1998) Preserved and stabilized natural latex, with water soluble carboxylic acid salts. U.S. Patent. 5 773 499.

9
References

Abraham, P. D., Taylor, R. S. (1967) Stimulation of latex flow in *Hevea brasiliensis*, *Exp. Agric.* **3**, 1–12.

Abraham, P. A., Wycherley, P. R., Pakianathan, S. W. (1968) Stimulation of latex flow in *Hevea brasiliensis* by 4-amino-3, 5,6-trichloropicolinic acid and 2-chloroethane phosphonic acid, *J. Rubb. Res. Inst. Malaya* **20**, 291–305.

Ahlroth, M., Alenius, H., Makinen-Kiljunen, S., Turjanmaa, K., Reunala, Palosuo, T. (1994) Cross-reacting allergens in natural rubber latex and avocado [abstract.], *J. Allergy Clin. Immunol.* **93**, 299.

Alenius, H., Turjanmaa, K., Palosuo, T., Makinen-Kiljumem, S., Reunala, T. (1991) Surgical latex glove allergy: Characterization of rubber protein allergens by immunoblotting, *Int. Arch. Allergy Appl. Immunol.* **96**, 376–380.

Alenius, H., Reunala, T., Turjanmaa, K. (1992) Detection of IgG_4 and IgE antibodies to rubber proteins by immunoblotting in latex allergy, *Allergy Proc.* **13**, 75–79.

Alenius, H., Kurup, V., Kelly, K., Palosuo, T., Turjanmaa, K., Fink, J. (1994a) Latex allergy: Frequent occurrence of IgE antibodies to a cluster of 11 latex proteins in patients with spina bifida and histories of anaphylaxis, *J. Lab. Clin. Med.* **123**, 712–720.

Alenius, H., Makinen-Kiljunen, S., Turjanmaa, K., Palosuo, T., Reunala, Reunala, T. (1994b) Allergen and protein content of latex gloves, *Ann. Allergy* **75**, 315–320.

Allen, P. W., Jones, K. P. (1988) A historical perspective of the rubber industry, in: *Natural Rubber Science and Technology* (Roberts, A.D., Ed), pp. 1–34. New York: Oxford University Press.

Amalon, Z., Bangratz, J., Chrestin, H. (1992) Ethrel (ethylene releaser) induced increase in the adenylated pool and transtonoplast pH within latex cells, *Plant Physiol.* **98**, 1270–1276.

Anabarro, B., Garcia-Ara, M. C., Pascual, C. (1993) Associated sensitization to latex and chestnut, *Allergy* **48**, 130–131.

Apelbaum, A., Winkler, C., Seakiotakia, E. (1984) Increased mitochondrial DNA and RNA polymerase activity in ethylene-treated potato tubers, *Plant Physiol.* **76**, 461–465.

Archer, B. L. (1960) The protein of *Hevea brasiliensis* latex. 4. Isolation and characterization of crystalline hevein, *Biochem. J.* **75**, 236–240.

Archer, B. L. (1980) Polyisoprene, in: *Encyclopedia of Plant Physiology New Series, Vol. 8, Plant Products* (Bell, E.A., Charlwood, B.W., Eds.), pp. 316–327. Basel: Springer-Verlag.

Archer, B. L. (1983) An alkaline protease inhibitor from *Hevea brasiliensis* latex, *Phytochemistry* **22**, 633–639.

Archer, B. L., Audley, B. G. (1967) Biosynthesis of Rubber, in: *Advances in Enzymology. Vol. 29* (Nord, F.F., Ed.), pp. 221–257. New York: Interscience.

Archer, B. L., Audley, B. G. (1973) Rubber, gutta percha and chicle, in: *Phytochemistry Vol. 2* (Miller, L.P., Ed), pp. 310–343. New York: Van Nostrand Reinhold.

Archer, B. L., Audley, B. G. (1987) New aspects of rubber biosynthesis, *Bot. J. Linnean Soc.* **94**, 309–332.

Archer, B. L., Barnard, G., Cockbain, E. G. C., Dickenson, P. B., McMullen, A. I. (1963a) Structure, composition and biochemistry of *Hevea* latex, in: *Chemistry and Physics of Rubber-Like Substances* (Bateman, L., Ed.), pp. 41–72. London: McLaren and Sons.

Archer, B. L., Audley, B. G., Cockbain, E. G., McSweeney, G. P.(1963b) Biosynthesis of rubber, *Biochem. J.* **89**, 565–574.

Archer, B. L., Cockbain, E. G., McSweeney, G. P., Hong, T. C. (1969) Studies on the composition of

latex serum and bottom fraction, *J. Rubb. Res. Inst. Malaya* **21**, 560–569.
Archer, B. L., Audley, B. G., Bealing, F. L. (1982) Biosynthesis of rubber in *Hevea brasiliensis, Plast. Rubb. Intl* **7**, 109–111.
Attanyaka, D. P. S. T. G., Kekwick, R. G. O., Franklin, F. H. C. (1991) Molecular cloning and nucleotide sequencing of rubber elongation factor from *Hevea brasiliensis, Plant Mol. Biol.* **16**, 1079–1081.
Audley, B. G. (1965) Studies of organelles in *Hevea* latex containing helical protein microfibrils, in: *Proc. Natl: Rubber Prod. Res. Assoc., Jubilee Conf.* (Mullins, L., Ed.), pp. 67–72. London: McLaren & Sons.
Audley, B. G. (1966) The isolation and chemical composition of helical protein microfibrils from *Hevea brasiliensis* latex, *Biochem. J.* **98**, 335–341.
Audley, B. G., Archer, B. L. (1988) Biosynthesis of rubber, in: *Natural Rubber Science and Technology* (Roberts, A. D., Ed.), pp. 35–62. London: Oxford University Press.
Baba, T., Allen, C. M. (1980) Prenyl transferase from *Micrococcus luteus, Arch. Biochem. Biophys.* **200**, 474–484.
Bach, T. J. (1986) Hydroxylmethylglutaryl CoA reductase, a key enzyme in phytosterol synthesis, *Lipids* **21**, 82–88.
Backhaus, R. A. (1985) Rubber formation in plants: a mini-review, *Israel J. Bot.* **34**, 283–293.
Backhaus, R. A., Walsh, S. (1983) The ontogeny of rubber formation in guayule, *Parthenium argentatum* Gray, *Bot. Gaz.* **144**, 391–400.
Baulkwill, W. J. (1989) The history of natural rubber production, in: *Rubber* (Webster, C. C., Baulkwill, W. J., Eds), pp. 1–56. Essex: Longman.
Bealing, F. J. (1965) Role of rubber and other terpenoids in plant metabolism, in: *Proc. Nat. Rubber Prod. Res. Assoc., Jubilee Conf.* (Mullins, L., Ed), pp. 113–122. London: McLaren & Sons.
Bealing, F. J. (1969) Carbohydrate metabolism in *Hevea* latex, availability and utilization of substrates, *J. Rubb. Res. Inst. Malaya* **21**, 445–455.
Bealing, F. J. (1976) Quantitative aspects of latex metabolism: possible involvement of precursors other than sucrose in the biosynthesis of *Hevea* rubber, in: *Proc. Intl. Rubb. Conf. Vol. 2*, pp. 543–565. Kuala Lumpur: RRIM.
Benedict, C. R. (1983) The biosynthesis of rubber, in: *Biosynthesis of Isoprenoid Compounds* (Porter, J.W., Spurgeon, S.L., Eds), pp. 355–369. New York: John Wiley & Sons.
Bonner, J. (1967) Rubber biosynthesis, in: *Biogenesis of Natural Compounds* (Bernfield, P., Ed.), pp. 491–552. Oxford: Pergamon.
Bonner, J. (1975) Physiology and chemistry of guayule, in: *An International Conference on the Utilization of Guayule* (McGinnies, N.G., Haas, E.F., Eds.), pp. 78–83. Tucson: University of Arizona.
Bonner, J., Galston, A. (1947) Rubber formation in plants - Review, *Bot. Rev.* **13**, 543–596.
Broekaert, W., Lee, H., Kush, A., Chua, N-.H., Raikhel, N. (1990) Wound-induced accumulation of mRNA containing a hevein sequence in laticifers of rubber tree (*Hevea brasiliensis*), *Proc. Natl: Acad. Sci. USA* **87**, 7633–7637.
Brown, M. S. Goldstein, J. L. (1980) Multivalent feedback regulation of HMG CoA reductase, a control mechanism co-ordinating isoprenoid synthesis and cell growth, *J. Lipid. Res.* **21**, 505–517.
Brydson, J. A. (1978) The historical development of rubber chemistry, in: *Rubber Chemistry*, pp. 2–9. London: Applied Science.
Bucks, D. A., Nakayama, F. S., French, O. F. (1984) Water management for guayule rubber production, *Trans. ASAE* **27**, 1763–1770.
Bucks, D. A., Nakayama, F. S., French, O. F., Legard, W. W., Alexander, W. L. (1985a) Irrigated guyule-production and water use relationships, *Agric. Water Manag.* **10**, 95–102.
Bucks, D. A., Roth, R. L., Nakayama, F. S., Gardner, B. R. (1985b) Irrigation water, nitrogen, and bioregulation for guayule production, *Trans. ASAE* **28**, 1196–1205.
Chambeyron, C., Dry, J., Leynadier, F. (1992) Study of allergenic fractions of latex allergy, *Allergy* **47**, 92–97.
Chrestin, H. (1989) Biochemical aspects of bark dryness induced by over-stimulation of rubber trees with Ethrel, in: *Physiology of rubber tree latex* (d'Auzac J., Jacob, J.-L., Chrestin, H., Eds.), pp. 431–4421. Florida: CRC Press.
Chrestin, H., Gidrol, X. (1986) Contribution of lutoidic tonoplast in regulation of cytosolic pH of latex from *Hevea brasiliensis*, in: *Proc. Int. Rubber Conf. Vol. 3*, pp. 66–87. Kuala Lumpur: RRIM.
Chrestin, H., Jacob, J.-L., d'Auzac, J. (1986) Biochemical basis for the cessation of latex flow and occurrence of physiological bark dryness, in: *Proc. Int. Rubber Conf. Vol. 3*, pp. 20–42. Kuala Lumpur: RRIM.
Chrispeels, M. J., Raikhel, N. V. (1991) Lectins, Lectin genes and their role in plant disease, *Plant Cell* **3**, 1–9.
Christofferson, R. E., Laties, G. G. (1982) Ethylene regulation of gene expression in carrots, *Proc. Natl: Acad. Sci. USA* **79**, 4060–4064.

Churngchow, N., Suntaro, A., Wititsuwannakul, R. (1995) β-1,3-Glucanase isozymes from latex of *Hevea brasiliensis*, *Phytochemistry* **39**, 505–509.

Chye, M. L., Kush, A., Tan, T. C., Chua, N. H. (1991) Characterization of cDNA and genomic clones encoding HMG CoA reductase from *Hevea brasiliensis*, *Plant Mol. Biol.* **16**, 567–577.

Clausen, M. K., Christansen K., Jensen, P. K., Behnke, O. (1974) Isolation of lipid particles from baker's yeast, *FEBS Lett.* **43**, 176–179.

Cornforth, J. W., Cornforth, R. H., Donninger, C., Popjak, G. (1966) Biosynthesis of cholesterol, steric course of hydrogen eliminations and C-C bond formation in squalene biosynthesis, *Proc. R. Soc. B London.* **163**, 429–432.

Cornforth, J. W., Clifford, K., Mallaby, R., Phillips, G. T. (1972) Stereochemistry of isopentenyl pyrophosphate isomerase, *Proc. R. Soc. B London.* **182**, 277–281.

Cornish, K. (1993) The separate role of plant *cis* and *trans* prenyl transferases in *cis*-1,4-polyisoprene biosynthesis, *Eur. J. Biochem.* **218**, 267–271.

Coupe, M., Chrestin, H. (1989) Physico-chemical and biochemical mechanisms of hormonal (ethylene) stimulation, in: *Physiology of rubber tree latex* (d'Auzac J., Jacob, J.-L., Chrestin, H., Eds.), pp. 295–319. Florida: CRC Press.

Cox, A. D., Der, C. J. (1992) Protein prenylation: more than just glue? *Curr. Opin. Cell Biol.* **4**, 1008–1016.

Czuppens, J. L., Van Durme, P., Dooms-Goosens, A. (1992) Latex allergy in patient with allergy to fruit [letter], *Lancet* **339**, 493.

Czuppon, A., Chen, Z., Baur, X. (1993a) Chemical synthesis of a peptide representing a major latex allergen, *Chest* **104**, 159S.

Czuppon, A. B., Chen, Z., Rennert, S., Engelke, T., Meyer, H. E., Heber, M., Baur, X. (1993b) The rubber elongation factor of rubber trees (*Hevea brasiliensis*) is the major allergen in latex, *J. Allergy Clin. Immunol.* **92**, 690–697.

d'Auzac, J. (1989a) Factors involved in the stopping of flow after tapping, in: *Physiology of rubber tree latex* (d'Auzac J., Jacob, J.-L., Chrestin, H., Eds.), pp. 257–285. Florida: CRC Press.

d'Auzac, J. (1989b) Historical account, in: *Physiology of rubber tree latex* (d'Auzac J., Jacob, J.-L., Chrestin, H., Eds.), pp. 289–293. Florida: CRC Press.

d'Auzac, J., Jacob, J.-L. (1969) Regulation of glycolysis in latex of *Hevea brasiliensis*, *J. Rubb. Res. Inst. Malaya* **21**, 417–444.

d'Auzac, J., Jacob, J.-L. (1989) The composition of latex from *Hevea brasiliensis* as a laticiferous cytoplasm, in: *Physiology of rubber tree latex* (d'Auzac J., Jacob, J.-L., Chrestin, H., Eds.), pp. 59–96. Florida: CRC Press.

d'Auzac, J., Sanier, C., Chrestin, H. (1986) Study of a NADH-quinone-reductase producing toxic oxygen from *Hevea* latex, in: *Proc. Int. Rubb. Conf., Vol. 3*, pp. 102–126. Kuala Lumpur: RRIM.

Dalrymple, S. J., Audley, B. G. (1992) Allergenic proteins in dipped products: Factors influencing extractable protein levels. *Rubb. Dev.*. **45**, 51–60.

de Camargo, A. P., Schmidt, N. C., Cardoso, R. M. G. (1976) South American leaf blight epidemics and rubber phenology in Sao Paulo, in: *Proc. Intl. Rubb. Conf. Vol. 3*, pp. 251–265. Kuala Lumpur: RRIM.

De Corres, L., Moneo, I., Minoz, D. Bernada, G., Fernandez, E., Auaicana, M., Urrutia, T. (1993) Sensitization from chestnuts and bananas in patients with urticaria and anaphylaxis from contact with latex, *Ann. Allergy* **70**, 35–39.

de Fay, E., Hebant, Ch., Jacob, J. L. (1989) Cytology and cytochemistry of the laticiferous system, in: *Physiology of rubber tree latex* (d'Auzac J., Jacob, J.-L., Chrestin, H., Eds.), pp.15–30. Florida: CRC Press.

Dennis, M. S., Henzel, W. J., Bell, J., Kohr, W., Light, D. R. (1989) Amino acid sequence of rubber elongation factor protein associated with rubber particles in *Hevea* latex, *J. Biol. Chem.* **264**, 18618–18626.

Dickenson, P. B. (1965) The ultrastructure of the latex vessel of *Hevea brasiliensis*, in: *Proc. Natl. Rubber Prod. Res. Assoc., Jubilee Conf.* (Mullins, L., Ed.), pp. 52–66. London: McLaren & Sons.

Dickenson, P. B. (1969) Electron microscopical studies of latex vessel system of *Hevea brasiliensis*, *J. Rubb. Res. Inst. Malaya* **21**, 543–559.

Dooms-Goossens, A. (1988) Contact urticaria caused by rubber gloves [letter], *J. Am. Acad. Dermatol.* **18**, 1360–1361.

Dupont, J., Moreau, F., Lance, C., Jacob, J. L. (1976) Phospholipid composition of the membrane of lutoids of *Hevea brasiliensis* latex, *Phytochemistry* **15**, 1215–1217.

Fisher, A.A. (1986) Contact dermatitis from foods and food additives, In: *Contact Dermatitis*, pp. 582–586. Philadelphia: Lea & Febiger.

Fisher, A. A. (1987) Contact urticaria and anaphylactoid reaction due to cornstarch surgical glove power, *Contact Dermatitis* **16**, 224–225.

Fisher, A. A. (1993) Association of latex and food allergy, *Cutis* **52**, 70–71.

Foegeding, P. M., Busta F. F. (1991) Disinfection, sterilization and preservation, in: *Chemical Food Preservatives* (Block, S. S., Ed.), pp. 816–820. Philadelphia: Lea & Febiger.

Gidrol, X., Chrestin, H., Tan, H. L., Kush, A. (1994) Hevein, a lectin like protein from *Hevea brasiliensis* (rubber tree) is involved in the coagulation of latex, *J. Biol. Chem.* **269**, 9278–9283.

Goldstein J. L., Brown, M. S. (1990) Regulation of mevalonate pathway, *Nature* **343**, 425–430.

Gomez, J. B. (1976) Comparative ultracytology of young and native vessel in *Hevea brasiliensis*, in: *Proc. Int. Rubber Conf. Vol. 2*, pp. 143–164. Kuala Lumpur: RRIM.

Gomez, J. B., Hamzah, S. (1989) Frey-Wyssling complex in *Hevea latex* – Uniqueness of the organelle, *J. Natl: Rubb. Res.* **4**, 75–85.

Gomez, J. B., Moir, G. F. J. (1979) The ultracytology of latex vessels in *Hevea brasiliensis, Malaysian Rubber Research and Development Board.* Monograph No. **8.**

Gomez, J. B., Southorn, W. A. (1969) Studies in lutoid membrane ultrastructure, *J. Rubb. Res. Inst. Malaya* **21**, 513–523.

Gomez, J. B., Tata, S. J. (1977) Further studies on the occurrence and distribution of microhelices in clones of *Hevea, J. Rubb. Res. Inst. Malaya* **25**, 120–124.

Goren, R., Huberman, M. (1976) Effects of ethylene and 2,4-D on the activity of cellulose isozymes in abscission zones of the developing orange fruit, *Plant Physiol.* **37**, 123–127.

Goss, R. A., Benedict, C. R., Keithly, J. H., Nessler, C. L., Stipanovic, R. D. (1984) cis-Polyisoprene synthesis in guayule plants (*Parthenium argentatum* Gray) exposed to low, non-freezing temperatures, *Plant Physiol.* **74**, 534–537.

Goyvaerts, E., Dennis, M., Light, D., Chua, N. H. (1991) Cloning and sequencing of the cDNA encoding the rubber elongation factor of *Hevea brasiliensis, Plant Physiol.* **97**, 317–321.

Granady, L. C., Slater, J. E. (1995) The history and diagnosis of latex allergy, in: *Immunology and Allergy Clinics of North America; Latex Allergy, Vol. 15* (Fink, J. N., Ed), pp. 21–29. Philadelphia: W. B. Saunders.

Hamann, C. P. (1993) Natural rubber latex protein sensitivity in review, *Am. J. Contact Dermatitis* **4**, 4–29.

Hasma, H. (1987) Proteolipids of the natural rubber particles, *J. Natl: Rubb. Res.* **2**, 129–133.

Hasma, H., Subramaniam, A. (1978) The occurrence of furanoid fatty acid in *Hevea brasiliensis* latex, *Lipids* **13**, 905–908.

Ho, C. C., Subramaniam, A., Yong, W. M. (1976) Lipids associated with the particles in *Hevea* latex, in: *Proc. Int. Rubber Conf.* pp. 441–456. Kuala Lumpur: RRIM.

Jacob, J. L., Moreau, F., Dupont, J., Lance, C. (1976) Some characteristics of the lutoids in *Hevea brasiliensis* latex, in: *Proc. Int. Rubber Conf.* pp. 470–483. Kuala Lumpur: RRIM.

Jacob, J. L., Eschbach, J. M., Prevot, J. C., Roussel, D., Lacrotte, R., Chrestin, H., d'Auzac, J. (1986) Physiological basis for latex diagnosis of the functioning of the laticiferous system in rubber-tree, in: *Proc. Int. Rubber Conf. Vol. 3.* pp. 43–65. Kualar Lumpur: RRIM.

Jacob, J. L., Prevot, J. C., Kekwick, R. G. O. (1989) General metabolism of *Hevea brasiliensis* latex (with the exception of isoprenoid anabolism), in: *Physiology of rubber tree latex* (d'Auzac J., Jacob, J.-L., Chrestin, H., Eds.), pp.101–144, Florida: CRC Press.

John, P. (1993) Rubber, in: *Biosynthesis of the Major Crop Products*, pp. 114–126. New York: John Wiley & Sons.

Keenan, M. V., Allen, C. M. (1974) Phospholipid activation of *Lactobacillus plantarum* undecaprenyl pyrophosphate synthetase. *Biochem. Biophys. Res. Commun.* **61**, 338–342.

Kekwick, R. G. O. (1989) The formation of polyisoprenoids in *Hevea* latex, in: *Physiology of rubber tree latex* (d'Auzac J., Jacob, J. L., Chrestin, H., Eds.), Florida: CRC Press.

Kekwick, R. G. O., Archer, B. L., Barnard, D., Higgins, G. M. C., McSweeney, G. P., Moore, C. G. (1959) Incorporation of DL-(2-^{14}C)-mevalonic acid lactone into polyisoprene, *Nature* **184**, 268–270.

Kleinig, H. (1989) The role of plastids in isoprenoid biosynthesis, *Annu. Rev. Plant Physiol. Plant Mol. Biol.* **40**, 39–59.

Koiwa, H., Bressan, R. A., Hasegawa, P. M. (1997) Regulation of protease inhibitors and plant defense, *Trends Plant Sci.* **2**, 379–384.

Koyama, T., Wititsuwannakul, D., Wititsuwannakul, R., Ogura, K. (1995) Analysis of prenyltransferase products from C-serum of *Hevea brasiliensis*, in: *Biopolymers and Bioproducts* (Svasti, J., Ed.), pp. 608–612. Bangkok: IUBMB.

Koyama, T., Wititsuwannakul, D., Asawatreratanakul, K., Wititsuwannakul, R., Ohya, N., Tanaka, Y., Ogura, K. (1996) Isopentenyl diphosphate isomerase in rubber latex, *Phytochemistry* **43**, 769–772.

Kurup, V.P., Kely, K.J., Turjanmaa, K. (1992) Characterization of latex antigen and demonstration of latex-specific antibodies by enzyme-linked immunosorbent assay in patients with latex hypersensitivity, *Allergy Proc.* **6**, 329–333.

Kurup, V. P., Kelly, K. J., Fink, J. N. (1993a) Characterization of monoclonal antibody against

latex protein associated with the latex allergy, *J. Allergy Clin. Immunol.* **92**, 638–643.
Kurup, V. P., Kelly, K. J., Turjanmaa, K. (1993b) Immunoglobulin E reactivity to latex antigens in the sera of patients from Finland and the United States, *J. Allergy Clin. Immunol.* **91**, 1128–1134.
Kurup, V. P., Murali, P. S., Kelly, K. J. (1995) Latex antigens, in: *Immunology and Allergy Clinics of North America; Latex Allergy, Vol. 15* (Fink, J. N., Ed.), pp. 45–59. Philadelphia: W. B. Saunders.
Kush, A. (1994) Isoprenoid biosynthesis: the *Hevea* factory, *Plant Physiol. Biochem.* **32**, 761–767.
Kush, A., Goyvaerts, E., Chye, M. L., Chua, N. H. (1990) Laticifer specific gene expression in *Hevea brasiliensis* (rubber tree), *Proc. Natl. Acad. Sci. USA* **87**, 1787–1790.
Lau, C. M., Gomez, J. B., Subramaniam, A. (1986) An electron microscopy study of the epoxidation of natural rubber particles, in: *Proc. Int. Rubb. Conf. Vol. 2*. pp. 525–529. Kuala Lumpur: RRIM.
Lavaud, F., Cossart, C., Reiter, V., Bernard, J., Deltour, G., Holmquist, I. (1992) Latex allergy in patient with allergy to fruit, *Lancet* **339**, 492–493.
Lee, H., Broekaert, W. F., Raikhel, N. V. (1991) Co- and post-translational processing of the hevein preproprotein of latex of the rubber tree (*Hevea brasiliensis*), *J. Biol. Chem.* **266**, 15944–15948.
Light, D. R., Dennis, M. S. (1989) Purification of prenyltransferase that elongates cis-polyisoprene rubber from the latex of *Hevea brasiliensis, J. Biol. Chem.* **264**, 18589–18597.
Light, D. R., Lazarus, R. A., Dennis, M. S. (1989) Rubber elongation by farnesyl pyrophosphate synthases involves a novel switch in enzyme stereospecificity, *J. Biol. Chem.* **264**, 8598–8607.
Low, F. C. (1978) Induction and control of flowing in *Hevea, J. Rubb. Res. Inst. Malaya* **26**, 21–32.
Low, F. C., Yeang, H. Y. (1985) Effect of ethephon stimulation on latex invertase in *Hevea, J. Rubb. Res. Inst. Malaya* **33**, 37–47.
Lynen, F. (1969) Biochemical problems of rubber synthesis, *J. Rubb. Res. Inst. Malaya* **21**, 389–406.
Mahlberg, P. G. (1993) Laticifers: An historical prospective, *Bot. Rev.* **59**, 1–23.
Makinen-Kilijunen, S., Turjanmaa, K., Palosuo, T., Reunala, T. (1992) Characterization of latex antigens and allergens in surgical gloves and natural rubber by immunoelectrophoretic methods, *J. Allergy Clin. Immunol.* **90**, 230–235.
Makinen-Kilijunen, S., Alenius, H., Palosuo, T., Reunala, T. (1993) Immunoblot inhibition detects several common allergens in rubber latex and banana [Abstract], *J. Allergy Clin. Immunol.* **91**, 242–242.
Martin, M. N. (1991) The latex of *Hevea brasiliensis* contains high levels of both chitinases and chitinases/lysozymes, *Plant Physiol.* **95**, 469–476.
McMullen, A. I. (1962) Particulate ribonucleoprotein components of *Hevea brasiliensis* latex, *Biochem. J.* **85**, 491–495.
McMullen, A. I., McSweeney, G. P. (1966) Biosynthesis of rubber, *Biochem. J.* **101**, 42–47.
Mehta, I. J. (1982) Stem anatomy of *Parthenium argentatum, P. incanum* and their natural hybrids, *Am. J. Bot.* **69**, 503–512.
Mehta, I. J., Hanson, G. P. (1983) Distribution of rubber and comparative stem anatomy of high and low rubber-bearing guayule (*Parthenium argentatum*) from Mexico, in: *Proceedings of the Third International Guayule Conference*, pp. 181–197. Tucson: University of Arizona.
Mernagh, L. R. (1986) Rubber, in: *The New Encyclopedia Britannica, Vol. 21* (15th edn), pp. 282–285. London.
Miyamoto, S., Piela, K., Davis, J. (1984) Water use, growth and rubber yields of guayule selections as related to irrigation regimes. *Irrig. Sci.* **5**, 95–103.
Moir, G. F. J. (1959) Ultracentrifugation and staining of *Hevea* latex, *Nature* **184**, 1626–1628.
Moir, G. F. J., Tata, S. J. (1960) The proteins of *Hevea brasiliensis* latex. III. The soluble proteins of bottom fraction, *J. Rubb. Res. Inst. Malaya.* **16**, 155–159.
Moreau, F., Jacob, J. L., Dupont, J., Lance, C. (1975) Electron transport in the membrane of lutoids from the latex of *Hevea brasiliensis, Biochim. Biophys. Acta* **396**, 116–124.
Nutter, A. F. (1979) Contact urticaria to rubber, *Br. J. Dermatol.* **101**, 597–598.
Oh, S. K., Kang, H., Shin, D. H., Yang, J., Chow, K-S., Yeang, H.Y., Wagner, B., Breiteneder, H., Han, K-H. (1999) Isolation, characterization, and functional analysis of a novel cDNA clone encoding a small rubber particle protein from *Hevea brasiliensis, J. Biol. Chem.* **274**, 17132–17138.
Paardekooper, E. C. (1989) Exploitation of rubber tree, in: *Rubber* (Webster, C. C., Baulkwill, W. J., Eds), pp. 319–414. Essex: Longman.
Pakianathan, S. W., Milford, G. F. J. (1973) Changes in the bottom fraction contents of the latex during flow in *Hevea brasiliensis, J. Rubb. Res. Inst. Malaya* **23**, 391–400.
Pakianathan, S. W., Boatman, S. G., Taysum, D. H. (1966) Particle aggregation following dilution of *Hevea* latex: a possible mechanism for the closure of latex vessels after tapping, *J. Rubb. Res. Inst. Malaya* **19**, 259–271.

Pakianathan, S. W., Wain, R. L., Ng, E. K. (1976) Studies on displacement area on tapping in mature *Hevea* trees, in: *Proc. Int. Rubber Conf. Vol. 3*. pp. 225–246. Kuala Lumpur: RRIM.

Park, R. B., Bonner, J. (1958) The enzymatic synthesis of rubber from mevalonic acid, *J. Biol. Chem.* **233**, 340–342.

Philpott, M. W., Wesgarth, D. R. (1953) Stability and mineral composition of *Hevea* latex, *J. Rubb. Res. Inst. Malaya* **14**, 133–148.

Pujade-Renaud, V., Clement, A., Perrot, C., Prevot, J. C., Chrestin, H., Jacob, J. L., Guern, J. (1994) Ethylene induced increase in glutamine synthetase activity and mRNA levels in *Hevea brasiliensis* latex cells, *Plant Physiol.* **105**, 127–132.

Randall, S. K., Marshall, M. S., Crowell, D. N. (1993) Protein isoprenylation in suspension-cultured tobacco cells, *Plant Cell* **5**, 433–442.

Ritter, F. J. (1954) Biosynthesis of rubber and other isoprenoid compounds, *Rubber J.* **126**, 55–71.

Ross, B. D., McCullough, J., Owenby, D. R. (1992) Partial cross-reactivity between latex and banana allergens, *J. Allergy Clin. Immunol.* **90**, 409–410.

Ryan, C. A. (1984) Defense response in plants, in: *Genes Involved in Microbe Plant Interaction* (Verma, D., Hihn, T., Eds), pp. 377–386. Berlin: Springer-Verlag.

Sherief, P. M., Sethuraj, M. R. (1978) The role of lipids and proteins in the mechanism of latex vessel plugging in *Hevea brasiliensis, Physiologia Plantarum* **42**, 351–355.

Southorn, W. A. (1961) Microscopy of *Hevea* latex, in: *Proc. Nat. Rubb. Conf. 1960*. pp. 766–776. Kuala Lumpur: RRIM.

Southorn, W. A. (1969) Physiology of *Hevea* latex flow, *J. Rubb. Res. Inst. Malaya* **21**, 494–512.

Southorn, W. A., Edwin, E. E. (1968) Latex flow studies, II. Influence of lutoids on the stability and flow of *Hevea* latex, *J. Rubb. Res. Inst. Malaya* **20**, 187–200.

Southorn, W. A., Yip, E. (1968) Latex flow studies. III. Electrostatic considerations in the colloidal stability of fresh *Hevea* latex, *J. Rubb. Res. Inst. Malaya* **20**, 201–215.

Stewart, W. D., Wachtel, J. J., Shipman, J. J., Yanks, J. A. (1955) Synthesis of rubber by fungi, *Science* **122**, 1271–1272.

Subramaniam, A. (1976) Molecular weight and other properties of natural rubber: a study of clonal variation, in: *Proc. Int. Rubber Conf. 1975, Vol. 4*. pp. 3–11. Kuala Lumpur: RRIM.

Subramaniam, A. (1995) The chemistry of natural rubber latex, in: *Immunology and Allergy Clinics of North America; Latex Allergy, Vol. 15* (Fink, J. N., Ed.), pp. 1–20. Philadelphia: W. B. Saunders.

Suvachittanont, W., Wititsuwannakul, R. (1995) 3-Hydroxy-methylglutaryl-Coenzyme A synthase in *Hevea brasiliensis, Phytochemistry* **40**, 757–761.

Tanaka, Y. (1986) Structural characterization of *cis*-polyisoprene from sunflower, *Hevea* and guayule, in: *Proc. Int. Rubb. Conf. Vol. 2*, pp. 1–9. Kuala Lumpur: RRIM.

Tanaka, Y. (1989) Structure and biosynthesis mechanism of natural polyisoprene, *Prog. Polym. Sci.* **14**, 339–371.

Tanaka, Y. (1991) Recent advances in structural characterization of elastomers, *Rubber Chem. Technol.* **64**, 325–385.

Tanaka, Y., Mori, M., Ute, K., Hatada, K. (1990) Structure and biosynthesis mechanism of rubber from fungi, *Rubber Chem. Technol.* **63**, 39–45.

Tanaka, Y., Eng, A. H., Ohya, N., Nishiyama, N., Tangpakdee, J., Kawahara, S., Wititsuwannakul, R. (1996) Initiation of rubber biosynthesis in *Hevea brasiliensis*: Characterization of initiating species by structural analysis, *Phytochemistry* **41**, 1501–1505.

Tangpakdee, J., Tanaka, Y., Ogura, K., Koyama, T., Wititsuwannakul, R., Wititsuwannakul, D. (1997a) Isopentenyl diphosphate isomerase and prenyl transferase activities in bottom fraction and C-serum from *Hevea* latex, *Phytochemistry* **45**, 261–267.

Tangpakdee, J., Tanaka, Y., Ogura, K., Koyama, T., Wititsuwannakul, R., Wititsuwannakul, D. (1997b) Rubber formation by fresh bottom fraction of *Hevea* latex, *Phytochemistry* **45**, 267–274.

Tangpakdee, J., Tanaka, Y., Jacob, J. L., d'Auzac, J. (1999) Characterization of *Hevea brasiliensis* rubber from virgin trees: A possible role of *cis*-polyisoprene in unexploited tree, *Rubber Chem. Technol.* **72**, 299–307.

Tata, S. J. (1976) Hevein: its isolation, purification and some structural aspects in: *Proc. Int. Rubber Conf. 1975, Vol. 2*. pp. 449–517. Kuala Lumpur: RRIM.

Tata, S. J., Edwin, E. E. (1970) *Hevea* latex enzymes detected by zymogram technique after starch gel electrophoresis, *J. Rubb. Res. Inst. Malaya* **23**, 1–12.

Tata, S. J., Moir, G. F. J. (1964) The proteins of *Hevea brasiliensis* latex, V. Starch gel electrophoresis of C-serum proteins, *J. Rubb. Res. Inst. Malaya* **18**, 97–101.

Tata, S. J., Boyce, A. N., Archer, B. L., Audley, B. G. (1976) Lysozymes: major component of the sedimentable phase of *Hevea brasiliensis* latex, *J. Rubb. Res. Inst. Malaya* **24**, 233–240.

Tata, S. J., Beintema, J. J., Balabaskaran, S. (1983) The lysozyme of *Hevea brasiliensis* latex, isolation purification: enzyme kinetics and a partial amino acid sequence, *J. Rubb. Res. Inst. Malaya* **31**, 35–48.

Tateyama, S., Wititsuwannakul, R., Wititsuwannakul, D., Sagami, H., Ogura, K. (1999) Dolicols of rubber plant, ginkgo and pine, *Phytochemistry* **51**, 11–15.

Taylor, F. R., Parks, L. W. (1978) Metabolic conversion of free sterols and steryl esters in *Saccharomyces cerevisiae*, *J. Bacteriol.* **126**, 531–537.

Truscott, W. (1995) The industry perspective on latex, in: *Immunology and Allergy Clinics of North America; Latex Allergy, Vol. 15* (Fink, J. N., Ed.), pp. 89–121. Philadelphia: W. B. Saunders.

Tupy, J. (1969) Stimulatory effects of 2,4-dichlorophenoxyacetic acid and 1-naphthyacetic acid on sucrose level invertase activity and sucrose utilization in the latex of *Hevea brasiliensis*, *Planta* **88**, 144–148.

Tupy, J. (1973) The regulation of invertase activity in latex of *Hevea brasiliensis*, *Exp. Bot.* **24**, 515–524.

Tupy, J. (1989) Sucrose supply and utilization for latex production, in: *Physiology of rubber tree latex* (d'Auzac J., Jacob, J.-L., Chrestin, H., Eds) pp. 179–199, Florida: CRC Press.

Tupy, J., Resing, W. L. (1969) Substrate and metabolism of carbon dioxide formation in *Hevea* latex *in vitro*, *J. Rubb. Res. Inst. Malaya* **21**, 456–460.

Turjanmaa, K., Laurila, K., Makinen-Kiljunen, S., Reunala, T. (1988) Rubber contact urticaria; Allergenic properties of 19 brands of latex gloves, *Contact Dermatitis* **19**, 362–367.

Turjanmaa, K., Makinen-Kiljunen, S., Reunala, T, Alenius, H., Palosuo, T. (1995) Natural rubber latex allergy: The European experience, in: *Immunology and Allergy Clinics of North America; Latex Allergy, Vol. 15* (Fink, J. N., Ed.), pp. 71–88. Philadelphia: W. B. Saunders.

Vehaar, G. (1959) Natural latex as a colloidal system, *Rubber Chem. Technol.* **32**, 1622–1627.

Vreugdenhil, D., Oerlemans, A. P. C., Steeghs, M. H. G. (1984) Hormonal regulation of tuber induction in radishes (*Raphanus sativus*): The role of ethylene, *Plant Physiol.* **62**, 175–179.

Walujono, K., Schiolma, R. A., Beintema, J. J., Mariono, A., Hahv, A. M. (1975) Amino acid sequence of Hevein, in: *Proc. Int. Rubber Conf. 1975, Vol. 2*. pp. 518–531. Kuala Lumpur: RRIM.

Wititsuwannakul, D., Wititsuwannakul, R., Pasitkul, P. (1997) Lectin binding protein in C-serum of *Hevea brasiliensis*, *Plant Physiol. (Suppl.)* **114**, 72.

Wititsuwannakul, R. (1986) Diurnal variation of 3-hydroxy-3-methylglutaryl coenzyme A reductase in latex of *Hevea brasiliensis* and its relationship to rubber content, *Experientia* **42**, 44–45.

Wititsuwannakul, R., Wititsuwannakul, D., Sothibanhhu, R., Suvachithanont, W., Sukonrat, W. (1988) Correlation studies on 3-hydroxy-3-methylglutaryl Coenzyme A reductase activity and dry rubber yield in *Hevea brasiliensis*, in: *C.R. Coll. Expl. Physiol. Amel. Hevea, IRCA-CIRAD, France.* 2–7 Nov. pp. 161–172. Paris: IRCA-CIRAD.

Wititsuwannakul, R., Wititsuwannakul, D., Suwanmanee, P. (1990a) 3-Hydroxy-3-methyglutaryl Coenzyme A reductase from the latex of *Hevea brasiliensis*, *Phytochemistry* **29**, 1401–1403.

Wititsuwannakul, R., Wititsuwannakul, D., Dumkong, S. (1990b) *Hevea* calmodulin: Regulation of the activity of latex 3-hydroxy-3-methylglutaryl coenzyme A reductase, *Phytochemistry* **29**, 1755–1758.

Wititsuwannakul, R., Wititsuwannakul, D., Pasitkul, P. (1997) Rubber latex coagulation by lutoidic lectin in *Hevea brasiliensis*, *Plant Physiol. (Suppl.)* **114**, 71.

Wititsuwannakul, R., Wititsuwannakul, D., Sakulborirug, C. (1998) A lectin from the bark of the rubber tree (*Hevea brasiliensis*), *Phytochemistry* **47**, 183–187.

Wrangsjo, K., Mellstrom, G., Axelsson, G. (1986) Discomfort from rubber gloves indicating contact urticaria, *Contact Dermatitis* **15**, 70–84.

Yeang, H. Y., David, M. N. (1984) Quantitation of latex vessels plugging by the intensity of plugging, *J. Rubb. Res. Inst. Malaya* **32**, 164–169.

Yeang, H. Y., Paranjothy, K. (1982) Initial physiological changes in *Hevea* latex flow characteristics associated with intensive tapping, *J. Rubb. Res. Inst. Malaya* **30**, 31–36.

Yeang, H. Y., Low, F. C., Gomez, J. B., Paranjothy, K., Sivakumaran, S. (1984) A preliminary investigation into the relationship between latex invertase and latex vessel plugging in *Hevea brasiliensis*, *J. Rubb. Res. Inst. Malaya* **32**, 50–54.

Yeang, H. Y., Jacob, J. L., Prevot, J. C., Vidal, A. (1986) Invertase activity in *Hevea* latex serum: Interaction between pH and serum concentration, *J. Natl. Rubb. Res.* **1**, 16–22.

Yeang, H. Y., Yusof, F., Abdullah, L. (1995) Precipitation of *Hevea brasiliensis* latex proteins with trichloroacetic acid and phosphotungstic acid in preparation for the Lowry protein assay, *Anal. Biochem.* **226**, 35–43.

Yeang, H. Y., Cheong, K. F., Sunderasan, E., Hamzah, S., Chew, N. P., Hamid, S., Hamilton, R.

G., Cardosa, M. J. (1996) The 14.6 kD rubber elongation factor (Hev b1) and 24 kD (Hev b3) rubber particle proteins are recognized by IgE from patients with spina bifida and latex allergy, *J. Allergy Clin. Immunol.* **98**, 628–639.

Yeang, H. Y., Ward, M. A., Zamri, A. S. M., Dennis, M. S., Light, D. R. (1998) Amino-acid sequence similarity of Hev. b3 to 2 previously reported 27 kDa and 23 kDa latex proteins allergenic to spina bifida patients, *Allergy* **53**, 513–519.

Yip, E., Gomez, J. B. (1984) Characterization of cell sap of *Hevea* and its influence on cessation of latex flow, *J. Rubb. Res. Inst. Malaya* **32**, 1–19.

Yip, E., Southern, W. A. (1968) Latex flow studies, VI. Effects of high pressure gradients on flow of fresh latex in narrow bore capillaries, *J. Rubb. Res. Inst. Malaya* **20**, 248–256.

Yokoyama, H., Hayman., W. J., Hsu, W. J., Poling, S. M., Bauman, A. J. (1977) Chemical bioinduction of rubber in guayule, *Science* **197**, 1076–1077.

Yong, W. M., Singh, M. M. (1976) Thin-layer chromatographic resolution of free amino acids in clonal latices of natural rubber, in: *Proc. Int. Rubber Conf. Vol. 2*. pp. 484–498. Kuala Lumpur: RRIM.

Zhao, X-Q. (1987) The significance of the structure of laticifer with relation to the exudation of latex in *Hevea brasiliensis*, *J. Natl. Rubb. Res.* **2**, 94–98.

Zhu, J.-K., Bressan, R. A., Hasegawa, P. M. (1993) Isoprenylation of the plant molecular chaperone ANJI facilitates membrane association and function at high temperature, *Proc. Natl. Acad Sci. USA* **90**, 8557–8561.

5 Biotechnological Processes for Recycling of Rubber Products

Dr. Katarina Bredberg[1], Dr. Magdalena Christiansson[2], Dr. Bengt Stenberg[3], Dr. Olle Holst[4]

[1] Department of Biotechnology, Center for Chemistry and Chemical Engineering, Lund University, P.O. Box 124, 221 00 Lund, Sweden; Tel: +46-462224626, Fax: +46-462224713, E-mail: katarina.bredberg@biotek.lu.se

[2] Department of Biotechnology, Center for Chemistry and Chemical Engineering, Lund University, P.O. Box 124, 221 00 Lund, Sweden; Tel: +46-462224626, Fax: +46-462224713

[3] Polymer Technology, Royal Institute of Technology, 100 44 Stockholm, Sweden; Tel: +46-87908269, Fax: +46-8208856, E-mail: stenberg@polymer.kth.se

[4] Department of Biotechnology, Center for Chemistry and Chemical Engineering, Lund University, P.O. Box 124, 221 00 Lund, Sweden; Tel: +46-462229844, Fax: +46-462224713, E-mail: olle.holst@biotek.lu.se

Biotechnology of Biopolymers. From Synthesis to Patents. Edited by A. Steinbüchel and Y. Doi

ISBN: 3-527-31110-6

CGTR	cryo-ground tire rubber
CBS	cyclohexylbenzothiazole sulfenamide
SSSE	solid state shear extrusion

1 Introduction

Rubber material as we know it is a material with properties that differ widely from those of the viscous sap of the rubber tree, *Hevea brasiliensis*. Elasticity and extensibility are the features most characteristic for rubber (Blow and Hepburn, 1982; Brydson, 1988; Franta, 1989; Gent, 1992). To gain these desirable rubbery properties – which make it possible to produce tires, hoses, and a wide range of other rubber products – it is necessary to vulcanize the raw rubber. By adding a few percent of sulfur and heating the mixture under pressure, sulfur bridges will be formed between the hydrocarbon chains (Figure 1). This irreversible reaction will produce a material with unique mechanical and thermoelastic properties, which is stable over a wide temperature range. Unfortunately, the material will also be almost impossible to reshape.

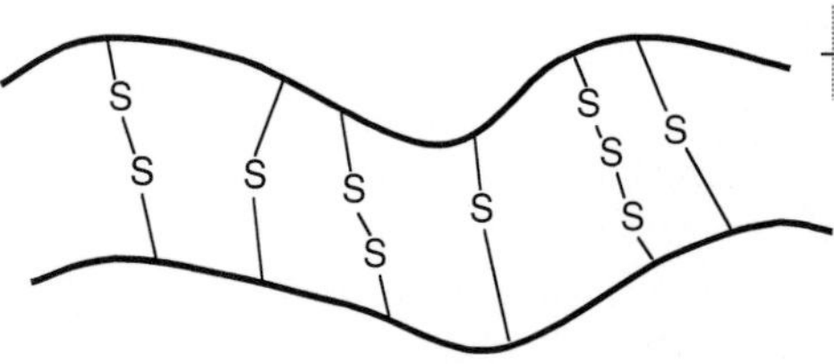

Fig. 1 Sulfur cross-links between rubber molecules. The exact structures of the sulfur bridges are not known, though several different suggestions have been presented (Roberts, 1988). The molecular weights between the cross-links (Mc) vary between 500 and 20,000, depending on the cross-link density.

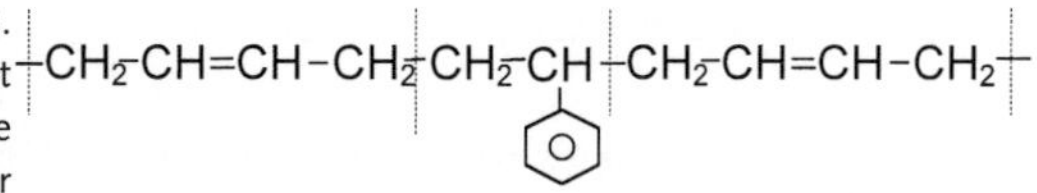

The synthetic rubbers differ from natural rubber, and from each other in chemical structure as well as physical and chemical properties. The molecular weight of natural rubber is several millions, while the molecular weight of synthetic rubbers is much lower. Rubber materials can also gain different properties by adding for example accelerators, activators, retarders, and fillers. In this way, a wide range of properties can be obtained. The chemical structures of natural and synthetic rubbers are shown in Figures 2–4.

In 1999, the natural rubber consumption worldwide was 6.7 million tons, while synthetic rubber consumption was 10 million

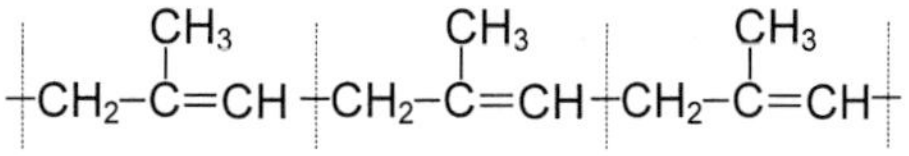

$$-CH_2-C(CH_3)=CH-CH_2-C(CH_3)=CH-CH_2-C(CH_3)=CH-$$

Fig. 2 Chemical structure of natural rubber, poly-*cis*-1,4-isoprene.

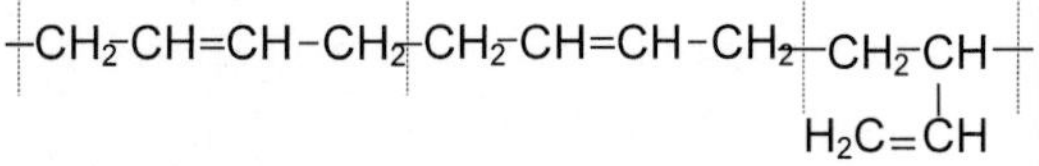

$$-CH_2-CH=CH-CH_2-CH_2-CH=CH-CH_2-CH_2-CH(CH=CH_2)-$$

Fig. 3 Chemical structure of butadiene rubber consisting of 1,4-butadiene monomers and 1,2-butadiene monomers.

$$-CH_2-CH=CH-CH_2-CH_2-CH(C_6H_5)-CH_2-CH=CH-CH_2-$$

Fig. 4 Chemical structure of styrene-butadiene rubber consisting of styrene and butadiene monomers.

tons (IRSG, 1999). The ratio between natural and synthetic rubber consumption has increased by a few percent during the past decade. The major part of the produced rubber is used in the manufacture of tires, but different areas of rubber use are shown in Figure 5.

As a consequence of the difficulties in reusing rubber material and of the widespread use of rubber products (mainly in the form of tires), huge numbers of used tires are stockpiled all over the world, with 2 – 3 billion tires currently estimated to be stockpiled in the US alone (Jang et al., 1998). Those tons of material could, instead of being environmental and health threats, provide a great asset of raw material, if efficient methods for recycling rubber could be developed. The problem with the reuse of rubber is caused by the strong sulfur bridges that occur between the hydrocarbon chains, these making it impossible to melt and reshape the material as can be done with thermoplastic and thermoelastic materials. In general, a priority order for decreasing the amount of waste should be: (1) reduction of consumption; (2) reuse of the product; (3) recycling of materials; (4) energy recovery; and (5) as a last possibility, deposition of the waste. In practice, the most environmentally friendly way to reuse scrap tires is of course dependent upon the costs, energy demands and pollution that such reuse creates. In part, the reduction of consumption and reuse of the product is, in the case of scrap tires, obtained by retreading.

A wide range of different methods for reusing and recycling tires and rubber material in general has been developed over the years (Figure 6). The most common method

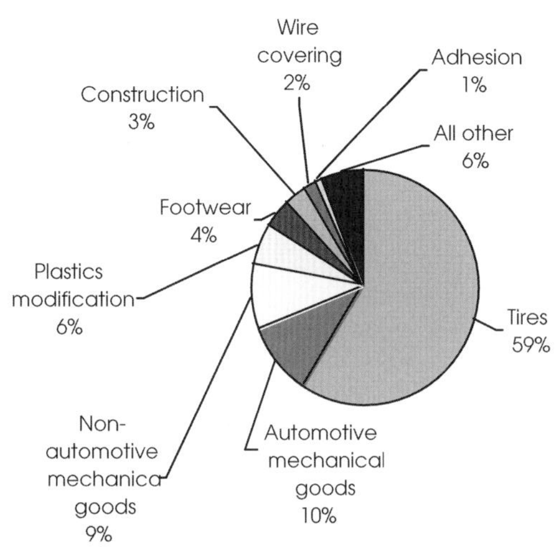

Fig. 5 Global synthetic rubber use in 1990. Total synthetic rubber consumption in 1990 was 7.1 million tons (Schnecko, 1998).

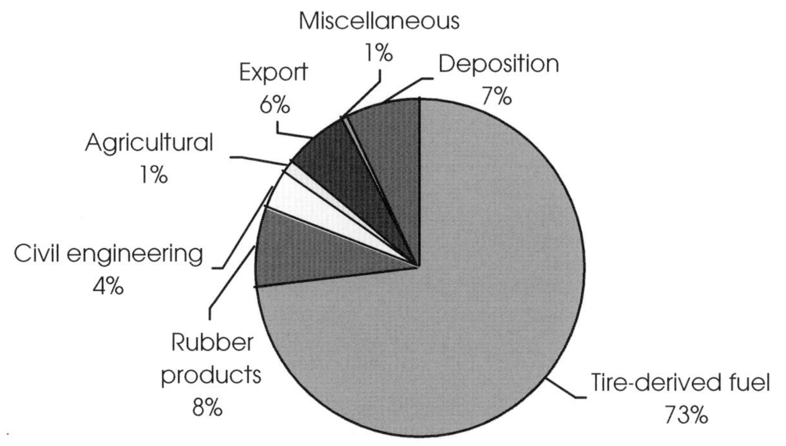

Fig. 6 Use of scrap tires worldwide in 1996 (Scheirs, 1998).

is to burn the tires in cement kilns and power plants as tire-derived fuel, or to use them as land-fill or as artificial reefs, for example. Ground rubber powder is used as filling material in a wide range of applications, together with different matrices such as new rubber material, plastics and cement. Shoe soles, plant pots and swings for children's playgrounds are other possible application areas, though at present these are limited in number.

1.1 Grinding of Rubber Material

Some type of size reduction is a prerequisite for most applications of waste rubber material, such as revulcanization with fresh rubber material, blends with other plastic materials, applications in civil engineering, soil improvements, running tracks and even energy recovery. Several such methods have been described (Astafan, 1995; Klingensmith, 1991; Kohler and O'Neill, 1996; Kohler and Jackson, 1998). In general, several steps are needed in the grinding process. The tires are cut or chopped into large pieces, and these are then ground to particles of millimeter size during the removal of textile, dirt and metals. For special applications, crumbs or particles <1 mm are produced. The requirements of energy increase drastically with decreasing particle size. The energy consumption constitutes the major part of the cost in the rubber grinding process. When using crumb rubber in the manufacture of new rubber products, the price of the ground rubber should be around half the price of raw rubber material or less, in order to make the process cost-efficient.

Cutting and grinding of the rubber material requires the use of special equipment due to its elasticity. Most processes are based on a combination of cutting and tearing. However, a general problem of grinding technology is the heat build-up in the rubber material, which means that cooling has to be applied in all grinding techniques. Cutting with hydraulically agitated knives is used to produce rubber chips of a large size. Shredding equipment is mainly used for rough grinding, the device consisting of low-speed rotating shafts with overlapping knives (Astafan, 1995). Granulators or hammer mills are equipped with rotor knives or rotating hammers, the counterparts being stationary knives and a grate, respectively. These can generate further size reduction. The waste rubber is exposed to large shear forces during grinding, and these – together with the heat build-up – leads to the problem of surface degradation. Ground particles tend to re-agglomerate, which further limits throughput. Water slurries of rubber particles and a flour-grinding mill can be used to produce a wet ground rubber, an advantage of which is its uniformity and cleanliness (Klingensmith and Baranwal, 1998).

Another way to avoid degradation is to cool the rubber under the glass transition temperature and then apply impact forces; this leads to cracking of the brittle material. Liquid nitrogen has to be used as cooling medium as the temperature required is –70 °C to –130 °C. Cryo-grinding (as this procedure is called) generates a powder with particle sizes down to 0.2 mm which can be produced without degradation and oxidation of the surface. Easier fiber and metal liberation is also reported (Astafan, 1995; Klingensmith, 1991; Kohler and O'Neill, 1996; Kohler and Jackson, 1998). The surface structure of ground rubber differs between cryo-grinding and ambient milling. Cryo-ground rubber has a smooth surface with relatively little surface area as compared with the 'cauliflower' structure obtained with ambient milling processes (Kohler and Jackson, 1998). Considerations on how to handle spent material and a cryo-grinding process have been described by Liaskos (1994). Cryo-grinding is a relatively expensive method, due to the cost of the liquid nitrogen. Another approach of interest is solid state shear extrusion (SSSE). Waste rubber chips are fed to a cooled co-rotating twin-screw extruder, which permits pulverized material of controlled particle size (down to 0.05 mm) to be obtained (Khait, 1997).

1.2 Recycling of Rubber Material

One way to recycle rubber material is to shred it, grind it, mix the crumbs with virgin rubber, and then to vulcanize it to new rubber products. Rubber materials for certain applications, e.g. the tread on off-the-road tires, contain a few percent of recycled ground rubber which, in addition to giving a cost-reducing effect, also increases stiffness and tear strength. Furthermore, the ground rubber improves mixing and processing characteristics, facilitates the removal of enclosed air, and is an efficient way to control die swell in extrusion operations (Franta, 1989; Manuel and Dierkes, 1997). However, the material's mechanical properties, such as tensile strength and elongation at break, decline drastically when the concentration of ground rubber is higher than a few percent (Asplund, 1996). The reason for this decline is probably the low cross-linking in the interface between the virgin material and the ground spent material, that is due to a low molecular interaction between the polymer chains in the interface. Only a few new sulfur bonds are then formed during the vulcanization process (Myhre and MacKillop, 1996). Another reason could be that sulfur migrates from the new rubber matrix in to the ground rubber particles during vulcanization. This makes the new rubber matrix less cross-linked (Gibala et al., 1999). The fracture areas of test pieces of rubber containing cryo-ground tire rubber (CGTR) have been studied with scanning electron microscopy (SEM). Those pictures clearly show that the rubber crumbs are discrete particles in a surrounding matrix (Figure 7).

Different methods for surface activation of the rubber crumbs have been developed to enhance the interactions between the crumbs and the new rubber matrix, and in this way improve the mechanical properties. Those include mechanical, chemical and a few biotechnological methods.

Other ways to recycle rubber include, for example, biotechnological processes such as microbial degradation of rubber into products that are in one way or another useful. There are generally some specific advantages

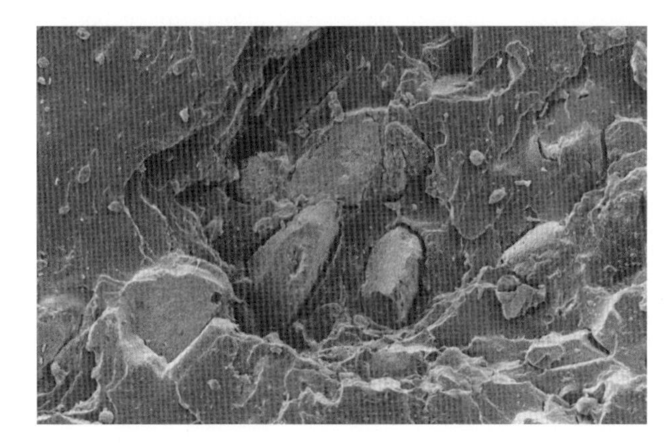

Fig. 7 Scanning electron micrograph of fracture area of rubber material containing 15% cryo-ground tire rubber (original magnification, ×150).

with biotechnological processes compared with chemical and physical processes. Biotechnology does not include any harmful or toxic chemicals, and is normally not energy intensive. Furthermore, the specificity of microorganisms and enzymes results in less unwanted degradation of the material. However, there are also some drawbacks, most notably that microorganisms are sensitive towards many chemical substances, including rubber additives. When manufacturing a tire for example, many different substances are added to the polymer to obtain a product that is stable and functional over a wide temperature range and in different applications. Except for sulfur, a vulcanization system also consists of accelerators, activators (e.g. zinc oxide and stearic acid), and retarders. Antidegradants, processing aids, mineral oil, pigments and fungicides are also added. Most of those additives are toxic towards many microorganisms. Some additives, for example fungicides, are added because of just that property, i.e. to prevent microbial attack of the product during use.

Zyska (1981) suggests that, in the manufacture of vulcanized rubbers, compounding ingredients which have a satisfactory level of microbiological resistance must be selected. Finding ingredients that are resistant to microorganisms implies fewer difficulties than finding an adequate fungicide for rubber. Such compounding ingredients will cause problems when biotechnology is to be used for rubber recycling.

1.3 Biotechnological Processes

Biotechnological processes can be defined very generally as "controlled and deliberate application of simple biological agents – living or dead cells or cell components – in technologically useful operations" (Bu'Lock and Kristiansen, 1987). They have also been described as "the integration of natural science and engineering science in order to achieve the application of organisms, cells, parts thereof and molecular analogues for products and services" (European Federation of Biotechnology, 1995). These kinds of processes include such diverse operations as the production of animal feed, citric acid and antibiotics, waste treatment, water purification and agricultural plant improvement. When designing a biotechnological process, one must consider a variety of different methods and reactor designs, for example whether to use an aerobic or anaerobic procedure; to use whole cells or pure enzymes; to use a continuous reactor or a batch-reactor. These different procedures have their specific advantages and drawbacks depending on the process to be accomplished.

2 Historical Outline

Rubber has been used by man for thousands of years (Blow and Hepburn, 1982). Columbus and his crew were the first Europeans to encounter rubber material when they discovered South America in the 1490s. At that time, the material was already well known and used by the Indians, not only as rubber balls for different games but also to make bottles, shoes and waterproof hats. Despite this early encounter, it took many years before rubber became widespread outside South America, and it was not until the middle of the nineteenth century that scientists first began to experiment with rubber. This led to the development of a variety of different rubber products, including the rubber eraser, which was invented by Joseph Priestly in 1770 and considered at that time to be an expensive luxury. In fact, the name 'rubber' was derived from its use as an eraser.

The problem with rubber at this time was that the material was sticky and impervious. This made the use of rubber products relatively limited until Charles Goodyear happened to leave a piece of rubber sprinkled with sulfur on a warm oven in 1839. The result of this was, to his surprise, a stable non-sticky material, which in the next few years became one of the most used materials in the world. Fifty years later, Dunlop invented the pneumatic tire, which is at present the most important rubber product. During World War I, Germany began to produce synthetic rubber, and they developed the technique further during the 1930s. Ten years later, during World War II, the USA was forced to start producing synthetic rubber due to the Japanese occupation of the rubber plantations. Today, about 60% of the manufactured rubber is synthetic.

When it comes to the recycling of polymers, rubber lags significantly behind the much more recently developed thermoplastic materials. This is in spite of the fact that rubber has been widely used, and recycled, for over 150 years (Rader et al., 1995). The use of recycled rubber was more prevalent 30 years ago than it is now. During the 1960s, recycled rubber constituted about 20% of the rubber industry's raw material. However, due to the more widespread use of synthetic rubber, production liability concerns, and the introduction of the steel-belted radial tires, the use of recycled rubber has fallen to ~2%. In recent years, significant technical innovations have been made in the tire-recycling field, and it was only between 1994 and 1996 that the use of tire-derived fuel increased by 70%. The use of ground tires in the production of new rubber products has increased from 4.5 million tires to 10 million tires, while the number of tires used for civil engineering purposes has been constant over those years (Scheirs, 1998). During the years of expansive use of rubber products, there has always been the problem of microbial deterioration, and extensive studies have been carried out aimed at its prevention. One idea is to use microbial deterioration to dispose of rubber material, and perhaps even to gain a useful end product as a result.

3 Rubber Products to which Biotechnological Recycling is Applied

The number of biotechnological processes for recycling of rubber material is still very limited. Recycling through microbial desulfurization is today probably the most investigated option in this area, but is yet not in commercial use (see Chapter 10, 'Biotechnological Desulfurization of Rubber Products'). Biotechnological methods in general present promising prospects for finding solutions to the future demands of increased rubber recycling.

3.1 Microbial Degradation of Rubber

Microbial deterioration of rubber products has attracted much interest, and many studies have been carried out on the degradation of both pure rubber elastomers and vulcanized rubber products. The main purpose of those studies has been to reduce rubber deterioration during use of rubber products, but the degradation of rubber products might represent a possible opening in the search for efficient biotechnological methods to recycle rubber products.

In 1955, Rook wrote a survey of the principal publications on microbial attack on rubber. He concluded that the data concerning degradation of vulcanized rubber in most cases are very vague and should be interpreted with great care (Rook, 1955). The same conclusions were drawn by Heap and

Morrel (1968) some years later. Several different rubber-degrading microorganisms have been isolated. Heisey and Papadatos (1995) isolated 10 strains belonging to the genera *Streptomyces*, *Amycolatopsis* and *Nocardia*, which reduced the weight of rubber latex gloves by more than 10% in 6 weeks. Jendrossek and co-workers (1997) used natural rubber latex as sole source of carbon and energy to find 50 isolates of rubber- degrading bacteria. A total of 1220 strains from different culture collections was also screened, and revealed 46 positive strains; both their own isolates and strains from the culture collections were Actinomycetes. Linos and Steinbüchel (1998) isolated a bacterium belonging to the genus *Gordona* that used rubber as sole carbon source. Tsuchii and coworkers made extensive studies on rubber degradation and studied the degradation products from simple natural rubber materials after treatment with Nocardia 835 A. They also found that an increased amount of carbon black, sulfur and addition of the accelerator cyclohexylbenzothiazole sulfenamide (CBS) increased the resistance towards microbial degradation, and that the shape of the particles is of importance for degradation. With Nocardia strain 835 A, a 51% weight loss of truck tire was obtained when pieces of truck tires were incubated with pieces of rubber gloves as cosubstrate, but the degradation rate was slow (Tsuchii et al., 1985, 1990, 1996, 1997). The common observations reported in the different studies are that the isolates capable of rubber degradation generally belong to Actinomycetes, and that natural rubber is more easily degraded than synthetic rubbers.

Linos and Steinbüchel (1996) studied the growth of adapted isolates on natural and synthetic rubber elastomers. The growth behavior of one isolate indicates production of biosurfactants. Tsuchii and Takeda (1990) reported on extracellular rubber-degrading enzymes from the Gram-negative bacterium *Xanthomonas* sp. strain 35 Y. The enzymes were shown to degrade natural rubber latex to create the same degradation products as *Nocardia* strain 835 A. When more complex rubber materials were tested, little success was obtained. Linos and coworkers (2000) isolated the first Gram-negative bacterium exhibiting strong rubber-decomposing properties. Interestingly, the most efficient degradation has been obtained with natural rubber and almost only with low-wear products or rubbers with low complexity. There are probably two explanations for this: (1) that microorganisms find it difficult to degrade complex high-molecular weight compounds; and (2) the addition of toxic compounds to the rubber materials. Cundell and Mulcock (1973) investigated the effect of curing agent concentrations on the microbial deterioration of vulcanized rubber, and concluded that the cross-link densities did not affect the deterioration rate, while the nature of the cross-links did affect the rate of microbial deterioration.

Loomis and coworkers (1991) used controlled microbial chemostats to evaluate the polymer degradation with emphasis on plastics, while Steinbüchel and Linos (1997) used natural and synthetic polyisoprenes as model compounds for studies on microbial action on rubber materials. Keursten and Groenevelt (1996) concluded after a series of experiments with unknown microorganisms in soil-rubber mixtures, that the degradation of rubber particles in soil could be described by first-order kinetics. Williams (1986) has suggested that microorganisms preferentially attack the stearic acid content in vulcanized rubber, which is an interesting idea and must be considered in the efforts to prevent rubber deterioration.

The time scales for microbial deterioration in the reports mentioned above are weeks and longer. Tsuchii and coworkers (1985) reported a 100% rubber weight loss of unvulcanized natural rubber in 8 weeks, but only 7–17% weight loss of tire tread in the same time. Linos and Steinbüchel (1998) noted a weight loss of 20% after 20 weeks of incubation. This indicates that efficient methods for degradation of rubber products such as tires on a large scale are still far from being applicable on a regular basis.

3.2 Surface Modification

A wide range of different methods for surface modification of ground rubber particles has been suggested over the years. Most of these are chemical or physical methods, but some biotechnological approaches have been proposed. The purpose of surface modification is to increase the interactions between ground rubber particles and a new matrix when rubber crumbs are mixed with a new material. This is done to improve the properties of materials that consist partly of ground rubber material.

3.2.1 Cleavage of the Hydrocarbon Chains

A great number of microorganisms have been studied in the context of degradation of rubber (see above). Most studies were aimed at preventing rubber deterioration, though it might be possible that cleavage of the hydrocarbon chains on the surface of rubber crumbs can enhance the interaction with a surrounding matrix material. Limited cleavage of the hydrocarbon chains on the surface of vulcanized rubber could make the chains more flexible, and a larger number of unsaturated bonds would be available to form new bonds with the surrounding material. This could for example apply to CGTR in a new matrix of unvulcanized rubber material or other polymers. Romine suggested that surface-modified rubber can bind with polar species in asphalt to establish a rubber-asphalt phase (Romine and Snowden-Swan, 1993)

3.2.2 Desulfurization of the Rubber Material

The possibility of enhancing the interactions between old ground rubber crumbs and a new rubber matrix by desulfurization of the vulcanized rubber has been studied by several groups (Torma and Raghavan, 1990; Löffler et al., 1993; Christiansson et al., 1998; Romine and Romine, 1998; Kim and Park, 1999; Bredberg et al., 2001; Bredberg et al., in preparation). The principal effects would be about the same as for cleavage of the hydrocarbon chains, but possibly more efficient. Cleft sulfur bonds make the hydrocarbon chains more flexible and a produce a larger number of active sites available for new bonding, but do not affect the quality of the polymers. This topic is extensively discussed in Chapter 10.

3.3 Microbial Detoxification

The largest single rubber product – tires – normally contain a wide range of different rubber additives such as accelerators, retarders, and antioxidants, and this applies also to many other rubber products. Many of the rubber additives used in different applications have been shown to be toxic to microorganisms (Zyska et al., 1971; Christiansson et al., 2000; Bredberg et al., 2001). It has been shown that most toxic rubber additives can be leached from the rubber with organic solvents, but from an environmental viewpoint this is not the best solution. This situation has also become a problem in other applications of used tires. For example, there are reasons to believe that artificial reefs made from used tires pose a threat to the aquatic environment, while ground tires incorporated into roadways and athletic tracks may leak toxic substances (Evans, 1997; Galbraith and Burns, 1997). This is of course a major obstacle in the biotechnological processes that include rubber material, but it might be a problem that can be overcome with microbial detoxification. The possibility of using microorganisms capable of degrading for example aromatic substances in order to detoxify used rubber has immense appeal. *Rhodococcus*, for example, is reported to be able to metabolize aromatic compounds and a wide range of other compounds (Warhurst and Fewson, 1994). Extensive studies have been carried out with different species of fungi in the area of soil remediation, but there are no reports about this application in the context of rubber recycling.

Some possible ways to reuse rubber material are shown in Figure 8.

4
Current State

At present, a very limited number of biotechnological processes exist for the reuse of waste rubber material. A number of patents exist for the biotechnological desulfurization of rubber material, but to our knowledge none of these processes is currently in commercial

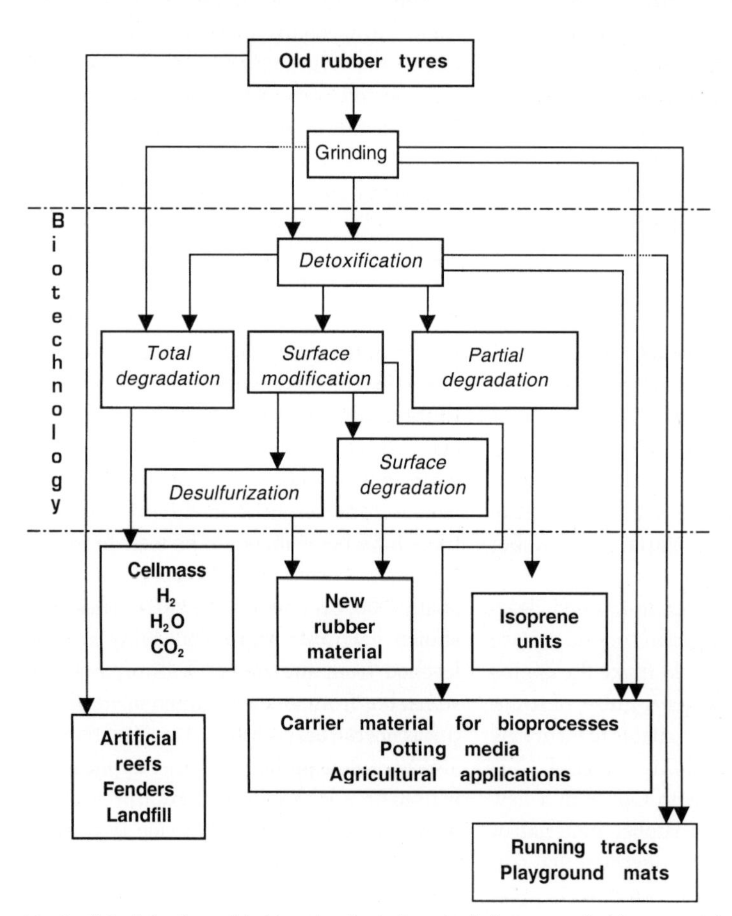

Fig. 8 Schedule of possible biotechnological methods for reuse of rubber material.

use. Although discarded rubber tires are widely used today for landfill, incorporation into asphalt, playground and athletic mats and artificial reefs, knowledge of the environmental impact of these procedures is very limited. There are reports that tire leachates are acutely toxic to rainbow trout (Mayer and Ellersieck, 1968), and that exposure of lake water to rubber significantly lowered the concentrations of bacteria, picophytoplankton and some species of zooplankton (Galbraith and Burns, 1997). From Evans' report of the toxicity from tire leachates, it is evident that these leachates exert toxicity that affects aquatic animals both lethally and nonlethally (Evans, 1997). It is not known to what degree tire leachates from urban areas, landfills, playgrounds, athletic surfaces and tire reefs alter and impair aquatic organisms. Further investigations on the impact of rubber material on ground and water environment are obviously needed if the use of rubber material as land fill and in other similar applications is to continue.

5 Rubber as a Carbon Source for Fermentative Processes

Although waste rubber material is a putative carbon source for microbial growth, very few reports exist on this subject. Crane has suggested the possibility of using discarded tires as raw material for microbial production of single cell protein for food supplement use (Crane et al., 1978). The production of biomass or different metabolic products such as methanol and methane gas might represent another interesting, albeit as yet unexplored, concept.

6 Putative Products

The use of different types of microorganisms to cleave hydrocarbon chains to smaller hydrocarbon units, might be a possible application of the biotechnological recycling of rubber. Tsuchii and Doi have each reported on microbial rubber degradation processes producing isoprene oligomeres (Tsuchii et al., 1985; Doi et al., 1986). Species of the genus Rhodococcus display a diverse range of metabolic capabilities, in that they are able to degrade short-chain, long-chain, and halogenated hydrocarbons. Tolerance to starvation, the frequent lack of catabolite repression, and their environmental persistence makes these bacteria excellent candidates for bioremediation treatments. *Rhodococcus rhodochrous* IGTS8 is able to use dibenzothiophene as a sole sulfur source, but not as a carbon source; thus it may be possible to use this strain for desulfurization processes (Warhurst and Fewson, 1994). Harris and Ghandimathi (1998) used natural rubber coagulum to immobilize yeast cells in the production of ethanol; hence it may be possible to use discarded rubber particles to immobilize microorganisms in biotechnological processes. The use of reclaimed rubber as a matrix for slow-release fertilizer has also been suggested (Martyniak, 1970; Han and Ma, 1996). The use of shredded tires as a bulking agent for the composting of sewage sludge and to support the biofilm in anaerobic wastewater treatment has also been suggested (Higgins et al., 1987; Reyes et al., 1990). Nickerson and Faber (1975) indicated that fermented rubber particles might be valuable as a soil conditioner as they possess ion-exchange capacity, while ground rubber material has also been proposed as a soil-less potting medium (Handreck, 1996). Microbial detoxification may offer the possibility of overcoming the problems of toxicity that have been reported. Rubber latex effluents from rubber production factories could be used as a carbon source for microbial processes, or as a source for the isolation of the antifungal protein hevein, which is enriched in these effluents. On the fringe of the area of biotechnological processing, Lehman and coworkers (1998) have suggested the reprocessing of waste tires to produce activated-carbon adsorbents for air-control applications.

7 Outlook and Developments

The increasing use of rubber material worldwide implies great demands upon recycling possibilities. The production of rubber material that is easy to reclaim and recycle is an important step in the progress towards sustainable rubber recycling. The exclusion of certain rubber additives from tire recipes to facilitate recycling of the rubber material is a possibility, though this has not yet been explored to any great extent. Bioplastics and polyesters produced by microorganisms might be an option in the future, as these types of polymer are (by definition) biodegradable, with some having been shown to possess rubber-like properties (Poirier et al., 1995, Steinbüchel, 1996). De Koning et al. (1994) cross-linked poly(hydroxyalkanoate) from *Pseudomonas oleovorans* and produced a biodegradable rubber. Although the research on biotechnological recycling of rubber material and biodegradable rubbers and plastics has been continuing for many years, very few practical applications have emerged. Except for the economic and environmental aspects, this is also an issue of legislation.

Acknowledgements

Financial support has been provided by the Swedish Foundation for Strategic Environmental Research (MISTRA) and CF Environmental Foundation.

8 References

Asplund, J. (1996) *RAPRA in the Environmental age – Progress in recycling*, Shropshire, UK.

Astafan, C. G. (1995) Machines for processing whole tires to specifically sized pieces to be used as feed material in crumb rubber processing systems, Rubber Division 148th Fall Technical Meeting, Cleveland, Ohio.

Blow, C. M., Hepburn, C. (1982) *Rubber Technology and Manufacture*, 2nd edn. London: Butterworth.

Bredberg, K., Persson, J., Christiansson, M., Stenberg, B., Holst, O. (2001) Anaerobic desulphurisation of ground rubber with the thermophilic archaea *Pyrococcus furious* – A new method for rubber recycling, *Appl. Microbiol. Biotechnol.* **55**, 43–48.

Bredberg, K., Christiansson, M., Bellander, M., Stenberg, B., Holst, O. (in preparation). Properties of rubber materials containing microbial devulcanized cryo-ground tire rubber.

Brydson, J. A. (1988) *Rubbery Materials and their Compounds*, London: Elsevier Applied Science.

Bu'Lock, J., Kristiansen, B. (1987) *Basic Biotechnology*, London: Academic Press.

Christiansson, M., Stenberg, B., Wallenberg, L. R., Holst, O. (1998) Reduction of surface sulphur upon microbial devulcanization of rubber materials, *Biotechnol. Lett.* **20**, 637–642.

Crane, G., Elefritz, R. A., Kay, E. L., Laman, J. R. (1978) Scrap tire disposal procedures, *Rubber Chem. Technol.* **51**, 577–599.

Christiansson, M., Stenberg, B., Holst, O. (2000) Toxic additives – a problem for microbial waste rubber desulphurisation, *Resour. Environ. Biotechnol.* **3**, 11–21.

Crane, G., Elefritz, R. A., Kay, E. L., Laman, J. R. (1978) Scrap tire disposal procedures, *Rubber Chem. Technol.* **51**, 577–599.

Cundell, A. M., Mulcock, A. P. (1973) The effect of curing agent concentration on the microbiological deterioration of vulcanized natural rubber, *Int. Biodeterior. Bull.* **9**, 91–94.

de Koning, G. J. M., van Bilsen, H. M. M., Lemstra, P. J., Hazenberg, W., Witholt, B., van der Galien, J. G., Schirmer, A., Jendrossek, D. (1994) A biodegradable rubber by crosslinking poly(hydroxyalkanoate) from *Pseudomonas oleovorans*, *Polymer* **35**, 2090–2096.

Doi, A., Suzuki, T., Takeda, K., Dazai, M. (1986) Isoprene oligomeres by microbial degradation of rubber, (abstract) Kokai Tokkyo Koho, Japan.

European Federation of Biotechnology (1995) *EFB Newsletter*, October. Porter South: Macmillan Magazines Ltd.

Evans, J. J. (1997) Rubber tire leachates in the aquatic environment. *Rev. Environ. Contamin. Toxicol.* **151**, 67–115.

Franta, I. (1989) Elastomers and rubber compounding materials, in: *Studies in Polymer Science* (Franta, I., Ed.), Amsterdam: Elsevier.

Galbraith, L., Burns, C. W. (1997) The toxic effects of rubber contaminants on microbial food webs and zooplankton of two lakes of different trophic status. *Hydrobiologia* **353**, 29–38.

Gent, A. G. (1992) *Engineering with Rubber. How to Design Rubber Components*, Munich: Hanser Publishers.

Gibala, D., Thomas, D., Hamed, G. R. (1999) Cure and mechanical behavior of rubber compounds containing ground vulcanizates: Part III tensile and tear strength, *Rubber Chem. Technol.* **72**, 357–360.

Han, H., Ma, X. (1996) Study on slow release fertilizer in reclaimed-rubber matrix, (abstract) *Beijing Res. Design Inst. Rubber Ind.* **43**, 143–147.

Handreck, K. A. (1996) Zinc toxicity from tire rubber in soilless potting media (abstract), *Commun. Soil Sci. Plant Anal.* **27**, 2615–2623.

Harris, E. M., Ghandimathi, H. (1998) New biotechnological use of natural rubber. Immobilization of yeast cells for alcohol production, (abstract) *Kautsch. Gummi Kunstst.* **51**, 804–807.

Heap, W. M., Morrell, S. H. (1968) Microbiological deterioration of rubbers and plastics, *J. Appl. Chem.* **18**, 189–193.

Heisey, R. M., Papadatos, S. (1995) Isolation of microorganisms able to metabolize purified natural rubber. *Appl. Environ. Microbiol.* **61**, 3092–3097.

Higgins, A. J., Suhr, J. L., Rahman, M. S., Singley, M. E., Rajput, V. (1987) Shredded rubber tires as a bulking agent for composting sewage sludge (abstract), *Gov. Rep.* Announce. Index (U.S.) **87** (15).

IRSG (1999) Rubber output, demand to grow, IRSG predicts, (abstract) *Rubber Plast. News* **29**, 6.

Jang, J.-W., Yoo, T.-S., Oh, J.-H., Iwasaki, I. (1998) Discarded tire recycling practices in the United States, Japan and Korea. *Resour. Conserv. Recycl.* **22**, 1–14.

Jendrossek, D., Tomasi, G., Kroppenstedt, R. M. (1997) Bacterial degradation of natural rubber: a privilege of actinomycetes? *FEMS Microbiol. Lett.* **150**, 179–188.

Keursten, G. T. G., Groenevelt, P. H. (1996) Biodegradation of rubber particles in soil, *Biodegradation* **7**, 329–333.

Khait, K. (1997) New solid-state shear extrusion pulverization process for used tire rubber recovery. *Rubber World* **216**, 38–39.

Kim, J. K., Park, J. W. (1999) The biological and chemical desulfurization of crumb rubber for the rubber compounding. *J. Appl. Polym. Sci.* **72**, 1543–1549.

Klingensmith, B. (1991) Recycling, production and use of reprocessed rubbers. *Rubber World* **203**, 16–21.

Klingensmith, W., Baranwal, K. (1998) Recycling of rubber: an overview, *Rubber World* **218**, 41–46.

Kohler, R., Jackson, J. D. (1998) Enhancements in cryogenic fine grinding, *Rubber Plast. News (June)*, 12–13.

Kohler, R., O'Neill, J. (1996) Rubber Division 150th Fall Technical Meeting, Louisville.

Lehman, C. M. B., Rostam-Abadi, M., Rood, M. J., Sun, J. (1998) Reprocessing of waste tire rubber to solve air quality problems, (abstract) *Energy Fuels* **12**, 1095–1099.

Liaskos, J. (1994) Rubber tyre recycling, *UNEP Ind. Environ.* (July-September), 26–29.

Linos, A., Steinbüchel, A. (1996) Investigations on the microbial breakdown of natural and synthetic rubbers, in: *Biodeterioration and Biodegradation* (Kreysa, G., Ed.), pp. 279–286. Veinheim: VCH-Verlagsgeselleschaft.

Linos, A., Steinbüchel, A. (1998) Microbial degradation of natural and synthetic rubbers by novel bacteria belonging to the genus *Gordona*, *Kautsch. Gummi Kunstst.* **51**, 496–499.

Linos, A., Reichelt, R., Keller, U., Steinbüchel, A. (2000), A Gram-negative bacterium, identified as *Pseudomonas aeruginosa* AL98, is a potent degrader of natural rubber and synthetic cis-1,4-polyisoprene, *FEMS Microbiol. Lett.* **182**, 155–161.

Löffler, M., Straube, G., Straube, E. (1993) Desulfurization of rubber by thiobacilli, in: *Biohydrometallurgical Technologies* (Torma, A. E., Apel, M. L., Brierley, C. L., Eds.), vol. **II**, pp. 673–680. Jackson Hole, Wyoming, USA.

Loomis, G. L., Romesser, J. A., Jewell, W. J. (1991) Evaluation of polymer degradation in controlled microbial chemostats, *Emerging Technologies in Plastics Recycling, Philadelphia*, **8**, 163–169.

Manuel, H. J., Dierkes, W. (1997) Recycling of rubber, Rapra report 99, ISBN: 1–85957-129-8.

Martyniak, D. (1970) Fertilizer compositions, Patent nr. FR 19681230, France.

Mayer, F. L., Ellersieck, M. R. (1968) Manual of acute toxicity and data base for 410 chemicals and 66 species of fresh water animals, (abstract) U. S. Department of the Interior, Fish and Wildlife Service, Resource, Publication 106.

Myhre, M. J., MacKillop, D. A. (1996) Modification of crumb rubber to enhance physical properties of recycled rubber products, *Rubber World* **214**, 42–46.

Nickerson, W. J., Faber, M. D. (1975) Microbial degradation and transformation of natural and synthetic insoluble polymeric substances, (abstract) *Dev. Ind. Microbiol.* **16**, 111–118.

Poirier, Y., Nawrath, C., Sommerville, C. (1995) Production of polyhydroxyalkanoates, a family of biodegradable plastics and elastomers, in bacteria and plants, *Bio/Technology* **13**, 142–150.

Rader, C. P., Baldwin, S. D., Cornell, D. D., Sadler, G. D., Stockel, R. F., (1995) Plastics, Rubber, and Paper Recycling. A Pragmatic Approach. ACS symposium series 609. Washington: American Chemical Society.

Reyes, O., Sánchez, E., Rovirosa, N., Borja, R., Cruz, M., Colmenarejo, M. F., Escobedo, R., Ruiz, M., Rodríguez, X., Correa, O. (1990) Low-strength wastewater treatment by a multistage anaerobic filter packed with waste tyre rubber, *Bioresour. Technol.* **70**, 55–60.

Roberts, A. D. (1988) *Natural Rubber Science and Technology*, Oxford: Oxford University Press.

Romine, R. A., Romine, M. F. (1998) Rubbercycle: a bioprocess for surface modification of waste tyre rubber, *Polym. Degrad. Stab.* **59**, 353–358.

Romine, R. A., Snowden-Swan, L. (1993) Microbial processing of waste tire rubber: A project review, Pacific Northwest Laboratory, 1–3 Dec.

Rook, J. J. (1955) Microbiological deterioration of vulcanized rubber, *Appl. Microbiol.* **3**, 302–309.

Scheirs, J. (1998) Polymer recycling, in: *Polymer Science*. Chichester: John Wiley & Sons.

Schnecko, H. (1998) Rubber recycling, *Macromolecular Symposium* **135**, 327–343.

Steinbüchel, A. (1996) Synthese und Produktion biologisch abbaubarer Thermoplaste und Elastomere: gegenwärtiger Stand und Ausblick, *Kautsch. Gummi Kunstst.* **49**,120–124.

Steinbüchel, A., Linos, A. (1997) Microbial degradation of natural and synthetic rubber, International Rubber Conference, Nürnberg, 305–307.

Torma, A. E., Raghavan, D. (1990) Biodesulfurization of rubber materials, *Bioproc. Eng. Symp.* **16**, 81–87.

Tsuchii, A., Takeda, K. (1990) Rubber-degrading enzyme from a bacterial culture, *Appl. Environ. Microbiol.* **56**, 269–274.

Tsuchii, A., Suzuki, T., Takeda, K. (1985) Microbial degradation of natural rubber vulcanizates, *Appl. Environ. Microbiol.* **50**, 965–970.

Tsuchii, A., Hayashi, K., Hironiwa, T., Matsunaka, H., Takeda, K. (1990) The effect of compounding ingredients on microbial degradation of vulcanized natural rubber, *J. Appl. Polym. Sci.* **41**, 1181–1187.

Tsuchii, A., Takeda, K., Suzuki, T., Tokiwa, Y. (1996) Colonization and degradation of rubber pieces by *Nocardia* sp., *Biodegradation* **7**, 41–48.

Tsuchii, A., Takeda, K., Tokiwa, Y. (1997) Degradation of the rubber in truck tires by a strain of *Nocardia*, *Biodegradation* **7**, 405–413.

Warhurst, A. M., Fewson, C. A. (1994) Biotransformations catalysed by the genus *Rhodococcus*, *Crit. Rev. Biotechnol.* **14**, 29–73.

Williams, G. R. (1986) The biodeterioration of vulcanized rubbers, *Int. Biodeterior.* **22**, 307–311.

Zyska, B. J. (1981) Rubber, in: *Microbial Biodeterioration* (Rose, A. H., Ed.), Vol. 6, Economic Microbiology, pp. 323–385. New York: Academic Press Inc.

Zyska, B. J., Rytych, B. J., Zankowicz, L. P., Fudalej, D. S. (1971) Microbial deterioration of rubber cables in deep mines and the evaluation of some fungicides in rubber, 2nd International Biodeterioration Symposium, Lunteren, Netherlands, 256–267.

6
Biotechnological Processes for Desulfurization of Rubber Products

Dr. Katarina Bredberg[1], Dr. Magdalena Christiansson[2], Dr. Bengt Stenberg[3], Dr. Olle Holst[4]

[1] Department of Biotechnology, Center for Chemistry and Chemical Engineering, Lund University, P.O. Box 124, 221 00 Lund, Sweden; Tel: +46-462224626; Fax: +46-462224713; E-mail: katarina.bredberg@biotek.lu.se

[2] Department of Biotechnology, Center for Chemistry and Chemical Engineering, Lund University, P.O. Box 124, 221 00 Lund, Sweden; Tel: +46-462224626; Fax: +46-462224713

[3] Polymer Technology, Royal Institute of Technology, 100 44 Stockholm, Sweden; Tel: +46-8-790-82-69; Fax: +46-8208856; E-mail: stenberg@polymer.kth.se

[4] Department of Biotechnology, Center for Chemistry and Chemical Engineering, Lund University, P.O. Box 124, 221 00 Lund, Sweden; Tel: +46-462229844; Fax: +46-462224713; E-mail: olle.holst@biotek.lu.se

Biotechnology of Biopolymers. From Synthesis to Patents. Edited by A. Steinbüchel and Y. Doi

ISBN: 3-527-31110-6

BR	butadiene rubber
CGTR	cryo-ground tire rubber
EDS	energy-dispersive X-ray spectrometry
ESCA	electron spectroscopy for chemical analysis
FTIR	Fourier transformed infra-red
NR	natural rubber
SBR	styrene/butadiene rubber
XANES	X-ray for near-edge surfaces

1 Introduction

Traditional rubber materials, such as tire rubber, consist of sulfur cross-linked polymers reinforced with carbon black. During the process of vulcanization, covalent C–S bonds are formed between the hydrocarbon chains under the influence of heat and pressure. These irreversible reactions create a material with exclusive and desirable properties, but with few possibilities for easy recycling. Because of the large production and usage of rubber materials – primarily in the form of tires – the difficult task of solving the recycling problem is of great priority. The annual worldwide production of rubber material amounts to several millions of tons, and millions of tires are stock-piled around the world due to lack of proper recycling methods. The exploration of biotechnological methods for desulfurization of rubber might offer a possible means of promoting rubber recycling in the future.

Tab. 1 The general order of priority for decreasing the amount of rubber waste

1.	Reduction of consumption
2.	Reuse of the product
3.	Recycling of the material
4.	Energy recovery
5.	Deposition of the waste

The general order of priority for reducing the amount of rubber waste is outlined in Table 1. This chapter will focus on the third alternative – recycling of the material.

1.1 Rubber Material

Rubber materials have been known and used by man for many hundreds of years. Although the material was first brought to Europe by Columbus in the fifteenth century, the major breakthrough for this exotic material did not occur until 1839 when Goodyear, by accident, discovered the secret of vulcanization. Today, rubber is used almost everywhere, and the application areas are expanding day by day due to the developemnt of an ever-increasing variety of properties.

When the temperature is raised on an amorphous rigid polymeric material such as rubber, the physical properties of the material change drastically, and the material becomes elastic and soft. At the point of the glass transition temperature (T_g), the polymer chains become partly mobile – which gives the material its rubbery properties. When the temperature is raised further, increased freedom and movement of the polymer chains finally results in the formation of a viscous liquid. The temperature range of the rubbery plateau is dependent on the molecular structure (Table 2).

Tab. 2 Glass transition temperatures (T_g) for different materials (Barnes et al., 1989; Zarzycki 1991)

Material	*T_g (°C)*
Natural rubber (NR)	–73
Butadiene rubber (BR)	–73
Styrene-butadiene rubber (SBR)	–70 to +20*
Polypropylene (PP)	–15
Polyethylene terephthalate (PET)	70
Nylon-66	40
Polystyrene	100
Window glass	540
Glucose	20
Glycerol	–80

* T_g decreases with increasing butadiene concentration.

The requirements for rubber properties are large, flexible polymer chains, a certain regularity in the structure, and cross-linking of the polymer chains in a loose network (Blow and Hepburn, 1982; Brydson, 1988; Mark et al., 1993). The chain entanglement of the large molecules (i.e. those with a molecular weight >1 million) is claimed to contribute as equally to the modulus as does the cross-linking (Ferry, 1970).

The pure rubber elastomer must be vulcanized, i.e. cross-linked, to become strong and elastic over a wide temperature range. Unsaturations in the polymer provide sites for vulcanization. By adding a few percent of elemental sulfur to a rubber elastomer with suitable filler and heating the material under pressure, the original tacky and viscous material is transformed into a non-tacky and elastic compound. Apart from sulfur, the vulcanization process also involves the use of accelerators, activators (e.g. zinc oxide and stearic acid), and retarders (Blow and Hepburn, 1982). In order to fulfil the different demands on rubber materials for various applications, a number of additives are used. Among these are antidegradants, which protect the material from influence of sunlight, oxygen and ozone, processing aids, mineral oil, pigments, and fungicides. Depending on the formulation of the rubber recipe and vulcanization parameters, the cross-links formed can be either polysulfidic, disulfidic or monosulfidic (Figure 1) (Roberts, 1988). Furthermore, pendant groups carrying accelerator fragments and carbon–carbon bonds can be formed during vulcanization.

The distance between bonds or entanglements in a typical rubber material is 15 to 1500 monomers or, expressed as cross-link density, 10^{-3}–10^{-5} mole cm^{-3} (Cornan, 1994). A higher sulfur/accelerator ratio in the vulcanizing system often increases the number of sulfur units in the cross-links. Increased addition of sulfur normally results in a higher cross-linking density, while an extremely low sulfur content is used, together with sulfur donors such as tetramethylthiuram-disulfide, in order to achieve carbon–carbon bonds, and which results in good heat and compression set properties. The nature of the bonds has an impact on the properties of the material, which can be understood from the bond dissociation energy of different cross-links. The sulfur–carbon bond dissociation energies are 272 kJ mol^{-1}, whereas the dissociation energy of carbon–carbon bonds is much higher, 346 kJ mol^{-1} and the dissociation energy of sulfur–sulfur bonds slightly lower, 270 kJ mol^{-1} (Brydson, 1981).

Tires are normally made of vulcanized natural rubber (NR), styrene butadiene rub-

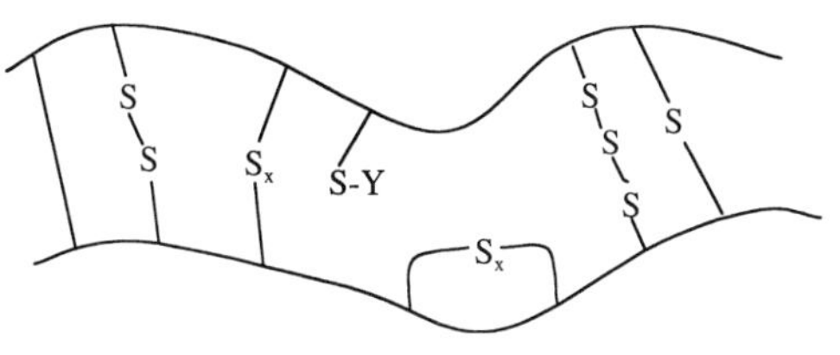

Fig. 1 Different types of possible cross-links in a sulfur-cured polymer (Y is the accelerator fragment).

ber (SBR) and/or butadiene rubber (BR) in different proportions. A car tire for winter use consists primarily of NR, whereas in tires for summer use SBR dominates. BR is used in the side-walls of tires, since it has excellent fatigue properties. An example of a tire recipe is presented in Table 3.

1.2 Grinding of Rubber Material

One way of recycling rubber material is to shred it, grind it, mix the crumbs with virgin rubber, and vulcanize it to new rubber products. In a standard procedure of grinding rubber products (i.e. tires), the rubber item is cut to pieces, and metals and textiles are removed. The pieces are then ground at ambient temperatures with or without the addition of water, or cryogenically after refrigerating with liquid nitrogen. For further descriptions of rubber grinding, see Chapter 9, 'Biotechnological Processes for Recycling of Rubber'. Rubber materials for certain applications, for example the tread on off-road tires, contain a few percent of recycled ground rubber which, in addition to giving a cost-reducing effect, also increases stiffness and tear strength. Furthermore, the ground rubber improves mixing and processing characteristics, facilitates the removal of enclosed air, and is an efficient way to control die swell in extrusion operations (Manuel and Dierkes, 1997). However, the material's mechanical properties, such as tensile strength and elongation at break, decline drastically when the concentration of ground rubber is higher than a few percent (Asplund, 1996). The reason for this decline is probably the low cross-linking in the interface between the virgin material and the ground spent material, that is due to a low molecular interaction between the polymer chains in the interface. Only a few new sulfur bonds are then formed during vulcanization (Myhre and MacKillop, 1996). The fracture areas of test pieces of rubber containing cryo-ground tire rubber (CGTR) have been studied with scanning electron microscopy (SEM). Those pictures clearly show that the rubber crumbs are discrete particles in a surrounding matrix (Figure 2).

If the polymer chains on the surface of ground tire rubber can be made more mobile, the mixing between polymer chains is likely to increase, and more sites for cross-links will be exposed. The main goal of the rubber desulfurization research is to improve the interactions between the vulcanized rubber crumb and the new rubber matrix.

Sulfur-utilizing microorganisms can be used to activate or modify the surface of ground rubber material by oxidizing or reducing sulfur cross-links. With microbial desulfurization, it is possible to create a transformation of the outermost layer of the ground rubber particles from an elastic to a viscous, sticky state. The material obtained this way could form a processable and curable material that can be mixed with virgin

Tab. 3 An example of a rubber 'recipe'

Compound	*% (w/w)*	*Compound*	*% (w/w)*
Natural rubber	65	Retarder	0.1
Carbon black N330	26	Antioxidizer	0.6
Sulfur	1.5	Antiozonants	0.6
Accelerator	0.3	High-aromatic oils	2.6
Zinc oxide	1.9	Anti-sun checking agent	0.3
Stearic acid	1.6		

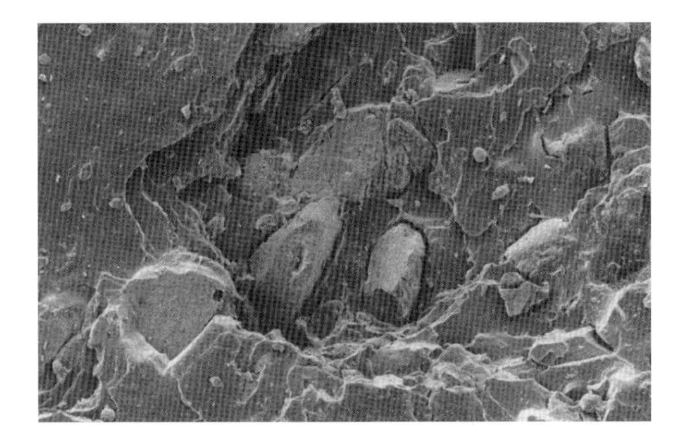

Fig. 2 Scanning electron micrograph of the fracture area of rubber material containing 15% (w/w) cryo-ground tire rubber.

material and used in new, high-quality rubber products.

Several methods for chemical and physical modification of crumb rubber have been developed. Several reviews have been prepared on the topic, and new methods are constantly suggested (Beckman et al., 1974; Crane et al., 1978; Manuel and Dierkes, 1997; Warner, 1994). The mechanism of chemical methods is to add a compound that either modifies the surface (Bauman, 1995; Dierkes, 1996) or reverses the cross-linking completely (Datta and Ivany, 1995; Myhre and MacKillop, 1996; Kohler and O'Neill, 1997). Some of the methods are available commercially, and so is the modified rubber. The drawback with chemical modification techniques is that they often require the use of hazardous chemicals. Physical methods are not impaired in this way, but are energy-intensive. An example of a physical method is the ultrasonication technique developed by Isayev and coworkers (Levin et al., 1996; Isayev et al., 1997; Johnston et al., 1997). Microbial devulcanization has the potential of being an inexpensive and environmentally friendly alternative (Holst et al., 1998).

2 Historical Outline

In 1945, when operational problems were found in fire hoses in England, a closer examination revealed that there was sulfuric acid found in the residual water inside the hoses, and that this had been produced by microorganisms. Within the hose, sulfur-oxidizing microorganisms had oxidized sulfur in the rubber to sulfate, which appeared as sulfuric acid (Thaysen et al., 1945). Almost half a century later, this situation was confirmed by Torma and Raghavan (1990), who studied the utilization by *Thiobacillus ferrooxidans* and *Thiobacillus thiooxidans* of sulfur from newly made pulverized rubber materials, their aim being to facilitate recycling of spent rubber. A number of subsequent studies have been conducted to investigate this same possibility (Raghavan et al., 1990; Löffler et al., 1993; Romine and Romine, 1998; Bredberg et al., in preparation).

The experience and knowledge obtained from the microbial desulfurization of minerals, oil and coal is utilized in the development of processes for microbial desulfurization of rubber. There are many advantages of this kind of technology such as low energy demands, and also that the industrial process is similar to the global biochemical sulfur-cycles that occur in the biosphere (Brombacher et al., 1997).

While microbial desulfurization is useful in the removal of sulfur and iron, i.e. pyrites, from low-grade clay (Ryu et al., 1995), uncontrolled leaching due to microbial activity in mine-dumps represents an environmental problem (Johnson et al., 1993; Tichy et al., 1998). Environmental problems caused by sulfur pollution resulting from the combustion of coal can be prevented by desulfurization. The sulfur contained in oil, coal and ore is mostly inorganic, i.e. pyrites, but organically bound sulfur also exists. The major problem with the organic desulfurization processes that have been designed so far is to make them economically viable (Shennan, 1996).

During the 1990s, techniques to treat sulfur-rich gas and liquid streams were developed (Buisman et al., 1990). Sulfide could be removed by partial oxidation to elemental sulfur using a *Thiobacillus* spp. under controlled oxygen concentrations (Buisman et al., 1990); the product, termed 'bio-sulfur', is hydrophilic in character and easily removed by settling (Janssen et al., 1996). Partially oxidized sulfur species in liquid streams and gases can be removed by reduction to H_2S by sulfate-reducing bacteria, followed by oxidation to elemental sulfur. Industrial waste streams containing sulfate are treated in this way (Hulshoff Pol et al., 1998; Rintala and Lepistö, 1998), the technique being applied on an industrial scale at the zinc refinery Budelco in the Netherlands (Scheeren et al., 1993). This type of process can provide a feasible approach to handling waste water from rubber desulfurization schemes.

During recent years, many studies have been conducted on sulfur metabolism in microorganisms, and the possibile use of such technology in industrial applications. However, very few studies have been conducted on the biotechnological desulfurization of rubber.

3
Rubber Products to which Desulfurization is Applied

Interest in rubber desulfurization lies mainly in the area of recycling, and as tires are the largest single rubber product, research has focused mainly on recycling of this commodity. Due to its actual size, the microorganism involved cannot diffuse into the rubber material, and hence desulfurization is a surface phenomenon. Consequently, the best effect is obtained when using very small rubber particles, as confirmed by the studies of Löffler et al. (1995), the extent of desulfurization decreasing as the particle size increased. Calculations based on a simple spherical model showed that the finest fraction (i.e. 0.1–0.2 mm particles) was desulfurized to a depth of 1 – 2 µm.

Desulfurization studies have been carried out both with sulfur-oxidizing and sulfur-reducing microorganisms (Table 4). These different microorganisms are discussed more fully in Sections 3.1 and 3.2, respectively.

Table 5 presents different possibilities for rubber desulfurization processes in terms of temperature and presence of oxygen (see Sectios 3.1 and 3.2). As can be seen, mesophilic sulfur reduction is still unexplored, but might represent a possible process condition in the future.

3.1
Sulfur-Oxidizing Microorganisms

Several different sulfur-oxidizing microorganisms have been tested for desulfurization of rubber material. Those strains oxidize the sulfur to sulfate according to the following reaction:

$$S_x + O_2 \rightarrow SO_4^{2-}$$

Tab. 4 Sulfur-utilizing strains investigated in the context of rubber desulfurization. The references provide general microbiological information about the strains

Strain	*Aerobe/ Anaerobe*	*Optimum growth temperature*	*Optimum pH*	*References*
Thiobacillus ferrooxidans	Aerobe	30–35 °C	2–4	(Temple and Colmer 1951)
Thiobacillus thiooxidans	Aerobe	28–30	2–3	(Waksman and Joffe 1922)
Thiobacillus thioparus	Aerobe	25–30 °C	6–8	(Kelly et al. 1989)
Sulfolobus solfataricus	Aerobe	87 °C	3,5–5	(Zillig et al. 1980)
Sulfolobus acidocaldarius	Aerobe	70–75 °C	2–3	(Brock et al. 1972)
Rhodococcus rhodochrous	Aerobe	30 °C	7	(Lechevalier et al. 1986)
Pyrococcus furiosus	Anaerobe	100 °C	7	(Fiala and Stetter 1986)
Acidianus brierleyi	Aerobe/ anaerobe	70 °C	1,5–2	(Zillig et al. 1980)

Tab. 5 Different possibilities for rubber desulfurization processes. The asterisks represent the investigated options

	Oxidation	*Reduction*
Mesophilic	*	–
Thermophilic	*	*

Torma and Raghavan (1990) studied eight synthetic rubber materials of unknown composition, and sulfur contents between 1.2% and 15.5%. The highest concentration of sulfate was obtained when a mixed culture of *T. ferrooxidans* and *T. thiooxidans* was added to the rubber material containing 15.5% sulfur. *T. thioparus* was reported to utilize more of the sulfur from powder accumulated from recapping of truck tires than did either *T. ferrooxidans* or *T. thiooxidans* in pure or mixed cultures, i.e. 4.7% for *T. thioparus* and 3.96% for the mixed culture (Löffler et al., 1993). It is favorable that *T. thioparus* grows at neutral pH, as this is also the pH of suspended rubber particles. Ground tire rubber of 200 mesh (74 μm) was used to screen six cultures of microorganisms for desulfurization activity (Romine and Romine, 1998). After 7 days of treatment with *Sulfolobus acidocaldarius*, 13.4% (w/w) of the sulfur was removed compared with 10% (w/w) for the mixed culture of *T. ferrooxidans* and *T. thiooxidans*. There was no release of sulfate detected for either *Rhodococcus rhodochrous* or the mixed culture ATCC 39327 (Romine and Romine, 1998).

Studies on desulfurization with *Thiobacillus ferrooxidans, T. thioparus,* and *Acidianus brierleyi* have also been performed (Christiansson et al., 1998), with ground spent car tires being the target material for rubber desulfurization. When *T. ferrooxidans* was growing on cryo-ground spent car tire rubber for 20 days the increase in sulfate concentration in the medium corresponded to approximately 8% of the total sulfur content in the spent rubber. In the corresponding controls the sulfur content in the rubber decreased by ~3%. Neither *A. brierleyi* nor *T. thioparus* were growing in the presence of this particular cryo-ground spent rubber material. When *Sulfolobus solfataricus* was tested for desulfurization, 13% (w/w) of the sulfur in the spent rubber was released during 7 days (Romine et al., 1998)

Further analyses on the post-desulfurization product must be performed at a molecular level. Hopefully, most of the sulfur will react upon vulcanization and form cross-links, but traces of dissolved unreacted elemental sulfur will remain in the rubber after vulcanization. Sulfur adsorbed to the surface of carbon black will not contribute to the vulcanization. Sometimes, vicinal cross-links are formed that act as single links but demand the double amount of sulfur. None of the published studies concerning microbial desulfurization has distinguished between elemental sulfur and the sulfur that contributes to cross-links (Raghavan et al., 1990; Löffler et al., 1993, 1995; Romine and Romine, 1998; Torma and Raghavan, 1990). It is likely that microorganisms first consume the unreacted elemental sulfur, after which the polysulfidic and disulfidic cross-links could provide energy for the microorganisms by oxidation. Except for *R. rhodochrous* and the unidentified mixed bacteria culture ATCC 39327, there are no reports of strains oxidizing organically bound sulfur. However, rubber desulfurization could not be detected when those were tested on vulcanized rubber (Romine and Romine, 1998). The formation of sulfone and sulfoxide groups on microbial treatment (Christiansson et al., 1998; Romine and Romine, 1998) provides partial evidence for the fact that oxidation of poly- and disulfides actually has occurred, and that the microbial growth is not only due to remains of unreacted elemental sulfur. The origin of the released sulfate must be analyzed, and quantification of different sulfur species in the rubber has to be carried out to find the optimal strain and cultivation conditions for efficient rubber desulfurization.

The reports on rubber desulfurization deal only with axenic cultures of sulfur-oxidizing microorganisms. The advantage of axenic cultures is that those strains can be chosen which cause little chain-scission and oxidation of the polymer backbone. Chain-scission and oxidation reduces the performance of the rubber material, and makes the material more hydrophilic. Hydrophilic rubber adsorbs more water, and this can cause blistering when the rubber is revulcanized.

3.2 Sulfur-Reducing Microorganisms

Only a very limited number of studies have been performed on microbial desulfurization of rubber with sulfur-reducing microorganisms. Sulfur-reducing microorganisms reduce sulfur to hydrogen sulfide according to the following reaction (see Sect. 3.1):

$$S_x + H_2 \rightarrow H_2S$$

In 1991, Borman suggested the possibility of using thermophilic bacteria capable of reducing sulfur to hydrogen sulfide in the coal desulfurization processes (Borman, 1991). *Pyrococcus furiosus*, an anaerobic archaeon, was able to utilize sulfur in rubber material (Bredberg et al., 2001). Although the growth of *P. furiosus* is inhibited by hydrogen gas produced during growth, this inhibition can be prevented by the addition of sulfur, whereupon hydrogen sulfide is formed (Malik et al., 1989; Adams, 1990). This reaction can be applied to vulcanized rubber material. CGTR treated with *P. furiosus* for 10 days was shown to have altered and improved properties when mixed with a new rubber matrix (Bredberg et al., 2001).

The anaerobic treatment has the advantage that no oxidation of the hydrocarbon chains occurs, which limits the deterioration of the rubber. However, *P. furiosus* demands higher temperatures than the sulfur-oxidizing strains. There is no formation of active sulfone and sulfoxide on the surface in sulfur-reducing treatment, but how this affects the outcome of the treatment, and whether other active groups that can facilitate the revulcanization are formed has still to be investigated.

3.3 Bioreactors

A wide range of different bioreactors might be used for biological desulfurization, designed and adapted to the growth conditions of the microorganisms. Bredberg and coworkers used 2 L, airlift reactors for desulfurization with *Thiobacillus ferrooxidans* in laboratory-scale experiments (Bredberg et al., in preparation). By contrast, others used shake-flasks with volumes ranging from 250 mL to 2 L (Torma and Raghavan, 1990; Löffler et al., 1995; Romine and Romine, 1998; Kim and Park, 1999). Romine and Snowden-Swan (1997) have patented a method for breaking the sulfur cross-links in rubber with the aid of purified enzymes, but did not reveal the size and construction of their bioreactor, as did Kilbane (1991a,b). Neumann and coworkers patented a batch or semi-continuous bioreactor for the reprocessing of waste rubber with chemolithotrophic bacteria (Neumann et al., 1992), and showed that slurry density and particle size each influenced the desulfurization process. Torma and Raghavan (1990) showed that the increase in sulfate concentration was proportional to the slurry density of rubber material up to about 8%. At higher concentrations the rate of utilization decreased, and this was suggested to be an effect of toxic compounds exceeding a critical concentration. Löffler and coworkers (1995) studied the influence of pulp concentration and found a linear behavior up to 20%. However, at higher concentrations the slurry became viscous, and shear forces were believed to influence the microorganisms in a negative manner. Clearly, it is essential that these processes are further optimized before being established on a commercial basis. An outline for a tentative biotechnological rubber recycling process is shown in Figure 3.

3.4 Analytical Tools and Techniques

In principle, three types of analyses are possible for the evaluation of microbial devulcanization methods:

1) Analyses of the liquid part of the rubber slurry, e.g. released sulfate, pH and microbial growth.
2) Analyses of the surface of the treated rubber.
3) Testing of a cured rubber material containing desulfurized crumb rubber.

The mechanisms behind desulfurization are oxidation or reduction of the cross-links to sulfate and sulfide, respectively. The simplest method to monitor the process is to measure the released sulfate or sulfide concentration and the pH in the growth medium. Several simple ways to measure released sulfur compounds exist, inclduing precipitation of the sulfate with barium chloride, or detection of hydrogen sulfide in the gas phase using gas chromatography (Christiansson et al., 1998; Bredberg et al., 2001).

Measuring dissolved sulfate and sulfide in the medium has two major drawbacks. First, the sulfide can cause precipitation with the rubber additives and may thereby not be detectable by these types of analysis. Second, this method does not detect partially oxidized sulfur bridges, and consequently no conclusions can be drawn regarding the sulfur left on the surface of the rubber material. Cross-links can be broken without the formation of sulfate, the same being true of course for sulfide production in sulfur-reducing processes. Romine and Romine (1998) investigated this by following the formation of sulfone and sulfoxide groups with Fourier transformed infrared spectroscopy (FTIR) and X-ray for near-edge surfaces (XANES). The FTIR absorbency was strongest after 2–3 days, and then slowly disappeared. After 7 days, the ground rubber was completely stripped of surface sulfur. Christiansson and coworkers (1998) used two analytical surface methods to detect the decreases in sulfur content, i.e. SEM, in which the elemental constituents were analyzed by energy-

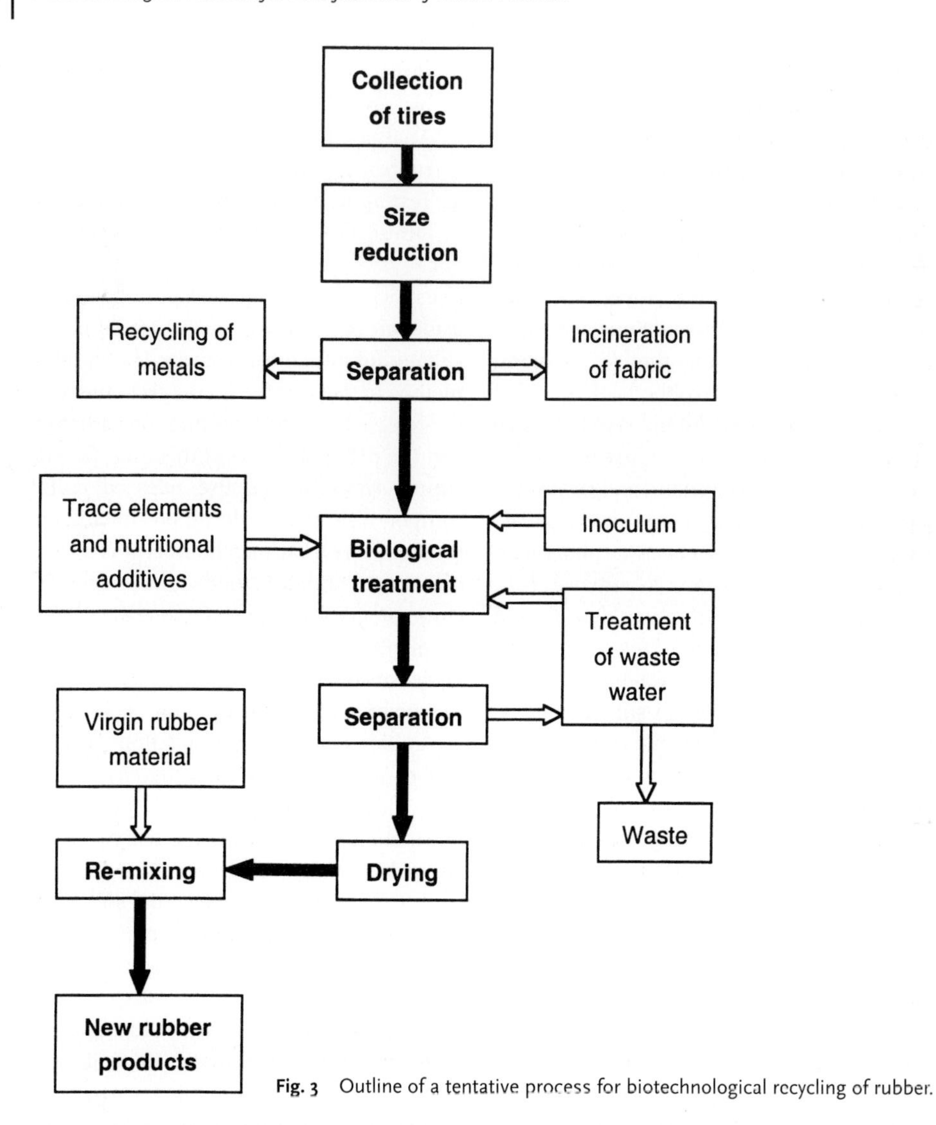

Fig. 3 Outline of a tentative process for biotechnological recycling of rubber.

dispersive X-ray spectrometry (EDS) and electron spectroscopy for chemical analysis (ESCA). EDS measures the relative amount of an element on a scanned surface. Using a model with titanium oxide as internal standard, sheets of the material were cut into flakes, desulfurized with *T. ferrooxidans* and analyzed with SEM/EDS. The flakes were scanned over areas of 10 μm × 10 μm, and the sulfur contents measured as arbitrary units based on the titanium content of the two surfaces. The sulfur was not completely depleted, but the amount had decreased on the surface of rubber flakes treated with *T. ferrooxidans*. The same rubber material was also analyzed with ESCA. This technique is based on measurements of differences in bond energies between elements, and it is thereby possible to distinguish oxidized sulfur from elemental sulfur or sulfides. A higher content of oxidized sulfur species was detected on the surface of the microbe-treated sample than on controls and interior of the sample. However, the total sulfur content was low (< 2%), and thus close to the detection limit; consequently, quantitative measurements could not be made. Nevertheless, ESCA is a technique that makes it possible to follow changes in sulfur species upon desulfurization.

Physical testing of rubber material containing microbially treated rubber crumbs provides an indication of the properties of the recycled rubber. The properties of rubber containing different concentrations of microbially treated and untreated rubber crumbs have been evaluated (Löffler et al., 1995; Romine et al., 1995; Kim and Park, 1999; Bredberg et al., 2001; Bredberg et al., in preparation). Several physical properties, such as stress at break, stress relaxation rate and swelling properties were tested. The swelling properties indicate the degree of cross-linking in the rubber. Efficient desulfurization results in more flexible hydrocarbon chains on the rubber surface, and a higher degree of linking to the rubber matrix. This is shown as a reduced swelling.

3.5 Toxic Effects

Ground tire rubber is a heterogeneous material. Due to the complex and varied rubber recipes, ground spent rubber is a mixture of different materials of a variety of origins. As the composition of the polymer backbone is unknown, and the mixture can also contain sand and soil, physical testing of rubber blends containing reused material is difficult to interpret. Moreover, tire rubber also contains a variety of additives that can be toxic to microorganisms. During the development of rubber products, there has been ongoing discussion on how to prevent rubber products from deterioration. Microbial deterioration is a major problem, especially for applications in close contact with soil and water (Cundell and Mulcock, 1973; Cundell et al., 1973; Rook, 1955; Williams, 1986; Zyska, 1981). Tests to determine the toxicity of rubber additives on sulfur-utilizing microorganisms (Christiansson et al., 2000) have shown that different strains were sensitive to different additives. That is, there were no specific substances that killed all types of microorganism (Table 6).

One possibility is to remove additives from the rubber by leaching with organic solvents. Christiansson and coworkers (2000) showed that microorganisms incapable of growth on used tire rubber could grow on the same material after had been leached with ethanol. Another approach was to use microorganisms capable of degrading organic substances and thereby detoxifying the rubber material (see Chapter 9). Desulfurizing microorganisms that are less sensitive to rubber additives might be an even more promising approach. Clearly, it would be interesting to evaluate the possibility of substituting the original rubber additives with new, less toxic compounds.

3.6 Properties of Rubber Material Containing Microbially Treated Rubber

Little has been written about the final results of microbial desulfurization, that is the properties of rubber materials containing microbe-treated spent rubber. Romine and Romine (1998) have evaluated desulfurization with *S. acidocaldarius* and *T. ferrooxidans* by tensile and elongation testing. Ground tire rubber was desulfurized for 7 days, blended (20%, w/w) with virgin rubber material, and vulcanized. No positive effect of the desulfurization was detected when comparing the

Tab. 6 The effects of rubber chemicals on the growth of different microorganisms in concentrations corresponding to a 10% rubber slurry (* growth in presence of the substance, † no growth) (Bredberg *et al.*, 2001; Christiansson *et al.*, 2000).

	Thiobacillus ferrooxidans	***Rhodococcus rhodochrous***	***Acidianus brierleyi***	***Sulfolobus solfataricus***	***Pyrococcus furiosus***
Natural rubber	*	*	*	*	*
Butadiene rubber	*	*	*	*	*
Styrene-butadiene rubber 1500	†	*	†	†	†
Styrene-butadiene rubber 1712	†	*	†	†	†
Carbon black N375	†	*	*	*	*
Carbon black N660	*	*	*	*	*
Paraffin	*	*	*	*	*
Mineral oil	*	*	†	†	*
Stearic acid	*	*	†	†	†
ZnO	*	†	†	*	†
Zn-salt	*	*	†	†	†
Sulfur	*	*	*	*	*
Tetramethyl-thiuram-monosulfide	†	†	†	†	*
Tetramethyl-thiuram-disulfide	†	*	†	†	*
Dibenzothiazyl-disulfide	*	*	*	*	*
Cyclohexyl benzo-thiazole-sulfenamide	*	*	†	*	*
N-oxydiethen-2-benzo thiazyl-sulfenamid	*	*	*	†	†
Dimethyl-phenyl-*p*-phenylene diamine	†	*	†	†	*
Cyclohexyl thiophthalimide	*	†	†	†	*
Trimethyl-dihydro quinoline	*	*	*	*	*
Anti-sunchecking agent	*	*	*	*	*

results with those performed on material containing untreated crumb. However, improved properties were obtained when analyzing ground tire rubber desulfurized with *S. acidocaldarius* for 72 h. With a 15% (w/w) loading of treated rubber the elastic modulus was 15% higher than with the untreated control. The authors suggested that these differences in results were due to sulfone and sufoxide functionalities, both of which are easily reduced in the presence of elemental sulfur and may thus contribute to the revulcanization. According to this theory, the process must be optimized to produce as many sulfones and sulfoxides as possible in the oxidizing processes. However, sulfur bridges in rubber material are quite dispersed, with approximately 100 monomers between the different sulfur functionalities. Every unsaturation in the hydrocarbon backbone can be accessible during a second revulcanization, provided that appropriate mixing between virgin and old polymer chains is achieved. By creating a looser surface, the flexibility of the chains increases and a higher degree of mixing can be expected. This is probably the most important advantage of techniques for splitting crosslinks in vulcanized rubber materials.

Rubber material containing 15% ground tire rubber treated with a mixed *Thiobacillus* culture had a lower elastic modulus than the control, but the reasons were not further identified (Romine and Romine, 1998). This was in contrast to the results obtained by Bredberg and coworkers (in preparation) when ethanol-treated rubber was desulfurized in batch reactors for 7 days. The material was then mixed (15%, w/w) with virgin rubber and revulcanized. Stress at break, swell and stress relaxation rate was measured. The material treated with *T. ferrooxidans* had better properties than the corresponding control, but the quality of virgin rubber was not regained except for the stress relaxation rate (Bredberg et al., in preparation). Extraction with ethanol had a detrimental effect on the ground crumb rubber, but the loss in properties was recovered after the bioreactor. The results of mixing CGTR treated anaerobically with *P. furiosus* for 10 days gave the same indications as did the results of Bredberg and coworkers (Bredberg et al., 2001). The mechanical properties were affected, some of them are positively, but it must be borne in mind that research in this area is still in its infancy.

4 Current State

Numerous methods have been reported on rubber devulcanization, most of them involving chemical or mechanical processes. The commercial uses are somewhat limited, but a few devulcanization agents, e.g. DeLink, and devulcanized rubber materials, e.g. Surcrum, are available commercially (Ishiaku et al., 1999).

The situation is similar for microbial desulfurization of oil, coal and gas, i.e. numerous reports but few commercial applications. However, the situation with microbial devulcanization of rubber materials is quite different. The reports are very limited, the patents number less than ten, and there are no practical applications – to our knowledge – yet in commercial use, though this is partly an issue of economical and environmental legislation.

Some major questions remain to be answered on the way to commercialization, one of which is the microorganism to be used. Romine and Snowden-Swan (1997) have patented a process with different *Thiobacillus* strains, while Kilbane (1991a,b) suggested the use of mutants of *Bacillus* and *Rhodococcus* for selective cleavage of organic C–S bonds. The question of whether to use purified enzymes from microorganisms, or possibly to modify microorganisms to enhance their efficiency, must also be considered. The methods and apparatus for microbial devulcanization are other important issues: Neumann and coworkers (1992) have patented one bioreaction process for chemolithotrophic bacteria with controlled addition of oxygen. The choice of equipment is obviously dependent on the type of microorganism or enzyme that is used.

5 Outlook and Developments

Important aspects of forthcoming research towards an efficient use of biotechnological desulfurization of rubber in a rubber recycling perspective include the following:

- new catalysts (microorganisms and enzymes)
- reactor constructions
- time optimization in the reactor
- partial degradation of the polymer backbone to increase accessibility
- detoxification of the rubber
- waste water handling
- new products from recycled rubber
- secondary reuse of recycled rubber
- altering of the rubber recipes to improve recyclability.

The search for more efficient microorganisms and enzymes will continue, and there is great potential for screening in several different natural and man-made environments for new strains. Also, the areas of genetic as well as protein engineering remain unexplored. It is however, very inconvenient to employ sterile processes when working with discarded rubber tires. One must rely on selective pressure for the microorganisms in question, which places certain demands on the choice of microorganism. Robust, insensitive microorganisms with reasonable growth rates are needed, while those that demand extreme environments might provide some advantages in the sense of less risk for growth of other strains that might degrade the hydrocarbon structure of the rubber material. In addition, the influence of extreme environments on the rubber has still to be investigated.

The development of processes and apparatus remains an important issue. The development of efficient bioreactors is a major step in the process, but the entire process-chain from cutting and grinding to microbial treatment and revulcanization must also be considered. One important aspect to consider is the waste from the desulfurization process, as the process water will contain sulfate or, in the case of anaerobic microorganisms, hydrogen sulfide. Using strains with the capacity to reduce sulfate to elemental sulfur is one possible way to overcome this problem. The processing water will also contain microorganisms, and some kind of treatment and/or recycling of this may become necessary (Mao et al., 1991).

A two-stage process is a possible way to enhance desulfurization. In the first step, microorganisms cut the hydrocarbon chains on the surface slightly to facilitate access of the sulfur-metabolizing microorganisms. This could decrease the quality of the rubber but, in the right balance, this effect would be counteracted by the increased desulfurization.

Microbial detoxification of the rubber might be necessary in order to avoid problems with the sensitivity of the microorganism towards rubber additives. Another possibility is to screen for microorganisms that are less sensitive towards rubber additives.

Different possible applications for a desulfurization process must also be considered. For example, it might be more economic to concentrate on more expensive and complex rubbers for special uses, e.g. the tread on off-road tires. It might also be necessary to look at the developments of new rubber products to find a market for recycled rubber.

The question of how to recycle or reuse recycled rubber on more than one occasion should also be considered, and indeed whether it is possible to use rubber more than once. The possibilty must be considered that if rubber cannot be recycled a second time, then alternative end products may exist to which this 'waste' may be converted.

A long-term more environmentally friendly approach would be to develop rubber materials that are easily recycled. This could include, for example, exchange of toxic rubber additives to more environmentally friendly ones, in addition to a classification system for different types of rubber materials to facilitate the recycling system.

6
Relevant Patents

Several processes and microorganisms for the desulfurization of rubber have been patented during recent years:

- In 1991, Kilbane patented mutant microorganisms capable of cleaving organic C–S bonds, the intention being to use them for desulfurization of fossil fuels and rubber materials (Kilbane, 1991a,b).
- Romine and Snowden-Swan have patented a method for surface activation with enzymes from *Sulfolobus acidocaldarius*. The sulfur bridges on the surface of rubber crumbs were oxidized to sulfoxide or sulfone (Romine and Snowden-Swan, 1997).
- Neumann and coworkers (1999) patented a biotechnological method for modification of sulfur cross-linked rubber particles with biologically active material. Oxidation of the sulfur on the surface of the rubber resulted in hydroxyl-, epoxy-, and carboxyl-groups.
- Darzins and coworkers (1997) have patented a method for the production of desulfurization enzymes in recombinant pseudomonads.
- A bioreactor system with chemolithotrophic bacteria for desulfurization of rubber wastes has been patented by Neumann and coworkers (1992).

Acknowledgements
Financial support has been provided by Swedish Foundation for Strategic Environmental Research (MISTRA) and CF Environmental Foundation.

7
References

Adams, M. W. W. (1990) The metabolism of hydrogen by extremely thermophilic, sulfur dependent bacteria, *FEMS Microbiol. Rev.* **75**, 219–238.

Asplund, J. (1996) *RAPRA in the Environmental age – Progress in recycling*, Shropshire, UK.

Barnes, H. A., Hutton, J. F., Walters, K. (1989) An introduction to rheology, in: *Rheology series*, 1st edn. Amsterdam: Elsevier Science Publishers B.V.

Bauman, B. D. (1995) High-value engineering materials from scrap rubber, *Rubber World* **212**, 30–33.

Beckman, J. A., Crane, G., Kay, E. L., Laman, J. R. (1974) Scrap tire disposal, *Rubber Chem. Technol.* **47**, 597–624.

Blow, C. M., Hepburn, C. (1982), *Rubber Technology and Manufacture*, 2nd edn. London: Butterworth.

Borman, S. (1991) Bacteria that flourish above 100°C could benefit industrial processing, *Chem. Eng. News* **69**, 31–34.

Bredberg, K., Persson, J., Christiansson, M., Stenberg, B., Holst, O. (2001) Anaerobic desulphurisation of ground rubber with the thermophilic archaea *Pyrococcus furiosus* – a new method for rubber recycling, *Appl. Microbiol. Biotechnol.* **55**, 43–48.

Bredberg, K., Christiansson, M., Bellander, M., Stenberg, B., Holst, O. (in preparation). Properties of rubber materials containing microbially devulcanized cryo-ground tire rubber.

Brock, T. D., Brock, K. M., Belly, R. T., Weiss, R. L. (1972) *Sulfolobus*: a new genus of sulfur-oxidizing bacteria living at low pH and high temperature, *Arch. Microbiol.* **84**, 54–68.

Brombacher, C., Bachofen, R., Brandl, H. (1997) Biohydrometallurgical processing of solids: a patent review, *Appl. Microbiol. Biotechnol.* **48**, 577–587.

Brydson, J. A. (1981) *Flow Properties of Polymer Melts*, 2nd edn., London: George Godwin Limited.

Brydson, J. A. (1988) *Rubbery Materials and their Compounds*, 1st edn., London: Elsevier applied science.

Buisman, C. J. N., Geraats, B. G., Ijspeert, P., Lettinga, G. (1990) Optimization of sulphur production in a biotechnological sulphide-removing reactor, *Biotechnol. Bioeng.* **35**, 50–56.

Christiansson, M., Stenberg, B., Wallenberg, L. R., Holst, O. (1998), Reduction of surface sulphur upon microbial devulcanization of rubber materials, *Biotechnol. Lett.* **20**, 637–642.

Christiansson, M., Stenberg, B., Holst, O. (2000) Toxic additives – a problem for microbial waste rubber desulphurisation, *Resour. Environ. Biotechnol.* **3**, 11–21.

Cornan, A. Y. (1994) *Science and Technology of Rubber*, 2nd Edn. Chapter 7 Vulcanization (Mark, J. E., Erman, B., Eirich, F. R., Eds.), San Diego: Academic Press.

Crane, G., Elefritz, R. A., Kay, E. L., Laman, J. R. (1978) Scrap tire disposal procedures, *Rubber Chem. Technol.* **51**, 577–599.

Cundell, A. M., Mulcock, A. P. (1973) The effect of curing agent concentration on the microbiological deterioration of vulcanized natural rubber, *Int. Biodeterior. Bull.* **9**, 91–94.

Cundell, A. M., Mulcock, A. P., Hills, D. A. (1973) The influence of antioxidants and sulphur level on the microbiological deterioration of vulcanised NR, *Rubber J.* **155**, 22–35.

Datta, R. N., Ivany, M. S. (1995). A chemical for reversion resistant compounding, *Rubber World* **213**, 24–29, 93.

Darzins, A., Xi, L., Childs, J. D., Monticello, D. J., Squires C. H. (1997) Dsz gene expression in *Pseudomonas* hosts, United States Patenet: 5,952,208, Application No.: 851088

Dierkes, I. W. (1996) Solutions to the rubber waste problem incorporating the use of recycled rubber, *Rubber World* **214**, 25–28.

Ferry, J. P. (1970) *Viscoelastic Properties of Polymers*, 2nd edn., New York: John Wiley & sons inc.

Fiala, G., Stetter, K. O. (1986) *Pyrococcus furiosus* sp. nov. represents a novel genus of marine heterotropic archaebacteria growing optimally at 100°C, *Arch. Microbiol.* **145**, 56–61.

Holst, O., Stenberg, B., Christiansson, M. (1998) Biotechnological possibilities for waste tyre-rubber treatment, *Biodegradation* **9**, 301–310.

Hulshoff Pol, L. W., Lens, P. N. L., Stams, A. J. M., Lettinga, G. (1998) Anaerobic treatment of sulphate-rich waste water, *Biodegradation* **9**, 213–224.

Isayev, A. I., Kim, S. H., Levin, V. Y. (1997) Superior mechanical properties of reclaimed SBR with bimodal network, *Rubber Chem. Technol.* **70**, 194–201.

Ishiaku, U. S., Chong, C. S., Ismail, H. (1999) Determination of optimum De-Link R concentration in a recycled rubber compound, *Polym. Test.* **18**, 621–633.

Janssen, A. J. H., De Keizer, A., Van Aelst, A., Fokking, R., Yangling, H., Lettinga, G. (1996) Surface characteristics and aggregation of microbially produced sulphur particles, *Colloids Surf. B* **9**, 115–129.

Johnson, D. B., Nicolau, P., Bridge, T. A. M. (1993) Manipulation and monitoring of acidophilic bacterial population, *Int. Biohydrometallurgy Symposium, Jackson Hole, Wyoming, USA.*

Johnston, S. T., Massey, J., Meerwall, E., Kim, S. H., Levin, V. Y., Isayev, A. I. (1997) Ultrasound devulcanization of SBR: molecular mobility of gel and sol, *Rubber Chem. Technol.* **70**, 183–193.

Kelly, D. P., Watson, S., Zavarzin, G. A. (1989) Aerobic chemolithotrophic bacteria and associated organisms, in: *Bergey's Manual of Systematic Bacteriology* (Staley, J. T., Bryant, M. P., Pfenning, N., Holt, J. G., Eds.), pp. 1854–1858. Williams & Wilkins.

Kilbane, J. J. (1991a) Mutant microorganisms useful for cleavage of organic C–S bonds, Europe, Patent Appl. No. 91250005.5, Publication No. 0 441 462 A2.

Kilbane, J. J. (1991b) Mutant microorganisms useful for cleavage of organic C–S bonds, USA, Patent No. 5.002.888.

Kim, J. K., Park, J. W. (1999) The biological and chemical desulfurization of crumb rubber for the rubber compounding, *J. Appl. Polym. Sci.* **72**, 1543–1549.

Kohler, R., O'Neill, J. (1997) New technology for the devulcanization of sulfur-cured scrap elastomers, *Rubber World* **216**, 32–36.

Lechevalier, H. A. (1986) Nocardioforms, in: *Bergey's Manual of Systematic Bacteriology* (Sneath, P. H. A., Mair, N. S., Sharpe, N. S., Holt, J. G., Eds.), pp. 1477–1481. Williams & Wilkins.

Levin, V. Y., Kim, S. H., Isayev, A. I. (1996) Ultrasound devulcanization of sulfur vulcanized SBR: Crosslink density and molecular mobility, *Rubber Chem. Technol.* **69**, 104–114.

Löffler, M., Straube, G., Straube, E. (1993) Desulfurization of rubber by *Thiobacilli*, in: *Biohydrometallurgical Technologies* (Torma, A. E., Apel, M. L., Brierley, C. L., Eds.) Jackson Hole, Wyoming, USA, pp. 673–680.

Löffler, M., Neuman, W., Straube, E., Straube, G. (1995) Mikrobielle Oberflächenentschwefelung von Altgummigranulat – ein Beitrag zur stofflichen Wiederverwertung von Altgummi, *Kautsch. Gummi Kunstst.* **48**, 454–457.

Malik, B., Su, W.-w., Wald, H. L., Blumentals, I. I., Kelly, R. M. (1989) Growth and gas production for hyperthermophilic archaebacterium, *Pyrococcus furiosus, Biotechnol. Bioeng.* **34**, 1050–1057.

Manuel, H. J., Dierkes, W. (1997) Recycling of rubber, *Rapra report 99*, ISBN: 1–85957-129-8.

Mao, H., Guoliang, T., Shenwei, C. (1991) Treatment of regenerated rubber desulfurization waste water, *Shanghai Huanjing Kexue* **10**(4), 6–9 (abstract).

Mark, J. E., Eisenberg, A., Graessley, W. W., Mandelkern, L., Samulski, E. T., Koenig, J. L., Wignall, G. D. (1993) *Physical Properties of Polymers*, 2nd edn. Washington DC: ACS Professional Reference Book.

Myhre, M. J., MacKillop, D. A. (1996) Modification of crumb rubber to enhance physical properties of recycled rubber products, *Rubber World* **214**, 42–46.

Neumann, W., Straube, G., Rueckauf, H., Forkmann, R. (1992) Method and apparatus for reprocessing of waste rubber, Germany, Patent No. DE 40 42 009 A1.

Neumann, W., Loeffler, M., Hölzemann, J. (1999) Surface activation and modification of sulphur-vulcanised rubber particles, Germany, Patent No. DE 197 28 036 A1.

Raghavan, D., Guay, R., Torma, A. E. (1990) A study of biodegradation of polyethylene and biodesulfurization of rubber, *Appl. Biochem. Biotechnol.* **24/25**, 387–396.

Rintala, J. A., Lepistö, S. S. (1998) Thermophilic anaerobic treatment of sulphur rich forest industry wastewater, *Biodegradation* **9**, 225–232.

Roberts, A. D. (1988) *Natural Rubber Science and Technology*, 1st edn. Oxford: Oxford University Press.

Romine, R. A., Romine, M. F. (1998), Rubbercycle: a bioprocess for surface modification of waste tyre rubber, *Polym. Degrad. Stab.* **59**, 353–358.

Romine, R. A., Snowden-Swan, L. J. (1997) Method for the addition of vulcanized waste rubber to virgin rubber products, USA. Patent Appl. No. 528076.

Romine, R. A., Romine, M. F., Snowden-Swan, L. (1995) Microbial processing of waste tire rubber, Rubber Division 148th Fall Technical Meeting, Clevland, Ohio.

Rook, J. J. (1955) Microbiological deterioration of vulcanized rubber, *Appl. Microbiol.* **3**, 302–309.

Ryu, H. W., Cho, K. S., Chang, Y. K., Kim, S. D., Mori, T. (1995) Refinement of low-grade clay by microbial removal of sulfur and iron compounds using *Thiobacillus ferrooxidans*, *J. Ferment. Bioeng.* **80**, 46–52.

Scheeren, P. J. H., Koch, R. O., Buisman, C. J. N. (1993) Geohydrological contaminant systems and microbial water treatment plant for metal contaminated groundwater at Budelco, *Int. Symp. World Zink '93, Hobart.*

Shennan, J. L. (1996) Microbial attack on sulphur-containing hydrocarbons: Implications for the biodesulphurisation of oils and coals, *J. Chem. Technol. Biotechnol.* **67**, 109–123.

Temple, K. L., Colmer, A. R. (1951) The autotrophic oxidation of iron by a new bacterium: *Thiobacillus ferrooxidans*, *J. Bacteriol.* **62**, 605–611.

Thaysen, A. C., Bunker, H. J., Adams, M. E. (1945) 'Rubber acid' damage in fire hoses, *Nature* **155**, 322–325.

Tichy, R., Lens, P., Grotenhuis, J. T. C., Bos, P. (1998) Solid-state reduced sulfur compounds: Environmental aspects and bio-remediation, *Crit. Rev. Environ. Sci. Technol.* **28**, 1–40.

Torma, A. E., Raghavan, D. (1990) Biodesulfurization of rubber materials, *Bioproc. Eng. Symp.* **16**, 81–87.

Waksman, S. A. & Joffe, J. S. (1922) Microorganisms concerned in the oxidation of sulfur in the soil, II, *Thiobacillus thiooxidans*, a new sulfur-oxidizing organism isolated from the soil. *J. Bacteriol.* **7**, 239–256.

Warner, W. C. (1994) Methods of devulcanisation, *Rubber Chem. Technol.* **67**, 559–566.

Williams, G. R. (1986) The Biodeterioration of vulcanized rubbers, *Int. Biodeter.* **22**, 307–311.

Zarzycki, J. (1991) Glasses and the vitreous state, in: *Cambridge Solid State Science Series* (Cahn, R. W., Davies, E. A., Ward, I. M., Eds.), pp. 12–13. Cambridge University Press.

Zillig, W., Stetter, K. O., Wunderl, S., Schulz, W., Priess, H., Scholz, I. (1980) The *Sulfolobus-"Caldariella"* group: Taxonomy on the basis of the structure of DNA-dependent RNA polymerase, *Arch. Microbiol.* **125**, 259–269.

Zyska, B. J. (1981) Rubber, in: *Microbiall Biodeterioration* (Rose, A. H., Ed.), vol. 6 Economic Microbiology, pp. 323–385. New York: Academic Press Inc.

III
Polyesters

Biotechnology of Biopolymers. From Synthesis to Patents. Edited by A. Steinbüchel and Y. Doi

ISBN: 3-527-31110-6

7 Metabolic Pathways and Engineering of PHA Biosynthesis

Dr. Kazunori Taguchi[1], Dr. Seiichi Taguchi[2], Dr. Kumar Sudesh[3], Dr. Akira Maehara[4], Dr. Takeharu Tsuge[5], Prof. Dr. Yoshiharu Doi[6]

[1] Polymer Chemistry Laboratory, RIKEN Institute, Hirosawa 2-1, Wako-shi, Saitama 351-0198, Japan; Tel.: +81-48-467-9403; Fax: +81-48-462-4667; E-mail: kaztag@postman.riken.go.jp

[2] Polymer Chemistry Laboratory, RIKEN Institute, Hirosawa 2-1, Wako-shi, Saitama 351-0198, Japan; Tel.: +81-48-467-9404; Fax: +81-48-462-4667; E-mail: staguchi@postman.riken.go.jp

[3] Polymer Chemistry Laboratory, RIKEN Institute, Hirosawa 2-1, Wako-shi, Saitama 351-0198, Japan; Tel.: +81-48-467-9403; Fax: +81-48-462-4667; E-mail: sudesh@postman.riken.go.jp

[4] Polymer Chemistry Laboratory, RIKEN Institute, Hirosawa 2-1, Wako-shi, Saitama 351-0198, Japan; Tel.: +81-48-467-9404; Fax: +81-48-462-4667; E-mail: amaehara@postman.riken.go.jp

[5] Polymer Chemistry Laboratory, RIKEN Institute, Hirosawa 2-1, Wako-shi, Saitama 351-0198, Japan; Tel.: +81-48-467-9404; Fax: +81-48-462-4667; E-mail: ttsuge@postman.riken.go.jp

[6] Polymer Chemistry Laboratory, RIKEN Institute, Hirosawa 2-1, Wako-shi, Saitama 351-0198, Japan; Tel.: +81-48-467-9402; Fax: +81-48-462-4667; E-mail: ydoi@postman.riken.go.jp

Biotechnology of Biopolymers. From Synthesis to Patents. Edited by A. Steinbüchel and Y. Doi

ISBN: 3-527-31110-6

ACP	acyl carrier protein
Buk	butyrate kinase
CoA	co-enzyme A
GAP	granule-associated protein
(*R*)-3HA, 3HA	(*R*)-3-hydroxyalkanoate
(*R*)-3HB, 3HB	(*R*)-3-hydroxybutyrate
(*R*)-3HD, 3HD	(*R*)-3-hydroxydecanoate
(*R*)-3HDD, 3HDD	(*R*)-3-hydroxydodecanoate
(*R*)-3HHp, 3HHp	(*R*)-3-hydroxyheptanoate
(*R*)-3HHx, 3HHx	(*R*)-3-hydroxyhexanoate
(*R*)-3HO, 3HO	(*R*)-3-hydroxyoctanoate
(*R*)-3HV, 3HV	(*R*)-3-hydroxyvalerate
(*R*)-4HB, 4HB	(*R*)-4-hydroxybutyrate
4HbD	4-hydroxybutyrate dehydrogenase
Mcl	medium-chain-length
OrfZ	4-hydroxybutyric acid-CoA transferase
PHA	polyhydroxyalkanoate
PhaA, BktB, PhbA	β-ketothiolase
PhaB, PhbB	acetoacetyl-CoA reductase
PhaC, PhbC	PHA synthase
PhaD	regulator involved in mcl-PHA synthesis
PhaE	subunit of PHA synthase
PhaF	Mcl-PHA GAP 1
PhaG	(*R*)-3-hydroxyacyl-ACP-CoA transferase
PhaI	Mcl-PHA GAP 2
PhaJ	(*R*)-specific enoyl-CoA hydratase
PhaP	phasin (scl-PHA GAP)
PhaR	putative regulator involved in scl-PHA synthesis
PhaZ	intracellular PHA depolymerase
Ptb	phosphotransbutyrylase
Scl	short-chain-length

1
Introduction

Since polyhydroxyalkanoates (PHAs) were first identified as environmentally friendly biological plastics, their production from renewable carbon sources in either microbes or plants has become a valuable commercial prospect due lower carbon feedstock costs. The biosynthesis of PHA comprises two enzymatic reaction steps: (1) supply of the substrate monomer; and (2) polymerization of the generated monomer units. For efficient PHA production, it is first necessary to explore the metabolic pathways by which the monomers can be supplied to the PHA biosynthetic process. In this chapter, metabolic pathways that are capable of channeling substrates into PHA biosynthesis are described, and the constituent enzymes involved in the pathways, together with the organization of their genes, are summarized. To date, the substrate monomers for polymerization are all derived from fatty acid metabolism pathways involved in both biosynthesis and degradation, except for one acetyl-CoA dimerization pathway. In particular, the relevant monomer-supplying enzymes are described with respect to their substrate specificity.

The combination of genetic engineering methods with fermentation technology of naturally occurring or recombinant PHA producers has led to the high-level production of poly-[(*R*)-3-hydroxybutyrate] (poly(3HB), or PHB). It has been shown that copolymerization of 3HB with longer side-chains (*R*)-3-hydroxyalkanoate (3HA), poly(3HB-*co*-3HA), confers the improved physical properties – namely, increased ductility and strength compared with the PHB homopolyester. Therefore, it is important to consider how the compositional variation of incorporated units forming the copolyester should be controlled, based on the metabolic pathways utilized for monomer supply. A representative bacterium, *Ralstonia eutropha* (formerly *Alcaligenes eutrophus*), when supplied with different precursors, is able to synthesize not only PHB but also various PHA copolymers. Metabolic engineering of the synthesis of polymers should provide greater control over the properties of the polymer produced, and should also offer alternative organisms for use in the production processes. For this study, the PHA nonproducer *Escherichia coli*, which has a rich genetic background, together with PHA-negative mutants of PHA-producing organisms, have been useful hosts, and these will be described later in the chapter. Over 100 examples of metabolic engineering are included, and the importance of host selection and gene selection has been emphasized through the performance of many experimental trials.

2
Metabolic Pathways for PHA Biosynthesis

In recent years it has become evident that various metabolic pathways can contribute to the generation of hydroxyalkanoate monomers for PHA biosynthesis. The biosynthesis of PHA by microorganisms is largely dependent on the type of carbon source available. In general, the carbon sources are referred to as 'related' and 'unrelated': the former gives rise to hydroxyalkanoate monomers with a chemical structure similar or related to the carbon sources provided, whilst the latter involves the generation of hydroxyalkanoate monomers that, structurally, are completely unrelated to the given carbon source. The reason for this difference lies in the metabolic pathways that are functioning in a particular microorganism.

The three best-known pathways involved in PHA biosynthesis are summarized diagrammatically in Figure 1. Pathway I – which generates 3HB monomers from the condensation of acetyl-CoA – is probably the most common, and is found in a wide range of microorganisms. Regardless of the type of carbon source, acetyl-CoA is generated in all living organisms, which explains the widespread presence of poly(3HB) in different environmental samples. In some bacteria, for example *Rhodospirillum rubrum*, an additional step involving two stereospecific enoyl-CoA hydratases (Moskowitz and Merrick, 1969) is needed to convert the (*S*)-isomer of 3-hydroxybutyryl-CoA to the (*R*)-isomer, which is the only isomeric form usually accepted by the PHA synthase. Unlike the NADPH-dependent acetoacetyl-CoA reductase of *R. eutropha* ($PhaB_{Re}$), which generates (*R*)-3-hydroxybutyryl-CoA, the respective enzyme of *R. rubrum* is NADH-dependent. It must be noted, however, that the corresponding reductase of *Allochromatium vinosum* (formerly *Chromatium vinosum*) is also NADH-dependent, but it yields the (*R*)-isomer of 3-hydroxybutyryl-CoA (Steinbüchel and Füchtenbusch, 1998). A new development which should be highlighted is the ability of $PhaB_{Re}$ to carry out the conversion of 3-ketoacyl-CoA intermediates in pathway II to the corresponding (*R*)-3-hydroxyacyl-CoA in *E. coli* (Ren et al., 2000).

Based on the finding that fatty acids are a suitable carbon source for the biosynthesis of PHA, it has been shown that the pathways involved in fatty acid metabolism can generate various hydroxyalkanoate monomers for PHA biosynthesis (Lageveen et al., 1988). The utilization of fatty acids by bacteria requires coordinated induction of the β-oxidation enzymes, plus a fatty acid transport system (Weeks et al., 1969). Since the PHA synthase is only active towards the (*R*) isomers that are linked to CoA, the intermediates of the fatty acid β-oxidation (pathway II) must undergo appropriate conversion. Some of the enzymes involved in this conversion have been identified and characterized at the genetic level. Of significant interest is the PhaJ protein [(*R*)-specific enoyl-CoA hydratase], the gene for which was found to be clustered with other genes involved in PHA biosynthesis in *Aeromonas caviae* (Fukui and Doi, 1997; Fukui et al., 1998).

It was also noted that certain bacteria capable of producing medium-chain-length PHA, poly(HA_{mcl}), from fatty acids could also supply (*R*)-3-hydroxyacyl-CoAs from glucose and other unrelated carbon sources (Huijberts et al., 1992). Similar observations were made previously by others (Haywood et al., 1990; Timm and Steinbüchel, 1990). Later, it became clear that PHA biosynthesis from glucose in these bacteria is linked to fatty acid biosynthesis (pathway III) (Huijberts et al., 1994), these authors demonstrating that several *Pseudomonas* strains, when cultivated on unrelated carbon sources, accumulate PHAs consisting of predominantly (*R*)-3-hydroxydecanoate monomers. Minor constituents of these polymers were 3-hydroxyhexanoate, 3-hydroxyoctanoate, and 3-hydroxydodecanoate. The similarity in composition of PHA formed on different unrelated carbon sources indicates that a common intermediate in the metabolism of these substrates is also a precursor in the synthesis of PHA monomers, the most likely candidate being acetyl-CoA. The generation of many different monomers for PHA biosynthesis from structurally unrelated and simple carbon sources is considered an important factor for the cost-effective production of PHAs. Among the various microorganisms studied, the pseudomonads are probably the most versatile in their ability to produce PHAs from a variety of related and unrelated carbon sources, including many different *n*-alkanes, *n*-alkenes, alkanoic acids, and alkenoic acids. Based on

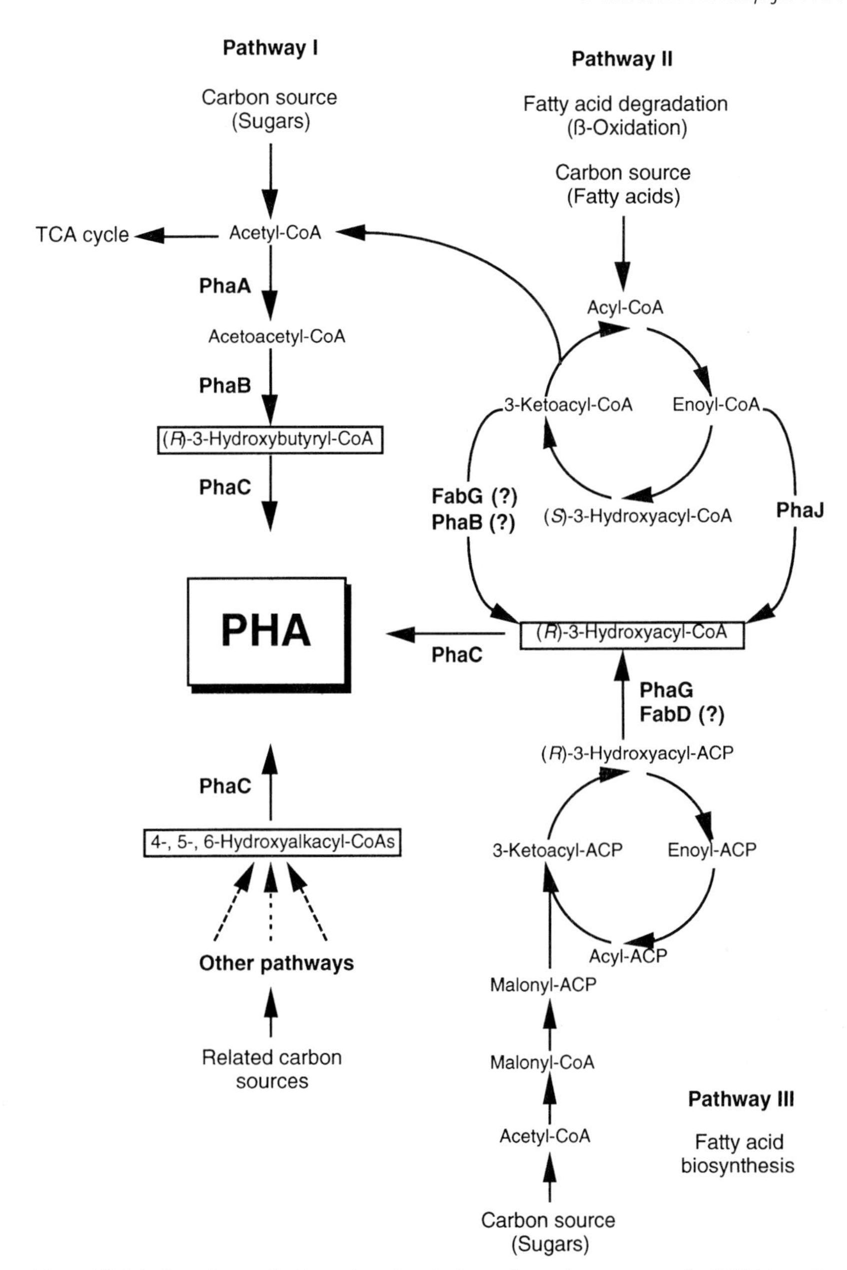

Fig. 1 Metabolic pathways that supply various hydroxyalkanoate monomers for PHA biosynthesis. PhaA, β-ketothiolase; PhaB, NADPH-dependent acetoacetyl-CoA reductase; PhaC, PHA synthase; PhaG, 3-hydroxyacyl-ACP-CoA transferase; PhaJ, (*R*)-enoyl-CoA hydratase; FabD, malonyl-CoA-ACP transacylase; FabG, 3-ketoacyl-ACP reductase.

this finding, the fatty acid biosynthesis pathway is considered to be of great interest. In order to channel suitable intermediates of fatty acid biosynthesis to PHA biosynthesis, the (*R*)-3-hydroxyacyl intermediates have to be converted from the acyl carrier protein (ACP) to the CoA form. The existence of enzyme(s) to carry out this conversion has long been anticipated, and a gene (*phaG*) has been identified recently whose product is involved in such a conversion. Discovery of the acyl-ACP–CoA transacylase thus established the metabolic link between fatty acid synthesis and PHA biosynthesis (Rehm et al., 1998).

Besides the three major pathways described above, others are also present (or have been constructed) which generate monomers such as 4-hydroxybutyrate (4HB) from related and unrelated carbon sources (Valentin et al., 1995, 2000; Valentin and Dennis, 1997). The wealth of knowledge concerning the various monomer-supplying pathways undoubtedly prove valuable in the construction of recombinant systems that can produce a desired type of PHA from low-cost, readily available carbon sources.

3 Organization of the Genes Involved in PHA Biosynthesis

A list of bacteria harboring the genes which are related to PHA biosynthesis is presented in Table 1. (These are listed in alphabetical order of PHA-producing microbes.) To date, 41 bacteria are known to be actual PHA producers, or to be producer candidates based on the database of proteins. Accession numbers are available for addressing the PHA synthesis genes of interest. The patterns of the gene organization can be divided into five groups in terms of arrangements of the *phaC* gene with respect to other genes related to PHA synthesis, as follows (refer to the detailed schematic diagrams in Figure 2): (1) The *phaC* gene exists by forming a cluster with *phaA* and *phaB* genes (*R. eutropha* H16 is a representative bacterium). (2) The *phaC* gene exists alone in a different locus where the *phaA* gene or *phaB* gene exists, or coexists with either the *phaA* gene or *phaB* gene; in both cases, there are two gene clusters. (3) Two homologous *phaC* genes (generally termed *phaC1* and *phaC2*) are tandemly arranged and are separated by the *phaZ* gene, which encodes an intracellular PHA depolymerase. This arrangement is mostly found in pseudomonads, but exceptionally, *Pseudomonas* sp. 61–3 also possesses a *phaBAC* operon at a different locus. At present, the *phaG* gene for (*R*)-3-hydroxyacyl-ACP-CoA transacylase is distributed specifically in pseudomonads. (4) The *phaC* gene exists with the *phaJ* gene (encodes (*R*)-specific enoyl-CoA hydratase) by forming the same cluster (as the case of *A. caviae*) or possibly not (*R. rubrum*). (5) At present, distinct grouping cannot be carried out, but current genome-sequencing projects have provided clues to find new genes that are related to the metabolic pathways for PHA biosynthesis, even though PHA production has not been observed, as in the cases of *Rickettsia prowazekii* Madrid E and *Vibrio cholerae*.

4 Monomer-supplying Enzymes for PHA Biosynthesis

In this section, the four monomer-supplying enzymes for PHA biosynthesis are presented with respect not only to their catalytic properties but also in particular to their substrate specificities.

Tab. 1 Bacteria in which PHA synthesis genes were cloned and sequenced

Organism	*Type of PHA*	*Accession No.*	*Designation*	*Organization pattern*	*References*
Acinetobacter sp. RA3849	S	U04848	*phaC*	(1)	Schembri et al. (1994)
		L37761	*phaBPCA*		Schembri et al. (1995)
Aeromonas caviae FA440	S, M	D88825	*phaPCJ*	(4)	Fukui and Doi (1997); Fukui et al. (1998)
Alcaligenes latus DSM1123	S	AF078795	*phaCAB*	(1)	Choi et al. (1998)
Alcaligenes latus DSM1124	S	U47026	*phaCAB*	(1)	Genser et al. (1998)
		AF004933	*phaC*		Hong et al. (1999)
Alcaligenes sp. SH-69	S	U78047	*phaC*	(2)	Lee, I. et al. (1996)
		AF002013	*phaA*		Lee et al. (1998)
		AF002014	*phaB*		Lee et al. (1998)
Allochromatium vinosum D	S	L01112	*phaCEARPB*	(1)	Liebergesell and Steinbüchel (1992, 1996)
Azorhizobium caulinodans ORS571	S	AJ006237	*phaC*	(5)	Mandon et al. (1998)
Bacillus megaterium ATCC11561	S	AF109909	*phaPBC*	(2)	McCool and Cannon (1999)
Burkholderia sp. DSM9242	S	AF153086	*phaCABR*	(1)	Rodrigues et al. (2000)
Caulobacter crescentus	N.O.	AY007313	*phaC*	(5)	Qi et al. (2000)
Chromobacterium violaceum DSM30191	S	AF061446	*phaCA*	(2)	Kolibachuk et al. (1999)
Comamonas acidovorans DS-17	S	AB009273	*phaCA*	(2)	Sudesh et al. (1998)
Ectothiorhodospira shaposhnikovii	N.O.	AF307334	*phaCEAPB*	(1)	Zhang et al. (2000)
Methylobacterium extorquens IBT6	S	L07893	*phaC*	(5)	Valentin and Steinbüchel (1993)
Nocardia corallina	S, M	AF019964	*phaC*	(5)	Hall et al. (1998)
Paracoccus denitrificans	S	D49362	*phaAB*	(2)	Yabutani et al. (1995)
		D43764	*phaC*		Ueda et al. (1996)
		AB017045	*phaZCPR*		Maehara et al. (1999)
Pseudomonas acidophila	S	-	*phaCABR*	(1)	Umeda et al. (1998)
Pseudomonas aeruginosa DSM1707	M	X66592	*phaC1ZC2D*	(3)	Timm and Steinbüchel (1992)
		AE004919	*phaC1ZC2DFI*		Stover et al. (2000)
		AB040025	*phaJ1*		Tsuge et al. (2000)
		AB040026	*phaJ2*		Tsuge et al. (2000)
		AF209711	*phaG*		Hoffmann et al. (2000a)

Tab. 1 (cont.)

Organism	*Type of PHA*	*Accession No.*	*Designation*	*Organization pattern*	*References*
Pseudomonas oleovorans	M	M58445	*phaC1ZC2D*	(3)	Huisman et al. (1991); Klinke et al. (2000)
		AJ010393	*phaFI*		Prieto et al. (1999a)
		AF169252	*phaG*		Hoffmann et al. (2000b)
Pseudomonas putida BM01	M	AF042276	*phaC2DFI*	(3)	Valentin et al. (1998)
Pseudomonas putida KT2440	M	AF052507	*phaG*	(5)	Rehm et al. (1998)
Pseudomonas putida U	M	AF150670	*phaC1ZC2*	(3)	García et al. (1999)
Pseudomonas resinovorans	M	AF129396	*phaC1ZC2D*	(3)	Solaiman (2000)
Pseudomonas sp. 61–3	S, M	AB014757	*phbRBAC*	(1), (3)	Matsusaki et al. (1998)
		AB014758	*phaC1ZC2*		Matsusaki et al. (1998)
		AB047080	*phaG*		Matsumoto et al. (2001)
Ralstonia eutropha H16	S	J05003	*phaC*	(1)	Peoples and Sinskey (1989b); Slater et al. (1988)
		J04987	*phaAB*		Peoples and Sinskey (1989a)
		M64341	*phaC*		Schubert et al. (1988, 1991)
		X85729	*phaP*		Wieczorek et al. (1995); Hanley et al. (1999)
		AF026544	*phaR,bktB*		Slater et al. (1998)
		AB017612	*phaZ*		Saegusa et al. (2001); Handrick et al. (2000)
Rhizobium etli	S	U30612	*phaC*	(5)	Cevallos et al. (1996)
Rhodobacter capsulatus	S	-	*phaC, phaAB*	(2)	Kranz et al. (1997)
Rhodobacter sphaeroides ATCC17023	S	L17049	*phaZCP*	(5)	Hustede and Steinbüchel (1993)
Rhodobacter sphaeroides 2.4.1	S	AF098459	*phaZC*	(5)	Kim and Lee (1997)
Rhodobacter sphaeroides RV	N.O.	X97200	*phaC*	(5)	Franchi et al. (1997)
Rhodococcus ruber NCIMB40126	S	X66407	*phaCP*	(5)	Pieper and Steinbüchel (1992); Pieper-Furst et al. (1994)
Rhodospirillum rubrum ATCC25903	S, M	AF178117	*phaC*	(4)	Clemente et al. (2000)
		AF156879	*phaJ*		Reiser et al. (2000)
Rhodospirillum rubrum Ha	S, M	AJ245888	*phaC*	(5)	Clemente et al. (2000)
Rickettsia prowazekii Madrid E	N.O.	AJ235273	*phaCA, phaC, phaB*	(2)	Andersson et al. (1998)
Sinorhizobium meliloti 41	S	U17226	*phaAB*	(2)	Tombolini et al. (1995)
		U17227	*phaC*		Tombolini et al. (1995)
Sinorhizobium meliloti Rm1021	S	AF031938	*phaC*	(5)	Willis and Walker (1998)

Tab. 1 (cont.)

Organism	*Type of PHA*	*Accession No.*	*Designation*	*Organization pattern*	*References*
Synechocystis sp. PCC6803	S	D90906	*phaCE*	(2)	Kaneko et al. (1996); Hein et al. (1998);
		D90910	*phaAB*		Taroncher-Oldenburg et al. (2000)
Thiococcus pfennigii	S, M	A49465	*phaCE*	(5)	Liebergesell et al. (1993)
Thiocystis violacea 2311	S	L01113	*phaCE*	(2)	Liebergesell and Steinbüchel (1993)
Vibrio cholerae	N.O.	AE004398	*phaBAPC*	(1)	Heidelberg et al. (2000)
Zoogloea ramigera	S	U66242	*phaC*	(2)	Lee, S. P. et al. (1996)
		J02631	*phaA(B)*		Peoples et al. (1987)

S, short-chain-length (C_3 to C_5); M, medium-chain-length (C_6 to C_{14}); N.O., not opened; Explanation of the gene organization patterns is given in the text. Abbreviations of the genes are as follows; *phaA* and *bktB*, β-ketothiolase; *phaB*, acetoacetyl-CoA reductase; *phaC*, PHA synthase; *phaD*, regulator involved in mcl-PHA synthesis; *phaE*, subunit of PHA synthase; *phaF*, mcl-PHA granule associated protein (GAP) 1; *phaI*, mcl-PHA GAP 2; *phaG*, (*R*)-3-hydroxyacyl-ACP-CoA transferase; *phaJ*, (*R*)-specific enoyl-CoA hydratase; *phaP*, phasin (scl-PHA GAP); *phaR*, putative regulator involved in scl-PHA synthesis; *phaZ*, intracellular PHA depolymerase. Bacteria in which the presence of PHA synthesis genes is obvious are listed in this table, even though the nucleotide sequences of genes are not available in database.

4.1 β-Ketothiolase (PhaA)

As the first reaction step in poly(3HB) formation, this enzyme catalyzes the condensation of two acetyl-CoA molecules to generate acetoacetyl-CoA. Although many genes encoding PhaA (over 40) have been identified (Table 2), enzymatic studies have been mostly limited to *R. eutropha* $PhaA_{Re}$ and *Z. ramigera* $PhaA_{Zr}$. *R. eutropha* is known to produce at least three similar enzymes, including $PhaA_{Re}$ and another (termed BktB), that are primarily responsible for generating 3-ketovaleryl-CoA during growth on fructose and propionate (Slater et al., 1998). Both enzymes exist as a homotetramer in the native state (Haywood et al., 1988a). *In vitro* analysis has revealed that $PhaA_{Re}$ shows thiolysis activity towards only acetoacetyl-CoA (preferably) and 3-ketopentanoyl-CoA, while BktB possesses broad substrate specificity (C_4 to C_{10}). The detailed investigation of $PhaA_{Zr}$ serves a mechanistic role of two active-site cysteines in the condensation of acetyl-CoA with either acetyl-CoA or acyl-CoA (Thompson et al., 1989; Palmer et al., 1991).

4.2 Acetoacetyl-CoA Reductase (PhaB)

This enzyme catalyzes the second step in the poly(3HB) biosynthesis pathway I by converting acetoacetyl-CoA into (*R*)-3-hydroxybutyryl-CoA. Table 3 shows a compilation of the PhaB genes identified to date. Although there are two types of reductase which catalyze in either NADH- or NADPH-dependent manner, most have been categorized to the NADPH-dependent enzyme family, except for the NADH-dependent reductase from *A. vinosum* (Liebergesell and Steinbüchel, 1992). *Z. ramigera* $PhaB_{Zr}$ is a homotetramer with identical 25 kDa

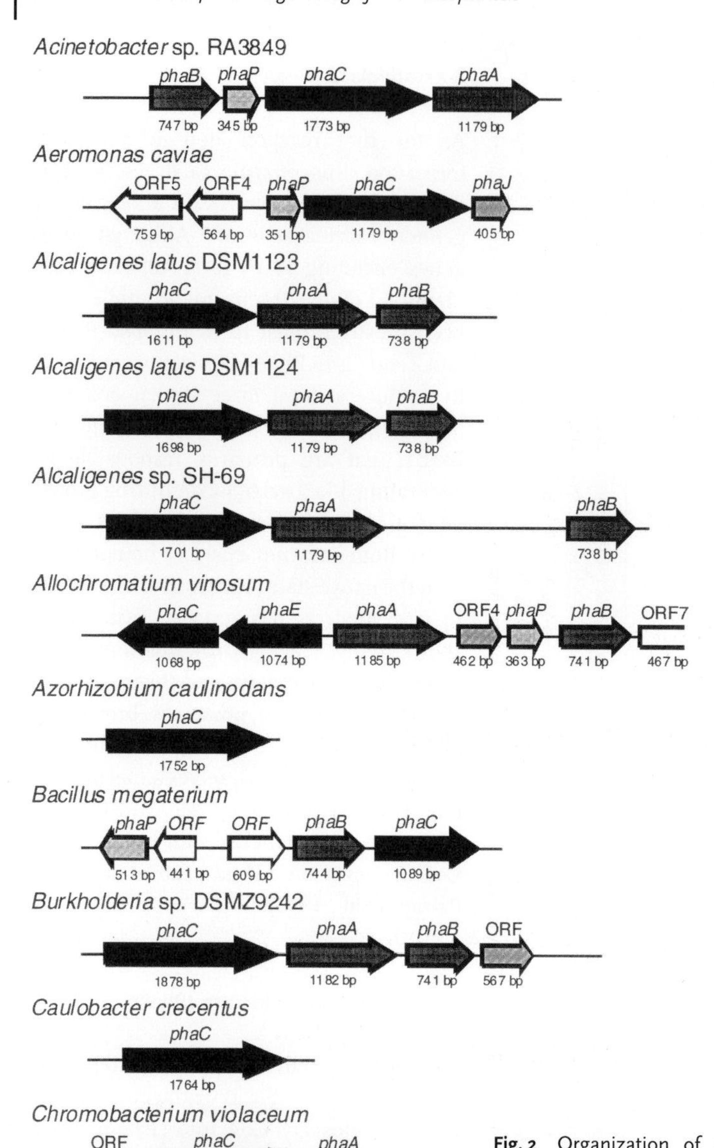

Fig. 2 Organization of genes involved in PHA synthesis. The size of genes includes the termination codon.

subunits and a NADPH-dependent type reductase (Fukui et al., 1987; Ploux et al., 1988). In the case of *R. eutropha*, both NADPH- and NADH-dependent reductase activities have been observed, but only the NADPH-dependent enzyme is responsible for poly(3HB) biosynthesis (Haywood et al., 1988b). NADPH-dependent reductases of *Z. ramigera* and *R. eutropha* show the substrate specificities of C_4 to C_5 (for $PhaB_{Re}$) and C_4 to C_6 (for $PhaB_{Zr}$), respectively.

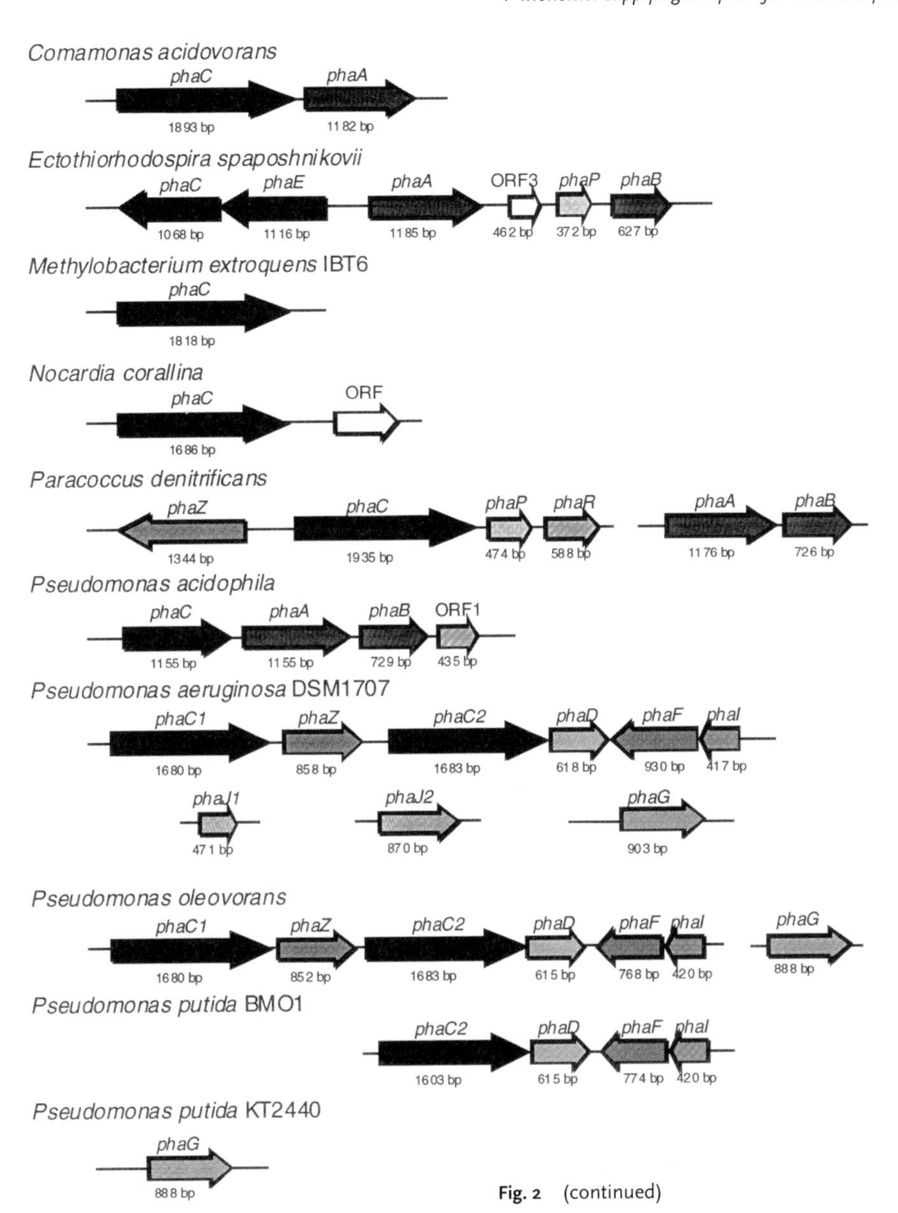

Fig. 2 (continued)

4.3
(*R*)-3-Hydroxyacyl-ACP–CoA transferase (PhaG)

By using PHA nonproducing mutants of *Pseudomonas putida* KT2440, a gene specific for transacylase $PhaG_{Pp}$ was identified as a metabolic link between *de novo* fatty acid biosynthesis and PHA biosynthesis (Rehm et al., 1998). This enzyme catalyzes the conversion reaction of (*R*)-3-hydroxyacyl-ACP intermediates of fatty acid synthesis

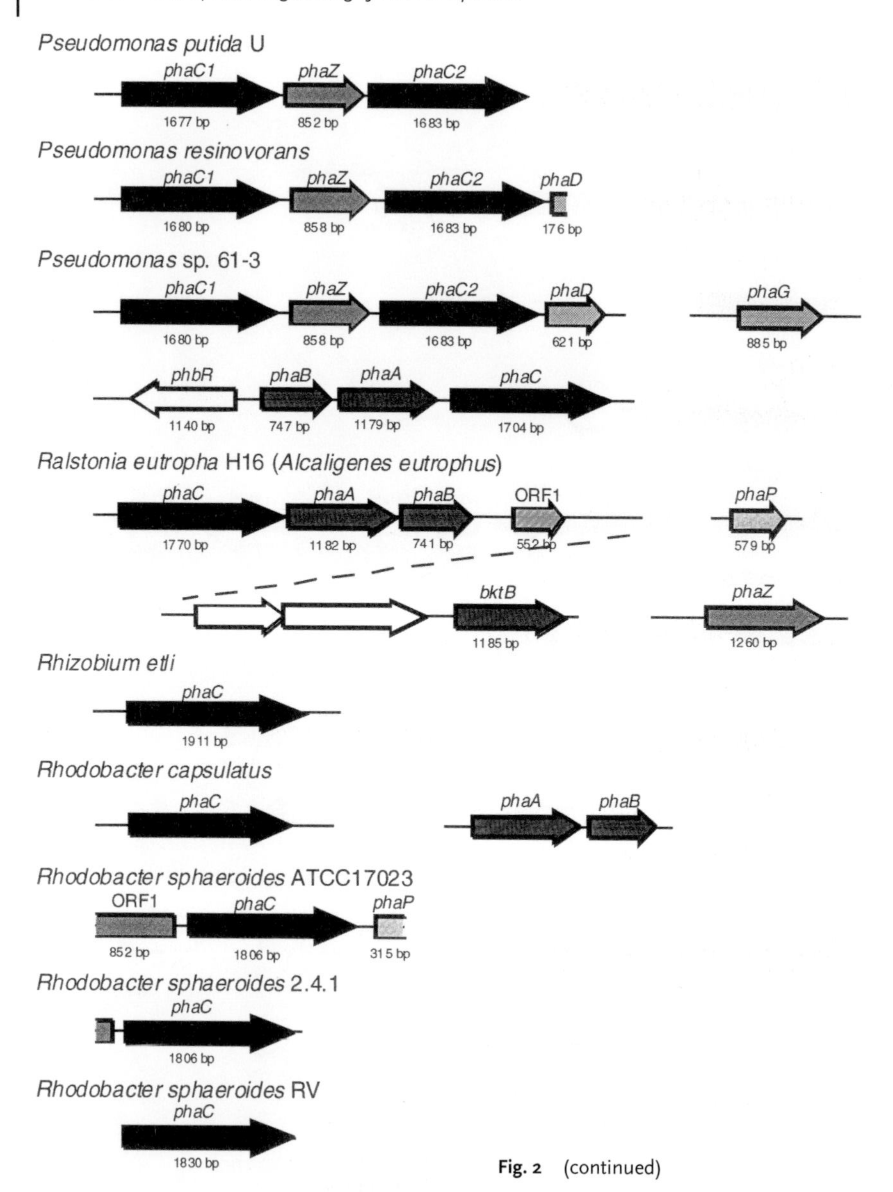

Fig. 2 (continued)

to the corresponding CoA derivatives. Consequently, the generated (*R*)-3-hydroxyacyl-CoA can be utilized as a substrate for PHA biosynthesis. Among substrates tested, a transfer reaction of the 3-hydroxydecanoyl moiety from CoA to ACP was obtained at the highest rate by purified $PhaG_{Pp}$ in the presence of $MgCl_2$. Later, as shown in Table 4, three other genes encoding PhaG were cloned from *P. aeruginosa* (Hoffmann et al., 2000b), *P. oleovorans* (Hoffmann et al., 2000a) and *Pseudomonas* sp. 61–3 (Matsu-

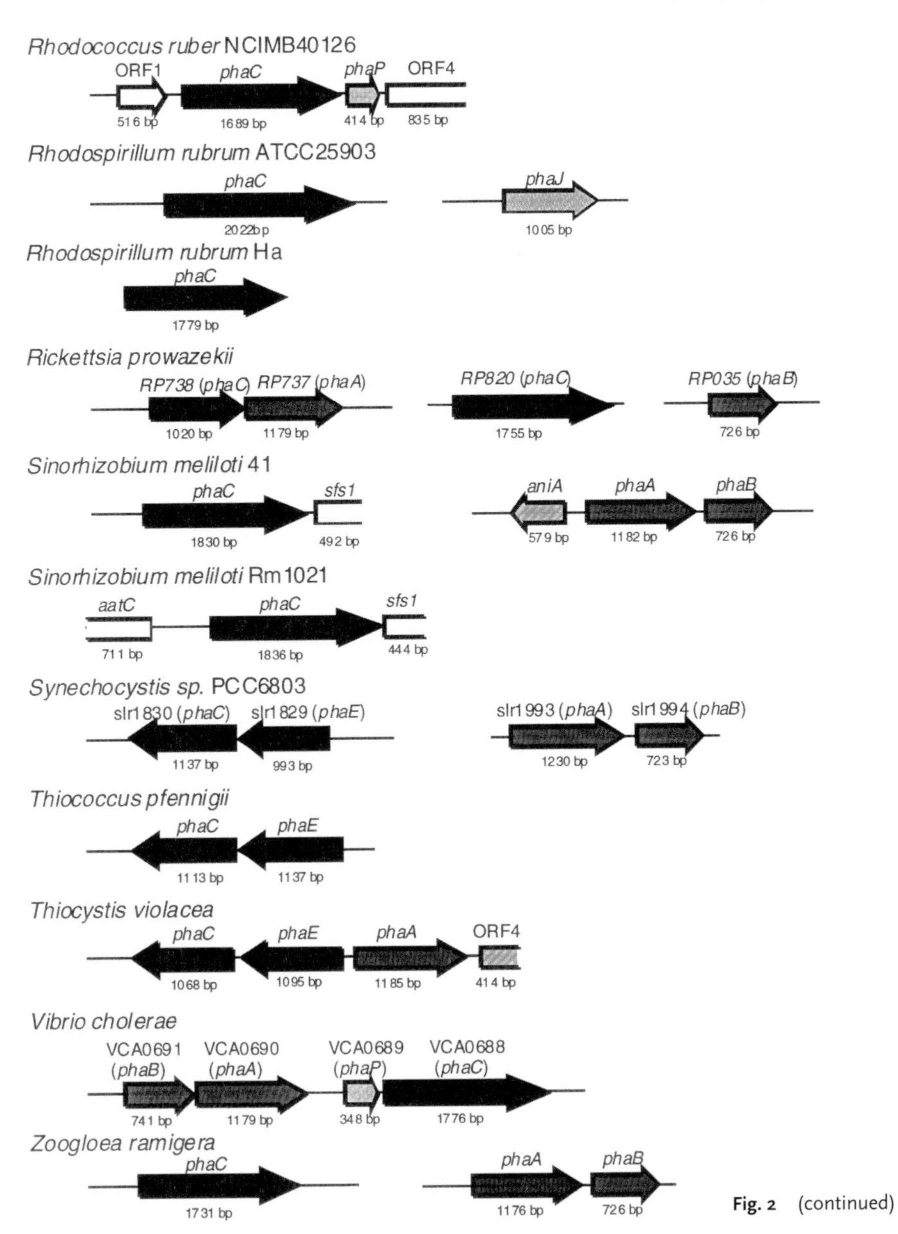

Fig. 2 (continued)

moto et al., 2001). The average molecular mass of the four gene products was calculated to be approximately 34 kDa by deducing their nucleotide sequences. A HX4D or similar motif-encoding sequence is found in all four of the *phaG* gene products, this configuration being commonly shared with a variety of glycerolipid acyltransferases involved in rhamnolipid biosynthesis. Site-directed mutagenesis revealed that a His residue in the motif plays an important role in $PhaG_{Ps}$ activity (Matsumoto et al., 2001). As a case study, $PhaG_{Pp}$-mediated synthesis of PHA consisting of medium-chain-length

Tab. 2 β-Ketothiolases (PhaA) involved in PHA synthesis

Organism	*Substrate specificity*	*Calculated molecular mass*	*Reference or accession No.*
Acinetobacter sp. RA3849	S	40,631	Schembri et al. (1995)
Alcaligenes latus ATCC29714	S	40,522	Choi et al. (1998)
Alcaligenes latus ATCC29713	S	40,284	Genser et al. (1998)
Alcaligenes latus SH-69	S	40,165	AF002013; Lee et al. (1997)
Allochromatium vinosum	S	40,996	Liebergesell and Steinbüchel (1992)
Burkholderia sp. DSMZ9242	S	40,413	Rodrigues et al. (2000)
Chromobacterium violaceum	S	N-terminal 143 aa	Kolibachuk et al. (1999)
Comamonas acidvorans	S	40,238	Sudesh et al. (1998)
Ectothiorhodospira spaposhnikovii	N.T.	40,800	AF307334; Zhang et al. (2000)
Paracoccus denitrificans	S	40,751	Yabutani et al. (1995)
Pseudomonas acidophila	S	40,200	Umeda et al. (1998)
Pseudomonas sp. 61–3	S	40,590	Matsusaki et al. (1998)
Ralstonia eutropha H16 (*phaA*)	S	40,549	Schubert et al. (1991)
Ralstonia eutropha H16 (*bktB*)	S, M	40,904	Haywood et al. (1988a); Slater et al. (1998)
Rhodobacter capsulatus	S	N.O.	Kranz et al. (1997)
Rickettsia prowazekii Madrid E	N.T.	41,723	AJ235273; Andersson et al. (1998)
Sinorhizobium meliloti 41	S	40,851	Tombolini et al. (1995)
Synechocystis sp. PCC6803	S	43,281	Taroncher-Oldenburg et al. (2000)
Thiocystis violacea 2311	S	40,897	Liebergesell and Steinbüchel (1993)
Vibrio cholerae	N.T.	41,400	AE004398; Heidelberg et al. (2000)
Zoogloea ramigera	S	40,416	Thompson et al. (1989); Peoples et al. (1987)

bktB, homologous gene of *phaA* gene; S, short-chain-length (C_3 to C_5); M, medium-chain-length (C_6 to C_{14}); N.T., not tested; N.O., not opened.

constituents was achieved from nonrelated carbon sources in recombinant *P. fragi*, which intrinsically is not a PHA producer (Fiedler et al., 2000).

4.4 (*R*)-Specific Enoyl-CoA Hydratase (PhaJ)

This enzyme was identified as a molecular converter of β-oxidation intermediates to (*R*)-3-hydroxyacyl-CoA monomer units for PHA biosynthesis in *A. caviae* (Fukui et al., 1998). The $PhaJ_{Ac}$ functionally overexpressed by *E. coli* forms a homodimer, and shows (*R*)-specific hydration activity towards *trans*-2-enoyl-CoA with four to six carbon atoms. Most recently, the tertiary structure of the $PhaJ_{Ac}$ was resolved at 1.5 Å resolution (Hisano et al., 2001). The recombinant product of a gene encoding the hydratase homolog ($PhaJ_{Rr}$) found in *R. rubrum* shows similar substrate specificity to $PhaJ_{Ac}$, but is characteristic in that $PhaJ_{Rr}$ possesses a C-terminal extension of amino acid residues and exists as a homotetramer in the native state (Reiser et al., 2000). Two *P. aeruginosa* genes homologous to the $phaJ_{Ac}$, referred to as $phaJ1_{Pa}$ and $phaJ2_{Pa}$, were also cloned and investigated as to whether their recombinant gene products conferred the ability to supply monomer units for PHA (Tsuge et al., 2000). Different substrate specificity was found for

Tab. 3 Acetoacetyl-CoA reductases (PhaB) involved in PHA synthesis

Organism	*Substrate specificity*	*Calculated molecular mass*	*Reference*
Acinetobacter sp. RA3849	S	26,727	Schembri et al. (1995)
Alcaligenes latus ATCC29714	S	26,134	Choi et al. (1998)
Alcaligenes latus ATCC29713	S	26,235	Genser et al. (1998)
Alcaligenes latus SH-69	S	26,223	AF002014; Lee et al. (1997)
Allochromatium vinosum	S	26,123	Liebergesell and Steinbüchel (1992)
Bacillus megaterium	S	26,095	McCool and Cannon (1999)
Burkholderia sp. DSMZ9242	S	26,315	Rodrigues et al. (2000)
Ectothiorhodospira spaposhnikovii	N.T.	21,999	AF307334; Zhang et al. (2000)
Paracoccus denitrificans	S	25,536	Yabutani et al. (1995)
Pseudomonas acidophila	S	25,860	Umeda et al. (1998)
Pseudomonas sp. 61–3	S	26,693	Matsusaki et al. (1998)
Ralstonia eutropha H16	S, M	26,370	Schubert et al. (1991)
Rhodobacter capsulatus	S	N.O.	Kranz et al. (1997)
Rickettsia prowazekii	N.T.	26,225	AJ235273; Andersson et al. (1998)
Sinorhizobium meliloti 41	S	25,370	Tombolini et al. (1995)
Synechocystis sp. PCC6803	S	25,333	Taroncher-Oldenburg et al. (2000)
Vibrio cholerae	M.T.	26,785	AE004398; Heidelberg et al. (2000)
Zoogloea ramigera	S	25,285	Peoples et al. (1987)

S, short-chain-length (C_3 to C_5); M, medium-chain-length (C_6 to C_{14}); N.T., not tested; N.O., not opened.

Tab. 4 (*R*)-3-hydroxyacyl-ACP-CoA transacylases (PhaG) and (*R*)-specific enoyl-CoA hydratases (PhaJ) involved in PHA synthesis

Organism	*Substrate specificity*	*Calculated molecular mass*	*Reference*
(PhaG)			
Pseudomonas putida	M	33,893	Rehm et al. (1998)
Pseudomonas aeruginosa	M	34,225	Hoffmann et al. (2000b)
*Pseudomonas oleovorans**	M	33,696	Hoffmann et al. (2000a)
Pseudomonas sp. 61–3	M	33,550	Matsumoto et al. (2001)
(PhaJ)			
Aeromonas caviae	S, M	14,086	Fukui et al. (1998)
*Rhodospirillum rubrum**	S, M	15,353	Reiser et al. (2000)
*Pseudomonas aeruginosa (phaJ1)**	S, M	16,901	Tsuge et al. (2000)
*Pseudomonas aeruginosa (phaJ2)**	M	31,631	Tsuge et al. (2000)

S, short-chain-length (C_3 to C_5); M, medium-chain-length (C_6 to C_{14}); Asterisk indicates not demonstrated gene responsible for PHA synthesis in the original cells.

both hydratases; that is, $PhaJ1_{Pa}$ was specific for short-chain-length enoyl-CoAs as well as $PhaJ_{Ac}$, while $PhaJ2_{Pa}$ preferred medium-chain-length enoyl-CoAs.

5 Metabolic Engineering and Improvements in PHA Biosynthesis

A number of wild-type bacteria producing PHA have been isolated. For a successful commercial implementation of PHA production, it is essential to investigate many physical and biological parameters such as optimal temperature, pH, substrate, and medium for growth. In addition, it is necessary to determine fermentation conditions for PHA accumulation within bacterial cells.

Although *E. coli* has no ability to synthesize and degrade PHA, it has been well characterized genetically, and is now established as an excellent host for genetic and metabolic engineering techniques facilitating the 'construction' of recombinant bacteria for PHA production. However, in addition to *E. coli*, a number of different recombinant bacteria have been utilized for the same purpose.

The following section outlines the metabolic engineering and improvements for PHA production not only in *E. coli* strains, but also in a *R. eutropha* PHA-negative strain and pseudomonads, by the use of genetic recombination techniques.

5.1 Metabolic Engineering for PHA Production in Recombinant *Escherichia coli*

A number of PHA synthetic genes have been cloned from PHA-producing bacteria. These genes facilitate the construction of recombinant *E. coli* for PHA biosynthesis. Although *R. eutropha* is a typical bacterium for PHA production, its natural strain synthesizes only short-chain-length PHA, poly(HA_{scl}), consisting of C_3–C_5 3HA monomer units, and the organism is not well characterized genetically. Recombinant *E. coli* can be used to solve these problems, and representative examples of metabolic engineering for PHA production in recombinant *E. coli* employing genetic techniques are listed in Table 5.

The first example of PHA production by a recombinant *E. coli* was achieved by the introduction of PHA synthetic genes from *R. eutropha*. During the 1980s, the $phaA_{Re}$, $phaB_{Re}$, and $phaC_{Re}$ genes from *R. eutropha* were cloned in *E. coli* and directed poly(3HB) accumulation from glucose or gluconate (Schubert et al., 1988; Slater et al., 1988; Peoples and Sinskey, 1989b). The content of poly(3HB) produced in the recombinant host was up to 54 wt.% of the dry cell matter. However, for even greater poly(3HB) accumulation, Lee and Chang (1995) constructed several *E. coli* strains equipped with a *parB*-stabilized plasmid containing $phaA_{Re}$, $phaB_{Re}$, and $phaC_{Re}$. Such plasmid stabilization and a defined cultivation resulted in the accumulation of more than 80 wt.% poly(3HB). Kusaka et al. (1997) found that these recombinants synthesized ultra-high-molecular weight poly(3HB) (number-average molecular weight up to 2×10^7) on pH6-stat cultivation. Further genetic improvement of the gene promoter and ribosome-binding sequence resulted in a maximal accumulation of up to 95 wt.% poly(3HB) of dry cell weight (Kalousek and Lubitz, 1995). By contrast, Wong and Lee (1998) investigated pH-stat, fed-batch cultivation using whey as a carbon source and showed that, by overexpressing the *ftsZ* gene, biomass production and poly(3HB) accumulation were each improved. Other recombinant *E. coli* harboring *phaA*, *phaB*, and *phaC* from *Alcaligenes latus* or photosynthetic bacteria

Tab. 5 Metabolic engineering for PHA production in recombinant *Escherichia coli*

Host	Introduced genes	Carbon source	PHA content (wt.%)	Composition (mol.%)								Reference
				3HB	4HB	3HV	4HV	3HHx	3HO	3HD	3HDD	
LE392	$phaA_{Re}$, $phaB_{Re}$, $phaC_{Re}$	Gluconate (LB)	20	100								Slater et al. (1988)
DH5α			54	100								
LM83	$phaA_{Re}$, $phaB_{Re}$, $phaC_{Re}$	Glucose (LB)	26	100								Schubert et al. (1988)
K12			30	100								
S17-1			18	100								
DH5α	$phaA_{Re}$, $phaB_{Re}$, $phaC_{Re}$	Glucose	50	100								Peoples and Sinskey (1989b)
LS5218 [*fadR*, *atoC* (Con)]	$phaA_{Re}$, $phaB_{Re}$, $phaC_{Re}$	Glucose + C3 + acetate	1.8 mg/ml culture	73		27						Slater et al. (1992)
K-12	$phaA_{Re}$, $phaB_{Re}$, $phaC_{Re}$	Glucose + C3	30	99			1					Rhie and Dennis (1995)
JMU170 (*fadR*)	$phaB_{Re}$, $phaC_{Re}$	Glucose	4	100								
		Glucose + C3	6	71		29						
	$phaA_{Re}$, $phaB_{Re}$, $phaC_{Re}$	Glucose + C3	13	85		15						
XL1-Blue	$phaA_{Re}$, $phaB_{Re}$, $phaC_{Re}$	Glucose (LB)	81	100								Lee and Chang (1995)
B	$phaA_{Re}$, $phaB_{Re}$, $phaC_{Re}$		76	100								
JM109	$phaA_{Re}$, $phaB_{Re}$, $phaC_{Re}$		85	100								
HB101	$phaA_{Re}$, $phaB_{Re}$, $phaC_{Re}$		75	100								
MC4100	$phaA_{Re}$, $phaB_{Re}$, $phaC_{Re}$	Glucose	95	100								Kalousek and Lubitz (1995)
XL1-Blue	$phaA_{Re}$, $phaB_{Re}$, $phaC_{Re}$	Glucose (LB, pH.7)	30	100								Kusaka et al. (1997)
		Glucose (LB, pH.6)	30*	100								
LS12998 (*fadB*)	$phaC1_{Pa}$	C10 (LB)	21					5	19	73	3	Langenbach et al. (1997)
XL1-Blue	$phaC_{Re}$, $orfZ_{Ck}$ (co-linear to *lac* promoter)	Glucose + 4-HB (LB)	15	64	36							Hein et al. (1997)
		4-HB (LB)	22	66	36							
DH5α	$phaA_{Re}$, $phaB_{Re}$, $phaC_{Re}$, $4hbD_{Ck}$, $sucD_{Ck}$, $orfZ_{Ck}$	Glucose	45	99	1							Valentin and Dennis (1997)
GCSC6576	$phaA_{Re}$, $phaB_{Re}$, $phaC_{Re}$	Lactose + whey	80	100								Wong and Lee (1998)

Tab. 5 (cont.)

Host	Introduced genes	Carbon source	PHA content (wt.%)	Composition (mol.%)								Reference
				3HB	4HB	3HV	4HV	3HHx	3HO	3HD	3HDD	
S17-1	$phaE_{Syn}$, $phaC_{Syn}$, $phaA_{Re}$, $phaB_{Re}$	Glucose (LB)	13	100								Hein and Steinbüchel (1998)
BL21	$phaA_{Al}$, $phaB_{Al}$, $phaC_{Al}$	Glucose	52	100								Genser et al. (1998)
XL1-Blue	$phaA_{Al}$, $phaB_{Al}$, $phaC_{Al}$	Glucose	73	100								Choi et al. (1998)
LS5218 [*fadR*, *atoC* (Con)]	$phaP_{Ac}$, $phaC_{Ac}$, $phaJ_{Ac}$	C12	38	83				17				Fukui et al. (1999)
JMU193 (*fadR*, *fabB*)	$phaC1_{Po}$, $alkS_{Po}$, 'tes_{Ec}	Gluconate	2.0					27	67	6		Klinke et al. (1999)
XL1-Blue	$phaA_{Pd}$, $phaB_{Pd}$, $phaC_{Pd}$, $phaP_{Pd}$	Lactate (LB)	43	100								Maehara et al. (1999)
	$phaA_{Pd}$, $phaB_{Pd}$, $phaC_{Pd}$, $phaP_{Pd}$, $phaR_{Pd}$		44	100								
XL1-Blue	$phaA_{Al}$, $phaB_{Al}$, $phaC_{Al}$	Glucose	77	100								Lee et al. (1999)
HB101	$phaC_{Ac}$, $fabG_{Ec}$	C12	8.4	86				14				Taguchi et al. (1999)
	$phaC1_{Ps6}$, $fabG_{Ec}$		8.1					12	80	8		
193MC1 (xyl/Pm::$phaC1_{Po}$)	($phaC1_{Po}$ in chromosomal DNA)	Palmitic acid	12					14	66	20		Prieto et al. (1999b)
JM109	$phaE_{Tp}$, $phaC_{Tp}$, buk_{Ca}, ptb_{Ca}	Glucose + 3HB + 4-HB	68.7	88	12							Liu and Steinbüchel (2000)
		Glucose + 3HB + 4-HV	64.0	94			6					
		Glucose + 3-HB + 4-HB + 4-HV	68.4	85	13		2					
LS5218	$phaC_{Ac}$, $phaJ1_{Pa}$	C12	28	91				9				Tsuge et al. (2000)
	$phaC_{Ac}$, $phaJ2_{Pa}$		25	90				10				
	$phaC1_{Ps6}$, $phaJ1_{Pa}$		29	10				78	7	3	2	
	$phaC1_{Ps6}$, $phaJ2_{Pa}$		14	8				45	30	11	6	
DH5α	$phaE_{Tp}$, $phaC_{Tp}$, $phaJ_{Rsr}$	Oleic acid	4.5	96		2		2				Reiser et al. (2000)

Tab. 5 (cont.)

Host	Introduced genes	Carbon source	PHA content (wt.%)	Composition (mol.%)								Reference
				3HB	4HB	3HV	4HV	3HHx	3HO	3HD	3HDD	
XL1-Blue	$phaA_{Re}$, $phaB_{Re}$, $phaC_{Re}$	Soy waste	28	100								Hong et al. (2000)
CGSC4401	$phaA_{Al}$, $phaB_{Al}$, $phaC_{Al}$	Lactose + whey	76	100								Ahn et al. (2000)
XL10	$phaE_{Syn}$, $phaC_{Syn}$, $phaA_{Syn}$, $phaB_{Syn}$	glucose	13	100								Taroncher-Oldenburg et al. (2000)

*, High-molecular weight PHB (number average molecular weight $= 2 \times 10^7$). Al, *Alcaligenes latus*; Ca, *Clostridium acetobutylicum*; Ck, *Clostridium kluyveri*; Nc, *Nocardia corallina*; Pd, *Paracoccus denitrificans*; Pa, *Pseudomonas aeruginosa*; Po, *Pseudomonas oleovorans*; Ps6, *Pseudomonas* sp. 61 – 3; Re, *Ralstonia eutropha*; Rsr, *Rhodospirillum rubrum*; Tp, *Thicapsa pfennigii*; Syn, *Synechocystis* PCC6803 are indicated as hosts. Abbreviations of the genes are as follows: *4hbD*, 4-hydroxybutyrate dehydrogenase; *buk*, butyrate kinase; *orfZ*, 4-hydroxybutyric acid-CoA transferase; *phaA*, β-ketothiolase; *phaB*, acetoacetyl-CoA reductase; *phaC*, PHA synthase; *phaD*, regulator involved in mcl-PHA synthesis; *phaE*, subunit of PHA synthase; *phaF*, mcl-PHA GAP 1; *phaI*, mcl-PHA GAP 2; *phaG*, (*R*)-3-hydroxyacyl-ACP-CoA transferase; *phaJ*, (*R*)-specific enoyl-CoA hydratase; *phaP*, phasin (scl-PHA GAP); *phaR*, putative regulator involved in scl-PHA synthesis; *ptb*, phosphotransbutyrylase; *sucD*, succinate semialdehyde dehydrogenase. CST, corn steep liquor; LB, Luria-Bertani medium; C3, propionate; C12, dodecanoate; 3-HB, 3-hydroxybutyrate; 4-HB, 4-hydroxybutyrate; 4-HHx, 4-hydroxyhexanoate are indicated as carbon sources. 3HB, 3-hydroxybutyrate; 4HB, 4-hydroxybutyrate; 3HV, 3-hydroxyvalerate; 4HV, 4-hydroxyvalerate; 3HHx, 3-hydroxyhexanoate; 3HO, 3-hydroxyoctanoate; 3HD, 3-hydroxydecanoate; 3HDD, 3-hydroxydodecanoate are indicated as monomer composition.

have also been constructed to produce poly(3HB) homopolymer (Choi et al., 1998; Genser et al., 1998; Taroncher-Oldenburg et al., 2000). PhaR, which in *Paracoccus denitrificans* plays the role of a putative regulator involved in PHA synthesis, was also studied for PHA accumulation by expression in *E. coli* (Maehara et al., 1999).

Several examples of metabolic engineering have been carried out to produce copolymers consisting of (*R*)-3-hydroxyalkanoates (3HAs) with different chain lengths in recombinant *E. coli*. The first example of PHA copolymer production in recombinant *E. coli* was demonstrated by Slater et al. (1992). The *E. coli* LS5218 strain that lacked *fadR*, a regulator gene of β-oxidation, and harbored $phaA_{Re}$, $phaB_{Re}$, and $phaC_{Re}$, possessed an ability to incorporate 3HB and 3-hydroxyvalerate (3HV) to yield a poly(3HB-*co*-3HV) copolymer. The monomer composition of PHA in bacteria depends not only on the specificity of the monomer-supplying enzyme, but also on the specificity of the PHA synthase. *E. coli* LS1298 which contained *phaC1* from *Pseudomonas aeruginosa*, accumulated poly(HA_{mcl}) from fatty acids. Copolymer synthesis in *fad* system mutants which harbored only the *phaC* gene indicated that *E. coli* possesses the capability of supplying (*R*)-3HA-CoA towards PHA synthase.

For the incorporation of 4HB into PHA, *phaC* from *R. eutropha* and *orfZ* encoding 4-hydroxybutyric acid-CoA transferase from *Clostridium kluyveri* were introduced into *E. coli* XL1-Blue. The recombinant was cultivated in a medium containing 4HB as the sole carbon source, and this resulted in the accumulation of poly(3HB-*co*-36 mol.% 4HB) (Hein et al., 1997). In addition, a recombinant *E. coli* produced poly(3HB-*co*-4HB) from glucose (Valentin and Dennis, 1997), this being achieved by introducing the succinate degradation pathway from *C. kluyveri* with the PHB synthesis pathway from *R. eutropha*. Liu and Steinbüchel (2000) designed a new pathway for PHA synthesis from hydroxyfatty acids by employing the genes for butyrate kinase (Buk) and phosphotransbutyrylase (Ptb) from *Clostridium acetobutylicum*, and the genes for PHA synthase from *Thiocapsa pfennigii*.

Two remarkable enzymes were identified as molecular converters of intermediates (via the β-oxidation pathway) into (*R*)-3HA-CoA. One enzyme was the (*R*)-specific enoyl-CoA hydratase encoded by *phaJ*. The *phaJ* genes were isolated from *A. caviae*, *P. aeruginosa*, and *R. rubrum* (Fukui et al., 1999; Reiser et al., 2000; Tsuge et al., 2000). The other enzyme proved to be one involved in fatty acid biosynthesis, FabG, which reduces 3-ketoacyl-CoA with NADPH into (*R*)-3HA-CoA (Taguchi et al., 1999). The FabG enzyme was shown to be an isomorph of the NADPH-dependent acetoacetyl-CoA reductase encoded by *phaB*.

5.2 Metabolic Improvement for PHA Production in *Ralstonia eutropha* PHB-4 and Pseudomonads

R. eutropha and several *Pseudomonas* strains are natural producers of PHA, and possess both PHA synthases and a monomer-supplying system. Within these bacteria, metabolic improvements related to PHA production have been carried out using a range of genetic engineering techniques.

The metabolic improvements for PHA production in *R. eutropha* with a disrupted *phaC* gene, together with that for some *Pseudomonas* strains, are shown in Tables 6 and 7. The PHA-negative mutants (i.e., *R. eutropha* PHB-4; *P. putida* GPp104) are convenient hosts for the isolation of PHA synthase genes from various PHA-accumulating bacteria. In fact, many examples listed

Tab. 6 Metabolic improvements for PHA production in *Ralstonia eutropha* PHB-4

Introduced genes	Carbon source	PHA content (wt%)	Composition (mol.%)									Reference
			3HB	4HB	3HV	3HHx	4HHx	3HHp	3HO	3HD	3HDD	
$phaC_{Rcr}$	Fructose	18	100									Pieper and Steinbüchel (1992)
$phaC_{Rs}$	Gluconate	80	100									Hustede et al. (1992)
$phaC_{Rsr}$	Gluconate	70	100									
$PhaC_{Av}$	Fructose	38	100									Liebergesell and Steinbüchel (1992)
$phaC1_{Pa}$, $phaC2_{Pa}$, $phaD_{Pa}$*	C5	4	9		81			10				Timm and Steinbüchel (1992)
	C8	4			7			93				
$phaA_{Tv}$, $phaC_{Tv}$, $phaE_{Tv}$	C3	34	42		58							Liebergesell et al. (1993)
$phaC_{Lr}$*	C3	41	44		56							
$PhaC_{Av}$*	C3	9	56		44							
$phaC_{Tp}$*	C3	26	30		70							
$phaC_{Rsr}$*	C3	2	71		29							
$phaC_{Rs}$*	C3	11	63		37							
$phaC_{Tp}$	4-HHx	49	96			2	2					Valentin et al. (1994)
$phaC_{PsG}$*	Gluconate	25	100									Timm et al. (1994)
	C8	70	100									
$phaC_{PsA}$*	C4	17	99						1			Lee et al. (1995)
	4-HB	6	79						4	17		
	C8	6	15			9			72	4		
	Gluconate	8	90							10		
$phaC_{Ac}$	C5	52	38		62							Fukui et al. (1997)
	C7	70	42		48			10				
	C9	78	66		30			4				
$phaC_{Ac}$	Palm oil	81	96			4						Fukui et al. (1998)
	Oleate	70	96			4						
$phaB_{Re}$, $phaC_{Re}$	C6	49	90			10						Dennis et al. (1998)
	C7	36	12		88							
	C8	46	90			10						

Tab. 6 (cont.)

Introduced genes	*Carbon source*	*PHA content (wt%)*	*Composition (mol.%)*									*Reference*
			3HB	*4HB*	*3HV*	*3HHx*	*4HHx*	*3HHp*	*3HO*	*3HD*	*3HDD*	
phaA$_{Ca}$, *phaC*$_{Ca}$	C5	27	56		44							Sudesh et al. (1998)
	4-HB	5	86	14								
phaC1$_{Ps6}$, *phaC2*$_{Ps6}$, *phaD*$_{Ps6}$, *phaZ*$_{Ps6}$	C8	6	92						8			Matsusaki et al. (1998)
	C12	4	91						3	3	3	
phaC$_{Ac}$, *phaJ*$_{Ac}$, *phaP*$_{Ac}$	C4	45	99			1						Kichise et al. (1999)
	C5	15	5		95							
	C6	11	89			11						
	C7	56	33		61			6				

*Introduced fragment was not sequenced completely. Ac, *Aeromonas cavia*; Av, *Allochromatium vinosum*; Lr, *Lamprocystis roseopersicina*; Nc, *Nocardia corallina*; Re, *Ralstonia eutropha*; Rsr, *Rhodospirillum rubrum*; Rs, *Rhodobactor shaeroides*; Rcr, *Rhodococcus ruber*; Pa, *Pseudomonas aeruginosa*; Pc, *Pseudomonas citronellolis*; Pm, *Pseudomonas mendocina*; Po, *Pseudomonas oleovorans*; PsA, *Pseudomonas* sp. A33; PsG; *Pseudomonas* sp. GP4BH1; Ps6, *Pseudomonas* sp. 61–3; Tp, *Thiocapsa pfennigii*; Tv, *Thiocystis violacea*. Abbreviations of the genes are as follows: *phaA*, β-ketothiolase; *phaB*, acetoacetyl-CoA reductase; *phaC*, PHA synthase; *phaD*, regulator involved in mcl-PHA synthesis; *phaE*, subunit of PHA synthase; *phaF*, mcl-PHA GAP 1; *phaI*, mcl-PHA GAP 2; *phaG*, (*R*)-3-hydroxyacyl-ACP-CoA transferase; *phaJ*, (*R*)-specific enoyl-CoA hydratase; *phaP*, phasin (scl-PHA GAP); *phaR*, putative regulator involved in scl-PHA synthesis *phaZ*, intracellular PHA depolymerase. C3, propionate; C4, butyrate; C5, valerate; C6, hexanoate; C7, heptanoate; C8, octanoate; C9, nonanoate; C12, dodecanoate; 4-HB, 4-hydroxybutyrate; 4-HHx, 4-hydroxyhexanoate are indicated as carbon sources. 3HB, 3-hydroxybutyrate; 4HB, 4-hydroxybutyrate; 3HV, 3-hydroxyvalerate; 3HHx, 3-hydroxyhexanoate; 4HHx, 4-hydroxyhexanoate; 3HHp, 3-hydroxyheptanoate; 3HO, 3-hydroxyoctanoate; 3HD, 3-hydroxydecanoate; 3HDD, 3-hydroxydodecanoate are indicated as monomer composition.

Tab. 7 Metabolic improvements for PHA production in recombinant pseudomonads

Host	*Introduced genes*	*Carbon source*	*PHA content (wt%)*	*Composition (mol.%)*					*Reference*
				3HB	*3HHx*	*3HO*	*3HD*	*3HDD*	
Pseudomonas putida GPp104	*phaC1*$_{Po}$	C10	29		7	49	45		Huisman et al. (1991)
(PHA-negative mutant)	*phaC2*$_{Po}$	C10	36		4	51	45		
	phaC1$_{Pa}$, *phaD*$_{Pa}$*	Gluconate	32		3	54	43		Timm and Steinbüchel (1992)
		C8	41		7	93			
	phaC2$_{Pa}$*	Gluconate	26		5	34	56	5	
		C8	16		5	95			
	phaA$_{Tv}$, *phaC*$_{Tv}$, *phaE*$_{Tv}$	C8	14	99	1				Liebergesell et al. (1993)
	phaC$_{Tp}$*	C8	2	100					
	phaC$_{Av}$*	C8	19	85	15				
	phaC$_{Lr}$*	C8	22	49	47	4			
	phaC$_{Rrub}$*	C8	1	100					
	phaC$_{Rsh}$*	C8	16	96	4				
	phaC1$_{Pc}$*	Gluconate	24		3	18	69	10	Timm et al. (1994)
		C8	47		5	92	3		
	phaC1$_{Pm}$*	Gluconate	0						
		C8	40		6	94			
	phaC1$_{PsG}$*	Gluconate	21		5	12	70	13	
		C8	62		7	90	3		
	phaC1$_{PsA}$*	Gluconate	29	2	1	6	64	28	Lee et al. (1995)
		C8	89		6	90	3	1	
	phaC$_{Ac}$, *phaJ*$_{Ac}$, *phaP*$_{Ac}$	Gluconate	4	71	29				Fukui and Doi (1997)
		C6	38	60	40				
		C8	48	69	31				
	phaC$_{Nc}$	C4	2	93	6	1			Hall et al. (1998)
		C6	31	13	87				
		C8	32	25	71	4			

Tab. 7 (cont.)

Host	*Introduced genes*	*Carbon source*	*PHA content (wt%)*	*Composition (mol.%)*					*Reference*
				3HB	*3HHx*	*3HO*	*3HD*	*3HDD*	
Pseudomonas putida GPp104 (PHA-negative mutant)	$phbA_{\mathrm{Ps6}}$, $phbB_{\mathrm{Ps6}}$, $phbC_{\mathrm{Ps6}}$, $phbR_{\mathrm{Ps6}}$	Gluconate	20	100					Matsusaki et al. (1998)
	$phaC1_{\mathrm{Ps6}}$	Gluconate	9		2	15	58	25	
		C8	43	3	16	77	2		
	$phaC2_{\mathrm{Ps6}}$, $phaD_{\mathrm{Ps6}}$	Gluconate	10		1	13	62	24	
		C8	6		10	72	18		
	$phaC1_{\mathrm{Ps6}}$, $phbA_{\mathrm{Re}}$, $phbB_{\mathrm{Re}}$	Gluconate	10	49	14	6	30	15	Matsusaki et al. (2000b)
		C8	18	9	9	75	1	1	
Pseudomonas oleovorans	$phbC_{\mathrm{Re}}$	Gluconate	11	100					Timm and Steinbüchel (1990)
		C8	77	47	5	48	1		
Pseudomonas sp. 61 – 3 (*phbC*$^-$)	$phbA_{\mathrm{Re}}$, $phbB_{\mathrm{Re}}$, $phaC1_{\mathrm{Ps6}}$	Gluconate	45	92		1	3	4	Matsusaki et al. (2000a)
Pseudomonas fragi	$phaC1_{\mathrm{Pa}}$, $phaG_{\mathrm{Pp}}$	Gluconate	9		2	22	63	13	Fiedler et al. (2000)
		Oleate	6		11	43	33	13	

*Introduced fragment was not sequenced completely. Ac, *Aeromonas cavia*; Av, *Allochromatium vinosum*; Lr, *Lamprocystis roseopersicina*; Nc, *Nocardia corallina*; Re, *Ralstonia eutropha*; Rsr, *Rhodospirillum rubrum*; Rs, *Rhodobactor shaeroides*; Pa, *Pseudomonas aeruginosa*; Pc, *Pseudomonas citronellolis*; Pm, *Pseudomonas mendocina*; Po, *Pseudomonas oleovorans*; PsA, *Pseudomonas* sp. A33; PsG; *Pseudomonas* sp. GP4BH1; Ps6, *Pseudomonas* sp. 61 – 3; Tp, *Thiocapsa pfennigii*; Tv, *Thiocystis violacea*. Abbreviations of the genes are as follows: *phaA*, β-ketothiolase; *phaB*, acetoacetyl-CoA reductase; *phaC*, PHA synthase; *phaD*, regulator involved in mcl-PHA synthesis; *phaE*, subunit of PHA synthase; *phaF*, mcl-PHA GAP 1; *phaI*, mcl-PHA GAP 2; *phaG*, (*R*)-3-hydroxyacyl-ACP-CoA transferase; *phaJ*, (*R*)-specific enoyl-CoA hydratase; *phaP*, phasin (scl-PHA GAP); *phaR*, putative regulator involved in scl-PHA synthesis; *phaZ*, intracellular PHA depolymerase. C4, butyrate; C6, hexanoate; C8, octanoate; C10, decanoate; C12, dodecanoate are indicated as carbon sources.3HB, 3-hydroxybutyrate; 4HB, 4-hydroxybutyrate; 3HV, 3-hydroxyvalerate; 3HHx, 3-hydroxyhexanoate; 4HHx, 4-hydroxyhexanoate; 3HHp, 3-hydroxyheptanoate; 3HO, 3-hydroxyoctanoate; 3HD, 3-hydroxydecanoate; 3HDD, 3-hydroxydodecanoate are indicated as monomer composition.

in Tables 6 and 7 are used to clone *phaC* genes, and to confirm the specificity of the gene products.

Overexpression of $phaB_{Re}$ and $phaC_{Re}$ was carried out in *R. eutropha* PHB-4 (Dennis et al., 1998). The recombinant $PHB^{-}4$ strain produced PHA copolymers consisting of 3HB and other 3HAs from various fatty acids, suggesting that $PhaC_{Re}$ is capable of incorporating various 3HAs and that its specificity is relatively broad. Kichise et al. (1999) investigated the composition of PHA produced in *R. eutropha* PHB-4 harboring $phaC_{Ac}$, $phaJ_{Ac}$, and $phaP_{Ac}$, and found that introduction of the *phaJ* gene led to an increase in the content of 3-hydroxyvalerate and 3-hydroxyhexanoate monomer units in the accumulated PHA. The introduction of *A. caviae* PHA synthetic genes into *P. putida* GPp104 strain achieved the production of poly(3HB-*co*-mcl-3HA), not only from fatty acids but also from gluconate (Fukui et al., 1998).

Pseudomonas sp. 61–3 was isolated as a novel bacterium which accumulated both poly(3HB) homopolymer and poly(3HB-*co*-mcl-3HA) copolymer (Abe et al., 1994), and this in turn led to the genes involved in poly(3HB) and poly(3HB-*co*-3HA) synthesis being cloned (Matsusaki et al., 1998). Matsusaki and colleagues also constructed a recombinant *Pseudomonas* sp. 61–3 harboring its PHA synthetic gene and *R. eutropha* PHA synthetic genes to incorporate more than 90 mol.% 3HB into PHA, and characterized the mechanical properties of the resulting poly(3HB-*co*-6 mol.% mcl-3HA) (Matsusaki et al., 2000a,b). The mechanical properties of the PHA were the same as low-density polyethylene. Recently, the first example of *phaG* gene expression in a nonPHA-producing bacterium, *Pseudomonas fragi*, was reported, the production of poly(HA_{mcl}) in this organism being accomplished by the introduction of *P. putida phaG* and *P. aeruginosa phaC1* (Fiedler et al., 2000).

6 Outlook and Perspectives

In recent years a number of PHA synthesis genes have been isolated from natural PHA producers, and introduced into *E. coli*, *R. eutropha*, and pseudomonads in order to investigate metabolic engineering techniques with regard to PHA production. Although these genes are essential for PHA formation, PHA synthesis activity is also seen to be affected by other factors such as the enzymes of the central pathways, global metabolic regulation, and control of the surface of PHA granules. Hence, a survey of the molecular dynamics, together with metabolic profiling of PHA-accumulating bacteria, will be necessary in order to create the next generation of PHA-producing bacteria.

In this respect, the use of new biotechnological techniques will permit the controlled expression of PHA synthesis genes in recombinant organisms, and consequently it will also be possible to modify the specificities of the PHA synthesis enzymes. Moreover, in future the information derived from genetically engineered, PHA-producing bacteria may be applied to the development of PHA production in transgenic plants, and this approach holds great promise as a potential (and highly cost-effective) means of producing PHA from carbon dioxide.

7 References

Abe, H., Doi, Y., Fukushima, T., Eya, H. (1994) Biosynthesis from gluconate of a random copolyester consisting of 3-hydroxybutyrate and medium-chain-length 3-hydroxyalkanoates by *Pseudomonas* sp. 61–3, *Int. J. Biol. Macromol.* **16**, 115–119.

Ahn, W. S., Park, S. J., Lee, S. Y. (2000) Production of poly(3-hydroxybutyrate) by fed-batch culture of recombinant *Escherichia coli* with a highly concentrated whey solution, *Appl. Environ. Microbiol.* **66**, 3624–3627.

Andersson, S. G., Zomorodipour, A., Andersson, J. O., Sicheritz-Ponten, T., Alsmark, U. C., Podowski, R. M., Naslund, A. K., Eriksson, A. S., Winkler, H. H., Kurland, C. G. (1998) The genome sequence of *Rickettsia prowazekii* and the origin of mitochondria, *Nature* **396**, 133–140.

Cevallos, M. A., Encarnacion, S., Leija, A., Mora, Y., Mora, J. (1996) Genetic and physiological characterization of a *Rhizobium etli* mutant strain unable to synthesize poly-beta-hydroxybutyrate, *J. Bacteriol.* **178**, 1646–1654.

Choi, J. I., Lee, S. Y., Han, K. (1998) Cloning of the *Alcaligenes latus* polyhydroxyalkanoate biosynthesis genes and use of these genes for enhanced production of poly(3-hydroxybutyrate) in *Escherichia coli, Appl. Environ. Microbiol.* **64**, 4897–4903.

Clemente, T., Shah, D., Tran, M., Stark, D., Padgette, S., Dennis, D., Bruckener, K., Steinbüchel, A., Mitsky, T. (2000) Sequence of PHA synthase gene from two strains of *Rhodospirillum rubrum* and in vivo substrate specificity of four PHA synthases across two heterologous expression systems, *Appl. Microbiol. Biotechnol.* **53**, 420–429.

Dennis, D., McCoy, M., Stangl, A., Valentin, H. E., Wu, Z. (1998) Formation of poly(3-hydroxybutyrate-*co*-3-hydroxyhexanoate) by PHA synthase from *Ralstonia eutropha, J. Biotechnol.* **64**, 177–186.

Fiedler, S., Steinbüchel, A., Rehm, B. H. (2000) PhaG-mediated synthesis of poly(3-hydroxyalkanoates) consisting of medium-chain-length constituents from nonrelated carbon sources in recombinant *Pseudomonas fragi, Appl. Environ. Microbiol.* **66**, 2117–2124.

Franchi, E., Tosi, C., Pedroni, P. (1997) GenBank accession No. X97200.

Fukui, T., Doi, Y. (1997) Cloning and analysis of the poly(3-hydroxybutyrate-*co*-3-hydroxyhexanoate) biosynthesis genes of *Aeromonas caviae, J. Bacteriol.* **179**, 4821–4830.

Fukui, T., Ito, M., Saito, T., Tomita, K. (1987) Purification and characterization of NADP-linked acetoacetyl-CoA reductase from *Zoogloea ramigera* I-16-M, *Biochim. Biophys. Acta* **917**, 365–371.

Fukui, T., Kichise, T., Yoshida, Y., Doi, Y. (1997) Biosynthesis of poly(3-hydroxybutyrate-*co*-3-hydroxyvalerate-*co*-3-hydroxyheptanoate) terpolymers by recombinant *Alcaligenes eutrophus, Biotechnol. Lett.* **19**, 1093–1097.

Fukui, T., Shiomi, N., Doi, Y. (1998) Expression and characterization of (*R*)-specific enoyl coenzyme A hydratase involved in polyhydroxyalkanoate biosynthesis by *Aeromonas caviae, J. Bacteriol.* **180**, 667–673.

Fukui, T., Yokomizu, S., Kobayashi, S., Doi, Y. (1999) Co-expression of polyhydroxyalkanoate synthase and (*R*)-enoyl-CoA hydratase genes of *Aeromonas caviae* establishes copolyester biosynthesis pathway in *Escherichia coli, FEMS Microbiol. Lett.* **170**, 69–75.

García, B., Olivera, E. R., Minambres, B., Fernandez-Valverde, M., Canedo, L. M., Prieto, M. A., Garcia, J. L., Martinez, M., Luengo, J. M. (1999) Novel biodegradable aromatic plastics from a bacterial source. Genetic and biochemical studies on a route of the phenylacetyl-CoA catabolon, *J. Biol. Chem.* **274**, 29228–29241.

Genser, K. F., Renner, G., Schwab, H. (1998) Molecular cloning, sequencing and expression in *Escherichia coli* of the poly(3-hydroxyalkanoate) synthesis genes from *Alcaligenes latus* DSM1124, *J. Biotechnol.* **64**, 125–135.

Hall, B., Baldwin, J., Rhie, H. G., Dennis, D. (1998) Cloning of the *Nocardia corallina* polyhydroxyalkanoate synthase gene and production of poly-(3-hydroxybutyrate-*co*-3-hydroxyhexanoate) and poly-(3-hydroxyvalerate-*co*-3-hydroxyheptanoate), *Can. J. Microbiol.* **44**, 687–691.

Handrick, R., Reinhardt, S., Jendrossek, D. (2000) Mobilization of poly(3-hydroxybutyrate) in *Ralstonia eutropha*, *J. Bacteriol.* **182**, 5916–5918.

Hanley, S. Z., Pappin, D. J. C., Rahman, D., White, A. J., Elborough, K. M., Slabas, A. R. (1999) Re-evaluation of the primary structure of *Ralstonia eutropha* phasin and implications for polyhydroxyalkanoic acid granule binding, *FEBS Lett.* **447**, 99–105.

Haywood, G. W., Anderson, A. J., Chu, L., Dawes, E. A. (1988a) Characterization of two 3-ketothiolases in the polyhydroxyalkanoate synthesizing organism *Alcaligenes eutrophus*, *FEMS Microbiol. Lett.* **52**, 91–96.

Haywood, G. W., Anderson, A. J., Chu, L., Dawes, E. A. (1988b) The role of NADH- and NADPH-linked acetoacetyl-CoA reductases in the poly-3-hydroxyalkanoate synthesizing organism *Alcaligenes eutrophus*, *FEMS Microbiol. Lett.* **52**, 259–264.

Haywood, G. W., Anderson, A. J., Ewing, D. F., Dawes, E. A. (1990) Accumulation of a polyhydroxyalkanoate containing primarily 3-hydroxydecanoate from simple carbohydrate substrates by *Pseudomonas* sp. strain NCIMB 40135, *Appl. Environ. Microbiol.* **56**, 3354–3359.

Heidelberg, J. F., Eisen, J. A., Nelson, W. C., Clayton, R. A., Gwinn, M. L., Dodson, R. J., Haft, D. H., Hickey, E. K., Peterson, J. D., Umayam, L., Gill, S. R., Nelson, K. E., Read, T. D., Tettelin, H., Richardson, D., Ermolaeva, M. D., Vamathevan, J., Bass, S., Qin, H., Dragoi, I., Sellers, P., McDonald, L., Utterback, T., Fleishmann, R. D., Nierman, W. C., White, O. (2000) DNA sequence of both chromosomes of the cholera pathogen *Vibrio cholerae*, *Nature* **406**, 477–483.

Hein, S., Söhling, B., Gottschalk, G., Steinbüchel, A. (1997) Biosynthesis of poly(4-hydroxybutyric acid) by recombinant strains of *Escherichia coli*, *FEMS Microbiol. Lett.* **153**, 411–418.

Hein, S., Tran, H., Steinbüchel, A. (1998) *Synechocystis* sp. PCC6803 possesses a two-component polyhydroxyalkanoic acid synthase similar to that of anoxygenic purple sulfur bacteria, *Arch. Microbiol.* **170**, 162–170.

Hisano, T., Fukui, T., Iwata, T., Doi, Y. (2001) Crystallization and preliminary X-ray analysis of (*R*)-specific enoyl-CoA hydratase from *Aeromonas caviae* involved in polyhydroxyalkanoate biosynthesis, *Acta Crystallogr. Section D* **57**, 145–147.

Hoffmann, N., Steinbüchel, A., Rehm, B. H. A. (2000a) Homologous functional expression of cryptic *phaG* from *Pseudomonas oleovorans* establishes the transacylase-mediated polyhydroxyalkanoate biosynthetic pathway, *Appl. Microbiol. Biotechnol.* **54**, 665–670.

Hoffmann, N., Steinbüchel, A., Rehm, B. H. A. (2000b) The *Pseudomonas aeruginosa phaG* gene product is involved in the synthesis of polyhydroxyalkanoic acid consisting of medium-chain-length constituents from non-related carbon sources, *FEMS Microbiol. Lett.* **184**, 253–259.

Hong, K., Leung, Y. C., Kwok, S. Y., Law, K. H., Lo, W. H., Chua, H., Yu, P. H. (2000) Construction of recombinant *Escherichia coli* strains for polyhydroxybutyrate production using soy waste as nutrient, *Appl. Biochem. Biotechnol.*, **84-86**, 381–390.

Hong, S. K., Park, J. S., Park, H. C., Lee, Y. H., Huh, T. L. (1999) Cloning and characterization of the gene for polyhydroxyalkanoate synthase from *Alcaligenes latus*, GenBank accession No. AF004933.

Huijberts, G. N., Eggink, G., de Waard, P., Huisman, G. W., Witholt, B. (1992) *Pseudomonas putida* KT2442 cultivated on glucose accumulates poly(hydroxyalkanoates) consisting of saturated and unsaturated monomers, *Appl. Environ. Microbiol.* **58**, 536–544.

Huijberts, G. N., de Rijk, T. C., de Waard, P., Eggink, G. (1994) ^{13}C nuclear magnetic resonance studies of *Pseudomonas putida* fatty acid metabolic routes involved in poly(3-hydroxyalkanoate) synthesis, *J. Bacteriol.* **176**, 1661–1666.

Huisman, G. W., Wonink, E. W., Meima, R., Kazemier, B., Tersptra, P., Witholt, B. (1991) Metabolism of poly(3-hydroxyalkanoates) (PHAs) by *Pseudomonas oleovorans*. Identification and sequences of genes and function of the encoded proteins in the synthesis and degradation of PHA, *J. Biol. Chem.* **266**, 2191–2198.

Hustede, E., Steinbüchel, A. (1993) Characterization of the polyhydroxyalkanoate synthase gene locus of *Rhodobacter sphaeroides*, *Biotechnol. Lett.* **15**, 709–714.

Hustede, E., Steinbüchel, A., Schlegel, H. G. (1992) Cloning of poly(3-hydroxybutyric acid) synthase genes of *Rhodobacter sphaeroides* and *Rhodospirillum rubrum* and heterologous expression in *Alcaligenes eutrophus*, *FEMS Microbiol. Lett.* **72**, 285–290.

Kalousek, S., Lubitz, W. (1995) High-level poly(beta-hydroxybutyrate) production in recombinant *Escherichia coli* in sugar-free, complex medium, *Can. J. Microbiol.* **41**, 216–221.

Kaneko, T., Sato, S., Kotani, H., Tanaka, A., Asamizu, E., Nakamura, Y., Miyajima, N., Hirosawa, M., Sugiura, M., Sasamoto, S., Kimura, T., Hosouchi, T., Matsuno, A., Muraki, A., Nakazaki, N., Naruo, K., Okumura, S., Shimpo, S., Takeuchi, C., Wada, T., Watanabe, A., Yamada, M., Yasuda, M., Tabata, S. (1996) Sequence analysis of the genome of the unicellular cyanobacterium *Synechocystis* sp. strain PCC6803. II. Sequence determination of the entire genome and assignment of potential protein-coding regions, *DNA Res.* **3**, 109–136.

Kichise, T., Fukui, T., Yoshida, Y., Doi, Y. (1999) Biosynthesis of polyhydroxyalkanoates (PHA) by recombinant *Ralstonia eutropha* and effects of PHA synthase activity on in vivo PHA biosynthesis, *Int. J. Biol. Macromol.* **25**, 69–77.

Kim, J.-H., Lee, J. K. (1997) Cloning, nucleotide sequence and expression of gene coding for poly-3-hydroxybutyric acid (PHB) synthase of *Rhodobacter sphaeroides* 2.4.1, *J. Microbiol. Biotechnol.* **7**, 229–236.

Klinke, S., Ren, Q., Witholt, B., Kessler, B. (1999) Production of medium-chain-length poly(3-hydroxyalkanoates) from gluconate by recombinant *Escherichia coli*, *Appl. Environ. Microbiol.* **65**, 540–548.

Klinke, S., de Roo, G., Witholt, B., Kessler, B. (2000) Role of *phaD* in accumulation of medium-chain-length poly(3-hydroxyalkanoates) in *Pseudomonas oleovorans*, *Appl. Environ. Microbiol.* **66**, 3705–3710.

Kolibachuk, D., Miller, A., Dennis, D. (1999) Cloning, molecular analysis, and expression of the polyhydroxyalkanoic acid synthase (*phaC*) gene from *Chromobacterium violaceum*, *Appl. Environ. Microbiol.* **65**, 3561–3565.

Kranz, R. G., Gabbert, K. K., Locke, T. A., Madigan, M. T. (1997) Polyhydroxyalkanoate production in *Rhodobacter capsulatus*: genes, mutants, expression, and physiology, *Appl. Environ. Microbiol.* **63**, 3003–3009.

Kusaka, S., Abe, H., Lee, S. Y., Doi, Y. (1997) Molecular mass of poly[(*R*)-3-hydroxybutyric acid] produced in a recombinant *Escherichia coli*, *Appl. Microbiol. Biotechnol.* **47**, 140–143.

Lageveen, R. G., Huisman, G. W., Preusting, H., Ketelaar, P., Eggink, G., Witholt, B. (1988) Formation of polyesters by *Pseudomonas oleovorans*: effect of substrates on formation and composition of poly-(*R*)-3-hydroxyalkanoates and poly-(*R*)-3-hydroxyalkenoates, *Appl. Environ. Microbiol.* **54**, 2924–2932.

Langenbach, S., Rehm, B. H. A., Steinbüchel, A. (1997) Functional expression of the PHA synthase gene *phaC1* from *Pseudomonas aeruginosa* in *Escherichia coli* results in poly(3-hydroxyalkanoate) synthesis, *FEMS Microbiol. Lett.* **150**, 303–309.

Lee, E. Y., Jendrossek, D., Schirmer, A., Choi, C. Y., Steinbüchel, A. (1995) Biosynthesis of copolyesters consisting of 3-hydroxybutyric acid and medium-chain-length 3-hydroxyalkanoic acids from 1,3-butanediol or from 3-hydroxybutyrate by *Pseudomonas* sp. A33, *Appl. Microbiol. Biotechnol.* **42**, 901–909.

Lee, I., Rhee, Y. H., Kim, J.-Y. (1996) Cloning and functional expression in *Escherichia coli* of the polyhydroxyalkanoate synthase (*phaC*) gene from *Alcaligenes* sp. SH-69, *J. Microbiol. Biotechnol.* **6**, 309–314.

Lee, I., Rhee, Y. H., Kim, J.-Y. (1997) Cloning and sequencing of phaB from *Alcaligenes* sp. SH-69, GenBank accession No. AF002014.

Lee, I., Rhu, H. M., Kim, M. W., Rhee, Y. H., Kim, J.-Y. (1998) Identification and heterologous expression of the polyhydroxyalkanoate biosynthesis genes from *Alcaligenes* sp. SH-69, *Biotechnol. Lett.* **20**, 969–975.

Lee, S. P., Do, V., Huisman, G. W., Peoples, O. P. (1996) PHB polymerase from *Zoogloea ramigera*, GenBank accession No. U66242.

Lee, S. Y., Chang, H. N. (1995) Production of poly(3-hydroxybutyric acid) by recombinant *Escherichia coli* strains: genetic and fermentation studies, *Can. J. Microbiol.* **41**, 207–215.

Lee, S. Y., Choi, J., Han, K., Song, J. Y. (1999) Removal of endotoxin during purification of poly(3-hydroxybutyrate) from gram-negative bacteria, *Appl. Environ. Microbiol.* **65**, 2762–2764.

Liebergesell, M., Steinbüchel, A. (1992) Cloning and nucleotide sequences of genes relevant for biosynthesis of poly(3-hydroxybutyric acid) in *Chromatium vinosum* strain D., *Eur. J. Biochem.* **209**, 135–150.

Liebergesell, M., Steinbüchel, A. (1993) Cloning and molecular analysis of the poly(3-hydroxybutyric acid) biosynthetic genes of *Thiocystis violacea*, *Appl. Microbiol. Biotechnol.* **38**, 493–501.

Liebergesell, M., Steinbüchel, A. (1996) New knowledge about the PHA-locus and P(3HB) granule-associated proteins in *Chromatium vinosum*, *Biotechnol. Lett.* **18**, 719–724.

Liebergesell, M., Mayer, F., Steinbüchel, A. (1993) Analysis of polyhydroxyalkanoic acid-biosynthesis genes of anoxygenic phototrophic bacteria reveals synthesis of a polyester exhibiting an unusual composition, *Appl. Microbiol. Biotechnol.* **40**, 292–300.

Liu, S. J., Steinbüchel, A. (2000) A novel genetically engineered pathway for synthesis of poly(hydroxyalkanoic acids) in *Escherichia coli*, *Appl. Environ. Microbiol.* **66**, 739–743.

Maehara, A., Ueda, S., Nakano, H., Yamane, T. (1999) Analyses of a polyhydroxyalkanoic acid granule-associated 16-kilodalton protein and its putative regulator in the *pha* locus of *Paracoccus denitrificans*, *J. Bacteriol.* **181**, 2914–2921.

Mandon, K., Michel-Reydellet, N., Encarnacion, S., Kaminski, P. A., Leija, A., Cevallos, M. A., Elmerich, C., Mora, J. (1998) Poly-beta-hydroxybutyrate turnover in *Azorhizobium caulinodans* is required for growth and affects *nifA* expression, *J. Bacteriol.* **180**, 5070–5076.

Matsumoto, K., Matsusaki, H., Taguchi, S., Seki, M., Doi, Y. (2001) Cloning and characterization of the *Pseudomonas* sp. 61–3 *phaG* gene involved in polyhydroxyalkanoate biosynthesis, *Biomacromolecules* **2**, 142–147.

Matsusaki, H., Manji, S., Taguchi, K., Kato, M., Fukui, T., Doi, Y. (1998) Cloning and molecular analysis of the poly(3-hydroxybutyrate) and poly(3-hydroxybutyrate-*co*-3-hydroxyalkanoate) biosynthesis genes in *Pseudomonas* sp. strain 61–3, *J. Bacteriol.* **180**, 6459–6467.

Matsusaki, H., Abe, H., Doi, Y. (2000a) Biosynthesis and properties of poly(3-hydroxybutyrate-*co*-3-hydroxyalkanoates) by recombinant strains of *Pseudomonas* sp. 61–3, *Biomacromolecules* **1**, 17–22.

Matsusaki, H., Abe, H., Taguchi, K., Fukui, T., Doi, Y. (2000b) Biosynthesis of poly(3-hydroxybutyrate-*co*-3-hydroxyalkanoates) by recombinant bacteria expressing the PHA synthase gene *phaC1* from *Pseudomonas* sp. 61–3, *Appl. Microbiol. Biotechnol.* **53**, 401–409.

McCool, G. J., Cannon, M. C. (1999) Polyhydroxyalkanoate inclusion body-associated proteins and coding region in *Bacillus megaterium*, *J. Bacteriol.* **181**, 585–592.

Moskowitz, G. J., Merrick, J. M. (1969) Metabolism of poly-β-hydroxybutyrate. Enzymatic synthesis of D(-)-β-hydroxybutyryl-Coenzyme A by an enoyl hydrase from *Rhodospirillum rubrum*, *Biochemistry* **8**, 2748–2755.

Palmer, M. A., Differding, E., Gamboni, R., Williams, S. F., Peoples, O. P., Walsh, C. T., Sinskey, A. J., Masamune, S. (1991) Biosynthetic thiolase from *Zoogloea ramigera*. Evidence for a mechanism involving Cys-378 as the active site base, *J. Biol. Chem.* **266**, 8369–8375.

Peoples, O. P., Sinskey, A. J. (1989a) Poly-β-hydroxybutyrate biosynthesis in *Alcaligenes eutrophus* H16. Characterization of the genes encoding β-ketothiolase and acetoacetyl-CoA reductase, *J. Biol. Chem.* **264**, 15293–15297.

Peoples, O. P., Sinskey, A. J. (1989b) Poly-β-hydroxybutyrate biosynthesis in *Alcaligenes eutrophus* H16. Identification and characterization of the PHB polymerase gene (*phbC*), *J. Biol. Chem.* **264**, 15298–15303.

Peoples, O. P., Masamune, S., Walsh, C. T., Sinskey, A. J. (1987) Biosynthetic thiolase from *Zoogloea ramigera*. III. Isolation and characterization of the structural gene, *J. Biol. Chem.* **262**, 97–102.

Pieper, U., Steinbuchel, A. (1992) Identification, cloning and sequence analysis of the poly(3-hydroxyalkanoic acid) synthase gene of the gram-positive bacterium *Rhodococcus ruber*, *FEMS Microbiol. Lett.* **75**, 73–79.

Pieper-Furst, U., Madkour, M. H., Mayer, F., Steinbuchel, A. (1994) Purification and characterization of a 14-kilodalton protein that is bound to the surface of polyhydroxyalkanoic acid granules in *Rhodococcus ruber*, *J. Bacteriol.* **176**, 4328–4337.

Ploux, O., Masamune, S., Walsh, C. T. (1988) The NADPH-linked acetoacetyl-CoA reductase from *Zoogloea ramigera*. Characterization and mechanistic studies of the cloned enzyme over-produced in *Escherichia coli*, *Eur. J. Biochem.* **174**, 177–182.

Prieto, M. A., Buhler, B., Jung, K., Witholt, B., Kessler, B. (1999a) PhaF, a polyhydroxyalkanoate-granule-associated protein of *Pseudomonas oleovorans* GPo1 involved in the regulatory expression system for *pha* genes, *J. Bacteriol.* **181**, 858–868.

Prieto, M. A., Kellerhals, M. B., Bozzato, G. B., Radnovic, D., Witholt, B., Kessler, B. (1999b) Engineering of stable recombinant bacteria for production of chiral medium-chain-length poly-3-hydroxyalkanoates, *Appl. Environ. Microbiol.* **65**, 3265–3271.

Qi, Q., Rehm, B., Steinbüchel, A. (2000) Assignment and characterization of PHB synthase in

Caulobacter crescentus, GenBank accession No. AY007313.

Rehm, B. H., Kruger, N., Steinbuchel, A. (1998) A new metabolic link between fatty acid de novo synthesis and polyhydroxyalkanoic acid synthesis. The PHAG gene from *Pseudomonas putida* KT2440 encodes a 3-hydroxyacyl-acyl carrier protein-coenzyme a transferase, *J. Biol. Chem.* **273**, 24044–24051.

Reiser, S. E., Mitsky, T. A., Gruys, K. J. (2000) Characterization and cloning of an (*R*)-specific trans-2,3-enoylacyl-CoA hydratase from *Rhodospirillum rubrum* and use of this enzyme for PHA production in *Escherichia coli*, *Appl. Microbiol. Biotechnol.* **53**, 209–218.

Ren, Q., Sierro, N., Kellerhals, M., Kessler, B., Witholt, B. (2000) Properties of engineered poly-3-hydroxyalkanoates produced in recombinant *Escherichia coli* strains, *Appl. Environ. Microbiol.* **66**, 1311–1312.

Rodrigues, M. F., Valentin, H. E., Berger, P. A., Tran, M., Asrar, J., Gruys, K. J., Steinbüchel, A. (2000) Polyhydroxyalkanoate accumulation in *Burkholderia* sp.: a molecular approach to elucidate the genes involved in the formation of two homopolymers consisting of short-chain-length 3-hydroxyalkanoic acids, *Appl. Microbiol. Biotechnol.* **53**, 453–460.

Saegusa, H., Shiraki, M., Kanai, C., Saito, T. (2001) Cloning of an intracellular poly[D(-)-3-hydroxybutyrate] depolymerase gene from *Ralstonia eutropha* H16 and characterization of the gene product, *J. Bacteriol.* **183**, 94–100.

Schembri, M. A., Bayly, R. C., Davies, J. K. (1994) Cloning and analysis of the polyhydroxyalkanoic acid synthase gene from an *Acinetobacter* sp.: evidence that the gene is both plasmid and chromosomally located, *FEMS Microbiol. Lett.* **118**, 145–152.

Schembri, M. A., Woods, A. A., Bayly, R. C., Davies, J. K. (1995) Identification of a 13-kDa protein associated with the polyhydroxyalkanoic acid granules from *Acinetobacter* spp., *FEMS Microbiol. Lett.* **133**, 277–283.

Schubert, P., Steinbüchel, A., Schlegel, H. G. (1988) Cloning of the *Alcaligenes eutrophus* genes for synthesis of poly-β-hydroxybutyric acid (PHB) and synthesis of PHB in *Escherichia coli*, *J. Bacteriol.* **170**, 5837–5847.

Schubert, P., Krüger, N., Steinbüchel, A. (1991) Molecular analysis of the *Alcaligenes eutrophus* poly(3-hydroxybutyrate) biosynthetic operon: identification of the N terminus of poly-(3-hydroxybutyrate) synthase and identification of the promoter, *J. Bacteriol.* **173**, 168–175.

Slater, S. C., Voige, W. H., Dennis, D. E. (1988) Cloning and expression in *Escherichia coli* of the *Alcaligenes eutrophus* H16 poly-β-hydroxybutyrate biosynthetic pathway, *J. Bacteriol.* **170**, 4431–4436.

Slater, S., Gallaher, T., Dennis, D. (1992) Production of poly-(3-hydroxybutyrate-co-3-hydroxyvalerate) in a recombinant *Escherichia coli* strain, *Appl. Environ. Microbiol.* **58**, 1089–1094.

Slater, S., Houmiel, K. L., Tran, M., Mitsky, T. A., Taylor, N. B., Padgette, S. R., Gruys, K. J. (1998) Multiple β-ketothiolases mediate poly(β-hydroxyalkanoate) copolymer synthesis in *Ralstonia eutropha*, *J. Bacteriol.* **180**, 1979–1987.

Solaiman, D. K. Y. (2000) PCR cloning of *Pseudomonas resinovorans* polyhydroxyalkanoate biosynthesis genes and expression in *Escherichia coli*, *Biotechnol. Lett.* **22**, 789–794.

Steinbüchel, A., Füchtenbusch, B. (1998) Bacterial and other biological systems for polyester production, *Trends Biotechnol.* **16**, 419–427.

Stover, C. K., Pham, X. Q., Erwin, A. L., Mizoguchi, S. D., Warrener, P., Hickey, M. J., Brinkman, F. S., Hufnagle, W. O., Kowalik, D. J., Lagrou, M., Garber, R. L., Goltry, L., Tolentino, E., Westbrock-Wadman, S., Yuan, Y., Brody, L. L., Coulter, S. N., Folger, K. R., Kas, A., Larbig, K., Lim, R., Smith, K., Spencer, D., Wong, G. K., Wu, Z., Paulsen, I. T. (2000) Complete genome sequence of *Pseudomonas aeruginosa* PA01, an opportunistic pathogen, *Nature* **406**, 959–964.

Sudesh, K., Fukui, T., Doi, Y. (1998) Genetic analysis of *Comamonas acidovorans* polyhydroxyalkanoate synthase and factors affecting the incorporation of 4-hydroxybutyrate monomer, *Appl. Environ. Microbiol.* **64**, 3437–3443.

Taguchi, K., Aoyagi, Y., Matsusaki, H., Fukui, T., Doi, Y. (1999) Co-expression of 3-ketoacyl-ACP reductase and polyhydroxyalkanoate synthase genes induces PHA production in *Escherichia coli* HB101 strain, *FEMS Microbiol. Lett.* **176**, 183–190.

Taroncher-Oldenburg, G., Nishina, K., Stephanopoulos, G. (2000) Identification and analysis of the polyhydroxyalkanoate-specific beta-ketothiolase and acetoacetyl coenzyme A reductase genes in the cyanobacterium *Synechocystis* sp. strain PCC6803, *Appl. Environ. Microbiol.* **66**, 4440–4448.

Thompson, S., Mayerl, F., Peoples, O. P., Masamune, S., Sinskey, A. J., Walsh, C. T. (1989) Mechanistic studies on beta-ketoacyl thiolase from *Zoogloea ramigera*: identification of the active-site nucleophile as Cys89, its mutation to Ser89, and kinetic and thermodynamic characterization of wild-type and mutant enzymes, *Biochemistry* **28**, 5735–5742.

Timm, A., Steinbüchel, A. (1990) Formation of polyesters consisting of medium-chain-length 3-hydroxyalkanoic acids from gluconate by *Pseudomonas aeruginosa* and other fluorescent pseudomonads, *Appl. Environ. Microbiol.* **56**, 3360–3367.

Timm, A., Steinbüchel, A. (1992) Cloning and molecular analysis of the poly(3-hydroxyalkanoic acid) gene locus of *Pseudomonas aeruginosa* PAO1, *Eur. J. Biochem.* **209**, 15–30.

Timm, A., Wiese, S., Steinbüchel, A. (1994) A general method for identification of polyhydroxyalkanoic acid synthase genes from pseudomonads belonging to the rRNA homology group I, *Appl. Microbiol. Biotechnol.* **40**, 669–675.

Tombolini, R., Povolo, S., Buson, A., Squartini, A., Nuti, M. P. (1995) Poly-beta-hydroxybutyrate (PHB) biosynthetic genes in *Rhizobium meliloti* 41, *Microbiology* **141**, 2553–2559.

Tsuge, T., Fukui, T., Matsusaki, H., Taguchi, S., Kobayashi, G., Ishizaki, A., Doi, Y. (2000) Molecular cloning of two (*R*)-specific enoyl-CoA hydratase genes from *Pseudomonas aeruginosa* and their use for polyhydroxyalkanoate synthesis, *FEMS Microbiol. Lett.* **184**, 193–198.

Ueda, S., Yabutani, T., Maehara, A., Yamane, T. (1996) Molecular analysis of the poly(3-hydroxyalkanoate) synthase gene from a methylotrophic bacterium, *Paracoccus denitrificans, J. Bacteriol.* **178**, 774–779.

Umeda, F., Kitano, Y., Murakami, Y., Yagi, K., Miura, Y., Mizoguchi, T. (1998) Cloning and sequence analysis of the poly (3-hydroxyalkanoic acid)-synthesis genes of *Pseudomonas acidophila, Appl. Biochem. Biotechnol.* **70-72**, 341–352.

Valentin, H. E., Dennis, D. (1997) Production of poly(3-hydroxybutyrate-*co*-4-hydroxybutyrate in recombinant *Escherichia coli* grown on glucose, *J. Biotechnol.* **58**, 33–38.

Valentin, H. E., Steinbüchel, A. (1993) Cloning and characterization of the *Methylobacterium extorquens* polyhydroxyalkanoic-acid-synthase structural gene, *Appl. Microbiol. Biotechnol.* **39**, 309–317.

Valentin, H. E., Lee, E. Y., Choi, C. Y., Steinbüchel, A. (1994) Identification of 4-hydroxyhexanoic acid as a new constituent of biosynthetic polyhydroxyalkanoic acids from bacteria, *Appl. Microbiol. Biotechnol.* **40**, 710–716.

Valentin, H. E., Zwingmann, G., Schönebaum, A., Steinbüchel, A. (1995) Metabolic pathway for biosynthesis of poly(3-hydroxybutyrate-*co*-4-hydroxybutyrate) from 4-hydroxybutyrate by *Alcaligenes eutrophus, Eur. J. Biochem.* **227**, 43–60.

Valentin, H. E., Stuart, E. S., Fuller, R. C., Lenz, R. W., Dennis, D. (1998) Investigation of the function of proteins associated to polyhydroxyalkanoate inclusions in *Pseudomonas putida* BMO1, *J. Biotechnol.* **64**, 145–157.

Valentin, H. E., Reiser, S., Gruys, K. J. (2000) Poly(3-hydroxybutyrate-*co*-4-hydroxybutyrate formation from γ-aminobutyrate and glutamate, *Biotechnol. Bioeng.* **67**, 291–299.

Weeks, G., Shapiro, M., Burns, R. O., Wakil, S. J. (1969) Control of fatty acid metabolism, *J. Bacteriol.* **97**, 827–836.

Wieczorek, R., Pries, A., Steinbüchel, A., Mayer, F. (1995) Analysis of a 24-kilodalton protein associated with the polyhydroxyalkanoic acid granules in *Alcaligenes eutrophus, J. Bacteriol.* **177**, 2425–2435.

Willis, L. B., Walker, G. C. (1998) The *phbC* (poly-beta-hydroxybutyrate synthase) gene of *Rhizobium (Sinorhizobium) meliloti* and characterization of *phbC* mutants, *Can J. Microbiol.* **44**, 554–564.

Wong, H. H., Lee, S. Y. (1998) Poly-(3-hydroxybutyrate) production from whey by high-density cultivation of recombinant *Escherichia coli, Appl. Microbiol. Biotechnol.* **50**, 30–33.

Yabutani, T., Maehara, A., Ueda, S., Yamane, T. (1995) Analysis of beta-ketothiolase and acetoacetyl-CoA reductase genes of a methylotrophic bacterium, *Paracoccus denitrificans*, and their expression in *Escherichia coli, FEMS Microbiol. Lett.* **133**, 85–90.

Zhang, S., Kolvek, S., Lenz, R. W., Goodwin, S. (2000) Cloning of the genes relevant for polyhydroxyalkanoic acid (PHA) biosynthesis and characterization of the PHA synthase from *Ectothiorhodospira shaposhnikovii*, GenBank accession No. AF307334.

8
Metabolic Flux Analysis on the Production of Poly(3-hydroxybutyrate)

Prof. Dr. Sang Yup Lee[1], M. Eng. Soon Ho Hong[2], M. Eng. Si Jae Park[3], Dr. Richard van Wegen[4], Dr. Anton P. J. Middelberg[5]

[1] Metabolic and Biomolecular Engineering National Research Laboratory, Department of Chemical Engineering and BioProcess Engineering Research Center, Korea Advanced Institute of Science and Technology 373-1 Kusong-dong, Yusong-gu, Taejon 305-701, Korea; Tel.: +82-42-869-3930; Fax: +82-42-869-3910; E-mail: leesy@mail.kaist.ac.kr

[2] Metabolic and Biomolecular Engineering National Research Laboratory, Department of Chemical Engineering and BioProcess Engineering Research Center, Korea Advanced Institute of Science and Technology 373-1 Kusong-dong, Yusong-gu, Taejon 305-701, Korea; Tel.: +82-42-869-5970; Fax: +82-42-869-3910; E-mail: soonho@mail.kaist.ac.kr

[3] Metabolic and Biomolecular Engineering National Research Laboratory, Department of Chemical Engineering and BioProcess Engineering Research Center, Korea Advanced Institute of Science and Technology 373-1 Kusong-dong, Yusong-gu, Taejon 305-701, Korea; Tel.: +82-42-869-5970; Fax: +82-42-869-3910; E-mail: parksj@mail.kaist.ac.kr

[4] Department of Chemical Engineering, University of Adelaide, SA 5005, Australia; Tel.: +43(01) 796-6362-311; Fax: +43(01) 796-6362-333; E-mail: richard.vanwegen@priorsep.com

[5] Department of Chemical Engineering, University of Cambridge, Pembroke Street, Cambridge, CB2 3RA, UK; Tel.: +44-1223-335245; Fax: +44-1223-334796; E-mail: antonm@cheng.cam.ac.uk

Biotechnology of Biopolymers. From Synthesis to Patents. Edited by A. Steinbüchel and Y. Doi

ISBN: 3-527-31110-6

3HB	3-hydroxybutyrate
CoA	coenzyme A
DOC	dissolved oxygen concentration
MCA	metabolic control analysis
MFA	metabolic flux analysis
NADH	nicotinamide adenine dinucleotide (reduced form)
NADPH	nicotinamide adenine dinucleotide phosphate (reduced form)
PHA	polyhydroxyalkanoate
SCL	short-chain-length
TCA	tricarboxylic acid cycle

1 Introduction

Polyhydroxyalkanoates (PHAs) have been considered to be good candidates as alternatives to synthetic nondegradable polymers due to their similar mechanical properties to petroleum-derived polymers and their complete biodegradability (Steinbüchel, 1992; Brandl et al., 1995; Lee, 1996a). A number of microorganisms have been found to accumulate PHAs under unfavorable growth conditions and in the presence of excess carbon source (Anderson and Dawes, 1990; Doi, 1990; Steinbüchel, 1991; Lee, 1996b; Steinbüchel and Füchtenbusch, 1998; Madison and Huisman, 1999). Among the various members of PHAs, poly(3-hydroxybutyrate), poly(3HB), a member of short-chain-length (SCL) PHAs, is best characterized and has been produced on a semicommercial scale. One of the major drawbacks in the commercialization of PHAs is the much higher production cost of PHAs compared with petrochemical-based polymers. Therefore, much effort has been devoted to reduce the production costs of PHAs by developing better bacterial strains and downstream processes such as more efficient fermentation and more economical recovery processes (Lee, 1996a,b; Choi et al., 1998; Choi and Lee, 1999a,b). Process design and economic analysis of SCL-PHA production by various bacteria have been reported, which provided the guidelines for designing an efficient means of PHA production (Choi and Lee, 1997, 1999c, 2000; Lee and Choi, 1998). Several factors affecting the production cost of PHA, including PHA productivity, PHA content and yield on carbon substrate, the cost of carbon substrate, and the recovery yield of PHA have been examined in detail. Based on the results of economic evaluation, *Ralstonia eutropha*, *Alcaligenes latus*, and recombinant *Escherichia coli* have been suggested as good candidates for the efficient production of SCL-PHAs (Lee, 1996a,b).

By using a metabolic flux analysis (MFA) technique, the intracellular metabolic fluxes can be quantified by the measurement of extracellular metabolite concentrations in combination with the stoichiometry of intracellular reactions (Nielsen and Villadsen, 1994; Edwards et al., 1999). MFA is based on the pseudo-steady-state assumption, which means that there is no net accumulation of intermediates (Stephanopoulos, 1999). To analyze the system using the MFA technique, the system should be determined or over-determined, which means that the number of constraints is equal to or greater than that of the reactions. To analyze the under-determined system, more constraints are required. This is often replaced by setting up an objective function, such as maximum growth or maximum metabolite production, and solving by linear programing (Varma and Palsson, 1994). MFA has been applied to calculate the maximum theoretical yield of a desired metabolite to be produced. Another application is to identify the rigidity of branch points in metabolic pathways. The rigidity of a branch point is important, as a rigid branch point resists changes in flux split ratios, while a flexible branch point tends to be more accommodating (Stephanopoulos and Vallino, 1991). The third possible application is the identification of alternative metabolic pathways. The detailed theories and applications of the MFA can be found in recent reviews (Edwards et al., 1999; Stephanopoulos, 1999).

Metabolic control analysis (MCA) is a statistical modeling technique that can be used to understand the control of metabolic pathways and pathway regulations. MCA allows us to understand how metabolic fluxes are controlled by certain enzyme activities and metabolite concentrations (Kacser and Burns, 1973; Heinrich and Rapoport, 1974). The responses of small changes in enzyme activities and metabolite concentrations to metabolic flux distribution can be predicted by MCA (Nielsen and Villadsen, 1994). If we consider a linear chain of N enzymatic reactions, there are N − 1 intermediates X_j, j = 1, ..., N − 1. The response coefficient of a certain metabolite is defined as the ratio of the relative change in the reaction rate brought about by the change in the metabolite concentration, viz:

$$\varepsilon_{ji} = \frac{X_j}{r_i}\frac{\partial r_i}{\partial X_j}; \quad i = 1, .., N \text{ and } j = 1, .., N - 1$$

where r_i is the net rate of the i-th enzymatic reaction and X_j is the size of the j-th metabolite pool. The flux control coefficient of a certain enzyme is the relative change in the steady-state flux resulting from the change in the activity of an enzyme of the pathway, viz:

$$C_i^r = \frac{X_{ei}}{r}\frac{\partial r}{\partial X_{ei}}; \quad i = 1, ..., N$$

where X_{ei} is the activity of i-th enzyme and r is the overall steady-state flux. By analyzing the flux control coefficients and response coefficients, one can propose which enzymatic reaction step is rate controlling (e.g., a reaction step with high flux control coefficient is the rate-controlling step), and also predict the results of deviations in certain enzymatic reactions. Readers are encouraged to refer to an excellent monograph on MCA by Fell (1997).

In this chapter, we review the applications of MFA and MCA on the production of poly(3HB) by various bacterial strains. The effects of various environmental conditions on poly(3HB) production are evaluated, and the important factors such as intracellular metabolite concentrations and enzyme activities on poly(3HB) biosynthesis are examined by MFA and MCA.

2 Historical Outline

A brief historical outline of the commercial production of poly(3HB) is shown below:

Year	History	Strain
1925	Production of poly(3HB) first discovered	*Bacillus megaterium*
1973	MCA first proposed	
1988	The first *phb* full operon cloned	*Ralstonia eutropha*
1988	Poly(3HB) first produced in *E. coli*	*Escherichia coli*
1991	Metabolic engineering first proposed	
1993	Effect of nitrogen limitation analyzed	*Ralstonia eutropha*
1997	Effect of carbon sources analyzed	*Ralstonia eutropha*
1998	Poly(3HB)-producing mechanism identified	
1999	Effect of oxygen limitation analyzed	*Escherichia coli*
2001	Control factors of poly(3HB) production identified	*Escherichia coli*

3 Production of Poly(3HB) by *Ralstonia eutropha*

Ralstonia eutropha (formerly *Alcaligenes eutrophus*), has been intensively examined for the efficient production of SCL-PHAs. The PHA biosynthetic pathway in *R. eutropha* has been well characterized: two acetyl-CoA moieties are condensed to form acetoacetyl-CoA by β-ketothiolase. Acetoacetyl-CoA is then reduced to (*R*)-3-hydroxybutyryl-CoA by an NADPH-dependent reductase. PHA synthase finally links (*R*)-3-hydroxybutyryl-CoA to the growing chain of poly(3HB). The genes coding for the three enzymes were found to form a *phb* operon in the order of PHA synthase, β-ketothiolase, and reductase (Schubert et al., 1988; Slater et al., 1988; Peoples and Sinskey, 1989). The three-step poly(3HB) biosynthesis pathway from acetyl-CoA is shown in Figure 1. Fed-batch culture strategies to achieve high productivity of PHAs have also been developed (Kim et al., 1994a,b; Ryu et al., 1997). An optimal fed-batch culture strategy resulted in both high poly(3HB) concentration and productivity of 232 g L^{-1} and 3.14 g poly(3HB) L^{-1} h^{-1}, respectively (Ryu et al., 1997).

The cost of raw materials (especially of the carbon source) is a very important factor affecting the overall production cost of poly(3HB). Therefore, the effect of different carbon sources on the production of poly(3HB) was extensively analyzed using MFA (Shi et al., 1997). *R. eutropha* can utilize various organic acids as carbon sources which, from a practical stand-point, is important for the production of poly(3HB) from organic wastes, for example food waste. Consequently, butyrate, lactate, and acetate were examined as carbon sources. When *R. eutropha* was cultivated on a mixture of three carbon sources, lactate was consumed first as a large amount of ATP is needed for the transport of acetate and butyrate. The central metabolic pathways for the utilization of three carbon sources are very similar, except for the anaplerotic pathway which replenishes carbon intermediates to the tricarboxylic acid (TCA) cycle. As shown in Figure 2, carbon dioxide is wasted during the conversion of pyruvate to acetyl-CoA when lactate is used as a carbon source. To decide which carbon source is better for the production of poly(3HB), simulations were carried out at different specific growth rates with three carbon sources. The maximum poly(3HB) yields (g g^{-1}) without cell growth obtainable with acetate, lactate, and butyrate were 0.33, 0.33, and 0.67, respectively. The poly(3HB) yield was limited by NADPH regeneration, as it is required for the conversion of acetoacetyl-CoA to (*R*)-3-

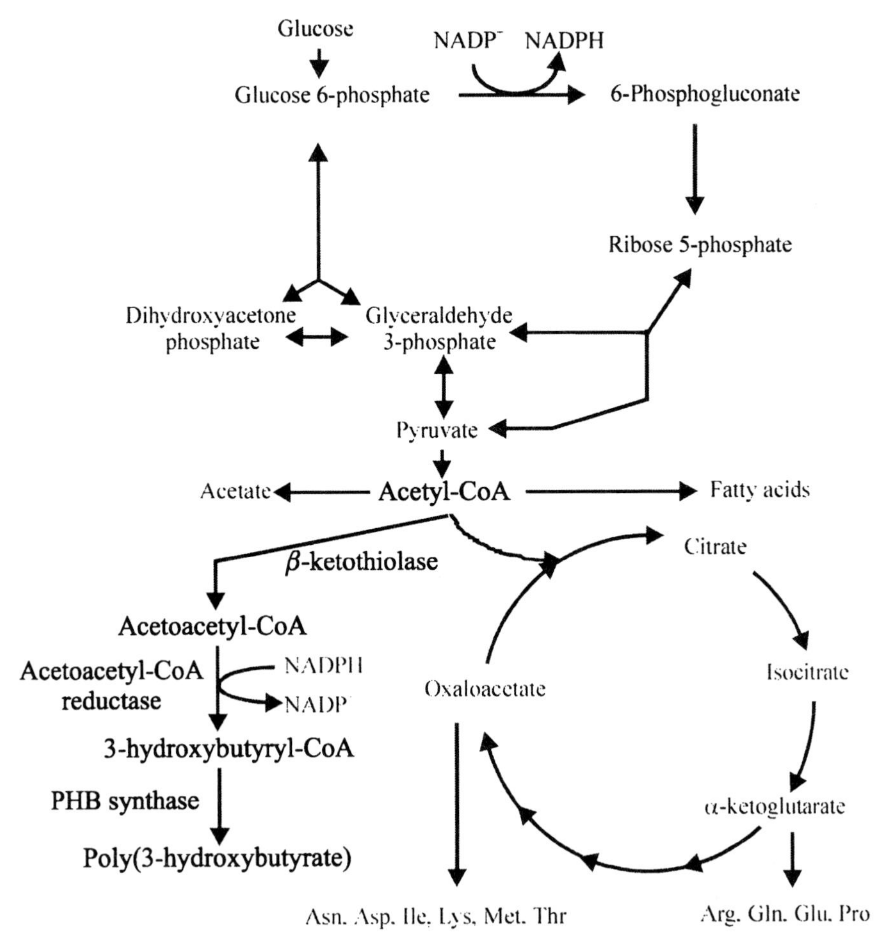

Fig. 1 Simplified central metabolic pathway of *E. coli* and PHB biosynthesis pathway.

hydroxybutyryl-CoA. NADPH is regenerated solely by isocitrate dehydrogenase in the TCA cycle. Butyrate was found to be more efficient for the production of poly(3HB), because about 67% of butyrate was directed to the TCA cycle, in which the required NADPH was regenerated, while only 33% of acetate and lactate were metabolized through the TCA cycle (Figure 2B, C, and D).

Poly(3HB) production by *R. eutropha* H16 (ATCC 17699) was also examined by fed-batch cultivation using butyrate as a carbon source (Shimizu et al., 1993). The maximum poly(3HB) yield of 0.85 (gg^{-1}) was obtained at a butyrate concentration of 3 g L^{-1} and at pH 8.0. The final poly(3HB) content was 75 wt.% dry cell weight. The changing profiles of the intracellular metabolic flux distribution during the fermentation were evaluated by MFA (Shi et al., 1997). The accumulation of poly(3HB) was induced by depletion of the nitrogen source. When the strain was shifted from the growth phase to the poly(3HB) production phase, flux into the poly(3HB) biosynthetic pathway increased, whilst flux into the glyoxylate bypass decreased (Figure 2A, B). There was no significant change in flux to the TCA cycle through

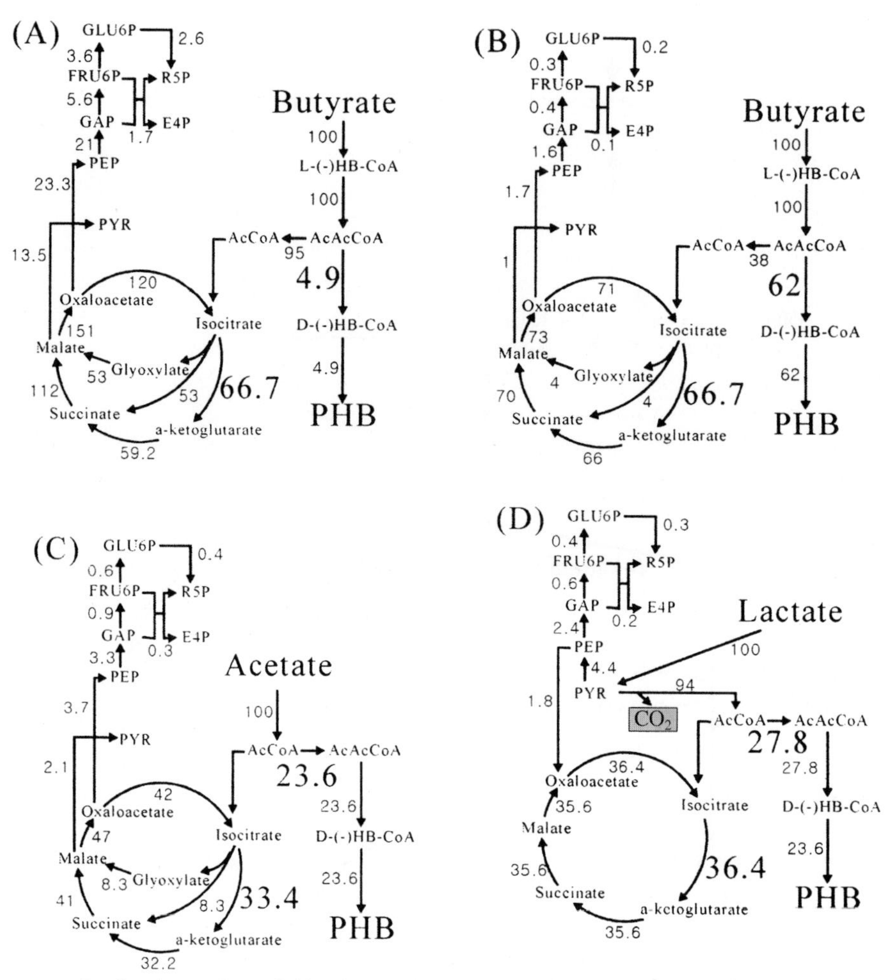

Fig. 2 Flux distribution during fed-batch culture on (A) butyrate at cell growth phase and (B) at poly(3HB) production phase, and on (C) acetate and (D) lactate at poly(3HB) production phase. Fluxes were normalized by substrates uptake rates. [Redrawn from Shi et al. (1997) with permission.]

isocitrate dehydrogenase. It can be concluded that glyoxylate bypass is required for the growth on butyrate, and this competes with the poly(3HB) biosynthetic pathway. Since flux into the TCA cycle was almost constant throughout the cultivation, the amount of NADPH produced during cultivation was not significantly changed. The NADPH consumption rate, however, was decreased because limitation of the nitrogen source blocked the amino acid synthesis pathways, especially the reaction from α-ketoglutarate to glutamate, which assimilates ammonium ions into the cell. Therefore, the residual NADPH was accumulated, which in turn enhanced poly(3HB) biosynthesis.

As mentioned earlier, the cost of raw materials is one of the most important factors affecting the economics of PHA production. Since the cost of glucose is lower than that of butyrate, glucose is better for the industrial production of PHAs. To use glucose directly as a carbon source, a mixed culture system was examined, in which glucose was converted to lactate by *Lactobacillus delbrueckii* and the lactate was converted to poly(3HB) by *R. eutropha* (Katoh et al., 1999). To evaluate the effects of environmental conditions on cell growth and poly(3HB) production, MFA was carried out under various conditions. It was found that NH_3 plays an important role in the production of poly(3HB). When NH_3 was sufficient, most of the NADPH was utilized for amino acids biosynthesis, and cell growth was activated. NADPH was used as a coenzyme of acetoacetyl-CoA reductase for the conversion of acetoacetyl-CoA to (*R*)-3-hydroxybutyryl-CoA under nitrogen-limiting conditions. From these results, it can be said that poly(3HB) production can be enhanced by providing a condition in which NADPH is in excess.

4 Production of Poly(3HB) by Recombinant *Escherichia coli*

Since the first demonstration that a large amount of poly(3HB) could be synthesized in recombinant *E. coli* harboring the PHA biosynthesis genes of *R. eutropha* (Schubert et al., 1988), recombinant *E. coli* has been intensively examined for the production of poly(3HB). Recombinant *E. coli* has been considered to be a good candidate as a producer for SCL-PHAs as it has several advantages over wild-type PHA producers such as *R. eutropha* and *A. latus* (Lee, 1997). Various metabolically engineered *E. coli* strains have been developed for the efficient production of poly(3HB), with high productivity of up to 4.63 g L^{-1} h^{-1} (Wang and Lee, 1997, 1998; Choi et al., 1998; Ahn et al., 2000). Recombinant *E. coli* equipped with a heterologous PHA biosynthetic pathway produces PHA in a growth-associated manner. Since poly(3HB) is a new intracellular metabolite of *E. coli*, the effect of poly(3HB) biosynthesis on the alteration of metabolic flux distribution should be evaluated in order to understand the physiological consequences of poly(3HB) accumulation in the cells. The dissolved oxygen concentration (DOC) is one of the important factors in poly(3HB) production by recombinant *E. coli*, since oxygen often becomes limited during high cell density cultivation (Wang and Lee, 1997; Wong et al., 1999). It was reported recently that poly(3HB) production by recombinant *E. coli* could be enhanced under oxygen-limiting conditions (Wang and Lee, 1997). To understand this phenomenon, the effects of DOC on metabolic flux distribution were examined using the MFA technique. The result of simulation using this approach suggested that 100 mol and 67 mol of poly(3HB) were produced from 100 mol of glucose under oxygen-sufficient and oxygen-limiting conditions, respectively. It was also found that the flux through pyruvate formate-lyase increased without any change in pyruvate dehydrogenase flux under oxygen-limiting conditions, causing the accumulation of intracellular acetyl-CoA from 350 μg g^{-1} RCM to 600 μg g^{-1} RCM (van Wegen et al., 2001). This accumulated acetyl-CoA was efficiently channeled to the poly(3HB) synthetic pathway, and more poly(3HB) could be accumulated under the oxygen-limited condition.

The amount of acetyl-CoA and NADPH available have also been found to be important factors in poly(3HB) production by recombinant *E. coli*, and this agrees well with the results of MFA that acetyl-CoA and NADPH play important roles in poly(3HB) biosynthesis by *R. eutropha* (Lee et al., 1996). The regulatory effects of NADPH and enzyme activities on poly(3HB) biosynthesis were examined in recombinant *E. coli* XL1-

Blue harboring the *R. eutropha* PHA biosynthesis genes. Cells were grown in various culture media including complex Luria-Bertani (LB) medium, LB supplemented with glucose, and chemically defined medium. Among these, poly(3HB) was most favorably accumulated in LB + glucose medium, which supported the highest NADPH/NADH ratio. The activity of citrate synthase, which competes with β-ketothiolase for acetyl-CoA, was also much lower when cells were cultured in LB + glucose medium (Lee S. Y. et al., 1995; Lee I. Y. et al., 1996) also indicated the importance of the availability of NADPH on PHA biosynthesis in recombinant *E. coli*. Supplementation of complex nitrogen sources, oleic acid, or amino acids to the chemically defined medium significantly enhanced poly(3HB) production. Because the biosynthesis of amino acids and oleic acid requires large amounts of reducing equivalents, poly(3HB) production by recombinant *E. coli* in chemically defined media was inefficient compared with poly(3HB) production in complex media (Lee et al., 1995). These experimental findings were supported by studies on the effects of acetyl-CoA and NADPH on the intracellular metabolic flux distribution of recombinant *E. coli* (Shi et al., 1999). The results of MFA suggested that in order to achieve the maximum poly(3HB) yield, about one-half of the carbon flux should be directed to the pentose phosphate (PP) pathway, and flux to the TCA cycle should be shut down. These two pathways affect the availability of two substrates for poly(3HB) synthesis, NADPH and acetyl-CoA.

MCA was also carried out in order to evaluate thoroughly the effects of NADPH and acetyl-CoA on poly(3HB) biosynthesis (van Wegen et al., 2001). The kinetics data of related enzymes are required for the calculation of flux control and elasticity coefficients. The β-ketothiolase catalyzes the conversion of acetyl-CoA to acetoacetyl-CoA, and it follows a ping-pong, Bi-Bi mechanism (Leaf and Srienc, 1998). The rate equation is:

$$\frac{r}{r_{max}} = \frac{AB - PQ/K_{eq}}{K_bA + K_aB + AB + K_qV_1P/V_2K_{eq} + K_pV_1Q/V_2K_{eq} + V_1PQ/V_2K_{eq} + K_qV_1AP/V_2K_{ia}K_{eq} + K_aBQ/K_{iq}}$$

where

r = forward reaction rate (μM substrate min^{-1})
r_{max} = maximum forward reaction rate (μM substrate min^{-1})
A = first substrate (acetyl-CoA) concentration (μM)
B = second substrate (also acetyl-CoA) concentration (μM)
P = first product (CoA) concentration (μM)
Q = second product (acetoacetyl-CoA) concentration (μM)
V_1/V_2 = ratio of maximum forward reaction rate to maximum reverse reaction rate
K_{eq} = equilibrium constant
K_a, K_b, K_p, K_q, K_{ia}, K_{ib}, K_{ip}, K_{iq}, = various kinetic constants (μM)

The kinetics of acetoacetyl-CoA reductase has been studied to a lesser extent, but most likely follows a rapid equilibrium, random, Bi-Bi mechanism. The rate equation is:

$$\frac{r}{r_{max}} = \frac{AB - PQ/K_{eq}}{K_{ia}K_b + K_bA + K_aB + AB + K_qV_1P/V_2K_{eq} + K_pV_1Q/V_2K_{eq} + V_1PQ/V_2K_{eq}}$$

where

A = first substrate (acetoacetyl-CoA) concentration (μM)
B = second substrate (NADPH) concentration (μM)
P = first product ((R)-3-hydroxybutyryl-CoA) concentration (μM)
Q = second product (NADP) concentration (μM)

The last enzyme, PHA synthase, catalyzes the polymerization of (*R*)-3-hydroxybutyryl-CoA, and the reaction is considered to be diffusion-limited.

$$\frac{r}{r_{\max}} = \frac{[(R)\text{-3-hydroxybutyryl-CoA}]}{K_m + [(R)\text{-3-3-hydroxybutyryl-CoA}]}$$

The kinetic constants for three enzymes are summarized in Table 1 (Leaf and Srienc, 1998).

The results of MCA suggested that the poly(3HB) biosynthesis flux was highly sensitive to the acetyl-CoA/CoA ratio (response coefficient 0.8) and total acetyl-CoA + CoA concentration (response coefficient 0.7), while it is less sensitive to the NADPH/NADP ratio (response coefficient 0.25) (Table 2). This means that to increase the flux to the same extent, the NADPH/NADP ratio needs to be increased more than the acetyl-CoA/CoA ratio. Finally, it was proposed that the overexpression of acetoacetyl-CoA reductase seems to be the most efficient way to enhance poly(3HB) productivity, as the flux control coefficients were 0.6, 0.25, and 0.15 for acetoacetyl-CoA reductase, PHA synthase, and β-ketothiolase, respectively (see Table 2).

From the results of MFA and MCA, it was found that acetyl-CoA and NADPH are the most important factors affecting poly(3HB) production. To identify whether these two actually affect poly(3HB) biosynthesis, two different mutant *E. coli* strains were examined as host strains for the production of poly(3HB). The first strain was TA3516 harboring pJM9131 and containing the *R. eutropha phb* operon. In this strain, acetyl-CoA was expected to be overproduced because of the inactivation of phosphotransacetylase and acetate kinase (Shi et al., 1999). The production rate of lactic acid decreased significantly, whilst those of pyruvate and acetic acid decreased only slightly. However,

Tab. 1 Kinetic constants for β-ketothiolase, acetoacetyl-CoA reductase and PHA synthase[a]

β-Ketothiolase			
K_{eq}	4×10^{-5}	K_q	64.6 μM
V_1/V_2	2.5×10^{-4}	K_{ia}	5.96 μM
K_a	3.78×10^{-3} μM	K_{ib}	841 μM
K_b	840 μM	K_{ip}	12.4 μM
K_p	31.4 μM	K_{iq}	1.62×10^{-2} μM
Acetoacetyl-CoA reductase			
K_{eq}	500	K_p, K_{ip}	16.5 μM
K_a, K_{ia}	5 μM	K_q, K_{iq}	31 μM
K_b, K_{ib}	19 μM		
PHA synthase			
K_m	720 μM		

[a]Data taken from Leaf and Srienc (1998).

Tab. 2 Response and flux control coefficients[a]

	Acetyl-CoA/CoA ratio	*Acetyl-CoA conc.*	*NADPH/NADP ratio*	*NADPH + NADP conc.*	*pH*
Metabolite response coefficient	0.8	0.7	0.2–0.3	0.06–0.09	–1
	β-ketothiolase		*Acetoacetyl-CoA reductase*		*PHA synthase*
Flux control coefficient	0.1–0.15		0.6		0.25

[a]Data from van Wegen et al. (2001).

no significant increase in poly(3HB) concentration was observed, and the final poly(3HB) content was only 10% of the dry cell weight. This result suggested that acetyl-CoA may not be accumulated due to the robustness of the primary metabolic network.

The effect of NADPH availability on the production of poly(3HB) was also investigated. The increase in NADPH flux could be achieved by rerouting the carbon fluxes to the PP pathway. The mutant strain DF11, which has lost phosphoglucose isomerase activity, was transformed with pAeKG1 containing the *R. eutropha phb* operon. This recombinant strain allowed enhanced production of poly(3HB), which again suggested the importance of NADPH. The conclusion which can be drawn from these results (and from the MCA results) is that both acetyl-CoA and NADPH are important in poly(3HB) biosynthesis, although it appears much easier to increase the amount of NADPH than the amount of acetyl-CoA.

Recently, proteome analysis of recombinant *E. coli* producing poly(3HB) has been carried out (Han et al., 2001). Several heat shock proteins including GroEL, GroES, and DnaK were up-regulated when poly(3HB) was accumulated to a great extent. It was also found that some enzymes in the glycolytic and Entner–Doudoroff pathways were overexpressed during the poly(3HB) production phase to support more acetyl-CoA and NADPH. Thus, it can be concluded that a sufficient supply of acetyl-CoA and NADPH is important for the efficient biosynthesis of poly(3HB).

5
Conclusions and Prospects

Systematic analyses on the accumulation of poly(3HB) – a member of the SCL-PHAs – have been carried out in *R. eutropha* and recombinant *E. coli* using MFA. The importance of acetyl-CoA and NADPH availability on poly(3HB) biosynthesis has been indicated by MFA, and this was confirmed totally by MCA. Proteome analysis also revealed that the Entner–Doudoroff pathway, which generally is not very active under normal growth conditions, was activated to supply more NADPH in recombinant *E. coli* producing poly(3HB). As yet, metabolic engineering strategies have been based on the simple deletion, amplification, or modification of some pathways. However, this strategy often failed because the global metabolic network is much more resistant to any modification of pathways than was originally thought. In order to redirect the metabolic fluxes towards the desired direction in an efficient manner, the effects of modifying the levels of certain enzymes and pathways on the global metabolic pathways should be carefully evaluated by MFA and MCA. Important data such as the rate-controlling steps which affect production of the desired materials can be identified from the results of MFA and MCA. This type of *in-silico* metabolic analysis will in turn reduce the time, money, and effort required to develop metabolically engineered strains for the enhanced production of various bioproducts, including PHAs.

Acknowledgments
The studies described in this chapter were supported by the National Research Laboratory program of Korean Ministry of Science and Technology (MOST).

6 Patents

Patent number (date)	Inventors	Title	Assignee
US4358583 (1982.11.9)	Walker and Whitton	Extraction of poly(beta-hydroxy butyric acid)	ICI
US4786598 (1988.11.22)	Lafferty and Braunegg	Process for the biotechnological preparation of poly-d-(-)-3-hydroxybutyric acid	Danubia Petrochemie
US5229279 (1993.7.20)	Peoples and Sinskey	Method for producing novel polyester biopolymers	MIT
US5250430 (1993.8.5)	Peoples and Sinskey	Polyhydroxyalkanoate polymerase	MIT
US5334520 (1994.8.2)	Dennis, D. E	Production of poly-beta-hydroxybutyrate in transformed *Escherichia coli*	CIT
US5512669 (1996.4.3)	Peoples and Sinskey	Gene encoding bacterial acetoacetyl-CoA reductase	MIT
US5518907 (1996.5.21)	Dennis, D. E	Cloning and expression in *Escherichia coli* of the *Alcaligenes eutrophus* H16 poly-beta-hydroxybutyrate biosynthetic pathway	CIT
US5661026 (1997.8.26)	Peoples and Sinskey	Gene encoding bacterial beta-ketothiolase	MIT
US5811272 (1998.9.2)	Snell et al.	Method for controlling molecular weight of polyhydroxyalkanoates	MIT
US595874 (1999.9.28)	Gruys et al.	Methods of optimizing substrate pools and biosynthesis of poly- beta -hydroxybutyrate-*co*-poly-beta-hydroxyvalerate in bacteria and plants	Monsanto Co.
US6043063 (2000.3.28)	Asrar et al.	Methods of PHA extraction and recovery using nonhalogenated solvents	Monsanto Co.

7
References

Ahn, W. S., Park, S. J., Lee, S. Y. (2000) Production of poly(3-hydroxybutyrate) by fed-batch culture of recombinant *Escherichia coli* with a highly concentrated whey solution, *Appl. Environ. Microbiol.* **66**, 3624–3627.

Anderson, A. J., Dawes, E. A. (1990) Occurrence, metabolism, metabolic role, and industrial uses of bacterial polyhydroxyalkanoates, *Microbiol. Rev.* **54**, 450–472.

Brandl, H., Bachofen, R., Mayer, J., Wintermantel, E. (1995) Degradation and applications of polyhydroxyalkanoates, *Can. J. Microbiol.* **41**(Suppl. 1), 143–153.

Choi, J., Lee, S. Y. (1997) Process analysis and economic evaluation for poly(3-hydroxybutyrate) production by fermentation, *Bioprocess. Eng.* **17**, 335–342.

Choi, J., Lee, S. Y. (1999a) Efficient and economical recovery of poly(3-hydroxybutyrate) from recombinant *Escherichia coli* by simple digestion with chemicals, *Biotechnol. Bioeng.* **62**, 546–553.

Choi, J., Lee, S. Y. (1999b) High-level production of poly(3-hydroxybutyrate-*co*-3-hydroxyvalerate) by fed-batch culture of recombinant *Escherichia coli*, *Appl. Environ. Microbiol.* **65**, 4363–4368.

Choi, J., Lee, S. Y. (1999c) Factors affecting the economics of polyhydroxyalkanoate production by bacterial fermentation, *Appl. Microbiol. Biotechnol.* **51**, 13–21.

Choi, J., Lee, S. Y. (2000) Economic consideration in the production of poly(3-hydroxybutyrate-*co*-3-hydroxyvalerate) by bacterial fermentation, *Appl. Microbiol. Biotechnol.* **53**, 646–649.

Choi, J., Lee, S. Y., Han, K. (1998) Cloning of the *Alcaligenes latus* polyhydroxyalkanoate biosynthesis genes and use of these genes for enhanced production of poly(3-hydroxybutyrate) in *Escherichia coli*, *Appl. Environ. Microbiol.* *64*, 4897–4903.

Doi, Y. (1990) *Microbial Polyesters*. New York: VCH.

Edwards, J. S., Ramakrishna, R., Schilling, C. H., Palsson, B. O. (1999) Metabolic flux balance analysis, in: *Metabolic Engineering* (Lee, S. Y., Papoutsakis, E. T., Eds), New York: Marcel Dekker, 13–57.

Fell, D. (1997) Metabolic control analysis, in: *Understanding the Control of Metabolism* (Fell, D., Ed.), London, UK: Portland Press, 101–132.

Han, M. J., Yoon, S. S., Lee, S. Y. (2001) Proteome analysis of metabolically engineered *Escherichia coli* producing poly(3-hydroxybutyrate), *J. Bacteriol.* **183**, 301–308.

Heinrich, R., Rapoport, T. A. (1974) A linear steady-state treatment of enzymatic chains. General properties, control and effector strength, *Eur. J. Biochem.* **42**, 89–95.

Kacser, H., Burns, J. A. (1973) The control of flux, *Symp. Soc. Exp. Biol.* **27**, 65–104

Katoh, T., Yuguchi, D., Yoshii, H., Shi, H., Shimizu, K. (1999) Dynamic and modeling on fermentative production of poly(β-hydroxybutyric acid) from sugars via lactate by a mixed culture of *Lactobacillus delbrueckii* and *Alcaligenes eutrophus*, *J. Biotechnol.* **67**, 113–134.

Kim, B. S., Lee, S. C., Lee, S. Y., Chang, H. N., Chang, Y. K., Woo, S. I. (1994a) Production of poly(3-hydroxybutyric acid) by fed-batch culture of *Alcaligenes eutrophus* with glucose concentration control, *Biotechnol. Bioeng.* **43**, 892–898.

Kim, B. S., Lee, S. C., Lee, S. Y., Chang, H. N., Chang, Y. K., Woo, S. I. (1994b) Production of poly(3-hydroxybutyric-*co*-3-hydroxyvaleric acid) by fed-batch culture of *Alcaligenes eutrophus* with substrate control using on-line glucose analyzer, *Enzyme Microbiol. Technol.* **16**, 556–561.

Leaf, T. A., Srienc, F. (1998) Metabolic modelling of polyhydroxybutyrate biosynthesis, *Biotechnol. Bioeng.* **57**, 557–570.

Lee, I. Y., Kim, M. K., Park, Y. H., Lee, S. Y. (1996) Regulatory effects of cellular nicotinamide nucleotide and enzyme activities on poly(3-hydroxybutyrate) synthesis in recombinant *Escherichia coli*, *Biotechnol. Bioeng.* **52**, 707–712.

Lee, S. Y. (1996a) Plastic bacteria? Progress and prospects for polyhydroxyalkanoate production in bacteria, *Trends Biotechnol.* **14**, 431–438.

Lee, S. Y. (1996b) Bacterial polyhydroxyalkanoates, *Biotechnol. Bioeng.* **49**, 1–14.

Lee, S. Y. (1997) *E. coli* moves into the plastic age, *Nature Biotechnol.* **15**, 17–18.

Lee, S. Y., Choi, J. (1998) Effect of fermentation performance on the economics of poly(3-hydroxybutyrate) production by *Alcaligenes latus*, *Polymer Degrad. Stabil.* **59**, 387–393.

Lee, S. Y., Lee, Y. K., Chang, H. N. (1995) Stimulatory effects of amino acids and oleic acid on poly(3-hydroxybutyric acid) synthesis by recombinant *Escherichia coli*, *J. Ferment. Bioeng.* **79**, 177–180.

Madison, L. L., Huisman, G. W. (1999) Metabolic engineering of poly(3-hydroxyalkanoates): from DNA to plastic, *Microbiol. Mol. Biol. Rev.* **63**, 21–53.

Nielsen, J., Villadsen, J. (1994) Analysis of reaction rates, in: *Bioreaction Engineering Principles* (Nielsen, J., Villadsen, J., Eds), New York: Plenum Press, 97–161.

Peoples, O. P., Sinskey, A. J. (1989) Poly-β-hydroxybutyrate biosynthesis in *Alcaligenes eutrophus* H16. Identification and characterization of the PHB polymerase gene (*phbC*), *J. Biol. Chem.* **264**, 15298–15303.

Ryu, H. W., Hahn, S. K., Chang, Y. K., Chang, H. N. (1997) Production of poly(3-hydroxybutyrate) by high cell density fed-batch culture of *Alcaligenes eutrophus* with phosphate limitation, *Biotechnol. Bioeng.* **55**, 28–32.

Schubert, P., Steinbüchel, A., Schlegel, H. G. (1988) Cloning of the *Alcaligenes eutrophus* poly-β-hydroxybutyrate synthetic pathway and synthesis of PHB in *Escherichia coli*. *J. Bacteriol.* **170**, 5837–5847.

Shi, H., Shiraishi, M., Shimizu, K. (1997) Metabolic flux analysis for biosynthesis of poly(β-hydroxybutyric acid) in *Alcaligenes eutrophus* from various carbon sources, *J. Ferment. Bioeng.* **84**, 579–587.

Shi, H., Nikawa, J., Shimizu, K. (1999) Effect of modifying metabolic network on poly-3-hydroxybutyrate biosynthesis in recombinant *Escherichia coli*, *J. Biosci. Bioeng.* **87**(5), 667–677.

Shimizu, H., Tamura, S., Shioya, S., Suga, K. (1993) Kinetic study of poly-D(-)-3-hydroxybutyric acid (PHB) production and its molecular weight distribution control in a fed-batch culture of *Alcaligenes eutrophus*, *J. Ferment. Bioeng.* **76**, 465–469.

Slater, S. C., Voige, W. H., Dennis, D. (1988) Cloning and expression in *Escherichia coli* of the *Alcaligenes eutrophus* H16 poly-β-hydroxybutyrate biosynthetic pathway, *J. Bacteriol.* **170**, 4431–4436.

Steinbüchel, A. (1991) Polyhydroxyalkanoic acids, in: *Biomaterials: Novel Materials from Biological Sources* (Byrom, D., Ed.), New York: Stockton, 124–213.

Steinbüchel, A. (1992) Biodegradable plastics, *Curr. Opin. Biotechnol.* **3**, 291–297.

Steinbüchel, A., Füchtenbusch, B. (1998) Bacterial and other biological systems for polyester production, *Trends Biotechnol.* **16**, 419–427.

Stephanopoulos, G. (1999) Metabolic fluxes and metabolic engineering, *Metab. Eng.* **1**, 1–11.

Stephanopoulos, G., Vallino, J. J. (1991) Network rigidity and metabolic engineering in metabolite overproduction, *Science* **252**, 1675–1681.

van Wegen, R. J., Lee, S. Y., Middelberg, A. P. J. (2001) Metabolic and kinetic analysis of poly(3-hydroxybutyrate) production by recombinant *Escherichia coli*, *Biotechnol. Bioeng.* **74**, 70–81.

Varma, A., Palsson, B. O. (1994) Metabolic flux balancing: basic concepts, scientific and practical use, *Nature Biotechnol.* **12**, 994–998.

Wang, F., Lee, S. Y. (1997) Production of poly(3-hydroxybutyrate) by fed-batch culture of filamentation-suppressed recombinant *Escherichia coli*, *Appl. Environ. Microbiol.* **63**, 4765–4769.

Wang, F., Lee, S. Y. (1998) High cell density culture of metabolically engineered *Escherichia coli* for the production of poly(3-hydroxybutyrate) in a defined medium, *Biotechnol. Bioeng.* **58**, 325–328.

Wong, H. H., van Wegen, R. J., Choi, J., Lee, S. Y., Middelberg, A. P. J. (1999) Metabolic analysis of poly(3-hydroxybutyrate) production by recombinant *Escherichia coli*, *J. Microbiol. Biotechnol.* **9**, 593–603.

9
Fermentative Production of Short-chain-length PHAs

Prof. Dr. Sang Yup Lee[1], **M. Eng. Si Jae Park**[2]
[1] Metabolic and Biomolecular Engineering National Research Laboratory, Department of Chemical Engineering and BioProcess Engineering Research Center, Korea Advanced Institute of Science and Technology 373-1 Kusong-dong, Yusong-gu, Taejon 305-701, Korea; Tel.: +82-42-869-3930; Fax: +82-42-869-3910; E-mail: leesy@mail.kaist.ac.kr
[2] Metabolic and Biomolecular Engineering National Research Laboratory, Department of Chemical Engineering and BioProcess Engineering Research Center, Korea Advanced Institute of Science and Technology 373-1 Kusong-dong, Yusong-gu, Taejon 305-701, Korea; Tel.: +82-42-869-5970; Fax: +82-42-869-8800; E-mail: parksj@mail.kaist.ac.kr

Biotechnology of Biopolymers. From Synthesis to Patents. Edited by A. Steinbüchel and Y. Doi

ISBN: 3-527-31110-6

3HB	3-hydroxybutyrate
3HP	3-hydroxypropionate
3HV	3-hydroxyvalerate
4HB	4-hydroxybutyrate
4HV	4-hydroxyvalerate
CoA	coenzyme A
DCW	dry cell weight
DO	dissolved oxygen
DOC	dissolved oxygen concentration
IPTG	isopropyl-β-D-thiogalactopyranoside
MCL	medium-chain-length
NADH	nicotinamide adenine dinucleotide (reduced form)
NADPH	nicotinamide adenine dinucleotide phosphate (reduced form)
PHA	polyhydroxyalkanoate
SCL	short-chain-length

1 Introduction

Polyhydroxyalkanoates (PHAs) are intracellular carbon and energy storage material that are accumulated in various microorganisms, usually under the unfavorable growth conditions in the presence of excess carbon source (Anderson and Dawes, 1990; Doi, 1990; Steinbüchel, 1991; Lee, 1996c; Steinbüchel and Füchtenbusch, 1998; Madison and Huisman, 1999). During recent years, environmental problems caused by the accumulation of nondegradable plastics have increasingly become recognized as crucial. Consequently, attention has focused on PHAs as substitutes for chemically synthesized polymers, due not only to their similar mechanical properties but also to their complete biodegradability (Brandl et al., 1995; Lee, 1996b; Steinbüchel, 1992). Poly(3-hydroxybutyrate), poly(3HB), was first discovered in the bacterium *Bacillus megaterium* in 1926 (Lemoigne, 1926), since when many different PHAs containing various numbers of main chain carbon atoms and (*R*)-pending groups have been reported (Steinbüchel and Valentin, 1995). According to the number of carbon atoms constituting the monomer units, bacterial PHAs can be classified into two groups: short-chain-length (SCL) PHAs consisting of three to five carbon atoms; and medium-chain-length (MCL) PHAs consisting of 6–14 carbon atoms (Lee, 1996c; Lee and Choi, 1999; Steinbüchel and Füchtenbusch, 1998). Recently, a number of reports have been made of PHAs consisting of both SCL and MCL monomer units (Brandl et al., 1989; Haywood et al., 1991; Steinbüchel and Wiese, 1992; Kobayashi et al., 1994; Kato et al., 1996). The physical properties of PHAs are highly dependent upon their monomer units; hence, biodegradable polymers having a wide range of properties can be produced by incorporating different monomer units. The commercialization of PHAs has been hampered by the high production costs of PHA compared with petrochemical-based polymers. Consequently, much effort has been expended to reduce the production costs of PHAs by developing better bacterial strains and downstream processes, for example more efficient fermentation and more economical recovery processes (Choi et al., 1998; Choi and Lee, 1999a,b; Lee, 1996b,c).

In this chapter, fermentation and metabolic engineering strategies employed for the efficient production of SCL-PHAs in various wild-type and recombinant bacterial strains are reviewed.

2 Historical Outline

A brief historical outline of the production of SCL-PHAs is shown in the following chart:

1926 Discovery of poly(3HB) in *Bacillus* sp.
1988 Cloning of poly(3HB) biosynthesis genes from *Ralstonia eutropha*
1992 Fed-batch cultivation of recombinant *Escherichia coli* harboring *R. eutropha* poly(3HB) biosynthesis genes in complex media for the production of poly(3HB)
1992 Production of poly(3HB) from cheap carbon sources such as whey, by recombinant *E. coli*
1994 Fed-batch cultivation of *R. eutropha* glucose-utilizing mutant for the production of poly(3HB) or poly(3HB-*co*-3HV) under nitrogen limitation
1997 Fed-batch cultivation of *R. eutropha* glucose-utilizing mutant for the production of poly(3HB) under phosphorus limitation
1997 Fed-batch cultivation of *Alcaligenes latus* for the production of poly(3HB) under nitrogen limitation
1998 Fed-batch cultivation of recombinant *E. coli* harboring *R. eutropha* poly(3HB) biosynthesis genes in chemically defined media for the production of poly(3HB)
1998 Cloning of poly(3HB) biosynthesis genes from *A. latus*
1998 Fed-batch cultivation of recombinant *E. coli* harboring *A. latus* poly(3HB) biosynthesis genes in chemically defined media for the production of poly(3HB)
1999 Fed-batch cultivation of recombinant *E. coli* harboring *A. latus* poly(3HB) biosynthesis genes in chemically defined media for the production of poly(3HB-*co*-3HV)
2001 Proteome analysis of recombinant *E. coli* producing poly(3HB)

3 Production of SCL-PHAs by Bacterial Fermentation

In general, PHA-producing bacteria can be divided into two groups by the characteristics of their PHA accumulation (Lee and Chang, 1995; Lee, 1996b). One group of bacteria starts to accumulate PHA when their cell growth is impaired by the limitation of an essential nutrient such as N, P, Mg, K, O, or S in the presence of excess carbon source. The other group produces PHA in a growth-associated manner. Many bacteria, including *Ralstonia eutropha* and methylotrophic bacteria, belong to the first group, whilst *Alcaligenes latus, Azotobacter beijerinckii* and recombinant *Escherichia coli* belong to the second group. Therefore, it is important to develop a suitable fermentation strategy for each of these microorganisms in order to enhance the production of PHAs. For the fed-batch culture of bacteria, which accumulate PHA under nutrient limitation, a two-step fed-batch culture method has been widely employed. The cells are first grown to a desired cell concentration, whereupon PHA biosynthesis is triggered by the nutrient limitation (Kim et al., 1994a,b; Ryu et al., 1997). During the active PHA biosynthesis phase, cell concentration increases mainly due to the concomitant PHA accumulation. For the growth-associated PHA producers, an optimal nutrient feeding strategy, which supports both cell growth and PHA accumulation, needs to be employed. However, there have been several reports that suitable nutrient limitation might significantly enhance PHA accumulation in growth-associated PHA producers such as *A. latus* and recombinant *E. coli* (Wang and Lee, 1997a,b, 1998; Ahn et al., 2000, 2001). Therefore, it is important to identify the time point at which to apply nutrient limitation in order to achieve high PHA productivity.

Among the various PHAs, poly(3HB) and poly(3-hydroxybutyrate-*co*-3-hydroxyvalerate), poly(3HB-*co*-3HV), were the only members produced on a semi-commercial scale, though their high production cost remains a major problem hampering the commercialization of PHAs. Although transgenic plants may offer an attractive economical solution for the production of PHAs in future, bacterial fermentation is currently the only feasible means by which PHAs may be produced efficiently. Moreover, issues on the use of genetically modified plants in the open field need to be carefully examined. Several fermentation strategies, ranging from high cell density fed-batch culture to continuous cell culture, have been developed for the production of PHAs, each with high productivity (Lee, 1996b). Process analysis and economic evaluation for the production of SCL-PHAs based on the previously reported fermentation data have been reported, and have permitted the design of an efficient system of PHA production (Choi and Lee, 1997, 1999c, 2000; Lee and Choi, 1998). Several factors affecting the production cost of PHA have been examined in detail. PHA productivity, PHA content and yield, the cost of carbon substrate, and the recovery yield of PHA have all been shown to be important factors, and should be optimized. Among these factors, PHA content was the most important as it has multiple effects on PHA yield and recovery efficiency. Based on economic considerations, *R. eutropha, A. latus*, and recombinant *E. coli* have been suggested as strong candidates for the efficient production of SCL-PHAs (Lee, 1996b,c).

3.1
Ralstonia eutropha

Until recently, *Ralstonia eutropha* – formerly known as *Alcaligenes eutrophus* – has been used for the semi-commercial production of poly(3HB-*co*-3HV), first by Zeneca BioProducts (Billingham, UK), and later by Monsanto (St. Louis, MO, USA). The PHA biosynthesis pathway in *R. eutropha* has been well characterized. Two acetyl-CoA moieties being condensed to form acetoacetyl-CoA by β-ketothiolase. The acetoacetyl-CoA is then reduced to (*R*)-3-hydroxybutyryl-CoA by an NADPH-dependent reductase. PHA synthase finally links (*R*)-3-hydroxybutyryl-CoA to the growing chain of poly(3HB) (Schubert et al., 1988; Slater et al., 1988; Peoples and Sinskey, 1989). In general, the microorganisms which accumulate SCL-PHAs from sugars (except for *Rhodospirillum rubrum*) have been known to follow the same PHA biosynthesis pathway as *R. eutropha* (Lee, 1996c; Madison and Huisman, 1999). *R. eutropha* grows well in a relatively inexpensive minimal medium, and accumulates a large amount of poly(3HB) under unbalanced growth conditions, and using several carbon sources (except glucose). As glucose is often the most favorable carbon substrate among a number of sugars, a glucose-utilizing mutant strain of *R. eutropha* was developed and used for the production of poly(3HB) and poly(3HB-*co*-3HV). Various *R. eutropha* strains used in PHA research are detailed in Table 1. The fed-batch culture strategy, which is the most popular method to achieve high cell density

Tab. 1 Various *R. eutropha* strains used in PHA research

R. eutropha strain	*Characteristics*	*Sources and References*
H16	Wild-type, prototrophic	ATCC17699; DSM428; NCIMB 10442
NCIMB 11599	Glucose-utilizing mutant of H16	ICI; Kim et al. (1994a,b); Ryu et al. (1997)
PHB^{-4}	PHA-negative mutant of H16	DSM541
F11-1-116-EtOH	Alcohol-utilizing mutant of NCIMB 11599	ICI; Park and Damodaran (1994)
DSM 545	Glucose-utilizing mutant of H1(DSM 529)	DSM 545; Ramsay et al. (1990a)
DSM 11348	Mutant from DSM 531, Increased growth on succinate and glucose	DSM 11348; Bormann et al. (1998b)

and consequently to achieve high productivity of a desired product, has been applied for the efficient production of PHAs. Examination of conditions for the cultivation of *R. eutropha* has suggested that it is important to maintain the concentration of carbon source at an optimal value, and to apply the nutrient limitation triggering of PHA accumulation at an optimal point (Kim et al., 1994a; Ryu et al., 1997). Fed-batch culture of *R. eutropha* DSM 545 using glucose resulted in the production of poly(3HB) up to 24 g L^{-1} under ammonium limitation conditions (Ramsay et al., 1990a). During the poly(3HB) accumulation phase, the glucose concentration was maintained at between 5 and 16 g L^{-1} by the constant addition of 50 wt.% glucose solution. Also, 17 g L^{-1} of poly(3HB-*co*-3HV) could be produced when the glucose feed was replaced with a mixture of glucose and propionic acid during the polymer accumulation phase. The mole fraction of 3HV in the copolymer was 5 mol.% (Ramsay et al., 1990a).

A high cell density culture of *R. eutropha* NCIMB 11599 has been studied extensively. In order to maintain the glucose concentration within the optimal range, several feeding strategies for fed-batch cultures were developed (Kim et al., 1994a; Ryu et al., 1997). Because the increase of pH in response to carbon depletion was found to be slow, a dissolved oxygen (DO)-stat feeding strategy was applied in order to improve poly(3HB) production. When nitrogen limitation was applied at a cell concentration of 70 g dry cell weight (DCW) L^{-1}, a poly(3HB) concentration of 121 g L^{-1} was obtained in 50 h, and this resulted in a poly(3HB) productivity of 2.42 g poly(3HB) L^{-1} h^{-1} (Kim et al., 1994a). Because the pH was controlled by adding NaOH under nitrogen-limiting conditions, the toxicity of highly concentrated hydroxide had a negative effect on cell growth at high cell density. Consequently, a fed-batch culture of *R. eutropha* NCIMB 11599 with DO-stat feeding strategy under phosphorus limitation was examined (Ryu et al., 1997). The pH was maintained by adding NH_4OH instead of NaOH, whereupon it was found that maintaining phosphate and magnesium concentrations above 0.35 g L^{-1} and 10 mg L^{-1}, respectively, was important to obtain a high poly(3HB) concentration. Under these conditions, a final cell concentration, poly(3HB) concentration and poly(3HB) content of 281 g DCW/L, 232 g L^{-1} and 80 wt.%, respectively, were obtained, and this resulted in a high productivity of 3.14 g poly(3HB) L^{-1} h^{-1} (Ryu et al., 1997) (Table 2).

A two-stage fed-batch culture of *R. eutropha* NCIMB 11599 for the production of poly(3HB-*co*-3HV) from glucose and pro-

Tab. 2 Production of poly(3HB) by fed-batch culture of wild-type microorganisms

Strain	*Feeding strategy*	*Limiting nutrient*	*Substrate*	*Time* [h]	*Cell conc.* [g L^{-1}]	*Poly(3HB) conc.* [g L^{-1}]	*Poly(3HB) content* [%]	*Productivity* [g L^{-1}h^{-1}]	*Reference*
R. eutropha	Glucose control	N	Glucose	50	164	121	76	2.42	Kim et al. (1994a)
R. eutropha	DO-stat	P	Glucose	74	281	232	82	3.14	Ryu et al. (1997)
A. latus	pH-stat	No	Sucrose	18	142	68.4	50	3.97	Yamane et al. (1996)
A. latus	DO-stat	N	Sucrose	20	111.7	98.7	88	4.94	Wang and Lee (1997a)
M. organophilum	Methanol control	P	Methanol	70	250	130	52	1.86	Kim et al. (1996)

pionic acid has been reported (Kim et al., 1994b, Lee I. Y. et al., 1995a). The glucose concentration of the culture broth was controlled at 10–20 g L^{-1} using an on-line glucose analyzer. Cells were first grown to 60–70 g L^{-1} using only glucose, after which nitrogen limitation was applied and the glucose feed replaced by a mixture of glucose and propionic acid. As the ratio of propionic acid to glucose (P/G ratio) was increased, less poly(3HB-*co*-3HV) was produced; however, the mole fraction of 3HV in the copolymer increased. When the P/G ratios in the feed were 0.17, 0.35, and 0.52, the final poly(3HB-*co*-3HV) concentrations of 117, 74, and 64 g L^{-1} and poly(3HB-*co*-3HV) contents of 74, 57, and 56.5 wt.% were obtained, respectively. The mole fraction of 3HV in the copolymer was increased to 14.3 mol.% at a P/G ratio of 0.52 (Kim et al., 1994b). Lee I. Y. et al. (1995a) also reported the production of poly(3HB-*co*-3HV) using *R. eutropha* NCIMB 11599 by controlling the concentration of propionic acid. When glucose addition was sufficient and the feeding rate of propionic acid decreased, the poly(3HB-*co*-3HV) concentration increased, whilst the 3HV fraction in the copolymer decreased. By controlling the propionic acid concentration at 1–4 g L^{-1}, poly(3HB-*co*-3HV) was produced at a high concentration of 85.6 g L^{-1}, with a 3HV fraction of 11.4 mol.% in 59 h. Based on these results, the concentration of propionic acid was suggested to be one of the most important parameters for the production of poly(3HB-*co*-3HV) having a high 3HV fraction. A mixture of glucose and valeric acid as cosubstrates was also examined for the production of poly(3HB-*co*-3HV) by the fed-batch culture of *R. eutropha* NCIMB 11599 (Lee I. Y. et al., 1995b). Under nitrogen-limited conditions, 90.4 g L^{-1} of poly(3HB-*co*-3HV) containing 20.4 mol.% of 3HV was obtained in 50 h. The use of valeric acid as a cosubstrate allowed a higher copolymer production rate, yield, and 3HV fraction of copolymer than were obtained with propionic acid. Recently, Choi and Lee (2000) reported the production cost of poly(3HB-*co*-3HV) to be highly dependent on the 3HV fraction in copolymer, the cost of the cosubstrate, and the 3HV yield on the carbon substrate. Hence, a careful examination should be made as to which cosubstrate is economically beneficial for the production of poly(3HB-*co*-3HV).

R. eutropha NCIMB 11599 was engineered to use an alcohol as a substrate at Imperial

Chemical Industries (ICI, Billingham, UK), and subsequently examined for the production of poly(3HB-*co*-3HV) in fed-batch cultures (Park and Damodaran, 1994). Ethanol and propanol were used as substrates, and various feeding methods after phosphate limitation were investigated to enhance the polymer yield on alcohols. When propanol was used as a sole carbon source, 24 g L^{-1} of poly(3HB-*co*-3HV) containing 36.5 mol.% of 3HV was produced with a PHA yield of 0.41 g PHA g^{-1} alcohol. When alcohol feeding was changed from a mixture of 50% ethanol + 50% propanol to 100% propanol, and from 100% ethanol to a mixture of 65% ethanol + 35% propanol, the PHA yields on alcohol were increased to 0.51 g PHA g^{-1} alcohol, and 0.46 g PHA g^{-1} alcohol, respectively. However, the fraction of 3HV in copolymer was lower than that obtained with propanol alone. When ethanol was changed to propanol directly, cell growth and polymer production were severely inhibited. The polymer production rate and yield were highly dependent upon the alcohol feeding mode.

The production of poly(3HB-*co*-3HV) was examined in continuous culture under nitrogen-limited conditions using *R. eutropha* H16 (Koyama and Doi, 1995). Concentrations of fructose and valeric acid were maintained at 17.5 g L^{-1} and 2.5 g L^{-1}, respectively, and the maximum poly(3HB-*co*-3HV) productivity of 0.31 g L^{-1} h^{-1} was obtained at a dilution rate of 0.17 h^{-1}. As the dilution rate was increased from 0.06 to 0.32 h^{-1}, the 3HV fraction in copolymer increased from 11 to 79 mol.%.

R. eutropha DSM 11348 was used for the production of poly(3HB) from casein peptone or casamino acids (Bormann et al., 1998b). The cell concentration of 65 g DCW L^{-1} and the poly(3HB) content of 60–80 wt.% were achieved in this medium. The highest poly(3HB) productivity was 1.2 g poly(3HB) L^{-1} h^{-1}. However, the cost of these complex nitrogen sources, and the reproducibility of the fermentation results may be problematic in a general sense.

3.2 *Alcaligenes latus*

A. latus has been considered to be a good candidate for the production of poly(3HB) because it grows rapidly, produces poly(3HB) during the actively growing stage, and can use sucrose as a carbon substrate (see Table 2). Since it can utilize sucrose as a carbon source, cheap substrates such as raw sugar, beet, or cane molasses can also be used for the production of poly(3HB) (Hangii, 1990). However, the poly(3HB) content obtained was typically about 50 wt.% of DCW (Yamane et al., 1996), which made the recovery process inefficient, and resulted in a low yield of poly(3HB) on carbon substrate (Choi and Lee, 1997; Lee and Choi, 1998). A report was made on the pH-stat fed-batch culture of *A. latus*, which resulted in a poly(3HB) concentration of 68.4 g L^{-1} in 18 h with a high poly(3HB) productivity of 3.97 g poly(3HB) L^{-1} h^{-1} by using a high inoculum size of 13.7 g L^{-1} (Yamane et al., 1996). However, the poly(3HB) content obtained was rather low at 50 wt.%, and needs to be considerably increased in order to make this process industrially attractive. Wang and Lee (1997a) examined the strategy of nutrient limitation to increase poly(3HB) content. After examining the effect of limiting nitrogen, phosphorus, magnesium, or sulfur in flask cultures, nitrogen limitation was chosen as the best strategy as it allowed the greatest enhancement of poly(3HB) production. Cell growth under nitrogen-sufficient conditions was apparent from the increase of residual cell concentration. In contrast, residual cell concentration did not increase under nitrogen-limited conditions, and only poly(3HB) concentration was increased. During fed-batch culture of *A. latus*, two feeding strategies were applied. Since the DO response was better than the pH response on carbon depletion, the DO-stat was used during the actively growing stage. Following the application of nitrogen limitation, the feeding strategy was changed from the DO-stat method to the optimally determined feeding profile. This was because there was no apparent DO increase upon carbon depletion under nitrogen-limited conditions, though this was to be expected from the fact that there was no further significant cell growth under this condition. The feeding profile was determined experimentally by calculating the rate of sucrose consumption during the nitrogen-limited period. Nitrogen limitation was applied at an optimal cell concentration of 76 g L^{-1}, since it was revealed from the fed-batch cultures of *R. eutropha* that both the poly(3HB) concentration and content were highly dependent on the time point of application of nutrient limitation (Kim et al., 1994a). Sucrose concentration was maintained at 5–20 g L^{-1}, and within 8 h of nitrogen limitation, cell concentration, poly(3HB) concentration, and poly(3HB) content reached 111.7 g L^{-1}, 98.7 g L^{-1}, and 88 wt.%, respectively, resulting in the productivity of 4.94 g poly(3-HB) L^{-1} h^{-1} (Figure 1). The highest productivity of 5.13 g poly(3HB) L^{-1} h^{-1} was obtained after 16 h (Wang and Lee, 1997a).

Based on the two different fermentation results of *A. latus* for the production of poly(3HB), a semi-commercial process for poly(3HB) production (100,000 tonnes/year) by *A. latus* was designed and subsequently investigated (Lee and Choi, 1998). From the computer-aided process simulation, the achievement of a high poly(3HB) content and productivity were suggested to

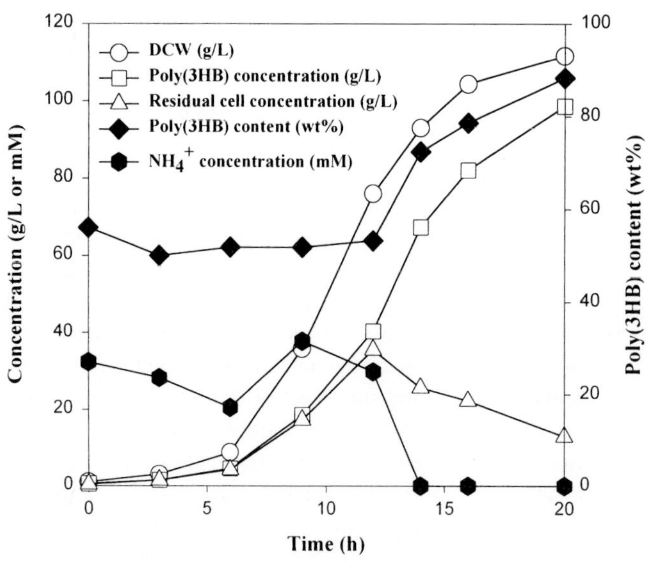

Fig. 1 Time profiles of the cell concentration (○), poly(3HB) concentration (□), residual cell concentration (△), and NH_4^+ concentration (●), and poly(3HB) content (◆) during the fed-batch culture of *A. latus* DSM1123. NH_4OH was replaced with NaOH when cell concentration reached 76 g L^{-1} at 12 h. Nitrogen depletion occurred at 12.5 h. [Reproduced from Wang and Lee (1997a), with permission.]

be important in order to reduce the final poly(3HB) production cost. In particular, the poly(3HB) content showed the multiple effects on production cost of poly(3HB) because a high poly(3HB) content resulted in an efficient recovery process, lower waste treatment costs, and lower fermentation-related operating costs (mainly the cost of carbon source). By developing an efficient fermentation strategy using *A. latus*, which resulted in a high poly(3HB) content of 88.3 wt.% and poly(3HB) productivity of 4.94 g poly(3HB) L^{-1} h^{-1} (Wang and Lee, 1997a), the production cost of poly(3HB) could be reduced to US\$2.6 kg^{-1} (Lee and Choi, 1998). The detailed analysis of process economics will be described in Section 5.

The single-stage chemostat cultivation of *A. latus* ATCC 29714 under nitrogen-limited conditions resulted in a poly(3HB) content of 40 wt.% at a dilution rate of 0.15 h^{-1} (Ramsay et al., 1990a). Two-stage chemostat cultivation of *A. latus* ATCC 29714, in which the second stage was fed only with the effluent from the first stage, was also carried out (Ramsay et al., 1990a). When propionic acid was added to the first stage of the two-stage chemostat, *A. latus* produced poly(3HB-*co*-3HV) up to 43 wt.% of DCW containing 18.5 mol.% of 3HV. In the second stage, poly(3HB-*co*-3HV) was accumulated up to 58 wt.% of DCW with a 3HV fraction of 11 mol.%, without any further addition of carbon substrate. The 3HV composition could be regulated by the concentration of propionic acid in the feed (Ramsay et al., 1990a).

3.3 Other Bacteria

Methylotrophic bacteria have also been considered as potential PHA producers, as they are able to use cheap (and currently abundant) methanol as a carbon source. However, the lower PHA productivity and PHA yield on carbon substrate of this strain, compared with those of other PHA producers such as *R. eutropha* and *A. latus*, are the major drawbacks for the commercialization of this process.

Suzuki et al. (1986a,b,c, 1988) carried out an extensive examination of the effects of physiological parameters such as temperature, pH, and methanol concentration on the growth and accumulation of poly(3HB) by *Protomonas extorquens*. By using a fed-batch culture of *P. extorquens* sp. Strain K, a poly(3HB) concentration of 136 g L^{-1} was obtained in 175 h. By optimizing the medium composition, 149 g L^{-1} of poly(3HB) was produced in 170 h (Suzuki et al., 1986a,b). There was another report on the fed-batch culture of methylotroph, with slow feeding of methanol to prevent oxygen limitation, and which resulted in a poly-(3HB) content of 45 wt.% and poly(3HB) productivity of 0.56 g poly(3HB) L^{-1} h^{-1} (Bourque et al., 1995). As a result of the slow feeding of methanol, a molecular mass of $>$ 1,000,000 Da could be obtained, which was much higher than that typically obtainable in other methylotrophic bacteria. Suzuki et al. (1988) suggested that methanol concentration during the fed-batch culture was an important parameter for molecular weight control. The production of poly(3HB) by microcomputer-aided automatic fed-batch culture of *Methylobacterium organophilum* under potassium-limited conditions has also been reported (Kim et al., 1996). The concentration of methanol was kept within the range of 2–3 g L^{-1}, which had no inhibitory effect on cell growth. Poly(3HB) accumulation was stimulated when the potassium concentration in the culture broth fell below 25 mg L^{-1}. After 70 h of cultivation, the cell concentration and poly(3HB) concentration reached 250 and 130 g L^{-1}, respectively, which corresponded to a poly(3HB) productivity of 1.8 g poly(3-HB) L^{-1} h^{-1}. The poly(3HB) content was 56 wt.% of DCW, and the poly(3HB) yield was 0.19 g poly(3HB) g^{-1} methanol.

Poly(3HB-*co*-3HV) was also produced by newly isolated *M. extorquens* (Bourque et al., 1992) and *Methylobacterium* sp. KCTC 0048 (Kang et al., 1993). In a fed-batch culture of *M. extorquens* using a methanol-valeric acid mixture as a carbon source and supplemented with complex nitrogen sources, poly(3HB-*co*-3HV) containing 20 mol.% of 3HV was accumulated to 30 wt.% of DCW (Bourque et al., 1992). *Methylobacterium* sp. KCTC 0048 accumulated poly(3HB-*co*-3HV) in a nitrogen-limited medium containing methanol when valeric acid, pentanol, or heptanoic acid was used as a cosubstrate (Kang et al., 1993). Poly(3-hydroxybutyrate-*co*-4-hydroxybutyrate), poly(3HB-*co*-4HB), was also produced when a mixture of methanol and 4-hydroxybutyrate, 1,4-butanediol, or γ-butyrolactone was used as a carbon source. Furthermore, poly(3-hydroxybutyrate-*co*-3-hydroxypropionate), poly(3HB-*co*-3HP), was produced up to 30 wt.% of DCW from a mixture of methanol and 3-hydroxypropionate (Kang et al., 1993). Poly(3HB-*co*-3HV) was also produced by a fed-batch culture of *M. organophilum* NCIB 11278 (Kim et al., 1999) when valeric acid was supplied as an auxiliary carbon source, with methanol as a main carbon source. Among the six nutritional components tested (NH_4^+, K^+, Mg^{2+}, SO_4^{2-}, PO_4^{2-}, and dissolved oxygen), the synthesis of poly(3HB-*co*-3HV) was most favored under potassium-limited conditions. When the volume ratio of valeric acid to methanol in the feed was 1:10 (1 mL valeric acid per 10 mL methanol), a copolymer content of 41 wt.% of DCW and a 3HV fraction of 14 mol.% were obtained. During 78 h of cultivation under these optimized conditions, poly(3HB-*co*-3HV) was produced up to 99 g L^{-1} with a 3HV fraction of 11.0 mol.%. Poly(3HB-*co*-3HV) production by *Paracoccus denitrificans* and *M. extorquens* from methanol and *n*-amyl alcohol was also investigated (Ueda et al., 1992). The copolymer synthesized in *P. denitrificans* was rich in 3HV (up to 91.5 mol.%) with increasing concentration of *n*-amyl alcohol, but in *M. extorquens* the maximum 3HV content was limited to 38.2 mol.%. Using a fed-batch culture of *P. denitrificans* under nitrogen-limited conditions, a cell concentration of 9 g DCW L^{-1} and a poly(3HB-*co*-3HV) content of 26 wt.% were obtained in 120 h. The mole fraction of 3HV in copolymer was increased up to 60 mol.% (Ueda et al., 1992).

Azotobacter beijerinckii produces poly(3HB) in a growth-associated manner when cultured in media containing casein peptone, yeast extract, casamino acids and urea, combined with glucose or sucrose (Bormann et al., 1998a). A poly(3HB) content of up to 50 wt.% was obtained by fermentation on casein peptone + glucose or urea + glucose media, the poly(3HB) productivity being 0.8–1.1 g poly(3HB) L^{-1} h^{-1}. Oxygen limitation was essential for efficient poly(3HB) production in this strain. *A. vinelandii* UWD was known to produce a poly(3HB) from complex carbon sources such as corn syrup, cane molasses, beet molasses or malt extract (Page, 1989) in a growth-associated manner due to a defective respiratory NADH oxidase (Page and Knosp, 1989). This strain was considered ideal for the fermentation process because it can produce poly(3HB) from an unpurified carbon substrate during growth.

A. vinelandii UWD accumulates poly(3HB-*co*-3HV) when grown in a medium containing glucose and valeric acid (Page et al., 1992). Copolymer was not produced when propionic acid was added to the glucose medium, but was produced when heptanoic acid, nonanoic acid, or *trans*-2-propionic acid was present. The fed-batch cultures of *A. vinelandii* in beet molasses with varying concentrations of valeric acid allowed the production of poly(3HB-*co*-3HV) containing 8.5 to 23 mol.% of 3HV up to 22 g L^{-1} in 38–40 h; the PHA content was 59–71 wt.% (Page et al., 1992). *A. vinelandii* UWD was also reported to produce poly(3HB-*co*-3HV) from swine waste liquor consisting primarily of acetic acid, propionic acid, butyric acid, and valeric acid (Cho et al., 1997). By using a two-fold dilution of swine waste liquor, *A. vinelandii* UWD accumulated poly(3HB-*co*-3HV) containing 7.9 mol.% of 3HV up to 34 wt.% of DCW. Supplementation of glucose increased the cell growth and PHA production four-fold, but decreased the 3HV fraction in the copolymer (Cho et al., 1997).

A bacterial strain able to synthesize poly(3HB-*co*-3HV) copolymer from glucose only has also been reported. A fed-batch culture of *Alcaligenes* sp. SH-69 was used for the production of poly(3HB-*co*-3HV) with glucose as a sole carbon source (Rhee et al., 1993). Poly(3HB-*co*-3HV) was favorably synthesized at a dissolved oxygen concentration (DOC) of 20% and a C/N ratio of 23.1. Under optimal conditions, poly(3HB-*co*-3HV) containing 3 mol.% of 3HV was produced up to 36 g L^{-1}. Reducing the C/N ratio resulted in an increase of 3HV fraction in the copolymer up to 6.3 mol.%. During the growth of *Alcaligenes* sp. SH-69 on glucose, a relatively low level of levulinic acid was detected (Jang and Rogers, 1996). The effects of added levulinic acid on cell growth and PHA synthesis in the presence of excess glucose were examined in both batch and continuous cultures. When levulinic acid was added at 0.5 g L^{-1} h^{-1}, a maximal PHA content of 38.3 wt.% and a 3HV fraction of 23.5 mol.% were obtained in a nitrogen-limited, two-stage continuous culture at a dilution rate of 0.078 h^{-1}. The stimulatory effect of levulinic acid on PHA synthesis was greater than that of propionic acid.

A fed-batch culture process to produce poly(3HV) homopolymer using *Chromobacterium violaceum* has also been developed (Schmack and Steinbüchel, 1995). When valeric acid and nutrient broth were used as substrates, a cell concentration and a poly(3HV) content of 40 g DCW L^{-1} and 70 wt.%, respectively, were obtained in 90 h at 10-L fermentation scale. The material properties of poly(3HV) were shown to be adequate for applications with low strength requirements.

4 Biosynthesis and Production of PHAs by Metabolically Engineered Bacteria

Various bacteria have been metabolically engineered to produce PHAs efficiently by the amplification of heterologous or homologous PHA biosynthesis genes, or by the modification of inherent PHA metabolic pathways. Among these bacteria, recombinant *E. coli* and *R. eutropha* have been intensively investigated for the efficient production of SCL-PHAs.

4.1 Recombinant *E. coli*

E. coli has most often been used as a host strain for the production of various metabolites and recombinant proteins as it is the most extensively studied bacterium in all aspects (Lee, 1996a). Following the first demonstration that a large amount of poly(3HB) could be synthesized in *E. coli* harboring the PHA biosynthesis genes of *R. eutropha* (Schubert et al., 1988), which were constitutively expressed in *E. coli*, recombinant *E. coli* has been extensively investigated for the production of poly(3HB). Detailed studies on the structure and organization of PHA biosynthesis genes and the metabolic pathways for PHA biosynthesis have allowed the construction of metabolically engineered *E. coli* for efficient PHA production. Recombinant *E. coli* has long been considered to be a strong candidate as a poly(3HB) producer due to several advantages over wild-type poly(3HB) producers such as *R. eutropha* and *A. latus*: these advantages include a wide range of utilizable carbon sources, the accumulation of poly(3HB) to a high content with high productivity, and easy polymer recovery due to fragility of the cells after having accumulated a large amount of poly(3HB) (Lee, 1997). In order to achieve an efficient production of poly(3HB) in recombinant *E. coli*, it was found to be important to maintain the stable expression of the PHA biosynthesis genes during cultivation. A series of reports have been published describing the development of host-plasmid systems and the strategies for the production of poly(3HB) to a high concentration with high productivity (Kim et al., 1992; Lee and Chang, 1994, 1996; Lee S. Y. et al., 1994a,b,c, 1995; Wang and Lee, 1997b, 1998; Choi et al., 1998) (Table 3).

Initial studies with recombinant *E. coli* were carried out in a complex medium (Luria-Bertani, LB). Poly(3HB) could be produced to >80 g L^{-1} by a fed-batch culture of recombinant *E. coli* with high productivity of >2 g poly(3HB) L^{-1} h^{-1} in LB medium containing 20 g L^{-1} glucose (Kim et al., 1992). Unlike the PHA synthesis by natural PHA-producing bacteria, PHA production by recombinant *E. coli* was not apparently triggered by nutrient limitation. The PHA biosynthesis operon was expressed using the native promoter of *R. eutropha*, which allowed constitutive expression of these genes in *E. coli*. This is important because expensive inducers such as isopropyl-β-D-thiogalactopyranoside (IPTG) cannot be used for the production of bulk materials such as PHA.

It was found that poly(3HB) production by recombinant *E. coli* was dependent on the amount of acetyl-CoA and NADPH available (Lee I. Y. et al., 1996b). For this reason, poly(3HB) production by recombinant *E. coli* in chemically defined media was inefficient compared with poly(3HB) production in complex media. The addition of various complex nitrogen sources, amino acids, or oleic acid to the chemically defined media significantly enhanced poly(3HB) production in recombinant *E. coli*. This was explained by the availability of more NADPH for poly(3HB) production because the biosynthesis of amino acids and oleic acid require substantial amounts of reducing

Tab. 3 Production of P(3HB) by fed-batch culture of recombinant *E. coli*

Feeding strategy	*Substrate*	*Time [h]*	*Cell conc. [g L^{-1}]*	*PHA conc. [g L^{-1}]*	*PHA content [%]*	*Productivity [g L^{-1} h^{-1}]*	*Reference*
PH-stat	Glucose	49	204.3	157.1	77	3.2	Wang and Lee (1997a)
PH-stat	Glucose	30.6	194.1	141.6	73	4.63	Choi et al. (1998)
PH-stat	Glucose	36	153.7	101.3	65.9	2.8	Wang and Lee (1998)
PH-stat	Whey	49	87	69	80	1.4	Wong and Lee (1998)
PH-stat	Whey	37.5	119.5	96.2	80	2.57	Ahn et al. (2000)
PH-stat +cell recycle	Whey	36.5	194	168	87	4.61	Ahn et al. (2001)

equivalents (Lee S. Y. et al., 1995; Lee I. Y. et al., 1996b).

When using fed-batch cultures of recombinant *E. coli* XL1-Blue in defined media, growth was found to be impaired by considerable cell filamentation during the polymer synthesis and accumulation stage, which resulted in low poly(3HB) productivity. Cell filamentation could be successfully suppressed by the overexpression of the *ftsZ* gene, which is involved in cell division, and twice the amount of poly(3HB) could be produced by using the filamentation-suppressed recombinant *E. coli* XL1-Blue (Lee, 1994). The fed-batch culture of this strain allowed the production of poly(3HB) up to 104 g L^{-1} with a poly(3HB) content of 70 wt.%. The poly(3HB) productivity of 2 g poly(3HB) L^{-1} h^{-1} could be achieved in a chemically defined medium by employing this strain (Wang and Lee, 1998).

During high cell density fed-batch culture of recombinant *E. coli*, a large amount of oxygen was needed to maintain the DOC above 10% of air saturation. Since supplementation with pure oxygen is economically undesirable, and oxygen transfer is also generally poor in a large-scale fermentor, the effect of limited oxygen supply on poly(3HB) production was examined (Wang and Lee, 1997b). Based on the results of fed-batch cultures, Wang and Lee (1997b) suggested that fed-batch cultivation could be divided into two phases: (1) an active growth phase during which poly(3HB) content is kept constant and at a low level; and (2) an active poly(3HB) synthesis phase during which poly(3HB) is actively accumulated, with a concomitant increase in poly(3HB) content. When oxygen was supplied insufficiently to maintain the DOC at 1–3% of air saturation during the actively growing cell stage, cell growth stopped and poly(3HB) accumulation was not enhanced. However, cell growth and poly(3HB) accumulation were not inhibited when oxygen was limited in the poly(3HB) synthesis phase, during which poly(3HB) began to accumulate actively, resulting in a concomitant increase in poly(3HB) content. The poly(3HB) concentration and poly(3HB) productivity of 157 g L^{-1} and 3.2 g poly(3HB) L^{-1} h^{-1}, respectively, could be obtained (Wang and Lee, 1997b). It was also important to maintain sufficient DOC during the active growth phase in order to achieve high final cell and poly(3HB) concentrations. For the scale-up studies, fed-batch cultures were carried out in a poorly aerated 50-L fermenter, in which the DOC decreased to zero when cell concentration reached 50 g L^{-1}. However, a relatively high poly(3HB) concentration of 101 g L^{-1} and poly(3HB) productivity of 2.8 g poly(3HB) L^{-1} h^{-1} could still be obtained, which demonstrated the strong possibility of an industrial production of poly(3HB) in a chemically defined medium by employing recombinant *E. coli* (Wang and Lee, 1997b). The effect of oxygen limitation on poly(3HB) production was further examined by metabolic flux analysis, which revealed that metabolic flux to the poly(3HB) biosynthetic pathway increased upon oxygen limitation at high cell density (Wong et al., 1999).

Recently, the PHA biosynthesis genes were cloned from *A. latus* (Choi et al., 1998). Several plasmids containing the *A. latus* PHA biosynthesis genes were constructed and examined for the efficient production of poly(3HB). Using the fed-batch culture of recombinant *E. coli* harboring an optimally designed plasmid containing the *A. latus* PHA biosynthesis genes, the higher poly(3HB) productivity of 4.63 g poly(3HB) L^{-1} h^{-1} could be obtained, which allowed more economic production of poly(3HB) (Figure 2) (Choi et al., 1998).

One of the advantages of recombinant *E. coli* over other natural PHA producers is

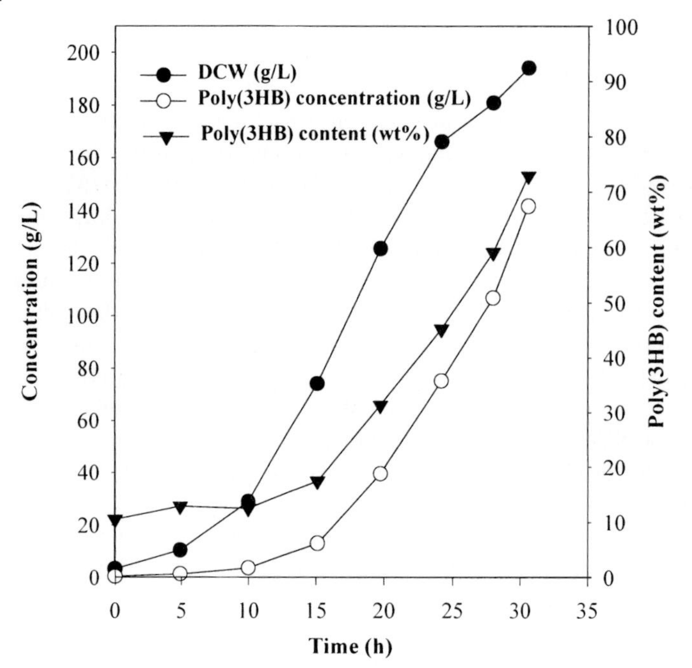

Fig. 2 Time profile of cell concentration (•), poly(3HB) concentration (○) and poly(3HB) content (▼) during the fed-batch culture of *E. coli* XL1-Blue (pJC4) in a chemically defined medium. [Reproduced from Choi et al. (1998), with permission.]

its ability to use inexpensive carbon sources. Therefore, it is possible to produce PHA from cheap carbon sources, such as whey, hemicellulose hydrolyzates, and molasses (Fidler and Dennis, 1992; Lee S. Y. et al., 1997; Liu et al., 1998; Wong and Lee, 1998; Ahn et al., 2000, 2001). A number of reports have been made that poly(3HB) could be produced from a whey-based medium using lactose utilizing recombinant *E. coli* strains in flask culture (Fidler and Dennis, 1992; Lee S. Y. et al., 1997). By using a fed-batch culture of recombinant *E. coli* strain harboring the *R. eutropha* PHA biosynthesis genes, 69 g L^{-1} of poly(3HB) could be produced from whey solution, resulting in a poly(3HB) productivity of 1.4 g poly(3HB) L^{-1} h^{-1} (Wong and Lee, 1998). A major problem encountered during the production of poly(3HB) from whey was the volumetric limitation of the fermentor due to the low amounts of lactose in the feeding solution as a result of the low solubility of lactose (ca. 200 g L^{-1}) in water and low poly(3HB) productivity compared with that obtained from glucose (Wong and Lee, 1998; Ahn et al., 2000).

In order to solve these inherent problems, Ahn et al. (2000) examined new fermentation strategies for the production of poly(3HB) from whey by recombinant *E. coli* strain CGSC 4401 harboring plasmid pJC4 containing the *A. latus* PHA biosynthesis genes. The pH-stat fed-batch cultures of recombinant *E. coli* CGSC 4401 were carried out using a highly concentrated whey solution containing 280 g L^{-1} lactose equivalent. By applying an optimal DOC reduction strategy, final cell and poly(3HB) concentrations of 119.5 and 96.2 g L^{-1}, respectively, were obtained in

37.5 h, which resulted in the poly(3HB) productivity of 2.57 g poly(3HB) L^{-1} h^{-1}. The productivity of poly(3HB) could be further increased by the cell recycle fed-batch culture using an external membrane unit (Ahn et al., 2001). Generally, membrane-based cell recycle culture is used to remove product or byproduct inhibition in a continuous stirred tank reactor for achieving high-density culture and high productivity of a desired product (Chang et al., 1994). In this fermentation, the purpose of cell recycle culture was to concentrate cells and to maintain the culture volume constant. By using the cell recycle fed-batch culture, and a whey solution containing 280 g L^{-1} lactose equivalent, a poly(3HB) concentration of 168 g L^{-1} was obtained in 36.5 h, resulting in a very high poly(3HB) productivity of 4.6 g poly(3HB) L^{-1} h^{-1} (Figure 3) (Ahn et al., 2001).

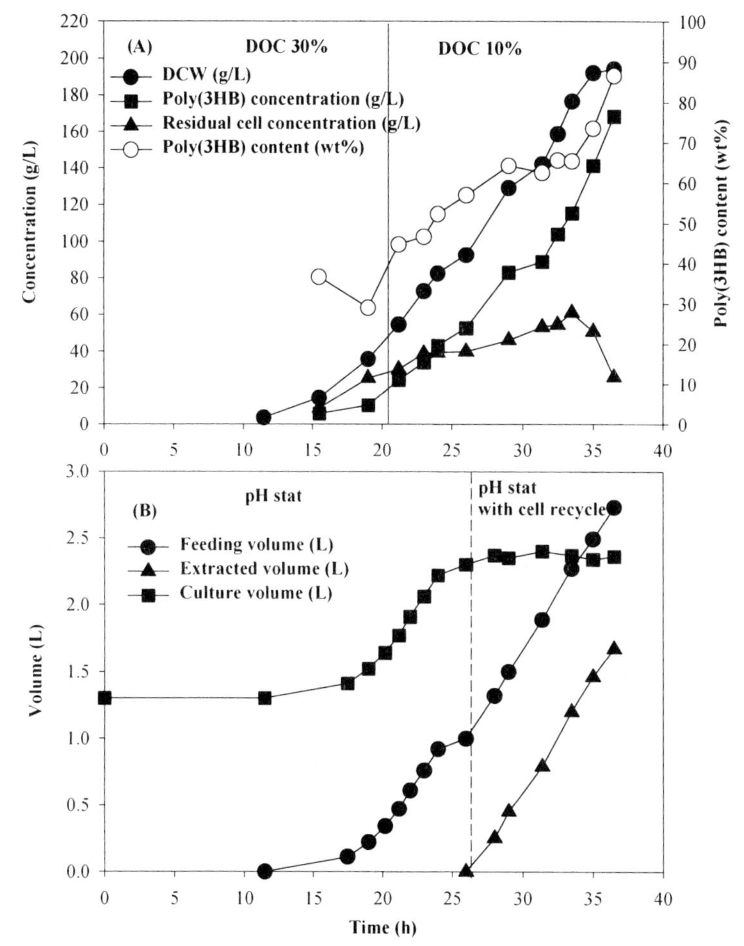

Fig. 3 Time profiles of cell concentration (●), poly(3HB) concentration (■), residual cell concentration (▲) and poly(3HB) content (●) (A) and feeding volume (○) and extracted volume (▲) and culture volume (■) (B) during the fed-batch culture of *E. coli* CGSC 4401 (pJC4) with the feeding solution containing lactose (280 g L^{-1}). The dissolved oxygen concentration (DOC) was maintained at 30% of air saturation during actively growing stage (up to the OD_{600} of 150) and was decreased to 10%. [Reproduced from Ahn et al. (2001), with permission.]

A sucrose-utilizing *E. coli* strain has also been metabolically engineered to produce poly(3HB) (Lee and Chang, 1993; Zhang et al., 1994). Even though recombinant *E. coli* W harboring the *R. eutropha* PHA biosynthesis genes could be cultured to a relatively high concentration of 125 g L^{-1} in 48 h, the poly(3HB) content of 27.5 wt.% of DCW was rather low compared with those obtained in recombinant *E. coli* XL1-Blue cultured in medium containing glucose (Lee and Chang, 1993). However, this result shows the possibility that poly(3HB) can be produced efficiently using cheap carbon sources. Also, by using a fed-batch culture of recombinant *E. coli* strain harboring the *R. eutropha* PHA biosynthesis genes, poly(3HB) was produced from hydrolyzed molasses with a poly(3HB) productivity of 1 g poly(3HB) L^{-1} h^{-1} (Liu et al., 1998).

Several recombinant *E. coli* strains harboring the *R. eutropha* PHA biosynthesis genes, which can produce poly(3HB) from xylose were also constructed (Lee, 1998). The supplementation of a small amount of cotton seed hydrolysate or soybean hydrolysate could enhance poly(3HB) accumulation up to 74 wt.% of DCW.

The production of poly(3HB-*co*-3HV) in recombinant *E. coli* has also been demonstrated by the alteration of metabolism, and by the supplementation of propionic acid or valeric acid, which is similar to the strategy used in *R. eutropha*. Because the uptake of propionic acid in *E. coli* is not efficient, the cells were first grown and adapted in a medium containing acetic acid, after which a glucose-propionic acid mixture was added (Slater et al., 1992). The *E. coli* strain used was LS5218 (*fadR atoC* (Con)) harboring the *R. eutropha* PHA biosynthesis genes. The *fadR atoC* (Con) genotype allows constitutive expression of the enzymes involved in the use of fatty acids to improve the uptake and utilization of propionic acid. The 3HV fraction in the copolymer could be altered with varying concentration of propionic acid in the medium.

Propionyl-CoA formation was studied using *E. coli* LS5218 with further mutations in *ackA* and *pta*, which encode acetate kinase and phosphotransacetylase, respectively (Rhie and Dennis, 1995). In the absence of either of these enzymes, 3HV incorporation was reduced significantly. Overexpression of the *ackA* gene led to an increase in 3HV formation. It was shown that *E. coli* requires the Ack and Pta activities for the efficient incorporation of 3HV. Non-*fadR atoC* (Con) *E. coli* strains were also examined for the production of poly(3HB-*co*-3HV) (Yim et al., 1996). Several recombinant *E. coli* strains, including XL1-Blue, JM109, HB101 and DH5α harboring the *R. eutropha* PHA biosynthesis genes were examined for this purpose. All recombinant *E. coli* strains could synthesize poly(3HB-*co*-3HV) from glucose and propionic acid, but poly(3HB) homopolymer was accumulated from glucose and valeric acid. The PHA concentration and 3HV fraction could be increased by inducing cells with acetic acid, propionic acid, and/or oleic acid. When supplemented with oleic acid, the 3HV fraction was increased four-fold compared with that obtained without induction. Among the strains examined, *E. coli* XL1-Blue harboring the *R. eutropha* PHA biosynthesis genes accumulated poly(3HB-*co*-3HV) containing 33 mol.% of 3HV up to 63 wt.% of DCW from glucose and propionic acid in flask culture when induced by propionic acid.

Fermentation strategies for the high-level production of poly(3HB-*co*-3HV) were developed using recombinant *E. coli* XL1-Blue harboring the *A. latus* PHA biosynthesis genes (Choi and Lee, 1999b). Acetic acid or oleic acid induction was used to activate the propionic acid uptake into the cells. Poly(3HB-*co*-3HV) having a 3HV fraction of 3 to 20 mol.% could be produced simply by varying the concentration of propionic acid in the feeding solution. When an optimized feeding strategy was used with acetic acid induction and oleic acid supplementation, poly(3HB-*co*-3HV) containing 10.6 mol.% of 3HV could be produced to a high concentration of 158.8 g L^{-1}, and with a high productivity of 2.88 g PHA L^{-1} h^{-1} (Figure 4).

4.2 Recombinant *R. eutropha*

Since *R. eutropha* can accumulate a significant amount of PHA (up to 80 wt.%), several

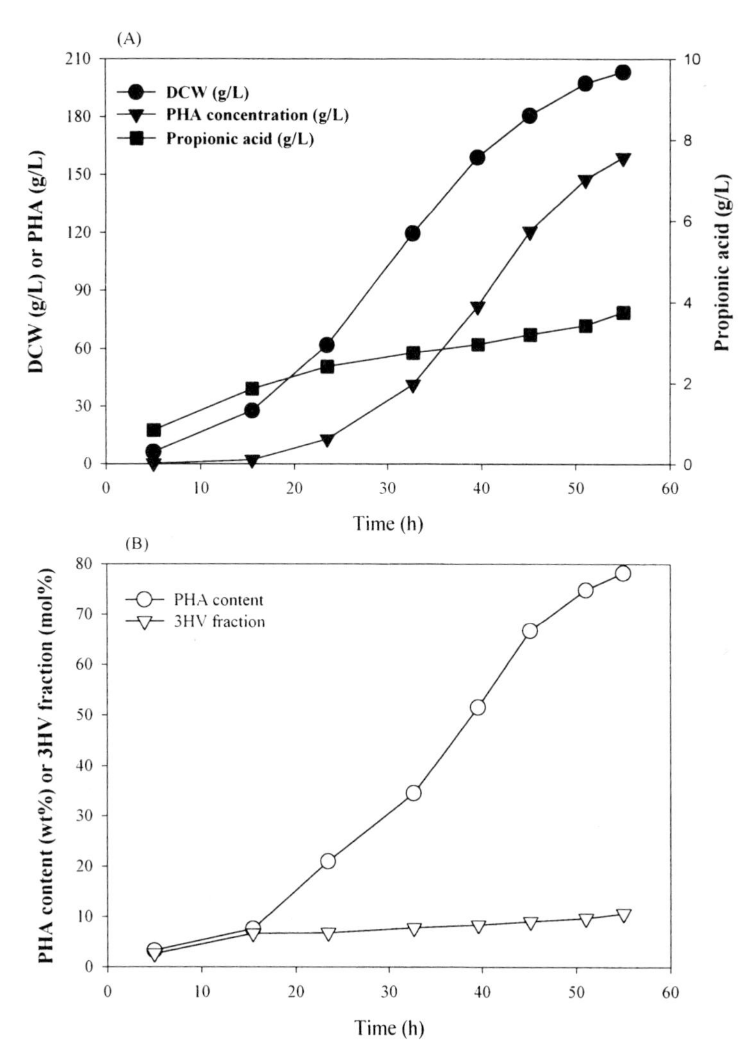

Fig. 4 Time profiles of (A) cell concentration (●), PHA concentration (▼) and residual propionic acid concentration in the medium (■), and (B) PHA content (wt.%) (○) and 3HV fraction in PHA (mol.%) (▽) during the fed-batch culture of *E. coli* XL1-Blue (pJC4) with oleic acid supplementation after acetic acid induction. The feeding solution was added to increase the concentrations of glucose and propionic acid to 20 g L^{-1} and to 5 mM, respectively, after each feeding. [Reproduced from Choi and Lee (1999b), with permission.]

metabolic engineering studies have focused on increasing the PHA synthesis rate and/or yield on the carbon substrate, rather than increasing the PHA content (which is already high). *R. eutropha* was transformed with a broad-host range plasmid containing its own PHA biosynthesis genes (*phaCAB*$_{Re}$) for the homologous amplification of the enzyme activities involved in PHA biosynthesis (Park et al., 1995). The specific activities of three enzymes in recombinant *R. eutropha* were all increased. However, in the batch culture for the production of poly(3HB), the final cell concentrations of the recombinant *R. eutropha* strains did not differ greatly from that of the parent strain. Recombinant *R. eutropha* strains harboring the *phaCAB*$_{Re}$ or *phaC*$_{Re}$ genes produced a slightly higher poly(3HB) concentration and content compared with the wild-type (Park et al., 1995, 1997). The poly(3HB) concentration and content were increased by 18–22% and 8–10%, respectively. However, the reported poly(3HB) contents were only 33 wt.% for the wild-type and 36–40 wt.% for the recombinants – values which were too low considering that the poly(3HB) content typically obtainable with *R. eutropha* was 60–80 wt.%.

The *in-vivo* regulatory mechanisms of the biosynthesis of poly(3HB) and poly(3HB-*co*-3HV) of *R. eutropha* were investigated by using various transformants containing their own PHA biosynthesis genes (*phaCAB*$_{Re}$) (Jung et al., 2000). The biosynthesis rates of poly(3HB) and poly(3HB-*co*-3HV) were found to be controlled by β-ketothiolase and acetoacetyl-CoA reductase, and especially by β-ketothiolase condensing acetyl-CoA or propionyl-CoA. The contents of poly(3HB) and poly(3HB-*co*-3HV) were controlled by poly(3HB) synthase. The mole fraction of 3HV in poly(3HB-*co*-3HV) was also closely correlated with the poly(3HB) synthase activity. The effect of modulating the copy number of the *phbCAB*$_{Re}$ operon on poly(3HB) synthesis rate in *R. eutropha* was also investigated (Jackson and Srienc, 1999). The gene dosage of the poly(3HB) biosynthesis operon in *R. eutropha* was increased by transformation of the broad-host-range vector pKT230 containing the native *R. eutropha* *phbCAB*$_{Re}$ operon. The specific activities of β-ketothiolase and acetoacetyl-CoA reductase were elevated 6.0- and 6.2-fold, respectively, as compared to the control strain with a single operon. However, the parent strain produced poly(3HB) more efficiently during exponential growth on fructose; this suggests that in order to enhance poly(3HB) productivity in *R. eutropha*, increasing the intracellular concentrations of poly(3HB) precursors may be a better strategy than raising the levels of poly(3HB) enzymes (Jackson and Srienc, 1999). There has been no report however on the production of poly(3HB) to a high concentration by fed-batch cultures of these recombinant *R. eutropha* strains (Table 4).

Use of the recombinant *R. eutropha* strains harboring the *phaC*$_{Re}$ gene also led to the production of poly(3HB-*co*-3HV) and poly(3HB-*co*-4HB) (Lee Y. H. et al., 1997). The cell concentration and the PHA content were not enhanced, but only the molar fractions of 3HV and 4HB increased in recombinant *R. eutropha*. Recombinant *R. eutropha* harboring multicopies of the *phaC*$_{Re}$ gene produced poly(3HB-*co*-3HV) with a 3HV mole fraction of 50.5 mol.% from 20 g L^{-1} of propionic acid, whilst the wild-type produced poly(3HB-*co*-3HV) with a 3HV mole fraction of 34.1 mol.%. When 20 g L^{-1} of 4-hydroxybutyric acid was used, the 4HV fraction in PHA was increased from 24.1 mol.% in the wild-type to 57.6 mol.% in the recombinant strain. Other examples have been reported showing that recombinant *R. eutropha* harboring extra copies of the PHA biosynthesis genes produced PHA with an increased

Tab. 4 Production of poly(3HB-*co*-3HV) by fed-batch culture of microorganisms

Strain	*PHA*	*Substrate*	*Time* [h]	*Cell conc.* [g L^{-1}]	*PHA conc.* [g L^{-1}]	*PHA content* [%]	*Productivity* [g L^{-1} h^{-1}]	*Reference*
R. eutropha	Poly(3HB-*co*-3HV)	Glucose +propionic acid	46	158	117	74	2.55	Kim et al. (1994b)
R. eutropha	Poly(3HB-*co*-3HV)	Ethanol +propanol	44	41.4	32	77	0.72	Park and Damodaran (1994)
P. denitrificans	Poly(3HB-*co*-3HV)	Methanol +*n*-amyl alcohol	120	9	2.34	26	0.02	Ueda et al. (1992)
M. organophilum	Poly(3HB-*co*-3HV)	Methanol + valeric acid	78	240	99	41.3	2.8	Kim et al. (1999)
Alcaligenes sp. SH-69	Poly(3HB-*co*-3HV)	Glucose + yeast extract	48	62	36	57	0.87	Rhee et el. (1993)
Recombinant *E. coli*	Poly(3HB-*co*-3HV)	Glucose +propionic acid	55.1	203.1	158.8	78.2	2.88	Choi and Lee (1999b)

PHA content and mole fraction of comonomers. When the *R. eutropha* mutant strain accumulating a terpolyester poly(3HB-*co*-3HV-*co*-4HV) with a 4HV mole fraction of up to 22.7 mol.% was transformed with the hybrid plasmid harboring the *R. eutropha* PHA biosynthesis genes, the molar fraction of 4HV in the polymer that accumulated in the recombinant was increased to 30 mol.%, but the PHA content was increased only slightly (Valentin and Steinbüchel, 1995).

Several mutant strains of *R. eutropha* possessing defective metabolic pathways that competed with the PHA biosynthetic pathway were developed for enhanced PHA production. The isocitrate dehydrogenase leaky mutant of *R. eutropha* accumulated poly(3HB) more favorably at a lower carbon/nitrogen molar ratio and at a lower carbon concentration than did the parent strain (Park and Lee, 1996). In batch culture, the final cell and poly(3HB) concentrations and yield on glucose were also slightly increased. Moreover, in the poly(3HB-*co*-3HV) biosynthesis, the molar fraction of 3HV and the 3HV yield on propionic acid was increased due to the enhanced conversion of propionic acid to 3-hydroxyvaleryl-CoA rather than to acetyl-CoA and CO_2 in this mutant. Another mutant *R. eutropha* strain which was unable to assimilate propionic acid for cell growth produced poly(3HB-*co*-3HV) with an increased molar fraction of 3HV and 3HV yield on propionic acid compared with the parent strain, but the concentration of poly(3HB-*co*-3HV) was decreased (Lee I. Y. et al., 1996a).

5 Economic Considerations in the Production of SCL-PHAs by Bacterial Fermentation

A number of microorganisms have been found to produce PHAs from renewable sources. However, economically feasible PHA producers are limited to several bacteria such as *R. eutropha, A. latus,* and recombinant *E. coli*. To determine which strain is most suitable for the economically feasible production of PHAs on a commercial scale, several factors which have considerable bearing on the production cost of PHAs – including PHA productivity, content, yield on carbon substrate, and the recovery method of PHA from bacterial cells

– have each been carefully examined (Choi and Lee, 1997, 1999c, 2000; Lee and Choi, 1998).

PHA productivity mainly affects the equipment-related cost such as direct-fixed capital dependent cost and labor-dependent cost. PHA content has multiple effects on the efficiency of the recovery process and the PHA yield on the carbon substrate. As the PHA content rises, then a higher recovery yield and purity of PHA can be obtained. This permits the use of less recovery chemicals and leads to a reduction in the cost of waste treatment. A high PHA content also allows the cost of the carbon source to be reduced, and this contributes significantly to the overall production cost of PHA. A good example showing the importance of PHA content in the production of PHA can be seen in the process employing methylotrophic bacteria (Choi and Lee, 1997). Even though this strain can utilize one of the cheapest carbon sources, methanol, this advantage could not compensate the increase in poly(3HB) production cost resulting from the low PHA content and subsequently low PHA yield on carbon source. Several carbon substrates, including cane and beet molasses, whey, plant oils, starch, cellulose, and hemicellulose, have been suggested as good carbon sources for the production of PHAs because of their low price. As mentioned repeatedly, the cost of carbon source has a significant effect on PHA production cost. As described earlier in this chapter, PHA productivity and content are generally low in natural PHA producers from these inexpensive carbon sources. However, as shown from the results of fed-batch cultures of recombinant *E. coli* from whey (Ahn et al., 2000, 2001), the efficient production of PHAs from cheap, renewable carbon sources might be realized by the use of metabolic engineering. The PHA yield on carbon substrate also has a significant effect on the production cost of PHAs, especially when the cosubstrate (e.g., propionic acid or valeric acid), more expensive than when a primary carbon source of glucose, is used to produce the copolymer, poly(3HB-*co*-3HV) (Choi and Lee, 2000). Using a cosubstrate such as propionic acid also has a harmful effect on cell growth, and results in the reduction of PHA productivity. As recently shown (Choi and Lee, 2000), the production cost of poly(3HB-*co*-3HV) is increased in line with the increase in the 3HV fraction in copolymer. Hence, to design an optimal process for the production of copolymer having desirable mechanical properties, the effect of the cosubstrate should be carefully examined.

The recovery of PHAs from bacterial cells has involved several methods, including solvent extraction, digestion of the cells by hypochlorite, enzymatic digestion and simple alkaline digestion (Berger et al., 1989; Holmes and Lim, 1990; Ramsay et al., 1990b, 1994; Hahn et al., 1993; Choi and Lee, 1999a). The recovery process of PHA should be considered on the basis of the efficiency of PHA recovery, the cost of recovery equipment and chemicals used, and waste treatment cost. From an economic standpoint, a simple digestion method using inexpensive chemicals such as NaOH appears to be the most cost-effective. However, this process has so far been successfully demonstrated only for the recovery of PHA from fragile cells such as recombinant *E. coli* with high PHA content (Choi and Lee, 1999a). The process for the production and recovery of PHA by recombinant *E. coli* is shown in Figure 5.

Until now, the lowest poly(3HB) production cost of \$ 2.5–3.0 kg^{-1} has been obtained in processes using *A. latus* and recombinant *E. coli*. From an economic consideration of poly(3HB) production, several factors have been shown to have a combined effect on the

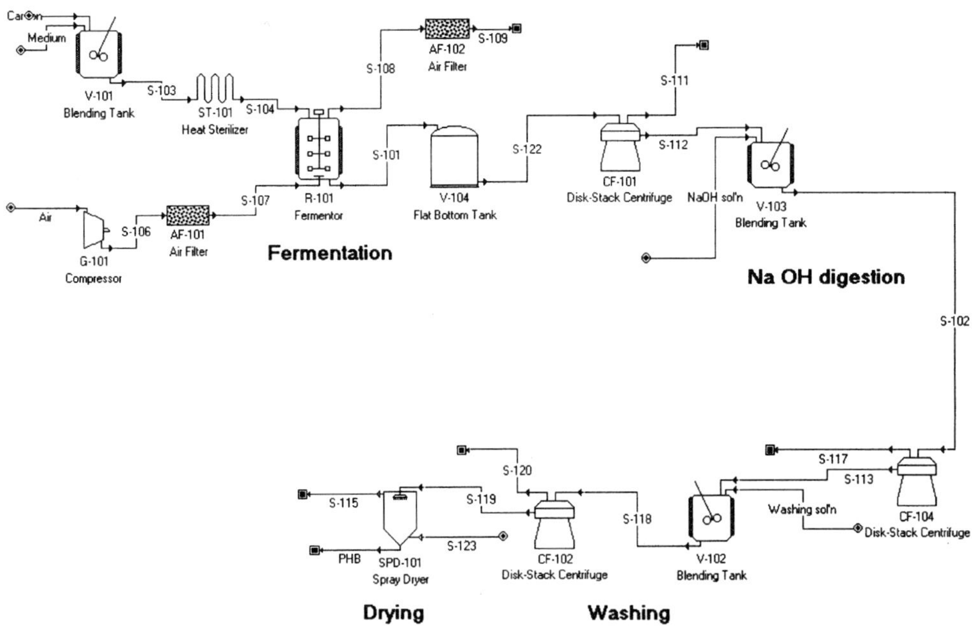

Fig. 5 Process flowsheet for poly(3HB) production by recombinant *E. coli* using the NaOH digestion method.

overall cost of poly(3HB). Recently, several metabolically engineered bacterial strains have been constructed for the efficient production of PHAs, and fermentation strategies for these strains have also been developed. The incorporation of PHA production processes using these strains into the waste treatment process operated in the agricultural industry is a good example of the cost-effective production of PHAs, namely the conversion of environmentally polluting material to environmentally friendly PHA polymers.

6 Outlook and Perspectives

Whilst PHAs have been considered to be good candidates as biodegradable polymers, it is essential to reduce their production costs if they are to be used in commercial applications. Intensive studies carried out on the mechanisms and metabolic pathways of PHA biosynthesis during past decade have allowed the construction of various metabolically engineered bacteria producing SCL-PHAs that are superior to those obtained from 'natural' PHA producers. Among these, recombinant *E. coli* appears to be the best candidate for PHA production, for several advantageous reasons. In recombinant *E. coli* or other bacteria, the method by which PHA precursors are efficiently channeled to PHA synthesis from the existing and/or newly created metabolic pathways is critical for efficient PHA production. This can be accomplished by metabolic engineering based on a thorough knowledge of the physiology, biochemistry, and molecular genetics of PHA biosynthesis. At present, only poly(3HB) and poly(3HB-*co*-3HV) can be produced with sufficiently high productivity. The production of SCL-PHAs using different bacteria is summarized in Tables 2, 3 and 4. Clearly, SCL-PHAs can be produced efficiently from various carbon sources. PHA productivity of almost 5 g L^{-1} h^{-1} is impressively high if it is appreciated that PHA is an intracellular polymeric product, and that the maximum cell productivity (which means simply culturing cells) in fed-batch culture is close to ca. 10 g L^{-1} h^{-1}. It can be seen from the tables that poly(3HB) and poly(3HB-*co*-3HV) are produced most efficiently by recombinant *E. coli* from glucose, though if sucrose (or molasses) is to be used as a carbon source, *A. latus* can be a good host strain for poly(3HB) production. The development of optimized fermentation strategies for the production of PHAs other than poly(3HB) remain to be further optimized. Advances in strain development and fermentation strategies will allow economically feasible production of PHAs, including SCL-PHAs, to be carried out.

7 Patents

Patent number (date)	Inventors	Title	Assignee
US4358583 (1982.11.9)	Walker and Whitton	Extraction of poly(beta-hydroxy butyric acid)	ICI
US4786598 (1988.11.22)	Lafferty and Braunegg	Process for the biotechnological preparation of poly-d-(–)-3-hydroxybutyric acid	Danubia Petrochemie
US5229279 (1993.7.20)	Peoples and Sinskey	Method for producing novel polyester biopolymers	MIT
US5250430 (1993.8.5)	Peoples and Sinskey	Polyhydroxyalkanoate polymerase	MIT
US5334520 (1994.8.2)	Dennis, D. E	Production of poly-beta-hydroxybutyrate in transformed *Escherichia coli*	CIT
US5512669 (1996.4.3)	Peoples and Sinskey	Gene encoding bacterial acetoacetyl-CoA reductase	MIT
US5518907 (1996.5.21)	Dennis, D. E	Cloning and expression in *Escherichia coli* of the *Alcaligenes eutrophus* H16 poly-beta-hydroxybutyrate biosynthetic pathway	CIT
US5661026 (1997.8.26)	Peoples and Sinskey	Gene encoding bacterial beta-ketothiolase	MIT
US5811272 (1998.9.2)	Snell et al.	Method for controlling molecular weight of polyhydroxyalkanoates	MIT
US595874 (1999.9.28)	Gruys et al.	Methods of optimizing substrate pools and biosynthesis of poly-beta-hydroxybutyrate-*co*-poly-beta-hydroxyvalerate in bacteria and plants	Monsanto Co.
US6043063 (2000.3.28)	Asrar et al.	Methods of PHA extraction and recovery using non-halogenated solvents	Monsanto Co.

Acknowledgments

The studies described in this chapter were supported by the Korea-Australia International Cooperative Research Project from the Korean Ministry of Science and Technology (MOST) and by LG Chemicals Ltd.

8
References

Ahn, W. S., Park, S. J., Lee, S. Y. (2000) Production of poly(3-hydroxybutyrate) by fed-batch culture of recombinant *Escherichia coli* with a highly concentrated whey solution, *Appl. Environ. Microbiol.* **66**, 3624–3627.

Ahn, W. S., Park, S. J., Lee, S. Y. (2001) Production of poly(3-hydroxybutyrate) from whey by cell recycle fed-batch culture of recombinant *Escherichia coli*, *Biotechnol. Lett.* **23**, 235–240.

Anderson, A. J., Dawes, E. A. (1990) Occurrence, metabolism, metabolic role, and industrial uses of bacterial polyhydroxyalkanoates, *Microbiol. Rev.* **54**, 450–472.

Berger, E., Ramsay, B. A., Ramsay, J. A., Chavarie, C., Braunegg, G. (1989) PHB recovery by hypochlorite digestion of non-PHB biomass, *Biotechnol. Tech.* **3**, 227–232.

Bormann, E. J., Leißner, M., Beer, B. (1998a) Growth associated production of poly(hydroxybutyric acids) by *Azotobacter beijerinckii* from organic nitrogen substrates, *Appl. Microbiol. Biotechnol.* **49**, 84–88.

Bormann, E. J., Leißner, M., Roth, M., Beer, B., Metzner, K. (1998b) Production of polyhydroxybutyrate by *Ralstonia eutropha* from protein hydrolysates, *Appl. Microbiol. Biotechnol.* **50**, 604–607.

Bourque, D., Ouellette, B., Andre, G., Groleau, D. (1992) Production of poly-β-hydroxybutyrate (PHB) from methanol: characterization of a new isolate of *Methylobacterium extorquens*, *Appl. Microbiol. Biotechnol.* **37**, 7–12.

Bourque, D., Pomerleau, Y., Groleau, D. (1995) High-cell-density production of poly-β-hydroxybutyrate (PHB) from methanol by *Methylobacterium extorquens*: production of high-molecular-mass PHB, *Appl. Microbiol. Biotechnol.* **44**, 367–376.

Brandl, H., Knee, E. J., Fuller, R.C., Gross, R.A., Renz, R. W. (1989) Ability of the phototrophic bacterium *Rhodospirillum rubrum* to produce various poly(β-hydroxyalkanoates): potential sources for biodegradable polyester, *Int. J. Biol. Macromol.* **11**, 49–55.

Brandl, H., Bachofen, R., Mayer, J., Wintermantel, E. (1995) Degradation and applications of polyhydroxyalkanoates, *Can. J. Microbiol.* **41**(Suppl. 1), 143–153.

Chang, H. N., Yoo, I., Kim, B. S. (1994) High density cell culture by membrane-based cell recycle. *Biotech. Adv.* 12, 467–487.

Cho, K.-S., Ryu, H. W., Park, C.-H, Goodrich, P. R. (1997) Poly(hydroxybutyrate-*co*-hydroxyvalerate) from swine waste liquor by *Azotobacter vinelandii* UWD, *Biotechnol. Lett.*, **19**, 7–10.

Choi, J., Lee, S. Y. (1997) Process analysis and economic evaluation for poly(3-hydroxybutyrate) production by fermentation, *Bioprocess Eng.* **17**, 335–342.

Choi, J., Lee, S. Y. (1999a) Efficient and economical recovery of poly(3-hydroxybutyrate) from recombinant *Escherichia coli* by simple digestion with chemicals, *Biotechnol. Bioeng.* **62**, 546–553.

Choi, J., Lee, S. Y. (1999b) High-level production of poly(3-hydroxybutyrate-*co*-3-hydroxyvalerate) by fed-batch culture of recombinant *Escherichia coli*, *Appl. Environ. Microbiol.* **65**, 4363–4368.

Choi, J., Lee, S. Y. (1999c) Factors affecting the economics of polyhydroxyalkanoate production by bacterial fermentation, *Appl. Microbiol. Biotechnol.* **51**, 13–21.

Choi, J., Lee, S. Y. (2000) Economic consideration in the production of poly(3-hydroxybutyrate-*co*-3-hydroxyvalerate) by bacterial fermentation, *Appl. Microbiol. Biotechnol.* **53**, 646–649.

Choi, J., Lee, S. Y., Han, K. (1998) Cloning of the *Alcaligenes latus* polyhydroxyalkanoate biosynthesis genes and use of these genes for enhanced production of poly(3-hydroxybutyrate) in *Esche-*

richia coli, *Appl. Environ. Microbiol.* **64**, 4897–4903.

Doi, Y. (1990) *Microbial Polyesters.* New York: VCH.

Fidler, S., Dennis, D. (1992) Polyhydroxyalkanoate production in recombinant *Escherichia coli, FEMS Microbiol. Rev.* **103**, 231–236.

Hahn, S. K., Chang, Y. K., Kim, B. S., Lee, K. M., Chang, H. N. (1993) The recovery of poly(3-hydroxybutyrate) by using dispersions of sodium hypochlorite solution and chloroform, *Biotechnol. Tech.* **7**, 209–212.

Hangii, U. J. (1990) Pilot scale production of PHB with *Alcaligenes latus*, in: *Novel Biodegradable Microbial Polymers* (Dawes, E. A., Ed.), Dordrecht: Kluwer Academic Publishers, 65–70.

Haywood, G. W., Anderson, A. J., Williams, G. A., Dawes, E. A., Ewing, D. F. (1991) Accumulation of a poly(hydroxyalkanoate) copolymer containing primarily 3-hydroxyvalerate from simple carbohydrate substrates by *Rhodococcus* sp. NCIMB 40126, *Int. J. Biol. Macromol.* **13**, 83–87.

Holmes, P. A., Lim, G. B. (1990) Separation process, US patent 4,910,145.

Jackson, J. K., Srienc, F. (1999) Effects of recombinant modulation of the *phbCAB* operon copy number on PHB synthesis rates in *Ralstonia eutropha, J. Biotechnol.* **68**, 49–60.

Jang, J.-H., and Rogers, P. L. (1996) Effect of levulinic acid on cell growth and poly-β-hydroxyalkanoate production by *Alcaligenes* sp. SH-69, *Biotechnol. Lett.* **18**, 219–224.

Jung, Y. M., Park, J. S., Lee, Y. H. (2000) Metabolic engineering of *Alcaligenes eutrophus* through the transformation of cloned *phbCAB* genes for the investigation of the regulatory mechanism of polyhydroxyalkanoate biosynthesis, *Enzyme Microb. Technol.* **26**, 201–208.

Kang, C. K., Lee, H. S., Kim, J. H. (1993) Accumulation of PHA and its copolyester by *Methylobacterium* sp. KCTC 0048, *Biotechnol. Lett.* **15**, 1017–1020.

Kato, M., Bao, H. J., Kang, C. K., Fukui, T., Doi, Y. (1996) Production of a novel copolyester of 3-hydroxybutyric acids and medium-chain-length 3-hydroxyalkanoic acids by *Pseudomonas* sp.61-3 from sugars, *Appl. Microbiol. Biotechnol.* **45**, 363–370.

Kim, B. S., Lee, S. Y., Chang, H. N. (1992) Production of poly-β-hydroxybutyrate by fed-batch culture of recombinant *Escherichia coli, Biotechnol. Lett.* **14**, 811–816.

Kim, B. S., Lee, S. C., Lee, S. Y., Chang, H. N., Chang, Y. K., Woo, S. I. (1994a) Production of poly(3-hydroxybutyric acid) by fed-batch culture of *Alcaligenes eutrophus* with glucose concentration control, *Biotechnol. Bioeng.* **43**, 892–898.

Kim, B. S., Lee, S. C., Lee, S. Y., Chang, H. N., Chang, Y. K., Woo, S. I. (1994b) Production of poly(3-hydroxybutyric-*co*-3-hydroxyvaleric acid) by fed-batch culture of *Alcaligenes eutrophus* with substrate control using on-line glucose analyzer, *Enzyme Microbiol. Technol.* **16**, 556–561.

Kim, S. W., Kim, P., Kim, J. H. (1996) High production of poly-β-hydroxybutyrate (PHB) from *Methylobacterium organophilum* under potassium limitation, *Biotechnol. Lett.* **18**, 25–30.

Kim, S. W., Kim, P., Kim, J. H. (1999) Production of poly(3-hydroxybutyrate-*co*-3-hydroxyvalerate) from *Methylobacterium organophilum* by potassium-limited fed-batch culture, *Enzyme Microb. Technol.* **24**, 555–560.

Kobayashi, G., Shiotani, T., Shima, Y., Doi, Y. (1994) Biosynthesis and characterization of poly(3-hydroxybutyrate-*co*-3-hydroxyhexanoate) from oils and fats by *Aeromonas* sp. OL-338 and *Aeromonas* sp. FA440, in: *Biodegradable Plastics and Polymers* (Doi, Y. and Fukuda, K. Eds), Amsterdam: Elsevier, 410–416.

Koyama, N., Doi, Y. (1995) Continuous production of poly(3-hydroxybutyrate-*co*-3-hydroxyvalerate) by *Alcaligenes eutrophus, Biotechnol. Lett.* **17**, 281–284.

Lee, I. Y., Kim, G. J. Shin, Y. C., Chang, H. N., Park, Y. H. (1995a) Production of poly(β-hydroxybutyrate-*co*-β-hydroxyvalerate) by two-stage fed-batch fermentation of *Alcaligenes eutrophus, J. Microbiol. Biotechnol.* **5**, 292–296.

Lee, I. Y., Kim, M. K., Kim, J. G., Chang, H. N., Park, Y. H. (1995b) Production of poly(β-hydroxybutyrate-*co*-β-hydroxyvalerate) from glucose and valerate in *Alcaligenes eutrophus, Biotechnol. Lett.* **17**, 571–574.

Lee, I. Y., Kim, G. J., Choi, D. K., Yeon, B. K., Park, Y. H. (1996a) Improvement of hydroxyvalerate fraction in poly(β-hydroxybutyrate-*co*-β-hydroxyvalerate) by a mutant strain of *Alcaligenes eutrophus, J. Ferment. Bioeng.* **81**, 255–258.

Lee, I. Y., Kim, M. K., Park, Y. H., Lee, S. Y. (1996b) Regulatory effects of cellular nicotinamide nucleotide and enzyme activities on poly(3-hydroxybutyrate) synthesis in recombinant *Escherichia coli, Biotechnol. Bioeng.* **52**, 707–712.

Lee, S. Y. (1994) Suppression of filamentation in recombinant *Escherichia coli* by amplified FtsZ activity, *Biotechnol. Lett.* **16**, 1247–1252.

Lee, S. Y. (1996a) High cell-density culture of *Escherichia coli*, Trends Biotechnol. **14**, 98–105.

Lee, S. Y. (1996b) Plastic bacteria? Progress and prospects for polyhydroxyalkanoate production in bacteria, *Trends Biotechnol.* **14**, 431–438.

Lee, S. Y. (1996c) Bacterial polyhydroxyalkanoates, *Biotechnol. Bioeng.* **49**, 1–14.

Lee, S. Y. (1997) *E. coli* moves into the plastic age, *Nature Biotechnol.* **15**, 17–18.

Lee, S. Y. (1998) Poly(3-hydroxyalkanoate) production from xylose by recombinant *Escherichia coli*, *Bioprocess Eng.* **18**, 397–399.

Lee, S. Y., Chang, H. N. (1993) High cell density cultivation of *Escherichia coli* W using sucrose as a carbon source, *Biotechnol. Lett.* **15**, 971–974.

Lee, S.Y., Chang, H. N. (1994) Effect of complex nitrogen source on the synthesis and accumulation of poly(3-hydroxybutyric acid) by recombinant *Escherichia coli* in flask and fed-batch cultures, *J. Environ. Polymer Degrad.* **2**, 169–176.

Lee, S. Y., Chang, H. N. (1995) Production of poly(hydroxyalkanoic acid), *Adv. Biochem. Eng. Biotechnol.* **52**, 27–58.

Lee, S. Y., Chang, H. N. (1996) Characteristics of poly(3-hydroxybutyric acid) synthesis by recombinant *Escherichia coli*, *Ann. N.Y. Acad. Sci.* **782**, 133–142.

Lee, S. Y., Choi, J. (1998) Effect of fermentation performance on the economics of poly(3-hydroxybutyrate) production by *Alcaligenes latus*, *Polymer Degrad. Stabil.* **59**, 387–393.

Lee, S. Y., Choi J. (1999) Polyhydroxyalkanoates: biodegradable polymer, in: *Manual of Industrial Microorganisms and Biotechnology* (Demain, A. L., Davies, J. E., Eds), Washington, DC: American Society for Microbiology, 616–627.

Lee, S. Y., Chang, H. N., Chang, Y. K. (1994a) Production of poly(β-hydroxybutyric acid) by recombinant *Escherichia coli*, *Ann. N.Y. Acad. Sci.* **721**, 43–53.

Lee, S. Y., Lee, K. M., Chang, H. N., Steinbüchel, A. (1994b) Comparison of *Escherichia coli* strains for synthesis and accumulation of poly-(3-hydroxybutyric acid) and morphological changes, *Biotechnol. Bioeng.* **44**, 1337–1347.

Lee, S. Y., Yim, K. S., Chang, H. N., Chang, Y. K. (1994c) Construction of plasmids, estimation of plasmid stability, and use of stable plasmids for the production of poly(3-hydroxybutyric acid) in *Escherichia coli*, *J. Biotechnol.* **32**, 203–211.

Lee, S. Y., Lee, Y. K., Chang, H. N. (1995) Stimulatory effects of amino acids and oleic acid on poly(3-hydroxybutyric acid) synthesis by recombinant *Escherichia coli*, *J. Ferment. Bioeng.* **79**, 177–180.

Lee, S. Y., Middelberg, A. P. J., Lee, Y. K. (1997) Poly(3-hydroxybutyrate) production from whey using recombinant *Escherichia coli*, *Biotechnol. Lett.* **19**, 1033–1035.

Lee, Y. H., Park, J. S., Huh, T. L. (1997) Enhanced biosynthesis of P(3HB-3HV) and P(3HB-4HB) by amplification of the cloned PHB biosynthesis genes in *Alcaligenes eutrophus*. *Biotechnol. Lett.* **19**, 771–774.

Lemoigne, M. (1926) Products of dehydration and of polymerization of β-hydroxybutyric acid, *Bull. Soc. Chem. Biol.* **8**, 770–782.

Liu, F., Li, W., Ridgway, D., Gu, T. (1998) Production of poly-β-hydroxybutyrate on molasses by recombinant *Escherichia coli*, *Biotechnol. Lett.* **20**, 345–348.

Madison, L. L, Huisman, G. W. (1999) Metabolic engineering of poly(3-hydroxyalkanoates): from DNA to plastic, *Microbiol. Mol. Biol. Rev.* **63**, 21–53.

Page, W. J. (1989) Production of poly-β-hydroxybutyrate by *Azotobacter vinelandii* strain UWD during growth on molasses and other complex carbon sources, *Appl. Microbiol. Biotechnol.* **31**, 329–333.

Page, W. J., Knosp, O. (1989) Hyperproduction of poly-β-hydroxybutyrate during exponential growth of *Azotobacter vinelandii* UWD, *Appl. Environ. Microbiol.* **55**, 1334–1339.

Page, W. J., Manchak, J., Rudy, B. (1992) Formation of poly(hydroxybutyrate-*co*-hydroxyvalerate) by *Azotobacter vinelandii* UWD, *Appl. Environ. Microbiol.* **58**, 2866–2873.

Park, C.-H., Damodaran, V. K. (1994) Effect of alcohol feeding mode on the biosynthesis of poly(3-hydroxybutyrate-*co*-3-hydroxyvalerate), *Biotechnol. Bioeng.* **44**, 1306–1314.

Park, H. C., Park, J. S., Lee, Y. H., Huh, T. L. (1995) Production of poly-β-hydroxybutyrate by *Alcaligenes eutrophus* transformants harboring cloned *phbCAB* genes, *Biotechnol. Lett.* **17**, 735–740.

Park, J. S., Lee, Y. H. (1996) Metabolic characteristics of isocitrate dehydrogenase leaky mutant of *Alcaligenes eutrophus* and its utilization for poly-β-hydroxybutyrate production, *J. Ferment. Bioeng.* **81**, 197–205.

Park, J. S., Huh, T. L., Lee, Y. H. (1997) Characteristics of cell growth and poly-β-hydroxybutyrate biosynthesis of *Alcaligenes eutrophus* transformants harbouring cloned *phbCAB* genes, *Enzyme Microb. Technol.* **21**, 85–90.

Peoples, O. P., Sinskey, A. J. (1989) Poly-β-hydroxybutyrate biosynthesis in *Alcaligenes eutrophus* H16. Identification and characterization of the PHB polymerase gene (*phbC*), *J. Biol. Chem.* **264**, 15298–15303.

Ramsay, B. A., Lomaliza, K., Chavarie, C., Dube, B., Bataille, P., Ramsay, J. A. (1990a) Production of poly-(β-hydroxybutyric-*co*-β-hydroxyvaleric) acids, *Appl. Environ. Microbiol.* **56**, 2093–2098.

Ramsay, J. A., Berger, E., Ramsay, B. A., Chavarie, C. (1990b) Recovery of poly-β-hydroxybutyric acid granules by a surfactant-hypochlorite treatment, *Biotechnol. Tech.* **4**, 221–226.

Ramsay, J. A., Berger, E., Voyer, R., Chavarie, C., Ramsay, B. A. (1994) Extraction of poly-β-hydroxybutyrate using chlorinated solvents, *Biotechnol. Tech.* **8**, 589–594.

Rhee, Y. H., Jang, J. H., Rogers, P. L.(1993) Production of copolymer consisting of 3-hydroxybutyrate and 3-hydroxyvalerate by fed-batch culture of *Alcaligenes* sp. SH-69, *Biotechnol. Lett.* **15**, 377–382.

Rhie, H. G., Dennis, D. (1995) The function of *ackA* and *pfa* genes is necessary for poly(3-hydroxybutyrate-*co*-3-hydroxyvalerate) synthesis in recombinant *pha*+ *Escherichia coli*, *Can. J. Microbiol.* **41**(Suppl. 1), 200–206.

Ryu, H. W., Hahn, S. K., Chang, Y. K., Chang, H. N. (1997) Production of poly(3-hydroxybutyrate) by high cell density fed-batch culture of *Alcaligenes eutrophus* with phosphate limitation, *Biotechnol. Bioeng.* **55**, 28–32.

Schmack, G., Steinbüchel, A. (1995) Large-scale production of poly(3-hydroxyvaleric acid) by fermentation of *Chromobacterium violaceum*, processing, and characterization of the homopolyester. *J. Environ. Polymer Degrad.* **3**, 243–258.

Schubert, P., Steinbüchel, A., Schlegel, H. G. (1988) Cloning of the *Alcaligenes eutrophus* poly-β-hydroxybutyrate synthetic pathway and synthesis of PHB in *Escherichia coli*, *J. Bacteriol.* **170**, 5837–5847.

Slater, S. C., Voige, W. H., Dennis, D. (1988) Cloning and expression in *Escherichia coli* of the *Alcaligenes eutrophus* H16 poly-β-hydroxybutyrate biosynthetic pathway, *J. Bacteriol.* **170**, 4431–4436.

Slater, S. C., Gallaher, T., Dennis, D. (1992) Production of poly-(3-hydroxybutyrate-*co*-3-hydroxyvalerate) in a recombinant *Escherichia coli* strain, *Appl. Environ. Microbiol.* **58**, 1089–1094.

Steinbüchel, A. (1991) Polyhydroxyalkanoic acids, in: *Biomaterials: Novel Materials from Biological Sources* (Byrom, D., Ed.), New York: Stockton, 124–213.

Steinbüchel, A. (1992) Biodegradable plastics, *Curr. Opin. Biotechnol.* **3**, 291–297.

Steinbüchel, A., Fuchtenbusch, B. (1998) Bacterial and other biological systems for polyester production, *Trends Biotechnol.* **16**, 419–427.

Steinbüchel, A., Valentin, H. E. (1995) Diversity of bacterial polyhydroxyalkanoic acid, *FEMS Microbiol. Lett.* **128**, 219–228.

Steinbüchel, A., Wiese, S. (1992) A *Pseudomonas* strain accumulating polyesters of 3-hydroxybutyric acid and medium-chain-length 3-hydroxyalkanoic acids, *Appl. Microbiol. Biotechnol.* **37**, 691–697.

Suzuki, T., Yamane, T., Shimizu, S. (1986a) Kinetics and effect of nitrogen source feeding on production of poly-β-hydroxybutyric acid by fed-batch culture, *Appl. Microbiol. Biotechnol.* **24**, 366–369.

Suzuki, T., Yamane, T., Shimizu, S. (1986b) Mass production of poly-β-hydroxybutyric acid by fed-batch culture with controlled carbon/nitrogen feeding, *Appl. Microbiol. Biotechnol.* **24**, 370–374.

Suzuki, T., Yamane, T., Shimizu, S. (1986c) Mass production of poly-β-hydroxybutyric acid by fully automatic fed-batch culture of a methylotroph, *Appl. Microbiol. Biotechnol.* **23**, 322–329.

Suzuki, T., Deguchi, H., Yamane, T., Shimizu, S., Gekko, K. (1988) Control of molecular weight of poly-β-hydroxybutyric acid produced in fed-batch culture of *Protomonas extorquens*, *Appl. Microbiol. Biotechnol.* **27**, 487–491.

Ueda, S., Matsumoto, S., Takagi, A., Yamane, T. (1992) Synthesis of poly(3-hydroxybutyrate-*co*-3-hydroxyvalerate) from methanol and n-amyl alcohol by the Methylotrophic bacteria *Paracoccus denitrificans* and *Methylobacterium extorquens*, *Appl. Environ. Microbiol.* **58**, 3574–3579.

Valentin, H. E., Steinbüchel, A. (1995) Accumulation of poly(3-hydroxybutyric-*co*-3-hydroxyvaleric acid-*co*-4-hydroxyvaleric acid) by mutants and recombinant strains of *Alcaligenes eutrophus*, *J. Environ. Polymer Degrad.* **3**, 169–175.

Wang, F., Lee, S. Y. (1997a) Poly(3-hydroxybutyrate) production with high polymer content by fed-batch culture of *Alcaligenes latus* under nitrogen limitation, *Appl. Environ. Microbiol.* **63**, 3703–3706.

Wang, F., Lee, S. Y. (1997b) Production of poly(3-hydroxybutyrate) by fed-batch culture of filamentation-suppressed recombinant *Escherichia coli*, *Appl. Environ. Microbiol.* **63**, 4765–4769.

Wang, F., Lee, S. Y. (1998) High cell density culture of metabolically engineered *Escherichia coli* for the production of poly(3-hydroxybutyrate) in a defined medium, *Biotechnol. Bioeng.* **58**, 325–328.

Wong, H. H., Lee, S. Y. (1998) Poly(3-hydroxybutyrate) production from whey by high-density cultivation of recombinant *Escherichia coli, Appl. Microbiol. Biotechnol.* **50**, 30–33.

Wong, H. H., van Wegen, R. J., Choi, J., Lee, S. Y., Middelberg, A. P. J. (1999) Metabolic analysis of poly(3-hydroxybutyrate) production by recombinant *Escherichia coli, J. Microbiol. Biotechnol.* **9**, 593–603.

Yamane, T., Fukunaga, M., Lee, Y. W. (1996) Increased PHB productivity by high-cell-density fed-batch culture of *Alcaligenes latus*, a growth-associated PHB producer, *Biotechnol. Bioeng.* **50**, 197–202.

Yim, K. S., Lee, S. Y., Chang, H. N. (1996) Synthesis of poly(3-hydroxybutyrate-*co*-3-hydroxyvalerate) by recombinant *Escherichia coli, Biotechnol. Bioeng.* **49**, 495–503.

Zhang, H., Obias, V., Gonyer, K., Dennis, D. (1994) Production of polyhydroxyalkanoates in sucrose-utilizing recombinant *Escherichia coli* and *Klebsiella* strains, *Appl. Environ. Microbiol.* **60**, 1198–1205.

10
Fermentative Production of Medium-chain-length Poly(3-hydroxyalkanoate)

Ruud A. Weusthuis[1], **Birgit Kessler**[2], **Marcia P. M. Dielissen**[3], **Bernard Witholt**[4], **Gerrit Eggink**[5]

[1] Agrotechnological Research Institute (ATO-DLO), Bornsesteeg 59, PO Box 17, 6700 AA Wageningen, The Netherlands; Tel.: +31-317-475300; Fax: +31-317-475347; E-mail: r.a.weusthuis@ato.wag-ur.nl

[2] Institute of Biotechnology, ETH Zürich, Hönggerberg HPT, 8093 Zürich, Switzerland; Tel.: +41-1-633-3286; Fax: +41-1-633-1051; E-mail: kessler@biotech.biol.ethz.ch

[3] Agrotechnological Research Institute (ATO-DLO), Bornsesteeg 59, PO Box 17, 6700 AA Wageningen, The Netherlands; Tel.: +31-317-475300; Fax: +31-317-475347; E-mail: m.p.m.dielissen@ato.wag-ur.nl

[4] Institute of Biotechnology, ETH Zürich, Hönggerberg HPT, 8093 Zürich, Switzerland; Tel.: +41-1-633-3286; Fax: +41-1-633-1051; E-mail: witholt@biotech.biol.ethz.ch

[5] Agrotechnological Research Institute (ATO-DLO), Bornsesteeg 59, PO Box 17, 6700 AA Wageningen, The Netherlands; Tel.: +31-317-475300; Fax: +31-317-475347; E-mail: g.eggink@ato.wag-ur.nl

Biotechnology of Biopolymers. From Synthesis to Patents. Edited by A. Steinbüchel and Y. Doi

ISBN: 3-527-31110-6

MCL	medium chain length
Poly(3HA)	poly(3-hydroxyalkanoate)
Poly(3HB)	poly(3-hydroxybutyrate)
Poly(3HB-*co*-3HV)	poly(3-hydroxybutyrate-*co*-3-hydroxyvalerate)
PSA	pressure-sensitive adhesive
SCL	short chain length

1 Introduction

Medium-chain-length poly-3-hydroxyalkanoate (MCL-Poly(3HA)) forms a large and versatile family of polyesters produced by various bacteria in nature. MCL-Poly(3HA)s are receiving considerable attention because of their potential as renewable and biodegradable plastics, and the monomers as a source of chiral synthons. A wide range of substituted hydroxyalkanoic acids can be incorporated into these polyesters in biotechnological processes. Various fermentation strategies have been developed and optimized in order to control the monomer composition of the polymer, enabling the tailoring of the material properties and the production of MCL-Poly(3HA)s in an economically efficient manner. Production processes of MCL-Poly(3HA) are presented in comparison to alternative production strategies. Furthermore, biosynthesis of MCL-Poly(3HA)s, including 'functionalized' Poly(3HA)s, is discussed.

2
Historical Outline

The first example of microbial Poly(3HA)s to be discovered was polyhydroxybutyrate (Poly(3HB)) in 1926 (Lemoigne, 1926). Since then Poly(3HB) accumulation was found in various microorganisms, representatives of Gram-negative and Gram-positive species (i.e. autotrophs, heterotrophs, phototrophs, aerobes, and anaerobes), and archaebacteria (as reviewed elsewhere: Steinbüchel, 1991; Lee, 1996; Sasikala and Ramana, 1996).

The discovery of a polyester consisting mainly of hydroxyoctanoate monomers by de Smet et al. (1983) was the first example of a new group, the so-called MCL-Poly(3HA)s which can contain a wide variety of different monomers.

The MCL-Poly(3HA)s are of interest for specific uses, where the chirality and elastomeric properties of the polymers are important. In addition, the monomers of Poly(3HA)s that contain different functional groups in their side chain are receiving more and more attention as source of chiral synthons (Ohashi and Hasegawa, 1992; Witholt et al., 1992). In this report we will focus on microbial production of these polyesters by fermentation and present economic considerations.

3
Occurrence

MCL-Poly(3HA) production is restricted to fluorescent Pseudomonads belonging to rRNA-homology group 1 (Huisman et al., 1989). Members of this group are, amongst others, *Pseudomonas aeruginosa, Pseudomonas fluorescens, Pseudomonas oleovorans, Pseudomonas lemonnieri, Pseudomonas testosteroni*, and *Pseudomonas putida*. MCL-Poly(3HA) is not just one single polymer, but a family of biopolyesters, which differ with respect to monomer composition. To date, more than 100 different monomers were found in the polymers (Steinbüchel and Valentin, 1995). Among those are 3-hydroxy acids of 6–16 carbon atoms with a large variety of saturated, unsaturated, straight, or branched chains containing aliphatic or aromatic side groups. Furthermore, monomers with various different functional groups in the side chain such as halogen atoms, hydroxy-, epoxy-, cyano-, carboxyl-, phenoxy-, cyanophenoxy-, nitrophenoxy-, and esterified carboxyl groups have been introduced into MCL-Poly(3HA)s (for review, see: Lenz et al., 1992; Steinbüchel and Valentin, 1995; Sasikala and Ramana, 1996). The 3-hydroxyalkanoic acid monomer units in these microbial polyesters are all in the *R*-configuration due to the stereospecificity of biosynthetic enzymes.

The molecular weights of the polymers range from 2×10^5 to 3×10^6, depending on the specific polymer, the microorganism, and the growth conditions.

4
Functions

MCL-Poly(3HA)s function as a reserve material for carbon and energy. They are formed when an excess carbon source is present.

Because MCL-Poly(3HA) is a polymer, a large amount of reserve material can be stored without affecting the osmotic pressure of the cell. When the supply of the carbon source becomes limiting, Poly(3HA) can be degraded by intracellular depolymerases and subsequently metabolized as carbon and energy source (Merrick and Doudoroff, 1964). The ability to convert excess substrate in the environment to reserve material is an advantage in the competition for survival because it limits the availability of the substrate for other microorganisms.

Another possible function of MCL-Poly-(3HA) is detoxification. Substrates such as alkanes, alkanols, and fatty acids are toxic to microorganisms at low concentrations. Fast removal of these substrates from the environment by conversion to MCL-Poly(3HA) would improve the viability of the microorganism (Kranz et al., 1997).

Apparently, different kinds of Poly(3HA)s have been developed during evolution. This makes one wonder what the functional differences between these Poly(3HA)s are. The calculated energetic efficiencies of MCL-Poly(3HA) and SCL-Poly(3HA) are compared below.

MCL-Poly(3HA) is especially effective as a storage material when aliphatic substrates are used as a carbon source. For example, the conversion of decanoic acid into acetyl-CoA via MCL-Poly(3HA) (poly(3-hydroxydecanoate)) costs only 1 additional ATP compared to the direct conversion of decanoic acid to acetyl-CoA, assuming that the Poly-(3HA) monomers are activated after depolymerization by means of a synthetase (Figure 1a). If SCL-Poly(3HA) (Poly(3HB)) is the storage material, 2.5 ATP has to be invested (Figure 1b). Also the efficiency in storage of the reducing power of MCL-Poly(3HA) with aliphatic substrates is higher. The conversion of decanoic acid into 3-hydroxydecanoic acid generates only 1 FADH; the remaining reducing power is stored in the polymer (Figure 1a). The conversion of decanoic acid in 3-hydroxybutyric acid, on the other hand, generates more reducing power equivalents, 1.5 NADH and 4 FADH, resulting in a lower reducing power storage capacity (Figure 1b).

SCL-Poly(3HA)s, on the other hand, are more efficient storage materials when carbohydrates are used as a carbon source. This is caused by the fact that production of MCL-Poly(3HA) by fatty acid synthesis requires more ATP and reducing equivalents than the degradation of MCL-Poly(3HA) by β-oxidation generates (Figures 1c and d).

Thus, MCL-Poly(3HA) is the more efficient storage material when aliphatic substrates are degraded by the β-oxidation pathway, whereas SCL-Poly(3HA) are more efficient with other substrates.

5 Biochemistry

The material properties of MCL-Poly(3HA) can be programmed during the fermentation phase. The most important tool to control the material properties is the monomer composition. The monomer composition of MCL-Poly(3HA) can be varied by using different substrates. The conversion of these substrates is specific for the substrate used and the metabolic pathway involved.

5.1 β-Oxidation

Lageveen et al. (1988) showed that the monomer composition of aliphatic saturated MCL-Poly(3HA) produced by *P. oleovorans* depended on the type of *n*-alkane used. It appeared that the *n*-alkanes were degraded by the subsequent removal of C2 units and therefore it was proposed that the β-oxida-

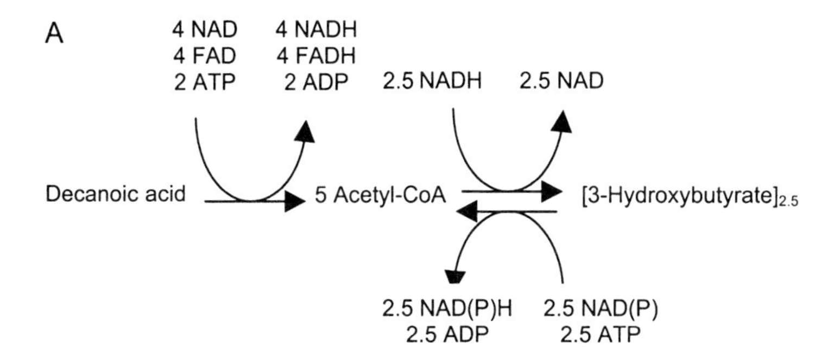

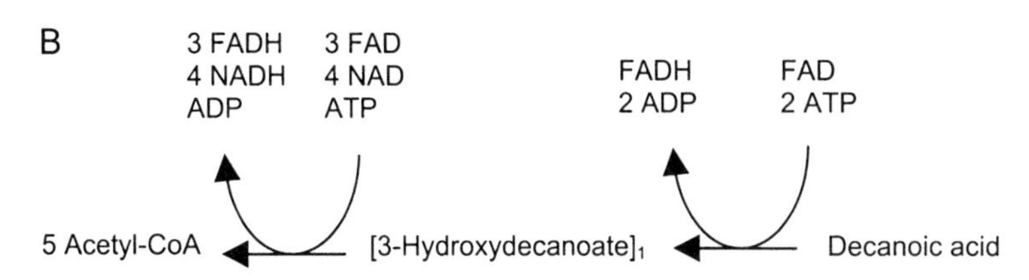

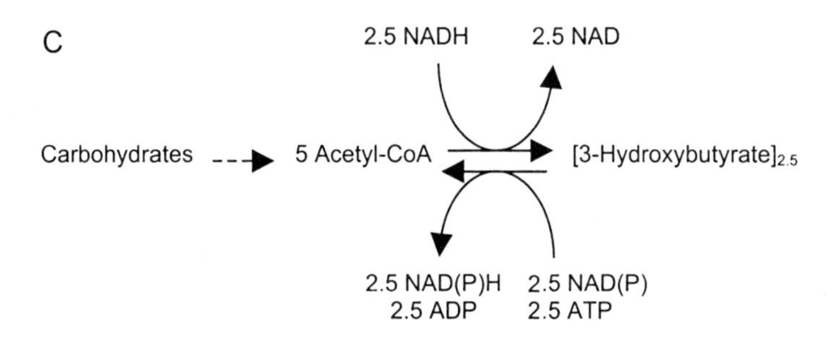

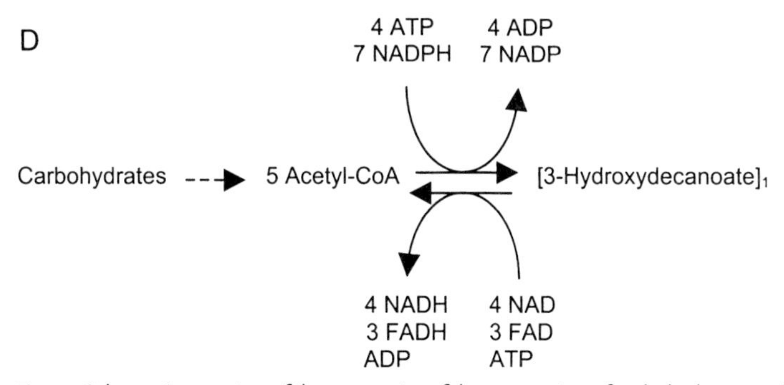

Fig. 1 Schematic overview of the energetics of the conversion of carbohydrates and fatty acids into SCL-PHA (PHB) and MCL-PHA (poly-3-hydroxydecanoate).

Tab. 1 Monomer composition of the MCL-poly(3HA)s produced by *P. oleovorans* grown on *n*-alkanes as the sole carbon and energy source (Preusting et al., 1990)

C source	*C4*	*C5*	*C6*	*C7*	*C8*	*C9*	*C10*
n-Hexane			83.1 ± 0.5	< 1.0	12.0 ± 0.2	< 1.0	4.9 ± 0.4
n-Heptane		2.5 ± 0.1		97.5 ± 0.1	< 1.0	< 1.0	
n-Octane	< 1.0		12.2 ± 0.2		87.8 ± 0.2		< 1.0
n-Nonane		2.3 ± 0.1		40.6 ± 0.4	1.4 ± 0.2	55.7 ± 0.4	
n-Decane	< 1.0		11.1 ± 0.4	1.1 ± 0.1	65.8 ± 0.2	1.2 ± 0.1	20.8 ± 0.7

C4: 3-hydroxybutyrate; C5: 3-hydroxyvalerate; C6: 3-hydroxyhexanoate; C7: 3-hydroxyheptanoate; C8: 3-hydroxyoctanoate; C9: 3-hydroxynonanoate; C10: 3-hydroxydecanoate.

tion pathway was involved in MCL-Poly(3HA) biosynthesis. Preusting et al. (1990) confirmed these results, but also showed that with hexane as substrate 3-hydroxyoctanoate and 3-hydroxydecanoate were produced, indicating that also other pathways were involved in MCL-Poly(3HA) biosynthesis (Table 1).

Comparable results were found with MCL-Poly(3HA) production by *P. putida* KT2442 using fatty acids as substrate (Huijberts et al., 1995). Studies with ^{13}C-labeled decanoic acid and inhibitors of β-oxidation and fatty acid synthesis showed that this substrate was converted into MCL-Poly(3HA) by the β-oxidation pathway exclusively.

5.2 Fatty Acid Synthesis

Experiments with ^{13}C-labeled hexanoic acid as a substrate for *P. putida* KT2442 showed that three pathways are involved in its conversion into MCL-Poly(3HA). Hexanoic acid can be incorporated directly into MCL-Poly(3HA) as 3-hydroxyhexanoic acid after half a cycle of β-oxidation. Further, it appeared that part of the hexanoic acid is partly or fully degraded by the β-oxidation cycle and that the generated acetyl-CoA is used for *de novo* fatty acid synthesis to produce C6 to C14 monomers. Also, the presence of unsaturated monomers suggests that *de novo* fatty acid synthesis is active. There was also evidence that hexanoic acid was elongated to 3-hydroxyoctanoic acid (Huijberts et al., 1994).

Non-MCL-Poly(3HA)-related substrates like glucose, fructose, and glycerol can be converted into MCL-Poly(3HA) (Haywood et al., 1990; Timm and Steinbuchel, 1990; Huijberts et al., 1992). This MCL-Poly(3HA) consists mainly of C8 and C10 monomers. The fatty acid synthesis inhibitor cerulenin stopped MCL-Poly(3HA) production. Also, the temperature-dependent presence of unsaturated monomers, which resembled the temperature-dependent production of unsaturated fatty acids, indicated that carbohydrates can be transformed into MCL-Poly(3HA) by means of fatty acid synthesis.

5.3 Unsaturated Fatty Acids

The 3-hydroxy fatty acids with functional groups can be incorporated in MCL-Poly(3HA). In particular, unsaturated 3-hydroxy fatty acids are readily integrated in MCL-Poly(3HA) by using aliphatic unsaturated substrates for growth. De Waard et al. (1993) used oleic acid and linoleic acid as substrates for MCL-Poly(3HA) production by *P. putida* KT2442. It was found that oleic acid was degraded via the enoyl-CoA isomerase-dependent route and linoleic acid via the dienoyl-CoA reductase-dependent route.

6 Physiology and Process Development

Process development of microbial MCL-Poly(3HA) production has been focussed on optimization of such process parameters as yield, productivity, and Poly(3HA) content of the biomass; on the dilemma of how to dose toxic substrates that are difficult to measure on-line at substrate excess concentrations; and on the control of the monomer composition and material characteristics of MCL-Poly(3HA) by adjustment of the feed composition.

6.1 Fermentation Process Development

Fermentation process development has recently been reviewed by Kessler et al. (2001). Of the fluorescent Pseudomonads, two species have been studied most extensively for MCL-Poly(3HA) production; *P. oleovorans* and *P. putida*. These microorganisms show a striking physiological dissimilarity with respect to MCL-Poly(3HA) production.

P. oleovorans is able to use alkanes and alkenes as a substrate due to the presence of the OCT1 plasmid (Kok, 1988), whereas *P. putida* is not able to oxidize alkanes/alkenes. *P. putida*, however, can, in contrast to *P. oleovorans*, use carbohydrates, such as glucose and fructose, for the production of MCL-Poly(3HA) (Haywood et al., 1990; Timm and Steinbuchel, 1990; Huijberts et al., 1992).

P. putida is able to produce MCL-Poly(3-HA) during exponential growth, when all nutrients are available in sufficient amounts. MCL-Poly(3HA) production in *P. oleovorans*, however, only occurs when the concentration of one of the nutrients is limiting growth.

6.1.1 *P. oleovorans*

The development of fermentation processes for the production of MCL-Poly(3HA) started with the experiments carried out by Preusting et al. (1993a). *P. oleovorans* was grown in two-phase fed-batch cultivation. The two phases consisted of a watery phase containing mineral nutrients and an organic phase of octane. Using an organic phase is convenient because this results, without extra addition during the process, in a constant availability of the carbon source for the microorganisms in the watery phase. The feed rate of the growth-limiting substrate was constant. After an initial batch period nitrogen became limited. A biomass concentration of 37.1 g L^{-1} was reached in 48 h, containing 33% of MCL-Poly(3HA), resulting in a productivity of 0.25 g L^{-1} h^{-1}.

With a comparable set-up, continuous cultivations were performed (Preusting et al., 1993b). The optimal growth rate was 0.05 h^{-1}. The maximum productivity was 0.58 g L^{-1} h^{-1}, with a maximum biomass concentration of 11.6 g L^{-1}. Compared with the fed-batch experiments, however, the MCL-Poly(3HA) content decreased to 20%. The restricted retention time of the microorganism in the culture appears to limit the maximal attainable Poly(3HA) content.

The medium composition used in the fed-batch process was optimized resulting in cell densities near 100 g L^{-1}. By applying an exponential feed rate resulting in a growth rate of 0.05 h^{-1}, the maximal biomass concentration increased further to 112 g L^{-1}, with a biomass productivity of 1.8 g L^{-1} h^{-1}. The MCL-Poly(3HA) productivity, however, was low, 0.34 g L^{-1} h^{-1}, caused by a steady decrease of the MCL-Poly(3HA) content during the last part of the fermentation

(Hazenberg, 1997). When this optimized medium composition was used in the chemostat set-up described above, a maximum biomass concentration of 18 g L^{-1} was reached. The MCL-Poly(3HA) content, however, remained low at approximately 10% (Hazenberg, 1997). It is still unclear what causes these low MCL-Poly(3HA) contents.

In order to develop a more efficient MCL-Poly(3HA) production process, a two-stage continuous culture system was set-up. In the first phase, biomass was produced,; in the second stage, MCL-Poly(3HA) was synthesized in the absence of a nitrogen source. A maximum polymer content of 63% was reached, at a productivity of 1.06 g L^{-1} h^{-1}. This polymer content is the highest reported for MCL-Poly(3HA) to date (Hazenberg, 1997; Jung et al., submitted).

Fed-batch fermentations with *P. oleovorans* have been carried out using octanol and octanoate as substrate (Lee and Chang, 1995). Pure oxygen was used to ensure high oxygen transfer rates. With octanoate as substrate, 41.8 g L^{-1} biomass with a cellular Poly(3HA) content of 37% and a productivity of 0.34 g L^{-1} were reached. Higher biomass concentrations could not be achieved due to accumulation of the toxic octanoate.

6.1.2
P. putida

In parallel, MCL-Poly(3HA) production processes with *P. putida* have been developed. *P. putida* does, in contrast to *P. oleovorans,* not have to be grown under nutrient-limited conditions to produce MCL-Poly(3HA). Another difference between both organisms is that *P. putida* is not able to use alkanes or alkenes as substrate. Instead, fatty acids have been used as a carbon source. These fatty acids cannot, however, be used as a second phase during fermentation because the resulting high concentrations of the fatty acids are toxic. In high-cell-density continuous culture *P. putida* has been grown to 30 g L^{-1} and 23% MCL-Poly(3HA) with oleic acid as substrate, corresponding to a productivity of 0.67 g L^{-1} h^{-1} (Huijberts and Eggink, 1996).

To perform fed-batch experiments with *P. putida* a method had to be developed to prevent carbon limitation and to prevent a build-up of the concentration of the fatty acids to inhibitory levels. High-performance liquid chromatography methods to measure the concentration of aliphatic substrates have been reported, also for octanoic acid (Kim et al., 1996, 1997), but these are not suitable for the detection of long-chain fatty acids in a watery phase due to their low solubility. Instead a method was developed in which the fatty acids were added pulse-wise to the cultures (Huijberts, 1996; Weusthuis et al., 1997). Substrate exhaustion was detected by a sudden increase in dissolved oxygen tension and this signal was used to pulse a further amount of fatty acids into the fermentor. In this way the time the culture was carbon-limited could be minimized and the maximum concentration of fatty acids could be controlled to prevent toxic levels. With coconut oil fatty acids as substrate, a maximal biomass concentration of 131 g L^{-1} after 36 h was reached containing 59% of MCL-Poly(3HA) resulting in a maximal productivity of 2.3 g MCL-Poly(3HA) L^{-1} h^{-1} (Figure 2). This is the highest productivity reported to date. The same experiment has also been performed with fatty acids derived from linseed oil, coconut oil, tall oil, rape seed oil, and mixtures of these with comparable results. This allows the production of MCL-Poly(3HA) with various monomer compositions.

These results show that, up to now, fed-batch cultivation is the method of choice for *P. putida*. The low Poly(3HA) content of the

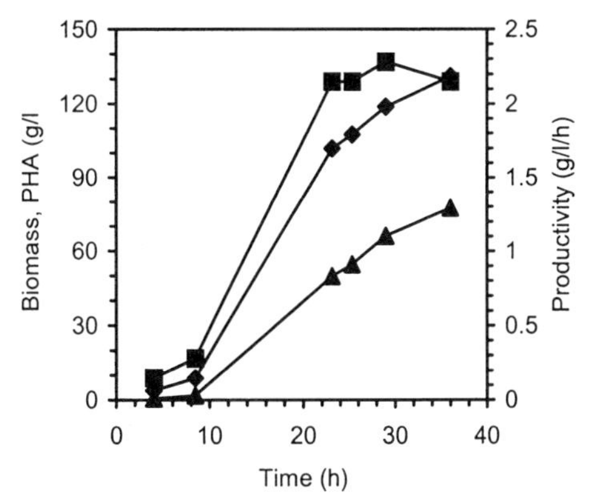

Fig. 2 MCL-Poly(3HA) production in a fed-batch fermentation with *P. putida* KT2442 and coconut oil fatty acids as substrate. ◆: Biomass; ▲: MCL-Poly(3HA); ■: MCL-Poly(3HA) productivity.

biomass grown in chemostat cultures renders this cultivation method unsuitable for large-scale production.

6.2 Control of MCL-Poly(3HA) Monomer Composition

Intermediates of the β-oxidation pathway are incorporated in MCL-Poly(3HA), as shown in Section 5.1. Both the β-oxidation pathway and the enzymes involved in MCL-Poly(3-HA) formation are highly aspecific. This opens the possibility to control the monomer composition of MCL-Poly(3HA) and to program material characteristics.

6.2.1 Length and Unsaturation of MCL-Poly(3HA) Monomers

With oleic acid, mono-unsaturated monomers were incorporated in MCL-Poly(3HA); with linoleic acid, 2-fold unsaturated monomers were also detected (De Waard et al., 1993). Casini et al. (1997) used hydrolyzed linseed oil as substrate for *P. putida* KT2442. The presence of the 3-fold unsaturated linoleic acid led to the incorporation of C14:3 and C16:3 3-hydroxy fatty acids in MCL-Poly(3HA). This was the first time that C16 3-hydroxy fatty acids were found to be also incorporated in MCL-Poly(3HA).

Furthermore, MCL-Poly(3HA)s were produced from free fatty acid mixtures derived from industrial byproducts, such as tall oil fatty acids, which showed an interesting potential as low-cost renewable resources. Isolation and analysis of the polymer allowed the identification of 16 different saturated, mono-unsaturated, and di-unsaturated monomers (Kellerhals, 1999). Except for the presence of diene-containing monomers and the large number of minor components, the monomer composition of the fatty acid mixture-derived MCL-Poly(3HA) did not differ significantly from oleic acid-derived Poly(3HA)s.

When a mixture of fatty acids or hydrocarbons is used as substrate, all compounds are simultaneously used for growth and MCL-Poly(3HA) production. In that way it is possible to control the monomer composition (length of carbon chain of monomer, number and type of unsaturations, and other functionalities) of MCL-Poly(3HA) to some extent, enabling the tailoring of the material properties to meet the demands of specific applications (Figure 3).

6.2.2 Production of MCL-Poly(3HA)s with other Functionalities

It has been shown that more than 60 different monomers can be incorporated into Poly(3HA) by Pseudomonads (Steinbüchel and Valentin, 1995). Poly(3HA)s containing a functional group in their side chain are generally called 'functional' Poly(3HA)s.

One strategy to produce MCL-Poly(3HA)s with a certain monomer content is co-feeding of two different substrates in a certain ratio. In principle three types of

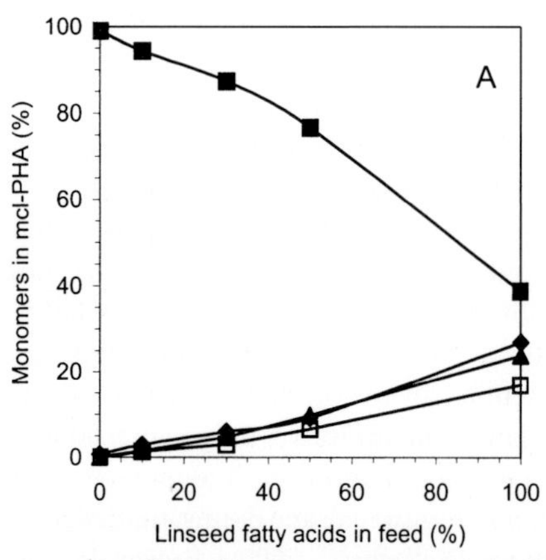

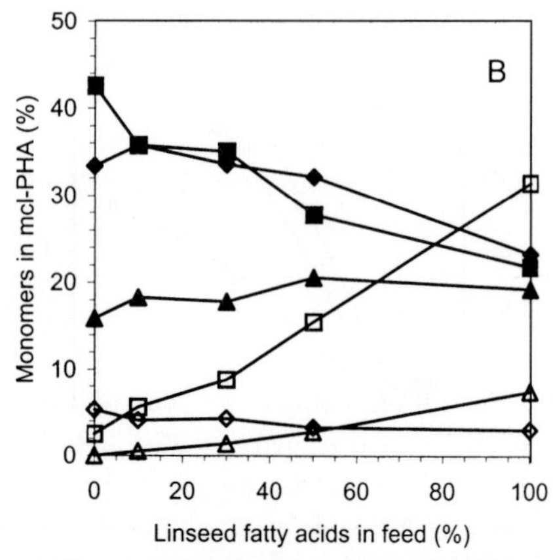

Fig. 3 The effect of the fatty acid composition of the substrate fed to a high-cell-density fed-batch cultivation of *P. putida* KT2442 on the degree of unsaturation (A) and the carbon chain length of the MCL-Poly(3HA) monomers. The substrates used were mixtures of coconut oil fatty acids and linseed oil fatty acids. (A) ■: saturated; ◆: monounsaturated; □: 2-fold unsaturated; ▲: 3-fold unsaturated. (B) □: 3-hydroxy C6 fatty acids; ■: 3-hydroxy C8 fatty acids; ◆: 3-hydroxy C10 fatty acids; ▲: 3-hydroxy C12 fatty acids; ◇↓: 3-hydroxy C14 fatty acids; Δ: 3-hydroxy C16 fatty acids.

substrates have to be differentiated: (1) substrates which support cell growth and Poly(3HA) production, (2) substrates which support growth but not Poly(3HA) production, and (3) substrates which do not support growth but do support Poly(3HA) production (Lenz et al., 1992). Therefore, dependent on the type of substrate, different cultivation processes and feeding strategies have to be used.

It has been shown that application of carbon source mixtures such as citrate/octanoate (Durner, 1998) or glucose/octanoic acid (Kim et al., 1996, 1997) which support cell growth and Poly(3HA) production, respectively, are utilized simultaneously in batch cultures, and that fatty acids were used for Poly(3HA) synthesis and carbohydrates were dissimilated to supply the maintenance energy. In general, support of bacterial cell growth by one substrate and Poly(3HA) formation, especially the incorporation of specific monomers, by the other substrate is a widespread technique for the production of 'functionalized' polymers (e.g. Scholtz et al., 1994; de Koning et al., 1994; Hori et al., 1994; Kim, O. Y., et al., 1995, 1996; Curley et al., 1996a; Gross et al., 1996; Song et al., 1996). Another possibility is to perform a two-stage cultivation process. In the first stage bacterial cell mass is produced and in the second stage Poly(3HA)-forming substrates are added to the culture, as has been reported for the production of Poly(3HA) containing multi-fluorinated, cyano, or nitrophenoxy side chain substituents (Kim, O. Y., et al., 1995, 1996).

In many cases co-feeding strategies are not only used to produce specific random copolymers – even block polymers or polymer blends can be obtained. Growth of *P. oleovorans* or *P. putida* on a mixture of 5-phenylvaleric acid (or other arylalkyl acids) and nonanoic acid results in a homopolymer poly-3-hydroxy-5-phenylvalerate, and a random copolymer consisting of 3-hydroxynonanoate and 3-hydroxyheptanoate (Kim et al., 1991; Curley et al., 1996b; Hazer et al.,

1996). It has been shown that both types of polyesters occur in the same granule (Curley, 1996b). Interestingly, it has even been proposed that by sequential feeding of nonanoic acid and 10-undecenoic acid, a physical mixture of two different polymers is produced; however, with small amounts of Poly(3HA) containing repeating units from both substrates (Kim, Y. B., et al., 1997), whereas co-feeding of octanoate and cyanophenoxyalkanoates resulted in Poly(3HA) block polymers containing chain segments that are enriched in 3-hydroxycyanophenoxyalkanoate monomers (Gross et al., 1996).

Production of MCL-Poly(3HA)s from toxic organic solvents requires other cultivation strategies. A cultivation method was developed to improve growth of *P. oleovorans* on toxic organic solvents, such as 1-hexene. This method includes dilution of 1-hexene with a non-metabolizable second organic phase to lower the toxic effect of the apolar carbon source and a long-term chemostat enrichment culture to increase the solvent tolerance and the specific growth rate (Jung et al., submitted). Furthermore, application of dual-carbon/nitrogen-limited conditions for cell growth and Poly(3HA) production on volatile and toxic substrates resulted in decreased cell lysis, side product formation, and biosurfactant production, and therefore higher cell and Poly(3HA) yields (Jung et al., submitted).

6.3
Oxygen Transfer and Heat Production

The importance of a good oxygen transfer is stressed in many publications concerning the biotechnological production of MCL-Poly(3HA) (e.g. Lee and Chang, 1995; Huijberts, 1996; Hazenberg, 1997). Oxygen uptake rates as high as 200 (Hazenberg, 1997) and 220 (Huijberts, 1996) mmol $L^{-1}\,h^{-1}$ have been described. By using reduced substrates as alkanes and fatty acids a lot of oxygen is necessary for the conversion of these aliphatic substrates into MCL-Poly(3HA) and, especially, into biomass.

In the fed-batch production process of MCL-Poly(3HA) by *P. putida* KT2442 as described above, the oxygen transfer limits the productivity and final biomass concentration. In addition, the Poly(3HA) content of biomass is positively affected by high oxygen transfer rates. At the end of the cultivation biomass production stops because all oxygen is used for maintenance processes (Figure 4a). The productivity of biomass and MCL-Poly(3HA), but also the final biomass concentration, final Poly(3-HA) concentration, and maximal Poly(3HA) content of the biomass, depend on the maximal oxygen transfer rate during the fermentation (Figure 4b).

The high oxygen transfer rates reached in laboratory fermentors are not easily reached at a production scale. The heat development by excessive oxygen consumption will also result in cooling problems. Methods to reduce the oxygen consumption rate have been mentioned. There are two promising possibilities. First, by increasing the Poly(3-HA) content of the biomass (thereby decreasing the amount of biomass) the oxygen consumption can be limited. Second, by using oxidated co-substrates the oxygen consumption can be decreased. Durner (1998) showed that citrate and octanoic acid can be used simultaneously in batch cultures of *P. oleovorans* and Kim et al. (1996, 1997) demonstrated the same for the combination of glucose and octanoic acid by high-cell-density fed-batch processes of *P. putida*. These findings indicate that Pseudomonads are able to use different unrelated substrates simultaneously, even under carbon excess conditions.

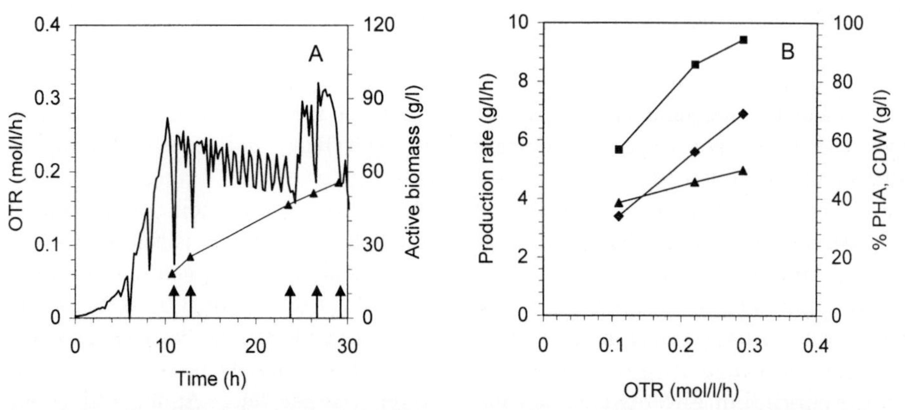

Fig. 4 (A) Relationship between the active biomass concentration (amount of biomass without MCL-Poly(3HA), ▲), endogenous respiration for maintenance purposes and the oxygen transfer rate (OTR, —) in a high-cell-density fed-batch fermentation of *P. putida* KT2442 cultivated on coconut oil fatty acids. The endogenous respiration was estimated at several time intervals by measuring the transient decrease in oxygen uptake rate going from substrate excess conditions to conditions where the substrate was fully consumed (indicated by arrows). There is a clear correlation between the endogenous respiration rate and the concentration of active biomass resulting in a complete utilization of the transferred oxygen for maintenance purposes after 24 h. After 24 h the oxygen transfer rate was increased by using a higher aeration rate, resulting in further growth. (B) The effect of the maximum oxygen transfer rate on the biomass concentration (■), MCL-Poly(3HA) content (▲) and biomass production rate (◆) of high-cell-density fed-batch fermentation of *P. putida* KT2442 using coconut oil fatty acids as substrate. The production rate is defined as the increase in biomass concentration during the linear growth phase divided by the duration of the linear growth phase.

6.4 Byproducts

Fluorescent Pseudomonads are known to produce several byproducts. Siderophores as pyoverdines (Hohnadel and Meyer, 1988) and pyochelin (Cox et al., 1981), antibiotics such as phenazine derivatives (Thomashow and Weller, 1988), pyrrolnitrin, pyoluteorin, and 2,4 diacetyl phloroglucinol (Dowling and O'Gara, 1994), and surfactants such as rhamnolipids (Fiechter, 1992) can be produced. No attempts have been made to determine the presence of these compounds in fermentation broth used for MCL-Poly(3-HA) production. Fluorescence and excessive foaming of fermentation broth samples of *P. putida* cultivated in high-cell-density fed-batch fermentations with fatty acids as substrate, however, indicate that in cultures used for MCL-Poly(3HA) production byproducts are also formed and that they influence the fermentation process (unpublished results).

7 Molecular Genetics

An alternative and additional approach to increase Poly(3HA) yield, productivity, and Poly(3HA) content of biomass has been by genetic modification. Poly(3HA) production of both recombinant fluorescent Pseudomonads and *Escherichia coli* has been studied.

7.1 Recombinant Pseudomonads

P. oleovorans GPo1, *P. putida* KT2442, and *P. aeruginosa* PAO1 are the best studied MCL-

Poly(3HA)-producing strains on a genetic level. These bacteria contain two Poly(3HA) polymerases (also called Poly(3HA) synthases) encoded by *phaC1* and *phaC2* of the *pha* gene cluster (Huisman et al., 1991; Timm and Steinbüchel, 1992). It has been shown for *P. oleovorans* that the two Poly(3HA) polymerases have a small difference in substrate specificity (Huisman et al., 1992). Moreover, it was demonstrated that both polymerases are functional proteins which are able to catalyze Poly(3HA) formation independently from each other, i.e. only one of the polymerase encoding genes is enough to produce MCL-Poly(3HA) in heterologous hosts (Huisman et al., 1992; Langenbach et al., 1997; Ren, 1997; Mittendorf et al., 1998). Introduction of additional copies of the Poly(3HA) polymerase-expressing genes resulted in a nearly 2-fold increase in Poly(3HA) when the strains were cultivated under non-limited conditions (Kraak et al., 1997). However, no significant increase in Poly(3HA) accumulation was observed when the recombinants were cultivated under nutrient-limited conditions. The only effect of additional copies of the Poly(3HA) polymerase-encoding genes was a slight change in the monomer composition of the polymer and a decrease in its molecular weight (Huisman et al., 1992; Kraak et al., 1997). Furthermore, it has been reported that GPp120, a chemical mutant of *P. putida* KT2442, produces higher levels of Poly(3-HA) in shaking flask experiments than the parental strain and that the mutant did not show any downregulation of Poly(3HA) formation under non-limiting conditions (Ren et al., 1998). However, a detailed analysis of the mutant showed that due to the reduced growth rate, the Poly(3HA) yield was only half of that found for the wild-type strain (unpublished results). In conclusion, all recombinant *Pseudomonas* strains or mutants tested so far cannot compete with respect to Poly(3HA) productivity with the wild-type organisms if the wild-type strains are cultivated under appropriate Poly(3HA)-forming conditions.

However, recombinant *Pseudomonas* strains have been used successfully for the production of Poly(3HA) polymers containing unusual monomers. For example, a Poly(3HA)-negative mutant of *P. putida* KT2442, called GPp104 (Huisman et al., 1991), expressing the Poly(3HA) synthase-encoding gene of *Thiocapsa pfennigii* has been cultivated in two-stage batch or fed-batch mode with 5-hydroxyhexanoic acid, 4-hydroxyheptanoic acid, or 4-hydroxyoctanoic acid as a carbon source in the second stage in order to produce polymers containing 5-hydroxyhexanoic acid, 4-hydroxyheptanoic acid, or 4-hydroxyoctanoic acid monomers, respectively (Valentin et al., 1996). A polyester containing Poly(3HA) with 4-hydroxyvaleric acid monomers has been produced in a 15-L scale two-stage aerobic fed-batch process using the recombinant GPp104 strain and octanoic and levulinic acid as carbon sources (Schmack et al., 1998). Cell densities of 20 g L^{-1} could be achieved and the Poly(3HA) content of these cells amounted to up to 50% of cell dry weight. Although the produced polymer consisted mainly of hydroxybutyric and hydroxyvaleric acid monomers, the polyesters showed a distinctly elastomeric behavior due to the low content of MCL monomers (15 mol.% hydroxyhexanoic acid and 2 mol.% hydroxyoctanoic acid) (Schmack et al., 1998).

In summary, recombinant *Pseudomonas* strains seem to be useful for the production of polymers containing certain unusual monomers, but less so for the production of the 'classical' MCL-Poly(3HA) polymers.

7.2 Recombinant *E. coli*

E. coli strains blocked in the 3-ketoacyl thiolase (FadA) or 3-hydroxyacyl-CoA dehydrogenase (FadB) enzyme activity of the β-oxidation pathway are able to accumulate MCL-Poly(3HA)s when only the *phaC1* or *phaC2* gene of *Pseudomonas* is expressed (Langenbach et al., 1997; Qi et al., 1997). It is assumed that the β-oxidation has to be slowed down in *E. coli* in order to accumulate specific intermediates, which can serve as precursors for Poly(3HA) synthesis. Various expression systems have been used and, depending on the carbon source and growth conditions Poly(3HA), amounts up to 33% of cell dry weight have been achieved (Table 2). The Poly(3HA) content in those β-oxidation-deficient *E. coli* strains could even be further increased to up to 50% of cell dry weight by using acrylic acid, a β-oxidation inhibitor (Qi et al., 1998). Recently, it has been shown that Poly(3HA) can also be produced in *E. coli* strains containing a functional, non-inhibited β-oxidation pathway. Overexpression of a 3-ketoacyl-CoA reductase-encoding gene of *P. aeruginosa* or *E. coli* in addition to one of the Poly(3HA) polymerase encoding genes resulted in the production of 3 or 8% Poly(3HA) per cell dry weight, respectively (Taguchi et al., 1999; Ren et al., 2000b). Moreover, *E. coli* recombinants containing in addition to a Poly(3HA) polymerase a *R*-specific enoyl-CoA hydratase of *P. aeruginosa* with a substrate specificity

Tab. 2 Production of MCL-poly(3HA) using recombinant *E. coli* strains

Strain	*Genes*	*Experimental system*	*C source*	*β-Oxidation inhibition*	*Percent poly(3HA) (w/w)*	*Monomer composition (mol.%)*					*Reference*
						C4	*C6*	*C8*	*C10*	*C12*	
LS1298	$phaC1_{Pa}$	P_{lac}	LB, C10	*fadB* mut.	21	ND	4.5	19	73	3.5	Langenbach et al. (1997)
LS1298	$phaC2_{Pa}$	P_{lac}	LB, C12	*fadB* mut.	15	ND	0	10	52	38	Qi et al. (1997)
JMU194	$phaC1_{Po}$	P_{alk}	YE, C16	*fadA* mut.	25	ND	10	90	ND	ND	Ren (1997)
JMU193	$phaC1_{Po}$	P_{alk}	YE, C16	*fadB* mut.	33	ND	40	60	ND	ND	Ren (1997)
JM109	$phaC1_{Pa}$	P_{lac}	LB, C10	acrylic acid	28	ND	ND	ND	ND	ND	Qi et al. (1998)
RS3097	$phaC1_{Pa}$	P_{lac}	LB, C10	acrylic acid	46	ND	2.5	20	72	5.5	Qi et al. (1998)
JMU194	$phaC2_{Po}$ $phbB_{Re}$	P_{tac} P_{lac}	YE, C16	*fadA* mut.	20	ND	42	38	20	ND	Ren et al. (2000a)
JMU194	$phaC2_{Po}$ $fabG_{Pa}$	P_{tac} P_{lac}	YE, C16	*fudA* mut.	21	ND	23	65	12	ND	Ren et al. (2000b)
RS3338	$phaC2_{Po}$ $fabG_{Pa}$	P_{tac} P_{lac}	C8	no	3	ND	9	91	0	ND	Ren et al. (2000b)
HB101	$phaC1_{Ps}$ $fabG_{Ec}$	P_{trc} P_{lac}	C12	no	8	0	12	80	8	0	Taguchi et al. (1999)
LS5218	$phaC1_{Ps}$ $phaJ1_{Pa}$	P_{trc} P_{lac}	C12	no	29	10	78	7	3	2	Tsuge et al. (1999)
LS5218	$phaC1_{Ps,}$ $phaJ2_{Pa}$	P_{trc} P_{lac}	C12	no	14	8	45	30	11	6	Tsuge et al. (1999)

C4: 3-hydroxybutyrate; C6: 3-hydroxyhexanoate; C8: 3-hydroxyoctanoate; C10: 3-hydroxydecanoate; C12: 3-hydroxydodecanoate; Pa: *P. aeruginosa*; Po: *P. oleovorans*; Re: *R. eutropha*; Ec: *E. coli*; LB: Luria Bertani medium; YE: Yeast Extract; ND: not determined.

towards SCL or MCL substrates produced 29 or 14% Poly(3HA) per cell dry weight, respectively (Tsuge et al., 1999). The existence of *R*-specific enoyl-CoA hydratases in Poly(3HA)-producing *Pseudomonas* strains clearly indicates that the Poly(3HA) synthesis pathway proceeds via a stereospecific hydratase reaction rather than the epimerase activity of the β-oxidation. Furthermore, it appears that the monomer composition of the Poly(3HA) produced by the different recombinants is determined by the substrate specificity of the introduced enoyl-CoA hydratase or 3-ketoacyl-CoA reductase (Table 2). Thus, specific *E. coli* recombinants can be engineered in order to produce polymers with desired monomer compositions. In addition, pathway engineering can be used to synthesize MCL-Poly(3HA) with altered physical properties. Introduction of the acetoacetyl-CoA reductase of *Ralstonia eutropha* and blockage of the ketoacyl-CoA degradation step of the β-oxidation not only caused significant changes in the monomer composition but also caused an increase of the molecular weight and loss of the melting point (Ren et al., 2000a). A high-molecular-weight peak of around 10^6 Da was observed that could be caused by the higher C6 monomer content of the polymer and which might alter the ratio of chain elongation to chain termination events, resulting in longer Poly(3HA) chains, comparable to Poly(3HB). Another possibility is that the high-molecular weight peak is due to the presence of C6 monomer stretches which facilitate strong non-covalent interactions among Poly(3HA) chains and thus result in the formation of microgels (Ren et al., 2000a).

In summary, it is now possible to produce not only significant amounts of MCL-Poly(3-HA) but also different types of MCL-Poly(3-HA) polymers in recombinant *E. coli*. However, the lack of stability of the recombinants is often a major drawback for the production of sufficient amounts of Poly(3HA) (Ren, 1997). In addition, a major problem in general in applying plasmid-containing recombinants in large-scale fermentation is plasmid maintenance and stability. The classical approach to maintain the phenotype of the recombinant strain is to add antibiotics to the culture medium. This can have a considerable effect on the reproducibility of the results and the final cost of the product. An attractive alternative using minitransposons for stable, regulated, and inexpensive *phaC* gene expression in recombinant bacteria has been developed by Prieto et al. (1999). The stability of the system to culture MCL-Poly(3HA)-producing recombinant *E. coli* in a bioreactor operated in batch or continuous cultivation mode in the absence of selection marker has been exploited (Prieto et al., 1999). The phenotype was 100% stable throughout the fermentation processes. Furthermore, it has been observed that the chain length of the polymer produced by the recombinants varies depending on the amount of inducer added to the medium. Reduction of inducer concentration caused an increase in the number of polymer molecules with longer chain length, which can only be explained by fewer molecules of Poly(3HA) polymerase (Prieto et al., 1999). This is in agreement with the hypothesis that higher enzyme levels could lead to an increased number of chain initiation events resulting in shorter polymer chain lengths (Huisman et al., 1992).

Taking together all the information gained so far from Poly(3HA)-producing recombinant *E. coli* and the advantage that the fermentation and downstream process technology is already established for *E. coli*, it seems likely that *E. coli* is an interesting candidate for the production of specific designed Poly(3HA) polymers in the future.

8 Downstream Processing

Recovery procedures for MCL-Poly(3HA) resemble those originally developed for the production of Poly(3HB). A number of solvent extraction processes have been assessed to separate MCL-Poly(3HA)s from biomass. These usually involve the use of a chlorinated solvent such as chloroform (Lageveen et al., 1988) or methylene chloride. Recently, it has been reported that MCL-Poly(3HA)s can be extracted with hexane or acetone instead of chlorinated solvents (Williams et al., 1999) and subsequently precipitated by the addition of a non-solvent for the Poly(3HA), such as methanol. Using this method, the resulting polymer can be obtained in high purity. An alternative, non-solvent-based extraction process was developed by de Koning et al. (1997ab) and further optimized to make the overall production process more attractive (Kellerhals, 1999). The biomass is separated from the medium by centrifugation, and treated with a protease cocktail and a detergent to solubilize all cell components. Removal of the solubilized cell material and concentration of the resulting Poly(3HA) suspension is achieved by crossflow microfiltration (Kellerhals, 1999) or continuous centrifugation (unpublished results). The submicron MCL-Poly(3HA) granules display a density close to that of water (Preusting et al., 1993), as a result of which a MCL-Poly(3HA) suspension does not settle (Marchessault et al., 1995), in fact it forms a highly stable polymer latex. The overall purity of the latex amounts to 95%. Furthermore, supercritical CO_2 is highly effective at extracting lipids and other hydrophobic contaminants from Poly(3HA)-containing cells and 100% purity can be reached in a single step (Williams et al., 1999).

Since the liberation of chromosomal DNA during lysis causes a dramatic increase in viscosity, a nuclease-encoding gene from *Staphylococcus aureus* was integrated into the genomes of several Poly(3HA) producers. The nuclease is directed to the periplasm, and occasionally to the culture medium, without affecting Poly(3HA) production or strain stability, and reducing the viscosity of the lysate significantly during the downstream process (Boynton et al., 1999)

9 Production

MCL-Poly(3HA) has, in contrast to Poly(3HB), not been produced on a commercial scale yet. There is sufficient material available for R&D purposes and several applications have been developed.

9.1 MCL-Poly(3HA) Production versus SCL-Poly(3HA) Production

In contrast to Poly(3HB), MCL-Poly(3HA) has not been produced on a commercial scale yet. The process development of Poly(3HB) has also received a lot more attention than processes for the production of MCL-Poly(3HA). It is therefore interesting to compare production parameters of MCL-Poly(3HA) production with those of Poly(3HB). The parameters of the best Poly(3HB) and MCL-Poly(3HA) processes are given in Table 3.

Looking at the fed-batch operated cultures the main difference concerning process parameters between Poly(3HB) and MCL-Poly(3HA) production seems to be the lower MCL-Poly(3HA) content. It is reported that a low MCL-Poly(3HA) content decreases the productivity and yield, and increases the costs for down-stream processing and waste disposal (Choi et al., 1999).

Tab. 3 Process parameters of poly(3HB) and MCL-poly(3HA) production

	Poly(3HB)	MCL-poly(3HA)	
Organism	*A. latus*	*P. putida*	*P. oleovorans*
Fermentation type	fed-batch	fed-batch	two-stage continuous
Substrate	sucrose	coconut oil fatty acids	octane
Culture time (h)	20	36	–
Cell concentration (g L^{-1})	111.7	131	17
Poly(3HA) content (%)	88	59	66
Productivity (g L^{-1} h^{-1})	4.94	2.3	1.06
Yield (g g^{-1})	0.42	0.3–0.4	not determined
Reference	Wang and Lee (1997)	see Figure 2	Hazenberg (1997)

Notable, the Poly(3HA) content is always expressed as the weight ratio between Poly(3HA) and total biomass weight. The density of Poly(3HB), however, is 1.24 g mL^{-1} (Marchessault et al., 1990) whereas the density of MCL-Poly(3HA), depending on the monomer composition, is close to 1.00 g mL^{-1}. On a volume basis, the Poly(3HA) content of 66% (Table 3) in *P. oleovorans* corresponds with a Poly(3HB) content of 82%! Also, in terms of applications, the volume of the material is more important than the weight. If Poly(3HB) and MCL-Poly(3HA) could be used for the same application, 24% more Poly(3HB) would be necessary on a weight basis.

9.2 Producers

MCL-Poly(3HA) is not produced at a commercial scale yet. It is being produced routinely at ATO (The Netherlands) with *P. putida* KT2442 using fatty acids as substrate. Typical fermentation process parameters are: 120 g L^{-1} biomass containing 50% MCL-Poly(3HA) produced in 24–35 h. MCL-Poly(3HA) batches (several kilogram amounts) with specific monomer compositions are available for research purposes.

9.3 Applications

The application of MCL-Poly(3HA) has been reviewed extensively by van der Walle et al. (2001).

The material properties of MCL-Poly(3-HA) are strongly related to the chemical characteristics (i.e. monomer composition) of the various polymers. Since the polyester structure can be tailored quite simply, the polymer properties therefore can be readily adjusted to meet the specific demands for a particular application. Moreover, the unsaturated MCL-Poly(3HA)s are chemically reactive and completely amorphous.

MCL-Poly(3HA)s can be manufactured to many different materials and shapes. Furthermore, they can be processed in latex (granules in water) or in solution with several different solvents. Together with the material properties of MCL-Poly(3HA)s, this opens up a whole field of feasible commercial applications to be explored and exploited.

In general, due to their biodegradability, water resistance, and oxygen impermeability, Poly(3HA)s can be used for all sorts of biodegradable packaging materials, including composting bags and food packaging. Also, the use of Poly(3HA)s in single-use sanitary articles like diapers is considered as economically feasible. In addition, in marine environments (fishing nets and other discarded objects that cause severe damage when made from non-degradable materials), construction materials (adhesives, laminates, foams and rubbers), and in agricultural industries, there is promising market potential for new biodegradable materials.

The potential for biomedical applications is very promising, since the added value to these special products is remarkably high (Hocking and Marchesault, 1994; Lafferty et al., 1988; Williams et al., 1999); although research in this field is of unique complexity, it is both technical and economical very compelling to succeed.

Several applications on basis of MCL-Poly(3HA) have been developed.

9.3.1 Pressure-sensitive Adhesives (PSAs)

Babu et al. (1997) described the development of a biodegradable PSA on the basis of MCL-Poly(3HA). Different Poly(3HA)s were tested, produced by cultivating *P. oleovorans* on octanoic acid, decanoic acid, mixtures of octanoic and nonanoic, or mixtures of octanoic and 11-undecenoic acid. Tackifiers were added to the Poly(3HA) to give a PSA with improved tack and the strength of the Poly(3HA) was increased by UV radiation crosslinking using a photosensitizer. All but the mixtures with octanoic acid gave PSAs with good properties. Biodegradation studies indicated that the PSA formulations were still biodegradable (Babu et al., 1997)

9.3.2 Biodegradable Rubbers

Biodegradable rubbers have been manufactured from unsaturated Poly(3HA)s by crosslinking of the biopolyesters. This has been accomplished by either chemical reaction with sulfur or peroxides (Gagnon et al., 1994ab), or by radiation curing using UV or an electron-beam source (De Koning et al., 1994; Ashby et al., 1998). The MCL-Poly(3-HA)-based rubbers are still biodegradable because the ester bond is still hydrolyzable. By choosing different types of starting material and varying the crosslinking conditions, material properties like mechanical strength, tear resistance, tensile set, and flexibility of the biorubbers were readily adjusted (De Koning et al., 1994; Gagnon et al., 1994a,b; Ashby et al., 1998).

9.3.3 Paint Binders

Recently, the development of environmentally friendly paints and coatings based on MCL-Poly(3HA) has been reported (van der Walle et al., 1999). Fatty acid mixtures derived from tall oil, linseed oil, and rape seed oil with unsaturated fatty acids have been used as a substrate for MCL-Poly(3HA) paint binders. Due to the relatively low molecular weight and narrow molecular weight distribution of MCL-Poly(3HA), the viscosity of the resulting paint is low compared to synthetic binders such as polyacrylates and polyurethanes. To adjust the viscosity of the MCL-Poly(3HA) paint to optimal values for paint applications, less organic solvents are necessary compared to the synthetic binders. This could have a significant potential, since organic solvents in DIY paints will be, and in some EU countries already are, further restricted by future legislation. Further studies are focussed on the application of MCL-Poly(3-HA) latexes in totally organic solvent-free paints. The application of such water-borne paint systems is a promising perspective in further reducing the use of organic solvents in paints and coatings (van de Walle et al., 1999).

9.3.4
Cheese Coatings

Cheeses are generally coated by a non-biodegradable, synthetic plastic-based latex, typically a copolymer of polyvinyl acetate and dibutyl maleic acid. This has prompted research towards the development of a fully biodegradable cheese coating.

The technical demands for a cheese coating are very comprehensive since it has to fulfill a large number of functions (Castle et al., 1993), such as mechanical and hygienic protection, semi-permeability for water, CO_2 and certain other flavoring components, easy applicability, long stability, etc.

A new biodegradable cheese coating has been developed on the basis of a MCL-Poly(3HA) latex derived from saturated fatty acids. An extensive test program showed that the functional aspects of the Poly(3HA)-based cheese coatings, like ripening control and mechanical and bacterial protection, are equivalent to the current generation of plastic coatings (van der Walle et al., 2001).

9.4
Patents

There are many patents concerning Poly(3-HA)s in general; many of them also valid for MCL-Poly(3HA)s. There are only of few patents specifically for microbial MCL-Poly-(3HA) production and applications on the basis of MCL-Poly(3HA)s (Table 4).

There are two key patents on the fermentative production of MCL-Poly(3HA) and its monomers. In WO9012104A1 (Witholt et al., 1992) the production of MCL-Poly(3-HA) and its monomers by fluorescent Pseudomonads from aliphatic substrates is claimed. The production of MCL-Poly(3HA) and its monomers by transformed *E. coli* is claimed in WO9854329 (Witholt et al., 1998).

The applications mentioned above (medical applications, paints, cheese coatings and adhesives) are patented (Table 4).

10
Outlook and Perspectives

MCL-Poly(3HA) is a unique (bio)polymer due to such properties as biodegradability, biocompatibility, water insensitivity, and chemical reactivity. Due to these characteristics MCL-Poly(3HA)s have their own niche in application development.

MCL-Poly(3HA) is not one polymer, but a class of biopolyesters. The monomeric composition is variable and can be easily controlled by simply changing the aliphatic fermentation feedstock. In this way it is possible to produce a whole range of bioplastics with distinctive material properties, allowing the tailoring of the material characteristics to meet the demands of several applications. This increases the applicability of MCL-Poly(3HA); it cannot only be used for bulk applications but also for specialties. Different types of MCL-Poly(3HA) can all be produced using the same or similar fermentation process by simply changing the type of substrate(s) used. In that way it is possible to produce tailor-made MCL-Poly(3HA) variants for specific applications – in other words, it is possible to produce high added-value specialties at a low-cost, bulk scale.

The costs of fermentative MCL-Poly(3HA) production are mainly caused by costs for feedstock, but also for a significant part by costs for waste disposal and cooling. Further optimization of MCL-Poly(3HA) fermentation processes has to focus on these three items. A further increase in MCL-Poly(3HA) content of the microbial biomass is the best solution, since it will decrease costs for feedstock, downstream processing, cooling,

Tab. 4 Patents concerning fermentation processes and applications specific for MCL-poly(3HA)

Number	*Holder*	*Inventors*	*Title*	*Date of publication*
WO9012104	Rijksuniversiteit Groningen	Witholt, B., Eggink, G., Huisman, G.W.	Microbiological production of polyesters	October 1990
JP5000159	Terumo Corp.	Ishikawa Kenji	Medical soft member	January 1993
JP6038739	Kobe Steel Ltd.	Masaichirou, N., Morio, M., Yoshimasa, T	Microorganism capable of producing poly(3-hydroxyalkanoate)	February 1994
WO9600263	Stichting Ontwikkeling en Onderzoek Noord Nederland (SOONN)	Eggink, G. and Northolt, M. D.	Method for the production of biologically degradable polyhydroxyalkanoate coating with the aid of an aqueous dispersion of polyhydroxyalkanoate	January 1996
US5614576	Minnesota Mining & Manufacturing Co.	Rutherford, D. R., Hammar, J., Babu, G. N.	Poly(β-hydroxyorganoate) pressure sensitive adhesive composition	March 1997
WO9851812	Metabolix Inc.	Williams, S. F., Martin, D. P., Gerngross, T., Horowitz, D. M	Polyhydroxyalkanoates for *in vivo* applications	November 1998
EP0881293	ETH	Kessler, B., Witholt, B.	Production of medium chain length poly-3-hydroxyalkanoates in *E. coli*, and monomers derived therefrom	December 1988
WO9939588	W. M. Wrigley Jr. Co.	Li, W., Orfan, C., Liu, J., Foster, J.W.	Environmentally friendly chewing gum bases including polyhydroxyalkanoates	August 1999
US6024784	Institute for Agrotechnological Research (ATO)	Buisman, G. J. H, Cuperus, F. P., Weusthuis, R. A., Eggink, G.	Poly(3-hydroxyalkanoate) paint and method for the preparation thereof	February 2000

and waste disposal simultaneously (Choi and Lee, 1999).

An alternative for fermentative production of MCL-Poly(3HA) is by means of genetically engineered plants (Poirier, 1999; Van der Leij and Witholt, 1995). The production of MCL-Poly(3HA) by *Arabidopsis thaliana* has been investigated (Mittendorf et al., 1998). A polymer content of 0.4% has been reached. The time-to-market of these materials is estimated at 10–15 years from now. The production costs of MCL-Poly(3HA) in plants is potentially lower than when MCL-Poly(3HA) is produced in fermentation processes. It is likely that the flexibility to incorporate various monomers will decrease when genetically modified plants are used for MCL-Poly(3HA) production. Even then, however, the possibility to (bio)chemically modify the polymers makes it possible to adapt the material characteristics to meet the demands of a multitude of applications.

MCL-Poly(3HA) is not available on the market yet. There are two main reasons to consider.

First of all, the economics of fermentative MCL-Poly(3HA) production is often compared with that of SCL-Poly(3HA). However, this is not justified since these are completely different materials. SCL-Poly(3HA)s have to compete with commodity plastics such as polyethylene and polypropylene. The costs of these commodity plastics are so low (0.96–1.10 and 0.84 $/kg, respectively; *Chemical Market Reporter*, January 2001) that fermentatively produced SCL-Poly(3HA) will not be able to compete. MCL-Poly(3HA), on the other hand, as a specialty polymer, has to compete with materials such as polyurethanes, isoprenes, styrene-butadienes, and chloroprenes. The price of these materials varies between 2–5 $ kg^{-1}. In a state-of-the-art fermentation process MCL-Poly(3-HA) can be produced with production costs ranging between 3.5 and 6.0 $/kg MCL-Poly(3HA), indicating that fermentatively produced MCL-Poly(3HA) indeed could compete cost-wise with its synthetic counterparts.

Secondly, the obvious advantage that the material characteristics of MCL-Poly(3HA) can be programmed also has an important negative side-effect. In order to adjust the material characteristics of MCL-Poly(3HA) to meet the demands of the application, the application developer has to work in close cooperation with the MCL-Poly(3HA) producer. This also implies that a potential MCL-Poly(3HA) producer has to have a wide network of application developers, to establish a sufficient market for bulk-scale production of MCL-Poly(3HA). On the other hand, this reduces the risk for the producer since its products are used for several applications and bought by several clients.

To introduce MCL-Poly(3HA) on the market in a short term it is therefore important to establish a network of (a) potential MCL-Poly(3HA) producer(s) and application developers and the availability of significant amounts of tailor-made MCL-Poly(3HA) to allow small-scale application development, field-trials, and market introduction of specific high-end MCL-Poly(3HA) products.

Acknowledgements
Parts of the research presented in this chapter was performed in an EET project (EETK960008) sponsored by the Dutch Ministries of Economische Zaken (EZ), Onderwijs, Cultuur en Wetenschappen (OCenW), and Volkshuisvesting, Ruimtelijke Ordening en Milieubeheer (VROM).

11
References

Ashby, R. D., Cromwick A.-M., Foglia, T. A. (1998) Radiation crosslinking of a bacterial medium-chain-length poly(hydroxyalkanoate) elastomer from tallow, *Int. J. Biochem. Macromol.* **23**, 61–72.

Babu, G. N., Hammar, W. J., Rutherford, D. R, Lenz, R. W., Richards, R, Goodwin, S. D. (1997) Poly-3-hydroxyalkanoates as pressure sensitive adhesives. in: *1996 International Symposium on Bacterial Polyhydroxyalkanoates* (Eggink, G., Steinbüchel, A., Poirier, Y., Witholt, B., Eds.), Ottawa: NRC Research Press, 48–56.

Boynton, Z. L., Koon, J. J., Brennan, E. M., Clouart, J. D., Horowitz, D. M., Gerngross, T. U., Huisman, G. W. (1999) Reduction of cell lysate viscosity during processing of poly(3-hydroxyalkanoates) by chromosomal integration of the staphylococcal nuclease gene in *Pseudomonas putida, Appl. Environ. Microbiol.* **65**, 1524–1529.

Casini, E., De Rijk, T. C., De Waard, P., Eggink, G. (1997) Synthesis of MCL-poly(hydroxyalkanoate) from hydrolysed linseed oil, *J. Environ. Polym. Degrad.* **5**, 153–158.

Castle, L., Kelly, M., Gilbert, J. (1993) Migration of mineral hydrocarbons into foods. 3. Cheese coatings and temporary casings for skinless sausages, *Food Additives and Contaminants* **10**, 175–184.

Choi, J., Lee, S. Y. (1999) Factors affecting the economics of polyhydroxyalkanoate production by bacterial fermentation, *Appl. Microbiol. Biotechnol.* **51**, 13–21.

Cox, C. D., Rinehart, K. L., Moore, M. L., Cook, J. C. (1981) Pyochelin: novel structure of an iron-chelating growth promoter for *Pseudomonas aeruginosa, Proc. Natl. Acad. Sci. USA* **78**, 4256–4260.

Curley, J. M., Hazer, B., Lenz, R. W., Fuller, R. C. (1996a) Production of poly(3-hydroxyalkanoates) containing aromatic substituents by *Pseudomonas oleovorans, Macromolecules* **29**, 1762–1766.

Curley, J. M., Lenz, R. W., Fuller, R. C. (1996b) Sequential production of two different polyesters in the inclusion bodies of *Pseudomonas oleovorans, Int. J. Biol. Macromol.* **19**, 29–34.

de Koning, G. J. M., Witholt, B. (1997) A process for the recovery of poly(hydroxyalkanoates) from Pseudomonads. 1. Solubilization, *Bioprocess Eng.* **17**, 7–13.

de Koning, G. J. M., van Bilsen, H. H. M., Lemstra, P. J., Hazenberg, W., Witholt, B, Preusting, H., van der Galiën, J. G., Schirmer, A., Jendrossek, D. (1994) A biodegradable rubber by crosslinking poly(hydroxyalkanoate) from *Pseudomonas oleovorans, Polymer* **35**, 2090–2097.

de Koning, G. J. M., Kellerhals, M., van Meurs, C., Witholt, B. (1997) A process for the recovery of poly(hydroxyalkanoates from Pseudomonads. 2. Process development and economic evaluation, *Bioprocess Eng.* **17**, 15–21.

de Smet, M. J., Eggink, G., Witholt, B., Kingma, J., Wynberg, H. (1983) Characterization of intracellular inclusions formed by *Pseudomonas oleovorans* during growth on octane, *J. Bacteriol.* **154**, 870–878.

de Waard, P., van der Wal, H., Huijberts, G. N. M., Eggink, G. (1993) Heteronuclear NMR analysis of unsaturated fatty acids in poly(3-hydroxyalkanoates), *J. Biol. Chem.* **268**, 157–163.

Dowling, N., O'Gara, F. (1994) Metabolites of *Pseudomonas* involved in the biocontrol of plant disease, *Trends Biotechnol.* **12**, 133–141.

Durner, R. A. (1998) *Feast and Starvation: Accumulation of Bioplastics in Pseudomonas oleovorans.* Zürich: ETH.

Eggink, G., de Waard, P., Huijberts, G. N. M. (1995) Formation of novel poly(hydroxyalkanoates) from long-chain fatty acids, *Can. J. Microbiol.* **41** (Suppl. 1), 14–21.

Fiechter, A. (1992) Biosurfactant: moving towards industrial application, *Trends Biotechnol.* **10**, 208–217.

Gagnon, K. D., Lenz, R. W., Farris, R. J., Fuller, R. C. (1994a) Chemical modification of bacterial elastomers: 1. Peroxide crosslinking, *Polymer* **35**, 4358–4367.

Gagnon, K. D., Lenz, R. W., Farris, R. J., Fuller, R. C. (1994b) Chemical modification of bacterial elastomers: 2. Sulfur vulcanisation, *Polymer* **35**, 4368–4375.

Gross, R. A., Kim, O., Rutherford, D. R., Newmark, R. A. (1996) Cyanophenoxy- containing microbial polyesters: structural analysis, thermal properties, second harmonic generation and *in vivo* biodegradability, *Polym. Int.* **39**, 205–213.

Haywood. G. W. A., Anderson, A. J., Ewing, D. F., Dawes, E. A. (1990) Accumulation of a polyhydroxyalkanoate containing primarily 3-hydroxydecanoate from simple carbohydrate substrates by *Pseudomonas* sp. strain NCIMB 40135, *Appl. Environ. Microbiol.* **56**, 3354–3359.

Hazenberg, W. M. (1997) *Production of Poly(3-hydroxyalkanoates) by Pseudomonas oleovorans in Two-Liquid-Phase Media*. Zürich: ETH.

Hazer, B., Lenz, R. W., Fuller, R. C. (1996) Bacterial production of poly-3-hydroxyalkanoates containing arylalkyl substituent groups, *Polymer* **37**, 5951–5957.

Hocking, P. J., Marchessault, R. H. (1994) Biopolyesters, in: *Chemistry and Technology of Biodegradable Polymers* (Griffin, G. J. L, Ed.), London: Chapman & Hall, 48.

Hohnadel, D., Meyer, J. M. (1988) Specificity of pyoverdine-mediated iron uptake among fluorescent *Pseudomonas* strains, *J. Bacteriol.* **170**, 4865–4873.

Hori, K., Soga, K., Doi, Y. (1994) Production of poly(3-hydroxyalkanoates-*co*-3-hydroxy-omega-fluoroalkanoates) by *Pseudomonas oleovorans* from 1-fluorononane and gluconate, *Biotechnol. Lett.* **16**, 501–506.

Huijberts, G. N. M. (1996) *Microbial Formation of Poly(3-hydroxyalkanoates)*. Groningen: Rijksuniversiteit.

Huijberts, G. N. M., Eggink, G. (1996) Production of poly(3-hydroxyalkanoates) by *Pseudomonas putida* KT2442 in continuous cultures, *Appl. Microbiol. Biotechnol.* **46**, 33–239.

Huijberts, G. N. M., Eggink, G., de Waard, P., Huisman, G. W., Witholt, B. (1992) *Pseudomonas putida* KT2442 cultivated on glucose accumulates poly(3-hydroxyalkanoates) consisting of saturated and unsaturated monomers, *Appl. Environ. Microbiol.* **58**, 536–544.

Huijberts, G. N. M., de Rijk, T. C., de Waard, P., Eggink, G. (1994) ^{13}C Nuclear magnetic resonance studies of *Pseudomonas putida* fatty acid metabolic routes involved in poly(3-hydroxyalkanoate) synthesis, *J. Bacteriol.* **176**, 1661–1666.

Huisman, G. W., de Leeuw, O., Eggink, G., Witholt, B. (1989). Synthesis of poly-3-hydroxyalkanoates is a common feature of fluorescent pseudomonads, *Appl. Environ. Microbiol.* **55**, 1949–1954.

Huisman, G. W., Wonink, E., Meima, R., Kazemier, B., Terpstra, P., Witholt, B. (1991) Metabolism of poly(3-hydroxyalkanoates) (PHAs) by *Pseudomonas oleovorans*, *J. Biol. Chem.* **266**, 2191–2198.

Huisman, G. W., Wonink, E., de Koning, G. J. M., Preusting, H., Witholt, B. (1992) Synthesis of poly (3-hydroxyalkanoates) by mutant and recombinant *Pseudomonas* strains, *Appl. Microbiol. Biotechnol.* **38**, 1–5.

Jung, K., Hazenberg, W., Prieto, M. A., Witholt, B. (2001) Two stage chemostat process development for the effective production of medium-chain-length poly(3-hydroxyalkanoates), *Biotechnol. Bioeng.* **72**, 19–24.

Jung, K., Kessler, B., Schmid, A., Witholt, B. (2001) Strategy to produce MCL-PHA on toxic substrates.

Jung, K., Sierro, N., Egli, T., Kessler, B., Witholt, B. (2001) Reduced toxicity and accumulation of low and high molecular weight polyhydroxyalkanoate (PHA) in *Pseudomonas oleovorans* during dual-nutrient-limited growth in solvent based bioprocess.

Kellerhals, M. B. (1999) *Microbial Polyesters for the 21st Century: Production of Poly(*R*-3-hydroxyalkanoates) by Pseudomonas putida KT2442*. Zürich: ETH.

Kellerhals, M. B., Kessler, B., Witholt, B. (1999). Development of a closed-loop control system for production of medium-chain-length poly(3-hydroxyalkanoates) (MCL-PHAs) from bacteria, *Macromol. Symp.* **144**, 385–389.

Kessler, B., Weusthuis, R. A., Witholt, B., Eggink, G. (2001) Production of microbial polyesters: fermentation and downstream processes, in: *Advances in Biochemical Engineering/Biotechnology, vol. 71, Biopolyesters* (Steinbüchel, A., Babel. W., Eds.), New York: Springer-Verlag, 159–182.

Kim, G. J., Lee, I. Y. Yoon, S. C., Shin, Y. C., Park Y. H. (1997) Enhanced yield and a high production of medium chain length poly(3-hydroxyalkanoates in a two step fed-batch cultivation of *Pseudomonas putida* by combined use of glucose and octanoate, *Enzyme Microb. Technol.* **20** (7), 500–505.

Kim, O. Y., Gross, R. A., Rutherford, D. R. (1995) Bioengineering of poly(beta-hydroxyalkanoates) for advanced material applications: incorporation of cyano and nitrophenoxy side chain substituents, *Can. J. Microbiol.* **41** (Suppl. 1), 32–43.

Kim, O. Y., Gross, R. A., Hammar, W. J., Newmark, R. A. (1996) Microbial synthesis of poly(beta-hydroxyalkanoates) containing fluorinated side-chain substituents, *Macromolecules* **29**, 4572–4581.

Kim, S. W., Kim, P., Lee, H. S., Kim, J. H. (1996) High production of poly-beta-hydroxybutyrate (Poly(3HB)) from *Methylobacterium organophilum* under potassium limitation, *Biotechnol. Lett.* **18**, 25–30.

Kim, Y. B., Lenz, R. B., Fuller, R. C. (1991) Preparation and characterisation of poly(β-hydroxyalkanoates) obtained from *Pseudomonas oleovorans* grown with mixtures of 5-phenylvaleric acid and n-alkanoic acids, *Macromolecules* **24**, 5256–5260.

Kim, Y. B., Rhee, Y. H., Lenz, R. W. (1997) Poly(3-hydroxyalkanoate)s produced by *Pseudomonas oleovorans* grown by feeding nonanoic and 10-undecenoic acids in sequence, *Polym. J.* **29**, 894–898.

Kok, M. (1988) Alkane utilization by *Pseudomonas oleovorans.* Taking the lid of the oil bar rel. Groningen: Ryksuniversiteit.

Kraak, M. N., Smits, T. H. M., Kessler, B., Witholt, B. (1997) Polymerase C1 levels and poly(*R*-3-hydroxyalkanoate) synthesis in wild-type and recombinant *Pseudomonas* strains, *J. Bacteriol.* **179**, 4985–4991.

Kranz, R. G., Gabbert, K. K., Madigan, M. T. (1997) Positive selection systems for discovery of novel polyester biosynthesis genes based on fatty acid detoxification, *Appl. Environ. Microbiol.* **63**, 3010–3013.

Lafferty, R. M., Korsatko, B., Korstako, W. (1988) *Biotechnology*, Vol. 6b (Rehm, H.-J., Reed, G., Eds.). Weinheim: VCH.

Lageveen, R. G., Huisman, G. W., Preusting, H., Ketelaar, P., Eggink, G., Witholt, B. (1988) Formation of polyesters by *Pseudomonas oleovorans*: effect of substrates on formation and composition of poly-(*R*)-3-hydroxyalkanoates and poly-(*R*)-3-hydroxyalkenoates, *Appl. Environ. Microbiol.* **54**, 2924–2932.

Langenbach, S., Rehm, B. H. A., Steinbüchel, A. (1997) Functional expression of the PHA synthase gene *phaC1* from *Pseudomonas aeruginosa* in *Escherichia coli* results in poly(3-hydroxyalkanoate) synthesis, *FEMS Microbiol. Lett.* **150**, 303–309.

Lee, S. Y. (1996) Bacterial polyhydroxyalkanoates, *Biotechnol. Bioeng.* **49**, 1–14.

Lee, S. Y., Chang, H. N. (1995) Production of poly(hydroxyalkanoic acid), *Adv. Biochem. Eng. Biotechnol.* **52**, 27–58.

Lemoigne, M. (1926) Produits de deshydration et de polymerisation de l'acide β-oxybutyric, *Bull. Soc. Chem. Biol.* **8**, 770–782.

Lenz, R. W., Kim, Y. B., Fuller, R. C. (1992) Production of unusual bacterial polyesters by *Pseudomonas oleovorans* through cometabolism, *FEMS Microbiol. Rev.* **103**, 207–214.

Marchessault, R. H., Monasterios, C. J., Morin, F. G., Sundararajan, P. R. (1990). Chiral poly(beta-hydroxyalkanoates): an adaptable helix influenced by the alkane side-chain, *Int. J. Biol. Macromol.* **12**, 158–165.

Marchessault, R. H., Morin, F. G., Wong, S., Saracovan, I. (1995) Artificial granule suspensions of long side chain poly(3-hydroxyalkanoate), *Can. J. Microbiol.* **41** (Suppl. 1), 138–142.

Merrick, J. M., Doudoroff, M. (1964) Depolymerization of poly-beta-hydroxybutyrate by an intracellular enzyme system, *J. Bacteriol.* **88**, 60–71.

Mittendorf, V., Robertson, E. J., Leech, R. M., Krüger, N., Steinbüchel, A., Poirier, Y. (1998) Synthesis of medium-chain-length polyhydroxyalkanoates in *Arabidopsis thaliana* using intermediates of peroxisomal fatty acid beta-oxidation, *Proc. Natl. Acad. Sci. USA* **95**, 13397–13402.

Ohashi, T., Hasegawa, J. (1992) D(–)-β-Hydroxycarboxylic acids as raw materials for captopril and β-lactams, in: *Chirality in Industry* (Collins, A. N., Sheldrake, G. N., Crosby, J., Eds.), Manchester: Zeneca Specialties, 269–278.

Poirier, Y. (1999) Production of new polymeric compounds in plants, *Curr. Opin. Biotechnol.* **10**, 181–185.

Preusting, H., Hazenberg, W., Witholt, B. (1993) Continuous production of poly(3-hydroxyalkanoates) by *Pseudomonas oleovorans* in a high-cell-density, two-liquid-phase chemostat, *Enzyme Microb. Technol.* **15**, 311–316.

Preusting, H., Kingma, J., Huisman, G., Steinbüchel, A., Witholt, B. (1993) Formation of polyester blends by a recombinant strain of *Pseudomonas oleovorans*: different poly(3-hydroxyalksnoates) are stored in separate granules, *J. Environ. Polym. Degrad.* **1**, 11–21.

Preusting, H., Nijenhuis, A., Witholt, B. (1990) Physical characteristics of poly(3-hydroxyalkanoates) and poly(3-hydroxyalkenoates) produced by *Pseudomonas oleovorans* grown on aliphatic hydrocarbons, *Macromolecules* **23**, 4220–4224.

Preusting, H., van Houten, R., Hoefs, A., Kool van Langenberghe, E., Favre-Bulle, O., Witholt, B. (1993) High cell density cultivation of *Pseudomonas oleovorans*: growth and production of poly (3-hydroxyalkanoates) in two-liquid phase batch and fed-batch systems, *Biotechnol. Bioeng.* **41**, 550–556.

Prieto, M. A., Kellerhals, M. B., Bozzato, G. B., Radnovic, D., Witholt, B., Kessler, B. (1999). Engineering of stable recombinant bacteria for production of chiral medium-chain-length poly-3-hydroxyalkanoates, *Appl. Environ. Microbiol.* **65**, 265–3271.

Qi, Q. S., Rehm, B. H. A., Steinbüchel, A. (1997) Synthesis of poly(3-hydroxyalkanoates) in *Escherichia coli* expressing the PHA synthase gene *phaC2* from *Pseudomonas aeruginosa*: comparison of PhaC1 and PhaC2, *FEMS Microbiol. Lett.* **157**, 155–162.

Qi, Q. S., Steinbüchel, A., Rehm, B. H. A. (1998) Metabolic routing towards polyhydroxyalkanoic acid synthesis in recombinant *Escherichia coli* (*fadR*): Inhibition of fatty acid beta-oxidation by acrylic acid, *FEMS Microbiol. Lett.* **167**, 89–94.

Ren, Q. (1997) Biosynthesis of medium chain length poly-3-hydroxyalkanoates: From *Pseudomonas* to *Escherichia coli*. Zürich: ETH.

Ren, Q., Kessler, B., van der Leij, F., Witholt, B. (1998) Mutants of *Pseudomonas putida* affected in poly-3-hydroxyalkanoate synthesis, *Appl. Microbiol. Biotechnol.* **49**, 743–750.

Ren, Q., Sierro, N., Kellerhals, M, Kessler, B., Witholt, B. (2000a) Properties of engineered poly-3-hydroxyalkanoates produced in recombinant *Escherichia coli* strains, *Appl. Environ. Microbiol.* **66**, 1311–1320.

Ren, Q., Sierro, N., Witholt, B., Kessler, B. (2000b) FabG, an NADPH-dependent 3-ketoacyl reductase of *Pseudomonas aeruginosa*, provides precursors for medium-chain-length poly-3-hydroxyalkanoate biosynthesis in *Escherichia coli*, *J. Bacteriol.* **182**, 2978–2981.

Sasikala, C., Ramana, C. V. (1996) Biodegradable polyesters, *Adv. Appl. Microbiol.* **42**, 97–218.

Schmack, G., Gorenflo, V., Steinbüchel, A. (1998) Biotechnological production and characterization of polyesters containing 4-hydroxyvaleric acid and medium-chain-length hydroxyalkanoic acids, *Macromolecules* **31**, 644–649.

Scholtz, C., Fuller, R. C., Lenz, R. W. (1994) Production of poly(beta-hydroxyalkanoates) with beta-substituents containing terminal ester groups by *Pseudomonas oleovorans*, *Macromol. Chem. Phys.* **195**, 1405–1421.

Song, J. J., Yoon, S. C. (1996) Biosynthesis of novel aromatic copolyesters from insoluble 11-phenoxyundecanoic acid by *Pseudomonas putida* BMo, *Appl. Environ. Microbiol.* **62**, 536–544.

Steinbüchel, A. (1991) Polyhydroxyalkanoic Acid, in: *Biomaterials. Novel Materials from Biological Sources.* (Byrom, D., Ed.), Basingstoke: Macmillan, 123–213.

Steinbüchel, A., Valentin, H. E. (1995) Diversity of bacterial polyhydroxyalkanoic acids, *FEMS Microbiol. Lett.* **128**,219–228.

Taguchi, K., Aoyagi, Y., Matsusaki, H., Fukui, T., Doi, Y. (1999) Co-expression of 3-ketoacyl-ACP reductase and polyhydroxyalkanoate synthase genes induces PHA production in *Escherichia coli* HB101 strain, *FEMS Microbiol. Lett.* **176**, 183–190.

Timm, A., Steinbüchel, A. (1990) Formation of polyesters consisting of medium-chain-length 3-hydroxyalkanoic acids from gluconate by *Pseudomonas aeruginosa* and other fluorescent pseudomonads, *Appl. Environ. Microbiol.* **56**, 3360–3367.

Timm, A., Steinbüchel, A. (1992) Cloning and molecular analysis of the poly(3-hydroxyalkanoic acid) gene locus of *Pseudomonas aeruginosa* PAO1, *Eur. J. Biochem.* **209**, 15–30.

Tomashow, L. S., Weller, D. M. (1988) Role of a phenazine antibiotic from *Pseudomonas fluorescens* in biological control of *Gaeumannomyces graminis* var. tritici. *J. Bacteriol.* **170**, 3499–3508.

Tsuge, T., Fukui, T., Matsusaki, H., Taguchi, S., Kobayashi, G., Ishizaki, A., Doi, Y. (1999) Molecular cloning of two (*R*)-specific enoyl-CoA hydratase genes from *Pseudomonas aeruginosa* and their use for polyhydroxyalkanoate synthesis, *FEMS Microbiol. Lett.* **184**, 193–198.

Valentin, H. E., Schönebaum, A., Steinbüchel, A. (1996) Identification of 5-hydroxyhexanoic acid, 4-hydroxyheptanoic acid and 4-hydroxyoctanoic acid as new constituents of bacterial polyhydroxyalkanoic acids, *Appl. Microbiol. Biotechnol.* **46**, 261–267.

Van der Leij, F. R., Witholt, B. (1995) Strategies for the sustainable production of new biodegradable polyesters in plants: a review, *Can. J. Microbiol.* **41** (Suppl. 1), 222–238.

van der Walle, G. A. M., Buisman, G. J. H., Weusthuis, R. A., Eggink, G. (1999) Development of environmentally friendly coatings and paints using medium-chain-length poly(3-hydroxyalkanoates) as the polymer binder, *Int. J. Biol. Macromol.* **25**, 123–128.

van der Walle, G. A. M., de Koning, G. J. M., Weusthuis, R. A., Eggink. G. (2001) Properties, modifications and applications of biopolyesters, in: *Advances in Biochemical Engineering/Biotechnology, vol. 71, Biopolyesters* (Steinbüchel, A., Babel, W., Eds.), New York: Springer-Verlag, 263–292.

Weusthuis, R. A., Huijberts, G. N. M., Eggink, G. (1997) Production of MCL-poly(hydroxyalkanoates) (Review), in: *1996 International Symposium on Bacterial Polyhydroxyalkanoates* (Eggink, G., Steinbüchel, A., Poirier, Y., Witholt, B., Eds.), Ottawa: NRC Research Press, 102–110.

Williams, S. F., Martin, D. P., Horowitz, D. M., Peoples, O. P. (1999). PHA applications: addressing the price performance issue. I. Tissue engineering, *Int. J. Biol. Macromol.* **25**, 111–121.

11
Biosynthesis and Fermentative Production of Short-chain-length Medium-chain-length PHAs

Prof. Dr. Sang Yup Lee[1], M. Eng. Si Jae Park[2]

[1] Metabolic and Biomolecular Engineering National Research Laboratory, Department of Chemical Engineering and BioProcess Engineering Research Center, Korea Advanced Institute of Science and Technology, 373-1 Kusong-dong, Yusong-gu, Taejon 305-701, Korea; Tel.: +82-42-869-3930; Fax: +82-42-869-3910; E-mail: leesy@mail.kaist.ac.kr

[2] Metabolic and Biomolecular Engineering National Research Laboratory, Department of Chemical Engineering and BioProcess Engineering Research Center, Korea Advanced Institute of Science and Technology, 373-1 Kusong-dong, Yusong-gu, Taejon 305-701, Korea; Tel.: +82-42-869-5970; Fax: +82-42-869-8800; E-mail: parksj@mail.kaist.ac.kr

Biotechnology of Biopolymers. From Synthesis to Patents. Edited by A. Steinbüchel and Y. Doi

ISBN: 3-527-31110-6

ACP	acyl carrier protein
CoA	coenzyme A
DCW	dry cell weight
DO	dissolved oxygen
DOC	dissolved oxygen concentration
3HA	3-hydroxyalkanoate
3HB	3-hydroxybutyrate
3HD	3-hydroxydecanoate
3HDD	3-hydroxydodecanoate
3HHp	3-hydroxyheptanoate
3HHx	3-hydroxyhexanoate
3HO	3-hydroxyoctanoate
3HV	3-hydroxyvalerate
LDPE	low-density polyethylene
MCL	medium-chain-length
NMR	nuclear magnetic resonance
PHA	polyhydroxyalkanoate
SCL	short-chain-length

1 Introduction

Polyhydroxyalkanoates (PHAs) are polyesters of various hydroxyalkanoates, which are accumulated as a carbon and energy storage material in various microorganisms, usually under conditions of limiting nutritional elements such as N, P, S, O, or Mg and in the presence of excess carbon source (Anderson and Dawes, 1990; Doi, 1990; Lee, 1996c; Steinbüchel, 1991; Steinbüchel and Füchtenbusch, 1998; Madison and Huisman, 1999). Since poly(3-hydroxybutyrate), poly(3HB), was first discovered in the bacterium *Bacillus megaterium* in 1926 (Lemoigne, 1926), many different PHAs consisting of different numbers of main-chain carbon atoms and various R-pendant groups have been reported (Steinbüchel and Valentin, 1995). Generally, PHAs can be divided into two groups depending on the number of carbon atoms in the monomer units: short-chain-length (SCL) PHA consisting of three to five carbon atoms; and medium-chain-length (MCL) PHA consisting of 6–14 carbon atoms (Lee, 1996c; Steinbüchel and Füchtenbusch, 1998; Lee and Choi, 1999). Recently, several bacteria capable of producing PHAs containing both SCL- and MCL-monomer units have been isolated (Brandl et al., 1989; Haywood et al., 1991; Liebergesell et al., 1991; Steinbüchel and Wiese, 1992; Kobayashi et al., 1994; Kato et al., 1996).

In recent years, PHAs have received much attention as effective, biodegradable substitutes for petroleum-derived synthetic polymers, not only due to their similar material properties but also to their complete biodegradability after disposal (Brandl et al., 1995; Lee, 1996b; Steinbüchel, 1992). The physical properties of PHAs are dependent on the monomer composition and molecular weight. Poly(3HB) homopolymer, the most ubiquitous biopolyester, is rather stiff and brittle (Madison and Huisman, 1999) and has a high melting temperature of ca. 170 °C that makes its processing difficult. Poly(3-hydroxybutyrate-*co*-3-hydroxyvalerate), poly(3HB-*co*-3HV) copolymer is less stiff and tougher. Poly(3HB-*co*-3HV) copolymer also shows higher elongation to break depending on the polymer composition (0–25 mol.% 3HV) and reduced melting point, which ranges from 160 to 100 °C. MCL-PHAs possess a much lower crystallinity and higher elasticity, and have a different range of applications compared with SCL-PHA. It has been suggested that MCL-PHAs could be used for biodegradable rubber and coating materials. Recently, reports have been made on the synthesis of SCL-MCL-PHA copolymers, including poly(3-hydroxybutyrate-*co*-3-hydroxyhexanoate), poly(3HB-*co*-3HHx), and poly(3-hydroxybutyrate-*co*-3-hydroxyalkanoate), poly(3HB-*co*-3HA) (Kobayashi et al., 1994; Kato et al., 1996). Poly(3HB-*co*-3HHx) is a flexible material, and films of this copolymer show a high degree of elongation to break that resembles the properties of low-density polyethylene (LDPE) (Doi et al., 1995). Currently, the superior properties of SCL-MCL-PHA copolymers are attracting various industrial applications.

The commercialization of PHAs has been hampered due to the high production cost of PHAs compared with petrochemical-based polymers. Hence, much effort has been devoted to reducing the production cost of PHAs by developing better bacterial strains and downstream processes, for example more efficient fermentation and more economic recovery processes (Lee, 1996b,c, 1997; Choi et al., 1998; Choi and Lee, 1999a,b). Metabolic engineering allows us to modify the existing metabolic pathways and/or to introduce new metabolic pathways to increase PHA biosynthesis capability, to broaden the utilizable substrate range, and to produce novel polymers (Lee, 1996b). In this chapter, fermentation and metabolic engineering strategies employed for the production of SCL-MCL-PHAs in various wild-type and recombinant strains are reviewed.

2 Historical Outline

A brief historical outline of the biosynthesis and fermentative production of MCL- and SCL-PHAs is illustrated in the following chart:

- 1983 Discovery of MCL-PHA in *Pesudomonas oleovorans*
- 1989 Discovery of PHA consisting of both SCL and MCL monomers
- 1992 Cloning of MCL-PHA biosynthesis genes from *Pseudomonas aeruginosa*
- 1994 Screening *Aeromonas* strains accumulating poly(3HB-*co*-3HHx)
- 1996 Production of PHA consisting of 3HB and MCL-3HA from glucose by *Pseudomonas* sp. 61–3
- 1997 Cloning of PHA biosynthesis genes from *Aeromonas caviae*
- 1997 Production of MCL-PHA from recombinant *Escherichia coli* harboring *P. aeruginosa* PHA synthase
- 1998 Cloning of PHA biosynthesis genes from *Pseudomonas* sp. 61–3
- 1999 Production of poly(3HB-*co*-3HHx) by recombinant *E. coli*
- 1999 Production of poly(3HB-*co*-3HHx) by recombinant *Ralstonia eutropha*
- 2000 Fed-batch cultivation of *A. hydrophila* for the production of poly(3HB-*co*-3HHx)
- 2000 Production of SCL-MCL-PHA consisting of high mole fraction of 3HB in recombinant *Pseudomonas* strain
- 2001 Fed-batch cultivation of recombinant *E. coli* harboring artificial *Aeromonas* PHA biosynthesis genes for the production of poly(3HB-*co*-3HHx)

3
General Metabolic Pathways for the Synthesis of PHAs

To date, three major different pathways involved in the synthesis of PHA have been elucidated (Figures 1 and 2). In *Ralstonia eutropha*, two acetyl-CoA moieties are condensed to form acetoacetyl-CoA by β-ketothiolase, after which the acetoacetyl-CoA is reduced to (*R*)-3-hydroxybutyryl-CoA by an NADPH-dependent reductase. PHA synthase finally links (*R*)-3-hydroxybutyryl-CoA to the growing chain of poly(3HB) (Schubert et al., 1988; Slater et al., 1988; Peoples and Sinskey, 1989). *Pseudomonas oleovorans* and most pseudomonads belonging to the rRNA homology group I can accumulate MCL-PHAs using 3-hydroxyacyl-CoA intermediates channeled from the β-oxidation pathway when cultured on various MCL-alkanes, alkanols, or alkanoates (Steinbüchel, 1992; Lee and Chang, 1995; Lee, 1996c; Madison and Huisman, 1999; Witholt and Kessler, 1999). Three putative enzymes including hydratase, epimerase, and 3-ketoacyl-CoA reductase have been suggested to supply 3-hydroxyacyl-CoA from the β-oxidation pathway (Madison and Huisman, 1999). The exact mechanisms of which enzyme is truly involved and how it is channeled in the PHA biosynthesis have not been elucidated. According to reports on cloning of the (*R*)-specific enoyl-CoA hydratase genes from *Aeromonas caviae* and *P. aeruginosa*, it is more plausible to suppose that the hydratase is a major candidate supplying hydroxyacyl-CoA from the β-oxidation pathway (Fukui and Doi, 1997, 1998a; Tsuge et al., 2000). Most pseudomonads belonging to the rRNA homology group I

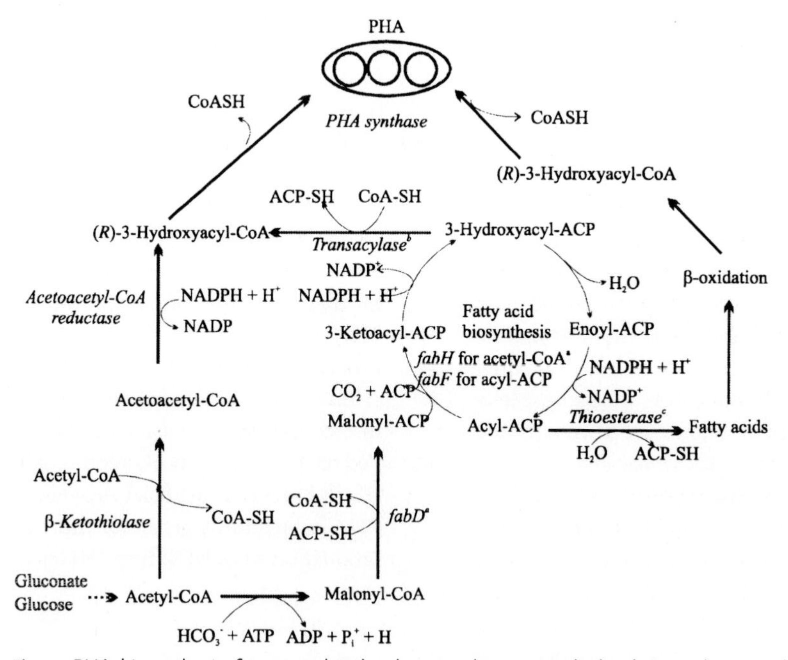

Fig. 1 PHA biosynthesis from unrelated substrates by a central glycolytic pathway and the fatty acid biosynthesis pathway. (Lee, 1996c; [a]Taguchi et al., 1999b; [b]Fiedler et al., 2000; Hoffman et al., 2000a,b; Matsumoto et al., 2001; Rehm et al., 1998, [c]Klinke et al., 1999).

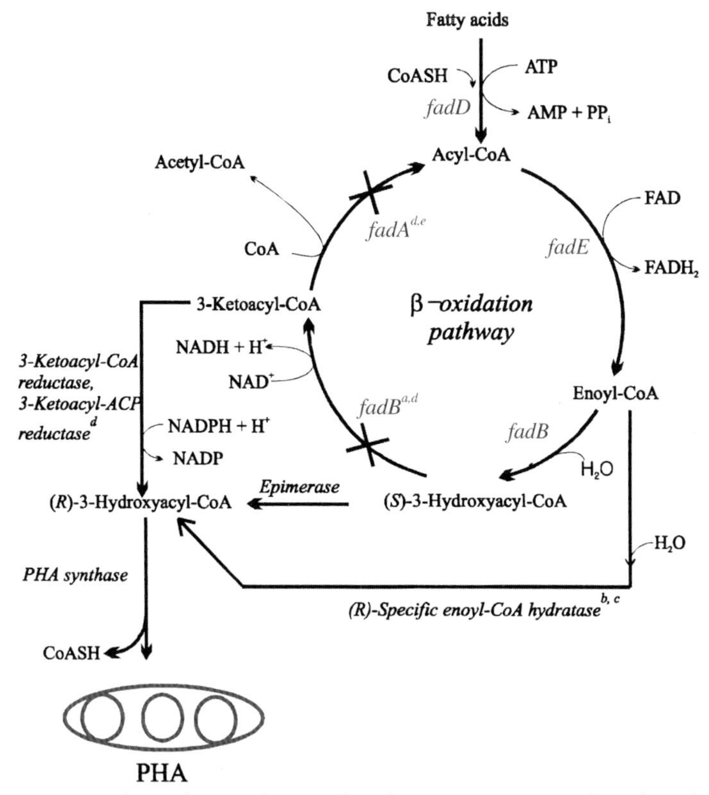

Fig. 2 PHA biosynthesis pathway via β-oxidation. Various intermediates from the β-oxidation pathway can be channeled to PHA biosynthesis by metabolic engineering (Lee, 1996c; [a, e]Qi et al., 1997, 1998. [b,c]Fukui et al., 1999; Tsuge et al., 2000; Park et al., 2001a,b; [d]Ren et al., 2000; Taguchi et al., 1999a).

(except *P. oleovorans*) can also accumulate MCL-PHAs from the β-oxidation pathway or from acetyl-CoA via fatty acid biosynthetic pathway (Huijberts et al., 1992; Lee, 1996c; Madison and Huisman, 1999). The composition of the PHA formed by pseudomonads is highly dependent on the structure of the carbon substrate used (Brandl et al., 1988; Lageveen et al., 1988; Huisman et al., 1989). A ^{13}C-NMR study of *P. putida* revealed that fatty acid biosynthesis and the β-oxidation pathway can independently provide PHA precursors (Huijberts et al., 1994). Since the time when PHA biosynthesis genes were first cloned from *R. eutropha*, a number of PHA biosynthesis-related genes have been cloned from various bacteria, and these have been recently reviewed (Steinbüchel and Füchtenbusch, 1998; Madison and Huisman, 1999). The PHA biosynthesis genes of *A. caviae* and *Aeromonas salmonisida*, which can produce a copolymer poly(3HB-*co*-3HHx), were cloned and characterized in detail (Fukui and Doi, 1997, 1998a; Fukui et al., 2001; Park et al., 2001a). The PHA biosynthesis genes of *A. caviae* consisted of $phaP_{Ac}$, $phaC_{Ac}$ and $phaJ_{Ac}$ genes encoding phasin, PHA synthase, and (*R*)-specific enoyl-CoA hydratase, respectively (Fukui and Doi, 1997, 1998a; Fukui et al., 2001). In certain *Pseudomonas* spp., the key enzyme 3-hydroxydecanoyl-ACP:CoA transacylase encoded by the *phaG* gene was found to play an important role in channeling the intermediates of fatty acid biosynthesis pathway to PHA production (see Figure 1; Rehm et al., 1998; Hoffman et al., 2000a, b; Matsumoto et al., 2001). The elucidation of various genes which are either directly or indirectly involved in PHA biosynthesis has permitted us to develop new metabolic engineering strategies for the efficient production of PHAs. Even though, ultimately, the entire metabolic pathways in the cell are involved in PHA production, several enzymes and proteins are more directly involved than others. The genes involved more directly in PHA biosynthesis discovered to date are summarized in Table 1.

4 Production of SCL-MCL-PHAs by Wild-type Bacteria

Several bacteria which produce PHAs containing both SCL and MCL monomers were first identified during the late 1980s and early 1990s. *Rhodospirillum rubrum* (Brandl et al., 1989), *Rhodocyclus gelatinosus* (Liebergesell et al., 1991), and *Rhodococcus ruber* (Haywood et al., 1991) were reported to produce terpolymers consisting of 3HB, 3HV, and 3HHx from hexanoate. Kobayashi et al. (1994) reported that two strains belonging to the aeromonads can accumulate poly(3HB-*co*-3HHx) when fatty acids comprising more than 12 carbons are used as a

Tab. 1 Key enzymes involved in the PHA biosynthesis

Enzyme (gene)	*Characteristic*	*Representative sources*	*Metabolic pathway*	*Reference*
β-ketothiolase (*phaA*)	Condensation of acetyl-CoA to acetoacetyl-CoA	*R. eutropha*	Central glycolysis	Peoples and Sinskey (1989); Schubert et al. (1988); Slater et al. (1988)
Acetoacetyl-CoA reductase (*phaB*)	Reduction of acetoacetyl-CoA to (*R*)-3-hydroxybutyryl-CoA	*R. eutropha*	Central glycolysis	Peoples and Sinskey (1989); Schubert et al. (1988); Slater et al. (1988)
3-hydroxydecanoyl-ACP:CoA transacylase (*phaG*)	Conversion of 3-hydroxyacyl-ACP to 3-hydroxyacyl-CoA	*P. putida*	Fatty acid biosynthesis	Rehm et al. (1998)
(*R*)-specific enoyl-CoA hydratase (*phaJ*)	Conversion of enoyl-CoA to (*R*)-3-hydroxyacyl-CoA	*A. caviae*	β-oxidation	Fukui and Doi (1998a)
3-ketoacyl-ACP synthase III (*fabH*)	Conversion of SCL-3-hydroxyacyl-ACP to SCL-3-hydroxyacyl-CoA	*E. coli*	Fatty acid biosynthesis	Taguchi et al. (1999b)
Malonyl-CoA-ACP transacylase (*fabD*)	Conversion of SCL-3-hydroxyacyl-ACP to SCL-3-hydroxyacyl-CoA	*Pseudomonas* sp. 61 – 3	Fatty acid biosynthesis	Taguchi et al. (1999b)
3-ketoacyl-ACP reductase (*fabG*)	Reduction of 3-ketoacyl-CoA to 3-hydroxyacyl-CoA	*E. coli, P. aeruginosa*	β-oxidation	Taguchi et al. (1999a); Ren et al. (2000)
Thioesterase I (*tesA*)	Conversion of acyl-ACP to free fatty acid	*E. coli*	Link between fatty acid biosynthesis and β-oxidation	Klinke et al. (1999)

carbon source. When triaurin, coconut oil, and olive oil were used as a carbon source in two-stage batch cultures, two *Aeromonas* strains produced poly(3HB-*co*-3HHx) containing different 3HHx fraction in the copolymer. The 3HHx fraction was dependent on the concentration of carbon source and culture conditions. The PHA biosynthesis genes of one of these strains, *A. caviae*, were cloned and characterized at the molecular level (Fukui and Doi, 1997, 1998a; Fukui et al., 2001).

Lee et al. (2000) reported the production of poly(3HB-*co*-3HHx) by a fed-batch culture of newly isolated *A. hydrophila*. The characteristics of cell growth and polymer accumulation were examined using various carbon sources including glucose, sodium gluconate, hexanoic acid, heptanoic acid, octanoic acid, lauric acid, and oleic acid, although poly(3HB-*co*-3HHx) was produced only from lauric acid and oleic acid. In flask cultures the poly(3HB-*co*-3HHx) content could be increased by applying phosphorus limitation. Even though the poly(3HB-*co*-3HHx) content was highest when lauric acid was used as a sole carbon source, oleic acid was used in fed-batch culture as a carbon source because lauric acid caused several problems during fed-batch cultures, for example difficulty of feeding and severe foam formation. Feeding strategies were developed based on the characteristics of cell growth, because the pH and the dissolved oxygen concentration (DOC) did not respond sensitively to the depletion of oleic acid during the early stage of fermentation. Manual feeding based on the cell yield on oleic acid, DO-stat feeding, and feeding by the response of the pH on the depletion of oleic acid were applied sequentially during the fed-batch culture. By applying an optimally designed nutrient feeding strategy using oleic acid as a carbon source, cell concentration, poly(3HB-*co*-3HHx) content, and 3HHx fraction obtained in 43 h were 95.7 g dry cell weight (DCW) L^{-1}, 45.2 wt.%, and 17 mol.%, respectively, resulting in a PHA productivity of 1.01 g PHA L^{-1} h^{-1} (Figure 3). This process was developed at KAIST (Taejon, Korea) and transferred to the Jiangmen Biotechnology Institute (Guangdong, China) in collaboration with G. Chen (Tsinghua University, Beijing, China) and Procter & Gamble Co. (Cincinnati, OH, USA) for the semi-commercial scale (4,000 L and 20,000 L) production of poly(3HB-*co*-3HHx).

The production of PHA copolymer consisting of 3HB and MCL-3-hydroxyalkanoates (3HA) of C_6, C_8, C_{10}, and C_{12} by *Pseudomonas* sp. 61–3 was also reported (Kato et al., 1996). When sugars (including glucose, fructose, and mannose) were used as the sole carbon source, a blend of poly(3HB) homopolymer and poly(3HB-*co*-3HA) copolymer composed of 44 mol.% 3HB, 5 mol.% 3HHx, 21 mol.% 3-hydroxyoctanoate (3HO), 25 mol.% 3-hydroxydecanoate (3HD), 2 mol.% 3-hydroxydodecanoate (3HDD), and 3 mol.% 3-hydroxy-5-*cis*-dodecanoate was produced under nitrogen-limited conditions. The separate poly(3HB) and poly(3HB-*co*-3HA) biosynthesis genes were recently cloned from *Pseudomonas* sp. 61–3 and analyzed at the molecular level (Matsusaki et al., 1998).

5
Production of SCL-MCL-PHAs by Metabolically Engineered Bacteria

Metabolic engineering has led to the introduction of new PHA metabolic pathways in bacteria capable of the efficient production of novel PHAs consisting of various monomers. Among these bacteria, recombinant *E. coli*, *R. eutropha* and *Pseudomonas* sp. have undergone intense investigation for the production of SCL-MCL-PHAs.

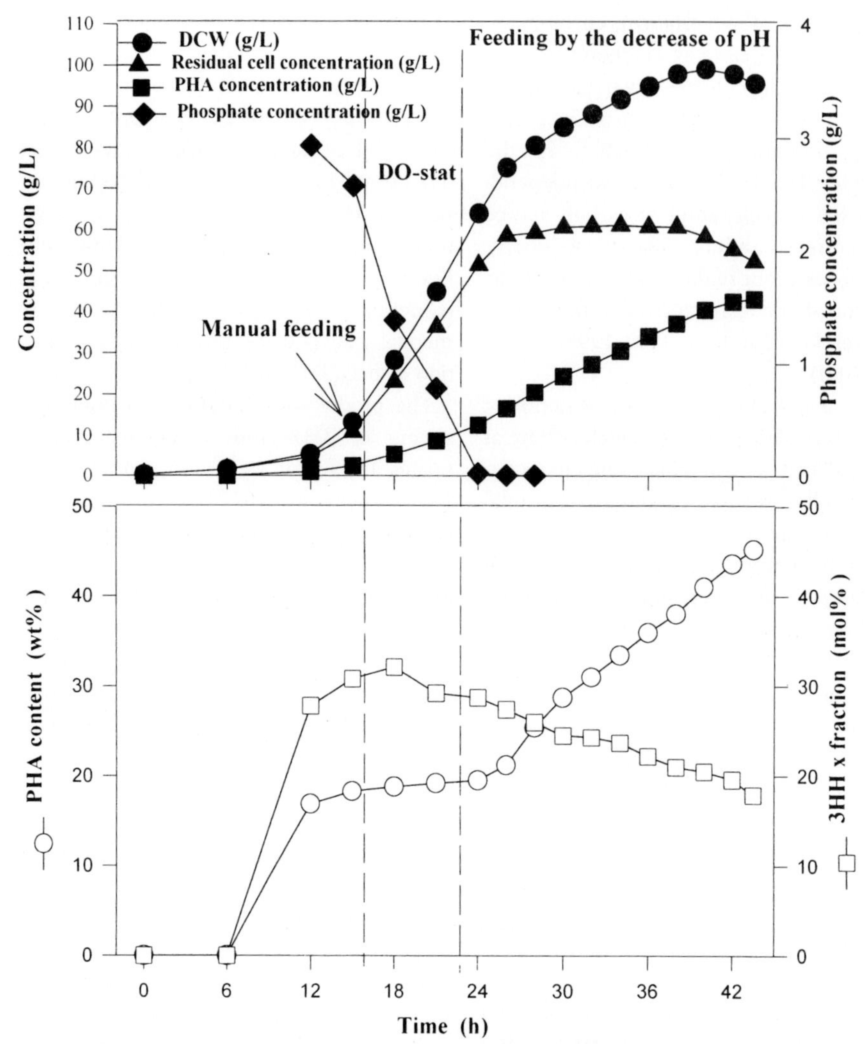

Fig. 3 Time profiles of cell concentration (●), residual cell concentration (▲), PHA concentration (■), phosphate concentration (◆), PHA content (○), and 3HHx fraction (□) during fed-batch cultivation of *A. hydrophila* in chemically defined medium. Oleic acid was fed by manual feeding, DO-stat, feeding on the decrease of pH responding to carbon depletion, sequentially. [Reproduced from Lee et al. (2000) with permission.]

5.1
Biosynthesis of MCL- and SCL-MCL-PHAs in Recombinant *E. coli*

E. coli has been regarded as the most promising strain for the production of SCL-PHAs (Lee, 1996a,b), and has also been closely examined with regard to the production of MCL-PHAs and SCL-MCL-PHAs. However, the polymer contents of MCL-PHAs and SCL-MCL-PHAs produced in recombinant *E. coli* were much lower compared with that of SCL-PHAs and the fermentation strategies were not developed further for the efficient production of these PHAs. In general, metabolic intermediates of the β-oxidation and fatty acid biosynthesis pathways can be modified to provide substrates for PHA synthase, and (*R*)-3-hydroxyacyl-CoA for the production of MCL-and SCL-MCL-PHAs (see Figures 1 and 2).

Recombinant *E. coli* harboring the *Pseudomonas* MCL-PHA synthase gene has been shown to produce MCL-PHA from 3-hydroxyacyl-CoA intermediates generated from the β-oxidation pathway (Langenbach et al., 1997; Qi et al., 1997, 1998; Prieto et al., 1999). Functional expression of the *P. aeruginosa* $phaC1_{Pa}$ gene in *E. coli* LS1298 (*fadB*), which is defective in β-oxidation, allowed the accumulation of MCL-PHA up to 21 wt.% DCW when cells were grown in Luria-Bertani (LB) medium supplemented with 5 g L^{-1} decanoate (Langenbach et al., 1997). The PHA obtained consisted of 2.5 mol.% 3HHx, 20 mol.% 3HO, 72.5 mol.% 3HD, and 5 mol.% 3HDD. It is clear that, to some extent, intermediates of the fatty acid biosynthesis pathway were incorporated into the polymer. The function of the *P. aeruginosa* $phaC1_{Pa}$ gene was compared with that of the $phaC2_{Pa}$ gene in *E. coli* LS1298 (Qi et al., 1997). When recombinant *E. coli* LS1298 harboring either the $phaC1_{Pa}$ or $phaC2_{Pa}$ gene (or both genes) was grown in LB medium supplemented with 5 g L^{-1} dodecanoate, MCL-PHA was accumulated up to 15 wt.% DCW in all cases, and there was little difference in the PHA composition. Inhibition of the β-oxidation pathway by the addition of acrylic acid successfully supported the 3-hydroxyacyl-CoA formation for the MCL-PHA production in recombinant *E. coli* strains (Qi et al., 1998), among which *E. coli* RS3097 (*fadR*41) accumulated MCL-PHA up to 60 wt.% DCW from decanoate.

E. coli can also be engineered to produce PHA via the fatty acid biosynthesis pathway (see Figure 1). Recently, the *phaG* gene encoding 3-hydroxydecanoyl-ACP:CoA transacylase was cloned from *P. putida* (Rhem et al., 1998). If this gene is functional in *E. coli*, it may be possible to synthesize PHA from unrelated carbon sources such as glucose or gluconate. However, during cultivation of *E. coli* JM109 harboring both the *P. aeruginosa* $phaC1_{Pa}$ and *P. putida* $phaG_{Pp}$ genes in two plasmids, no PHA accumulation was observed from glucose (Rhem et al., 1998). It was recently demonstrated that expression of these two genes in the non-PHA-accumulating bacterium *P. fragi* allowed PHA accumulation, which showed that PhaG (transacylase) directly links *de-novo* fatty acid biosynthesis with PHA biosynthesis (Fiedler et al., 2000). This role of PhaG was further confirmed by cloning and expression of the $phaG_{Pa}$ gene from *P. aeruginosa* (Hoffmann et al., 2000a). It was found later that PhaG is present in all *Pseudomonas* strains producing MCL-PHAs from unrelated carbon sources (Hoffmann et al., 2000b). It was of interest to find that *P. oleovorans*, which is not able to synthesize MCL-PHAs from unrelated carbon sources, also possesses the $phaG_{Po}$ gene. However, the gene is not expressed in *P. oleovorans*, and hence MCL-PHA synthesis from glucose or gluconate is not possible in this bacterium (Hoffmann et al., 2000b).

The metabolic link between the fatty acid β-oxidation and fatty acid biosynthesis pathways was realized by overexpression of cytosolic thioesterase I in *E. coli* (see Figure 1; Klinke et al., 1999). Thioesterase I mediates channeling of acyl-ACP intermediates from the fatty acid biosynthesis pathway to free fatty acids. The *fadR fadB* mutant *E. coli* strain expressing the *P. oleovorans* PHA synthase and the thioesterase I genes accumulated MCL-PHA from gluconate up to 2.3 wt.% DCW, the polymer being composed of 3HHx, 3HO, and 3HD. The amount of

PHA and the thioesterase I activity were found to be directly correlated. A stable recombinant *E. coli* strain for MCL-PHA production was developed by using chromosomal insertion of the *P. oleovorans* $phaC1_{Po}$ gene using a mini-Tn5 system (Prieto et al., 1999). PHA synthase expression was controlled by the xylS/Pm regulatory system, in which XylS is the regulatory protein that turns on the Pm promoter when it is activated by benzoates or toluates. Recombinant *E. coli* accumulated MCL-PHA from palmitic acid up to 11 wt.% DCW when induced by 3-methylbenzoic acid.

SCL-MCL copolymers could also be produced in recombinant *E. coli* using the β-oxidation pathway with functional expression of different PHA biosynthesis-related genes. Recombinant *E. coli* LS1298 harboring the *R. eutropha* PHA synthase gene could produce poly(3HB-*co*-3HO) and poly(3HB-*co*-3HO-*co*-3HDD) when octanoate and decanoate or dodecanoate was provided as a carbon source in LB medium. Even though the molar fractions of 3HO and 3HDD in copolymers were rather low (<4 mol.%), these results show that the *R. eutropha* PHA synthase possesses much broader substrate specificity than was previously thought (Antonio et al., 2000). Recently, the PHA biosynthesis operon was cloned from *A. caviae* and characterized (Fukui and Doi, 1997). Recombinant *E. coli* LS5218 harboring only the *A. caviae* $phaC_{Ac}$ gene did not accumulate PHA from dodecanoate (Fukui et al., 1999). However, coexpression of the *A. caviae* $phaC_{Ac}$ and $phaJ_{Ac}$ genes under the *A. caviae* native promoter resulted in poly(3HB-*co*-3HHx) accumulation up to 11 wt.% DCW from octanoate and dodecanoate. The high-level expression of both genes under the *lac* promoter enhanced PHA accumulation up to 38 wt.% DCW from dodecanoate. Enzyme assays suggested that it is essential to channel PHA precursor efficiently from β-oxidation by the high-level expression of $phaJ_{Ac}$ for achieving a high content of PHA (Fukui et al., 1999). Recently, the genes for two (*R*)-specific enoyl-CoA hydratases (encoded by $phaJ1_{Pa}$ and $phaJ2_{Pa}$), which are homologous to *A. caviae* (*R*)-specific enoyl-CoA hydratase but have different substrate specificity, were cloned from *P. aeruginosa* using the polymerase chain reaction (PCR) (Tsuge et al., 2000). The $phaJ1_{Pa}$ and $phaJ2_{Pa}$ products were found to be specific for SCL-enoyl-CoA and MCL-enoyl-CoA, respectively. Coexpression of these two hydratase genes along with the PHA synthase gene in *E. coli* LS5218 resulted in the accumulation of PHA up to 29 wt.% DCW from dodecanoate as sole carbon source. The composition of PHA was dependent on the type of PHA synthase and hydratase employed.

A novel strategy for the fed-batch culture of recombinant *E. coli* for the production of poly(3HB-*co*-3HHx) from fatty acids was recently reported (Park et al., 2001a) (Figure 4). A recombinant *E. coli* LS5218 (*fadR atoC* (Con)) was constructed which harbored the *Aeromonas* PHA synthase gene, (*R*)-specific enoyl-CoA hydratase gene and *R. eutropha* reductase gene. During the fermentation, the DOC was reduced stepwise (from 40% to 10%, then to 5%) to supply efficiently the intermediates of the β-oxidation pathway. By using a fed-batch culture combined with a DO-stat feeding strategy, *E. coli* LS5218 harboring the artificial PHA operon could produce poly(3HB-*co*-3HHx) from dodecanoic acid up to 21.5 g L^{-1} in 40.8 h, resulting in a PHA productivity of 0.53 g PHA L^{-1} h^{-1} (Figure 4). Recombinant *E. coli* LS5218 harboring the *Aeromonas* PHA synthase, (*R*)-specific enoyl-CoA hydratase gene and *R. eutropha* reductase gene was also examined for the production of terpolymer poly(3HB-*co*-3HV-*co*-3HHx) (Park et al., 2001b). Several reports have been made on the production of poly(3HB-*co*-3HV-*co*-

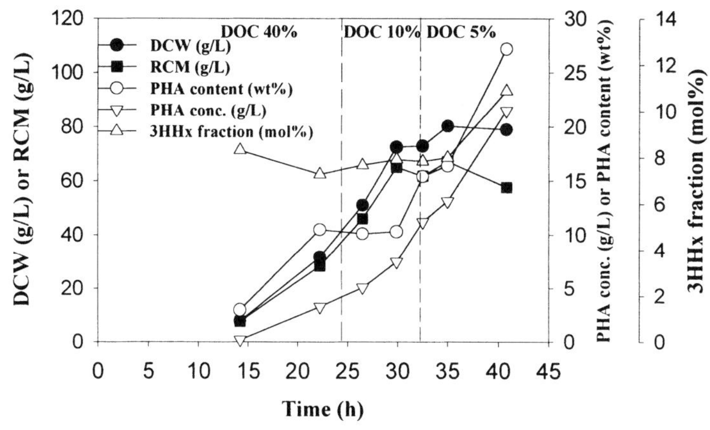

Fig. 4 Time profiles of cell concentration (●), residual cell concentration (■), PHA content (○), PHA concentration (▽) and 3HHx fraction in polymer (△) during the fed-batch culture of *E. coli* LS5218 harboring *Aeromonas* PHA synthase gene, (*R*)-specific enoyl-CoA hydratase gene and *R. eutropha* reductase gene. [Reproduced from Park et al. (2001a), with permission.]

3HHx) in wild-type and recombinant bacteria harboring the *Nocardia corallina* PHA synthase gene from various fatty acids (Brandl et al., 1989; Haywood et al., 1991; Liebergesell et al., 1991; Hall et al., 1998). However, the monomer fraction in the terpolymer could not be controlled in previous studies. Park et al. (2001b) reported that the metabolically engineered *E. coli* strain could produce a terpolymer having different monomer fractions by simple supplementation with various odd carbon-number fatty acid and dodecanoic acid. In particular, the phasin was found to play a critical role in modulating the monomer fraction in the terpolymer. This unusual behavior of phasin has been reported in various recombinant bacteria (Kichise et al., 1999; Fukui et al., 2001; Park et al., 2001a). Coexpression of the *Aeromonas* sp. phasin gene led to the 3HHx fraction in poly(3HB-*co*-3HHx) being significantly increased, though the reason for this phenomenon has still to be identified.

Coexpression of the *E. coli* 3-ketoacyl-ACP reductase gene (*fabG*$_{Ec}$) and the *A. caviae* *phaC*$_{Ac}$ gene or *Pseudomonas* sp. 61–3 *phaC1*$_{Ps}$ gene also allowed accumulation of PHA via fatty acid β-oxidation (see Figure 2; Taguchi et al., 1999a). Using a two-stage cultivation, and with dodecanoate as sole carbon source, recombinant *E. coli* HB101 harboring the *fabG*$_{Ec}$ gene along with the *A. caviae phaC*$_{Ac}$ or the *Pseudomonas* sp. 61–3 *phaC1*$_{Ps}$ gene accumulated poly(3HB-*co*-3HHx) or MCL-PHA consisting of 3HHx, 3HO, 3HD, and 3HDD, the monomer composition of PHA being dependent on the PHA synthase employed. These results suggested that overexpression of the *fabG*$_{Ec}$ gene allowed 3-hydroxyacyl-CoA to be channeled for PHA synthesis via the β-oxidation pathway. In addition, functional expression of the *P. aeruginosa fabG*$_{Pa}$ gene and the *P. oleovorans phaC1*$_{Po}$ gene in mutant *E. coli* strains, which are deficient in steps downstream or upstream of the 3-ketoacyl-CoA formation steps during β-oxidation, might support the formation of MCL-PHA from octanoic acid. A higher level of PHA could be produced in *E. coli* JMU194 (*fadR fadA*) harboring the *phaC1*$_{Po}$ and *fabG*$_{Pa}$ genes, whereas a similar level of PHA was accumulated in *E. coli* JMU193 (*fadR fadB*), compared with those obtained with the corresponding mutant strains carrying the *phaC1*$_{Po}$ gene alone (see Figure 2; Ren et al., 2000).

The production of poly(3HB) from acetyl-CoA via the fatty acid biosynthesis pathway was also demonstrated by coexpression of the *E. coli fabH*$_{Ec}$ gene encoding 3-ketoacyl-ACP synthase III, and the *A. caviae phaC*$_{Ac}$ or the *Pseudomonas* sp. 61–3 *fabD*$_{Ps}$ gene encoding malonyl-CoA-ACP transacylase and the *A. caviae phaC*$_{Ac}$ gene (Taguchi et al., 1999b). FabH and FabD were suggested to have transacylase activity with narrow specificity towards 3-hydroxyacyl-ACP, based on the finding that only poly(3HB) homopolymer was accumulated, even though the *A. caviae phaC*$_{Ac}$ has broad substrate specificity from 3HB to 3-hydroxyheptanoate (3HHp) (Fukui et al., 1997).

5.2
Biosynthesis of MCL-PHAs and SCL-MCL-PHAs in Recombinant *Pseudomonas* sp.

Various recombinant *Pseudomonas* strains have been examined intensively for the efficient production of MCL-PHAs and SCL-MCL-PHAs. Metabolic engineering strategies have also been applied to improve MCL-PHA production by *Pseudomonas* spp. Overexpression of the *P. oleovorans phaC1*$_{Po}$ gene in the parent strain resulted in an increase of PHA content up to 64 wt.% DCW under non-nitrogen-limiting conditions, whilst the PHA content in the wild-type *P. oleovorans* was 34 wt.% under the same conditions. Under nitrogen-limiting conditions, however, the PHA content increased only slightly from 44 wt.% in the parent strain to 47 wt.% in the recombinant strain (Kraak et al., 1997). In addition, recombinant *Pseudomonas* strains containing an additional copy of the PHA synthase gene showed only small changes in the PHA composition (Kraak et al., 1997). Overexpression of either of the two PHA synthase genes of *P. oleovorans* (*phaC1*$_{Po}$ or *phaC2*$_{Po}$) in *P. putida* resulted in a decrease in the molecular weight of PHAs (Huisman et al., 1992). Production of PHA in the PHA-negative mutant strain *P. putida* GPp104 harboring the heterologous PHA biosynthesis genes has also been reported. Such a bacterium, harboring the *P. oleovorans* PHA synthase gene, produced poly(3HHx-*co*-3HO) with a different 3HHx fraction in the copolymer depending on the *P. oleovorans* PHA synthase gene employed (Huisman et al., 1992). When the PHA-negative mutant *P. putida* GPp104 harboring the *Thiocapsa pfennigii* PHA synthase genes was cultured on octanoate, PHA copolymer consisting of almost equimolar amounts of 3HB and 3HHx, plus small amounts of 3HO, was produced (Liebergesell et al., 1993).

Recombinant *P. oleovorans* harboring the *R. eutropha* PHA biosynthesis genes accumulated PHA composed of 3HB, 3HHx, and 3HO from sodium octanoate. However, poly(3HB) homopolymer and poly(3HHx-*co*-3HO) copolymer were accumulated in separate granules, indicating that two different PHA synthases have the exclusive substrate specificities (Timm et al., 1990). Indeed, it was found later that the *R. eutropha* PHA synthase can incorporate C_3-C_{12} monomers (Dennis et al., 1998; Antonio et al., 2000), suggesting that the exclusive behavior observed previously seems to be due to differences in enzyme specificity and the amounts of metabolic precursors available. As mentioned earlier, heterologous expression of the *P. putida phaG*$_{Pp}$ gene in *P. oleovorans*, which is not capable of synthesizing PHA from gluconate, enabled MCL-PHA synthesis on gluconate up to 46 wt.% DCW. A PHA biosynthetic pathway could also be newly established in the nonPHA-accumulating bacterium *P.*

fragi by coexpression of the *P. putida* $phaG_{Pp}$ gene and the *P. aeruginosa* $phaC1_{Pa}$ gene. The cultivation of recombinant *P. fragi* on gluconate resulted in the accumulation of MCL-PHA, mainly composed of 3HD (60 mol.%), up to 14 wt.% DCW (Fiedler et al., 2000). The role of the *phaG* gene in providing MCL-PHA precursors was also examined in wild-type *Pseudomonas* sp. 61–3, and in *Pseudomonas* sp. 61–3 that was deficient in either the $phaG_{Ps}$ gene, or the $phaG_{Ps}$ and $phbC_{Ps}$ genes. When the $phaG_{Ps}$ gene was disrupted, the 3HB fraction in poly(3HB-*co*-3HA) copolymer was increased to 92 mol.%. Homologous amplification of the $phaG_{Ps}$ gene under the *lac* or native promoter further increased the fraction of MCL-monomer units (Matsumoto et al., 2001).

Expression of the *A. caviae* PHA biosynthesis genes in the PHA-negative mutant strain of *P. putida* GPp104 resulted in the accumulation of poly(3HB-*co*-3HHx) from gluconate, hexanoate, or octanoate up to 48 wt.% DCW (Fukui and Doi, 1997). The 3HHx fraction in the copolymer varied depending on the carbon source used. The PHA-negative mutant strain *P. putida* GPp104 harboring the *Pseudomonas* sp. 61–3 PHA biosynthesis genes accumulated poly(3HB-*co*-3HA) consisting of 3HB, 3HHx, 3HO, 3HD, 3HDD, and 3-hydroxy-*cis*-5-dodecenoate up to 43 wt.% DCW from gluconate and various alkanoates including octanoate, dodecanoate, and tetradecanoate (Matsusaki et al., 1998). The 3HB fraction in the copolymer varied from 1 mol.% to 92 mol.% depending on the carbon sources. To increase the 3HB fraction in the poly(3HB-*co*-3HA) copolymer, the *R. eutropha phbAB* genes were coexpressed under the control of *Pseudomonas* sp. 61–3 *pha* promoter or *R. eutropha pha* promoter, together with the *Pseudomonas* sp. 61–3 $phaC1_{Ps}$ gene in *P. putida* GPp104 (Matsusaki et al., 2000a). The 3HB fraction in the poly(3HB-*co*-3HA) copolymer produced from gluconate or alkanoates could be increased to 49 mol.% by the coexpression of *R. eutropha* $phaAB_{Re}$ genes. This can be compared with the results that 3HB monomer was not incorporated at all, or at a very low fraction (3 mol.%), into the copolymer in *P. putida* GPp104 carrying the *Pseudomonas* sp. 61–3 $phaC1_{Ps}$ gene only. The effect of co-expressing *R. eutropha phbAB* genes along with the *Pseudomonas* sp. 61–3 $phaC1_{Ps}$ gene was also examined in *Pseudomonas* sp. 61–3 (*phbC::tet*), which is a $phbC_{Ps}$-disrupted mutant (Matsusaki et al., 2000b). The *Pseudomonas* sp. 61–3 (*phbC::tet*) strain harboring these three genes synthesized poly(3HB-*co*-3HA) copolymers from glucose with a high 3HB fraction up to 94 mol.%. This poly(3HB-*co*-3HA) copolymer was found to possess good physical properties, and to be a flexible material with moderate toughness, similar to LDPE (Matsusaki et al., 2000b).

5.3 Biosynthesis of SCL-MCL-PHAs in Recombinant *R. eutropha*

It has been reported that the heterologous expression of various PHA biosynthesis genes in *R. eutropha* resulted in the accumulation of PHA consisting of different monomers compared with that found in the parent strain. It was shown recently that a recombinant strain of a PHA-negative *R. eutropha* mutant expressing its own $phaB_{Re}$ and $phaC_{Re}$ genes was able to accumulate poly(3HB-*co*-3HHx) with a PHA content of 81 wt.% when even-numbered carbon chain fatty acids (chain length $\geq C_6$) were supplied as carbon sources (Dennis et al., 1998). The molar fraction of 3HHx in copolymer was 8 mol.%. These studies also demonstrated that the substrate specificity of the *R. eutropha* PHA synthase is broader than C_3–C_5, as previously thought.

The PHA-negative mutant strain of *R. eutropha* harboring the *A. caviae* PHA biosynthesis genes accumulated poly(3HB-*co*-3HHx) up to 96 wt.% DCW from hexanoate or octanoate in flask cultures (Fukui and Doi, 1997). In addition, a PHA-negative *R. eutropha* mutant strain expressing the *A. caviae* PHA synthase gene produced a copolymer of poly(3HB-*co*-3HHx) from plant oils approximately up to 80 wt.% DCW (Fukui and Doi, 1998b). The molar fraction of 3HHx was 3–5 mol.%, regardless of the structure of the triglycerides fed. Furthermore, a recombinant strain of a PHA-negative *R. eutropha* mutant harboring the *A. caviae* PHA synthase genes synthesized poly(3HB-*co*-3HV-*co*-3HHp) terpolymer up to 78 wt.% DCW from alkanoic acids of odd-number carbon (Fukui et al., 1997). A PHA-negative *R. eutropha* mutant harboring the *A. caviae* PHA biosynthesis genes under the control of *R. eutropha pha* promoter was examined for the PHA production from various alkanoates (Kichise et al., 1999). These recombinant strains produced poly(3HB-*co*-3HHx) from hexanoate or octanoate, and poly(3HB-*co*-3HV-*co*-3HHp) from pentanoate or nonanoate up to 87 wt.% DCW. A PHA-negative mutant strain of *R. eutropha* harboring the *Pseudomonas* sp. 61–3 PHA biosynthesis genes accumulated poly(3HB-*co*-3HA) consisting of 3HB, 3HHx, 3HO, 3HD, and 3HDD up to 14 wt.% DCW from various alkanoates, including octanoate, dodecanoate, and tetradecanoate. However, only poly(3HB) was accumulated from gluconate (Matsusaki et al., 1998). The coexpression of *R. eutropha* $phaAB_{Re}$ genes under the control of *Pseudomonas* sp. 61–3 *pha* promoter or *R. eutropha pha* promoter, together with *Pseudomonas* sp. 61–3 $phaC1_{Ps}$ gene in a PHA-negative mutant *R. eutropha*, were examined possibly to increase the 3HB fraction in poly(3HB-*co*-3HA) (Matsusaki et al., 2000a). Recombinant *R. eutropha* produced poly(3HB-*co*-3HA) with higher 3HB fractions from alkanoates and plant oils. One of the recombinant strains, *R. eutropha* harboring the $phaC1_{Ps}$ and $phaAB_{Re}$ genes under the control of *Pseudomonas* sp. 61–3 *pha* promoter accumulated poly(3HB-*co*-3HA) copolyester with a very high 3HB fraction (85 mol.%) from palm oil (Matsusaki et al., 2000a).

6
Outlook and Perspectives

Intensive studies on PHA carried out during the past decade have provided us with ample knowledge on the metabolism and regulation of PHA biosynthesis, and degradation. Various pathways involved in providing the precursors of PHA have been elucidated, and further discoveries are expected. In addition, novel metabolic engineering strategies have been developed in order to produce PHAs consisting of desirable monomers, and also to enhance polymer production. It is now possible to produce large amounts of SCL-MCL PHA copolymers by employing various wild- type and metabolically engineered bacteria, though the efficiency of MCL-PHA or SCL-MCL-PHA production is still much lower than that for SCL-PHA. The reason for this discrepancy appears to be that the glycolytic flux providing SCL-PHA precursors is much higher than that providing MCL-PHA precursors by fatty acid degradation (β-oxidation) and/or fatty acid biosynthesis. As has been recognized from metabolic engineering studies involving the production of other primary and secondary metabolites, the simple amplification, deletion or modification of some pathways will often fail to provide the desired production profiles (Lee and Papoutsakis, 1999). Quantitative as well as qualitative analysis of the metabolic network is necessary in order to understand the steps controlling the fluxes towards PHA, and subsequently to design an optimal pathway. Henceforth, it will be necessary to understand the global metabolic network in order to enhance PHA biosynthesis and modulate polymer composition. Global transcriptional analysis by using DNA chip (DNA microarray) and proteome analysis by two-dimensional gel electrophoresis will allow us to understand better, and comparatively, the cellular metabolism/physiology under various conditions. This was recently demonstrated by the proteome analysis of metabolically engineered *E. coli* producing poly(3HB), which showed certain physiological changes caused by poly(3HB) accumulation (Han et al., 2001). Many of the changes were not obvious as they were not directly related to the PHA biosynthetic pathways. Our further understanding on global metabolism during PHA biosynthesis and degradation will allow us to design better engineered bacterial strains, and also to develop more efficient fermentation strategies. All of these efforts will undoubtedly in time lead to the development of a group of "plastic bacteria" (Lee, 1996b), which can be used for the efficient production of not only SCL-PHAs but also MCL- and SCL-MCL-PHAs from renewable feedstocks.

7
Patents

Patent number (date)	Inventors	Title	Assignee
US6143952 (2000.11.7)	Jackson et al.	Modified *Pseudomonas oleovorans phaC1* nucleic acids encoding bispecific polyhydroxyalkanoate polymerase	Minnesota Univ.
US6027787 (2000.2.22)	I. Noda	Films and absorbent articles comprising a biodegradable polyhydroxyalkanoate comprising 3-hydroxybutyrate and 3-hydroxyhexanoate comonomer units	P & G
US6011144 (2000.1.4)	Pries et al.	PHA E and PHA C components of poly(hydroxy fatty acid) synthase from *Thiocapsa pfennigii*	Monsanto Co.
US5849894 (1998.12.15)	Clemente et al.	*Rhodospirillum rubrum* poly-beta-hydroxyalkanoate synthase	Monsanto Co
US5618855 (1997.4.8)	I. Noda	Biodegradable copolymers and plastic articles comprising biodegradable copolymers	P & G
US5536564 (1996.7.16)	I. Noda	Biodegradable copolymers and plastic articles comprising biodegradable copolymers of 3-hydroxyhexanoate	P & G
US6077931 (2000.6.20)	I. Noda	Biodegradable PHA copolymers P & G	
US6013590 (2000.1.11)	I. Noda	Fibers, nonwoven fabrics, and absorbent articles comprising a biodegradable polyhydroxyalkanoate comprising 3-hydroxybutyrate and 3-hydroxyhexanoate	P & G

Acknowledgments

The studies described in this chapter were supported by Procter & Gamble Company and by the National Research Laboratory program of the Korean Ministry of Science and Technology (MOST).

8 References

Anderson, A. J., Dawes, E. A. (1990) Occurrence, metabolism, metabolic role, and industrial uses of bacterial polyhydroxyalkanoates, *Microbiol. Rev.* **54**, 450–472.

Antonio, R. V., Steinbüchel, A., Rehm, B. H. A. (2000) Analysis of in vivo substrate specificity of the PHA synthase from *Ralstonia eutropha*: formation of novel copolyesters in recombinant *Escherichia coli, FEMS Microbiol. Lett.* **182**, 111–117.

Brandl, H., Gross, R. A., Lenz, R. W., Fuller, R. C. (1988) *Pseudomonas oleovorans* as a source of poly(β-hydroxyalkanoates) for potential applications as biodegradable polyesters, *Appl. Environ. Microbiol.* **54**, 1977–1982.

Brandl, H., Knee, E. J., Fuller, R.C., Gross, R.A., Renz, R. W. (1989) Ability of the phototrophic bacterium *Rhodospirillum rubrum* to produce various poly(β-hydroxyalkanoates):potential sources for biodegradable polyester, *Int. J. Biol. Macromol.* **11**, 49–55.

Brandl, H., Bachofen, R., Mayer, J., Wintermantel, E. (1995) Degradation and applications of polyhydroxyalkanoates, *Can. J. Microbiol.* **41**(Suppl. 1), 143–153.

Choi, J., Lee, S. Y. (1999a) Efficient and economical recovery of poly(3-hydroxybutyrate) from recombinant *Escherichia coli* by simple digestion with chemicals, *Biotechnol. Bioeng.* **62**, 546–553.

Choi, J., Lee, S. Y. (1999b) High-level production of poly(3-hydroxybutyrate-*co*-3-hydroxyvalerate) by fed-batch culture of recombinant *Escherichia coli, Appl. Environ. Microbiol.* **65**, 4363–4368.

Choi, J., Lee, S. Y., Han, K. (1998) Cloning of the *Alcaligenes latus* polyhydroxyalkanoate biosynthesis genes and use of these genes for enhanced production of poly(3-hydroxybutyrate) in *Escherichia coli, Appl. Environ. Microbiol.* **64**, 4897–4903.

Dennis, D., McCoy, M., Stangl, A., Valentin, H. E., Wu, Z. (1998) Formation of poly(3-hydroxybutyrate-*co*-3-hydroxyhexanoate) by PHA synthase from *Ralstonia eutropha, J. Biotechnol.* **64**, 177–186.

Doi, Y. (1990) *Microbial Polyesters.* New York: VCH.

Doi, Y., Kitamura, S., Abe, H. (1995) Microbial synthesis and characterization of poly(3-hydroxybutyrate-*co*-3-hydroxyhexanoate, *Macromolecules* **28**, 4822–4828.

Fiedler, S., Steinbüchel, A., Rehm, B. H. A. (2000) PhaG-mediated synthesis of poly(3-hydroxyalkanoates) consisting of medium-chain-length constituents from nonrelated carbon sources in recombinant *Pseudomonas fragi, Appl. Environ. Microbiol.* **66**, 2117–2124.

Fukui, T., Doi, Y. (1997) Cloning and analysis of the poly(3-hydroxybutyrate-*co*-3-hydroxyhexanoate) biosynthesis genes of *Aeromonas caviae, J. Bacteriol.* **179**, 4821–4830.

Fukui, T., Doi, Y. (1998a) Expression and characterization of (R)-specific enoyl coenzyme A hydratase involved in polyhydroxyalkanoate biosynthesis by *Aeromonas caviae, J. Bacteriol.* **180**, 667–673.

Fukui, T., Doi, Y. (1998b) Efficient production of polyhydroxyalkanoates from plant oils by *Alcaligenes eutrophus* and its recombinant strain, *Appl. Microbiol. Biotechnol.* **49**, 333–336.

Fukui, T., Kichise, T., Yoshida, Y., Doi, Y. (1997) Biosynthesis of poly(3-hydroxybutyrate-*co*-3-hydroxyvalerate-*co*-3-hydroxyheptanoate) terpolymers by recombinant *Alcaligenes eutrophus, Biotechnol. Lett.* **19**, 1093–1097.

Fukui, T., Yokomizo, S., Kobayashi, G., Doi, Y. (1999) Co-expression of polyhydroxyalkanoate synthase and (R)-enoyl-CoA hydratase genes of *Aeromonas caviae* establishes copolyester biosynthesis pathway in *Escherichia coli, FEMS Microbiol. Lett.* **170**, 69–75.

Fukui, T., Kichise, T., Iwata, T., Doi, Y. (2001) Characterization of 13 kDa granule-associated protein in *Aeromonas caviae* and biosynthesis of polyhydroxyalkanoates with altered molar composition by recombinant bacteria, *Biomacromolecules* **2**, 148–153.

Hall, B., Baldwin, J., Rhie, H. G., Dennis, D. (1998) Cloning of the *Nocardia corallina* polyhydroxyalkanoate synthase gene and production of poly-(3-hydroxybutyrate-*co*-3-hydroxyhexanoate) and poly-(3-hydroxyvalerate-*co*-3-hydroxyheptanoate), *Can. J. Microbiol.* **44**, 687–691.

Han, M.-J., Yoon, S. S., Lee, S. Y. (2001) Proteome analysis of metabolically engineered *Escherichia coli* producing poly(3-hydroxybutyrate), *J. Bacteriol.* **183**, 301–308.

Haywood, G. W., Anderson, A. J., Williams, G. A., Dawes, E. A., Ewing, D. F. (1991) Accumulation of a poly(hydroxyalkanoate) copolymer containing primarily 3-hydroxyvalerate from simple carbohydrate substrates by *Rhodococcus* sp. NCIMB 40126, *Int. J. Biol. Macromol.* **13**, 83–87.

Hoffmann, N. Steinbüchel, A., Rehm, B. H. A. (2000a) The *Pseudomonas aeruginosa phaG* gene product is involved in the synthesis of polyhydroxyalkanoic acid consisting of medium-chain-length constituents from non-related carbon sources, *Microbiol. Lett.* **184**, 253–259.

Hoffmann, N. Steinbüchel, A., Rehm, B. H. A. (2000b) Homologous functional expression of cryptic *phaG* from *Pseudomonas olevorans* establishes the transacylase-mediated polyhydroxyalkanoate biosynthetic pathway, *Appl. Microbiol. Biotechnol.* **54**, 665–670.

Huijberts, G. N., Eggink, G., de Waard, P., Huisman, G. W., Witholt, B. (1992) *Pseudomonas putida* KT2442 cultivated on glucose accumulates poly(3-hydroxyalkanoates) consisting of saturated and unsaturated monomers, *Appl. Environ. Microbiol.* **58**, 536–544.

Huijberts, G. N., de Rijk, T. C., de Waard, P., Eggink, G. (1994) ^{13}C nuclear magnetic resonance studies of *Pseudomonas putida* fatty acid metabolic routes involved in poly(3-hydroxyalkanoate) synthesis, *J. Bacteriol.* **176**, 1661–1666.

Huisman, G. W., de Leeuw, O., Eggink, G., Witholt, B. (1989) Synthesis of poly-3-hydroxyalkanoates is a common feature of fluorescent pseudomonads. *Appl. Environ. Microbiol.* **55**, 1949–1954.

Huisman, G. W., Wonink, E., de Koning. G., Preusting, H., Witholt, B. (1992) Synthesis of poly(3-hydroxyalkanates) by mutant and recombinant *Pseudomonas* strains, *Appl. Microbiol. Biotechnol.* **38**, 1–5.

Kato, M., Bao, H. J., Kang, C. K., Fukui, T., Doi, Y. (1996) Production of a novel copolyester of 3-hydroxybutyric acids and medium-chain-length 3-hydroxyalkanoic acids by *Pseudomonas* sp. 61–3 from sugars, *Appl. Microbiol. Biotechnol.* **45**, 363–370.

Kichise, T., Fukui, T., Yoshida, Y., Doi, Y. (1999) Biosynthesis of polyhydroxyalkanoates (PHA) by recombinant *Ralstonia eutropha* and effects of PHA synthase activity on in vivo PHA biosynthesis, *Int. J. Biol. Macromol.* **25**, 69–77.

Klinke, S., Ren, Q., Witholt, B., Kessler, B. (1999) Production of medium-chain-length poly(3-hydroxyalkanoates) from gluconate by recombinant *Escherichia coli*, *Appl. Environ. Microbiol.* **65**, 540–548.

Kobayashi, G., Shiotani, T., Shima, Y., Doi, Y. (1994) Biosynthesis and characterization of poly(3-hydroxybutyrate-*co*-3-hydroxyhexanoate) from oils and fats by *Aeromonas* sp. OL-338 and *Aeromonas* sp. FA440, in: *Biodegradable Plastics and Polymers* (Doi, Y., Fukuda, K., Eds), Amsterdam: Elsevier, 410–416.

Kraak, M. N., Smits, T. H., Kessler, B., Witholt, B. (1997) Polymerase C1 levels and poly(R-3-hydroxyalkanoate) synthesis in wild-type and recombinant *Pseudomonas* strains, *J. Bacteriol.* **179**, 4985–4991.

Lageveen, R. G., Huisman, G. W., Preusting, H., Ketelaar, P., Eggink, G., Witholt, B. (1988) Formation of polyesters by *Pseudomonas oleovorans*: effect of substrates on formation and composition of poly-(*R*)-3-hydroxyalkanoates and poly-(*R*)-3-hydroxyalkenoates, *Appl. Environ. Microbiol.* **54**, 2924–2932.

Langenbach, S., Rehm, B. H. A., Steinbüchel, A. (1997) Functional expression of the PHA synthase gene *phaC1* from *Pseudomonas aeruginosa* in *Escherichia coli* results in poly(3-hydroxyalkanoate) synthesis, *FEMS Microbiol. Lett.* **150**, 303–309.

Lee, S. H., Oh, D. H., Ahn, W. S., Lee, Y., Choi, J., Lee, S. Y. (2000) Production of poly(3-hydroxybutyrate-*co*-3-hydroxyhexanoate) by high-cell-density cultivation of *Aeromonas hydrophila*, *Biotechnol. Bioeng.* **67**, 240–244.

Lee, S. Y. (1996a) High cell-density culture of *Escherichia coli*, *Trends Biotechnol.* **14**, 98–105.

Lee, S. Y. (1996b) Plastic bacteria? Progress and prospects for polyhydroxyalkanoate production in bacteria, *Trends Biotechnol.* **14**, 431–438.

Lee, S. Y. (1996c) Bacterial polyhydroxyalkanoates, *Biotechnol. Bioeng.* **49**, 1–14.

Lee, S. Y. (1997) *E. coli* moves into the plastic age, *Nature Biotechnol.* **15**, 17–18.

Lee, S. Y., Chang, H. N. (1995) Production of poly(hydroxyalkanoic acid), *Adv. Biochem. Eng. Biotechnol.* **52**, 27–58.

Lee, S. Y., Choi J. (1999) Polyhydroxyalkanoates: biodegradable polymer, in: *Manual of Industrial Microorganisms and Biotechnology* (Demain, A. L., Davies, J. E., Eds), Washington, DC: American Society for Microbiology, 616–627.

Lee, S. Y., Papoutsakis, E. T. (1999) *Metabolic Engineering*. New York: Marcel Dekker, Inc.

Lemoigne, M. (1926) Products of dehydration and of polymerization of β-hydroxybutyric acid, *Bull. Soc. Chem. Biol.* **8**, 770–782.

Liebergesell, M., Hustede, E., Timm, A., Steinbüchel, A., Fuller, R. C., Lenz, R. W., Schlegel, H. G. (1991) Formation of poly(3-hydroxyalkanoates) by phototrophic and chemolithotrophic bacteria, *Arch. Microbiol.* **155**, 415–421.

Liebergesell, M., Mayer, F., Steinbüchel, A. (1993) Analysis of polyhydroxyalkanoic acid biosynthesis genes of anoxygenic phototrophic bacteria reveals synthesis of a polyester exhibiting an unusual composition, *Appl. Microbiol. Biotechnol.* **40**, 292–300

Madison, L. L, Huisman, G. W. (1999) Metabolic engineering of poly(3-hydroxyalkanoates): from DNA to plastic, *Microbiol. Mol. Biol. Rev.* **63**, 21–53.

Matsumoto, K., Matsusaki, H., Taguchi, S., Seki, M., Doi, Y. (2001) Cloning and characterization of the *Pseudomonas* sp. 61–3 *phaG* gene involved in polyhydroxyalkanoate biosynthesis, *Biomacromolecules* **2**, 142–147.

Matsusaki, H., Manji, S., Taguchi, K., Kato, M., Fukui, T., Doi, Y. (1998) Cloning and molecular analysis of the poly(3-hydroxybutyrate) and poly(3-hydroxybutyrate-*co*-3-hydroxyalkanoate) biosynthesis genes in *Pseudomonas* sp. strain 61–3, *J. Bacteriol.* **180**, 6459–6467.

Matsusaki, H., Abe, H., Taguchi, K., Fukui, T., Doi, Y. (2000a) Biosynthesis of poly(3-hydroxybutyrate-*co*-3-hydroxyalkanoates) by recombinant bacteria expressing the PHA synthase gene *phaC1* from *Pseudomonas* sp. 61–3, *Appl. Microbiol. Biotechnol.* **53**, 401–419.

Matsusaki, H., Abe, H., Doi, Y. (2000b) Biosynthesis and properties of poly(3-hydroxybutyrate-*co*-3-hydroxyalkanoates) by recombinant strains of *Pseudomonas* sp. 61–3, *Biomacromolecules* **1**, 17–22.

Park, S. J., Ahn, W. S., Green P. R., Lee, S. Y. (2001a) Production of poly(3-hydroxybutyrate-*co*-3-hydroxyhexanoate) by metabolically engineered *Escherichia coli* strains, *Biomacromolecules* **2**, 248–254.

Park, S. J., Ahn, W. S., Green P. R., Lee, S. Y. (2001b) Biosynthesis of poly(3-hydroxybutyrate-*co*-3-hydroxyvalerate-*co*-3-hydroxyhexanoate) by metabolically engineered *Escherichia coli* strains, *Biotechnol. Bioeng.* **74**, 82–87.

Peoples, O. P., Sinskey, A. J. (1989) Poly-β-hydroxybutyrate biosynthesis in *Alcaligenes eutrophus* H16. Identification and characterization of the PHB polymerase gene (*phbC*), *J. Biol. Chem.* **264**, 15298–15303.

Prieto, M. A., Kellerhals, M. B., Bozzato, G. B., Radnovic, A., Witholt, B., Kessler, B. (1999) Engineering of stable recombinant bacteria for production of chiral medium-chain-length poly-3-hydroxyalkanoates, *Appl. Environ. Microbiol.* **65**, 3265–3271.

Qi, Q., Rehm, B. H. A., Steinbüchel, A. (1997) Synthesis of poly(3-hydroxyalkanoates) in *Escherichia coli* expressing the PHA synthase gene *phaC2* from *Pseudomonas aeruginosa*: comparison of PhaC1 and PhaC2, *FEMS Microbiol. Lett.* **157**, 155–162.

Qi, Q., Steinbüchel, A., Rehm, B. H. A. (1998) Metabolic routing towards polyhydroxyalkanoic acid synthesis in recombinant *Escherichia coli* (*fadR*): inhibition of fatty acid beta-oxidation by acrylic acid, *FEMS Microbiol. Lett.* **167**, 89–94.

Rehm, B. H. A., Kruger, N., Steinbüchel, A. (1998) A new metabolic link between fatty acid de novo synthesis and polyhydroxyalkanoic acid synthesis. The *phaG* gene from *Pseudomonas putida* KT2440 encodes a 3-hydroxyacyl-acyl carrier protein-coenzyme a transferase, *J. Biol. Chem.* **273**, 24044–24051.

Ren, Q., Sierro, N., Witholt, B., Kessler, B. (2000) FabG, an NADPH-dependent 3-ketoacyl reductase of *Pseudomonas aeruginosa*, provides precursors for medium-chain-length poly-3-hydroxyalkanoate biosynthesis in *Escherichia coli*, *J. Bacteriol.* **182**, 2978–2981.

Schubert, P., Steinbüchel, A, Schlegel, H. G. (1988) Cloning of the *Alcaligenes eutrophus* poly-β-hydroxybutyrate synthetic pathway and synthesis of PHB in *Escherichia coli*. *J. Bacteriol.* **170**, 5837–5847.

Slater, S. C., Voige, W. H., Dennis, D. (1988) Cloning and expression in *Escherichia coli* of the *Alcaligenes eutrophus* H16 poly-β-hydroxybutyrate biosynthetic pathway, *J. Bacteriol.* **170**, 4431–4436.

Steinbüchel, A. (1991) Polyhydroxyalkanoic acids, in: *Biomaterials: Novel Materials from Biological Sources* (Byrom, D., Ed.), New York: Stockton, 124–213.

Steinbüchel, A. (1992) Biodegradable plastics, *Curr. Opin. Biotechnol.* **3**, 291–297.

Steinbüchel, A., Füchtenbusch, B. (1998) Bacterial and other biological systems for polyester production, *Trends Biotechnol.* **16**, 419–427.

Steinbüchel, A., Valentin, H. E. (1995) Diversity of bacterial polyhydroxyalkanoic acid, *FEMS Microbiol. Lett.* **128**, 219–228.

Steinbüchel, A., Wiese, S. (1992) A *Pseudomonas* strain accumulating polyesters of 3-hydroxybutyric acid and medium-chain-length 3-hydroxyalkanoic acids, *Appl. Microbiol. Biotechnol.* **37**, 691–697.

Taguchi, K., Aoyagi, Y., Matsusaki, H., Fukui, T., Doi, Y (1999a) Co-expression of 3-ketoacyl-ACP reductase and polyhydroxyalkanoate synthase genes induces PHA production in *Escherichia coli* HB101 strain. *FEMS Microbiol. Lett.* **176**, 183–190.

Taguchi, K., Aoyagi, Y., Matsusaki, H., Fukui, T., Doi, Y. (1999b) Over-expression of 3-ketoacyl-ACP synthase III or malonyl-CoA-ACP transacylase gene induces monomer supply for the polyhydroxybutyrate production in *Escherichia coli* HB101, *Biotechnol. Lett.* **21**, 579–584.

Timm, A., Byrom, D., Steinbüchel, A. (1990) Formation of blends of various poly(3-hydroxyalkanoic acid) by a recombinant strain of *Pseudomonas oleovorans*, *Appl. Microbiol. Biotechnol.* **33**, 296–301.

Tsuge, T., Fukui, T., Matsusaki, H., Taguchi, S., Kobayashi, G., Ishizaki, A., Doi, Y. (2000) Molecular cloning of two (R)-specific enoyl-CoA hydratase genes from *Pseudomonas aeruginosa* and their use for polyhydroxyalkanoate synthesis, *FEMS Microbiol. Lett.* **184**, 193–198.

Witholt, B., Kessler, B. (1999) Perspectives of medium chain length poly(hydroxyalkanoates), a versatile set of bacterial bioplastics, *Curr. Opin. Biotechnol.* **10**, 279–285.

12 Production of Polyhydroxyalkanoates in Transgenic Plants

Prof. Dr. Yves Poirier[1], **Dr. Kenneth J. Gruys**[2]
[1] Institut d'Écologie-Laboratoire de Biotechnologie Végétale, Bâtiment de Biologie, Université de Lausanne, 1015 Lausanne, Switzerland; Tel: +41-21-692-4222; Fax: +41-21-692-4195; yves.poirier@ie-bpv.unil.ch
[2] Monsanto, 700 Chesterfield Parway North, Chesterfield, MO 63198, USA; Tel: +01-636-737-7345; Fax: +01-636-737-7015; E-mail: kenneth.j.gruys@monsanto.com

ACP	acyl carrier protein;
CaMV	cauliflower mosaic virus
DAGAT	diacylglycerol acyltransferase
dwt	dry weight

Biotechnology of Biopolymers. From Synthesis to Patents. Edited by A. Steinbüchel and Y. Doi

ISBN: 3-527-31110-6

fwt	fresh weight
FMV	figwort mosaic virus
GC	gas chromatography
GC-MS	gas chromatography–mass spectrometry
MCL-PHA	medium-chain-length polyhydroxyalkanoate
MFP	multifunctional protein
NMR	nuclear magnetic resonance
PDC	puruvate dehydrogenase complex
PE	polyethylene
PHA	polyhydroxyalkanoate
phaA	3-ketothiolase
phaB	acetoacetyl-CoA reductase
phaC	poly(3HB) synthase
Poly(3HB)	poly(3-hydroxybutyrate)
Poly(3HB-*co*-3HV)	poly(3-hydroxybutyrate-*co*-3-hydroxyvalerate)
Poly(3HO)	poly(3-hydroxyoctanoate)

1 Introduction

Crop plants cannot only be harvested as a source of food but can also be regarded as a factory capable of producing millions of tons of chemicals at a price comparable to many petroleum-derived commodities. Good examples of present-day commodity chemicals produced in plants are starch and oils, which both can be used for food and non-food purposes. Recent progress in molecular biology and plant transformation has enabled the creation of plants that have the capacity to produce new industrially useful products that are not naturally found in plants (Goddijn and Pen, 1995). Examples of such products are polyesters of the family of polyhydroxyalkanoates (PHAs). These polyesters, naturally synthesized in bacteria, have attracted considerable interest in the past 20 years because of their potential to act as a source of biodegradable thermoplastics and elastomers (Anderson and Dawes, 1990). The main rational for the synthesis of PHA in plants is the potential for producing the polymer on a large scale and at a cost lower than bacterial fermentation, making PHA competitive with petroleum-derived plastics (Poirier et al., 1995a).

This chapter will focus on the synthesis of poly(3-hydroxybutyrate) (Poly(3HB)) and PHA copolymers in plants and is complementary with Volume 4, Chapter 3 by Asrar and Gruys, which addresses in greater depth the production of poly(3-hydroxybutyrate-*co*-3-hydroxyvalerate) (Poly(3HB-*co*-3HV)). Furthermore, since most of the knowledge on the genes and enzymes involved in PHA have been obtained from studies on bacteria, we recommend the readers to refer to other chapters of this volume as well to reviews on bacterial PHAs (Anderson and Dawes, 1990; Doi, 1990; Steinbüchel, 1991; Steinbüchel and Schlegel, 1991; Poirier et al., 1995a; Braunegg et al., 1998; Steinbüchel and Füchtenbusch, 1998) to integrate the topic of PHA synthesis in plants within the larger topic of biological PHAs.

2 Historical Perspective

Although bacterial Poly(3HB) was initially described by M. Lemoigne in 1926, it was not until 1962 that the value of Poly(3HB) as a thermoplastic was recognized in a US patent and until 1982 when the first PHA produced by bacterial fermentation was commercialized. Synthesis of PHA in plants is a relatively young science since it was first demonstrated in 1992 in the plant *Arabidopsis thaliana* (Poirier et al., 1992a). Although of no agronomic value, this plant was chosen because it could easily be transformed and its short life-cycle (approximately 2 months from seed to seed) enabled the rapid creation of hybrid plants expressing the various genes necessary for PHA biosynthesis. This plant also has the advantage of being a model organism for which a number of mutants affected in various aspects of metabolism were available and could be used to study PHA synthesis (Meyerowitz, 1987; Meyerowitz and Somerville, 1994). Finally, *A. thaliana* is a plant which accumulates approximately 40% lipids in its seeds and thus can be used as a good model for the synthesis of PHA in the major oil crops, such as rapeseed or soybean. The first bacterial genes involved in the synthesis of PHA were cloned from *Ralstonia eutropha* (formerly *Alcaligenes eutrophus*) and were involved in the synthesis of Poly(3HB) (Schubert et al., 1988; Slater et al., 1988; Peoples and Sinskey, 1989a,b). Thus, although Poly(3HB) has relatively poor physical properties, being too stiff and brittle for its use in consumer products (de Koning, 1995), it was the *R. eutropha* genes for Poly(3HB) biosynthesis that were first expressed in plants. Expression of the Poly(3HB) biosynthetic pathway in the cytoplasm of cells of *A. thaliana* resulted in the accumulation of 0.1% dry weight (dwt) Poly(3HB) (Poirier et al., 1992a). Although successful, these experiments quickly pointed out the potential limitation of using the cytoplasm as a site of Poly(3HB) synthesis. Two years later, Nawrath et al. (1994a,b) demonstrated that 100-fold more Poly(3HB) could be synthesized in *A. thaliana* by targeting the Poly(3HB) biosynthetic pathway to the chloroplast. Since then, synthesis of Poly(3HB) in the cytoplasm, plastids, and peroxisomes of leaves or seeds has been demonstrated in a number of plants, including rape, tobacco, potato, corn, and cotton (John and Keller, 1996; Hahn et al., 1999; Houmiel et al., 1999; Nakashita et al., 1999; Poirier, 1999; Bohmert et al., 2000). The highest amount of Poly(3HB) produced in plants is presently 40% dwt (Bohmert et al., 2000).

A further milestone was set in 1998 and 1999 when synthesis of PHA copolymers was reported in plants. Mittendorf et al. (1998a,b) first demonstrated the synthesis of low amounts (0.4% dwt or less) of medium-chain-length PHA (MCL-PHA) copolymers in the peroxisomes of *A. thaliana*. One year later, the same group improved the quantity and quality of the polymer through additional genetic engineering of the fatty acid biosynthetic pathway (Mittendorf et al., 1999; Poirier et al., 1999). Slater et al. (1999) at Monsanto reported the synthesis of the copolymer Poly(3HB-*co*-3HV) in the plastids of both *A. thaliana* and seeds of rape. This later study was important since Poly(3HB-*co*-3HV) had been a commercial target for bacterial fermentation for nearly 20 years.

A spectrum of PHAs ranging from the stiff Poly(3HB) to the more flexible Poly(3HB-*co*-3HV) plastics and MCL-PHA elastomers have thus been successfully synthesized in plants (Poirier, 1999). Table 1 provides a summary of the status of PHA synthesis in plants, which will be described in detail in the following sections. The

Tab. 1 Summary of transgenic plants producing PHAs

Sub-cellular compart-ment	*Species*	*Tissue*	*PHA type*	*Bacterial gene*	*PHA quantity (% dwt)*	*Reference*
Cytoplasm	*A. thaliana*	shoot	Poly(3HB)	*R. eutropha phaB, phaC*	0.1	Poirier et al., 1992a
	rapeseed	shoot	Poly(3HB)	*R. eutropha phaB, phaC*	0.1	P. Fentem, pers. commun.
	tobacco	shoot	Poly(3HB)	*R. eutropha phaB, A. cariae phaC*	0.01	Nakashita et al., 1999
	cotton	fiber	Poly(3HB)	*R. eutropha phaB, phaC*	0.3	John and Keller, 1996
Plastid	*A. thaliana*	shoot	Poly(3HB)	*R. eutropha phaA, phaB, phaC*	14–40	Nawrath et al., 1994a; Bohmert et al., 2000; K. J. Gruys, pers. commun.
	rapeseed	seeds	Poly(3HB)	*R. eutropha phaA, phaB, phaC*	8	Houmiel et al., 1999
	cotton	fibers	Poly(3HB)	*R. eutropha phaA, phaB, phaC*	0.05	John, 1997
	A. thaliana	shoot	Poly(3HB-*co*-3HV)	*R. eutropha bktA, phaB, phaC; E. coli ilvA*	1.6	Slater et al., 1999
	rapeseed	seed	Poly(3HB-*co*-3HV)	*R. eutropha bktA, phaB, phaC; E. coli ilvA*	2.3	Slater et al., 1999
	potato	shoot	Poly(3HB)	*R. eutropha phaA, phaB, phaC*	0.02	K. Bohmert, pers. commun.
	tobacco	shoot	Poly(3HB)	*R. eutropha phaA, phaB, phaC*	0.04	K. Bohmert, pers. commun.
	alfalfa	shoot	Poly(3HB)	*R. eutropha phaA, phaB, phaC*	0.2	D.P. Saruul, pers. commun.
	corn	shoot	Poly(3HB)	*R. eutropha phaA, phaB, phaC*	6	Mitsky et al., 2000
Peroxisome	*A. thaliana*	whole plants	MCL-PHA	*P. aeruginosa phaC1*	0.6	Mittendorf et al., 1998b
	maize	cell suspension	Poly(3HB)	*R. eutropha phaA, phaB, phaC*	2	Hahn et al., 1999

challenge for the future is to succeed in high level production (15% dwt or greater) of a limited number of useful PHAs without a decrease in crop yield. This must also be accompanied by the development of protocols for the efficient, ecological, and economical extraction of PHAs from plants.

3
Synthesis of PHA in Plants

This section will first present in detail the knowledge gathered over the last 10 years on the synthesis of Poly(3HB) in *A. thaliana* and other plants. It will then present the most recent data on the synthesis of various PHA copolymers in plants.

3.1 Synthesis of Poly(3HB) in *Arabidopsis thaliana*

Most of the knowledge on the genes and enzymes involved in Poly(3HB) synthesis have been obtained from the bacteria *R. eutropha,* although a similar pathway exists in most bacteria synthesizing Poly(3HB). Poly(3HB) is synthesized from acetyl-CoA via a three-step reaction (Figure 1). The first enzyme of the pathway, 3-ketothiolase, catalyses the condensation of two molecules of acetyl-CoA to form acetoacetyl-CoA. The enzyme acetoacetyl-CoA reductase subsequently reduces acetoacetyl-CoA to *R-3-*hydroxybutyryl-CoA. The last enzyme, the PHA synthase, polymerizes *R-3*-hydroxybutyryl-CoA to produce Poly(3HB). Since acetyl-CoA is present in plant cells in the cytosol, plastid, mitochondrion, and peroxisome, the synthesis of Poly(3HB) in plants could, in theory, be achieved in any of these subcellular compartments. However, the cytoplasm was targeted as the first site for Poly(3HB) synthesis because it had the advantage that the bacterial enzymes could be directly expressed in this compartment without any modification of the proteins. Furthermore, an endogenous plant 3-ketothiolase is present in the cytoplasm as part of mevalonate pathway. Thus, creation of the Poly(3HB) biosynthetic pathway in the cytoplasm required only the expression of two additional enzymes, the reductase and synthase (Figure 1). The *R. eutropha* genes encoding the acetoacetyl-CoA reductase (*phaB*) and PHA synthase (*phaC*) were expressed in plants under the control of the cauliflower mosaic virus (CaMV) 35S promoter, allowing a relatively high expression of the enzymes in a broad range of tissues (Poirier et al., 1992a). Transgenic *A. thaliana* expressing the Poly(3HB) synthase or acetoacetyl-CoA reductase were cross-pollinated to obtain hybrids having all enzyme activities necessary for Poly(3HB) synthesis. The highest amount of Poly(3HB) measured in the shoots of these hybrids was approximately 0.1% dwt (Poirier et al., 1992a). Staining of the tissue with Nile Blue A and visualization by epifluorescence microscopy revealed the presence of Poly(3-HB) inclusions in all organs of the hybrid transgenic plants, including root, leaf, cotyledon, and seed. Transmission electron microscopy revealed that Poly(3HB) accumulated in the form of agglomerations of electron-lucent inclusions surrounded by an electron-dense layer (Poirier et al., 1992a,b). The size (0.2 – 1 μm) and general appearance of these inclusions were similar to PHA inclusions found in bacteria. Surprisingly, even though the Poly(3HB) metabolic pathway was expressed in the cytoplasm, Poly-(3HB) agglomerations were found in several subcellular compartments, i.e. cytosol, vacuole, and nucleus (Poirier et al., 1992a,b). No Poly(3HB) inclusions were found in plastids and mitochondria. From these results, it was hypothesized that the hydrophobic nature of Poly(3HB) inclusions might allow them to pass through the single membrane of the vacuole but not the double membrane of organelles such as the plastid and mitochondrion. The nuclear localization of Poly(3HB) granules may be explained by some affinity of the inclusions to nuclear constituents, leading to their entrapment in the nucleus during cell division.

Poly(3HB) synthesized in *A. thaliana* was purified from a plant suspension culture in sufficient quantity to enable a detailed analysis of its chemical structure, confirming that that polymer was isotactic poly(*R*-(–)-3-hydroxybutyrate) and that the thermal properties of plant Poly(3HB) were similar

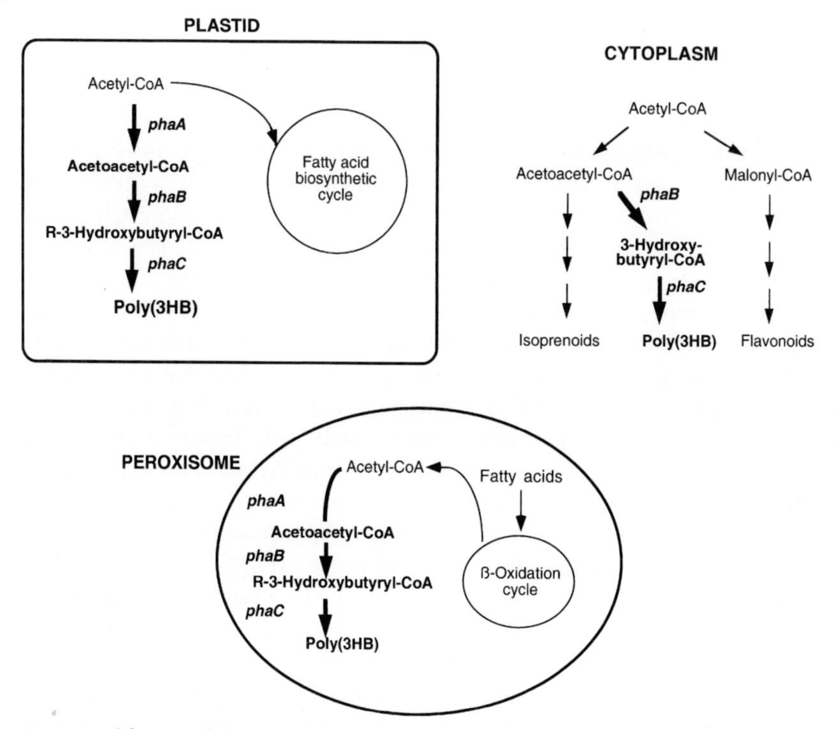

Fig. 1 Modification of plant metabolic pathways for the synthesis of Poly(3HB). Endogenous plant pathways are in plain letters while pathways created by the expression of transgenes are highlighted in bold. The transgenes expressed in plants are indicated in italics. The *phaA*, *phaB*, and *phaC* genes encode the 3-ketothiolase, acetoacetyl-CoA reductase, and PHA synthase from *R. eutropha*.

to bacterial Poly(3HB) (Poirier et al., 1995b). The only difference noted was that plant Poly(3HB) had a broader molecular weight distribution, ranging from 10^4 to 10^6, compared to bacterial Poly(3HB) with a narrow molecular weight distribution near 10^6 (Poirier et al., 1995b). These results demonstrated that although other proteins, such as phasins, may be found on the surface of PHA granules in bacteria (Pieper-Fürst et al., 1995), expression of only the acetoacetyl-CoA reductase and PHA synthase were sufficient for the synthesis of high molecular weight PHA accumulating in the form of inclusions. Heterogeneity in the size of the Poly(3HB) inclusions present in the various organelles was noted, with inclusions found in the nucleus being smaller than inclusions found in the cytoplasm or vacuole (Poirier et al., 1992a,b). It is likely that, depending on the subcellular compartment, different plant amphiphatic proteins could be absorbed on the surface of inclusions and that these proteins may affect the sizes of the granules through the promotion or prevention of inclusion fusion, in a manner analogous to the phasin proteins found associated with bacterial PHA inclusions (Pieper-Fürst et al., 1995; Wieczorek et al., 1995).

Plants expressing the Poly(3HB) pathway in the cytoplasm accumulated polymer to 0.1% (Poirier et al., 1992a,b), which is approximately 200–400 times lower than lipid accumulation in seeds of oil crops (20–40% dwt in soybean or rape, respectively) and 800–900 times lower than Poly(3HB) accumulation in *R. eutropha*. Furthermore, plants expressing both acetoacetyl-CoA reductase and Poly(3HB) synthase were small in comparison to wild-type plants (Poirier et al., 1992a,b). Although the reasons for the dwarf phenotype has not been unambiguously determined, it is hypothesized that the diversion of cytoplasmic acetyl-CoA and acetoacetyl-CoA away from the endogenous isoprenoid and flavonoid pathways might lead to a depletion of essential metabolites which may affect growth. Since the plant isoprenoid pathway contributes to the synthesis of three classes of plant hormones, i.e. cytokinins, gibberelins, and brassinosteroids, it is likely that even a small imbalance in the synthesis of these hormones may strongly affects plant growth. Cytoplasmic acetyl-CoA and acetoacetyl-CoA are also implicated in the synthesis of sterols, which are essential components of membranes. A decrease in the carbon flux towards the flavonoid pathway was indicated by the fact that plants expressing high levels of acetoacetyl-CoA reductase have a decreased amount of anthocyanins in the seed coat (Poirier et al., 1992b).

In view of the hypothesis that the limited supply of cytoplasmic acetyl-CoA and the depletion of metabolites derived from it was thought to be the main factors limiting Poly(3HB) accumulation and reduced plant growth in these first generation transgenic plants, expression of the Poly(3HB) pathway in a compartment with a higher flux through acetyl-CoA was thought to be a potential solution. Fatty acid biosynthesis in plants occurs primarily in the plastid using acetyl-CoA as precursor (Figure 1). The plastid is therefore a site with a large flux of carbon through acetyl-CoA in tissues having a high proportion of lipids. This is particularly true in seeds of plants accumulating triacylglycerides as the main carbon reserve. For example, *A. thaliana* synthesizes up to 40% of the seed dry weight as triacylglycerides during a period of 5–7 days in seed development. The large flux of acetyl-CoA in the plastids was thus hypothesized to allow a significantly higher production of Poly(3HB) while minimizing potential deleterious effects on plant growth. Plastids are also the site of starch accumulation. In a manner analogous to bacterial PHA inclusions, starch is synthesized as osmotically inert inclusions and can accumulate to high levels in photosynthetic chloroplasts and amyloplasts of storage tissues. The plastid can, therefore, accommodate inclusions without disruption of organelle function. In addition, the absence of Poly(3HB) granules in plastids of transgenic plants expressing the Poly(3HB) enzymes in the cytoplasm raised the possibility that the plastid double membrane may be impervious to penetration by Poly(3HB) inclusions. Expression of the Poly(3HB) biosynthetic pathway in plastids was therefore hypothesized to lead to the accumulation of inclusions exclusively in the plastid, thus preventing potential disruption of other subcellular structures by the migrating inclusions.

In order to express the Poly(3HB) biosynthetic pathway in plastids of *A. thaliana*, the *R. eutropha pha*A, *pha*B, and *pha*C proteins were modified by the addition of a plastid targeting sequence derived from the chloroplast small subunit of the ribulose bisphosphate carboxylase from pea (Nawrath et al., 1994a,b). The modified 3-ketothiolase, acetoacetyl-CoA reductase and Poly(3HB) synthase were individually expressed in plants under the control of the constitutive CaMV 35S promoter. Transgenic plants expressing a high level of plastid-targeted reductase were crossed with plants expressing a high level of plastid-targeted PHA synthase. The resulting hybrids did not produce detectable Poly(3HB), since 3-ketothiolase activity is most likely not available in the plastid (Nawrath et al., 1994a). These double hybrids were subsequently cross-pollinated with transgenic plants expressing the plastid-targeted 3-ketothiolase to obtain triple hybrids expressing all three PHA enzymes in the organelle. These triple hybrids produced Poly(3HB) as detected by gas chromatography–mass spectroscopy (GC-MS) (Nawrath et al., 1994a). Transmission electron microscopy revealed that Poly(3HB) inclusions accumulated exclusively in the plastids. The size and general appearance of these inclusions were similar to bacterial PHA inclusions (Figure 2) (Nawrath et al., 1994a). The quantity of Poly(3HB) in these plants was found to gradually increase over the life span of the plant, with fully expanded pre-senescing leaves typically accumulating 10 times more Poly(3HB) than young expanding leaves of the same plant. The maximal amount of Poly(3HB) detected in pre-senescing leaves was 10 mg/g fresh weight (fwt), representing approximately 14% dwt. Synthesis of Poly(3HB) in these plants was not accompanied by a significant reduction in growth. However, slight chlorosis of fully expanded leaves accumulating more than 3 mg/g fwt could be detected, indicating an alteration in some of the chloroplast functions (Nawrath et al., 1994a). Transmission electron micrographs indicated that in these leaves a sizeable fraction of plastids were filled with Poly(3-HB) inclusions, leaving little space for thylakoid membranes.

In an effort to avoid the difficulties associated with the combination of three independent transgenes via cross-pollination and the selection of triple hybrids, a strategy was devise to combine all three genes for Poly(3HB) biosynthesis on a single vector (Bohmert et al., 2000), along the lines described by work done at Monsanto in *Arabidopsis*, rapeseed, and corn (Houmiel et al., 1999; Slater et al., 1999; Mitskey et al., 2000; Gruys et al., unpublished). This vector was made using the same *pha*A, *pha*B, and *pha*C that had been modified for targeting the protein to the plastids (Nawrath et al., 1994b). Following transformation of this complex vector in *A. thaliana*, a GC-MS method was used to rapidly screen transgenic lines accumulating high amounts of Poly(3HB). By this approach, a number of lines were identified which accumulated between 3 and 40% dwt Poly(3HB). Transmission electron microscopy confirmed the

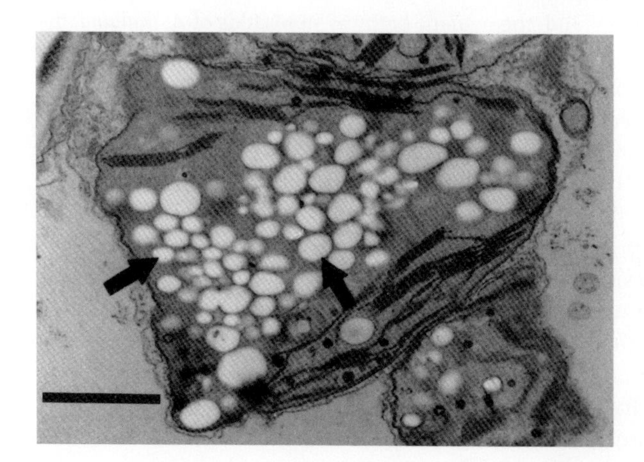

Fig. 2 Accumulation of Poly(3HB) inclusions in transgenic *Arabidopsis*. Transgenic cell expressing the Poly(3HB) pathway in the plastid showing accumulation of polymer inclusions (arrows) in the chloroplast of a leaf mesophyll cell. Bar represents 1 μm.

accumulation of Poly(3HB) in plastids as inclusions of 0.1 – 0.7 μm in diameter. While in a line accumulating 3% dwt most of the plastids contained some Poly(3HB) inclusions, all plastids of mesophyll cells were packed with inclusion in the line containing 40% dwt Poly(3HB). Analysis of the expression of all three *pha* genes by Northern analysis failed to identify a clear correlation between Poly(3HB) amount and expression of a particular *pha* gene. Interestingly, these transgenic plants showed a strong negative correlation between Poly(3HB) accumulation and plant growth. While plants containing 3% dwt showed only a relatively small reduction in growth, plants accumulating between 30 and 40% dwt Poly(3HB) were dwarf and produced no seeds. As previously observed by Nawrath et al. (1994a), all plants producing above 3% dwt Poly(3HB) showed some chlorosis. Analysis of over 60 metabolites (15 amino acids, 12 organic acids, 16 fatty acids, inorganic phosphate, and 16 sugars and sugar alcohols) was performed on these transgenic lines in order to gain an understanding of the metabolic changes occurring as a result of Poly(3HB) synthesis. Surprisingly, no changes in fatty acids were observed. There was, however, a correlation between an increase in Poly(3HB) with a decrease in levels of fumarate and isocitrate. This may indicate a reduction in tricarboxylic acid cycle activity, leading perhaps to a reduction in pools of acetyl-CoA which may result in growth retardation. There was also a positive correlation between Poly(3HB) accumulation and levels of several sugars such as mannitol, glucose, fructose, and sucrose. Together, these data indicate that high amount of accumulation of Poly(3HB) in chloroplasts has a negative effect on plant metabolism. While some of these effects can be explained, such has the decrease in tricarboxylic acid intermediates due to the demand on acetyl-CoA, others are more difficult to interpret. Nevertheless, these recent results indicate that plants are capable of accumulating higher amounts of PHA then previously described.

At Monsanto, somewhat similar results in *Arabidopsis* to that described by Nawrath et al. (1994a) and Bohmert et al. (2000) were obtained through transformation strategies that included:

(1) the crossing of two homozygous independent transformation events (one harboring the β-ketothiolase, the other the reductase and the synthase),

(2) simultaneous cotransformation of two vectors that together contained all three genes,

(3) re-transformation of plants that already harbored gene(s) for part of the pathway with a vector that contained the gene(s) that were missing, and

(4) the use of multigene vectors that contained the entire Poly(3HB) biosynthetic pathway (Mitsky et al., 2000).

In all these constructs the gene products were directed to the chloroplasts using plastid targeting sequences.

For all transformation strategies, excluding the use of multigene vectors, Poly(3HB) production was modest and never exceeded 2% dwt. This is despite the fact that the β-ketothiolase and reductase activities for the *Arabidopsis* events containing the full pathway were found to be in a similar range as that described by Nawrath et al. (1994a). The reasons for the smaller levels of polymer production are unclear, but it is known that the plants generated by Nawrath et al. (1994a) contained multiple insertions of some transgenes, whereas in most cases single insertions were obtained in the plants generated at Monsanto. It is possible that this difference affected the timing and maintenance of expression throughout the life of the plant, and this in turn impacted polymer levels. It is worth noting that the

Arabidopsis plants generated by Nawrath et al. (1994a) did not maintain a high Poly(3HB)-producing phenotype in the progeny from the original parents (Mitsky et al., 2000) This may be due to cosuppression (Finnegan and McElroy, 1994) or to segregation of high-producing insertions in the progeny.

Similar to the reasons described by Bohmert et al. (2000), the desire to increase the speed with which to generate Poly(3HB)-producing plants and significantly simplify the genetic analysis of transgenic events was the driving force at Monsanto for the transformation of *Arabidopsis* with multigene constructs that contained the entire Poly(3-HB) biosynthetic pathway. This strategy also allowed the creation of many more independent polymer-producing events relative to other transformation strategies. The results of this effort proved to be very fruitful in that a number of heterozygous lines containing dwt levels of Poly(3HB) up to 7–8% were generated (Valentin et al., 1999; Mitsky et al., 2000). More importantly, the trait was stable in many homozygous progeny and generally gave increased levels of Poly(3HB) compared to the heterozygous parents, some in the 10–14% range (Gruys et al., unpublished results). While there were cases of plants with more than 5% dwt Poly(3HB) showing no negative phenotype, there generally was a reduction in growth and the appearance of chlorotic tissue that correlated with Poly(3HB) levels higher than 4% dwt, similar to results described above by Bohmert et al. (2000) and Nawrath et al. (1994a). In addition, while many lines gave an increase in Poly(3HB) levels once brought to homozygosity, there were some examples where there was a dramatic decrease in polymer content at the homozygosity stage. This could be due to cosuppression since there is significant redundancy in the genetic elements present in the construct (e.g. three copies of the CaMV 35S promoter, the transit peptide sequence, etc.).

It is somewhat puzzling that the results from the Monsanto laboratories did not produce some events with the extraordinary high levels of Poly(3HB) as reported by Bohmert et al. (2000) considering that the multigene constructs used were virtually identical. Perhaps the differences can be related to the particular *A. thaliana* biotypes that were transformed in each laboratory and/or the fact that the highest producers reported by Bohmert et al. (2000) contained the construct at more than one insertion site. One striking difference noted in the two independent efforts was the cellular pattern of Poly(3HB) accumulation. Bohmert et al. (2000) reported that in the higher producing lines, the chloroplasts in all leaf cell types, including mesophyll, were full of Poly(3HB) inclusions. Results produced at Monsanto, however, showed by transmission electron microscopy that there was differential production of Poly(3HB), with epidermal cells and cells associated with the vascular tissue producing the highest amounts (Gruys et al., unpublished results). This was especially clear when the analysis was done during mid-vegetative development. In this case, little if any Poly(3HB) was visible in mesophyll cells compared to epidermal cells or cells associated with vascular tissue, which showed visible inclusions. However, by senescence, some Poly(3HB) inclusions were evident in mesophyll cells, although still visibly less than the other cell types. The reasons for the differential production is not clear, but may be associated with known different metabolic functions for various cell types. For example, epidermal cells produce wax, implying that fatty acid biosynthesis is robust and requires that there be sufficient steady-state levels of acetyl-CoA. On the other hand, mesophyll cells are mainly involved with producing starch and may not need high levels of acetyl-CoA to the same degree as epidermal cells. Alternatively, gene expression using the CaMV 35S promoter may not be equal in all the cell types in the leaf. The reasons why this particular Poly(3HB) distribution was not observed by Bohmert et al. (2000) are unclear at present. It is possible that in the study of Bohmert et al. (2000) the electron microscope observations were only done once the plant reached senescence, making the difference in Poly(3HB) between cell types less evident. It is worth noting that this differential production of Poly(3HB) was not exclusive to *Arabidopsis*, as will be shown and discussed in Section 3.2 describing the biosynthesis of Poly(3HB) in corn.

There are several possible explanations for the increased levels of polymer present in plants transformed with multigene vectors compared to transgenic plants generated by other transformation and crossing strategies. As mentioned above, one explanation derives from the fact that it was possible to generate more independent lines with the multigene vectors and the screening of more plants allowed detection of the relatively rare high-producing lines. This is one clear advantage of having the entire pathway on a single vector. However, the distribution of polymer synthesized in plants produced by the various methods suggests that numbers alone do not account for the increased polymer production of multigene vectors. It is also possible that having a metabolic pathway genetically linked at a single integration locus is more metabolically favorable due to some level of concerted gene expression and/or mRNA metabolism. This phenomenon is common in bacteria, but there are not many examples of clustering genes in plants for concerted gene expression. Another possibility is that the high local concentration of promoters may lead to locally high levels of transcription factors. Still another possibility is that having the genes tightly linked may reduce gene silencing, or cosuppression, in certain cases. Regardless, it is clearly evident that the difficulty in the initial construction of such complex vectors was rewarded with plants producing high levels of Poly(3HB).

Collectively, all experiments done in *A. thaliana* demonstrated that redirecting the Poly(3HB) biosynthetic pathway from the cytoplasm to the plastid resulted in an approximate 100- to 400-fold increase in Poly(3HB) production (Nawrath et al., 1994a; Bohmert et al., 2000; Mitsky et al., 2000; Gruys et al., unpublished). These results indicate a difference either in the quantity and/or availability of the plastidial acetyl-CoA to be diverted from endogenous metabolic pathways as compared to cytoplasmic acetyl-CoA. However, the chlorosis and reduction in growth observed in plants expressing more then 3–4% dwt Poly(3HB) in leaves indicate that Poly(3HB) accumulation, even in the plastid, has its limits and can affect metabolism directly through competition for primary metabolites, and/or indirectly by disruption of normal subcellular processes such as photosynthesis.

3.2 Synthesis of Poly(3HB) in Other Plants

Synthesis of Poly(3HB) has also been demonstrated in a few plants other than *Arabidopsis* by expression of the Poly(3HB) biosynthetic pathway in either the cytoplasm, plastid, or peroxisome. Expression of the Poly(3HB) biosynthetic pathway in the cytosol of cells of *Brassica napus* gave results similar to experiments in *A. thaliana*. Cross-pollination of transgenic rapeseed expressing the acetoacetyl-CoA reductase with plants expressing the Poly(3HB) synthase, both genes expressed under the CaMV 35S promoter, led to F_1 hybrids producing Poly(3HB) in the range of 0.02–0.1% dwt (P. A. Fentem, unpublished data). Some transgenic rapeseeds were also stunted in growth, similar to *A. thaliana* producing Poly(3HB) in the cytoplasm. Interestingly, overexpression of the bacterial 3-ketothiolase in plants expressing the reductase and Poly(3HB) synthase did not lead to an increase in Poly(3HB) production. This result indicated that 3-ketothiolase activity was not limiting Poly(3HB) synthesis in the cytoplasm, but rather that other factors, such as the low flux of acetyl-CoA, may be important.

Synthesis of Poly(3HB) has also been demonstrated in tobacco through the coexpression of the *phaB* gene from *R. eutropha* and the PHA synthase from *Aeromonas caviae* (Nakashita et al., 1999). Although the bacterial genes were expressed under the strong constitutive CaMV 35S promoter, expression of both proteins was relatively low. Low expression of protein derived from the transcription of *R. eutropha* genes has also been previously observed in tobacco (Poirier and Nawrath, unpublished data). The reason for this is unknown, but may be related to inefficient translation of the mRNA in tobacco due to secondary structure or the different codon usage between tobacco and *R. eutropha* genes, or to an unexplained instability of the protein. The maximal amount of Poly(3HB) detected in leaves was 10 $\mu g\ g^{-1}$ fwt. Analysis of the polymer revealed that the number-average molecular weights and polydispersity were 32,000 and 1.9, respectively (Nakashita et al., 1999).

Scientists at Monsanto recently reported the successful production of Poly(3HB) in oilseed leukoplasts of *B. napus* (Houmiel et al., 1999; Valentin et al., 1999). This was done to critically evaluate the possibility of using oilseed crops as a commercial production system for PHAs. Since acetyl-CoA is the primary metabolite for both Poly(3HB) and fatty acid biosynthesis, and *B. napus* seeds are very efficient in the commercial production of oil, this crop was hypothesized to be an ideal system. The generation of *B. napus* producing Poly(3HB) was accomplished through *Agrobacterium*-mediated transformation using multigene vectors that contained the entire Poly(3HB) biosynthetic pathway. Each gene was placed under the control of the fatty acid hydroxylase promoter from *Lesquerella fendeleri* which directs gene expression to the developing seed (Broun et al., 1997) and the expressed proteins were fused to a plastid transit peptide for transport to the seed leukoplast. Previous to this published work, Monsanto had generated Poly(3HB)-containing *B. napus* seeds using the 7S seed promoter driving the Poly(3HB) biosynthetic genes and through the crossing of homozygous plants containing parts of the pathway (Mitsky et al., 2000). The polymer production in these crosses was modest (below 2% dwt), which was similar to the findings in *Arabidopsis* using crossing strategies. However, as discussed for *Arabidopsis*, greater levels of Poly(3HB) were obtained from rapeseed transformed with the multigene constructs.

Results from using multigene pathway constructs showed an 87% frequency of transgene coexpression in *B. napus* seeds from more than 400 transformants analyzed. Polymer levels up to 7.7% fwt of mature seeds were found and averaged from 1–1.7% dwt from two different multigene vectors (Houmiel et al., 1999). Electron micrographs of a high-producing line revealed that Poly(3HB) accumulated within the leukoplast and that every visible plastid contained the polymer. Interestingly, compared to nontransformed seed, the size of the leukoplast was expanded to accommodate Poly(3HB) granules. This is not unlike the behavior observed with amyloplasts that accumulate starch during seed development and suggests that seed plastids adjust in size to accommodate any granular product. Seeds from these transformed plants appeared normal and germinated at the same rates as nontransformed seeds. These results demonstrated that *B. napus* is a possible production system for PHAs, although approximately twice the levels of PHA would need to be generated in the seed relative to the levels observed in the highest producing lines. Also, for *B. napus* to be viable as a PHA commercial system, little to no loss in the amount of oil will be required since this coproduct, in addition to the protein meal, would still need to be extracted and sold for economical considerations.

Monsanto has also demonstrated Poly(3-HB) production in corn. Some of the results from this work are described by Mitsky et al. (2000). The strategy for this crop was for production to occur not in the kernel, but in the stover (leaves and stock) which is normally left as waste biomass in fields after harvest. Four multigene vectors were generated, similar to those constructed for other plants, testing the effectiveness of four different promoters to drive gene expression. The vectors contained the plant viral CaMV 35S or figwort mosaic virus (FMV) promoters, a rice actin promoter (rACT), or the promoter for maize chlorophyll A/B binding protein (P-ChlA/B), and included the HSP70 intron designed to enhance expression in monocots. All enzymes were fused to the transit peptide of the *Arabidopsis* ribulose bisphosphate carboxylase small subunit for protein translocation to the chloroplasts.

The experiments demonstrated that maize can also produce moderate levels of Poly(3-HB), though there were differences associated with the promoter utilized in the construct (Table 2). The maize chlorophyll A/B binding protein promoter gave the highest levels and the greatest number of positive Poly(3HB) producing transformants. The CaMV 35S promoter gave a low frequency of Poly(3HB)-producing plants relative to the other promoters, but some of this may be due to cosuppression of gene expression in these plants with a frequency greater than that seen with plants transformed with the other vectors. This is partially supported by the observation that some of the CaMV 35S lines stopped producing Poly(3HB), as measured by quantitative GC, early during vegetative growth concurrent with a drop in gene expression. This is not a characteristic normally associated with the activity of the CaMV 35S promoter in maize. The rice actin promoter gave plants with low Poly(3HB) production and this was consistent with gene expression also being low relative to the other promoter constructs. Although the vectors contained constitutive promoters, the polymer distribution in corn plants was highest in the leaf. The seed contained little detectable Poly(3-HB). During vegetative development, the older leaves contained the greatest Poly(3-HB) levels, resembling the findings in *Arabidopsis*. Also, like *Arabidopsis*, there was a general correlation with a chlorotic phenotype with plants producing the higher levels of Poly(3HB). For corn this was independent of the promoter construct.

Similar to Monsanto's findings in *Arabidopsis*, the production level of Poly(3HB) in various cell types in the maize leaf was also found to be variable. Interestingly, the pattern was quite comparable to *Arabidopsis* in that mesophyll cells showed few Poly(3-HB) granules during vegetative growth relative to cells associated with the vascular tissue (bundle sheath cells). These latter cells were packed with Poly(3HB) granules. This is shown quite dramatically in the electron micrograph shown in Figure 3. Again, the exact reason for this differential Poly(3HB) production could be related to either differences in gene expression or levels of available acetyl-CoA, or both. How-

Tab. 2 Polymer production in the plastids of corn leaves using multigene vectors

Plant construct description	*No. of plants assayed*	*No. of plants positive*	*Poly(3HB) (% dwt)*
P-35S phbC *P-35S phbA* *P-35S phbB*	90	16	0.10–4.81 average: 1.53 SD: 1.09
P-FMV phbC *P-FMV phbA* *P-FMV phbB*	113	53	0.10–4.84 average: 0.7 SD: 1.06
P-ChlA/B, phbC *P-ChlA/B, phbA* *P-ChlA/B, phbB*	132	78	0.10–5.73 average: 1.75 SD: 1.26
P-rACT phbC *P-rACT phbA* *P-rACT phbB*	130	72	0.10–0.80 average: 0.22 SD: 0.14

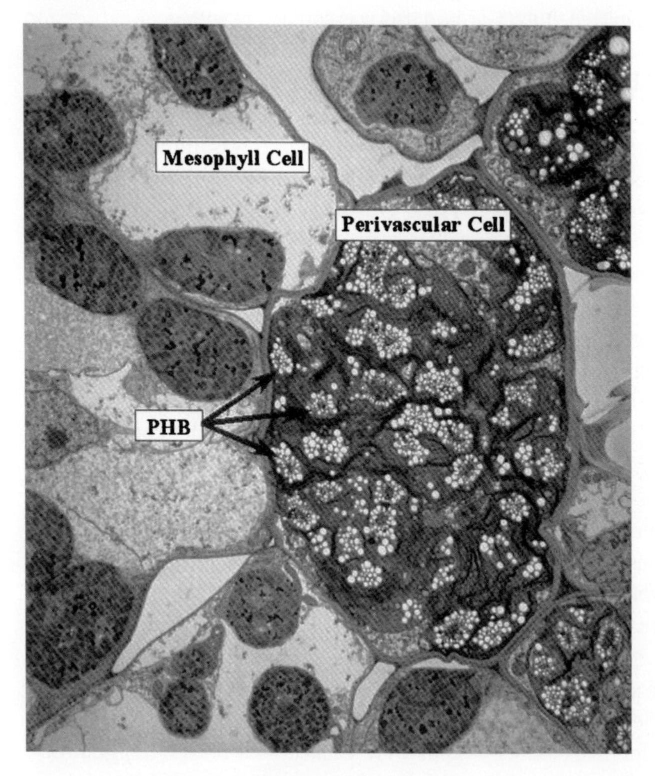

Fig. 3 Accumulation of Poly(3HB) inclusions in transgenic corn expressing the biosynthetic pathway in plastids. Poly(3HB) inclusions are seen as electron-lucent granules (arrows). Note the greater abundance of Poly(3HB) inclusions in the plastids of the perivascular cell compared to the surrounding mesophyll cells.

ever, this pattern of Poly(3HB) production was seen for plants transformed with either the CaMV 35S and P-ChlA/B promoters, and the latter is known to be a strong promoter for the production of chlorophyll A/B-binding protein in mesophyll cells. This suggests that the availability of acetyl-CoA in these cell types differs and that Poly(3HB) production may be a useful tool for measuring such differences (see Section 4.3).

Efforts have also been made to transfer the Poly(3HB) pathway in the plastids of potato. Transformation of potato using the same vector containing all three *R. eutropha pha* genes that was used for transformation of *A. thaliana* (Bohmert et al., 2000) failed to yield any transgenic plants (Bohmert, unpublished data). Further investigation revealed that constitutive expression of the bacterial 3-ketothiolase prevented the recovery of transformants. From these data, a novel vector was designed where the constitutive CaMV 35S promoter was replaced by the salicylic acid-inducible promoter Prp1, while the *phaB* and *phaC* genes remained under the control of the CaMV 35S promoter. Using this vector, potato transformants accumulating very low amounts of Poly(3-HB) (0.02% dwt) were isolated (Bohmert et al., unpublished). Transformation of tobacco with the same vector also yielded only transgenic plants accumulating similar low amounts of Poly(3HB) (0.04% dwt). Since both tobacco and potato are members of the Solanaceae family, it is possible that the factors limiting Poly(3HB) synthesis in both plants may be the same.

Alfalfa has been recently transformed with the *R. eutropha phaA*, *phaB*, and *phaC* genes engineered for plastid targeting, and put under the control of the CaMV 35S promoter. GC analysis of 100 plants revealed an accumulation of Poly(3HB) between 0.02–0.2% dwt (D. P. Saruul et al., unpublished). Analysis of the chloroplast by electron microscopy revealed the presence of numerous typical Poly(3HB) inclusions. Growth of these plants was not significantly different from wild-type.

Acetyl-CoA, the central building block of PHA, is found not only in the cytoplasm and plastids, but also in the mitochondria and peroxisomes, being implicated in these organelles in the tricarboxylic acid and β-oxidation cycles, respectively. No conclusive demonstration of Poly(3HB) in plant mitochondria has yet been reported. In contrast, synthesis of Poly(3HB) was reported in transgenic Black Mexican sweet corn suspension cell cultures expressing the Poly(3-HB) biosynthetic pathway in the peroxisomes (Hahn et al., 1999) (Figure 1). In these experiments, the *phaA*, *phaB*, and *phaC* genes from *R. eutropha* were modified in order to add at the carboxy-terminal end of each protein a 6-amino-acid peptide RAV-ARL, which has previously been shown to localize the enzyme glycolate oxidase to the peroxisome of tobacco (Volokita, 1991). Biolistic transformation of maize suspension culture with a mixture of all three genes lead to the isolation of a transformant expressing all three enzyme activities and accumulating 2% dwt Poly(3HB). As no transgenic plants have been obtained from these transformed cells, it is difficult at this point to evaluate the potential effects of Poly(3HB) synthesis in peroxisome on growth and metabolism.

3.3 Synthesis of Poly(3HB-*co*-3HV) Copolymer in Plants

The successful use of crop plants as a production vector for PHA not only depends on the amount of polymer accumulated in plants but also on the type and quality of the PHA synthesized. Since Poly(3HB) is a polymer with relatively poor physical characteristics, being too stiff and brittle for use in most consumer products (de Koning, 1995), it was important to engineer plants for the synthesis of PHA copolymers with better physical characteristics. Poly(3HB-*co*-3HV) is one of the best studied PHA copolymers. Poly(3HB-*co*-3HV) has lower crystallinity, is more flexible, and is less brittle than Poly(3-HB) homopolymer (de Koning, 1995). Synthesis of Poly(3HB-*co*-3HV) in bacteria was first achieved by fermentation of *R. eutropha* on glucose and propionic acid as discussed by Anderson and Dawes (1990), and seen in a number of patents of Zeneca from the 1980s. For a number of years, bacterial production of Poly(3HB-*co*-3HV), known also under the trade name Biopol™, has been central to the marketing and production strategies of PHA by Zeneca, who had strong intellectual property rights covering polymer composition and synthesis. It was therefore natural that efforts on PHA copolymer production in plants would also be initially focused on Poly(3HB-*co*-3HV) synthesis.

Production of Poly(3HB-*co*-3HV) copolymer in the plastids of plants has been demonstrated by the PHA group of Monsanto (Slater et al., 1999), which acquired the Biopol™ business from Zeneca in 1996. The readers are referred to the Volume 4, Chapter 3 by Asrar and Gruys for a detailed account of the Monsanto strategy to synthesize Poly(3HB-*co*-3HV) copolymer in plants, including the testing of the metabolic pathway in *Escherichia coli*. In this chapter, we wish to give a brief summary of the experiments and results demonstrating Poly(3HB-*co*-3HV) synthesis in plants.

In the commercial production of Poly(3-HB-*co*-3HV) from *R. eutropha*, propionate is added to the growth media in order to create an intracellular pool of propionyl-CoA which is condensed with acetyl-CoA to form 3-ketovaleryl-CoA. The 3-ketovaleryl-CoA is then reduced by the acetoacetyl-CoA reductase to give *R*-3-hydroxyvaleryl-CoA, which is subsequently copolymerized with *R*-3-hydroxybutyryl-CoA to form Poly(3HB-*co*-3HV). For the synthesis of Poly(3HB-*co*-3HV) in plants, it was thus necessary to create an endogenous pool of propionyl-CoA which could be used by the PHA pathway. Although several metabolic pathways exist in prokaryotes and eukaryotes that can generate propionyl-CoA, the strategy adopted by Slater et al. (1999) was the conversion of 2-ketobutyrate to propionyl-CoA by the pyruvate dehydrogenase complex (PDC), an enzyme naturally located in the plastid. Although PDC normally decarboxylates pyruvate to give acetyl-CoA, experiments had previously shown that PDC can also decarboxylate 2-ketobutyrate, albeit at low efficiency, to give propionyl-CoA (Camp and Randall, 1985) (Figure 4). Since 2-ketobutyrate is also found in the plastid as an intermediate in the synthesis of isoleucine from threonine, both the substrate and the enzyme complex required for the generation of propionyl-CoA are present in this organelle. However, since PDC would have to compete for the 2-ketobutyrate with the acetolactate synthase, an enzyme involved in isoleucine biosynthesis, the quantity of 2-ketobutyrate present in the plastid was enhanced through the expression of the *E. coli ilvA* gene, which encodes a threonine deaminase (Figure 4).

The genes encoding the *E. coli ilvA*, the *R. eutropha phaB* and *phaC*, as well as the *bktB* gene from *R. eutropha* encoding a novel 3-thiolase having high affinity for both acetyl-CoA and propionyl-CoA (Slater et al., 1998), were all modified to add a plastid-targeting signal to the proteins. All genes were expressed under the control of the CaMV 35S promoter. Constitutive expression of the *ilvA* protein along with *bktB*, *phaB*, and *phaC* proteins in the plastids of *A. thaliana* lead to the synthesis of Poly(3HB-*co*-3HV) in the range of 0.1 – 1.6% dwt, and with a HV level between 2 and 17 mol% (Slater et al., 1999). These plants were generated through coinfiltration using two distinct vectors. Coexpression of an isoleucine insensitive mutant of the *ilvA* gene along with the *bktB*, *phaA*, and *phaC* genes, all under the control of the seed-specific promoter from the *Lesquerella* hydroxylase gene, in the developing seed of

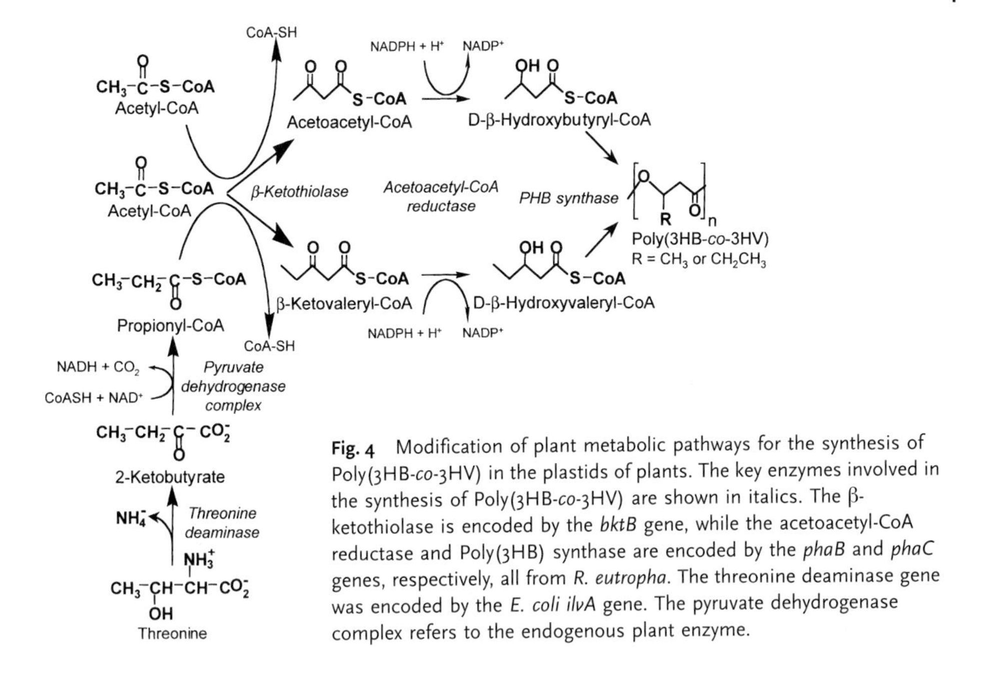

Fig. 4 Modification of plant metabolic pathways for the synthesis of Poly(3HB-*co*-3HV) in the plastids of plants. The key enzymes involved in the synthesis of Poly(3HB-*co*-3HV) are shown in italics. The β-ketothiolase is encoded by the *bktB* gene, while the acetoacetyl-CoA reductase and Poly(3HB) synthase are encoded by the *phaB* and *phaC* genes, respectively, all from *R. eutropha*. The threonine deaminase gene was encoded by the *E. coli ilvA* gene. The pyruvate dehydrogenase complex refers to the endogenous plant enzyme.

the oil crop *B. napus* lead to the accumulation of Poly(3HB-*co*-3HV) in the range of 0.7–2.3% dwt, with a HV content of 2.3-6.4 mol% (Slater et al., 1999). These plants were generated utilizing a multigene construct that contained all four genes. In all these transgenic plants, the presence of the HV units in the polymer was confirmed by GC-MS as well as nuclear magnetic resonance (NMR). The average molecular mass of the polymer produced ranged from 0.5×10^6 to 1.2×10^6 and polydispersity ranged from 1.8 to 2.4. These characteristics, combined with the significant level of HV monomer in a number of the *Arabidopsis* and *B. napus* lines, makes this polymer quite suitable for commercial applications. However, additional metabolic engineering would be required to reach plant polymer levels of 15% dwt while at the same time maintaining the HV content at a minimum of 5 mol%.

The biosynthesis of Poly(3HB-*co*-3HV) in plants demonstrated that specific intermediates involved in amino acid and fatty acid biosynthesis could be utilized to successfully produce this copolymer. This was an intriguing result, but also of interest was the information generated from measuring the impact on amino acid metabolism when a wild-type or partially deregulated threonine deaminase (IlvA) was overexpressed to enhance the concentration of 2-ketobutyrate (Figure 4). As mentioned briefly above, this strategy worked in both *Arabidopsis* and *Brassica*, with the former showing an approximately 20-fold increase in 2-ketobutyrate concentration when the wild-type *E. coli IlvA* was expressed and targeted to the chloroplast (Slater et al., 1999). Not too surprisingly, the levels of free isoleucine also increased from 5- to 10-fold in both plants due to the fact that threonine deaminase is the committed and regulated step in isoleucine biosynthesis. In addition, a high concentration of 2-aminobutyrate, 17-fold greater than *Arabidopsis* control plants, was generated as a result of transamination of 2-ketobutyrate. As a consequence, the large majority of carbon generated through the action of IlvA ended up in isoleucine or 2-aminobutyrate rather than the targeted 2-ketobutyrate. Interestingly, this did not occur at the expense of threonine or aspartate, the latter being the anabolic precursor of threonine and isoleucine. The levels of these two amino acids were similar to control plants. This suggests that threonine biosynthesis in plants is highly robust and can compensate for a large increase in flux created by the enhanced levels of IlvA.

In addition to these experimental results, Daae et al. (1999) used a mathematical simulation model to address the theoretical prospects of producing the Poly(3HB-*co*-3HV) copolymer in plant plastids. This model was generated using information on various metabolite levels essential to the pathway plus detailed knowledge on the kinetics of the enzymatic steps. The model suggested that both the HV/HB ratio and the copolymer production rate vary considerably between dark and light conditions of plant growth. This has to do with the difference in plastid pH under these two conditions, which impacts the equilibrium constant for the reductase reaction (Figure 4). The results generated using metabolic control analysis suggested that the β-ketothiolase predominately controls the copolymer production rate, but that the activity of all three PHA biosynthetic enzymes influence the copolymer ratio. Dynamic simulations further suggested that controlled expression of the three enzymes at different levels may lead to the desired changes in both the copolymer production rate and the HV/HB ratio. Not surprisingly, the model showed that natural variations in substrate and cofactor levels will have a considerable impact on both the production rate and the HV/HB ratio.

3.4 Synthesis of MCL-PHA in Plants

Approximately 150 different hydroxyacid monomers have been found to be included into PHAs (Steinbüchel and Valentin, 1995). One large group of PHAs, defined as MCL-PHAs, represents polyesters containing 3-hydroxyacids ranging from 6 to 16 carbons. Although these PHAs are typically described as elastomers, their actual physical properties are very diverse and are dependent on the monomer composition (de Koning, 1995). Monomers present in MCL-PHA may have a number of functional groups, such as unsaturated bonds and phenoxy groups (Steinbüchel and Valentin, 1995). The presence of reactive groups in the side chain offers opportunities to polymer chemists to modify the structure and physical properties of PHAs after extraction. For example, the inclusion of unsaturated monomers into MCL-PHA enabled the creation of a cross-linked polymer after electron-beam irradiation, resulting in the formation of a true rubber with constant physical properties over a wide range of temperatures (de Koning et al., 1994).

There are two main routes for the synthesis of MCL-PHA in bacteria (Poirier et al., 1995a; Steinbüchel, 1991; Steinbüchel and Füchtenbusch, 1998). The first is the synthesis of PHA using intermediates of fatty acid β-oxidation. This pathway is found in bacteria, such as *Pseudomonas oleovorans*, which can synthesize MCL-PHA from a source of alkanoic acids or fatty acids. In these bacteria, the type of PHAs produced is directly influenced by the carbon source added to the growth media, being composed of monomers which are 2*n* carbons shorter than the substrates used. For example, growth of *P. oleovorans* on octanoate (C8) generates a PHA copolymer containing C8 and C6 monomers, whereas growth on dodecanoate (C12) generates a PHA containing C12, C10, C8, and C6 monomers (Lageveen et al., 1988). Alkanoic acids present in the media are converted to CoA esters by an acyl-CoA synthetase and then channeled to the β-oxidation pathway where a number of 3-hydroxyacyl-CoA intermediates can be synthesized. Since the PHA synthase accepts only the *R* isomer of 3-hydroxyacyl-CoAs and β-oxidation of saturated fatty acids generates only the *S* isomer of 3-hydroxyacyl-CoAs (Gerhard, 1993; Schulz, 1991), bacteria must have enzymes capable of generating *R*-3-hydroxyacyl-CoAs. One enzyme is a 3-hydroxyacyl-CoA epimerase, mediating the reversible conversion of the *S* and *R* isomers of 3-hydroxyacyl-CoA. This enzyme activity is found as a part of the multifunctional protein (MFP), an enzyme participating in the core β-oxidation cycle and which possess, in addition to the 3-hydroxyacyl-CoA epimerase, an enoyl-CoA hydratase I, *S*-3-hydroxyacyl-CoA dehydrogenase and a Δ^3-Δ^2-enoyl-CoA isomerase (Hiltunen et al., 1996). Furthermore, several monofunctional enoyl-CoA hydratase II enzymes, converting directly enoyl-CoA to *R*-3-hydroxyacyl-CoA have been identified in *A. cavea*, *Rhodospirillum rubrum* and *Pseudomonas aeruginosa* (Fukui et al., 1998, 1999; Reiser et al., 2000; Tsuge et al., 2000). Finally, it is also possible that a 3-ketoacyl-CoA reductase that could specifically generate *R*-3-hydroxyacyl-CoA may exist in bacteria, although such an enzyme has not yet been unambiguously identified. However, it has been shown that the enzyme 3-ketoacyl-acyl carrier protein (ACP) reductase, participating normally in the fatty acid biosynthetic pathway, may also act on the 3-ketoacyl-CoA to generate *R*-3-hydroxyacyl-CoA (Taguchi et al., 1999).

While *P. oleovorans* and *P. fragi* can only synthesize MCL-PHA from related alkanoic acids present in the growth media, other Pseudomonads, such as *Pseudomonas putida*, can synthesize MCL-PHAs when grown on unrelated substrates, such as glucose (Haywood et al., 1990; Timm and Steinbüchel, 1990). Analysis of the composition of PHA produced by *P. putida* grown on glucose revealed the presence of the monomers 3-hydroxy-5-*cis*-dodecenoic acid and 3-hydroxy-7-*cis*-tetradecenoic acid (Huijbert et al., 1992). Since these monomers are structurally identical to the acyl moieties of the 3-hydroxyacyl-ACP intermediates of the *de novo* fatty acid biosynthesis, it was hypothesized that intermediates from the *de novo* fatty acid biosynthetic pathway could be used to form PHAs. This conclusion was also supported by studies using ^{13}C-labeled acetate (Saito and Doi, 1993; Huijbert et al., 1994). A key gene linking fatty acid biosynthesis and PHA synthesis was first identified and cloned in the bacteria *P. putida*. This gene, named *phaG*, was shown to have a 3-hydroxyacyl-CoA-ACP transferase (Rehm et al., 1998). Expression of this gene in *P. oleovorans* and *P. fragii* confers the capacity to synthesize PHA from glucose (Rehm et al., 1998; Fiedler et al., 2000; Hoffman et al., 2000a). Interestingly, a homologue of *phaG* has also been identified in *P. oleovorans*, but was found to be transcriptionally inactive, explaining the inability of this organism to synthesize PHA from fatty acid biosynthetic intermediates (Hoffmann et al., 2000b).

Synthesis of MCL-PHA in plants has recently been demonstrated in *A. thaliana* (Mittendorf et al., 1998a,b). The approach was to use the 3-hydroxyacyl-CoA intermediates of the β-oxidation of endogenous fatty acids for MCL-PHA production. Since in plants β-oxidation occurs principally in the peroxisomes, PHA biosynthetic proteins needed to be targeted to this organelle. The phaC1 synthase from *P. aeruginosa* was thus modified for peroxisome targeting by the addition of the last 34 amino acids from the peroxisomal protein isocitrate lyase of *B.*

napus. The modified gene was expressed under the control of the CaMV 35S promoter, allowing constitutive expression in a wide spectrum of tissues. Immunolocalization of the PHA synthase demonstrated the appropriate targeting of the PHA synthase in plant peroxisomes (Mittendorf et al., 1998b). Furthermore, plants expressing the PHA synthase showed the presence of electron-lucent inclusions within the peroxisomes, which was enlarged compared to peroxisomes of wild-type plants. GC-MS and NMR of the purified polymer confirmed synthesis of MCL-PHA in these transgenic plants. The PHA had a weight-average molecular weight of 23,700, a number-average molecular weight of 5500, and a polydispersity of 4.3, indicating that, similar to Poly(3HB) synthesized in the cytoplasm of *A. thaliana* (Poirier et al., 1995b), the PHA produced in plants had a significantly lower molecular weight and a broader distribution compared to bacterial PHA (Mittendorf et al., 1998b). The monomer composition was fairly complex, including saturated and unsaturated monomers ranging from 6 to 16 carbons. The production of peroxisomal MCL-PHA was relatively low, reaching 0.4% dwt in 7-day-old germinating seedlings and then decreasing as the plant matured to reach approximately 0.02% dwt. This decrease is not thought to reflect PHA degradation but rather the fact that in expanding green tissues the plant weight increases faster then the rate of PHA synthesis. Interestingly, there was a 2- to 3-fold increase in PHA during leaf senescence. These data support the link between β-oxidation and PHA synthesis, since these pathways, in association with the glyoxylate cycle, are most active during germination and senescence where they are involved in the conversion of fatty acids to carbohydrates. No effects of MCL-PHA accumulation on plant growth or seed germination were observed.

The wide range of monomers found into plant MCL-PHA suggests that, as with bacteria, plants also have enzymes capable of converting the β-oxidation intermediates *S*-3-hydroxyacyl-CoA to the *R* isomer (Figure 5). Such enzymes could be either the 3-hydroxyacyl-CoA epimerase present on the plant MFP (Hiltunen et al., 1996) or an enoyl-CoA hydratase II activity which is specific for the generation of *R*-3-hydroxyacyl-CoA from 2-*trans*-enoyl-CoA (Engeland and Kindl, 1991). A third route for the synthesis of *R*-3-hydroxyacyl-CoA is the hydration of 2-*cis*-enoyl-CoA by the enoyl-CoA hydratase I activity (Schulz, 1991). The substrate 2-*cis*-enoyl-CoA is derived from the β-oxidation of unsaturated fatty acids having a *cis* double bond at an even position, such as found in linoleic (18:2 Δ9*cis* 12*cis*) and linonelic acid (18:3 Δ9*cis* 12*cis* 15 *cis*), two fatty acids which are abundant in plants.

Modulation of the quantity and monomer composition of the MCL-PHA synthesized in plant peroxisomes has been demonstrated in various experiments aimed primarily at influencing the quantity and nature of the fatty acids targeted to the β-oxidation cycle (Mittendorf et al., 1999). Growth of transgenic plants in liquid media supplemented with detergents containing various fatty acids resulted in an increased accumulation of MCL-PHA containing monomers derived from the β-oxidation of these external fatty acids (Mittendorf et al., 1999). For example, addition to the media of the detergent polyoxyethylenesorbitan esterified to lauric acid (Tween 20) resulted in a 8- to 10-fold increase in the amount of PHA synthesized in 14-day-old plants compared to plants growing in the same media without detergent. The monomer composition of the MCL-PHA synthesized in media containing Tween 20 showed a large increase in the proportion of saturated monomers with 12 carbons and lower, and a corresponding

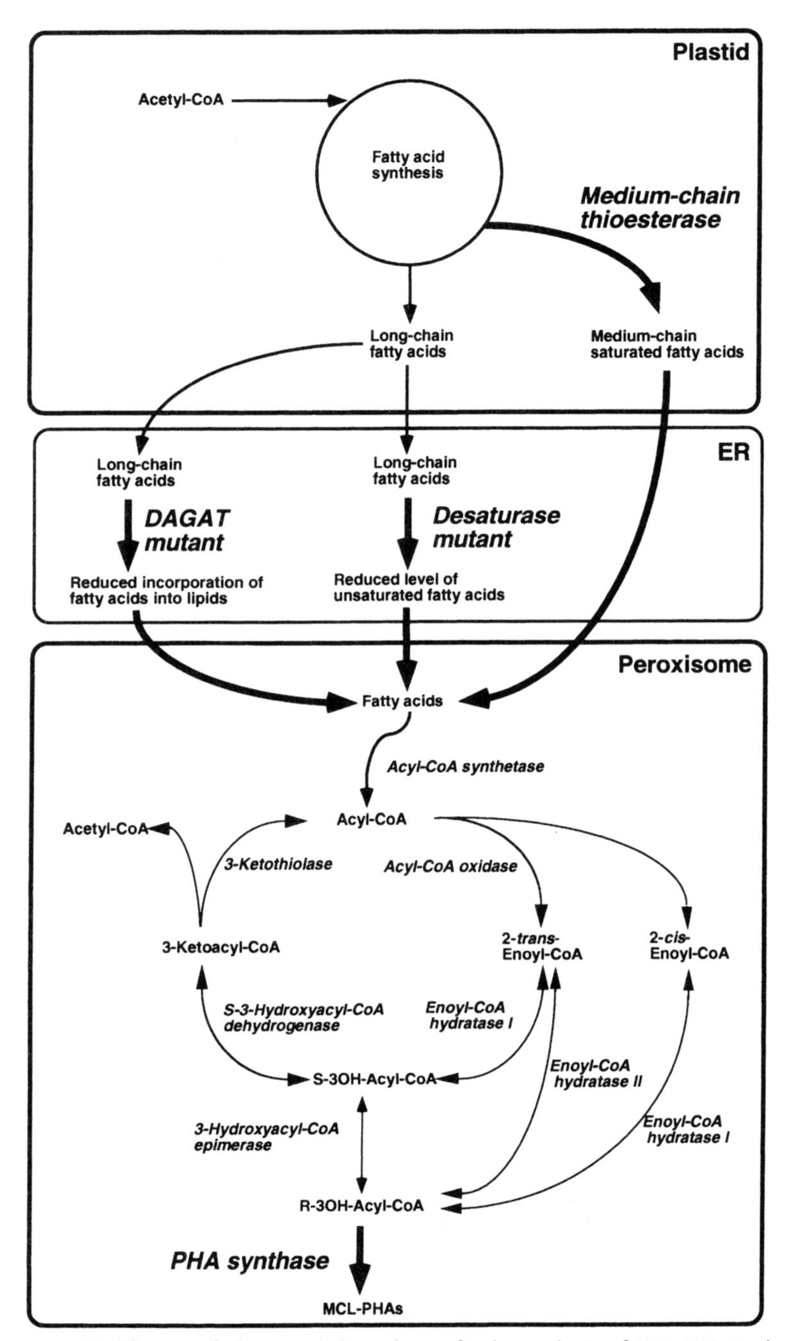

Fig. 5 Modification of plant metabolic pathways for the synthesis of MCL-PHAs in the peroxisomes of plants. The pathways created or enhanced by the expression of transgenes from bacteria (PHA synthase) or plant (medium-chain thioesterase) or the use of mutant genes (DAGAT and fatty acid desaturates) are highlighted in bold.

decrease in the proportion of all unsaturated monomers. This shift in monomer composition is accounted for by the fact that β-oxidation of lauric acid, a 12 carbon saturated fatty acid, gives saturated 3-hydroxyacyl-CoA intermediates of 12 carbons and lower. Further experiments have shown that addition in the plant growth media of either tridecanoic acid, tridecenoic acid (C13:1 Δ12) or 8-methyl-nonanoic acid resulted in the production of MCL-PHA containing mainly saturated odd-chain, unsaturated odd-chain, or branched-chain 3-hydroxyacid monomers, respectively (Mittendorf et al., 1999). These results demonstrated that the plant β-oxidation cycle was capable of generating a large spectrum of monomers that can be included in MCL-PHA even from fatty acids which are not present to significant quantities in plants.

Modulation of the quantity and/or monomer composition of MCL-PHA synthesized in peroxisomes was also achieved by modifying the endogenous fatty acid biosynthetic pathway (Mittendorf et al., 1999). For example, expression of the peroxisomal PHA synthase in an *A. thaliana* mutant deficient in the synthesis of triunsaturated fatty acids (McConn and Browse, 1996) resulted in the synthesis of a PHA having an almost complete absence of all 3-hydroxyacids derived from the degradation of these fatty acids, including triunsaturated 3-hydroxyacid monomers (Mittendorf et al., 1999). Since numerous fatty acid desaturases have been cloned, and the number and position of unsaturated bonds in fatty acids can be controlled to a significant extend in transgenic plants, this approach could be extended to modulate the proportion of a number of 3-hydroxyacid monomers in PHAs.

In a different strategy aimed at modulating PHA synthesis by influencing the flux of fatty acids targeted to the β-oxidation cycle, expression of a fatty acyl-ACP thioesterase in the plastid was combined with the expression of a peroxisomal PHA synthase (Mittendorf et al., 1999). Fatty acyl-ACP thioesterases are enzymes catalyzing the release of the fatty acyl-ACP intermediate from the fatty acid synthase complex and are thus responsible for stopping the elongation of fatty acids. Numerous thioesterases having different fatty acid chain-length specificities have been cloned in the last decade (Voelker et al., 1992; Jones et al., 1995; Facciotti and Yuan, 1998). Typically, expression in the seeds of transgenic *B. napus* of a thioesterase specific for medium-chain fatty acyl-ACPs results in the accumulation of medium-chain fatty acids in the seed reserve lipids (Jones et al., 1995; Voelker et al., 1992). Interestingly, it was also shown that expression of the same enzyme in roots or leaves does not result in the presence of measurable medium-chain fatty acids, suggesting that they are specifically channeled towards peroxisomal β-oxidation in these vegetative tissues (Eccleston et al., 1996). These results indicated that expression of a thioesterase might be a way of increasing the carbon flux towards β-oxidation and peroxisomal PHA biosynthesis. This hypothesis was tested by combining the constitutive expression of the peroxisomal PHA synthase with the caproyl-ACP thioesterase from *Cuphea lanceolata* (Martini et al., 1999) in the plastid of *A. thaliana*. Expression of both enzymes lead to a 7- to 8-fold increase in the amount of MCL-PHA synthesized in adult plants as compared to plants expressing only the PHA synthase (Mittendorf et al., 1999). Furthermore, the composition of the MCL-PHA in the thioesterase/PHA synthase double-transgenic plant was shifted towards saturated 3-hydroxyacid monomers containing 10 carbons and less, in agreement with an increase in the flux of decanoic acid towards β-oxidation triggered by the expression of the caproyl-ACP thioesterase (Mittendorf et al.,

1999). From these results, a working hypothesis has been developed where enzymes and genes involved in the synthesis of unusual fatty acids in plants can be used to modulate the quantity and quality of substrates channeled towards MCL-PHA.

Synthesis of MCL-PHA has also been demonstrated in seeds of *A. thaliana* by putting the PHA synthase gene modified for peroxisomal targeting under the control of the seed-specific napin promoter. In such transgenic plants MCL-PHAs accumulated to 0.006% dwt in mature seeds and the monomer composition was relatively similar to the PHA synthesized in germinating seedlings. However, combination of the PHA synthase with the seed-specific expression of the caproyl-ACP thioesterase from *C. lanceolata* in the plastid resulted in a nearly 20-fold increase in seed PHA, reaching 0.1% dwt in mature seeds. Furthermore, as found with the expression of PHA synthase and caproyl-ACP thioesterase in whole plants, expression of these two enzymes in seeds resulted in a large increase in the proportion of 3-hydroxyacid monomers containing 10 carbons and less in the PHA. These data clearly indicate that even though expression of the caproyl-ACP thioesterase in seeds leads to the accumulation of medium-chain fatty acids in triacylglycerides, there is still a significant proportion of these fatty acids that are channeled towards β-oxidation. These results are very significant, considering that there is only a 4-fold difference between the maximal amount of PHA synthesized in germinating seedlings (0.4% dwt), where β-oxidation is thought to be maximal, and the PHA synthesized in the developing seeds expressing the thioesterase (0.1% dwt), where metabolism should be mainly devoted to the synthesis of fatty acid synthesis.

The absolute amount of MCL-PHA accumulating in leaves or seeds of transgenic plants expressing the PHA synthase, with or without the thioesterase, remains relatively low (0.6% dwt or less) compared to Poly(3HB) synthesized in the plastids (10–40% dwt). The reasons for this difference could be multiple, including different activities of bacterial enzymes in various plant subcellular compartments (peroxisome versus plastids), and the relative ability of the created and endogenous metabolic pathways to compete for the same substrates (acetyl-CoA for Poly(3HB) and 3-hydroxyacyl-CoAs for MCL-PHA). It is thus expected that further metabolic engineering will be necessary in order to increase the amount of MCL-PHA synthesized in plants to commercially viable amounts.

The discovery of the *P. putida phaG* gene encoding a *R*-3-hydroxyacyl-ACP-CoA transacylase opened the possibility of synthesizing MCL-PHAs from intermediates of fatty acid biosynthesis in the plastid of plants. Constitutive expression in *A. thaliana* of only *phaG* modified for plastid targeting leads to a marked deleterious effect on plant growth, the plants being dwarf with crinkly leaves, and the seed set being strongly reduced (Mittendorf, unpublished results). The reasons for this phenotype are not known, but are thought to be related to modification in fatty acid biosynthesis. Co-expression of the *P. aeruginosa* PHA synthase along with phaG protein in the plastid did not conclusively lead to MCL-PHA accumulation (V. Mittendorf, unpublished results). Thus, despite the advantages of the plastid as a location for Poly(3HB) and Poly(3HB-*co*-3HV) synthesis, the synthesis of MCL-PHAs in this organelle using fatty acid biosynthetic intermediates appears problematic at present.

4 Novel Uses for PHA in Plants

Although the primary interest in the production of PHAs has been as a source of biodegradable plastics and elastomers, synthesis of PHAs in plants has opened novel avenues for the use of these polymers in both plant biotechnology and basic research.

4.1 PHA to Modify Fiber Properties

An inventive and novel perspective on the use of PHA in plants was achieved by the expression of the Poly(3HB) biosynthetic pathway in the cytoplasm of cotton fiber cells (John and Keller, 1996; John, 1997). In this system, PHA is produced not as a source of polyester to be extracted from the plant and used in the plastic industries, but is rather used as an intracellular agent that modifies the physical properties of the fiber. The genes from *R. eutropha* encoding the 3-ketothiolase, acetoacetyl-CoA reductase and Poly(3HB) synthase were expressed in transgenic cotton under the control of a fiber-specific promoter (John and Keller, 1996; John, 1997). Poly(3HB) accumulated in the cytoplasm to 0.3% dwt of the mature fiber. The polymer was in the molecular mass range of 0.6×10^6 to 1.8×10^6 and thus similar to bacterial Poly(3HB). Interestingly, even if the amount of Poly(3HB) accumulated was relatively low, significant changes in the thermal properties of the fiber were measured. The transgenic fibers conducted less heat, cooled down more slowly, and took up more heat than conventional fibers, thus improving the overall property of the fiber. In an effort to increase the amount of Poly(3HB) in the fibers, the Poly(3HB) pathway was also expressed in the plastids of the fiber cells. However, in contrast to similar studies done in *A. thaliana* and other plants, expression of the Poly(3HB) in plastids of fiber cells resulted in a lower level of Poly(3HB) as compared to cytoplasmic expression (John, 1997). These results suggest a difference either in the number and/or the metabolic activity of the plastids in cotton fiber cells as compared to the chloroplast of leaves or leukoplasts of seeds. Nevertheless, these experiments are interesting as they opened the range of uses of PHA in plants. In this light, it is tempting to speculate whether the properties of wood, rubber, or starch could also be positively modified through the coaccumulation of PHAs.

4.2 PHA as a Feed Supplement

The biodegradation of PHA by a wide variety of bacteria and fungi present in our environment, as well as the limited hydrolysis of PHA under acidic conditions, raised the possibility that PHA could be digested in the gut of animals and that the released monomers or oligomers could be used as an energy source. The concept of using PHA as a supplement in animal feed has recently been examined. In a study reported by Peoples et al. (1999), broiler chicks were fed with meals supplemented with 3% (w/w) purified bacterial Poly(3HB) or poly(3-hydroxyoctanoate) (Poly(3HO)) granules, as well as soy oil. Addition of soy oil to the meal lead to an increase in the energy content of the meal of 4.5% compared to an increase of 3.16% for Poly(3HB) and 2.95% for Poly(3HO). The authors concluded that the available energy released from the bacterial PHA in broiler chick was in the range between carbohydrates and oil. It was thus suggested that plants accumulating PHA, either in seeds or leaves, could be used improve the energy content of the meal. In this context, the PHA present in the meal could either represent the full amount of polymer accumulated in the tissues or the residues left after extraction of the polymer for its use as a bioplastic.

Similar feeding studies done in other monogastric animals or ruminants has, however, not confirmed the digestibility of PHA in the digestive tract. In two studies done in sheep and pigs, bacterial Poly(3HB-*co*-3HV) added to the feed was found to go through the animal digestive system largely undegraded and did not positively or negatively affect the energy content of the meal (Forni et al., 1999a,b,c). Pre-treatment of the Poly(3HB-*co*-3HV) with sodium hydroxide did lead to an increase in digestibility of the polymer (Forni et al., 1999b,c). However, digestibility was directly correlated with the fraction of monomers released from the polymer by the pretreatment, indicating that only hydroxyacid monomers found in the meal are absorbed by the animals and used as an energy source, while the remaining polymer remained largely undigested. The reasons for the discrepancies between these studies done in sheep and pig and the one done in broiler chicks are unclear at present. Clearly, further studies are needed in order to measure the value of PHA present in plant material as a feed supplement.

4.3
PHA as a Tool to Study Plant Metabolic Pathways

Synthesis of PHAs in plants cannot only be used directly in biotechnology to produce bioplastics, but can also be utilized as a unique novel tool in the basic studies of plant biochemistry. This is because PHA synthesized in plants acts as a terminal carbon sink, since plants do not appear to have enzymes, such as PHA depolymerases (Jendrossek et al., 1996), required for degradation of the polymer. The quantity and composition of PHA can thus be used to monitor the quantity and quality of the carbon flux to different pathways that can influence PHA biosynthesis.

Synthesis of PHA in various subcellular compartments could be used to study how plants adjust gene expression and metabolic pathways to accommodate the production of a new sink, and how carbon flux through one pathway can affect carbon flux through another. For example, one could study how modification of the carbon flux to starch or lipid in the plastid may affect the flux of carbon to acetyl-CoA and Poly(3HB). In the context of Poly(3HB) synthesis in the cytoplasm, one could study how plants adjust the activity of genes and proteins involved in isoprenoid and flavonoid biosynthesis, since these three pathways compete for the same building block, i.e. acetyl-CoA. In addition, the results previously described from the Monsanto group regarding the production of Poly(3HB) in *Arabidopsis* and corn demonstrated how Poly(3HB), through visible inclusions in the plastid, might be an indicator that can qualitatively differentiate acetyl-CoA levels or availability in different cell types.

Very powerful tools have been developed in genomics, and in particular the DNA microarray technology, that would allow a global view of changes in gene expression resulting from PHA synthesis in plants. Likewise, the utilization of powerful biochemical analytical tools, such as liquid chromatography/MS, will advance our understanding of changes resulting from PHA synthesis at the level of metabolite concentrations, as shown for Poly(3HB) in the plastids of *A. thaliana* (Bohmert et al., 2000). Such studies would help determine the potential bottlenecks limiting PHA accumulation as well as identify potential solutions to the growth defect observed in some transgenic plants synthesizing PHAs.

Synthesis of MCL-PHAs in peroxisomes has also brought novel insights into the field of fatty acid degradation. For example, as discussed in Section 3.4, the increase in

MCL-PHA synthesized in transgenic plants expressing the caproyl-ACP thioesterase clearly showed the presence of a futile cycling of medium-chain fatty acids towards β-oxidation in leaves as well as in developing seeds (Mittendorf et al., 1999; Poirier et al., 1999). Synthesis of MCL-PHA in the peroxisomes of developing seeds has also demonstrated the presence of a large increase in the flux of fatty acids towards β-oxidation in plants having a deficiency in the enzyme diacylglycerol acyltransferase (DAGAT) (Poirier et al., 1999). The *tag1* mutant of *A. thaliana* was shown by Katavic et al. (1995) to be deficient in DAGAT activity in developing seeds, resulting in a decrease accumulation of triacylglycerides, and corresponding increase in diacylglycerides and free fatty acids in mature seeds. It was hypothesized that the imbalance created between that capacity of the plastid to synthesize fatty acids and the capacity of the lipid biosynthetic machinery of the endoplasmic reticulum to include these fatty acids into triacylglycerides might have two basic consequences. These would be that either fatty acid biosynthesis would be reduced (feedback inhibited) in order to match it with triacylglyceride biosynthesis, or that excess fatty acids which cannot be included in triacylglycerides would be channeled towards β-oxidation. Expression of the peroxisomal PHA synthase in the *tag1* mutant resulted in a 10-fold increase in the amount of MCL-PHA accumulating in mature seeds compared to expression of the transgene in wild-type plants. Although these results cannot address whether fatty acid biosynthesis is decreased in the *tag1* mutant, they nevertheless clearly indicate that a decrease in triacylglyceride biosynthesis results in a large increase in the flux of fatty acids towards β-oxidation. Thus, carbon flux to the β-oxidation cycle can be modulated to a great extent and appears to play an important role in lipid homeostasis in plants even in tissues which are primarily devoted to lipid biosynthesis, such as the developing seeds.

As discussed in Section 3.4, changes detected in the monomer composition of peroxisomal MCL-PHAs in plants that are either fed externally with various fatty acids, deficient in the synthesis of particular unsaturated fatty acid, or expressing a medium-chain acyl-ACP thioesterase, clearly reflects the nature of the fatty acid being degraded and how this fatty acid is degraded by the β-oxidation pathway (Mittendorf et al., 1999). It is, thus, possible to use peroxisomal PHA to elucidate the pathways involved in the degradation of unsaturated and unusual fatty acids. Such studies could have an important impact in our ability to create transgenic plants accumulating novel valuable fatty acids, such as ricinoleic acid and vernolic acid (van de Loo et al., 1993). This is because it is expected in some cases that an oilseed crop accumulating a novel fatty acid may show poor germination if the specialized enzymes required to handle the presence of novel groups in the fatty acids, such as epoxy or hydroxy groups, are missing in the transgenic plants (Gerhardt, 1992). Thus, it may important to know what are the enzymes and genes required for the β-oxidation of unusual fatty acids. Furthermore, as revealed by the work with the caproyl-ACP thioesterase, the quantity of peroxisomal MCL-PHA can be used as a tool to study the extent of a loss of unusual fatty acid in seeds through futile cycling.

The usefulness of using peroxisomal MCL-PHAs to study the pathway of fatty acid degradation in plants has recently been demonstrated for the β-oxidation of unsaturated fatty acids (Allenbach and Poirier, 2000). Degradation of fatty acids having *cis* double bonds on even-numbered carbons requires the presence of auxiliary enzymes in addition to the enzymes of the core β-oxidation cycle. This is because hydration of *cis*-2-enoyl-CoA by the 2-enoyl-CoA hydratase I generates the *R* isomer of 3-hydroxyacyl-CoA which is not a substrate for the S-3-hydroxyacyl-CoA dehydrogenase. Two alternative pathways have been described to degrade these fatty acids. One pathway involves the participation of the enzymes 2,4-dienoyl-CoA reductase and Δ^3-Δ^2-enoyl-CoA isomerase, while the second involves the epimerization of *R*-3-hydroxyacyl-CoA via a 3-hydroxyacyl-CoA epimerase or the action of two stereo-specific enoyl-CoA hydratases. Whereas degradation of these fatty acids in bacteria and mammalian peroxisomes was shown to involve mainly the reductase–isomerase pathway (Yang et al., 1986), previous analysis of the relative activity of the enoyl-CoA hydratase II and 2,4-dienoyl-CoA reductase in plants indicated that degradation occurred mainly, if not exclusively, through the epimerase pathway (Engeland and Kindl, 1991). The relative contribution of the isomerase and reductase pathway could be examined in transgenic plants synthesizing peroxisomal PHA since the degradation of *cis*-10-heptadecenoic or *cis*-10-pentadecenoic acids via the epimerase or reductase–isomerase pathways results in the introduction of some distinctive 3-hydroxyacid monomers in PHA. Analysis of the PHA produced from transgenic plants fed with these different fatty acids revealed that a significant proportion of fatty acid were degraded via the reductase–isomerase pathway in addition of the epimerase pathway (Allenbach and Poirier, 2000).

5 Patents on PHA in plants

In view of the potentially important role that production of PHA in agricultural crops may play in the large-scale commercialization of PHA, it is not surprising to find several patents relating to the synthesis of various PHAs in plants. One of the first patent application to be specifically directed towards synthesis of PHA in plants was filed in 1993 by Zeneca (WO 94/11519) (Table 3). This patent was largely prophetic, since it described the principle of the transformation of plants with the genes involved in PHA synthesis without providing any experimental data to support the claims. The first proof-of-concept for PHA synthesis in plants was provided in 1992 by the transformation of the plant *A. thaliana* with the *R. eutropha* genes involved in Poly(3HB) synthesis (Poirier et al., 1992). Since then, it has become common to include claims dealing with PHA in plants in patents based on such diverse aspects of PHA production as the cloning of bacterial genes involved in the synthesis of PHA, the creation of novel metabolic pathways for PHA production in bacteria, or the isolation of PHA. There are also several patents focussing on various aspects of plant transformation and transgene expression that mention the synthesis of PHA in plants. Thus, over 50 patents can be found in the US patent database alone since 1992 that have some links with PHA in plants. However, for the majority of them, direct demonstration of PHA synthesis in plants is not given.

Table 3 gives a summary of the major published or issued patents which either bring experimental proof of PHA production in plants and/or have synthesis of PHA in plants as the primary focus.

6 Perspective on PHA Production in Agricultural Crops

A major barrier to the wide-spread use of biodegradable plastics in consumer products

Tab. 3 Major patents on PHA synthesis in plants

Patent no.	*Holder*	*Inventor*	*Title*	*Comments*[a]	*Date of publication*
WO94/11519	Zeneca	P. A. Fentem	Production of polyhydroxyalkanoate in plants	Description of the concept of PHA synthesis in plants PHA–	26-5-94
US5502273; US6175061	Zeneca	P. A. Fentem	Production of polyhydroxyalkanoate in plants	Expression of *R. eutropha phaA* and *phbB* genes in plants PHA–	26-3-96; 16–1-01
US5650555	Michigan State University	C. R. Somerville, Y. Poirier, D. E. Dennis	Transgenic plants producing polyhydroxyalkanoates	First demonstration of PHA synthesis in plants PHA+	22-7-97
US5610041	Michigan State University	C. R. Somerville, C. Nawrath, Y. Poirier	Processes for producing polyhydroxybutyrate and related polyhydroxyalkanoates in the plastids of higher plants	Improvement of PHA synthesis by targeting the pathway to the plastid PHA+	11-3-97
US5602321	Monsanto	M. John	Transgenic cotton plants producing heterologous polyhydroxybutyrate bioplastic	Synthesis of PHB in cotton fibers PHA+	11-2-97
WO98/00557; US5942660	Monsanto	K. J. Gruys, T. A. Mitsky, G. M. Kishore, S. C. Slater, S. R. Padgette, D. M. Stark, M. A. Hinchee, T. E. Clemente, D. V. Connor-Ward, M. J. Fedele	Methods of optimizing substrate pools and biosynthesis of poly-β-hydroxybutyrate-co-poly- β-hydroxyvalerate in bacteria and plants	Demonstration of Poly(3HB-*co*-3HV) synthesis in plants PHA+	8-1-98 24–8-99
WO99/45122	Metabolix	O. P. Peoples, L. Boynton, G. W. Huiisman, M. Moloney, N. Patterson, K. Snell	Modification of fatty acid metabolism in plants	Expression of bacterial genes involved in β-oxidation PHA–	10-9-99
WO99/35278	Monsanto	Y. Poirier, V. Mittendorf	Biosynthesis of medium-chain-length polyhydroxyalkanoates	Demonstration of MCL-PHA in plant peroxisome PHA+	15-7-99
WO00/52183	Monsanto	T. A. Mitsky, S. C. Slater, S. E. Reiser, M. Hao, K. L. Hourniel,	Multigene expression vectors for the biosynthesis of products via multienzyme biological pathways	Improvement of PHA synthesis in plants through the creation of multigene vectors PHA+	8-9-00
US6103956	University of Minnesota	F. Srienc, D. A. Somers, J. J. Hahn, A. C. Eschenlauer	Polyhydroxyalkanoate synthesis in plants	Demonstration of Poly(3HB) in plant peroxisome PHA+	15-8-00

[a] PHA+ denotes patents where synthesis of PHA in plants was shown experimentally, while PHA– denotes patents where no demonstration of PHA is reported.

is their higher production cost compared to petroleum-derived plastics. For example, whereas the cost of polypropylene is well below 1 US$/kg, the costs of some of the cheapest biodegradable plastics are 3–6 US$/kg. Several reviews have discussed the economics of PHA synthesis by bacterial fermentation (de Koning et al., 1997a,b; Lee, 1996; Lee et al., 1997; Page, 1997). The most optimistic production cost for bacterial Poly(3HB) is approximately 4 US$/kg, considering a production scale of one million tons per year. The cost could be lower at higher production scale, but is not expected to reach below 2–3 US$/kg. One key component of the cost of bacterial PHA is the cost of the feedstock. Considering that for the synthesis of Poly(3HB) from *R. eutropha* 3.3 g of glucose is used to synthesize 1 g of polymer and that the price of glucose is around 0.5 US$/kg, the cost of carbon is at least 1.65 US$/kg of Poly(3HB). Although other cheaper carbon sources have been suggested, such as methanol or cane molasses (Page, 1997), the PHA production yield from bacteria using these carbon sources is typically lower, thus again affecting production costs.

The main advantage of plants is that the carbon feedstock used to make PHA would be very cheap, being essentially derived from CO_2 and sunlight through the process of photosynthesis. The main question is whether this fact alone is sufficient to make the production of PHA in plants as cheap as the production of vegetable oils (0.5–1 US$/kg) or starch (0.25 US$/kg).

The final cost of producing PHA in plants will depend on a number of factors. One important factor will be whether PHA can be produced in high quantity in plants without affecting the overall yield of other plant products, such as oils. In contrast to the production of PHA by bacterial fermentation, where the system is designed to produce only PHA, agricultural production of PHA is likely to be only viable through the recovery of not only PHA, but also all other useful components of the crop. For example, in the case of an oil crop such as *B. napus*, one must be able to recover PHA and the oil, as well as still being able to use the delipidized protein-rich meal for animal feed. In the case of a carbohydrate-producing crop such as sugar beet or sugar cane, both the sucrose and PHA would have to be recovered. It is thus important that synthesis of PHA in plants is achieved without decreasing too significantly the yield in oil, proteins, or carbohydrates. We know thus far that Poly(3HB) can be produced in the seed of rape to 8% dwt without obvious deleterious effects on plant growth (Houmiel et al., 1999). It remains, however, to be determined to what extend PHA can accumulate in seeds without affecting too significantly the production of oils and proteins. Synthesis of PHA in the plastids of leaves appears to be more limiting since production levels higher then 3–4% dwt leads to chlorosis (Nawrath et al., 1994a) and growth reduction (Bohmert et al., 2000).

A second important factor in the production cost of PHA will be the extraction process. The extraction of PHA from bacteria and plants has recently been reviewed (Poirier, 2001). Although a number of strategies have been described in the literature, there is yet little actual experience in validating these extraction processes for the large-scale production of PHA from plants. The process must not only ensure that extraction of PHA is efficient, economic, and ecological, but also that the quality of the other plant products, such as oil, carbohydrates, and protein, is not negatively affected by the extraction of PHA. It is likely that the extraction costs will be higher for plants than for bacteria, if it is only for the fact that PHA synthesis in plants will be lower than the 80–85% dwt achievable by bacterial fermentation.

All of these factors mean that production of PHA in plants will likely be more expensive than starch. However, considering that starch costs about 0.25 US$/kg, even tripling the production cost of PHA compared to starch would make PHA in plants cheaper than PHA obtained from bacterial fermentation and most likely the cheapest biodegradable plastic made from renewable resources.

An interesting aspect beyond the economics of PHA production is the relative energy and CO_2 costs associated with the synthesis of PHA by bacterial fermentation or in plants compared to the synthesis of petroleum-derived plastics, such as polyethylene (PE). An evaluation performed by Gerngross (1999) has indicated that the energy and CO_2 cost associated with the synthesis of PHA in bacteria is significantly higher than PE production. Similarly, Coulon et al. (2000) and Gerngross and Slater (2000) have evaluated the greenhouse gas profile of PHA made in corn stover compared to PE production. It was concluded that synthesis of PHA in plants would only give a better greenhouse gas profile than PE if the plant biomass was also used as an energy source for the processing of the polymer. In the comparison of PE versus biological PHA, one must remember that PE is derived from petroleum, which is a finite resource that is expected to become more limiting and expensive in the future. It is, thus, likely that the renewability of biological PHA will become an increasingly important aspect as petroleum reserves do become limiting.

Davies, H. M. (1992) Fatty acid biosynthesis redirected to medium chains in transgenic oilseed plants, *Science* **257**, 72–74.
Yang, S. Y., Cuebas, D., Schultz, H. (1986) 3-Hydrpoxyacyl-CoA epimerase of rat liver peroxisomes and *Escherichia coli* function as auxiliary enzymes in the β-oxidation of polyunsaturated fatty acids, *J. Biol. Chem.* **261**, 12238–12243.
Wieczorek, R., Pries, A., Steinbüchel, A., Mayer, F. (1995) Analysis of a 24-kilodalton protein associated with the polyhydroxyalkanoic acid granules in *Alcaligenes eutrophus*, *J. Bacteriol.* **177**, 2425–2435.

13
Fermentative Production of Building Blocks for Chemical Synthesis of Polyesters

Dr. Sang Yup Lee[1], Dr. Sang Hyun Park[2,3], M. Eng. Soon Ho Hong[4], M. Eng. Young Lee[5,6], M. Eng. Seung Hwan Lee[7]

[1] Metabolic and Biomolecular Engineering National Research Laboratory, Department of Chemical Engineering and BioProcess Engineering Research Center, Korea Advanced Institute of Science and Technology, 373-1 Kusong-dong, Yusong-gu, Taejon 305-701, Korea; Tel.: +82-42-869-3930; Fax: +82-42-869-3910; E-mail: "leesy@mail.kaist.ac.kr

[2] Metabolic and Biomolecular Engineering National Research Laboratory, Department of Chemical Engineering and BioProcess Engineering Research Center, Korea Advanced Institute of Science and Technology, 373-1 Kusong-dong, Yusong-gu, Taejon 305-701, Korea; Tel.: +82-42-869-3906; Fax: +82-42-869-3910; E-mail: parksh@mail.kaist.ac.kr"

[3] ChiroBio Inc., #2324 Undergraduate Building 2, KAIST, 373-1 Kusong-dong, Yusong-gu, Taejon 305-701, Korea; Tel. +82-42-863-8609; Fax +82-42-869-8800; E-mail: parksh@mail.kaist.ac.kr

[4] Metabolic and Biomolecular Engineering National Research Laboratory, Department of Chemical Engineering and BioProcess Engineering Research Center, Korea Advanced Institute of Science and Technology, 373-1 Kusong-dong, Yusong-gu, Taejon 305-701, Korea; Tel.: +82-42-869-3970; Fax: +82-42-869-3910; E-mail: soonho@kaist.ac.kr

[5] Metabolic and biomolecular Engineering National Research Laboratory, Department of chemical Engineering and BioProcess Engineering Research Center, Korea Advanced Institute of Science and Technology, 373-1 Kusong-dong, Yusong-gu, Taejon 305-701, Korea; Tel: +82-42-869-3970; Fax: +82-42-869-3910; E-mail: s_ylee@kaist.ac.kr

[6] ChiroBio Inc., #2324 Undergraduate Building 2, KAIST, 373-1 Kusong-dong, Yusong-gu, Taejon 305-701, Korea; Tel: +82-42-869-5990; Fax: +82-42-869-8800; E-mail: chirobio@mail.kaist.ac.kr

[7] Metabolic and Biomolecular Engineering National Research Laboratory, Department of Chemical Engineering and bioProcess Engineering Research Center, Korea Advanced Institute of Science and Technology, 373-1 Kusong-dong, Yusong-gu, Taejon 305-701, Korea; Tel.: +82-42-869-3970; Fax: +82-42-869-3910; E-mail: sively@kaist.ac.kr

Biotechnology of Biopolymers. From Synthesis to Patents. Edited by A. Steinbüchel and Y. Doi

ISBN: 3-527-31110-6

ADP adenosine diphosphate
ATP adenosine triphosphate
EDTA ethylenediaminetetra-acetic acid
$FADH_2$ reduced form of flavin adenine dinucleotide
GC gas chromatography
MCL medium-chain-length
MFA metabolic flux analysis
NAD nicotinamide adenine dinucleotide
NADH reduced form of nicotinamide adenine dinucleotide
NADP nicotinamide adenine dinucleotide phosphate
NADPH reduced form of nicotinamide adenine dinucleotide phosphate
PBT poly(butylene terephthalate)

PEP	phosphoenolpyruvate
PET	poly(ethylene terephthalate)
PHAs	polyhydroxyalkanoates
Poly(3HB)	poly[(*R*)-3-hydroxybutyric acid]
Ppc	phosphoenolpyruvate carboxylase
PPT	poly(propylene terephthalate)
Pta	phosphotransacetylase
PTS	phosphotransferase system
SCL	short-chain-length
TCA	tricarboxylic acid

1 Introduction

Polymers are used in our everyday life for various purposes and in a wide variety of forms, including packaging materials, films, lubricants, and coating agents. The reason for diversity is that polymers possess excellent properties of economy, lightness, durability, non-degradability, and non-corrosivity to acids and alkalis. Polymers also have certain disadvantages however, and in many countries plastic wastes have become a serious environmental problem because they cannot be degraded after use. In an attempt to solve this problem, extensive research has led to the development of several different waste treatment technologies. Indeed, in the near future the development of biodegradable plastics in order to resolve the problems associated with such waste will surely lead to a host of highly innovative materials.

Many biodegradable polymers have been identified, including starch and cellulose (and their derivatives), polyhydroxyalkanoates, poly(vinyl alcohol), polycaprolactone, and poly(lactic acid) (Mayer and Kaplan, 1994). Amylose, cellulose acetate, and amylopectin are polysaccharide-based biodegradable polymers which can be produced from corn, potato, wheat, rice, and wood pulp. Poly(vinyl alcohol) is prepared chemically by the hydrolysis of poly(vinyl acetate), and is used as a wrapping film not only for agricultural chemicals but also for food products (especially where direct food-wrapper contact is unavoidable). Polyhydroxyalkanoates (PHAs), polycaprolactone and poly(lactic acid) all belong to this group of biodegradable polyesters. Polycaprolactone, a synthetic polyester of ε-caprolactone, has been used for orthopedic casts, adhesive mold release agents and pigment dispersants. Poly(lactic acid) is produced commercially in two forms: L-(-)-lactide-based poly(lactic acid); and DL-lactide-based poly(lactic acid), each having significantly different properties. Poly(lactic acid) has mostly been used for medical applications such as sutures, drug delivery, vascular grafts, and artificial skin. By contrast, PHAs are energy and/or reducing power storage materials that are accumulated intracellularly by many bacteria under conditions of either nitrogen, phosphorus, sulfur, oxygen or magnesium limitations and in the presence of excess carbon source (Anderson and Dawes, 1990; Doi, 1990; Lee, 1996). Since the discovery of poly (3-hydroxybutyric acid), poly(3HB), produced by the bacterium *Bacillus megaterium* (Lemoigne, 1927), PHAs with monomers of different numbers of main-chain carbon atoms and different types of pendent groups have been reported (Steinbüchel and Valentin, 1995). These bacterial polyesters have been divided tentatively into two groups on the basis of their structures: short-chain-length PHAs (SCL-PHAs), which consist of three to five carbon atoms; and medium-chain-length PHAs (MCL-PHAs), which consist of 6 to 14 carbon atoms (Lee and Chang, 1995; Steinbüchel and Valentin, 1995). Recently, some bacteria were shown to accumulate SCL-*co*-MCL-PHA copolymers.

In addition to these polymers, many others consisting of oxalic, succinic, malic, mandelic, fumaric, ketomalonic, ketoglutaric, diglycolic, itaconic, allylmalonic, adipic, and phthalic acids, butanediol, β-butyrolactone, γ-valerolactone, and higher lactones have been described (Pitt, 1992). Among these polymers, extensive investigations have revealed the presence of several monomers, including succinic acid, propanediol, and lactic acid.

In this chapter, the strategies employed for the fermentative production and/or simple chemical transformation after fermentation of various building blocks including succinic acid, adipic acid, 1,2-propanediol, 1,3-propanediol, 1,4-butanediol, lactic acid, and lactones are reviewed. The use of these building blocks in the synthesis of polymers is reviewed elsewhere in this book, and so will not be described here.

2 Historical Outline

A brief historical outline of the development of polyesters is given in the following table:

1881 Detection of 1,3-propanediol in the glycerol-fermenting mixed culture
1940 solation of 1,3-propanediol-producing *Aerobacter*
1976 Production of (*S*)-1,2-propanediol by *Escherichia coli*
1976 Isolation of *Anaerobiospirillum succiniciproducens*
1983 Biological synthesis of adipic acid
1985 Isolation of *Pseudomonas* producing 1,4-butanediol
1993 Biological synthesis of chiral 1,2-propanediol by resolution
1996 Production of adipic acid by recombinant system
1997 Production of optically pure 1,2-propanediol by recombinant system
1997 Metabolic flux analysis for propanediol producing *E. coli*
1997 Cloning of the PEP carboxykinase gene of *Anaerobiospirillum succiniciproducens*
1997 Construction of *E. coli* strain NZN111
1998 Isolation of *E. coli* mutant strain AFP111, a spontaneous *ptsG* gene mutant of NZN111
1999 Homofermentative production of lactic acid by *E. coli*
1999 Isolation of *Actinobacillus succinogenes*
2000 Metabolic flux analysis of *E. coli* NZN111

3 Dicarboxylic Acids

The dicarboxylic acid family can be used as a monomer in the production of various polyesters. Two such compounds – succinic acid and adipic acid – can be produced by biological methods. The various strategies for their production will be reviewed in this section.

3.1 Succinic Acid

Succinic acid is a member of the C_4-dicarboxylic acid family, and is manufactured by hydrogenation of maleic anhydride to succinic anhydride, followed by hydration to succinic acid. Currently, more than 15,000 tons of succinic acid is manufactured yearly, and is sold at US$ 5.90–8.80/kg, depending on its purity. Succinic acid has been used in many industrial applications (Figure 1), and has been used as a surfactant, a detergent extender, a foaming agent, and an ion chelator. It has also been used as a food additive, a flavoring agent, and a supplement to pharmaceuticals, antibiotics, and vitamins. At present, succinic acid produced by microbial fermentation is used only in the food and pharmaceutical industries. Recently, a major effort has been made to produce succinic acid and its derivatives by microbial fermentation, using renewable feedstocks (Zeikus, 1980), in time it is expected that fermentatively produced succinic acid will replace much of the commodity chemicals currently produced from petrochemicals.

Succinic acid can also be used as a precursor of several important chemicals, for example 1,4-butanediol, tetrahydrofuran, γ-butyrolactone, and other four-carbon compounds. Important chemicals that can be produced from succinic acid include *N*-methylpyrrolidone, adipic acid, and linear aliphatic esters (Sado and Tajima, 1980; Dake et al., 1987; Jain et al., 1989). *N*-methylpyrrolidone has been recommended as a replacement for the common solvent methylene chloride, mainly because it is less volatile than methylene chloride and so can be recycled without (toxic) loss to the atmosphere. 1,4-Butanediol is an important building block which can be used as a precursor of polybutylene terephthalate resin, while adipic acid is the precursor of nylon 6,6 and can also be used as a raw material for lubricants, foams, and food products. Tetra-

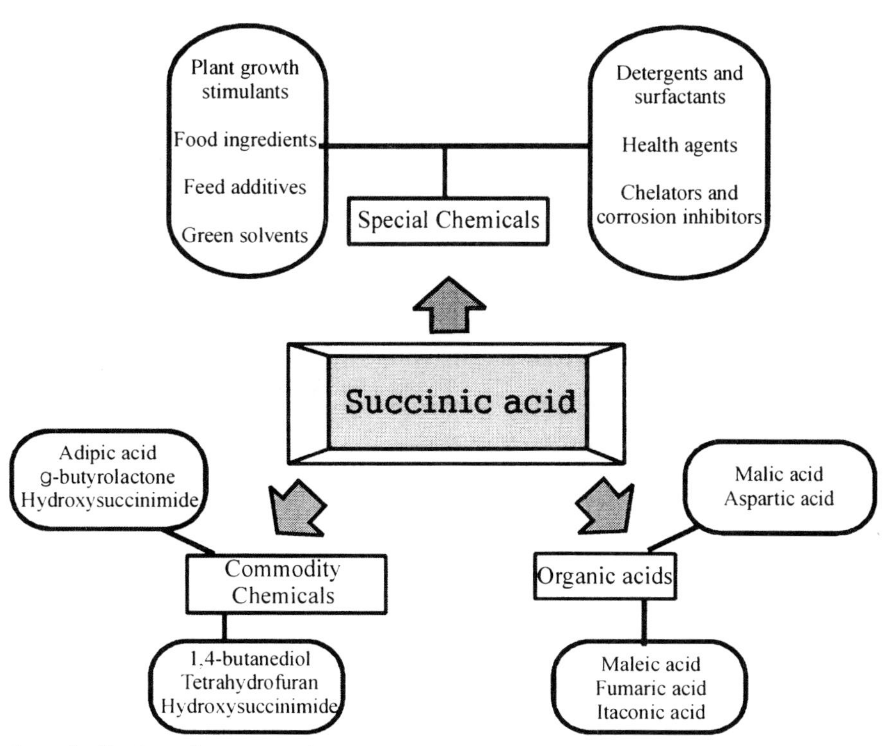

Fig. 1 Applications of succinic acid.

hydrofuran has widespread use as a solvent, and as an ingredient of adhesive, printing and magnetic tapes.

In addition to these many applications, new uses for succinic acid are undergoing rapid development. Diethyl succinate can be used to clean metal surfaces, while ethylenediaminedisuccinate can be used as a substitute for ethylenediaminetetra-acetate (EDTA) (Zwicker et al., 1997). Sodium succinate can replace monosodium glutamate as a flavor enhancer (Turk, 1993), while succinic acid can act as an intermediate in several environmentally friendly ("green") chemicals and materials; for example, the ester of succinic acid and 1,4-butanediol is used to make the biodegradable polymer, Bionelle® (Showa Highpolymer Co., Ltd., Tokyo, Japan). Succinic acid can also be used for the succinylation of lysine residues, the production of modified succinimides that can be used as a fuel constituent (Wollenberg and Frank, 1988), the succinylation of cellulose to improve water absorbtivity, and the succinylation of starch to create thickening agents (Diamantoglon and Meyer, 1988).

3.1.1 Production of Succinic Acid by Non-Recombinant Microorganisms

Succinic acid is a common metabolite of several anaerobic and facultative microorganisms. The best-known succinic acid-producing microorganisms are *Anaerobiospirillum succiniciproducens*, propionate-producing bacteria such as *Propionibacterium* species, gastrointestinal bacteria such as *Escherichia coli*, *Pectinatus* sp., *Bacteroides* sp., rumen bacteria such as *Ruminococcus flavefaciens*, *Actinobacillus succinogenes*, *Bacteroides amylophilus*, *Prevotella ruminicola*, *Succcinimonas amylolytica*, *Succinivibrio dextrinisolvens*, *Wolinella succinogenes*, and *Cytophaga succinicans* (Bryant and Small, 1956; Bryant et al., 1958; Scheifinger and Wolin, 1973; Davis et al., 1976; Van der Werf et al., 1997; Guettler et al., 1999).

A. succiniciproducens and *A. succinogenes* have been shown to be the most efficient succinic acid-producing strains, with both bacteria producing succinic acid via four reaction steps catalyzed respectively by phosphoenolpyruvate (PEP) carboxykinase, malate dehydrogenase, fumarase, and fumarate dehydrogenase (Van der Werf et al., 1997). It was reported that the PEP carboxykinase pathway, a common succinic acid production pathway in *A. succiniciproducens* and *A. succinogenes*, is regulated by CO_2 concentration (Samuelov et al., 1991; Van der Werf et al., 1997; Lee et al., 1999a). In these strains, each mol of PEP, ADP and CO_2 is converted to 1 mol of oxaloacetate and ATP by PEP carboxykinase. At low CO_2 concentration (10 mol CO_2/100 mol glucose), lactic acid and ethanol are produced as major products by both *A. succiniciproducens* and *A. succinogenes*, but if sufficient CO_2 is supplied (100 mol CO_2/100 mol glucose) then both strains produce succinic acid as a major end product, with only small amounts of lactic acid and/or ethanol. CO_2 was shown to regulate the expression levels of key enzymes of anaerobic pathways in *A. succinogenes* (Van der Werf et al., 1997). At high CO_2 concentrations, the PEP carboxykinase level is elevated, while expression of the lactate dehydrogenase and alcohol dehydrogenase genes is not observed. Thus, it can be concluded that anaerobic dissimilation of PEP is regulated by CO_2 concentration.

Continuous culture of *A. succiniciproducens* has been carried out over a wide range of dilution rates (0.056 to 0.636 h^{-1}) to obtain valuable steady-state fermentation parameters (Lee et al., 2000b). During continuous culture in a medium containing 19 and 38 g L^{-1} glucose, the experimental succinic acid yields ($Y_{SA/S}$) were maintained at between 0.83 and 0.88 g g^{-1} over the entire range of dilution rates. The volumetric productivity of succinic acid increased with increasing dilution rates (D), since succinic acid was found to be a growth-associated product (Figure 2). The maximum volumetric productivity was 6.1 g succinic acid L^{-1} h^{-1}.

A. succiniciproducens, which was isolated from the mouth of beagle dogs, produces a mixture of succinic acid, acetic acid and lactic acid from glucose and lactose, and a mixture of succinic acid and acetic acid at a molar ratio of 4:1 from glucose under strictly anaerobic conditions (Davis et al., 1976). Since the purification of succinic acid from a mixture of other acids is inefficient and costly, much effort has been devoted to developing a mutant strain and/or modifying the cultivation conditions which will allow succinic acid production but with minimal byproduct formation. A fluoroacetate-resistant variant of *A. succiniciproducens* was found that could produce succinic acid and only trace amounts of acetic acid (Guettler and Jain, 1996).

In order to evaluate the effects of carbon source on the growth and production of succinic acid, various carbon substrates such as glucose, fructose, mannose, rhamnose, arabinose, inositol, mannitol, sucrose, maltose, lactose and glycerol were investigated (Lee et al., 1999b, 2000c). When cells were cultured anaerobically in a medium containing 6.5 g L^{-1} glycerol, a high succinic acid yield (1.3 g g^{-1} glycerol) was obtained with significantly reduced byproduct formation (Lee et al., 2000a). The weight ratio of succinic acid to acetic acid was 25.8:1, some 6.5-fold higher than was obtained with glucose (ca. 4:1). When glucose and glycerol were cofermented with an increasing ratio of glucose to glycerol, the ratio of succinic acid

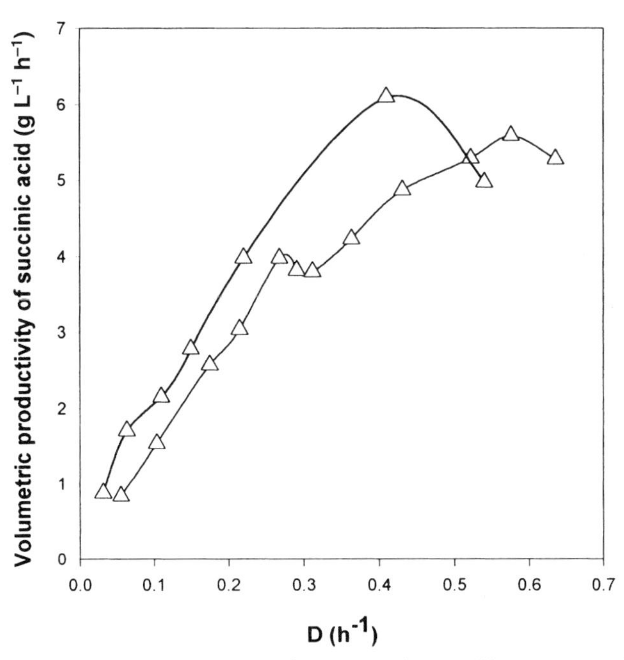

Fig. 2 Volumetric productivity of succinic acid versus dilution rate (D) during the continuous cultivation of *Anaerobiospillum succiniciproducens* in a medium containing 19 g L^{-1} glucose (p) and 38 g L^{-1} glucose (r) (reproduced from Lee et al., 2000b with permission).

to acetic acid and succinic acid yield decreased, suggesting that glucose enhanced acetic acid formation irrespective of the presence of glycerol. Based on these results, it is clear that glycerol is a good candidate carbon source for the production of succinic acid because it will reduce the separation and hence total production costs of succinic acid.

Whey, a byproduct produced during cheese making, was also investigated as a carbon source, as it contains ca. 4.5% lactose, which can be used by *A. succiniciproducens* to produce succinic acid (Samuelov et al., 1999). As mentioned earlier, succinic acid productivity is strongly affected by CO_2 concentration, and at high CO_2 concentration, >90% of lactose was consumed, the ratio of succinic acid to acetic acid being 3.6–4.3. Under CO_2-limiting conditions, only small amounts of lactose were consumed, and lactic acid was produced as a major product. A succinic acid yield of 0.8 g succinic acid/g whey lactose was obtained in batch cultivation, while the highest succinic acid yield of 0.9 g succinic acid/g whey lactose was obtained in fed-batch cultivation (Samuelov et al., 1999).

Productivity of up to 1.35 g succinic acid L^{-1} h^{-1} could be obtained by continuous culture of *A. succiniciproducens* on whey (Lee et al., 2000b), the ratio of succinic acid to acetic acid (g g^{-1}) being 5.1–5.8. The effect of dilution rate on the conversion of whey lactose to succinic acid was also evaluated in continuous culture where, at a dilution rate of 0.15 h^{-1}, a succinic acid productivity of 3.0 g L^{-1} h^{-1} and a yield of 0.64 g succinic acid/g whey lactose were obtained. When succinic acid produced by fermentation was added to rumen fluid, it was completely consumed by the mixed rumen microbial population, and was 90% decarboxylated to propionate on a molar basis. Therefore, the whey fermentation product formed under excess CO_2, which contained mainly organic acids and cells, might (potentially) be used as an animal feed supplement (Lee et al., 2000b).

The rumen bacterium *A. succinogenes* 130Z produces succinic acid at high concentration from various carbon substrates such as glucose, fructose, lactose, maltose, mannitol, mannose, sucrose, xylose, and cellobiose (Van der Werf et al., 1997). In contrast to *A. succiniciproducens*, *A. succinogenes* is a facultative bacterium (Guettler et al., 1999), and a fluoroacetate-resistant mutant of this was isolated which could grow and produce succinic acid in a medium containing 96 g L^{-1} sodium succinate or 130 g L^{-1} magnesium succinate (Guettler et al., 1996). This mutant strain was able to produce 110 g L^{-1} succinic acid with an apparent yield of 120 mol succinic acid/100 mol glucose. With supplementation of hydrogen, the ratio of succinic acid to acetic acid was increased from 1 to 1.5 during glucose fermentation by *A. succinogenes* (Van der Werf et al., 1997). For the conversion of fumaric acid to succinic acid, a low redox potential electron (which can be provided by pyruvate oxidation) is needed, and this is why an equivalent amount of acetic acid must be produced during glucose fermentation. If hydrogen is supplied, *A. succinogenes* can generate low redox potential electrons by the action of the hydrogenase, and hence more PEP can be directed towards succinic acid.

3.1.2
Production of Succinic Acid by Metabolically Engineered *E. coli*

E. coli produces succinic acid as a minor fermentation product (typically, 7.8% of total) under anaerobic conditions (Neijssel et al., 1996). The anaerobic metabolic pathway of *E. coli* is shown in Figure 3.

Several metabolic engineering strategies have been proposed for the enhanced production of succinic acid by *E. coli*. Metabolic engineering can be defined as directed

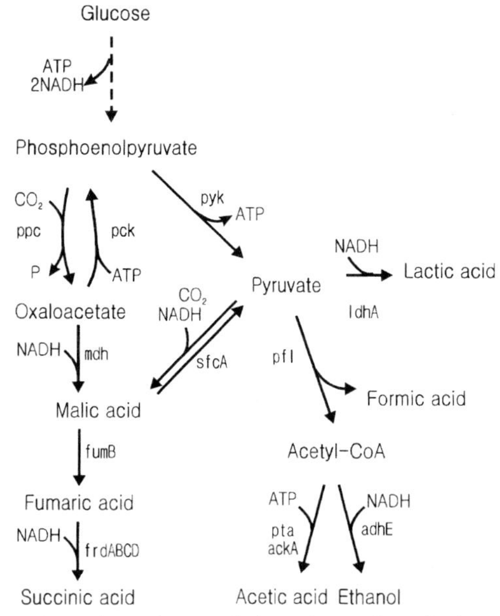

Fig. 3 Anaerobic metabolic pathway of *E. coli*. The genes shown are: *ppc*, PEP carboxylase; *pck*, PEP carboxykinase; *mdh*, malate dehydrogenase; *pyk*, pyruvate kinase; *ldhA*, lactate dehydrogenase; *pfl*, pyruvate formate-lyase; *pta*, phosphotransacetylase; *ackA*, acetate kinase; *adhE*, alcohol/acetaldehyde dehydrogenase; *sfcA*, malic enzyme; *fumB*, fumarase B; *frdABCD*, fumarate reductase.

modification of cellular metabolism and properties through the introduction, deletion, and modification of metabolic pathways by using recombinant DNA and other molecular biological tools (Bailey, 1991; Lee and Papoutsakis, 1999).

In one example, fumarate reductase (FrdABCD) was overexpressed in *E. coli* for the conversion of fumaric acid to succinic acid (Goldberg et al., 1983; Wang et al., 1998). Fumarate reductase catalyzes the reduction of fumaric acid to succinic acid, and is a key enzyme under anaerobic condition when fumaric acid is the terminal electron acceptor. The amino acid sequence of fumarate reductase is similar to that of the tricarboxylic acid (TCA) cycle enzyme succinate dehydrogenase (Blattner et al., 1997). Indeed, the two enzymes have distinct functional characteristics (e.g., substrate affinities and reaction rates), and both catalyze the interconversion of fumaric acid and succinic acid. The recombinant *E. coli* strain DH5α harboring pGC1002, which contains the *E. coli* fumarate reductase gene *frdABCD* and the ampicillin resistance (β-lactamase) gene, was cultured in Luria-Bertani (LB) medium at 37 °C. Glucose and fumaric acid were used as carbon sources, since glucose was required for the conversion of fumaric acid to succinic acid (see below). In the absence of glucose, or in cultures of low cell density, malic acid was accumulated, while the recombinant strain produced 47.5 g L^{-1} succinic acid from 50.8 g L^{-1} fumaric acid and 23.7 g L^{-1} glucose after 32 h of cultivation (Wang et al., 1998). The weight yield of succinic acid, based on the quantity of fumaric acid consumed, was 0.93. The effect of initial fumaric acid concentration on the production of succinic acid was also examined. When 90 g L^{-1} fumaric acid was used, substrate inhibition occurred and malic acid was produced as the major product rather than succinic acid. It was also found that glucose, when used without fumaric acid, was not converted to succinic acid. Glucose was seen to be used to supplement the reducing power and to regenerate the cofactor $FADH_2$, which is required in the conversion of fumaric acid to succinic acid.

In another study, overexpression of the *E. coli* PEP carboxylase gene was examined for the possibility of enhancing succinic acid production (Millard et al., 1996). This enzyme catalyzes the carboxylation of PEP to oxaloacetate, and is important in the anaplerotic pathway since it functions to replenish oxaloacetate consumed in biosynthetic pathways under aerobic conditions. Under anaerobic conditions, the enzyme functions to direct PEP to succinic acid. PEP carboxykinase also plays an important role in gluconeogenesis by catalyzing the phosphorylation of oxaloacetate to PEP, with concomitant consumption of ATP. It was reported that PEP carboxylase and PEP carboxykinase formed an energy-consuming futile cycle when the PEP carboxykinase gene was overexpressed. Overexpression of PEP carboxylase gene provided a significant increase in the amount of succinic acid produced, while overexpression of PEP carboxykinase gene did not. The amount of succinic acid produced was increased from 3.27 to 4.44 g L^{-1} when PEP carboxylase was overexpressed.

The anaerobic PEP carboxykinase gene (*pckA*) of *A. succiniciproducens* was recently cloned (Laivenieks et al., 1997). PEP carboxykinase of *A. succiniciproducens* is more suitable for the production of succinic acid since it produces ATP and conserves the free energy of PEP, while the PEP carboxylase of *E. coli* dissipates the free energy of PEP. The sequence of *A. succiniciproducens* PEP carboxykinase was found to be similar to those of all known ATP/ADP-dependent PEP carboxykinases. In particular, the amino acid sequence of *A. succiniciproducens* PEP carboxykinase was 67.3% identical and 79.2% similar to that of *E. coli* PEP carboxykinase. The PEP carboxykinase of *A. succiniciproducens* was found to comprise 532 amino acids, with a calculated molecular mass of 58.7 kDa. The enzyme is oxygen-stable, has an optimum pH of 6.5–7.1, and requires Mn^{2+}, Co^{2+} and Mg^{2+} as cofactors. Overexpression of the *A. succiniciproducens* PEP carboxykinase gene in *E. coli*, however, did not lead to any increase in the amount of succinic acid produced (Laivenieks et al., 1997).

In *E. coli*, PEP is converted to malic acid in two reaction steps catalyzed by PEP carboxylase and malate dehydrogenase (Mdh). As mentioned above, the free energy of PEP is dissipated through the conversion of PEP to oxaloacetate. On the other hand, one ATP is produced when PEP is directly converted to pyruvate by pyruvate kinase (Pyk). Therefore, a new route – conversion of PEP to malic acid by pyruvate kinase and malic enzyme – was proposed to conserve the free energy of PEP. In *E. coli*, two types of malic enzyme exist: NAD^+-dependent and $NADP^+$-dependent (Stols and Donnelly, 1997). The NAD^+-dependent malic enzyme encoded by the *sfcA* gene functions to convert malic acid to pyruvate. The K_m values of NAD^+-dependent malic enzyme in *E. coli* are 16 mM for pyruvate and 0.26 mM for malic acid. Because of this large difference in K_m values, malic acid is not produced from pyruvate under normal cultivation conditions. This is why a mutant *E. coli* strain NZN111, in which pyruvate-formate lyase (PFL) and lactate dehydrogenase (LDH) are partially blocked, was used. By double mutation of *pfl* and *ldhA* genes, anaerobic pyruvate dissipation pathways are blocked and pyruvate accumulates (Stols and Donnelly, 1997). The NAD^+-dependent malic enzyme gene was overexpressed under the control of *trc* promoter in NZN111 to convert accumulated pyruvate to malic acid, the latter then being further converted to succinic acid. At the end of fermentation, 12.8 g L^{-1} succinic acid was produced, and the apparent yield of succinic acid during anaerobic cultivation was 1.2 g succinic acid g^{-1} glucose.

In the fermentor-scale studies, however, a considerable amount of malic acid (the precursor of succinic acid) was also produced. In order to analyze the involvement of each of the steps in the overall metabolic pathways, the process of metabolic flux analysis (MFA) was carried out (Hong and Lee, 2001). In using MFA, intracellular metabolic fluxes can be quantified by measuring extracellular metabolite concentrations in combination with the stoichiometry of intracellular reactions (Nielsen and Villadsen, 1994; Edwards et al., 1999). As mentioned previously, intracellular pyruvate accumulation is required to change the direction of the reaction catalyzed by malic enzyme; therefore, fluxes to the intracellular pools of pyruvate and succinic acid were introduced in order to mimic the accumulation of intracellular metabolites (especially, pyruvate and succinic acid). The results of MFA showed that about 80% of pyruvate was redirected to malic acid by the malic enzyme, and the intracellular pyruvate concentration was maintained at a low level (<1 mM) throughout the anaerobic cultivation period. It was also proposed from MFA results that fumarase became a bottleneck in this engineered pathway (Hong and Lee, 2001), but new strategies are needed to solve this problem and to enhance succinic acid production. Overexpression of fumarase gene or the adjustment of redox balance using more reduced carbon substrate may provide a solution to this problem.

The malic enzyme gene of *Ascaris suum* was also overexpressed in *E. coli* NZN111 (Stols et al., 1997). The K_m values of this enzyme are 45 mM for pyruvate and 0.4 mM for malic acid, the enzyme being overexpressed under the control of *trc* promoter. As a result, production of succinic acid (the major product) was increased from 2.06 to 7.07 g L^{-1}.

Recently, a mutant *E. coli* strain AFP111, a spontaneous *ptsG* gene mutant of NZN111, was isolated (Donnelly et al., 1998; Chatterjee et al., 2001). The AFP111 mutant was able to grow on glucose minimal medium (although its parent strain NZN111 failed to grow) and produced 2 mol of succinic acid, 1 mol of acetic acid and 1 mol of ethanol from 2 mol of glucose. The protein $EIICB^{glc}$, which is encoded by the *ptsG* gene, is the glucose-specific permease of the phosphotransferase system (PTS). To evaluate the effect of mutation in the *ptsG* gene on metabolite production, the disrupted *ptsG* gene was introduced to a *ldhA pfl* double mutant strain DC1327, which contained a disrupted *ptsG* gene; this restored the glucose-fermenting ability and led to the production of succinic acid, acetic acid and ethanol in a molar ratio of 2:1:1, which is equivalent to those obtained with AFP111. To examine the effect of the *ptsG* gene in more detail, the inactivated *ptsG* gene was introduced into several *E. coli* strains; in all cases, succinic acid was produced in threefold greater quantities than was obtained with the parent strains. On the basis of these results it appears that inactivation of the *ptsG* gene redirects metabolic fluxes towards succinic acid production, although the mechanism of this effect is not clear.

As reviewed above, succinic acid can be produced at high concentrations, and at high yield and productivity, by using several different wild-type and metabolically engineered bacteria. Future improvements of microbial strains and fermentation/recovery processes will permit the cost-effective production of succinic acid from renewable resources, and hence succinic acid will be made available for use in a wide range of applications, including the synthesis of biodegradable polyesters.

3.2
Adipic Acid

Adipic acid is a member of the C_6-dicarboxylic acid family, and is an important intermediate which is used mainly (about 90%) in the manufacture of nylon. Other uses of adipic acid include plasticizers, lubricants and food additives (Beziat et al., 1996; Frost and Draths, 1996).

The industrial-scale production of adipic acid is usually carried out by multi-step chemical synthesis starting from the oxidation of cyclohexane (Beziat et al., 1996). Cyclohexane is oxidized to form cyclohexanone and cyclohexanol, which are subsequently oxidized with nitric acid to form a mixture of carboxylic acids. This process is regarded as poor because the yield is low and the chemicals used are not environmentally friendly; consequently, extensive research has been directed towards developing an alternative process. Adipic acid can also be produced by the hydrogenation and carbonylation of succinic acid, produced by fermentation with microorganisms (see Section 3.1; Zeikus et al., 1999). Many chemicals such as butadiene, five- and six-carbon lactones, six-carbon diacids, 1,4-disubstituted-2-butene, and pentenoic ester are used in the multi-step chemical synthesis of adipic acid (Isogai et al., 1983; Atadan and Bruner, 1994; Bruner et al., 1998). 1,3-Butadiene can be used as a starting material for producing adipic acid via a hydroesterification reaction with a cobalt carbonyl catalyst. Another metallic catalyst, rhodium, has been developed for the production of adipic acid using 1,3-butadiene or allylic butenols or their esters as starting materials (Bruner, 1992). Burke (1994) reported the production of adipic acid by reacting the lactones with carbon monoxide and water in the presence of a homogeneous rhodium catalyst at high temperature (190–250 °C) and pressure (100–2000 psi).

In addition to chemical transformation, a biological approach has also been investigated (Frost and Draths, 1996). *Cis,cis*-muconic acid, which can be hydrogenated to produce adipic acid, is produced by *Pseudomonas putida* using toluene as a carbon source (Maxwell, 1982). Strains of *Acinetobacter* and *Nocardia* can also produce adipic acid as a metabolic intermediate from cyclohexanol as a sole carbon source (Norris and Trudgill, 1971; Donoghue and Trudgill, 1975). Alkane-utilizing *Pseudomonas* spp. were also reported to produce adipic acid from hexanoate, 6-hydroxyhexanoate, and 6-oxohexanoate (Kunz and Weimer, 1983). However, due to the high cost of starting materials, these processes are not commercially viable. A process combining chemical and biocatalytical methods has also been reported and which involves first, biomass compounds (e.g., cellulose, wood, corn stalks) being converted to 1,6-hexanediol by multi-step chemical reactions. The 1,6-hexanediol is then oxidized to adipic acid by *Gluconobacter oxydans* (Faber, 1983). Recombinant bacteria have also been developed for the production of adipic acid (Frost and Draths, 1996). A genetically engineered *E. coli* has been developed for the single-step conversion of nontoxic, renewable carbon sources such as glucose, starch, cellulose and other biomass resources into adipic acid. Figure 4 shows the metabolic pathway of recombinant *E. coli* producing adipic acid from D-glucose. Recombinant *E. coli* equipped with a plasmid containing the *Klebsiella pneumoniae* 3-dehydroshikimate dehydratase and protocatechuate decarboxylase genes, and the *Acinetobacter calcoaceiens* catechol 1,2-dioxygenase gene was capable of converting several carbon sources to adipic acid via catechol and *cis,cis*-muconic acid (Draths and Frost, 1994). The final concentration of *cis,cis*-muconic acid and the yield on D-glucose after 48 h were 1.75 g L^{-1} and 0.3 mol mol^{-1}, respectively. The final yield of adipic acid obtained by the reduction of *cis,cis*-muconic acid was 0.9 mol mol^{-1}.

Until recently, the production of adipic acid relied on chemical synthesis rather than biological conversion, because the fermentative production of adipic acid by natural strains is inefficient and substrates are expensive. From the biosynthesis pathway of adipic acid in *Pseudomonas* sp., it was suggested that adipic acid could be synthe-

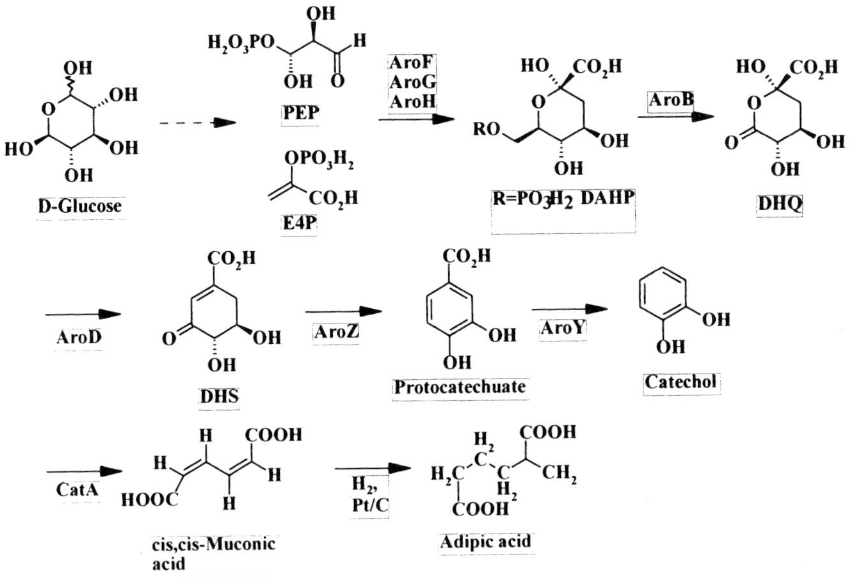

Fig. 4 Metabolic pathway of recombinant *E. coli* producing adipic acid from D-glucose. PEP, phosphoenolpyruvate; E4P, D-erythrose 4-phosphate; AroF, tyrosine-sensitive 3-deoxy-D-arabino-heptulosonate 7-phosphate (DAHP) synthase; AroG, phenylalanine-sensitive DAHP synthase; AroH, tryptophan-sensitive DAHP synthase; DAH, AroB 3-dehydroquinate (DHQ) synthase; AroD, DHQ dehydratase; AroZ, 3-dehydroshikimate (DHS) dehydratase; AroY, protocatechuate decarboxylase; CatA, catechol 1,2-dioxygenase (redrawn from Frost and Draths, 1996).

sized from fatty acids. Also, metabolically engineered *E. coli* demonstrated the possibility of adipic acid production from renewable carbon sources such as glucose. Further studies on the enzymes, pathways and genes involved in adipic acid metabolism should be carried out for the development of superior microbial strains efficiently producing adipic acid from renewable resources.

4 Diols

Several diols have been produced fermentatively by various microorganisms. Metabolic engineering strategies have also been applied for the development of improved strains producing these diols more efficiently. The structures of three diols reviewed in this section are shown in Figure 5.

4.1 1,2-Propanediol

1,2-Propanediol (propylene glycol) is a three-carbon diol with a chiral center at the central carbon atom. Racemic 1,2-propanediol is a commodity chemical with an annual production of over 0.5 million tonnes in the United States (Cameron et al., 1998). The commercial route to racemic 1,2-propanediol is by the hydration of propylene oxide, which is derived from propylene. Several methods exist for the production of 1,2-propanediol from renewable feedstocks. Hydrogenolysis of sugars at high temperature and pressure, and in the presence of a metal catalyst, results in the production of a mixture of 1,2-propanediol and other polyols.

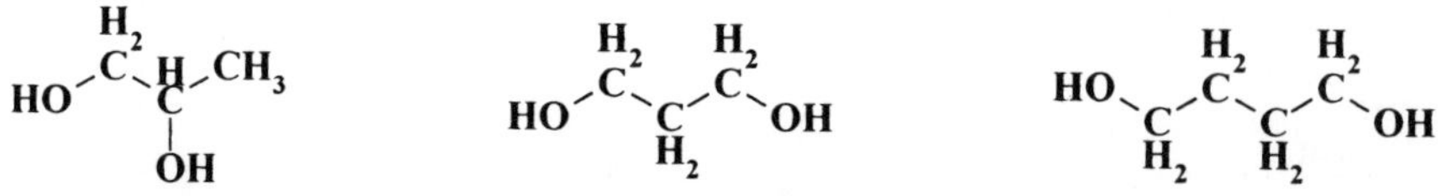

1,2-Propanediol **1,3-Propanediol** **1,4-Butanediol**

Fig. 5 The structural formulas of diols.

Optically pure 1,2-propanediol can be produced using several methods (Levene and Walti; 1943; Fryzuk and Bosnich, 1978; Kometani et al., 1993), each of which has both advantages and disadvantages. The fermentative production of enantiomerically pure 1,2-propanediol has also been reported (Boronat and Aguilar, 1981; Cameron et al., 1998). The first method is conversion of 6-deoxyhexose to (*S*)-1,2-propanediol by *E. coli.*, though this route is not economical as the least expensive sugar (L-rhamnose) used costs over $300 per kg (Pfanstiehl Laboratories, 1996). The second method to produce (*R*)-1,2-propanediol from glucose, xylose and several other common sugars using various microorganisms. *Clostridium sphenoides* DSM614, *Thermoanaerobacterium thermosaccharolyticum*, and *Thermoanaerobacter ethanolicus* ATCC 33223 are known to produce (*R*)-1,2- propanediol from such sugars (Cameron et al., 1998).

4.1.1
Microbial Production of 1,2-Propanediol by Wild-type Bacteria

Several reports have been made on the fermentative production of 1,2- propanediol from common sugars. In *C. sphenoides* fermentation, less than 2 g L^{-1} of 1,2-propanediol was produced from glucose, xylose, mannose, and cellobiose under phosphorus-limited conditions (Tran-Din and Gottschalk, 1985). The metabolic pathway to 1,2-propanediol from dihydroxyacetone is shown in Figure 6.

Cameron and Cooney (1986) reported that *T. thermosaccharolyticum* produces up to

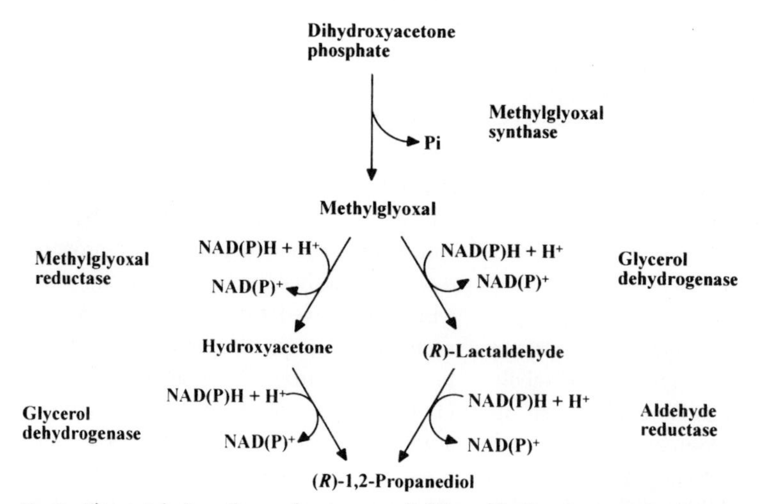

Fig. 6 The metabolic pathway of 1,2-propanediol from dihydroxyacetone phosphate (modified from Altaras and Cameron, 1999).

7.9 g L^{-1} of (*R*)-1,2 propanediol from glucose, xylose, mannose, and cellobiose, but no cell growth or propanediol production was observed with either fructose, galactose, or lactose. Batch culture of *T. thermosaccharolyticum* resulted in the production of 9.0 g L^{-1} of (*R*)-1,2 propanediol in 25 h, with a yield of 0.2 g propanediol g^{-1} glucose (Sanchez-Riera et al., 1987). Continuous cultures of *T. thermosaccharolyticum* have been carried out for the production of (*R*)-1,2 propanediol from either glucose or galactose. Using glucose as a carbon source, the maximum concentration of (*R*)-1,2-propanediol obtained was only 0.7 g L^{-1} under phosphorus- or potassium-limited conditions. However, when galactose was used as a carbon source, (*R*)-1,2-propanediol concentration was increased to 3.5 g L^{-1} under phosphorus-limited conditions (Wrobel, 1997). It should be appreciated that galactose is more expensive than glucose, however.

4.1.2
Production of 1,2-Propanediol by Metabolically Engineered Bacteria

Wild-type *E. coli* does not normally produce 1,2-propanediol from common sugars. Since *E. coli* naturally synthesizes some methylglyoxal (an intermediate of 1,2-propanediol), 1,2-propanediol can be produced in *E. coli* if an appropriate enzyme and/or reducing power is provided.

The use of aldose reductase, an NADPH-linked enzyme was examined initially (Cameron et al., 1998). Rat lens aldose reductase gene was overexpressed under the inducible *trc* promoter. By the overexpression of this gene, 0.1 g L^{-1} of (*R*)-1,2-propanediol could be produced, while less than 0.05 g L^{-1} of (*R*)-1,2-propanediol was produced in the absence of the gene. The next enzyme to be investigated was glycerol dehydrogenase, an NADH-linked reductase that has the primary activity of oxidizing glycerol to dihydroxyacetone. However, this enzyme is known to have broad substrate specificities (Tang et al., 1979), and can also reduce hydroxyacetone (Lee and Whitesides, 1986). Overexpression of the *E. coli* glycerol dehydrogenase gene resulted in the production of 1,2-propanediol to ca. 1.5 g L^{-1}, which is slightly higher than that obtained by overexpression of the aldose reductase gene. The product was analyzed using chiral gas chromatography (GC) and found to consist of 100% of (*R*)-1,2-propanediol (Cameron et al., 1998).

Recently, the *E. coli* methylglyoxal synthase gene was cloned (Percy and Harrison, 1996). Altaras and Cameron (1999) reported the production of optically pure (*R*)-1,2-propanediol from glucose in *E. coli* expressing the NADH-linked glycerol dehydrogenase genes (*E. coli gldA* or *K. pneumoniae dhaD*) and the methylglyoxal synthase gene (*mgs*). The expression of either glycerol dehydrogenase or methylglyoxal synthase gene resulted in the anaerobic production of ~0.25 g L^{-1} of (*R*)-1,2-propanediol. However, when both the methylglyoxal synthase and glycerol dehydrogenase genes were coexpressed, (*R*)-1,2-propanediol production was increased to 0.7 g L^{-1}. *In vitro* studies revealed that the route to (*R*)-1,2-propanediol involved reduction of methylglyoxal to (*R*)-lactaldehyde by the recombinant glycerol dehydrogenase and reduction of (*R*)-lactaldehyde to (*R*)-1,2-propanediol by the native *E. coli* aldehyde reductase (Altaras and Cameron, 1999). Huang et al. (1999) have reported that (*R*)-1,2-propanediol can be produced by coexpression of the *E. coli* glycerol dehydrogenase gene and the newly cloned methylglyoxal synthase gene from *Clostridium acetobutylicum* in *E. coli*. About 0.3 g L^{-1} of 1,2-propanediol was produced by cultivation of this recombinant *E. coli* in M9 medium supplemented with fructose and yeast extract.

Recently, it has been reported that the yield and product concentration of (*R*)-1,2-propanediol can be increased by introducing methylglyoxal synthase and glycerol dehydrogenase genes, and either yeast alcohol dehydrogenase gene or *E. coli* 1,2-propanediol oxidoreductase gene (Altaras and Cameron, 2000). In this report, the final concentration of (*R*)-1,2-propanediol obtained was 4.5 g L^{-1}, with a final yield of 0.19 g of 1,2-propanediol g^{-1} glucose.

4.2 1,3-Propanediol

1,3-Propanediol is one of oldest known anaerobic fermentation products, and is produced by numerous bacteria from glycerol as a fermentation substrate. It was identified as early as 1881 by August Freund, in a glycerol-fermenting mixed culture that clearly contained *Clostridium pasteurianum* as an active organism (Freund, 1881). 1,3-Propanediol has been considered to be a good starting chemical for the production of polymers and other organic chemicals (Biebl et al., 1999). Having two functional groups, 1,3-propanediol can be used as a monomer for polycondensation reactions to produce polyesters, polyethers and polyurethanes. At present, the use of 1,3-propanediol produced by fermentation as a starting material for the synthesis of novel polymers is receiving worldwide attention, mainly as DuPont (Wilmington, DE, USA) commenced this project several years ago. The polymer developed at DuPont is based on reacting the 1,3-propanediol monomer with terephthalic acid (Potera, 1997). This polyester, called poly(propylene terephthalate) (PPT), was found to be suitable for use in carpeting and textiles, and had good recyclability. Even though poly(ethylene terephthalate) (PET), which uses ethylene glycol as a monomer, and poly(butylene terephthalate) (PBT), which uses 1,4-butanediol as a monomer, are composed of similar monomers to PPT, PPT shows unique properties due to the presence of 1,3-propanediol (Smith et al., 1966).

4.2.1 Production of 1,3-Propanediol by Wild-type Bacteria

Numerous microorganisms are able to convert glycerol to 1,3-propanediol during fermentation, including the genera *Klebsiella* (*K. pneumoniae*), *Enterobacter* (*E. agglomerans*), *Citrobacter* (*C. freundii*), *Lactobacillus* (*L. brevis* and *L. buchneri*) and *Clostridium* (*C. butyricum* and *C. pasteurianum*). Figure 7 shows the metabolic route leading to the formation of 1,3-propanediol from glycerol. Both enterobacteria and clostridia are suitable for the fermentative production of 1,3-propanediol. Among these, *K. pneumoniae* and *C. freundii*, as facultative anaerobes, are usually selected for easy handling. However, these strains are opportunistic pathogens, and thus special precautions are required for their cultivation (Homann et al., 1990).

The 1,3-propanediol yield obtainable with *K. pneumoniae* (0.53 mol mol^{-1}) was found to be generally lower than that with *Citrobacter* (0.65 mol mol^{-1}). However, *K. pneumoniae* grows much more rapidly, and therefore 1,3-propanediol productivity is

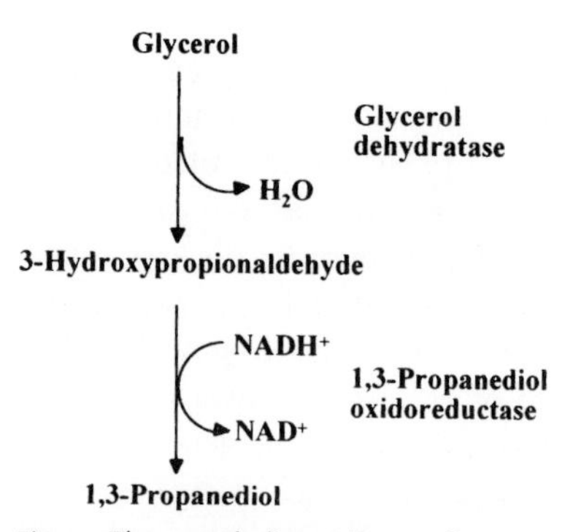

Fig. 7 The metabolic pathway of 1,3-propanediol biosynthesis from glycerol.

considerably higher when this bacterium is used (Deckwer, 1995).

Batch fermentations of *K. pneumoniae* at various glycerol concentrations (from 21.8 to 201.7 g L^{-1}) have been reported (Tag, 1990). At a concentration of up to 120 g L^{-1}, almost 100% of the substrate was consumed, the highest 1,3-propanediol concentration obtained being 57.7 g L^{-1} at a glycerol concentration of 154.1 g L^{-1}. The highest 1,3-propanediol productivity of 2.3 g L^{-1} h^{-1} was obtained at a glycerol concentration of 122 g L^{-1}.

Fed-batch culture of *K. pneumoniae* was carried out by Held (1996), the final yield of 1,3-propanediol on glycerol being 0.39 g g^{-1} (0.48 mol mol^{-1}). Glycerol concentration was maintained at around 20 g L^{-1}, and additional yeast extract was added during the culture. The final 1,3-propanediol concentration and the maximum 1,3-propanediol productivity obtained in 20 h were 73.3 g L^{-1} and 2.5 g L^{-1} h^{-1}, respectively. 2,3-Butanediol, lactate, acetate, ethanol, formate, and succinate were produced as byproducts, the major byproduct (2,3-butanediol) being produced at up to 20 g L^{-1}. The formation of these byproducts can be reduced by the development of an engineered strain and/or by changing the fermentation strategy.

Continuous cultures of *K. pneumoniae* have also been reported. By optimizing the various inlet flow rates and glycerol concentrations, a maximum 1,3-propanediol productivity of 8.1 g L^{-1} h^{-1} could be obtained, though the highest concentration was only ~30 g L^{-1} (Tag, 1990). Menzel et al. (1997) reported that a final 1,3-propanediol concentration of 35.2–48.5 g L^{-1} and a productivity of 4.9–8.8 g L^{-1} h^{-1} could be obtained at dilution rates between 0.1 and 0.25 h^{-1} in a continuous fermentation of *K. pneumoniae* using glycerol as the carbon source. The highest 1,3-propanediol concentration of 48.5 g L^{-1} was obtained at a dilution rate of 0.1 h^{-1}. The highest 1,3-propanediol concentration obtained by continuous culture was somewhat lower than was obtained in batch and fed-batch cultures (70 g L^{-1}), for obvious reasons. Productivity obtained by continuous culture was, however, up to 3.5-fold higher than that obtained by the best fed-batch culture.

The production of 1,3-propanediol by a fed-batch culture of *C. butyricum* DSM 5431 using glycerol as a carbon source has been reported, the highest concentration and productivity obtained being 58 g L^{-1} and 2.7 g L^{-1} h^{-1}, respectively. During the cultivation, substrate inhibition was observed at the initial glycerol concentration of 50 g L^{-1}; therefore, fed-batch operation led to better results for this strain. To investigate the effect of reactor type and size, two reactors with different sizes, an air-lift reactor (up to 1.2 m^3) and a stirred vessel (up to 2 m^3), were compared, but no difference was found during the course of the fermentation. These results show that reactor type and size did not seriously affect microbial 1,3-propanediol production (Gunzel et al., 1991). Better results were obtained by carefully controlling nutrient supply during the fed-batch cultivation, however. Saint-Amans et al. (1994) used CO_2 as the control parameter, while Reimann and Biebl (1996) combined pH correction by KOH with nutrient addition. Glycerol concentration in the fermentation broth could be maintained at a constant to a slightly excessive level, which was necessary to reduce butyrate formation.

The process for the recovery and purification of 1,3-propanediol is based on mechanical and thermal operations (Deckwer, 1995), the 1,3-propanediol concentration in the fermentation broth being the most important factor relating to downstream processing costs.

Continuous fermentation of 1,3-propanediol from glycerol by cell recycle culture of *C. butyricum* has also been reported. The performance of the culture system was examined at a retention ratio (dilution rate/bleed rate) of 5, dilution rates between 0.2 and 1.0 h^{-1}, and inlet glycerol concentrations of 32 or 56 g L^{-1}. The 1,3-propanediol concentration was maintained 26.5 g L^{-1} up to a dilution rate of 0.5 h^{-1}, when the inlet glycerol concentration was 56 g L^{-1}. With a further increase in dilution rate, the 1,3-propanediol concentration was reduced, while the productivity was highest at a dilution rate of 0.7 h^{-1}. Although productivity could be increased four-fold by cell recycle cultivation, the maximum product concentration could not be increased significantly using this system (Reimann, 1997).

Continuous cultures of *C. butyricum* were carried out using industrial glycerol as a carbon source, this being the major by-product of the biodiesel production process. The highest 1,3-propanediol concentration of 48 g L^{-1} and the glycerol yield of 0.55 g 1,3-propanediol g^{-1} glycerol were obtained in single-stage continuous culture. The maximum 1,3-propanediol productivity obtained was 5.5 g L^{-1} h^{-1}. A two-stage continuous fermentation was also carried out in which, by optimizing the culture parameters (including the dilution rate), a 1,3-propanediol concentration of 46 g L^{-1} and a productivity of 3.4 g L^{-1} h^{-1} could be achieved (Papanikolaou et al., 2000). Another report on glycerol fermentation by *C. freundii* was made in which cell retention was achieved by immobilization on a polyurethane carrier (Pflugmacher and Gottschalk 1994). In this culture, a productivity of 8.2 g L^{-1} h^{-1} and a 1,3-propanediol concentration of 16.3 g L^{-1} were achieved, and at a dilution rate of 0.5 h^{-1}.

Production of 1,3-propanediol was investigated using a newly screened *Enterobacter agglomerans*, a facultatively anaerobic Gram-negative bacterium. A 20 g L^{-1} glycerol solution was converted mainly to 1,3-propanediol (0.51 mol mol^{-1}) and acetate (0.18 mol mol^{-1}) in batch culture. The 1,3-propanediol yield could be increased to 0.61 mol mol^{-1} at high initial glycerol concentrations (71 g L^{-1} and 100 g L^{-1}). Some 3-hydroxypropionaldehyde was also accumulated in the culture medium, which reduced cell growth and 1,3-propanediol production due to its toxicity (Barbirato et al., 1995). Therefore, the formation and/or accumulation of this metabolite should be eliminated by strain engineering and/or by employing a different cultivation condition.

4.2.2
Production of 1,3-Propanediol by Metabolically Engineered Bacteria

The microbial production of 1,3-propanediol might be an attractive alternative to chemical synthesis, and its fermentative production on an industrial scale is gradually being realized. Many researchers have suggested a number of strategies for making the biotechnological process economically feasible: (1) To increase the 1,3-propanediol formation by amplifying the genes in the rate-controlling steps and/or deletion of genes involved in the formation of undesirable by-products; and (2) To reduce the cost of carbon substrates by using less expensive carbon sources. Abbad-Andaloussi et al. (1996) and Ahrens et al. (1998) reported that glycerol dehydratase is a rate-limiting enzyme for 1,3-propanediol production in *C. butyricum* and *K. pneumoniae*. However, by overexpression of a gene coding for this enzyme, a higher 1,3-propanediol production and less byproduct formation could be achieved.

Although glucose and some other sugars are considerably cheaper than glycerol in many countries, no known wild-type microorganism produces 1,3-propanediol directly from sugar. One approach to ferment sugar to 1,3-propanediol is the co-cultivation of two microorganisms, whereby one bacterium converts the sugar to glycerol, and the other ferments the glycerol to 1,3-propanediol. By applying a mixed culture or two-stage fermentation, this approach could be successfully demonstrated (Haynie and Wagner, 1996).

Another approach to produce 1,3-propanediol from sugar is to use a metabolically engineered strain. A gene that allows formation of glycerol from sugars or intermediates of glycolysis could be introduced to a bacterial strain producing 1,3-propanediol from glycerol. Alternatively, the genes that allow conversion of glycerol to 1,3-propanediol could be introduced to a bacterium producing glycerol from sugar. If the need to engineer a bacterium were greater, then genes that allow the formation of both glycerol and 1,3-propanediol could be introduced.

The first strategy was applied by expression of the glycerol-3-phosphatase gene. This enzyme catalyzes the conversion of glycerol-3-phosphate to glycerol in glycerol-producing microorganisms. Two different groups cloned the genes coding for glycerol-3-phosphatase from *Saccharomyces cerevisiae* and *Bacillus licheniformis* (Norbeck et al., 1996; Skraly and Cameron, 1998). Recently, Nakamura et al. (2000) reported that 1,3-propanediol can be produced by recombinant microorganisms by applying this strategy. In that report, various sugars such as glucose, fructose, lactose, sucrose, maltose and mannose were converted to 1,3-propanediol by cultivation of recombinant *K. pneumoniae* expressing glycerol-3-phosphate dehydrogenase and glycerol-3-phosphate phosphatase genes of *S. cerevisiae* and/or glycerol dehydrogenase and 1,3-propanediol oxidoreductase genes of various microorganisms. When these microorganisms were cultivated using glucose as a carbon source, 10.9 g L^{-1} of 1,3-propanediol was produced, with a yield of 0.12 g 1,3-propanediol g^{-1} glucose. Based on this patent and further developments which have been made, DuPont (Wilmington, DE, USA) is currently commercializing the process for the fermentative production of 1,3-propanediol.

The second strategy was demonstrated by overexpressing glycerol hydratase and 1,3-propanediol oxidoreductase genes in an organism which produces glycerol naturally, such as *S. cerevisiae* (Nevoigt and Stahl, 1997). For the expression in yeasts, bacterial genes were expressed under the control of eucaryotic promoters. However, only a small amount of 1,3-propanediol (<0.1 g L^{-1}) was produced (Laffend et al., 1996; Cameron et al., 1998). To overcome this problem, recombinant *S. cerevisiae* has recently been developed by integrating the glycerol hydratase and 1,3-propanediol oxidoreductase genes of *K. pneumoniae* into the chromosome (Nakamura et al., 2000). In this way, a 1,3-propanediol concentration of 0.53 g L^{-1} was achieved from glucose.

The final strategy was demonstrated by expressing the genes of the 1,3-propanediol pathway heterologously. Forage and Lin (1982) reported the genes coding for the enzymes of glycerol metabolism, glycerol dehydratase (*dhaB, C, E*), 1,3-propanediol oxidoreductase (*dhaT*), glycerol dehydrogenase (*dhaD*), dihydroxyacetone kinase (*dhaK*) and a putative regulatory gene (*dhaR*). All these genes belonging to the *dha* regulon have been cloned and sequenced from *K. pneumoniae* and *C. freundii* (Sprenger et al., 1989; Tong et al., 1991; Daniel and Gottschalk, 1992; Daniel et al., 1995a,b; Seyfried et al., 1996; Tobimatsu et al., 1996). By culturing a recombinant *E. coli* harboring all these genes, 6.5 g L^{-1} of 1,3-propanediol could be produced (Tong and Cameron, 1992). Skraly and Cameron (1998) recently constructed several synthetic operons containing the *K. pneumoniae dhaB* and *dhaT* genes under the control of a single *trc* promoter, which can be expressed in several different hosts. Fed-batch culture of *E. coli* AG1 harboring these synthetic operons resulted in a final 1,3-propanediol concentration of 6.33 g L^{-1} using glycerol and glucose as carbon sources.

4.3 1,4-Butanediol

1,4-Butanediol (1,4-butylene glycol) is an industrially important chemical which is used as a solvent, a humectant (moisture-retaining agent), an intermediate for plasticizers and pharmaceuticals, a cross-linking agent for polyurethane elastomers, and a precursor in the synthesis of tetrahydrofuran. 1,4-Butanediol has also received much attention as a raw material for the preparation of thermoplastic polyester in large quantities (Budge et al., 1999).

Both aliphatic acid and aromatic dicarboxylic acids can be used for condensation with 1,4-butanediol. The polyesters prepared by condensation of 1,4-butanediol and aliphatic dicarboxylic acids possess low melting points, and some of these polymers are commercially valuable, biodegradable polymers (Nagata et al., 2000).

1,4-Butanediol has been produced by the chemical transformation of various chemicals, for example by the hydrogenation of 1,4-butynediol, succinic acid, succinic anhydride, maleic acid, and maleic anhydride, and by the hydrolysis and transesterification of the acetic ester of 1,4-butanediol (Broecker and Schwarzmann, 1977; Coates and Newark, 1980; Yoshida and Oka, 1981).

Broecker and Schwarzmann (1977) reported that 1,4-butanediol could be produced from maleic or succinic anhydride via γ-butyrolactone using nickel and copper catalysts. However, in employing this type of catalyst, the reaction conditions were rather severe and the catalytic activities inadequate; hence, more effective catalysts were required. By using a ruthenium-type catalyst composed of ruthenium, an organic phosphine and a phosphorus compound, 1,4-butanediol could be produced from succinic acid, succinic anhydride, and γ-butyrolactone under mild conditions (Hara and Inagaki, 1991). Subsequently, these catalysts have been further developed to minimize byproduct formation. For example, Budge et al. (1999) reported a metal catalyst for the hydrogenation of maleic acid or other precursors to 1,4-butanediol; this was composed of palladium, silver and rhenium, and at least one part of iron, aluminum, cobalt and mixtures thereof on a carbon support. Using this catalyst, a higher conversion to and yield of butanediol could be obtained.

It was reported recently that 1,4-butanediol could be produced by the hydrogenation of 1,3-butadiene diepoxide in the presence of hydrogen and a hydrogenation catalyst (Fischer and Sigwart, 2000). Selective hydrogenation of 1,3-butadiene epoxide took place to form 1,4-butanediol only. By reacting allyl alcohol with CO and H_2 under a single catalyst system, 1,4-butanediol could also be produced (Zajacek and Shum, 2000). Initially, the reaction was carried out under relatively mild conditions using a catalyst system that consisted of a rhodium compound and a trialkyl phosphine. Subsequently, the reaction was combined with the same catalyst system (and preferably also the same reaction gas mixture) under more severe reaction conditions to form mainly 1,4-butanediol.

It has been reported that 1,4-butanediol could be produced by ω-oxidation of *n*-butanol by using several *Pseudomonas* strains (Stieglitz and Weimer, 1985). It can also be produced by the enzymatic conversion of 1,4-butanediol diester, using the carboxylesterase from *Brevibacterium linens*

IFO 12171 (Sakai et al., 1999), though further studies on this product are required.

Currently, the fermentative production of diols has received much attention because of their environmentally friendly processing from renewable resources. A biological process can also be applied for the production of optically active chemicals, which have been used as precursors for valued products, notably pharmaceuticals. Until recently, the yield and productivity of diols produced by fermentation have been rather low compared with those achieved by chemical processes. Metabolic engineering strategies have permitted an improved production of these chemicals by rearranging the existing metabolic pathways, or by generating new ones. However, the yield and productivity still need to be improved in order for these biological processes to remain competitive with their chemical counterparts. To overcome these drawbacks, it is essential to optimize metabolic pathways and to develop suitable fermentation strategies. In this sense, the development by DuPont and Genencor of an industrial process for the production of 1,3-propanediol is truly encouraging.

5 Hydroxy Acids

Another chemical group that can be used for the monomer of a polyester is that of hydroxy acids. In this section, the various production strategies of hydroxy acids such as lactic acid, lactone, and cyclic esters by biological systems will be discussed.

5.1 Lactic Acid

Lactic acid can be easily polymerized to polyesters because it contains both hydroxyl and carboxyl groups. Recently, lactate polymers have received an increasing amount of attention due to their various advantages, which include their high strength, thermoplasticity, ease of fabrication, biodegradability, and bioenvironmental compatibility.

Lactic acid can be produced by either chemical or biological routes. There are two enantiomeric forms that are designated L, *S*, or (+) (dextrorotary) and D, *R*, or (–) (levorotary). Racemic DL-lactic acid is obtained by chemical synthesis from lactonitrile (Holten et al., 1971), while optically pure L- or D-lactic acid can be produced by microbial fermentation. Some members of the lactic acid bacteria possess both D- and L-lactic acid dehydrogenases, resulting in the production of racemic lactic acid (Dennis et al., 1965). Recently, the fermentative production of lactic acid has attracted great interest, mainly as it can be produced from renewable resources.

The fermentation of lactic acid bacteria has a long history of uses in the pharmaceutical, chemical and food industries. The traditional applications of lactic acid and its derivatives are as food additives, for example as an acidulant and a preservative. Pharmaceutical applications of lactic acid and its derivatives include uses as drug intermediates, particularly optically pure methyl, ethyl, and isopropyl lactate esters for the synthesis of chiral molecules. For example, sodium lactate is used in parenteral and kidney dialysis solutions, while calcium and magnesium lactates are used in the treatment of mineral deficiencies (Benninga, 1990; Purac, 1993).

Due to this wide range of applications, approximately 50,000 tons of lactic acid is produced annually worldwide (Datta and Tsai, 1997). The commercial price of lactic acid ranges from US\$1.40 kg^{-1} for general uses to \$1.90 kg^{-1} for food grade (Chemical Market Report, 1999). However, the price of lactic acid is still too high for its economical use in the production of general-purpose polymers.

The traditional processes for lactic acid production consist of pretreatment of cheap carbon substrates, batch fermentation and separation/purification. In these processes, separation/purification is the most costly step. The efficiency and economics of the separation/purification step are mainly affected by the status of product solution after fermentation, including the concentrations of lactic acid and the presence of other contaminants (especially other acids and salts). The development of an efficient fermentation method, as well as efficient separation/purification techniques, is important if the overall production cost of lactic acid is to be reduced.

Much research effort has therefore been focused on optimizing several factors affecting the overall economics of fermentative lactic acid production, including the microorganisms employed, the carbon and nitrogen sources, pH, temperature, and various cell culture methods (Litchfield, 1996; Hofvendahl and Hahn-Hägerdal, 2000).

5.1.1 Microorganisms

Numerous bacteria exist which can convert carbohydrates to lactic acid as a major fermentation product: *Carnobacterium, Enterococcus, Lactobacillus, Lactococcus, Leuconostoc, Oenococcus, Pediococcus, Streptococcus, Tetragenococcus, Vagococcus,* and *Weissella* (Stiles and Holzapfel, 1997). Lactic acid bacteria lack aerobic or oxidative respiratory pathways, and synthesize ATP by glycolysis, the lactic acid being produced as a major end product to recycle the NAD^+ required to continue glycolysis. Most of lactic acid bacteria are anaerobic, catalase-negative, nonmotile, and do not form spores. They are also generally tolerant to low pH (< 5.0). Except for some pathogenic strains such as streptococci, most lactic acid bacteria are considered GRAS (generally regarded as safe).

Complex nitrogen sources providing amino acids and vitamins are required for most lactic acid bacteria to grow and to produce acids (Chopin, 1993), and this leads to an increase in the production costs of lactic acid.

Lactic acid bacteria are classified as: (1) homofermentative, producing only lactic acid as the end product of glucose metabolism (Figure 8A); or (2) heterofermentative, producing equimolar amounts of lactic acid, carbon dioxide and ethanol or acetate (Figure 8B). Fermentation of sugars by homofermentative strains takes place via the Embden–Meyerhof pathway, with the theoretical conversion of 1 mol of glucose to 2 mol of lactic acid (Smith et al., 1975; Thomas et al., 1979). The phosphoketolase pathway is used by facultative heterofermentative strains for the fermentation of pentoses, and for the fermentation of hexoses and pentoses by obligate heterofermentative strains (Kandler, 1983). The ratio of ethanol and acetate formed is dependent on the redox potential of the system (Garvie, 1980; Kandler, 1983). According to Kandler (1983), all lactic acid bacteria except for the type I lactobacilli (e.g. *Lactobacillus delbrueckii*) are able to ferment pentoses.

The yields of lactic acid from glucose by homofermentative strains are generally 90% or greater of the theoretical yield. From an economic point of view, homofermentative characteristics are regarded as an important criterion in strain selection. Even for the homofermentative bacteria, other metabolites such as ethanol, acetate and formate are formed under some conditions such as during glucose limitation, during growth on other sugars, and under high pH or low temperature (Figure 8C) (Fordyce et al., 1984; Sjöberg et al., 1995; Garrigues et al., 1997; Hofvendahl and Hahn-Hägerdal, 1997; Hof-

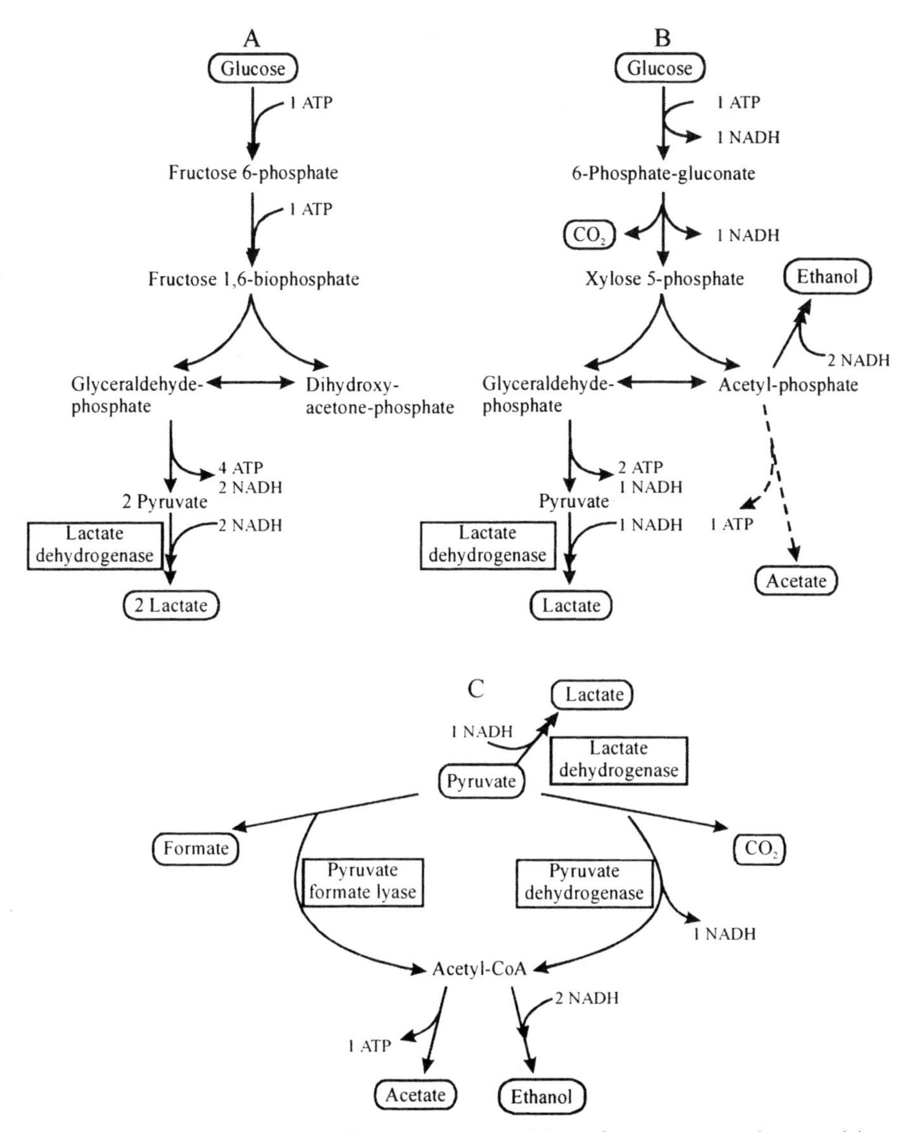

Fig. 8 Catabolic pathways in homofermentative (A) and heterofermentative (B) lactic acid bacteria. Production of ethanol, acetate and formate by homofermentative lactic acid bacteria under certain condition is also shown (C).

vendahl, 1998). In this case, pyruvate is also metabolized into formate and acetyl-CoA by pyruvate formate lyase under anaerobic conditions, or into carbon dioxide, acetyl-CoA and NADH in the presence of oxygen (Takahashi et al., 1982; de Vos, 1996).

Various bacterial strains produce L(+)-, D(−)-, or DL-lactic acid. The development of specialty polymers from lactic acid mentioned previously has led to an increased interest in the production of specific enantiomers. The dimers, LL, DD, and DL, can be used as building blocks for polylactic acid possessing different physical properties. The selection of a strain can be based on the desired lactic acid enantiomer, the substrate to be used, tolerance to temperature, pH and lactic acid, and yield and productivity. In addition to the usual desirable characteristics for the selection of strain, resistance to bacteriophage is another important issue which must be considered, and *Lactococcus* spp., *Lactobacillus* spp., and *Streptococcus* spp. are each subject to bacteriophage infection that can cause fermentations to fail. The various types of bacteriophage infection and the development of phage-resistant strains have been reviewed by Klaenhammer and Fitzgerald (1994).

Very recently, homofermentative lactic acid production using a metabolically engineered *E. coli* was reported (Chang et al., 1999). A *pta* mutant of *E. coli* RR1, which is deficient in the phosphotransacetylase (Pta) of the Pta-acetate kinase pathway, was found to metabolize glucose to D-lactic acid with only small amounts of succinic acid as a byproduct, under anaerobic conditions. An additional mutation in *ppc* coding for phosphoenolpyruvate carboxylase made the mutant produce D-lactic acid, as would a homofermentative bacterium. A nonindigenous fermentation product, L-lactic acid, could also be produced by introducing an L-lactate dehydrogenase gene from *Lactobacillus casei* into a *pta ldhA* mutant strain of *E. coli*.

5.1.2
Carbon and Nitrogen Sources

Various refined, unrefined, and even waste carbohydrates have been examined as carbon substrates for lactic acid production. When a refined carbohydrate was used, the product obtained was also pure, and this resulted in lower purification costs. However, due to the high price of refined sugars, less expensive waste products from agriculture such as whey and molasses, starch from crops or wastes, and lignocellulosic materials have received much attention for the production of lactic acid.

The most commonly used cheap substrate for fermentative production of lactic acid is whey (a waste product from cheese production), which contains proteins, salts, and lactose. Whey lactose can be hydrolyzed to glucose and galactose (Amrane and Prigent, 1994, 1996), deproteinized by ultrafiltration (Roy et al., 1986, 1987a,b; Aeschlimann et al., 1990; Krischke et al., 1991; Norton et al., 1994a), and demineralized (Linko et al., 1984). The strains that have often been employed for the production of lactic acid from whey are *L. delbrueckii* ssp. *bulgaricus* (Stieber and Gerhardt, 1981; Tewari et al., 1985; Podlech et al., 1990), *Lactobacillus helveticus* (Boyaval et al., 1987; Aeschlimann and von Stockar, 1989, 1991) and *L. casei* (Tuli et al., 1985; Roukas and Kotzekidou, 1991).

Molasses, a byproduct from the sugar manufacturing process, can also be used for lactic acid production (Aksu and Kutal, 1986; Göksungur and Güvenç, 1997; Monteagudo et al., 1997). *L. delbrueckii* has often been used for the production of lactic acid from molasses.

Starch from crops or wastes is another common substrate (Giraud et al., 1994; Zhang and Cheryan, 1994; Javanainen and Linko, 1995; Hofvendahl and Hahn-Hägerdal, 1997). Various lactic acid bacteria, including *L. casei* (Javanainen and Linko, 1995), *Lactobacillus plantarum* (Shamala and Sreekantiah, 1987), *L. delbrueckii* (Ray et al., 1991; Khan et al., 1995), *Lactobacillus helveticus* (Tsai and Millard, 1994) and *Lactococcus lactis* (Kurosawa et al., 1988; Hofvendahl and Hahn-Hägerdal, 1997) have been employed for lactic acid production from starch. Starch must be hydrolyzed to glucose, and maltose must be fermentable by lactic acid bacteria. To reduce the pretreatment costs, several amylase-producing organisms including *Lactobacillus fermentum* (Chatterjee et al., 1997), *Lactobacillus amylovorus* (Cheng et al., 1991; Zhang and Cheryan, 1991), and *Lactobacillus amylophilus* (Mercier et al., 1992; Yumoto and Ikeda, 1995) have been investigated for the direct production of lactic acid from untreated or liquefied/gelatinized starch. Amylase can be added separately to hydrolyze starch, after which the lactic acid bacteria can ferment glucose to produce lactic acid (Richter and Träger, 1994; Khan et al., 1995; Hofvendahl et al., 1999). A mixed-culture system containing amylase-producing *Aspergillus awamori* and *L. lactis* has also been reported (Kurosawa et al., 1988).

Lignocellulosic materials, including waste paper (McCaskey et al., 1994), plant materials (Chen and Lee, 1997; Melzoch et al., 1997) and wood (Linko et al., 1984) have also been used for the production of lactic acid in similar manner to starch. The main sugar components are three hexoses (glucose, galactose, and mannose) and two pentoses (xylose and arabinose). Simultaneous saccharification and fermentation of cellulose has been studied with *L. delbrueckii* (Abe and Takagi, 1991) and *Lactobacillus rhamnosus* (Parajó et al., 1997; Schmidt and Padukone, 1997). A mixed culture with the cellulase-producing fungus *Trichoderma reesei* has also been reported (Venkatesh et al., 1994).

Lactic acid bacteria require complex nutrients due to their limited ability to synthesize amino acids and vitamins B (Chopin, 1993). The addition of a complex nitrogen source generally has a positive effect on lactic acid production. The most effective complex nutrient for the production of lactic acid is yeast extract (Aeschlimann and von Stockar, 1990; Olmos-Dichara et al., 1997; Amrane and Prigent, 1998; Payot et al., 1999), though due to its high price (\$3.3–10 kg^{-1}, depending on its purification method), yeast extract is not cost-effective for the production of commodity chemicals such as lactic acid. In an economic analysis for the production of lactic acid (Mulligan et al., 1991; Tejayadi and Cheryan, 1995), the estimated cost contribution of the yeast extract was more than 30% of the total production cost. Many studies have been conducted to find a cheaper alternative to yeast extract (Lund et al., 1992; Hujanen and Linko, 1996; Yoo et al., 1997; Demirci et al., 1998). Barley malt sprouts (Hujanen and Linko, 1996) and soybean hydrolyzate with the addition of five vitamin B compounds (Yoo et al., 1997) might reduce the amount of yeast extract used to the level of 4 and 5 g L^{-1}, respectively, which is much less than that (10–20 g L^{-1}) used normally. Whey protein was also used to replace yeast extract, by using a newly screened strain (Börgardts et al., 1998). Recently, Kwon et al. (2000) reported the total replacement of yeast extract by soytone (an enzymatic digest of soybean mill) supplemented with seven selected vitamins for the production of lactic acid by *L. rhamnosus*.

The results of fermentations using various bacteria with different carbon and complex nitrogen sources are summarized in Table 1.

5.1.3 Cell Culture Modes

In large-scale production, the cell culture mode has a major effect on the final price of the product. Usually, batch mode has been employed in industrial-scale fermentations, but for the efficient production of lactic acid, a number of alternative, continuous culture models have been studied.

Batch Processes

Batch fermentation is most commonly used for lactic acid production. In an earlier commercial-scale process of American

Tab. 1 Summary of lactic acid production by various bacteria

Organism	*Carbon source and treatment*	*Nitrogen source*	*LA [g L⁻¹]*	$Y_{LA/tot}$ [g g⁻¹]	Q_v [g L⁻¹ h⁻¹]	*Ref.*
Lactobacillus amylophius ATCC 49845	Glucose		21	0.95	0.6	1
	Hydr corn starch		23	0.73	0.88	
L. amylovorus ATCC 33620	Cassava stach		4.8	0.48	0.69	2
	Corn starch		10	1.0	1.2	
	Potato starch		4.2	0.42	0.14	
	Rice starch		7.9	0.79	0.86	
	Wheat starch		7.8	0.78	1.2	
L. amylovorus ATCC 33622	Raw corn starch		45	0.82	8.6	3
	Liq corn starch		55	1.0	20	
L. casei NRRL B-441	Liq barley starch + glucoam		112	0.68		4
	Liq barley starch + glucoam + alpha		162	0.87		
L. casei NRRL B-441 +	Barley flour		36	0.20		5
L. amylovorus NRRL B-4542	Barley flour + glucoam		114	0.63		
L. delbrueckii IFO3534	Hydr newspaper		24	0.48		6
	Hydr pure cellulose		53	0.53		
L. delbrueckii sowjeskij	Sucrose	20% YE	21	1.1	1.0	7
	Sucrose	1% YE +0.5% pep	18	0.90	0.83	
L. delbrueckii sp. *bulgaricus* CNRZ369	Glucose		56	2.8		8
	Cellobiose		32	1.6		
	Xylose		41	2.1		
L. delbrueckii sp. *delbrueckii*	Glucose		87	0.87	5.5	9
	Fructose + glucose		94	0.94	5.5	
	Sucrose		85	0.85	6.2	
L. delbrueckii sp. *delbrueckii* ATCC 9649	Glucose		58	0.85		10
	Glucose	MRS +1% YE	58	0.48	0.72	11
	Glucose	MRS +3% YE	67	0.56	1.4	
	Lactose		40	0.75		10
	Hydr wheat flour, SSF	YE	106	0.82	1.6	12
	Hydr wheat flour, SSF	YE	109	0.91	3.6	
	Whey permeate		36	0.75	5.8	13
L. helveticus Milano	Whey permeate	YE higher conc	36	0.75	9.4	13
	Whey permeate	YE + pep	40	0.83	12	

Tab. 1 (cont.)

Organism	Carbon source and treatment	Nitrogen source	LA [g L^{-1}]	$Y_{LA/tot}$ [g g^{-1}]	Q_v [g L^{-1} h^{-1}]	Ref.
L. helveticus sp. *milano*	Glucose		18	0.36	4.2	14
	Maltose		42	0.84	5.0	
	Hydr whey	YE	44		5.5	15
	Hydr whey, clarified	CSL	41		4.4	
	Whey, UF	CSL	37		2.7	
L. paracasei No 8	Glucose		95	0.95	5.6	16
	Sweet sorghum		106	0.79	10	
	Sweet sorghum	YE + pep	91	0.91	10	
L. pentosus	Glucose	MRS	46	0.92	2.4	17
	Glucose	YE	45	0.90	2.3	
	Xylose		27	0.54	0.59	
	Glucose + xylose		90	1.8	4.0	
	Hydr wood		40	0.70	1.3	
L. pentosus NRRL B-227	Glucose		6.0	0.60		18
	Galactose		734	0.74		
	Mannose		6.8	0.68		
	Hydr cellulose: glu, mam, xyl, gal			0.51		
L. pentosus NRRL B-473	Glucose		6.9	0.69		18
	Galactose		5.9	0.59		
	Mannose		7.4	0.74		
	Xylose		1.4	0.14		
	Hydr cellulose: glu, mam, xyl, gal					
L. plantarum	Hydr soluble starch		15	0.30		19
	Hydr tapioca starch		15	0.30		
	Hydr tapioca flour		17	0.35		
L. plantarum ATCC 14917	Sorghum	5% vetch juice			2.2	20
	Sorghum	15% vetch juice			2.0	
	Sorghum	25% vetch juice			2.8	
L. plantarum NRRL B-787	Glucose		6.2	0.62		18
	Galactose		4.0	0.40		
	Mannose		6.6	0.66		
	Hydra cellulose: glu, man, xyl, gal			0.42		

Tab. 1 (cont.)

Organism	*Carbon source and treatment*	*Nitrogen source*	*LA [g L^{-1}]*	*$Y_{LA/tot}$ [g g^{-1}]*	*Q_v [g L^{-1} h^{-1}]*	*Ref.*
L. plantarum NRRL B-788	Glucose		6.0	0.60		18
	Galactose		4.9	0.49		
	Hydra cellulose: glu, man, xyl, gal			0.46		
L. plantarum NRRL B-813	Glucose		7.3	0.73		18
	Galactose		4.7	0.47		
	Mannose		8.3	0.83		
	Hydr cellulose: glu, man, xyl, gal			0.43		
L. plantarum USDA 422	Glucose		5.2	0.82		18
	Galactose		3.1	0.31		
	Mannose		6.2	0.62		
	Xylose		1.3	0.13		
L. plantarum USDA 422	Hydr cellulose: glu, man, xyl, gal			0.43		18
L. rhamnosus ATCC 10863	Glucose		17	0.86		21
	Glucose	0.25% YE +0.5% trp	57	0.81		22
	Glucose	0.5% YE +1% trp	58	0.96		
	Glucose	1.93% soy +7 vit	125	0.92		23
	Fructose		14	0.71		21
	Fructose + glucose		16	0.81		
	Sucrose		15	0.73		
	Hydr molasses, SSF		16	0.81		
	Hydr molasses, SSF	YE + pep	14	0.70		
	Alpha-cellulose		45		0.96	24
	Switch grass cellulose	YE + pep	28	0.96	0.50	
	Hydr wood	YE + pep	27	1.0	2.3	25
	Hydr wood, SSF		29		1.5	
LBM5 = mix of five Lactobacilli strains	Glucose		99	0.90		26
	Gtarch		90	0.82		
Lactococcus lactis sp. *lactis* ATCC 13673	Glucose		36	1.0	3.6	28
	Xylose		13	0.42	0.37	
L. lactis sp. *lactis* ATCC 19435	Glucose		4.9	0.86	2.5	29
	Maltose		3.2	0.70	1.0	
L. lactis sp. *lactis* ATCC 19435	Hydr wheat flour, 3 μL enz g^{-1}		75	0.78	1.2	30

Tab. 1 (cont.)

Organism	*Carbon source and treatment*	*Nitrogen source*	*LA [g L⁻¹]*	$Y_{LA/tot}$ *[g g⁻¹]*	Q_v *[g L⁻¹ h⁻¹]*	*Ref.*
	starch		75	0.83	0.85	
	Hydr wheat flour, 5 μL enz g^{-1}		90	0.98	1.5	
	starch		87	0.93	1.7	
	Hydr wheat flour, 6 μL enz g^{-1}		96	0.76	3.0	12
	starch	YE	106	0.88	3.3	
	Hydr wheat flour, 8 μL enz g^{-1}		90	0.98	1.5	
	starch	YE	87	0.96	3.3	30
	Hydr wheat flour, SSF		43		1.5	
	Hydr wheat flour, SSF					
	Hydr wheat flour, SSF					
	Hydr wheat flour, SSF					
	Hydr wheat flour + protease					
L. lactis sp. *lactis* ATCC 19435	Hydr wheat flour + protease	vitamins	46		2.4	30
	Hydr wheat flour + protease	amino acids	53		2.8	
	Hydr wheat flour + protease	peptides	44		2.2	
	Hydr wheat flour		17		0.23	
L. lactis sp. *lactis* NRRL B-4449	Glucose		6.6	0.66		18
	Galactose		2.8	0.28		
	Mannose		5.8	0.58		
	Xylose		1.8	0.18		
	Hydr cellulose: glu, man, xyl, gal			0.16		

Abbreviations: LA = lactic acid; Y_{LA}/tot = yield of g LA per g substrate provided; Qv = maximum volumetric LA productivity in g LA L^{-1} h^{-1}; Ref = reference; hydr = hydrolyzed; liq = liquefied; glucoam = glucoamylase; alpha = alpha-amylase; glu = glucose; man = mannose; xyl = xylose; gal = galactose; enz = enzyme; YE = yeast extract; pep = peptone; soy = soytone; vit = vitamin; CSL = corn steep liquor; trp = tryptone; SSF = simultaneous saccharification and fermentation; UF = ultrafiltration.

References: 1, Mercier et al., 1992; 2, Xiaodong et al., 1997; 3, Zhang and Cheryan, 1991; 4, Linko and Javanainen, 1996; 5, Javanainen and Linko, 1995; 6, Abe and Takagi, 1991; 7, Milko et al., 1966; 8, El Sabaeny, 1996; 9, Suskovic et al., 1991; 10, Ohleyer et al., 1985; 11, Demirci and Pometto, 1992; 12, Smith et al., 1975; 13, Amrane and Prigent, 1997; 14, Roy et al., 1987b; 15, Amrane and Prigent, 1993; 16, Richter and Träger, 1994; 17, Linko et al., 1984; 18, McCaskey et al., 1994; 19, Shamala and Sreekantiah, 1988; 20, Samuel et al., 1980; 21, Aksu and Kutal, 1986; 22, Olmos-Dichara et al., 1997; 23, Kwon et al., 2000; 24, Chen and Lee, 1997; 25, Parajó et al., 1997; 26, Tsai and Millard, 1994; 27, Ishizaki et al., 1992; 28, Tyree et al., 1990; 29, Hofvendahl, 1998; 30, Hofvendahl et al., 1999.

Maize Products Co. (Hammond, IN, USA), lactic acid concentrations reached 120–135 g L^{-1} from 150 g L^{-1} glucose, with yields in the range of 0.80 to 0.90 of lactic acid per gram substrate utilized (Litchfield, 1996). In this case, the lactic acid concentration was limited basically by the solubility of the calcium lactate formed.

Only a few laboratory-scale batch fermentation reported in the scientific literature have achieved lactic acid concentrations above 100 g L^{-1}, owing to the relatively low substrate concentrations used, and these are generally in the range of 20 to 100 g L^{-1}. Mehaia and Cheryan (1987a) were able to produce 115 g L^{-1} of lactic acid by employing *L. delbrueckii* ssp. *bulgaricus* using 150 g L^{-1} of whey permeate. A lactic acid concentration of 120 g L^{-1} was obtained by employing *L. amylovorus* using a medium containing 100 g L^{-1} of enzyme-thinned starch and 30 g L^{-1} of yeast extract (Cheng et al., 1991).

During lactic acid fermentation, end-product inhibition occurs as the (undissociated) lactic acid concentration increases (Friedman and Gaden, 1970; Bibal et al., 1989; Gätje and Gottschalk, 1991; Ohara et al., 1992; Dutta et al., 1996). Furthermore, Gonçalves et al. (1991) observed that increasing glucose concentrations inhibited growth of and lactic acid production by *L. rhamnosus* ATCC 10863.

The presence of substrate- and end-product inhibition implies that there is a limitation to the improvement of efficiency of lactic acid production by batch fermentation. Generally, substrate inhibition can be solved by using fed-batch fermentation, but end-product inhibition may result in a much less successful fermentation. Consequently, investigations have been made into the development of continuous fermentation processes.

Continuous Processes

Generally, a lactic acid concentration higher than 100 g L^{-1}, a yield >90%, and a lactic acid productivity of more than 5 g L^{-1} h^{-1} can be achieved in batch fermentation. Productivities obtained by batch fermentations are lower than might be desired from an economic standpoint, however, and consequently considerable effort has been made to develop continuous fermentation processes to improve system productivity. Continuous fermentation provides generally higher productivities than batch fermentations, but unfortunately in general has a low yield. As a result, cell immobilization and cell recycle systems have been investigated to overcome these disadvantages.

Immobilized cell systems. Cell immobilization avoids the need to separate cells from the fermentation medium, and also avoids the membrane fouling problems that are often encountered in recycled systems. Immobilized *Lactobacillus* systems have been used on a recycle basis for periods of up to 157 days (Linko et al., 1984). A number of investigators have described a method of entrapping living cells in polymeric beads, including agar (Tuli et al., 1985), calcium or sodium alginates (Guoqiang et al., 1991; Roukas and Kuzekidou, 1991), carrageenan combined with locust bean gum (Norton et al., 1994b), and polyacrylamide or polyvinyl alcohol gels (Tsutomu, 1982). Unfortunately, cell immobilization has not been very successful in terms of increasing the lactic acid yield and productivity, mainly due to problems of oxygen and substrate transport (Yahannavar and Wang, 1991; Wang et al., 1995). Indeed, in many cases better results were obtained using free cells.

Continuous membrane cell recycle systems. The beneficial effect of high cell concentrations on lactic acid productivity obtained with this type of bioreactor is clear from the results of Vickroy (1985). A continuous-stirred tank reactor (CSTR) equipped with an ultrafiltration unit (molecular weight cut-off 100,000 Da) allowed production of lactic acid to a concentration of 35 g L^{-1} and productivity of 76 g L^{-1} h^{-1} from glucose. The *L. delbrueckii* cell concentration was 54 g dry cell weight (DCW) L^{-1} under these conditions. Lactic acid and cell mass yields (g g^{-1} glucose) were 0.96 and 0.09, which can be compared with batch fermentation values of 0.90 and 0.16, respectively. Using the same strain, Major and Bull (1989) employed a CSTR-hollow-fiber membrane cell recycle system, and found that the molar ratio of lactate to ethanol and acetate was decreased as a result of glucose limitation. A higher lactic acid productivity could be obtained by cell recycle culture (10.1 – 12.0 g L^{-1} h^{-1}) compared with that obtained by chemostat (8.3 g L^{-1} h^{-1}). Mehaia and Cheryan (1987a,b) obtained an impressively high lactic acid productivity of 84.2 g L^{-1} h^{-1} from whey permeate, using a hollow-fiber membrane cell recycle system. Tejayadi and Cheryan (1995) reported a continuous membrane cell recycle process that produced lactic acid to a concentration of 90 g L^{-1} and with productivity of 22 g L^{-1} h^{-1} from whey permeate and yeast extract by using an ultrafiltration unit (molecular weight cut-off 30,000 Da). By employing this process, lactic acid production cost was estimated to be \$1.2 kg^{-1}.

5.1.4 Separation/Purification

In the traditional process of lactic acid production, the separation/purification step generated a large amount of calcium sulfate as a byproduct. In order to prevent the generation of such undesirable byproducts, and thereby to reduce production costs, electrodialysis has been suggested to replace the traditional recovery process. Desalting electrodialysis requires only a small amount of electric power for the recovery, purification, and concentration of lactic acid from the fermentation products. Electrodialysis can split lactic acid and alkali from the lactate salt, whilst the alkali produced can be reused to control pH during fermentation (Datta, 1989). By using two electrodialysis units and an ion-exchange unit, lactic acid could be purified with only 0.1% of protein contaminant. In this process, the electric power required was ~ 1 kWh kg^{-1} lactic acid (Datta, 1989; de Gooijer et al., 1996).

Other methods for the separation and purification of lactic acid have been outlined in a previous review (Litchfield, 1996; and references cited therein).

As described above, the fermentative production of lactic acid has attracted wide scientific and industrial interest, and continuous fermentation with cell recycling appears to be the most promising replacement for traditional batch fermentation. Electrodialysis is also becoming recognized as the standard method for the separation/purification of lactic acid produced by fermentation as its use, in combination with continuous fermentation, means that cells should be resistant to high osmotic pressure. Further research on metabolic and cellular engineering of lactic acid bacteria should be focused on the development of a lactic acid-tolerant strain, a strain that can directly ferment polymeric raw material such as starch and cellulose, and a strain that does not require expensive complex nitrogen sources.

Cargill Dow (Minnetonka, MN, USA) is constructing a plant for the annual production of 140,000 tons of lactic acid, which will be used to make polylactide (NatureWorks™). This development represents an exciting phase in commercial lactic acid production, and will undoubtedly trigger greater interest in the production of chemicals from renewable resources.

5.2 Lactones and Other Cyclic Esters

Lactones are cyclic esters formed by the intramolecular reaction of hydroxy acids in which the two functional groups, the hydroxyl and carboxyl groups, are separated by three or more carbon atoms. Often, γ-butyrolactone and its derivatives form spontaneously, especially from γ-hydroxy acids. Lactides are cyclic diesters formed by the intermolecular reaction of α-hydroxy acids (Figure 9).

Lactones and other cyclic esters have been used as monomers for the chemical synthesis of polyesters, the major advantage being that no small molecule is produced as a byproduct. Generally, polyesters can be produced by esterification of the carboxyl group of organic acids with the hydroxyl group of alcohols, or by transesterification between esters. During esterification and transesterification, one small molecule (e.g., water or alcohol) is produced as a byproduct, and this causes a shift in equilibrium, resulting in both a low yield and a low molecular-weight product. Since the problem of byproduct formation can be solved by ring-opening polyesterification of lactones or cyclic esters, high molecular-weight polyesters can be produced with high yields when these are used as building blocks.

It has been reported that several microorganisms produce lactones as secondary metabolites. These microbial lactones show pharmacological effects such as antibiotic activity, and hence research has focused on these materials for pharmaceutical purposes. These microbial lactones were found to be not suitable for the chemical synthesis of polyesters, however. Lactones and cyclic esters for the chemical synthesis of polyesters can be obtained initially by the fermentative production of an intermediate compound, followed by chemical transformation as described below.

Numerous microorganisms produce various hydroxy acids and their polyesters, including lactic acid and PHAs. Cargill Inc. (Minneapolis, MN, USA) has the patent on the continuous production of lactide from esters of lactic acid (Gruber et al., 1993), whilst DuPont (Wilmington, DE, USA) has the patent on preparing cyclic esters from the corresponding hydroxy esters or their oligomers (Bhatia, 1989). Lactic acid and PHAs can be produced efficiently by fermentative methods, as described in previous sections and in other chapters of this book. In turn, various cyclic esters can be prepared from these fermentative products, and these can be used subsequently in the synthesis of high molecular-weight polymers.

Several lactones have been investigated as building blocks for the synthesis of copolymers with lactides. For example, β-butyrolactone (Hori et al., 1993; Reeve et al., 1993), γ-butyrolactone (Nakayama et al., 1998), δ-valerolactone (Kurcok et al., 1992) and ε-caprolactone (Bero et al., 1993; Perego et al., 1993; Zhang et al., 1993) have been used as

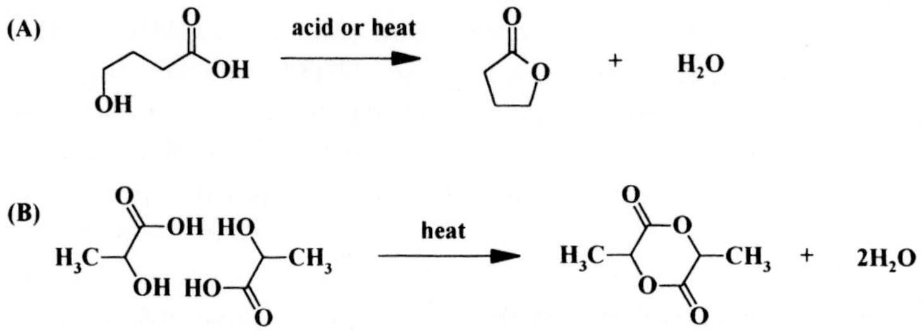

(A) acid or heat → + H_2O

(B) heat → + $2H_2O$

Fig. 9 The structural formulas of γ-butyrolactone (A) and a lactide (B).

reactants in the polymerization reaction with lactide, resulting in random or block copolyesters.

Reeve et al. (1993) synthesized diblock poly(lactide-*co*-β-butyrolactone) using natural low molecular-weight poly[(*R*)-3-hydroxybutyrate] [poly(3HB)] as a macroinitiator for lactide. Random poly(lactide-*co*-β-butyrolactone) was prepared by using a distannoxane catalyst (Hori et al., 1993), but due to the higher reactivity of L-lactide, its fraction in the copolymer (17%) was greater than that in the initial mixture (10%).

γ-Butyrolactone has a steadily growing demand because of its major use as an important reactant in the synthesis of other compounds, as well as its direct use as a dielectric material or as a solvent. It is currently produced by the hydrogenation of 1,4-butanediol, but can also be prepared by the hydrogenation of maleic anhydride (Hara et al., 2000). Due to its poor polymerizing ability, only 28 mol.% of γ-butyrolactone can be incorporated into the copolymer, poly(lactide-*co*-γ-butyrolactone), when its fraction in the feed mixture was 66% (Nakayama et al., 1998). Kurcok et al. (1992) reported the preparation of block copolymers of L-lactide and δ-valerolactone by the addition of L-lactide to a living poly(δ-valerolactone) macroinitiator. The copolymer of L-lactide and ε-caprolactone has attracted the interest of polymer scientists because its properties differ widely from those of polylactide and polycaprolactone homopolymers (Bero et al., 1993; Perego et al., 1993; Zhang et al., 1993).

β- and γ-butyrolactone can also be produced by fermentation combined with additional chemical treatment and/or transformation. β-Butyrolactone can be synthesized from the β-hydroxybutyric acid monomer of poly(3HB) (Seebach et al., 1987). An orthocarbonate derived from β-hydroxybutyric acid can be easily transformed to β-butyrolactone and diethyl carbonate by pyrolysis at 150–150 °C/60 Torr. The production of poly(3HB) and β-hydroxybutyric acid is described in other chapters of this book.

γ-Butyrolactone can be prepared by the hydrogenation of succinate, maleic anhydride or its esters, and by the traditional dehydrogenation of 1,4-butanediol (Castiglioni et al., 1996; Schlander and Turek, 1999; Hara et al., 2000). Gas-phase hydrogenation of succinate, maleic anhydride or its esters is a generally used process for the production of 1,4-butanediol (see above). Castiglioni et al. (1996) developed an aluminum-containing catalyst system, and γ-butyrolactone was produced with a yield of 96% by hydrogenation of maleic anhydride using a Cu/Zn/Al catalyst (25.3:50.7:24.0, atomic ratio) at 265 °C. Schlander and Turek (1999) used copper/zinc oxide catalysts for gas-phase hydrogenolysis of dimethyl maleate to 1,4-butanediol and γ-butyrolactone, and suggested that the equilibrium constant for the reaction between γ-butyrolactone and 1,4-butanediol was a function of temperature and pressure. In this copper/zinc oxide catalyst system, dimethyl maleate could be converted to γ-butyrolactone at 270 °C and 10 bar, without formation of 1,4-butanediol.

Recently, a hydrogenation system for the production of γ-butyrolactone catalyzed by ruthenium complexes consisting of $Ru(acac)_3$, $P(sctyl)_3$ and *p*-toluenesulfonic acid was demonstrated by Hara et al. (2000). The selectivity of the reaction towards γ-butyrolactone was as high as 97%, and over 90% of the catalysts could be recovered.

Although no report has been made on the fermentative production of lactones and cyclic esters of current interest, a combination of metabolic pathway engineering and enzyme evolution might allow this in the future. However, if ultimately this is unsuccessful, the possibility remains of combining catalysis and biocatalysis methods in order to improve the existing processes.

6 Outlook and Perspectives

Although it is possible to produce building blocks for the synthesis of polyesters by chemical transformations, these processes are not only based on fossil fuels (which have limited availability) – they also often generate waste streams containing environmental pollutants. Therefore, the production of chemical building blocks by using biological systems has received much attention on the basis of the renewable carbon sources used, the mild process conditions and the environmental benignancy. Moreover, it has been shown that many bulk chemicals produced by chemical processes can also be prepared using microbial fermentation.

For example, the biological production of succinic acid has been investigated extensively with regard to the so-called "Renewable Vision". At present, succinic acid is produced at almost maximum theoretical yield by both natural and recombinant strains of *E. coli*. However, the price of biologically produced succinic acid is still higher than its chemically produced counterpart, and the cost of separation processes remaining a major factor for production costs. In particular, the separation of succinic acid from other organic acids (including lactic and acetic acids) is extremely difficult and expensive. In succinic acid production, the target therefore is to reduce the amount of byproduct(s) formed, which in turn will lead to an overall reduction in production costs. Recently, adipic acid has been produced successfully from renewable carbon sources such as glucose, by using recombinant *E. coli*. Although (at present) the yield and productivity were not sufficiently high to generate interest from an industrial standpoint, further improvements in both bacterial strains and their cultivation strategies should render this process viable.

The production of diols, including 1,2-propanediol, 1,3-propanediol and 1,4-butanediol, has also received much attention, especially as these compounds can be used as monomers to generate superior polyesters such as PET, PPT, and PBT. The diols can be produced by either natural and/or recombinant bacteria, using a variety of carbon sources. Consequently, studies on diol production from renewable carbon sources have been extensive, and this has led to the development of several metabolically engineered microorganisms such as *E. coli*, *Saccharomyces cerevisiae* and *Klebsiella* spp. that are able to produce diols on an industrial scale.

The economic and competitive bulk production of lactic acid has been studied in great depth, and in this respect many natural and mutant bacterial strains have been used in fermentation processes. Many factors, including substrate, desired enantiomer and tolerance to temperature and pH, together with yield and productivity, have been considered when selecting such bacterial strains. The use of cheap carbon sources is an important factor in reducing lactic acid production costs; consequently, waste materials such as whey, molasses, starch and lignocellulosic materials will undoubtedly be investigated as possible substrates for this process. Ultimately, the further development of bacterial strains and fermentation/purification processes will result in lower production costs, and will also permit the fermentative production of lactic acid to become an indispensable industrial process in the future.

In the synthesis of polyesters, cyclic esters (including lactones and lactides) have been used as important monomers because they do not generate byproduct(s) during the polymerization reaction. Hence, a major research effort is likely to be directed towards improving the yield of these monomers by both chemical and biological means.

The commercial processes for the production of 1,3-propanediol have been developed by DuPont and Genencor, employing recombinant *K. pneumoniae* and *E. coli* and renewable carbon sources such as glucose. Indeed, a plant for the production of large amounts of lactic acid is currently under construction by Cargill Dow. These are all excellent examples of the fermentative production of commercial bulk chemicals, aided by metabolic engineering. It is clear that in the future, as our understanding of the metabolism of various microorganisms increases, and our experience of mid- and downstream processes gains momentum, an increasing number of industrial processes will be created for the production of chemicals from renewable resources.

Acknowledgements
The studies described in this chapter were supported by the Korean Ministry of Commerce, Industry and Energy (MOCIE) and by the National Research Laboratory program of the Korean Ministry of Science and Technology (MOST).

7 Patents

Publication number (Date)	*Assignee* / *Title of Patent*	*Inventors*
US6159738 (2000-12-12)	University of Chicago (IL)	Mark Donnelly (IL); Cynthia Sanville-Millard (IL); Ranjini Chatterjee (IL)
	Method for construction of bacterial strains with increased succinic acid production	
US5958744 (1999-09-28)	Applied Carbochemicals (MI)	Kris Berglund (MI); Sanjay Yedur (CA); Dilum Dunuwila (MI)
	Succinic acid production and purification	
US5168055 (1992-12-01)		Rathin Datta (IL); David Glassner (MI); Mahendra Jain (MI); John Vick Roy (MI)
	Fermentation and purification process for succinic acid	
US5143834 (1992-09-01)		David Glassner (MI); Rathin Datta (IL)
	Process for the production and purification of succinic acid	
US5770435 (1998-06-23)	University of Chicago (IL)	Mark Donnelly (IL); Cynthia Millard (IL); Lucy Stols (IL)
	Mutant *E. coli* strain with increased succinic acid production	
US5573931 (1996-11-12)	Michigan Biotechnology Institute (MI)	Michael Guettler (MI); Mahendra Jain (MI); Denise Rumler (MI)
	Method for making succinic acid, bacterial variants for use in the process, and methods for obtaining variants	

(Continued)

Publication number (Date)	*Assignee* *Title of Patent*	*Inventors*
US5521075 (1999-05-28)	Michigan Biotechnology Institute (MI)	Michael Guettler (MI); Mahendra Jain (MI)
	Method for making succinic acid, *anaerobiospirillum succiniciproducens* variants for use in process and methods for obtaining variants	
US5143833 (1992-09-01)		Rathin Datta (IL)
	Process for the production of succinic acid by anaerobic fermentation	
US5504004 (1996-04-02)	Michigan Biotechnology Institute (MI)	Michael Guettler (MI); Mahendra Jain (MI); Bhupendra Soni (IL)
	Process for making succinic acid, microorganisms for use in the process and methods of obtaining the microorganisms	
US5475031 (1995-12-12)		William Livingston
	Discovery of a valuable property for succinic acid and other intermediary metabolites	
US5202335 (1993-04-13)	Kissei Pharmaceutical Co., Ltd., (JP)	Fumiyasu Sato (JP); Atsushi Tsubaki (JP); Hiroshi Hokari (JP); Nobuyuki Tanaka (JP); Masaru Saito (JP); Kenji Akahane (JP); Michihiro Kobayashi (JP)
	Succinic acid compounds	
US6087140 (2000-07-11)	Wisconsin Alumni Research Foundation (US)	Douglas C. Cameron (US); Anita J. Shaw (US); Nedim E. Altaras (US)
	Microbial production of 1,2-propanediol from sugar	
US5254467 (1993-10-19)	Henkel Kommanditgesellschaft auf Aktien, Fesellschaft fuer Biotechnologische Forschung mbH (DE)	Kretschmann Josef (DE); Carduck Franz-Josef (DE); Deckwer Wolf-Dieter (DE); Tag Carmen (DE); Biebl Hanno (DE)
	Fermentive production of 1,3-propanediol	
WO9958686 (1999-11-18)	Du Pont (US); Genencor Int (US);	Bulthuis Ben (NL); Gatenby Anthony A (US); Trimbur Donald E (US); Whited Gregory M (US)
	Method for the production of 1,3-propanediol by recombinant organisms comprising genes for vitamin B12 transport	
US4400468 (1983-08-23)	Hydrocarbon Research Inc (US)	Faber Marcel (US)
	Process for producing adipic acid from biomass	
WO9821339 (1998-05-22)	Du Pont (Us); Genencor Int (Us)	Dias-Torres Maria (US); Haynie Sharon Loretta (US); Hsu Amy Kuang-Hua (US); Nair Ramesh V (US); Gatenby Anthony Arthur (US); Lareau Richard D (US); Nagarajan Vasantha (US); Nakamura Charles E (US); Payne Mark Scott (US); Picataggio Stephen Kenneth (US); Trimbur Donald E (US); Whited Gregory M (US)
	Method for the production of 1,3-propanediol by recombinant organisms	

(Continued)

Publication number (Date)	Assignee	Inventors	Title of Patent
US5821092 (1998-10-13)	Du Pont (US);	Nagarajan Vasantha (US); Nakamura Charles Edwin (US)	Production of 1,3-propanediol from glycerol by recombinant bacteria expressing recombinant diol dehydratase
US5487987 (1996-01-30)	Frost John W (US); Draths Karen M (US)	Purdue Research Foundation (US)	Synthesis of adipic acid from biomass-derived carbon sources
US6013494 (2000-01-11)	Du Pont (US); Genencor Int (US)	Diaz-Torres Maria (US); Gatenby Anthony A (US); Hsu Amy Kuang-Hua (US); La Reau Richard D (US); Nair Ramesh V (US); Haynie Sharon L (US); Payne Mark S (US); Nagarajan Vasantha (US); Nakamura Charles E (US); Picataggio Stephen K (US); Trimbur Donald E (US); Whited Gregory M (US)	Method for the production of 1,3-propanediol by recombinant microorganisms
US6214967 (1998-08-20)	Selin Johan Fredrik (FI); Jansson Kari (FI); Koskinen Jukka (FI); Neste OY (FI)	Selin Johan-Fredrik (FI); Jansson Kari (FI); Koskinen Jukka (FI)	Process for the polymerization of lactide
US6201072 (2001-03-13)	Macromed Inc (US)	Rathi Ramesh C (US); Jeong Byeongmoon (US); Zentner Gaylen M (US)	Biodegradable low molecular weight triblock poly(lactide-co-glycolide) polyethylene glycol copolymers having reverse thermal gelation properties
US6187951 (2001-02-13)	Cargill Inc (US)	Eyal Aharon M (IL); Hazan Betty (IL); Mizrahi Joseph (IL); Baniel Avraham M (IL); Fisher Rod R (US)	Lactic acid production, separation and/or recovery process
US6187901 (1998-08-20)	Selin Johan Fredrik (FI); Koskinen Jukka (FI); Neste OY (FI); Kaeaeriaeinen Kari (FI); Katsaras Nikitas (FI)	Selin Johan-Fredrik (FI); Koskinen Jukka (FI); Kaeaeriaeinen Kari (FI); Katsaras Nikitas (FI)	Method for the removal and recovery of lactide from polylactide
US6160173 (2000-12-12)	Cargill Inc (US)	Eyal Aharon Meir (IL); Mcwilliams Paul (US); Witzke David R (US); Gruber Patrick R (US)	Process for the recovery of lactic acid esters and amides from aqueous solutions of lactic acid and/or salts thereof
US6159724 (2000-12-12)	Agrano Ag (CH)	Ehret Aloyse (FR)	Process for preparing culture mediums for culturing yeasts and lactic acid bacteria
US6156353 (2000-12-05)	Nestec Sa (CH)	Beutler Ernst (CH); Illi Johann (CH); Favre-Galliand Leuka (CH); Sutter Andreas (US)	Strains of *Lactobacillus helveticus* for forming exclusively L(+) lactic acid in milk

(Continued)

Publication number (Date)	Assignee Title of Patent	Inventors
US6117949 (2000-09-12)	Macromed Inc (US) Biodegradable low molecular weight triblock poly (lactide-co-glycolide) polyethylene glycol copolymers having reverse thermal gelation properties	Rathi Ramesh C (US); Jeong Byeongmoon (US); Zentner Gaylen M (US)
US6111137 (1998-06-24)	Mitsui Chemicals Inc (JP) Purification process of lactic acid	Suizu Hiroshi (JP); Ajioka Masanobu (JP)
US6087532 (2000-07-11)	Cargill Inc. (US) Process for isolating lactic acid	Eyal Aharon M (IL); Hazan Betty (IL); Mizrahi Joseph (IL); Baniel Avraham M (IL); Kolstad Jeffrey J (US); Fisher Rod R (US); Stewart Brenda F (US)
US6077504 (2000-06-20)	Enteral dietary compositions comprising a mixture of live lactic bacteria consisting of *Streptococcus thermophilus, Bifidobacterium longum* and *Bifidobacterium infantis*	De Simone Claudio (IT); Cavaliere Ved Vesley Renata Ma (IT)
US6060622 (2000-05-09)	Shimadzu Corp. (JP) Methods of reproducing lactic acid components by using lactic acid based by-products	Kawamoto Tatsushi (JP); Ogaito Makoto (JP); Horibe Yasumasa (JP); Kawabe Takashi (JP); Okuyama Hisashi (JP)
US6054262 (1998-07-01)	Yakult Honsha Kk (JP) Method for controlling culture of lactic bacteria	Hayakawa Kazuhito (JP); Takeuchi Sogo (JP); Miyagi Akihiko (JP); Shibata Shinya (JP); Harada Katsuhisa (JP)
US6051663 (2000-04-18)	Basf Ag (DE) Lactic acid polymers	Yamamoto Motonori (DE); Witt Uwe (DE)
US6005067 (1999-12-21)	Cargill Inc. (US) Continuous process for manufacture of lactide polymers with controlled optical purity	Hall Eric Stanley (US); Benson Richard Douglas (US); Borchardt Ronald Leo (US); Gruber Patrick Richard (US); Iwen Matthew Lee (US); Kolstad Jeffrey John (US)
US6004573 (1999-12-21)	Macromed Inc (US) Biodegradable low molecular weight triblock poly(lactide-co-glycolide) polyethylene glycol copolymers having reverse thermal gelation properties	Rathi Ramesh C (US); Zentner Gaylen M (US)
US5959144 (1999-09-28)	Staley Mfg Co A E (US) Process for the recovery of lactic acid	Baniel Avraham (IL)
US5932455 (1999-08-03)	Cultor OY (FI) Method for preparing pure lactic acid	Viljava Tapio (FI); Koivikko Hannu (FI)
US5914248 (1996-05-22)	Stichting NL I Zuivelonderzoek (NL) Method for controlling the gene expression in lactic acid bacteria	De Vos Willem Meindert (NL); Kuipers Oscar Paul (NL)

(Continued)

Publication number (Date)	Assignee / Title of Patent	Inventors
US5900266 (1999-05-04)	Univ Missouri (US)	Iannotti Eugene L (US); Mueller Richard E (US); Jin Zhonglin (US)
	Heat-treated lactic and/or glycolic acid compositions and methods of use	
US5892109 (1999-04-06)	Cargill Inc (US)	Eyal Aharon M (IL); Hazan Betty (IL); Mizrahi Joseph (IL); Baniel Avraham M (IL); Kolstad Jeffrey J (US); Fisher Rod R (US); Stewart Brenda F (US)
	Lactic acid production, separation and/or recovery process	
US5866677 (1999-02-02)	Shimadzu Corp (JP); Kobe Steel Ltd (JP)	Kawada Eiichi (JP); Maeda Hiroshi (JP); Yamamoto Koji (JP); Fujii Yasuhiro (JP); Ito Masahiro (JP); Miyakawa Yutaka (JP); Ohara Hitomi (JP); Sawa Seiji (JP); Shimizu Kunihiko (JP)
	Method and system for producing poly (lactic acid)	
US5852117 (1998-12-22)	Nat Starch Chem Invest (US)	Harlan Robert D (US); Schoenberg Jules E (US)
	Process for making lactide graft copolymers	
US5849565 (1998-12-15)	Agrano Ag (CH)	Ehret Aloyse (FR)
	Panification ferment containing *Saccharomyces cerevisiae steineri* DSM 9211 and lactic acid bacteria	
US5844066 (1998-12-01)	Dainippon Ink & Chemicals (JP)	Kakizawa Yasutoshi (JP)
	Process for the preparation of lactic acid-based polyester	
US5837509 (1998-11-17)	Bioteknologisk Inst; Hansens Lab (DK)	Madsen Soeren Michael (DK); Johansen Eric (DK); Nilsson Dan (DK); Israelsen Hans (DK); Vrang Astrid (DK); Hansen Egon Bech (DK)
	Recombinant lactic acid bacterium containing an inserted promoter and method of constructing same	
US5833977 (1998-11-10)	Lahden Polttimo Ab Oy (FI)	Relander Harald (FI)
	Method of improving the quality of plant seeds using lactic acid producing microorganisms	
US5801255 (1998-09-01)	Shimadzu Corp (JP)	Ogaito Makoto (JP); Ohara Hitomi (JP)
	Method for producing lactide and apparatus used therefor	
US5801025 (1997-05-02)	Shimadzu Corp (JP)	Ohara Hitomi (JP); Yahata Masahito (JP)
	Method for producing L-lactic acid with high optical purity using bacillus strains	
US5798237 (1998-08-25)	Midwest Research Inst (US)	Mc Millan James D (US); Zhang Min (US); Finkelstein Mark (US); Franden Mary Ann (US); Picataggio Stephen K (US)
	Recombinant lactobacillus for fermentation of xylose to lactic acid and lactate	
US5786185 (1998-07-28)	Reilly Ind Inc (US)	Seo Jin-Ho (Kr); Tsao George T (US); Lee Seo Ju (US); Tsai Gow-Jen (US); Vorhies Susan L (US); Iyer Ganeshkumar (US); Mcquigg Donald W (US)
	Process for producing and recovering lactic acid	

(Continued)

Publication number (Date)	*Assignee* *Title of Patent*	*Inventors*
US5783725 (1998-07-21)	Haarmann & Reimer GmbH (DE) Stabilized lactic acid methyl ester	Hagena Detlef (DE); Kuhn Walter (DE); Koerber Alfred (DE); Langner Roland (DE)
US5780678 (1998-07-14)	Cargill Inc (US) Lactic acid production, separation and/or recovery process	Eyal Aharon M (IL); Hazan Betty (IL); Mizrahi Joseph (IL); Baniel Avraham M (IL); Kolstad Jeffrey J (US); Fisher Rod R (US); Stewart Brenda F (US)
US5759583 (1997-03-12)	Hoffmann La Roche (CH) Sustained release poly (lactic/glycolic) matrices	Kimura Akio (JP); Iwamoto Taro (JP); Ohyama Takehiko (JP); Takahashi Yasuyuki (JP)
US5728847 (1998-03-17)	Shimadzu Corp (JP) Method for recovering lactide from high-molecular weight polylactic acid	Ohara Hitomi (JP); Okamoto Toshio (JP)
US5525671 (1995-07-05)	Dainippon Ink & Chemicals (JP) Continuous production process for lactide copolymer	Ebato Hiroshi (JP); Imamura Shoji (JP)
US5521278 (1996-05-28)	Ecological Chemical Products (US) Integrated process for the manufacture of lactide	O'Brien William G (US); Cariello Lisa A (US); Wells Theodore F (US)
US5714618 (1998-02-03)	Toyo Boseki (JP) Polymer containing lactic acid as its constituting unit and method for producing the same	Arichi Minako (JP); Hotta Kiyoshi (JP); Ito Takeshi (JP); Uno Keiichi (JP); Aoyama Tomohiro (JP); Kimura Kunio (JP)
US5713788 (1998-02-03)	Univ North Carolina (US) Automated system for preparing animal carcasses for lactic acid fermentation and/or further processing	Ferket Peter R (US); Mckeithan Jr Jerry R (US); Stikeleather Larry F (US)
US5712152 (1998-01-27)	Agronomique Inst Nat Rech (FR) Yeast strains expressing the lactic lacticodehydrogenase gene and vectors useful in producing said strains	Dequin Sylvie (FR); Barre Pierre (FR)
US5707854 (1998-01-13)	Calpis Food Ind Co Ltd (JP) Lactic acid bacteria of the genus lactobacillus	Mizutani Jun (JP); Saito Yoshio (JP)
US5691424 (1997-11-25)	Mitsui Toatsu Chemicals (JP) Heat-resistant molded article of lactic acid-base polymer	Kitahara Yasuhiro (JP); Watanabe Takayuki (JP); Ajioka Masanobu (JP); Suzuki Kazuhiko (JP)
US5686540 (1997-04-02)	Dainippon Ink & Chemicals (JP) Process for the preparation of lactic acid-based polyester	Kakizawa Yasutoshi (JP)

(Continued)

Publication number (Date)	*Assignee* *Title of Patent*	*Inventors*
US5641406 (1997-06-24)	Vogelbusch GmbH (AT) Lactic acid extraction and purification process	Sarhaddar Schahroch (AT); Scheibl Anton (AT); Berghofer Emmerich (AT); Cramer Adalbert (AT)
US5635368 (1997-06-03)	Cultor Oy (FI) Bioreactor with immobilized lactic acid bacteria and the use thereof	Lommi Heikki (FI); Swinkels Wilhelmus J P M (NL); Viljava Timo T (FI); Hammond Roger C (FI)
US5618911 (1997-04-08)	Toyo Boseki (JP) Polymer containing lactic acid as its constituting unit and method for producing the same	Kimura Kunio (JP); Ito Takeshi (JP); Aoyama Tomohiro (JP); Uno Keiichi (JP); Hotta Kiyoshi (JP); Arichi Minako (JP)
US5616657 (1996-02-21)	Dainippon Ink & Chemicals (JP) Process for the preparation of high molecular lactic copolymer polyester	Ebato Hiroshi (JP); Imamura Shoji (JP)
US5605981 (1997-02-25)	Dainippon Ink & Chemicals (JP) Process for the preparation of high molecular lactic copolymer polyester	Imamura Shoji (JP); Ebato Hiroshi (JP)
US5605833 (1997-02-25)	Ind Tech Res Inst (TW) Process for preparation of D-lactic acid from d,l lactic acid ester using wheat germ or pancreatic lipase	Hsieh Chun-Lung (TW); Houng Jer-Yiing (TW)
US5597716 (1995-05-24)	Mitsubishi Rayon Co (JP) Process for producing D-lactic acid and L-lactamide	Iida Chinami C O Central Resea (JP); Kobayashi Yoshimasa C O Mitsub (JP); Ozaki Eiji C O Central Researc (JP); Sakimae Akihiro C O Central Re (JP); Sato Eiji C O Mitsubishi Rayon (JP)
US5574180 (1996-11-12)	Reilly Ind Inc (US) Process for recovering phytic acid, lactic acid or inositol	Mcquigg Donald (US); Marston Charles (US); Fitzpatrick Gina (US); Crowe Ernest (US); Vorhies Susan (US); Murugan Ramiah (US); Bailey Thomas D (US)
US5574129 (1996-11-12)	Japan Steel Works Ltd (JP) Process for producing lactic acid polymers and a process for the direct production of shaped articles from lactic acid polymers	Miyoshi Rika (JP); Sakai Tadamoto (JP); Hashimoto Noriaki (JP); Sumihiro Yukihiro (JP); Yokota Kayoko (JP); Koyanagi Kunihiko (JP)
US5543494 (1996-08-06)	Ministero Dell Univerita E Del (IT) Process for the production of poly(lactic acid)	Perego Gabriele (IT); Bastioli Catia (IT); Grzebieniak Karolina (Pl); Niekraszewicz Antoni (Pl)

(Continued)

Publication number (Date)	Assignee Title of Patent	Inventors
US5541111 (1995-05-10)	Calpis Food Ind Co Ltd (JP) *Lactobacillus helveticus* mutants having low increase in acidity of lactic acid during storage	Takano Toshiaki (JP); Masujima Yoshiko (JP); Yamamoto Naoyuki (JP)
US5512653 (1996-04-30)	Mitsui Toatsu Chemicals (JP) Lactic acid containing hydroxycarboxylic acid for the preparation of polyhydroxycarboxylic acid	Ohta Masahiro (JP); Obuchi Shoji (JP); Yoshida Yasunori (JP)
US5510526 (1996-04-23)	Cargill Inc (US) Lactic acid production, separation and/or recovery process	Baniel Abraham M (IL); Eyal Aharon M (IL); Mizrahi Joseph (IL); Hazan Betty (US); Fisher Rod R (US); Kolstad Jeffrey J (US); Stewart Brenda F (US)
US5506123 (1996-04-09)	Controlled Environment Syst (US) Municipal solid waste processing facility and commercial lactic acid production process	Chieffalo Rodger (US); Lightsey George R (US)
US5503750 (1996-04-02)	Membrane-based process for the recovery of lactic acid by fermentation of carbohydrate substrates containing sugars	Russo Jr Lawrence J (US); Kim Hyung S (US)
US5502215 (1995-06-14)	Musashino Kagaku Kenkyusho (JP) Method for purification of lactide	Yamaguchi Yoshiaki (JP); Arimura Tomohiro (JP)
US5498650 (1996-03-12)	Ecological Chemical Products (US) Poly(lactic acid) composition having improved physical properties	Flexman Edmund A (US); Kelly Jr William E (US)
US5488156 (1996-01-30)	Uop Inc (US) Preparation of a heat-stable lactic acid	Kulprathipanja Santi (US); Maher Gregory F (US); Lorsbach Thomas W (US)
US5482723 (1996-01-09)	Snow Brand Milk Prod Co Ltd (JP) Lactic acid bacteria, antibacterial substance produced by the bacteria, fermented milk starter containing the bacteria, and process for producing fermented milk by using the starter	Sasaki Masahiro (JP); Ishii Satoshi (JP); Yamauchi Yoshihiko (JP); Kitamura Katsushi (JP); Toyoda Shuji (JP); Ahiko Kenkichi (JP)
US5470340 (1995-04-12)	Ethicon Inc (US) Copolymers of (p-dioxanone/glycolide and/or lactide) and p-dioxanone	Shalaby Shalaby W (US); Bezwada Rao S (US); Koelmel Donald F (US)
US5466588 (1995-11-14)	Daicel Chem (JP) Production of high optical purity D-lactic acid	Kosaki Michio (JP); Kawai Kimitoshi (JP)

(Continued)

Publication number (Date)	Assignee Title of Patent	Inventors
US5464760 (1995-11-07)	Univ Chicago (US) Fermentation and recovery process for lactic acid production	Tsai Shih-Perng (US); Moon Seung H (US); Coleman Robert (US)
US5459053 (1995-10-17)	US Army (US) Use of rumen contents from slaughter cattle for the production of lactic acid	Rasmussen Mark A (US)
US5442033 (1995-08-15)	Ethicon Inc (US) Liquid copolymers of epsilon-caprolactone and lactide	Bezwada Rao S (US)
US5439985 (1995-08-08)	Univ Massachusetts Lowell (US) Biodegradable and hydrodegradable diblock copolymers composed of poly(beta-hydroxyalkanoates) and poly(lactones) or poly(lactide) chain segments	Gross Richard A (US); McCarthy Stephen P (US); Reeve Michael S (US)
US5434241 (1995-07-18)	Korea Inst Science Technology (K/R) Biodegradable poly(lactic acid)s having improved physical properties and process for their preparation	Kim Young H (KR); Ahn Kwang D (KR); Han Yang K (KR); Kim Soo H (KR); Kim Jeong B (KR)
US5416020 (1995-05-16)	Bio Tech Resources (US) *Lactobacillus delbrueckii* ssp. *bulgaricus* strain and fermentation process for producing L-(+)-lactic acid	Severson David K (US); Barrett Cheryl L (US)
US5403897 (1994-10-05)	Dainippon Ink & Chemicals (JP) Process for producing lactic acid-based copolyester and packaging material	Arai Kosuke (JP); Ebato Hiroshi (JP); Oya Satoshi (JP); Furuta Hideyuki (JP); Kakizawa Yasutoshi (JP)
US5401773 (1995-03-28)	Roussel Uclaf (FR) Lactic acid acylates	Noel Hugues (FR)
US5397572 (1995-03-14)	Univ Texas (US) Resorbable materials based on independently gelling polymers of a single enantiomeric lactide	Coombes Allan G A (US); Boyan Barbara D (US); Heckman James D (US)
US5389679 (1995-02-14)	Method of treating small mouth ulcers with lactic acid	Alliger Howard (US)
US5382617 (1995-01-17)	Du Pont (US) Stabilization of poly(hydroxy acids) derived from lactic or glycolic acid	Kelly Jr William E (US); Baird Richard L (US)
US5380813 (1994-04-20)	Neste OY (FI) Method for producing lactic acid based polyurethane	Su Tao (CN); Selin Johan-Fredrik (FI); Seppaelae Jukka (FI)
US5380525 (1995-01-10)	Upjohn Co (US) Ruminal bacterium for preventing acute lactic acidosis	Leedle Jane A Z (US); Greening Richard C (US); Smolenski Walter J (US)

(Continued)

Publication number (Date)	*Assignee* *Title of Patent*	*Inventors*
US5374743 (1994-12-20)	Du Pont (US) Process for the synthesis of lactide or glycolide from lactic acid or glycolide acid oligomers	Thayer Chester A (US); Bellis Harold E (US)
US5359027 (1994-10-25)	Himont Inc (US) Process for the synthesis of lactic acid polymers in the solid state and products thus obtained	Perego Gabriele (IT); Albizzati Enrico (IT)
US5359026 (1994-10-25)	Cargill Inc (US) Poly(lactide) copolymer and process for manufacture thereof	Gruber Patrick R (US); Kolstad Jeffrey J (US); Witzke David R (US)
US5357035 (1994-10-18)	Cargill Inc (US) Continuous process for manufacture of lactide polymers with purification by distillation	Kolstad Jeffrey J (US); Hall Eric S (US); Iwen Matthew L (US); Benson Richard D (US); Borchardt Ronald L (US); Gruber Patrick R (US)
US5357034 (1994-10-18)	Camelot Technologies Inc (US) Lactide polymerization	Kwok John (US); Downey Ronald J (US); Fridman Israel D (US); Nemphos Speros P (US)
US5338682 (1994-08-16)	Snow Brand Milk Prod Co Ltd (JP) Lactic acid bacteria, antibacterial substance produced by the bacteria, fermented milk starter containing the bacteria, and process for producing fermented milk by using the starter	Kitamura Katsushi (JP); Ishii Satoshi (JP); Ahiko Kenkichi (JP); Sasaki Masahiro (JP); Toyoda Shuji (JP); Yamauchi Yoshihiko (JP)
US5332839 (1994-07-26)	Biopak Technology Ltd (US) Catalytic production of lactide directly from lactic acid	Benecke Herman P (US); Markle Richard A (US); Sinclair Richard G (US)
US5331045 (1994-07-19)	Du Pont (US) Polyvinyl alcohol esterified with lactic acid and process therefor	Spinu Maria (US)
US5326744 (1994-07-05)	Basf Ag (DE) Glycol aldehyde and lactic acid derivatives and the preparation and use thereof	Rheinheimer Joachim (DE); Baumann Ernst (DE); Vogelbacher Uwe J (DE); Saupe Thomas (DE); Bratz Matthias (DE); Meyer Norbert (DE); Gerber Matthias (DE); Westphalen Karl-Otto (DE); Walter Helmut (DE); Kardorff Uwe (DE)
US5322781 (1994-06-21)	Cooperatieve Weiproduktenfabri (NL) Procedure for the preparation of D-(−)-lactic acid with *Lactobacillus bulgaricus*	Veringa Hubertus A (NL)

(Continued)

Publication number (Date)	Assignee Title of Patent	Inventors
US5320624 (1994-06-14)	United States Surgical Corp (US) Blends of glycolide and/or lactide polymers and caprolactone and/or trimethylene carbonate polymers and absorbable surgical devices made therefrom	Muth Ross R (US); Hermes Matthew (US); Kaplan Donald S (US); Kennedy John (US)
US5302693 (1994-04-12)	Boehringer Sohn Ingelheim (DE) Process for preparing poly-D,L-lactide and the use thereof as a carrier for active substances	Stricker Herbert (DE); Bendix Dieter (DE)
US5274127 (1993-12-28)	Biopak Technology Ltd (US) Lactide production from dehydration of aqueous lactic acid feed	Smith Russell K (US); Markle Richard A (US); Sinclair Richard G (US)
US5274073 (1993-12-28)	Cargill Inc (US) Continuous process for manufacture of a purified lactide	Kolstad Jeffrey J (US); Hall Eric S (US); Iwen Matthew L (US); Benson Richard D (US); Borchardt Ronald L (US); Gruber Patrick R (US)
US5266706 (1993-11-30)	Du Pont (US) Solvent scrubbing recovery of lactide and other dimeric cyclic esters	Bhatia Kamlesh K (US)
US5264592 (1993-11-23)	Camelot Technologies Inc (US) Lactide melt recrystallization	Fridman Israel D (US); Kwok John (US)
US5258488 (1993-11-02)	Cargill Inc (US) Continuous process for manufacture of lactide polymers with controlled optical purity	Kolstad Jeffrey J (US); Hall Eric S (US); Iwen Matthew L (US); Benson Richard D (US); Borchardt Ronald L (US); Gruber Patrick R (US)
US5252473 (1993-10-12)	Battelle Memorial Institute (US) Production of esters of lactic acid, esters of acrylic acid, lactic acid, and acrylic acid	Walkup Paul C (US); Rohrmann Charles A (US); Hallen Richard T (US); Eakin David E (US)
US5250182 (1993-10-05)	Zenon Environmental Inc (CA) Membrane-based process for the recovery of lactic acid and glycerol from a "corn thin stillage" stream	Bento John M A (US); Fleming Hubert L (US)
US5247059 (1993-09-21)	Cargill Inc (US) Continuous process for the manufacture of a purified lactide from esters of lactic acid	Kolstad Jeffrey J (US); Hall Eric S (US); Iwen Matthew L (US); Benson Richard D (US); Borchardt Ronald L (US); Gruber Patrick R (US)
US5247058 (1993-09-21)	Cargill Inc (US) Continuous process for manufacture of lactide polymers with controlled optical purity	Kolstad Jeffrey J (US); Hall Eric S (US); Iwan Matthew L (US); Benson Richard D (US); Borchardt Ronald L (US); Gruber Patrick R (US)

(Continued)

Publication number (Date)	*Assignee* *Title of Patent*	*Inventors*
US5235031 (1993-08-10)	Du Pont (US) Polymerization of lactide	Drysdale Neville E (US); Ford Thomas M (US); Mclain Stephan J (US)
US5234826 (1992-03-04)	Nitto Chemical Industry Co Ltd (JP) Biological process for preparing optically active lactic acid	Yamagami Tomohide (JP); Kobayashi Etsuko (JP); Endo Takakazu (JP)
US5219597 (1993-06-15)	Korea Food Res Inst (KR) Method for producing highly concentrated, lactic-acid fermented product utilizing unground grainy rice and improving qualities thereof by the secondary, enzymatic treatment at fermentation	Kim Young-Jin (KR); Mok Chul-Kyoon (KR); Nam Young-Jung (KR)
US5214159 (1993-05-25)	Boehringer Sohn Ingelheim (DE) Meso-lactide	Entenmann Gunther (DE); Schnell Wilhem-Gustav (DE); Bendix Dieter (DE); Hess Joachim (DE); Muller Manfred (DE)
US5210296 (1993-05-11)	Du Pont (US) Recovery of lactate esters and lactic acid from fermentation broth	Cockrem Michael C M (US); Johnson Pride D (US)
US5210294 (1992-12-09)	Himont Inc (US) Process for the production of purified lactic acid aqueous solutions starting from fermentation broths	Mantovani Giorgio (IT); Vaccari Giuseppe (IT); Campi Anna Lisa (IT)
US5196551 (1993-03-23)	Du Pont (US) Co-vaporization process for preparing lactide	Tarbell James V (US); Bhatia Kamlesh K (US)
US5177009 (1993-01-05)	Kampen Willem H (US) Process for manufacturing ethanol and for recovering glycerol, succinic acid, lactic acid, betaine, potassium sulfate, and free flowing distiller's dry grain and solubles or a solid fertilizer therefrom	Kampen Willem H (US)
US5177008 (1993-01-05)	Kampen Willem H (US) Process for manufacturing ethanol and for recovering glycerol, succinic acid, lactic acid, betaine, potassium sulfate, and free flowing distiller's dry grain and solubles or a solid fertilizer therefrom	Kampen Willem H (US)
US5149833 (1992-09-22)	Boehringer Sohn Ingelheim (DE) Process for preparing D,L-Lactide	Hess Joachim (DE); Muller Klaus R (DE); Muller Manfred (DE)
US5147668 (1992-09-15)	Munk Werner Georg (DE) Process of producing a reconstitutable solid lactic acid dried product	Munk Werner G (DE)
US5143845 (1992-09-01)	Toa Pharmaceutical Co Ltd (JP) Mixture of saccharifying lactic acid producing and butyric acid producing bacteria	Masuda Takashi (JP)

(Continued)

Publication number (Date)	Assignee Title of Patent	Inventors
US5142023 (1992-08-25)	Cargill Inc (US) Continuous process for manufacture of lactide polymers with controlled optical purity	Kolstad Jeffrey J (US); Hall Eric S (US); Iwen Matthew L (US); Benson Richard D (US); Borchardt Ronald L (US); Gruber Patrick R (US)
US5136057 (1992-08-04)	Du Pont (US) High yield recycle process for lactide	Bhatia Kamlesh K (US)
US5136017 (1992-08-26)	Novacor Chem Int (CH) Continuous lactide polymerization	Fridman Israel David (US); Kharas Gregory B (US); Nemphos Speros P (US)
US5117008 (1992-05-26)	Process for preparing D,L-Lactide	Kosak John R (US); Bhatia Kamlesh K (US); Drysdale Neville E (US)
US5114613 (1989-11-21)	Canon Kk (JP) Lactic acid derivative and liquid crystal composition containing same	Yoshinaga Kazuo (JP); Katagiri Kazuharu (JP); Tsuboyama Akira (JP); Kitayama Hiroyuki (JP); Shinjo Kenji (JP); Hioki Chieko (JP)
US5089664 (1990-06-27)	Donegani Guido Ist (IT) Process for recovering lactic acid from solutions which contain it	Dalcanale Enrico; Bonsignore Stefanio; Du Vosel Annick
US5084553 (1992-01-28)	Boehringer Ingelheim KG (DE) Copolymers of lactic acid and tartaric acid, the production and the use thereof	Hess Joachim (DE); Mueller Klaus R (DE)
US5077063 (1991-12-31)	Nikitenko Vyacheslav I (SU) Process for preparing lactic-acid products	Nikitenko Vyacheslav I (SU)
US5076807 (1991-02-06)	Random copolymers of p-dioxanone, lactide and/or glycolide as coating polymers for surgical filaments	Bezwada Rao S (US); Kronenthal Richard L (US)
US5075115 (1991-12-24)	FMC Corp (US) Process for polymerizing poly(lactic acid)	Brine Charles J (US)
US5071754 (1991-12-10)	Battelle Memorial Institute (US) Production of esters of lactic acid, esters of acrylic acid, lactic acid, and acrylic acid	Eakin David E (US); Walkup Paul C (US); Hallen Richard T (US); Rohrmann Charles A (US)
US5068418 (1991-11-26)	Uop Inc (US) Separation of lactic acid from fermentation broth with an anionic polymeric absorbent	Kulprathipanja Santi (US); Oroskar Anil R (US)
US5053522 (1991-10-01)	Boehringer Ingelheim Kg (DE) Process for the preparation of lactide	Muller Manfred (DE)

(Continued)

Publication number (Date)	Assignee Title of Patent	Inventors
US5053485 (1989-05-03)	CCA Biochem BV (NL) Polymer lactide, method for preparing it and a composition containing it	Mol Arie Cornelis; Nieuwenhuis Jan
US5011946 (1991-04-30)	Boehringer Sohn Ingelheim (DE) Process for preparing D,L-Lactide	Hess Joachim (DE); Muller Klaus R (DE); Muller Manfred (DE)
US5007939 (1991-04-16)	Solvay (BE) Article made of lactic acid polymer capable of being employed particularly as a biodegradable prosthesis and process for its manufacture	Ghyselinck Philippe (BE); Delcommune Luc (IT)
US5007923 (1991-08-07)	Ethicon Inc (US) Crystalline copolyesters of amorphous (lactide/glycolide) and p-dioxanone	Shalaby Shalaby W (US); Bezwada Rao S (US); Newman Hugh D Jr (US)
US4983745 (1991-01-08)	Boehringer Ingelheim KG (DE) Meso-lactide, processes for preparing it and polymers and copolymers produced therefrom	Entenmann Gunther (DE); Schnell Wilhem-Gustav (DE); Bendix Dieter (DE); Hess Joachim (DE); Muller Manfred (DE)
US4966982 (1989-11-02)	Mitsui Toatsu Chemicals (JP) Process for the production of lactide	Ono Hiroshi; Phala Heng
US4963486 (1990-10-16)	Cornell Res Foundation Inc (US) Direct fermentation of corn to L(+)-lactic acid by *Rhizopus oryzae*	Hang Yong D (US)
US4954450 (1990-09-04)	Miles Lab (US) Method for controlling the concurrent growth of two or more lactic acid producing bacteria	Brothersen Carl F (US); Knoespel Willard R W (US)
US4859763 (1989-01-18)	Mitsui Toatsu Chemicals (JP) Preparation process of DL-lactic acid-glycolic acid-copolymer	Takayanagi Hiroshi (JP); Kobayashi Tadashi (JP); Masuda Takayoshi (JP); Shinoda Hosei Azanishiakatsuch (JP)
US4855147 (1989-03-22)	Kagome KK (JP) Beverages by lactic acid fermentation and methods of producing same	Yokota Tetsuya (JP); Sakamoto Hideki (JP); Takahashi Naoto (JP)
US4846991 (1988-08-17)	Kao Corp (JP) Novel fatty acid-lactic acid ester	Suzue Shigetoshi (JP); Kimura Akio (JP); Tsukada Kiyoshi (JP); Noda Kozo (JP); Ogino Hidekazu (JP); Kamegai Jun (JP)
US4816267 (1989-03-28)	Nagano Miso KK (JP) Process for the production of a nutritional lactic acid fermentation product	Oka Hideki (JP)
US4808585 (1987-11-10)	Bayer AG (DE) Solutions of lactic acid salts of piperazinylquinolone- and piperazinyl-azaquinolone-carboxylic acids	Grohe Klaus (DE); Lammens Robert (DE)

(Continued)

Publication number (Date)	Assignee Title of Patent	Inventors
US4808583 (1987-11-10)	Bayer AG (DE) Solutions of lactic acid salts of piperazinylquinolone- and piperazinyl-azaquinolone-carboxylic acids	Grohe Klaus (DE); Lammens Robert (DE)
US4797468 (1988-07-27)	Akzo NV (NL) Preparation of polylactic acid and copolymers of lactic acids	De Vries Klaas Sybren (NL)
US4786756 (1988-11-22)	Standard Oil Co Ohio (US) Catalytic conversion of lactic acid and ammonium lactate to acrylic acid	Paparizos Christos (US); Dolhyj Serge R (US); Shaw Wilfrid G (US)
US4771001 (1988-09-13)	Neurex Corp (US) Production of lactic acid by continuous fermentation using an inexpensive raw material and a simplified method of lactic acid purification	Bailey Richard B (US); Joshi Dilip K (US); Michaels Stephen L (US); Wisdom Richard A (US)
US4769329 (1988-09-06)	Basf AG (DE) Preparation of optically pure D- and L- lactic acid	Cooper Bryan (DE); Kuesters Werner (DE); Martin Christoph (DE); Siegel Hardo (DE)
US4749652 (1988-06-07)	Texaco Inc (US) Lactic acid process	Robison Peter D (US)
US4705789 (1987-11-10)	Bayer AG (DE) Solutions of lactic acid salts of piperazinylquinolone- and piperazinyl-azaquinolone-carboxylic acids	Grohe Klaus (DE); Lammens Robert (DE)
US4698303 (1987-10-06)	Engenics Inc (US) Production of lactic acid by continuous fermentation using an inexpensive raw material and a simplified method of lactic acid purification	Bailey Richard B (US); Joshi Dilip K (US); Michaels Stephen L (US); Wisdom Richard A (US)
US4664919 (1987-05-12)	Taishi Foods (JP) Method of producing lactic-acid fermented soy milk	Yan Huang Y (CN); Peng Wang D (CN)
US4643734 (1987-02-17)	Hexcel Corp (US) Lactide/caprolactone polymer, method of making the same, composites thereof, and prostheses produced therefrom	Lin Steve (US)
US4643191 (1987-02-17)	Ethicon Inc (US) Crystalline copolymers of p-dioxanone and lactide and surgical devices made therefrom	Shalaby Shalaby W (US); Bezwada Rao S (US); Kafrawy Adel (US); Newman Jr Hugh (US)
US4637905 (1987-01-20)	Battelle Development Corp (US) Process of preparing microcapsules of lactides or lactide copolymers with glycolides and/or epsilon-caprolactones	Gardner David L (US)
US4596889 (1986-06-24)	Basf AG (DE) Preparation of alkenyl-lactic acid esters and the novel esters obtained	Kroener Michael (DE); Goetze Walter (DE)

(Continued)

Publication number (Date)	*Assignee* *Title of Patent*	*Inventors*
US4568760 (1986-02-04)	Bayer AG (DE)	Gallenkamp Bernd (DE); Krall Hermann D (DE)
	Process for the preparation of lactic acid silyl esters	
US4563356 (1986-01-07)	Sumitomo Chemical Co (JP)	Fujisawa Koichi (JP); Yokoyama Akiko (JP); Suzukamo Gohfu (JP)
	Lactic acid fermentation products of sunflower seed milk	
US4470416 (1984-09-11)	Ethicon Inc (US)	Shalaby Shalaby W (US); Kafrawy Adel (US); Mattei Frank V (US)
	Copolymers of lactide and/or glycolide with 1,5-dioxepan-2-one	
US4467034 (1984-08-21)	Hoechst AG (DE)	Voelskow Hartmut (DE); Sukatsch Dieter (DE)
	Process for the production of D-lactic acid with the use of *Lactobacillus bulgaricus* DSM 2129	
US4292339 (1981-09-29)	Boehringer Sohn Ingelheim (DE)	Bisle Hans E (DE)
	Stable concentrated lactic acid containing mixture	
US4258131 (1981-03-24)	Oriental Yeast Co Ltd (JP)	Takagahara Isamu (JP); Yamauti Juniti (JP); Yoshimura Setsuko (JP); Fujii Katsumi (JP); Horio Takekazu (JP)
	Method of fractional quantitative determination of isoenzyme of lactic dehydrogenase	
US4254222 (1981-03-03)	Owen Oliver E (US)	Owen Oliver E (US)
	Semi-quantitative assay of lactic acid and beta-hydroxybutyrate	
US4217419 (1980-08-12)	Ajinomoto KK (JP)	Suzuki Tadao (JP)
	Dried lactic acid bacteria composition	
US4190585 (1980-02-26)	Lonza AG (CH)	Tenud Leander (CH)
	Process for the production of indolyl lactic acid	
US4169102 (1979-09-25)	Goldschmidt AG TH (DE)	Hameyer Peter (DE); Tomczak Theodor (DE)
	Process for the manufacture of partially neutralized mixed esters of lactic acid, citric acid and partial glycerides of fatty acids	
US4157437 (1979-02-06)	Ethicon Inc (US)	Okuzumi Yuzi (US); Wasserman David (US); Mellon A Darline (US)
	Addition copolymers of lactide and glycolide and method of preparation	
US4147807 (1979-04-03)	Microlife Technics (US)	Gryczka Alfred J (US); Shah Ramesh B (US)
	Process for the treatment of meat with compositions including *Micrococcus varians* and a lactic acid producing bacteria	
US4137921 (1979-02-06)	Ethicon Inc (US)	Okuzumi Yuzi (US); Wasserman David (US); Mellon A Darline (US)
	Addition copolymers of lactide and glycolide and method of preparation	
US4072709 (1978-02-07)	Monsanto Co (US)	Tinker Harold Burnham (US)
	Production of lactic acid	

(Continued)

Publication number (Date)	Assignee Title of Patent	Inventors
US4057537 (1977-11-08)	Gulf Oil Corp (US) Copolymers of L-(-)-lactide and epsilon caprolactone	Sinclair Richard G (US)
US4045418 (1977-08-30)	Gulf Oil Corp Copolymers of D,L-lactide and epsilon caprolactone	Sinclair Richard G
US3846479 (1974-11-05)	ICI America Inc (US) Esters of lactic acid and fatty alcohols	Zech J (US)
US3839297 (1974-10-01)	Ethicon Inc (US) Use of stannous octoate catalyst in the manufacture of L(−)lactide-glycolide copolymer sutures	Wasserman D (US); Versfelt C (US)
US3835169 (1974-09-10)	Schlossman M (US); Kraft E (US) Lanolin derivatives essentially comprising esters of lanolin alcohol with lactic acid	Kraft E (US); Schlossman M (US)
US3716584 (1973-02-13)	Rhone Poulenc Sa (US) Process for the purification of lactic acid	Chaintron G (US)
US3619397 (1971-11-09)	Rhone Poulenc Sa (FR) Process for the purification of lactic acid	Jacquemet Jean-Claude (FR)
US1459395 (1923-06-19)	Chem Fab Vorm Goldenberg Gerom (US) Process of purifying lactic acid	Toni Hamburger (US)
US1447252 (1923-03-06)	Ward Kitchen Method of and means for treating lactic fluid	Joseph Moses (US)
US1401278 (1921-12-27)	Thomas J Coster (US) Process for producing lactic ferment culture for milk	Peter Petersen (US)

8
References

Abbad-Andaloussi, S., Guedon, E., Spiesser, E., Petitdemange, H. (1996) Glycerol dehydratase activity: the limiting step for 1,3-propanediol production by *Clostridium butyricum*, *Lett. Appl. Microbiol.* **22**, 311–314.

Abe, S. I., Takagi, M. (1991) Simultaneous saccharification and fermentation of cellulose to lactic acid, *Biotechnol. Bioeng.* **37**, 93–96.

Aeschlimann, A., von Stockar, U. (1989) The production of lactic acid from whey permeate by *Lactobacillus helveticus*, *Biotechnol. Lett.* **11**, 195–200.

Aeschlimann, A., von Stockar, U. (1990) The effect of yeast extract supplementation on the production of lactic acid from whey permeate by *Lactobacillus helveticus*, *Appl. Microbiol. Biotechnol.* **32**, 398–402.

Aeschlimann, A., von Stockar, U. (1991) Continuous production of lactic acid from whey permeate by *Lactobacillus helveticus* in a cell-recycle reactor, *Enzyme Microb. Technol.* **13**, 811–816.

Aeschlimann, A., Di Stasi, L., von Stockar, U. (1990) Continuous production of lactic acid from whey permeate by *Lactobacillus helveticus* in two chemostats in series, *Enzyme Microb. Technol.* **12**, 926–932.

Ahrens, K., Menzel, K., Zeng, A.-P., Deckwer, W.-D. (1998) Kinetic, dynamic, and pathway studies of glycerol metabolism by *Klebsiella pneumoniae* in anaerobic continuous culture. III. Enzymes and fluxes of glycerol dissimilation and 1,3-propanediol formation, *Biotechnol. Bioeng.* **59**, 544–552.

Aksu, Z., Kutal, T. (1986) Lactic acid production from molasses utilizing *Lactobacillus delbrueckii* and invertase together, *Biotechnol. Lett.* **8**, 157–160.

Altaras, N. E., Cameron D. C. (1999) Metabolic engineering of a 1,2-propanediol pathway in *Escherichia coli*, *Appl. Environ. Microbiol.* **65**, 1180–1185.

Altaras, N. E., Cameron D. C. (2000) Enhanced production of (*R*)-1,2-propanediol by metabolically engineered *Escherichia coli*, *Biotechnol. Prog.* **16**, 940–946.

Amrane, A., Prigent, Y. (1993) Influence of media composition on lactic acid production rate from whey by *Lactobacillus helveticus*, *Biotechnol. Lett.* **15**, 239–244.

Amrane, A., Prigent, Y. (1994) Lactic acid production from lactose in batch culture: analysis of the data with the help of a mathematical model; relevance for nitrogen source and preculture assessment, *Appl. Microbiol. Biotechnol.* **40**, 644–649.

Amrane, A., Prigent, Y. (1996) A novel concept of bioreactor: specialized function two-stage continuous reactor, and its application to lactose conversion into lactic acid, *J. Biotechnol.* **45**, 195–203.

Amrane, A., Prigent, Y. (1997) Growth and lactic acid production coupling for *Lactobacillus helveticus* cultivated on supplemented whey: influence of peptidic nitrogen deficiency, *J. Biotechnol.* **55**, 1–8.

Amrane, A., Prigent, Y. (1998) Influence of yeast extract concentration on batch cultures of *Lactobacillus helveticus*: growth and production coupling, *World J. Microbiol. Biotechnol.* **14**, 529–534.

Anderson, A. J., Dawes, E. A., (1990) Occurrence, metabolism, metabolic role, and industrial uses of bacterial polyhydroxyalkanoates, *Microbiol. Rev.* **54**, 450–472.

Atadan, E. M., Bruner, Jr. H. S. (1994) Process for the preparation of adipic acid or pentenoic acid, U.S. Patent No. 5,292,944.

Bailey, J. E. (1991) Towards a science of metabolic engineering, *Science* **252**, 1668–1674.

Barbirato, F., Bories, A., Camarasa-Claret, C., Grivet, J. P. (1995) Glycerol fermentation by a new 1,3-propanediol producing microorganism:

Enterobacter agglomerans, *Appl. Microbiol. Biotechnol.* **43**, 786–793.

Benninga, H. (1990) *A History of Lactic Acid Making*. Kluwer Academic: Dordrecht/Norwell, MA.

Bero, M., Kasperczyk, J., Adamus, G. (1993) Coordination polymerization of lactides. 3. Copolymerization of L, L-lactide and epsilon-caprolactone in the presence of initiators containing Zn and Al, *Makromol. Chem.* **194**, 907–912.

Beziat, J.C., Besson, M., Gallezot, P. (1996) Liquid phase oxidation of cylohexanol to adipic acid with molecular oxygen on metal catalysts, *Appl. Catal. A: General* **135**, L7–L11.

Bhatia, K. K. (1989) Atmospheric pressure process for preparing cyclic esters, U.S. Patent No. 4,835,293.

Bibal, B., Kapp, C., Goma, G., Pareilleux, A. (1989) Continuous culture of *Streptococcus cremoris* on lactose using various medium conditions, *Appl. Microbiol. Biotechnol.* 32, 155–159.

Biebl, H., Menzel, K., Zeng, A.-P., Deckwer, W.-D. (1999) Microbial production of 1,3-propanediol, *Appl. Microbiol. Biotechnol.* **52**, 289–297.

Blattner, F. R., Plunkett, G., Bloch, C. A., Perna, N. T., Burland, V., Riley, M., Collado-Vides, J., Glasner, J. D., Rode, K., Mayhew, G. F., Gregor, J., Davis, N. W., Kirkpatrick, H. A., Goeden, M. A., Rose, D. J., Mau, B., Shao, Y. (1997) The complete genome sequence of *Escherichia coli* K-12, *Science* **277**, 1453–1462.

Börgardts, P., Krischke, W., Trösch, W., Brunner, H. (1998) Integrated bioprocess for the simultaneous production of lactic acid and dairy sewage treatment, *Bioprocess Eng.* **19**, 321–329.

Boronat, A., Aguilar, J. (1981) Metabolism of L-fucose and L-rhamnose in *Escherichia coli*: differences in induction of propanediol oxidoreductase, *J. Bacteriol.* **147**, 181–185.

Boyaval, P., Corre, C., Terre, S. (1987) Continuous lactic acid fermentation with concentrated product recovery by ultrafiltration and electrodialysis, *Biotechnol. Lett.* **9**, 207–212.

Broecker, R. J., Schwarzmann, M. (1977) Manufacture of butanediol and/or tetrahydrofuran from maleic acid/or succinic anhydride via γ-butyrolactone, U.S. Patent No. 4,048,196.

Bruner, H. S., Jr. (1992) Process for the manufacture of adipic acid, U.S. Patent No. 5,166,421.

Bruner, H. S., Jr., Lane, S. L., Murphree, B. E. (1998) Manufacture of adipic acid, U.S. Patent No. 5,710,325.

Bryant, M. P., Small, N. (1956) Characteristics of two new genera of anaerobic curved rods isolated from the rumen of cattle, *J. Bacteriol.* **72**, 22–26.

Bryant, M. P., Bouma, C., Chu, H. (1958) *Bacteroides ruminicola* n. sp. and the new species *Succinomonas amylolytica*. Species of succinic acid producing anaerobic bacteria by the bovine rumen, *J. Bacteriol.* **76**, 15–23.

Budge, J. R., Attig, T. G., Dubbert, R. A. (1999) Catalysts for the hydrogenation of maleic acid to 1,4-butanediol, U.S. Patent No. 5969164.

Burke, P. M. (1994) Preparation of adipic acid from lactones, U.S. Patent No. 5,359,137.

Cameron, D. C., Cooney, C. L. (1986) A novel fermentation: the production of (*R*)-1,2-propanediol and acetol by *Clostridium thermosaccharolyticum*, *Bio/Technology* **4**, 651–654.

Cameron, D. C., Altaras, N. E., Hoffman, M. L., Shaw, A. J. (1998) Metabolic engineering of propanediol pathways, *Biotechnol. Progress* **14**, 116–125.

Castiglioni, G. L., Ferrari, M., Guercio, A., Vaccari, A., Lanci, R., Fumagalli, C. (1996) Chromium-free catalysts for selective vapor phase hydrogenation of maleic anhydride to γ-butyrolactone, *Catalysis Today* **27**, 181–186.

Chang, D. E., Jung, H. C., Rhee, J. S., Pan J. G. (1999) Homofermentative production of D- or L-lactate in metabolically engineered *Escherichia coli* RR1, *Appl. Environ. Microbiol.* **65**, 1384–1389.

Chatterjee, M., Chakrabarty, S. L., Chattopadhyay, B. D., Mandal, R. K. (1997) Production of lactic acid by direct fermentation of starchy wastes by an amylase-producing *Lactobacillus*, *Biotechnol. Lett.* **19**, 873–874.

Chatterjee, R., Millard, C. S., Champion, K., Clark, D. P., Donnelly, M. I. (2001) Mutation of the ptsG gene results in increased production of succinate in fermentation of glucose by *Escherichia coli*, *Appl. Environ. Microbiol.* **67**, 148–154.

Chemical Market Reporter (1999) **255**, 34–41.

Chen, R., Lee, Y. Y. (1997) Membrane mediated extractive fermentation for lactic acid production from cellulosic biomass, *Appl. Biochem. Biotechnol.* **63–65**, 435–448.

Cheng, P., Mueller, R. E., Jaeger, S., Bajpai, R., Iannotti, E. L. (1991) Lactic acid production from enzyme-thinned corn starch using *Lactobacillus amylovorus*, *J. Ind. Microbiol.* **7**, 27–34.

Chopin, A. (1993) Organization and regulation of genes for amino acid biosynthesis in lactic acid bacteria, *FEMS Microbiol. Rev.* **12**, 21–38.

Coates, J. S. Newark, D. E. (1980) Reducing color formers in 1,4-butanediol, U.S. Patent No. 4,213,000.

Dake, S., Gholap, R. V., Chaudhuri, R. V. (1987) Carbonylation of 1,4-butanediol diacetate using rhodminum complex catalyst: a kinetic study, *Ind. Eng. Chem. Res.* **26**, 1513–1518.

Daniel, R., Gottschalk, G. (1992) Growth temperature-dependent activity of glycerol dehydratase in *Escherichia coli* expressing the *Citrobacter freundii* regulon, *FEMS Microbiol. Lett.* **100**, 281–286.

Daniel, R., Boenigk, R., Gottschalk, G. (1995a) Purification of 1,3-propanediol dehydrogenase from *Citrobacter freundii* and cloning, sequencing, and overexpression of the corresponding gene in *Escherichia coli, J. Bacteriol.* **177**, 2151–2156.

Daniel, R., Stuertz, K., Gottschalk, G. (1995b) Biochemical and molecular characterization of the oxidative branch of glycerol utilization by *Citrobacter freundii, J. Bacteriol.* **177**, 4392–4401.

Datta, R. (1989) Recovery and purification of lactate salts from whole fermentation broth by electrodialysis, U.S. Patent No. 4,885,247.

Datta, R., Tsai, S.-P. (1997) Lactic acid production and potential uses: a technology and economics assessment, *ACS Symp. Ser.* **666**, 224–236.

Davis, C. P., Cleven, D., Brown, J., Balish, E. (1976) *Anaerobiospirillum*, a new genus of spiral-shaped bacteria, *Int. J. Syst. Bacteriol.* **26**, 498–504.

Deckwer, W.-D. (1995) Microbial conversion of glycerol to 1,3-propanediol, *FEMS Microbiol. Rev.* **16**, 143–149.

de Gooijer, C. D., Bakker, W. A. M., Beeftink, H. H., Tramper, J. (1996) Bioreactors in series: an overview of design procedures and practical applications, in: *International Congress on Chemicals from Biotechnology*, Hannover, Germany, Oct 18–20.

Demirci, A., Pometto, A. L., III (1992) Enhanced production of D(-)-lactic acid by mutants of *Lactobacillus delbrueckii* ATCC 9649, *J. Ind. Microbiol.* **11**, 23–28.

Demirci, A., Pometto, A. L., III, Lee, B., Hinz, P. N. (1998) Media evaluation of lactic acid repeated-batch fermentation with *Lactobacillus plantarum* and *Lactobacillus casei* subsp. *rhamnosus, J. Agric. Food. Chem.* **46**, 4771–4774.

Dennis, D., Reichlin, M., Kaplan, N. O. (1965) Lactic acid racemization, *Ann. NY Acad. Sci.* **119**, 868–876.

de Vos, W. M. (1996) Metabolic engineering of sugar catabolism in lactic acid bacteria, *Antonie van Leeuwenhoek* **70**, 223–242.

Diamantoglon, M., Meyer, G. (1988) Process for the production of water-insoluble fibers of cellulose monoesters of maleic acid, water-insoluble fibers of cellulose monoesters of maleic acid, succinic acid and phthalic acid, having an extremely high absorbability for water and physiological liquids, U.S. Patent No. 4,734,239.

Doi, Y. (1990) *Microbial Polyesters*, New York: VCH.

Donnelly, M. I., Millard, C. S., Clark, D. P., Chen, M. J., Rathke, J. W. (1998) A novel fermentation pathway in an *Escherichia coli* mutant producing succinic acid, acetic acid, and ethanol, *Appl. Biochem. Biotechnol.* **70-72**,187–198.

Donoghue, N. A., Trudgill, P. W. (1975) The metabolism of cyclohexanol by *Acinetobacter* NCIB 9871, *Eur. J. Biochem.* **60**, 1–7.

Draths, K. M., Frost, J. W. (1994) Environmentally compatible synthesis of adipic acid from D-glucose, *J. Am. Chem. Soc.* **116**, 399–400.

Dutta, S. K., Mukherjee, A., Chakraborty, P. (1996) Effect of product inhibition on lactic acid fermentation: simulation and modeling, *Appl. Microbiol. Biotechnol.* **46**, 410–413.

Edwards, J. S., Ramakrishna, R., Schilling, C. H., Palsson, B. O. (1999) Metabolic flux balance analysis, in: *Metabolic Engineering* (Lee, S. Y., Papoutsakis, E. T., Eds.), Marcel Dekker: New York, 13–57.

El Sabaeny, A. H. (1996) Influence of medium composition on lactic acid production from dried whey by *Lactobacillus delbrueckii, Microbiologia* **12**, 411–416.

Faber, M. (1983) Process for producing adipic acid from biomass, U.S. Patent No. 4,400,486.

Fischer, R., Sigwart, C. (2000) Preparation of 1,4-butanediol, U.S. Patent No. 6,103,941.

Forage, R. G., Lin, E. C. C. (1982) dha systems mediating aerobic and anaerobic dissimilation of glycerol in *Klebsiella pneumoniae* NCIB 418, *J. Bacteriol.* **151**, 591–599.

Fordyce, A. M., Crow, V. L., Thomas, T. D. (1984) Regulation of product formation during glucose or lactose limitation in nongrowing cells of *Streptococcus lactis, Appl. Environ. Microbiol.* **48**, 332–337.

Freund, A. (1881) Über die Bildung und Darstellung von Trimethylenalkohol aus Glycerin, *Monatsh. Chem.* **2**, 636–641.

Friedman, M. R., Gaden, E. L. J. (1970) Growth and acid production by *Lactobacillus delbrueckii* in a

dialysis culture system, *Biotechnol. Bioeng.* **12**, 961–974.
Frost, J.W., Draths, K. M. (1996) Synthesis of adipic acid from biomass-derived carbon source, U.S. Patent No. 5,487,987.
Fryzuk, M. D., Bosnich, B. (1978) Asymmetric synthesis. An asymmetric homogeneous hydrogenation catalyst which breeds its own chirality, *J. Am. Chem. Soc.* **100**, 5491–5494.
Garrigues, C., Loubiere, P., Lindley, ND., Cocaign-Bousquet, M. (1997) Control of shift from homolactic acid to mixed-acid fermentation in *Lactococcus lactis*: predominant role of the NADH/NAD^+ ratio, *J. Bacteriol.* **179**, 5282–5287.
Garvie, E. I. (1980) Bacterial lactate dehydrogenases, *Microbiol. Rev.* **44**, 106–139.
Gätje, G., Gottschalk, G. (1991) Limitation of growth and lactic acid production in batch and continuous cultures of *Lactobacillus helveticus*, *Appl. Microbiol. Biotechnol.* **34**, 446–449.
Giraud, E., Champailler, A., Raimbault, M. (1994) Degradation of raw starch by a wild amylotic strain of *Lactobacillus plantarum*, *Appl. Environ. Microbiol.* **60**, 4319–4323.
Göksungur, Y. and Güvenç, U. (1997) Batch and continuous production of lactic acid from beet molasses by *Lactobacillus delbrueckii* IFO 3202. *J. Chem. Technol. Biotechnol.* **69**, 399–404.
Goldberg, I., Lonberg-Holm, K., Bagley, E. A., Stieglitz, B. (1983) Improved conversion of fumarate to succinate by *Escherichia coli* strains amplified for fumarate reductase, *Appl. Environ. Microbiol.* **45**,1834–1847.
Gonçalves, L. M. D., Xavier, A. M. R., Almeida, J. S., Carrondo, M. J. T. (1991) Concomitant substrate and product inhibition kinetics in lactic acid production, *Enzyme Microb. Technol.* **13**, 314–319.
Gruber, P. R., Hall, E. S., Kolstad, J. J., Iwen, M. L., Benson, R. D., Borchardt, R. L. (1993) Continuous process for the manufacture of a purified lactide from esters of lactic acid, U.S. Patent No. 5,247,059.
Guettler, M. V., Jain, M. K. (1996) Method for making succinic acid, *Anaerobiospirillum succiniciproducens* variants for use in the process and methods for obtaining variants, U.S. Patent No. 5,521,075.
Guettler, M. V., Jain, M. K., Rumler, D. (1996) Method for making succinic acid, bacterial variants for use in the process, and methods for obtaining variants, U.S. Patent No. 5,573,931.
Guettler, M. V., Rumler, D., Jain, M. K. (1999) *Actinobacillus succinogenes* sp. nov., a novel succinic acid producing strain from the bovine rumen, *Int. J. Syst. Bacteriol.* **49**, 207–216.
Gunzel, B., Yonsel, S., Deckwer, W.-D. (1991) Fermentative production of 1,3-propanediol from glycerol by *Clostridium butyricum* up to a scale of 2 m^3. *Appl. Microbiol. Biotechnol.* **36**, 289–295.
Guoqiang, D., Kaul, R., Mattiasson, B. (1991) Evaluation of alginate-immobilized *Lactobacillus casei* for lactate production, *Appl. Microbiol. Biotechnol.* **36**, 309–314.
Hara, Y., Inagaki, H. (1991) Method for producing 1,4-butanediol, U.S. Patent No. 5,077,442.
Hara, Y., Kusaka, H., Inagaki, H., Takahashi, K., Wada, K. (2000) A novel production of γ-butyrolactone catalyzed by ruthenium complexes, *J. Catalysis* **194**, 188–197.
Haynie, S. L., Wagner, L. W. (1996) Process for making 1,3-propanediol from carbohydrates mixed microbial culture, U.S. Patent No. 5,599,689.
Held, A. M. (1996) *The fermentation of glycerol to 1,3-propanediol by Klebsiella pneumonia*, Master's thesis, University of Wisconsin-Madison, USA.
Hofvendahl, K. (1998) *Fermentation of wheat starch hydrolysate by Lactococcus lactis: factors affecting product formation*, Ph.D. Thesis, Lund University, Lund, Sweden.
Hofvendahl, K. and Hahn-Hägerdal, B. (1997) L-lactic acid production from whole wheat flour hydrolysate using strains of Lactobacilli and Lactococci, *Enzyme Microb. Technol.* **20**, 301–307.
Hofvendahl, K. and Hahn-Hägerdal, B. (2000) Factors affecting the fermentative lactic acid production from renewable resources. *Enzyme Microb. Technol.* **26**, 87–107.
Hofvendahl, K., van Niel, E. W. J., Hahn-Hägerdal, B. (1999) Effect of temperature and pH on growth and product formation of *Lactococcus lactis* spp. *lactis* ATCC 19435 growing on maltose. *Appl. Microbiol. Biotechnol.* **51**, 669–672.
Holten, C. H., Müller, A., Rehbinder, D. (1971) *Lactic Acid – Properties and Chemistry of Lactic Acid and Derivatives*, Weinheim, Germany: VCH.
Homann, T., Tag, C., Biebl, H., Deckwer, W.-D., Schink, B. (1990) Fermentation of glycerol to 1,3-propanediol by *Klebsiella* and *Citrobacter* strains, *Appl. Microbiol. Biotechnol.* **33**, 121–126.
Hong, S. H., Lee, S. Y. (2001) Metabolic flux analysis for succinic acid production by recombinant *Escherichia coli* with amplified malic enzyme activity, *Biotechnol. Bioeng.* **74**, 89–95.
Hori, Y., Takahashi, Y., Yamaguchi, A., Nishishita, T. (1993) Ring-opening copolymerization of optically-active beta-butyrolactone with several

lactones catalyzed by distannoxane complexes – Synthesis of new biodegradable polyesters, *Macromolecules* **26**, 4388–4390.

Huang, K., Rudolph, F. B., Bennett, G. N. (1999) Characterization of methylglyoxal synthase from *Clostridium acetobutylicum* ATCC 824 and its use in the formation of 1,2-propanediol, *Appl. Environ. Microbiol.* **65**, 3244–3247.

Hujanen, M., Linko, Y.-Y. (1996) Effect of temperature and various nitrogen sources on L(+)-lactic acid production by *Lactobacillus casei, Appl. Microbiol. Biotechnol.* **45**, 307–313.

Ishizaki, A., Ueda, T., Tanaka, K., Stanbury, P. F. (1992) L-Lactate production from xylose employing *Lactococcus lactis* IO-1, *Biotechnol. Lett.* **14**, 599–604.

Isogai, N., Hosokawa, M., Okawa, T., Wakui, N., Watanabe, T. Process for producing adipic acid diester, U.S. Patent No. 4,404,394.

Jain, M. K., Datta, R., Zeikus, J. G. (1989) High-value organic acids fermentation – emerging processes and products, in: *Bioprocess Engineering: The First Generation* (Ghosh, T. K., Ed.), Chichester: Ellis Harwood, 366–398.

Javanainen, P., Linko, Y.-Y. (1995) Lactic acid fermentation on barely flour without additional nutrients. *Biotechnol. Tech.* **9**, 543–548.

Kandler, O. (1983) Carbohydrate metabolism in lactic acid bacteria, *Antonie van Leeuwenhoek* **49**, 209–224.

Khan, J., Baig, M. A., Ehtehsamuddin, A. F. M. (1995) Production of lactic acid from potato by *Lactobacillus delbrueckii, Sarhad, J. Ag.* **11**, 13–18.

Klaenhammer, T. R., Fitzgerald, G. F. (1994) Bacteriophages and bacteriophage resistance, in: *Genetics and Biotechnology of Lactic Acid Bacteria* (Gasson, M. J., de Vos, W. M., Eds.), London: Blackie, 106–168.

Kometani, T., Morita, Y., Furui, H., Yoshii, H., Matsuno, R. (1993) Preparation of chiral 1,2-alkanediols with baker's yeast-mediated oxidation, *Chem. Lett.* **12**, 2123–2124.

Krischke, W., Schöder, M., Trösch, W. (1991) Continuous production of L-lactic acid from whey permeate by immobilized *Lactobacillus casei* spp. *casei, Appl. Microbiol. Biotechnol.* **34**, 573–578.

Kunz, D. A., Weimer, P. J. (1983) Bacterial formation and metabolism of 6-hydroxyhexanoate: evidence of a potential role for ω-oxidation, *J. Bacteriol.* **156**, 567–575.

Kurcok, P., Penczek, J., Franek, J., Jedliski, Z. (1992) Anionic-polymerization of lactones. 14. Anionic block copolymerization of delta-valerolactone and L-lactide initiated with potassium methoxide, *Macromolecules* **25**, 2285–2289.

Kurosawa, H. Ishikawa, H., Tanaka, H. (1988) L-lactic acid production from starch by coimmobilized mixed culture system of *Aspergillus awamori* and *Streptococcus lactis, Biotechnol. Bioeng.* **31**, 183–187.

Kwon, S., Lee, P. C., Lee, E. G., Chang, Y. K., Chang, H. N. (2000) Production of lactic acid by *Lactobacillus rhamnosus* with vitamin-supplemented soybean hydrolysate, *Enzyme Microb. Technol.* **26**, 209–215.

Laffend, L. A., Nagarajan, V., Nakamura, C. E. (1996) Bioconversion of a fermentable carbon source to 1,3-propanediol by a single microorganism, WO96/53796.

Laivenieks, M., Vieille, C., Zeikus, J. G. (1997) Cloning, sequencing, and overexpression of the *Anaerobiospirillum succiniciproducens* phosphoenolpyruvate carboxykinase (*pckA*) gene, *Appl. Environ. Microbiol.* **63**, 2273–2280.

Lee, L. G., Whitesides, G. M. (1986) Preparation of optically active 1,2-diols and R-hydroxy ketones using glycerol dehydrogenase as catalyst: limits to enzyme-catalyzed synthesis due to noncompetitive and mixed inhibition byproduct, *J. Org. Chem.* **51**, 25–36.

Lee, P. C., Lee, W. G., Kwon, S., Lee, S. Y., Chang, H. N. (1999a) Succinic acid production by *Anaerobiospirillum succiniciproducens*: effects of the H_2/CO_2 supplying and glucose concentration, *Enzyme. Microb. Technol.* **24**, 549–554.

Lee, P. C., Lee, W. G., Lee, S. Y., Chang, H. N. (1999b) Effects of medium components on the growth of *Anaerobiospirillum succiniciproducens* and succinic acid production, *Process Biochem.* **35**, 49–55.

Lee, P. C., Lee, W. G., Lee, S. Y., Chang, H. N. (2000a) Succinic acid production with reduced byproduct formation in the fermentation of *Anaerobiospirillum succiniproducens* using glycerol as a carbon source, *Biotechnol. Bioeng.* **72**, 41–48.

Lee, P. C., Lee, W. G., Kwon, S. H., Lee, S. Y., Chang, H. N. (2000b) Batch and Continuous fermentation of *Anaerobiospirillum succiniproducens* for the production of succinic acid from whey, *Appl. Microbiol. Biotechnol.* **54**, 23–27.

Lee, P. C., Lee, W. G., Lee, S. Y., Chang, Y. K., Chang, H. N. (2000c) Fermentative production of succinic acid from glucose and corn steep liquor by *Anaerobiospirillum succiniproducens, Biotechnol. Bioprocess Eng.* **5**, 379–381.

Lee, S. Y. (1996) Bacterial polyhydroxyalkanoates, *Biotechnol. Bioeng.* **49**, 1–14.

Lee, S. Y., Chang, H. N. (1995) Production of poly-(hydroxyalkanoic acid), *Adv. Biochem. Eng. Biotechnol.* **52**, 27–58.

Lee, S. Y., Papoutsakis, E. T. (1999) The challenges and promise of metabolic engineering, in: *Metabolic Engineering* (Lee, S. Y., Papoutsakis, E. T., Eds.), New York: Marcel Dekker, 1–12.

Lemoigne, M. (1927) *Ann. Inst. Pasteur* **41**, 148–165.

Levene, P. A., Walti, A. (1943) l-Propylene glycol, in: *Organic Syntheses Collective* (Blatt, A. H., Ed.), New York: John Wiley & Sons, 545–547.

Linko, P., Stenroos, S.-L., Linko, Y.-Y., Koistinen, T., Harju, M., Heikonen, M. (1984) Applications of immobilized lactic acid bacteria. *Ann. NY Acad. Sci.* **434**, 406–417.

Linko, Y.-Y., Javanainen, P. (1996) Simultaneous liquefaction, saccharification, and lactic acid fermentation on barely starch, *Enzyme Microb. Technol.* **19**, 118–123.

Litchfield, J. H. (1996) Microbiological production of lactic acid, in: *Advances in Applied Microbiology* (Neidleman, S. L., Laskin, A. I., Eds.), New York: Academic Press, 45–95, vol. 42.

Lund, B., Norddahl, B., Ahring, B. (1992) Production of lactic acid from whey using hydrolysed whey protein as nitrogen source, *Biotechnol. Lett.* **14**, 851–856.

Major, N. C., Bull, A. T. (1989) The physiology of lactate production by *Lactobacillus delbrueckii* in a chemostat with cell recycle, *Biotechnol. Bioeng.* **34**, 592–599.

Maxwell, P. C. (1982) Production of muconic acid, U.S. Patent No. 4,355,107.

Mayer, J.M. Kaplan, D. (1994) Biodegradable materials: balancing degradability and performance, *Trends Polym. Sci.* **2**, 227–235.

McCaskey, T. A., Zhou, S. D., Britt, S. N., Strickland, R. (1994) Bioconversion of municipal solid waste to lactic acid by *Lactobacillus* species. *Appl. Biochem. Biotechnol.* **45-46**, 555–563.

Mehaia, M. A., Cheryan, M. (1987a) Production of lactic acid from sweet whey permeate concentrates, *Process Biochem.* **22**, 185–188.

Mehaia, M. A., Cheryan, M. (1987b) Immobilization of *Lactobacillus bulgaricus* in a hollow-fiber bioreactor for production of lactic acid from whey permeate, *Appl. Biochem. Biotechnol.* **14**, 21–27.

Melzoch, K., Votruva, J., Habova, V., Rychtera, M. (1997) Lactic acid production in a cell retention continuous culture using lignocellulosic hydrolysate as a substrate. *J. Biotechnol.* **56**, 25–31.

Menzel, K., Zeng, A.-P., Deckwer, W.-D. (1997) High concentration and productivity of 1,3-propanediol from continuous fermentation of glycerol by *Klebsiella pneumoniae, Enzyme Microbiol. Technol.* **20**, 82–86.

Mercier, P., Yerushalmi, L., Rouleau, D., Dochain, D. (1992) Kinetics of lactic acid fermentation on glucose and corn by *Lactobacillus amylophilus, J. Chem. Technol. Biotechnol.* **55**, 111–121.

Milko, E. S., Sperelup, O. V., Rabotnova, I. L. (1966) Die Milchsäuregärung von *Lactobacterium delbrueckii* bei kontinuerlicher Kultivierung, *Z. Allg. Mikrobiol.* **6**, 297–301.

Millard, C. S., Chao, Y., Liao, J. C., Donnelly, M. I. (1996) Enhanced production of succinic acid by overexpression of phosphoenolpyruvate carboxylase in *Escherichia coli, Appl. Environ. Microbiol.* **62**, 1808–1810.

Monteagudo, J. M., Rodriguez, L., Rincon, J., Fuertes, J. (1997) Kinetics of lactic acid fermentation by *Lactobacillus delbrueckii* grown on beet molasses, *J. Chem. Technol. Biotechnol.* **68**, 271–276.

Mulligan, C. N., Safi, B. F., Groleau, D. (1991) Continuous production of ammonium lactate by *Streptococcus cremoris* in a three-stage reactor, *Biotechnol. Bioeng.* **38**, 1173–1181.

Nagata, M., Goto, H., Sakai, W., Tsutsumi, T. (2000) Synthesis and enzymatic degradation of poly(tetramethylene succinate) copolymers with ererphthalic acid, *Polymer* **41**, 4373–4376.

Nakamura, C. E., Gatenby, A. A., Hsu, A. K., Reau, R. D., Haynie, S. L., Diaz-Torres, M., Trimbur, D. E., Whited, G. M., Nagarajan, V., Payne, M. S., Picataggio, S. K., Nair, R. V. (2000) Method for the production of 1,3-propanediol by recombinant microorganisms, U.S. Patent No. 6,013,494.

Nakayama, A., Kawasaki, N., Aiba, S., Maeda, Y., Arvanitoyannis, I., Yamamoto, N. (1998) Synthesis and biodegradability of novel copolyesters containing gamma-butyrolactone units, *Polymer* **39**, 1213–1222.

Neijssel, O. M., de Mattos, M. J., Tempest, D. W. (1996) Growth yield and energy distribution, in: *Escherichia coli and Salmonella* (Neidhardt, F. C., Ed.), Washington D.C.: ASM Press, 1683–1692.

Nevoigt, E., Stahl, U. (1997) Osmoregulation and glycerol metabolism in the yeast *Saccharomyces cerevisiae, FEMS Microbiol. Rev.* **21**, 231–241.

Nielsen, J., Villadsen, J. (1994) Analysis of reaction rates, in: *Bioreaction Engineering Principles* (Nielsen, J., Villadsen, J., Eds.), New York: Plenum Press, 97–161.

Norbeck, J., Påhlman, A.-K., Akhtar, N., Blomberg, A., Adler, L. (1996) Purification and characterization of two isoenzymes of DL-glycerol-3-phosphatase from *Saccharomyces cerevisiae*, *J. Biol. Chem.* **271**, 13875–13881.

Norris, D. B., Trudgill, P. W. (1971) The metabolism of cyclohexanol by *Nocardia globerula* CL1, *Biochem. J.* **121**, 363–370.

Norton S., Lacroix, C., Vuillemard, J.-C. (1994a) Reduction of yeast extract supplementation in lactic acid fermentation of whey permeate by immobilized cell technology, *J. Dairy Sci.* **77**, 2494–2508.

Norton S., Lacroix, C., Vuillemard, J.-C. (1994b) Kinetic study of continuous whey permeate fermentation by immobilized *Lactobacillus helveticus* for lactic acid production, *Enzyme Microb. Technol.* **16**, 457–466.

Ohara, H., Hiyama, K., Yoshida T. (1992) Kinetics of growth and lactic acid production in continuous and batch culture, *Appl. Microbiol. Biotechnol.* **37**, 544–548.

Ohleyer, E., Blanch, H. W., Wilke, C. R. (1985) Continuous production of lactic acid in a cell recycle reactor, *Appl. Biochem. Biotechnol.* **11**, 317–332.

Olmos-Dichara, A., Ampe, F., Uribelarrea, J.-L., Pareilleux, A., Goma, G. (1997) Growth and lactic acid production by *Lactobacillus rhamnosus* in batch and membrane bioreactor: influence of east extract and tryptone enrichment, *Biotechnol. Lett.* **19**, 709–714.

Papanikolaou, S., Ruiz-Sanchez, P., Pariset, B., Blanchard, F., Fick, M. (2000) High production of 1,3-propanediol from industrial glycerol by a newly isolated *Clostridium butyricum* strain, *J. Biotechnol.* **77**, 191–208.

Parajó, J. C., Alonso, J. L., Moldes, A. B. (1997) Production of lactic acid from lignocellulose in a single stage of hydrolysis and fermentation. *Food Biotechnol.* **11**, 45–48.

Payot, T., Chemaly, Z., Fick, M. (1999) Lactic acid production by *Bacillus coagulans* –kinetic studies and optimization of culture medium for batch and continuous fermentations, *Enzyme Microb. Technol.* **24**, 191–199.

Percy, D. S., Harrison, D. H. T. (1996) Abstracts of Papers, Annual Meeting of the American Society for Biochemistry and Molecular Biology, New Orleans; American Society for Biochemistry and Molecular Biology: Bethesda, Abstract 1367.

Perego, G., Vercelio, T., Balbontin, G. (1993) Copolymers of L-lactide and D,L-lactide with 6-caprolactone – Synthesis and characterization, *Makromol. Chem.* **194**, 2463–2469.

Pflugmacher, U., Gottschalk, G. (1994) Development of an immobilized cell reactor for the production of 1,3-propanediol by *Citrobacter freundii*, *Appl. Microbiol. Biotechnol.* **41**, 313–316.

Pitt, C. G. (1992) Non-microbial degradation of polyester mechanisms and modifications, in: *Biodegradable Polymers and Plastics* (Vert, M., Feijen, J., Albertsson, A., Scott, G., Chiellini, E., Eds.), Wiltshire: Redwood Press, 7–19.

Podlech, P.A.S., Luna, M.F., Jerke, P.R., De Souza Neto, C.A.C, Dos Passos, R.F., Souza, O., Borzani, W. (1990) Semicontinuous lactic fermentation of whey by *Lactobacillus bulgaricus*. I. Experimental results, *Biotechnol. Lett.* **12**, 531–534.

Potera, C. (1997) Genencor & Dupont create "green" polyester, *Gen. Eng. News* **12**, 17.

Purac, L. (1993) *Natural Lactic Acid*, Lincolnshire, IL.

Ray, L., Mukherjee, G., Majumdar, S. K. (1991) Production of lactic acid from potato fermentation, *Ind. J. Exp. Biol.* **29**, 681–685.

Reeve, M. S., McCarthy, S. P., Gross, R. A. (1993) Preparation and characterization of (*R*)-poly(beta-hydroxybutyrate) poly(epsilon-caprolactone) and (*R*)-poly(beta-hydroxybutyrate) poly(lactide) degradable diblock copolymers, *Macromolecules* **26**, 888–894.

Reimann, A. (1997) Produktion von 1,3-Propandiol aus Glycerin durch *Clostridium butyricum* DSM 5431 und produkttolerante Mutanten, Dissertation, University of Braunschweig, Germany.

Reimann, A., Biebl, H. (1996) Production of 1,3-propanediol by *Clostridium* DSM5431 and product tolerant mutants in fed-batch culture: feeding strategy for glycerol and ammonium, *Biotechnol. Lett.* **18**, 827–832.

Richter, K. and Träger, A. (1994) L(+)-Lactic acid from sweet sorghum by submerged and solid-state fermentations. *Acta Biotechnol.* **14**, 367–378.

Roukas, T. and Kotzekidou, P. (1991) Production of lactic acid from deproteinized whey by coimmobilized *Lactobacillus casei* and *Lactococcus lactis* cells. *Enzyme Microb. Technol.* **13**, 33–38.

Roukas, T. and Kotzekidou, P. (1998) Lactic acid production from deproteinized whey by mixed cultures of free and coimmobilized *Lactobacillus casei* and *Lactococcus lactis* cells using fedbatch culture. *Enzyme Microb. Technol.* **22**, 199–204.

Roy, D., Goulet, J., LeDuy, A. (1986) Batch fermentation of whey ultrafiltrate by *Lactobacillus helveticus* for lactic acid production. *Appl. Microbiol. Biotechnol.* **24**, 206–213.

Roy, D., Goulet, J., LeDuy, A. (1987a) Continuous production of lactic acid from whey permeate by free and calcium alginate entrapped *Lactobacillus helveticus. J. Dairy Sci.* **70**, 506–513.

Roy, D., LeDuy, A., Goulet, J. (1987b) Kinetics of growth and lactic acid production from whey permeate by *Lactobacillus helveticus, Can. J. Chem. Eng.* **65**, 597–603.

Sado, T., Tajima, M., Noguchi Research Foundation. (1980) Carbonylation of 1,4-butanediol by homogeneous catalyst, Japan Kakai Tokkyo Koho 8,051,037.

Saint-Amans, S., Perlot, P., Goma, G., Soucaille, P. (1994) High production of 1,3-propanediol from glycerol by *Clostridium butyricum* VPI 3266 in a simply controlled fed-batch system, *Biotechnol. Lett.* **16**, 832–836.

Sakai, Y., Ishikawa, J., Fukasaka, S., Urimoto, H., Mitsui, R., Yanase, H., Kato, N. (1999) A new carboxylesterase from *Brevibacterium linens* IFO 12171 responsible for the conversion of 1,4-butanediol diacrylate to 4-hydroxybutyl adcrylate: purification, characterization, gene cloning, and gene expression in *Escherichia coli, Biosci. Biotechnol. Biochem.* **63**, 688–697.

Samuel, W. A., Lee, Y. Y., Anthony, W. B. (1980) Lactic acid fermentation of crude sorghum extract, *Biotechnol. Bioeng.* **22**, 757–758.

Samuelov, N. S., Lamed, R., Lowe, S., Zeikus, J. G. (1991) Influence of CO_2-HCO_3^- level and pH on growth, succinate production, and enzyme activities of *Anaerobiospirillum succiniciproducens, Appl. Environ. Microbiol.* **57**, 3013–3019.

Samuelov, N. S., Datta, R., Jain, M. K., Zeikus, J. G. (1999) Whey fermentation by *Anaerobiospirillum succiniciproducens* for production of a succinate-based animal feed additive, *Appl. Environ. Microbiol.* **65**, 2260–2263.

Sanchez-Riera, F., Cameron, D. C., Cooney, C. L. (1987) Influence of environmental factors in the production of *(R)*-(-)-1,2-propanediol by *Clostridium thermosaccharolyticum, Biotechnol. Lett.* **9**, 449–454.

Scheifinger, C. C., Wolin, M. J. (1973) Propionate formation from cellulose and soluble sugars by combined cultures of *Bacteroides succinogenes* and *Selenomonas ruminantium, Appl. Microbiol.* **26**, 789–795.

Schlander, J. H., Turek, T. (1999) Gas-phase hydrogenolysis of dimethyl maleate to 1,4-butanediol and γ-butyrolactone over copper/zinc oxide catalysts, *Ind. Eng. Chem. Res.* **38**, 1264–1270.

Schmidt, S. and Padukone, N. (1997) Production of lactic acid from wastepaper as a cellulosic feedback. *J. Ind. Microb. Biotechnol.* **18**, 10–14.

Seebach, D., Roggo, S., Zimmermann, J. (1987) Biological-chemical preparation of 3-hydroxycarboxylic acids and their use in EPC-syntheses, in *Stereochemistry of Organic and Bioorganic Transformation*, Workshop Conferences Hoechst, Vol. 17, (Bartmann, W., Sharpless, K.B., Eds.), Weinheim, Germany: VCH, 85–126.

Seyfried, M., Daniel, R., Gottschalk, G. (1996) Cloning, sequencing and overexpression of the genes encoding coenzyme B12-dependent glycerol dehydratase of *Citrobacter freundii, J. Bacteriol.* **178**, 5793–5796.

Shamala, T. R., Sreekantiah, K. R. (1987) Degradation of starchy substrates by a crude enzyme preparation and utilization of the hydrolysates for lactic fermentation. *Enzyme Microb. Technol.* **9**, 726–729.

Shamala, T. R., Sreekantiah, K. R. (1988) Fermentation of starch hydrolysates by *Lactobacillus plantarum, J. Ind. Microbiol.* **3**, 175–178.

Sjöberg, A., Persson, I., Quednau, M., Hahn-Hägerdal, B. (1995) The influence of limiting and non-limiting growth conditions on glucose and maltose metabolism in *Lactococcus lactis* spp. *lactis* strain, *Appl. Microbiol. Biotechnol.* **42**, 931–938.

Skraly, F. A., Cameron, D. C. (1998) Purification and characterization of a *Bacillus licheniformis* phosphatase specific for D-(*R*)-glycerophosphate, *Arch. Biochem. Biophys.* **349**, 27–35.

Smith, J. G., Kibler, C. J., Sublett, B. J., (1966) Preparation and properties of poly(methylene terephthalates), *J. Polym. Sci. Part A-1* **4**, 1851–1859.

Smith, J. S., Hillier, A. J., Lees, G. J. (1975) The nature of the stimulation of the growth of *Streptococcus lactis* by yeast extract, *J. Dairy Res.* **42**, 123–138.

Sprenger, G. A., Hammer, G. A., Johnson, E. A., Lin, E. C. C. (1989) Anaerobic growth of *Escherichia coli* on glycerol by importing genes of the *dha* regulon from *Klebsiella pneumoniae, J. Gen. Microbiol.* **135**, 1255–1262.

Steinbüchel, A., Valentin, H. E. (1995) Diversity of bacterial polyhydroxyalkanoic acid, *FEMS Microbiol. Lett.* **128**, 219–228.

Stieber, R. W., Gerhardt, P. (1981) Dialysis continuous process for ammonium lactate fermentation: simulated and experimental dialysate-feed, immobilized-cell systems, *Biotechnol. Bioeng.* **23**, 535–550.

Stieglitz, B., Weimer, P. J. (1985) Novel microbial screen for detection of 1,4-butanediol, ethylene glycol, and adipic acid, *Appl. Environ. Microbiol.* **49**, 593–598.

Stiles, M. E., Holzapfel, W. H. (1997) Lactic acid bacteria of foods and their current taxonomy, *Int. J. Food Microbiol.* **36**, 1–29.

Stols, L., Donnelly, M. I. (1997) Production of succinic acid through overexpression of NAD^+-dependent malic enzyme in an *Escherichia coli* mutant, *Appl. Environ. Microbiol.* **63**, 2695–2701.

Stols, L., Kulkarni, G., Harris, B. G., Donnelly, M. I. (1997) Expression of *Ascaris suum* enzyme in a mutant *Escherichia coli* allows production of succinic acid from glucose, *Appl. Biochem. Biotechnol.* **63–65**, 153–158.

Suskovic, J., Novak, S., Maric, V., Matosic, S. (1991) Lactic acid fermentation kinetics on different carbon sources, *Prehrambeno-Technol. Biotechnol. Rev.* **29**, 155–158.

Tag, C.G. (1990) *Mikrobielle Herstellung von 1,3-Propandiol*, Thesis, Universitat Oldenburg, Oldenburg, Germany.

Takahashi, S., Abbe, K., Yamada, T. (1982) Purification of pyruvate formate lyase from *Streptococcus mutans* and its regulatory properties, *J. Bacteriol.* **149**, 1034–1040.

Tang, C.-T. Ruch, F. E. Lin, E. C. C. (1979) Purification and properties of nicotinamide adenine dinucleotide-linked dehydrogenase that serves an *Escherichia coli* mutant for glycerol catabolism, *J. Bacteriol.* **140**, 182–187.

Tejayadi, S., Cheryan, M. (1995) Lactic acid from cheese whey permeate. Productivity and economics of a continuous membrane bioreactor, *Appl. Microbiol. Biotechnol.* **43**, 242–248.

Tewari, H. K., Sethi, R. P., Sood, A., Singh, L. (1985) Lactic acid production from paneer whey by *Lactobacillus bulgaricus*, *J. Res. Punjab Agric. Univ.* **22**, 89–98.

Thomas, T. D., Ellwood, D. C., Longyear, M. C. (1979) Change from homo- to heterolactic fermentation by *Streptococcus lactis* resulting from glucose limitation in anaerobic chemostat cultures, *J. Bacteriol.* **138**, 109–117.

Tobimatsu, T., Azuma, M., Matsubara, H., Tskatori, H., Niida, T., Nishimoto, K., Satoh, H., Hayashi, R., Toraya, T. (1996) Cloning, sequencing, and high level expression of the genes encoding adenosylcobalamin-dependent glycerol-dehydratase of *Klebsiella pneumoniae*, *J. Biol. Chem.* **271**, 22352–22357.

Tong, I.-T., Cameron, D. C. (1992) Enhancement of 1,3-propanediol production by cofermentation in *Escherichia coli* expressing *Klebsiella pneumoniae* *dha* regulon, *Appl. Biochem. Biotechnol.* 34/35, 149–159.

Tong, I.-T., Liao, H. H., Cameron, D. C. (1991) 1,3-Propanediol production by *Escherichia coli* expressing genes from the *Klebsiella pneumoniae* *dha* regulon, *Appl. Environ. Microbiol.* **57**, 3541–3546.

Tran-Din, K., Gottschalk, G. (1985) Formation of D-(-)-1,2-propanediol and D-(-)-lactate from glucose by *Clostridium sphenoides* under phosphate limitation, *Arch. Microbiol.* **142**, 87–92.

Tsai, T. S. and Millard, C. S. (1994) Improved pretreatment process for lactic acid production. PCT Int. Appl. Patent. WO 94/13826:PCT/US93/11759.

Tsutomu, O. (1982) Preparation of L-lactic acid, JP57-110192A.

Tuli, A., Sethi, R. P., Khanna, P. K., Maarwaha, S. S. (1985) Lactic acid production from whey permeate by immobilized *Lactobacillus casei*, *Enzyme Microb. Technol.* **7**, 164–168.

Turk, R. (1993) Metal free and low metal salt substitutes containing lysine, U.S. Patent No. 5,229,161.

Tyree, R. W., Clausen, E. C., Gaddy, J. L. (1990) The fermentative characteristics of *Lactobacillus xylosus* on glucose and xylose, *Biotechnol. Lett.* **14**, 599–604.

Van der Werf, M. J., Guettler, M. V., Jain, M. K., Zeikus, J. G. (1997) Environmental and physiological factors affecting the succinate product ratio during carbohydrate fermentation by *Actinobacillus* sp. 130Z, *Arch. Microbiol.* **167**, 332–342.

Venkatesh, K.V., Wankat, P. C., Okos, M. R. (1994) Kinetic model for lactic acid production from cellulose by simultaneous fermentation and saccharification. *AIChE Symp. Ser.* **300**, 80–87.

Vickroy, T. B. (1985) Lactic acid, in: *The Practice of Biotechnology: Commodity Products* (Blanch, H. W., Drew, S., Wang, D. I. C., Eds.), New York: Vickroy, 761–776.

Wang, H., Seki, M., Furusaki, S. (1995) Mathematical model for analysis of mass transfer for immobilized cells in lactic acid fermentation, *Biotechnol. Prog.* **11**, 558–564.

Wang, X., Gong, C. S., Tsao, G. T. (1998) Bioconversion of fumaric acid to succinic acid by recombinant *E. coli*, *Appl. Biochem. Biotechnol.* **70-72**, 919–928.

Wollenberg, R. H., Frank, P. (1988) Modified succinimides in fuel composition, U.S. Patent No. 4,767,850.

Wrobel, S. A. (1997) Continuous culture of *Thermoanaerobacterium thermosaccharolyticum* and the characterization of 1,2-propanediol production, M.S. Thesis, University of Wisconsin, Madison.

Xiaodong, W., Xuan, G., Rakshit, S. K. (1997) Direct fermentative production of lactic acid on cassava and other starch substrates, *Biotechnol. Lett.* **19**, 841–843.

Yahannavar, V. M., Wang, D. I. C. (1991) Analysis of mass transfer for immobilized cells in an extractive lactic acid fermentation, *Biotechnol. Bioeng.* **37**, 544–550.

Yoo, I. K., Chang, H. N., Lee, E. G., Chang, Y. K., Moon, S. H. (1997) Effect of B vitamin supplementation on lactic acid production by *Lactobacillus casei*, *J. Ferment. Bioeng.* **84**, 172–175.

Yoshida, Y., Oka, H. (1981) Process for producing tetrahydrofuran and 1,4-butanediol, U.S. Patent No. 4,268,447.

Yumoto, I. and Ikeda, K. (1995) Direct fermentation of starch to L-(+)-lactic acid using *Lactobacillus amylophilus*, *Biotechnol. Lett.* **17**, 543–546.

Zajacek, J. G., Shum, W. P. (2000) Butanediol production, U.S. Patent No. 6,127,584.

Zeikus, J. G. (1980) Chemical and fuel production by anaerobic bacteria, *Annu. Rev. Microbiol.* **34**, 423–464.

Zeikus, J. G., Jain, M. K., Elankovan, P. (1999) Biotechnology of succinic acid production and markets for derived industrial products, *Appl. Microbiol. Biotechnol.* **51**. 545–552.

Zhang, D. X., Cheryan, M. (1991) Direct fermentation of starch to lactic acid by *Lactobacillus amylovorus*, *Biotechnol. Lett.* **13**, 733–738.

Zhang, D. X., Cheryan, M. (1994) Starch to lactic acid in a continuous membrane bioreactor, *Process Biochem.* **29**, 145–150.

Zhang, X., Wyss, U. P., Pichora, D., Goosen, M. F. A. (1993) Biodegradable polymers for orthopedic applications – Synthesis and processability of poly(L-lactide) and poly(lactide-*co*-epsilon-caprolactone), *J. Macromol. Sci. - Pure Appl. Chem.* **A30**, 933–947.

Zwicker, N., Theobald, U., Zahner, H., Fiedler, H. P. (1997) Optimization of fermentation conditions for the production of ethylene-diamine-disuccinic acid by *Amycolatopsis orientalis*, *J. Ind. Microbiol. Biotechnol.* **19**, 280–285.

IV
Polysaccharides

Biotechnology of Biopolymers. From Synthesis to Patents. Edited by A. Steinbüchel and Y. Doi

ISBN: 3-527-31110-6

14
Bacterial Cellulose

Prof. Dr. Eng. Stanislaw Bielecki[1], Dr. Eng. Alina Krystynowicz[2], Prof. Dr. Marianna Turkiewicz[3], Dr. Eng. Halina Kalinowska[4]

[1] Institute of Technical Biochemistry, Technical University of Lódz, Stefanowskiego 4/10, 90-924 Lódz, Poland; Tel: +48-4263-13442; Fax: +48-4263-402; E-mail: biochem@ck-sg.p.lodz.pl

[2] Institute of Technical Biochemistry, Technical University of Lódz, Stefanowskiego 4/10, 90-924 Lódz, Poland; Tel: +48-4263-13442; Fax: +48-4263-402; E-mail: biochem@ck-sg.p.lodz.pl

[3] Institute of Technical Biochemistry, Technical University of Lódz, Stefanowskiego 4/10, 90-924 Lódz, Poland; Tel: +48-4263-13442; Fax: +48-4263-402; E-mail: biochem@ck-sg.p.lodz.pl

[4] Institute of Technical Biochemistry, Technical University of Lódz, Stefanowskiego 4/10, 90-924 Lódz, Poland; Tel: +48-4263-13442; Fax: +48-4263-402; E-mail: biochem@ck-sg.p.lodz.pl

Biotechnology of Biopolymers. From Synthesis to Patents. Edited by A. Steinbüchel and Y. Doi

ISBN: 3-527-31110-6

A-BC	bacterial cellulose from agitated culture
ATP	adenosine triphosphate
BC	bacterial cellulose
CBH	cellobiohydrolase
c-di-GMP	cyclic diguanosine monophosphate
Cel^-	cellulose-negative mutant
Cel6A, Cel7A	cellobiohydrolases belonging to 6A and 7A families, respectively
CM	carboxymethyl-
CS	cellulose synthase
CSL	corn steep liquor
D	aspartic acid

DMSO	dimethyl sulfoxide
DP	degree of polymerization
E	glutamic acid
FBP	fructose-1,6-biphosphate phosphatase
FK	fructokinase
Fru-bi-P	fructose-1,6-biphosphate
Fru-6-P	fructose-6-phosphate
G	guanine
GK	glucokinase
Glc	glucose
Glc-6(1)-P	glucose-6(1)-phosphate
G6PDH	glucose-6-phosphate dehydrogenase
H-S medium	Hestrin and Schramm medium (1954)
IS	insertion sequence
Lip	lipid
LP	UDPGPT: lipid pyrophosphate: UDPGlc- phosphotransferase
LPP	lipid pyrophosphate phosphohydrolase
Man	mannose
NMR	nuclear magnetic resonance
PC	plant cellulose
PDEA	phosphodiesterase A
PDEB	phosphodiesterase B
Pel^-	pellicle non-forming
1PFK	fructose-1-phosphate kinase
PGA	phosphogluconic acid
PGI	phosphoglucoisomerase
PMG	phosphoglucomutase
PTS	system of phosphotransferases
Q	glutamine
R	arginine
Rha	rhamnose
Rib	D-ribose
S	serine
S-BC	bacterial cellulose from static culture
TC	terminal complexe
U	uridine
UDP	uridine diphosphate
UDPGlc	uridine diphosphoglucose
UGP	pyrophosphorylase uridine diphosphoglucose
UMP	uridine monophosphate
v.v.m.	volume per volume per minute
W	tryptophan

1
Introduction

Cellulose is the most abundant biopolymer on earth, recognized as the major component of plant biomass, but also a representative of microbial extracellular polymers. Bacterial cellulose (BC) belongs to specific products of primary metabolism and is mainly a protective coating, whereas plant cellulose (PC) plays a structural role.

Cellulose is synthesized by bacteria belonging to the genera *Acetobacter, Rhizobium, Agrobacterium,* and *Sarcina* (Jonas and Farah, 1998). Its most efficient producers are Gram-negative, acetic acid bacteria *Acetobacter xylinum* (reclassified as *Gluconacetobacter xylinus,* Yamada et al., 1997; Yamada, 2000), which have been applied as model microorganisms for basic and applied studies on cellulose (Cannon and Anderson, 1991). Investigations have been focused on the mechanism of biopolymer synthesis, as well as on its structure and properties, which determine practical use (Legge, 1990; Ross et al., 1991). One of the most important features of BC is its chemical purity, which distinguishes this cellulose from that from plants, usually associated with hemicelluloses and lignin, removal of which is inherently difficult.

Because of the unique properties, resulting from the ultrafine reticulated structure, BC has found a multitude of applications in paper, textile, and food industries, and as a biomaterial in cosmetics and medicine (Ring et al., 1986). Wider application of this polysaccharide is obviously dependent on the scale of production and its cost. Therefore, basic studies run together with intensive research on strain improvement and production process development.

2
Historical Outline

Although synthesis of an extracellular gelatinous mat by *A. xylinum* was reported for the first time in 1886 by A. J. Brown, BC attracted more attention in the second half of the 20th century. Intensive studies on BC synthesis, using *A. xylinum* as a model bacterium, were started by Hestrin et al. (1947, 1954), who proved that resting and lyophilized *Acetobacter* cells synthesized cellulose in the presence of glucose and oxygen. Next, Colvin (1957) detected cellulose synthesis in samples containing cell-free extract of *A. xylinum,* glucose, and ATP. Further milestones in studies on BC synthesis, presented in this review, contributed to the elucidation of mechanisms governing not only the biogenesis of the bacterial polymer, but also that of plants, thus leading to the understanding of one of the most important processes in nature. The true historical outline is presented throughout all the paragraphs below, including the references.

3
Structure of BC

Cellulose is an unbranched polymer of β-1,4-linked glucopyranose residues. Extensive research on BC revealed that it is chemically identical to PC, but its macromolecular structure and properties differ from the latter (Figure 1). Nascent chains of BC aggregate to form subfibrils, which have a width of approximately 1.5 nm and belong to the thinnest naturally occurring fibers, comparable only to subelemental fibers of cellulose detected in the cambium of some plants and in quinee mucous (Kudlicka, 1989). BC subfibrils are crystallized into microfibrils (Jonas and Farah, 1998), these into bundles, and the latter into ribbons (Yamanaka et al.,

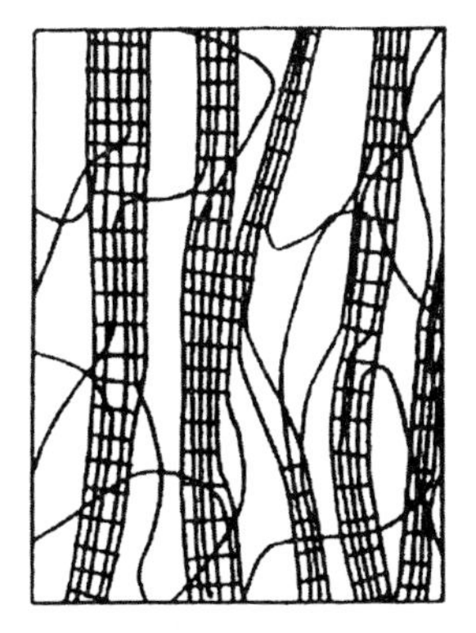
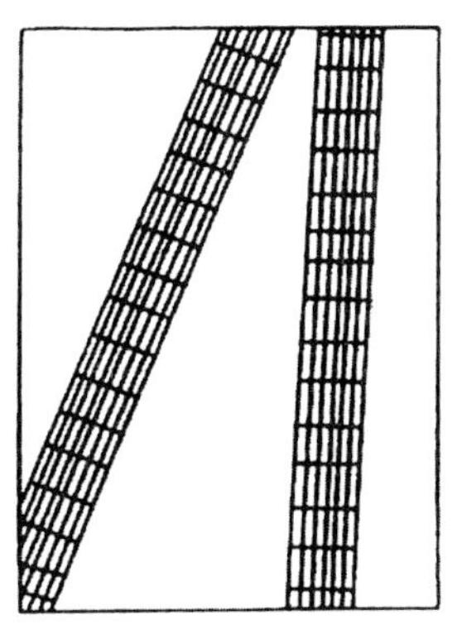

Fig. 1 Schematic model of BC microfibrils (right) drawn in comparison with the 'fringed micelles'; of PC fibrils (Iguchi et al., 2000; with kind permission).

2000). Dimensions of the ribbons are 3–4 (thickness) × 70–80 nm (width), according to Zaar (1977), 3.2 × 133 nm, according to Brown et al. (1976), or 4.1 × 117 nm, according to Yamanaka et al. (2000), whereas the width of cellulose fibers produced by pulping of birch or pine wood is two orders of magnitude larger ($1.4-4.0 \times 10^{-2}$ and $3.0-7.5 \times 10^{-2}$ mm, respectively). The ultrafine ribbons of microbial cellulose, the length of which ranges from 1 to 9 μm, form a dense reticulated structure (Figure 2), stabilized by extensive hydrogen bonding (Figure 3). BC is also distinguished from its plant counterpart by a high crystallinity index (above 60%) and different degree of polymerization (DP), usually between 2000 and 6000 (Jonas and Farah, 1998), but in some cases reaching even 16,000 or 20,000 (Watanabe et al., 1998b), whereas the average DP of plant polymer varies from 13,000 to 14,000 (Teeri, 1997).

Macroscopic morphology of BC strictly depends on culture conditions (Watanabe et al., 1998a; Yamanaka et al., 2000). In static conditions (Figure 4), bacteria accumulate cellulose mats (S-BC) on the surface of nutrient broth, at the oxygen-rich air–liquid interface. The subfibrils of cellulose are continuously extruded from linearly ordered pores at the surface of the bacterial cell, crystallized into microfibrils, and forced deeper into the growth medium. Therefore, the leather-like pellicle, supporting the population of *A. xylinum* cells, consists of overlapping and intertwisted cellulose rib-

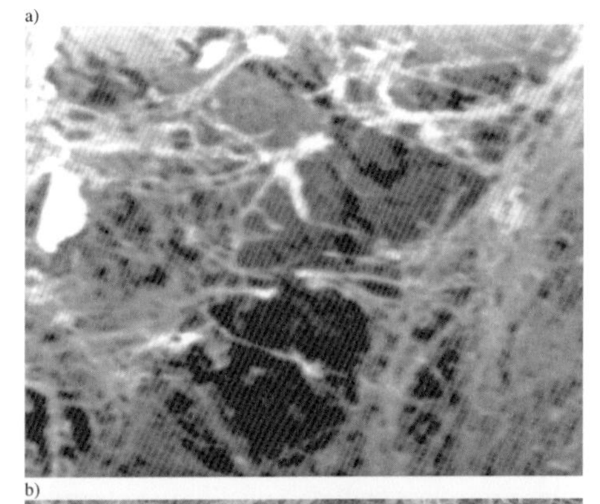

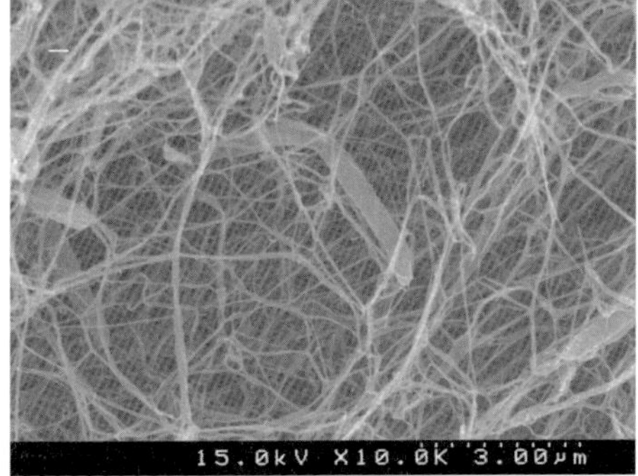

Fig. 2 Scanning electron microscopy images of BC membrane from static culture of *A. xylinum* (a) and bacterial cell with attached cellulose ribbons (b).

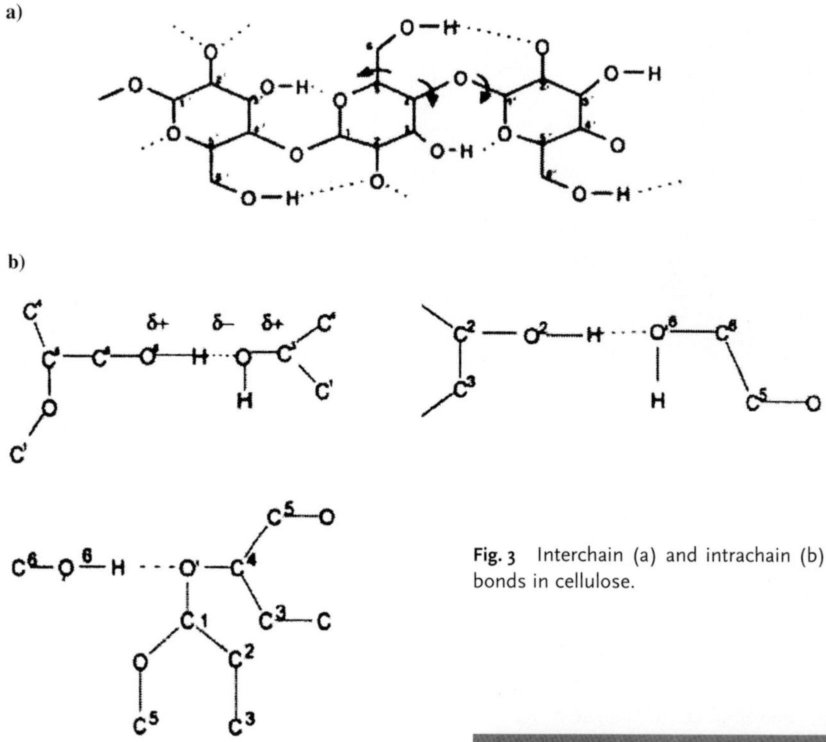

a)

b)

Fig. 3 Interchain (a) and intrachain (b) hydrogen bonds in cellulose.

bons, forming parallel but disorganized planes (Jonas and Farah, 1998). The adjacent S-BC strands branch and interconnect less frequently than these in BC produced in agitated culture (A-BC), in a form of irregular granules, stellate and fibrous strands, well-dispersed in culture broth (Figure 5) (Vandamme et al., 1998). The strands of reticulated A-BC interconnect to form a grid-like pattern, and have both roughly perpendicular and roughly parallel orientations (Watanabe et al., 1998a).

Differences in three-dimensional structure of A-BC and S-BC are noticeable in their scanning electron micrographs. The S-BC fibrils are more extended and piled above one another in a criss-crossing manner. Strands of A-BC are entangled and curved (Johnson and Neogi, 1989). Besides, they

Fig. 4 BC pellicle formed in static culture.

Fig. 5 BC pellets formed in agitated culture.

have a larger cross-sectional width (0.1–0.2 μm) than S-BC fibrils (usually 0.05–0.10 μm). Morphological differences between S-BC and A-BC contribute to varying degrees of crystallinity, different crystallite size and I_{α} cellulose content.

Two common crystalline forms of cellulose, designated as I and II, are distinguishable by X-ray, nuclear magnetic resonance (NMR), Raman spectroscopy, and infrared analysis (Johnson and Neogi, 1989). It is known that in the metastable cellulose I, which is synthesized by the majority of plants and also by *A. xylinum* in static culture, parallel β-1,4-glucan chains are arranged uniaxially, whereas β-1,4-glucan chains of cellulose II are arranged in a random manner. They are mostly antiparallel and linked with a larger number of hydrogen bonds that results in higher thermodynamic stability of the cellulose II.

A-BC has a lower crystallinity index and a smaller crystallite size than S-BC (Watanabe et al., 1998a). It was also observed that a significant portion of cellulose II occurred in BC synthesized in agitated culture. In nature, cellulose II is synthesized by only a few organisms (some algae, molds, and bacteria, such as *Sarcina ventriculi*) (Jonas and Farah, 1998); the industrial production of this kind of cellulose is based on chemical conversion of PC.

Using CP/MAS ^{13}C-NMR it is possible to reveal the presence of cellulose I_{α} and I_{β} – two distinct forms of cellulose (Watanabe et al., 1998a). These forms occur in algae-, bacteria-, and plant-derived cellulose. The latter one contains less I_{α} cellulose than BC (Johnson and Neogi, 1989). The irreversible crystal transformation from cellulose I_{α} to I_{β} shifts the X-ray and CP/MAS ^{13}C-NMR spectra because of the difference in the unit cell. S-BC contains more cellulose I_{α} than A-BC. It was reported that the difference in cellulose I_{α} content between A-BC and S-BC exceeded that in crystallinity index (Watanabe et al., 1998a), and the mass fraction of cellulose I_{α} was closely related to the crystallite size.

4
Chemical Analysis and Detection

For the detection of either crystalline or amorphous cellulose, several direct dyes, specific for the linear β-1,4-glucan, are used (Mondal and Kai, 2000). All of them are fluorescent brightening agents and form dye–cellulose complexes, stabilized by van der Waals and/or hydrogen bonding. One of these dyes is the fluorescent brightener Calcofluor. The direct dyes do not only enable visualization of cellulose chains, but also have been intensively applied for studies on nascent cellulose chains association and crystallization (see Section 7.3.2).

The weight-average DP of cellulose and the DP distribution are determined by high-performance gel-permeation chromatography (Watanabe et al., 1998a) of nitrated cellulose samples.

The differences between the reticulated structure of microbial cellulose, produced under agitated culture conditions, and the disorganized layered structure of cellulose pellicle, formed in static culture, are noticeable in scanning electron microscopy images (Johnson and Neogi, 1989).

To distinguish the parallel chain crystalline lattice of cellulose I from the antiparallel one of cellulose II, X-ray diffraction, Raman spectroscopy, infrared analysis, and NMR are applied (Johnson and Neogi, 1989). The crystallinity index and crystallite size are calculated based on X-ray diffraction measurements (Watanabe et al., 1998a).

Two distinct forms of cellulose I, i.e. cellulose I_{α} and I_{β}, are not distinguishable by X-ray diffraction, and therefore CP/MAS ^{13}C-NMR analysis, carried out on freeze-dried cellulose samples, has to be performed to determine their mass fractions (Watanabe et al., 1998a).

The physicochemical properties of cellulose such as water holding capacity, viscosity of disintegrated cellulose suspension, and the Young's modulus of dried sheets are determined using conventional methods (Watanabe et al., 1998a, Iguchi et al., 2000).

5 Occurrence

BC is synthesized by several bacterial genera, of which *Acetobacter* strains are best known. An overview of BC producers is presented in Table 1 (Jonas and Farah, 1998). The polymer structure depends on the organism, although the pathway of biosynthesis and mechanism of its regulation are probably common for the majority of BC-producing bacteria (Ross et al., 1991; Jonas and Farah, 1998).

Tab. 1 BC producers (Jonas and Farah, 1998, modified)

Genus	*Cellulose structure*
Acetobacter	extracellular pellicle composed of ribbons
Achromobacter	fibrils
Aerobacter	fibrils
Agrobacterium	short fibrils
Alcaligenes	fibrils
Pseudomonas	no distinct fibrils
Rhizobium	short fibrils
Sarcina	amorphous cellulose
Zoogloea	not well defined

A. xylinum (synonyms *A. aceti* ssp. xylinum, *A. xylinus*), which is the most efficient producer of cellulose, has been recently reclassified and included within the novel genus *Gluconacetobacter*, as *G. xylinus* (Yamada et al., 1998, 2000) together with some other species (*G. hansenii*, *G. europaeus*, *G. oboediens*, and *G. intermedius*).

6 Physiological Function

In natural habitats, the majority of bacteria synthesize extracellular polysaccharides, which form envelopes around the cells (Costeron, 1999). BC is an example of such a substance. Cells of cellulose-producing bacteria are entrapped in the polymer network, frequently supporting the population at the liquid–air interface (Wiliams and Cannon, 1989). Therefore, BC-forming strains can inhabit sewage (Jonas and Farah, 1998). The polymer matrix takes part in adhesion of the cells onto any accessible surface and facilitates nutrient supply, since their concentration in the polymer lattice is markedly enhanced due to its adsorptive properties, in comparison to the surrounding aqueous environment (Jonas and Farah, 1998; Costeron, 1999). Some authors suppose that cellulose synthesized by *A. xylinum* also plays a storage role and can be utilized by the starving microorganisms. Its decomposition would be then catalyzed by exo- and endoglucanases, the co-presence of which was detected in the culture broth of some cellulose-producing *A. xylinum* strains (Okamoto et al., 1994).

Because of the viscosity and hydrophilic properties of the cellulose layer, the resistance of producing bacterial cells against unfavorable changes (a decrease in water content, variations in pH, appearance of toxic substances, pathogenic organisms, etc.) in an habitat is increased, and they can further grow and develop inside the envelope. It was also found that cellulose placed over bacterial cells protects them from ultraviolet radiation. As much as 23% of the acetic acid bacteria cells covered with BC survived a 1 h treatment with ultraviolet irradiation. Removal of the protective polysaccharide brought about a drastic decrease in their viability (3% only) (Ross et al., 1991).

7 Biosynthesis of BC

Synthesis of BC is a precisely and specifically regulated multi-step process, involving a large number of both individual enzymes and complexes of catalytic and regulatory proteins, whose supramolecular structure has not yet been well defined. The process includes the synthesis of uridine diphosphoglucose (UDPGlc), which is the cellulose precursor, followed by glucose polymerization into the β-1,4-glucan chain, and nascent chain association into characteristic ribbon-like structure, formed by hundreds or even thousands of individual cellulose chains. Pathways and mechanisms of UDPGlc synthesis are relatively well known, whereas molecular mechanisms of glucose polymerization into long and unbranched chains, their extrusion outside the cell, and self-assembly into fibrils require further elucidation.

Moreover, studies on BC synthesis may contribute to better understanding of PC biogenesis.

7.1 Synthesis of the Cellulose Precursor

Cellulose synthesized by *A. xylinum* is a final product of carbon metabolism, which depending on the physiological state of the cell involves either the pentose phosphate cycle or the Krebs cycle, coupled with gluconeogenesis (Figure 6) (Ross et al., 1991; Tonouchi et al., 1996). Glycolysis does not operate in acetic acid bacteria since they do not synthesize the crucial enzyme of this pathway – phosphofructose kinase (EC 2.7.1.56) (Ross et al., 1991). In *A. xylinum*, cellulose synthesis is tightly associated with catabolic processes of oxidation and consumes as much as 10% of energy derived from catabolic reactions (Weinhouse, 1977). BC production does not interfere with other anabolic processes, including protein synthesis (Ross et al., 1991).

A. xylinum converts various carbon compounds, such as hexoses, glycerol, dihydroxyacetone, pyruvate, and dicarboxylic acids, into cellulose, usually with about 50% efficiency. The latter compounds enter the Krebs cycle and due to oxalacetate decarboxylation to pyruvate undergo conversion to hexoses via gluconeogenesis, similarly to glycerol, dihydroxyacetone, and intermediates of the pentose phosphate cycle (Figure 6).

The direct cellulose precursor is UDPGlc, which is a product of a conventional pathway, common of many organisms, including plants, and involving glucose phosphoryla-

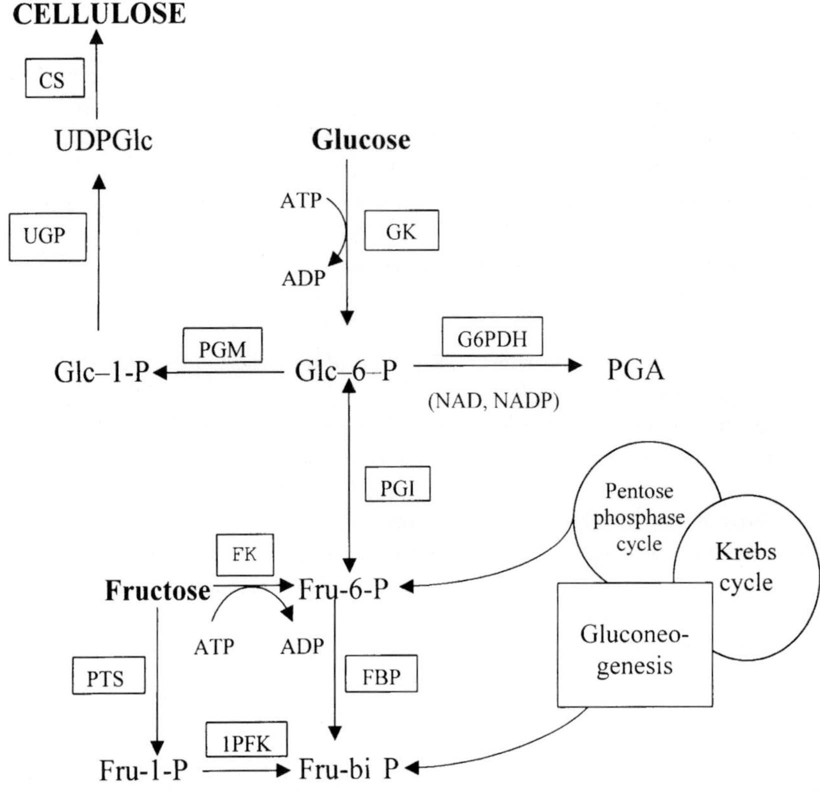

Fig. 6 Pathways of carbon metabolism in *A. xylinum*. CS, cellulose synthase (EC 2.4.2.12); FBP, fructose-1,6-biphosphate phosphatase (EC 3.1.3.11); FK, glucokinase (EC 2.7.1.2); G6PDH, glucose-6-phosphate dehydrogenase (EC 1.1.1.49); 1PFK, fructose-1-phosphate kinase (EC 2.7.1.56); PGI, phosphoglucoisomerase; PMG, phosphoglucomutase (EC 5.3.1.9); PTS, system of phosphotransferases; UGP, pyrophosphorylase UDPGlc (EC 2.7.7.9); Fru-bi-P, fructose-1,6-bi-phosphate; Fru-6-P, fructose-6-phosphate; Glc-6(1)-P, glucose-6(1)-phosphate; PGA, phosphogluconic acid; UDPGlc, uridine diphosphoglucose.

tion to glucose-6-phosphate (Glc-6-P), catalyzed by glucokinase, followed by isomerization of this intermediate to Glc-α-1-P, catalyzed by phosphoglucomutase, and conversion of the latter metabolite to UDPGlc by UDPGlc pyrophosphorylase. This last enzyme seems to be the crucial one involved in cellulose synthesis, since some phenotypic cellulose-negative mutants (Cel$^-$) are specifically deficient in this enzyme (Valla et al., 1989), though they display cellulose synthase (CS) activity, this was confirmed *in vitro* by means of observation of cellulose synthesis, catalyzed by cell-free extracts of Cel$^-$ strains (Saxena et al., 1989). Furthermore, the pyrophosphorylase activity varies between different *A. xylinum* strains and the highest activity was detected in the most effective cellulose producers, such as *A. xylinum* ssp. sucrofermentans BPR2001. The latter strain prefers fructose as a carbon source, displays high activity of phosphoglucoisomerase, and possesses a system of phosphotransferases, dependent on phosphoenolpyruvate. The system catalyzes conversion of fructose to fructose-1-phosphate and further to fructose-1,6-biphosphate (Figure 6).

7.2 Cellulose Synthase

Cellulose biosynthesis both in plants and in prokaryotes is catalyzed by the uridine diphosphate (UDP)-forming CS which is basically a processing 4-β-glycosyltransferase (EC 2.4.1.12, UDPGlc: 1,4-β-glucan 4-β-D-glucosyltransferase), since it transfers consecutive glucopyranose residues from UDPGlc to the newly formed polysaccharide chain and is all the time linked with this chain. Oligomeric CS complexes are frequently called terminal complexes (TCs). It is presumed that TCs are responsible first for β-1,4-glucan chains synthesis, extrusion through the outer membrane (if the globular, catalytic domain of CS is localized on the cytoplasmic side of the cell membrane), as well as association and crystallization into defined supramolecular structures, that follow the first two processes.

Cellulose synthase of *A. xylinum* is a typical membrane-anchored protein, having a molecular mass of 400–500 kDa (Lin and Brown, 1989), and is tightly bound to the cytoplasmic membrane. Because of this localization, purification of CS was extremely difficult, and isolation of the membrane fraction was necessary before CS solubilization and purification. Furthermore, *A. xylinum* CS appeared to be a very unstable protein (Lin and Brown, 1989). CS isolation from membranes was carried out using digitonin (Lin and Brown, 1989), or detergents (Triton X-100) and treatment with trypsin (Saxena et al., 1989), followed by CS entrapment on cellulose. According to Lin and Brown (1989), the purified CS preparation contained three different types of subunits, having molecular masses of 90, 67, and 54 kDa. Saxena et al. (1989) found only two types of polypeptides (83 and 93 kDa). The latter result seems to be more probable, since the mass of both subunits corresponds to the size of two genes *cesA* and *cesB*, detected in the CS operon by Saxena et al. (1990b, 1991), and the genes *bcsA* and *bcsB*, reported by Wong et al. (1990) and Ben-Bassat et al. (1993) for the same operon (see Section 7.4).

Photolabeling affinity studies indicate that the 83-kDa polypeptide is a catalytic subunit, displaying high affinity towards UDPGlc (Lin et al., 1990). According to the gene sequence, it contains 723 amino acid residues, of which the majority are hydrophobic. This subunit is probably synthesized as a proprotein, which contains a signal sequence, composed of 24 amino acid residues, which is cut off during the maturation process (Wong et al., 1990). The mature protein is anchored in the cell membrane by means of a transmembrane helix, close to the N-terminus of the polypeptide chain. The protein contains five more transmembrane helices, which can interact both with each other and with a large, catalytic site-comprising globular domain submerged in cytoplasma. Brown and Saxena (2000) claim that the catalytic subunit of *A. xylinum* CS operates in the same manner as processing glycosyltransferases of the second family, i.e. it catalyzes the direct glycoside bond synthesis in cellulose, assisted by simultaneous inversion of the configuration on the anomeric carbon, from the α-configuration in the UDPGlc, which is the monomer donor, to the β-configuration in the polysaccharide. A catalytically active aspartic acid residue (D), and short sequences DXD and QXXRW were found in the globular fragment of the numerous processing glycosyltransferases (Saxena and Brown, 1997). The same motives were detected in *A. xylinum* CS. This globular fragment has a two-domain character (i.e., comprises two domains A and B) in many of glycoside synthases (Saxena et al., 1995). The domain A includes the D residue, crucial for catalysis, and the DXD motive, which probably binds the nucleotide–sugar substrate. Participation of this motive in UDPGlc binding by *A. xylinum* CS was recently confirmed by Brown and Saxena (2000), who investigated enzyme mutants obtained using site-directed mutagenesis. It is not yet clear if the globular fragment of *A. xylinum* CS comprises one or two domains, although this second possibility is more probable. However, it is known that the catalytic subunit of *A. xylinum* CS is glycosylated. Two potential sites of glycosylation were found in its primary structure, deduced based on the gene sequence (Saxena et al., 1990a).

The 93-kDa polypeptide is tightly bound to the catalytic subunit. It contains the signal sequence close to its N-terminus and one transmembrane helix close to the C-terminus. The helix enables anchoring in the cell membrane. The polypeptide is probably a regulatory subunit, which interacts with the CS activator – cyclic diguanosine monophospahte (c-di-GMP) (see Section 7.5). The 93-kDa polypeptide does not combine with the 83-kDa subunit antibodies. However, it is not yet clear if CS is composed of these two subunits only, since this oligomeric complex plays multiple roles, i.e. β-1,4-glucan polymerization, extrusion of the subfibrils outside the cell, as well as self-assembly and crystallization of cellulose ribbons, composed of microfibrils. Therefore, the CS complex is probably comprised of other polypeptides involved in transmembrane pore formation. The analysis of microscopic images of the purified CS indicates that the enzyme tends to form tetrameric or octameric aggregates (Lin et al., 1989).

7.3 Mechanism of Biosynthesis

Synthesis of the metastable cellulose I allomorph, in *A. xylinum* and other cellulose-producing organisms, including plants, involves at least two intermediary steps, i.e. (1) polymerization of glucose molecules to the linear 1,4-β-glucan, and (2) assembly and crystallization of individual nascent polymer chains into supramolecular structures, characteristic for each cellulose-producing organism.

Assuming that the CS globular domain is localized on the cytoplasmic side of the cell membrane (see Section 7.2), the transfer of 1,4-β-glucan chains through the membrane, outside the cell is also required. All three steps are tightly coupled, and the rate of polymerization is limited by the rate of assembly and crystallization.

7.3.1 Mechanism of 1,4-β-Glucan Polymerization

Formation of BC is catalyzed by the CS complexes aligned linearly in the *A. xylinum* cytoplasmic membrane. Since these complexes can polymerize up to 200,000 molecules of Glc s^{-1} into the β-1,4-glucan chain (Ross et al., 1991), the process must run with high intensity. The mechanism of the reaction has not yet been definitely clarified. Currently, two different hypotheses for this mechanism in *A. xylinum* have been proposed.

The first of them, developed in Brown's laboratory (Brown, 1996; Brown and Saxena, 2000), assumes that 1,4-β-glucan polymerization does not involve a lipid intermediate, which transfers glucose from UDPGlc to the newly synthesized polymer chain. This hypothesis agrees with the fact that glycosyltransferases responsible for synthesis of unbranched homopolysaccharides are processing enzymes. The scheme of 1,4-β-glucan polymerization, proposed by Brown and his colleagues, is presented in Figure 7.

According to this model, there are three catalytic sites in the globular fragment of the CS catalytic subunit, similar to other proc-

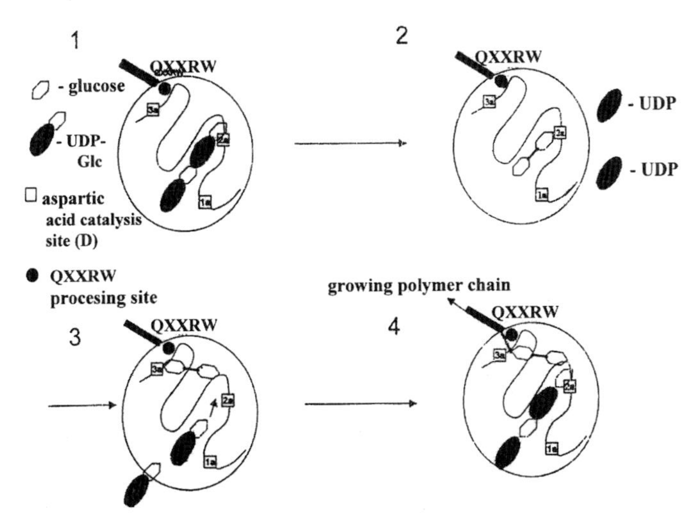

Fig. 7 Generalized concept for the polymerization reactions leading to β-1,4-glucan chain biosynthesis (Brown and Saxena, 2000; with kind permission).

essing glycosyltransferases. One of the catalytic sites (2a), comprising the DXD motive, binds two UDPGlc molecules (see Figure 7.1). The second catalytic site (1a), containing the crucial acidic aspartic acid (D) residue, catalyzes formation of the β-1,4-linkage between the two Glc residues docked in the pocket (see Figure 7.2), accompanied by releasing two UDP molecules. The third catalytic site (3a), containing the QXXRW motif, pulls the reducing end of the synthesized cellobiose (see Figure 7.3). The disaccharide leaves the area 1a–2a, which may bind two subsequent UDPGlc molecules (see Figure 7.4), and forms the second cellobiose molecule. In the next step, involving the QXXRW motif, the reducing end of the first cellobiose molecule is forced to the site of extrusion. At the same moment, the second cellobiose molecule is linked (with the aid of the D residue in the site 1a) to the nonreducing end of the first one, thus giving cellotetraose. The emptied 1a–2a area may bind two subsequent UDPGlc molecules, to repeat the cycle of reactions, attaching two more glucose residues to the chain. The simplified scheme of the proposed model of polymerization is depicted in Figure 8. According to this model, extrusion of the newly synthesized β-1,4-glucan chain starts from its reducing end.

Studies by Koyama et al. (1997) confirm this model. The authors proved that glucose residues are added to the nonreducing end of the polysaccharide and that reducing ends of nascent polymer chains are situated away from the cells. Furthermore, the torsion angle between two adjacent glucose residues in cellulose molecule is 180°, thus adding cellobiose residues (rather than single glucose moieties) to the growing chain favors maintaining the 2-fold screw axis of the β-1,4-glucan (Brown, 1996). Also Kuga and Brown (1988), who applied silver labeling, proved that cellulose chains were extruded outside the cell, starting from their reducing ends.

Han and Robyt (1998) who studied BC synthesis by the *A. xylinum* ATCC 10821 strain (resting cells and membrane preparations) using the ^{14}C pulse and chase reaction with D-glucose and UDPGlc, respectively,

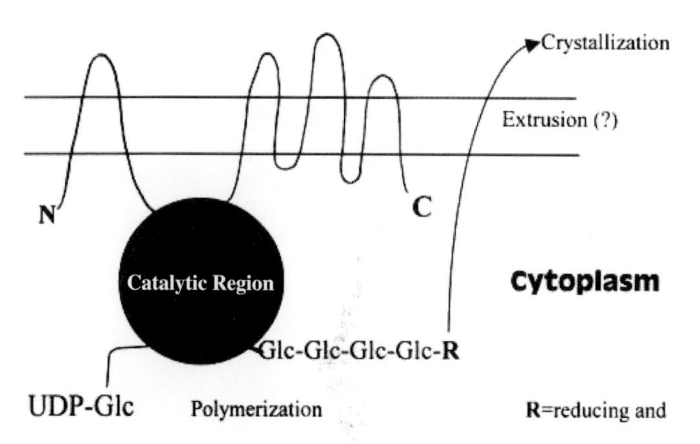

N and C – N and C-terminus of CS subunit of 83 kDa, respectively

Fig. 8 Simplified scheme of BC biosynthesis according to the Brown's model (Brown and Saxena, 2000; with kind permission).

proposed a second, different molecular mechanism of this process. They believe that the consecutive residues of the activated glucopyranoses (from UDPGlc) are linked to the reducing end of the growing cellulose chain, because the ratio of D-[^{14}C]glucitol obtained from the extruded reducing end of the cellulose chain to D-[^{14}C]glucose was decreasing with time. Han and Robyt assumed that a lipid pyrophosphate with a polyisoprenoid character takes part in the biosynthesis. According to them, formation of the lipid intermediate during BC synthesis was confirmed by an extraction of as much as 33% of the pulsed ^{14}C label with a mixture of chloroform and methanol. Its quantitative extraction was impossible, since when the polysaccharide chain length exceeded 8–10 sugar moieties, the complex with lipid became insoluble in the extraction mixture.

Han and Robyt proposed that BC biosynthesis involved three enzymes embedded in the cytoplasmic membrane: CS, lipid pyrophosphate: UDPGlc phosphotransferase (LipPP: UDPGlc-PT), and lipid pyrophosphate phosphohydrolase (LPP). The reaction mechanism, called by the authors the insertion reaction, is presented in the Figure 9.

The first enzyme transfers Glc-α-1-P from UDPGlc onto the lipid monophosphate (Lip-P), thus giving the lipid pyrophosphate-α-D-Glc (LipPP-α-Glc) (reaction 1, Figure 9). The α configuration on the anomeric carbon, involved in the Glc phosphoester bond, remains the same as in the substrate molecule. The second product of this reaction is UMP (according to Brown's model, UDP is released).

In the next reaction (reaction 2, Figure 9), catalyzed by CS, the glucose residue is transferred from one LipPP-α-Glc molecule onto another one and the β-1,4-glycosidic linkage between the two glucose residues is formed, due to the attack of the C-4 hydroxyl group of one of them onto C-1 hydroxyl group of the second Glc (from the second LipPP-α-Glc).

In the third reaction (reaction 3, Figure 9), hydrolysis of the lipid pyrophosphate formed in the previous step occurs and

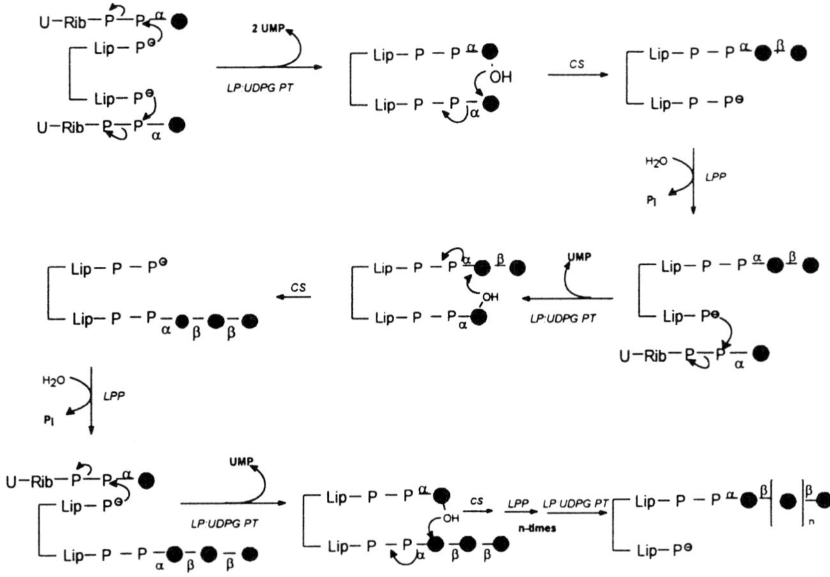

Fig. 9 Mechanism of BC biosynthesis involving lipid intermediates (Han and Robyt, 1998; with kind permission; modified). Lip, lipid; P, phosphoryl; Rib, D-ribose; U, uridine, •, glucosyl residue; LP: UDPGPT, lipid pyrophosphate: UDPGlc-phosphotransferase; CS; cellulose synthase; LPP, lipid pyrophosphate phosphohydrolase.

another Glc-α-1P from UDPGlc can be attached to the LipP, released in this reaction. The cycle of these three reactions is continued to give the β-1,4-glucan chain of an appropriate length. The mechanism includes the inversion of a configuration on the anomeric carbon in the Glc residue transferred either from one LipPP-α-Glc molecule to another, or to cellobiose, -triose, -tetraose, and subsequently to *n*-ose. An acceptor of the elongated chain of β-1,4-glucan is always a single α-D-Glc residue, linked with one of two LipPP molecules present in the polymerizing system. It means that the chain elongation occurs from its reducing end and nonreducing ends of nascent cellulose chains are situated away from the cells; this is contradictory to Brown's model. According to Han and Robyt's model, both the two cooperating LipPP molecules and the CS complex, which is not a processing glycosyltransferase, are embedded in the cytoplasmic membrane of the bacterial cell.

Participation of a lipid intermediate in cellulose biosynthesis was confirmed for *Agrobacterium tumefaciens* (Matthysse et al., 1995), as well as for synthesis of *Salmonella* O-antigen polysaccharide (Bray and Robbins, 1967) and *Xanthomonas campestris* xanthan (Ielpi et al., 1981). Furthermore, the lipid intermediate is probably also involved in the synthesis of acetan (De Lannino et al., 1988), which is a soluble polysaccharide produced together with cellulose by numerous *A. xylinum* strains (see Section 7.6). Further studies have to be carried out to elucidate whether the lipid intermediate plays in cellulose formation, the role postulated by Han and Robyt (1998). On the contrary, Brown's hypothesis (1996, 2000) is confirmed by cellulose synthesis *in vitro*, catalyzed by *A. xylinum* membrane preparations solubilized with digitonin (Lin et al., 1985; Bureau and Brown, 1987), and by polymer synthesis by purified CS subunits (Lin and Brown, 1989), which did not contain the lipid component.

According to both proposed hypotheses, BC biosynthesis does not require any primer, in agreement with earlier deductions (Canale-Parda and Wolte, 1964).

7.3.2
Assembly and Crystallization of Cellulose Chains

Polymerization of glucopyranose molecules to the β-1,4-glucan, whatever the mechanism, is the least complicated stage of cellulose synthesis, also in bacteria. The unique structure and properties of cellulose, which are dependent on its origin, result from the course of further stages: it means extrusion of the chains and their assembly outside the cell, giving supramolecular structures. Most important in these processes is the specific organization of multimeric complexes of CS, which are anchored in cytoplasmic membrane. In *A. xylinum* cells these complexes are ordered linearly along the long axis of the cell and one cell contains 50–80 TCs (Brown et al., 1976), whereas in vascular plants TCs form characteristic, 6-fold symmetrical rosettes (Brown and Saxena, 2000).

Stronger aeration, e.g., during agitated *A. xylinum* cultures, or the presence of certain substances, that cannot penetrate inside the cells, but can form competitive hydrogen bonds with the β-1,4-glucan chains (e.g., carboxymethyl-(CM)-cellulose, fluorescent brightener Calcofluor white), bring about significant changes in the supramolecular organization of cellulose chains, and instead of the ribbon-like polymer, i.e. the metastable cellulose I (Figure 10), which fibrils are colinear to CS complexes and gathered in a row, the second, thermodynamically more stable allomorph amorphous cellulose II is formed. This phenomenon is accompanied by disorganization of the TC' linear order. Discussing this problem, Brown and Saxena (2000) state that the conditions-dependent form of the final product may somehow determine the TC' arrangement in the cytoplasmic membrane, though it is believed that in *A. xylinum* these complexes have a stationary character, as opposed to some algae (Kudlicka, 1989). Worth mentioning is the well-known phenomenon of *A. xylinum* cell motion during BC synthesis, in the direction opposite to cellulose chain extrusion (Brown et al., 1976). Reversal movement of *A. xylinum* cells enabled calculation

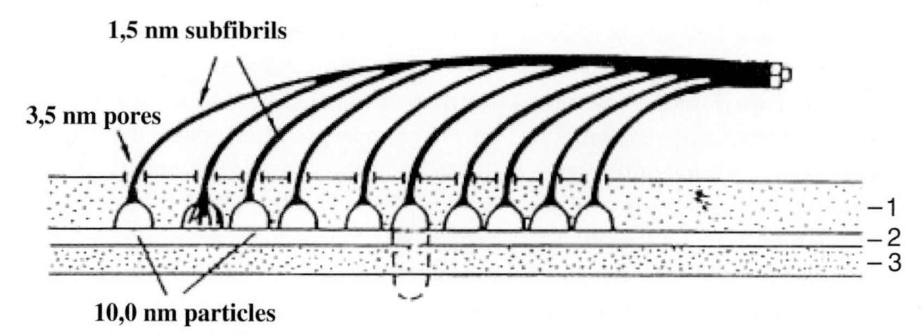

Fig. 10 Model of *A. xylinum* cellulose subfibrils formation: 1, lipopolysaccharide layer; 2, periplasmic space; 3, plasmalemma; 10 nm particles, CS subunits (Kudlicka, 1989; with kind permission).

of the typical chain elongation rate, which is equal to 2 μm min^{-1}, and corresponds to a polymerization of more than 10^8 glucose molecules into the β-1,4-glucan per hour and per single bacterial cell. Thus if forces associated with subfibrils assembly to bundles and ribbons, are strong enough to shift the motion of whole *A. xylinum* cells, these forces may as well influence the spatial arrangement of CS subunits, localized in the semi-liquid lipid bilayer of the membrane.

The spatial assembly of the β-1,4-glucan chains has a hierarchical character (Figure 10). In the first stage, 10–15 nascent β-1,4-glucan chains form a subfibril (also called a protomicrofibril), 1.5 nm in diameter, which is not a rod-like structure, but a left-handed triple helix (Ross et al., 1991). The subfibrils gather to form (also twisted) microfibrils composed of numerous parallel chains. The next structure in hierarchy is a bundle of microfibrils, followed by loosely wound ribbon, comprising about 1000 individual chains of the β-1,4-glucan (Haigler, 1985). In the presence of Calcofluor, which penetrates the outer membrane of the cell envelope, and forms complexes with protomicrofibrils, their aggregation to microfibrils is stopped. In the presence of much larger molecules such as CM-cellulose and xyloglucan (Yamamoto et al., 1996), which are too large to traverse the membrane, assembly of the bundles of microfibrils and ribbons is disrupted, thus indicating that formation of these two latter structures occurs exocellularly. Ross et al. (1991), who analyzed the mechanism of *A. xylinum* cellulose chain association, emphasized the important role of specific sites of adhesion in the inner and outer membranes of cell envelope, whose presence in bacterial cells was first reported by Bayer (1979). Their occurrence would enable an export of cellulose microfibrils without any interactions with the peptidoglycan gel, which fills the periplasmic space. Such interactions would prevent formation of correct hydrogen bonding between the protomicrofibrils.

The molecular organization of pores, through which *A. xylinum* cellulose microfibrils migrate outside the cell, remains unknown. The participation of expression products of the *bcsC* and *bcsD* genes, belonging to CS operon (see Section 7.4), in their formation cannot be excluded (Wong et al., 1990), all the more since these expression products appeared to be essential for cellulose synthesis *in vivo* (Ben-Bassat et al., 1993). Also the function of three proteins: CSAP20, 54, and 59 (CS-associated proteins), has to be scrutinized. They bind to cellulose together with the CS catalytic subunit, as revealed by experiments on CS purification (Benziman and Tal, 1995). It seems probable that isolation and purification of other components of the complex machinery responsible for cellulose synthesis, in concert with direct mutations of genes from *A. xylinum* CS operon, may help to elucidate the sequence of events during this process on the molecular level.

The process of assembly of nascent cellulose chains, not separated from parental *A. xylinum* cells, displays a unique and remarkable dynamics. Extrusion of these chains outside and their assembly into ordered structures is assisted by reversal of the motion of the cells. Throughout the whole process of the β-1,4-glucan synthesis and secretion, *A. xylinum* cell simultaneously turns around its own axis (uncoiling motion) (Brown et al., 1976). Also, this movement is driven by forming of twists of the ribbons, which are anchored in extrusion loci, localized on the side surface of the elongated cell. These twists are noticeable under the electron microscope.

7.4 Genetic Basis of Cellulose Biosynthesis

Cellulose biosynthesis involves several enzymes. Their function starts from the synthesis of the direct cellulose precursor UDPGlc, followed by polymerization of the monomer. It was found that the first reactions catalyzed by glucokinase, phosphoglucomutase and UDPGlc pyrophosphorylase, do not limit the final rate of BC synthesis, since *A. xylinum* synthesizes an excess of UDPGlc (Ben-Basat et al.,1993). The rate of BC synthesis is limited by diguanylate cyclase and oligomeric, regulatory complexes of CS. Therefore construction of more efficient cellulose-producing *A. xylinum* mutants, requires determination of organization of the genes coding for these enzymes. This has been studied by Wong et al. (1990) and Ben-Bassat et al. (1993), who proved that BC biosynthesis involves four coupled genes: *bcsA* (2261 bp), *bcsB* (2405 bp), *bcsC* (3956 bp), and *bcsD* (467 bp), forming the CS operon, which is 9217 bp in length, and is transcribed simultaneously as polycistron mRNA.

Although the function of the proteins encoded by each of these genes has not yet been precisely determined, some information on their role was derived from genetic complementation experiments (Wong et al., 1990). Results of these studies indicate that CS-deficient strains can be complemented by the gene *bcsB*, which codes for the 85-kDa protein (802 amino acid residues), and that *A. xylinum* mutants defective in both CS and diguanylate cyclase are complemented by the gene *bcsA*, coding for the 84-kDa protein (754 amino acid residues) (Ben-Bassat et al., 1993). Based on different mutations it was concluded that protein A takes part in the interaction of the CS complex with c-di-GMP, which is an allosteric activator of CS, earlier postulated by Saxena et al. (1991). The protein B is capable of binding UDPGlc and synthesizing β-1,4-glucan chains. A gene coding for the catalytic subunit (*cesA*) and analogous to the gene *bcsB* (Wong et al., 1990; Ben-Bassat et al., 1993) was also described by Saxena et al. (1990a), who one year later identified the gene coding for the regulatory subunit (*cesB*). However, Saxena et al. (1990a, 1991) suppose that the position of the first two genes in the CS operon is opposite and that the first gene codes for the CS catalytic subunit.

The role of the expression products of both *bcsC* and *bcsD* genes has not been determined so precisely. These genes probably also code for proteins (molecular mass of 141 and 17 kDa, respectively), anchored in the membrane, and both essential for cellulose synthesis *in vivo* and for protein secretion from the cells. Ben-Bassat et al. (1993) reported that splitting or deletion of the *bcsD* gene markedly reduced synthesis of cellulose. The presence of genes, which govern the assembly of cellulose chains, in the CS operon was also predicted by Saxena et al. (1990a) and Ross et al. (1991).

It is not clear if plasmid DNA participates in BC synthesis and, if so, what its role is. However it was proven that the composition and size of plasmids detected in 60% of the non-reverting Cel^{-} mutants of the *A. xylinum* ATCC 10245 (currently 17005) strain, that were obtained by means of mutagenesis with *N*-methyl-*N'*-nitro-*N*-nitrosoguanidine, were markedly different from that of the wild strain. This observation suggests the plausible relationship between plasmid DNA structure and BC synthesis.

Mobile DNA fragments, including insertion sequences (IS), particularly widespread among prokaryotes, and 750–2500 bp in length (Galas and Chandler, 1989), are important factors of regulation of many genes. The ISs may activate or inactivate genes, and start DNA rearrangement such as deletions, inversions, and cointegrations. It was revealed that ISs participated in regulation of extracellular polysaccharides synthesis in *Pseudomonas atlantica* (Bartlett and Silverman, 1989), *Xanthomonas campestris* (Hotte et al., 1990), and *Zoogloea ramigera* (Easson et al., 1987). Their participation in BC synthesis regulation is also probable.

The best recognized insertion sequence in *A. xylinum* is the IS 1031 (950 bp, with known nucleotide sequence). Alteration in the IS 1031 profile was detected in the majority of Cel^- mutants, in comparison to the wild strain. Some recombinants contained two or more IS 1031 fragments. Coucheron (1991) observed even more significant changes in the IS pattern, pointing to DNA rearrangement and tried to obtain revertants by splitting off the additional ISs from the inactivated gene and obtained pseudorevertants, which produced a wax-like substance on the surface of nutrient broth, but the capability of cellulose synthesis was not restored. These experiments indicate that the presence of ISs contributes to the genetic instability of *A. xylinum*.

7.5
Regulation of Bacterial Cellulose Synthesis

Cyclo-3,6′:3′6 diguanosine monophosphate (c-di-GMP, see Figure 11) is a reversible allosteric activator of *A. xylinum* CS and plays a crucial role in regulation of the whole β-1,4-glucan biogenesis (Ross et al., 1987).

This compound binds to the enzyme regulatory subunit, and induces conformational changes that facilitate association of the CS protomers and lead to the enhancement of its reactivity (Ross et al., 1987).

The concentration of c-di-GMP in *A. xylinum* cells is regulated by 3 enzymes: diguanyl cyclase (CDG), phosphodiesterase A (PDEA) and phosphodiesterase B (PDEB)

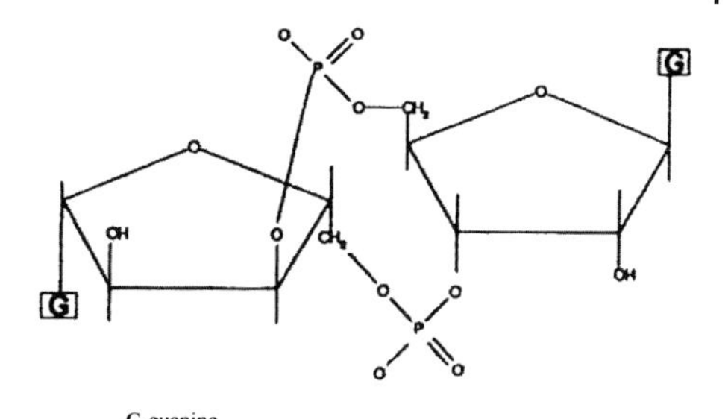

Fig. 11 Structure of c-di-GMP – the allosteric activator of *A. xylinum* CS.

(Figure 12) (Ross et al., 1987). PDEA and PDEB are anchored in the cytoplasmic membrane, and CDG has two forms. One of them is anchored in the cytoplasmic membrane and the second one operates in the cytoplasma (Ross et al., 1991). Recently, Weinhouse et al. (1997) reported on a new c-di-GMP-binding protein, which probably also participates in the intracellular regulation of free c-di-GMP concentration. Tal et al. (1998) proved that the cellular turnover of c-di-GMP in *A. xylinum* is controlled by three operons.

CDG, the key regulatory enzyme in the *A. xylinum* cellulose synthesis system, is composed of two polypeptide chains and encoded by two genes (Nichols and Singletary, 1998). CDG is activated by Mg^{2+} ions and specifically inhibited by saponin (Ohana et al., 1998), a glycosylated terpenoid. CDG converts two GTP molecules at first into the linear (pppGpG) and then into the cyclic (c-di-GMP) diguanosine monophosphate, which activates CS. PDEA splits active c-di-GMP into a linear inactive dimer (pGpG, di-GMP), further decomposed by PDEB into two molecules of 5′GMP (Figure 12). PDEA is inhibited by Ca^{2+} ions, which do not influence PDEB. Therefore, the Ca^{2+} concentration indirectly affects the rate of cellulose synthesis. High concentration of these ions enhances this rate, since the

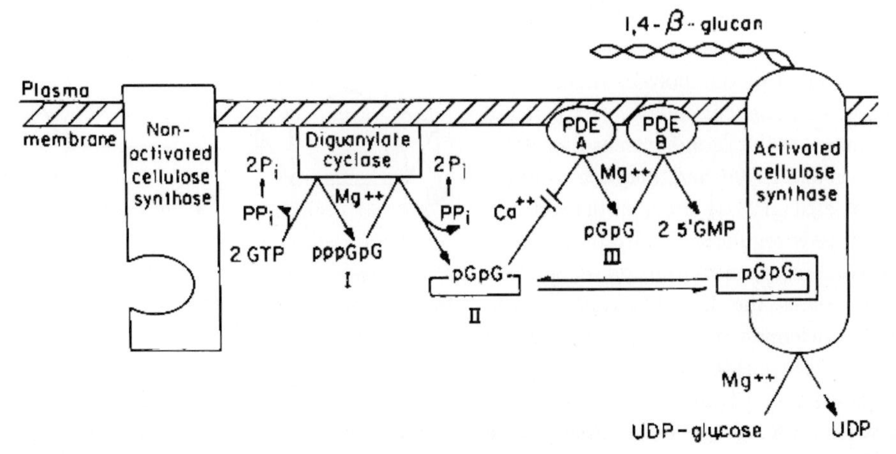

Fig. 12 Proposed model for cellulose synthesis in *A. xylinum*. For simplicity, the synthesis of a single β-1,4-glucan chain is depicted, although a more complex form of CS, polymerizing several chains simultaneously, might be the active enzyme unit in cellulose biogenesis (Ross et al., 1991; with kind permission). PDEA, phosphodiesterase A; PDEB, phosphodiesterase B; pppGpG, diphospho-di-GMP; pGpG (II), c-di-GMP (cyclic-di-GMP); pGpG, di-GMP (linear di-GMP).

conversion of c-di-GMP to the inactive linear form is inhibited.

It is believed that both the molecular background of the CS activation by c-di-GMP as well as the regulation of its synthesis, involving a positive influence of CDG, and a negative effect of PDEA and PDEB, have a unique character. To date, such a mechanism of cellulose synthesis regulation has only been detected in *A. xylinum*.

7.6 Soluble Polysaccharides Synthesized by *A. xylinum*

Apart from cellulose, some related soluble polysaccharides are also synthesized by *A. xylinum*. In 1970s, a soluble β-homoglucan was detected. Its main backbone was composed of β-1,4-linked glucose residues and every third glucose moiety of this chain was substituted with another glucose via β-1,2 linkage (Colvin and Leppard, 1977; Colvin et al., 1979). *A. xylinum* cellulose pellicle non-forming (Pel^-) strains synthesize α-glucan linked with the cell envelope (Dekker et al., 1977). Valla and Kjosbakken (1981) isolated a soluble polysaccharide containing Glc, Man, Rha, and glucuronic acid residues (molar ratio of 3:1:1:1, respectively) from the culture broth of the Cel^- strain. The related polymer, composed of the same moieties (altered molar ratio of 6:2:1:1, respectively), synthesized by another Cel^- strain, was detected by Minakami et al. (1984). Probably some of the residues in these polymers are acetylated (Tayama et al., 1985). Polysaccharides of this type have received the common name acetan (De Lannino et al., 1988). Moreover, some wild *A. xylinum* strains synthesize a similar soluble polysaccharide (Glc, Man, Rha, and glucuronic acid; 6:1:1:1, respectively) together with cellulose (Savidge and Colvin, 1985).

The question is whether cellulose-assisting soluble polysaccharides play any role in its biogenesis. It seems that they do not play any direct role, although kinetics of their biosynthesis are the same as that of cellulose (Marx-Figini and Pion, 1974) and, moreover,

UDPGlc was preferentially used for production of these soluble saccharides instead of the production of cellulose (Delmer, 1982). Some authors state that *in vivo* nascent cellulose microfibrils are coated with amorphous material (Leppard et al., 1975), which may be composed of the above-mentioned soluble polysaccharides. The absence of these soluble polymers in the liquid fraction of static culture broths and their presence in the cellulose mat confirms that assumption (Valla and Kjosbakken, 1981).

Much earlier, Ben-Hayyim and Ohad revealed in 1965 that CM-cellulose present in the medium, used for BC synthesis, resulted in incorporation of this soluble derivative of cellulose into microfibrils of the polymer. However, incorporation of the *A. xylinum* soluble exopolysaccharides into cellulose microfibrils has not been observed.

7.7 Role of Endo- and Exocellulases Synthesized by *A. xylinum*

The first reports on *A. xylinum* cellulases, detected independently by Okamoto et al. (1994) and Standal et al. (1994), included description of their genes. The first authors selected from *A. xylinum* IFO 3288 DNA gene libraries one gene coding for a 24-kDa protein (218 amino acid residues), which displayed CM cellulase activity. Standal et al. (1994) found another endocellulase gene in the *A. xylinum* ATCC 23769 Cel^- mutant, localized upstream the CS operon. They revealed that the loss of cellulase synthesis capability resulted from gene splitting. The conclusions of Tonouchi et al. (1997), who investigated the localization of the *A. xylinum* BPR 2001 endo-1,4-β-glucanase gene, were similar. The latter gene was localized upstream the CS operon, whereas the gene of the second cellulolytic enzyme, produced by this strain, i.e. an exo-1,4-β-glucosidase, was found downstream this operon.

A. xylinum BPR 2001 endo- and exocellulase were purified and characterized (Oikawa et al., 1997; Tahara et al., 1998). The studies of Tahara et al. (1997) also revealed that at pH 5.0, optimal for growth and BC synthesis, the activity of both cellulases is several times higher than at pH 4.0, at which the BC production is only slightly declined. It was also observed that at pH 5.0, the average DP was decreasing from 16,800 to 11,000 with the time of cultivation, whereas at pH 4.0 these changes were negligible. The cellulose obtained at pH 5.0 had inferior physical properties, i.e. a lower tensile strength (a lower Young's modulus value). These results and colocalization of genes of cellulases and CS operon, suggest that the endo-1,4-β-glucanases and exo-1,4-β-glucosidase are involved in *A. xylinum* cellulose biogenesis (Tahara, 1998).

More particularly, it relates to possible degradation of the nascent β-glucan chains, synthesized in the later phase of growth. Furthermore, the *A. xylinum* BPR 2001 exo-1,4-β-glucosidase displays both hydrolytic and transglycosylating activity (Tahara et al., 1998), similarly to other glycosidases, which do not cause inversion of configuration on the anomeric carbon. The latter activity may be responsible *in vivo* for changes in the average DP of the polymer; however, further research is necessary to explain this hypothesis. Tentative evidence for such a possibility might be the behavior of soybean cells cultured *in vitro*. The activity of their β-glucosidase, which also has transglycosylating activity, increases together with the length of the cells, since this enzyme probably participates in the transfer of glycosidic residues of hemicelluloses precursors to the growing terminus of the cell wall, in which intensive synthesis of these polysaccharides is taking place (Nari et al., 1983).

Whether *A. xylinum* cellulases are involved in releasing the energy stored in the form of cellulose, in periods of starvation, is still a question to answer, although some studies confirm this thesis (Okamoto et al., 1994). It is known that some *Sclerotium* endo-1,3-β-glucanases play such a role (Jones et al., 1974).

Current observations indicate that modification of the physical properties and DP of BC can be achieved either by using compounds that influence its biosynthesis, or by exploiting the activity and specificity of *A. xylinum* cellulases.

8
Biodegradation of BC

Independent of its origin, cellulose undergoes total biodegradation in nature. However, in comparison to PC, associated both with polymers susceptible to degradation, like hemicelluloses and pectin, and with lignin, which is the most resistant plant polymer, BC is relatively pure and *a priori* more susceptible to attack by cellulolytic enzymes, which are produced mainly by fungi and numerous bacterial species. In this respect, commercial application of BC is friendly for the environment.

Furthermore, the structure and properties of BC (large accessible surface area) make it the superior model substrate for studies on cellulases (1,4-β-glucan 4-glucanohydrolases, formerly endocellulases, EC 3.2.1.4) and cellobiohydrolases (cellulose 1,4-β-cellobiosidases, EC 3.2.1.91), that are the main components of cellulosomes, the specialized multienzymatic particles from cellulolytic bacteria, as well as fungal systems for cellulose decomposition.

Recently, Boisset et al. (1999) proved that *Clostridium thermocellum* cellulosomes completely decompose *A. xylinum* cellulose microfibrils faster than microcrystalline *Valonia ventricosa* cellulose. Ultrastructural observations of the hydrolysis process, using transmission electron microscopy, infrared spectroscopy, and X-ray diffraction analysis, indicated that the rapid hydrolysis of BC resulted from very efficient synergistic action of the various enzymic components, present in the cellulosome scaffolding structure. Further studies of Boisset et al. (2000) revealed that *Humicola insolvens* cellobiohydrolase Cel7A (previously CBH I) brought about thinning of dispersed BC ribbons, whereas the mixture of Cel6A (CBH II) and Cel7A (in the ratio of 2:1, respectively) cut the ribbons to shorter pieces, thus suggesting the partly endo-manner of CBH II attack. The phenomenon of low inherent endoactivity of some exoglucanases has been known for several years and explains more efficient synergistic action of cellobiohydrolases in comparison to endoglucanases (Teeri, 1997). According to current opinions, the system of multiple exo- and endocellulases, both in bacterial cellulosomes, and in fungal cellulolytic complexes, represents the perfectly balanced continuum of activity, providing efficient degradation of various cellulosic substrates.

BC susceptibility to cellulases was also observed by Samejima et al. (1997), who found that the mixture of *Trichoderma viride* CBH I and endoglucanase II, drastically disintegrated the twisted and bent ribbon-like structure of microfibril bundles to linear needle-like microcrystallites, and also caused rapid polymer fragmentation. Transformation of the coiled structure of BC ribbons to the needle-like one is driven by the remarkable twisting motion of the substrate, which is probably a result of a tension – released when the microfibrils are being decomposed to shorter fragments by cellulases. Samejima et al. (1997) also detected that the acid-treated BC, containing many microfibril aggregates, was not susceptible to attack of both enzymes.

Similar results were achieved by Srisodsuk et al. (1998), who digested microcrystalline BC with *Trichoderma reesei* CBH I and observed rapid solubilization of the polymer, but slow decrease in its DP. The endocellulase I from this organism attacked the substrate in the opposite manner. Both enzymes hydrolyzed cotton cellulose more slowly than BC, though cotton cellulose exhibits relatively high purity as compared to other plants.

One of the reasons of the susceptibility of BC to the attack of cellulases is its large accessible surface area, that even increases throughout hydrolysis and facilitates effective binding of cellulolytic enzymes, the crucial step in the process of degradation. It was concluded based on studies of Palonen et al. (1999), who found that the number of *T. reesei* CBH I and CBH II molecules linked to BC, increased during hydrolysis. Furthermore, the CBH II binding was markedly stronger (the adsorption/desorption of the enzyme from its substrate was assisted by 60–70% hysteresis), pointing to a distinctly different processing character of this enzyme.

Stalbrandt et al. (1998) compared the mode of attack of four *Cellulomonas fimi* endo- and exocellulases against microcrystalline BC and acid-swollen Avicel cellulose. The latter was decomposed more effectively by all the enzymes. In most cases 45–65% solubilization and a decrease in DP were observed, except for Cbh B, which caused 27% solubilization. A high degree of bacterial polymer solubilization (67%) was achieved for Cbh A, known as the processing enzyme. Results of Stalbrandt et al. (1998) proved that only the external surface of BC fibrils is accessible for *C. fimi* cellulases. Since Cbh A is the processing enzyme, it can remove external fibrils much faster than the other three cellulases and attack the deeper internal surface of the polymer. Therefore, the other enzymes (two nonprocessing endocellulases and Cbh B, which displays weaker processing properties) could not solubilize BC so effectively, whereas amorphous soluble cellulose was equally available to all of them.

Current results indicate that complete digestion of the highly crystalline bacterial polymer will not cause any problems, all the more since organisms responsible for BC biodegradation in natural environments produce a multitude of various endo- and exocellulases, whose activities complement each other. Furthermore, as opposed to cellulose of plant origin, which requires pretreatment to provide easier enzymatic conversion to sugars, BC can be directly digested by cellulases.

9
Biotechnological Production

To achieve high productivity and yields of BC and to reduce cost of its production, special emphasis should be given to the following aspects:

- development of screening methods providing selection of *A. xylinum* strains, which can efficiently produce cellulose from various inexpensive waste carbon sources – wild strains derived from the screening could be improved using genetic engineering methods;
- optimization of *A. xylinum* culture conditions (static or agitated culture, fermentor type), determining both a form (a pellicle or an amorphous gel) and properties (resilience, elasticity, mechanical strength, absorbency) of BC, which have to be tailored to the further polymer application;

- optimization of culture broth composition (carbon and nitrogen sources, biosynthesis stimulating compounds, microelements, etc.) and process conditions (pH, temperature, aeration).

These are discussed in the next sections.

9.1 Isolation from Natural Sources and Improvement of BC-producing Strains

One of the methods enabling selection of proper *A. xylinum* strains, is the screening for strains, which cannot oxidize glucose via gluconic acid to 2-, 5-, or 2,5-ketogluconate (Winkelman and Clark, 1984; Johnson and Neogi, 1989; De Wulf et al., 1996; Vandamme, 1998). In this respect, De Wulf et al. (1996) applied an agar nutrient medium containing Br^- and BrO_3^- ions, that combine at acidic pH to release molecular bromine, toxic to *A. xylinum* cells. Only those mutants, which do not convert glucose into gluconate or its derivatives, can survive in this medium. Vandamme et al. (1998) successfully used this method to isolate *A. xylinum* KJ33 strain, producing 3.3 g cellulose L^{-1}. *A. xylinum* strains, which do not metabolize glucose to gluconic acids, were also selected on $CaCO_3$ containing medium (Johnson and Neogi, 1989). Another approach to avoid conversion of glucose to organic acids, was screening for mutants defective in glucose-6-phosphate dehydrogenase. The mentioned increase in final concentration, volumetric productivity and total cellulose yield, was achieved in agitated cultures. For the same purpose, i.e. selection of the best cellulose producers, preparations of cellulases, added into the growth medium, have also been used (Brown, 1989c).

Mutation with chemical compounds, such as *N*-methyl-*N'*-nitro-*N*-nitrosoguanidine or ethylmethanesulphate, as well as with ultraviolet irradiation gave mutants that displayed higher cellulose productivity, reduced synthesis of the soluble polysaccharide acetane, and a much lower degree of glucose conversion into organic acids (Johnson and Noegi, 1989).

Screening of *A. xylinum* strains was also aimed at isolation of spontaneous or induced cellulose II-producing mutants. This cellulose is synthesized by bacteria that form smooth colonies and do not produce a pellicle on the surface of the nutrient broth. Instead, the polymer is dispersed in the whole volume of the medium.

Selection of the best BC producers has been also performed traditionally, by determination of the synthetic activity of a plethora of individual monocultures. For example, Toyosaki et al. (1995) isolated 2096 *Acetobacter* strains from a large number of plant samples (fruits, flowers) and 412 strains forming a mat on the surface of nutrient medium were subjected to further investigations. The procedure yielded *A. xylinum* ssp. sucrofermentans BPR2001 – the strain which efficiently synthesized BC in agitated culture.

9.1.1 Improvement of Cellulose-producing Strains by Genetic Engineering

Expression of genes coding for various carbohydrases from organisms other than *A. xylinum* enhances the scope of carbon sources available, that may increase BC yield and decrease the nutrient broth price. An example is the expression of sucrose phosphorylase (EC 2.4.1.7) gene from *Leuconostoc mesenteroides* in *Acetobacter* sp. (Tonouchi et al., 1998), that resulted in utilization of sucrose as a carbon source and increased cellulose production. In addition, Nakai et al. (1999) obtained double cellulose yield by means of expression of the mutant sucrose synthase (EC 2.4.1.13) gene from mung bean (*Vigna radiata*, Wilczek) in *A. xylinum* strain. In the mutant enzyme, the eleventh serine residue was replaced by glutamic acid (S11E); this caused higher affinity of the engineered enzyme toward sucrose and favored cleavage of this sugar for the synthesis of UDPGlc. Introduction of the mutant gene into *A. xylinum* not only changed sucrose metabolism in the recombinant strain, but also created a new metabolic pathway of direct UDPGlc synthesis (Figure 13). The energy driving this process is derived only from the cleavage of the glycosidic bond in sucrose without any other energy input required, which is the case when UDPGlc is synthesized via the conventional pathway. Furthermore, the UDP molecules resulting from glucose polymerization process can be directly recycled by the sucrose synthase activity, which increases the rate of the coupled reactions and a higher rate of cellulose synthesis is observed. It should be emphasized that higher plants have both systems of UDPGlc synthesis.

9.2 Fermentation Production

Growth on the surface of liquid media and synthesis of the gelatinous, leather-like mat, are natural properties of *A. xylinum* strains. Therefore static conditions seemed to be optimal for BC synthesis. However, the main drawbacks of static culture, such as the synthesis of the polymer only in the form of a sheet and relatively low productivity, have contributed to the development of new fermentation processes.

9.2.1 Carbon and Nitrogen Sources

In studies on factors affecting the BC production yield, much attention has been paid to carbon sources. Numerous mono-, di-, and polysaccharides, alcohols, organic acids, and other compounds were compared by Jonas and Farah (1998), who found out that the preferred carbon sources were D-arabitol and D-mannitol, which resulted in a 6.2- or 3.8-fold higher cellulose production, respectively, in comparison to glucose. Both sugar alcohols provided stabilization of the pH throughout the culture period, since they were not converted to gluconic acids.

Tonouchi et al. (1996), who used a strain of *A. xylinum* to obtain cellulose from glucose and fructose, found that fructose stimulated the activities of phosphoglucose isomerase and UDPGlc pyrophosphorylase, thus enhancing cellulose yield.

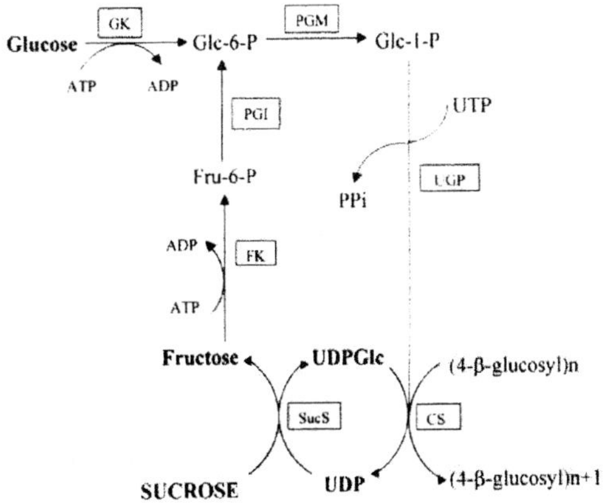

Fig. 13 UDPGlc biosynthesis pathways in a recombinant strain of *A. xylinum* containing the sucrose synthase (SucS) gene. Other abbreviations are the same as in Figure 6 (see Section 7.1).

When maltose was the sole carbon source in the nutrient medium (Masaoka, 1993) BC production was even 10 times lower than in glucose-containing medium and the polymer was markedly shorter (DP = 4000–5000) than in the presence of glucose (DP = 11,500).

Matsuoka et al. (1996) investigated BC synthesis by *A. xylinum* ssp. sucrofermentous BPR2001 in agitated culture and found out that the presence of lactate in the growth medium stimulated bacterial growth and enhanced the cellulose yield 4–5 times. The source of lactate is usually corn steep liquor (CSL), which is one of the main medium components applied for BC production, especially in agitated cultures. Contrary to glucose and fructose, lactate is not converted to UDPGlc, but is metabolized via pyruvate and oxalacetate in the Krebs cycle, being a source of energy for cellulose production. To provide lactate in the growth medium, cultivation of mixed cultures of lactic acid and acetic acid bacteria has been carried out. The best results gave strains of *Lactobacillus, Leuconostoc, Pediococcus,* and *Streptococcus.* The lactic and acetic acid bacteria were grown also in the presence of sucrose-hydrolyzing *Saccharomyces* yeast (β-fructofuranosidase producers). This principle provided a high BC yield, since after 14 days of agitated culture, as much as 8.1 g of cellulose L^{-1} was obtained, when in the absence of *Lactobacillus* strain, only 6.4 g L^{-1} (Seto et al., 1997).

One of the cellulose synthesis-stimulating compounds appeared to be ethanol (Naritomi et al., 1998). Continuous culture of *A. xylinum* in fructose-containing medium, revealed that 10 g ethanol L^{-1} enhanced a BC yield, but a concentration of 15 g L^{-1} prevented polymer synthesis. These results suggest that similarly to lactate, ethanol is also a source of energy (accumulated as ATP) and not a substrate for cellulose synthesis. ATP activates fructose kinase and inhibits glucose-6-phosphate dehydrogenase, thus halting conversion of 6-P-glucose to 6-phosphogluconate. It can be concluded that – due to these coupled reactions – the amount of fructose further isomerized to glucose-6-phosphate is increased.

Each cellulose-producing strain requires a specific complex nitrogen source, providing not only amino acids, but vitamins and mineral salts as well. These requirements are met with yeast extract, CSL as well as hydrolyzates of casein and other proteins. The preferred nitrogen sources are yeast extract and peptone, which are basic components of the model medium developed by Hestrin and Schramm (1954; H-S medium), applied in numerous studies on BC synthesis. However the most recommended nitrogen source for agitated cultures is CSL (Johnson and Noegi, 1989).

It was also found that a significant part of the expensive medium components, i.e., yeast extract and bactopeptone, can be replaced with CSL or even white cabbage juice. Also waste plant materials such as sugar beet molasses, spent liquors after glucose separation from starch hydrolyzates, as well as whey and some fermentation industry wastes (e.g., spent liquors after dextran precipitation with ethanol) were appropriate medium components (Krystynowicz et al., 2000).

Studies on the influence of vitamins on BC synthesis (Ishikawa et al., 1995, 1996b) revealed that the most stimulating ones were pyridoxine, nicotinic acid, *p*-aminobenzoic acid and biotin, and even CSL media should be fortified with these substances. Some other compounds, strongly stimulating cellulose production by *A. xylinum* strains, such as derivatives of choline, betaine, and fatty acids (salts and esters), were also identified (Hikawu et al., 1996).

High BC production (up to 25 g L^{-1}) can be achieved using optimum growth medium composition, designed by mathematical methods and computer analysis (Joris et al., 1990; Embuscado et al., 1994; Vandamme et al., 1998; Galas et al., 1999).

9.2.2 Effect of pH and Temperature

Analysis of the influence of pH on *A. xylinum* cellulose yield and properties, indicates that optimum pH depends on the strain, and usually varies between 4.0 and 7.0 (Johnson and Neogi, 1989; Galas et al., 1999). For instance, Ishikawa et al. (1996a) and Tahara et al. (1997), who applied two different *A. xylinum* strains, observed the highest polymer yield at pH 5.0. The same pH was found to be optimal by Krystynowicz et al. (1997). Comparison of adsorptive properties of the polymer obtained at different pH, learned that cellulose, accumulated in the S-H medium pH 4.8–6.0, displayed the highest water-binding capacity (Wlochowicz, 2001).

In addition to the pH of the nutrient broth, also the temperature influences BC yield and properties. In majority of reported experiments, the temperature ranged from 28 to 30 °C, and its variations caused changes of cellulose DP and water-binding capacity. For instance, BC synthesized at 30 °C had a lower DP (approximately 10,000) and a higher water-binding capacity (164%) in comparison to that produced at 25 and 35 °C (Wlochowicz, 2001).

9.2.3 Static and Agitated Cultures; Fermentor Types

Synthesis of BC is run either in static culture or in submerged conditions, providing proper agitation and aeration, necessary for medium homogeneity and effective mass transfer. The choice of culture conditions strictly depends on the polymer use and destination.

BC yield in static cultures is mostly dependent on the surface/volume ratio. Optimum surface/volume ratio protects from either too high (unnecessary) or too low aeration (cell growth and BC synthesis termination). Reported values of surface:volume ratio vary from 2.2 cm^{-1} (Joris et al., 1990) to 0.7 cm^{-1} (Krystynowicz et al., 1997). BC synthesis in static conditions can be achieved either in a one-step (medium inoculated with 5–10% cell suspension) or a two-step procedure. The latter one starts from agitated fermentation, followed by the static culture (Okiyama et al., 1992). Krystynowicz et al. (1997) modified the two-step procedure and applied two consecutive static cultures. Pellicles containing entrapped *A. xylinum* E_{25} cells, obtained in the first step (24 h), were cut to uniform pieces and used as an inoculum to start the next culture, run for 4–5 days. The method provided uniform bacterial growth and production of homogeneous pellicles.

The control of BC synthesis in static culture is very difficult since the pellicle limits access to the liquid medium. A particularly important parameter, which requires continuous control, is the pH. Accumulation of keto-gluconic acids in the culture broth brings about a decrease in pH far below the value that is optimum for growth and polysaccharide synthesis. Because conventional methods of pH adjustment cannot be used in static cultures, Vandamme et al. (1998) applied an *in situ* pH control via an optimized fermentation medium design, based on introducing acetic acid as an additional substrate for *Acetobacter* sp. LMG1518. Products of acetic acid catabolism counteracted the pH decrease caused by keto-gluconate formation, and provided a constant pH of 5.5 of the growth medium, throughout the whole process.

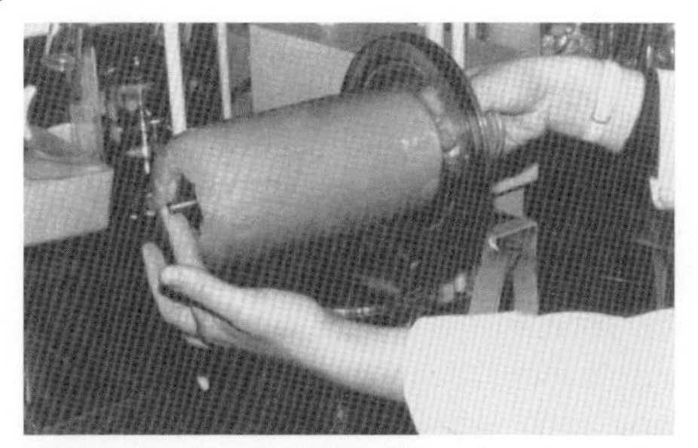

Fig. 14 *A. xylinum* cellulose formed on the surface of a roller in horizontal fermentor.

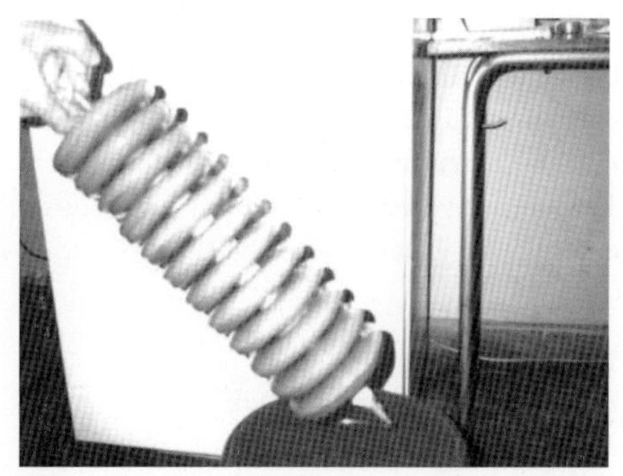

Fig. 15 *A. xylinum* cellulose deposited on disks in rotating disk fermentor.

Better control of BC synthesis can be achieved in special fermentors. Cellulose production in horizontal fermentors provides a combination of stationary and submerged cultures. The polymer is deposited on the surface of rollers or disks, rotating around the long axis (Figure 14). Part of their surface temporarily dips in the liquid medium or is above its surface (in the air). The advantages of this method include a larger polymer surface, synthesis of cellulose in a form of hollow fibers, different in diameter, as well as good process control, easy scale up, appropriate accessible surface for adhesion of bacteria and product deposition, better substrate utilization, higher rate of cellulose production, possibility of nutrient supply, additional aeration during the process, etc. (Sattler and Fiedler, 1990). Bungay and Serafica (1997) produced BC in a 1-L disk (12 cm in diameter) fermentor and revealed that optimum sugar (sucrose or glucose) concentration was 10 g L^{-1}, disks rotation rate 12 r.p.m., and constant pH 5.0. Krystynowicz et al. (1997) successfully applied a 11-L disk reactor containing 3 l of the H-S medium for BC synthesis by the *A. xylinum* E_{25} strain (Figure 15). The optimum rate of rotation of its 11 disks (12 cm in diameter each) appeared to be 3 r.p.m. As much as 4.2 g L^{-1} of dry cellulose was obtained after 7 d of growth.

BC production in submerged cultures usually requires common fermentors, equipped with some static parts (baffles, blades, etc.), enabling cell adhesion, since cellulose synthesis in the free aqueous phase usually drops. In analogy, various water-insoluble microparticles, such as sand, diatomaceous earth, or glass beads, added to the culture medium, enhanced BC productivity, since the biofilm formed by bacteria on the particles probably limited oxygen transfer and stopped glucose oxidation to gluconic acids (Vandamme et al., 1998).

Large-scale cellulose production, in fermentors with continuous agitation and aeration, encounters many problems, including spontaneous appearance of Cel^- mutants, which contribute to a decline in cellulose production. Optimized agitation and aeration prevent turbulence, which negatively effects cellulose polymerization and crystallization, thus reducing the polysaccharide yield. For instance, a rate of agitation equal to 60 r.p.m. and aeration of 0.6 volume per volume per minute (v.v.m.) were optimum for the *A. aceti* ssp. xylinum ATCC 2178 strain cultured in a 300-L fermentor for 45 h at 30 °C, and 10 g BC L^{-1} d^{-1} was obtained (Laboureur, 1988). Ben-Bassat et al. (1989) investigated BC production by the *A. xylinum* 1306–21 strain in 14-L Chemap fermentor, cultured in CSL medium containing glucose and CSL, 2% of each. The rate of agitation was equal to 900 r.p.m. and the dissolved oxygen concentration corresponded to 30% air saturation. The polymer yield amounted to 5.1 g L^{-1} d^{-1}. Chao et al. (2000) applied a 50-L internal loop airlift reactor for BC synthesis with the *A. xylinum* sp. BPR2001 strain. Aeration with oxygen-enriched air enlarged cellulose yield from 3.8 to 8 g L^{-1} after 67 h.

Production of BC in fermentors encounters similar agitation problems as cultivation of fungi or streptomycetes, since most of these organisms grow in pellet or filament form and their culture broths become non-Newtonian fluids. A high concentration of mycelium in the form of a dense suspension of diversely shaped particles limits agitation and gas transfer. Increasing the rate of agitation in order to improve aeration leads to damage to the product structure due to shearing forces. Bauer et al. (1992) took this aspect into consideration and introduced a polyacrylamide-protecting agent into the growth medium, applied for BC synthesis in fermentors. The protector reduced the shear damage and affected the specific productivity at high densities negatively since the cell growth rate was lower.

Accumulation of some metabolites can also affect production of cellulose. For instance, studies of Kouda et al. (1998) revealed that a high partial pressure of CO_2 negatively affected *A. xylinum* growth and reduced BC yield.

Laboureur (1998) developed a method for BC production by *A. xylinum* sp. ATCC 21780 strain in 300- and 500-L fermentors in a medium containing 5% sucrose, 0.05% yeast extract, 0.2% citric acid, nitrogen salts, Mg^{2+}, and phosphates. The medium was inoculated with 12% of inoculum, cultivated for 160 h. The process was run for 45 h at 30 °C and pH 4.8, and an aeration rate of 0.6 v.v.m. BC synthesis yield reached 18 g L^{-1} (10 g L^{-1} d^{-1}). Another strain, *Acetobacter* sp. ATCC 8303, was grown at 28 °C and pH 4.6, in a modified medium, containing 0.28% glucose, 0.07% maltose, 0.03% CSL, and 0.03% yeast extract. The aeration was 1 v.v.m. for the first 33 h and 0.5 v.v.m. for the last 12 h of the process, and the BC yield was 13 g L^{-1} d^{-1}.

Preparing an inoculum of appropriate cell density for large-volume fermentors is also a problem, mainly because the cells are entrapped in cellulose. To liberate the cells and increase their density, Brown (1989c) applied preparations of cellulases for partial cellulose hydrolysis. In the presence of these enzymes, the cell density reached 10^8 mL^{-1}, whereas in their absence it was only $1.12 \cdot 10^7$ mL^{-1}, after 30 h of growth.

In addition to the above-mentioned methods, the production of cellulose in the form of hollow fibrils of various diameters has also been tried (Yamanaka et al., 1990). Such material could be especially useful for the production of small-diameter blood vessels. The hollow BC fibril is obtained by growing the BC-producing bacteria on the inner and/or outer surface of an oxygen-permeable hollow carrier, produced from, for example, cellophane, Teflon, silicone, ceramics, etc.

9.2.4 Continuous Cultivation

Microbial cellulose can be also produced in a static continuous culture (Sakair et al., 1998). *A. xylinum* strain was cultured on trays, in the S-H medium, and after 2 d the pellicle produced on the surface was picked up, passed through a sodium dodecylsulfate bath to kill the cells, and set on a winding roller. The process was continued for a couple of weeks at a winding rate of 35 mm h^{-1} and fresh S-H medium was added into the trays every 8–12 h to keep its optimum level. A cellulose filament of more than 5 m was collected using this method, indicating its industrial application potential.

9.3
In vitro Biosynthesis

In vitro synthesis of cellulose has been one of the most difficult topics in the area of cellulose research. For the first time, cellulose was synthesized utilizing the 'reversed action' of cellulase as a catalyst, by Kobayashi et al. (1991), even though chemical synthetic methods have been applied before. Several monomers and catalysts have been used (Nakatsubo et al., 1989), but none of them gave the desired product, the stereoregular polymer of β-1,4-linked glucopyranoses. The idea of BC synthesis *in vitro* originated from experiments using *A. xylinum* cell-free extracts (Glaser, 1958), raw preparations of membranes (Colvin, 1980; Swissa et al., 1980; Aloni et al., 1982), and membranes solubilized with digitonin (Aloni et al., 1983). Such investigations proved that UDPGlc is the substrate for *A. xylinum* cellulose synthesis, c-di-GMP is the specific activator (Ross et al., 1987), and Ca^{2+} and Mg^{2+} ions are essential for this process (Swissa et al., 1980). Synthesis *in vitro*, carried out using *A. xylinum* membranes solubilized with digitonin (Lin et al., 1985), gave cellulose fibers 17 ± 2 Å in diameter. Their morphology and size resembled *A. xylinum* cellulose fibrils formed in the presence of factors disturbing crystallization of the nascent polymer (see Section 7.3.2). Further research, including X-ray diffraction analysis (Bureau and Brown, 1987) of cellulose produced by the *A. xylinum* membrane fractions, revealed that synthesis *in vitro* led to the amorphous cellulose II allomorph, whereas *in vivo* the highly crystalline cellulose I was obtained. Lin and Brown (1989) proved that purified CS also catalyzed cellulose synthesis *in vitro*. All these experiments were run on the microscale level and it is hard to believe today that commercial *in vitro* production of microbial polymer is possible, all the more since in the 1990s no progress in this area has been made.

9.4
Chemo-enzymatic Synthesis

Experiments on the chemical synthesis of cellulose have not given the expected results. Branched and low-molecular-weight glucans (Husemann and Muller, 1966) or polymers of β-1,4- and α-1,4-linked glucopyranoses (Micheel and Brodde, 1974) were obtained.

Successful experiments of Kobayashi et al. (1991) indicate that using coupled chemical and enzymatic methods may lead to the development of commercial *in vitro* synthesis of cellulose. The authors applied β-D-cellobiosyl fluoride, obtained by means of chemical synthesis, as a substrate and *T. reesei* cellulase as a catalyst in an aqueous–organic solvent system (a mixture of acetic buffer pH 5.0 and acetonitrile, 1:5 v/v) and achieved water-insoluble 'synthetic cellulose' having a DP > 22. X-ray and ^{13}C-NMR analyzes revealed that the product was the cellulose II. The authors stated that the DP of the product depends on reaction conditions, especially on the aqueous–organic solvent composition. They found that in the presence of higher concentrations of the substrate or acetonitrile, the cellulase – which acts as a glycosyltransferase – produces mainly soluble cellooligosaccharides (DP ≤ 8), which may also find numerous uses. Although these studies have not been continued, the method proposed by Kobayashi might be a promising alternative for total enzymatic *in vitro* cellulose production.

9.5 Production Processes Expected to be Applied in the Future

Recently, an original method of BC production on a large-scale was proposed by Nichols and Singletary (1998), who patented the idea of the construction of transgenic plants capable of this polymer synthesis. The authors intend to express three *A. xylinum* CS genes (*bcsA*, *bcsB*, and *bcsC*, see Section 7.4) and two *A. xylinum* diguanyl cyclase genes in storage tissues (roots, tubers, grains) of crops (potato, maize, oat, sorghum, millet, wheat, rice, sugarcane, etc.). They suppose to obtain masses of the pure polymer, easy and cheap to recover. According to their method, its production would be less expensive than by means of fermentation. The authors also emphasize the ecological aspect of their procedure, since an additional benefit of breeding of pure cellulose-producing transgenic plants would be a more economical use of forest resources.

9.6 Recovery and Purification

Microbial cellulose obtained via stationary or agitated culture is not completely pure and contains some impurities, such as culture broth components and whole *A. xylinum* cells. Prior to use in medicine, food production, or even the papermaking industry, all these impurities must be removed.

One of the most widely used purification methods is based on treatment of BC with a solution of hydroxides (mainly sodium and potassium), sodium chlorate or hypochlorate, H_2O_2, diluted acids, organic solvents, or hot water. The reagents can be used alone or in combination (Yamanaka et al., 1990). Immersing BC in such solutions (for 14–18 h, in some cases up to 24 h) at elevated temperature (55–65 °C) markedly reduces the number of cells and coloration degree.

Boiling in 2% NaOH solution after preliminary rinsing with running tap water was also reported (Yamanaka et al., 1989). Watanabe et al. (1998a) immersed the polymer in 0.1 NaOH at 80 °C for 20 min and then washed it with distilled water. Takai et al. (1997) treated BC with distilled water and 2% NaOH, and neutralized the mat with 2% acetic acid.

Krystynowicz et al. (1997) developed a procedure starting from washing crude BC in running tap water (overnight), followed by boiling in 1% NaOH solution for 2 h, washing in tap water to accomplish NaOH removal (1 d), neutralization with 5% acetic acid, and its removal with tap water. The final BC preparations obtained contained less than 3% protein, and were suitable for certain food and medical purposes.

Medical application of BC requires special procedures to remove bacterial cells and toxins, which can cause pyrogenic reactions. One of the most effective protocols begins with gentle pressing of the cellulose pellicle between absorbent sheets to expel about 80% of the liquid phase and then immersing the mat in 3% NaOH for 12 h. This procedure is repeated 3 times, and after that the pellicle is incubated in 3% HCl solution, pressed, and thoroughly washed in distilled water. The purified pellicle is sterilized in an autoclave or by ^{60}Co irradiation. It performs excellently as a wound dressing since it contains only 1–50 ng of lipopolysaccharide endotoxins, whereas BC purified using conventional methods usually contains 30 μg or more of these substances (Ring et al., 1986).

10
Properties

BC is an extremely insoluble, resilient, and elastic polymer, having high tensile strength. It has a reticulated structure, in which numerous ribbon-shaped fibrils are composed of highly crystalline and highly uniaxially oriented cellulose subfibrils. This three-dimensional structure, not found in plant-originating cellulose, brings about a higher crystallinity index (60–70%) of BC. Furthermore, fibers of the plant polymer are about 100 times thicker than BC microfibrils and therefore the bacterial polymer has an about 200 times larger accessible surface. In comparison to *A. xylinum* cellulose from agitated cultures, the polymer synthesized under static conditions has a higher DP (14,400 and 10,900, respectively), crystallinity index (71 and 63%, respectively), tensile strength (Young's modulus of 33.3 and 28.3, respectively), but lower water holding-capacity (45 and 170 g water g BC^{-1}, respectively) and suspension viscosity (0.04 and 0.52 Pa·s, respectively) in its disintegrated form (Watanabe et al., 1998a).

Microbial cellulose appears to be gelatinous, since its liquid component (usually water), present in voids among very fine ribbons, amounts to at least 95% by weight. The bacterial polysaccharide has a high water-binding capacity, but the majority of the water is not bound to the polymer and it can be squeezed out by gentle pressing. Drying of BC leads to paper-like sheets, having a thickness of 0.01–0.5 mm (Yamanaka et al., 1990; Krystynowicz et al., 1995, 1997) and good absorptive properties. In addition to a high Young's modulus, BC also displays high sonic velocity and, because of these unique mechanical properties, it can be applied as acoustic membranes (Vandamme et al., 1998).

The properties of BC can be modified either during its synthesis or when the culture is completed. Some compounds, like cellulose derivatives, sulfonic acids, alkylphosphates, or other polysaccharides (starch, dextran), introduced into the nutrient broth alter the macroscopic morphology, the tensile strength, the optical density, and the absorptive properties of the final product (Yamanaka et al., 2000). BC can also be combined with other substances, added either to wet or dried cellulose, thus giving composites of desired physicochemical properties (Yamanaka, 1990). The auxiliary materials used for this purpose comprise granules and fibers of various inorganic and organic compounds, such as alumina, glass, agar, alginates, carragenan, pullulan, dextran, polyacrylamide, heparin, polyhydroxylalcohols, gelatin, collagen, etc. They are combined with BC sheets by impregnation, lamination, or adsorption, or mixed with the disintegrated polymer. The composites are further subjected to a shaping treatment, thus giving various products.

Recently, Kim et al. (1999) reported an enzymatic method of BC modification using *L. mesenteroides* dextransucrase and alternansucrase. In their presence *A. xylinum* ATCC 10821 synthesized 'soluble cellulose', which was a glucan composed of 1,4-, 1,6-, and 1,3-linked monomers.

The basic BC properties are summarized in Table 2.

11
Applications

BC belongs to the generally recognized as safe (GRAS) polysaccharides and therefore it has already been put to a multitude of different uses. Commercial application of this polymer results from its unique properties and developments in effective technologies of production, based on growth of improved bacterial strains on cheap waste materials. The advantage of BC is its chem-

Tab. 2 Properties of BC

High purity
High degree of crystallinity
Greater surface area than that of conventional wood pulp
Sheet density from 300 to 900 kg m^{-3}
High tensile strength even at low sheet density (below 500 kg m^{-3})
High absorbency
High water-binding capacity
High elasticity, resilience, and durability
Nontoxicity
Metabolic inertness
Biocompatibility
Susceptibility to biodegradation
Good shape retention
Easy tailoring of physicochemical properties

ical purity and the absence of substances usually occurring in the plant polysaccharide, which requires laborious purification. In addition to the shape of BC sheets, their area and thickness can be tailored by means of culture conditions. Relatively easy BC modification during its biosynthesis enables regulation of such properties as molecular mass, elasticity, resilience, water-holding capacity, crystallinity index, etc. BC microfibrils may bind both low-molecular-weight substances and polymers, added for instance to the growth medium, thus giving novel commodities. BC can be also a raw material for further chemical modifications.

Based on the assumption that as much as 10,000 kg of the bacterial polymer can be obtained per year from static culture having 1 hectare surface area and only 600 kg of cotton is harvested in the same period of time from a field of the same area, the perspectives for a wider use of BC become more apparent (Kudlicka, 1989).

The main potential BC applications are summarized in the Table 3.

11.1 Technical Applications

Compared to PC sheets, BC has satisfactory tensile strength even at low sheet density (300–500 kg m^{-3}) (Johnson and Neogi,

Tab. 3 Applications of bacterial cellulose

Sector	*Application*
Cosmetics	Stabilizer of emulsions such as creams, tonics, nail conditioners, and polishes; component of artificial nails
Textile industry	Artificial skin and textiles; highly adsorptive materials
Tourism and sports	Sport clothes, tents, and camping equipment
Mining and refinery waste treatment	Spilt oils collecting sponges, materials for toxins adsorption, and recycling of minerals and oils
Sewage purification	Municipal sewage purification and water ultrafiltration
Broadcasting	Sensitive diaphragms for microphones and stereo headphones
Forestry	Artificial wood replacer, multi-layer plywood and heavy-duty containers
Paper industry	Specialty papers, archival documents repair, more durable banknotes, diapers, and napkins;
Machine industry	Car bodies, airplane parts, and sealing of cracks in rocket casings
Food production	Edible cellulose and 'nata de coco'
Medicine	Temporary artificial skin for therapy of decubitus, burns and ulcers; component of dental and arterial implants
Laboratory/research	Immobilization of proteins, chromatographic techniques, and medium component of *in vitro* tissue cultures

1989). Therefore, BC is an excellent component of papers, providing better mechanical properties. Microfibrils of the bacterial polymer form a great number of hydrogen bonds when the paper is subjected to drying, thus giving improved chemical adhesion and tensile strength. BC-containing paper not only shows better retention of solid additives, such as fillers and pigments, but is also more elastic, air permeable, resistant to tearing and bursting forces, and binds more water (Iguchi et al., 2000).

A beneficial effect such as an improved aging resistance was achieved by adding small amounts of BC to cotton fibers to obtain hand-made paper, used for old document repair. The paper displayed appropriate ink receptivity and specific snap (Krystynowicz et al., 1997). Paper containing 1% of BC meets the international standard ISO 9706:1994 for information and documentation papers, and also has a specific snap comparable to that of rag paper. It is possible to use BC for pressboard making, as well as production of paperboard used as an electro-insulating material and for bookbinding.

BC can also be used as a surface coating for specialty papers. For this purpose, a filtered suspension of BC (homogenized and mixed with other components) is added using special applicators, either within wet sheet formation, or to a partially or completely dried sheet. The coating improves properties such as gloss, brightness, smoothness, porosity, ink receptivity, and tensile strength. Other additives like starch, organic polymers, including CM-cellulose, organic, or inorganic pigments may also be used. Johnson and Neogi (1989) claim that paper coated with 3% of BC (solid matter) displays gloss properties and surface strength similar to rotogravure paper having 20% traditional coating. The authors also state that coating with a mixture of BC and CM-cellulose gives even better properties, since they act synergistically.

BC is also a valuable component of synthetic paper (Iguchi et al., 2000) since nonpolar polypropylene and polyethylene fibers, providing insulation, heat resistance, and fire-retarding properties, cannot form hydrogen bonds. The amount of wood pulp in this type of paper is usually from 20 to 50% to achieve good quality. Using BC enables us to decrease the amount of the additives without any effect on the synthetic paper properties.

BC also appeared to be a good binder in nonwoven fabric-like products (Yamanaka et al., 1990) commonly used in surgical drapes and gowns, and containing various hydrophilic and hydrophobic, natural and synthetic fibers, such as cellulose esters, polyolefin, nylon, acrylic glass, or metal fibers. Even small amounts of BC improve tensile and tear strength of the fabric, e.g. 10% of the bacterial polymer is equivalent to 20–30% of latex binder.

The scope of BC uses can be even wider since it is modifiable during synthesis (Brown, 1989a,b). For this purpose, CM-cellulose or copolymers of saccharides and dicarboxylic acids are added directly to the culture medium (Yamanaka et al., 1989). Cellulose obtained in the presence of CM-cellulose and dried with organic solvents has a resilient and elastic character, as well as higher water binding capacity (adsorbs faster more water). The optimum CM-cellulose concentrations range from 0.1 to 5% (w/v). The polarity of the organic solvent also influences the BC features. After treatment with acetone the polymer is elastic and rubber-like, whereas after drying with absolute ethanol, BC resembles leather.

Since BC is susceptible to enzymatic digestion, it is modified to obtain various composites of satisfactory biodegrability and strength. Their production is based either on chemical reaction between cellulose and the copolymer or on culturing the BC-producing strain in the copolymer-containing medium.

11.2 Medical Applications

A cellulose pad from a static culture is a ready-to-use, naturally 'woven' wound dressing material that meets the standards of modern wound dressings (Figure 16). It is sterilizable, biocompatible, porous, elastic, easy to handle and store, adsorbs exudation, provides optimum humidity, which is essential for fast wound healing, protects from secondary infection and mechanical injury, does not stick to the newly regenerated tissue, and alleviates pain by heat adsorption from burns. BC sheets are also excellent carriers for immobilization of medicinal preparations, which speed up the healing process.

Since BC pellicles can have various dimensions, it is relatively easy to produce dressings for extensive wounds. Because of recent problems with products of animal origin, collagen dressings can be replaced with BC ones. This thesis is additionally supported by undoubtedly positive results of clinical tests. For instance, the BC preparation, Prima Cel™, produced by Xylos Corp., according to the Rensselaer Polytechnic Institute (USA), has been applied as a wound dressing in clinical tests to heal ulcers. The results obtained were satisfactory, since after 8 weeks 54% of the patients recovered and the remaining ulcers were almost healed (Jonas and Farah, 1998). Other commercial preparations of *A. xylinum* cellulose, such as Biofil® and Bioprocess®, appeared to be excellent as skin transplants, and in the treatment of third-degree burns, ulcers, and decubitus. Another preparation, Gengiflex®, found application in recovering periodontal tissues (Jonas and Farah, 1998). Investigations on *A. xylinum* cellulose pads prepared by Krystynowicz et al. (2000) (Figure 17) indicate that general use as wound dressings is possible.

Results of investigations on BC hollow fibers use as artificial blood vessels and ureters are also promising (Yamanaka et al., 1990). Antithrombic BC property (blood

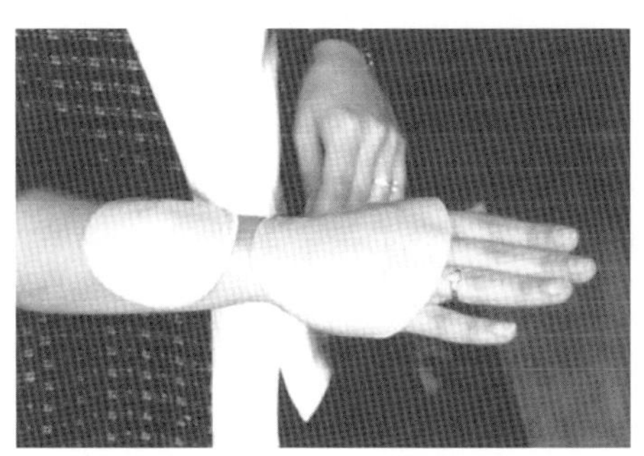

Fig. 16 Cellulose pellicle as a material for wound dressings.

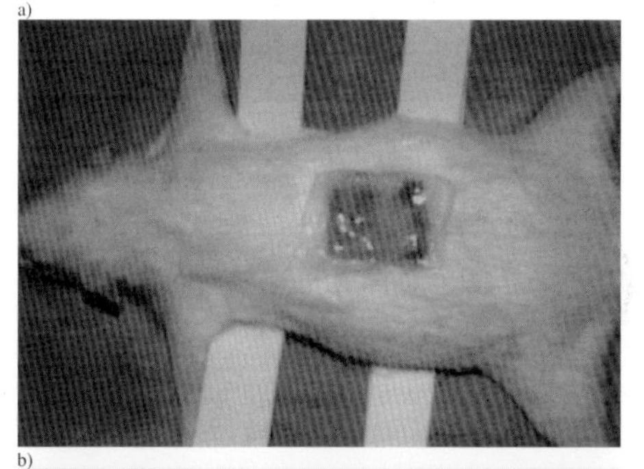

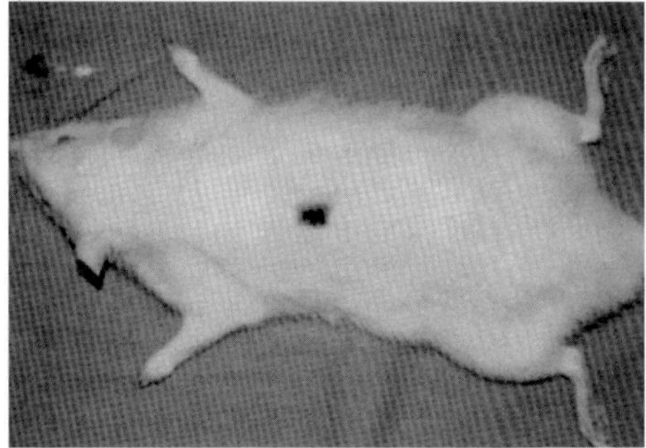

Fig. 17 Burns treated with a BC dressing, before (a) and after (b) healing (Krystynowicz et al., 2000).

compatibility) was evaluated by a test based on replacement of adult dog blood vessels (parts of the descending aorta and jugular vein) with the artificial counterpart made of BC. One month later the artificial vessel was removed and the state of adhesion of thrombi on its inner surface was examined. A good open state of the BC blood vessel was maintained.

Another successful experiment showing the application of biosynthetic cellulose was substitution of the dog dura matter in the brain (Jonas and Farah, 1998).

Because of its high tensile strength, elasticity, and permeability to liquids and gases, dried BC was applied as an additional membrane to protect immobilized glucose oxidase in biosensors used for assays of blood glucose levels. This BC membrane enhanced the electrode stability in 10 times diluted human blood solution, up to 200 h. Other commercial protecting membranes, such as cuprophan (AKZO, England), provided an electrode stability up to 30 h only. In undiluted human blood the biosensor coated with cuprophan was stable for 3–4 h, whereas the BC membrane prolonged its stability up to 24 h.

11.3
Food Applications

Chemically pure and metabolically inert BC has been applied as a noncaloric bulking and stabilizing agent in processed food. Similar to pure preparations of the plant polysaccharide and its derivatives, it is used for stabilization of foams, pectin and starch gels, emulsions, e.g. canned chocolate drinks and soups, texture modification, e.g. improvement of the consistency of pulps, enhancement of adhesion, replacement of lipids, including oils, and dietary fiber supplements (Ang and Miller, 1991; Kent et al., 1991; Krystynowicz et al., 1999).

The first successful commercial application of BC in food production is 'nata de coco' (Sutherland, 1998). It is a traditional dessert from the Philippines, prepared from coconut milk or coconut water with sucrose, which serves as a growth medium for BC-producing bacteria. Consumption of the pellicle is believed to protect against bowel cancer, artheriosclerosis, and coronary thrombosis, and prevents a sudden rise of glucose in the urine. Therefore 'nata de coco' is becoming increasingly popular, not only in Asia.

Another popular BC-containing food product is Chinese Kombucha (teakvass or tea-fungus), obtained by growing yeast and acetic acid bacteria on tea and sugar extract. The pellicle formed on the surface contains both cellulose and enzymes healthy for humans. Their abiotic activity is especially stimulating for the large bowel and the whole alimentary tract. Kombucha is believed to protect against certain cancers (Iguchi et al., 2000).

Results of studies carried out by the authors, who applied BC pellicles synthesized by *A. xylinum* E_{25} for wine and juice filtration, and for immobilization of polyphenols, are promising (Krystynowicz et al., 1999). Preparations of bioactive anthocyanins, enriched in dietary fiber, are excellent for functional food production. BC also appeared to be an attractive component of bakery products, since it plays a role of dietary fiber, is taste and odorless, and prolongs the shelf-life.

11.4
Miscellaneous Uses

The large accessible surface area, high durability, and superior adsorptive properties as well as possibility of modification by means of physical or chemical methods, means that BC can be applied as a carrier for immobilization of biocatalysts. Cellulose gels containing immobilized animal cells were used for their culture to produce interferon, interleukin-1, cytostatics, and monoclonal antibodies (Iguchi et al., 2000).

BC was also applied for adsorption of cells of *Gluconobacter oxydans, Acetobacter methanolyticus* and *Saccharomyces cerevisiae.* The immobilized strains appeared to be effective producers of gluconate (84–92% yield), dihydroxyacetone (90–98% yield), and ethanol (88–92% yield), respectively, and displayed better operational and thermal stability.

Purified BC can be a raw material for synthesis of cellulose acetate, nitrocellulose, CM-cellulose, hydroxymethylcellulose, methylcellulose, and hydroxycellulose (Yamanaka, 1990). If BC is produced in the presence of compounds that interfere with regular fibril assembly and which influence the β-1,4-glucan structure, such as CM-cellulose or other cellulose derivatives, and other carbohydrates (starch, dextran), sulfonates, and alkylphosphates, the resulting microbial polymer has novel, additive-dependent, and useful properties, including optical transparency or higher water-binding capacity, even after repeated soaking and drying.

The potential application of BC in the chemical, paper, and textile industries depends on its price and accessibility. To meet these demands, BC has to be produced by highly efficient strains, growing on cheap substrates, in sophisticated surface, solid-state, or submerged fermentors (Vandamme et al., 1998).

12 Patents

Growing rapidly, the interest in BC is reflected in a number of patents (annually about 20 since the 1980s) and publications (20–40 per year for the last 10 years) devoted to this unique polymer (Iguchi et al., 2000). Some basic patents concerning the biosynthesis, properties, and applications of BC are presented in Table 4. These patents that are cited in chapter are listed in the references.

13 Outlook and Perspectives

The first scientific paper reporting on an unusual substance formed by acetic acid bacteria and known for ages in many countries as 'vinegar plant' (Yamanaka et al., 1989) was published 115 years ago (Brown, 1886). Further studies revealed that the substance is a super-pure cellulose. Metabolic pathways and the complicated molecular machinery of the polysaccharide biosynthesis, as well as the intriguing dynamics of its nascent chains association into a structure with unique properties, has been elucidated. Although progress and limited commercialization of BC have undoubtedly taken place, the relevant biotechnology, competitive to modern industrial technologies for PC production, has not yet been developed.

So, what to do, to achieve success and accomplish commercialization of BC production? First of all, stable overproducer strains of the polymer have to be constructed, using recent achievements in molecular genetics and biology. These strains have to assimilate a wide range of carbon sources, as well as display lower tendency towards spontaneous mutation to $Cell^-$ mutants and effectively synthesize cellulose (10–15 $g\ L^{-1}\ d^{-1}$) under agitated culture conditions. Construction of new bioreactors for both stationary and submerged culture is necessary. The considerable reduction of BC production cost could be attained by replacing expensive nutrient media components

Tab. 4 Selected patents concerning BC

Patent number	*Patent holder*	*Inventors*	*Title*	*Date of publication*	*Major claims*
WO 0125470	Novozymes A/S, Denmark	Herbert, W., Chanzy, H. D., Ernst, S., Schulein, M., Husum, T. L., Kongsbak, L.	Cellulose films for screening	2001	BC microfibril films containing fluorescein-labeled hemoglobin or galactomannan can be used to detect proteases and mannanases, respectively
WO 0105838	Pharmacia Corp., USA	Yang, Z. F., Sharma, S., Mohan, C., Kobzeff, J.	Process for drying reticulated bacterial cellulose without co-agents	2001	The reticulated *Acetobacter* cellulose subjected to dispersing in a solvent, e.g. hydrocarbon (hexane), aliphatic alcohol, and/or alkyl sulfoxide (DMSO), separation from the solvent, and drying, may be rehydrated to provide uniform dispersions having high viscosity
JP 11255806 A2	Bio-Polymer Research Co. Ltd, Japan	Watanabe, O.	Freeze drying method for microfibrous cellulose concentrate	1999	Freeze drying of *Acetobacter* cellulose gives dry microfibrous cellulose with good retention of its original properties after redispersing in water
WO 9940153	Monsanto	Smith, B. A., Colegrove, G. T., Rakitsky, W. G.	Acid-stable, cationic compatible cellulose compositions useful as rheology modifiers	1999	Cationic co-agents, acids, and/or cationic surfactants are used to form acid-stable *Acetobacter* cellulose compositions, which are useful as rheology modifiers
WO 9943748	Sony	Uryu, M., Tokura, K.	Biodegradable composite polymer material	1999	A new composite material has been obtained by drying the *A. xylinum* cellulose, its pulverization, and blending with a biodegradable polymer material
JP 0056669 A2	Canon Co., Japan	Minami, M., Mihara, C., Takeda, T., Kikuchi, Y.	Composite, for use in thermoformed articles, comprises cellulose and saccharide ester derivative	1999	A composite comprising BC and a saccharide ester can be used for production of biodegradable, thermoformed articles, which have improved processability, mechanical strength and flexibility
JP 11172115 A2	Ajinomoto Co. Inc., Japan	Suzuki, O., Kitamura, N., Matsumoto, R.	Bacterial cellulose-containing composite absorbents with high liquid absorption	1999	Dispersing highly water-absorbing polymer particles, dissolved in an organic solvent, in an aqueous solution of defibrillated BC, followed by organic solvent removal and partial drying, gives excellent and stable absorbents

Tab. 4 (cont.)

Patent number	Patent holder	Inventors	Title	Date of publication	Major claims
JP 11246602 A2	Bio-Polymer Research Co. Ltd, Japan	Tahara, N., Hagamida, T., Miyashita, H., Watanabe, O.	Preservation of wet bacterial cellulose	1999	Wet BC can be preserved with alkyl sulfate salts or NaOH and/or KOH
JP 11187896 A2	Bio-Polymer Research Co. Ltd, Japan	Tabata, T., Toyosaki, H., Tsuchida, T., Yoshinaga, F.	A method for screening cellulose-producing bacteria using cellulase	1999	A rapid and convenient method for screening a large number of cellulose membrane-producing strains is presented; the metabolic peculiarities of the strains and conditions needed for them to produce cellulose pellicle are explained
JP 11092502 A2	Bio-Polymer Research Co. Ltd, Japan	Shoda, M., Kanno, Y., Koda, T., Yoshinaga, F.	Manufacture of bacterial cellulose under oxygen-enriched conditions	1999	Passing more than 21% oxygen-containing air through *A. xylinum* ssp. sucrofermentans culture in an air-lift fermentor, enables accumulation of 6.5 g L^{-1} of BC after 75 h
JP 11117120 A2	Toray Industries Inc., Japan	Hara, T., Amano, J.	Fibers made from blends of bacterial cellulose and polymers having flexible main backbone	1999	The blends contain polyvinyl alcohol-type polymers at weight ratio of 2–50%
JP 11181001 A2	Bio-Polymer Research Co. Ltd, Japan	Matsuoka, M., Toyosaki, H., Matsumura, T., Ougiya, H., Tsuchida, T., Yoshinaga, F.	Production of bacterial cellulose	1999	BC, useful for filler retention aids for paper-making, can be produced in agitated cultures of *Acetobacter* strains, in the presence of water-soluble polysaccharides, e.g. CM-cellulose
JP 11137163 A2	Shikishima Seipan Co. Ltd, Japan	Kondo, M., Yamada, M., Inoue, S.	Manufacture of bread from dough containing bacterial cellulose	1999	Addition of BC increases water absorption of dough, thus giving bread with high water-holding capacity
JP 11221072 A2	Bio-Polymer Research Co. Ltd, Japan	Ishikawa, A., Tsuchida, T., Yoshinaga, F.	Bacterial cellulose production enhancement	1999	An *A. xylinum* mutant having higher cellular levels of UDP, UTP, and UDPGlc as well as higher carbamoyl phosphate synthetase activity is an excellent cellulose producer

Tab. 4 (cont.)

Patent number	*Patent holder*	*Inventors*	*Title*	*Date of publication*	*Major claims*
JP 11269797 A2	Toppan Printing Co. Ltd, Nakano Vinegar Co. Ltd, Japan	Yamawaki, K., Tomita, T., Harasawa, A., Kaminaga, J., Kawasaki, K., Matsuo, R., Sasaki, N., Fukagai, M., Tsukamoto, Y.	Impregnated paper with good water resistance and stiffness	1999	Paper impregnated with silane coupler-grafted BC derivatives is moisture resistant
PP 299907	Technical University of Lodz, Poland	Krystynowicz, A., Czaja, W., Bielecki, S.	Biosynthesis and application of bacterial cellulose	1999	Production of BC by wild and mutant *A. xylinum* strains, cultured in static or agitated cultures, and under various conditions as well as an influence of culture conditions on the cellulose properties are described
JP 11018758 A2	Bio-Polymer Research Co. Ltd, Japan	Yamamoto, T., Yano, H., Yoshinaga, F.	Horizontal type spinner culture vessel, having high oxygen-supplying efficiency	1999	BC can be produced in the spinner culture tank, equipped with mixing impellers, and providing high efficiency of oxygen supply
JP 10298204 A2	Ajinomoto Co. Inc., Japan	Ishikara, M., Yamanaka, S.	Bacterial cellulose with ribbon-like microfibril shape	1998	Cellulose fibrils having a short axis 10–100 nm and a long axis 160–1000 nm are produced extracellularly by cellulose-generating bacteria, e.g. *Acetobacter pasteurianus*, in a culture containing cell division inhibitor
US 005723764	Pioneer Hi-Bred International Inc., USA	Nichols, S. E., Singletary, G. W.	Cellulose synthesis in the storage tissue of transgenic plants	1998	Introducing the genes for cellulose biosynthesis from the species *A. xylinum* into a given plant provides a method of synthesizing cellulose in the storage tissue of transgenic plants
JP 10077302	Bio-Polymer Research Co. Ltd, Japan	Tabuchi, M., Watanabe, K., Morinaga, Y.	Solubilized bacterial cellulose and its compositions or composites for moldings and coatings.	1998	Cellulose synthesized by *A. xylinum* in stationary culture was solubilized by stirring in a mixture of DMSO and paraformaldehyde (25:5) at 100 °C for 3 h

Tab. 4 (cont.)

Patent number	*Patent holder*	*Inventors*	*Title*	*Date of publication*	*Major claims*
WO 97/05271	Rensselaer Polytechnic Institute, USA	Bungay, H. R., Serafica, G.	Production of microbial cellulose using a rotating disc film bioreactor	1997	BC can be deposited by *A. xylinum*, cultured in a liquid medium inside the horizontal fermentor, on a plurality of disks, which rotate around the long axis of the fermentor and are partly submerged in the culture medium
JP 09025302 A2	Bio-Polymer Research Co. Ltd, Japan	Hioki, S., Watanabe, K., Ogya, H., Morinaga, Y.	Preparation of disintegrated bacterial cellulose for improved additive retention in paper manufacture	1997	BC which helps additive retention and causes no harm to freeness during paper manufacturing can be obtained by disintegration using a self-excited ultrasonic pulverizer
JP 09107892 A	Nakano Vinegar Co. Ltd, Japan	Furukawa, H., Maruyama, Y., Fukaya, M., Tsukamoto, Y., Kawamura, K.	Composition for stabilizing dispersion used in food	1997	A fine particulate (210 μm average particle diameter) BC, obtained by hydrolyzing *A. xylinum* cellulose with a mineral acid, is of low viscosity and is a sufficient dispersion stability agent in food products
WO 9744477	Bio-Polymer Research Co. Ltd, Japan	Naritomi, T., Kouda, T., Naritomi, M., Yano, H., Yoshinaga, F.	Continuous preparation of bacterial cellulose having a high production rate and yield	1997	BC obtained in continuous culture of *A. xylinum* (production rate at least 0.4 $g L^{-1} h^{-1}$) in a medium containing a substance enhancing the apparent substrate affinity to sugar (e.g. lactic acid) is useful as a thickener, humectant or stabilizer for production of food, cosmetics or paints, etc.
WO 9740135	Bio-Polymer Research Co. Ltd, Japan	Tsuchida, T., Tonouchi, N., Seto, A., Kojima, Y., Matsuoka, M., Yoshinaga, F.	Novel cellulose-producing bacteria and a process of producing it	1997	Production of cellulose with a new *A. xylinum* ssp. nonactoxidans, which lacks an ability to oxidize acetates and lactates, yields odor- and taste-less products, having excellent dispersability in water
WO 9712987	Bio-Polymer Research Co. Ltd, Japan	Kouda, T., Naritomi, T., Yano, H., Yoshinaga, F.	Production process for bacterial cellulose which is useful as material in various fields	1997	A new process for BC production involving maintaining a certain pressure inside a fermentor, reduces power required for agitation, and elevates production rate and yield
JP 09056392 A	Kikkoman Corp., Japan	Fukazawa, K., Imai, H., Kijima, T., Kikuchi, T.	Production of microorganism cellulose	1997	A cellulose pellicle synthesized by *A. xylinum* strain precultured in stationary conditions can be homogenized and used to inoculate the fresh culture medium

Tab. 4 (cont.)

Patent number	*Patent holder*	*Inventors*	*Title*	*Date of publication*	*Major claims*
PL 171952 B1	Technical University of Lodz, Poland	Galas, E., Krystynowicz, A.	Method of obtaining bacterial cellulose	1997	A method of BC production in static culture using *A. xylinum* P23 strain is described
US 0824096	Bio-Polymer Research Co. Ltd, Japan	Kouda, T., Nagata, Y., Yano, H., Yoshinaga, F.	Production of bacterial cellulose through cultivation of cellulose-producing bacteria under specified conditions in aerated and agitated culture	1997	Cellulose produced by species of *Acetobacter, Agrobacterium, Rhizobium, Sarcina, Pseudomonas, Achromobacter, Alcaligenes, Aerobacter, Azotobacter,* and *Zoogloea*, in aerated and agitated cultures can be applied in production of food and cosmetics
JP 97–21905	Bio-Polymer Research Co. Ltd, Japan	Seto, H., Tsuchida, T., Yoshinaga, F.	Manufacture of bacterial cellulose by mixed culture of micro-organisms	1997	To provide lactate and split sucrose, cellulose-producing *A. xylinum* strain can be grown together with lactic acid bacteria and *Saccharomyces* yeast; their presence in the culture broth markedly enhances BC yield
JP 08127601 A	Bio-Polymer Research Co. Ltd, Japan	Hioki, S., Watabe, O., Hori, S., Morinaga, Y., Yoshinaga, F.	Freeness regulating agent	1996	Production of a freeness regulating agent, comprising a defiberized BC, synthesized by *A. xylinum* ssp. sucrofermentans is described
JP 96316922	Bio-Polymer Research Co. Ltd, Japan	Hikawu, S., Hiroshi, T., Takayasu, T., Yoshinaga, F.	Manufacture of bacterial cellulose by addition of cellulose formation stimulators	1996	Compounds such as derivatives of choline, betaine, and fatty acids (salts and esters) appeared to stimulate cellulose production by *A. xylinum* strains
JP 08056689 A	Bio-Polymer Research Co. Ltd, Japan	Seto, H., Tsuchida, T., Yoshinaga, F.	Production of bacterial cellulose	1996	To obtain an edible BC, excellent in aqueous dispersibility, useful for retaining the viscosity of foods, cosmetics, coatings, etc., and enrichment of foods, *Acetobacter* strains can be cultured in saponin-containing medium
JP 08033494 A	Bio-Polymer Research Co. Ltd, Japan	Tawara, N., Koda, T., Hagamida, T., Morinaga, Y., Yano, H.	Method for circulating continuous production and separation of bacterial cellulose	1996	Circulating a culture liquid containing *A. xylinum* cells between a culture and a separation apparatus enables separation of the produced cellulose; also, a flotation separator or an edge filter can be applied for this purpose

Tab. 4 (cont.)

Patent number	*Patent holder*	*Inventors*	*Title*	*Date of publication*	*Major claims*
JP 08034802 A	Gun Ei Chemical Industries Co. Ltd, Japan	Hirooka, S., Hamano, T., Miyashita, Y., Hanaue, K., Yamazaki, K., Shiichi, K., Shiichi, F.	Bacterial cellulose, production thereof and processed product made therefrom	1996	*A. xylinum* can synthesize cellulose in culture broths containing difficult to ferment, branched oligosaccharides
JP 08000260 A	Bio-Polymer Research Co. Ltd, Japan	Ishikawa, A., Tsuchida, T., Yoshinaga, F.	Production of bacterial cellulose with pyrimidine analogue-resistant strain	1996	A method of production of BC, in a high yield and at a low cost, using a pyrimidine analog-resistant *Acetobacter* mutant is presented
JP 08325301 A	Bio-Polymer Research Co. Ltd, Japan	Hori, S., Watabe, O., Morinaga, Y., Yoshinaga, F.	Cellulose having high dispersibility and its production	1996	BC synthesized by *A. xylinum* ssp. sucrofermentans, subjected after harvesting to partial hydrolysis with HCl, yields a fraction exhibiting high birefringence
JP 08276126	Bio-Polymer Research Co. Ltd, Japan	Ogiya, H., Watabe, O., Shibata, A., Hioki, S., Morinaga, Y., Yoshinaga, F.	Emulsification stabilizer	1996	The method of production of the emulsification stabilizer containing BC obtained from agitated culture of *A. xylinum* ssp. sucrofermentans and having low index of crystallization, is presented
JP 08009965 A	Bio-Polymer Research Co. Ltd, Japan	Ishikawa, A., Tsuchida, T., Yoshinaga, F.	Production of bacterial cellulose using microbial strain resistant to inhibitor of DHO-dehydrogenase	1996	A method of efficient production of BC, using *Acetobacter* strains resistant either to an inhibitor of DHO-dehydrogenase or to DNP is presented
US 005382656	Weyerhauser Co., USA	Benziman, M., Tal, R.	Cellulose synthase associated proteins	1995	CS-associated proteins, which have molecular weights of 20, 54, and 59 kDa are not encoded by CS operon genes *bcsA*, *B*, *C*, or *D*
JP 07313181 A	Bio-Polymer Research Co. Ltd, Japan	Takemura, H., Tsuchida, T., Yoshinaga, F., Matsuoka, M.	Production of bacterial cellulose using PQQ-unproductive strain	1995	Pyrroloquinolinequinone-unproductive *Acetobacter* mutant strain enables high-yield production of BC
JP 07184675 A	Bio-Polymer Research Co. Ltd, Japan	Matsuoka, M., Tsuchida, T., Yoshinaga, F.	Production of bacterial cellulose	1995	To obtain BC useful for retaining the viscosity of foods, cosmetics, or coatings, *Acetobacter* strains can be cultured in a methionine-containing medium

Tab. 4 (cont.)

Patent number	Patent holder	Inventors	Title	Date of publication	Major claims
JP 07268128	Fujitsuko Co. Ltd, Japan	Kiriyama, S., Fukui, H., Toda, T., Yamagishi, H.	Dried material of cellulose derived from microorganism and its production	1995	Water-soluble stabilizer (preferably glucose or gelatin) added to cellulose gel obtained by culturing *A. xylinum* strain before drying of the material provides excellent and stable physical properties
WO 95/32279	Bio-Polymer Research Co. Ltd, Japan	Tonouchi, N., Tsuchida, T., Yoshinaga, F., Horinouchi, S., Beppu, T.	Cellulose-producing bacterium transformed with gene coding for enzyme related to sucrose metabolism	1995	*A. xylinum* transformant with a gene coding for invertase accumulates cellulose in a sucrose-containing medium
JP 07184677 A	Bio-Polymer Research Co. Ltd, Japan	Seto, H., Tsuchida, T., Yoshinaga, F.	Production of bacterial cellulose	1995	High-yield production of an edible BC useful for retaining the viscosity of cosmetics or coatings can be achieved by culturing an *Acetobacter* strain in an invertase-added medium containing sucrose as a carbon source
JP 07039386 A	Bio-Polymer Research Co. Ltd, Japan	Matsuoka, M., Takemura, H., Tsuchida, T., Yoshinaga, F.	Production of bacterial cellulose	1995	BC, useful for retaining the viscosity of a food, a cosmetic, a coating, etc., and usable as a food additive, an emulsion stabilizer, etc., can be obtained by culturing *A. xylinum* in a culture medium containing a carboxylic acid (e.g. lactic acid) salt
JP 07274987 A	Bio-Polymer Research Co. Ltd, Japan	Toda K., Asakura, T.	Production of bacterial cellulose	1995	To obtain BC in high yield, the cellulose-producing *Acetobacter* strain is cultured in a cylindrical container and oxygen is fed through an oxygen-permeable membrane at the bottom
JP 06125780	Nakano Vinegar Co. Ltd, Japan	Fukaya, M., Okumura, H., Kawamura, K.	Production of cellulosic substance of microorganism	1994	A method used to improve production efficiency of cellulose by an *Acetobacter* strain is presented; the method is based on adding a protein having affinity to the cellulose to the culture medium
JP 06248594 A	Mitsubishi Paper Mills Ltd, Japan	Katsura, T., Okafuro, K.	Low-density paper having high smoothness	1994	Excellent, low-density paper having high smoothness contains BC and the broad-leaved pulp in the ratio of 1/99:1/1

Tab. 4 (cont.)

Patent number	*Patent holder*	*Inventors*	*Title*	*Date of publication*	*Major claims*
JP 06206904	Shin Etsu Chemical Co. Ltd, Japan	Horii, F., Yamamoto, H.	Bacterial cellulose, its production and method for controlling crystal structure thereof	1994	Xanthan gum or sodium CM-cellulose added to the culture broth of *A. xylinum* enable control of the crystal structure of BC
JP 06113873 A	Nippon Paper Industries Co. Ltd, Japan	Samejima, K., Mamoto, K.	Production of microbial cellulose	1994	Adding a sulfite pulp waste liquor and/or its permeate from ultrafiltration into a culture medium of a cellulose-producing microorganism enhances the cellulose yield and lowers the cost of its production
WO 94/20626	Bio-Polymer Research Co. Ltd, Japan	Beppu, T., Tonouchi, N., Horinouchi, S., Tsuchida, T.	*Acetobacter*, plasmid originating therein, and shuttle vector constructed from said plasmid	1994	Genetic recombination of cellulose-producing *Acetobacter* strains is performed using an *Acetobacter* strain, its endogenous plasmid, and a shuttle vector constructed from the latter plasmid and a plasmid from *Escherichia coli*
JP 06001647 A	Shimizu Corp., Japan	Yano, H., Narutomi, T., Okamura, K., Kawai, T., Minami, S.	Concrete and coating material	1994	The concrete or coating material containing disaggregated BC displays better dispersibility of cement or pigment particles and an enhanced fluidity
US 005268274 A	Cetus Corp., USA	Ben-Bassat, A., Calhoon, R. D., Fear, A. L., Gelfand, D. H., Meade, J. H., Tal, R., Wong, H., Benziman, M.	Methods and nucleic acid sequences for the expression of the cellulose synthase operon	1993	Nucleic acid sequences encoding the BC synthase from *A. xylinum*, and methods for isolating the genes and their expression in hosts are presented
EP 0396344 A2	Ajinomoto Co. Inc., Japan	Yamanaka, S., Ono, E., Watanabe, K., Kusakabe, M., Suzuki, Y.	Hollow microbial cellulose, process for preparation thereof, and artificial blood vessel formed of said cellulose	1990	BC prepared by culturing a cellulose-producing strain on one or both surfaces of an oxygen-permeable hollow carrier is useful as a substitute for a blood vessel or another internal hollow organ; the cellulose can be impregnated with a medium, cured, and cut if necessary

Tab. 4 (cont.)

Patent number	Patent holder	Inventors	Title	Date of publication	Major claims
US 004950597	University of Texas, USA	Saxena, I. M., Roberts, E. M., Brown, R. M.	Modification of cellulose normally synthesized by cellulose-producing micro-organisms	1990	Mutants of *A. xylinum* that do not form a pellicle in liquid culture and synthesize cellulose almost exclusively as the allomorph cellulose II, that arise spontaneously or by nitrosoguanidine mutagenesis are described and the cellulose they produce is characterized
JP 02182194 A	Asahi Chemical Industries Co. Ltd, Japan	Matsuda, Y., Kamiide, K.	Production of cellulose with acetic acid bacterium	1990	Acetic acid bacteria having a synchronized cell cycle enable efficient production of BC, excellent in water holding properties, tensile strength, and purity, and displaying a relatively low DP
WO 89/12107	Brown, R. M.	Brown, R. M.	Microbial cellulose as a building block resource for specialty products and processes thereof	1989	A novel process for manufacturing BC using different bacterial species belonging to *Acetobacter, Rhizobium, Agrobacterium*, and *Pseudomonas*, and production of various articles from this polymer are described
EP 0258038 A3	Brown, R. M.	Brown, R. M.	Use of cellulase preparations in the cultivation and use of cellulose-producing microorganisms	1989	To prepare an inoculum of appropriate cell density for large-volume fermentors, *A. xylinum* cells entrapped in cellulose can be liberated with cellulase preparation, which causes a partial cellulose hydrolysis
US 004863565	Weyerhauser Co., USA	Johnson, D. C., Neogi, A. M.	Sheeted products formed from reticulated microbial cellulose	1989	Strains of *Acetobacter* that are stable under agitated culture conditions and that exhibit reduced gluconic acids production, synthesize unique reticulated cellulose sheets, characterized by resistance to densification and great tensile strength
WO 89/11783	Brown, R. M.	Brown, R. M.	Microbial cellulose composites and processes for producing same	1989	Methods enabling production of various objects utilizing BC produced *in situ* or applied as a film are presented; a process for manufacturing currency from BC is described

Tab. 4 (cont.)

Patent number	*Patent holder*	*Inventors*	*Title*	*Date of publication*	*Major claims*
EP 0323717 A3	ICI Plc, UK	Byrom, D.	Process for the production of microbial cellulose	1988	A process for the production of BC using a novel strain of the genus *Acetobacter* is described
EP 0 289993 A3	Weyerhaeuser Co., USA	Johnson, D. C., LeBlanc, H. A., Neogi, A. N.	Bacterial cellulose as surface treatment for fibrous web	1988	BC applied at relatively low concentrations, singularly or in combination, to at least one surface of a fibrous web gives excellent properties of gloss, smoothness, ink receptivity and holdout, and surface strength
WO 88/09381	Financial Union for Agricultural Development, France	Labourer, P. F.	Process for producing bacterial cellulose from material of plant origin	1988	Culturing of *A. xylinum* strain in a plant polysaccharide-containing medium enables efficient cellulose production
US 004655758	Johnson & Johnson products, Inc., USA	Ring, D. F., Nashed, W., Dow, T.	Microbial polysaccharide articles and methods of production	1987	After removal of excess liquid and bacterial cells, the cellulose pellicle can be impregnated and used for various purposes
WO 86/02095	Bio-Fill Industria e Comercio de Produtos Medico Hospitalares, Ltd, Brazil	Farah, L. F. X.	Process for the preparation of cellulose film, cellulose film produced thereby, artificial skin graft and its use	1986	BC film preparation, including optimal conditions of *A. xylinum* culturing, and methods of removing the formed film are described; the film appeared to be suitable for use as an artificial skin graft, a separating membrane, or artificial leather
EP 0200409 A3	Ajinomoto Co. Inc., Japan	Iguchi, M., Mitsuhashi, S., Ichimura, K., Nishi, Y., Uryu, M., Yamanake, S., Watanabe, K.	Molded material comprising bacteria-produced cellulose	1986	BC is an excellent component of molded materials having high dynamic strength as compared to conventional molded materials

with industrial wastes, rich in proper carbon sources, such as spent liquors from crystalline glucose production, etc. Simultaneously, a significant environmental benefit would be obtained. However, the major advantage of mass production of BC would be the protection of forests, which are presently disappearing at an alarming rate, thus leading to soil eutrophication and global climate changes.

BC has already been put to numerous uses, presented in this review. The polysaccharide can not only be replaced by some animal polymers (collagen), but also carriers of substances having a positive impact on human health (e.g. antioxidants and prebiotics). The usefulness of the bacterial polymer in medicine (wound, burn and ulcer dressing materials, component of implants) is not longer questioned. Moreover, cellulose granulates can be an excellent matrix for the immobilization of medicinal preparations. For example, if specific substances (receptors) are adsorbed on BC, the resulting molecules can be scavengers of either toxins or of the pathogenic microflora inhabiting the alimentary tract. Recently, unique nanocrystals ($30 \times 600-800$ nm) of BC have been obtained, derived from its commercial preparation Prima Cel™ (Xylos). Selective modification (trimethyl silylation) of the surface of these nanocrystals while leaving their core intact has been achieved. Such modified crystals have great potential in several advanced technologies.

Deciphering of all the riddles of cellulose biosynthesis will lead to improvement and tailoring of BC supramolecular structure and properties; as a consequence, novel concepts for both its inexpensive production, and for its bulk and specialty applications will be developed.

14
References

Aloni, Y., Benziman, M. (1982) Intermediates of cellulose synthesis in *Acetobacter*, in: *Cellulose and Other Natural Polymers System* (Brown, R. M., Jr., Ed.), New York: Plenum Press, 341–361.

Aloni, Y., Cohen, Y., Benziman, M., Delmer, D. P. (1983) Solubilization of UDP-glucose: 1,4-β-glucan 4-β-D-glucosyl transferase (cellulose synthase) from *Acetobacter xylinum*, *J. Biol. Chem.* **258**, 4419–4423.

Ang, J. F., Miller, W. B. (1991) Multiple functions of powdered cellulose as a food ingredient, *J. Am. Ass. Cereal Chem.* **36**, 558–564.

Bartlett, D. H., Silverman, M. (1989) Nucleotide sequences of IS492, a novel insertion sequence causing variation of extracellular production in the marine bacterium *Pseudomonas atlantica*, *J. Bacteriol.* **171**, 1763–66.

Bauer, K., Codolington, K., Ben-Bassat, A. (1992) Methods for improving production of bacterial cellulose, *Abstr. Paper Am. Chem. Soc.*, **203** Meet., Pt 1, Biot. 94.

Bayer, M. E. (1979) The fusion sites between outer membrane and cytoplasmic membrane in bacteria: their role in membrane assembly and virus infection, in: *Bacterial Outer Membranes* (Inouye, M., Ed.), New York: John Wiley & Sons, 167–202.

Ben-Bassat, A., Bruner, R., Wong, H., Shoemaker, S., Aloni, Y. (1989) Production of bacterial cellulose by *Acetobacter*, *Abstr. Pap. Am. Chem. Soc.*, **198** Meet., MBTD20.

Ben-Bassat, A., Calhoon, R. D., Fear, A. L., Gelfand, D. H., Mead, J. H., Tal, R., Wong, H., Benziman, M. (1993) Methods and nucleic acid sequences for expression of the cellulose synthase operon, US patent 5 268 274.

Ben-Hayyim, G., Ohad, I. (1965) Synthesis of cellulose by *Acetobacter xylinum*; VIII. On the formation and orientation of bacterial cellulose fibrils in the presence of acidic polysaccharides, *J. Cell Biol.* **25**, 191–207.

Benziman, M., Tal, R. (1995) Cellulose synthase associated proteins. US patent 5 382 656.

Boisset, C., Chanzy, H., Henrissat, B., Lamed, R., Shoham, Y., Bayer, E. A. (1999) Digestion of crystalline cellulose substrates by the *Clostridium thermocellum* cellulosome: structural and morphological aspects, *Biochem. J.* **340**, 829–835.

Boisset, C., Fraschini, C., Schulein, M., Henrissat, B., Chanzy, H. (2000) Imaging the enzymatic digestion of bacterial cellulose ribbons reveals the endo character of the cellobiohydrolase Cel6A from *Humicola insolens* and its mode of synergy with cellobiohydrolase Cel7A, *Appl. Environ. Microbiol.* **66**, 1444–1452.

Brown, A. J. (1886) An acetic ferment which forms cellulose, *J. Chem. Soc.* **49**, 432–439.

Brown R. M., Jr. (1989a) Microbial cellulose composites and processes for producing same. WO 89/11783.

Brown, R. M., Jr. (1989b) Microbial cellulose as a building block resource for specialty products and processes thereof, WO 89/12107.

Brown, R. M., Jr. (1989c) Use of cellulase preparations in the cultivation and use of cellulose-producing microorganisms, European patent 0258038A3.

Brown, R. M., Jr. (1996) The biosynthesis of cellulose, *Pure Appl. Chem.* **A33**, 1345–1373.

Brown, R. M., Jr., Saxena, I. M. (2000) Cellulose biosynthesis: a model for understanding the assembly of biopolymers, *Plant Physiol. Biochem.* **38**, 57–60.

Brown, R. M., Jr., Willison, J. H. M., Richardson, C. L. (1976) Cellulose biosynthesis in *Acetobacter xylinum*: visualisation of the site of synthesis and direct measurement of the *in vivo* process, *Proc. Natl. Acad. Sci. USA* **73**, 4565–4569.

Bungay, H. R., Serafica, G. (1997) Production of microbial cellulose using a rotating disc film bioreactor, WO 97/05271.

Bureau, T. E., Brown, R. M., Jr. (1987) *In vitro* synthesis of cellulose II from a cytoplasmic membrane fraction of *Acetobacter xylinum*, *Proc. Natl. Acad. Sci. USA* **84**, 6985–6989.

Canale-Parda, E., Wolfe, R. S. (1964) Synthesis of cellulose by *Sarcina ventriculi*, *Biochim. Biophys. Acta* **82**, 403–405.

Cannon, R. E., Anderson, S. M. (1991) Biogenesis of bacterial cellulose, *Crit. Rev. Microbiol.* **17**, 435–439.

Chao, Y., Ishida, T., Sugano, Y., Shoda, M. (2000) Bacterial cellulose production by *Acetobacter xylinum* in a 50 L internal-loop airlift reactor, *Biotechnol. Bioeng.* **68**, 345–352.

Colvin, J. R. (1957) Formation of cellulose microfibrils in a homogenate of *Acetobacter xylinum*, *Arch. Biochem. Biophys.* **70**, 294–295.

Colvin, J. R. (1980) The biosynthesis of cellulose, in: *Plant Biochemistry* (Priess, J., Ed.), New York: Academic Press, 543–570, Vol. 3

Colvin, J. R., Leppard, G. G. (1977) The biosynthesis of cellulose by *Acetobacter xylinum* and *Acetobacter acetigenus*, *Can. J. Microbiol.* **23**, 701–709.

Colvin, J. R., Sowden, L. C., Daoust, V., Perry, M. (1979) Additional properties of a soluble polymer of glucose from cultures of *Acetobacter xylinum*, *Can. J. Biochem.* **57**, 1284–1288.

Costeron, J. W. (1999) The role of bacterial exopolysaccharides in nature and disease, *J. Ind. Microbiol. Biotechnol.* **22**, 551–563.

Coucheron, D. H. (1991) An *Acetobacter xylinum* insertion sequence element associated with inactivation of cellulose production, *J. Bacteriol.* **173**, 5723–2731.

Dekker, R. F. H., Rietschel, E. T., Sandermann, H. (1977) Isolation of α-glucan and lipopolysaccharide fractions from *Acetobacter xylinum*, *Arch. Microbiol.* **115**, 353–357.

De Lannino, N., Cuoso, R. O., Dankert, M. A. (1988) Lipid-linked intermediates and the synthesis of acetan in *A. xylinum*, *J. Gen. Microbiol.* **134**, 1731–1736.

Delmer, D. P. (1982) Biosynthesis of cellulose, *Adv. Carbohydr. Chem. Biochem.* **41**, 105–153.

De Wulf, P., Joris, K., Vandamme, E. J. (1996) Improved cellulose formation by an *Acetobacter xylinum* mutant limited in (keto)gluconate synthesis, *J. Chem. Tech. Biotechnol.* **67**, 376–380.

Easson, D. D., Jr., Sinskey, A. J., Peoples, O. P. (1987) Isolation of *Zoogloea ramigera* I-16 M exopolysaccharides biosynthetic genes and evidence for instability within this region, *J. Bacteriol.* **169**, 4518–4524.

Embuscado, M. E., Marks, J. S., Miller, J. N. (1994) Bacterial cellulose. II. Optimization of cellulose production by *Acetobacter xylinum* through response surface methodology, *Food Hydrocolloids* **8**, 419–430.

Galas, D. J., Chandler, M. (1989) Bacterial insertion sequences, in: *Mobile DNA* (Berg, D. E., Howe, M. M., Eds.) Washington, DC: ASM, 109–162.

Galas, E., Krystynowicz, A., Tarabasz-Szymanska, L., Pankiewicz, T., Rzyska, M. (1999) Optimization of the production of bacterial cellulose using multivariable linear regression analysis, *Acta Biotechnol.* **19**, 251–260.

Glaser, L. (1958) The synthesis of cellulose in cell-free extracts of *Acetobacter xylinum*, *J. Biol. Chem.* **232**, 627–636.

Haigler, C. H. (1985) The function and biogenesis of native cellulose, in: *Cellulose Chemistry and its Applications* (Nevel, R. P., Zeronian, S. H., Eds.), Chichester: Ellis Horwood, 30–83.

Han, N. S., Robyt, J. F. (1998) The mechanism of *Acetobacter xylinum* cellulose biosynthesis: direction of chain elongation and the role of lipid pyrophosphate intermediates in the cell membrane, *Carbohydr. Res.* **313**, 125–133.

Hestrin, S., Aschner, M., Mager J. (1947) Synthesis of cellulose by resting cells of *Acetobacter xylinum*, *Nature* **159**, 64–65.

Hestrin, S., Schramm, M. (1954) Synthesis of cellulose by *Acetobacter xylinum*, II. Preparation of freeze-dried cells capable of polymerizing glucose to cellulose, *Biochem. J.* **58**, 345–352.

Hikawu, S., Hiroshi, T., Takayasu, T., Yoshinaga, F. (1996) Manufacture of bacterial cellulose by addition of cellulose formation stimulators, Japanese patent 96316922.

Hotte, B., Roth-Arnold, I., Puhler, A., Simon, R. (1990) Cloning and analysis of a 35.3 kb DNA region involved in exopolysaccharide production by *Xanthomonas campestris*, *J. Bacteriol.* **172**, 2804–2807.

Husemann, E., Muller, G. J. M. (1966) Synthesis of unbranched polysaccharides, *Macromol. Chem.* **91**, 212–230.

Ielpi, L., Couso, R., Dankert, M. A. (1981) Lipid-linked intermediates in the biosynthesis of xanthan gum, *FEBS Lett.* **130**, 253–256.

Iguchi, M., Yamanaka, S., Budhioko, A. (2000) Bacterial cellulose – a masterpiece of nature's arts, *J. Mater. Sci.* **35**, 261–270.

Ishikawa, A., Matsuoka, M., Tsuchida, T., Yoshinaga, F. (1995) Increasing of bacterial cellulose

production by sulfoguanidine-resistant mutants derived from *Acetobacter xylinum* subsp. sucrofermentans BPR2001, *Biosci. Biotechnol. Biochem.* **59**, 2259–2263.
Ishikawa, A., Tsuchida, T., Yoshinaga, F. (1996a) Production of bacterial cellulose using microbial strain resistant to inhibitor of DHO-dehydrogenase, Japanese patent 08009965A.
Ishikawa, A., Tsuchida, T., Yoshinaga, F. (1996b) Production of bacterial cellulose with pyrimidine analogue-resistant strain, Japanese patent 08000260.
Johnson, D. C., Neogi, A. N. (1989) Sheeted products formed from reticulated microbial cellulose, US patent 4 863 565.
Jonas, R., Farah, L. F. (1998) Production and application of microbial cellulose, *Polym. Degrad. Stabil.* **59**, 101–106.
Jones, D., Gordon, A. H., Bacon, J. S. D. (1974) β-1,3-Glucanases from *Sclerotium rolfsii*, *Biochem. J.* **140**, 47–55.
Joris, K., Billiet, F., Drieghe, S., Brachx, D., Vandamme, E. (1990) Microbial production of β-1,4-glucan, *Meded. Fac. Landbouwwet Rijksuniv. Gent* **55**, 1563–1566.
Kenji, K., Yukiko, M., Hidehi, L., Kunihiko, O. (1990) Effect of culture conditions of acetic acid bacteria on cellulose biosynthesis, *Br. Polym. J.* **22**, 167–171.
Kent, R. A., Stephens, R. S., Westland, J. A. (1991) Bacterial cellulose fiber provides an alternative for thickening and coating, *Food Technol.* **45**, 108.
Kim, D., Kim, Y. M., Park, D. H. (1999) Modification of *Acetobacter xylinum* bacterial cellulose using dextransucrase and alternansucrase, *J. Microbiol. Biotechnol.* **9**, 704–708.
Kobayashi, S., Kashiwa, K., Kawasaki, T., Shoda, S. (1991) Novel method for polysaccharide synthesis using an enzyme: the first *in vitro* synthesis of cellulose via nonbiosynthetic path utilising cellulase as catalyst, *J. Am. Chem. Soc.* **113**, 3079–3084.
Kouda, T., Naritomi, T., Yano, H., Yoshinaga, F. (1998) Inhibitory effect of carbon dioxide on bacterial cellulose production by *Acetobacter* in agitated culture, *J. Ferment. Bioeng.* **85**, 318–321.
Koyama, M., Helbert, W., Imai, T., Sugiyama, J., Henrissat, B. (1997) Parallel-up structure evidence the molecular directionality during biosynthesis of bacterial cellulose, *Proc. Natl. Acad. Sci. USA* **94**, 9091–9095.
Krystynowicz, A., Turkiewicz, M., Drynska, E., Galas, E. (1995) Bacterial cellulose – biosynthesis and application, *Biotechnologia* **30**, 120–132.
Krystynowicz, A., Galas, E., Pawlak, E. (1997) Method of bacterial cellulose production. Polish patent P-299907.
Krystynowicz, A., Czaja, W., Bielecki, S. (1999) Biosynthesis and application of bacterial cellulose, *Zywnosc* **3**, 22–33.
Krystynowicz, A., Czaja, W., Pomorski, L., Kolodziejczyk, M., Bielecki, S. (2000) The evaluation of usefulness of microbial cellulose as a wound dressing material, 14th Forum for Applied Biotechnology, Gent, Belgium, *Meded. Fac. Landbouwwet Rijksuniv. Gent*, Proceedings Part I, 213–220.
Kudlicka K. (1989) Terminal complexes in cellulose synthesis, *Postepy biologii komórki* **16**, 197–212 (abstract in English).
Kuga, G., Brown, R. M., Jr. (1988) Silver labeling of the reducing ends of bacterial cellulose, *Carbohydr. Res.* **180**, 345–350.
Laboureur, P. (1988) Process for producing bacterial cellulose from material of plant origin, WO 88/09381.
Legge R. L. (1990) Microbial cellulose as a specialty chemical, *Biotechnol. Adv.* **8**, 303–319.
Leppard G. G., Sowden, L. C., Ross, C. J. (1975) Nascent stage of cellulose biosynthesis, *Science* **189**, 1094–1095.
Lin, F. C., Brown, R. M., Jr. (1989) Purification of cellulose synthase from *Acetobacter xylinum*, in: *Cellulose and Wood Chemistry and Technology* (Schmerck, C., Ed.), New York: John Wiley & Sons, 473–492.
Lin, F. C., Brown, R. M., Jr., Cooper, J. B., Delmer, D. P. (1985) Synthesis of fibrils *in vitro* by a solubilized cellulose synthase from *Acetobacter xylinum*, *Science* **230**, 822–825.
Lin, F. C., Brown, R. M., Jr., Drake, R. R., Jr., Haley. B. E. (1990) Identification of the uridine-5′-diphosphoglucose (UDPGlc) binding subunit of cellulose synthase in *Acetobacter xylinum* using the photoaffinity probe 5-azido-UDPGlc, *J. Biol. Chem.* **265**, 4782–4784.
Marx-Figini, M., Pion, B. G. (1974) Kinetic investigations on biosynthesis of cellulose by *Acetobacter xylinum*, *Biochim. Biophys. Acta* **338**, 382–393.
Masaoka, S., Ohe, T., Sakota, N. (1993) Production of cellulose from glucose by *Acetobacter xylinum*, *J. Ferment. Bioeng.* **75**, 18–22.
Matsuoka, M., Tsuchida, T., Matsuchita, K., Adachi, O., Yoshinaga, F. (1996) A synthetic medium for bacterial cellulose production by *Acetobacter xylinum* subsp. sucrofermentans, *Biosci. Biotechnol. Biochem.* **60**, 575–579.

Matthyse, A., Thomas, D. I., White, A. R. (1995) Mechanism of cellulose synthesis in *Agrobacterium tumefaciens*, *J. Bacteriol.* **177**, 1076–1081.

Micheel, F., Brodde, O. E. (1974) Polymerization of 1,4-anhydro-2,3,6 tri-*O*-benzyl-α-D-glucopyranose, *Liebigs Ann. Chem.* **124**, 702–708.

Minakami, H., Entani, K., Tayama, S., Fujiyama, S., Masai, H. (1984) Isolation and characterization of a new polysaccharide-producing *Acetobacter* sp. *Agric. Biol. Chem.* **48**, 2405–2414.

Mondal, I. H., Kai, A. (2000) Control of the crystal structure of microbial cellulose during nascent stage, *J. Appl. Polym. Sci.***79**, 1726–1734.

Nakai, T., Tonouchi, N., Konishi, T., Kojima, Y., Tsuchida, T., Yoshinaga, F., Sakai, F., Hayashi, T. (1999) Enhancement of cellulose production by expression of sucrose synthase in *Acetobacter xylinum, Proc. Natl. Acad. Sci. USA* **96**, 14–18.

Nakatsubo, F., Takano, T., Kawada, T., Murakami, K. (1989) Toward the synthesis of cellulose: synthesis of cellooligosaccharides, in: *Cellulose: Structural and Functional Aspects* (Kennedy, J. F., Philips, G. O., Williams, P. A., Eds.), New York: Ellis Horwood, 201–206.

Nari, J., Noat, G., Richard, J., Franchini, E., Monstacas, A. M. (1983) Catalytic properties and tentative function of a cell wall β-glucosyltransferase from soybean cells cultured *in vitro, Plant Sci. Lett.* **28**, 313–320.

Naritomi, T., Kouda, T., Yano, H., Yoshinaga, F. (1998) Effect of ethanol on bacterial cellulose production from fructose in continuous culture, *J. Ferment Bioeng.* **85**, 598–603.

Nichols, S. E., Singletary, G. W. (1998) Cellulose synthesis in the storage tissue of transgenic plants, US patent 5 723 764.

Ohana, P., Delmer, D. P., Volman, G., Benziman, M. (1998) Glycosylated triterpenoid saponin: a specific inhibitor of diguanylate cyclase from *Acetobacter xylinum, Plant Cell Physiol.* **39**, 153–159.

Oikawa, T., Kamatani, N., Kaimura, T., Ameyama, M., Soda, K. (1997) Endo-β-glucanase from *A. xylinum* – purification and characterization, *Curr. Microbiol.* **34**, 309–313.

Okamoto, T., Yamano, S., Ikeaga, H., Nakamura, K. (1994) Cloning of the *A. xylinum* cellulase gene and its expression in *E. coli* and *Zymomonas mobilis, Appl. Microbiol. Biotechnol.* **42**, 563–568.

Okiyama, A., Shirae, H., Kano, H., Yamanaka, S. (1992) Two-stage fermentation process for cellulose production by *Acetobacter aceti, Food Hydrocolloids* **6**, 471–477.

Palonen, H., Tenkanen, M., Linder, M. (1999) Dynamic interaction of *Trichoderma reesei* cellobiohydrolases Cel6A and Cel7A and cellulose at equilibrium and during hydrolysis, *Appl. Environ. Microbiol.* **65**, 5229–5233.

Ring, D. F., Nashed, W., Dow, T. (1986) Liquid loaded pad for medical applications, US patent 4 588 400.

Ross, P., Weinhouse, H., Aloni, Y., Michaeli, D., Ohana, P., Mayer, R., Braun, S., de Vroom, E., van der Marel, G. A., van Boom, J. H., Benziman, M. (1987) Regulation of cellulose synthesis in *Acetobacter xylinum* by cyclic diguanylic acid, *Nature* **325**, 279–281.

Ross, P., Mayer, R., Benziman, M. (1991) Cellulose biosynthesis and function in bacteria, *Microbiol. Rev.* **55**, 35–58.

Sakair, N., Asamo, H., Ogawa, M., Nishi, N., Tokura, S. (1998) A method for direct harvest of bacterial cellulose filaments during continuous cultivation of *Acetobacter xylinum, Carbohydr. Polym.* **35**, 233–237.

Samejima, M., Sugiyama, J., Igarashi, K., Eriksson, K. E. L. (1997) Enzymatic hydrolysis of bacterial cellulose, *Carbohydr. Res.* **305**, 281–288.

Sattler, K., Fiedler, S. (1990) Production and application of bacterial cellulose. II. Cultivation in a rotating drum fermentor, *Zbl. Microbiol.* **145**, 247–252.

Savidge, R. A., Colvin, J. R. (1985) Production of cellulose and soluble polysaccharides by *Acetobacter xylinum, Can. J. Microbiol.* **31**, 1019–1025.

Saxena, I. M., Brown, R. M., Jr. (1989) Cellulose biosynthesis in *Acetobacter xylinum*: a genetic approach, in: *Cellulose and Wood Chemistry and Technology* (Schnerck, C., Ed.). New York: John Wiley & Sons, 537–557.

Saxena, I. M., Brown, R. M., Jr. (1997) Identification of cellulose synthase(s) in higher plants: Sequence analysis of processive β-glycosyltransferases with common motif 'D, D, D 35Q (RQ)XRW', *Cellulose* **4**, 33–49.

Saxena, I. M., Brown, R. M., Jr. (2000) Cellulose synthases and related proteins, *Curr. Opin. Plant Biol.* **3**, 523–531.

Saxena, I. M., Lin, F. C., Brown, R. M., Jr. (1990a) Cloning and sequencing of the cellulase synthase catalytic subunit gene of *Acetobacter xylinum, Plant Mol. Biol.* **15**, 673–683.

Saxena, I. M., Roberts, E. M., Brown, R. M., Jr. (1990b) Modification of cellulose normally synthesised by cellulose-producing microorganisms, US patent 4 950 597.

Saxena, I. M., Lin, F. C., Brown, R. M., Jr. (1991) Identification of a new gene in an operon for

cellulose biosynthesis in *Acetobacter xylinum*, *Plant Mol. Biol.* **16**, 947–954.
Saxena, I. M., Brown, R. M., Jr., Fevre, M., Geremia, R. A., Henrissat, B. (1995) Multidomain architecture of β-glycosyl transferase: Implications for mechanism of action, *J. Bacteriol.* **177**, 1419–1424.
Seto, H., Tsuchida, T., Yoshinaga, F., Beppu, T., Horinouchi, S. (1996) Characterization of the biosynthetic pathway of cellulose from glucose and fructose in *Acetobacter xylinum*, *Biosci. Biotechnol. Biochem.* **60**, 1377–1379.
Seto, H., Tsuchida, T., Yoshinaga, F. (1997) Manufacture of bacterial cellulose by mixed culture of microorganisms, Japanese patent 9721905.
Srisodsuk, M., Kleman-Leyer, K., Keranen, S., Kirk, T. K., Teeri, T. T. (1998) Modes of action on cotton and bacterial cellulose of a homologous endoglucanase-exoglucanase pair from *Trichoderma reesei*, *Eur. J. Biochem.* **62**, 185–187.
Stalbrandt, H., Mansfield, S. D., Saddler, J. N., Kilburn, D. G., Warren, R. A. J., Gilkes, N. R. (1998) Analysis of molecular size of cellulose by recombinant *Cellulomonas fimi* β-1,4-glucanase, *Appl. Environ. Microbiol.* **64**, 2374–2379.
Standal, R., Iversen, T. G., Coucheron, D. H., Fjaervik, E., Blatny, J. M., Valla, S. (1994) A new gene required for cellulose production and gene encoding cellulolytic activity in *A. xylinum* are localized with *bcs* operon, *J. Bacteriol.* **176**, 665–672.
Sutherland, I. W. (1998) Novel and established applications of microbial polysaccharides, *TIBTECH* **16**, 41–46.
Swissa, M., Aloni, Y., Weinhouse, H., Benziman, M. (1980) Intermediary steps in cellulose synthesis in *Acetobacter xylinum*: studies with whole cells and cell-free preparation of the wild type and a cellulose-less mutant, *J. Bacteriol.* **143**, 1142–1150.
Tahara, N., Tabuchi, M., Watanabe, K., Yano, H., Morinaga, Y., Yoshinaga, F. (1997) Degree of polymerisation of cellulose from *A. xylinum* BPR 2001 decreased by cellulase producing strain, *Biosci. Biotechnol. Biochem.* **61**, 1862–1865.
Tahara, N., Tonouchi, M., Yano, H., Yoshinaga, F. (1998) Purification and characterization of exo-1,4-β-glucosidase from *A. xylinum* BPR 2001, *J. Ferment. Bioeng.* **85**, 589–594.
Takai, M., Tsuta, Y., Watanabe, S. (1997) Biosynthesis of cellulose by *Acetobacter xylinum* and characterization of bacterial cellulose, *Polym. J.* **7**, 137–146.
Tal, R., Wong, H. C., Calhon, R., Gelfand, D., Fear, A. L., Volman, G., Mayer, R., Ross, P., Amikam, D., Weinhouse, H., Cohen, A., Sapir, S., Ohana, P., Benziman, M. (1998) Three *cdg* operons control cellular turnover of c-di-GMP in *Acetobacter xylinum*: genetic organization and occurrence of conserved domain in isozymes, *J. Bacteriol.* **180**, 4416–4425.
Tayama, K., Minakami, H., Entani, E., Fujiyama, S., Masai, H. (1985) Structure of an acidic polysaccharide from *Acetobacter*, sp. NBI 1022, *Agric. Biol. Chem.* **49**, 959–966.
Teeri, T. T. (1997) Crystalline cellulose degradation: new insight into the function of cellobiohydrolases, *TIBTECH* **15**, 160–167.
Tonouchi, N., Tsuchida, T., Yoshinaga, F., Beppu, T. (1996) Characterization of the biosynthetic pathway of cellulose from glucose and fructose in *Acetobacter xylinum*, *Biosci. Biotechnol. Biochem.* **60**, 1377–1379.
Tonouchi, N., Tahara, N., Kojima, Y., Nakai, T., Sakai, F., Hayashi, T., Tsuchida, T., Yoshinaga, F. (1997) A β-glucosidase gene downstream of the cellulase synthase operon in cellulase producing *Acetobacter*, *Biosci. Biotechnol. Biochem.* **61**, 1789–1790.
Tonouchi, N., Hirinouchi, S., Tsuchida, T., Yoshinaga, F. (1998) Increased cellulose production by *Acetobacter* after introducing the sucrose phosphorylase gene, *Biosci. Biotechnol. Biochem.* **62**, 1778–1780.
Toyosaki, H., Naritomi, T., Seto, A., Matsuoka, M., Tsuchida, T., Yoshinga, F. (1995) Screening of bacterial cellulose- producing *Acetobacter* strains suitable for agitated culture, *Biosci. Biotechnol. Biochem.* **59**, 1498–1502.
Valla, S., Kjosbakken, J. (1981) Isolation and characterization of a new extracellular polysaccharide from a cellulose-negative strain of *Acetobacter xylinum*, *Can. J. Microbiol.* **27**, 599–603.
Valla, S., Coucheron, D. H., Fjaervik, E., Kjosbakken, J., Weinhose, H., Ross, P., Amikam, D., Benziman, M. (1989) Cloning of a gene involved in cellulose biosynthesis in *Acetobacter xylinum*: complementation of cellulose-negative mutant by the UDPG pyrophosphorylase structure gene, *Mol. Gen. Genet.* **217**, 26–30.
Vandamme, E. J., De Baets, S., Vanbaelen, A., Joris, K., De Wulf P. (1998) Improved production of bacterial cellulose and its application potential, *J. Polymer Degrad. Stabil.* **59**, 93–99.
Watanabe, K., Tabuchi, M., Morinaga, Y., Yoshinaga, F. (1998a) Structural features and properties of bacterial cellulose produced in agitated culture, *Cellulose* **5**, 187–200.
Watanabe, K., Tabuchi, M., Ischikawa, M., Takemura, H., Tsuchida, T., Morinaga, Y. (1998b)

Acetobacter xylinum mutant with high cellulose productivity and ordered structure, *Biosci. Biotechnol. Biochem.* **62**, 1290–1292.

Weinhouse, R. (1977) Regulation of carbohydrate metabolism in *Acetobacter xylinum*, PhD thesis, Hebrew University of Jerusalem. Jerusalem, Israel.

Weinhouse, H., Sapir, S., Amikam, D., Shiko, Y., Volman, G., Ohana, P., Benziman, M. (1997) c-di-GMP-binding protein, a new factor regulating cellulose synthesis in *Acetobacter xylinum, FEBS Lett.* **416**, 207–211.

Wiliams, W. S., Cannon, R. E. (1989) Alternative environmental roles for cellulose produced by *Acetobacter xylinum, Appl. Environ. Microbiol.* **55**, 2448–2452.

Winkelman, J. W., Clark, D. P. (1984) Proton suicide: general method for direct selection of sugar transport and fermentation-defective mutants, *J. Bacteriol.* **11**, 687–690.

Wlochowicz, A. (2001) Personal communication.

Wong, H. C., Fear, A. L., Calhoon, R. D., Eichinger, G. M., Mayer, R., Amikam, D., Benziman, M., Gelfand, D. H., Meade, J. H., Emerick, A. W., Bruner, R., BenBassat, A., Tal, R. (1990) Genetic organization of cellulose synthase operon in *A. xylinum, Proc. Natl. Acad. Sci. USA* **87**, 8130–8134.

Yamada, Y., Hoshino, K., Ishikawa, T. (1997) The phylogeny of acetic acid bacteria based on the partial sequences of 16 S ribosomal RNA: the elevation of the subgenus *Gluconacetobacter* to the generic level, *Biosci. Biotechnol. Biochem.* **61**, 1244–51.

Yamada, Y., Hoshino, K., Ishikawa, T. (1998) *Gluconacetobacter corrig.* (*Gluconoacetobacter* [sic]), in: Validation of publication of new names and new combinations previously effectively published outside the IJSB, List no 64, *Int. J. Syst. Bacteriol.* **48**, 327–328.

Yamada, Y. (2000) Transfer *Acetobacter oboediens* and *A. intermedius* to the genus *Gluconacetobacter* as *G. oboediens* comb. nov. and *G. intermedius* comb. nov., *Int. J. System. Evolut. Microbiol.* **50**, 2225–2227.

Yamamoto, H., Horii, F., Hirai, A. (1996) *In situ* crystallization of bacterial cellulose. 2. Influence of different polymeric additives on the formation of celluloses I_α and I_β at the early stage of incubation, *Cellulose* **3**, 229–242.

Yamanaka, S., Watanabe, K., Kitamura, N., Iguchi, M., Mitsuhashi, S., Nishi, Y., Uryu, M. (1989) The structure and mechanical properties of sheets prepared from bacterial cellulose, *J. Mater. Sci.* **24**, 3141–3145.

Yamanaka, S., Watanabe, K., Suzuki, Y. (1990) Hollow microbial cellulose, process for preparation thereof, and artificial blood vessel formed of said cellulose, European patent 0396344A2.

Yamanaka, S., Ishihara, M., Sugiyama, J. (2000) Structural modification of bacterial cellulose, *Cellulose* **7**, 213–225.

Zaar, K. (1977) The biogenesis of cellulose by *Acetobacter xylinum, Cytobiologie* **16**, 1–15.

15
Bioemulsans: Surface-active Polysaccharide-containing Complexes

Prof. Dr. Eugene Rosenberg[1], **Prof. Dr. Eliora Z. Ron**[2]

[1] Department of Molecular Microbiology and Biotechnology, Tel Aviv University, Ramat Aviv, Israel 69978; Tel.: +972-3-640 9838; Fax: +972-3-642 9377; E-mail: eros@post.tau.ac.il

[2] Department of Molecular Microbiology and Biotechnology, Tel Aviv University, Ramat Aviv, Israel 69978; Tel.: +972-3-640 9879; Fax: +972-3-641 4138; E-mail: eliora@post.tau.ac.il

Biotechnology of Biopolymers. From Synthesis to Patents. Edited by A. Steinbüchel and Y. Doi

ISBN: 3-527-31110-6

CMC critical micelle concentration
OmpA outer membrane protein A
PAHs polyaromatic hydrocarbons
pgi phosphoglucoisomerase gene
SCP single-cell protein

1 Introduction

In general, microorganisms are specialists, and in any particular ecological niche one microorganism or a limited number of strains will dominate. These microorganisms have adapted through evolution to be able to survive in this niche during the long periods when growth is impossible, but when nutrients do become available they can outgrow their competitors. The rapid growth of microorganisms depends largely on their high surface-to-volume ratio, which allows for the efficient uptake of nutrients and release of waste products. The price that the microorganism pays for the high surface-to-volume ratio is that it is totally exposed. All of the components on the outside of the cell must be able to function under the specific conditions of the ecological niche. It is probably for this reason that the "diversity of the microbial world" is best expressed on the outside of the cell. Because of their high surface-to-volume ratio and the diversity of their exocellular polymers, microorganisms are a rich source of potentially useful polymers. One group of such polymers, referred to as emulsans, is amphipathic polysaccharides and/or proteins that stabilize oil-in-water emulsions.

2 Historical Outline

Research on microbial bioemulsifiers can be traced to pioneering attempts to: (1) solve practical problems in the petroleum industry; and (2) understand how microorganisms grow on water-insoluble hydrocarbons.

2.1 The Petroleum Industry Connection

The two major problems of the petroleum industry that involved surface-active agents (surfactants) were the secondary recovery of oil, and the production of single-cell protein (SCP) from hydrocarbons. One of the earliest studies of the microbial production of surfactants was carried out by La Riviere (1955a,b), who showed that eight different microbial cultures, including *Aspergillus ni-*

ger, Psuedomonas aeruginosa, Candida lipolytica, Desulfovibrio desulfuricans and *Mycobacterium phlei*, lowered surface tension by 14 to 34 mN m^{-1}. Updegraff and Wren (1954) patented a process for the release of oil from petroleum-bearing materials using sulfate-reducing bacteria. However, subsequent studies demonstrated that only small increases (ca. 5%) in oil recovery were obtained with microbial surfactants (Dostalek and Spurny, 1958). The early work on the use of microorganisms for secondary recovery of oil has been reviewed by Davis (1967) and Zajic and Panchal (1976).

During the early 1960s, British Petroleum announced that it intended to manufacture microbial protein from hydrocarbons on a large scale. This stimulated many oil companies to support in-house as well as university research on the optimization of the microbial conversion of n-paraffins to what came to be called SCP. It soon became clear that the growth of microorganisms at high cell density in fermentors on n-paraffins was limited by the low solubility of hydrocarbons in water (Johnson, 1968; Erickson and Nakahara, 1975). For example, the solubilities of decane, dodecane and tetradecane in water at 25 °C are 3.1×10^{-7} M, 1.7×10^{-8} M, and 9.8×10^{-10} M, respectively (Goldberg, 1985). Accordingly, the growth rate of microorganisms on hydrocarbons is limited by the transfer rate of the substrate into solution, and then across the cell membrane. Numerous observations were reported indicating that the dispersion of hydrocarbons in oil increased during fermentation as a result of increased interfacial area between the oil drops and the aqueous culture broth. This led to the discovery that microorganisms, in this case yeast, produce extracellular bioemulsifiers, such as fatty acids and various polymers (Iguchi et al., 1969; Abbot and Gledhill, 1971).

2.2 The Emulsan Story

By means of the enrichment culture technique, a mixed population of microorganisms was obtained which catalyzed the emulsification of crude oil in supplemented seawater (Reisfeld et al., 1972). From this mixed culture, one strain (RAG-1) was isolated that degraded ca. 60% of the oil and efficiently dispersed the nondegraded oil into small droplets (2–5 μm diameter). The initial studies indicated that the material responsible for emulsification was present in the extracellular fluid of cultures grown on crude oil as the carbon and energy source. The emulsifier-producing RAG-1 strain was initially mis-classified as a member of the genus *Arthrobacter*. Subsequently, the strain was identified as an *Acinetobacter* (Pines and Gutnick, 1981) and referred to as *A. calcoaceticus* RAG-1 (ATCC 31012).

RAG-1 was initially isolated with the goal of using it to treat oil pollution in the sea. However, it soon became clear that this goal was impracticable because growth of this bacterium, as with any known microorganism, requires utilizable nitrogen (N) and phosphorus (P) sources in addition to the crude oil. The oceans are "deserts" when it comes to N and P compounds, and crude oil contains very little N and P. The addition of utilizable N and P compounds (e.g., urea, nitrates and phosphates) to an open system such as an ocean is of no use because these water-soluble compounds diffuse away from the oil and are not available to the bacteria adhering to the oil spill. Consequently, Rosenberg et al. (1974) used RAG-1 in a closed system, the cargo compartment of an oil tanker during its ballast voyage, in order to prevent oil pollution resulting from discharge of oily ballast water.

The oil tanker/RAG-1 experiment was performed in January, 1973 on a ballast voyage from Eilat in the Gulf of Aqaba to Khargh Island in the Persian Gulf. One of the slop tanks containing 100 m^3 of oily ballast water over a layer of sludge was supplemented with urea and phosphate. After inoculation with RAG-1, the slop tank was oxygenated with air at 3 $m^3 min^{-1}$. A nonaerated slop tank served as a control. Samples removed from the tanks during the first 3 days of the voyage showed that the number of bacteria in the tank that was supplemented with urea and phosphate, aerated and inoculated with RAG-1, reached over 10^8 bacteria per mL, whereas the nonaerated control tank contained ca. 10^4 bacteria per mL.

Furthermore, the nondegraded oil in the experimental tank was thoroughly emulsified. After 4 days, the tanks were emptied, whereupon the experimental tank was completely free of the thick layer of oily sludge that remained in the control tank. In this experiment, removal of the sludge was made possible by providing conditions that favored bioemulsification rather than optimum cell growth. This microbiological cleaning method has the advantage that cleaning takes place while the tank is full of ballast water, thereby decreasing the danger of explosion. Moreover, all submerged components of the tank are in contact with the cleaning agent.

The potential of using RAG-1 to clean the cargo compartments of oil tankers during their ballast voyage (Gutnick and Rosenberg, 1977) led to an initial patent on the process (U.S. patent No. 3,941,692) and more than 20 subsequent patents throughout the world on RAG-1 and emulsans. The oil tanker experiment reached the attention of Leslie Misrock, a patent attorney and pioneer biotechnology entrepreneur, who obtained the rights from Tel Aviv University and established a company, Petroferm USA, to exploit these discoveries. Between 1975 and 1990, most of the research on emulsans was supported by Petroferm. RAG-1 emulsan was the first of a number of emulsans that are produced by microorganisms. During the past few years, several reviews have been published on different aspects of emulsans: their role in bacterial adhesion (Neu, 1996); growth on hydrocarbons (Rosenberg and Ron, 1996); surface-active polymers from the genus *Acinetobacter* (Rosenberg and Ron, 1998a); production (Wang and Wand, 1990); molecular genetics (Sullivan, 1998); commercial applications (Fiechter, 1992); enhanced oil recovery (Banat, 1995); bioemulsans (Rosenberg and Ron, 1997); high and low molecular mass microbial surfactants (Rosenberg and Ron, 1999); and their natural roles (Ron and Rosenberg, 2001).

3
Occurrence and Chemical Properties

A large number of bacterial species from different genera produce exocellular polymeric surfactants composed of polysaccharides, protein, lipopolysaccharides, lipoproteins, or complex mixtures of these biopolymers (Table 1). Xanthan, although it has some surfactant properties, is primarily a thickener, and as such, will not be discussed in this chapter.

3.1
RAG-1 Emulsan

The best-studied polymeric bioemulsifiers are the emulsans produced by different species of *Acinetobacter* (Rosenberg and Ron, 1998a). RAG-1 emulsan is a complex of an anionic heteropolysaccharide and protein (Rosenberg and Kaplan, 1987). Its surface activity is due to the presence of fatty acids, comprising 15% of the emulsan dry weight, that are attached to the polysaccha-

Tab. 1 Selected microbially produced emulsans

Emulsan	*Producing microorganisms*	*References*
RAG-1 emuslan	*A. calcoaceticus* RAG-1	Rosenberg et al. (1979a,b)
BD4 emulsan	*A. calcoaceticus* BD413	Kaplan and Rosenberg (1982)
Alasan	*A. radioresistens* KA53	Navon-Venezia et al. (1995)
Biodispersan	*A. calcoaceticus* A2	Rosenberg et al. (1988a,b)
Mannan-lipid-protein	*C. tropicalis*	Kaeppeli et al. (1984)
Liposan	*C. lipolytica*	Cirigliano and Carman (1984)
Emulsan 378	*P. fluorescens*	Persson et al. (1988)
Protein complex	*M. thermoautotrophium*	De Acevedo and McInerney (1996)
Insecticide emulsifier	*P. tralucida*	Appaiah and Karanth (1991)
Thermophilic emulsifier	*B. stearothermophilus*	Gunjar et al. (1995)
Acetylheteropolysaccharide	*S. paucimobilis*	Ashtaputre and Shah (1995)
Food emulsifier	*C. utilis*	Shepherd et al. (1995)
Sulfated polysaccharide	*H. eurihalinia*	Calvo et al. (1998)
Undefined complex	*Rhodococcus* sp.	Bredholt and Eimhjellen (1999)

ride backbone via O-ester and N-acyl linkages (Belsky et al., 1979). The chemical and physical properties of RAG-1 emulsan are summarized in Table 2. RAG-1 cells excrete maximum amounts of emulsan in shake flasks when grown in a minimal medium containing 2% ethanol as the sole carbon and energy source. Under these conditions, approximately 80% of the emulsan produced is released when the cells are in stationary phase (Goldman et al., 1982). Emulsan is an effective emulsifier at low concentrations (0.01–0.001%), representing emulsan-to-hydrocarbon ratios of 1:100 to 1:1000, and exhibits considerable substrate specificity (Rosenberg et al., 1979a). RAG-1 emulsan does not emulsify pure aliphatic, aromatic, or cyclic hydrocarbons; however, all mixtures that contain an appropriate mixture of an aliphatic and an aromatic (or cyclic alkane) are emulsified efficiently. Maximum emulsifying activity occurs over a wide pH range and requires the presence of divalent cations (Rosenberg et al., 1979b).

Tab. 2 Chemical and physical properties of RAG-1 emulsan (Data from Rosenberg et al., 1979b)

Measurement	*Result*
Chemical composition	D-galactosamine, 25%
	L-galactosaminouronic acid, 25%
	Dideoxy-diaminohexose, 25%
	3-Hydroxydodecanoic acid, 10%
	2-Hydroxydodecanoic acid, 10%
	Water and ash, 10%
Intrinsic viscosity ($cm^3 g^{-1}$)	550
Diffusion constant ($cm^3 g^{-1}$)	5.3×10^{-8}
Partial molar volume ($cm^3 g^{-1}$)	0.71
Molecular weight (kDa)	980
Dimensions (nm)	3×200

3.2 BD4 Emulsan

A. calcoaceticus BD4, which was initially isolated and characterized by Taylor and Juni (1961), produces a large polysaccharide capsule. Under certain growth conditions (e.g., 2% ethanol as the carbon and energy source), an enhanced release of the capsules was obtained (Sar and Rosenberg, 1983). However, in contrast to RAG-1, no decrease in capsular polysaccharide was observed; rather, capsular polysaccharide levels remained constant when enhanced levels of extracellular polysaccharide were obtained. When the crude capsular material was applied to a Sepharose 4B column, it eluted in a single peak containing a polysaccharide–protein complex. The polysaccharide component was obtained by deproteinization and its chemical structure elucidated (Figure 1). The protein component was obtained from the extracellular supernatant fluid of strain BD4-R7, a capsule-negative mutant of BD4 that produces no extracellular polysaccharide. The polysaccharide and protein components had no emulsifying activity by themselves. However, mixing the protein and polysaccharide fractions led to a reconstitution of the emulsifying activity (Kaplan et al., 1987). Apparently, the protein (which is hydrophobic) binds to the hydrocarbon initially in a reversible fashion. The polysaccharide then attaches to the protein and stabilizes the oil-in-water emulsion.

3.3 Alasan

Alasan, produced by a strain of *Acinetobacter radioresistens*, is a complex of an anionic polysaccharide and protein with a molecular weight of approximately 10^3 kDa (Navon-Venezia et al., 1995). However, the polysaccharide component of alasan is unusual in that it contains covalently bound alanine. The protein component of alasan appears to play an important role in both the structure and activity of the complex (Navon-Venezia et al., 1998). Deproteinization of alasan with hot phenol or treatment with specific proteinases caused a loss in most of the emulsifying activity. When a solution of alasan was exposed to increasing temperatures, there were large changes in the viscosity and emulsifying activity of the complex (Figure 2). Between 30 and 50 °C, the viscosity increased 2.6-fold, but with no change in activity. However, between 50 and 90 °C the viscosity decreased 4.8-fold while the activity increased 5-fold. None of these changes occurred with the protein-free polysaccharide, indicating that they were due to the interactions of the protein and polysaccharide portions of the complex. Alasan lowers interfacial tension from 69 to 41 dynes cm^{-1} at 20 °C and has a critical micelle concentration (CMC) of 200 μg/ml (Barkay et al., 1999). Although alasan, as well as the other emulsans, is not as effective as some of the lower molecular-weight bioemulsifiers at lowering interfacial tension, the emulsans are extremely effective at stabilizing oil-in-water emulsions.

```
→3)-a-L-Rha-(1 → 3)-α-D-Man-(1 → 3)-α-L-Rha-(1 → 3)-α-L-Rha-(1 → 3)-β-D-Glc-(1→
                          2
                          ↑
                          1
α-L-Rha-(1 →4)-β-D-GlcUA
```

Fig. 1 The chemical structure of the *Acinetobacter calcoaceticus* BD4 capsular polysaccharide (Kaplan et al., 1985).

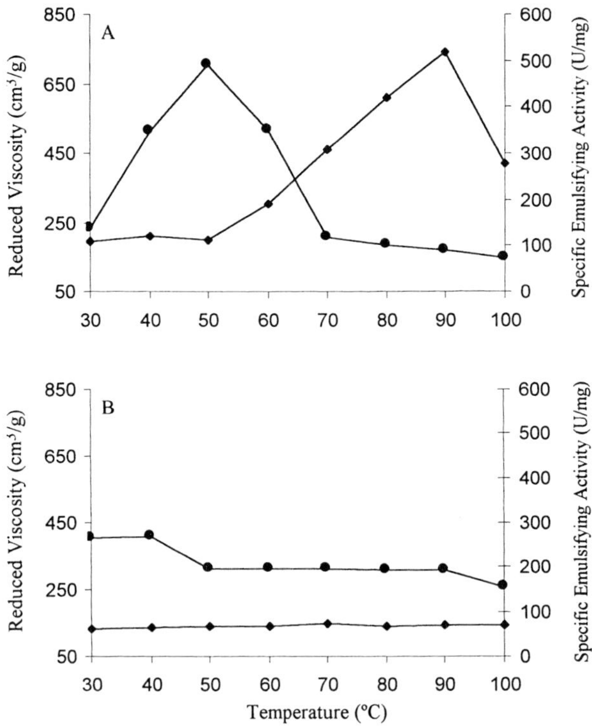

Fig. 2 Effect of temperature on viscosity and emulsifying activity of alasan (A) and apo-alasan (B). Solutions of alasan and apo-alasan (2 mg mL^{-1} in 20 mM Tris–HCl buffer, pH 7.0) were placed in a thermoblock and incubated for 10 min at 30 °C. After treatment, samples were taken for measuring emulsifying activity (squares) and reduced viscosity (circles). The solutions were then heated at the next highest temperature for 10 min. The procedure was repeated up to 100 °C. All measurements of viscosity and emulsifying activity were carried out at 30 °C.

Recently, alasan has been fractionated into three proteins (of 16, 31 and 45 kDa) and the alasan polysaccharide (Toren et al., 2001). Each of the three alasan proteins showed emulsifying activity: the 45-kDa protein had the highest specific activity, 11% higher than the intact alasan complex. The N-terminal amino acid sequence of the 45-kDa protein showed high similarity to the OmpA protein of several Gram-negative bacteria. The function of the alasan polysaccharide is not clear, but it may play a role in the release of the proteins into the medium and in protecting the protein complex against proteolytic activities. It was interesting to note that the purified 45-kDa protein was readily hydrolyzed by trypsin, whereas the protein that was bound to the polysaccharide was resistant.

3.4 Biodispersan

A. calcoaceticus A2 produces an extracellular anionic polysaccharide surfactant of molecular mass 51,400 Daltons that effectively disperses limestone and titanium dioxide (Rosenberg et al., 1988a,b). The biopolymer, referred to as biodispersan, binds to powdered calcium carbonate and changes its surface properties in a way that allows for better dispersion in water. In addition to being a surfactant, biodispersan acts also as a surfactant, aiding in the fracturing of limestone during the grinding process (Rosenberg et al., 1989).

3.5 Other High Molecular-mass Bioemulsifiers

Sar and Rosenberg (1983) reported that the majority of *Acinetobacter* strains produce extracellular nondialyzable emulsifiers. These strains included both soil and hospital isolates. Marin et al. (1996) have reported the isolation of a strain of *A. calcoaceticus* from contaminated heating oil which emulsifies that substrate. Neufeld and Zajic (1984) demonstrated that whole cells of *A. calcoaceticus* 2CA2 have the ability to act as emulsifiers, in addition to producing an extracellular emulsifier.

A large number of high molecular-weight complex bioemulsifiers have been reported, though in general little is known about these other than the producing organism and the overall chemical composition of the crude mixture. An alkane-oxidizing *Rhodococcus* sp. produces a high molecular-weight complex that when released into the medium stabilizes oil-in-water emulsions (Bredholt and Eimhjellen, 1999). *Halomonas eurihalina* produces an extracellular sulfated heteropolysaccharide (Calvo et al., 1998), while *Pseudomonas tralucida* produced an extracellular acetylated polysaccharide that was effective in emulsifying several insecticides (Appaiah and Karanth, 1991). Several recently reported bioemulsifiers are effective at high temperature, including the protein complex from *Methanobacterium thermoautotrophium* (De Acevedo and McInerney, 1996) and the protein–polysaccharide–lipid complex of *Bacillus stearothermophilus* ATCC 12980 (Gunjar et al., 1995).

Yeasts produce a number of emulsifiers, which is particularly interesting because of the food-grade status of several yeasts. Liposan is an extracellular emulsifier produced by *Candida lipolytica* (Cirigliano and Carman, 1985) and is composed of 83% carbohydrate and 17% protein. Mannanprotein emulsifiers are produced by *Saccharomyces cerevisiae* (Cameron et al., 1988), while a variety of polymeric bioemulsifiers for potential use in foods was reported by Shepard et al. (1995).

4 Natural Role of Emulsans

Although bioemulsifiers are produced by a large number of microorganisms and are clearly significant in many aspects of growth, it is difficult to generalize on their role in microbial physiology. To begin with, relevant experiments have been performed only with few emulsifier-producing microbes. In addition, with increasing numbers of identified microbial emulsifiers it has become clear that microbial surfactants have very different structures, are produced by a wide variety of microorganisms, and have very different surface properties. Thus, it is expected that bioemulsifiers have various roles, some of which are unique to the physiology and ecology of the producing microorganisms. At this stage it is impossible to draw any generalizations or to identify one or more functions that are clearly common to all microbial surfactants. Thus, most of the concepts have been derived from a consideration of the surface properties of the biosurfactants and experiments in which biosurfactants are added to microorganisms growing on water-insoluble substrates. In the following discussion, we will present a few natural roles for emulsans that have been suggested or demonstrated.

4.1 Increasing the Surface Area of Hydrophobic Water-insoluble Substrates

For bacteria growing on hydrocarbons the growth rate can be limited by the interfacial surface area between water and oil. When the surface area becomes limiting, biomass increases arithmetically rather than exponentially. The evidence that emulsification is a natural process brought about by extracellular agents is indirect, and there are certain conceptual difficulties in understanding how emulsification can provide an (evolutionary) advantage for the microorganism producing the emulsifier. Stated briefly, emulsification is a cell density-dependent phenomenon; that is, the greater the number of cells, the higher the concentration of extracellular product. The concentration of cells in an open system, such as an oil-polluted body of water, never reaches a high enough value to emulsify oil effectively. Furthermore, any emulsified oil would disperse in the water and not be any more available to the emulsifier-producing strain than to competing microorganisms. One way to reconcile the existing data with these theoretical considerations is to suggest that the emulsifying agents do play a natural role in oil degradation, but not in producing macroscopic emulsions in the bulk liquid. If emulsion occurs at or very close to the cell surface, and no mixing occurs at the microscopic level, then each cell creates its own micro-environment and no cell density dependence would be expected.

4.2 Increasing the Bioavailability of Hydrophobic Water-insoluble Substrates

One of the major reasons for the prolonged persistence of high molecular-weight hydrophobic compounds is their low water solubility that increases their sorption to surfaces and limits their availability to biodegrading microorganisms. When organic molecules are bound irreversibly to surfaces, biodegradation is inhibited (Van Delden et al., 1998). Biosurfactants can enhance growth on bound substrates by desorbing them from surfaces or by increasing their apparent water solubility (Deziel et al., 1996). Surfactants that lower interfacial tension are particularly effective in mobilizing bound hydrophobic molecules and making them available for biodegradation. Recently, it has been demonstrated that alasan increases the apparent solubilities of polycyclic aromatic hydrocarbons (PAHs) between 5- and 20-fold, and also significantly increases their rate of biodegradation (Barkay et al., 1999; Rosenberg et al., 1999).

4.3 Binding of Toxic Heavy Metals

Emulsans, like most anionic polysaccharides, bind cations. In some cases, the uronic acid residues of the polysaccharide are arranged so that divalent cations are bound avidly. RAG-1 emulsan is particularly efficient at binding uranyl ions, partially because of its negative charge and partially due to hydrophobic interactions (Zosim et al., 1983).

4.4 Regulating the Attachment-detachment of Microorganisms to and from Surfaces

One of the most fundamental survival strategies of microorganisms is their ability to locate themselves in an ecological niche where they can multiply. This is true not only for microbes that live in or on animals and plants, but also for those that inhabit soil and aquatic environments. The key elements in this strategy are cell surface structures that are responsible for the attachment of the microbes to the proper surface. Neu (1996) has reviewed how surfactants can affect the interaction between bacteria and interfaces. If a biosurfactant is excreted it can form a conditioning film on an interface, thereby stimulating certain microorganisms to attach to the interface while inhibiting the attachment of others. In the case where the substratum is also a water-insoluble substrate (e.g., sulfur and hydrocarbons), the biosurfactant stimulates growth (Beebe and Umbreit, 1971; Bunster et al., 1989). If the biosurfactant is cell-bound it can cause the microbial cell surface to become more hydrophobic, depending on its orientation. For example, the cell- surface hydrophobicity of *P. aeruginosa* was greatly increased by the presence of cell-bound rhamnolipid (Zhang and Miller, 1994), whereas the cell-surface hydrophobicity of *Acinetobacter* strains was reduced by the presence of its cell-bound emulsifier (Rosenberg et al., 1983). These data suggest that microorganisms can use their biosurfactants to regulate their cell-surface properties in order to attach or detach from surfaces according to need. This has been demonstrated for *A. calcoaceticus* RAG-1 growing on crude oil (Rosenberg and Ron, 1998a) when, during exponential growth, emulsan is cell-bound in the form of a minicapsule. This bacterium utilizes only relatively long-chain n-alkanes for growth. When these compounds have been utilized, RAG-1 becomes starved, despite still being attached to the oil droplet that is enriched in aromatics and cyclic paraffins. Starvation of RAG-1 causes release of the minicapsule of emulsan. It was shown that this released emulsan forms a polymeric film on the n-alkane-depleted oil droplet, thereby desorbing the starved cell (Rosenberg et al., 1983). In effect, the "emulsifier" frees the cell to find fresh substrate. At the same time, the depleted oil droplet has been "marked" as having been used, because it now has a hydrophilic outer surface to which the bacterium cannot attach.

5 Genetics and Regulation of Bioemulsan Production

The genetics of bioemulsifier production has been studied using mutants, both naturally occurring or those induced by transposition. The screening of such mutants is made difficult by the fact that the loss of ability to produce the emulsifier usually does not result in an easily selectable phenotype. In addition, the genetics of many emulsifier-producing bacteria has not been adequately elucidated, and important genetic tools (plasmids, transposons, gene libraries) are still to be developed.

The synthesis of high molecular-weight heteropolysaccharides requires a large number of genes, and the genetics is even more complex for polysaccharide–protein complexes. From the genetics point of view, the best-studied polysaccharide bioemulsifier is that produced by *A. calcoaceticus* BD4. The genes involved in its synthesis were identified in a cosmid library that was used to complement nonproducing mutants (Stark, 1996). These biosynthetic genes are organized in a cluster of about 60 kilobases. The first gene in the biosynthesis of the BD4 emulsifier (Accession no. X89900) was identified as a homologue of genes coding for phosphoglucoisomerase (*pgi*). The product of this gene is a protein of about 60 kDa molecular weight and carries out the bidirectional conversion of glucose-6-phosphate to fructose-6-phosphate. The gene is highly conserved from bacteria to mammals, with about 40% homology in amino acids. Additional genes in the cluster (*epsX* and *epsM*; Accession no. X81320) show homology to the genes coding for GDP-mannose pyrophosphorylase and phosphomannose isomerase of enteric bacteria (Stark, 1996). It is interesting to note that mutants in the *epsX* and *epsM* grow poorly under conditions that favor formation of the bioemulsifier, and these deleterious mutations can be overcome by an additional (suppressor) mutation in the first gene in the pathway – *pgi*. These results suggest that mutants in the biosynthesis of polysaccharide accumulate a toxic intermediate, probably fructose-6-phosphate (Fraenkel, 1992). The accumulation of the toxic substance, as well as the inhibition of growth, can be overcome by a mutation in a previous metabolic reaction, since it blocks the synthesis of this toxic intermediate. The finding that some mutants in capsule synthesis grow poorly under conditions that favor capsule synthesis may explain the difficulty often encountered in obtaining such mutants (as an example, mutants unable to synthesize alasan were not obtained after screening more than 7000 transposants of *A. radioresistens* KA53; Dahan, 1998). As screening for the mutants is usually performed on media that maximize the contrast between the capsule-producing wild-type and the mutant, i.e., media that favor capsule formation, it is possible that these conditions are strongly inhibitory – or even lethal – for many of the mutants.

In *Acinetobacter*, the production of polysaccharide bioemulsifiers is concurrent with stationary phase. It has also been suggested that UDP-glucose – one of the precursors in the synthesis of polysaccharide bioemulsifiers – is a signal molecule in the control of sigma S and sigma S-dependent genes (Bohringer et al., 1995). The bioemulsifier of *A. calcoaceticus* BD4 (Kaplan and Rosenberg, 1982), emulsan of *A. calcoaceticus* RAG-1 (Rubinovitz et al., 1982), biodispersan of *A. calcoaceticus* A2 (Rosenberg et al., 1988a), alasan of *A. radioresistens* (Navon Venezia et al., 1995; Dahan, 1998), and an uncharacterized bioemulsifier from *A. junii* (Goldenberg-Dvir, 1998) can be detected in cultures only after more than 10 h of growth, and maximal production occurs when the cultures have progressed well into the stationary phase. These results suggest the possibility that the high molecular-weight bioemulsifiers are also controlled by quorum sensing, although there is, as yet, no direct proof.

The results presented here suggest that production of bioemulsifiers by bacteria is correlated with high bacterial density. This finding may be fortuitous, or may reflect an indirect correlation with one or more physiological factors affected by high bacterial density such as availability of energy, nitrogen, or oxygen. However, it is possible that the production of bioemulsifiers at high bacterial density has a selective advantage. For emulsifiers produced by pathogens, it has been suggested (Sullivan, 1998) that – being virulence factors – they are produced when the cell density is high enough to cause a localized attack on the host. It is easier to explain the need for bioemulsifiers in bacteria growing on hydrocarbons. As these bacteria are growing at the oil–water interface, production of emulsifiers when the density is high will increase the surface area of the drops, allowing more bacteria to feed. Furthermore, when the utilizable fraction of the hydrocarbon is consumed, as in the case of oil that consists of many types of hydrocarbons, the production of the emulsifiers allows the bacteria to detach from the "used" droplet and find a new one (Rosenberg et al., 1983).

6 Biodegradation

Although it is generally assumed that bioemulsifiers are biodegradable, few studies have been conducted on the biodegradation of bioemulsans. A mixed bacterial culture was obtained by enrichment culture that was capable of degrading RAG-1 emulsan and using it as a carbon source (Shoham et al., 1983). From this mixed culture an aerobic, Gram-negative, rod-shaped bacterium was isolated which formed translucent plaques on *A. calcoaceticus* lawns. The plaques were due to the solubilization of the emulsan capsule of RAG-1. A pure culture of the capsule-degrading *Bacillus* depolymerized RAG-1 emulsan, and then used the breakdown products for growth. Subsequently it was shown that depolymerization of RAG-1 emulsan was due to the secretion of an extracellular emulsan depolymerase of molecular weight 89 000 daltons (Shoham and Rosenberg, 1983).

Emulsan depolymerase activity was the result of an eliminase reaction which split glycosidic linkages within the heteropolysaccharide backbone of emulsan to generate reducing groups and α,β-unsaturated uronides with an absorbance maximum of 233 nm. Deesterified emulsan was degraded by emulsan depolymerase at only 27% of the rate of the native polymer. The treatment of emulsan solutions with emulsan depolymerase for brief periods caused a rapid and parallel drop in viscosity and emulsifying activity. More than 75% of the viscosity and emulsifying activity was lost at a time when less than 0.5% of the glycosidic linkages were broken. These data indicate that: (1) emulsan depolymerase is an endoglycosidase; and (2) the higher the molecular weight of emulsan, the greater its emulsifying activity.

To study alasan degradation, an enrichment culture was used to isolate a bacterial strain that grew on alasan as the sole source of carbon and energy. A strain was obtained that degraded alasan, causing the loss of the protein portion of alasan, as well as the emulsifying activity (Navon-Venezia et al., 1998). The degradation was mediated by extracellular proteinases/alasanases. One of these enzymes, referred to as alasanase II, was purified to homogeneity. Alasanase II, as well as pronase, inactivated alasan, whereas a polysaccharide-degrading enzyme mixture, snail juice, had no effect on emulsifying activity.

7 Production

The production of both high and low molecular-weight bioemulsifiers was reviewed by Desai and Banat (1997).

7.1 Shake-flask Experiments

Initial studies on the production of RAG-1 emulsan was carried out using hexadecane and ethanol as the carbon sources (Rosenberg et al., 1979b). In both cases, extracellular emulsifier was produced during the exponential growth phase, but continued to accumulate during stationary phase. Using antibodies specific for emulsan, it was subsequently shown that cells in the early exponential growth phase exhibited relatively large amounts of cell-associated emulsan in the form of a mini-capsule (Goldman et al., 1982). During late exponential phase and early stationary phase, the cell-bound emulsan was released and was accompanied by a rise in the extracellular bioemulsifier concentration.

When exponentially growing cultures of *A. calcoaceticus* RAG-1 were either treated with inhibitors of protein synthesis or starved of a required amino acid, there was a stimulation in the production of emulsan (Rubinowitz et al., 1982). Emulsan synthesis in the presence of chloramphenicol was dependent on utilizable sources of carbon and nitrogen, and was inhibited by cyanide or azide or anaerobic conditions. Radioactive tracer experiments indicated that the enhanced production of emulsan after the addition of chloramphenicol was due to both the release of material synthesized before the addition of the antibiotic (40%) and to *de novo* synthesis of the polymer (60%). Chemical analysis of RAG-1 cells demonstrated the presence of large amounts of polymeric amino sugars; it was estimated that cell-associated emulsan comprised about 15% of the dry weight of growing cells. These data are consistent with the hypothesis that a polymeric precursor of emulsan accumulates on the cell surface during the exponential growth phase; in the stationary phase or during inhibition of protein synthesis, the polymer is released as a potent emulsifier.

The heavily encapsulated *A. calcoaceticus* BD4 and the "mini-encapsulated" single-step mutant *A. calcoaceticus* BD413 produced extracellular polysaccharides in addition to the capsular material (Kaplan and Rosenberg, 1982). The molar ratio of rhamnose to glucose (3:1) in the extracellular BD413 polysaccharide fraction was similar to the composition of the capsular material. In both strains, the increase in capsular polysaccharide was parallel to cell growth and remained constant in stationary phase. The extracellular polysaccharides were detected starting from mid-logarithmic phase and continued to accumulate in the growth medium for 5–8 h after the onset of stationary phase. Strain BD413 produced one-fourth the total rhamnose exopolysaccharide per cell that strain BD4 did. Depending on the growth medium, 32 to 63% of the rhamnose polysaccharide produced by strain BD413 was extracellular, whereas in strain BD4 only 7–14% was extracellular. In all cases, strain BD413 produced more extracellular rhamnose polysaccharide than did strain BD4. In glucose medium, strain BD413 also produced approximately 10-fold more extracellular emulsifying activity than did strain BD4. The isolated capsular polysaccharide obtained after shearing of BD4 cells showed no emulsifying activity. Thus, strain BD413 either produces a modified extracellular polysaccharide or excretes an additional substance(s) that is responsible for the emulsifying activity.

7.2
Production in Small Fermentors

The growth of *A. calcoaceticus* RAG-1 and emulsan production was studied in a 5-L jar fermentor at 30 °C, operating at 500 r.p.m. and 0.5 v.v.m. (volume air per volume medium per min) (Choi et al., 1996). Optimum emulsan production was achieved at an ethanol concentration of 6.5 g L^{-1} and a phosphate concentration of 12.1 g L^{-1}. In a fed-batch culture, high volumetric emulsan production was achieved by continuous ethanol and phosphate feeding to maintain their concentrations at optimum levels. During the exponential growth phase, *A. calcoaceticus* RAG-1 accumulates emulsan on the cell surface in the form of a minicapsule. When the cells are starved, the cell-bound emulsan is released into the extracellular medium. An exocellular esterase is required for the release of emulsan (Shabtai and Gutnick, 1985).

Mechanisms for RAG-1 emulsan accumulation in immobilized *A. calcoaceticus* RAG-1 cells were studied using a Celite support matrix (Wang and Wand, 1990). The ratio of cell-bound emulsan to cell dry weight was much higher for immobilized cells grown at low shear forces than for free cells or immobilized cells grown at shear stress values above 5 dynes cm^{-2}. The most efficient production of emulsan appears to be via self-cycling fermentation (Brown and Cooper, 1991). The specific emulsan productivity achieved by self-cycling fermentation was about 50-fold greater than that of a batch culture, and more than twice that of a chemostat.

The production of alasan was studied in a batch-fed 2.5-L fermentor (Navon-Venezia et al., 1995). *A. radioresistens* KA53 had an approximate doubling time of 2.4 h during the initial 20-h exponential phase of growth, while the viable count decreased slowly after reaching a maximum of 8.0×10^9 cells per mL at 68 h. The total dry weight (cell biomass and extracellular material) at the end of the fermentation was 23.6 g L^{-1}. The ratio of emulsifying activity to cell biomass increased from 5.3 to 7.3 to 11.6 at 24, 64, and 87 h, respectively. The extracellular activity was 190 U mL^{-1} after 87 h, indicating that the majority of emulsification activity was extracellular. The extracellular biopolymer as a fraction of the cell biomass increased during growth from 9.8% at 20 h to 24% at 87 h.

The production and secretion of biodispersan by *A. calcoaceticus* A2 was studied in protein secretion mutants (Elkeles et al., 1994), the active component being an extracellular anionic polysaccharide. Extracellular fluid also contains a high concentration of secreted proteins that create problems in the purification and application of biodispersan. Strains that were defective in protein secretion were selected following chemical mutagenesis. These mutants produced equal, or even higher, levels of extracellular biodispersan than the parent strain, suggesting that the production and release of polysaccharides proceeded independently of protein secretion. Thus, protein secretion mutants are potentially useful in the production of extracellular polysaccharides.

7.3 Large-scale Production of Emulsans

Several tons of RAG-1 emulsan was produced by Petroferm USA in California during the 1980s. No information is available on the fermentation conditions, other than that denatured ethanol was used as the carbon source. Part of the emulsan that was produced was used for cleaning the decks of aircraft carriers and oil separators by the U.S. Navy (J. Jones-Meehan, personal communication).

Alasan was produced in a 20 000-L fermentor at 30 °C using ethanol as the carbon source and maintaining the pH at 6.8–7.0 (by addition of NH_3). The ethanol concentration was maintained at 0.5–1.5% (v/v) by periodic addition, and the foam was controlled by Dow Corning 1520 Silicone antifoam. The yield was 120 kg. The alasan produced was used in grinding limestone for the production of paper by American-Israeli Paper Mills Ltd. (Rosenberg et al., 1989).

7.4 Patents

A number of material, process and production patents were issued on RAG-1 emulsan; the U.S. patents are listed in Table 3. Patents have also been issued on biodispersan (Rosenberg and Ron, 1990) and alasan (Rosenberg and Ron, 1998b).

8 Potential Applications

Surfactants are used widely in industry, agriculture and medicine. The materials currently in use commercially as surfactants are produced mainly by chemical synthesis or as a relatively inexpensive by-product of an industrial process (e.g., lignin sulfates from

Tab. 3 U.S. patents issued on RAG-1 emulsan, alasan and emulcyan

Authors	*Title*	*U.S. Patent No.*
Gutnick, D.L., Rosenberg, E. (1976)	Cleaning of cargo compartments	3,941,692
Gutnick, D.L., Rosenberg, E. (1980)	Production of emulsans	4,230,801
Gutnick, D.L., Rosenberg. E., Shoham, Y. (1980)	Production of α-emulsans	4,234,689
Gutnick, D.L., Rosenberg, E. (1981)	Cleaning oil-contaminated vessels with α-emulsans	4,276,094
Gutnick, D.L., Rosenberg, E., Belsky. I., Zosim, Z. (1982)	Apo-β-emulsans	4,311,829
Gutnick, D.L., Rosenberg, E., Belsky, I., Zosim, Z. (1982)	Apo-β-emulsans	4,311,830
Gutnick, D.L., Rosenberg, E., Belsky, I., Zosim, Z. (1982)	Apo-γ-emulsans	4,311,831
Gutnick, D.L., Rosenberg, E., Belsky, I., Zosim, Z. (1982)	Proemulsans	4,311,832
Gutnick, D.L., Rosenberg, E., Belsky, I., Zosim, Z. (1983)	γ-emulsans	4,380,504
Gutnick, D.L., Rosenberg, E., Belsky, I., Zosim, Z. (1983)	Polyanionic heteropolysaccharide biopolymers	4,395,353
Gutnick, D.L., Rosenberg, E, Belsky, I., Zosim, Z. (1983)	Emulsans	4,395,354
Shoham, Y., Gutnick, D.L., Rosenberg, E. (1987)	Enzymatic degradation of lipopolysaccharide bioemulsifiers	4,704,360
Sar, N., Rosenberg, E., Gutnick, D.L., Nestaas, E. (1984)	Bioemulsifier production by *Acinetobacter calacoaceticus* strain	4,676,916
Hayes, M.E. et al. (1990)	Bioemulsifier-stabilized hydrocarbosols	4,943,390
Gutnick, D.L. et al. (1989)	Bioemulsifier production by *Acinetobacter* strains	4,883,757
Hayes, M.E. (1989)	Bioemulsified-containing personal care products	4,870,010
Shilo, M., Fattom, A. (1986)	Cyanobacterial produced bioemulsifier composition	4,693,842
Rosenberg, E., Ron, E.Z. (1998)	Alasan	5,840,847

paper manufacturing). In order for a microbial polysaccharide surfactant to penetrate the market, it must provide a clear advantage over the existing competing materials. Special properties of polysaccharide surfactants that may provide added incentive for their commercialization include:

1) Biodegradability and controlled inactivation. Several chemically synthesized, commercial available surfactants (e.g., perfluorinated anionics) resist biodegradation, accumulate in nature, and cause ecological problems. Microbial polysaccharides, like all natural products, are susceptible to degradation by microorganisms in water and soil (Zajic et al., 1977; Shoham et al., 1983). Hydrocarbon-in-water emulsions stabilized by emulsan can be broken by minute quantities of the enzymes. In principle, it should be possible to isolate a specific enzyme that will reverse the effect of any particular polysaccharide surfactant. The ability rapidly to break a specific emulsion or dispersion under mild conditions by the addition of catalytic quantities of an enzyme may lead to new applications.
2) Selectivity for specific interfaces. Compared with chemically synthesized materials, one of the general features of polysaccharide molecules is their remarkable specificity. Two examples of microbial surfactants that exhibit substrate specificity are emulsan toward a mixture of aliphatic and aromatic hydrocarbons (Rosenberg et al., 1979a), and the solubilizing factor of *Pseudomonas* PG-1 toward pristane. Some of the best examples of specificity involve polysaccharide–lectin interactions.
3) Surface modifications. A polysaccharide emulsifying or dispersing agent not only causes a reduction in the average particle size, but also changes the surface properties of the particle in a fundamental manner (Neu and Poralla, 1990). For example, oil droplets that have been emulsified with emulsan contain an outer hydrophilic shell of spatially oriented carboxyl and hydroxyl groups. These emulsan-stabilized oil droplets bind large quantities of uranium (Zosim et al., 1983). In an elegant experiment, Pines et al. (1983) demonstrated that emulsan on the surface of hexadecane droplets acts as a receptor for *A. calcoaceticus* bacteriophage ap3. Clearly, each polysaccharide alters, in a characteristic manner, the interface to which it adheres. Usually, 0.1 – 1.0% of a polysaccharide bioemulsifier or biodispersant is sufficient to saturate the surface of the dispersed material. Thus, small quantities of a dispersant can alter dramatically a material's surface properties, such as its surface charge, its hydrophobicity and, most interestingly, its pattern recognition based on the three-dimensional structure of the adherent polymer.

At present, the use of bioemulsifiers is limited by the cost of production and insufficient experience in applications. However, since there is increasing awareness of water quality and environmental conservation, as well as an expanding demand for natural products, it appears inevitable that high-quality, microbially produced, bioemulsifiers will replace the currently used chemical emulsifiers in many applications.

9
Outlook and Perspectives

Although a large number of bioemulsans have been isolated, much of the research has been quite descriptive: characterization of the producing microorganism; flask conditions for producing the bioemulsifier; and a very superficial description of the macromolecules. With the exception of the BD4 polysaccharide (Kaplan et al., 1985) and one of the alasan proteins (Toren et al., 2001), none of the bioemulsan macromolecules has been purified to homogeneity, and neither has their chemical structure been elucidated.

The requirements for bioemulsans to find their rightful position in microbial physiology and biotechnology are threefold. First, the chemical structures of bioemulsans must be elucidated. In the case of polysaccharides, structures can be determined using modern NMR studies coupled to classical carbohydrate chemistry. With surface-active proteins, powerful molecular genetic techniques are now available to determine the precise amino acid sequence of the protein. Second, when structures have been determined, then it becomes possible to perform structure–function studies. The fundamental questions remaining to be answered include: (1) How do bioemulsans differ from other macromolecules? (2) What aspect of their three-dimensional structure makes them such effective stabilizers of oil-in-water emulsions, or what other surface-active properties do they express? (3) Are their activities a direct outcome of their monomeric sequences, or are they modified after polymerization? Using the power of modern structural biology, these questions should be answered within the next decade – at least for some bioemulsifiers.

The third question relates to the reason why bioemulsans have not made a greater penetration into the market place, despite their special properties. The two major problems are cost and lack of the necessary toxicity testing. The cost of existing commercial emulsifiers ranges from US$5 to US$15 per kg, compared with the current cost of producing RAG-1 emulsan of about US$50 per kg. It should be borne in mind however, that for many applications bioemulsans are 5- to 10-fold more effective on a weight basis than commercial emulsifiers. Furthermore, up-scaling and strain improvement should reduce the production cost of bioemulsans considerably. The greater obstacle to overcome before bioemulsans achieve their commercial potential is the cost of the toxicity testing required in order to introduce any new product. This is especially true in the food industry, but is equally applicable to the cosmetics and home care industries. Consequently, it is predicted that the company which chooses the correct bioemulsan and invests in lowering its production costs and obtaining the necessary product approvals, will reap the benefits of introducing this new biotechnology product.

10 References

Abbot, B. J., Gledhill, W. E. (1971) The extracellular accumulation of metabolic products by hydrocarbon-degrading microorganisms, *Adv. Appl. Microbiol.* **14**, 249–388.

Appaiah, A. K. A., Karanth, N. G. K. (1991) Insecticide specific emulsifier production by hexachlorocyclohexane-utilizing *Pseudomonas tralucida* Ptm+ strain, *Biotechnol. Lett.* **13**, 371–374.

Ashtaputre, A. A., Shah, A. K. (1995) Emulsifying property of a viscous exopolysaccharide from *Sphingomonas paucimobilis, World J. Microbiol. Biotechnol.* **11**, 219–222.

Banat, I. M. (1995) Biosurfactants production and possible use in microbial enhanced oil recovery and oil pollution remediation: a review, *Biosource Technol.* **51**, 1–12.

Barkay, T., Navon-Venezia, S., Ron, E. Z., Rosenberg, E. (1999) Enhancement of solubilization and biodegradation of polyaromatic hydrocarbons by the bioemulsifier alasan, *Appl. Environ. Microbiol.* **65**, 2697–2702.

Beebe, J. L., Umbreit, W. W. (1971) Extracellular lipid of *Thiobacillus thiooxidans, J Bacteriol.* **108**, 612–615.

Belsky, I., Gutnick, D. L., Rosenberg, E. (1979) Emulsifier of *Arthrobacter* RAG-1: determination of emulsifier-bound fatty acids, *FEBS Lett.* **101**, 175–178.

Bohringer, J., Fischer, D., Mosler, G., Hengge-Aronis, R. (1995) UDP-glucose is a potential intracellular signal molecule in the control of expression of sigma S and sigma S-dependent genes in *Escherichia coli, J. Bacteriol.* **177**, 413–422.

Bredholt, H., Eimhjellen, K. (1999) Induction and development of the oil emulsifying system in an alkane oxidizing *Rhodococcus* species, *Can. J. Microbiol.* **45**, 700–708.

Brown, W. A., Cooper, D. G. (1991) Self-cycling fermentation applied to *Acinetobacter calcoaceticus* RAG-1, *Appl. Environ. Microbiol.* **57**, 2901–2906.

Bunster, L., Fokkema, N. J., Shippers, B. (1989) Effect of surface-active *Pseudomonas* spp. on leaf wettability, *Appl. Environ. Microbiol.* **55**, 1434–1435.

Calvo, C., Martinez-Checa, F., Mota, A., Bejar, V., Quesada, E. (1998) Effect of cations, pH and sulfate content on the viscosity and emulsifying activity on the *Halomonas eurihalina, J. Ind. Microbiol. Biotechnol.* **20**, 205–209.

Cameron, D. R., Cooper, D. G., Neufeld, R. J. (1988) The mannoprotein of *Saccharomyces cerevisiae* is an effective bioemulsifier, *Appl. Environ. Microbiol.* **54**, 1420–1425.

Choi, J. W., Choi, H. G., Lee, W. H. (1996) Effects of ethanol and phosphate on emulsan production by *Acinetobacter calcoaceticus* RAG-1, *J. Biotechnol.* **45**, 217–225.

Cirigliano, M. C., Carman, G. M. (1984) Purification and characterization of liposan, a bioemulsifier from *Candida lipolytica, Appl. Environ. Microbiol.* **50**, 846–850.

Dahan, O. (1998) Isolation and characterization of alasan mutants in *Acinetobacter radioresistens,* MSc Thesis, Tel Aviv University.

Davis, J. B. (1967) *Petroleum Microbiology.* New York: Elsevier.

De Acevedo, G. T., McInerney, M. J. (1996) Emulsifying activity in thermophilic and extremely thermophilic microorganisms, *J. Ind. Microbiol.* **16**, 17–22.

Desai, J., Banat, I. (1997) Microbial production of surfactants and their commercial potential, *Microbiol. Mol. Biol. R.* **61**, 47–48.

Deziel, E., Paquette, G., Villemur, R., Lepine, F., Bisaillon, J. G. (1996) Biosurfactant production by a soil *Pseudomonas* strain growing on poly-

cyclic aromatic hydrocarbons, *Appl. Environ. Microbiol.* **62**, 1908–1912.
Dostalek, M., Spurny, M. (1958) Bacterial release of oil, *I. Folia Biol.* (Prague), **IV**, 166–171.
Elkeles, A., Rosenberg, E., Ron, E. Z. (1994) Production and secretion of the polysaccharide biodispersan of *Acinetobacter calcoaceticus* A1 in protein secretion mutants, *Appl. Environ. Microbiol.* **60**, 4642–4645.
Erickson, L. E., Nakahara, T. (1975) Growth in cultures with two liquid phases: hydrocarbon uptake and transport, *Process Biochem.* **10**, 9–13.
Fiechter, A. (1992) Biosurfactants: moving towards industrial application, *Trends Biotechnol.* **10**, 208–217.
Fraenkel, D. G. (1992) Genetics and intermediary metabolism, *Annu. Rev. Genet.* **26**, 159–177.
Goldberg, I. (1985) *Single Cell Protein.* Heidelberg: Springer-Verlag.
Goldenberg-Dvir, V. (1998) A new bioemulsifier produced by the oil-degrading *Acinetobacter junii* V-26, MSc thesis, Tel-Aviv University.
Goldman, S., Shabtai, Y., Rubinovitz, C., Rosenberg, E., Gutnick, D. L. (1982) Emulsan in *Acinetobacter calcoaceticus* RAG-1: distribution of cell-free and cell-associated cross-reacting materials, *Appl. Environ. Microbiol.* **44**, 165–170.
Gunjar, M., Khire, J. M., Khan, M. I. (1995) Bioemulsifier production by *Bacillus stearothermophilus* VR8 isolate, *Lett. Appl. Microbiol.* **21**, 83–86.
Gutnick, D. L., Rosenberg, E. (1977) Oil tankers and pollution: a microbiological approach, *Annu. Rev. Microbiol.* **31**, 379–396.
Iguchi, T., Takeda, I., Ohsawa, H. (1969) Emulsifying factor of hydrocarbon produced by a hydrocarbon-assimilating yeast, *Agric. Biol. Chem.* **33**, 1657–1658.
Johnson, M. J. (1968) Utilization of hydrocarbon by microorganisms, *Chem. Ind.* (London) **36**, 1532–1537.
Kaplan, N., Rosenberg E. (1982) Exopolysaccharide distribution and bioemulsifier production in *Acinetobacter calcoaceticus* BD4 and BD413, *Appl. Environ. Microbiol.* **44**, 1335–1341.
Kaplan, N., Jann, B., Jann, K. (1985) Structural studies on the capsular polysaccharide of *Acinetobacter calcoaceticus* BD4, *Eur. J. Biochem.* **152**, 453–458.
Kaplan, N., Zosim, Z., Rosenberg, E. (1987) *Acinetobacter calcoaceticus* BD4 emulsan: reconstitution of emulsifying activity with pure polysaccharide and protein, *Appl. Environ. Microbiol.* **53**, 440–446.
Kaeppeli, O., Walther, P., Mueller, M., Fiechter, A. (1984) Structure of cell surface of the yeast *Candida tropicalis* and its relation to hydrocarbon transport, *Arch. Microbiol.* **138**, 279–282.
La Riviere, J. W. M. (1955a) The production of surface active compounds by microorganisms and its possible significance in oil recovery. I. Some general observations on the change of surface tension in microbial cultures, *Antonie van Leeuwenhoek, J. Microbiol. Serol.* **21**, 1–8.
La Riviere, J. W. M. (1955b) The production of surface active compounds by microorganisms and its possible significance in oil recovery. II. On the release of oil from oil-sand mixtures with the aid of sulphate-reducing bacteria. *Antonie van Leeuwenhoek, J. Microbiol. Serol.* **21**, 9–16.
Marin, M., Pedregosa, A., Laborda, F. (1996) Emulsifier production and microscopical study of emulsions and biofilms formed by the hydrocarbon-utilizing bacteria *Acinetobacter calcoaceticus* MM5, *Appl. Microbiol. Biotechnol.* **44**, 660–667.
Navon-Venezia, S., Zosim, Z., Gottlieb, A., Legmann, R., Carmeli, S., Ron, E. Z., Rosenberg, E. (1995) Alasan, a new bioemulsifier from *Acinetobacter radioresistens*, *Appl. Environ. Microbiol.* **61**, 3240–3244.
Navon-Venezia, S., Banin, E., Ron, E. Z., Rosenberg, E. (1998) The bioemulsifier alasan: role of protein in maintaining structure and activity. *Appl. Microbiol. Biotechnol.* **49**, 382–384.
Neu, T. R. (1996) Significance of bacterial surface-active compounds in interaction of bacteria with interfaces, *Microbiol. Rev.* **60**, 151–166.
Neu, T. R., Poralla, K. (1990) Emulsifying agent from bacteria isolated during screening for cells with hydrophobic surfaces, *Appl. Microbiol. Biotechnol.* **32**, 521–525.
Neufeld, R. J., Zajic, J. E. (1984) The surface activity of *Acinetobacter calcoaceticus* sp. 2CA2, *Biotechnol. Bioeng.* **26**, 1108–1114.
Pines, O., Bayer, E.A., Gutnick, D. L. (1983) Localization of emulsan-like polymers associated with the cell surface of *Acinetobacter calcoaceticus, J. Bacteriol.* **154**, 893–905.
Pines, P., Gutnick, D. L. (1981) Relationship between phase resistance and emulsan production, interaction of phases with the cell-surface of *Acinetobacter calcoaceticus, Arch. Microbiol.* **130**, 129–133.
Reisfeld, A., Rosenberg, E., Gutnick, D. (1972) Microbial degradation of crude oil: factors affecting the dispersion in sea water by mixed and pure cultures, *Appl. Environ. Microbiol.* **24**, 363–368.

Ron, E.Z., Rosenberg, E. (2001) Natural roles of biosurfactants, *Environ. Microbiol.* **3**, 229–236.

Rosenberg, E., Kaplan, N. (1987) Surface-active properties of *Acinetobacter* expolysaccharides, in: *Bacterial Outer Membranes as Model Systems* (Inouye, M., Ed.), New York: John Wiley & Sons, 311–342.

Rosenberg, E., Ron, E. Z. (1990) Bacterial process for the production of biodispersants. Israel-Patent Application No. 76981.

Rosenberg, E., Ron, E. Z. (1996) Bioremediation of petroleum contamination, in: *Bioremediation: Principles and Applications* (Crawford, R. L., Crawford, D. L., Eds.), Cambridge University Press, 100–124.

Rosenberg, E., Ron, E. Z (1997) Bioemulsans: microbial polymeric emulsifiers, *Curr. Opin. Biotechnol.* **8**, 313–316.

Rosenberg, E., Ron, E. Z. (1998a) Surface active polymers from the genus *Acinetobacter*, in: *Biopolymers from Renewable Resources* (Kaplan, D. L., Ed.), New York: Springer-Verlag, 281–291.

Rosenberg, E., Ron, E. Z. (1998b) Novel bioemulsifiers. Israeli Patent Application No. 112,254, 5/1/95: U.S. Patent No. 5,840,547 (November 24, 1998).

Rosenberg, E., Ron, E. Z. (1999). High and low molecular mass microbial surfactants, *Appl. Microbiol. Biotechnol.* **52**, 154–162.

Rosenberg, E., Horowitz, A., Englander, E., Gutnick, D. L. (1974) Bacterial emulsion of crude oil in seawater, Symposium on Impact of Microorganisms on the Aquatic Environment, *USEPA Publication*, 157–168.

Rosenberg, E., Perry, A., Gibson, D. T., Gutnick, D. (1979a) Emulsifier of *Arthrobacter* RAG-1: specificity of hydrocarbon substrate, *Appl. Environ. Microbiol.* **37**, 409–413.

Rosenberg, E., Zuckerberg, A., Rubinovitz, C., Gutnick, D. L. (1979b) Emulsifier of *Arthrobacter* RAG-1: Isolation and emulsifying properties, *Appl. Environ. Microbiol.* **37**, 402–408.

Rosenberg, E., Gottlieb, A., Rosenberg, M. (1983) Inhibition of bacterial adherence to hydrocarbons and epithelial cells by emulsan, *Infect. Immun.* **39**, 1024–1028.

Rosenberg, E., Rubinovitz, C., Gottlieb, A., Rosenhak, S., Ron, E. Z. (1988a) Production of biodispersan by *Acinetobacter calcoaceticus* A2, *Appl. Environ. Microbiol.* **54**, 317–322.

Rosenberg, E., Rubinovitz, C., Legmann, R., Ron, E. Z. (1988b) Purification and chemical properties of *Acinetobacter calcoaceticus* A2 biodispersan, *Appl. Environ. Microbiol.* **54**, 323–326.

Rosenberg, E., Schwartz, Z., Tenenbaum, A., Rubinovitz, C., Legmann, R., Ron, E. Z. (1989) A microbial polymer that changes the surface properties of limestone: effect of biodispersan in grinding limestone and making paper, *J. Dispersion Sci. Technol.* **10**, 241–250.

Rosenberg, E., Barkay, T., Navon-Venezia, S., Ron, E. Z. (1999) Role of *Acinetobacter* bioemulsans in petroleum degradation, in: *Novel Approaches for Bioremediation of Organic Pollution* (Fass, R., Flasher, Y., Eds), New York: Kluwer Academic/Plenum Publishers.

Rubinovitz, C., Gutnick, D. L., Rosenberg, E. (1982) Emulsan production by *Acinetobacter calcoaceticus* in the presence of chloramphenicol, *J. Bacteriol.* **152**, 126–132

Sar, N., Rosenberg, E. (1983) Emulsifier production by *Acinetobacter calcoaceticus* strains, *Curr. Microbiol.* **9**, 309–314.

Shabtai, Y., Gutnick, D. L. (1985) Exocellular esterase and emulsan release from the cell surface of *Acinetobacter calcoaceticus*, *J. Bacteriol.* **161**, 1176–1181.

Shepard, R., Rockey, J., Sutherland, I. W., Roller, S. (1995) Novel bioemulsifiers from microorganisms for use in foods, *J. Biotechnol.* **40**, 207–217.

Shoham, Y., Rosenberg, E. (1983) Enzymatic depolymerization of emulsan, *J. Bacteriol.* **156**, 161–167.

Shoham, Y., Rosenberg, M., Rosenberg, E. (1983) Bacterial degradation of emulsan, *Appl. Environ. Microbiol.* **46**, 573–579.

Stark, M. (1996) Analysis of the exopolysaccharide gene cluster from *Acinetobacter calcoaceticus* BD4, PhD thesis, Tel-Aviv University.

Sullivan, E. R. (1998) Molecular genetics of biosurfactant production, *Curr. Opin. Biotechnol.* **9**, 263–269.

Taylor, W. H., Juni, E. (1961) Pathways for biosynthesis of a bacterial capsular polysaccharide. I. Characterization of the organism and polysaccharide, *J. Bacteriol.* **81**, 688–693.

Toren, A., Navon-Venezia, S., Ron, E. Z., Rosenberg, E. (2001) Emulsifying activities of purified alasan proteins from *Acinetobacter radioresistens* KA53, *Appl. Environ. Microbiol.* **67**, 1102–1106.

Updegraff, D. M., Wren, G. B. (1954) The release of oil from petroleum bearing materials by sulfate reducing bacteria, *Appl. Microbiol.* **2**, 309–316.

Van Delden, C., Pesci, E. C., Pearson, J. P., Iglewski, B. H. (1998) Starvation selection restores elastase and rhamnolipid production in a *Pseudomonas aeruginosa* quorum-sensing mutant, *Infect. Immun.* **66**, 4499–4502.

Wang, S. D., Wand, D. I. C. (1990) Mechanisms for biopolymer accumulation in immobilized *Acinetobacter calcoaceticus* system, *Biotechnol. Bioeng.* **36**, 402–410.
Zajic, J. E., Panchal, C. J. (1976) Bioemulsifiers, *CRC Crit. Rev. Microbiol.* **5**, 39–66.
Zajic, J. E., Guignard, H., Gerson, D. F. (1977) Properties and biodegradation of a bioemulsifier from *Corynebacterium hydrocarboclastus*, *Biotechnol. Bioeng.* **19**, 1303–1320.
Zhang, Y., Miller, R. M. (1994) Effect of a *Pseudomonas* rhamnolipid biosurfactant on cell hydrophobicity and biodegradation of octadecane, *Appl. Environ. Microbiol.* **60**, 2101–2106.
Zosim, Z., Gutnick, D. L., Rosenberg, E. (1983) Uranium binding by emulsan and emulsanosols, *Biotech. Bioeng.* **25**, 1725–1735.

16
Curdlan

Dr. In-Young Lee
Korea Research Institute of Bioscience and Biotechnology and DawMaJin Biotech Corp., P.O. Box 115, Daeduk Valley, Daejon 305-600, Korea; Tel.: +82-428620722; Fax: +82-428620702; E-mail: leeiy@dmj-biotech.com

Biotechnology of Biopolymers. From Synthesis to Patents. Edited by A. Steinbüchel and Y. Doi

ISBN: 3-527-31110-6

AIDS	acquired immunodeficiency syndrome
AMP	adenosine 5′-monophosphate
ATCC	American Type Culture Collection
DMSO	dimethylsulfoxide
DP	degree of polymerization
FDA	Food and Drug Administration
HIV	human immunodeficiency virus
MNNG (also NTG)	*N*-methyl-*N*-nitro-*N*-nitrosoguanidine
NMR	nuclear magnetic resonance
TEM	transmission electron microscopy
UDP-glucose	uridine 5′-diphosphate glucose
UMP	uridine 5′-monophosphate

1 Introduction

Curdlan, an insoluble microbial exopolymer is composed almost exclusively of β-(1,3)-glucosidic linkages. One of the unique features of curdlan is that aqueous suspensions can be thermally induced to produce high-set gels, which will not return to the liquid state upon reheating (Harada et al., 1968), and this has attracted the attention of the food industry. In addition to this, curdlan offers many health benefits, as the beta-glucan family is well known among the scientific community to have immunestimulatory effects. Many informative reviews have been produced on this subject (Harada, 1977; Harada et al., 1993); however, apart from offering a brief review of β-(1,3)-glucan, the present chapter article provides an updated overview of the production, properties, and application of curdlan.

2 Historical Outline

Curdlan was discovered in 1966 by Professor Harada and coworkers, and given its name because of its ability to "curdle" when heated (Harada et al., 1966). At this time, Harada and his colleagues were working on the identification of organisms capable of utilizing petrochemical materials, and isolated *Alcaligenes faecalis* var. *myxogenes* 10C3 from soil. This organism was found to be capable of growing on a medium containing 10% ethylene glycol as the sole carbon source

(Harada et al., 1965), and also produced a new β-(1,3)-glucan that contained about 10% succinic acid, and which was named succinoglucan (Harada 1965; Harada and Yoshimura, 1965). They were also able to derive a spontaneous mutant that mainly produced a water-insoluble neutral polysaccharide, β-(1,3)-glucan, and which did not contain succinoglucan.

Scientists at Takeda Chemical Industries Ltd. (Osaka, Japan) have played a pioneering role in both the research and development of curdlan. Thus, as early as 1989, curdlan was approved and commercialized for food usage in Korea, Taiwan, and Japan. Upon obtaining approval in December 1996, Pureglucan™ – the tradename of curdlan – was launched in the US market as a formulation aid, processing aid, stabilizer, and thickener or texture modifier for food use (Spicer et al., 1999). No evidence of any toxicity nor carcinogenicity of Pureglucan has been observed.

3 Structure

3.1 Beta-1,3-glucan

Both chemical and enzymatic analyses have confirmed that curdlan is a homopolymer of D-glucose linked in β-(1,3) fashion (Saito et al., 1968) (Figure 1). Curdlan has an average degree of polymerization (DP) of approximately 450, and is unbranched (Naganishi et al., 1976). Nakata et al. (1998) reported that the average molecular weight of curdlan in 0.3 N NaOH is in the range of 5.3×10^4 to 2.0×10^6 daltons. Within the class of polysaccharides classified as β-(1,3)-glucans, there are a number of structural variants. The sources of glucans and their structural differences are listed in Table 1. Mycelial fungi are an abundant source of β-(1,3)-glucans; grifolan, which is produced from *Grifola frondosa* and stimulates cytokine production from macrophages, is a β-(1,3)-glucan with a molecular weight of $> 4.5 \times 10^5$ daltons (Okazaki et al., 1995), while lentinan, from *Lentinus edodes*, has a molecular weight of 5×10^5 daltons and two glucose branches for every five β-(1,3)-glucosyl units in the backbone (Jong and Birmingham, 1993). The structure of schizophyllan is very similar to that of lentinan, but it has one glucose branch for every third glucose in the β-1,3-backbone and its molecular weight is 4.5×10^5 daltons (Misaki et al., 1993). Scleroglucan from *Sclerotium rolfsii* has one glucose branch for every third glucose unit (Farina et al., 2001), whereas SSG from *Sclerotinia sclerotiorum* is a highly branched β-(1,3)-glucan (Sakurai et al., 1991). Pachyman, from *Poria cocos*, has an average of 3.2 branch points per molecule of β-(1,3)-glucan (Okuyama et al., 1996; Zhang et al., 1997), whilst krestin is a protein-linked β-glucan with a molecular weight of c. 100,000 daltons, which can be extracted from the mycelia of *Coriolus versicolor* (Azuma, 1987). β-(1,3)-Glucan is also present in the inner cell wall of the bakers' yeast *Saccha-*

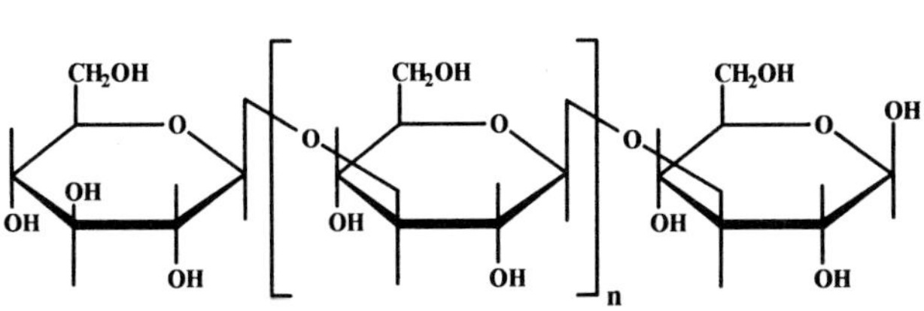

Fig. 1 Structure of curdlan.

Tab. 1 A variety of glucans having β-1,3 linkage in their backbones

Source	*Branch*	M_w	*Reference*
Bacteria			
Curdlan (*Agrobacterium sp. Alcaligens sp.*)	Exclusively β-(1,3)-glucosidic linkages	$5.3\times10^4 - 2.0\times10^6$	Nakata et al. (1998)
Fungi			
Grifolan (*Grifola frondosa*)	Branched β-1,3-gulcan	4.5×10^5	Okazaki et al. (1995)
Lentinan (*Lentinus eeodes*)	Two glucose branches for every five glucose unit	5×10^5	Jong and Birmingham (1993)
Schizophyllan (*Schizophyllum commune*)	One glucose branch for every third glucose unit	4.5×10^5	Misaki et al. (1993)
Scleroglucan (*Sclerotium glucanum*)	One glucose branch for every third glucose unit	$1.6-5.0\times10^6$	Farina et al. (2001)
SSG (*Sclerotinia sclerotiorum*)	Highly branched β-1,3-glucan	$2\times10^5-2\times10^6$	Okazaki et al. (1995)
Pachyman (*Poria cocos*)	Several β-1,6-linked branch points per molecule	2.06×10^4, 8.93×10^4	Zhang et al. (1997)
Krestin (*Coriolus versicolor*)	β-1,3-glucan	1.0×10^6	Azuma (1987)
Yeast (*Saccharomyces cerevisiae*)			
Soluble glucan	β(1,6) linkage to β-(1,3) backbone	2×10^5–2×10^6	Janusz et al. (1986)
Insoluble glucan	β-(1,6) linkage to β-(1,3) backbone	3.53×10^4, 4.57×10^6	Williams et al. (1994)
Brown algea			
Laminarin (*Laminaria digitata*)	β-1,3-glucan and β-1,6-glucan		Read et al. (1996)

romyces cerevisiae to support the structural strength of its cell wall. Unlike lentinan, schizophyllan and scleroglucan, the side branches of yeast β-(1,3)-glucans are chains of glucose molecules, and not single glucose residues. Depending on the extraction procedure used and their subsequent treatment, the yeast glucans may be either particulate water-insoluble or water-soluble macromolecules.

3.2 Conformation in Solution

Many researchers have investigated the molecular structures of curdlan in aqueous system. Three conformers of soluble curdlan have been reported, including single-helix, triple-helix, and random coil. Ogawa et al. (1972) studied the conformational behavior of curdlan in alkaline solution by measuring the optical rotatory dispersion, intrinsic viscosity and flow birefringence. At low concentrations of sodium hydroxide, curdlan has a helical (ordered) conformation, but a significant conformational change occurs at a NaOH concentration of 0.19–0.24 N NaOH. In alkaline solution >0.2 N NaOH, curdlan is completely soluble and exists as random coils, but upon neutralization the polymer adopts an 'ordered state', which is composed of a mixture of single and triple helices. A ^{13}C-NMR study supported this finding (Saito et al., 1977). Increasing the salt concentration shifts the point of conformational transition to a higher alkali concentration (Ogawa et al., 1973a), and addition of nonsolvents such as 2-chloroethanol, dioxan or water to dimethylsulfoxide (DMSO) solution also changes the conformation of curdlan to a rigid, ordered structure (Ogawa et al., 1973b). These workers also showed that the optical rotation was dependent upon the DP of the curdlan in 0.1 N sodium hydroxide (Ogawa et al., 1973c), and concluded that the content of the ordered form increases with DP until becoming constant at DP values of about 200. Electron microscopic comparison of the molecular structures of curdlan with different DPs showed that only curdlan with higher DP can form a gel when heated (Koreeda et al., 1974).

The conversion between triple-helix and single-helix conformers is mediated by different chemical or physical treatments. Treatment of the triple-helix schizophyllan with NaOH has been used to prepare single helix-rich forms (Ohno et al., 1995). Young et al. (2000) proposed a transition mechanism after an investigation using fluorescence resonance energy transfer spectroscopy, which showed that a partially opened triple-helix conformer was formed on treatment with NaOH, and that increasing degrees of strand opening were associated with increasing concentrations of NaOH. After neutralizing the NaOH, the partially opened conformers gradually reverted to the triple-helix.

3.3 Gel Structure

Some clarification of the fine structure of dispersed molecules and networks is necessary to understand the viscoelastic properties of curdlan, which forms two distinct types of gel. For both gels – described as low-set and high-set gels – transmission electron microscopy (TEM) showed them to be composed of three curdlan molecules that are associated to form a triple helix. Tada et al. (1997) proposed a mechanism of formation of the low-set gel using static light-scattering measurements. The molecular associates are formed at a NaOH concentration of 0.01–0.1 N at 25 °C, and this association progresses with as the NaOH concentration decreases. Consequently, the average molecular weight for the molecular associate in 0.01 N NaOH is higher than that in 0.1 N NaOH aqueous solution. The molecular associates consist of a dense core and hydrophilic surface at low NaOH concentrations. By contrast, Kasai and Harada (1980) proposed an annealing model to form a high-set, resilient gel upon heating to >80 °C. This annealing is associated with the irreversible loss of water, and resulted in a more tightly coiled triple helix. The structure crystallizes as a triplex of right-handed, six-fold helical chains in a hexagonal unit cell with a fiber repeating length of 18.78 Å (Chuah et al., 1983). Further removal of water from this structure, by drying under vacuum, results in further tightening of the six-fold triple helix and a decrease in the fiber period to only 5.87 Å (Deslandes et al., 1980).

4 Occurrence

As described in Section 2, β-(1-3)-D-glucans are present in a variety of living systems, including fungi, yeasts, algae, bacteria and higher plants. However, until now only bacteria belonging to the *Alcaligenes* and *Agrobacterium* species have been reported to produce the linear β-(1,3)-glucan type of homopolymer, curdlan. Figure 2 illustrates

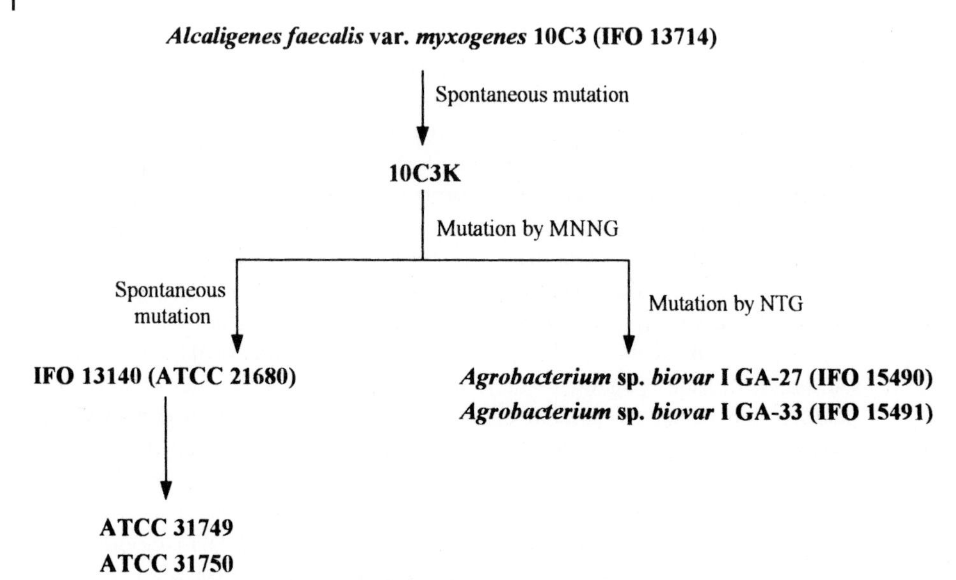

Fig. 2 Lineage of representative strains for curdlan production.

the lineage of the curdlan-producing strains since Harada et al. first isolated *Alcaligenes faecalis* var. *myxogenes* 10C3 during the screening of soil bacteria capable of metabolizing various petroleum fractions (Harada et al., 1965). The parent strain produced two different types of exopolysaccharide; a water-insoluble neutral homoglucan called 'curdlan', and a water-soluble acidic heteroglucan containing about 10% succinic acid, and referred to as 'succinoglucan' (Harada, 1965; Harada and Yoshimura 1965). Moreover, a mutant strain 10C3K was isolated from a stock culture of 10C3, which produced only curdlan. Strain 10C3K is a spontaneous mutant with a stable ability to produce exocellular polysaccharide. By inducing mutagenesis with *N*-methyl-*N*-nitro-*N*-nitrosoguanidine (MNNG), Takeda Chemical Industries Ltd. later isolated a uracil auxotrophic mutant from strain 10C3K, which was named *Alcaligenes faecalis* var. *myxogenes* IFO 13140 (ATCC 21680) and had improved gel-forming β-(1,3)-glucan-producing ability. Phillips and Lawford (1983) isolated a mutant strain from strain ATCC 21680 in a nitrogen-limited chemostat culture (the accession number was ATCC 31749). Unlike its auxotrophic parent strain, ATCC 31749 does not require uracil for its growth, and is not a revertant as it can be distinguished from 10C3K by its inability to hydrolyze starch and its ability to grow on citric acid as sole carbon source. ATCC 31750, which arose as a spontaneous variant of the parent ATCC 31749, produces only the water-insoluble curdlan-type glucan, while the parent strain produces both soluble and insoluble polysaccharides. All of these strains, which were formerly regarded as *Alcaligenes* species, have now been taxonomically reclassified as *Agrobacterium* species (IFO Research Communications, Vol. 15, pp. 57–75 (1991)). Takeda Chemical Industries Ltd. derived further mutant strains from strain 10C3K, which reduced the activity of the enzyme, phosphoenol pyruvic acid carboxykinase (Kanegae et al., 1996).

Naganishi et al. (1974) examined the occurrence of curdlan-type polysaccharides in microorganisms by using the water-soluble dye aniline blue, with which curdlan forms a blue complex. It was also shown that the rate of color complex formation was dependent both on the polymer concentration and DP; hence these findings provided an excellent tool for the screening of curdlan-producing bacteria. Naganishi et al. (1976) tested 687 strains of different genus of bacteria using the aniline blue staining technique. Among those examined, some strains of *Alcaligenes* and *Agrobacterium* species turned blue on agar plates containing aniline blue, and these have been used widely in the production of curdlan-type polysaccharides. Some strains of *Bacillus* formed blue complexes with aniline blue, but their polymeric constitution has not yet been studied.

5 Biosynthesis

Sutherland (1977, 1993) generalized the biosynthesis of extracellular polysaccharides into three major steps: (1) substrate uptake; (2) intracellular formation of polysaccharide; and (3) extrusion from cell. A metabolic pathway for exopolysaccharide biosynthesis is shown schematically in Figure 3. First, a carbohydrate substrate enters the cell by active transport and group translocation involving substrate phosphorylation. The substrate is then directed along either catabolic pathways, or those leading to polysaccharide synthesis. UDP-glucose, a key precursor, is synthesized by the UDP-glucose pyrophosphorylase-induced conversion of glucose-1-phosphate to UDP-glucose. Subsequently, polymer construction occurs together with the transfer of monosaccharides from UDP-glucose to a carrier lipid.

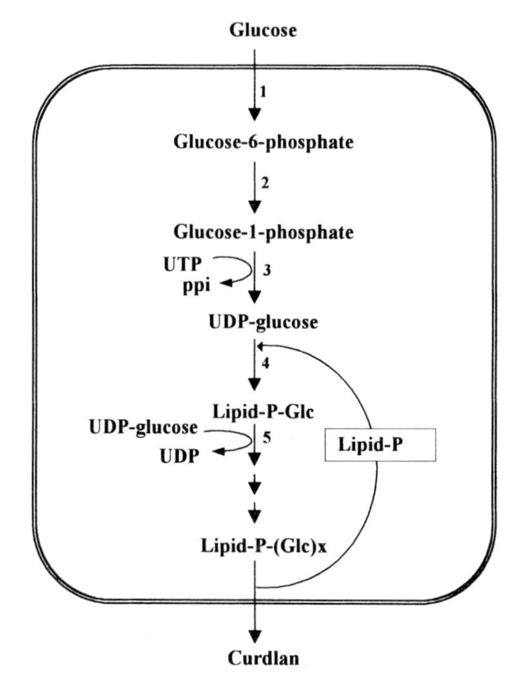

Fig. 3 Metabolic pathway for the synthesis of curdlan. 1, hexokinase; 2, phosphoglucomutase; 3, UDP-glucose pyrophosphorylase; 4, transferase; 5, polymerase. Lipid-P represents isoprenoid lipid phosphate.

After further chain elongation, the polymer is extruded from the cells.

Nitrogen-limited culture has also been employed for the production of curdlan, and has been generally explained by the roles of the carrier lipids (Sutherland, 1977; Ielpi et al., 1993). The availability of isoprenoid lipid may provide a way of regulating polysaccharide synthesis. Since curdlan biosynthesis takes place most extensively after cell growth has been stopped due to nitrogen exhaustion, isoprenoid lipids would be more available for carrying oligosaccharides instead of cellular liposaccharide and peptidoglycan.

The synthesis of precursor molecules is also of considerable importance in polysaccharide synthesis, in terms of the metabolic driving force. UDP-glucose serves as an activated precursor for glycosyl moieties in the synthesis of curdlan, a homopolysaccharide composed exclusively of β-1,3-linked glucose residues. In addition, cellular nucleotides not only play an important role in the synthesis of sugar nucleotides, but also have a widespread regulatory potential in cellular metabolism. Kim et al. (1999) examined the change of intracellular nucleotide levels and their stimulatory effects on curdlan synthesis in *Agrobacterium* species under different culture conditions. Under nitrogen-limited conditions where curdlan synthesis was stimulated, intracellular levels of UMP and AMP were at least twice as high as those occurring under nitrogen-sufficient conditions, though UDP-glucose levels were similar. The time profiles of curdlan synthesis and cellular nucleotide levels showed that curdlan synthesis is positively related with intracellular levels of UMP and AMP. *In vitro* enzyme reactions involved in the synthesis of UDP-glucose showed that a higher UMP concentration promotes the synthesis of UDP-glucose, while AMP neither inhibits nor facilitates the activity of UDP-glucose pyrophosphorylase. The addition of UMP to the medium also increased curdlan synthesis. From these results, these workers concluded that the higher intracellular UMP levels caused by nitrogen limitation enhance the metabolic flux of curdlan synthesis by promoting cellular UDP-glucose synthesis.

Kai et al. (1993) reported a study of the biosynthetic pathway of curdlan using ^{13}C-labeled glucose in *Agrobacterium* sp. ATCC 31749. By analyzing the labeled products, the biosynthesis of curdlan was interpreted as involving five routes: direct polymerization from glucose; rearrangement; isomerization of cleaved trioses; from fructose-6-phosphate; and from fructose fragments produced in various pathways of glycolysis. However, it was noted that more than 60% of curdlan is synthesized by direct polymerization, and that curdlan biosynthesis via glycolysis is comparatively low. This analysis also indicated that glycolysis occurs mainly via the pentose cycle and the Entner–Doudoroff pathway rather than the Embden–Meyerhof pathway.

6
Molecular Genetics

Little is known about the molecular genetics of bacterial curdlan biosynthesis, while there is growing information about the genes required for β-(1,3)-glucan synthesis in yeasts and filamentous fungi. Recently, Stasinopoulos et al. (1999) cloned genes that were essential for the production of curdlan and, by using comparative sequence analysis, identified them as putative curdlan synthase genes. Further genetic investigations will open up new avenues for curdlan synthesis, and these will doubtlessly be exploited to produce curdlan in higher yields.

7 Production

7.1 Carbon Source

Many crucial factors that affect curdlan production, including carbon, nitrogen, phosphate, oxygen supply, and pH have been investigated. High productivity using cheap carbon sources is important for the industrial production of curdlan. Lee, I.-Y. et al. (1997) reported that maltose and sucrose were efficient carbon sources for the production of curdlan by a strain of *Agrobacterium* species, with maximal production (60 g L^{-1}) being obtained from sucrose, with a productivity of 0.5 g L^{-1} h^{-1} when nitrogen was limited at a cell concentration of 16 g L^{-1}. Molasses, which contains large amounts of sucrose, might also be the substrate of choice, with up to 42 g L^{-1} of curdlan, at a yield of 0.35 g curdlan per gram total sugar, being obtained in 5-day cultivation. Sucrose is a less expensive substrate than glucose, and as sugar beet or sugar cane molasses are cheap byproducts widely available from the sugar industry, they are very attractive carbon sources from an economic point of view.

7.2 Nitrogen Effect

As previously described, relatively few strains of *Agrobacterium* and *Alcaligenes* species are known to produce curdlan. In such strains, curdlan production is associated with the poststationary phase of nitrogen depletion; thus, the operation involves an initial production of biomass, which is followed by curdlan production. Therefore, it is important to determine the initial concentration of the nitrogen source because it provides the limiting factor for cell growth during batch fermentation. Kim, M.-K. et al. (2000) reported that the cell growth rate decreased as the ammonium concentration increased. However, since higher cell concentrations produce more curdlan, an optimal ammonium concentration should be determined to provide an appropriate cell concentration while minimizing the inhibitory effect of the ammonium ion.

7.3 Oxygen Supply

Curdlan-producing stains are highly aerobic, and an adequate oxygen supply is therefore a key factor in production. Since curdlan is insoluble in water, the fermentation broth is of relatively low viscosity, and there is little resistance to oxygen transfer from gas to the liquid. However, a layer of insoluble exopolymer surrounding the cell mass offers resistance to oxygen transfer from the liquid into the cell, and therefore a high dissolved oxygen concentration is required for maximal productivity. Shake-flask fermentation results have shown that the specific production rate decreases as the volume of medium is increased, indicating that these cultures are limited by the relatively low oxygen transfer capacity of the system. Several investigations have been made into developing the process for curdlan production, especially with respect to reactor design (Lawford et al., 1986; Lawford and Rousseau, 1991, 1992). These workers employed two different types of impeller: a radial-flow, flat-blade impeller; and an axial-flow impeller. The radial-flow impeller was effective at providing high oxygen transfer rates to increase the production of curdlan, but the high shear characteristics of this design yielded a product of inferior quality in terms of tensile strength of the thermally induced gel. An axial-flow impeller typically produces less shear and more pumping. High volumetric oxygen transfer can be achieved by low shear designs equipped with sparging devices, which consist of microporous materials through which oxygen-enriched air is dispersed. The maximal specific production rate was 90 mg per g cells h^{-1} when 30% oxygen-enriched air was supplied in the low-shear system.

7.4 Phosphate Effect

Phosphate concentration must also be considered because it significantly influences cell growth and product formation. The production of rhamnose-containing polysaccharide by a *Klebsiella* strain was enhanced by a reduction in the phosphate content of the medium (Farres et al., 1997). In contrast, a sufficiency of phosphate resulted in good alginate yields in *Pseudomonas* strains and showed growth-associated production (Conti et al., 1994). Thus, the effect of phosphate on the production of polysaccharides is variable. Kim, M.-K. et al. (2000) investigated the influence of inorganic phosphate concentration on the production of curdlan by *Agrobacterium* species. Under nitrogen-limited conditions which allow curdlan production, the concentration of phosphate remains constant as it is not further utilized for cell growth. The optimal residual phosphate concentration for curdlan production was in the range 0.1–0.5 g L^{-1}. Relatively low concentrations appeared to be optimal for curdlan production, although without phosphate, curdlan production was extremely low. However, on increasing the cell phosphate concentration from 0.42 to 1.68 g L^{-1}, curdlan production increased from 0.44 to 2.80 g L^{-1}. Moreover, the optimal phosphate concentration range was not dependent upon cell concentration, and the specific production rate was about 70 mg curdlan per g cells h^{-1}, irrespective of cell concentration.

7.5 pH Effect

The pH of the culture is one of the most important factors because it significantly influences rates of both cell growth and product formation. High viscosity of the culture broth is often a critical problem in polysaccharide production. However, the viscosity problem can be obviated by operating the fermentation at a slightly acidic pH, since curdlan is insoluble under these conditions. Moreover, there appears to be more than one single optimal pH because fermentation of the culture for curdlan production is divided into two phases – the cell growth phase and the curdlan production phase. Lee, J.-H. et al. (1999) sought an optimal pH profile to maximize curdlan production in a batch fermentation of *Agrobacterium* species. The cell growth rate was maximal at pH 7.0, while curdlan production was maximal at pH 5.5. The pH profile provided a strategy to shift the culture pH from the optimal growth condition (pH 7.0) to the optimal production one (pH 5.5) at the time of ammonium exhaustion. By adopting the optimal pH profile in a batch process, these workers obtained a significant improvement in curdlan production (64 g L^{-1}) compared with that obtained using a constant-pH operation (36 g L^{-1}).

7.6 Batch Production

A high curdlan production was attempted by employing the optimal operation strategy with an *Agrobacterium* strain (Lee, I.-Y. et al., 1997, 1999a; Lee, J.-H. et al., 1999; Kim, M.-K. et al., 2000) that was tolerant of high concentrations of sucrose. This made batch operation possible, with an initial sucrose concentration of 140 g L^{-1}. An initial ammonium concentration 0.8 g L^{-1} was chosen, which produced 6.4 g L^{-1} of cells, the cell yield from ammonium being 8.0 g cells per g ammonium. A typical batch fermentation profile in a 300-L stirred tank reactor is shown in Figure 4. To produce a volumetric oxygen transfer coefficient of 146.5 h^{-1}, the agitation speed was set at 200 r.p.m. over the whole fermentation period, with an inner pressure of 0.2 bar. When the ammonium was exhausted in the culture broth at 20 h, the cell concentration had reached 6.8 g L^{-1}. The pH was controlled at 7.0 with 4 N NaOH/KOH at the cell growing stage. The culture pH was then shifted from 7.0 to 5.5 by adding 3 N HCl at the time of nitrogen limitation. Aeration rates were maintained at 0.5 vvm (volume $volume^{-1}$ min^{-1}). Curdlan production began from the onset of nitrogen exhaustion, and a maximum concentration of 58 g L^{-1} was obtained in 120 h cultivation. The dissolved oxygen level fell rapidly during cell growth, and increased immediately after nitrogen exhaustion, as cell growth ceased. Subsequently, as the curdlan concentration increased, the dissolved oxygen level became limited due to the increased viscosity of the culture broth.

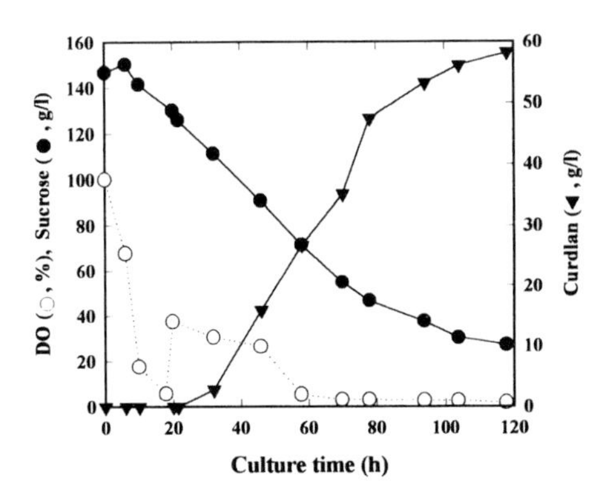

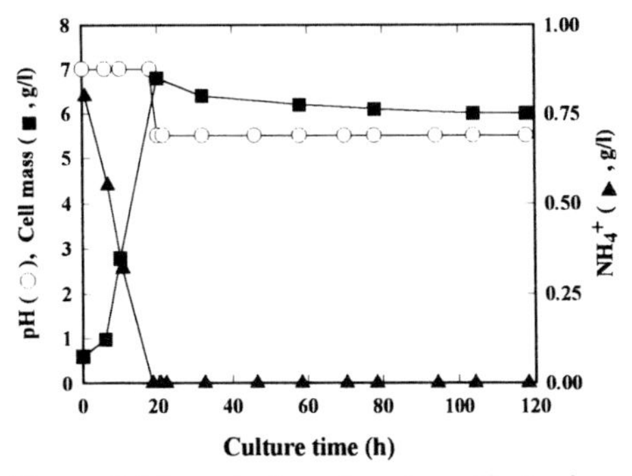

Fig. 4 Batch production of curdlan with *Agrobacterium sp.* ATCC 31750 in a 300-L jar fermenter (Lee, I.-Y. et al., 1999a).

7.7 Continuous Production

Continuous production of curdlan was attempted in a two-stage continuous process (Phillips and Lawford, 1983). In the first stage, the curdlan-producing strain *Alcaligenes* sp. ATCC 31749 was grown aerobically in a medium containing carbon and nitrogen. However, the amount of nitrogen in the first stage was so limited that the effluent contained substantially no inorganic nitrogen. The effluent was fed into the second stage in a constant-volume fermenter, where it was mixed with a nitrogen-free medium in order to induce curdlan production without further cell growth. Curdlan production was 7 g L^{-1} at a dilution rate of 0.02 h^{-1} in a steady-state culture, providing a curdlan productivity of 0.14 g L^{-1} h^{-1}.

7.8 Isolation Process

The recovery procedure is based on the conformational transition which occurs when the concentration of alkali exceeds 0.2 N (Ogawa et al., 1972). Under alkaline conditions, the biomass can be separated from the dissolved curdlan, which remains in the supernatant. Upon neutralization of the alkaline supernatant, the polymer forms an insoluble gel that can be recovered by centrifugation, the polymer being subsequently washed free of contaminating salt. Procedures for preserving curdlan in the dry state include both dehydration with organic solvents and spray drying.

8 Properties

8.1 Gel Formation

Curdlan is water-insoluble but soluble in alkali, and forms two-types of gels: high-set or low-set gels. The high-set gel is formed upon heating to around 80 °C, depending on its concentration in water (Maeda et al., 1967). The high-set curdlan gels are thermally irreversible and, unlike agar gel, further heating does not cause them to revert to the liquid state; they are also very elastic and resilient, whereas agar gels are brittle and relatively fragile. Curdlan forms a low-set reversible gel when its suspension is heated to 55–65 °C and then cooled. When a low-set gel is heated to 80 °C, however, it turns into a high-set gel. At the same curdlan concentration, a low-set gel has a weaker strength than a high-set gel. Changes in external factors such as pH, temperature, and ionic strength greatly affect gelation ability. Curdlan also forms a reversible gel when its alkaline solution is neutralized, and the addition of calcium or magnesium ions to a weakly alkaline solution of curdlan produces a gel with a bridged structure (Aizawa et al., 1974). Other methods to prepare curdlan gels include cooling DMSO solutions, or dialyzing its alkaline or DMSO solutions in water (Ogawa et al., 1972). The strength of the curdlan gel increases with temperature, concentration, and heating time (Maeda et al., 1967; Takeda technical report, 1997) (Figure 5). Curdlan, unlike other gelling agents, has a unique ability to form gels over a wide range of pH values (3.0–10.0).

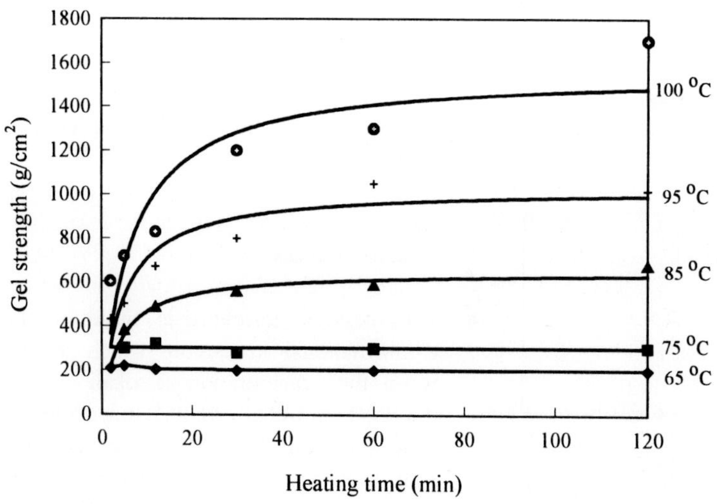

Fig. 5 Effect of heating temperature and time on curdlan gel strength (2% gel) (Takeda technical report, 1997).

8.2 Immunestimulatory Activity

β-(1,3)-Glucan has been found to be effective in stimulating macrophages (leukocytes or white blood cells). Macrophages form the immune system's first line of defense against foreign invaders, and can recognize and kill tumor cells, remove foreign debris resulting from oxidative and radiation damage, speed up the recovery of damage tissue, further activate cytokines, and initiate an immune cascade system to mobilize B and T cells (Di Luzio et al., 1979; Di Luzio, 1983; Suzuki et al., 1992; Hadden, 1993). Curdlan is a bacterial polysaccharide composed entirely of β-(1,3)-D-glucosidic linkages, and has been reported to show antitumor activity (Sasaki and Takasuka, 1976; Yoshioka et al., 1985). However, differences in biological activities seem to be dependent upon the degree of branching, molecular conformation, and molecular weight. Sasaki et al. (1978) examined the effect of chain length on antitumor activity by using different molecular weight glucans, which were obtained by the acid hydrolysis of curdlan. The results show that insoluble glucans with a number-average DP >50 have strong antitumor activity. As described in Section 2, β-glucans exist in three conformers: single-helix, triple-helix, and random coil. Among these, the ordered (helical) conformations are considered to be biologically active forms (Bohn and BeMiller, 1995). Some researchers reported that the triple-helical structure of schizophyllan, a β-glucan obtained from *Schizophyllum commune*, is essential for its anti-tumor activity (Yanaki et al., 1983; Kojima et al., 1986). However, other studies have suggested that the single-helix is more potent in this respect (Saito et al., 1991; Aketagawa et al., 1993). Ohno et al. (1995, 1996) showed that both the triple- and single-helix conformers of schizopyllan are active against tumors and leukopenia. In addition, the triple-helix conformer had a significant antagonistic effect upon zymosan-mediated hydrogen peroxide synthesis in peritoneal macrophages, whereas the single-helix conformer showed strong activity on the synthesis of tumor necrosis factor, nitric oxide, and hydrogen peroxide. It remains unclear as to which of the helical conformers is the most active, but it is likely that the structure–activity relationship of the β-glucan-mediated immunopharmacological activities vary, and are dependent upon the assay systems adopted.

9 Applications

9.1 Food Applications

Since the time of curdlan's original discovery, it has been mainly proposed for use in the food industry as a possible extender or substitute for natural plant gums in food preparations requiring a thickening or bodying additive. The food applications of curdlan are summarized in Table 2. These food additives and essential ingredients are generally divided in terms of how much curdlan is used; food additives require less than 1% curdlan, and essential ingredients more than 1%. Curdlan is useful as a gelling material to improve the textural quality, water-holding capacity and thermal stability of various foods. The polymer can be added during the production process, before heating, either as a powder, or as a suspension or slurry in either water or aqueous alcohol. The foods employing these functions include soy-bean curd (tofu), sweet bean paste jelly, boiled fish paste, noodles, sausages, jellies, and jams (Masayuki and Yukihiro, 1990; Taguchi et al., 1991; Ken and Akirou, 1992; Masanori et al., 1992; Shunsuke and Masatoshi, 1992; Hideaki et al., 1993; Hiroki and Etsuko, 1993; Masahiro and Masaru, 1993; Masatoshi et al., 1993; Yukihiro and Takahiro, 1993; Yasuhiro, 1994; Masaaki and Takaaki, 1995; Masataka et al., 1997; Akihiro and Takaaki, 1998; Hiroshi et al., 1999; Kazushi, 1999; Masao and Toshimi, 1999; Takaaki, 1999). Various tests have shown that this polymer is safe. The gel of curdlan has properties intermediate between the brittleness of agar gel and the elasticity of gelatin (Kimura et al., 1973). Furthermore, since the β-(1,3)-glucan polymer is not readily degraded by human digestive enzymes, it offers the possibility of new calorie-reduced products, such as those often referred to as "dietetic" foods. The fact that curdlan set-gels are known efficiently to absorb high concentrations of sugars from syrups and are relatively resistant to syneresis, suggests a use in sweet jellies and various other dessert-type foods. The resistance of curdlan gels to degradation by freezing and thawing also indicates potential in frozen food products. The pseudoplastic flow behavior of curdlan-containing fluids points to a possible role as thickeners and stabilizers in liquefied foods, such as salad dressings and spreads. Curdlan is also used to produce an edible casting film for foods with other water-soluble polymeric substances having heat-sealability (Akira and Atsushi, 1989).

9.2 Pharmaceutical Applications

Curdlan sulfate having a β-(1,3)-glucan backbone showed high anti-AIDS (acquired immunodeficiency syndrome) virus activity, with few adverse side effects; therefore, curdlan sulfate has a growing potential as an anti-AIDS drug (Jagodzinski et al., 1994; Takeda-Hirokawa et al., 1997). The Phase I/II trials (toxicity testing) of curdlan sulfate for human immunodeficiency virus (HIV) carriers have been carried out in the United States under the auspices of the Food and Drugs Administration (FDA) since 1992. A sulfuric acid ester of curdlan low molecular polymer, with an average DP ranging from 10 to 400 was found to be an active antiretrovirus ingredient useful for preventing and treating infection because of its high water solubility, low toxicity and relatively high inhibitory activity upon human retrovirus multiplication (Takashi and Junji, 1991). Mikio et al. (1995) disclosed that a

Tab. 2 Application of curdlan as food additives and essential ingredient (Spicer et al., 1999)

Function	*Objective foods*
Essential ingredient	
Gelling agent	Dessert, jelly, pudding, dry mixes
Bulking agent	Dietetic foods, diabetic foods
Low-calorie foods	Edible film, edible fiber casings
Food additive	
Improving viscoelasticity	Noodle, hamburger, sausage
Binding agent	Hamburger, starch jelly
Water-holding agent	Noodle, sausage, ham, starch jelly
Prevention of deterioration	Frozen egg products
Masking of malodors or aromas	Boiled rice
Retention of shape	Starch jelly, dry desert mixes
Thickeners and stabilizers	Salad dressing, low-calorie foods, frozen foods
Coating agent	Flavors

curdlan derivative modified by reaction with glycidol developed excellent antiviral activity whilst showing extremely low toxicity. Evans et al. (1998) subsequently reported that the low toxicity of curdlan and its marked anti-invasion activity on merozoites makes it a potential auxiliary treatment for severe malaria. Moreover, linear β-(1,3)-glucan sulfate has potential as a blood coagulation inhibitor useful for both the prevention and treatment of thrombosis, because linear β-(1,3)-glucan sulfate has been found to have a strong inhibitory action on blood coagulation (Tsuneo, 1990). Curdlan is also used as a sustained release suppository containing drug-active components, such as indomethacin, diclofenac sodium and ibuprofen (Toshiko, 1992). Kanke et al. (1992) investigated the *in vitro* release of a curdlan tablet containing theophylline; drug release from the tablet was the lowest tested, and was unaffected by pH and various other ions. Kim, B.-S. et al. (2000) proposed that hydroxyethyl derivatives of curdlan can be used as protein drug delivery vehicles. A curdlan hydrolysate with a number average molecular weight ranging from 340 to 4000 was used as an immunoactivator and did not cause any adverse side effects (Masahiro et al., 1998a). Hiroshi et al. (1997) have also reported that linear or branched β-(1,3)-glucans such as lentinan and curdlan are effective in the treatment of dementia.

9.3 Agricultural Applications

Fumio et al. (1989) reported that polysaccharides with the β-(1,3)-glucan structure, such as lentinan, schizophyllan, curdlan or laminarin are useful for the multiplication of *Bifidobacterium* bacteria and, when given to animals in the food, also inhibit the intestinal putrefying bacteria, thereby preventing the animals from aging. In addition, curdlan is incorporated into fish feed mixtures to improve immune activity (Masahiro et al., 1998b).

9.4 Other Industrial Applications

Curdlan has been studied as a support for immobilized enzymes (Murooka et al., 1977). For example, in the immobilization of an enzyme or a mold using curdlan gel, the curdlan is gelatinized by heating an aqueous suspension containing curdlan, an enzyme or a mold (Toshio, 1986). Other potential industrial applications include the use of curdlan films and fibers which are water-insoluble, biodegradable and impermeable to oxygen. Curdlan is also used to prevent nozzle clogging, and the electrostatic damage of electronic components when used as a seal in an ink-jet recoding head (Yoichi et al., 1996). Furthermore, curdlan functions as a concrete admixture to produce concrete with high fluidity but low separating properties (Toshiyuki et al., 1995; Lee, I.-Y. et al., 1999b). Applications of curdlan other than in the food industry are shown in Figure 6.

10 Outlook and Perspectives

Curdlan has been well accepted in the meat processing and noodle industries for its extremely unusual property of forming gels that are resistant to freezing or retorting. Takeda has been marketing this innovative product in Japan for several years, and sales are promising. Indeed, production has continued to grow by more than 10% each year during the past eight years such that, in recent years, Takeda Chemical Industries, Ltd. has produced 600–700 tonnes of curdlan annually.

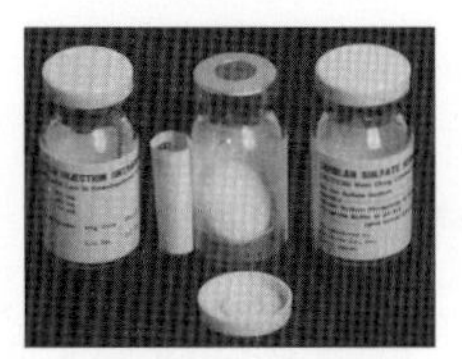

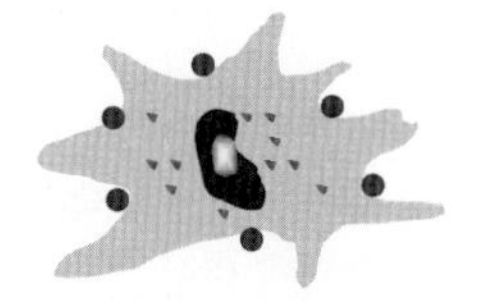

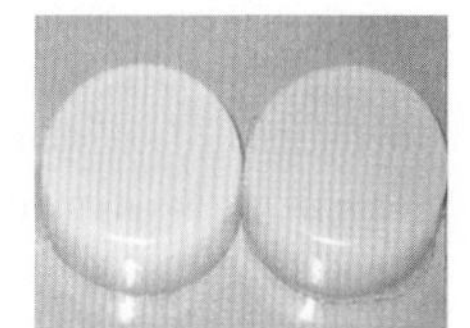

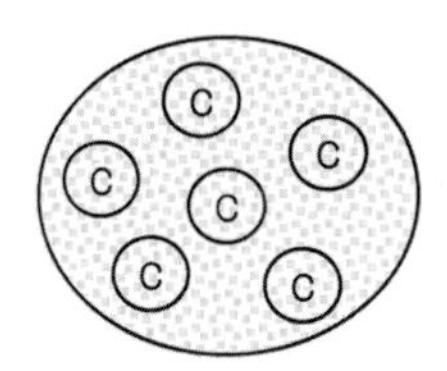

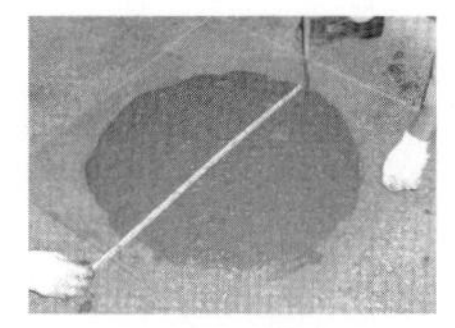

Fig. 6 Applications of curdlan in fields other than the food industry.

In addition, it should be noted that a variety of glucans with the β-1,3 linkage have immunestimulatory effects. Krestin, a β-glucan extracted from the mycelium of the basidiomycete *Coriolus versicolor*, has a general immunestimulatory effect in humans and has also been used as an anticancer agent (Kamisato and Nowakowski, 1988). Lentinan from *Lentinus edodes* is another glucan which inhibits tumor growth and increases resistance to infections by bacteria, viruses, and parasites (Jong and Birmingham, 1993). Schizophyllan (Numasaki et al., 1990) from *Schizophyllum commune* has the property of inducing the formation of cytotoxic macrophages and generating antitumor activity in the human body. In their preparation, these fungal glucans require extensive extraction operations to be performed in order to obtain comparatively small amounts of pure glucans, which makes the process very expensive. By contrast, curdlan as an exotype of β-(1,3)-glucan can be readily produced on a large scale by a submerged bacterial culture. Thus, if the material is correctly designed on a molecular

Tab. 3 Patent holders and publication year

Company	***Year of publication***													
	1988	***1989***	***1990***	***1991***	***1992***	***1993***	***1994***	***1995***	***1996***	***1997***	***1998***	***1999***	***2000***	***Total***
Ajinomoto	2		1							2				5
Asai							1							2
Chem Reizou	1													1
Dainippon	1	1							1	1	1			5
Daiichi		1												1
Daito										1				1
Endo Akira	1													1
Ezaki Glico														1
Fuji Oil	1				2	1							1	5
Hinoshoku												1	1	2
House foods	1							1						2
Hodaka						1								1
Ina						1					1		4	6
Japan Organo						2		2	2	7	2	2	2	19
Kanai Masako												1		1
Kanebo					2			1			1			4
Kanegafuchi				1			1				1			3
Kiteii					1									1
Kibun foods									1					1
Kyokuto				1										1
Meiji			2		1					1	1			5
Morigana												1		1
Nagai				1	1									2
Nikken									1					1
Nippon					2	1							1	4
Nitta Gelatin							1							1
Nokyo									1					1
Okada					2									2
Sanei		4		1			1		1					7
Sanyo	1													1
Shiseido							1		2					3
Snow Brand					1						2	1	2	6
Sumisho												1		1
Takeda		1	12	8	6	5	1	5	3	5	9	3	4	62
Taiyo Kagaku							1							1
Tsubuki								1						1
Tsukishima						1								1
Unie Colloid							2							2
Wako	5													5
Yamamoto										1				1
Total	13	7	15	12	18	12	8	13	9	19	19	11	15	171

basis to enhance its immunestimulatory activity, there is a large potential market in both the nutraceutical and pharmaceutical industries.

11 Patents

The Japanese patent literature between 1988 and 2000 was reviewed, as most of the patents relating to the application of curdlan were first filed in Japan. The companies holding patents, together with the annual number of patents filed, are listed in Table 3. Takeda and Japan Organo each hold about one-half of total 171 patents filed, indicating their major role in the development of curdlan application. It should also be noted that 122 patents relate to the application of curdlan as food additives or ingredients. The patents described in Section 8 are listed in Table 4.

Acknowledgments
The author thanks Mi-Kyoung Kim for her help in the preparation of the manuscript.

Tab. 4 A list of the patents of curdlan application

Patent no.	*Inventor(s)*	*Title*	*Year*
JP-61015688	Toshio, O.	Immobilized enzyme, or the like and preparation thereof	1986
JP-01289457	Akira, K., Atsushi, I.	Edible film	1989
JP-01137990A	Fumio, I., Kimikazu, I., Satoshi, S.	Polysaccharide having activity for multiplying *Bifidobacterium*	1989
JP-02124902	Tsuneo, A., Koichi, K., Junji, K.	Blood coagulation inhibitor	1990
JP-02249466	Masayuki, T., Yukihiro, N.	Noodle made of rice powder and producing method thereof	1990
JP-03218317	Takashi, T., Junji, K.	Anti-retrovirus agent	1991
JP-03157401	Taguchi, T., Hiroshi, K., Yukihiro, N.	Production of linear gel	1991
JP-04210507	Masanori, T., Tetsuya, T., Yukihiro, N.	Noodles	1992
JP-04330256	Shunsuke, O., Masatoshi, S.	Production of uncooked Chinese noodle and boiled Chinese noodle	1992
JP-04197148	Ken, O., Akirou, M.	Soybean protein-containing solid food	1992
JP-04074115	Toshiko, S.	Sustained release suppository	1992
JP-05184310	Masatoshi, H., Masami, F., Kenichi, H., Hiromi, K.	Devil's tongue for frozen food and its production	1993
JP-05207859	Hideaki, Y., Chieko, M., Takeshi, A.	Ganmodoki and its production	1993
JP-05023143	Yukihiro, N., Takahiro, F.	Ground fish meat or fish or cattle paste product and composition therefor	1993
JP-05076290	Hiroki, O., Etsuko, M.	Noodle-like soybean protein-containing food	1993
JP-05288909	Masahiro, K., Masaru, N.	Preparation of processed edible meat	1993
JP-06335370	Yasuhiro, S.	Formed jelly-containing liquid food and its production	1994
JP-07010624	Toshiyuki, U., Yoshio, T., Minoru, Y.	Additive for concrete	1995

Tab. 4 (cont.)

Patent no.	*Inventor(s)*	*Title*	*Year*
JP-07274876	Masaaki, A., Takaaki, I.	Quality improver for wheat flour product and wheat flour product having improved texture	1995
JP-07228601	Mikio, K., Yoshiro, O., Hirotomo, O., Yoshikazu, K., Hiroshi, I., Hisashi, M., Takeshi, K.	Water soluble β-1,3-glucan derivative and antiviral agent containing the derivative	1995
JP-0818787	Yoichi, T., Hiroshi, S., Makiko, K.	Biodegradable component for protecting ink jet recording head	1996
USP-5508191	Kanegae, Y. Yutani, A., Nakatsui, I.	Mutant strains of *Agrobacterium* for producing beta-1,3-glucan	1996
JP-09075017	Masataka, M., Hideo, Y., Yukihiro, N.	Devil's-tongue jelly and its production	1997
Patent number	Inventors	Title	Year
JP-09255579	Hiroshi, S., Nobuyoshi, N., Yutaro, K., Toru, M.	Medicine for treating dementia	1997
JP-10194977	Masahiro, K., Masaaki, K., Shinji, M., Hiroaki, K.	Immunoactivator	1998a
JP-10313794	Masahiro, K., Mikitomo, A., Akitomo, A., Tomonori, Y.	Feed composition for fish kind cultivation	1998b
JP-10014541	Akihiro, S., Takaaki, I.	Quality improving agent for fishery paste product and production of fishery paste product	1998
JP-11187819	Hiroshi, Y., Yuichi, S., Norio, I., Nana, I.	Frozen dessert food and its production	1999
JP-11075726	Masao, K., Toshimi, T.	Jelly-like food	1999
JP-11178533	Kazushi, M.	Production of bean curd including ingredient	1999
JP-11056246	Takaaki, I.	Quality improver for bean jam product and production of bean jam product	1999

12
References

Aizawa, M., Takahashi, M., Suzuki, S. (1974) Gel formation of curdlan-type polysaccharide in DMSO-H_2O mixed solvents, *Chem. Lett.* 193–196.

Aketagawa, J., Tanaka, S., Tamura, H., Shibata, Y., Saito, H. (1993) Activation of *Limulus* coagulation factor G by several (1 → 3)-β-D-glucans: comparison of the potency of glucans with identical degree of polymerization but different conformations, *J. Biochem.* **113**, 683–686.

Akihiro, S., Takaaki, I. (1998) Quality improving agent for fishery paste product and production of fishery paste product. Japanese patent 10014541.

Akira, K., Atsushi, I. (1989) Edible film. Japanese patent 01289457.

Azuma, I. (1987) Development of immunostimulants in Japan, in: *Immunostimulants: Now and Tomorrow* (Azuma, I., Jolles, G., Eds.), Tokyo: Japan Sci. Soc. Press/Berlin: Springer-Verlag, 41–45.

Bohn, J. A., BeMiller, J. N. (1995) (1 → 3)-β-D-glucans as biological response modifiers: a review of structure–functional activity relationships, *Carbohydr. Res.* **28**, 3–14.

Chuah, C. T., Sarko, A., Deslandes, Y., Marchessault, R. H. (1983) Triple-helical crystalline structure of curdlan and paramylon hydrates, *Macromolecules* **16**, 1375–1382.

Conti, E., Flaibani, A., O'Regan, M., Sutherland, I. W. (1994) Alginate from *Pseudomonas fluorescens* and *P. putida*: production and properties, *Microbiology* **140**, 1125–1132.

Deslandes, Y., Marchessault, R. H., Sarko, A. (1980) Triple-helical structure of (1 → 3)-β-D-glucan, *Macromolecules* **13**, 1466–1471.

Di Luzio, N. R. (1983) Immunopharmacology of glucan: a broad spectrum enhancer of host defence mechanisms, *Trends Pharmacol. Sci.* **4**, 344–347.

Di Luzio, N. R., Williams, D. L., McNamee, R. B., Edwards, B. F., Kitahama, A. (1979) Comparative tumor-inhibitory and anti-bacterial activity of soluble and particulate glucan, *Int. J. Cancer* **24**, 773–779.

Evans, S. G., Morrison, D., Kaneko, Y., Havlik, I. (1998) The effect of curdlan sulfate on development *in vitro* of *Plasmodium falciparum*, *Trans. R. Soc. Trop. Med. Hyg.* **92**, 87–89.

Farina, J. I., Sineriz, F., Molina, O. E., Perotti, N. I. (2001) Isolation and physico-chemical characterization of soluble scleroglucan from *Sclerotium rolfsii*. Rheological properties, molecular weight and conformational characteristics, *Carbohydr. Res.* **44**, 41–50.

Farres, J., Caminal, G., Lopez-Santin, J. (1997) Influence of phosphate on rhamnose-containing exopolysaccharide rheology and production by *Klebsiella* 1-714, *Appl. Microbiol. Biotechnol.* **48**, 522–527.

Fumio, I., Kimikazu, I., Satoshi, S. (1989) Polysaccharide having activity for multiplying *Bifidobacterium*. Japanese patent 01137990A.

Hadden, J. W. (1993) Immunostimulants, *Immunology Today* **14**, 275–280.

Harada, T. (1965) Succinoglucan 10C3: a new acidic polysaccharide of *Alcaligenes faecalis* var. *myxogenes*, *Arch. Biochem. Biophys.* **112**, 65–69.

Harada, T. (1977) Production, properties, and application of curdlan, in: *Extracellular Microbial Polysaccharides* (Sanford, P. A. Laskin, A., Eds.), Washington, DC: American Chemical Society, 265–283.

Harada, T., Yoshimura, T. (1965) Rheological properties of succinoglucan 10C3 from *Alcaligenes faecalis* var. *myxogenes*, *Agr. Biol. Chem.* **29**, 1027–1032.

Harada, T., Yoshimura, T., Hidaka, H., Koreeda, A. (1965) Production of a new acidic polysaccharide, succinoglucan by *Alcaligenes faecalis* var. *myxogenes*, *Agr. Biol. Chem.* **29**, 757–762.

Harada, T., Masada, M., Fujimori, K., Maeda, I. (1966) Production of a firm, resilient gel-forming polysaccharide by a mutant of *Alcaligenes faecalis* var. *myxogenes* 10C3, *Agr. Biol. Chem.* **30**, 196–198.

Harada, T., Misaki, A., Saito, H. (1968) Curdlan: a bacterial gel-forming β-1,3-glucan, *Arch. Biochem. Biophys.* **124**, 292–298.

Harada, T., Terasaki, M., Harada, A. (1993) Curdlan, in: *Industrial Gums* (Whistler, R. L., BeMiller, J. N., Eds.), San Diego, CA: Academic Press, Inc., 427–445.

Hideaki, Y., Chieko, M., Takeshi, A. (1993) Ganmodoki and its production. Japanese patent 05207859.

Hiroki, O., Etsuko, M. (1993) Noodle-like soybean protein-containing food. Japanese patent 05076290.

Hiroshi, S., Nobuyoshi, N., Yutaro, K., Toru, M. (1997) Medicine for treating dementia. Japanese patent 09255579.

Hiroshi, Y., Yuichi, S., Norio, I., Nana, I. (1999) Frozen dessert food and its production. Japanese patent 11187819.

Ielpi, L., Couso, R. O., Dankert, M. A. (1993) Sequential assembly and polymerization of the polyphenol-linked pentasaccharide repeating unit of the xanthan polysaccharide in *Xanthomonas campestris*, *J. Bacteriol.* **175**, 2490–2500.

Jagodzinski, P. P., Wiaderkiewicz, R., Kurzawski, G., Kloczewiak, M., Nakashima, H., Hyjek, E., Yamamoto, N., Uryu, T., Kaneko, Y., Posner, M. R., Kozbor, D. (1994) Mechanism of the inhibitory effect of curdlan sulfate on HIV-1 infection *in vitro*, *Virology* **202**, 735–745.

Janusz, M. J., Austen, K. F., Czop, J. K. (1986) Isolation of soluble yeast β-glucans that inhibit human monocyte phagocytosis mediated by β-glucan receptors, *J. Immunol.* **137**, 3270–3276.

Jong, S. C., Birmingham, J. M. (1993) Medicinal and therapeutic value of the Shiitake mushroom, *Adv. Appl. Microbiol.* **39**, 153–184.

Kai, A., Ishino, T., Arashida, T., Hatanaka, K., Akaike, Y., Matsuzaki, K., Kaneko, Y., Mimura, T. (1993) Biosynthesis of curdlan from culture media containing ^{13}C-labeled glucose as the carbon source, *Carbohydr. Res.* **240**, 153–159.

Kamisato, J. K., Nowakowski, M. (1988) Morphological and biochemical alterations of macrophages produced by a glucan, PSK, *Immunopharmacology* **16**, 88–96.

Kanegae, Y. Yutani, A., Nakatsui, I. (1996) Mutant strains of *Agrobacterium* for producing beta-1,3-glucan. US patent 5508191.

Kanke, M., Koda, K., Koda, Y., Katayama, H. (1992) Application of curdlan to controlled drug delivery. I. The preparation and evaluation of theophylline-containing curdlan tablets, *Pharmaceut. Res.* **9**, 414–418.

Kasai, N., Harada, T. (1980) Ultrastructure of curdlan. in: *Fiber Diffraction Methods*, (French, A. D., Gardner, K. H., Eds.), Washington, DC: American Chemical Society Symposium Series, 363–383.

Kazushi, M. (1999) Production of bean curd including ingredient. Japanese patent 11178533.

Ken, O., Akirou, M. (1992) Soybean protein-containing solid food. Japanese patent 04197148.

Kim, B.-S., Jung, I.-D., Kim, J.-S., Lee, J.-H., Lee, I.-Y., Lee, K.-B. (2000) Curdlan gels as protein drug delivery vehicles, *Biotechnol. Lett.* **22**, 1127–1130.

Kim, M.-K., Lee, I.-Y., Ko, J.-H., Rhee, Y.-H., Park, Y.-H., (1999) Higher intracellular levels of uridine monophosphate under nitrogen-limited conditions enhance metabolic flux of curdlan synthesis in *Agrobacterium* species, *Biotechnol. Bioeng.* **62**, 317–323.

Kim, M.-K., Lee, I.-Y., Lee, J.-H, Kim, K.-T., Rhee, Y.-H., Park, Y.-H. (2000) Residual phosphate concentration under nitrogen-limiting conditions regulates curdlan production in *Agrobacterium* species, *J. Ind. Microbiol. Biotechnol.* **25**, 180–183.

Kimura, H., Moritaka, S., Misaki, M. (1973) Polysaccharide 13140: a new thermo-gelable polysaccharide, *J. Food Sci.* **38**, 668–670.

Kojima, T., Tabata, K., Itoh, W., Yanaki, T. (1986) Molecular weight dependence of the antitumor activity of schizophyllan, *Agr. Biol. Chem.* **50**, 231–232.

Koreeda, A., Harada, T., Ogawa, K., Sato, S. Kasai, N. (1974) Study of the ultrastructure of gel-forming (1→3)-β-D-glucan (curdlan-type polysaccharide) by electron microscopy, *Carbohydr. Res.* **33**, 396–399.

Lawford, H. G., Rousseau, J. D. (1991) Bioreactor design considerations in the production of high-quality microbial exopolysaccharide, *Appl. Biochem. Biotechnol.* **28/29**, 667–684.

Lawford, H. G., Rousseau, J. D. (1992) Production of β-1,3-glucan exopolysaccharide in low shear systems, *Appl. Biochem. Biotechnol.* **34/35**, 597–612.

Lawford, H., Keenan, J., Phillips, K., Orts, W. (1986) Influence of bioreactor design on the rate and amount of curdlan-type exopolysaccharide production by *Alcaligenes faecalis*, *Biotechnol. Lett.* **8**, 145–150.

Lee, J.-H, Lee, I.-Y., Kim, M.-K., Park, Y.-H. (1999) Optimal pH control of batch processes for production of curdlan by *Agrobacterium* species, *J. Ind. Microbiol. Biotechnol.* **23**, 143–148.

Lee, I.-Y., Seo, W.-T., Kim, K.-G., Kim, M.-K., Park, C.-S., Park, Y.-H. (1997) Production of curdlan using sucrose or sugar cane molasses by two-step fed-batch cultivation of *Agrobacterium* species, *J. Ind. Microbiol. Biotechnol.* **18**, 255–259.

Lee, I.-Y, Kim, M.-K., Lee, J.-H., Seo, W.-T., Jung, J.-K., Lee, H.-W., Park, Y.-H., (1999a) Influence of agitation speed on production of curdlan by *Agrobacterium* species, *Bioprocess Eng.* **20**, 283–287.

Lee, I.-Y., Kim, S.-W., Lee, J.-H., Kim, M.-K., Cho, I.-S., Park, Y.-H. (1999b) A high viscosity of curdlan at alkaline pH increases segregational resistance of concrete, *Korean J. Biotechnol. Bioeng.* **14**, 1–5.

Maeda, I., Saito, H., Masda, M., Misaki, A., Harada, T. (1967) Properties of gels formed by heat treatment of curdlan, a bacterial β-1,3 glucan, *Agr. Biol. Chem.* **31**, 1184–1188.

Masaaki, A., Takaaki, I. (1995) Quality improver for wheat flour product and wheat flour product having improved texture. Japanese patent 07274876.

Masahiro, K., Masaru, N. (1993) Preparation of processed edible meat. Japanese patent 05288909.

Masahiro, K., Masaaki, K., Shinji, M., Hiroaki, K. (1998a) Immunoactivator. Japanese patent 10194977.

Masahiro, K., Mikitomo, A., Akitomo, A., Tomonori, Y. (1998b) Feed composition for fish kind cultivation. Japanese patent 10313794.

Masanori, T., Tetsuya, T., Yukihiro, N. (1992) Noodles. Japanese patent 04210507.

Masao, K., Toshimi, T. (1999) Jelly-like food. Japanese patent 11075726.

Masataka, M., Hideo, Y., Yukihiro, N. (1997) Devil's-tongue jelly and its production. Japanese patent 09075017.

Masatoshi, H., Masami, F., Kenichi, H., Hiromi, K. (1993) Devil's tongue for frozen food and its production. Japanese patent 05184310.

Masayuki, T., Yukihiro, N. (1990) Noodle made of rice powder and producing method thereof. Japanese patent 02249466.

Mikio, K., Yoshiro, O., Hirotomo, O., Yoshikazu, K., Hiroshi, I., Hisashi, M., Takeshi, K. (1995) Water soluble β-1,3-glucan derivative and antiviral agent containing the derivative. Japanese patent 07228601.

Misaki, A., Kishida, E., Kakuta, M., Tabata, K. (1993) Antitumor fungal (1→3)-β-D-glucans: structural diversity and effects of chemical modification. In: *Carbohydrates and Carbohydrate Polymers* (Yalpani, M., Ed.), Mount Prospect, IL: ATL Press, 116–129.

Murooka, Y., Yamada, T., Harada, T. (1977) Affinity chromatography of *Klebsiella* arylsulfatase on tyrosyl-hexamethylenediamine-β-1,3-glucan and immunoadsorbent, *Biochim. Biophys. Acta* **485**, 134–140.

Naganishi, I., Kimura, K., Kusui, S., Yamazaki, E. (1974) Complex formation of gel-forming bacterial (1→3)-β-D-glucan (curdlan type polysaccharide) with dyes in aqueous solution, *Carbohydr. Res.* **32**, 47–52.

Naganishi, I., Kimura, K., Suzuki, T., Ishikawa, M., Banno, I., Sakene, T., Harada, T. (1976) Demonstration of curdlan-type polysaccharide and some other β-1,3-glucan in microorganisms with aniline blue, *J. Gen. Appl. Microbiol.* **22**, 1–11.

Nakata, M., Kawaguchi, T., Kodama, Y., Konno, A. (1998) Characterization of curdlan in aqueous sodium hydroxide, *Polymer* **39**, 1475–1481.

Numasaki, Y., Kikuchi, M., Sugiyama, Y., Ohba, Y. (1990) A glucan Sizofiran: T cell adjuvant property and antitumor and cytotoxic macrophage inducing activities, *Oyo Yakuri Pharmacometrics* **39**, 39–48.

Ogawa, K., Watanabe, T., Tsurugi, J., Ono, S. (1972) Conformational behavior of a gel-forming (1→3)-β-D-glucan in alkaline solution, *Carbohydr. Res.* **23**, 399–405.

Ogawa, K., Tsurugi, J., Watanabe, T. (1973a) Effect of salt on the conformation of gel-forming β-1,3-D-glucan in alkaline solution, *Chem. Lett.* 95–98.

Ogawa, K., Miyagi, M., Fukumoto, T., Watanabe, T. (1973b) Effect of 2-chloroethanol, dioxane, or water on the conformation of a gel-forming β-1,3-glucan in DMSO, *Chem. Lett.* 943–946.

Ogawa, K., Tsurugi, J., Watanabe, T. (1973c) The dependence of the conformation of a (1→3)-β-D-glucan on chain length in alkaline solution, *Carbohydr. Res.* **29**, 397–403.

Ohno, N., Miura, N. N., Chiba, N., Adachi, Y., Yadomae, T. (1995) Comparison of the immunopharmacological activities of triple and single-helical schizophyllan in mice, *Biol. Pharm. Bull.* **18**, 1242–1247.

Ohno, N., Hashimoto, T., Adachi, Y., Yadomae, T. (1996) Conformation dependency of nitric oxide synthesis of murine peritoneal macrophages by β-glucans *in vitro*, *Immunol. Lett.* **52**, 1–7.

Okazaki, M., Adachi, Y., Ohno, N., Yadomae, T. (1995) Structure–activity relationship of (1→3)-β-D-glucans in the induction of cytokine production from macrophages, *in vitro*, *Biol. Pharm. Bull.* **18**, 1320–1327.

Okuyama, K., Obata, Y., Noguchi, K., Kusaba, T., Ito, Y., Ohno, S. (1996) Single structure of curdlan triacetate, *Biopolymers* **38**, 557–566.

Phillips, K. R., Lawford, H. G. (1983) Curdlan: Its properties and production in batch and continuous fermentations, in: *Progress Industrial Microbiology* (Bushell, D. E., Ed.), Amsterdam: Elsevier Scientific Publishing Co., 201–229.

Read, S. M., Currie, G., Bacic, A. (1996) Analysis of the structural heterogeneity of laminarin by electrospray-ionisation-mass spectrometry, *Carbohydr. Res.* **281**, 187–201.

Saito, H., Misaki, A., Harada, T. (1968) Comparison of structure of curdlan and pachyman, *Agr. Biol. Chem.* **32**, 1261–1269.

Saito, H., Ohki, T., Sasaki, T. (1977) A ^{13}C Nuclear Magnetic Resonance study of gel-forming (1 → 3)-β-D-glucans. Evidence of the presence of single-helical conformation in a resilient gel of a curdlan-type polysaccharide 13140 from *Alcaligenes faecalis* var. *myxogenes* IFO 13140, *Biochemistry* **16**, 908–914.

Saito, H., Yoshioka, Y., Uehara, N., Aketagawa, J., Tanaka, S., Shibata, Y. (1991) Relationship between conformation and biological response for (1 → 3)-β-D-glucans in the activation of coagulation factor G from limulus amebocyte lysate and host-mediated antitumor activity. Demonstration of single-helix conformation as a stimulant, *Carbohydr. Res.* **217**, 181–190.

Sakurai, T., Suzuki, I., Kinoshita, A., Oikawa, S., Masuda, A., Ohsawa, M., Tadomae, T. (1991) Effect of intraperitoneally administered β-1,3-glucan, SSG, obtained from *Sclerotinia sclerotiorum* IFO 9395 on the functions of murine alveolar macrophages, *Chem. Pharm. Bull.* **39**, 214–217.

Sasaki, T., Takasuka, N. (1976) Further study of the structure of lentinan, an anti-tumor polysaccharide from *Lentinus edodes*, *Carbohydr. Res.* **47**, 99–104.

Sasaki, T., Abiko, N., Sugino, Y., Nitta, K. (1978) Dependence on chain length of antitumor activity of (1 → 3)-β-D-glucan from *Alcaligenes faecalis* var. *myxogenes*, IFO 13140, and its acid-degraded products, *Cancer Res.* **38**, 379–383.

Shunsuke, O., Masatoshi, S. (1992) Production of uncooked Chinese noodle and boiled Chinese noodle. Japanese patent 04330256.

Spicer, E. J. F., Goldenthal, E. I., Ikeda, T. (1999) A toxicological assessment of curdlan. *Fd. Chem. Toxicol.* **37**, 455–479.

Stasinopoulos, S. J., Fisher, P. R., Stone, B. A., Stanisich, V. A. (1999) Detection of two loci involved in (1 → 3)-β-glucan (curdlan) biosynthesis by *Agrobacterium* sp. ATCC 31749, and comparative sequence analysis of the putative curdlan synthase gene, *Glycobiology* **9**, 21–31.

Sutherland, I. W. (1977) Microbial exopolysaccharide synthesis, in: *Extracellular Microbial Polysaccharides* (Sanford, P. A., Laskin, A., Eds.), Washington, DC: American Chemical Society, 40–57.

Sutherland, I. W. (1993) Biosynthesis of extracellular polysaccharides, in: *Industrial Gums* (Whistler, R. L., BeMiller, J. N., Eds.), San Diego, CA: Academic Press, Inc., 69–85.

Suzuki, T., Ohno, N., Saito, K., Yamdomae, T. (1992) Activation of the complement system by (1,3)-β-D-glucan having different degrees of branching and different ultrastructures, *J. Pharmacobiodyn.* **15**, 277–285.

Tada, T., Matsumoto, T., Masuda, T. (1997) Influence of alkaline concentration on molecular association structure and viscoelastic properties of curdlan aqueous systems, *Biopolymers* **42**, 479–487.

Taguchi, T., Hiroshi, K., Yukihiro, N. (1991) Production of linear gel. Japanese patent 03157401.

Takaaki, I. (1999) Quality improver for bean jam product and production of bean jam product. Japanese patent 11056246.

Takashi, T., Junji, K. (1991) Anti-retrovirus agent. Japanese patent 03218317.

Takeda technical report. (1997) Pureglucan: basic properties and food applications. Takeda Chemical Industries, Ltd. Japan.

Takeda-Hirokawa, N., Neoh, L. P., Akimoto, H., Kaneko, H., Hishikawa, T., Sekigawa, I., Hashimoto, H., Hirose, S.-I., Murakami, T., Yamamoto, N., Mimura, T., Kaneko, Y. (1997) Role of curdlan sulfate in the binding of HIV-1 gp120 to CD4 molecules and the production of gp120-mediated THF-α, *Microbiol. Immunol.* **41**, 741–745.

Toshio, O. (1986) Immobilized enzyme, or the like and preparation thereof. Japanese patent 61015688.

Toshiko, S. (1992) Sustained release suppository. Japanese patent 04074115.

Toshiyuki, U., Yoshio, T., Minoru, Y. (1995) Additive for concrete. Japanese patent 07010624.

Tsuneo, A., Koichi, K., Junji, K. (1990) Blood coagulation inhibitor. Japanese patent 02124902.

Williams, D. L., Pretus, H. A., Ensley, H. E., Browder, I. W. (1994) Molecular weight analysis of a water-insoluble, yeast derived (1 → 3)-β-D-

glucan by organic-phase size-exclusion chromatography, *Carbohydr. Res.* **253**, 293–298.
Yanaki, T., Ito, W., Tabata, K., Kojima, T., Norysuye, T., Takano, N., Fujita, H. (1983) Correlation between the antitumor activity of a polysaccharide schizophyllan and its triple-helical conformation in dilute aqueous solution, *Biophys. Chem.* **17**, 337–342.
Yasuhiro, S. (1994) Formed jelly-containing liquid food and its production. Japanese patent 06335370.
Yoichi, T., Hiroshi, S., Makiko, K. (1996) Biodegradable component for protecting ink jet recording head. Japanese patent 0818787.
Yoshioka, Y., Tabeta, R., Saito, R., Uehara, N., Fukuoka, F. (1985) Antitumor polysaccharides from *P. ostreatus* (Fr.) Quel: isolation and structure of a beta-glucan, *Carbohydr. Res.* **140**, 93–100.
Young, S.-H., Dong, W.-J., Jacobs, R. R. (2000) Observation of a partially opened triple-helix conformation in (1 → 3)-β-glucan by fluorescence resonance energy transfer spectroscopy, *J. Biol. Chem.* **275**, 11874–11879.
Yukihiro, N., Takahiro, F. (1993) Ground fish meat or fish or cattle paste product and composition therefor. Japanese patent 05023143.
Zhang, L., Ding, Q., Zhang, P., Zhu, R., Zhou, Y. (1997) Molecular weight and aggregation behaviour in solution of β-D-glucan from *Poria cocos* sclerotium, *Carbohydr. Res.* **303**, 193–197.

17
Succinoglycan

Dr. Miroslav Stredansky
Polytech, Area Science Park, Padriciano 99, 34012 Trieste, Italy;
Tel: +39-040-3756621; Fax: +39-040-9220016;
E-mail: miro@polytech3.area.trieste.it

Biotechnology of Biopolymers. From Synthesis to Patents. Edited by A. Steinbüchel and Y. Doi

ISBN: 3-527-31110-6

CoA	coenzyme A
EPS	exopolysaccharide
HMW	high molecular weight
LMW	low molecular weight
NMR	nuclear magnetic resonance
T_m	melting temperature
UDP	uridine 5′-diphosphate

1 Introduction

Succinoglycan is an acidic extracellular heteropolysaccharide produced by several strains of bacteria belonging to the genera *Rhizobium, Agrobacterium, Alcaligenes,* and *Pseudomonas* (Harada and Harada, 1996; Zevenhuizen, 1997). Succinoglycan chains are made up of octasaccharide repeating units modified with pyruvate, succinate, and acetate substituents. The repeat unit is branched, and contains one galactose and seven β-linked glucoses (Jansson et al., 1977; Chouly et al., 1995). The biosynthetic pathway of succinoglycan and the identity of the genes encoding the enzymes involved in it have been elucidated on the model symbiotic strain *R. meliloti* Rm 1201 and its mutants (Reuber and Walker, 1993a; Becker et al., 1993a; Gonzalez et al., 1996). The biosynthesis consists of the synthesis of nucleotide sugar precursors, stepwise assembly of the octasaccharide subunit, addition of decorations (succinate, pyruvate, and acetate), and finally polymerization followed by a transport to the cell surface.

Succinoglycan was found to play an important role in bacterium–plant symbiotic interactions (Long, 1989; Gonzalez et al., 1996). It was the first exopolysaccharide (EPS) to be recognized as being crucial for inducing the formation of nitrogen-fixing nodules on the roots of their leguminous host plants and it was considered to be a likely signal molecule. On the other hand, interesting rheological properties of succinoglycan together with satisfactory production processes led to its industrial applications as a thickening, suspending, emulsion-stabilizing, gel-forming, and precipitation agent (Sutherland, 1990).

2 Historical Outline

During studies on bacteria utilizable petrochemicals, Harada and Yoshimura (1964) isolated the strain *Alcaligenes faecalis* var. myxogenes, which grew on ethylene glycol and produced an acidic EPS containing succinate. The EPS was firstly named succinoglucan (Harada et al., 1965). Later the name was changed to succinoglycan because this polymer was found to contain also a minor portion of galactose besides glucose (Misaki et al., 1969). Subsequently some strains of the genera *Rhizobium, Agrobacterium,* and *Pseudomonas* were found to produce succinoglycans (Björndal et al., 1971; Zevenhuizen, 1971, 1997). The structure of the biopolymer was intensively studied and elucidated (Jansson et al., 1977; Åman et al.,

1981; Matulova et al., 1994; Chouly et al., 1995). Succinoglycans from all mentioned sources are composed of polymerized branched octasaccharide subunits, each of which consists of one galactose and seven glucoses (β-linked), and carries succinyl, pyruvyl, and acetyl modifications. Pyruvate is present in a stoichiometric ratio; the number and type of succinyl and acetyl groups vary, and depend on the bacterial origin and culture conditions. Only the biopolymer with this structure is called succinoglycan, although there are other polysaccharides containing succinate, e.g. marginalan (Osman and Fett, 1989). The biosynthetic pathway of succinoglycan and its genetic base have been fully determined (Reuber and Walker, 1993a; Gonzalez et al., 1996).

Succinoglycan is one of the best-understood symbiotically important EPSs. It is required for the successful invasion of nodules by bacteria in several *Rhizobium*–legume symbioses (Leigh and Walker, 1994). In addition to its great biological importance, succinoglycan has been produced industrially, because of its useful applications as an emulsion-stabilizing, suspending, and thickening agent (Sutherland, 1990). Various fermentation processes, such as batch, fed-batch, continuous, and solid state, have been developed for its production (Linton et al., 1987a; Stredansky and Conti, 1999a; Stredansky et al., 1999a; Knipper et al., 2000).

3
Chemical Structure

Succinoglycans from different sources have a common structure unit, a branched octasaccharide consisting of one galactose and seven glucoses with succinyl, pyruvyl, and acetyl modifications. The structure (Figure 1) was elucidated by various methods, such as sequential degradation of methylated saccharides followed by gas–liquid chromatography or high-performance liquid chromatography-mass spectroscopy (Jansson et al., 1977; Åman et al., 1981), enzymatic hydrolysis followed by paper or gas chromatography (Amemura and Harada,

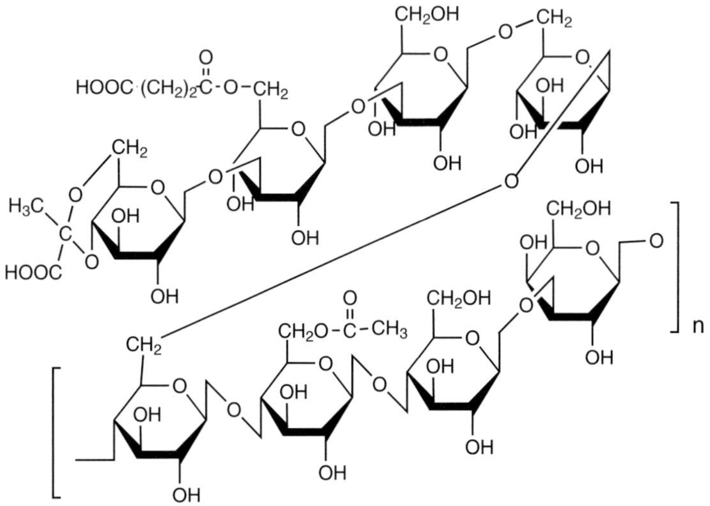

Fig. 1 Chemical structure of succinoglycan from *R. meliloti*.

1983), one- or two-dimensional nuclear magnetic resonance (NMR) studies (Matulova et al., 1994; Chouly et al., 1995; Evans et al., 2000), tandem mass spectroscopy with electrospray ionization, and collision-induced dissociation (Reinhold et al., 1994). The polymer chains have a neutral backbone with a regularly positioned tetrasaccharide side chain composed of β-D-glucose residues. The side chain, which contains pyruvate acetal linked at positions O^4 and O^6 on the terminal glucose residue, is connected to the main chain via a $\beta(1 \rightarrow 6)$ glycosidic linkage. Other glycosidic bonds in succinoglycan are $\beta(1 \rightarrow 3)$ and $\beta(1 \rightarrow 4)$. All hexosyl residues were determined to be β-pyranosidic from the low optical rotation and ^{1}H-NMR spectrum (Jansson et al., 1977). Each octasaccharide repeating unit (Figure 1) is further substituted by *O*-succinyl (side chain) and *O*-acetyl (main chain) groups. Their location was determined by detailed NMR studies (Matulova et al., 1994; Chouly et al., 1995) and by mass spectroscopy (Reinhold et al., 1994). The content of succinate and acetate per repeating unit is less than 1, and varies with the source. Molecular weight of succinoglycan is in the range from several hundred thousand to several million Daltons, and strongly depends on the used strain and culture conditions (Meade et al., 1995; Burova et al., 1996; Ridout et al., 1997; Stredansky et al., 1998).

4 Chemical Analysis and Detection

The detection and quantification of succinoglycan is important for both research work and monitoring production processes. However, there are no specific methods for succinoglycan determination. The only exception is a Calcofluor method used for the detection of succinoglycan-producing colonies of *Rhizobium* growing on agar plates (Leigh et al., 1985). The ability of succinoglycan to bind laundry whitener Calcofluor and fluoresce under UV light greatly facilitated subsequent genetic analysis of succinoglycan's biosynthesis, and its symbiotic roles, regulation, and degradation (Leigh and Walker, 1994). This method has never been applied for quantitative succinoglycan determination. Thus classical methods based on the EPS precipitation from cell-free supernatant with organic solvents (isopropanol, acetone, and ethanol) or quaternary ammonium compounds followed by gravimetry have been most frequently used (Meade et al., 1995; Linton et al., 1987a; Navarini et al., 1997; Stredansky et al., 1998). The precipitation with quaternary ammonium compounds is more specific for acidic EPS. The colorimetric determination of succinoglycan after the anthrone–sulfuric acid reaction was also reported (Zevenhuizen and Faleschini, 1991; Glucksmann et al., 1993b). The quantitative determination of nonsaccharide substituents was performed either directly by colorimetry (Zevenhuizen and Faleschini, 1991) and by NMR analysis (Matulova et al., 1994) or after mild hydrolysis of the EPS. Acetate and succinate are liberated by alkali hydrolysis and pyruvate by acidic hydrolysis (Reuber and Walker, 1993a). Subsequently, they could be determined by well-known enzymatic, spectrophotometric, or chromatographic methods.

5 Occurrence

Succinoglycan is produced and secreted by soil bacteria belonging to the genera *Rhizobium*, *Agrobacterium*, *Pseudomonas*, and *Alcaligenes*, which also often produce other EPSs (Harada and Harada, 1996; Zevenhuizen, 1997). Harada and Yoshimura (1964) isolated the soil bacterium *A. faecalis* var. myxogenes 10C3, which grew on ethylene glycol and produced succinoglycan. The production was significantly improved when sugars were used as a carbon source (Harada et al., 1965). The strain was later reclassified as belonging to the genera *Agrobacterium* (Yokota and Sakane, 1991). Thus the ability of *Alcaligenes* strains to produce succinoglycan is questionable, because no other strains of this genera were reported to synthesize it. On the other hand, a number of strains of *Rhizobium, Agrobacterium*, and *Pseudomonas* were found to produce succinoglycan. In the genera *Rhizobium* these include *R. meliloti* (Björndal et al., 1971), *R. trifolii, R. phaseoli, R. lupini* (Amemura and Harada, 1983), and *R. hedysari* (Navarini et al., 1997). Succinoglycan of *Rhizobium* strains is involved in symbiotic processes with legumes and is also termed EPS I or EPS a. In the genera *Agrobacterium* these include *A. radiobacter* (Harada et al., 1979; Linton et al., 1987a) and *A. tumefaciens* (Zevenhuizen, 1973; Stredansky et al., 1998). The *Pseudomonas* strains producing succinoglycan have been not fully classified (Cripps et al., 1981; Meade et al., 1995; Zevenhuizen, 1997).

6 Biological Function

A number of *Rhizobium* strains are capable of inducing the formation of nitrogen-fixing nodules on the roots of their leguminous host plants. This process involves specific recognition and progressive differentiation of both bacterial and host cells, where bacterial EPS play an important role. The *Rhizobium*–legume symbiosis is of a great economic importance, since legumes are widely cultivated as food or forage crops and actively used in agricultural rotations.

Nodulation of the leguminous plants by *Rhizobium* is a complex developmental process that requires a series of signal exchanges between the host and bacteria (Long, 1989). Plant flavonoids and bacterial Nod factors are among the best characterized signals involved in this interorganismal communication process. These specific flavonoids are secreted from the roots and induce Nod factor production by bacteria (van Rhijn and Vanderleyden, 1995). The Nod factor then induces root hair curling and initiates nodule development (Denarie and Cullimore, 1993). These are colonized by *Rhizobium* and form infection threads filled with bacterial cells. The infection threads elongate inside the root hairs, enter the root cortex, and release bacterial cells into newly formed plant cells in the nodule primordia. The bacterial cells then differentiate into bacteroids and fix nitrogen inside those plant cells (Kijne, 1992). In order to invade the nodules they elicit on alfalfa roots, *R. meliloti* cells must be able to synthesize at least one of the following polysaccharides: succinoglycan (EPS I), galactoglucan (EPS II), or a particular capsular polysaccharide (Leigh and Walker, 1994; Reuhs et al., 1995).

Succinoglycan was the first such EPS to be recognized as being crucial for bacterial invasion of nodules and was considered to be a likely signal. Although nodules formed in response to *exo* mutants (not producing succinoglycan), indicating that succinoglycan was neither the initiating signal nor the primary determinant of host specificity, they remained uncolonized, indicating that succinoglycan was required for nodule invasion (Leigh et al., 1985). Invasion appeared to be blocked at the stage where the infection thread penetrates beyond the epidermal cell layer. The structural features of the EPS required for nodule invasion suggest that its role is more specific than previously attributed to bacterial EPS. The first clue to this specificity was the observation that *exoH* mutants, which produce succinoglycan that lacks the succinyl moiety, have the same invasion phenotype as *exo* mutants that produce no succinoglycan: they fail to invade nodules (Leigh and Walker, 1994). The succinoglycan produced by *exoH* mutants also lacks the low-molecular-weight (LMW) fraction produced by the wild strain. This might be because the succinyl moiety is required for degradation of the EPS to LMW forms (York and Walker, 1998a). Thus either succinylated or LMW succinoglycan is required for stimulating the infection thread in the plant (Battisti et al., 1992; Cheng and Walker, 1998a). On the contrary, the acetyl substituent is not required (Reuber and Walker, 1993b). The specificity is further supported by the fact that the deficiency of *exo* mutants in nodule invasion can be partially suppressed by adding homologous EPS to the plant root at the time of inoculation (Djordjevic et al., 1987; Battisti et al., 1992; Urzainqui and Walker, 1992). It has been shown for three different *Rhizobium*–plant pairs that wild-type EPS from the usual nodulating strain will partially restore invasion of an *exo* mutant strain, while EPS from another species will not. Furthermore, invasion of alfalfa nodules by *exo* mutants of *R. meliloti* requires a particular LMW succinoglycan which is a trimer of the repeating subunit having succinyl and pyruvyl substituents (Gonzalez et al., 1998). The LMW succinoglycan is produced either by its direct biosynthesis or by expressing glycanases that cleave nascent HMW succinoglycan (York and Walker, 1998a).

The biological function of succinoglycan in *Pseudomonas* and *Agrobacterium* is unknown. It is apparently unnecessary in *A. tumefaciens*–plant pathogenic processes (Leigh and Coplin, 1992). However, the production and function of succinoglycan in bacterial ecosystems should not be restricted to a mere specific role during symbiosis of *Rhizobium* bacteria with their appropriate host plant; it points to a much more general functioning in bacterial systems, such as water-holding capacity, adhesive properties in bacterial flocs and biofilms of water organisms, and in aiding soil aggregation and stability (Zevenhuizen, 1997).

Tab. 1 The genes involved in the succinoglycan biosynthesis

Exo genes	*Function of encoded protein(s)*	*Reference*
C	phosphoglucomutase	Uttaro et al., 1990
N	UDP-pyrophosphorylase	Glucksmann et al., 1993b
B	UDP-glucose-4-epimerase	Buendia et al., 1991
Y, F	transfer of galactose onto the lipid carrier	Gray et al., 1990; Müller et al., 1993
A, L, M, O, U, W	glucosyl transferases	Glucksmann et al., 1993a
Z	acetylation	Buendia et al., 1991
H	succinylation	Becker et al., 1993b
V	pyruvylation	Glucksmann et al., 1993b
P, Q, T	polymerization/export of succinoglycan	Glucksmann et al., 1993b, Müller et al., 1993
K	cleavage of HMW succinoglycan	Becker et al., 1993b
X, R, S	regulation of biosynthesis	Gray et al., 1990; Reed et al., 1991a; Cheng and Walker, 1998b

7
Biosynthesis

The biosynthetic pathway of succinoglycan and the identity of the genes encoding the enzymes involved in it (Table 1) have been studied intensively on the model symbiotic strain *R. meliloti* Rm 1201 and its mutants (Reuber and Walker, 1993a; Becker et al., 1993a; Gonzalez et al., 1996). The pathway in strains of *A. tumefaciens* looks to be very similar (Cangelosi et al., 1987). Moreover many biosynthetic steps, enzymes, and gene sequences involved in succinoglycan synthesis showed a similarity to their counterparts in the biosynthesis of other bacterial EPSs (Reuber and Walker, 1993a; Coplin and Cook, 1990). Succinoglycan biosynthesis consists of the synthesis of sugar precursors, stepwise assembly of the octasaccharide subunit, addition of substituents, and finally polymerization followed by transport to the cell surface.

7.1
Biosynthetic Pathway

Succinoglycan is synthesized by a set of more than 20 *exo* gene products (Glucksmann et al., 1993a,b; Becker et al., 1995). The assembly of polymer subunits takes place on lipid carriers on the cytoplasmic face of the cytoplasmic membrane and polymerization takes place on the periplasmic side (Tolmasky et al., 1982; Reuber and Walker, 1993a). Sugar precursors, acetyl-coenzyme A (CoA), succinyl-CoA, and phosphoenolpyruvate are derived from the glucose catabolism pathway (Entner–Doudoroff and Krebs cycle) (Dussap et al., 1991). The scheme of succinoglycan synthesis is illustrated in Figure 2. The *exoC* gene encoding for phosphoglucomutase is the only one located on the chromosome. *exoN* protein is 42% identical to the *Acetobacter xylinum cel*A gene product, which codes for a uridine 5′-diphosphoglucose (UDP)-pyro-

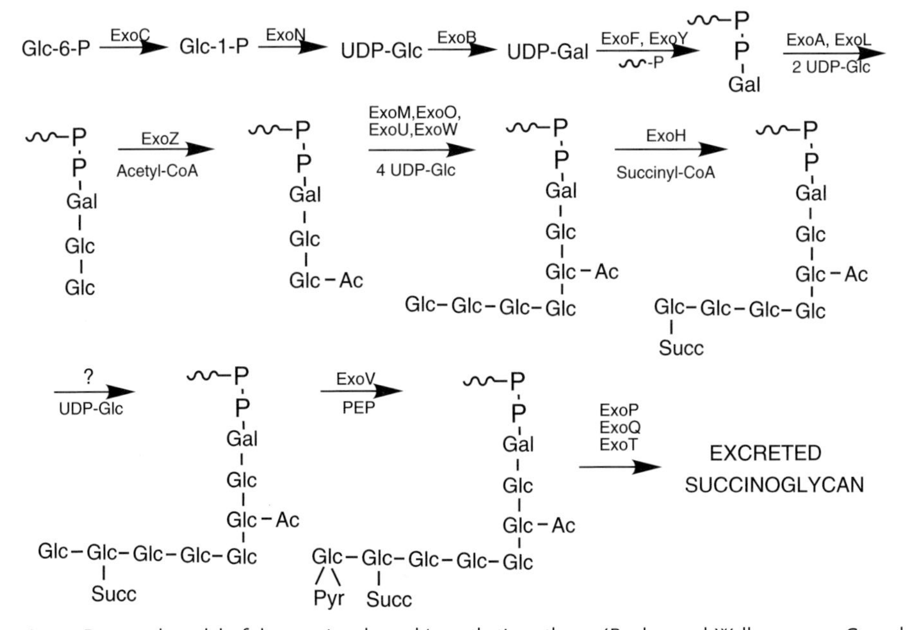

Fig. 2 Proposed model of the succinoglycan biosynthetic pathway (Reuber and Walker, 1993a; Gonzalez et al., 1996).

phosphorylase catalyzing formation of UDP-glucose (Glucksmann et al., 1993b). *exoB* functions as a UDP-glucose-4-epimerase (Buendia et al., 1991). The *exoY*, galactosyl-1-P-transferase, and *exoF* gene products are required for the transfer of galactose onto the lipid carrier. The *exoA, exoL, exoM, exoO, exoU,* and *exoW* gene products showed glucosyltransferase activities (Glucksmann et al., 1993a). The *exoZ* and *exoH* genes code for transmembrane proteins, which are needed for acetyl and succinyl modification, respectively. The *exoV* product is a ketalase responsible for transferring pyruvate to the terminal glucose (Reuber and Walker, 1993a; Glucksmann et al., 1993b). The membrane-associated proteins *exoP, exoQ,* and *exoT* are involved in polymerization and export of succinoglycan, where the *exoQ* produces high-molecular-weight (HMW) and the *exoT* produces LMW polymer (Gonzalez et al., 1996; Becker and Pühler, 1998). Another way to produce LMW succinoglycan, which is important in *Rhizobium*–legume interactions, is the cleavage of HMW succinoglycan by the *exoK* gene product showing endo-β-glycanase activity (Becker et al., 1993b).

7.2 Genetic Basis of Biosynthesis

The work of Gonzalez et al. (1996) showed that succinoglycan is synthesized by a set of 22 *exo* gene products, with 19 of them clustered in a 25-kb region on the *R. meliloti* megaplasmid 2 (Figure 3). Becker et al. (1995) extended it and showed that 21 *exo* and two *exs* genes are located in a 27-kb gene cluster on this megaplasmid. The *exp* gene cluster, responsible for the synthesis of EPS II, (galactoglucan) is localized on the same megaplasmid, separated from the *exo* cluster by about 200 kb (Charles and Finan, 1991). The DNA sequence of the *exo* region has been determined. Functional and evolutionary relatedness of genes for EPS synthesis in various *Rhizobium* strains were found (Zhan et al., 1990; Leigh and Coplin, 1992), and similarity to some genes encoding EPS in other bacteria is also apparent (Coplin and Cook, 1990).

7.3 Regulation of Biosynthesis

The succinoglycan synthesis in *Rhizobium* is regulated at both the transcriptional and posttranscriptional levels. The EPS production is regulated in part by the *exoS, chvI,* and *exoR* (a negative regulatory locus) genes located on the chromosome (Cheng and Walker, 1998b; Reed et al., 1991a). *exoS* and *chvI* proteins constitute a two-component regulatory system involved in controlling *exo* gene expression. *exoS* is the membrane sensor and *chvI* is the response regulator. The negative regulatory nature of *exoR* is supported by the existence of several independent insertion mutations, all of which caused increased EPS synthesis, and by the recessive nature of the mutations. Alkaline phosphatase activities of *exo–phoA* fusions indicate that *exoR* regulates the transcription or translation of most of the genes of the *exo*

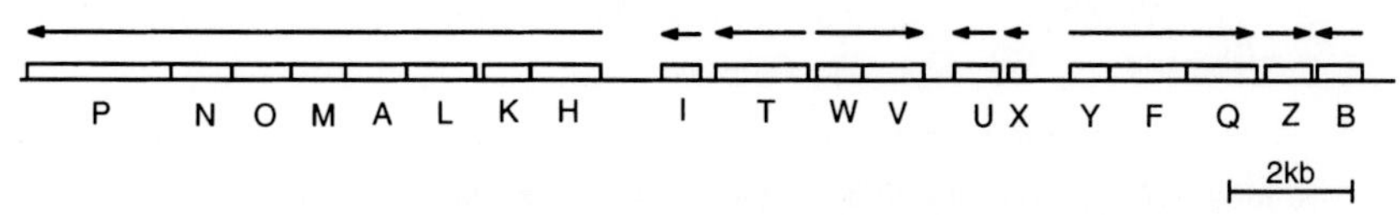

Fig. 3 Map of the *exo* gene cluster (Glucksmann et al., 1993b; Gonzalez et al., 1996).

cluster, with the exception of *exoB* (Reuber et al., 1991).

Another regulatory mechanism, mediated by the *exoX* and *exoY* genes, is conserved among several species of *Rhizobium* (Reed et al., 1991b; Zhan and Leigh, 1990). A multicopy of the small protein encoded by *exoX* inhibits EPS synthesis posttranscriptionally, because the gene dosage of *exoX* does not alter the expression of translational fusions to other *exo* genes. The inhibitory effect of *exoX* is counterbalanced in a copy number-dependent manner by *exoY*, which stimulates EPS synthesis.

The recently identified chromozomal *mucR* gene from *Rhizobium* codes for the regulatory protein that plays a key role in the control of the biosynthesis of both EPSs important in the nodulation process, i.e. succinoglycan and galactoglucan (Keller et al., 1995). The encoded protein exerts a positive effect on the transcription of the *exoK*, *exoY*, and *exoF* genes.

Succinoglycan synthesis has been shown to be regulated both in the free-living stage and during symbiosis. In the free-living stage, EPS production by *R. meliloti* is sensitive to the concentration of ammonia, phosphate, and sulfate present in the growth media. For the symbiotic stage, gene fusions have been used to show that the *exo* genes are expressed only in the invasion zone inside the nodule, not in the more mature nodule regions (Reuber et al., 1991; Mendrygal and Gonzalez, 2000).

8
Biodegradation

Succinoglycan is a fully biodegradable biopolymer. A partial cleavage of HMW to LMW succinoglycan in *R. meliloti* plays an important role in the nodulation process (York and Walker, 1998b). The *exoK* gene encodes a protein employed in this cleavage. This gene was sequenced and its high homology to β(1,3)–β(1,4)-glucanases genes present in various bacteria was found (Becker et al., 1993b; Glucksmann et al., 1993b).

In addition to succinoglycan-producing bacteria, other microorganisms also showed ability to degrade this polymer and enzymes from various sources were used in the structure elucidation (Harada et al., 1979; Amemura and Harada, 1983). *Cytophaga arvensicola* was capable growing on succinoglycan as the sole carbon source by means of extracellular succinoglycan depolymerase and intracellular endo-(1 → 6)β-D-glucanase. The former enzyme cleaves the polymer to octasaccharide repeating units and the latter enzyme hydrolyzes the octasaccharide to two tetrasaccharides. These could be further hydrolyzed by other enzymes of this bacterium or by glycosidases from other various sources to form oligo- and monosaccharides (Harada and Harada, 1996).

9
Solution and Rheological Properties

Succinoglycan is a polyelectrolyte and exhibits reversible pseudoplastic behavior in aqueous solutions. It does not form gels, but gives rise to extremely viscous solutions characterized under proper conditions by a viscoelastic behavior typical of ‘weak gels’. This behavior is exhibited by worm-like polysaccharides in aqueous solution at sufficiently high concentration and/or molecular weight and under conditions of an ordered conformation. The rheological properties are sensitive to the bacterial origin of the EPS, and hence to the level and type of noncarbohydrate substituents (Gravanis et al., 1990b). The viscosity of succinoglycan solutions is affected by the ionic strength, shear rate, nature of counterions, concentration, and molecular weight of the polymer (Gravanis et al., 1990b; Ridout et al., 1997). Succinoglycans show a sharp order–disorder transition on heating and cooling. The conformational transition was studied by various techniques, such as NMR, optical rotation, differential scanning calorimetry, conductivity and viscosity measurements, and atomic force microscopy (Dentini et al., 1989; Gravanis et al., 1990a; Burova et al., 1996; Ridout et al., 1997; Balnois et al., 2000). The ordered form is promoted by lowering the temperature and/or raising the ionic strength. Variation of the transition temperature (T_m, melting temperature) with ionic strength is consistent with the behavior expected for a single helix to stretched coil transition. Burova et al. (1996) showed the single helix–coil type of transition mechanism at low EPS concentrations in salt-free solution, and with increasing EPS and/or salt concentrations the transition included two stages: the cooperative dissociation of the helix dimer and subsequent two-state melting of the helix monomer. The reported T_ms are in the range from 60 to 75 °C depending on the ionic strength, EPS composition (presence or absence of acyl groups), and molecular weight. T_m could also be controlled by addition of various salts and alcohols, in particular polyols (Lau and Davies, 1992). The viscosity of succinoglycan samples falls on heating to above the T_m and increases on subsequent recooling. The first thermal cycle results in an irreversible decrease in viscosity as measured under ambient conditions and it is accompanied by a decrease in molecular weight (Dentini et al., 1989; Gravanis et al., 1990b; Stredansky et al., 1999a). Such changes have been attributed to chain breakage and/or disruption of aggregates. No further reductions occur on subsequent thermal recycling.

10
Biotechnological Production

Succinoglycan is a subject of commercial interest, thus various processes, such as batch, fed-batch, continuous, and solid-state fermentation, have been developed for its production using strains of the genera *Rhizobium*, *Agrobacterium*, and *Pseudomonas* (Cripps et al., 1981; Azoulay, 1983; Linton et al., 1987a; Stredansky and Conti, 1999a; Stredansky et al., 1999a; Knipper et al., 2000).

During the course of fermentation, the excretion of EPS results in a highly viscous shear-thinning broth. The rheological behavior of the viscous broth causes serious problems with regard to mixing, heat transfer, and oxygen supply, thus limiting both the maximum polymer concentration achievable and product quality. As in the case of many microbial EPS production processes, a specific design of bioreactors, in particular impeller systems, is necessary for improving the bioprocess performance. The design of the agitation system requires special attention to the correct distribution of power to ensure good culture homogeneity and turbulence, to minimize bubble coalescence, to promote small bubble formation, and to achieve adequate fluid movement at heat transfer surfaces (Margaritis and Pace, 1985).

Control of the cultivation medium composition, its addition, and other environmental parameters are critical in achieving the desired rates of synthesis and yields of succinoglycan. The media suited for succinoglycan production are relatively poor with a high carbon/nitrogen ratio (Linton et al., 1987a; Stredansky et al., 1999a; Knipper et al., 2000). The best carbon sources are sugars (glucose, sucrose, and maltose) and alditols (mannitol and sorbitol) in the concentration range from 1 to 5% with 50–70% conversion into polymer. A high sugar concentration had an inhibitory effect (Stredansky et al., 1998). Nitrogen sources, such as ammonium salts, nitrates, glutamate, and lysine, are used mostly as growth-limiting factors, thus the excess energy deriving from carbon substrate metabolism is committed to EPS synthesis. A low level of minerals and other growth factors is also important. The optimum cultivation temperature for succinoglycan production by most strains is around 30 °C. Maintenance of a neutral pH during the process is required. Oxygen availability depends on the performance of aeration and agitation systems (discussed above), and showed a significant effect. Under nonlimited oxygen conditions a better yield of succinoglycan with a HMW was reached (Stredansky et al., 1998, 1999a).

10.1 Continuous Process

The major advantages of the continuous process are a high production rate and overall bioreactor productivity. On the other hand, some problems arise connected with poor long-term stability of the strain, higher contamination risk, and diluted product, which increases downstream costs. The process is easily controlled through a single limiting nutrient and the dilution rate. Detailed studies of the continuous production of succinoglycan by *A. radiobacter* were reported by Linton and his coworkers (Linton et al., 1987a,b; Cornish et al., 1987). The maximum observed yield of EPS was 3.5 g/g O_2 and 0.67 g/g glucose in nitrogen-limiting chemostat culture at dilution rate 0.047 h^{-1} (steady-state) with a succinoglycan concentration of 9.8 g/L. When the observed yields were corrected for cell production, they came very close to the theoretical values. The yields of EPS decreased sharply when the carbon source was either more reduced or oxidized than glucose.

10.2 Batch and Fed-Batch Fermentation

The batch process is simpler than the continuous one; in particular, when performed on a large scale, the risk of strain instability and culture contamination is lower, and a higher final product concentration could be reached. The batch process looks to be favorable in the case of succinoglycan, which is formed prevailingly in the non-growth phase of cultivation. A rapid period of cell growth is followed by a stationary phase, during which the product is generated (Linton et al., 1985; Stredansky et al., 1998). Draw fill techniques may be applied to the batch process to increase overall bioreactor productivity. The best reported succinoglycan yields per substrate in batch process are high, more than 60%. On the other hand, the highest volumetric yields obtained in laboratory bioreactors are about 16 g/L (Dasinger and McArthur, 1992; Meade et al., 1995; Knipper et al., 2000), i.e. lower than those reported for some other EPS (Margaritis and Pace, 1985). However, improved yields could be expected in large-scale bioreactors with a suitable design and/or substrate feeding. Stredansky et al. (1999a) reported an increase of the succinoglycan yield of 47% using a simple fed-batch process, as compared to the batch one. The batch production allows also utilization of less-defined cheap substrates, e.g. wood hydrolyzates (Meade et al., 1995). The reported cultivation times are 3–4 days.

10.3 Solid-state Fermentation

Solid-state fermentation has been used for many centuries in Asian and African countries for the production of fermented food. It is a process in which microorganisms grow on a moist substrate in the absence of free water and which allows utilization of cheap raw materials and residues of agro and food industries as substrates (Pandey, 1992). Production of succinoglycan by *Agrobacterium* and *Rhizobium* strains and production of other EPS, e.g. xanthan, on spent malt grains impregnated with a nutrient solution have been recently reported (Stredansky et al., 1999b,c, Stredansky and Conti, 1999a,b). The solid-state fermentation process overcomes the problems connected to the highly viscous broth typical of the submerged cultivation. The highest succinoglycan yield was achieved with *A. tumefaciens* in a 4-day static cultivation, reaching 42 g/L of impregnating solution, corresponding to 30 g/kg of total wet solid substrate (Stredansky and Conti, 1999a). The increased EPS production in a solid-state fermentation system could be partially accounted for by a higher tolerance of cells to a high sugar concentration. The enhanced metabolic activity observed in microorganisms growing in solid-state fermentation might be stimulated by the formation of a concentration gradient of nutrients, causing a local drop in substrate concentration, which significantly minimizes catabolite repression. However, many scale-up problems of the solid-state fermentation process have to be solved before a laboratory scale process can be transferred to the large-scale level.

10.4 Recovery and Purification

The cost of recovery of EPS is a significant part of the total production cost due to the dilute nature of the stream leaving the bioreactor, the presence of cells and solutes in the stream, and its high viscosity. Cell removal is essential if succinoglycan is to be used not only in cosmetics or foods, but also in enhanced oil recovery where the presence of solids in the polymer flood causes plugging of the pores of the oil-bearing rock. The high viscosity of the fermentation broth results in conventional separation steps such as centrifugation, filtration, and flocculation being less effective and slow. However, some advantages may be derived during processing by heating the broth above the T_m of succinoglycan (60–75 °C) to lower its viscosity, but care must be taken not to degrade the product (Gravanis et al., 1990b, Stredansky et al., 1999a).

To obtain a low-grade product, the cell-free supernatant could be concentrated and/or dried; to obtain a high-grade product, the EPS must be precipitated. Suitable precipitation agents are organic solvents, such as isopropanol (Linton et al., 1987a), acetone (Navarini et al., 1997) and ethanol (Stredansky et al., 1998), and or quaternary ammonium compounds (Meade et al., 1995). Increasing the concentration of EPS and salts prior to precipitation will decrease the amount of solvent processed per unit weight of polymer recovered. The precipitate is subsequently dewatered, dried, and milled to a suitable particle size. A simple and effective isolation procedure was described by Stredansky et al. (1999a). The whole fermentation broth was heated to 90 °C after fermentation. Cell removal, deodorization, and decoloration of the broth were simultaneously achieved by addition of activated charcoal (1% w/v), followed by stirring and filtration through a layer of washed kieselguhr. The warm filtrate could be directly dried or concentrated without cooling to save additional energy costs. As the fermentation process led to almost complete substrate utilization and the medium was poor in salts, succinoglycan of a high purity was obtained.

10.5 Patents and Commercial Products

The production of succinoglycan with various strains of genera *Rhizobium, Agrobacterium,* and *Pseudomonas* is covered by a number of patents (Table 2).

Several succinoglycan preparations have been produced industrially and introduced into the market. The company Shell marketed two products under the trademarks Shellflo-S and Enorflo-S produced with the strains *Pseudomonas* sp. NCIB 11592 and *A. radiobacter* NCIB 11883 (Clarke-Sturman et al., 1986; Linton et al., 1987a). Succinoglycan (a powder form) from *A. tumefaciens* I-736 modified by partial hydrolysis of acidic groups is actually marketed by the company Rhodia under the trademark Rheozan (around US$25/kg). The Halliburton Company offers the liquid concentrate of succinoglycan under the trademark FLO-PAC (about US$90/gallon). All these products are claimed for applications mainly in the oil industry.

11 Applications

Succinoglycan is used as a thickening, suspending, emulsion-stabilizing, gel-forming, and precipitation agent. The main field of its application is in the oil industry, where various synthetic and natural polymers are employed in various phases of the oil production. Succinoglycan possesses a combination of desirable properties for the use in this field: shear-thinning rheology, high stability at acidic and neutral pH, compatibility with anionic and nonionic surfactants, ease of mixing, temperature-insensitive viscosity below its T_m, and adjustable T_m over a wide range of temperature. It is often applied as a component of drilling fluids in the drilling of exploratory or production wells, when its lower T_m than that of xanthan is required (Linton et al., 1985; Lau and Davies, 1992; Aubert et al., 1999). Succinoglycan applications in replacement fluids used in enhanced oil recovery have been also patented (Cripps et al., 1981; Azoulay, 1983; Linton et al., 1985; Lau and Davies, 1992; Sydansk, 1992). The biopolymer-containing fluids and foam fluids are utilized for completion, workover, and kill operations in wells, and in profile modifications (Dasinger and McArthur, 1992; Rae and Johnston, 1995; Sydansk, 1998; Carpenter et al., 1998). Succinoglycan is particularly suitable in completing operations such as gravel packing, because it is effective in situations where cheaper hydroxyethylcellulose and xanthan are not (Lau, 1993; Lau and Bernardi, 1993). Gravel packing is the most commonly used method of sand control.

In addition to oil production, succinoglycan is used also in other fields. It has been employed in the manufacture of cosmetics, e.g. shampoos, creams, emulsions, and hair

Tab. 2 List of patents on the succinoglycan production

Patent	*Publication date*	*Patent holder*	*Inventors*
EP 040445A1	25 November 1981	Shell	R. E. Cripps, R. N. Ruffel and A. J. Sturman
US 4400466	23 August 1983	Dumas & Inchnuspe	E. E. Y. Azoulay
EP 0138255A2	24 April 1985	Shell	J. D. Linton, M. W. Evans and A. R. Godley
EP 0251638B1	22 April 1992	Pfizer	B. Dasinger and , H. A. I. McArthur
IT 1293620	8 March 1999	Polytech	M. Stredansky, E. Conti and C. Bertocchi
EP 0527061B1	10 May 2000	Rhodia Chimie	M. Knipper, A. Senechal and M. Raffart

dye compositions, as a thickener, stabilizer, and humectant (Guerin et al., 1998; Yoshida, 1998; Yoshida and Suzuki, 1998; Yoshida, 1999). Succinoglycan use in the production of detergents (Guerin and Knipper, 1991; Guillou, 1995), inks (Kondo and Matsubara, 1995; Kitaoka, 1998; Sugimoto and Yamamoto, 1999), and agricultural compositions (Philips et al., 1991; Macri et al., 1996) was described. Its application as a precipitation or stabilizing agent in the manufacturing processes of solid catalysts, adsorbents, and ceramics was also claimed (Chopin et al., 1991; Besnard et al., 1994). Although a potential use of succinoglycan in food has been described (Fung et al., 1992; Fuisz, 1993), it is probably not used for these purposes.

12
References

Åman P., McNeil, M., Franzen, L.-E., Darvill, A. G., Albersheim, P. (1981) Structural elucidation, using HPLC-MS and GLC-MS, of the acidic polysaccharide secreted by *Rhizobium meliloti* strain 1021, *Carbohydr. Res.* **95**, 263–282.

Amemura, A., Harada, T. (1983) Structural studies on extracellular acidic polysaccharides secreted by three non-nodulating *Rhizobia, Carbohydr. Res.* **112**, 85–93.

Aubert, D., Frouin, L., Morvan, M., Vincent, M.-M. (1999) Gel of an apolar medium, its use for the preparation of water-based drilling fluids, US patent 5 858 928.

Azoulay, E. E. Y. (1983) Process for the preparation of viscous water by action of microorganisms, US patent 4 400 466.

Balnois, E., Stoll, S., Wilkinson, K. J., Buffle, J., Rinaudo, M., Milas, M. (2000) Conformation of succinoglycan as observed by atomic force microscopy, *Macromolecules* **33**, 7440–7447.

Battisti, L., Lara, J. C., Leigh, J. A. (1992) Specific oligosaccharide form of the *Rhizobium meliloti* exopolysaccharide promotes nodule invasion in alfalfa, *Proc. Natl. Acad. Sci. USA* **89**, 5625–5629.

Becker, A., Pühler, A. (1998) Specific amino acid substitution in the proline-rich motif of the *Rhizobium meliloti exoP* protein results in enhanced production of low-molecular-weight succinoglycan at the expense of high-molecular-weight succinoglycan, *J. Bacteriol.* **180**, 395–399.

Becker, A., Kleickmann, A., Küster, H., Keller, M., Arnold, W., Pühler, A. (1993a) Analysis of the *Rhizobium meliloti* genes *exoU, exoV, exoW, exoT,* and *exoI* involved in exopolysaccharide biosynthesis and nodule invasion: *exoU* and *exoW* probably encode glucosyltransferases, *Mol. Plant–Microbe Interact.* **6**, 735–744.

Becker, A., Kleickmann, A., Arnold, W., Pühler, A. (1993b) Analysis of the *Rhizobium meliloti exoH, exoK, exoL* fragment: *exoK* shows homology to excreted endo-β-1,3–1,4 glucanases and *exoH* resembles membrane proteins, *Mol. Gen. Genet.* **238**, 145–154.

Becker, A., Küster, H., Niehaus, K., Pühler, A. (1995) Extension of the *Rhizobium meliloti* succinoglycan biosynthesis gene cluster: identification of *exsA* gene encoding an ABC transporter protein, and the *exsB* gene which probably codes for a regulator of succinoglycan biosynthesis, *Mol. Gen. Genet.* **249**, 487–497.

Besnard, M.-M., David, C., Knipper, M. (1994) Mixed polysaccharide precipitating agents and insulating articles shaped therefrom, US patent 5 350 524.

Björndal, H., Erbing, C., Lindberg, B., Fähraeus, G., Ljunggren, H. (1971) Studies on an extracellular polysaccharide from *Rhizobium meliloti, Acta Chem. Scand.* **25**, 1281–1286.

Buendia, A. M., Enenkel, B., Köplin, R., Niehaus, K., Arnold, W., Pühler, A. (1991) The *Rhizobium meliloti exoZ/exoB* fragment of megaplasmid 2: *exoB* functions as a UDP-glucose-4-epimerase and *exoZ* shows homology to *NodX* of *Rhizobium leguminosarum* biovar. vicia strain TOM, *Mol. Microbiol.* **5**, 1519–1530.

Burova, T. V., Golubeva, I. A., Grinberg, N. V., Mashkevich, A. Y., Grinberg, V. Y., Usov, A. I., Navarini, L., Cesàro, A. (1996) Calorimetric study of the order-disorder conformational transition in succinoglycan, *Biopolymers* **39**, 517–529.

Cangelosi, G. A., Hung, L., Puvanesarajah, V., Stacey, G., Ozga, D. A., Leigh, J. A., Nester, E. W. (1987) Common loci for *Agrobacterium tumefaciens* and *Rhizobium meliloti* exopolysaccharide synthesis and their roles in plant interactions, *J. Bacteriol.* **169**, 2086–2091.

Carpenter, R. B., Wilson, J. M., Loughridge, B. W., Ravi, K. M., Jones, R. R. (1998) Well com-

pletion spacer fluids and methods, US patent 5 789 352.
Charles, T. C., Finan, T. M. (1991) Analysis of a 1600-kilobase *Rhizobium* megaplasmid using defined deletions generated *in vivo*, *Genetics* **127**, 5–20.
Cheng, H.-P., Walker, G. C. (1998a) Succinoglycan is required for initiating and elongation of infection threads during nodulation of alfalfa by *Rhizobium meliloti*, *J. Bacteriol.* **180**, 5183–5191.
Cheng, H.-P., Walker, G. C. (1998b) Succinoglycan production by *Rhizobium meliloti* is regulated through the *exoS–chvI* two-component regulatory system, *J. Bacteriol.* **180**, 20–26.
Chopin, T., Quemere, E., Nortier, P., Schuppiser, J.-L., Segaud, C. (1991) Mechanically improved shaped articles, US patent 4 983 563.
Chouly, C., Colquhoun, I. C., Jodelet, A., York, G., Walker, G. C. (1995) NMR studies of succinoglycan repeating-unit octasaccharides from *Rhizobium meliloti* and *Agrobacterium radiobacter*, *Int. J. Biol. Macromol.* **17**, 357–363.
Clarke-Sturman, A. J., Pedley, J. B., Sturla, P. L. (1986) Influence of anions on the properties of microbial polysaccharides in solution, *Int. J. Biol. Macromol.* **8**, 355–360.
Coplin, D. L., Cook, D. (1990) Molecular genetics of extracellular polysaccharide synthesis in vascular phytopathogenic bacteria, *Mol. Plant–Microbe Interact.* **3**, 271–279.
Cornish, A., Linton, J. D., Jones, C. W. (1987) The effect of growth conditions on the respiratory system of a succinoglucan-producing strain *Agrobacterium radiobacter*, *J. Gen. Microbiol.* **133**, 2971–2978.
Cripps, R. E., Ruffel, R. N., Sturman, A. J. (1981) Fluid displacement with heteropolysaccharide solutions, and the microbial production of heteropolysaccharides, European patent 040445A1.
Dasinger, B., McArthur, H. A. I. (1992) Aqueous gel composition derived from succinoglycan, European patent 0251638B1.
Denarie, J., Cullimore, J. (1993) Lipo-oligosaccharide nodulation factors: a new class of signalling molecules mediating recognition and morphogenesis, *Cell* **74**, 951–954.
Dentini, M., Crescenzi, V., Fidanza, M., Coviello, T. (1989) On the aggregation and conformational state in aqueous solution of a succinoglycan polysaccharide, *Macromolecules* **22**, 954–959.
Djordjevic, S. P., Chen, H., Batley, M., Redmond, J. W., Rolfe, B. G. (1987) Nitrogen fixation ability of exopolysaccharide mutants of *Rhizobium sp.* strain NGR234 and *Rhizobium trifolii* is restored by the addition of homologous exopolysaccharides, *J. Bacteriol.* **169**, 53–60.
Dussap, C. G., De Vita, D., Pons, A. (1991) Modeling growth and succinoglycan production by *Agrobacterium radiobacter* NCIB 9042 in batch cultures, *Biotechnol. Bioeng.* **38**, 65–74.
Evans, L. R., Linker, A., Impallomeni, G. (2000) Structure of succinoglycan from an infectious strain of *Agrobacterium radiobacter*, *Int. J. Biol. Macromol.* **27**, 319–326.
Fuisz, R. C. (1993) Method of preparing a frozen comestible, US patent 5 238 696.
Fung, F.-N., Miller, J. W., Wuesthoff, M. T. (1992) Low-calorie fat substitute, US patent 5 158 798.
Glucksmann, M. A., Reuber, T. L., Walker, G. C. (1993a) Family of glycosyl transferases needed for the synthesis of succinoglycan by *Rhizobium meliloti*, *J. Bacteriol.* **175**, 7033–7044.
Glucksmann, M. A., Reuber, T. L., Walker, G. C. (1993b) Genes needed for modification, polymerization, export, and processing of succinoglycan by *Rhizobium meliloti*: a model for succinoglycan biosynthesis, *J. Bacteriol.* **175**, 7045–7055.
Gonzalez, J. E., York, G. M., Walker, G. C. (1996) *Rhizobium meliloti* exopolysaccharides: synthesis and symbiotic function, *Gene* **179**, 141–146.
Gonzalez, J. E., Semino, C. E., Wang, L.-X., Castellano-Torres, L. E., Walker, G. C. (1998) Biosynthetic control of molecular weight in the polymerization of the octasaccharide subunits of succinoglycan, a symbiotically important exopolysaccharide of *Rhizobium meliloti*, *Proc. Natl. Acad. Sci. USA* **95**, 13477–13482.
Gravanis, G., Milas, M., Rinaudo, M., Clarke-Sturman, A. J. (1990a) Conformational transition and polyelectrolyte behaviour of a succinoglycan polysaccharide, *Int. J. Biol. Macromol.* **12**, 195–200.
Gravanis, G., Milas, M., Rinaudo, M., Clarke-Sturman, A. J. (1990b) Rheological behaviour of a succinoglycan polysaccharide in dilute and semi-dilute solutions, *Int. J. Biol. Macromol.* **12**, 201–206.
Gray, J. X., Djordjevic, M. A., Rolfe, B. G. (1990) Two genes that regulate exopolysaccharide production in *Rhizobium* sp. strain NGR234: DNA sequences and resultant phenotypes, *J. Bacteriol.* **172**, 193–203.
Guerin, G., Knipper, M. (1991) Stable zeolite/succinoglycan suspensions, US patent 5 104 566.
Guerin, G., Larrey, M.-D., Nabavi, M., Ricca, J.-M. (1998) Aqueous cosmetic compositions with base of non-volatile insoluble silicon, stabilized by a succinoglycan, WO 98/24408.

Guillou, V. (1995) Sulfamic acid cleaning/stripping compositions comprising heteropolysaccharide thickening agents, US patent 5 431 839.
Harada, T., Yoshimura, T. (1964) Production of new acidic polysaccharide containing succinic acid by a soil bacterium, *Biochim. Biophys. Acta* **83**, 374.
Harada, T., Harada, A. (1996) Curdlan and succinoglycan, in: *Polysaccharides in Medicinal Applications* (Dumitriu, S., Ed.), New York: Marcel Dekker, 21–58.
Harada, T., Yoshimura, T., Hidaka, H., Koreda, A. (1965) Production of new acidic polysaccharide, succinoglucan, by *Alcaligenes faecalis* var. myxogenes, *Agric. Biol. Chem.* **29**, 757.
Harada, T., Amemura, A., Jansson, P.-E., Lindberg, B. (1979) Comparative studies of polysaccharides elaborated by *Rhizobium, Alcaligenes,* and *Agrobacterium, Carbohydr. Res.* **77**, 285–288.
Jansson, P.-E., Kenne, L., Lindberg, B., Ljunggren, H., Lönngren, J., Ruden, U., Svensson, S. (1977) Demonstration of an octasaccharide repeating unit in the extracellular polysaccharide of *Rhizobium meliloti* by sequential degradation, *J. Am. Chem. Soc.* **99**, 3812–3815.
Keller, M., Roxlau, A., Weng, W. M., Schmidt, M., Quandt, J., Niehaus, K., Jording, D., Arnold, W., Puhler, A. (1995) Molecular analysis of the *Rhizobium meliloti MucR* regulating the biosynthesis of the exopolysaccharides succinoglycan and galactoglucan, *Mol. Plant–Microbe Interact.* **8**, 267–277.
Kijne, J. W. (1992) The *Rhizobium* infection process, in: *Biological Nitrogen Fixation* (Stacey, G., Burris, R. H., Evans H. J., Eds), New York: Chapman & Hall, 349–398.
Kitaoka, N. (1998) Aqueous ink for ball point pen, Japanese patent 10130563A.
Knipper, M., Besnard, M.-M., David, C. (1995) Composition comportant un succinoglycane, European patent 0469953B1.
Knipper, M., Senechal, A., Raffart, M. (2000) Composition dérivant d'un succinoglycane, son procédé de préparation et ses applications, European patent 0527061B1.
Kondo, M., Matsubara, N. (1995) Aqueous ink composition for writing instrument, US patent 5 466 283.
Lau, H. C. (1993) Gravel packing process, US patent 5184679.
Lau, H. C., Bernardi, L. A. (1993) Low-viscosity gravel packing process, US patent 5 251 699.
Lau, H. C., Davies, D. R. (1992) Aqueous polysaccharide compositions and their use, GB patent 2 250 761A.
Leigh, J. A., Coplin, D. L. (1992) Exopolysaccharides in plant–bacterial interactions, *Annu. Rev. Microbiol.* **46**, 307–346.
Leigh, J. A., Walker, G. C. (1994) Exopolysaccharides of *Rhizobium*: synthesis, regulation and symbiotic function, *Trends Genet.* **10**, 63–67.
Leigh, J. A., Signer, E. R., Walker, G. C. (1985) Exopolysaccharide deficient mutants of *Rhizobium meliloti* that form ineffective nodules, *Proc. Natl. Acad. Sci. USA* **82**, 6231–6235.
Linton, J. D., Evans, M. W., Godley, A. R. (1985) Process for preparing a heteropolysaccharide obtained thereby, its use, and strain NCIB 11883, European patent 0138255A2.
Linton, J. D., Evans, M. W., Jones, D. S., Gouldney, D. N. (1987a) Exocellular succinoglycan production by *Agrobacterium radiobacter* NCIB 11883, *J. Gen. Microbiol.* **133**, 2961–2969.
Linton, J. D., Jones, D. S., Woodard, S. (1987b) Factors that control the rate of exopolysaccharide production by *Agrobacterium radiobacter* NCIB 11883, *J. Gen. Microbiol.* **133**, 2979–2987.
Long, S. R. (1989) *Rhizobium*–legume nodulation: life together in the underground, *Cell* **56**, 203–214.
Macri, C. A., Miller, J. W., Sarges, D. A., Sprott, D. J. (1996) Agricultural foam composition, European patent 0400914B1.
Margaritis, A., Pace, G. W. (1985) Microbial polysaccharides, in: *Comprehensive Biotechnology* (Moo Young, M., Ed.), Oxford: Pergamon Press, 1005–1044, Vol. 3.
Matulova, M., Toffanin, R., Navarini, L., Gilli, R., Paoletti, S., Cesaro, A. (1994) NMR analysis of succinoglycans from different microbial sources: partial assignment of their ^{1}H and ^{13}C NMR spectra and location of the succinate and the acetate groups, *Carbohydr. Res.* **265**, 167–179.
Meade, M. J., Tanenbaum, S. W., Nakas, J. P. (1995) Production and rheological properties of a succinoglycan from *Pseudomonas* sp. 31260, grown on wood hydrolysates, *Can. J. Microbiol.* **41**, 1147–1152.
Mendrygal, K. E., Gonzalez, J. E. (2000) Environmental regulation of exopolysaccharide production in *Sinrhizobium meliloti, J. Bacteriol.* **182**, 599–606.
Misaki, A., Saito, H., Ito, T, Harada, T. (1969) Structure of succinoglycan, an extracellular acidic polysaccharide from *Alcaligenes faecalis* var. myxogenes, *Biochemistry* **8**, 4645–4650.
Müller, P., Weng, M. W. M., Quandt, J., Arnold, W., Pühler, A. (1993) Genetic analysis of *Rhizobium meliloti exoYFQ* operon: exoY is homologous to

sugar transferases and ExoQ represents a transmembrane protein, *Mol. Plant–Microbe Int.* **6**, 55–65.

Navarini, L., Stredansky, M., Matulova, M., Bertocchi, C. (1997) Production and characterization of an exopolysaccharide from *Rhizobium hedysari* HCNT1, *Biotechnol. Lett.* **19**, 1231–1234.

Osman, S. F., Fett, W. F. (1989) Structure of an acidic exopolysaccharide from *Pseudomonas marginalis* HT041B, *J. Bacteriol.* **171**, 1760–1762.

Pandey, A. (1992) Recent process development in solid state fermentation, *Process Biochem.* **27**, 109–117.

Philips, J. C., Hoyt IV, H. L., Macri, C. A., Miller, W. L., O'Neilll, J. J. (1991) Agricultural gel-forming compositions, US patent 5 077 314.

Rae, P., Johnston, N. (1995) Storable liquid cementitious slurries for cementing oil and gas wells, US patent 5 447 197.

Reed, J. W., Glazebrook, J., Walker, G. C. (1991a) The *exoR* gene of *Rhizobium meliloti* affects RNA levels of other *exo* genes but lacks homology to known transcriptional regulators, *J. Bacteriol.* **173**, 3789–3794.

Reed, J. W., Capage, M., Walker, G. C. (1991b) *Rhizobium meliloti exoG* and *exoJ* mutations affect the exoX–exoY system for modulation of exopolysaccharide production, *J. Bacteriol.* **173**, 3776–3788.

Reinhold, B. B., Chan, S. Y., Reuber, T. L., Marra, A., Walker, G. C., Reihold, V. N. (1994) Detailed structural characterization of succinoglycan, the major exopolysaccharide of *Rhizobium meliloti* Rm1201, *J. Bacteriol.* **176**, 1997–2002.

Reuber, T. L., Walker, G. C. (1993a) Biosynthesis of succinoglycan, a symbiotically important exopolysaccharide of *Rhizobium meliloti*, *Cell* **74**, 269–280.

Reuber, T. L., Walker, G. C. (1993b) The acetyl substituent of succinoglycan is not necessary for alfalfa nodule invasion by *Rhizobium meliloti* Rm1201, *J. Bacteriol.* **175**, 3653–3655.

Reuber, T. L., Long, S., Walker, G. C. (1991) Regulation of *Rhizobium meliloti exo* genes in free-living cells and in plants examined by using Tn*phoA* fusions, *J. Bacteriol.* **173**, 426–434.

Reuhs, B. L., Williams, M. N. V., Kim, J. S., Carlson, R. W., Côté, F. (1995) Suppression of the Fix$^-$ phenotype of *Rhizobium meliloti exoB* mutants by *lpsZ* is correlated to a modified expression of the K polysaccharide, *J. Bacteriol.* **177**, 6231–6235.

Ridout, M. J., Brownsey, G. J., York, G. M., Walker, G. C., Morris, V. J. (1997) Effect of o-acyl substituent on the functional behaviour of *Rhizobium meliloti* succinoglycan, *Int. J. Biol. Macromol.* **20**, 1–7.

Stredansky, M., Conti, E. (1999a) Succinoglycan production by solid-state fermentation with *Agrobacterium tumefaciens*, *Appl. Microbiol. Biotechnol.* **52**, 332–337.

Stredansky, M., Conti, E. (1999b) Xanthan production by solid state fermentation, *Process Biochem.* **34**, 581–587.

Stredansky, M., Conti, E., Bertocchi, C., Matulova, M., Zanetti, F. (1998) Succinoglycan production by *Agrobacterium tumefaciens*, *J. Ferm. Bioeng.* **85**, 398–403.

Stredansky, M., Conti, E., Bertocchi, C., Navarini, L., Matulova, M., Zanetti, F. (1999a) Fed-batch production and simple isolation of succinoglycan from *Agrobacterium tumefaciens*, *Biotechnol. Tech.* **13**, 7–10.

Stredansky, M., Conti, E., Navarini, L., Bertocchi, C. (1999b) Production of bacterial exopolysaccharides by solid substrate fermentation, *Process Biochem.* **34**, 11–16.

Stredansky, M., Conti, E., Bertocchi, C. (1999c) Processo per la produzione di esopolisaccaridi batterici mediante fermentazione aerobica su substrato solido, Italian patent 1 293 620.

Sutherland, I. W. (1990) *Biotechnology of Microbial Polysaccharides*. Cambridge: Cambridge University Press.

Sugimoto, Y., Yamamoto, T. (1999) Ink composition for water base ballpoint, and water-base ballpoint, JP 11256090A.

Sydansk, R. D. (1992) Enhanced liquid hydrocarbon recovery process, US patent 5 129 457.

Sydansk, R. D. (1998) Polymer enhanced foam workover, completion, and kill fluids, US patent 5 706 895.

Tolmasky, M. E., Staneloni, R. J., Leloir, L. F. (1982) Lipid-bound saccharides in *Rhizobium meliloti*, *J. Biol. Chem.* **257**, 6751–6757.

Urzainqui, A., Walker, G. C. (1992) Exogenous suppression of the symbiotic deficiencies of *Rhizobium meliloti exo* mutants, *J. Bacteriol.* **174**, 3403–3406.

Uttaro, A. D., Cangelosi, G. A., Geremia, R. A., Nester, E. W., Ugalde, R. A. (1990) Biochemical characterization of avirulent *exoC* mutants of *Agrobacterium tumefaciens*, *J. Bacteriol.* **172**, 1640–1646.

van Rhijn, P., Vanderleyden, J. (1995) The *Rhizobium*–plant symbiosis, *Microbiol. Rev.* **59**, 124–142.

Yokota, A., Sakane, T. (1991) Taxonomic significance of fatty acid compositions in whole cells and lipopolysaccharides in *Rhizobiaceae, Inst. Ferment. Osaka, Res. Commun.* **15**, 57.

York, G. M., Walker, G. C. (1998a) The succinyl and acetyl modifications of succinoglycan influence susceptibility of succinoglycan to cleavage by the *Rhizobium meliloti* glycanases *exoK* and *exsH*, *J. Bacteriol.* **180**, 4184–4191.

York, G. M., Walker, G. C. (1998b) The *Rhizobium meliloti exoK* and *exsH* glycanases specifically depolymerize nascent succinoglycan chains, *J. Bacteriol.* **180**, 4912–4917.

Yoshida, M. (1998) Cosmetic material, Japanese patent 10203953A.

Yoshida, M. (1999) Composition for acid hair dye, Japanese patent 11240824A.

Yoshida, M., Suzuki, K. (1998) Hairdye compositions, WO 98/31330A1.

Zevenhuizen, L. P. T. M. (1971) Chemical composition of exopolysaccharides of *Rhizobium* and *Agrobacterium*, *J. Gen. Microbiol.* **68**, 239–243.

Zevenhuizen, L. P. T. M. (1973) Methylation analysis of acidic exopolysaccharides of *Rhizobium* and *Agrobacterium*, *Carbohydr. Res.* **26**, 409–419.

Zevenhuizen, L. P. T. M. (1997) Succinoglycan and galactoglucan, *Carbohydr. Polymers.* **33**, 139–144.

Zevenhuizen, L. P. T. M., Faleschini, P. (1991) Effect of the concentration of sodium chloride in the medium on the relative proportions of poly- and oligo-saccharides excreted by *Rhizobium meliloti* strain YE-2SL, *Carbohydr. Res.* **209**, 203–209.

Zhan, H., Leigh, J. A. (1990) Two genes that regulate exopolysaccharide production in *Rhizobium meliloti*, *J. Bacteriol.* **172**, 5254–5259.

Zhan, H., Gray, J. X., Levery, S. B., Rolfe, B., Leigh, J. A. (1990) Functional and evolutionary relatedness of genes for exopolysaccharide synthesis in *Rhizobium meliloti* and *Rhizobium* sp. strain NGR234, *J. Bacteriol.* **172**, 5245–5253.

18
Alginates from Bacteria

Dr. Bernd H. A. Rehm
Institut für Mikrobiologie, Westfälische Wilhelms-Universität Münster, Correnstraβe 3, 48149 Münster, Germany; Tel.: +49-251-8339848; Fax: +49-251-8338388; E-mail: rehm@uni-muenster.de

Biotechnology of Biopolymers. From Synthesis to Patents. Edited by A. Steinbüchel and Y. Doi

ISBN: 3-527-31110-6

CF cystic fibrosis
EPS extracellular polysaccharide (exopolysaccharide)
HSL homoserine lactone
LPS lipopolysaccharide
PMI-GMP phosphomannose isomerase/guanosine diphosphomannose pyrophosphorylase
PMI-GMP phosphomannose isomerase-GDP-mannose pyrophosphorylase
PMM phosphomannomutase
PMM phosphomannomutase
SCLM scanning laser confocal microscopy
TCA cycle tricarboxylic acid cycle

1 Introduction

Alginates are a family of unbranched, non-repeating copolymers consisting of variable amounts of (1-4)-linked β-D-mannuronic acid and its epimer α-L-guluronic acid. The monomers are distributed in blocks of continuous mannuronate residues (M-blocks), guluronate residues (G-blocks), or alternating residues (MG-blocks) (Figure 1). Alginates isolated from different natural sources vary in the length and distribution of the different block types. Alginates are produced by bacteria and brown seaweeds, and the mannuronate residues of the bacterial–but not those of the seaweed polymers–are acetylated to a variable extent at positions O-2 and/or O-3 (Skjåk-Bræk et al., 1986). The variability in monomer block structures and acetylation strongly affects the physico-chemical and rheological properties of the polymer, and the biological basis for the variability is therefore of both scientific and

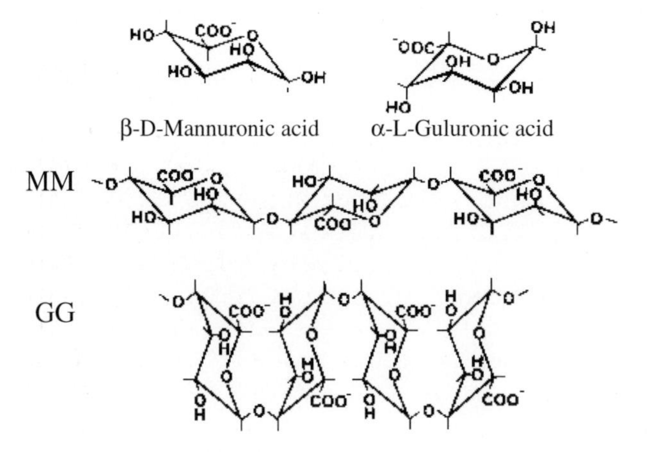

Fig. 1 The structure of alginate. MM: polymannuronate; GG: polyguluronate.

applied importance (Smidsrød and Draget, 1996). Alginate is used for a variety of industrial purposes, for example as a stabilizing, thickening and gelling agent in food production, or to immobilize cells in pharmaceutical and biotechnology industries (see below) (Onsøyen, 1996). The production is currently based exclusively on the harvesting of brown seaweeds. Several bacteria belonging to the genera *Pseudomonas* and *Azotobacter* also produce alginate (Gorin and Spencer, 1966; Linker and Jones, 1966; Govan et al., 1981; Cote and Krull, 1988), and the structures of the blocks of monomer residues are similar in alginates produced by seaweeds and those synthesized by *Azotobacter vinelandii*. In contrast, all *Pseudomonas* alginates lack G-blocks (Skjåk-Bræk et al., 1986). Most of our knowledge of the genetics of alginate biosynthesis originates from studies of *Pseudomonas aeruginosa*, mainly because of the medical relevance of this organism as an opportunistic human pathogen, particularly for patients suffering from cystic fibrosis (CF) (Govan and Harris, 1986; May and Chakrabarty, 1994). Alginate plays an important role as a virulence factor during the infectious process (Gacesa and Russell, 1990). The reason for this appears to be related to the alginate-mediated mode of biofilm growth, which causes resistant colonization of the lung. *A. vinelandii* and *P. aeruginosa* produce alginate as an extracellular polysaccharide (EPS) in vegetatively growing cells, whereas in *A. vinelandii* alginate is also involved in the differentiation process leading to a so-called cyst (Sadoff, 1975).

2 Historical Outline

The polysaccharide alginate was firstly isolated from marine macroalgae in the 19th century, but it was approximately 80 years later that a bacterial source (from *P. aeruginosa*) of the polysaccharide was identified (Linker and Jones, 1966). This alginate was later found to be similar to the commercially useful polymer obtained from marine algae (Lin and Hassid, 1966a; Linker and Jones, 1966) and also to the polysaccharide produced as a capsule by *Azotobacter vinelandii* (Gorin and Spencer, 1966). The association of mucoid *P. aeruginosa*, i.e., alginate-overproducing strains, with chronic lung infections in patients who suffer from the inherited disease CF has been well established, and is recognized as a major cause of pathogenesis in these individuals. Mucoid *P. aeruginosa* have also been isolated, albeit less frequently, from other patients, for example bronchiotatics and those with urinary tract or middle-ear infections (McAvoy et al., 1989), although not normally from individuals with infected burn sites. Although most of the mucoid isolates of *P. aeruginosa* have been obtained from clinical samples, it is clear that alginate production is important in a much wider context. Ten of 81 *P. aeruginosa* isolates from technical water systems showed a mucoid phenotype, which implies a more widespread occurrence of ecological niches for mucoid forms (Grobe et al., 1995). Alginate biosynthesis is a key factor in the establishment of stable mature biofilms of *P. aeruginosa* in a wide range of environmental situations (Nivens et al., 2001). Alginate production is fairly widespread amongst rRNA homology group I pseudomonads, as indicated by Southern hybridization experiments using various alginate biosynthesis genes as probes (Fialho et al., 1990; Fett et al., 1992; Rehm, 1996). Genomic DNA from representatives of groups II–IV gave very weak or no hybridization with the probes, except for *algC*, indicating that the ability to produce alginate is restricted to members of rRNA homology group I. This substantiates earlier physiological studies in which alginate was isolated from *Pseudomonas fluorescens, Pseudomonas putida, Pseudomonas mendocina* (Govan et al., 1981) and *P. syringae* (Fett et al., 1986; Gross and Rudolph, 1987). Alginate is also synthesized by *A. vinelandii* as part of the encystment process (Gorin and Spencer, 1966). The mature cysts are surrounded by two discrete alginate-rich layers (exine and intine) which enable the dormant cells to survive long periods of desiccation. Strains of *Azotobacter chroococcum* also produce alginate (Cote and Krull, 1988). In *Azotobacter,* abnormalities in alginate production resulted in impaired encystment, which indicated that alginate biosynthesis is required to survive and adapt to famine conditions, as was supposed for the establishment of biofilms by pseudomonads. Enzymological evidence for the biosynthesis pathway of alginate was obtained from the brown algae *Fucus gardneri* by Lin and Hassid (1966a) and from *A. vinelandii* about ten years later by Pindar and Bucke (1975). These studies indicated that fructose-6-phosphate is the first alginate precursor, which was derived from the Entner–Doudoroff pathway and from the fructose-1,6-bisphosphate aldolase reaction. The presence of the first alginate biosynthetic enzymes, phosphomannose isomerasc, GDP-mannose pyrophosphorylase and GDP-mannose dehydrogenase, was demonstrated by Piggot et al. (1981) in *P. aeruginosa*. Later, Padgett and Phibbs (1986) detected another alginate biosynthesis enzyme, phosphomannomutase. The medical relevance of mucoid *P. aeruginosa* stimulated research on the genetics of alginate biosynthesis. A stable mucoid mutant of *P. aeruginosa* (8830) had to be generated by chemical mutagenesis in order to obtain nonmucoid mutants (Darzins and Chakrabarty, 1984). Complementation studies of the various mutants allowed the identification and characterization of the alginate biosynthesis genes, as well as the respective regulatory genes (May and Chakrabarty, 1994). The identification and biochemical characterization of regulatory alginate biosynthesis proteins enabled a deeper understanding of the rather complex regulatory network (Govan and Deretic, 1996; Rehm and Valla, 1997). The extracellular Ca^{2+}-dependent C-5-epimerases were the first alginate-related enzymes from *A. vinelandii*, which were biochemically and genetically characterized (Ertesvåg et al., 1994, 1999). Meanwhile, hybridization of a lambda gene library of *A. vinelandii* with the outer membrane encoding the *algE* gene from *P. aeruginosa* as a probe, led to the identification of the second alginate biosynthesis gene cluster and the first gene cluster of a biotechnologically relevant microorganism (Rehm, 1996; Rehm et al., 1996). Comparative analysis of the genetics, biochemistry and regulation of alginate biosynthesis revealed strong similarities (Rehm and Valla, 1997; Gacesa, 1998). Moreover, the plant-pathogenic *P. syringae* pv. *syringae* alginate biosynthesis gene cluster was identified and characterized, exhibiting a virtually identical alginate gene arrangement (Penaloza-Vazquez et al., 1997).

3
Chemical Structures

Alginate is composed of the uronic acid β-D-mannuronate and its C-5 epimer α-L-guluronate. These monomers may be arranged in homopolymeric (poly-mannuronate or poly-guluronate) or heteropolymeric block structures (see Figure 1). In addition, bacterial alginates are normally O-acetylated on the 2 and/or 3 position(s) of the β-D-mannuronate residues. Consequently, bacteria produce a range of alginates with different block structures and degrees of O-acetylation.

The high molecular mass of bacterial alginate and the negative charge ensure that the polysaccharide is highly hydrated and viscous. It is well established that alginates from *P. aeruginosa* do not contain polyguluronate blocks (Sherbrock-Cox et al., 1984) but those from *A. vinelandii* may do so. The block structure and degree of O-acetylation, as well as the molecular weight, determine the physico-chemical properties of alginate. Alginates which contain polyguluronate form rigid gels in the presence of Ca^{2+}, and are therefore important in structural roles, for example the outer cyst wall (exine) of *A. vinelandii*. Conversely, an absence of polyguluronate, as in *P. aeruginosa*, produces relatively flexible gels in the presence of Ca^{2+}. Extensive O-acetylation of alginate increases the water-binding capacity of the polysaccharide, which may be significant in enhancing survival under desiccating conditions.

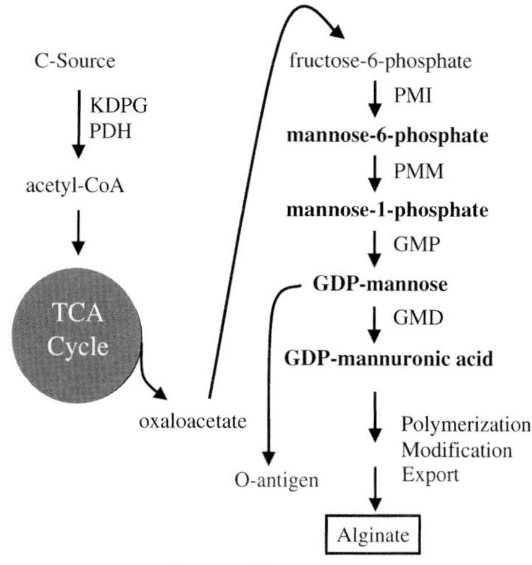

Fig. 2 Biosynthesis pathway of alginate. Oxaloacetate is converted to fructose-6-phosphate via gluconeogenesis. KDPG: ketodeoxyphosphogluconate pathway (Entner–Doudoroff pathway); PDH: pyruvate dehydrogenase; PMI-GMP: phosphomannose isomerase-GDP-mannose pyrophosphorylase; PMM: phosphomannomutase; GMD: GDP-mannose dehydrogenase. Intermediates shown in bold type are precursors of alginate.

4 Biosynthetic Pathway of the Alginate Precursor, GDP-Mannuronic Acid

A convincing pathway for alginate biosynthesis in *A. vinelandii* was first proposed by Pindar and Bucke (1975) based on the assay of individual enzyme activities. However, the corresponding enzymes in *P. aeruginosa* have proved more difficult to assay directly, and the pathway has been elucidated using a combination of complementation analyses and gene overexpression studies. Although the initial steps in the pathway are indisputable (Figure 2) (and the same applies to *A. vinelandii* and pseudomonads), there is still considerable debate about the final stages of biosynthesis and the export of alginate.

The alginate biosynthesis starts from fructose-6-phosphate in the cytosol. Radiolabeling studies have established that six-carbon growth substrates are oxidized via the Entner–Doudoroff pathway, and that the resultant pyruvate [1 mol (mol hexose)$^{-1}$] is ultimately channeled into alginate biosynthesis (Lynn and Sokatch, 1984). More detailed analyses of the labeling patterns in alginate indicate that the pyruvate derived from the oxidation of sugars is fed into the tricarboxylic acid (TCA) cycle prior to synthesis of fructose-6-phosphate and alginate (Narbad et al., 1988). Recent experiments using ^{13}C-NMR have clearly established the key role of the Entner–Doudoroff pathway in the oxidation of hexoses, and also the obligatory requirement of triose intermediates in alginate biosynthesis (Beale and Foster, 1996). The latter authors concluded that the pyruvate derived from the oxidation of hexoses feeds into alginate biosynthesis via the formation of oxaloacetate and subsequent gluconeogenesis (see Figure 2). This supports earlier data employing ^{13}C-labeled precursors, which suggested an obligatory role for the TCA cycle in the conversion of glucose to alginate (Narbad et al., 1988). This important role of the TCA cycle in alginate biosynthesis is supported by more recent genetic studies using nonmucoid mutants of *P. aeruginosa*. Levels of only the phosphorylated (active) form of the key TCA cycle enzyme succinyl-CoA synthetase are reduced in *algQ* mutants (Schlictman et al., 1994). The normally rare occurrence of mucoid forms of *P. aeruginosa* in culture can be significantly increased by growth on energy-poor media (Terry et al., 1991). However, there is a recent report that glucose can stimulate *algD* transcription and alginate production (Ma et al., 1997), though this contradicts earlier findings which proposed that glucose repression of *algD* occurs (Devault et al., 1991). Interestingly, alginate biosynthesis occurs in response to energy deprivation, despite biosynthesis of the polysaccharide being an energy-consuming process. It has been suggested that the AlgQ-controlled expression of succinyl-CoA synthetase and nucleoside-diphosphate kinase may result in a decreased pool size of GTP and hence less available GDP-mannose for alginate biosynthesis (Schlictman et al., 1994; Kim et al., 1998; Kapatral et al., 2000). However, quantification of the nucleotide sugars indicates that GDP-mannose is present in great excess (Tatnell et al., 1993), even when the GDP-mannose dehydrogenase gene was overexpressed (Tatnell et al., 1994). The initial steps in the alginate biosynthesis are related to general carbohydrate metabolism, and the intermediates are widely utilized. In particular, the intermediate GDP-mannose serves not only as a precursor for alginate biosynthesis but also for lipopolysaccharide (LPS) biosynthesis (Goldberg et al., 1993). Accordingly, the GDP-mannose dehydrogenase (AlgD) exhibits a key role in the biosynthesis of alginate. The *algD* gene is proximal to the promoter on the alginate biosynthesis gene cluster and expression is tightly controlled (Schurr et al., 1993). Analyses of nucleotide sugar pools and exopolysaccharide production clearly indicate that GDP-mannose dehydrogenase is the kinetic control point in the alginate pathway (Tatnell et al., 1994). However, the alginate biosynthesis enzyme phosphomannose isomerase/ guanosine-diphosphomannose pyrophosphorylase (PMI-GMP (AlgA)) is a bifunctional protein catalyzing the initial and third steps of alginate synthesis (see Figure 2). The PMI reaction pulls the fructose-6-phosphate out of the metabolic pool, leading to the first intermediate, mannose-6-phosphate. Phosphomannomutase (AlgC) then catalyzes the second step, resulting in the formation of mannose-1-phosphate. This is not the only reaction catalyzed by AlgC, which also exhibits phosphoglucomutase activity and which is evidently involved in rhamnolipid, LPS and alginate biosynthesis (Olvera et al., 1999). The GMP activity of PMI-GMP (AlgA) then, with concomitant GTP hydrolysis, converts mannose-1-phosphate to GDP-mannose. The enzyme favors the reverse reaction, but because of the efficient removal of the GDP-mannose in the next step, the entire pathway proceeds efficiently in the direction of alginate synthesis. The almost irreversible oxidation of GDP-mannose to GDP-mannuronic acid involves the enzyme guanosine-diphosphomannose dehydrogenase, and the reaction product is the immediate precursor for polymerization (see Figure 2). For further details on this pathway, readers are referred to the review on *P. aeruginosa* alginate synthesis by May and Chakrabarty (1994). Interestingly, the intracellular activities of key biosynthesis enzymes appear to be very low, even in extracts prepared from highly mucoid *P. aeruginosa* cells. It has been speculated that the enzymes phosphomannose isomerase-GDP-mannose pyrophosphorylase (PMI-GMP) and phosphomannomutase (PMM) may exist as an enzyme complex ("metabolon"), which allows the coupling of enzymatic reactions, as described for many metabolic pathways (Mathews, 1993).

5 Genetics of Alginate Biosynthesis

At least 24 genes have been directly implicated in alginate biosynthesis in *P. aeruginosa* (Figure 3; Table 1), and there is good evidence that others may also be involved, e.g., *glpM* (Schweizer et al., 1995). It is not possible to assign all *alg* genes identified so far as solely functioning in alginate biosynthesis, as it is now evident for example that some of the regulator genes act globally and encode proteins such as alternative sigma factors (Yu et al., 1995). Other "*alg*" genes, e.g., *algA, C,* are also required for LPS biosynthesis (Goldberg et al., 1993). The genes involved in the synthesis of the precursor GDP-mannuronic acid have all been identified and characterized, and they have been assigned the designations *algA* (encoding PMI-GMP), *algC* (encoding phosphomannomutase), and *algD* (encoding guanosine-diphosphomannose dehydrogenase). With the exception of *algC* (located at 10 min on the chromosome map; between 5,992,000 and 5,994,000 bp of the *P. aeruginosa* genome sequence), the other two genes and all other known structural genes involved in alginate biosynthesis in *P. aeruginosa* are clustered at 34 min (between positions 3,962,000 bp and 3,980,000 bp of the *P. aeruginosa* genome sequence) (Figure 3; Table 1). The biological functions of many of the gene products putatively involved in the polymerization process are poorly understood, mainly because the polymerase has

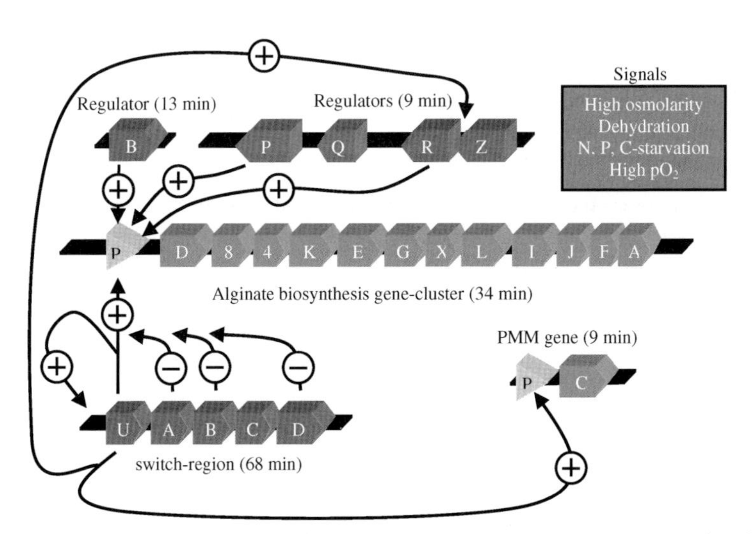

Fig. 3 Genetic organization of alginate genes and their regulation. Signals which induce alginate biosynthesis gene expression are given in the gray box. +: induction (positive effector); –: inhibition (negative effector); P: promoter; switch-region: muc-gene region (genotypic switch to alginate overproduction by mutation in *mucA, mucB* or *mucD*, respectively).

Tab. 1a Alginate genes from *Pseudomonas aeruginosa*

Gene	*Location*	*Gene product*
algD	34	GDP mannose dehydrogenase
alg8	34	Polymerase/export function?
alg44	34	Polymerase/export function?
*algK**	34	Polymerase/export function?
algE	34	Outer-membrane porin?
algG	34	Mannuronan C-5-epimerase
algX	34	Unidentified function but high sequence similarity to *algJ*
algL	34	Alginate lyase
algI	34	*O*-Acetylation
algJ	34	*O*-Acetylation
algF	34	*O*-Acetylation
algA	34	Phosphomannose isomerase/GDP mannose pyrophosphorylase
algB	13	Member of *ntrC* subclass of two-component regulators
algC	10	Phosphomannomutase
algH	?	Unknown function
algR1	9	Regulatory component of two-component sensory transduction system
algR2 (algQ)	9	Protein kinase or kinase regulator
algR3(algP)	9	Histone-like transcription regulator
algZ	9	AlgR cognate sensor
algU (algT)	68	Homologue of *E. coli* σ^E global stress response factor
algS (mucA)	68	Anti σ factor
algN (mucB)	68	Anti σ factor?
algM (mucC)	68	Regulator?
algW (mucD)	68	Homologue of serine protease (HtrA)

Modified according to Rehm and Valla (1997)

not been purified and no *in vitro* alginate synthesis assay has been established. However, the gene products of *alg8, alg44, algX* (formerly *alg60*) and *algK* are candidates for being subunits of the alginate polymerase. The deduced amino acid sequences of these proteins contain hydrophobic regions, suggesting a localization in the cytoplasmic membrane (Wang et al., 1987; Maharaj et al., 1993; Rehm and Valla, 1997). The *algK* gene has been identified located directly downstream of *alg44* in *P. aeruginosa* (Aarons et al., 1997). Evidence was obtained that AlgK is entirely periplasmic and is probably anchored in the cytoplasmic membrane. AlgK might be also involved in polymerization and/or export of alginate. Although most of the genes are essential for alginate biosynthesis, those encoding the epimerase and O-acetyltransferase(s) can be inactivated, provided that essential genes downstream of the mutation are expressed in trans, without abolishing alginate biosynthesis. The role of a gene (*algL*) encoding an alginate lyase in the biosynthesis operon (Boyd et al., 1993; Schiller et al., 1993; Monday and Schiller, 1996) is unknown. Contradictory data were published regarding the role of AlgL in alginate biosynthesis. Boyd and coworkers (1993) showed that a mutation in the *algL* gene had no effect on alginate biosynthesis. However, Monday and Schiller (1996) demonstrated that an *algL* mutation strongly impaired alginate biosynthesis, as

Tab. 1b Alginate genes from *Azotobacter vinelandii*

Gene	*Gene product*
alg8	Polymerase?
alg44	Polymerase/export function?
algA	Phosphomannose isomerase
	GDP mannose pyrophosphorylase
algD	GDP mannose dehydrogenase
algE1-7	Mannuronan C-5-epimerases
algG	Mannuronan C-5-epimerase
algJ	Export of alginate?
algL	Alginate lyase
algU	Homologue of *E. coli* σ^E global stress response factor
mucA	Anti σ factor
mucB	Anti σ factor
mucC	Regulator?
mucD	Homologue of serine protease (HtrA)

Modified according to Rehm and Valla (1997).

was recently confirmed by an *algL* mutant of *P. syringae* pv. *syringae* (Penaloza-Vazquez et al., 1997). The *P. aeruginosa* gene cluster at 34 min also contains the *algE* gene, encoding an outer membrane protein (Chu et al., 1991; Rehm et al., 1994a). Production of this protein is strictly correlated with the mucoid phenotype of *P. aeruginosa* (Grabert et al., 1990; Rehm et al., 1994a).

In *A. vinelandii*, our understanding of the genetics of alginate synthesis has improved greatly over the past few years. The first genes involved in alginate synthesis to be cloned and characterized encode a set of seven strongly related Ca^{2+}-dependent mannuronan C-5-epimerases, designated AlgE1 to AlgE7 (Ertesvåg et al., 1999). These proteins are structurally unusual, since they can all be described as repeats of two types of protein modules, designated A (385 amino acids) and R (153 amino acids). Each protein contains a short motif designated S at the carboxy-terminal end. The A modules are present once or twice in each protein, while the R modules are represented one to seven times. Each R module contains four to seven repeats of a nine-amino-acid sequence repeat putatively involved in the binding of Ca^{2+} ions (Ertesvåg et al., 1999). These epimerase genes (*algE1–algE7*) are clustered in the *A. vinelandii* chromosome, and they share no significant sequence homology to the *P. aeruginosa algG* gene.

It was shown recently that the *A. vinelandii* genome encodes a gene (*algD*) corresponding to the *P. aeruginosa algD* gene, and these two genes share 73% identity with each other at the protein level (Campos et al., 1996). An *A. vinelandii* gene (*algJ*) apparently corresponding to the *P. aeruginosa algE* gene was also recently identified (Rehm, 1996). AlgJ shares about 52% sequence identity with AlgE from *P. aeruginosa*, and topological modeling suggests that this protein is structurally similar to AlgE. It is also believed to be functionally equivalent, forming a pore which is involved in alginate export (Rehm, 1996). Surprisingly, the *A. vinelandii* genome also encodes a mannuronan C-5-epimerase (AlgG), which belongs to a different class from AlgE1–E7, and this epimerase seems to represent the equivalent of AlgG in *P. aeruginosa* (Rehm et al., 1996). In addition *A. vinelandii* encodes a protein

(AlgY) containing one A and one R module, but this protein appears to display no epimerase activity after expression in *Escherichia coli* (Svanem et al., 1999). Recently, the entire alginate biosynthesis gene cluster of *A. vinelandii* was identified, which revealed a similar physical organization of the *alg* genes as has been found in *P. aeruginosa* (Lloret et al., 1996; Rehm et al., 1996). A similar arrangement of *alg* genes has been also described in *Pseudomonas syringae* pv. *syringae* (Penaloza-Vazquez et al., 1997), and is also evident for *P. fluorescens* based on genome sequence analysis (B. H. A. Rehm, unpublished results). These data suggested that the physical arrangement of alginate genes in bacteria is conserved.

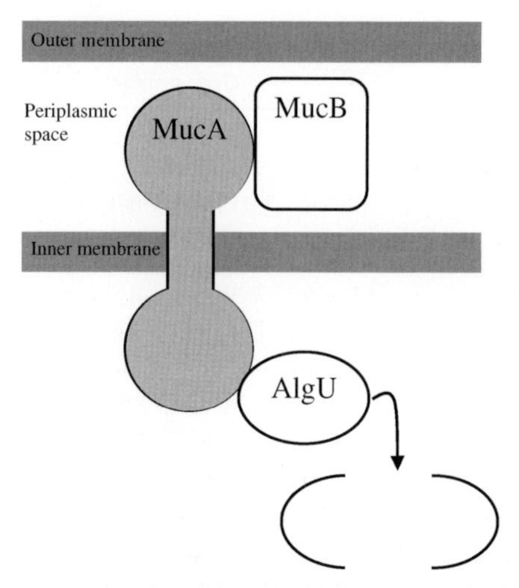

Fig. 4 Model of the action of anti-sigma factors MucA and MucB. AlgU is the alternative sigma-factor, which is required for *alg* gene expression. Binding of AlgU to MucA might expose AlgU to proteolytic digestion by MucD (Mathee et al., 1997).

6 Regulation of Alginate Biosynthesis

The regulation of alginate biosynthesis is complex, and involves specific gene products and those that act more globally (Figure 3 and Figure 4). Expression of the entire alginate biosynthesis gene cluster of *P. aeruginosa* is under the control of the *algD* promoter, and in essence this region (*algD-algA*) acts as an operon (Chitnis and Ohman, 1993), although there is sequence-based evidence for weak promoters within the gene cluster. One of the alginate biosynthesis genes, *algC*, is located at 10 min on the PAO1 map and is transcribed independently of the 34-min region, but is coordinately induced with the biosynthesis gene cluster. The regulatory genes in *P. aeruginosa* map at 9 min and 13 min, and genes responsible for a genotypic switch to alginate overproduction are found at 68 min (see Figure 3). The alginate structural genes are controlled via the positively regulated *algD* promoter (Deretic et al., 1989). In *A. vinelandii*, on the other hand, there seems to exist two additional *algD*-independent promoters, from which *alg8-algJ* and *algG-algA* are transcribed, respectively. These promoters were found to be independently regulated (Lloret et al., 1996; Vazquez et al., 1999). The genotypic switch region comprising *algU*, *mucA*, *mucB*, *mucC* and *mucD* (831,000–835,000, positions relative to the *P. aeruginosa* genome sequence) is found in both bacteria, and the corresponding gene sequences are highly homologous (see Figure 3). The genes have also been found to be biologically active and play a similar role in the two species (Martinez-Salazar et al., 1996). All the known *alg* genes and their corresponding proteins are listed in Table 1.

6.1 Environmentally Induced Activation of *alg* Genes

Conditions of high osmolarity, N, P or carbon starvation, dehydration, and the presence of phosphorylcholine activate a cascade of regulatory proteins in *P. aeruginosa* involved in the activation of the *algD*-promoter (Gacesa and Russell, 1990; Terry et al., 1991). Genes located at a region spanning 9 and 13 min on the *P. aeruginosa* chromosome, *algR*(*algR1*), *algQ*(*algR2*) and *algP*(*algR3*), and *algB* modulate the production of alginate and have been described as auxiliary regulators of mucoidy (Govan and Deretic, 1996). A two-component signal-transduction pathway comprising the putative sensor proteins AlgQ (kinase) and AlgZ, interacting with the cognate response regulator proteins such as AlgR and AlgB, were identified (see Figure 3) (May and Chakrabarty, 1994; Yu et al., 1997). Analysis of sequence data indicates that *algZ* encodes a sensory component of a signal transducer system, but that it lacks several expected motifs typical of histidine protein kinases (Yu et al., 1997). The best characterized of these regulators is AlgR, which gene is transcribed in response to the protein AlgU. AlgR binds to three sites upstream of the *algD* promoter and, in conjunction with AlgU, up-regulates transcription of *algD* and the downstream genes. AlgR also promotes expression of *algC*. The efficiency of AlgR is increased by phosphorylation by the cognate kinase AlgQ. AlgB also modulates *algD* expression and, based on sequence analysis, is a member of the NtrC subclass of two-component regulators (Wozniak and Ohman, 1991); however, the *algB* and *algR* regulatory systems appear to operate independently of each other (Wozniak and Ohman, 1994). Interestingly, AlgB and AlgR showed phosphorylation-independent activity on the induction of alginate biosynthesis (Ma et al., 1998). Binding of these positive regulators upstream of the *algD* promoter, presumably leads to the formation of a suprahelical structure with the aid of the histone-like AlgP protein, causing activation of transcription (Deretic and Konyecsni, 1989; Konyecsni and Deretic, 1990; Deretic et al., 1994). A comprehensive account of the inter-relationships of the regulators has been reviewed (Govan and Deretic, 1996). Furthermore, the recently identified sigma-like factor AlgU (AlgT) is responsible for the initiation of *algD* transcription (se Figure 3) (Hershberger et al., 1995). On the basis of sequence analysis (Martin et al., 1994), AlgU is a member of the σ^E class of sigma factors, i.e., analogous to RpoE of *Escherichia coli*, and is essential for alginate production. Subsequent studies have established that AlgU and RpoE are functionally equivalent (Yu et al., 1995), and that AlgU forms complexes with RNA polymerase (Schurr et al., 1995). AlgU causes an increase in alginate biosynthesis by a direct action on the *algD* promoter (see below), and indirectly by up-regulating transcription of another regulatory gene, *algR* (Martin et al., 1994). This environmentally induced transcription of the *alg* cluster and the resulting production of alginate occur only at a rather low level.

6.2 Genotypic Switch

Copious amounts of alginate are only produced in combination with inactivation (mutations) of the negative regulators of the AlgU activity (anti-sigma factors) MucA (AlgS) or MucB (AlgN) (Martin et al., 1993a,b). Mutational inactivation (genotypic switch) of MucA, MucB (Schurr et al., 1996) or MucD (Boucher et al., 1996) leads to full activity of AlgU, which allows strong transcription of the *alg* operon (Figure 4). MucA is supposed to be located in the cytoplasmic membrane interacting with MucB in the periplasm upon perception of an unknown stimulus, and transducing a signal to the cytoplasm which mediates degradation of

AlgU presumably due to the action of MucD (Mathee et al., 1997) (see Figure 4). MucD is orthologous to the *E. coli* periplasmic protease and chaperone DegP. DegP homologues are known virulence factors that play a protective role in stress responses in various species. Recently, negative regulation of AlgU by anti-sigma factors MucA and MucB and the transcriptional regulation of the *algD* gene have been also described for *A. vinelandii* (Campos et al., 1996; Martinez-Salazar et al., 1996).

7
Polymerization and Export of the Alginate Chain

Since no undecaprenol-linked intermediate has been identified in either *P. aeruginosa* or *A. vinelandii*, the polymerase–which presumably is localized as a protein complex in the cytoplasmic membrane–might synthesize alginate by an undecaprenol-independent mechanism (Sutherland, 1982). Alginate synthesis might occur similarly to bacterial cellulose synthesis. The cellulose synthase of *Acetobacter xylinum* is localized in the cytoplasmic membrane, and appears to be a protein complex of 420 kDa. This enzyme catalyzes the processive polymerization of glucose residues (from UDP-glucose), and the nascent β-(1,4)-linked glucosan chains appear to remain attached to the synthase during polymerization (Ross et al., 1991). Correspondingly, the alginate polymerase in the cytoplasmic membrane might accept the GDP-mannuronic acid at the cytosolic site while simultaneously translocating the nascent alginate chain through the cytoplasmic membrane (Rehm and Winkler, 1996). Preliminary studies using ^{14}C-GDP-mannuronic acid as substrate and defined oligomannuronates as primer have revealed that the envelope fraction of mucoid *P. aeruginosa* exhibited *in vitro* alginate polymerase activity (B. H. A. Rehm, unpublished results). Biochemical and electrophysiological characterization of AlgE revealed that it forms an anion-selective pore in the outer membrane (Rehm et al., 1994b). This pore could be partially blocked by GDP-mannuronic acid in lipid bilayer experiments. In addition, a topological model of AlgE has been developed and, according to this model, the protein is a β-barrel consisting of 18 β-strands (Rehm et al., 1994b) (Figure 5). A three-dimensional model of AlgE was developed by homology modeling, indicating pore diameters eligible for alginate export (see Figure 5). These data are consistent with the hypothesis that AlgE forms an alginate-specific pore that enables export of the nascent alginate chain through the outer membrane. Figure 6 summarizes all the findings regarding polymerization, modification and export in a model.

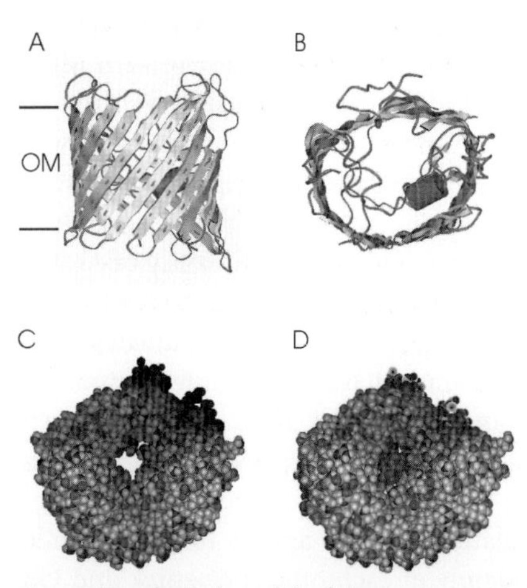

Fig. 5 Topological model of the outer membrane protein AlgE from *P. aeruginosa*. Experimental evidence was obtained that AlgE is involved in export of alginate through the outer membrane. (A) Side-view of the AlgE model (the bottom is exposed to periplasm, whereas the top is cell-surface-exposed. OM: outer membrane. (B) Top view of the AlgE model from outside the cell. (C) Top view of the AlgE model (in CPK format) from outside the cell. D, C with inserted alginate chain (dark gray).

8 Alginate-Modifying Enzymes

The alginate-modifying enzymes (the transacetylase and the C-5-mannuronan epimerase) carry N-terminal signal sequences and are mainly found in the periplasm in *P. aeruginosa* (Franklin et al., 1994). The corresponding alginate-modification reactions occur at the polymer level, presumably in the periplasm (see Figure 6). The genes encoding the transacetylase and other proteins involved in transacetylation (*algI, algJ, algF*), the epimerase (*algG*), and the lyase (*algL*) have been cloned, and the gene products have been characterized (Franklin and Ohman, 1993, 1996; Shinabarger et al., 1993; Franklin et al., 1994; Monday and Schiller, 1996). Transacetylation occurs at position(s) O-2 and/or O-3 of the mannuronic acid residue, preventing these residues from being epimerized to guluronic residues by AlgG and from degradation by AlgL (Franklin and Ohman, 1993; Franklin et al., 1994; Wong et al., 2000).

Thus, the periplasmic acetylase indirectly controls the periplasmic epimerase and lyase activity on the alginate polymer. The increasing degree of acetylation also causes the alginate polymer to have an enhanced water-binding capacity. This might be particularly important under dehydrating conditions, for example during infection and colonization of the lungs of patients with CF. The alginate lyase presumably functions as an editing protein to control the length of the polymer, or it might also serve the polymerase with alginate oligomers to prime synthesis. The lyase is not involved in providing a carbon source (Boyd et al., 1993).

8.1 Mannuronan C-5-epimerases

The G residues in alginates originate from a post-polymerization reaction catalyzed by mannuronan C-5-epimerases. In *P. aeruginosa*, and presumably also in other species belonging to this genus, there appears to be only one such enzyme encoded by the gene *algG* (Franklin et al., 1994). Like the proteins necessary for acetylation (Franklin and Ohman, 1996), the Ca^{2+}-independent AlgG is also probably located in the periplasm. The *A. vinelandii* genome also contains an *algG* homologue (Rehm et al., 1996), but this species in addition modifies its alginates by a family of extracellular Ca^{2+}-dependent epimerases that according to sequence-alignment studies are unrelated to AlgG. Seven

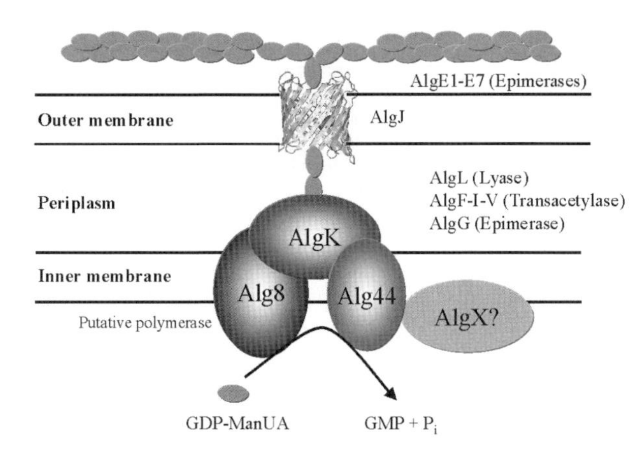

Fig. 6 Model of alginate polymerization, modification and export in *A. vinelandii* (Modified according to Rehm and Valla, 1997).

such enzymes (AlgE1–7) are now known (Ertesvåg et al., 1994, 1995; Svanem et al., 1999), and they can all be seen as composites of two structurally distinct modules, designated A and R. The A modules (about 385 amino acids) are present in one or two copies in each enzyme, while the R modules (about 155 amino acids) are present in one (AlgE4) and up to seven (AlgE3) copies. The N-terminal ends of each R module contain four to seven copies of a nine-amino-acids motif putatively involved in the binding of Ca^{2+}. In addition, *A. vinelandii* encodes a protein (AlgY) containing one A and one R module, but this protein appears to display no epimerase activity after expression in *E. coli* (Svanem et al., 1999).

8.1.1
Functional Differences

The A modules alone appeared to be sufficient both for catalyzing the epimerization reaction and for determining the epimerization pattern (Ertesvåg and Valla, 1999). It is, therefore, particularly important to understand the structure–function relationships in these modules. The epimerization patterns generated by all seven AlgE epimerases can be divided into two main groups: those which almost exclusively generate MG-blocks; and those which form G- blocks. From NMR spectroscopy analyses of the reaction products of all the enzymes, it is immediately obvious that only AlgE4 belongs to the first group. In addition, the C-terminal parts of AlgE1 and AlgE3 (AlgE1-2 and AlgE3-2) display this property, but it is not reasonable that these truncated forms are made *in vivo* in *A. vinelandii*. It is not known why the *A. vinelandii* genome encodes a specialized enzyme like AlgE4, but the physical properties of the alternating structure of its reaction product can be predicted to be quite different from the gel-forming G-block alginates (Smidsrød and Draget, 1996). All Ca^{2+}-dependent epimerases except AlgE4 are involved in the formation of G-blocks. AlgE3 and AlgE1 are composite enzymes, and due to the properties of each part they can both make long G-blocks and presumably put alternating structures between them. Comparison of the epimerization patterns of these two enzymes indicates that they share similar properties, but it was observed that less AlgE3 (measured as initial activity) compared to AlgE1, is needed to obtain high degrees of epimerization. For AlgE1 (Ertesvåg et al., 1998b) and AlgE3 the relative amount of G-blocks also increases with decreasing concentration of Ca^{2+}. This property, which was not observed for AlgE2 (Ramstad et al., 1999), probably reflects that the C-terminal part of AlgE3 displays less activity at low concentrations of Ca^{2+} than the N-terminal part, similar to what has been observed earlier for AlgE1 (Ertesvag et al., 1998b). It is also known that whole AlgE1 needs only 0.8 mM $CaCl_2$ for full activity (Ertesvåg et al., 1998b), while AlgE3 requires 3 mM $CaCl_2$, and displays less than 40% of full activity at 1 mM concentration of this cation. At low or moderate levels of epimerization, AlgE6 introduces more alternating structures than AlgE2 and AlgE5, although less than the composite enzymes AlgE1 and AlgE3. The average lengths of the G-blocks at about 40% epimerization, on the other hand, are similar to those made by AlgE2 and AlgE5. Alginate may, however, be epimerized to 90% G by AlgE6, and such highly epimerized alginate contain very long G-blocks and can be predicted to form very strong gels. This degree of epimerization has so far not been achieved using AlgE2 or AlgE5. These different epimerase specificities may indicate the requirement of the organism to generate alginate structures of importance for formation of the metabolically dormant and alginate-containing cysts which are generated under conditions of environmental stress in *A. vinelandii* (Sadoff, 1975). Furthermore, the significant alginate lyase activity of AlgE7 might be needed for the germination of the cysts (Wyss et al., 1961).

8.1.2
The Biotechnological Potential

Commercially, the alginates are harvested from different species of brown algae (Smidsrød and Draget, 1996), and it seems very likely that the composition of these products will not always meet industrial and biotechnological needs. In addition, alginate structures that are not available from seaweeds might have properties that would open the possibilities for totally new applications. By epimerizing algal alginates with the recombinantly produced epimerases, one might therefore be able either to upgrade their value or to generate completely new products. Since it is already clear that this is feasible, the questions are rather what prices are acceptable for each particular application, what the potential uses are, and how the market will react to products made using recombinant enzymes. A further extension of alginate modifications *in vitro* would be to design cells that directly synthesize the alginates of interest *in vivo*. This might lead to products of lower price, but it may prove to be more difficult to obtain the same level of control as achieved by in-vitro epimerization. In any case, the mannuronan C-5-epimerization system raises many interesting questions for basic science and applied biotechnology, and consequently the enzymes are likely to be the subject of active studies for many years to come.

8.2
O-Transacetylases

The products of *algI*, *algJ*, and *algF* from *P. aeruginosa* are required for the addition of O-acetyl groups to the alginate polymer, and mutations in *algI*, *algJ*, or *algF* resulted in production of an alginate polymer that was not O-acetylated (Shinabarger et al., 1993; Franklin and Ohman, 1996; Nivens et al., 2001). The mannuronate residues undergo modification by C-5 epimerization to form the L-guluronates and by the addition of acetyl groups at the O-2 and O-3 positions. By using genetic analysis, *algF* was identified and located upstream of *algA* in the 18-kb alginate biosynthesis operon, as a gene required for alginate acetylation (Franklin and Ohman, 1993; Shinabarger et al., 1993). An *algI*::*Tn501* mutant, which was defective in *algIJFA* because of the polar nature of the transposon insertion, produced alginate when *algA* was provided in *trans*. This indicated that the *algIJF* gene products were not required for polymer biosynthesis. To examine the potential role of these genes in alginate modification, mutants were constructed by gene replacement in which each gene (*algI*, *algJ*, or *algF*) was replaced by a polar gentamicin resistance cassette (Franklin and Ohman, 1996). Proton nuclear magnetic resonance (NMR) spectroscopy showed that polymers produced by strains deficient in *algIJF* still contained a mixture of D-mannuronate and L-guluronate, indicating that C-5 epimerization was not affected. Alginate acetylation was evaluated by a colorimetric assay and Fourier transform-infrared (FTIR) spectroscopy, and this analysis showed that strains deficient in *algIJF* produced nonacetylated alginate. Plasmids that supplied the downstream gene products affected by the polar mutations were introduced into each mutant. The strain defective only in *algF* expression produced an alginate that was not acetylated, confirming previous results. Strains missing only *algJ* or *algI* also produced nonacetylated alginates. Providing the respective missing gene (*algI, algJ*, or *algF*) *in trans* restored alginate acetylation. Mutants defective in *algI* or *algJ*, obtained by chemical and transposon mutagenesis, were also defective in their ability to acetylate alginate. Therefore, *algI* and *algJ* represent newly identified genes that, in addition to *algF*, are required for alginate acetylation. Once in the periplasmic space, alginate is O-acetylated by the combined action of the *algF, algI* and *algJ* gene products. Mutants deficient in any one of these three genes are unable to produce O-acetylated alginate, but the epimerization process and overall yields of alginate appear to be unaffected (Shinabarger et al., 1993; Franklin and Ohman, 1996). Acetyl-CoA is almost certainly the primary donor of O-acetyl groups for alginate modification; however, this metabolite is localized in the cytoplasm, whereas the O-acetylation process occurs in the periplasm. Therefore, at least one of the *algF, algI* or *algJ* gene products is likely to be involved in transport of O-acetyl groups across the cell membrane into the periplasmic space. Sequence data indicate that AlgI is probably a membrane-bound protein and may fulfil this role (Franklin and Ohman, 1996). AlgF has a signal peptide, which is processed by *E. coli*, indicating that it is a periplasmic protein (Shinabarger et al., 1993) and therefore could be the O-acetyltransferase. AlgJ shows remarkable similarity (30% identity, 69% similarity) to another gene product, AlgX (Monday and Schiller, 1996), which is of unknown function but is essential for alginate biosynthesis. However, neither AlgJ nor AlgX shows any significant similarity to other proteins in the databases and therefore their function in the O-acetylation process remains unresolved at this stage. The observation that a cell suspension of *P. syringae* is able to O-acetylate seaweed alginate suggests that this event is either periplasmic or extracellular (Lee and Day, 1995). It is likely that the O-acetylation of mannuronate is catalyzed by at least two enzymes each specific for either the 2- or 3-hydroxyl on the sugar ring. However, at this stage it is not known which gene products might be involved in determining these specific modifications.

8.3 Alginate Lyases

It is not clear why a degradative enzyme should be expressed concurrently with the enzymes involved in biosynthesis. One suggestion has been that the lyase may be involved in excising polysaccharide fragments from the biosynthesis complex, although there is no real evidence to support this contention. It is clear though that overexpression of the *algL* gene product (May and Chakrabarty, 1994) or of other alginate lyases (Gacesa and Goldberg, 1992) results in the release of planktonic bacteria from biofilms.

8.3.1 Reaction Mechanism

Alginate lyase catalyzes the degradation of alginate by a β-elimination mechanism, targeting the glycosidic 1,4 O-linkage between monomers. A double bond is formed between the C4 and C5 carbons of the six-membered ring from which the 4-O-glycosidic bond is eliminated, depolymerizing alginate and simultaneously yielding a product containing 4-deoxy-L-*erythro*-hex-4-enopyranosyluronic acid as the nonreducing terminal moiety (Haug et al., 1967). Gacesa (1987, 1992) proposed a catalytic mechanism for alginate lyase that described a three-step reaction to depolymerize alginate. This mechanism may also be shared with epimerase, another enzyme that acts on the alginate polymer. The two reactions differ only in the last step of the three-stage transformation of alginate. The three steps include: (1) removal of the negative charge on the carboxyl anion – essentially neutralizing the charge by a salt bridge (lysine or arginine may be the candidate residue); (2) a general base-catalyzed abstraction of the proton on C5 (aspartic acid, glutamic acid, histidine, lysine, and cysteine have been suggested for this role), where one residue may be required as the proton abstractor and another as the proton donor, although the proton may be derived from the solvent environment; and (3) a transfer of electrons from the carboxyl group to form a double bond between C4 and C5, resulting in the β-elimination of the 4-O-glycosidic bond. In the proposed mechanism for epimerase, the replacement of the proton at C5 (epimerization) takes place in step 3. The principle of the catalytic mechanism was confirmed by the identification of putative catalytic residues by structural analysis of the *Sphingomonas* lyase ALY1-III complexed with a trisaccharide (Yoon et al., 2001).

8.3.2
Function in Alginate-Producing Bacteria

Very few bacteria synthesize polysaccharides as well as the specific degrading enzymes (Kennedy et al., 1992; Sutherland, 1995). Periplasmically localized alginate lyases have been found in various species of *Pseudomonas* and *Azotobacter* that can synthesize an extracellular alginate but are unable to use alginate as a carbon or energy source. The extracellular poly(M)-rich alginate produced by these bacteria is O-acetylated at the C2 and/or C3 positions on the M residues to various degrees, which makes this polymer more resistant to degradation by the endogenously produced lyase (Nguyen and Schiller, 1989; Kennedy et al., 1992). Interestingly, alginate lyase genes in *Pseudomonas* and *Azotobacter* species are localized within their respective alginate biosynthesis gene clusters. This genetic organization has been reported in *P. aeruginosa* (Chitnis and Ohman, 1993), *P. syringae* pv. *syringae* (Penaloza-Vazquez et al., 1997), *A. vinelandii* (Rehm et al., 1996), and *A. chroococcum* (Pecina et al., 1999). This localization of the alginate lyase gene (*algL*) raises questions about the function of AlgL in alginate biosynthesis. It is possible that the lyase works as part of a polymerization complex within the periplasm, assembling the alginate exopolysaccharide for transport to the cell surface, or it may provide oligomeric alginate, which might serve as primer. Accordingly, the lyase-negative *P. aeruginosa* produced only small amounts of alginate (Monday and Schiller, 1996). However, Boyd et al. (1993) reported that lyase was not required for alginate production by *P. aeruginosa*. A similar study demonstrated that the absence of lyase activity reduced alginate production by *P. syringae* pv. *syringae* by ~50% (Penaloza-Vazquez et al., 1997). Hence, alginate lyase seems not to be essential for alginate biosynthesis, but for maximum production. May and Chakrabarty (1994) proposed that the lyase may function either as an editing protein to control the length of the alginate polymer or to provide short pieces of alginate to prime the polymerization reaction. The nascent polymannuronate in the periplasmic space would be most sensitive to endogenous lyase before acetylation and epimerization, whereas after modification, the polymer would be ready for export to the cell surface. The levels of Ca^{2+} and Mn^{2+} ions have been shown selectively to activate or inhibit alginate lyase and epimerization activity in algae (Madgwick et al., 1978). Since various cations have a strong influence on bacterial alginate lyase activity (Wong et al., 2000), this observation suggests that periplasmic ionic conditions could also regulate alginate modification in *P. aeruginosa*.

The production of lyase by *P. aeruginosa* may also be important in facilitating dissemination of the bacteria (Boyd and Chakrabarty, 1994). Alginate synthesis is increased upon attachment of the bacteria to a cell surface, resulting in stronger bacterial adhesion to the surface and colonization. However, overexpression of the lyase gene within a mucoid strain of *P. aeruginosa* led to a decrease in the length of alginate polymers and increased bacterial detachment from the surface (Boyd and Chakrabarty, 1994). Thus, cleavage of the alginate polymer within *P. aeruginosa* biofilms could enhance detachment of the bacteria, allowing them to spread and colonize new sites. In *Azotobacter* sp., alginate is produced by vegetatively growing cells as capsule, and by cells in the metabolically dormant-state in the cyst coat (Page and Sadoff, 1975, Sadoff, 1975). Both *A. vinelandii* and *A. chroococcum* strains produce M-specific endolytic lyases that are localized in the periplasmic space. These lyases may be biologically important in the differentiation of *Azotobacter* cells during encystment, when they most likely play a role in concert with epimerases to form the desiccation-resistant cyst capsule. It has now been reported that *A. vinelandii* has multicopy epimerase genes, making a gene family that is highly likely to be responsible for the synthesis of complex alginates of various polymeric composition in the cyst capsule (Svanem et al., 1999). One of these epimerases, AlgE7, also exhibited lyase activity. Although alginate-negative strains of *A. vinelandii* fail to encyst, it is not yet clear whether alginate lyase-negative strains can differentiate. An increase in alginate lyase activity just before cyst germination (Kennedy et al., 1992) suggests that lyase expression might support this process.

From the collective information on alginate and alginate lyases, it is concluded that alginate lyases are important enzymes in a broad spectrum of biological roles and applications. The lyases maintain a balance in the cell physiology of alginate-producers that efficiently use alginate as functional biopolymers and also in the natural environment, where the recycling of alginate is achieved through metabolic breakdown (for review, see Wong et al., 2000). In addition, the ability of lyases selectively to depolymerize alginate–which has become a very useful industrial polysaccharide–makes them important tools with great potential for advanced biotechnological uses.

8.3.3
Structure–Function Analysis

The cloning and sequencing of many alginate lyase genes have now enabled investigators to focus on structure–function analysis of this enzyme. DNA sequences and primary amino acid sequences are available for 23 alginate lyases (Wong et al., 2000). Various alginate lyases have been characterized with respect to enzyme properties (Wong et al., 2000; Table 2). Meanwhile, the coordinates of the three-dimensional structure of ALY1-III *Sphingomonas* sp. can be found in the Protein Data Bank (Brookhaven, NY) (Yoon et al., 1999). Based on sequence information, most alginate lyases appear to fall into three major classes according to their molecular mass: 20–35 kDa; ~40 kDa; and ~60 kDa. Analysis of amino acid sequences of alginate lyases revealed that they contain a hydrophilic central region and a hydrophobic sequence in the C terminus, with a slightly charged end. All alginate lyases share amino acid sequence similarities from 18 to 95%. Although several regions of similarity exist between these lyases, the core region residues of the lyases within the 40-kDa class

Tab. 2 Alginate lyases from Gram-negative bacteria: localization, characteristics, and sequence accession numbers (modified according to Wong et al., 2000)

Source	*Localization*	*Substrate specificity*	*Endo/exolytic*	*Cleavage site*	*Molecular mass (kDa)*	*pI*	*Opt. pH*	*Opt. T*	*Cations needed*	*GenBank accession No.*
A. chroococcum	Periplasmic	AlgL: M	Endolytic	–	43	–	–	30°C, pH 7.5	Na^+, K^+ Mg^{2+}	AJ223605
	Periplasmic	M			–	–	6.8	30°C		N/A
A. chroococcum 4A1M	Extracellular	PolyM	Endolytic	–	23–24	5.6	6	60°C	Ca^{2+}	N/A
A. vinelandii	Intracellular	M	–	–	~50		7.5	–	–	N/A
	Periplasmic	M			–	–	7.2	30°C		N/A
	Periplasmic	AlgL: M, acetyl'd (nonconsecutive M)	Endolytic not G-M	M-M and M-G,	39	5.1	8.1–8.4		Na^+, divalents not needed	AF037600
	Periplasmic	AlgE7 (epimerase and lyase): M and G	Endolytic	G-GM and G-MM	(384aa)	–	–	–	–	AF099800
A. vinelandii	N/A	M	N/A	N/A	(375aa)	N/A	N/A	N/A	N/A	AF027499
Enterobacte cloacae M-1	Extracellular	G	Endolytic	–	32–38	8.9	7.8	30°C	Ca^{2+}, Al^{3+}, Mn^{2+}	N/A
	Intracellular	G	Endolytic	7 subsites, cleaves G-G between subsites 2 and 3	31–39	8.9	7.5	40°C	–	N/A
K. aerogenes type 25	Intracellular	PolyG	Endolytic	–	28–31.6	–	7	37°C	Na^+ and K^+ (0.1–0.3 M)	N/A
	Extracellular	G	Endolytic	G-G, G-M	–	–	7	–	Mg^{2+} (0.05–0.1 M)	N/A
K. pneumoniae subsp. aerogenes	Extra/intra-celluar (R)	AlyA: polyG	–	–	28 8.9	–	–	–	N/A	25
	Extracellular (R)	AlyA: polyG	Endolytic		31.4 9.39 (calc.)	7.0 –7.6 h		50°C	Na^+	L19657
Pseudomonas sp. W7	N/A (R)	N/A	N/A	N/A	(345aa)	N/A	N/A	N/A	N/A	AF050114
P. aeruginosa	Intracellular	M, nonacetyl'd	–	–	–	–	6.2	20–40°C	–	N/A
P. aeruginosa	Intracellular	AlgL: M nonacetyl'd	–	–	53	–	8	–	Mg^{2+}, K^+, Na^+	N/A
P. aeruginosa CFl/Ml	Periplasmic	AlgL :M,	Endolytic	6 subsites, nonacetyl'd cleaves M-M between subsites 3 and 4	–	–	–	–	–	N/A
P. aeruginosa	Intracellular (R)	AlgL: M, nonacetyl'd	Endolytic	–	43.5	–	–	–	–	L14597
P. aeruginosa FRD1	Peripl.(N)/ intracell.(R)	AlgL :M, nonacetyl'd	Endolytic	M-M	39	9 (calc.)	7.0	–	Mg^{2+} , Na^+	U27829/L09724
P. aeruginosa	N/A (R)	AlgY :M	N/A	N/A	(685aa)	N/A	N/A	N/A	N/A	Z54213

Tab. 2 (cont.)

Source	*Localization*	*Substrate specificity*	*Endo/exolytic*	*Cleavage site*	*Molecular mass (kDa)*	*pI*	*Opt. pH*	*Opt. T*	*Cations needed*	*GenBank accession No.*
P. maltophilia and *P. putida*	Intracellular	M, acetyl'd	Endolytic	–	–	–	7.7–7.8	28–30°C	–	N/A
P. syringae pv. syringae	Periplasmic	PolyM (prefers deacetylated)	Endolytic		40	8.2	7	42°C	Cations not needed	AF22020
Pseudomonas	Intracellular	Multiple		–	–	90, 72, 60, 54	–	–	–	– N/A
sp. OS-ALG-9	Intra/ extracellular (R)	ALY or AlyP.M	Endolytic (preferred), G		46.3	–	–	–	AlyP:—	D10336
	Intracellular	(ALY or AlyP:—)	Endolytic	–	45	–	7.5	45°C	AlyP:—	N/A
	Intracellular (R)	ALYII : M	Endolytic	–	79.8 (calc.)	8.3	7	30°C	ALYII: None required; EDTA stimulates activity	AB003330
Sphingomonas sp.	Cytoplasmic	ALY1-I :M, nonacetyl'd, acetyl'd		Endolytic	–	60	9.03	7.5–8.5	70°C	ALY1-I, -II, -III: 2009330A
	Cytoplasmic	ALY1-II :G	Endolytic	–	25	6.82	7.5–8.5	70°C	None required;	sequenced
	Cytoplasmic	ALY1-III : M, acetyl'd (highly active) and nonacetyl'd	Endolytic	M-M, hetero MG	38	10.16	7.5–8.5	70°C	EDTA: no effect	1QAZ

N/A, not available

that share significant alignment are very well conserved; especially notable is the highly conserved six-amino-acid hydrophilic sequence "NNHSYW" in the center of the protein sequences. This region is also conserved in the alginate lyase ALY1-III from a *Sphingomonas* sp., which otherwise differs significantly in amino acid sequence from the other lyases in this grouping. Yoon et al. (1999) recently solved the three-dimensional structure for ALY1-III, revealing a structure with 12 α-helices, organized in a twisted α/α helix barrel composed of six inner and five outer helices. Analysis of this conformation identified a deep tunnel-like cleft, which was proposed as the catalytic site. The highly conserved NNHSYW sequence is located in the center of this cleft. As described below, studies suggest that alteration of the histidine residue in this site inactivates the lyase, providing evidence that this region is critical for enzyme activity. There is also a well-conserved nine-amino-acid hydrophobic sequence, "WLEPYCALY," in the C terminus of the lyases from *P. aeruginosa* and *Azotobacter* sp.. This nine-amino-acid block is weakly conserved in ALY1-III from *Sphingomonas* sp. Sequence and structural information indicates that this nine-amino-acid region is in H11, an inner α-helix of ALY1-III (Yoon et al., 1999).

The three-dimensional structure for alginate lyase from *Sphingomonas* sp. revealed an interesting feature of the enzyme, a disulfide bridge between Cys188 and Cys189, giving the structure a tight turn at the edge of the proposed active-site cleft (Yoon et al., 1999). This disulfide bond immediately precedes the conserved sequence of "NNHSYW" and is suggested to be very important in the maintenance of the active site conformation. This cysteine-cysteine pair in the alginate lyase of *Sphingomonas* sp. is a structural feature that is unique to this lyase, which is not surprising because the primary protein sequence of ALY1-III departs significantly from the general consensus of the majority of alginate lyases in the 40-kDa group. M-specific ALY1-III has an α-helix-rich structure and has no β-strand or sheet conformation (Yoon et al., 1999). The ALY1-III structure was used as a template to develop a threading model of the alginate lyase of *P. aeruginosa*, which exhibited 22% similarity to ALY1-III (B. H. A. Rehm, unpublished results). This model showed the tunnel-like cleft with the potential catalytic residues N171, H172, R219, Y226 closely arranged inside the cleft. Residues N171 and H172 are located in the conserved motif "NNHSYW" (Figure 7). Considering the many differences observed for the alginate lyases (primary sequence, M_W, substrate specificities, hydrophobicity profiles, etc.), it is important that additional enzymes will be biochemically analyzed as well as by X-ray crystallography. This will allow complete comparisons of structure and function between these enzymes, as well as related enzymes, and define their relationship with the structure of ALY1-III from *Sphingomonas* sp. (Yoon et al., 1999).

8.3.4
Substrate-Binding and Catalytic Sites

Experiments with defined alginate oligomers have indicated the specific substrate recognition sequences for alginate lyase action, highlighting the optimal capacity of the catalytic site for substrate units. Using a series of oligomannuronates (n D 2–9), Rehm (1998) reported that the poly(M) lyase from *P. aeruginosa* CF1/M1 would not accept oligomers that were smaller than five mannuronates for β-elimination, and that the highest enzyme activity was noted with the hexameric mannuronate oligomer, whereas the trimer was the most abundantly accumulating product. This study suggested that the catalytic site of alginate lyases probably

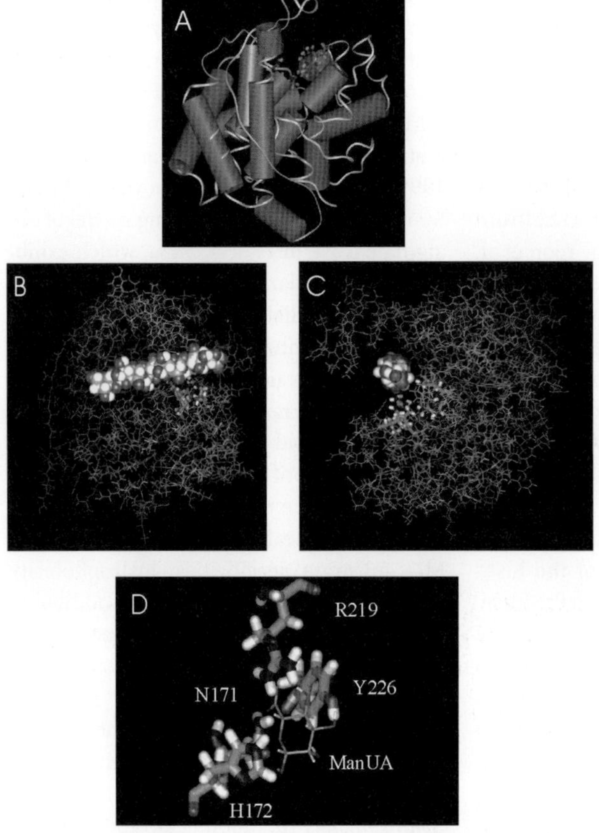

Fig. 7 Threading model of the alginate lyase from *P. aeruginosa* based on the structure of the *Sphingomonas* sp. ALY1-III alginate lyase. (A) Alginate lyase threading model indicating the secondary structure composition (cylinders represent α-helical segments) and amino acid side chain of the putative catalytic residues were demonstrated (see D). (B) The alginate lyase model in stick format complexed (docked) with the hexameric oligomannuronate and emphasis of the putative catalytic residues by thicker sticks in the tunnel-like cleft. (C) Model (B) turned around by 90°. (D) Spatial arrangement of the putative catalytic residues complexed (docked) with a mannuronic acid monomer (ManUA).

accommodates five to six residues. In the crystal structure of ALY1-III (Yoon et al., 1999), His192 is located in the highly conserved region NNHSYW, which is in the center of the proposed active-site cleft. Chemical modification of histidine residues in ALY1-III inactivated the enzyme (Yoon et al., 1999), which suggests strongly the role of histidine in the catalytic activity of ALY1-III. In ALY1-III, four conserved tryptophan residues, together with other aromatic residues, are located along the cleft of the active site, and two conserved arginine residues flank the entrance to the cleft. The charged arginine and lysine residues on both sides of the cleft and the aromatic side chains lining the active site have been suggested to be substrate-binding molecules. This active cleft can accommodate at least five residues of poly(M) along the curved surface (Yoon et al., 1999). The highly conserved sequence "INNHSY" located in the central region of the 40-kDa lyases is also highly conserved as "E/FNNVSY" in mannuronan C5-epimerases (Ertesvag et al., 1998b). This motif, found in alginate lyases and epimerases from *P. aeruginosa* (Boyd et al., 1993; Schiller et al., 1993; Franklin et al., 1994), *A. chroococcum* (Pecina et al., 1999), and *A. vinelandii* (Rehm et al., 1996; Ertesvag et al., 1998a; Svanem et al., 1999), appears to be mainly located within an average of 200 residues from the encoded N terminus of the proteins. However, this pattern of residues is not found in the G-lyases or the 30-kDa alginate lyases. Because both the alginate lyases and epimerases from *P. aeruginosa* and *A. vinelandii* are active on mannuronate units of alginate, it is highly possible that this pattern could be a binding motif for the poly(M) or the mannuronate and its glycosidic bond. The AlgG epimerases of the alginate biosynthesis operon in *A. vinelandii* (Rehm et al., 1996) and *P. aeruginosa* (Franklin et al., 1994) differ slightly from the epimerases found in the *A. vinelandii* epimerase gene family cluster (Rehm et al., 1996; Ertesvag et al., 1998a; Svanem et al., 1999). The AlgG proteins are smaller than most of the other epimerases (AlgE or AlgY), and the conserved "INNHSY" sequences from both AlgGs have greater homology between themselves than with the other epimerases. Also, the conserved sequences are located ~350 residues (instead of 200) from the encoded N termini of the respective AlgG. Through comparisons of the common "INNHSY" and "ENNVSY" motifs, it is clear that asparagine (N), serine (S), and tyrosine (Y) residues are the essential residues and are conserved throughout. A valine (V) or arginine (R) residue in the epimerases substitutes for the histidine (H) residue position in the M lyases. In the M lyases, the histidine (H) residue is highly important for catalytic activity, and this may be the residue that contributes to the main difference in catalytic mechanism between a lyase and an epimerase. However, only recently the crystal structure of ALY1-III complexed with trimeric mannuronate was obtained (Yoon et al., 2001). The binding of this substrate in the tunnel-like cleft, strongly suggested that the four residues – N191, H192, R239 and Y246 – are directly involved in substrate binding and catalysis. The following catalytic mechanism has been proposed: (1) The C5 carboxylate group is neutralized by R239 and N191; (2) subsequently, the C5 proton can more easily removed by Y246, the catalytic nucleophile, resulting in the formation of the carboxylate dianion intermediate; (3) Y246 then donates the proton to the oxygen of the glycosidic bond, which results in bond cleavage and formation of a C4–C5 double bond. H192 presumably stabilizes the carboxylate dianion intermediate during catalysis (Figure 8). A similar arrangement of the catalytic residues was obtained in the threading model of the alginate lyase from *P. aeruginosa*, supporting their role in catalysis (see Figure 7).

8.3.5
Future Applications

The therapeutic use of alginate lyase for the treatment of alginate in biofilms of *P. aeruginosa* colonizing the lungs of CF patients remains one of the most important goals of studying alginate lyase. Mrsny et al. (1996) described the complex distribution of DNA and alginate within the mucin matrix of CF sputa. The combination of alginate lyase and deoxyribonuclease demonstrated an additive reduction of the sputum viscoelasticity, which suggests that this approach deserves further study (Mrsny et al., 1994). Alginate lyases may also be used to generate defined products with potential applications in various fields. Alginates with low molecular weights act like oligosaccharides in their ability to regulate physiological processes in

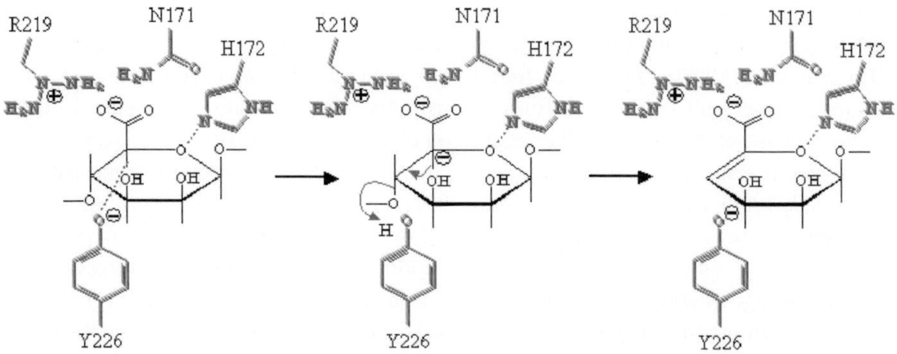

Fig. 8 The postulated catalytic mechanism of alginate lyases (see text for detailed description).

plants (Albersheim and Darvill, 1985). Oligomeric alginate (average M_W 2000 Daltons) obtained from lyase degradation of high-molecular weight alginate, can promote growth of *Bifidobacteria* sp. and thus has been proposed for use as a physiological food source (Murata et al., 1993; Akiyama et al., 1992). Furthermore, alginate lyase-degraded products (average M_W 1800 Daltons) greatly enhanced germination and shoot elongation in plants, despite repressing the growth of *Chlamydomonas* sp. and HeLa cells (Yonemoto et al., 1993). Trisaccharides (Natsume et al., 1994) or alginate lyase-lysate (Tomoda et al., 1994) has also been found to promote root growth in barley. In the presence of epidermal growth factor, dimers, trimers, and tetramers that possessed guluronic acid at the reduced ends highly induced the proliferation of keratinocytes (Kawada et al., 1997). Other alginate polymers (average M_W 230,000 Daltons) have antitumor effects and enhance the phagocytic activity of macrophages (Fujihara and Nagumo, 1993). Poly(M) block-rich alginate exhibited high antitumor activity (Fujihara and Nagumo, 1992) and could stimulate production of cytokines by human monocytes (Otterlei et al., 1991). The conformational properties of the macromolecules and the anionic character and molecular weight of the polysaccharide were suggested to be important in the effectiveness of their antitumor activity (Fujihara et al., 1984). Therefore, alginate lyases are crucial in the generation of such useful oligomeric products. The promising use of calcium alginate beads as a biomaterial in wide-ranging applications (Skjåk-Bræk and Martinsen, 1991) extends further the potential for alginates and alginate lyases. Calcium alginate, with or without a polylysine, polyarginine, or chitosan protective coating, has been used for the encapsulation of a variety of materials including drugs for controlled delivery; DNA and oligonucleotides for tumor development studies, gene delivery, gene therapy, and antisense oligonucleotide therapeutic agents; yeast cells coentrapped with lipase for the production of flavor esters; plant tissue, with the aim of developing artificial seed technology; *P. fluorescens* as a biocatalyst for phenolic-compound removal from wastewater; and entomopathogenic nematodes for agricultural biocontrol (for review, see Wong et al., 2000). Alginate in combination with other biomaterials such as hyaluronate and chitosan can be highly effective in many medical applications, including use in wound dressings impregnated with antibiotics and encapsulation of chondrocytes to engineer cartilage tissues *in vitro* for cartilage transplant and repair. Alginate with a G content >70% and average G-block length of >15, and with low polyphenol contamination, provides the most suitable characteristics for immobilization beads. Apart from generating alginate with a predictable monomeric sequence from native sources, it is possible to use a combination of D-mannuronan C5-epimerase and alginate lyase on a particular alginate substrate to engineer novel alginate polymers of defined composition (Skjåk-Bræk et al., 1986). The composition and properties of alginate gels are still being extensively studied, and the collective information will provide a guideline for the design of these novel polymers.

9
The Role of Alginate in Biofilm Formation

It has been proposed that contact with a surface may induce changes in gene expression, and there is evidence to support this idea in *P. aeruginosa*. Studies by Davies and coworkers showed that one of the genes required for the synthesis of the exopolysaccharide (EPS) alginate (*algC*) is up-regulated three- to five-fold in recently attached cells compared to their planktonic counterparts (Davies et al., 1993; Davies and Geesey, 1995). This result is not surprising because alginate – the regulation of which has been studied in depth (Govan and Fyfe, 1978; May et al., 1991) – has long been implicated as the extracellular matrix in biofilms of *P. aeruginosa*. These experiments were among the first to show surface contact-induced gene expression in *P. aeruginosa*. Recent studies in the laboratory of Wozniak have taken this observation a step further. Wozniak and colleagues noted that isolates of *P. aeruginosa* from the CF lung that produced large quantities of alginate (mucoid strains) were also nonmotile, and these authors suspected a link between the two phenotypes. In a series of genetic experiments, they showed that expression of a sigma factor (AlgU or σ^{22}) required for alginate synthesis resulted in down-regulation of a key flagellar biosynthetic gene (Garrett et al., 1999). These data suggest that, on contacting the surface, flagellar synthesis is down-regulated and alginate synthesis is up-regulated. Moreover, a recent study by Whiteley and coworkers (2001) employing DNA microarray analysis demonstrated that only about 1% of the *P. aeruginosa* genes were differentially regulated in mature biofilms as compared to planktonic cells. Interestingly, none of the *alg* genes appeared to be differentially regulated, whereas biosynthesis genes for pili IV and flagella were repressed. Scanning laser confocal microscopy (SCLM) analysis confirmed this result, demonstrating that nonmucoid strains formed densely packed biofilms that were generally less than 6 μm in depth. In contrast, *P. aeruginosa* FRD1 produced microcolonies that were approximately 40 μm in depth. An *algJ* mutant strain that produced alginate lacking O-acetyl groups produced only small microcolonies. After 44 h, the *algJ* mutant switched to the nonmucoid phenotype and formed uniform biofilms, similar to biofilms produced by the nonmucoid strains (Nivens et al., 2001). Results of both the ATR/FT-IR (attenuated total reflexion/Fourier transform infrared) and SCLM analyses of nonmucoid strains demonstrated that alginate was not required for *P. aeruginosa* biofilm formation, and therefore alginate did not act as a primary adhesin for the *P. aeruginosa* cells to these surfaces. Biofilm formation by *P. aeruginosa* FRD1131 that had a *Tn501* insertion in the alginate biosynthesis gene *algD* was still possible. Strain FRD1131 showed a delay in biofilm formation but ultimately formed biofilms, indicating that alginate biosynthesis is not essential for biofilm formation.

These results demonstrate that alginate, although not required for *P. aeruginosa* biofilm development, plays a role in the biofilm structure and may act as intercellular material, required for formation of thicker three-dimensional biofilms. The results also demonstrate the importance of alginate O-acetylation in *P. aeruginosa* biofilm architecture.

The portion of the biofilm developmental pathway that concerns detachment represents an important area of future research. One possible signal for detaching may be starvation, although this has not been investigated in detail. However, Boyd and Chakrabarty (1994) reported that the enzyme alginate lyase may play a role in the detachment phase in *P. aeruginosa*. These authors showed that overexpression of alginate lyase could speed detachment and cell sloughing from biofilms (Boyd and Chakrabarty, 1994). A recent study by Allison et al. (1998) showed that a *P. fluorescens* biofilm decreased after extended incubation, which was attributed – at least in part – to the loss of EPS. Furthermore, these authors presented evidence showing that acyl-HSLs and/or another factor present in stationary-phase culture supernatants mediated this effect (Allison et al., 1998). Little else is known about the functions or regulatory pathways involved in the release of bacteria from the biofilm.

10 The Applied Potential of Bacterial Alginates

At present, all alginates used for commercial purposes are produced from harvested brown seaweeds. The prices of such alginates are generally low, and it seems to be a difficult task to establish a competitive bacterial production process in this price range. However, there are at least two factors that now make it more probable that bacterial alginates may become commercial products. The first is related to the environmental concerns associated with seaweed harvesting and processing, and the second is related to the quality of the final polymer product. The environmental impact is not the topic of this review, but the possibility of producing bacterial alginates with improved qualities will be discussed on the basis of recent scientific discoveries. Alginates with unique qualitative properties have the obvious advantage that they may potentially be sold at higher prices than the bulk materials, and such new polymer products may at least theoretically also open new kinds of markets for this polymer. The properties of pure alginates are determined by their degree of polymerization and acetylation, and by their monomer composition and sequence (Moe et al., 1995). It appears likely that it will become possible to control these three parameters in bacterial alginates. The corresponding possibility does not seem realistic for alginates obtained by the harvesting of oceanic seaweeds, and producers of such products will therefore be limited by the need to fractionate the polymer mixtures produced by these organisms in their natural environments. The degree of polymerization affects the viscosity of alginates, and it will most likely be possible – at least to some extent – to control this parameter in bacteria by strict control of the fermentation conditions and/or genetic manipulations. One possible target for genetic modifications might be the alginate lyase found in both *Pseudomonas* and *Azotobacter* species (Kennedy et al., 1992; Lloret et al., 1996; Monday and Schiller, 1996; Rehm et al., 1996). Hence, bacterial alginates might in principle provide a particular viscosity if reduced quantities of material (on a weight basis) are used. Extreme viscosities would, on the other hand, also create problems with oxygen transfer in fermentors. From a commercial point of view it is not clear if this is a fruitful approach. It now seems obvious that the most interesting possibilities offered by bacterial alginates relate to the control of their monomer composition and sequential structures. Both seaweed and bacterial alginates are heterogeneous mixtures of molecules, and their sequential monomer composition can be described by only statistical models (Stokke et al., 1991), in contrast to accurately defined protein and nucleic acid sequences. If alginate structures are to be controlled, one must therefore first understand the origin of the structural variability observed in nature. On the basis of current knowledge it appears that *Pseudomonas* species have only one mannuronan C-5-epimerase, encoded by the *algG* gene, and this enzyme is able to introduce only single guluronic acid residues into the mannuronan chain. Consequently, these alginates cannot form the divalent-cation-dependent (typically Ca^{2+}) gels formed by alginates which contain G-blocks. The potential for in-vivo manipulations of monomer structures in *Pseudomonas* alginates therefore appears to be limited. However, by knocking out the *algG* gene, it may be possible to produce pure mannuronan, and evidence for this has been presented (Franklin et al., 1994). Mannuronan is known to be a strong immunostimulant, and might have a commercial potential in applications where this property is of interest (Skjåk-Bræk and Espevik, 1996). In *A. vinelandii*, the epimerization system (AlgE1–E7) is complicated, as described above. The known secreted epimerases and their different substrate specificities might allow the *in vitro* design of alginates (see above). Specialized alginates might be made *in vitro* by first producing deacetylated poly(mannuronic acid), possibly from a *Pseudomonas* species. Deacetylated alginates are needed because early studies indicate that acetylation seems to protect them from epimerization (Skjåk-Bræk et al., 1985). The acetyl groups might be prevented from being introduced by the use of an acetylation-deficient mutant of the production strain (Franklin and Ohman, 1993, 1996; Shinabarger et al., 1993), or they might be removed later using standard chemical deacetylation procedures. Mannuronan produced in this way could then be epimerized by one particular recombinantly produced epimerase, or by combinations of such enzymes. Alternatively, one might homogenize commercially produced alginates from brown seaweeds using recombinant epimerases. It is now very probable that almost any alginate structure of applied interest can be produced by such procedures. The problems of marketing these products are therefore most likely more related to parameters such as price, and to the technical advantages of the products compared with those already available from brown seaweeds. For food and pharmaceutical applications, approval by legal authorities may also be a major obstacle. In order to evaluate the technological properties of the alginates modified *in vitro*, the enzymes must be produced in sufficient quantities to allow physical and functional studies of the modified polymers. Alginates produced as described above would probably be quite expensive and, if the price were to become a hindrance to certain applications, then the production of similar products *in vivo* using metabolic engineering techniques might well be considered.

The first applications of bacterial alginates are likely to involve their use either as immunostimulants (e.g., mannuronan, see above) or as gel-forming agents for the immobilization of cells (e.g., in tissue engineering) (Gutowska et al., 2001). Immobilized cells might be used for a variety of biotechnological production processes (see

Skjåk-Bræk and Espevik, 1996 for mini review), or in medical transplantation technologies. A variety of proposals have been suggested by different groups (Skjåk-Bræk and Espevik, 1996), and one of the most interesting examples involves the reversal of type I diabetes by immobilizing insulin-producing cells in alginate capsules. These capsules have been implanted into the body of whole animals and even humans, and the biological effects of this kind of approach are currently being evaluated (Soon-Shiong, 1995).

Acknowledgments

These studies were supported by the Deutsche Forschungsgemeinschaft, the Deutsche Fördergesellschaft der Mukoviszidose-Forschung e.V. and the Minister für Wissenschaft und Forschung des Landes Nordrhein-Westfalen.

11 References

Aarons, S. J., Sutherland, I. W., Chakrabarty, A. M., Gallagher, M. P. (1997) A novel gene, *algK*, from the alginate biosynthesis cluster of *Pseudomonas aeruginosa*, *Microbiology* **143**, 641–652.

Akiyama, H., Endo, T., Nakakita, R., Murata, K., Yonemoto, Y., Okayama, K. (1992) Effect of depolymerized alginates on the growth of bifidobacteria, *Biosci. Biotechnol. Biochem.* **56**, 355–356.

Albersheim, P., Darvill, A. G. (1985) Oligosaccharins, *Sci. Am.* **253**, 58–64.

Allison, D. G., Ruiz, B., SanJose, C., Jaspe, A., Gilbert, P. (1998) Extracellular products as mediators of the formation and detachment of *Pseudomonas fluorescens* biofilms, *FEMS Microbiol. Lett.* **167**, 179–184.

Beale, J. M., Foster, J. L. (1996) Carbohydrate fluxes into alginate biosynthesis in *Azotobacter vinelandii* NCIB 8789 – nmr investigations of the triose pools, *Biochemistry* **35**, 4492–4501.

Boucher, J. C., Martinez-Salazar, J., Schurr, M. J., Mudd, M. H., Yu, H., Deretic, V. (1996) Two distinct loci affecting conversion to mucoidy in *Pseudomonas aeruginosa* in cystic fibrosis encode homologs of the serine-protease HtrA, *J. Bacteriol.* **178**, 511–523.

Boyd, A., Chakrabarty, A. M. (1994) Role of alginate lyase in cell detachment of *Pseudomonas aeruginosa*, *Appl. Environ. Microbiol.* **60**, 2355–2359.

Boyd, A., Gosh, M., May, T. B., Shinabarger, D., Keogh, R., Chakrabarty, A. M. (1993) Sequence of the *algL* gene from *Pseudomonas aeruginosa* and purification of its alginate lyase product, *Gene* **131**, 1–8.

Campos, M.-E., Martinez-Salazar, J. M., Lloret, L., Moreno, S., Nunez, C., Espin, G., Soberon-Chavez, G. (1996) Characterization of the gene coding for GDP-mannose dehydrogenase (*algD*) from *Azotobacter vinelandii*, *J. Bacteriol.* **178**, 1793–1799.

Chitnis, C. E., Ohman, D. E. (1993) Genetic analysis of the alginate biosynthetic gene cluster of *Pseudomonas aeruginosa* shows evidence of an operonic structure, *Mol. Microbiol.* **8**, 583–590.

Chu, L., May, T. B., Chakrabarty, A. M., Misra, T. K. (1991) Nucleotide sequence and expression of the *alg*E gene involved in alginate biosynthesis by *Pseudomonas aeruginosa*, *Gene* **107**, 1–10.

Cote, G. L., Krull, L. H. (1988) Characterization of the exocellular polysaccharides from *Azotobacter chroococcum*, *Carbohydr. Res.* **181**, 143–152.

Darzins, A., Chakrabarty, A. M. (1984) Cloning of genes controlling alginate biosynthesis from a mucoid cystic fibrosis isolate of *Pseudomonas aeruginosa*, *J. Bacteriol.* **159**, 9–18.

Davies, D. G., Geesey, G. G. (1995) Regulation of the alginate biosynthesis gene *algC* in *Pseudomonas aeruginosa* during biofilm development in continuous culture, *Appl. Environ. Microbiol.* **61**, 860–867.

Davies, D. G., Chakrabarty, A. M., Geesey, G. G. (1993) Exopolysaccharide production in biofilms – substratum activation of alginate gene-expression by *Pseudomonas aeruginosa*, *Appl. Environ. Microbiol.* **59**, 1181–1186.

Deretic, V., Konyecsni, W. M. (1989) Control of mucoidy in *Pseudomonas aeruginosa*: transcriptional regulation of *alg*R and identification of the second regulatory gene *alg*Q, *J. Bacteriol.* **171**, 3680–3688.

Deretic, V., Dikshit, R., Konyecsni, W. M., Chakrabarty, A. M., Misra, T. K. (1989) The *algR* gene, which regulates mucoidy in *Pseudomonas aeruginosa*, belongs to a class of environmentally response genes, *J. Bacteriol.* **171**, 1278–1283.

Deretic, V., Schurr, M. J., Boucher, J. C., Martin, D. W. (1994) Conversion of *Pseudomonas aeruginosa* to mucoidy in cystic fibrosis: environmental stress and regulation of bacterial virulence by alternative sigma factors, *J. Bacteriol.* **176**, 2773–2780.

Devault, J. D., Hendrickson, W., Kato, J., Chakrabarty, A. M. (1991) Environmentally regulated *algD* promoter is responsive to the cAMP receptor protein in *Escherichia coli*, *Mol. Microbiol.* **5**, 2503–2509.

Ertesvåg, H., Valla, S. (1999) The A modules of the *Azotobacter vinelandii* mannuronan-C-5-epimerase AlgE1 are sufficient for both epimerization and binding of Ca^{2+}, *J. Bacteriol.* **181**, 3033–3038.

Ertesvåg, H., Doseth, B., Larson, B., Skjåk-Bræk, G., Valla, S. (1994) Cloning and expression of an *Azotobacter vinelandii* mannonuran C-5-epimerase gene, *J. Bacteriol.* **176**, 2846–2853.

Ertesvåg, H., Hoidal, H. K., Hals, I. K., Rian, A., Doseth, B., Valla, S. (1995) A family of modular type mannuronan C-5-epimerase genes controls alginate structure in *Azotobacter vinelandii*, *Mol. Microbiol.* **9**, 719–731.

Ertesvåg, H., Frode, E., Skjåk-Bræk, G., Rehm, B. H. A., Valla, S. (1998a) Biochemical properties and substrate specificities of a recombinantly produced *Azotobacter vinelandii* alginate lyase, *J. Bacteriol.* **180**, 3779–3784.

Ertesvåg, H., Hoidal, H. K., Skjåk-Bræk, G., Valla, S. (1998b) The *Azotobacter vinelandii* mannuronan C-5-epimerase AlgE1 consists of two separate catalytic domains, *J. Biol. Chem.* **273**, 30927–30932.

Ertesvåg, H., Hoidal, H. K., Schjerven, H., Svanem, B. I., Valla, S. (1999) Mannuronan C-5-epimerases and their application for *in vitro* and *in vivo* design of new alginates useful in biotechnology, *Metab. Eng.* **1**, 262–269.

Fett, W. F., Osman, S. F., Fishman, M. L., Siebles, T. S. (1986) Alginate production by plant-pathogenic Pseudomonads, *Appl. Environ. Microbiol.* **52**, 466–473.

Fett, W. F., Wijey, C., Lifson, E. R. (1992) Occurrence of alginate gene-sequences among members of the Pseudomonad ribosomal-RNA homology groups I-IV, *FEMS Microbiol. Lett.* **99**, 151–157.

Fialho, A. M., Zielinski, N. A., Fett, W. F., Chakrabarty, A. M., Berry, A. (1990) Distribution of alginate gene-sequences in the *Pseudomonas* ribosomal-RNA homology group I-*Azomonas*-*Azotobacter* lineage of superfamily-B procaryotes, *Appl. Environ. Microbiol.* **56**, 436–443.

Franklin, M. J., Ohman, D. E. (1993) Identification of *alg*F in the alginate biosynthetic gene cluster of *Pseudomonas aeruginosa* which is required for alginate acetylation, *J. Bacteriol.* **175**, 5057–5065.

Franklin, M. J., Ohman, D. E. (1996) Identification of *algI* and *algJ* in the *Pseudomonas aeruginosa* alginate biosynthetic gene cluster which are required for alginate acetylation, *J. Bacteriol* **178**, 2186–2195.

Franklin, M. J., Chitnis, C. E., Gacesa, P., Sonesson, A., White, D. C., Ohman, D. E. (1994) *Pseudomonas aeruginosa* AlgG is a polymer level alginate C5-mannuronan epimerase, *J. Bacteriol.* **176**, 1821–1830.

Fujihara, M., Nagumo, T. (1992) The effect of the content of D-mannuronic acid and L-guluronic acid blocks in alginates on antitumor activity, *Carbohydr. Res.* **224**, 343–347.

Fujihara, M., Nagumo, T. (1993) An influence of the structure of alginate on the chemotactic activity of macrophages and the antitumor activity, *Carbohydr. Res.* **243**, 211–216.

Fujihara, M., Iizima, N., Yamamoto, I., Nagumo, T. (1984) Purification and chemical and physical characterization of an antitumor polysaccharide from the brown seaweed *Sargassum fulvellum*, *Carbohydr. Res.* **125**, 97–106.

Gacesa, P. (1987) Alginate-modifying enzymes: a proposed unified mechanism of action for the lyases and epimerases, *FEBS Lett.* **212**, 199–202.

Gacesa, P. (1992) Enzymic degradation of alginates, *Int. J. Biochem.* **24**, 545–552.

Gacesa, P. (1998) Bacterial alginate biosynthesis – recent progress and future prospects, *Microbiology* **144**, 1133–1143.

Gacesa, P., Goldberg, J. B. (1992) Heterologous expression of an alginate lyase gene in mucoid and non-mucoid strains of *Pseudomonas aeruginosa*, *J. Gen. Microbiol.* **138**, 1665–1670.

Gacesa, P., Russell, N. J. (1990) The structure and property of alginate, in: *Pseudomonas* Infection and Alginates (Gacesa, P., Russell, N. J., Eds.), London: Chapman & Hall, 29–49.

Garrett, E. S., Perlegas, D., Wozniak, D. J. (1999) Negative control of flagellum synthesis in *Pseudomonas aeruginosa* is modulated by the alternative sigma factor AlgT (AlgU), *J. Bacteriol.* **181**, 7401–7404.

Goldberg, J. B., Hatano, K., Pier, G. B. (1993). Synthesis of lipopolysaccharide-O side-chains by *Pseudomonas aeruginosa* PAO1 requires the enzyme phosphomannomutase, *J. Bacteriol.* **175**, 1605–1611.

Gorin, P. A., Spencer, J. F. T. (1966) Exocellular alginic acid from *Azotobacter vinelandii*, *Can. J. Chem.* **44**, 993–998.

Govan, J. R. W., Deretic, V. (1996) Microbial pathogenesis in cystic fibrosis: mucoid *Pseudo-*

monas aeruginosa and *Burkholderia cepacia*, *Microbiol. Rev.* **60**, 539–574.

Govan, J. R. W., Fyfe, J. F. M. (1978) Mucoid *Pseudomonas aeruginosa* and cystic fibrosis: resistance of the mucoid form to carbenicillin, flucloxacillin and tobramycin and the isolation of mucoid variants *in vitro*, *J. Antimicrob. Chemother.* **4**, 233–240.

Govan, J. R. W., Harris, G. S. (1986) *Pseudomonas aeruginosa* and cystic fibrosis: unusual bacterial adaptation and pathogenesis, *Microbiol. Sci.* **3**, 302–308.

Govan, J. R. W., Fyfe, J. F. M., Jarman, T. R. (1981) Isolation of alginate-producing mutants of *Pseudomonas fluorescens, Pseudomonas putida*, and *Pseudomonas mendocina*, *J. Gen. Microbiol.* **125**, 217–220.

Grabert, E., Wingender, J., Winkler, U. K. (1990) An outer membrane protein characteristic of mucoid strains of *Pseudomonas aeruginosa*, *FEMS Microbiol. Lett.* **68**, 83–88.

Grobe, S., Wingender, J., Trüper, H. G. (1995) Characterization of mucoid *Pseudomonas aeruginosa* strains isolated from technical water systems, *J. Appl. Bacteriol.* **79**, 94–102.

Gross, M., Rudolph, K. (1987) Demonstration of levan and alginate in bean-plants (*Phaseolus vulgaris*) infected by *Pseudomonas syringae* pv. *phaseolicola*, *J. Phytopathol.* **120**, 9–19.

Gutowska, A., Jeong, B., Jasionowski, M. (2001) Injectable gels for tissue engineering, *Anat. Rec.* **263**, 342–349.

Haug, A., Larsen, B., Smidsrød, O. (1967) Studies on the sequence of uronic acid residues in alginic acid, *Acta Chem. Scand.* **21**, 691–704.

Hershberger, C. D., Ye, R. W., Parsek, M. R., Xie, Z. D., Chakrabarty, A. M. (1995) The *algT* (*algU*) gene of *Pseudomonas aeruginosa*, a key regulator involved in alginate biosynthesis, encodes an alternative sigma factor (sigma^{E}), *Proc. Natl. Acad. Sci. USA* **92**, 7941–7945.

Kapatral, V., Bina, X., Chakrabarty, A. M. (2000) Succinyl coenzyme A synthetase of *Pseudomonas aeruginosa* with a broad specificity for nucleoside triphosphate (NTP) synthesis modulates specificity for NTP synthesis by the 12-kilodalton form of nucleoside diphosphate kinase, *J. Bacteriol.* **182**, 1333–1339.

Kawada, A., Hiura, N., Shiraiwa, M., Tajima, S., Hiruma, M. (1997) Stimulation of human keratinocyte growth by alginate oligosaccharides, a possible co-factor for epidermal growth factor in cell culture, *FEBS Lett.* **408**, 43–46.

Kennedy, L., McDowell, K., Sutherland, I. W. (1992) Alginases from *Azotobacter* species, *J. Gen. Microbiol.* **138**, 2465–2471.

Kim, H. Y., Schlictman, D., Shankar, S., Xie, Z., Chakrabarty, A. M., Kornberg, A. (1998) Alginate, inorganic polyphosphate, GTP and ppGpp synthesis co-regulated in *Pseudomonas aeruginosa*: implications for stationary phase survival and synthesis of RNA/DNA precursors, *Mol. Microbiol.* **27**, 717–725.

Konyecsni, W. M., Deretic, V. (1990) DNA sequence and expression analysis of *alg*P and *alg*Q, components of the multigene system transcriptionally regulating mucoidy in *Pseudomonas aeruginosa*: *algP* contains multiple direct repeats, *J. Bacteriol.* **172**, 2511–2520.

Lee, J. W., Day, D. F. (1995) Bioacetylation of seaweed alginate, *Appl. Environ. Microbiol.* **61**, 650–655.

Lin, T.-Y., Hassid, W. Z. (1966a) Pathway of alginic acid synthesis in the marine brown alga, *Fucus gardneri* Silva, *J. Biol. Chem.* **241**, 5284–5297.

Lin, T.-Y., Hassid, W. Z. (1966b) Isolation of guanosine diphosphate uronic acids from a marine brown alga, *Fucus gardneri* Silva, *J. Biol. Chem.* **241**, 3283–3293.

Linker, A., Jones, R. S. (1966) A new polysaccharide resembling alginic acid isolated from pseudomonads, *J. Biol. Chem.* **241**, 3845–3851.

Lloret, L., Barreto, R., Leon, R., Moreno, S., Martínez-Salazar, J., Espín, G., Soberón-Chávez, G. (1996) Genetic analysis of the transcriptional arrangement of *Azotobacter vinelandii* alginate biosynthetic genes: identification of two independent promoters, *Mol. Microbiol.* **21**, 449–457.

Lynn, A. R., Sokatch, J. R. (1984) Incorporation of isotope from specifically labelled glucose into alginates of *Pseudomonas aeruginosa* and *Azotobacter vinelandii*, *J. Bacteriol.* **158**, 1161–1162.

Ma, J. F., Phibbs, P. V., Hassett, D. J. (1997) Glucose stimulates alginate production and *algD* transcription in *Pseudomonas aeruginosa*, *FEMS Microbiol. Lett.* **148**, 217–221.

Ma, S., Selvaraj, U., Ohmann, D. E. Quarless, R., Hassett, D. J., Wozniak, D. J. (1998) Phsophorylation-independent activity of the response regulators AlgB and AlgR in promoting alginate biosynthesis in mucoid *Pseudomonas aeruginosa*, *J. Bacteriol.* **180**, 956–968.

Madgwick, J., Haug, A., Larsen, B. (1978) Ionic requirements of alginate-modifying enzymes in the marine alga *Pelvetia canaliculata* (L.) Dcne. et Thur, *Bot. Mar.* **21**, 1–3.

Maharaj, R., May, T. B., Wang, S. K., Chakrabarty, A. M. (1993) Sequence of the *alg8* and alg44 genes involved in the synthesis of alginate by *Pseudomonas aeruginosa*, *Gene* **136**, 267–269.

Martin, D. W., Schurr, M. J., Mudd, M. H., Deretic, V. (1993a) Differentiation of *Pseudomonas aeruginosa* into the alginate-producing form: inactivation of *mucB* causes conversion to mucoidy, *Mol. Microbiol.* **9**, 495–506.

Martin, D. W., Schurr, M. J., Mudd, M. H., Govan, J. R. W., Holloway, B. W., Deretic, V. (1993b) Mechanism of conversion to mucoidy in *Pseudomonas aeruginosa* infecting cystic fibrosis patients, *Proc. Natl. Acad. Sci. USA* **90**, 8377–8381.

Martin, D. W., Schurr, M. J., Yu, H., Deretic, V. (1994) Analysis of promoters controlled by the putative sigma-factor AlgU regulating conversion to mucoidy in *Pseudomonas aeruginosa* relationship to sigma(e) and stress-response, *J. Bacteriol.* **176**, 6688–6696.

Martínez-Salazar, J. M., Moreno, S., Najera, R., Boucher, J. C., Espín, G., Soberón-Chávez, G., Deretic, V. (1996) Characterization of genes coding for the putative sigma factor AlgU and its regulators MucA, MucB, MucC, and MucD in *Azotobacter vinelandii* and evaluation of their roles in alginate biosynthesis, *J. Bacteriol.* **178**, 1800–1808.

Mathee, K., McPherson, C. J., Ohmann, D. E. (1977) Posttranslational control of the algT (algU)-encoded sigma22 for expression of the alginate regulon in *Pseudomonas aeruginosa* and localization of its antagonist proteins MucA and MucB (AlgN), *J. Bacteriol.* **179**, 3711–3720.

Mathews, C. K. (1993) Enzyme organization in DNA precursor biosynthesis, *Prog. Nucleic Acid Res. Mol. Biol.* **44**, 167–203

May, T. B., Chakrabarty, A. M. (1994) *Pseudomonas aeruginosa*: genes and enzymes of alginate biosynthesis, *Trends Microbiol.* **2**, 151–157.

May, T. B., Shinabarger, D., Maharaj, R., Kato, J., Chu, L., Devault, J. D., Roychoudhury, S., Zielinski, N. A., Berry, A., Rothmel, R. K., Misra, T. K., Chakrabarty, A. M. (1991) Alginate synthesis by *Pseudomonas aeruginosa*: a key pathogenic factor in chronic pulmonary infections of cystic fibrosis patients, *Clin. Microbiol. Rev.* **4**, 191–206.

McAvoy, M. J., Newton, V., Paull, A., Morgan, J., Gacesa, P., Russell, N. J. (1989) Isolation of mucoid strains of *Pseudomonas aeruginosa* from non-cystic-fibrosis patients and characterization of the structure of their secreted alginate, *J. Med. Microbiol.* **28**, 183–189.

Moe, S. T., Draget, K. I., Skjåk-Bræk, G., Smidsrød, O. (1995) Alginates, in: *Food Polysaccharides and Applications* (Stephen, A. M., Ed.), New York: Marcel Dekker, Inc., 245–286.

Monday, S. R., Schiller, N. L. (1996) Alginate synthesis in *Pseudomonas aeruginosa*: the role of AlgL (alginate lyase) and AlgX, *J. Bacteriol.* **178**, 625–632.

Mrsny, R. J., Lazazzera, B. A., Daugherty, A. L., Schiller, N. L., Patapoff, T. W. (1994) Addition of a bacterial alginate lyase to purulent CF sputum *in vitro* can result in the disruption of alginate and modification of sputum viscoelasticity, *Pulm. Pharmacol.* **7**, 357–366.

Mrsny, R. J., Daugherty, A. L., Short, S. M., Widmer, R., Siegel, M. W., Keller, G.-A. (1996) Distribution of DNA and alginate in purulent cystic fibrosis sputum: implications to pulmonary targeting strategies, *J. Drug Target.* **4**, 233–243.

Murata, K., Inose, T., Hisano, T., Abe, S., Yonemoto, Y. (1993) Bacterial alginate lyase: enzymology, genetics and application, *J. Ferment. Bioeng.* **76**, 427–437.

Narbad, A., Russell, N. J., Gacesa, P. (1988) Radiolabelling patterns in alginate of *Pseudomonas aeruginosa* synthesized from specifically-labelled ^{13}C monosaccharide precursors, *Microbios* **54**, 171–179.

Natsume, M., Kamo, Y., Hirayama, M., Adachi, T. (1994) Isolation and characterization of alginate-derived oligosaccharides with root growth-promoting activities, *Carbohydr. Res.* **258**, 187–197.

Nivens, D. E., Ohman, D. E., Williams, J., Franklin, M. J. (2001) Role of alginate and its O acetylation in formation of *Pseudomonas aeruginosa* microcolonies and biofilms, *J. Bacteriol.* **183**, 1047–1057.

Nguyen L. K., Schiller, N. L. (1989) Identification of a slime exopolysaccharide de-polymerase in mucoid strains of *Pseudomonas aeruginosa*, *Curr. Microbiol.* **18**, 323–329.

Olvera, C., Goldberg, J. B., Sanchez, R., Soberon-Chavez, G. (1999) The *Pseudomonas aeruginosa algC* gene product participates in rhamnolipid biosynthesis, *FEMS Microbiol. Lett.* **179**, 85–90.

Onsøyen, E. (1996) Commercial applications of alginates. *Carbohydr. Eur.* **14**, 26–31.

Otterlei, M., Østgaard, K., Skjåk-Bræk, G., Smidsrød, O., Soon-Shoing, Espevik, T. (1991) Induction of cytokine production from human monocytes stimulated with alginate, *J. Immunother.* **10**, 286–291.

Padgett, P. J., Phibbs, P. V., Jr. (1986) Phosphomannomutase activity in wild-type and alginate-producing strains of *Pseudomonas aeruginosa*, *Curr. Microbiol.* **14**, 187–192.

Page, W. J., Sadoff, H. L. (1975) Relationship between calcium and uronic acids in the encystment of *Azotobacter vinelandii*, *J. Bacteriol.* **122**, 145–151.

Pecina, A., Pascual, A., Paneque, A. (1999) Cloning and expression of the *algL* gene, encoding the

Azotobacter chroococcum alginate lyase: purification and characterization of the enzyme, *J. Bacteriol.* **181**, 1409–1414.

Penaloza-Vazquez, A., Kidambi, S. P., Chakrabarty, A. M., Bender, C. L. (1997). Characterisation of the alginate biosynthetic gene cluster in *Pseudomonas syringae* pv. *syringae*, *J. Bacteriol.* **179**, 4464–4472.

Piggott, N. H., Sutherland, I. W., Jarman, T. R. (1981) Enzymes involved in the biosynthesis of alginate by *Pseudomonas aeruginosa*, *Eur. J. Appl. Microbiol. Biotechnol.* **13**, 179–183.

Pindar, D. F., Bucke, C. (1975) The biosynthesis of alginic acid by *Azotobacter vinelandii*, *Biochem. J.* **152**, 617–622.

Ramstad, M. V., Ellingsen, T. E., Josefsen, K. D., Hoidal, H. K., Valla, S., Skjåk-Bræk, G., Levine, D. W. (1999) Properties and action pattern of the recombinant mannuronan C-5-epimerase AlgE2, *Enzyme Microbiol. Technol.* **24**, 636–646.

Rehm, B. H. A. (1996) The *Azotobacter vinelandii* gene *algJ* encodes an outer membrane protein presumably involved in export of alginate, *Microbiology* **142**, 873–880.

Rehm, B. H. A. (1998) The alginate lyase from *Pseudomonas aeruginosa* CF1/M1 prefers the hexameric oligomannuronate as substrate, *FEMS Microbiol. Lett.* **165**, 175–180.

Rehm, B. H. A., Valla, S. (1997) Bacterial alginates: biosynthesis and applications, *Appl. Microbiol. Biotechnol.* **48**, 281–288.

Rehm, B. H. A., Winkler, U. K. (1996) Alginatbiosynthese bei *Pseudomonas aeruginosa* und *Azotobacter vinelandii*: Molekularbiologie und Bedeutung, *BIOspektrum* **4**, 31–36.

Rehm, B. H. A., Grabert, G., Hein, J., Winkler, U. K. (1994a) Antibody response of rabbits and cystic fibrosis patients to alginate-specific outer membrane protein of a mucoid strain of *Pseudomonas aeruginosa*, *Microb. Pathog.* **16**, 43–51.

Rehm, B. H. A., Boheim, G., Tommassen, J., Winkler, U. K. (1994b) Overexpression of *algE* in *Escherichia coli*: subcellular localization, purification, and ion channel properties, *J. Bacteriol.* **176**, 5639–5647.

Rehm, B. H. A., Ertesvåg, H., Valla, S. (1996) A new *Azotobacter vinelandii* mannuronan C-5-epimerase gene (*algG*) is part of an *alg* gene cluster physically organized in a manner similar to that in *Pseudomonas aeruginosa*, *J. Bacteriol.* **178**, 5884–5889.

Ross, P., Meyer, R., Benziman, M. (1991) Cellulose biosynthesis and function in bacteria, *Microbiol. Rev.* **55**, 35–58.

Sadoff, H. L. (1975) Encystment and germination in *Azotobacter vinelandii*, *Bacteriol. Rev.* **39**, 516–539.

Schiller, N. L., Monday, S. R., Boyd, C. M., Keen, N. T., Ohman, D. E. (1993) Characterization of the *Pseudomonas aeruginosa* alginate lyase gene (*algL*) – cloning, sequencing, and expression in *Escherichia coli*, *J. Bacteriol.* **175**, 4780–4789.

Schlictman, D., Kavanaughblack, A., Shankar, S., Chakrabarty, A. M. (1994) Energy metabolism and alginate biosynthesis in *Pseudomonas aeruginosa* – role of the tricarboxylic acid cycle, *J. Bacteriol.* **176**, 6023–6029.

Schurr, M. J., Martin, D. W., Mudd, M. H., Hibler, N. S., Boucher, J. C., Deretic, V. (1993) The *algD* promoter – regulation of alginate production by *Pseudomonas aeruginosa* in cystic fibrosis, *Cell. Mol. Biol. Res.* **39**, 371–376.

Schurr, M. J., Yu, H., Martinez-Salazar, J. M., Hibler, N. S., Deretic, V. (1995) Biochemical characterisation and posttranslational modification of AlgU, a regulator of stress response in *Pseudomonas aeruginosa*, *Biochem. Biophys. Res. Commun.* **216**, 874–880.

Schurr, M. J., Yu, H., Martinez-Salazar, J. M., Boucher, J. C., Deretic, V. (1996) Control of *algU*, a member of the sigma(e)-like family of stress sigma factors, by the negative regulators MucA and MucB and *Pseudomonas aeruginosa* conversion to mucoidy in cystic fibrosis, *J. Bacteriol.* **178**, 4997–5004.

Schweizer, H. P., Po, C., Bacic, M. K. (1995) Identification of *Pseudomonas aeruginosa glpM*, whose gene-product is required for efficient alginate biosynthesis from various carbon sources, *J. Bacteriol.* **177**, 4801–4804.

Sherbrock-Cox, V., Russell, N. J., Gacesa, P. (1984) The purification and chemical characterization of the alginate present in extracellular material produced by mucoid strains of *Pseudomonas aeruginosa*, *Carbohydr. Res.* **135**, 147–154.

Shinabarger, D., May, T. B., Boyd, A., Ghosh, M., Chakrabarty, A. M. (1993) Nucleotide sequence and expression of the *Pseudomonas aeruginosa algF* gene controlling acetylation of alginate, *Mol. Microbiol.* **9**, 1027–1035.

Skjåk-Bræk, G., Espevik, T. (1996) Application of alginate gels in biotechnology and medicine, *Carbohydr. Eur.* **14**, 19–25.

Skjåk-Bræk, G., Martinsen, A. (1991) Applications of some algal polysaccharides in biotechnology, in: *Seaweed Resources in Europe: Uses and Potential* (Guiry, M.D., Blunden, G., Eds.), New York: John Wiley & Sons, 219–257.

Skjåk-Bræk G., Larsen, B. Grasdalen, H., (1985) The role of O-acetyl groups in the biosynthesis of alginate by *Azotobacter vinelandii*, *Carbohydr. Res.* **145**, 169–174.

Skjåk-Bræk G., Grasdalen, H., Larsen, B. (1986) Monomer sequence and acetylation pattern in some bacterial alginates, *Carbohydr. Res.* **154**, 239–250.

Smidsrød, O., Draget, K. I. (1996) Chemistry and physical properties of alginates, *Carbohydr. Eur.* **14**, 6–13.

Svanem, B. J., Skjåk-Bræk G., Ertesvåg, H., Valla, S. (1999) Cloning and expression of three new *Azotobacter vinelandii* genes closely related to a previously described gene family encoding mannuronan C-5-epimerases, *J. Bacteriol.* **181**, 68–77

Soon-Shiong, P. (1995) Encapsulated islet cell therapy for the treatment of diabetes: intraperitoneal injection of islets, *J. Controlled Release* **39**, 399–409.

Stokke, B. T., Smidsrød, O., Bruheim, P., Skjåk-Bræk, B. (1991) Distribution of uronate residues in alginate chains in relation to alginate gelling properties, *Macromolecules* **24**, 4637–4645.

Sutherland, J. W. (1982) Biosynthesis of microbial exopolysaccharides, *Adv. Microb. Physiol.* **23**, 79–150.

Sutherland, I. W. (1995) Polysaccharide lyases, *FEMS Microbiol. Rev.* **16**, 323–347.

Tatnell, P. J., Russell, N. J., Gacesa, P. (1993) A metabolic study of the activity of GDP-mannose dehydrogenase and concentrations of activated intermediates of alginate biosynthesis in *Pseudomonas aeruginosa*, *J. Gen. Microbiol* **139**, 119–127.

Tatnell, P. J., Russell, N. J., Gacesa, P. (1994). GDP-mannose dehydrogenase is the key regulatory enzyme in alginate biosynthesis in *Pseudomonas aeruginosa*: evidence from metabolite studies, *Microbiology* **140**, 1745–1754.

Terry, J. M., Pina, S. E., Mattingly, S. J. (1991) Environmental conditions which influence mucoid conversion in *Pseudomonas aeruginosa* PAO1, *Infect. Immun.* **59**, 471–477.

Tomoda, Y., Umemura, K., Adachi, T.(1994) Promotion of barley root elongation under hypoxic conditions by alginate lyase-lysate, *Biosci. Biotechnol. Biochem.* **58**, 202–203.

Vazquez, A., Soledad, M., Guzman, J., Alvarado, A., Espin, G. (1999) Transcriptional organization of the *Azotobacter vinelandii algGXLVIFA* genes: characterization of *algF* mutants, *Gene* **232**, 217–222.

Wang, S. K., Sa-Correia, I., Darzins, A., Chakarabarty, A. M. (1987) Characterization of *Pseudomonas aeruginosa* alginate (*alg*) gene region II, *J. Gen. Microbiol.* **133**, 2303–2314.

Whiteley, M., Bangera, M. G., Bumgarner, R. E., Parsek, M. R., Teitzel, G. M., Lory, S., Greenberg, E. P. (2001) Gene expression in *Pseudomonas aeruginosa* biofilms, *Nature* **413**, 860–864.

Wong, T. Y., Preston, L. A., Schiller, N. L. (2000) Alginate lyase: review of major sources and enzyme characteristics, structure–function analysis, biological roles, and applications, *Annu. Rev. Microbiol.* **54**, 289–340.

Wozniak, D. J., Ohman, D. E. (1991) *Pseudomonas aeruginosa* AlgB, a 2-component response regulator of the NtrC family, is required for *algD* transcription, *J. Bacteriol.* **173**, 1406–1413.

Wozniak, D. J., Ohman, D. E. (1994) Transcriptional analysis of the *Pseudomonas aeruginosa* genes *algR*, *algB*, and *algD* reveals a hierarchy of alginate gene expression which is modulated by *algT*, *J. Bacteriol.* **176**, 6007–6014.

Wyss, O., Neumann, M. G., Socolofsky, M. D. (1961). Development and germination of the *Azotobacter* cyst, *J. Biophys. Biochem. Cytol.* **10**, 555–565.

Yonemoto, Y., Tanaka, H., Yamashita, T., Kitabatake, N., Ishida, Y. (1993) Promotion of germination and shoot elongation of some plants by alginate oligomers prepared with bacterial alginate lyase, *J. Ferment. Bioeng.* **75**, 68–70.

Yoon, H.-J., Mikami, B., Hashimoto, W., Murata, K. (1999) Crystal structure of alginate lyase A1-III from *Sphingomonas* species A1 at 1.78 Å resolution, *J. Mol. Biol.* **290**, 505–514.

Yoon, H.-J., Hashimoto, W., Miyake, O., Murata, K., Mikami, B. (2001) Crystal structure of alginate lyase A1-III complexed with trisaccharide product at 2.0 Å resolution, *J. Mol. Biol.* **307**, 9–16.

Yu, H., Schurr, M. J., Deretic, V. (1995) Functional equivalence of *Escherichia coli* sigma(e) and *Pseudomonas aeruginosa* AlgU. *Escherichia coli* RpoE restores mucoidy and reduces sensitivity to reactive oxygen intermediates in AlgU mutants of *Pseudomonas aeruginosa*, *J. Bacteriol.* **177**, 3259–3268.

Yu, H., Mudd, M., Boucher, J. C., Schurr, M. J., Deretic, V. (1997) Identification of the *algZ* gene upstream of the response regulator *algR* and its participation in control of alginate production in *Pseudomonas aeruginosa*, *J. Bacteriol.* **179**, 187–193.

19
Xanthan

Dr. Karin Born[1], Dr. Virginie Langendorff[2], Dr. Patrick Boulenguer[3]
[1] DEGUSSA Texturant Systems France SAS, Research Center,
F-50500 Baupte, France; Tel.: +33-2-33-713433; Fax: +33-2-33-713492;
E-mail: karin.born@degussa.com
[2] DEGUSSA Texturant Systems France SAS, Research Center,
F-50500 Baupte, France; Tel.: +33-2-33-713433; Fax: +33-2-33-713492;
E-mail: virginie.langendorff@degussa.com
[3] DEGUSSA Texturant Systems France SAS, Research Center,
F-50500 Baupte, France; Tel.: +33-2-33-713433; Fax: +33-2-33-713492;
E-mail: patrick.boulenguer@degussa.com

Biotechnology of Biopolymers. From Synthesis to Patents. Edited by A. Steinbüchel and Y. Doi

ISBN: 3-527-31110-6

ADI	acceptable daily intake
AFFF	asymmetric flow field fractionation
AFM	atomic force microscopy
EPS	exopolysaccharide
GM	genetically modified
GPC	gel-permeation chromatography
LALLS	low-angle laser light scattering
LBG	locust bean gum
MALLS	multi-angle laser light scattering
M_w	molecular weight
O/W	oil in water
SEC	size-exclusion chromatography
W/O	water in oil

1 Introduction

During the second half of the 20th century, many new and useful polysaccharides of scientific and commercial interest have been discovered which can be obtained from microbial fermentations. Microorganisms such as bacteria and fungi produce three distinct types of carbohydrate polymers: (1) extracellular polysaccharides, which can be found either as a capsule that envelops the microbial cell or as an amorphous mass secreted into the surrounding medium; (2) structural polysaccharides, which can be part of the cell wall; or (3) intracellular storage polysaccharides. The scientific and industrial success of polysaccharides of microbial origin is due to several factors. First, they can be produced under controlled conditions with selected species; second, they usually present a high structural regularity; and third, different microorganisms can synthesize a wide range of very specific ionic and neutral polysaccharides with widely varying compositions and properties. Such variety is not found among plant polysaccharides and, perhaps more importantly, it cannot be imitated by means of synthetic chemistry.

The usefulness of water-soluble carbohydrate polymers in industry relies on their wide range of functional properties. The most important characteristic is their ability to modify the properties of aqueous environments, that is their capacity to thicken, emulsify, stabilize, flocculate, swell and suspend or to form gels, films and membranes. Another very important aspect is that polysaccharides obtained from natural, renewable sources are both biocompatible and biodegradable.

Xanthan, a microbial biopolymer produced by the *Xanthomonas* bacterium, has provoked great scientific and industrial interest since its discovery in the late 1950s. In 1999 alone, more than 300 references of articles or patents dealing with xanthan are listed in *Chemical Abstracts*. Since 1990, more than 2000 patents have been listed in *Derwent World Patents Index*. This interest is due to the extraordinary properties of xanthan as well as to the successful establishment of an industrial process for its production.

2 Historical Outline

Xanthan gum was discovered in the 1950s by scientists of the Northern Regional Research Laboratory of the U.S. Department of Agriculture in the course of a screening which aimed at identifying microorganisms that produced water-soluble gums of commercial interest. The first industrial production of xanthan was carried out in 1960, and the product first became available commercially in 1964. Toxicology and safety studies showed that xanthan caused no acute toxicity, had no growth-inhibiting activity, and did not alter organ weights, hematological values or tumors when fed to rats or dogs, neither in short-term, nor in long-term feeding studies. The approval for food use was given by the U.S. Food and Drug administration in 1969, and the FAO/OMS specification followed in 1974. Authorization in France was given in March 1978, and approval in Europe was obtained in 1982, under the E number E415. The official definition of the EU food regulations for E415 is: "Xanthan gum is a high molecular-weight polysaccharide gum produced by a pure culture fermentation of a carbohydrate with natural strains of *Xanthomonas campestris*, purified by recovery with ethanol or propane-2-ol, dried and milled. It contains D-glucose and D-mannose as the dominant hexose units, along with D-glucuronic acid and pyruvic acid, and is prepared as the sodium, potassium or calcium salt. Its solutions are neutral. The molecular weight must be approximately 1 MDa and the color must be cream". Xanthan gum is approved as food additive with an acceptable daily intake (ADI) "not specified"; that is, no limit for ADI is defined and the gum may be used *quantum satis*, which means with just the quantity useful for the application. Today, xanthan is produced commercially by several companies, such as Monsanto/Kelco, Rhodia, Jungbunzlauer, Archer Daniels Midland, and SKW Biosystems. For the past few years xanthan gum has also been produced in China. Annual volumes worldwide are estimated to be about 35,000 tons in 2001.

3 Structure

3.1 Chemical Structure

Xanthan is a heteropolysaccharide with a very high molecular weight, consisting of repeating units (Figure 1). The sugars present in xanthan are D-glucose, D-mannose, and D-glucuronic acid. The glucoses are linked to form a β-1,4-D-glucan cellulosic backbone, and alternate glucoses have a short branch consisting of a glucuronic acid sandwiched between two mannose units. The side chain consists therefore of β-D-mannose-(1,4)-β-D-glucuronic acid-(1,2)-α-D-mannose. The terminal mannose moiety may carry pyruvate residues linked to the 4- and 6-positions. The internal mannose unit is acetylated at C-6. Acetyl and pyruvate substituents are linked in variable amounts to the side chains, depending upon which X. *campestris* strain the xanthan is isolated from. The pyruvic acid content also varies with the fermentation conditions. On average, about half of the terminal mannoses carry a pyruvate, with the number and positioning of the pyruvate and acetate residues conferring a certain irregularity to the otherwise very regular structure. Usually, the degree of substitution for pyruvate varies between 30 and 40%, whereas for acetate the degree of substitution is as high as 60–70%. Some of the repeating units may be devoid of the trisaccharide side chain.

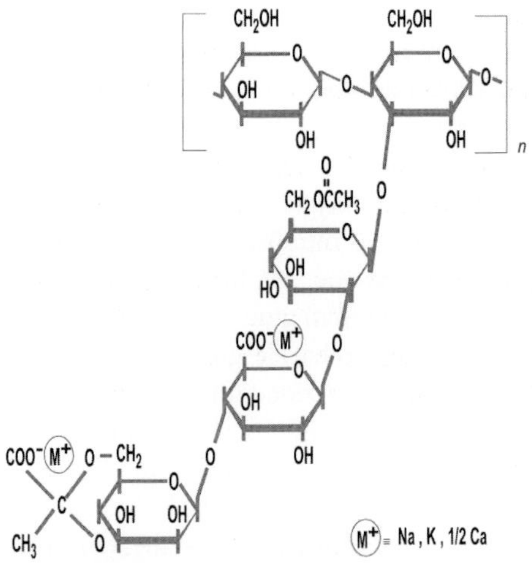

Fig. 1 Chemical structure of xanthan.

3.2 Superstructure/Secondary Structure

The secondary structure of xanthan depends on the conditions under which the molecule is characterized. The molecule may be in an ordered or in a disordered conformation. The ordered confirmation can be either native or renatured; in the native form the conformation is present at temperatures below the melting point of the molecule, a temperature which depends on the ionic strength of the medium in which xanthan is dissolved. The secondary structure of xanthan and the methods to analyze it have been recently reviewed in detail by Stokke et al. (1998). X-ray scattering results indicate that native xanthan in the ordered conformation exists as a right-hand helix with five-fold symmetry with a pitch of 4.7 nm and a diameter of 1.9 nm (Moorhouse, 1977). Two models, a single-strand helix and a double-strand or multi-strand helix, have been proposed, though most authors currently support the idea of a double helix. The helix is stabilized by noncovalent bonds, such as hydrogen bonds, electrostatic interaction, and steric effects; its structure can be described as rigid rod. In aqueous solution, the molecule may undergo a conformational transition which can be driven by changes in temperature and ionic strength, and which depends on the degree of ionization of the carboxyl groups and acetyl contents. The temperature-induced transition from an ordered to a disordered conformation is generally attributed to a complete or partial separation of the double-strand form. Renaturation may occur under favorable conditions, which means temperatures below the transition temperature and high salt concentrations. The transition from the denatured to the renatured state is reversible, whereas that from the native to the denatured state is irreversible. The model of a double-strand structure has been supported by several studies. Capron et al. (1997) demonstrated that upon heating to temperatures above the order–disorder transition temperature, denaturation of the native ordered conformation occurred, together with a reduction in molecular weight. The molecular weight is roughly halved, which supports the model of a double-strand conformation for the native form. The molecular weight was found to be invariant after renaturation on cooling. The renaturation probably occurs as an intramolecular process, which means that the restoration of the ordered form of xanthan seems to take place within a single molecule. Most likely, the conformation for the renatured form of xanthan is that of an anti-parallel, double-stranded structure consisting of one chain folded as a hairpin loop (Liu et al., 1987). The viscosity of renatured xanthan is higher than the viscosity of native xanthan however, thereby supporting the hypothesis that single-stranded xanthan molecules associate during renaturation to form supramolecular structures.

4 Occurrence

Xanthan is produced by *X. campestris*, a plant-associated bacterium that is generally pathogenic for plants belonging to the family Brassicaceae. *Xanthomonas* causes a variety of disease symptoms such as necrosis, gummosis and vascular parenchymatous diseases on leaves, stems or fruits; an example is "black rot" of crucifers such as cabbage, cauliflower or broccoli. *Xanthomonas* does not form spores, but it is very resistant to desiccation during relatively long periods. Survival at room temperature for 25 years has been reported by Leach et al. (1957). The resistance is usually due to the protective effect of the xanthan gum produced and exuded by the bacteria. Xanthan also protects the bacteria from the effects of light, and generally causes wilting of the leaves by blocking water movements (Leach et al., 1957). Exopolysaccharides, like xanthan, are also known to provide protection against bacteriophages by building a physical barrier against the phage attack (MacNeely, 1973). The polysaccharide is not a reserve energy source because in general the bacterium is not able to catabolize its own extracellular polysaccharide.

5 Physiological Function

The physiological function of the exopolysaccharide xanthan has received little attention compared with investigations into the molecule's production, properties, and applications. The bacteria (*Xanthomonas* sp.) that produce xanthan gum as a secondary metabolite are usually phytopathogenic, or may live in asymptomatic association with plant tissues or epiphytes. *Xanthomonas* infections have been observed in over 120 monocotyledonous and over 150 dicotyledonous plant species.

Xanthan is the predominant component of the bacterial slime. The physiological function of xanthan has been deduced as being analogous to the functions of other exopolysaccharides (Yang and Tseng, 1988; De Crecy et al., 1990; Daniels and Leach, 1993; Chan and Goodwin, 1999).

Enclosure of bacterial cells in the exopolysaccharide (EPS) results in prolonged survival, and increased resistance to both temperature and ultraviolet (UV) light (Leach et al., 1957). In rice, the wilting induced by EPS seems to play a role in pathogenesis (Kuo et al., 1970). EPS may increase cell membrane leakage, which in turn leads to wilting (Vidhyasekaran et al., 1989). In general, bacterial EPS induce water-soaking of the intercellular space which is important for bacterial colonization. It is also possible that xanthan forms a gel-like slime in the intercellular space as a result of synergy with other plant polysaccharides. This gel may then promote bacterial colonization of plant tissue by retarding the desiccation of the bacterial colony, by protecting the bacteria from bacteriostatic compounds, and by preventing close morphological contact of the colony with the cell wall, thus preventing the triggering of plant defense reactions. The amount of EPS produced by *Xanthomonas* is correlated to the organism's virulence; strains with attenuated virulence usually produce less EPS (Ramirez et al., 1988) and the distribution of the polysaccharide in infected leafs coincides with that of bacteria. This was seen in a study in rice, where EPS and bacteria were distributed in both the xylem and transverse veins (Watabe et al., 1993).

6 Analysis and Detection

In order to characterize xanthan, different parameters must be taken into consideration, such as chemical structure, acetate and pyruvate contents, molecular weight, secondary structure, and rheological behavior.

6.1 Chemical Characterization

By using chemical analysis, the sugar composition of the molecule as well as the nature and the degree of substituent content can be ascertained.

6.1.1 Sugar Composition

The sugar composition of xanthan is difficult to obtain as the cellulosic backbone is highly resistant to hydrolysis. Moreover, in the side chains the presence of uronic acid prevents complete hydrolysis of the aldobiouronic (β-D-GlcAp-(1 → 2)-D-Manp) acid without degradation of the glucuronic acid. Well-documented reports of this situation have been made (Tait and Sutherland, 1989; Tait et al., 1990) in which the most suitable conditions to determine the neutral sugars, the aldobiouronic (β-D-GlcAp-(1 → 2)-D-Manp) acid and the substituents are described. A single condition with one form of hydrolysis is insufficient to characterize all the constituents of xanthan quantitatively. Due to these problems, the official description of xanthan, e.g., by the JECFA (Joint Expert Committee for Food Additives), does not refer to its chemical composition, but only to its ability to gellify in the presence of locust bean gum (LBG). In the official description, there is no reference to acetyl groups, only to pyruvic acid.

6.1.2 Pyruvic Acid Determination

After hydrolysis (Cheetham and Punruckvong, 1985; Tait et al., 1990), the pyruvic acid content of xanthan can be determined in several ways. The oldest method described is a colorimetric procedure using 2,4-dinitrophenylhydrazine (DNPH) (Slonecker and Orentas, 1962), and this is still the reference method for the JECFA. Duckworth and Yaphe (1970), have developed an enzymatic method using lactate dehydrogenase (LDH), the reaction being as follows:

$$\text{Pyruvate} + \text{NADH} + \text{H}^+ \xrightarrow{\text{LDH/LD}} \text{lactate} + \text{NAD}^+ \quad (1)$$

The amount of NAD released is measured at 340 nm. More recent determinations use high-pressure liquid chromatography (HPLC) (Cheetham and Punruckvong, 1985; Tait et al., 1990) or nuclear magnetic resonance (NMR) methods, both of which permit the simultaneous detection of both pyruvate and acetate.

6.1.3 Acetate Determination

In 1949, Hestrin published a method for the determination of acetate which uses hydroxamic acid, but today (see Section 6.1.2) NMR and HPLC methods are more often used (Cheetham and Punruckvong, 1985; Tait et al., 1990).

6.2 Physical Characterization

Besides the chemical composition of xanthan, its physical characteristics such as molecular weight, secondary structure and rheological properties are the most important determinants of the behavior of this molecule in its final application.

6.2.1 Molecular Weight

Values reported in the literature for the molecular weight (M_w) of xanthan are usually between 4 and 12×10^6 g mol^{-1}. Accurate determination of the M_w of xanthan is difficult for several reasons, including: (1) the very high molecular weight; (2) the stiffness of the molecule; and (3) the presence of aggregates. Nonetheless, several techniques have been used to determine the molecular weight of xanthan, including GPC-MALLS (gel- permeation chromatography with multi-angle laser light scattering), AFFF-MALLS (asymmetrical flow field fractionation combined with multiangle laser light scattering) and electron microscopy.

GPC-MALLS

GPC-MALLS is a technique that permits the estimation of absolute molecular weight and gyration radius of polysaccharides, without the need for column calibration methods or standards. Often used in the field of polymer analysis, this technique is constituted by a GPC system which allows molecules to be separated as a function of their molecular size, and also by MALLS, which allows information to be obtained on the molecular weight of the fraction eluted from the column. With xanthan, the GPC technique presents several difficulties: first, the high molecular weight of the xanthan combined with its rod-like structure causes the xanthan molecules to have a very high hydrodynamic volume. The columns which are used today are unable to separate molecules with such high hydrodynamic volumes, and so xanthan appears as a monodisperse molecule. Second, for the same reason – and also due to the tight stationary phase of the column – the xanthan is submitted to a very high shear rate when eluted across the column, and this can degrade the molecule. In addition, the MALLS detection for xanthan analysis is problematic as, due to its high molecular weight, extrapolation to zero angle is not easy.

There are two classical methods to determine the molecular weight: the Zimm method; and the Debye method. In the Zimm method, we express $Kc/\Delta R_\theta = f(\sin \theta/2)$, whereby K is an optical parameter and R is the Rayleigh ratio. Since $1/M$ is obtained by extrapolation to zero angle, this method is not suitable for xanthan. The values obtained for $1/M$ are in the order of 10^{-7}, which leads to a significant variation of the value calculated for the molecular weight. In the Debye method, we express $\Delta R_0/Kc = f(\sin \theta/2)$; hence, by extrapolation to zero angle, M can be determined directly. This method is preferable for xanthan, but even in this case the very great angular dependence prevents the linear extrapolation and a polynomial of 3rd or 4th order is needed in order to obtain reproducible results. Some authors (Capron et al., 1997) prefer to use LALLS (low-angle laser light scattering), which provides a measure at a very low angle (5 °) and avoids this problem, but this type of apparatus does not provide any information on the gyration radius. Another problem is the presence of aggregates in the xanthan solution, and these are probably responsible for the very high molecular weight values reported in the literature. Such aggregates, even when present in very low quantities only, give very important signals in light scattering.

AFFF-MALLS

In order to avoid the problems which occur with GPC, recent investigations have used AFFF (Janca, 1988). AFFF uses a narrow channel in which a solvent flows, and a field is applied perpendicularly to this channel. Usually, the perpendicular field is created by a perpendicular flow. The sample is injected into the inlet of the channel and eluted by the solvent. At the same time, the field applied across the channel presses the sample against the wall (accumulation wall) of the channel. Due to the gradient of velocity across the channel coming from the laminar flow, the particles – depending on their distances from the wall – have different velocities. In AFFF, the separation is governed by the diffusion coefficient, and so this technique allows the separation of molecules on the basis of their hydrodynamic radius up to a size of several micrometers. An advantage of this method is that no packing material is needed, and this also avoids the problem of shear. The problem is that the different parameters – the two flows, the injected volume, and the solvent – must be carefully adjusted in order to obtain good results.

Electron Microscopy

This technique allows direct measurements of the xanthan molecule to be made after vacuum drying in the presence of glycerol and covering the molecule with a platinum film (Stokke et al., 1998). The contour length L of individual xanthan chains can be visualized by electron microscopy, and the average value of L reflects the molecular weight. Electron microscopy can also be used to detect the formation of microgels.

6.2.2 Secondary Structure

The physico-chemical properties of xanthan in aqueous solutions can be studied by means of various experimental techniques, such as light scattering measurements, hydrodynamic measurements, thermodynamic properties such as ion activity, dependence of the transition temperature on the ionic strength and calorimetric measurements. A relatively new method for studying the superstructure of xanthan is atomic force microscopy (AFM) (Capron et al., 1998a; Morris et al., 1999). AFM allows visualization of the surface of a sample, with a resolution close to the atomic scale. The mechanical properties of both native and denatured xanthan can be measured on the molecular scale (Li et al., 1999; Morris et al., 1999). AFM measures the interaction between the sample and the tip of the measuring device. A force exists between the atom of the tip of the microscope and those of the sample, separated by only a few Angstroms. By moving the tip, it is possible to follow the variation of this force on the surface of the sample and so obtain an image of the sample and estimate its shape. No modification of the sample is needed for the measurement, and this technique can be applied to both conducting and nonconducting samples. AFM avoids the drying step of the sample, and can provide images of individual xanthan molecules as well as of aggregated molecules. Molecules and molecular interactions can be imaged in the liquid environment.

Capron et al. (1997) have studied the size and conformation changes associated with the temperature-induced denaturation and renaturation of native xanthan under different salt conditions. The different methods used were LALLS, size- exclusion chromatography coupled with multi-angle light scattering (SEC-MALLS), low shear intrinsic viscosity measurement, and circular dichroism. The conformational transition can be monitored using NMR, optical rotation or calorimetric measurements. The specific optical rotation changes suddenly at the melting temperature of the molecule, from about –120 to almost zero. Circular dichroism studies near 200 nm show a decrease in overall ellipticity when passing through the transition region.

6.2.3 Rheology

Depending on the medium conditions, that is polymer concentration, salts or addition of other hydrocolloids, xanthan systems can be a Newtonian or pseudo-plastic solution or a gel. The rheological behavior can be determined using viscometers by applying shear rate and measuring shear stress and viscosity or using controlled shear stress or deformation rheometers to perform dynamic viscoelastic or flow measurements. Xanthan solutions can be characterized by classical rheological parameters, such as intrinsic viscosity $[\eta]$. The intrinsic viscosity corresponds to the hydrodynamic volume of the polymer chain in a given solvent. Classically, it can be obtained by measuring the viscosity at different low concentrations in the Newtonian domain, where the overlap between hydrodynamic volume of individual polymer chains is negligible and applying the equations of Huggins (Eq. 2) and Kraemer (Eq. 3) (Launay et al., 1986):

$$\eta_{sp}/C = [\eta] + \lambda_H[\eta]^2 C \quad (2)$$

$$\ln(\eta_r)/C = [\eta] + \lambda_k[\eta]^2 C \text{ with} \quad (3)$$

$$\eta_r = \eta/\eta_0 \quad (4)$$

$$\eta_{sp} = (\eta - \eta_0)/\eta_0 \quad (5)$$

where η is the solution viscosity, η_0 the solvent viscosity, η_r the relative viscosity, η_{sp} the specific viscosity, $[\eta]$ the intrinsic viscosity and λ_H and λ_k are constants which are functions of the hydrodynamic interaction between molecules. In dynamic measurement, the sample is submitted to a deformation $\gamma^*(t)$ or a stress $\sigma^*(t)$ which are sinusoidal function of time. When the system is viscoelastic linear, the stress $\sigma^*(t)$ is a sinusoidal function of time with the same frequency ω as $\gamma^*(t)$ and a phase angle gap δ.

$$\gamma^*(t) = \gamma_0(\cos \omega t + i \cdot \sin \omega t) \quad (6)$$

$$\sigma^*(t) = \sigma_0(\cos (\omega t + \delta) + i \cdot \sin (\omega t + \delta)) \quad (7)$$

If $\delta = 0$, stress and deformation are proportional at every moment, which means that the behavior is elastic linear. If $\delta = \pi/2$, stress and deformation speed are proportional at every moment, the behavior is Newtonian. If $0 < \delta < \pi/2$ the behavior is viscoelastic. The complex modulus G^* can be defined as:

$$G^* = \sigma^*/\gamma^* = G' + i\, G'' \text{ with} \quad (8)$$

$$G' = (\sigma_0 / \gamma_0) \cdot \cos \delta \quad (9)$$

$$G'' = (\sigma_0 / \gamma_0) \cdot \sin \delta \quad (10)$$

$$\tan \delta = G''/G' \quad (11)$$

G' is the storage modulus which corresponds to the elastic component of the system, and G'' is the loss modulus which corresponds to the viscous component of the system. We can also define the complex viscosity:

$$\eta^* = \sigma^*/\acute{y}^* = \eta' + i\, \eta' \text{ with} \quad (12)$$

$$\eta' = G''/\omega \quad (13)$$

$$\eta'' = G'/\omega \quad (14)$$

The gel-like structure of xanthan solutions can be characterized by deformation measurements. Small deformation measurements characterize the intact network, whilst large deformation measurements destroy the network and therefore give lower values. It is possible to measure the static yield point of a xanthan solution by applying an increasing deformation at constant shear rate and measuring the shear stress. Initially, the stress generated in resistance to the deformation increases linearly with the deformation, as in an elastic solid. Ultimately, the resistance reaches a maximum corresponding to the breaking point of the network (i.e., the yield point) and then drops again, settling down at a constant value which defines the steady shear viscosity. Xanthan gum solutions can also be characterized using dynamic light scattering. This technique allows characterization of the boundary between dilute and semi-dilute solutions. The degree of dilution is important because in a truly diluted state, the xanthan coils occupy a defined hydrodynamic volume, whereas above a critical concentration the molecules interact. Dynamic light scattering experiments can demonstrate the onset of molecular interaction and the onset of anisotropic aggregation (Rodd et al., 2000).

7 Biosynthesis

The path of xanthan biosynthesis has been described and reviewed by several authors (Leigh and Coplin, 1992; Becker et al., 1998). Xanthan synthesis starts with the assembly of the pentasaccharide repeating units, and these are then polymerized to produce the macromolecule. The oligosaccharide repeating units of xanthan are formed by the sequential addition of monosaccharides from energy-rich sugar nucleotides, involving acetyl-CoA and phosphoenolpyruvate. A polyisoprenol phosphate from the inner membrane functions as an acceptor (Ielpi et al., 1993). The first step of the pentasaccharide assembly is the transfer of glycosyl-1-phosphate from UDP-glucose to polyisoprenol phosphate. This transfer is followed by sequential transfer of the other sugar residues, D-mannose and D-glucuronic acid from GDP-mannose and UDP-glucuronic acid, respectively, which gives the complete lipid-linked repeating pentasaccharide unit. Acetyl and pyruvyl residues are added at this lipid-linked pentasaccharide stage; these are donated by acetyl-CoA and phosphoenolpyruvate, respectively. Depending on the strain and on the fermentation conditions, O-acetyl groups are attached in varying quantities to the internal mannose residue, and pyruvate is added to the terminal mannose. The xanthan chains grow at the reducing end (Figure 2). The final steps of the biosynthesis, which means the secretion from the cytoplasmic membrane, the passage across the periplasm and outer membrane and the excretion into the extracellular environment has not yet been entirely elucidated. The process requires energy, and it is probably accomplished via a specific transport system, which ensures the release of the polymer from the lipid carrier and the transport across the outer membrane (Daniels and Leach, 1993).

Many genes which are involved in xanthan biosynthesis have been identified, isolated and characterized (Figure 3). In *Xanthomonas campestris* pv. *campestris*, the biosynthesis is directed by a cluster of 12 genes, *gumB* to *gumM* (Vanderslice et al., 1989; Vojnov et al., 1998). Seven gene products are required for the transfer of the monosaccharides and for the acylation at the lipid intermediate level to form the complete acylated repeating unit. This gene cluster is not linked to the genes which are required for the synthesis of the sugar nucleotide precursors. The 12 genes of the cluster are

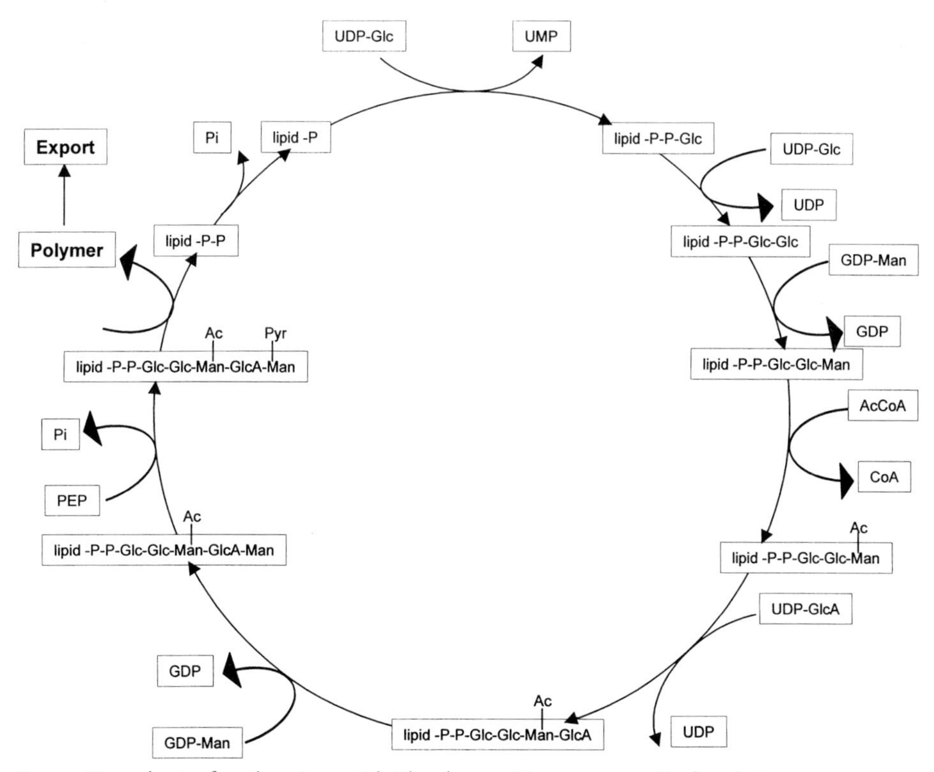

Fig. 2 Biosynthesis of xanthan. Ac: acetyl; Glc: glucose; Man: mannose; P: phosphate; PEP: phosphoenolpyruvate; Pyr: pyruvate (Adapted from Sutherland, 1993).

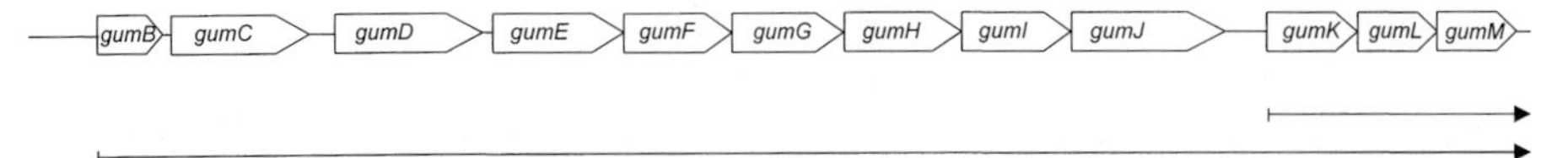

Fig. 3 Genetic map of the *Xanthomonas campestris* pv. *campestris gum* operon (adapted from Becker et al., 1998).

expressed as a single operon from a promoter which is located upstream of the first gene. Xanthan synthesis does not seem to be specifically controlled but is activated by the gene products of a cluster of at least five genes (Tang et al., 1990). The initial stages of xanthan biosynthesis appear to be regulated as part of a pathogenicity regulon (Coplin and Cook, 1990), and some promotors and transcription start sites have been characterized (Knoop et al., 1991; Kingsley et al., 1993; Katzen et al., 1998). It has been shown that a transposon insertion, located 15 bp upstream of the translational start site of the *gumB* gene leads to a defect in xanthan synthesis (Vojnov et al., 1998). Cells of the mutant strain were able to synthesize the lipid-linked pentasaccharide repeating unit of xanthan from the three nucleotide sugar donors (UDP-glucose, GDP-mannose, and UDP-glucuronic acid), but were unable to polymerize the pentasaccharide into mature xanthan. A subclone of the gum gene cluster carrying *gumB* and *gumC* restored xanthan production of the mutant strain to levels of almost 30% of the wild type. In contrast, subclones carrying *gumB* or *gumC* alone were not effective. These results indicate that *gumB* and *gumC* are both involved in the translocation of xanthan across the bacterial membranes. Tseng et al. (1999) have constructed a physical map of the *Xanthomonas campestris* pv. *campestris* chromosome, and the locations of eight loci involved in xanthan gum synthesis were determined. Furthermore, four loci were determined which may be involved in gum polymerization, secretion and regulation. Rodrigues and Aguilar (1997) have identified several mutants of *Xanthomonas campestris* showing increased viscosity and/or gum production after UV treatment. Xanthan solutions of the different mutants showed different intrinsic viscosity values, and no relationship was found between pyruvate or acetate contents and viscosifying ability of the xanthan. It is also possible to produce xanthan using genetically modified (GM) bacteria other than *Xanthomonas*. Pollock et al. (1997) have cloned 12 genes coding for assembly, acetylation, pyruvylation, polymerization and secretion of the polysaccharide xanthan gum in *Sphingomonas*, and these genes were sufficient for synthesis of xanthan gum. The polysaccharide from the recombinant microorganism was largely indistinguishable, both structurally and functionally, from the native xanthan gum.

8
Degradation

The xanthan molecule is very stable as long as it is in the ordered, double-stranded conformation. The double helix confers a good resistance to the molecule against degradation by free radicals, acid, enzymes or repeated freeze–thaw cycles. The stability of xanthan is significantly lowered if the molecule is in a disordered conformation, which is the case at temperatures above the transition temperature or at low ionic strength, for example in salt-free solutions at temperatures around 40–50 °C. Xanthan in solution will not degrade in a pH range from about 2 to 11, but at pH 9 xanthan gum starts to deacetylate. Temperatures up to

90 °C do not affect xanthan, and even after heat treatment at 120 °C for over 30 min the viscosity remained at about 90% of the original value. Chemical degradation of xanthan can be achieved through free radicals which are generated by strong oxidants such as hypochlorite and persulfate, at high temperature, or by a combination of H_2O_2 and Fe^{2+}. Oxidative–reductive depolymerization leads to a cleavage of the polymeric backbone (Herp, 1980). Acid hydrolysis leads preferentially to the hydrolysis of the terminal β-D-mannose in the side chain. Mechanical degradation is possible using high shear or ultrasound; however, the shear to be applied must be very high, in the order of $10^6\ s^{-1}$. Enzymatic degradation of xanthan is usually not very efficient, but if the molecules are in the disordered state they can be degraded by a class of enzymes called xanthanases. This group of enzymes comprises endo-1,4-β-glucanases and xanthan lyases. The glucanases can partially degrade the cellulosic backbone of xanthan (Rinaudo and Milas, 1980), and they are also able to degrade different celluloses. Xanthan lyases are able to cleave the β-D-mannosyl-D-glucuronic acid linkage of the trisaccharide side chain. The presence of acetate and pyruvate hampers the action of the cellulases, whereas the activity of xanthan lyase does not seem to be affected by either pyruvate or acetyl groups. It is likely that xanthan contains cellulosic regions that do not carry side chains and are thus preferred regions for an attack by cellulases. Christensen and Smidsrød (1996) have shown that acid hydrolysis prior to enzymatic degradation increases the depolymerization of xanthan by several orders of magnitude. Removal of the side chains from xanthan with a limited degree of backbone degradation can be obtained by partial hydrolysis. The terminal β-mannose residue linked to O-4 of the glucuronic acid is hydrolyzed quite rapidly (Christensen and Smidsrød, 1991; Christensen et al., 1993a,b). The α-mannose linked to the backbone is hydrolyzed about 10 times more slowly, the hydrolysis obeying pseudo first-order kinetics. Ruijssenaars et al. (1999) have identified several xanthan-degrading enzymes in a *Paenibacillus* strain; this strain is able to degrade about one-third of the xanthan molecule without attacking the backbone of the molecule. Nankai et al. (1999) have reported in detail an enzymatic route for the depolymerization of xanthan by a *Bacillus* strain. The enzymes which are necessary for the degradation of xanthan are xanthan lysase, glucanase, glucosidase, glucuronyl hydrolase and mannosidase. The degradation starts with the cleavage of the glycosidic bond between pyruvylated mannosyl and glucuronyl residues in xanthan side chains due to the action of an extracellular xanthan lyase. The modified xanthan can then be attacked by an extracellular β-D-glucanase, which produces a tetrasaccharide, representing the repeating unit of xanthan without the terminal mannosyl residue. The tetrasaccharide is then taken into the bacterial cell and is subsequently converted by β-D-glucosidase to yield the trisaccharide unsaturated glucuronyl-acetylated mannosyl-glucose. Afterwards, a glucuronyl hydrolase generates an unsaturated glucuronic acid and the disaccharide mannosyl-glucose. The last step in the degradation is the hydrolysis of this disaccharide to mannose and glucose by α-D-mannosidase. For more details, see Chapter 23.

9 Biotechnological Production

Xanthan gum is produced by fermentation; this is a very efficient and reliable process, but it does present some intrinsic problems.

9.1 General Description of the Process

Today, xanthan gum is the most successful industrial biopolymer produced by fermentation. Xanthan gum is produced by the aerobic fermentation of *Xanthomonas campestris*, and many different strains of this bacterium have been screened for their ability to produce xanthan gum (Gupte and Kamat, 1997; De Andrade Lima et al., 1997). Xanthan fermentation can be either batch-wise, semi-batch-wise, or continuous. Industrial production is usually carried out by a batch-wise, submersed fermentation in an aerated and agitated fermenter. The different steps of an industrial process are batch fermentation, alcoholic precipitation, first drying, rinsing with an alcohol/water mixture, final drying, grinding, quality control of the batch, and packaging (Figure 4), with the production strain usually preserved in a freeze-dried state. Galindo et al. (1994) and Salcedo et al. (1992) have reported the preservation of the production strain on agar slopes as well as preservation in sterile seeds. The strain is activated by inoculation into a nutrient medium containing a carbohydrate source, a nitrogen source and mineral salts. After growth, the culture can be used to inoculate successive fermenters through to the industrial scale. Throughout the fermentation process, pH, aeration, temperature and agitation are monitored and controlled. The optimal temperature for cell growth is between 24 and 27 °C, and the best temperature for xanthan production is 30–33 °C. A pH of 6–7.5 is most suitable for growth, while for xanthan production the pH optimum range is 7–8. During the fermentation, pH decreases; however, in an industrial fermentation process the pH is maintained close to neutral in order to allow the process to continue until complete exhaustion of the carbohydrate substrate. Xanthan is produced during the bacterial growth phase as well as during the stationary phase, though maximum production is seen during the exponential growth phase. The pyruvate and acetate contents of xanthan are highest in polymers synthesized immediately after the end of exponential growth (Sutherland, 1993). The quantity of polymer increases until about 30 h after inoculation, after which time a steady state is reached with termination of bacterial growth and polymer production, at which point the fermentation is stopped. The achievable productivity and concentrations reported in the literature range from 15 to 30 g L^{-1} (sometimes higher) and up to 0.7 g $L^{-1} \cdot h$) (mostly lower). Productivity in industrial fermentations may be significantly higher, however. At the end of the fermentation, the fermenter is emptied, cleaned and sterilized before the next fermentation takes place. The post-fermentation steps include a pasteurization by thermal treatment, and sometimes also

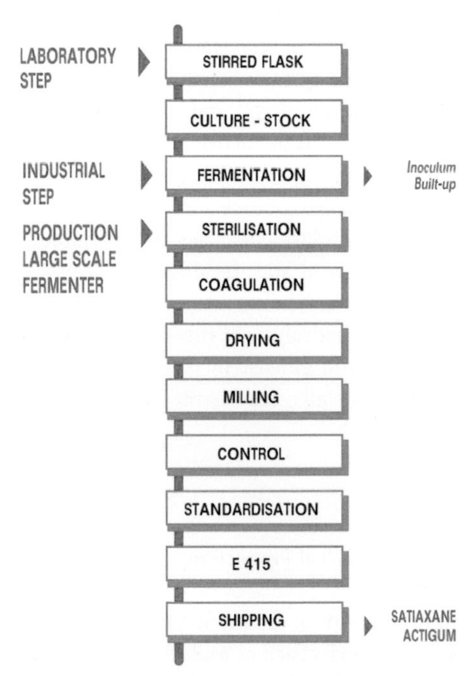

Fig. 4 Industrial production of xanthan.

purification. The heat treatment serves to kill the bacteria. Since *Xanthomonas* do not form spores, the fermentation can be stopped by heat treatment. Pasteurization also improves the properties of the gum, as it leads to a (partial) transformation of the native to the renatured state. Xanthan gum is recovered by precipitation in isopropyl alcohol or ethanol. Usually, the whole broth including biomass is precipitated, but for specific applications, purification may be necessary (see Section 8.4). In the different growth phases, xanthan molecules with different molecular weight and different degree of substitution are produced; hence, the xanthan recovered at the end of a batch fermentation is a mixture of structurally different molecules. The coagulum obtained is isolated by solid–liquid separation, rinsed with alcohol, pressed and dried. The dried xanthan is ground to obtain a white or cream-colored powder with the desired particle size (usually 80–250 μm). Xanthan powder may also be granulated in order to obtain products with specific dispersion and dissolution properties. The final steps in the industrial production chain are quality control, including chemical, physical, rheological and microbiological analysis, packaging and shipping.

9.2 Process Improvement

Although the process for xanthan production is well established, numerous and also recent publications deal with the improvement of xanthan production. Due to the rheological properties of the xanthan molecule, the production of xanthan by fermentation is technically challenging. During the course of the fermentation, the excretion of the polysaccharide results in a highly viscous and shear-thinning broth. As the fermentation broth becomes non-Newtonian, the rheology of the broth causes serious problems of mixing, oxygen supply and heat transfer, thereby limiting the maximum gum concentration achievable, as well as the product quality. Changes in viscosity during culture exceed four orders of magnitude. At the beginning of the fermentation the broth is more or less like water, whereas at the end of the process it is highly viscous and pseudoplastic, exhibiting yield stress. Besides the technical difficulties which always demand improvement, economic aspects of the fermentation process are also important, and improvement of the fermentation costs has a high priority for industrial xanthan producers.

9.2.1 General Improvement

Instead of batch fermentation, other approaches such as continuous fermentation, fed-batch or emulsion fermentation are possible. Continuous fermentation allows high productivity to be obtained, but the process is not easy to set up, the equipment needed is not standard, and stirring in the continuous process is often insufficient. Furthermore, the sterility of the process is difficult to maintain (Silman and Rogovin, 1972). Hence, continuous processes are usually not suitable for industrial production.

Fed-batch fermentation is more successful, and towards the end of the production nutrients are recharged and very good productivity and process efficiency can be obtained. Fermentation in emulsions or in water/oil dispersions can result in high yield in the water phase, but low yields in the total volume.

Emulsion fermentation can reduce the viscosity of the broth by dispersing it into an organic liquid, such as isoparaffin, with the help of an emulsifier, for example fatty acid alkanolamide. The water-in-oil emulsions obtained develop less viscosity and show less pseudoplastic behavior than normal fermentation broth. High oxygen transfer rates can be achieved, and mixing is efficient even at low power input (Schumpe et al., 1991). However, downstream processing is much more difficult than in one-phase submerged cultivation, and the gum may be contaminated with the emulsifier. The separation of cell growth and biopolymer production in a two-stage process can improve productivity (Banks and Browning, 1983; DeVuyst et al., 1987). Xanthan production by immobilized cells has also been described (Lebrun et al., 1994, Yang et al., 1996).

9.2.2
Oxygen Supply

Xanthomonas campestris is a strictly aerobic bacterium, and so oxygen is required for growth and for xanthan production. Sufficient oxygen supply is a prerequisite for efficient polymer production and good productivity. Oxygen is provided by agitation of the culture broth, but this becomes increasingly difficult during the process as the viscosity of the broth increases. Agitation is much affected by vessel geometry, liquid volume, the method of agitation and impeller properties. The productivity of *Xanthomonas* can be improved through the use of impeller systems and suitable reactor configurations which lead to increased oxygen supply for the bacteria. The various reactor configurations which have been proposed to increase oxygen transfer into the medium include stirred tank reactors, external circulation loop reactors, bubble column or air lift reactors. The impeller types which have been proposed are helical stirrers, which cause axial motion, and anchor and multiple rod stirrers which have the tangential action of flat-blade turbines (Galindo, 1994). Amanullah et al. (1998b) have investigated the agitator speed and dissolved oxygen effects in xanthan fermentation. Changes of shear stress in the vicinity of the microorganism which are caused by changes of agitation speed did not influence xanthan production up to a xanthan concentration of 20 g L^{-1}. At higher gum concentrations, xanthan production was enhanced at the higher agitation speed, due to higher microbial oxygen uptake rates.

9.2.3
Nutrients

A standard medium for xanthan production contains a nitrogen source, a carbon source, phosphate and magnesium ions, and trace elements. A typical composition of a medium is 0.06% ammonium nitrate, 0.5% potassium dihydrogen phosphate, 0.01% magnesium sulfate heptahydrate, 2.25% glucose and 97.18% water (Daniel et al., 1994). As carbon substrates, *Xanthomonas* can use different sources such as carbohydrates, amino acids and intermediates of the citric acid cycle. The most commonly used carbon source is glucose, with concentrations around 40 g L^{-1}. In fact, glucose concentrations above 50 g L^{-1} usually lead to poorer conversion into the polysaccharide. Nitrogen sources which allow growth and efficient xanthan production are ammonium salts or amino acids. The degree of substitution with pyruvate increases if $(NH_4)_2HPO_4$ is used as nitrogen source (US Secretary of Agriculture, 1978). The polymer production is promoted by a high ratio of carbon to nitrogen in the culture medium. The mode of glucose feeding can influence the productivity of the fermentation (Amanullah et al., 1998a). An improved performance cannot be achieved by increasing the initial glucose concentration above 50 g L^{-1}, nor by a single 10 g L^{-1} pulse addition with an initial glucose concentration of 40 g L^{-1}, while significant nitrogen is still present. However, a simple pulse and continuous feeding strategy, after nitrogen has been essentially exhausted and under conditions of nonlimiting dissolved oxygen, can result in a greatly enhanced performance compared with batch fermentations. A xanthan gum concentration > 60 g L^{-1} and a productivity of 0.72 g $L^{-1} \cdot$ h) were obtained, which is higher than the values usually reported in the literature for conventional stirred bioreactors.

Between 20–30% of the total process cost can be attributed to the culture medium. In order to reduce the costs of xanthan production, several cheap substrates – many of which are agricultural waste or byproducts of the agronomic industry – have been proposed for xanthan fermentation. Cost reduction is only one aspect in the selection of a nutrient, however; another important aspect is that cheap nutrients must allow both satisfactory productivity and good product quality to be obtained. Among the proposed alternative substrates are whey, corn-steep liquor, molasses, glucose syrup, and olive oil waste waters (De Vuyst and Vermeire, 1994). The cultivation of *Xanthomonas campestris* on solid substrates, such as spent malt grains, apple pomace, grape pomace and citrus peels is also possible. With most of these solid substrates, xanthan yields were comparable with those obtained in conventional submerged cultivation, the yield achieved ranging from 33 to 57 g L^{-1} (Stredansky and Conti, 1999; Stredansky et al., 1999). The biopolymers obtained were chemically and structurally similar to commercial xanthan. Moreno et al. (1998) have reported successful fermentations using different acid-hydrolyzed wastes – from melon, watermelon, cucumber, and tomato – in the culture medium for xanthan production. Melon acid-hydrolyzed waste was the best substrate for xanthan production, allowing an exopolysaccharide to be obtained the chemical composition of which was very similar to that of commercial xanthan. Sugar beet pulp added as a supplement to the xanthan culture medium can increase the yield of xanthan fermentation and allows the production of nonfood-grade xanthan gum (Yoo and Harcum, 1999). Chestnut flour can also be used as a nutrient in xanthan fermentation (Liakopoulou Kyriakides et al., 1999), with a maximum concentration of 33 g L^{-1} xanthan being obtained after 45 h of fermentation. Albergaria et al. (1999) have used LBG extracts as a growth medium in the fermentation; the maximum xanthan concentration obtained was 14 g L^{-1}, and the specific xanthan yield was 7 g g^{-1} cells. This substrate is more of academic interest however, since the yield is rather low and LBG extracts are not a cheap raw material source.

9.3 Modeling the Fermentation Process

Controlling a difficult fermentation process, especially of a highly viscous biopolymer such as xanthan gum, carries several intrinsic difficulties. The process is complex, varies over time, and is nonlinear; moreover, on-line measurements of many important variables is often not possible or is too difficult for industrial routine. A helpful approach for controlling the process can be mathematical modeling, and models for the control of a rheologically complex fermentation have been developed by several groups (Abraham et al., 1995; Garcia Ochoa et al., 1995). The proposed models are able to describe the process parameters including growth, polymer production, consumption of nutrients and evolution of dissolved oxygen, and can be used to improve significantly the production efficiency. Garcia Ochoa et al. (1998) have developed a kinetic model describing xanthan production by *Xanthomonas campestris* NRRL B-1459 in a batch stirred-tank. Parameters described include biomass production, carbon, nitrogen and oxygen consumption, xanthan production and temperature effects. The model can accurately describe fermentations over the temperature range 25–34 °C. Kuttuva et al. (1998) have simulated W/O xanthan fermentation by integrating the microbial kinetic behavior and the multiple-phase process characteristics. Two different models are proposed, one of which assumes a uniform redistribution of cells, substrates and product by frequent droplet breakup and coalescence, and another model which simulates the system of viscous aqueous phase with minimal droplet breakup and component redistribution. The evaluated parameters were the xanthan concentration in the aqueous phase and the volumetric productivity achieved at 200 h. According to the model, W/O fermentation can lead to very high xanthan concentrations in the aqueous phase of > 200 g L^{-1}, and a much increased volumetric productivity of > 0.8 g $L^{-1} \cdot h$ compared with conventional fermentations where the xanthan concentration reaches ~50 g L^{-1} and a volumetric productivity of ~0.5 g $L^{-1} \cdot h$). Serrano Carreon et al. (1998) have modeled xanthan gum fermentation through the interactions between the kinetics of growth and product formation and mixing, which is related to the rheological behavior of the broth. The mixing was linked to the power drawn, as a function of the agitation rate and xanthan content.

9.4 Post-fermentation Treatment

Isolation and purification of xanthan are costly procedures, with post-fermentation treatments accounting for up to 50% of the total production costs. Industrial xanthan is usually not purified. Following pasteurization, which takes place at the end of the fermentation process, the entire fermentation broth is precipitated with isopropanol or a similar alcohol – which means that the bacterial cells remain in the product after processing. The final product contains about 85% pure xanthan gum, 5% biomass, and 10% moisture. The alcohol is recovered by distillation. The biomass imparts turbidity of the final product, and so removal of the bacterial cells is required when high transparency or clarity of the product is required. The fermentation broth can be clarified by filtration, though Yang et al. (1998) removed the cells present in xanthan fermentation broth by adsorption of the bacterial cells onto fibers. These authors were able to show that rough-surfaced cotton fibers were better than smooth-surfaced polyester fibers in the cell adsorption process. Cell adsorption was facilitated by the anionic xanthan gum present in the solution, as in the absence of xanthan gum the cell adsorption to fibers was poor. Alternatively, proteolytic enzymes which degrade the cellular debris from *Xanthomonas* can be used for clarification; this is possible for example using extracellular enzymes secreted by *Trichoderma koningii* (Triveni and Shamala, 1999).

The thermal denaturation and renaturation of a fermentation broth of xanthan can affect the rheological properties of xanthan (Capron et al., 1998b). Depending on the concentration and degree of purification, thermal treatment will have different effects. Changes in both the viscoelastic properties and molecular weight are observed after heating above the melting order–disorder temperature, these being related to the order–disorder conformational transition of the xanthan molecules. At concentrations below 10 g L^{-1}, heat denaturation occurs with dissociation of the native double-stranded structure into two single strands. Upon cooling, the single strands will fold back on themselves to form renatured helices, but at higher concentrations no complete dissociation occurs. Xanthan, which is renatured in concentrated conditions (>10 g L^{-1}) has a higher viscosity than that of the native sample, and displays more gel-like properties. In order to optimize the economy of the post-fermentation treatment of the fermentation broth, Lo et al. (1997) proposed that ultrafiltration be carried out before the alcoholic precipitation. In this way, the xanthan solution could be concentrated to up to 15% (w/v). Mixing modes can also influence the recovery of xanthan gum by precipitation. Zhao et al. (1997) measured the rheological properties of xanthan solution at varying alcohol concentration, thereby introducing the concepts of mixing alcohol concentration and precipitation alcohol concentration. The best design for the precipitation was a continuous precipitation system, as this gave the highest yield and the least amount of impurities. Parlin (1997) proposed a scalable vibrating membrane filter system designed for the separation of solid–liquid foods. Instead of recovering the polymer by alcohol precipitation it is also possible to recover the gum by direct drying with drum dryers or spray dryers (Harrison et al., 1999). Alternative agents for precipitation which have been studied include quaternary ammonium salts or calcium salts. Commercial xanthan is usually dried until it reaches a moisture content of 8–12%, after which it is ground to a particle size usually between 80 and 250 μm. Easily-dispersible commercial products may also be granulated.

10 Properties

Xanthan has very interesting rheological properties. Generally speaking, xanthan gum is a thickener, and produces extremely viscous solutions in water, even at very low concentrations, though under certain conditions xanthan may also gellify.

The properties of pure xanthan gum and xanthan gum blends will be described in this section. As detailed above, xanthan molecules can exist in either an ordered or a disordered conformation. The most interesting properties of xanthan are due to the ordered form, in which the macromolecules adopt a helical conformation whereby the secondary structure resembles rigid rods that do not have any tendency to associate. The form and the rigidity of the xanthan macromolecules determine the rheology of xanthan solutions. The organized state is stabilized in the presence of electrolytes. Parker et al. (1999) determined the dissolution kinetics of xanthan powder, and showed the solubilization of xanthan powder to depend on the size and distribution of the particles; for example, a polydisperse powder will have tendency to form lumps. The dissolution behavior is different in pure water and salt solutions, however. In pure water, dissolution begins with a burst of aggregates, whereas the addition of salt diminishes the aggregate population. The presence of anionic side chains on xanthan gum molecules enhances hydration and renders it soluble in cold water.

10.1 Viscosity

At rest, xanthan solutions develop high viscosity values at concentrations as low as 1%, this being due to the high M_w of the molecules and to their secondary structure. The rigid double-stranded helix confers stability to the molecule; this is because the rigid rods cannot gyrate freely and, at rest, the macromolecules adopt an equilibrium position, stabilized by low interactions. Xanthan solutions are quite resistant to hydrolysis, temperature, and electrolytes.

Salts usually do not reduce the viscosity, though xanthan might precipitate in the presence of polyvalent cations under alkaline conditions. The nature of the counterion in single counterion xanthan influences xanthan viscosity. Solutions of xanthan gum are generally not affected by changes in pH value, and xanthan gum dissolves in most acids or bases. The viscosity of xanthan gum is stable at pH values ranging from 2 to 11 for long periods of time, whereas other hydrocolloids lose their viscosity under the same conditions. The viscosity varies only slightly with temperature, and xanthan solutions remain viscous at temperatures even exceeding 100 °C as long as the ordered conformation is maintained. Stability is enhanced by high gum concentrations.

10.2 Flow Behavior

At very low shear rate, xanthan solutions may present a Newtonian behavior. Milas et al. (1990) showed that transition from the Newtonian regime to viscoelastic behavior was characterized by a critical value of the shear rate, $\acute{y}_r$, i.e., a relaxation time $\acute{y}_r^{-1}$. Above a certain low polymer concentration, this critical shear rate depends on the concentration: the higher the concentration, the lower the shear rate, which can be explained by the change in the entanglement degree of the system. For $\acute{y} > \acute{y}_r$, the xanthan solutions present a shear-thinning behavior which, together with a high yield value, give xanthan solutions exceptional suspension properties. The shear-thinning effect is due to the orientation of the polymer. At low shear, the molecules are randomly oriented, and the flowing properties are bad. This state is relatively stable until the shear-stress exceeds a certain value, called the yield-value. At high shear, the molecules align in the direction of the applied force and the xanthan solution flows easily. Above the yield value, the shear-thinning behavior is more pronounced than with other gums with a random coil conformation. It has been reported (Stokke et al., 1998) that the dynamic viscosity $\eta^*(\omega)$ of a xanthan solution is higher than the steady flow viscosity $\eta(\acute{y})$ at a shear rate equal to ω, which means that xanthan does not follow the empirical Cox–Mertz rule, except at low concentration (<1 g L^{-1}), in the absence of salt and in low shear rate and frequency regions (Milas et al., 1990).

10.3 Weak Network Formation

Xanthan chains in solution form a continuous three-dimensional network, with weak cross-links. Therefore, xanthan solutions may also be characterized as weak gels. The chains separate easily under shearing, which allows the xanthan solution to flow and also accentuates the shear-thinning behavior. The mechanical spectra of xanthan solutions resembles the spectra of gels with $G' > G''$, with little frequency dependence in either modulus. Association of the negatively charged xanthan is promoted by metal ions; the order of effectiveness in inducing weak gels is $Ca^{2+} > K^+ > Na^+$.

10.4 Gelation

Under certain conditions, xanthan may gellify. This may occur in the presence of certain metal ions, or as a synergistic effect with other polymers. Heavy metal ions, such as Cr^{3+}, Al^{3+} or Fe^{3+} cross-link xanthan; in the case of Cr^{3+}, the cross-link occurs through a ligand exchange reaction, where the water molecule which is bound to the Cr^{3+} is exchanged with a carboxylic group on the xanthan molecule.

10.5 Interaction of Xanthan with Other Macromolecules

Xanthan gum can develop additive, synergistic and/or antagonistic effects with other molecules, such as galactomannans and glucomannans. Xanthan molecules in their rigid rod helical form can be cross-linked via the smooth zones of galactomannans. Xanthan shows a synergistic viscosity increase with the galactomannan guar, and it can form a strong thermoreversible gel in the presence of the galactomannan LBG, the glucomannans konjac mannan, the galactomannan tara, or modified guar (Morris, 1990). With LBG, the maximum gel strength is obtained at a xanthan:LBG ratio of 1:1. The formation of thermoreversible gels by synergistic mixtures of xanthan with certain galacto- and glucomannans has been ascribed to intermolecular binding through cocrystallization of denatured xanthan chains within the mannan crystallite (Morris, 1990). The cellulosic backbone of xanthan and the stereochemically similar mannan backbone both form ribbon-like structures. The mixed crystallites probably act as strong junction zones to consolidate the weak xanthan network. The xanthan–galactomannan system is interesting because each polysaccharide alone does not form a gel, but mixing results in a synergistic gelation. The mannose on galactose (M:G) ratio in galactomannans controls the mechanism and the temperature at which gelation occurs.

11 Applications

Xanthan is soluble in both hot and cold water, it develops high viscosity at low concentrations, is compatible with many salts even at elevated concentrations, and is stable at both acid and alkaline pH. Xanthan solutions are also quite resistant to high temperatures, even above 100 °C. These properties, in addition to the concept of a yield stress, and along with the shear-thinning behavior and water-binding capacity of xanthan gum, make it a highly valuable texturizing agent, and consequently it is used in a wide variety of applications. In the food industry, xanthan can act as thickener, as a suspending agent, as an emulsion stabilizer, and as a foam enhancer. Technical applications of xanthan make use of the same properties which are also important for the food industry, and include oil drilling emulsions, paints and glues, as well as fire-fighting formulations. Xanthan applications have been reviewed recently by Kang and Pettitt (1993) and Nussinovitch (1997). From the total volume of xanthan produced worldwide, about 65% is used in food applications, 15% in the petroleum industry, and about 20% in technical applications other than oil drilling.

11.1 Food Applications

Some examples of the multitude of potential food applications include:

- Use in dry mixes for products such as sauces, dressing, gravies, or desserts. Xanthan is used in these applications because it dissolves quite easily in hot or cold water, and provides a rapid build-up of viscosity. In dressings, xanthan is particularly suitable for the suspension of herbs, spices and vegetables, of which it assures an even distribution throughout the bottle over a long period of time (Figure 5). Xanthan provides the necessary high yield value and strong pseudoplasticity, and its properties are not hampered by acid pH, high salt concentra-

Fig. 5 Application of xanthan for the stabilization of herbs in salad dressing. Left: dressing without xanthan. Right: dressing containing 0.125% (w/v) xanthan.

tions, or heat treatment. Due to the shear-thinning behavior, the dressing flows easily from the bottle; however, once poured onto the salad, the sauce clings to the food, which is important for visual presentation.

- In syrup or chocolate toppings, xanthan confers a good consistency and flow properties due to the high yield value and high at-rest viscosity.
- In beverages, xanthan gives body and a good mouthfeel to the liquid and can also stabilize pulps, especially in combination with other polysaccharides such as carboxymethylcellulose.
- Applications in dairy products are often in combination with LBG or guar gum. These combinations stabilize cream emulsions, prevent whey-off and improve the physical and organoleptic properties of pasteurized products. Xanthan stabilizes air cells as well as particles, which makes it useful for mousses, whipped creams and pourable aerated desserts (Sanofi, 1993).
- In frozen desserts, xanthan gum confers heat-shock resistance, and it can also protect foods from freeze–thaw instability (MacNeely and Kang, 1973; Vafiadis, 1999). Xanthan gum is able to increase perceived creaminess and minimize the effects of temperature variation during storage. Freeze–thaw cycles often destroy the delicate texture of frozen desserts, as under unfavorable storing conditions small ice crystals will migrate towards large ice crystals during the thaw stage. During refreezing they will attach to large ice crystals, causing them to grow. Xanthan gum, together with other ingredients such as starch, slows down the recrystallization of smaller ice crystals and prevents the formation of large ice crystals.
- The heat stability and the stabilizing and suspending properties of xanthan are also used widely in canned foods.
- In baking, xanthan is helpful in dough preparation by preventing lump formation during kneading and improving dough homogeneity and volume (Collar et al., 1999). Xanthan also facilitates pumping of the dough during production. In baking, xanthan is also used to suspend larger solid particles such as fruits or nuts.

Other applications, as described by Tilly (1991) include:

- the stabilization of chocolate milk and yogurt-based beverages;
- in combination with other polysaccharides for low-fat spreads;
- the production of fat-reduced biscuits (Conforti et al., 1997); and
- reduced-calorie grape juice jellies which will show similar texture characteristics, for example gel hardness, cohesiveness and springiness, as a reduced-calorie grape juice jelly texturized with low methoxyl pectin (Gaspar et al., 1998).

The application of xanthan in combination with LBG the production of "melt-in-the-mouth" polysaccharide gelling systems for foods has also been described (Marrs, 1997).

11.2
Non-food Applications

The most important technical application for xanthan gum is its use oil drilling. In this application, xanthan is particularly useful due to its pseudoplastic behavior, temperature stability, and salt tolerance. The use of xanthan in petroleum production has been reviewed by Kang and Pettitt (1993). During oil drilling, a low viscosity is required at the drill bit, whereas a high viscosity is required in the annulus. The pseudoplasticity of xanthan solutions meets these requirements as xanthan develops low viscosity at the drill bit where the shear is high, and high viscosity in the annulus where shear is low. Therefore, drilling fluids containing xanthan allow a rapid penetration at the bit, and a suspension of cuttings in the annulus.

In textile printing, xanthan is used to provide the specific rheological properties needed in the production of sharp and clean patterns by preventing migration of the dye. Xanthan is compatible with most components of printing pastes, and it is also removed easily by washing. Xanthan is also used in ceramic glazes where it prevents agglomeration of the different components. In cleaning liquids, xanthan can be useful by providing a high viscosity to the solution at low shear, thereby allowing the cleaner to cling to inclined surfaces. Other technical applications described for xanthan gum include paint and ink, where it can stabilize and suspend pigments, wallpaper adhesives, formulation of pesticides and toothpastes, and industrial emulsions. The use of xanthan gum as a support for enzyme and cell immobilization has been described by Dumitriu and Chornet (1998). Xanthan gum can also be useful as controlled-release agent for pharmaceuticals (Philipon, 1997), with microspheres of gellified xanthan encapsulating active ingredients. The spheres are swallowed in a dry state and swell in the stomach, thereby gradually releasing the active ingredient. The active molecule can also be linked covalently to the polysaccharide and then released in the body by enzymatic hydrolysis.

12
Relevant Patents

As long ago as 1959, Esso Research patented the use of xanthan gum for the displacement of oil from partially depleted reservoirs. Later, several patents followed concerning processes for fermentation (1963, 1966), recovery (1963) and application (1975) of xanthan. In 1960, several patents of Jersey Production Research Co. appeared, covering a thickening agent and the process for its production, a substituted heteropolysaccharide, and a process for synthesizing polysaccharides. Kelco Biospecialities Ltd. has patented processes for xanthan production and recovery as well as xanthan application, since the early 1960s. This includes several patents concerning processes for producing the polysaccharides deposited in 1966, a process for xanthan gum production (1981), an inoculation procedure for xanthan gum production (1981), or the production of xanthan gum by emulsion fermentation (1981). Patents concerning the post-fermentation treatment include patents regarding the treatment of a *Xanthomonas* hydrophilic colloid and the resulting product (1964, 1968) or the precipitation of xanthan gum (1981). Special xanthans such as etherified *Xanthomonas* hydrophilic colloids (1965), polymeric derivatives of a cationic *Xanthomonas* colloid (1964), cationic ethers of *Xanthomonas* hydrophilic colloids (1963) or xanthan having a low pyruvate content (1981) have also been patented by Kelco. Applications patented by Kelco include edible compositions comprising oil-in-water emulsions (1960), joint-filling compositions (1963), and a dehydrated food product (1971). Patents of Rhone-Poulenc appeared mainly in the 1970s and 1980s; those of the 1970s covered improvements of the fermentation process (1973, 1976, 1978) and applications as a thickening agent (1975) and in oil drilling (1977). During the 1980s, other patents relating to the fermentation process were deposited (1984, 1987, 1988). Merck began working on xanthan in the 1970s, and this resulted in several patents concerning improvements in the process to obtain xanthan solutions with increased viscosity (1975), the preparation of a xanthan copolymer (1975), a deacetylated borate-biosynthetic gum composition (1979), a process for producing low-calcium xanthan gum by fermentation (1978), a process to prepare a xanthan gum which does not contain any cellulases (1979), the production of low-calcium, smooth-flow xanthan gum (1979), and a process for improved recovery of xanthan gum with high viscosity (1979). In the 1980s, the Merck patents covered a dispersible xanthan gum composite (1980) and a heteropolysaccharide and its production and use (1985). In 1986, Merck patented a recombinant DNA plasmid for xanthan gum synthesis containing some essential genetic material for xanthan gum synthesis and in 1991, a patent about low-ash xanthan gum (< 2%) appeared which is claimed to be particularly useful in the preparation of ceramics. Patents of Standard Oil Co. about xanthan gum appeared mainly in 1980/1981, and referred to different strains and the corresponding processes, such as *Xanthomonas campestris* ATCC 31600 (1980), *Xanthomonas campestris* NRRL B-12075 and NRRL B-12074 (1980), *Xanthomonas campestris* ATCC 31602 (1980) and *Xanthomonas campestris* ATCC 31601 (1980). Patents relating to the production process appeared in 1980; these included the production of xanthan gum from a chemically defined medium, a method for improving xanthan yield, and also for improving specific xanthan productivity during continuous fermentation. Standard Oil Co. has also patented a semicontinuous method for the production of xanthan gum using different *Xanthomonas* strains (1981). A method of producing a low-viscosity xanthan gum was patented in 1981. Sanofi has patented production processes (1986, 1987) and the application of xanthan (1993), as well as a mutant strain which produces a xanthan that does not develop any viscosity (1990). Oil companies other than Esso which have worked on xanthan development include Shell, which has patented the treatment of polysaccharide solutions (1980), a process for preparing *Xanthomonas* heteropolysaccharide, the heteropolysaccharide as prepared by this process and its use (1983), as well as a method for improving the filterability of a microbial broth and its use (1984). Mobil Oil Corporation, with patents relating to a waterflood process employing thickened water (1966) and a method for clarifying polysaccharide solutions (1973), and Phillips Petroleum Co., with a patent about recovery of a microbial polysaccharide (1971). Other companies which have patented xanthan-related processes or applications include: Archer Daniels Midland Co., who claim the biochemical synthesis of industrial gums (1962); Pfizer, who have patented a batch process (1979) and a continuous production of *Xanthomonas* biopolymer (1982); Stauffer Chemical Co., who have patented the fermentation of whey to produce a thickening polymer (1981); and Hoechst AG, who claim the lowering of viscosity of fermentation broths (1983). In addition, Celanese Corp. patented concentrated xanthan gum solutions obtained by ultrafiltration (1980), and Jungbunzlauer AG claimed an improved fermentation process (1982) and a process for obtaining polysaccharides as grains (1989). Several xanthan-related patents have been deposited by General Mills Chemicals Inc. about *Xanthomonas* gum amine salts (1974), about flash-drying of xanthan gum (1976), or about a cationic polysaccharide obtained by reacting a xanthan with isopropanol and NaOH (1986). Hercules Inc. has patented a xanthan recovery process (1976), and Henkel KGaA has patented some biopolymers obtained from *Xanthomonas* (1981) as well as a process for the preparation of exocellular biopolymers (1986). The Akademie der Wissenschaften der DDR of the former East Germany has worked on a method for microbial production of xanthan (1986). Getty Scientific Development Co. has patented a polysaccharide polymer made by *Xanthomonas* in 1985, and the recombinant-DNA mediated production of xanthan gum in 1986. The latter patent claims a gene cluster encoding enzymes necessary for the biosynthesis of xanthan gum which was isolated from *Xanthomonas campestris*. The U.S. Secretary of Agriculture has patented a method for producing an atypically salt-responsive alkali-deacetylated polysaccharide (1959), a method of recovering microbial polysaccharides from their fermentation broths (1962), a continuous process for producing *Xanthomonas* heteropolysaccharide (1967) as well as a nitrogen source for improved production of microbial polysaccharides (1967). The Institut Français du Petrole has patented the clarification of xanthan in 1980, and a process for the production of an improved xanthan for the oil drilling application in 1993. Cerestar Holding B.V. patented a fermentation feedstock containing simple sugars including mannose and glucose (1993). One of the most recent patents has been from Rhodia Inc. (1997), and claims the use of a liquid carbohydrate fermentation product in food. The claimed product can be either food grade or pharmaceutical grade, and it is delivered in a carbohydrate medium other than dairy whey, without any drying steps prior to use (Table 1).

Tab. 1 Relevant patents relating to xanthan products

Publication date	*Patent holder*	*Inventors*	*Patent No.*	*Patent title*
1961	U.S. Secretary of Agriculture	Jeanes, A. R., Sloneker, J. H.	US 3 000 790	Method of producing an atypically salt-responsive alkali-deacetylated polysaccharide.
1962	Jersey Production Research Co.	Patton, J. T., Lindblom, G. P.	US 3 020 206	Process for synthesizing polysaccharides.
1962	Jersey Production Research Co.	Patton, J. T.	US 3 020 207	Thickening agent and process for producing same.
1962	Kelco Co.	O'Connell, J. J.	US 3 067 038	Edible compositions comprising oil in water emulsions.
1964	Jersey Production Research Co.	Lindblom, G. P., Patton, J. T.	US 3 163 602	Substituted heteropolysaccharide.
1964	U.S. Secretary of Agriculture	Rogovin, S. P., Albrecht, W. J.	US 3 119 812	Method of recovering microbial polysaccharides from their fermentation broths.

Tab. 1 (cont.)

Publication date	*Patent holder*	*Inventors*	*Patent No.*	*Patent title*
1966	Archer Daniels Midland Co.	Weber, R. O., Horan, F. E.	US 3 271 267	Biochemical synthesis of industrial gums.
1966	Esso Production Research Co.	Lipps, B. J.	US 3 251 749	Fermentation process for preparing polysaccharides.
1966	Esso Production Research Co.	Lipps, B. J.	US 3 281 329	Fermentation process for producing a heteropolysaccharide.
1966	Kelco Co.	Schweiger, R. G.	US 3 244 695	Cationic ethers of *Xanthomonas* hydrophilic colloids.
1966	Kelco Co.	Schuppner, H. R.	US 3 279 934	Joint filling composition.
1967	Esso Production Research Co.	Lindblom, G. P., Ortloff, G. D., Patton, J. T.	US 3 305 016	Displacement of oil from partially depleted reservoirs.
1967	Esso Production Research Co.	Lindblom, G. P., Patton, J. T.	US 3 328 262	Heteropolysaccharide fermentation process.
1967	Kelco Co.	O'Connell, J. J.	US 3 355 447	Treatment of *Xanthomonas* hydrophilic colloid and resulting product.
1967	Kelco Co.	Schweiger, R. G.	US 3 349 077	Etherified *Xanthomonas* hydrophilic colloids and process of preparation.
1968	Esso Production Research Co.	Patton, J. T., Holman, W. E.	US 3 382 229	Polysaccharide recovery process.
1968	Kelco Co.	Schweiger, R. G.	US 3 376 282	Polymeric derivatives of cationic *Xanthomonas* colloid derivatives.
1968	Kelco Co.	MacNeely, W. H.	US 3 391 060	Process for producing polysaccharides.
1968	Kelco Co.	MacNeely, W. H.	US 3 391 061	Process for producing polysaccharides.
1968	Mobil Oil Corp.	Sherrod, A. W.	US 3 373 810	Waterflood process employing thickened water.
1969	General Mills Inc.	Nordgren, R., Wittcoff, H. A.	DE 1 919 790	Polysaccharide B1459 cationique.
1969	Kelco Co.	Macneely, W. H.	US 3 427 226	Process for preparing polysaccharide.
1969	U.S. Secretary of Agriculture	Rogovin, S. P.	US 3 485 719	Continuous process for producing *Xanthomonas* heteropolysaccharide.
1970	Kelco Co.	Colegrove, G. T.	US 3 516 983	Treatment of *Xanthomonas* hydrophilic colloid and resulting product.
1970	Mobil Oil Corp.	Abdo, M. K.	US 3 711 462	Method of clarifying polysaccharide solutions.
1972	Kelco Co.	Edlin, R. L.	US 3 694 236	Method of producing a dehydrated food product.
1973	Philips Petroleum Co.	Buchanan, B. B., Cottle, J. E.	US 3 773 752	Recovery of microbial polysaccharide.
1975	General Mills Chemicals Inc.	Jordan, W. A., Carter, W. H.	US 3 928 316	*Xanthomonas* gum amine salts.
1975	Société des Usines Chimiques Rhone-Poulenc		FR 2 251 620	Perfectionnement à la production de polysaccharides par fermentation.
1976	Rhone-Poulenc Industries	Falcoz, P., Celle, P., Campagne, J. C.	FR 2 299 366	Nouvelles compositions épaissisantes à base d'hétéropolysaccharides.

Tab. 1 (cont.)

Publication date	*Patent holder*	*Inventors*	*Patent No.*	*Patent title*
1977	Exxon Research & Engineering Co.	Naslund, L. A., Laskin, A. I.	FR 2 330 697	Procédé de traitement d'un hétéropolysaccharide et son application.
1977	General Mills Chemicals Inc.	Cahalan, P. T., Peterson, J. A., Arndt, D. A.	US 4 053 699	Flash drying of xanthan gum and product produced thereby.
1977	Hercules Inc.	Towle, G. A.	US 4 051 317	Xanthan recovery process.
1977	Merck & Co. Inc.	Kang, K. S., Burnett, D. B.	FR 2 318 926	Perfectionnement aux procédés pour accroitre la viscosité de solutions aqueuses de gomme xanthane et aux compositions obtenues.
1977	Merck & Co. Inc.	Cottrell, I. W.	FR 2 325 666	Copolymère greffe d'un colloide hydrophile de *Xanthomonas* et procédé pour sa préparation.
1977	Rhone-Poulenc Industries	Campagne, J. C.	FR 2 342 339	Procédé de production de polysaccharides par fermentation.
1979	Institut Français du Pétrole des Carburants & Lubrifiants - Rhone-Poulenc Industries	Ballerini, D., Claude, O., Chauveteau, G., Kohler, N., Vandecasteele, J. P.	FR 2 398 874	Utilisation de mouts de fermentation pour la récupération assistée du pétrole.
1979	Rhone-Poulenc Industries	Contat, F., Lartigau, G., Nocolas, O.	FR 2 414 555	Procédé de production de polysaccharides par fermentation.
1980	Merck & Co. Inc.	Empey, R., Dominik, J. G.	EP 20 097	Production of low-calcium smooth-flow xanthan gum.
1980	Merck & Co. Inc.	Roche, R. E.	FR 2 457 322	Procédé de récupération pour améliorer la viscosité de la gomme xanthane.
1980	Merck & Co. Inc.	Racciato, J. S., Cottrell, I. W.	US 4 214 912	Deacetylated borate-biosynthetic gum composition.
1981	Celanese Corp.	Lee, H. L.	US 4 299 825	Concentrated xanthan gum solutions.
1981	Merck & Co. Inc.	Kang, K. S.	FR 2 458 589	Procédé de préparation de gomme xanthane exempte de cellulase.
1981	Pfizer Inc.	Wernau, W. C.	US 4 282 321	Fermentation process for production of xanthan.
1981	Standard Oil Co.	Weisrock, W. P.	US 4 301 247	Method for improving xanthan yield.
1982	Kelco Biospecialities Ltd.	Jarman, T. R.	EP 66 957	Inoculation procedure for xanthan gum production.
1982	Henkel KGaA	Bahn, M., Engelskirchen, K., Schieferstein, L., Schindler, J., Schmid, R.	EP 58 364	Biopolymères à partir de *Xanthomonas*.

Tab. 1 (cont.)

Publication date	*Patent holder*	*Inventors*	*Patent No.*	*Patent title*
1982	Institut Français du Pétrole	Rinaudo, M., Milas, M., Kohler, N.	FR 2 491 494	Procédé enzymatique de clarification de gommes de xanthane permettant d'améliorer leur injectivité et leur filtrabilité.
1982	Kelco Biospecialities Ltd.	Jarman, T. R., Pace, G. W.	EP 66 961	Production of xanthan having a low pyruvate content.
1982	Kelco Biospecialties Ltd.	Jarman, T. R.	EP 66 377	Process for xanthan gum production.
1982	Kelco Biospecialties Ltd.	Maury, L. G.	US 4 352 882	Production of xanthan gum by emulsion fermentation.
1982	Merck & Co. Inc.	Sandford, P. A., Baird, J. K.	US 4 357 260	Dispersible xanthan gum composite.
1982	Shell Internationale Research Maatschappij B.V.	Van Lookeren Campagne, C. J., Roest, J. B.	EP 49 012	Treatment of polysaccharide solutions.
1982	Standard Oil Co.	Weisrock, W. P., MacCarthy, E. F.	EP 46 007	*Xanthomonas campestris* ATCC 31601 and process for use.
1982	Standard Oil Co.	Weisrock, W. P.	US 4 311 796	Method for improving specific xanthan productivity during continuous fermentation.
1982	Standard Oil Co.	Weisrock, W. P.	US 4 328 310	Semi-continuous method for production of xanthan gum using *Xanthomonas campestris* ATCC 31601.
1983	Kelco Biospecialties Ltd.	Smith, I. H.	EP 68 706	Precipitation of xanthan gum.
1983	Merck & Co. Inc.	Richmon, J. B.	US 4 375 512	Process for producing low calcium xanthan gum by fermentation.
1983	Standard Oil Co.	Weisrock, W. P., Klein, H. S.	US 4 374 929	Production of xanthan gum from a chemically defined medium introduction.
1983	Standard Oil Co.	Bauer, K. A., Khosrovi, B.	US 4 400 467	Process for using *Xanthomonas campestris* NRRL B-12075 and NRRL B-12074 for making heteropolysaccharide.
1983	Standard Oil Co.	Weisrock, W. P., MacCarthy, E. F.	US 4 407 950	*Xanthomonas campestris* ATCC 31602 and process for use.
1983	Standard Oil Co.	Weisrock, W. P., MacCarthy, E. F.	US 4 407 951	*Xanthomonas campestris* ATCC 31600 and process for use.
1983	Standard Oil Co.	Weisrock, W. P.	US 4 377 637	Method of producing a low viscosity xanthan gum.
1983	U.S. Secretary of Agriculture	Cadmus, M. C., Knutson, C. A.	US 4 394 447	Production of high-pyruvate xanthan gum on synthetic medium.
1984	Jungbunzlauer A.G.	Kirkovits, A. E., Waltenberger, I.	AT 373 916	Xanthan.

Tab. 1 (cont.)

Publication date	*Patent holder*	*Inventors*	*Patent No.*	*Patent title*
1984	Pfizer Inc.	Young, T. B.	EP 115 154	Continuous production of *Xanthomonas* biopolymer.
1984	Stauffer Chemical Co.	Schwartz, R. D., Bodie, E. A.	US 4 444 792	Fermentation of whey to produce a thickening polymer.
1985	Hoechst A.G.	Voelskow, H., Keller, R., Schlingmann, M.	DE 3 330 328	Lowering the viscosity of fermentation broths.
1985	Shell International Research Maatschappij B.V.	Downs, J. D.	EP 130 647	Process for preparing *Xanthomonas* heteropolysaccharide; heteropolysaccharide as prepared by the latter process and its use.
1986	Rhone-Poulenc Specialités Chimiques	Leproux, V., Peignier, M., Cros, P., Beucherie, J., Kennel, Y.	EP 187 092	Procédé de production de polysaccharides de type xanthane.
1986	Shell International Research Maatschappij B.V.	Drozd, J. W., Rye, A. J.	EP 184 882	Method for improving the filtrability broth and the use.
1987	Akademie der Wissenschaften der DDR	Behrens, U., Stottmeister, U.	DD 250720	Method for microbial synthesis of polysaccharides.
1987	Getty Scientific Development Co.	Vanderslice, R. W., Shanon, P.	EP 211 288	A polysaccharide polymer made by *Xanthomonas.*
1987	Getty Scientific Development Co.	Capage, M. A., Doherty, D. H., Betlach, M. R., Vanderslice, R. W.	WO 87/05 938	Recombinant-DNA mediated production of xanthan gum.
1987	Merck & Co. Inc.	Peik, J. A., Steenbergen, S. M., Veeder, G. T.	EP 209 277	Heteropolysaccharide and its production and use.
1987	Merck & Co. Inc.	Cleary, J. M., Rosen, I. G., Harding, N. E., Cabanas, D. K.	EP 233 019	Recombinant DNA plasmid for xanthan gum synthesis.
1988	Sanofi - Méro Rousselot Satia S.A.	Eyssautier, B.	EP 296 965	Procédé de fermentation pour l'obtention d'une polysaccharide de type xanthane.
1988	Sanofi Elf Bio-Industries	Eyssautier, B.	FR 2 606 423	Procédé d'obtention d'un xanthane à fort pourvoir épaississant et application de ce xanthane.
1989	Henkel KGaA	Viehweg, H.	US 4 871 665	Process for the preparation of exocellular biopolymers.
1989	Rhone-Poulenc Chimie	Tavernier, C.	FR 2 624 135	Procédé de production de polysaccharides.

Tab. 1 (cont.)

Publication date	*Patent holder*	*Inventors*	*Patent No.*	*Patent title*
1990	Jungbunzlauer A.G.	Westermayer, R., Stojan, O., Eder, J.	FR 2 646 857	Procédé pour obtenir sous forme grenue des polysaccharides formés par les bactéries du genre *Xanthomonas* ou *Arthrobacter*.
1990	Rhone-Poulenc chimie	Nicolas, O.	EP 365 390	Procédé de production de polysaccharides par fermentation d'une source carbonée à l'aide de microorganismes.
1992	Merck & Co. Inc.	Talashek, T., Cleary, J. M.	EP 511 784	Low-ash xanthan gum.
1992	Sanofi - Société Nationale Elf Aquitaine	Salome, M.	FR 2 671 097	Souche mutante de *Xanthomonas campestris*. Procédé d'obtention de xanthane et xanthane non visqueux.
1994	Cerestar Holding B.V.	De Troostemberghm J. C., Beck, R. H. F., De Wannemaeker, B. L. T.	EP 609 995	Fermentation feedstock.
1994	Institut Français du Pétrole	Monot, F., Noik, C., Ballerini, D.	FR 2 701 490	Procédé de production d'un mout de xanthane ayant une propriété améliorée. Composition obtenue et application de la composition dans une boue de forage de puits.
1995	Sanofi	Tilly, G.	EP 649 599	Stabilizer composition enabling the production of a pourable aerated dairy dessert.
1999	Rhodia Inc.	Hoppe, C. A., Lawrence, J., Shaheed, A.	WO 99/25 208	Use of liquid carbohydrate fermentation product in foods.

13 Current Problems and Limitations

As a food additive which is produced by fermentation, xanthan is affected by the current debate regarding the use and danger of GM organisms. Even though the xanthan production strains used today are not GM, the consumer is skeptical about a product which is obtained via a biotechnological process. Indeed, the consumer demands that the whole process from the very beginning is accomplished without using any substrate that may have a GM source. The culture medium used for xanthan production must contain carbon and nitrogen sources; in order for the production to be cost-competitive, these sources must be cheap and efficient, easy to handle during the fermentation, and lead to a product of good quality. These requirements are fulfilled by sources such as corn-steep water as a carbon source and soy protein as a nitrogen source. However, a significant part of the world production of soy and corn today is obtained from GM plants. Although it seems very unlikely that GM carbon and nitrogen sources, after having been metabolized by the bacterium to yield the biopolymer, will confer any risk to the final xanthan product, the customer – especially the European customer – is not accepting GM sources in the culture medium. One of the reasons is certainly that standard commercial xanthans are not 100% pure, but contain about 5% of biomass. The components of the culture medium – especially glucose and nitrogen – should be used completely at the end of the production and no GM material should be left in the final product, though no guarantee can be given for this. Hence, the culture medium must be adapted to contain only non-GM sources, yet costs, productivity and product quality should remain unchanged. This is a major challenge for today's xanthan producers.

The current refusal of genetic modification for food additives by the consumer also limits innovation concerning xanthan. GM *Xanthomonas* strains could be developed in order to increase productivity, to improve the use of nutrients, to enable the metabolism of cheap substrates and their conversion into xanthan gum. It has been proved in the past that this is possible. Fu and Tseng (1990) have constructed a *Xanthomonas campestris* strain which carries a recombinant β-galactosidase-encoding gene and which is able to convert whey into xanthan. GM strains could also be modified in order to produce biopolymers with different substitution patterns, different properties in terms of rheology, dissolution and dispersion behavior, or to provide new or different synergetic interactions with other molecules. For the oil drilling industry, xanthan with increased temperature stability would be valuable. However, the current GM debate makes such development highly unlikely in the near future.

14 Outlook and Perspectives

Today, xanthan gum is a very successful biopolymer, and this success is likely to continue as no other biopolymer with similar properties is available commercially at a similar price. The world market for xanthan gum is growing and new markets are emerging for xanthan consumption as well as for xanthan production; for example, China is currently estimated to produce 5–10% of the world's xanthan. As discussed above, it appears highly unlikely that any future development of xanthan gum will be based on genetic modification, and developments will rather focus on specialty xanthans with improved handling properties. However, some perspectives based on genetic modification and improvement will clearly be required, and these are discussed in the following section.

Common genetic methods such as conjugation, electroporation, chemical and transposon mutagenesis and site-directed mutagenesis can be used with *Xanthomonas campestris*. By modifying the biosynthetic pathway for xanthan production, the carbohydrate structure and substitution pattern of the polymer can be genetically controlled to produce polysaccharides with quite different properties (Betlach et al., 1987). Mutants that lack glucuronic acid and pyruvate residues have been constructed, as well as strains producing xanthan with an increased pyruvate content. Tait and Sutherland (1989) have constructed a strain producing truncated xanthan. Until today, strains for xanthan production have been selected and improved by conventional methods. Attempts to construct strains with improved xanthan yield have been unsuccessful in the past, and are not very likely in the future since the conversion rate of carbon to xanthan is very high (Linton, 1990). Improvement of yield by genetic methods seems less promising than improvement by a more efficient fermenter design and better culture media. Another perspective for the development of xanthan is the improvement of the molecule by chemical modification. Potential modifications might include oxidation, reduction, changing of the substitution pattern, altering the side chains, or grafting other molecules onto the xanthan molecule. However, any such modification would lead to products that today are not food-approved. Controlled degradation of xanthan might lead to oligosaccharides with bioactive properties.

Acknowledgements

The authors thank Annick Bourdais and Patricia Poutrel, without whom this chapter would never have been accomplished.

15 References

Abraham, N. H., Kent, C. A., Satti, S. M. (1995) Modeling for control of poorly-mixed bioreactors, *I. Chem E. Research Event* **2**, 1049–1051.

Akademie der Wissenschaften der DDR (1986) Method for microbial synthesis of polysaccharides, DD 250720.

Albergaria, H., Roseiro, J. C., Amaral Collaco, M. T. (1999) Technological aspects and kinetics analysis of microbial gum production in carob, *Agro-Food-Ind. Hi-Tech* **10**, 24–26.

Amanullah, A., Satti, S., Nienow, A. W. (1998a) Enhancing xanthan fermentations by different modes of glucose feeding, *Biotechnology* **14**, 265–269.

Amanullah, A., Tuttiet, B., Nienow, A. W. (1998b) Agitator speed and dissolved oxygen effects in xanthan fermentation, *Biotechnol. Bioeng.* **57**, 198–210.

Archer-Daniels-Midland Co. (1962) Biochemical synthesis of industrial gums, U.S. Patent No. 3 271 267.

Banks, G., Browning, F. (1983) The development of a two stage xanthan gum fermentation, *Process Biochem. Suppl. Proc. Conf. Adv. Ferment.* 163–170.

Becker, A., Katzen, F., Puhler, A., Ielpi, L. (1998) Xanthan gum biosynthesis and application: a biochemical/genetic perspective, *Appl. Microbiol. Biotechnol.* **50**, 145–152.

Betlach, M. R., Capage, M. A., Doherty, D. H., Hassler, R. A., Henderson, N. M., Vanderslice, R. W., Marreli, J. D., Ward, M. B. (1987) Genetically engineered polymers: manipulation of xanthan biosynthesis, in: *Industrial Polysaccharides: Genetic Engineering, Structure/Property Relations and Applications* (Yalpani, M., Ed.), Amsterdam: Elsevier, 35–50.

Bih-Ying, Y., Tseng, Y. H. (1988) Production of exopolysaccharide and levels of protease and pectinase activity in pathogenic and non-pathogenic strains of *Xanthomonas campestris* pv. *campestris*, *Bot. Bull. Acad. Sinica* **29**, 93–99.

Capron, I., Brigand, G., Muller, G. (1997) About the native and renatured conformation of xanthan exopolysaccharide, *Polymer* **38**, 5289–5295.

Capron, I., Alexandre, S., Muller, G. (1998a) An atomic force microscopy study of the molecular organisation of xanthan, *Polymer* **39**, 5725–5730.

Capron, I., Brigand, G., Muller, G. (1998b) Thermal denaturation and renaturation of a fermentation broth of xanthan: rheological consequences, *Int. J. Biol. Macromol.* **23**, 215–225.

Celanese Corp. (1980) Concentrated xanthan gum solutions, U.S. Patent No. 4 299 825.

Cerestar Holding B.V. (1993) Fermentation feedstock, EP 609 995.

Chan, J. W. Y. F., Goodwin, P. H. (1999) The molecular genetics of virulence of *Xanthomonas campestris*, *Biotechnol. Adv.* **17**, 489–508.

Cheetham, N. W. H., Punruckvong, A. (1985) An HPLC method for the determination of acetyl and pyruvyl groups in polysaccharides, *Carbohydr. Polym.* **5**, 399–406.

Christensen, B. E., Smidsrød, O. (1991) Hydrolysis of xanthan in dilute acid: effects on chemical composition, conformation, and intrinsic viscosity, *Carbohydr. Res.* **214**, 55–69.

Christensen, B. E., Smidsrød, O. (1996) Dependence of the content of unsubstituted (cellulosic) regions in prehydrolysed xanthans on the rate of hydrolysis by *Trichoderma reesei* endoglucanase, *Int. J. Biol. Macromol.* **18**, 93–99.

Christensen, B. E., Smidsrød, O., Elgsaeter, A., Stokke, B. T. (1993a) Depolymerization of double-stranded xanthan by acid hydrolysis: characterization of partially degraded double strands and single-stranded oligomers released from the ordered structures, *Macromolecules* **26**, 6111–6120.

Christensen, B. E., Smidsrød, O., Stokke, B. T. (1993b) Xanthans with partially hydrolysed side chains: conformation and transitions, in: *Carbohydrates and Carbohydrate Polymers, Analysis, Biotechnology, Modification, Antiviral, Biomedical and Other Applications* (Yalpani, M., Ed.), ATL Press, 166–173.

Collar, C., Andreu, P., Martinez, J. C., Armero, E. (1999) Optimisation of hydrocolloid addition to improve wheat bread dough functionality: a response surface methodology study, *Food Hydrocolloids* **13**, 467–475.

Conforti, F. D., Charles, S. A., Duncan, S. E. (1997) Evaluation of a carbohydrate-based fat replacer in a fat-reduced baking powder biscuit, *J. Food Qual.* **20**, 247–256.

Coplin, D. L., Cook, D. (1990) Molecular genetics of extracellular polysaccharide biosynthesis in vascular phytopathogenic bacteria, *Mol. Plant-Microbe Interact.* **3**, 271–279.

Daniel, J. R., Whistler, R. L., Voragen, A. C. J., Pilnik W. (1994) Starch and other polysaccharides, in: *Ullmann's Encyclopedia of Industrial Chemistry* (Elvers, B., Hawkins, S., Russey, W., Eds.), Weinheim: VCH, 1–62.

Daniels, M. J., Leach, J. E. (1993) Genetics of *Xanthomonas* in: *Xanthomonas* (Swings, J. G., Civerolo, E. L., Eds.), London: Chapman & Hall, 301–339.

De Andrade Lima, M. A. G., De Araujo, J. M., De Franca, F. P. (1997) The evaluation of different parameters to characterize xanthan gum-producing strains of *Xanthomonas* pv. *campestris*, *Arq. Biol. Tecnol.* **40**, 179–187.

De Crecy Lagard, V., Glaser, P., Lejeune, P., Sismeiro, O., Barber, C. E., Daniels, M. J., Danchin, A. (1990) A *Xanthomonas campestris* pv. *campestris* protein similar to catabolite activation factor is involved in regulation of phytopathogenicity, *J. Bacteriol.* **172**, 5877–5883.

De Vuyst, L., Vermeire, A. (1994) Use of industrial medium components for xanthan production by *Xanthomonas campestris* NRRL-B-1459, *Appl. Microbiol. Biotechnol.* **42**, 187–191.

De Vuyst, L., Van Loo, J., Vandamme, E. J. (1987) Two stage fermentation process for improved xanthan production by *Xanthomonas campestris* NRRL B-1459, *J. Chem. Technol. Biotechnol.* **39**, 263–273.

Duckworth, M., Yaphe, W. (1970) Definitive assay for pyruvic acid in agar and other algal polysaccharide, *Chem. Ind.* **23**, 747–748.

Dumitriu, S., Chornet, E. (1998) Polysaccharides as support for enzyme and cell immobilisation, in: *Polysaccharides. Structural Diversity and Functional Versatility* (Dumitriu, S., Ed.), New York: Marcel Dekker, 629–748.

Esso Production Research Co. (1959) Displacement of oil from partially depleted reservoirs, U.S. Patent No. 3 305 016.

Esso Production Research Co. (1963) Fermentation process for preparing polysaccharides, U.S. Patent No. 3 251 749.

Esso Production Research Co. (1963) Fermentation process for producing a heteropolysaccharide, U.S. Patent No. 3 281 329.

Esso Production Research Co. (1963) Polysaccharide recovery process, U.S. Patent No. 3 382 229.

Esso Production Research Co. (1966) Heteropolysaccharide fermentation process, U.S. Patent No. 3 328 262.

Exxon Research & Engineering Co. (1975) Procédé de traitement d'un hétéropolysaccharide et son application, FR 2 330 697.

Fu, J. F., Tseng, Y. H. (1990) Construction of lactose-utilizing *Xanthomonas campestris* and production of xanthan gum, *Appl. Environ. Microbiol.* **56**, 919–923.

Galindo, E. (1994) Aspects of the process for xanthan production, *Trans. I. Chem. E* **72**, 227–237.

Galindo, E., Salcedo, G., Ramirez, M. E. (1994) Preservation of *Xanthomonas campestris* on agar slopes: effects on xanthan production, *Appl. Microbiol. Biotechnol.* **40**, 634–637.

Garcia Ochoa, F., Santos, V. E., Alcon, A. (1995) Xanthan gum production: an unstructured kinetic model, *Enzyme Microb. Technol.* **17**, 206–217.

Garcia Ochoa, F., Santos, V. E., Alcon, A. (1998) Metabolic structured kinetic model for xanthan production, *Enzyme Microb. Technol.* **23**, 75–82.

Gaspar, C., Laureano, O., Sousa, I. (1998) Production of reduced-calorie grape juice jelly with gellan, xanthan and locust bean gums: sensory and objective analysis of texture, *Z. Lebensm. Unters. Forsch.* **206**, 169–174.

General Mills Chemicals Inc. (1974) *Xanthomonas* gum amine salts, U.S. Patent No. 3 928 316.

General Mills Chemicals Inc. (1976) Flash drying of xanthan gum and product produced thereby, U.S. Patent No. 4 053 699.

General Mills Inc. (1968) Polysaccharide B1459 cationique, DE 1 919 790.

Getty Scientific Development Co. (1985) A polysaccharide polymer made by *Xanthomonas*, EP 211 288.

Getty Scientific Development Co. (1986) Recombinant-DNA mediated production of xanthan gum, WO 87/05 938.

Gupte, M. D., Kamat, M. Y. (1997) Isolation of wild *Xanthomonas* strains from agricultural produce, their characterisation and potential related to polysaccharide production, *Folia Microbiol.* **42**, 621–628.

Harris, P. J., Fergusson, L. R., (1999) Dietary fibres may protect or enhance carcinogenesis, *Mutat. Res.* **443**, 95–110.

Harrison, G. M., Mun, R., Cooper, G., Boger, D. V. (1999) A note on the effect of polymer rigidity and concentration on spray atomisation, *J. Non-Newtonian Fluid Mech.* **85**, 93–104.

Henkel KGaA (1981) Biopolymères à partir de *Xanthomonas*, EP 58 364.

Henkel KGaA (1986) Process for the preparation of exocellular biopolymers, U.S. Patent No. 4 871 665.

Hercules Inc. (1976) Xanthan recovery process, U.S. Patent No. 4 051 317.

Herp, A. (1980) Oxidative-reductive depolymerization of polysaccharides, in: *The Carbohydrates*, Vol. Ib (Pigman, W., Horton, D., Eds.), New York: Academic Press, 1276–1297.

Hestrin, S. (1949) Reaction of acetylcholine and other carboxylic acid derivatives with hydroxylamine, and its analytical application, *J. Biol. Chem.* **180**, 249–261.

Hoechst A.G. (1983) Lowering the viscosity of fermentation broths, DE 3 330 328.

Ielpi, L., Couso, R. O., Dankert, M. A. (1993) Sequential assembly and polymerization of the polyprenol linked pentasaccharide repeating unit of the xanthan polysaccharide in *Xanthomonas campestris*, *J. Bacteriol.* **175**, 2490–2500.

Institut Français du Pétrole des Carburants & Lubrifiants - Rhone-Poulenc Industries (1977) Utilisation de mouts de fermentation pour la récupération assistée du pétrole, FR 2 398 874.

Institut Français du Pétrole (1980) Procédé enzymatique de clarification de gommes de xanthane permettant d'améliorer leur injectivité et leur filtrabilité, FR 2 491 494.

Institut Français du Pétrole (1993) Procédé de production d'un mout de xanthane ayant une propriété améliorée. Composition obtenue et application de la composition dans une boue de forage de puits, FR 2 701 490.

Janca, J. (Ed.) (1988) Field-Flow Fractionation: analysis of macromolecules and particles, New York: Marcel Dekker.

Jarman, T. R., Pace, G. W. (1984) Energy requirement for microbial expolysaccharide synthesis, *Arch. Microbiol.* **137**, 231–235.

Jersey Production Research Co. (1960) Process for synthesizing polysaccharides, U.S. Patent No. 3 020 206.

Jersey Production Research Co. (1960) Thickening agent and process for producing same, U.S. Patent No. 3 020 207.

Jersey Production Research Co. (1960) Substituted heteropolysaccharide, U.S. Patent No. 3 163 602.

Jungbunzlauer A.G. (1982) Xanthan, AT 373 916.

Jungbunzlauer A.G. (1989) Procédé pour obtenir sous forme grenue des polysaccharides formés par les bactéries du genre *Xanthomonas* ou *Arthrobacter*, FR 2 646 857.

Kang, K. S., Pettitt, D. L. (1993) Xanthan, gellan, welan and rhamsan, in: *Industrial Gums*, 3rd edn (Whistler, R. L., BeMiller, J. N., Eds.), San Diego, CA: Academic Press, 341–397.

Katzen, F., Ferreiro, D. U., Oddo, C. G., Ielmini, M. V., Becker, A., Puhler, A., Ielpi, L. (1998) *Xanthomonas campestris* pv. *campestris* gum mutants: effects on xanthan biosynthesis and plant virulence, *J. Bacteriol.* **180**, 1607–1617.

Kelco Biospecialities Ltd. (1981) Inoculation procedure for xanthan gum production, EP 66 957.

Kelco Biospecialities Ltd. (1981) Production of xanthan having a low pyruvate content, EP 66 961.

Kelco Biospecialties Ltd. (1981) Process for xanthan gum production, EP 66 377.

Kelco Biospecialties Ltd. (1981) Precipitation of xanthan gum, EP 68706.

Kelco Biospecialties Ltd. (1981) Production of xanthan gum by emulsion fermentation, U.S. Patent No. 4 352 882.

Kelco Co. (1960) Edible compositions comprising oil in water emulsions, U.S. Patent No. 3 067 038.

Kelco Co. (1963) Cationic ethers of *Xanthomonas* hydrophilic colloids, U.S. Patent No. 3 244 695.

Kelco Co. (1963) Joint filling composition, U.S. Patent No. 3 279 934.

Kelco Co. (1964) Treatment of *Xanthomonas* hydrophilic colloid and resulting product, U.S. Patent No. 3 355 447.

Kelco Co. (1964) Polymeric derivatives of cationic *Xanthomonas* colloid derivatives, U.S. Patent No. 3 376 282.

Kelco Co. (1965) Etherified *Xanthomonas* hydrophilic colloids and process of preparation, U.S. Patent No. 3 349 077.

Kelco Co. (1966) Process for producing polysaccharides, U.S. Patent No. 3 391 060.
Kelco Co. (1966) Process for producing polysaccharides, U.S. Patent No. 3 391 061.
Kelco Co. (1966) Process for preparing polysaccharide, U.S. Patent No. 3 427 226.
Kelco Co. (1968) Treatment of *Xanthomonas* hydrophilic colloid and resulting product, U.S. Patent No. 3 516 983.
Kelco Co. (1971) Method of producing a dehydrated food product, U.S. Patent No. 3 694 236.
Kingsley, M., Gabriel, D., Marlow, G., Roberts, P. (1993) The *ops*X locus of *Xanthomonas campestris* affects host range and biosynthesis of lipopolysaccharide and extracellular polysaccharide, *J. Bacteriol.* **175**, 5839–5850.
Knoop, V., Staskawicz, B., Bonas, U. (1991) Expression of the avirulence gene *avr*Bs3 from *Xanthomonas campestris* pv. *vesicatoria* is not under the control of *hrp* genes and is independent of plant factors, *J. Bacteriol.* **173**, 7142–7150.
Kuo, T. T., Lin, B. C., Li, C. C. (1970) Bacterial leaf blight of rice plant. III – Phytotoxic polysaccharides produced by *Xanthomonas oryzae*, *Bot. Bull. Acad. Sinica* **11**, 46–54.
Kuttuva, S. G., Sundararajan, A., Ju, L. K. (1998) Model simulation of water-in-oil xanthan fermentation, *J. Dispersion Sci. Technol.* **19**, 1003–1029.
Launay, B., Doublier, J.L., Cuvelier, G. (1986) Flow properties of aqueous solutions and dispersion of polysaccharides, in: *Functional Properties of Food Macromolecules* (Mitchell, J. R., Ledward, D. A., Eds.), London, New York: Elsevier Applied Science Publisher, 1–78.
Leach, J. G., Lilly, V. G., Wilson, H. A., Purvis, M. R. (1957) Bacterial polysaccharides: the nature and function of the exudate produced by *Xanthomonas phaseoli*, *Phytopathology* **47**, 113–120.
Lebrun, L., Junter, G. A., Jouenne, T., Mignot, L. (1994) Exopolysaccharide production by free and immobilized microbial cultures, *Enzyme Microb. Technol.* **16**, 1048–1054.
Leigh, J. A., Coplin, D. L. (1992) Exopolysaccharides in plant–bacterial interactions, *Annu. Rev. Microbiol.* **46**, 307–346, 1048–1054.
Li, H., Rief, M., Oesterhelt, F., Gaub, H. E. (1999) Force spectroscopy on single xanthan molecules, *Appl. Physics A* **68**, 407–410.
Liakopoulou Kyriakides, M., Psomas, S. K., Kyriakidis, D. A. (1999) Xanthan gum production by *Xanthomonas campestris* w.t. fermentation from chestnut extract, *Appl. Biochem. Biotechnol.* **82**, 175–183.
Linton, J. D. (1990) The relationship between metabolite production and the growth efficiency of the producing organism, *FEMS Microbiol. Rev.* **75**, 1–18.
Liu, W., Sato, T., Norisuye, T., Fujita, H. (1987) Thermally induced conformational change of xanthan in 0.01M aqueous sodium chloride, *Carbohydr. Res.* **160**, 267–281
Lo, Y. M., Yang, S. T., Min, D. B. (1997) Ultrafiltration of xanthan gum fermentation broth: process and economic analyses, *J. Food Eng.* **31**, 219–236.
Marrs, M. (1997) Melt-in-mouth gels, *World Ingr.* June, 39–40.
MacNeely, W. H., Kang, K. S. (1973) Xanthan and some other biosynthetic gums, in: *Industrial Gums*, 2nd edn (Whistler, R. L., BeMiller, J. N., Eds.), New York: Academic Press, 473–497.
Merck & Co. Inc. (1975) Copolymère greffe d'un colloide hydrophile de *Xanthomonas* et procédé pour sa préparation, FR 2 325 666.
Merck & Co. Inc. (1975) Perfectionnement aux procédés pour accroitre la viscosité de solutions aqueuses de gomme xanthane et aux compositions obtenues, FR 2 318 926.
Merck & Co. Inc. (1978) Process for producing low calcium xanthan gum by fermentation, U.S. Patent No. 4 375 512.
Merck & Co. Inc. (1979) Production of low-calcium smooth-flow xanthan gum, EP 20 097.
Merck & Co. Inc. (1979) Procédé de préparation de gomme xanthane exempte de cellulase, FR 2 458 589.
Merck & Co. Inc. (1979) Deacetylated borate-biosynthetic gum composition, U.S. Patent No. 4 214 912.
Merck & Co. Inc. (1980) Dispersible xanthan gum composite, U.S. Patent No. 4 357 260.
Merck & Co. Inc. (1985) heteropolysaccharide and its production and use, EP 209 277.
Merck & Co. Inc. (1986) Recombinant DNA plasmid for xanthan gum synthesis, EP 233 019.
Merck & Co. Inc. (1991) Low-ash xanthan gum, EP 511 784.
Milas, M., Rinaudo, M., Knipper, M., Schuppiser, J.L. (1990) Flow and viscoelastic properties of xanthan gum solutions, *Macromolecules* **23**, 2506–2511.
Mobil Oil Corp. (1966) Waterflood process employing thickened water, U.S. Patent No. 3 373 810.
Mobil Oil Corp. (1973) Method of clarifying polysaccharide solutions, U.S. Patent No. 3 711 462.
Moorhouse, R., Walkinshaw, M. D., Arnott, S. (1977) Xanthan gum. Molecular conformation

and interactions, in: *ACS Symposium Series 45, Extracellular Microbial Polysaccharides* (Sandford, P. A., Laskin, A., Eds.), Washington, DC: American Chemical Society, 90–102.

Moreno, J., Lopez, M. J., Vargas Garcia, C., Vasquez, R. (1998) Use of agricultural wastes for xanthan production by *Xanthomonas campestris, J. Ind. Microbiol. Biotechnol.* **21**, 242–246.

Morris, V. J. (1990) Science, structure and applications of microbial polysaccharides, in: *Gums and Stabilisers for the Food Industry* 5 (Phillips, G. O., Wedlock, D. J., Williams, P. A., Eds.), New York: IRL Press, 315–328.

Morris, V. J., Kirby, A. R., Gunning, A. P. (1999) Using atomic force microscopy (AFM) to probe food biopolymer functionality, *Scanning* **21**, 287–292.

Nankai, H., Hashimoto, W., Miki, H., Kawai, S., Murata, K. (1999) Microbial system for polysaccharide depolymerisation: enzymatic route for xanthan depolymerisation by *Bacillus* sp. strain GL1, *Appl. Environ. Microbiol.* **65**, 2520–2526.

Nussinovitch, A. (Ed.) (1997) Xanthan gum, in: *Hydrocolloids Applications: Gum Technology in the Food and Other Industries*, London: Blackie Academic & Professional, 154–168.

Parker, A., Michel, R., Vigouroux, F., Reed, W. F. (1999) Dissolution kinetics of polymer powders, *Polym. Prep. Amer. Chem. Soc. Div. Polym. Chem.* **40**, 685–686.

Parlin, S. (1997) Good vibrations. New scalable vibrating membrane filter system separates liquids, solids, *Food Process* **58**, 106–107.

Pfizer Inc. (1979) Fermentation process for production of xanthan, U.S. Patent No. 4 282 321.

Pfizer Inc. (1982) Continuous production of *Xanthomonas* biopolymer, EP 115 154.

Philipon, P. (1997) Des médicaments libérés sur commande, *Biofutur* **171**, 25–27.

Philips Petroleum Co. (1971) Recovery of microbial polysaccharide, U.S. Patent No. 3 773 752.

Pollock, T. J., Mikolajczak, M., Yamazaki, M., Thorme, L., Armentrout, R. W. (1997) Production of xanthan gum by *Sphingomonas* bacteria carrying genes from *Xanthomonas campestris, J. Ind. Microbiol. Biotechnol.* **19**, 92–97.

Ramirez, M. E., Fucikovsky, L., Garcia-Jimenez, F., Quintero, R., Galindo, E. (1988) Xanthan gum production by altered pathogenicity variants of *Xanthomonas campestris, Appl. Microbiol. Biotechnol.* **29**, 5–10.

Rhodia Inc. (1997) Use of liquid carbohydrate fermentation product in foods, WO 99/25 208.

Rhone-Poulenc Chimie (1987) Procédé de production de polysaccharides, FR 2 624 135.

Rhone-Poulenc Chimie (1988) Procédé de production de polysaccharides par fermentation d'une source carbonée à l'aide de microorganismes, EP 365 390.

Rhone-Poulenc Industries (1975) Nouvelles compositions épaississantes à base d'hétéropolysaccharides, FR 2 299 366.

Rhone-Poulenc Industries (1976) Procédé de production de polysaccharides par fermentation, FR 2 342 339.

Rhone-Poulenc Industries (1978) Procédé de production de polysaccharides par fermentation, FR 2 414 555.

Rhone-Poulenc Specialités Chimiques (1984) Procédé de production de polysaccharides de type xanthane, EP 187 092.

Rinaudo, M., Milas, M. (1980) Enzymic hydrolysis of the bacterial polysaccharide xanthan by cellulase, *Int. J. Biol. Macromol.* **2**, 45–48.

Rodd, A. B., Dunstan, D. E., Boger, D. V. (2000) Characterisation of xanthan gum solutions using dynamic light scattering and rheology, *Carbohydr. Polym.* **42**, 159–174.

Rodriguez, H., Aguilar, L. (1997) Detection of *Xanthomonas campestris* mutants with increased xanthan production, *J. Ind. Microbiol. Biotechnol.* **18**, 232–234.

Ruijssenaars, H. J., De Bont, J. A. M., Hartmans, S. (1999) A pyruvated mannose-specific xanthan lyase involved in xanthan degradation by *Paenibacillus alginolyticus* XL-1, *Appl. Environ. Microbiol.* **65**, 2446–2452.

Salcedo, G., Ramirez, M. E., Flores, C., Galindo, E. (1992) Preservation of *Xanthomonas campestris* in *Brassica oleracea* seeds, *Appl. Microbiol. Biotechnol.* **37**, 723–727

Sanofi Elf Bio-Industries (1986) Procédé d'obtention d'un xanthane à fort pourvoir épaississant et application de ce xanthane, FR 2 606 423.

Sanofi (1993) Stabilizer composition enabling the production of a pourable aerated dairy dessert, EP 649 599.

Sanofi, Méro Rousselot Satia S.A. (1987) Procédé de fermentation pour l'obtention d'une polysaccharide de type xanthane, EP 296 965.

Sanofi, Société Nationale Elf Aquitaine (1990) Souche mutante de *Xanthomonas campestris*, procédé d'obtention de xanthane et xanthane non visqueux, FR 2 671 097.

Schumpe, A., Diedrichs, S., Hesselink, P. G. M., Nene, S., Deckwer, W. D. (1991) Xanthan production in emulsions, *Proceedings of the Second*

International Symposium on Biochemical Engineering, Stuttgart, 196–199.
Serrano Carreon, L., Corona, R. M., Sanchez, A., Galindo, E. (1998) Prediction of xanthan fermentation development by a model linking kinetics, power drawn and mixing, *Proc. Biochem.* **33**, 133–146.
Shell International Research Maatschappij B.V. (1983) Process for preparing *Xanthomonas* heteropolysaccharide, heteropolysaccharide as prepared by the latter process and its use, EP 130 647.
Shell International Research Maatschappij B.V. (1984) Method for improving the filterability broth and the use, EP 184 882.
Shell International Research Maatschappij B.V. (1980) Treatment of polysaccharide solutions, EP 49 012.
Silman, R. W., Rogovin, P. (1972) Continuous fermentation to produce xanthan biopolymer: effect of dilution rate, *Biotechnol. Bioeng.* **14**, 23–31.
Sloneker, J. H., Orentas, D. G. (1962) Exocellular bacterial polysaccharide from *Xanthomonas campestris* NRRL B61459. II – Linkage of the pyruvic acid, *Can. J. Chem.* **40**, 2188–2189.
Société des Usines Chimiques Rhone-Poulenc (1973) Perfectionnement à la production de polysaccharides par fermentation, FR 2 251 620.
Standard Oil Co. (1980) *Xanthomonas campestris* ATCC 31601 and process for use, EP 46 007.
Standard Oil Co. (1980) Method for improving xanthan yield, U.S. Patent No. 4 301 247.
Standard Oil Co. (1980) Method for improving specific xanthan productivity during continuous fermentation, U.S. Patent No. 4 311 796.
Standard Oil Co. (1980) Production of xanthan gum from a chemically defined medium introduction, U.S. Patent No. 4 374 929.
Standard Oil Co. (1980) Process for using *Xanthomonas campestris* NRRL B-12075 and NRRL B-12074 for making heteropolysaccharide, U.S. Patent No. 4 400 467.
Standard Oil Co. (1980) *Xanthomonas campestris* ATCC 31602 and process for use, U.S. Patent No. 4 407 950.
Standard Oil Co. (1980) *Xanthomonas campestris* ATCC 31600 and process for use, U.S. Patent No. 4 407 951.
Standard Oil Co. (1981) Semi-continuous method for production of xanthan gum using *Xanthomonas campestris* ATCC 31601, U.S. Patent No. 4 328 310.
Standard Oil Co. (1981) Method of producing a low viscosity xanthan gum, U.S. Patent No. 4 377 637.
Stauffer Chemical Co. (1981) Fermentation of whey to produce a thickening polymer, U.S. Patent No. 4 444 792.
Stokke, B. J., Christensen, B. E., Smidsrød, O. (1998) Macromolecular properties of xanthan, in: *Polysaccharides. Structural, Diversity and Functional Versatility* (Dumitriu, S., Ed.), New York: Marcel Dekker, 433–472.
Stredansky, M., Conti, E. (1999) Xanthan production by solid state fermentation, *Process Biochem.* **34**, 581–587.
Stredansky, M., Conti, E., Navarini, L., Bertocchi, C. (1999) Production of bacterial exopolysaccharides by solid substrate fermentation, *Process Biochem.* **34**, 11–16.
Sutherland, I. W. (1993) Xanthan, in: *Xanthomonas* (Swings, J. G., Civerolo, E. L., Eds.), London: Chapman & Hall, 363–388.
Tait, M. I., Sutherland, I. W. (1989) Synthesis and properties of a mutant type of xanthan, *J. Appl. Bacteriol.* **66**, 457–460.
Tait, M. I., Sutherland, I. W., Clarke-Sturman, A. J. (1990) Acid hydrolysis and high-performance liquid chromatography of xanthan, *Carbohydr. Polym.* **13**, 133–148.
Tang, J. L., Gough, C. L., Daniels, M. J. (1990) Cloning of genes involved in negative regulation of production of extracellular enzymes and polysaccharide of *Xanthomonas campestris* pathovar *campestris*, *Mol. Gen. Genet.* **222**, 157–160.
Tilly, G. (1991) Stabilization of dairy products by hydrocolloids, in: *Food Ingredients Europe: Conference Proceedings* (Van Zeijst, R., Ed.) Maarsen: Expoconsult Publishers, 105–121.
Triveni, R., Shamala, T. R., (1999) Clarification of xanthan gum with extracellular enzymes secreted by *Trichoderma koningii*, *Process Biochem.* **34**, 49–53.
Tseng, Y. H., Choy, K. T., Hung, C. H., Lin, N. T., Liu, J. Y., Lou, C. H., Yang, B. Y., Wen, F. S., Wu, J. R. (1999) Chromosome map of *Xanthomonas campestris* pv. *campestris* 17 with locations of genes involved in xanthan gum synthesis and yellow pigmentation, *J. Bacteriol.* **181**, 117–125.
U. S. Secretary Agriculture (1959) Method of producing an atypically salt-responsive alkali-deacetylated polysaccharide, U.S. Patent No. 3 000 790.
U. S. Secretary Agriculture (1962) Method of recovering microbial polysaccharides from their fermentation broths, U.S. Patent No. 3 119 812.
U. S. Secretary Agriculture (1967) Continuous process for producing *Xanthomonas* heteropolysaccharide, U.S. Patent No. 3 485 719.

U. S. Secretary Agriculture (1978) Production of high-pyruvate xanthan gum on synthetic medium, U.S. Patent No. 4 394 447.

Vafiadis, D. (1999) Anti-shock treatment (frozen dairy products formulation), *Dairy Field* **182**, 85–88.

Vanderslice, R. W., Doherty, D. H., Capage, M. A., Betlach, M. R., Hassler, R. A., Henderson, N. M., Ryan-Graniero, J., Tecklenburg, M. (1989) Genetic engineering of polysaccharide structure in *Xanthomonas campestris*, in: *Biomedical and Biotechnological Advances in Industrial Polysaccharides* (Crescenzi, V., Dea, I. C. M., Paoletti, S., Stivala, S. S., Sutherland, I. W., Eds.), New York: Gordon and Breach, 145–156.

Vidhyasekaran, P., Alvenda, M. E., Mew, T. W. (1989) Physiological changes in rice seedlings induced by extracellular polysaccharide produced by *Xanthomonas campestris* pv. *oryzae*, *Physiol. Mol. Plant Pathol.* **35**, 391–402.

Vojnov, A. A., Zorreguieta, A., Dow, J. M., Daniels, M. J., Dankert, M. A. (1998) Evidence for a role for the gumB and gumC gene products in the formation of xanthan from its pentasaccharide repeating unit by *Xanthomonas campestris*, *Microbiology* **144**, 1487–1493.

Watabe, M., Yamaguchi, M., Kitamura, S., Horino, O. (1993) Immunohistochemical studies on localization of the extracellular polysaccharide produced by *Xanthomonas oryzae* pv. *oryzae* in infected rice leaves, *Can. J. Microbiol.* **39**, 1120–1126.

Yang, B. Y., Tseng, Y. H. (1988) Production of exopolysaccharide and levels of protease and pectinase activity in pathogenic and non-pathogenic strains of *Xanthomonas campestris* pv. *campestris*, *Bot. Bull. Academia Sinica* **29**, 93–99.

Yang, S. T., Lo, Y. M., Min, D. B. (1996) Xanthan gum fermentation by *Xanthomonas campestris* immobilized in a novel centrifugal fibrous-bed bioreactor, *Biotechnol. Prog.* **12**, 630–637.

Yang, S. T., Lo, Y. M., Chattopadhyay, D. (1998) Production of cell-free xanthan fermentation broth by cell adsorption on fibers, *Biotechnol. Prog.* **14**, 259–264.

Yoo, S. D., Harcum, S. W. (1999) Xanthan gum production from paste waste sugar beet pulp, *Bioresource Technol.* **70**, 105–109.

Zhao, X. M., Li, X. H., Ban, R., Zhu, Y. (1997) The effects of mixing modes on recovery of xanthan gum by precipitation, *BHR Group Conf. Ser. Publ.* **25**, 3–8.

20
Dextran

Dr. Timothy D. Leathers
Fermentation Biochemistry Research Unit, National Center for Agricultural Utilization Research, Agricultural Research Service, United States Department of Agriculture;1815 N. University St.; Peoria, IL, 61604, USA; Tel. +1-309-681-6377; Fax: +1-309-681-6427; E-mail: leathetd@ncaur.usda.gov

Biotechnology of Biopolymers. From Synthesis to Patents. Edited by A. Steinbüchel and Y. Doi

ISBN: 3-527-31110-6

ATP	adenosine 5′-triphosphate
CM	carboxymethyl
DEAE	diethylaminoethyl
EC	enzyme commission
FTIR	fourier-transform infrared
GRAS	Generally Regarded as Safe
HIV	human immunodeficiency virus
IU	international units
K_m	Michaelis-Menten constant
MRI	magnetic resonance imaging
NMR	nuclear magnetic resonance
QAE	diethyl(2-hydroxypropyl)aminoethyl
SP	sulfopropyl

Names are necessary to report factually on available data; however, the USDA neither guarantees nor warrants the standard of the product, and the use of the name by USDA implies no approval of the product to the exclusion of others that also may be suitable.

1 Introduction

Dextrans are defined as homopolysaccharides of glucose that feature a substantial number of consecutive α-(1 → 6) linkages in their major chains, usually more than 50% of total linkages. These α-D-glucans also possess side chains stemming from α-(1 → 2), α-(1 → 3), or α-(1 → 4) branch linkages. The exact structure of each type of dextran depends on its specific microbial strain of origin. Dextrans are produced by certain lactic acid bacteria, particularly strains of *Streptococcus* species and *Leuconostoc mesenteroides*. Dextrans from oral *Streptococcus* species are of clinical interest as components of dental plaque. Dextrans from *L. mesenteroides* are of commercial interest, primarily as specialty chemicals for clinical, pharmaceutical, research, and industrial uses. Numerous reviews have appeared on dextrans, including those by Evans and Hibbert (1946), Neely (1960), Jeanes (1966, 1978), Murphy and Whistler (1973), Sidebotham (1974), Walker (1978), Alsop (1983), Robyt (1986, 1992, 1995), de Belder (1990, 1993), and Cote and Ahlgren (1995).

2 Historical Outline

Because dextrans are formed from sucrose, they have long been known as troublesome contaminants of food products and sugar refineries. Pasteur made an early applied study of dextran formation in wine, proving that this phenomenon was caused by microbial activity (Pasteur, 1861). Scheibler (1874) determined that dextran was a carbohydrate of the empirical formula $(C_6H_{10}O_6)_n$ having a positive optical rotation and thus coined the term “dextran”. Van Tieghem (1878) identified a dextran-forming bacterium and named it *Leuconostoc mesenteroides*. Beijerinck (1912) investigated the phenomenon, and Hehre (1941) demonstrated dextran synthesis by a cell-free culture filtrate. Scientific interest in dextrans was stimulated by studies suggesting its value as a blood-plasma volume expander (Gronwall and Ingelman, 1945, 1948). In the late 1940s, an extensive research program on dextrans was initiated at the Northern Regional Research Laboratory (now the National Center for Agricultural Utilization Research) of the Agricultural Research Service, U.S. Department of Agriculture, in Peoria, Illinois. Numerous dextran-producing strains were characterized, including *L. mesenteroides* strain NRRL B-512F used today for commercial dextran production in North America and Western Europe.

3 Chemical Structure

Within the general definition of dextran as a glucan in which α-(1 → 6) linkages predominate, chemical structures vary considerably as a function of the specific microbial strain of origin. The article of commerce is the product of a single strain of *L. mesenteroides*, NRRL B-512F. As shown in Figure 1, dextran from this strain features α-(1 → 6) linkages in the main chains with a relatively low level (about 5%) of α-(1 → 3) branch linkages (Van Cleve et al., 1956; Jeanes et al., 1954; Slodki et al., 1986). Larm et al. (1971) estimated that 40% of these side chains are one subunit long and 45% are two subunits long. The remaining side chains are probably greater than 30 subunits long, and branches appear to be distributed randomly (Bovery, 1959; Covacevich and Richards, 1977; Taylor et al., 1985; Kuge et al., 1987).

Dextrans are produced by numerous additional strains of bacteria, and the structures of these dextrans are diverse. In the classic study of Jeanes et al. (1954), 96 strains of *Leuconostoc* and *Streptococcus* were surveyed for the formation of polysaccharides from sucrose. In this study it was reported that α-(1 → 6) linkages in dextran varied from 50% to 97% of total linkages. The balance represented α-(1 → 2), α-(1 → 3), or α-(1 → 4) linkages, usually at branch points. Several isolates produced more than one type of polysaccharide, which were named on the basis of their greater (fraction *S*, for soluble) or lesser (fraction *L*) degree of solubility in water–ethanol mixtures.

Leuconostoc citreum strain NRRL B-742 (formerly *L. mesenteroides*; Takahashi et al., 1992) produces a fraction *L* dextran that contains approximately 15% α-(1 → 4) branch linkages and a fraction *S* dextran that contains a variable (30% to 45%) percentage of α-(1 → 3) branches (Jeanes et al., 1954; Seymour et al., 1979a; Côté and Robyt, 1983; Slodki et al., 1986). Dextran from *L. mesenteroides* strain NRRL B-1299 has a high percentage (27% to 35%) of α-(1 → 2) linked single-glucose branches (Jeanes et al., 1954; Kobayashi and Matsuda, 1977; Slodki et al., 1986). Soluble dextran produced by *Streptococcus sobrinus* strain 6715 (formerly *Streptococcus mutans* strain 6715) appears to be structurally similar to the *S* dextran from *L. citreum* strain NRRL B-742 in that it is an α-(1 → 6) glucan with a relatively high percentage of α-(1 → 3) branch linkages (Shimamura et al., 1982). Many strains of oral *Streptococcus* species, and some strains of *L. mesenteroides,* also make α-D-glucans containing linear sequences of consecutive α-(1 → 3) linkages. These low solubility glucans were once proposed to be "Class 3" dextrans (Seymour and Knapp, 1980) but now are considered to be forms of mutan, an important component of dental plaque.

L. mesenteroides strains NRRL B-1355, NRLL B-1498, and NRRL B-1501 produce an *S* fraction glucan with a unique backbone structure of regularly alternating α-(1 → 3) and α-(1 → 6) linkages (Côté and Robyt, 1982; Misaki et al., 1980; Seymour and

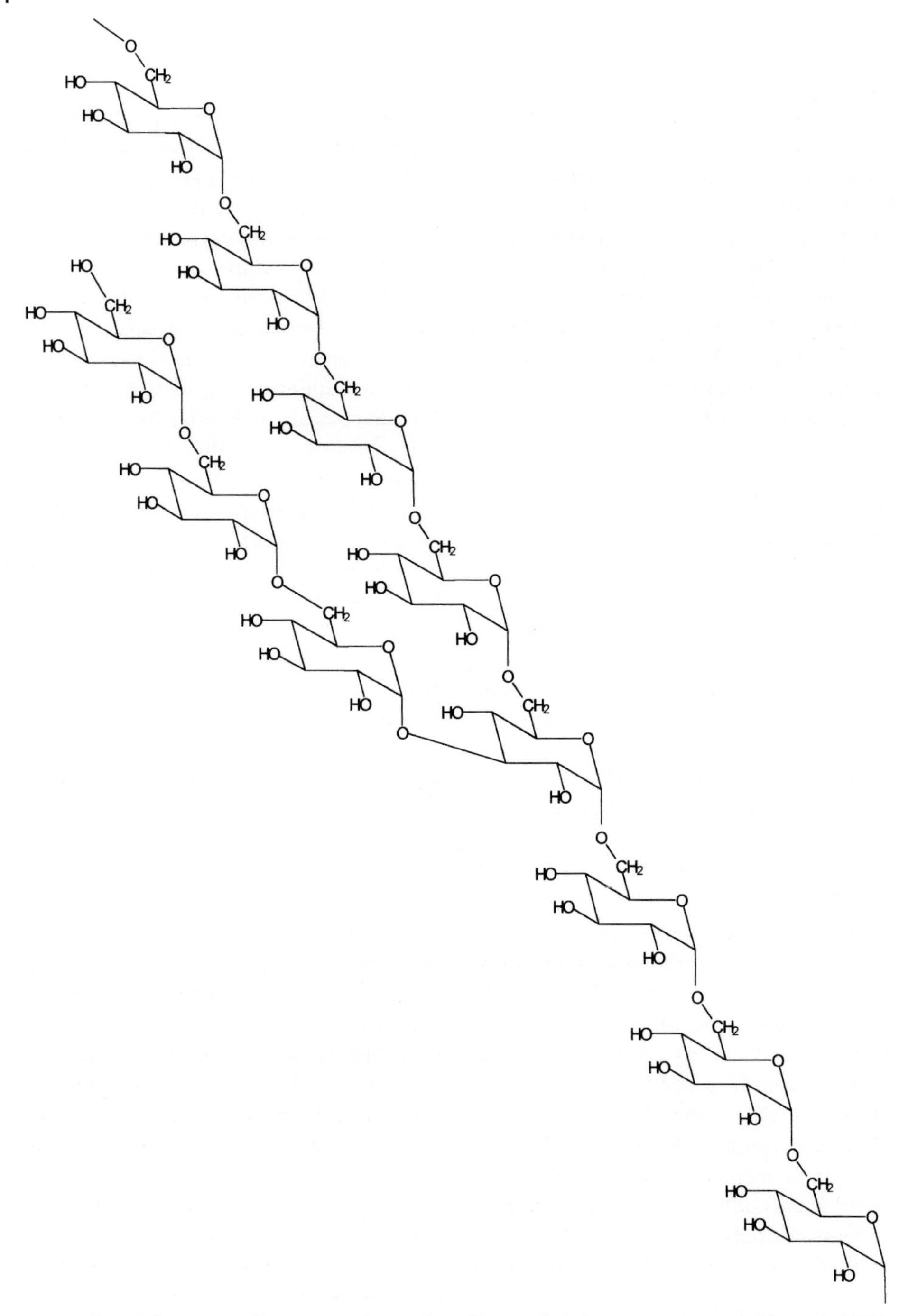

Fig. 1 Chemical structure of a representative portion of dextran from *Leuconostoc mesenteroides* strain NRRL B-512F. Figure courtesy of Dr. Gregory L. Côté.

Knapp, 1980). Because this polysaccharide does not contain significant regions of contiguous α-(1 → 6) linkages, it is not considered to be a true dextran and has been named alternan (Côté and Robyt, 1982). Alternan has distinctive properties of significant basic and applied interest, and it is the subject of Chapter 13, this volume. A mutant derivative of strain NRRL B-1355 recently was reported to produce a third polysaccharide, an insoluble α-D-glucan containing linear (1 → 3) and (1 → 6) linkages with (1 → 2) and (1 → 3) branch points (Smith et al., 1998; Côté et al., 1999).

4 Physiological Function

Oral *Streptococcus* species produce both dextran and mutan, an insoluble glucan in which α-(1 → 3) linkages predominate. These glucans comprise the matrix of dental plaque, which provides an environment for the proliferation of these bacteria (Hamada and Slade, 1980; Shimamura et al., 1982; Loesche, 1986). The physiological function of dextrans produced by *L. mesenteroides* is unknown. Since dextran-producing bacteria do not break down the polymers, dextrans presumably do not serve as storage materials. It is possible that these polysaccharides serve to protect cells from dessication or help them adhere to environmental substrates.

5 Chemical Analyses

General methods applicable to polysaccharides and polyglucans may be used to detect and quantitate dextrans. More specifically, dextrans show high positive specific optical rotation and exhibit characteristic infrared absorption bands at about 917, 840, and 768 cm^{-1} (Barker et al., 1956; Jeanes, 1966). Since dextrans are homopolysaccharides of glucose subunits, many studies have focused on the analysis of specific linkage patterns. Methods have included periodate analysis (Rankin and Jeanes, 1954; Dimler et al., 1955), methylation analysis (Van Cleve et al., 1956; Lindberg and Svensson, 1968; Seymour et al., 1977; Jeanes and Seymour, 1979; Seymour et al., 1979a; Slodki et al., 1986), nuclear magnetic resonance (NMR) spectroscopy (Seymour et al., 1976; Seymour et al., 1979b; Cheetham et al., 1991), and Fourier-transform infrared (FTIR) spectroscopy (Seymour and Julian, 1979). Enzymes that attack dextran in a specific fashion also have been exploited to reveal useful structural information (Covacevich and Richards, 1977; Sawai et al., 1978; Taylor et al., 1985; Pearce et al., 1990).

6 Occurrence

Several lactic acid bacteria have been reported to produce dextrans, principally including *Streptococcus* species and *L. mesenteroides* (Jeanes, 1966; Sidebotham, 1974; Cerning, 1990). *Leuconostoc* and *Streptococcus* are related genera, both composed of Gram-positive, facultatively anaerobic cocci. *L. mesenteroides* (incorporating as subspecies the former species *L. dextranicum* and *L. cremoris*) generally is found on plant materials, particularly on mature or harvested crops, and often plays a role in spoilage (Holzapfel and Schillinger, 1992; Stiles and Holzapfel, 1997). Because sucrose is the natural substrate for dextran synthesis, contamination problems are most evident in food products containing this sugar. *L. mesenteroides* can cause significant problems in sugar (particularly cane) refineries, where dextrans can clog filters and inhibit sugar crystallization (Jeanes, 1977). *Leuconostoc* species are considered Generally Regarded as Safe (GRAS) organisms because of their common appearance in natural fermented foods. In fact, certain strains are valued as starter cultures for buttermilk, cheese, and other dairy products (Holzapfel and Schillinger, 1992). *L. mesenteroides* also plays a role in the production of sauerkraut and other fermented vegetables. Recently, a strain of *L. mesenteroides* that produces both dextran and the related glucan alternan was isolated from the traditional fermented food Kim-Chi (Jung et al., 1999). Oral *Streptococcus* species produce dextran and the insoluble glucan mutan as components of dental plaque (Hamada and Slade, 1980; Shimamura et al., 1982). In addition, certain strains of the ubiquitous Gram-negative bacterium *Gluconobacter oxydans* (formerly *Acetobacter capsulatus*) produce dextran from starch-derived dextrins (Hehre and Hamilton, 1951; Kooi, 1958; Yamamoto et al., 1993a,b; Mountzouris et al., 1999).

7 Biosynthesis

Dextrans are produced extracellularly by secreted enzymes commonly referred to as glucansucrases or, more specifically, dextransucrases. These enzymes are glycosyltransferases (EC 2.4.1.5) that catalyze the transfer of D-glucopyranosyl subunits from sucrose to dextrans. Fructose is released and consumed by growing cells, if they are present. No adenosine 5′-triphosphate (ATP) or cofactors are required for these reactions, as the enzymes utilize energy available in the glycosidic bond between glucose and fructose. Glucansucrase synthesis in wild-type strains of *L. mesenteroides* is induced by growth on sucrose, while *Streptococcus* species produce this enzyme constitutively. Dextransucrase production by *L. mesenteroides* strain NRRL B-512F appears to be regulated at the transcriptional level (Quirasco et al., 1999).

Characterizations of dextransucrases have been complicated by the appearance of multiple enzyme species in culture fluids. Many of these species appear to be proteolytic-processing or degradation products, although some retain activity (Sanchez-Gonzalez et al., 1999). However, strains that produce more than one type of glucan appear to produce a separate glucansucrase for each. Furthermore, enzymes may exist as aggregates and typically are associated with their polysaccharide products, making purifications difficult. Despite these problems, a number of glucansucrases have been studied. Dextransucrase from *L. mesenteroides* strain NRRL B-512F has been purified and characterized (Robyt and Walseth, 1979; Kobayashi and Matsuda, 1980; Paul et al., 1984; Miller et al., 1986; Fu and Robyt, 1990; Kitaoka and Robyt, 1998a; Kim and Kim, 1999). The enzyme appears to have an initial molecular mass of 170 kDa, a pI value of 4.1, and a Michaelis-Menten constant (K_m) for sucrose of approximately 12 to 16 mM. Optimal reaction conditions are pH 5.0 to 5.5 and 30 °C. Kim and Kim (1999) reported a specific activity of up to 250 IU mg^{-1} protein for highly purified enzyme. Low levels of calcium are necessary for optimal enzyme production and activity. Various forms of dextransucrase have been described from *L. mesenteroides* strain NRRL B-1299, with molecular masses of 48 to 79 kDa, temperature optima of 35 °C to 45 °C, pH optima of 5.0 to 6.5, and K_m values for sucrose of 13 to 30 mM (Kobayashi and Matsuda, 1975, 1976; Dols et al., 1997). Multiple forms of dextransucrase have been described from *Streptococcus* species, having molecular masses of 94 to 170 kDa, pI values of approximately 4.0, temperature optima of

34 °C to 42 °C, pH optima of 5.0 to 5.7, and K_m values for sucrose of 2 to 9 mM (Chludzinski et al., 1974; Fukui et al., 1974; Shimamura et al., 1982; Furuta et al., 1985; McCabe, 1985). In addition, dextrandextrinase (also called dextrin dextranase) has been purified and characterized from *G. oxydans* strain ATTC 11894 (Yamamoto et al., 1992; Suzuki et al., 1999).

Robyt and colleagues have developed a model for the reaction mechanism of dextransucrase from *L. mesenteroides* strain NRRL B-512F (Robyt et al., 1974; Robyt, 1992, 1995; Su and Robyt, 1994). According to this model, two nucleophilic reaction sites exist in the catalytic domain of the enzyme. Sucrose is hydrolyzed at one or both sites, and the glucosyl residues are bound covalently to the enzyme in high-energy bonds conserved from sucrose. A dextran chain grows by successive glucosyl insertions between the enzyme and the reducing end of the chain, which remains bound to the enzyme. Branches are formed when glucosyl units or dextran chains are transferred to secondary hydroxyl positions on the dextran chains (Robyt and Taniguchi, 1976). Termination of chain extension occurs by transfer to an acceptor molecule.

Sucrose is strongly preferred as the glucosyl donor, although other natural and synthetic donors have been identified (Hehre and Suzuki, 1966; Binder and Robyt, 1983). However, a number of sugars and derivatives may function as alternative acceptors, including maltose, isomaltose, nigerose, α-methyl glucoside, and others (Robyt and Taniguchi, 1976; Robyt and Walseth, 1978; Robyt and Eklund, 1983; Fu et al., 1990). These acceptor reactions can be utilized to produce dextrans of lower average molecular weights, including clinical dextrans (Koepsell et al., 1955; Tsuchiya et al., 1955; Remaud et al., 1991; Robyt, 1992) and oligosaccharides of interest (Pelenc et al., 1991; Remaud et al., 1992; Remaud-Simeon et al., 1994; Dols et al., 1999). Maltose, the most effective alternative acceptor, accepts a glucosyl residue to form the trisaccharide panose (Killey et al., 1955; Heincke et al., 1999). The acceptor reaction with fructose, the natural co-product of dextran synthesis, has been studied for production of the disaccharide leucrose, a potential alternative sweetener and substrate for industrial conversions (Stodola et al., 1956; Swengers, 1991; Reh et al., 1996; Heincke et al., 1999). A minor product of this reaction is isomaltulose, also known as palatinose, likewise of interest as an alternative sweetener (Sharpe et al., 1960; Takazoe, 1989).

Robyt and Martin (1983) found evidence that a similar reaction mechanism exists for glucansucrases from *S. sobrinus* strain 6715. Alternative models for the glucansucrase reaction mechanism have been reviewed (Monchois et al., 1999). Because the glucansucrases from *Leuconostoc* and *Streptococcus* species appear to be closely related on a molecular level, it seems likely that they share a common reaction mechanism. If so, differences among the polymer structures might be determined by subtle differences in the stereochemistry of the reaction sites.

8 Genetics and Molecular Biology

L. mesenteroides strain NRRL B-512F, used for commercial production of dextran, has been described as a laboratory "substrain" that supplanted natural isolate NRRL B-512 in 1950 (Van Cleve et al., 1956). A dextransucrase hyperproducer mutant of NRRL B-512F was isolated as NRRL B-512FM (Miller and Robyt, 1984). Wild-type strains of *L. mesenteroides* form glucansucrases only when cultured on sucrose, and further mutations were obtained that allowed strain

B-512FMC to produce dextransucrase constitutively (Kim and Robyt, 1994). Further improvements in enzyme productivity were obtained through additional rounds of mutagenesis (Kim et al., 1997; Kitaoka and Robyt, 1998b). Although commercial dextran currently is produced by a fermentative process, such strains would be particularly valuable for dextran production by an enzymatic process. Similar glucansucrase mutants have been obtained for other strains of *Leuconostoc*, including NRRL B-742, NRRL B-1142, NRRL B-1299, and NRRL B-1355 (Kim and Robyt, 1994, 1995a,b, 1996; Kitaoka and Robyt, 1998b). Dextransucrase-deficient mutants of strain NRRL B-1355 also have been isolated for improved production of alternan (Smith et al., 1994; Leathers et al., 1995, 1997, 1998). Alternan is the subject of Chapter 13, this volume.

Because of clinical interest in developing anti-caries vaccines, a number of glucansucrase genes have been cloned from oral *Streptococcus* species (Shiroza et al., 1987; Ueda et al., 1988; Honda et al., 1990). Glucansucrase genes have been cloned and sequenced from *L. mesenteroides* NRRL B-512F (Wilke-Douglas et al., 1989; Bhatnagar and Singh, 1999; Arguello-Morales et al., 2000a; Funane et al., 2000; Ryu et al., 2000), *L. mesenteroides* strain NRRL B-1299 (Monchois et al., 1996, 1998), *L. citreum* strain NRRL B-742 (Kim et al., 2000), and *L. mesenteroides* strain NRRL B-1355 (Arguello-Morales et al., 2000b; Kossman et al., 2000). Interestingly, some of these genes apparently do not specify enzymes normally secreted *in vivo*. Glucansucrase genes appear to be closely related and exhibit a common organizational structure, with a conserved N-terminal catalytic domain and a C-terminal glucan-binding domain that contains a series of direct tandem repeat sequences (Monchois et al., 1999; Remaud-Simeon et al., 2000). Based on site-directed mutageneses and consensus sequences, potentially important catalytic sites have been proposed (Monchois et al., 1997; Arguello-Morales, 2000b; Monchois et al., 2000; Remaud-Simeon et al., 2000). On a broader scale, glucansucrases resemble enzymes in glycosyl hydrolase family 13, which includes α-amylases (Fujiwara et al., 1998; Janecek et al., 2000; Remaud-Simeon et al., 2000).

9 Biodegradation

A variety of fungi produce dextranases, including *Aspergillus* species (Carlson and Carlson, 1955b; Hiraoka et al., 1972), *Chaetomium gracile* (Hattori et al., 1981), *Fusarium* species (Simonson and Liberta, 1975; Shimizu et al., 1998), *Lipomyces starkeyi* (Webb and Spencer-Martins, 1983; Koenig and Day, 1988), *Paecilomyces lilacinus* (Lee and Fox, 1985; Sun et al., 1988; Galvez-Mariscal and Lopez-Munguia, 1991), and *Penicillium* species (Tsuchiya et al., 1956; Chaiet et al., 1970). These enzymes are endodextranases with specificity for internal α-(1 → 6) linkages, and they produce mainly isomaltose or isomaltotriose from dextran. Dextranases from *C. gracile* and *Penicillium* sp. are produced commercially and used for treatment of dextran contamination problems in sugar processing (Godfrey, 1983). Endodextranases have shown potential for the enzymatic production of specific molecular weight fractions of dextran (Carlson and Carlson, 1955a; Corman and Tsuchiya, 1957; Novak and Stoycos, 1958; Day and Kim, 1992; Kim and Day, 1995; Kim and Robyt, 1996). These enzymes also have been tested for the treatment of dental plaque (Fitzgerald et al., 1968; Caldwell et al., 1971), although the more highly branched dextrans are far less susceptible to endodextranase digestion. Limit endodextranase digestion of the branched dextran from *L. citreum* strain NRRL B-742 produces an interesting branched fraction with rheological characteristics similar to those of polydextrose (Cote et al., 1997).

Dextranases also have been reported from a number of bacteria. *Arthrobacter globiformis* produces isomaltodextranase, an exodextranase that successively releases isomaltose from the non-reducing ends of dextrans and oligosaccharides (Torii et al., 1976; Okada et al., 1988). This enzyme recognizes not only α-(1 → 6) linkages but also α-(1 → 2), α-(1 → 3), and α-(1 → 4) linkages. Unlike endodextranases, isomaltodextranase is able to partially hydrolyze alternan, producing an interesting limit alternan (Sawai et al., 1978; Cote, 1992). An isomaltodextranase from *Actinomadura* sp. exhibits slightly different specificities (Sawai et al., 1981). A dextran α-(1 → 2) debranching enzyme also has been described from a *Flavobacterium* sp. (Mitsuishi et al., 1979). Dextran from *L. mesenteroides* strain NRRL B-512F is degraded by intestinal bacteria and enzymes in mammalian tissues other than blood (Sery and Hehre, 1956; Fischer and Stein, 1960). Intravenously administered clinical dextrans are metabolized slowly and completely in the body.

10 Production

To date, commercial production of dextran has employed primarily simple batch fermentation methods, using live cultures grown on sucrose. Methods and conditions for dextran fermentation have been detailed (Tarr and Hibbert, 1931; Hehre et al., 1959; Jeanes, 1965b, 1966; Alsop, 1983; de Belder, 1993). *L. mesenteroides* is a fastidious organism, and its special nutritional requirements include glutamic acid, valine, biotin, nicotinic acid, thiamine, and pantothenic acid (Holzapfel and Schillinger, 1992). In dextran production, these needs are met by combinations of complex medium components, such as yeast extract, corn steep liquor, casamino acids, malt extract, peptone, and tryptone. Sucrose serves as a carbon source, inducer of dextransucrase, and substrate for dextran production. Low levels of calcium (e.g., 0.005%) are necessary for optimal enzyme and dextran yields, and other basal salts, including a source of phosphate, complete the medium. Operative production factors include initial pH (typically pH 6.7 to 7.2), temperature (about 25 °C), initial sucrose concentration (usually 2%), and time (usually 24 to 48 h). Dextran branching appears to increase at elevated temperatures (Sabatie et al., 1988). High levels of sucrose (10% to 50%) reduce the yield of high-molecular-weight dextran, and this observation has been exploited for the production of intermediate sized dextran (Tsuchiya et al., 1955; Alsop, 1983). The organism is facultatively anaerobic or microaerophilic, and fermentations are not aerated. During the first 20 h of fermentation, culture pH falls to approximately 5.0 because of the formation of organic acids, favorably near the optimal pH of dextran sucrase. Dextran may be recovered by precipitation with solvents, particularly alcohols (Hehre et al., 1959; Jeanes, 1965b).

It has long been recognized that dextran also can be produced enzymatically, using cell-free culture supernatants that contain dextransucrase (Hehre, 1941; Tsuchiya and Koepsell, 1954; Hellman et al., 1955; Behrens and Ringpfeil, 1962; Jeanes, 1965a). Accordingly, improved dextransucrase production and purification methods have been developed (Lawford et al., 1979; Paul et al., 1984; Miller et al., 1986; Fu and Robyt, 1990; Kim and Kim, 1999). Glucansucrases also have been immobilized, although this approach may be most useful for production of oligosaccharides (Kaboli and Reilly, 1980; Monsan et al., 1987; Cote and Ahlgren, 1994; Reh et al., 1996; Alcalde et al., 1999). Enzymatic synthesis offers advantages of product molecular weight and quality control, as well as the benefit of obtaining fructose as a valuable co-product. However, this approach has been largely ignored for commercial production, presumably for economic reasons. Dextran production from maltodextrins, using dextran-dextrinase from *Gluconobacter oxydans*, also has attracted interest (Hehre and Hamilton, 1951; Kooi, 1958; Yamamoto et al., 1993a,b; Mountzouris et al., 1999).

Clinical dextran fractions are produced primarily by simple methods of partial acid hydrolysis followed by differential fractionation in solvents (Wolff et al., 1955; Gronwall, 1957; de Belder, 1990). Attractive alternative methods to produce these fractions include the use of dextranases (Carlson and Carlson, 1955a; Corman and Tsuchiya, 1957; Novak and Stoycos, 1958; Day and Kim, 1992; Kim and Day, 1995; Kim and Robyt, 1996) and chain-terminating acceptor reactions (Koepsell et al., 1955; Tsuchiya et al., 1955; Remaud et al., 1991; Robyt, 1992).

Dextran has been produced commercially for many years and by a number of companies, including Dextran Products, Ltd., Toronto, Canada; Pfeifer und Langen, Dormagen, Germany; Pharmachem Corp., Bethlehem, Pennsylvania, USA; and Pharmacia, Uppsala, Sweden. Annual world production was recently estimated at 2000 tons per year (Vandamme et al., 1996). The wholesale price of dextran varies, but recently it has been near 3 USD per pound.

11 Properties and Applications

Purified dextrans are white, tasteless solids. Other physical and chemical properties vary depending on the specific chemical structure, which is determined by the microbial strain of origin and method of production. Dextrans with the highest percentages of α-$(1 \rightarrow 6)$ linkages are generally the most soluble in water. Dextran from *L. mesenteroides* strain NRRL B-512F is freely soluble in water and other solvents, including 6 M urea, 2 M glycine, formamide, glycerol, etc. (Jeanes, 1966; de Belder, 1990). Dextran solutions behave as Newtonian fluids, and their viscosity is a function of concentration, temperature, and average molecular weight (Granath, 1958; Gekko and Noguchi, 1971; Carrasco et al., 1989). Native dextran is polydisperse and typically of high average molecular weight (generally between 10^6 and 10^9 daltons). However, many of the current applications for dextran depend on the convenience with which it can be broken down to fractions of specific weight ranges. The relative linearity of dextran from strain NRRL B-512F is crucial for the production of such fractions. Dextrans exhibit characteristic serological reactions, apparently related to their molecular weight and degree of branching (Gronwall, 1957; Kabat and Bezer, 1958; Jeanes, 1986). However, intravenously administered clinical dextrans are of relatively low antigenicity, although individuals can exhibit hypersensitivity. The pharmacological properties of clinical dextrans have been reviewed recently (de Belder, 1996). Free hydroxyl groups in dextran are potential targets for chemical derivatizations, and dextran from strain NRRL B-512F is particularly suitable for these reactions because of its low level of branch linkages.

A number of bulk chemical applications have been demonstrated for dextran, including uses in oil-drilling operations, agriculture, food products, and the manufacture of photographic films and other products (Murphy and Whistler, 1973; Alsop, 1983; Glicksman, 1983). It should be noted that dextrans are not explicitly approved as food additives in the United States or Europe, although *L. mesenteroides* is a GRAS organism commonly found in fermented foods. Currently, dextran and dextran derivatives are used primarily as specialty chemicals in clinical, pharmaceutical, research, and industrial applications (Yalpani, 1986; de Belder, 1996; Vandamme et al., 1996). Early work by Gronwall and Ingelman (1945, 1948) established the potential of using a hydrolyzed dextran fraction as a blood-plasma volume expander (Figure 2). Clinical dextrans used today are Dextran 40 and Dextran 70, which are 40,000 and 70,000 dalton average molecular weight fractions, respectively. Dextrans are less expensive than the albumins and starch derivatives also used in plasma therapies (Lilley and Aucker, 1999). These colloids essentially replace normal blood proteins in providing osmotic pressure to pull fluid from the interstitial space into the plasma. This treatment is useful to prevent shock from hemorrhage, burns, surgery, or trauma and to reduce the risk of thrombosis and embolisms. Dextran 40 also improves blood flow and inhibits the aggregation of erythrocytes (de Belder, 1996).

Iron dextran is a colloidal preparation used especially in veterinary medicine for the treatment of anemia, particularly in newborn piglets. Special iron dextran preparations also have been developed to enhance magnetic resonance imaging (MRI) techniques (de Belder, 1996). Dextran sulfate has been used as a substitute for heparin in anticoagulant therapy, and, more recently, it is being studied as an antiviral agent, particularly in the treatment of human immunodeficiency virus (HIV) (Mitsuya et al., 1988; Piret et al., 2000). Dextran can be crosslinked by epichlorohydrin (Flodin and Porath, 1961) to form beads (Sephadex®) that have become widely used in

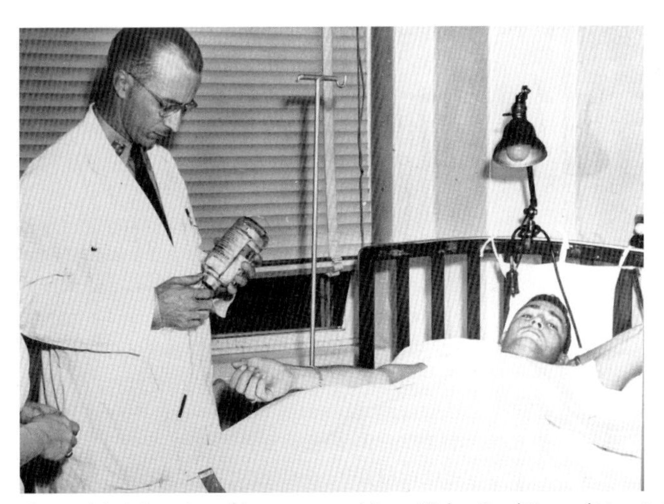

Fig. 2 Administration of dextran to a soldier at Walter Reed General Hospital, 1952. U.S. Dept. of Agriculture photograph.

research and industry for separations based on gel filtration. Anion and cation exchange resins based on Sephadex derivatives are widely used, including carboxymethyl (CM) Sephadex, diethylaminoethyl (DEAE) Sephadex, diethyl(2-hydroxypropyl)aminoethyl (QAE) Sephadex, and sulfopropyl (SP) Sephadex (Figure 3). Dextran is also an important component of many aqueous two-phase extraction systems, usually used in conjunction with polyethylene glycol (Tjerneld, 1992; Sinha et al., 2000).

Recently, oligosaccharides have received a great deal of attention as potential prebiotic compounds in food products, animal feeds, and cosmetics (Hidaka and Hirayama, 1991; Kohmoto et al., 1991; Monsan and Paul, 1995; Lamothe et al., 1996; Monsan et al., 2000). In contrast to clinical dextran preparations, prebiotic oligosaccharides must be resistant to digestion and preferentially utilized by beneficial bifidobacteria and lactic acid bacteria in the intestinal or skin microflora. Accordingly, dextran oligosaccharides of interest as prebiotics include the more branched varieties, containing α-(1 → 3) linkages from *L. citreum* strain NRRL B-742 (Remaud et al., 1992), α-(1 → 2) linkages from *L. mesenteroides* strain NRRL B-1299 (Remaud-Simeon et al., 1994; Dols et al., 1999), or the alternating α-(1 → 3) and α-(1 → 6) linkages from *L. mesenteroides* strain NRRL B-1355 (Pelenc et al., 1991).

Fig. 3 Sephadex ion-exchange media are used widely for process scale applications. Photograph courtesy of Amersham Pharmacia Biotech, Inc.

12 Patents

Numerous patents claim methods for the production of dextran and dextran derivatives. The following examples, also summarized in Table 1, are illustrative. An early patent by Gronwall and Ingelman (1948) suggested that dextran might be useful as a blood-plasma volume expander. Hehre et al. (1959) described methods for dextran production by fermentation. Enzymatic production of dextran using dextransucrase also has been claimed (Tsuchiya and Koepsell, 1954; Hellman et al., 1955; Behrens and Ringpfeil, 1962). Wolff et al. (1955) described partial acid hydrolysis and fractionation of dextran for clinical applications. Alternative methods for the production of clinical dextrans include the use of dextranases (Carlson and Carlson, 1955a; Corman and Tsuchiya, 1957; Novak and Stoycos, 1958; Day and Kim, 1992) and chain-terminating acceptor reactions (Koepsell et al., 1955). Flodin and Porath (1961) described the cross-linking of dextran to form beads (Sephadex®) useful for gel filtration. Methods have been patented for the production of therapeutic iron dextrans (London and Twigg, 1958; Herb, 1979) and dextran sulfate (Morii et al., 1964; Usher, 1989). Recently, dextran oligosaccharides have garnered interest as potential prebiotic compounds (Lamothe et al., 1996).

Tab. 1 Selected patents related to dextran

Patent number	*Holder*	*Inventors*	*Title*	*Date*
U.S. Patent 2,437, 518	Pharmacia AB, Sweden	A. Gronwall, B. Ingelman	Manufacture of infusion and injection fluids	1948
U.S. Patent 2,686,147	U.S. Dept. Agriculture	H. M. Tsuchiya, H. J. Koepsell	Production of dextransucrase	1954
U.S. Patent 2,709,150	Enzmatic Chemicals, Inc., Delaware, USA	V. W. Carlson, W. W. Carlson	Method of producing dextran material by bacteriological and enzymatic action	1955
U.S. Patent 2,712,007	U.S. Dept. Agriculture	I. A. Wolff, R. L. Mellies, C. E. Rist	Fractionation of dextran products	1955
U.S. Patent 2,726,190	U.S. Dept. Agriculture	H. J. Koepsell, N. N. Hellman, H. M. Tsuchiya	Modification of dextran synthesis by means of alternate glucosyl acceptors	1955
U.S. Patent 2,726,985	U.S. Dept. Agriculture	N. N Hellman, H. M. Tsuchiya, S. P. Rogovin, R. W. Jackson, F. R. Senti	Controlled enzymatic synthesis of dextran	1955
U.S. Patent 2,841,578	The Commonwealth Engineering Co. of Ohio	L. J. Novak, G. S. Stoycos	Method for producing clinical dextran	1958
U.S. Patent 2,776,925	U.S. Dept. Agriculture	J. Corman, H. M. Tsuchiya.	Enzymic production of dextran of intermediate molecular weights	1957
U.S. Patent 2,820, 740	Benger Laboratories Ltd., England	E. London, G. D. Twigg	Therapeutic preparation of iron	1958
U.S. Patent 2,906,669	U.S. Dept. Agriculture	E. J. Hehre, H. M. Tsuchiya, N. N. Hellman, F. R. Senti	Production of dextran	1959
U.S. Patent 3,002,823	Pharmacia AB, Sweden	P. G. M. Flodin, J. O. Porath	Process of separating materials having different molecular weights and dimensions	1961
U.S. Patent 3,044,940	VEB Serum-Werk Bernburg, Germany	U. Behrens, M. Ringpfeil	Process for enzymatic synthesis of dextran	1962
U.S. Patent 3,141,014	Meito Sangyo Kabushiki Kaishu, Japan	E. Morii, K. Iwata, H. Kokkoku	Sodium and potassium salts of the dextran sulfate acid ester having substantially no anticoagulant activity but having lipolytic activity and the method of preparation thereof	1964
U.S. Patent 4,180,567	Pharmachem Corp., USA	J. R. Herb	Iron preparations and methods of making and administering the same	1979

Tab. 1 (cont.)

Patent number	*Holder*	*Inventors*	*Title*	*Date*
U.S. Patent 4,855,416	Polydex Pharmaceuticals, Ltd., The Bahamas	T. C. Usher	Method for the manufacture of dextran sulfate and salts thereof	1989
U.S. Patent 5,229,277	Louisiana State Univ.	D. F. Day, D. Kim	Process for the production of dextran polymers of controlled molecular size and molecular size distributions	1992
U.S. Patent 5,518,733.	Bioeurope, France	J.-P. Lamothe, Y. G. Marchenay, P. F. Monsan, F. M. B. Paul, V. Pelenc	Cosmetic compositions containing oligosaccharides	1996

13
Outlook and Perspectives

Advances in the molecular biology of glucansucrases promise not only to resolve fundamental questions concerning enzyme structure, function, and regulation but also to open new avenues for dextran applications. Recombinant organisms that overproduce dextransucrases may reduce the cost of dextran production by enzymatic synthesis, making dextrans more competitive for bulk chemical applications. Alternatively, dextrans might be produced in transgenic crops, as has been demonstrated recently for fructans (Caimi et al., 1996; Pilon-Smits et al., 1996; Sevenier et al., 1998) and *Streptococcus* glucans (Nichols, 2000a,b,c). Novel dextransucrases might be created by site-directed mutagenesis, chimeric recombination, or shuffling of dextransucrase genes. At the same time, dextran oligosaccharides appear to have considerable potential to find new and expanded markets as prebiotic supplements in foods, cosmetics, and animal feeds.

14
References

Alcalde, M., Plou, F. J., Gomez de Segura, A., Remaud-Simeon, M. Willemot, R. M., Monsan, P., Ballesteros, A. (1999) Immobilization of native and dextran-free dextransucrases from *Leuconostoc mesenteroides* NRRL B-512F for the synthesis of glucooligosaccharides, *Biotechnol. Tech.* **13**, 749–755.

Alsop, R. M. (1983) Industrial production of dextrans, in: *Progress in Industrial Microbiology*, (Bushell, M. E., Ed.), London: Elsevier, 1–44, Vol. 18.

Arguello-Morales, M. A., Remaud-Simeon, M., Pizzut, S., Sarcabal, P., Willemot, R.-M., Monsan, P. (2000a) *Leuconostoc mesenteroides* NRRL B-1355 *dsrC* gene for dextransucrase. GenBank Accession No. AJ250172.

Arguello-Morales, M. A., Remaud-Simeon, M., Pizzut, S., Sarcabal, P., Willemot, R.-M., Monsan, P. (2000b) Sequence analysis of the gene encoding alternansucrase, a sucrose glucosyltransferase from *Leuconostoc mesenteroides* NRRL B-1355. *FEMS Microbiol. Lett.* **182**, 81–85.

Barker, S. A., Bourne, E. J., Whiffen, D. H. (1956) Use of infrared analysis in the determination of carbohydrate structure, in: *Methods of Biochemical Analysis*, (Glick, D., Ed.), New York: Interscience Publishers, Inc., 213–245, Vol. 3.

Behrens, U., Ringpfeil, M. (1962) Process for enzymatic synthesis of dextran. U.S. Patent 3,044,940.

Beijerinck, M. W. K. (1912) Mucilaginous substances of the cell wall produced from cane sugar by bacteria. *Folia Microbiol.* **1**, 377.

Bhatnagar, R., Singh, D. K. S. (1999) Cloning and characterization of dextransucrase gene from *Leuconostoc mesenteroides* NRRL B-512F. GenBank Accession No. U81374.

Binder, T. P., Robyt, J. F. (1983) *p*-nitrophenyl α-D-glucopyranoside, a new substrate for glucansucrases, *Carbohydr. Res.* **124**, 287–299.

Bovery, F. A. (1959) Enzymatic polymerization. I. Molecular weight and branching during the formation of dextran, *J. Polymer Sci.* **35**, 167–182.

Caimi, P. G., McCole, L. M., Klein, T. M., Kerr, P. S. (1996) Fructan accumulation and sucrose metabolism in transgenic maize endosperm expressing a *Bacillus amyloliquefaciens SacB* gene, *Plant Physiol.* **110**, 355–363.

Caldwell, R. C., Sandham, H. J., Mann, W. V., Finn, S. B., Formicola, A. J. (1971) The effect of a dextranase mouthwash on dental plaque in young adults and children, *J. Amer. Dent. Assoc.* **82**, 124–131.

Carlson, V. W., Carlson, W. W. (1955a) Method of producing dextran material by bacteriological and enzymatic action. U.S. Patent 2,709,150.

Carlson, V. W., Carlson, W. W. (1955b) Production of endodextranase by *Aspergillus wentii*. U.S. Patent 2,716,084.

Carrasco, F., Chornet, E., Overend, R. P., Costa, J. (1989) A generalized correlation for the viscosity of dextrans in aqueous solutions as a function of temperature, concentration, and molecular weight at low shear rates, *J. Appl. Polymer Sci.* **37**, 2087–2098.

Cerning, J. (1990) Exocellular polysaccharides produced by lactic acid bacteria, *FEMS Microbiol. Rev.* **87**,113–130.

Chaiet, L., Kempf, A. J., Harman, R., Kaczka, E., Weston, R., Nollstadt, K., Wolf, F. J. (1970) Isolation of a pure dextranase from *Penicillium funiculosum*, *Appl. Microbiol.* **20**, 421–426.

Cheetham, N. W. H., Fiala-Beer, E., Walker, G. J. (1991) Dextran structural details from high-field proton NMR spectroscopy, *Carbohydr. Polymers* **14**, 149–158.

Chludzinski, A. M., Germaine, G. R., Schachtele, C. F. (1974) Purification and properties of dextransucrase from *Streptococcus mutans, J. Bacteriol.* **118**, 1–7.

Corman, J., Tsuchiya, H. M. (1957) Enzymic production of dextran of intermediate molecular weights. U.S. Patent 2,776,925.

Côté, G. L. (1992) Low-viscosity α-D-glucan fractions derived from sucrose which are resistant to enzymatic digestion, *Carbohydr. Polym.* **19**, 249–252.

Côté, G. L., Ahlgren, J. A. (1994) Production, isolation, and immobilization of alternansucrase. Amer. Chem. Soc. 207 Natl. Meeting, Abstract CARB#12.

Côté, G. L., Ahlgren, J. A. (1995) Microbial polysaccharides, in: *Kirk-Othmer Encyclopedia of Chemical Technology* (Kroschvitz, J. I., Howe-Grant, M., Eds), New York: John Wiley & Sons, Inc., 578–612, 4th Ed., Vol. 16.

Côté, G. L., Robyt, J. F. (1982) Isolation and partial characterization of an extracellular glucansucrase from *L. mesenteroides* NRRL B-1355 that synthesizes an alternating (1 → 6), (1 → 3)-α-D-glucan, *Carbohyd. Res.* **101**, 57–74.

Côté, G. L., Robyt, J. F. (1983) The formation of α-D-(1 → 3) branch linkages by an exocellular glucansucrase from *Leuconostoc mesenteroides* NRRL B-742, *Carbohydr. Res.* **119**, 141–156.

Côté, G. L., Leathers, T. D., Ahlgren, J. A., Wyckoff, H. A., Hayman, G. T., Biely, P. (1997) Alternan and highly branched limit dextrans: Low-viscosity polysaccharides as potential new food ingredients, in: *Chemistry of Novel Foods* (Spanier, A. M., Tamura, M., Okai, H., Mills, O., Eds.), Carol Stream, IL: Allured Publishing Corp., 95–110.

Côté, G. L., Ahlgren, J. A., Smith, M. R. (1999) Some structural features of an insolube α-D-glucan from a mutant strain of *Leuconostoc mesenteroides* NRRL B-1355, *J. Ind. Microbiol. Biotechnol.* **23**, 656–660.

Covacevich, M. T., Richards, G. N. (1977) Frequency and distribution of branching in a dextran: an enzymic method, *Carbohydr. Res.* **54**, 311–315.

Day, D. F., Kim, D. (1992) Process for the production of dextran polymers of controlled molecular size and molecular size distributions. U.S. Patent 5,229,277.

de Belder, A. N. (1990) *Dextran.* Uppsala, Sweden: Pharmacia.

de Belder, A. N. (1993) Dextran, in: *Industrial Gums. Polysaccharides and Their Derivatives,* Third Edition (Whistler, R. L., BeMiller, J. N., Eds.), San Diego, CA: Academic Press, 399–425.

de Belder, A. N. (1996) Medical applications of dextran and its derivatives, in: *Polysaccharides. Medical Applications* (Dumitriu, S., Ed.), New York: Marcel Dekker, 505–523.

Dimler, R. J., Wolff, I. A., Sloan, J. W., Rist, C. E. (1955) Interpretation of periodate oxidation data on degraded dextran, *J. Am. Chem. Soc.* **77**, 6568–6573.

Dols, M., Remaud-Simeon, M., Willemot, R.-M., Vignon, M., Monsan, P. F. (1997) Characterization of dextransucrases from *Leuconostoc mesenteroides* NRRL B-1299, *Appl. Biochem. Biotechnol.* **62**, 47–59.

Dols, M., Remaud-Simeon, M., Willemot, R.-M., Demuth, B., Joerdening, H.-J., Buchholz, K., Monsan, P. (1999) Kinetic modeling of oligosaccharide synthesis catalyzed by *Leuconostoc mesenteroides* NRRL B-1299 dextransucrase, *Biotechnol. Bioeng.* **63**, 308–315.

Evans, T. H., Hibbert, H. (1946) Bacterial polysaccharides, in: *Adv. Carbohydr. Chem.* (Pigman, W. W., Wolfrom, M. L., Eds.). New York: Academic Press, 203–233, Vol. 2.

Fischer, E. H., Stein, E. A. (1960) Cleavage of O- and S-glycosidic bonds (survey), in: *The Enzymes* (Boyer, P. D., Lardy, H., Myrback, K., Eds.), New York: Academic Press, 301–312, Vol. 4.

Fitzgerald, R. J., Spinell, D. M., Stoudt, T. H. (1968) Enzymatic removal of artificial plaques, *Arch. Oral Biol.***13**, 125–128.

Flodin, P. G. M., Porath, J. O. (1961) Process of separating materials having different molecular weights and dimensions. U.S. Patent 3,002,823.

Fu, D., Robyt, J. F. (1990) A facile purification of *Leuconostoc mesenteroides* B- 512FM dextransucrase, *Prep. Biochem.* **20**, 93–106.

Fu, D., Slodki, M. E., Robyt, J. F. (1990) Specificity of acceptor binding to *Leuconostoc mesenteroides* B512F dextransucrase: binding and acceptor-product structure of α-methyl-D-glucopyranoside analogs modified at C-2, C-3, and C-4 by inversion of the hydroxyl and by replacement of the hydroxyl with hydrogen, *Arch. Biochem. Biophys.* **276**, 460–465.

Fujiwara, T. Terao, Y., Hoshino, T., Kawabata, S., Ooshima, T., Sobue, S., Kimura, S., Hamada, S. (1998) Molecular analyses of glucosyltransferase genes among strains of *Streptococcus mutans, FEMS Microbiol. Lett.* **161**, 331–336.

Fukui, K., Fukui, Y., Moriyama, T. (1974) Purification and properties of dextransucrase and invertase from *Streptococcus mutans, J. Bacteriol.* **118**, 796–804.

Funane, K., Mizuno, K., Takahara, H., Kobayashi, M. (2000) Gene encoding a dextransucrase-like

protein in *Leuconostoc mesenteroides* NRRL B-512F, *Biosci. Biotechnol. Biochem.* **64**, 29–38.
Furuta, T., Koga, T., Nisizawa, T., Okahashi, N., Hamada, S. (1985) Purification and characterization of glucosyltransferases from *Streptococcus mutans* 6715, *J. Gen. Microbiol.* **131**, 285–293.
Galvez-Mariscal, A., Lopez-Munguia, A. (1991) Production and characterization of a dextranase from an isolated *Paecilomyces lilacinus* strain, *Appl. Microbiol. Biotechnol.* **36**, 327–331.
Gekko, K., Noguchi, H. (1971) Physicochemical studies of oligodextran. I. Molecular weight dependence of intrinsic viscosity, partial specific compressibility and hydrated water, *Biopolymers* **10**, 1513–1524.
Glicksman, M. (1983) Dextran, in: *Food Hydrocolloids* (Glicksman, M., Ed.), Boca Raton: CRC Press 157–166.
Godfrey, T. (1983) Dextranase and sugar processing, in: *Industrial Enzymology. The Application of Enzymes in Industry* (Godfrey, T., Reichelt, T., Eds.), New York: Nature Press, 422–424.
Granath, K. A. (1958) Solution properties of branched dextrans, *J. Colloid Sci.* **13**, 308–328.
Gronwall, A. (1957) *Dextran and Its Use in Colloidal Infusion Solutions.* Stockholm: Almqvist & Wiksell.
Gronwall, A., Ingelman, B. (1945) Dextran as a substitute for plasma, *Nature* **155**, 45.
Gronwall, A., Ingelman, B. (1948) Manufacture of infusion and injection fluids. U.S. Patent 2,437, 518.
Hamada, S., Slade, H. D. (1980) Biology, immunology, and cariogenicity of *Streptococcus mutans*, *Microbiol. Rev.* **44**, 331–384.
Hattori, A., Ishibashi, K., Minato, S. (1981) The purification and characterization of the dextranase from *Chaetomium gracile*, *Agric. Biol. Chem.* **45**, 2409–2416.
Hehre, E. J. (1941) Production from sucrose of a serologically reactive polysaccharide by a sterile bacterial extract, *Science* **93**, 237–238.
Hehre, E. J., Hamilton, D. M. (1951) The biological synthesis of dextran from dextrins, *J. Biol. Chem.* **192**, 161–174.
Hehre, E. J., Tsuchiya, H. M., Hellman, N. N., Senti, F. R. (1959) Production of dextran. U.S. Patent 2,906,669.
Hehre, E. J., Suzuki, H. (1966) New reactions of dextransucrase: α-D-glucosyl transfers to and from the anomeric sites of lactulose and fructose, *Arch. Biochem. Biophys.* **113**, 675–683.
Heincke, K., Demuth, B., Jordening, H.-J., Buchholz, K. (1999) Kinetics of the dextransucrase acceptor reaction with maltose - experimental results and modeling, *Enzyme Microbial Technol.* **24**, 523–534.
Herb, J. R. (1979) Iron preparations and methods of making and administering the same. U.S. Patent 4,180,567.
Hellman, N. N., Tsuchiya, H. M., Rogovin, S. P., Jackson, R. W., Senti, F. R. (1955) Controlled enzymatic synthesis of dextran. U.S. Patent 2,726,985.
Hidaka, H., Hirayama, M. (1991) Useful characteristics and commercial applications of fructo-oligosaccharides, *Biochem. Soc. Trans.* **19**, 561–565.
Hiraoka, N., Fukumoto, J., Tsuru, D. (1972) Studies on mold dextranases: III. Purification and some enzymatic properties of *Aspergillus carneus* dextranase, *J. Biochem.* **71**, 57–64.
Holzapfel, W. H., Schillinger, U. (1992) The genus *Leuconostoc*, in: *The Procaryotes*, 2nd Ed., (Ballows, A., Truper, H. G., Dworkin, M., Harder, W., Schleifer, K.-H., Eds.), New York: Springer-Verlag, 1508–1534, Vol. 2.
Honda, O., Kato, C., Kuramitsu, H. K. (1990) Nucleotide sequence of the *Streptococcus mutans gtfD* gene encoding the glucosyltransferase-S enzyme, *J. Gen. Microbiol.* **136**, 2099–2105.
Janecek, S., Svensson, B, Russell, R. R. B. (2000) Location of repeat elements in glucansucrases of *Leuconostoc* and *Streptococcus* species, *FEMS Microbiol. Lett.* **192**, 53–57.
Jeanes, A. (1965a) Dextrans. Preparation of a water soluble dextran by enzymic synthesis, in: *Methods in Carbohydrate Chemistry* (Whistler, R. L., BeMiller, J. N., Eds.), New York: Academic Press, 127–132, Vol. 5.
Jeanes, A. (1965b) Dextrans. Preparation of dextrans from growing *Leuconostoc* cultures, in: *Methods in Carbohydrate Chemistry* (Whistler, R. L., BeMiller, J. N., Eds.), New York: Academic Press, 118–126, Vol. 5.
Jeanes, A. (1966) Dextran, in: *Encyclopedia of Polymer Science and Engineering*, (Mark, H. F.; Bikales, N. M.; Overberger, C. G.; Menges, G.; Kroschwitz, J. I., Eds.), New York: John Wiley & Sons, 752–767, Vol. 4.
Jeanes, A. (1977) Dextrans and pullulans: industrially significant α-D-glucans, in: *ACS Symp. Series No. 45, Extracellular Microbial Polysaccharides* (Sandford, P. A., Laskin, A., Eds.), Washington, D. C.: American Chemical Society, 284–298.
Jeanes, A. (1978) *Dextran Bibliography.* Washington, D. C.: U. S. Dept. Agriculture.

Jeanes, A. (1986) Immunochemical and related interactions with dextrans reviewed in terms of improved structural information. *Mol. Immun.* **23**, 999–1028.

Jeanes, A., Seymour, F. R. (1979) The α-D-glucopyranosidic linkages of dextrans: comparison of percentages from structural analysis by periodate oxidation and by methylation, *Carbohydr. Res.* **74**, 31–40.

Jeanes, A., Haynes, W. C., Wilham, C. A., Rankin, J. C., Melvin, E. H., Austin, M. J., Cluskey, J. E., Fisher, B. E., Tsuchiya, H. M., Rist, C. E. (1954) Characterization and classification of dextrans from ninety-six strains of bacteria, *J. Amer. Chem. Soc.* **76**, 5041–5052.

Jung, H-K., Kim, K-N., Lee, H-S., Jung, S-H. (1999) Production of alternan by *Leuconostoc mesenteroides* CBI-110, *Kor. J. Appl. Microbiol. Biotechnol.* **27**, 35–40.

Kabat, E. A., Bezer, A. E. (1958) The effect of variation in molecular weight on the antigenicity of dextran in man, *Arch. Biochem. Biophys.* **78**, 306–318.

Kaboli, H., Reilly, P. J. (1980) Immobilization and properties of *Leuconostoc mesenteroides* dextransucrase, *Biotechnol. Bioeng.* **22**, 1055–1069.

Killey, M., Dimler, R. J., Cluskey, J. E. (1955) Preparation of panose by the action of NRRL B-512 dextransucrase on a sucrose-maltose mixture, *J. Amer. Chem. Soc.* **77**, 3315–3318.

Kim, D., Day, D. F. (1995) Isolation of a dextranase constitutive mutant of *Lipomyces starkeyi* and its use for the production of clinical size dextran, *Lett. Appl. Microbiol.* **20**, 268–270.

Kim, D., Kim, D-W. (1999) Facile purification and characterization of dextransucrase from *Leuconostoc mesenteroides* B-512FMCM, *J. Microbiol. Biotechnol.* **9**, 219–222.

Kim, D., Robyt, J. F. (1994) Production and selection of mutants of *Leuconostoc mesenteroides* constitutive for glucansucrases, *Enzyme Microb. Technol.* **16**, 659–664.

Kim, D., Robyt, J. F. (1995a) Dextransucrase constitutive mutants of *Leuconostoc mesenteroides* B-1299, *Enzyme Microb. Technol.* **17**, 1050–1056.

Kim, D., Robyt, J. F. (1995b) Production, selection, and characteristics of mutants of *Leuconostoc mesenteroides* B-742 constitutive for dextransucrases, *Enzyme Microb. Technol.* **17**, 689–695.

Kim, D., Robyt, J. F. (1996) Properties and uses of dextransucrases elaborated by a new class of *Leuconostoc mesenteroides* mutants, *Prog. Biotechnol.* **12**, 125–144.

Kim, D., Kim, D-W., Lee, J-H., Park, K-H., Day, L. M., Day, D. F. (1997) Development of constitutive dextransucrase hyper-producing mutants of *Leuconostoc mesenteroides* using the synchrotron radiation in the 70–1000 eV region, *Biotechnol. Tech.* **11**, 319–321.

Kim, H., Kim, D., Ryu, W-H., Robyt, J. F. (2000) Cloning and sequencing of the α-1 → 6 dextransucrase gene from *Leuconostoc mesenteroides* B-742CB, *J. Microbiol. Biotechnol.* **10**, 559–563.

Kitaoka, M., Robyt, J. F. (1998a) Large-scale preparation of highly purified dextransucrase from a high-producing constitutive mutant of *Leuconostoc mesenteroides* B-512FMC, *Enzyme Microb. Technol.* **23**, 386–391.

Kitaoka, M., Robyt, J. F. (1998b) Use of a microtiter plate screening method for obtaining *Leuconostoc mesenteroides* mutants constitutive for glucansucrase, *Enzyme Microb. Technol.* **22**, 527–531.

Kobayashi, M., Matsuda, K. (1975) Purification and characterization of two activities of the intracellular dextransucrase from *Leuconostoc mesenteroides* NRRL B-1299, *Biochim. Biophys. Acta* **397**, 69–79.

Kobayashi, M., Matsuda, K. (1976) Purification and properties of the extracellular dextransucrase from *Leuconostoc mesenteroides* NRRL B-1299, *J. Biochem.* **79**, 1301–1308.

Kobayashi, M., Matsuda, K. (1977) Structural characteristics of dextrans synthesized by dextransucrases from *Leuconostoc mesenteroides* NRRL B-1299, *Agric. Biol. Chem.* **41**, 1931–1937.

Kobayashi, M., Matsuda, K. (1980) Characterization of the multiple forms and main component of dextransucrase from *Leuconostoc mesenteroides* NRRL B-512F, *Biochim. Biophys. Acta* **614**, 46–62.

Koenig, D. W., Day, D. F. (1988) Production of dextranase by *Lipomyces starkeyi*, *Biotechnol. Lett.* **10**, 117–122.

Koepsell, H. J., Hellman, N. N., Tsuchiya, H. M. (1955) Modification of dextran synthesis by means of alternate glucosyl acceptors. U.S. Patent 2,726,190.

Kohmoto, T., Fukui, F., Takaku, H., Mitsuoka, T. (1991) Dose-response test of isomaltooligosaccharides for increasing fecal bifidobacteria, *Agric. Biol. Chem.* **55**, 2157–2159.

Kooi, E. R. (1958) Production of dextran-dextrinase. U.S. Patent 2,833,695.

Kossman, J., Welsh, T., Quanz, M., Knuth, K. (2000) Nucleic acid molecules encoding alternansucrase. PCT Patent WO00/47727.

Kuge, T., Kobayashi, K., Kitamura, S., Tanahashi, H. (1987) Degrees of long-chain branching in dextrans, *Carbohydr. Res.* **160**, 205–214.

Lamothe, J.-P., Marchenay, Y. G., Monsan, P. F., Paul, F. M. B., Pelenc, V. (1996) Cosmetic compositions containing oligosaccharides. U.S. Patent 5,518,733.

Larm, O., Lindberg, B., Svensson, S. (1971) Studies on the length of the side chains of the dextran elaborated by *Leuconostoc mesenteroides* NRRL B-512, *Carbohydr. Res.* **20**, 39–48.

Lawford, G. R., Kligerman, A., Williams, T. (1979) Dextran biosynthesis and dextransucrase production by continuous culture of *Leuconostoc mesenteroides*, *Biotechnol. Bioeng.* **21**, 1121–1131.

Leathers, T. D., Hayman, G. T., Cote, G. L. (1995) Rapid screening of *Leuconostoc mesenteroides* mutants for elevated proportions of alternan to dextran, *Curr. Microbiol.* **31**, 19–22.

Leathers, T. D., Hayman, G. T., Cote, G. L. (1997) Microorganism strains that produce a high proportion of alternan to dextran. U.S. Patent 5,702,942.

Leathers, T. D., Hayman, G. T., Cote, G. L. (1998) Rapid screening method to select microorganism strains that produce a high proportion of alternan to dextran. U.S. Patent 5,789,209.

Lee, J. M., Fox, P. F. (1985) Purification and characterization of *Paecilomyces lilacinus* dextranase, *Enzyme Microb. Technol.* **7**, 573–577.

Lilley, L. L., Aucker, R. S. (1999) Fluids and electrolytes, in: *Pharmacology and the Nursing Process.* St. Louis, MO: Mosby, Inc., 335–348.

Lindberg, B., Svensson, S. (1968) Structural studies on dextran from *Leuconostoc mesenteroides* NRRL B-512, *Acta Chem. Scand.* **22**, 1907–1912.

Loesche, W. J. (1986) Role of *Streptococcus mutans* in human dental decay, *Microbiol. Rev.* **50**, 353–380.

London, E., Twigg, G. D. (1958) Therapeutic preparation of iron. U.S. Patent 2,820, 740.

McCabe, M. M. (1985) Purification and characterization of a primer-independent glucosyltransferase from *Streptococcus mutans* 6715-13 mutant 27, *Infect. Immun.* **50**, 771–777.

Miller, A. W., Robyt, J. F. (1984) Stabilization of dextransucrase from *Leuconostoc mesenteroides* NRRL B-512F by nonionic detergents, poly(ethylene glycol) and high-molecular-weight dextran, *Biochim. Biophys. Acta* **785**, 89–96.

Miller, A. W., Eklund, S. H., Robyt, J. F. (1986) Milligram to gram scale purification and characterization of dextransucrase from *Leuconostoc mesenteroides* NRRL B-512F, *Carbohydr. Res.* **147**, 119–133.

Misaki, A., Torii, M., Sawai, T., Goldstein, I. J. (1980) Structure of the dextran of *Leuconostoc mesenteroides* B-1355, *Carbohydr. Res.* **84**, 273–285.

Mitsuishi, Y., Kobayashi, M., Matsuda, K. (1979) Dextran α-1, 2 debranching enzyme from *Flavobacterium* sp. M-73: its production and purification, *Agric. Biol. Chem.* **43**, 2283–2290.

Mitsuya, H., Looney, D. J., Kuno, S., Ueno, R., Wong-Staal, F., Broder, S. (1988) Dextran sulfate suppression of viruses in the HIV family: inhibition of virion binding to $CD4^+$ cells, *Science* **240**, 646–649.

Monchois, V., Willemot, R.-M., Remaud-Simeon, M., Croux, C., Monsan, P. (1996) Cloning and sequencing of a gene coding for a novel dextransucrase from *Leuconostoc mesenteroides* NRRL B-1299 synthesizing only a α(1-6) and α(1-3) linkages, *Gene* **182**, 23–32.

Monchois, V., Remaud-Simeon, M., Russell, R. R. B., Monsan, P., Willemot, R.-M. (1997) Characterization of *Leuconostoc mesenteroides* NRRL B512F dextransucrase (DSRS) and identification of amino-acid residues playing a key role in enzyme activity, *Appl. Microbiol. Biotechnol.* **48**, 465–472.

Monchois, V., Remaud-Simeon, M., Monsan, P., Willemot, R.-M. (1998) Cloning and sequencing of a gene coding for an extracellular dextransucrase (DSRB) from *Leuconostoc mesenteroides* NRRL B-1299 synthesizing only α(1-6) glucan, *FEMS Microbiol. Lett.* **159**, 307–315.

Monchois, V., Willemot, R.-M., Monsan, P. (1999) Glucansucrases: mechanism of action and structure-function relationships, *FEMS Microbiol. Rev.* **23**, 131–151.

Monchois, V., Vignon, M., Russell, R. R. B. (2000) Mutagenesis of asp-569 of glucosyltransferase I glucansucrase modulates glucan and oligosaccharide synthesis, *Appl. Environ. Microbiol.* **66**, 1923–1927.

Monsan, P., Paul, F. (1995) Enzymatic synthesis of oligosaccharides, *FEMS Microbiol. Rev.* **16**, 187–192.

Monsan, P., Paul, F., Auriol, D., Lopez, A. (1987) Dextran synthesis using immobilized *Leuconostoc mesenteroides* dextransucrase, in: *Methods in Enzymology: Immobilized Enzymes and Cells* (Mosbach, K., Ed.), Orlando, FL: Academic Press, Inc, 239–254, Vol. 136.

Monsan, P., Potocki de Montalk, G., Sarcabal, P., Remaud-Simeon, M., Willemont, R.-M. (2000) Glucansucrases: efficient tools for the synthesis of oligosaccharides of nutritional interest, in: *Food Biotechnology* (Bielecki, S., Tramper, J., Polak, J., Eds.), Amsterdam: Elsevier Science B. V., 115–122.

Morii, E., Iwata, K., Kokkoku, H. (1964) Sodium and potassium salts of the dextran sulphate acid ester having substantially no anticoagulant activity but having lipolytic activity and the method of preparation thereof. U.S. Patent 3,141,014.

Mountzouris, K. C., Gilmour, S. G., Jay, A. J., Rastall, R. A. (1999) A study of dextran production from maltodextrin by cell suspensions of *Gluconobacter oxydans* NCIB 4943, *J. Appl. Microbiol.* **87**, 546–556.

Murphy, P.T., Whistler, R. L. (1973) Dextrans, in *Industrial Gums*, Second Ed. (Whistler, R. L., BeMiller, J. N., Eds.), New York: Academic Press, 513–542.

Neely, W. B. (1960) Dextran: structure and synthesis, in: *Adv. Carbohydr. Chem.* (Wolfrom, M. L., Tipson, R. S., Eds.), New York: Academic Press, 341–369, Vol. 15.

Nichols, S. E. (2000a) Plant cells and plants transformed with *Streptococcus mutans* gene encoding glucosyltransferase C enzyme. U.S. Patent 6,127,603.

Nichols, S. E. (2000b) Plant cells and plants transformed with *Streptococcus mutans* genes encoding wild-type or mutant glucosyltransferase B enzymes. U.S. Patent 6,087,559.

Nichols, S. E. (2000c) Plant cells and plants transformed with *Streptococcus mutans* genes encoding wild-type or mutant glucosyltransferase D enzymes. U.S. Patent 6,127,602.

Novak, L. J., Stoycos, G. S. (1958) Method for producing clinical dextran. U.S. Patent 2,841,578.

Okada, G., Takayanagi, T., Sawai, T. (1988) Improved purification and further characterization of an isomaltodextranase from *Arthrobacter globiformis* T6, *Agric. Biol. Chem.* **52**, 495–501.

Pasteur, L. (1861) On the viscous fermentation and the butyrous fermentation, *Bull. Soc. Chim. Paris*, 30–31.

Paul, F., Auriol, D., Oriol, E. Monsan, P. (1984) Production and purification of dextransucrase from *Leuconostoc mesenteroides* NRRL B-512(F). *Ann. N. Y. Acad. Sci.* **434**, 267–270.

Pearce, B. J., Walker, G. J., Slodki, M. E., Schuerch, C. (1990) Enzymic and methylation analysis of dextrans and (1-3)-α-D-glucans, *Carbohydr. Res.* **203**, 229–246.

Pelenc, V., Lopez-Munguia, A., Remaud, M., Biton, J., Michel, J. M., Paul, F., Monsan, P. (1991) Enzymatic synthesis of oligoalternans, *Sci. Aliments* **11**, 465–476.

Pilon-Smits, E. A. H., Ebskamp, M. J. M., Jeuken, M. J. W., van der Meer, I. M., Visser, R. G. F., Weisbeek, P. J., Smeekens, S. C. M. (1996) Microbial fructan production in transgenic potato plants and tubers, *Ind. Crops Products* **5**, 35–46.

Piret, J., Lamontagne, J., Bestman-Smith, J., Roy, S., Gourde, P., Desormeaux, A., Omar, R. F., Juhasz, J., Bergeron, M. G. (2000) *In vitro* and *in vivo* evaluations of sodium lauryl sulfate and dextran sulfate as microbicides against herpes simplex and human immunodeficiency viruses, *J. Clin. Microbiol.* **38**, 110–119.

Quirasco, M., Lopez-Munguia, A., Remaud-Simeon, M., Monsan, P., Farres, A. (1999) Induction and transcriptional studies of the dextransucrase gene in *Leuconostoc mesenteroides* NRRL B-512F, *Appl. Environ. Microbiol.* **65**, 5504–5509.

Rankin, J. C., Jeanes, A. (1954) Evaluation of periodate oxidation method for structural analysis of dextrans, *J. Am. Chem. Soc.* **76**, 4435–4441.

Reh, K.-D., Noll-Borchers, M., Buchholz, K. (1996) Productivity of immobilized dextransucrase for leucrose formation, *Enzyme Microb. Technol.* **19**, 518–524.

Remaud, M., Paul, F., Monsan, P., Heyraud, A., Rinaudo, M. (1991) Molecular weight characterization and structural properties of controlled molecular weight dextrans synthesized by acceptor reaction using highly purified dextransucrase, *J. Carbohydr. Chem.* **10**, 861–876.

Remaud, M., Paul, F., Monsan, P. (1992) Characterization of α-(1 → 3) branched oligosaccharides synthesized by acceptor reaction with the extracellular glucosyltransferases from *L. mesenteroides* NRRL B-742, *J. Carbohydr. Chem.* **11**, 359–378.

Remaud-Simeon, M., Lopez-Munguia, A., Pelenc, V., Paul, F., Monsan, P. (1994) Production and use of glucosyltransferases from *Leuconostoc mesenteroides* NRRL B-1299 for the synthesis of oligosaccharides containing α-(1 → 2) linkages, *Appl. Biochem. Biotechnol.* **44**, 101–117.

Remaud-Simeon, M., Willemot, R.-M., Sarcabal, P., Potocki de Montalk, G., Monsan, P. (2000) Glucansucrases: molecular engineering and oligosaccharide synthesis, *J. Mol. Catalysis B: Enzymatic* **10**, 117–128.

Robyt, J. F. (1986) Dextran, in: *Encyclopedia of Polymer Science and Engineering*, (Mark, H. F., Gaylord, N. G., Bikales, N. M., Eds.), New York: John Wiley & Sons, 752–767, Vol. 4.

Robyt, J. F. (1992) Structure, biosynthesis, and uses of nonstarch polysaccharides: dextran, alternan, pullulan, and algin, in: *Developments in Biochemistry and Biophysics* (Alexander, R. J., Zobel, H. F., Eds.), St. Paul: Amer. Assoc. Cereal Chemists, 261–292.

Robyt, J. F. (1995) Mechanisms in the glucansucrase synthesis of polysaccharides and oligosaccharides from sucrose, in: *Adv. Carbohydr. Chem. Biochem.* (Horton, D., Ed.), San Diego: Academic Press, 133–168, Vol. 51.

Robyt, J. F., Eklund, S. H. (1983) Relative, quantitative effects of acceptors in the reaction of *Leuconostoc mesenteroides* B-512F dextransucrase, *Carbohydr. Res.* **121**, 279–286.

Robyt, J. F., Martin, P. J. (1983) Mechanism of synthesis of D-glucans by D-glucosyltransferases from *Streptococcus mutans* 6715, *Carbohydr. Res.* **113**, 301–315.

Robyt, J. F., Taniguchi, H. (1976) The mechanism of dextransucrase action. Biosynthesis of branch linkages by acceptor reactions with dextran, *Arch. Biochem. Biophys.* **174**, 129–135.

Robyt, J. F., Walseth, T. F. (1978) The mechanism of acceptor reactions of *Leuconostoc mesenteroides* B-512F dextransucrase, *Carbohydr. Res.* **61**, 433–445.

Robyt, J. F., Walseth, T. F. (1979) Production, purification and properties of dextransucrase from *Leuconostoc mesenteroides* NRRL B-512F, *Carbohydr. Res.* **68**, 95–111.

Robyt, J. F., Kimble, B. K., Walseth, T. F. (1974) The mechanism of dextransucrase action. Direction of dextran biosynthesis, *Arch. Biochem. Biophys.* **165**, 634–640.

Ryu, H-J., Kim, D., Kim, D-W., Moon, Y-Y., Robyt, J. F. (2000) Cloning of a dextransucrase gene (*fmcmds*) from a constitutive dextransucrase hyper-producing *Leuconostoc mesenteroides* B-512FMCM developed using VUV, *Biotechnol. Lett.* **22**, 421–425.

Sabatie, J., Choplin, L., Moan, M., Doublier, J. L., Paul, F., Monsan, P. (1988) The effect of synthesis temperature on the structure of dextran NRRL B 512F, *Carbohydr. Polymers* **9**, 87–101.

Sanchez-Gonzalez, M., Alagon, A., Rodriguez-Sotres, R., Lopez-Munguia, A. (1999) Proteolytic processing of dextransucrase of *Leuconostoc mesenteroides, FEMS Microbiol. Lett.* **181**, 25–30.

Sawai, T., Tohyama, T., Natsume, T. (1978) Hydrolysis of fourteen native dextrans by *Arthrobacter* isomaltodextranase and correlation with dextran structure, *Carbohydr. Res.* **66**, 195–205.

Sawai, T., Ohara, S., Ichimi, Y., Okaji, S., Hisada, K., Fukaya, N. (1981) Purification and some properties of the isomaltodextranase of *Actinomadura* strain R10 and comparison with that of *Arthrobacter globiformis* T6, *Carbohydr. Res.* **89**, 289–299.

Scheibler, C. (1874) Investigation on the nature of the gelatinous excretion (so-called frog's spawn) which is observed in production of beet-sugar juices, *Z. Ver. Dtsch. Zucker-Ind.* **24**, 309–335.

Sery, T. W., Hehre, E. J. (1956) Degradation of dextrans by enzymes of intestinal bacteria, *J. Bacteriol.* **71**, 373–380.

Sevenier, R., Hall, R. D., van der Meer, I. M., Hakkert, H. J. C., van Tunen, A. J., Koops, A. J. (1998) High level fructan accumulation in a transgenic sugar beet, *Nature Biotechnol.* **16**, 843–846.

Seymour, F. R., Julian, R. L. (1979) Fourier-transform, infrared difference-spectrometry for structural analysis of dextrans, *Carbohydr. Res.* **74**, 63–75.

Seymour, F. R., Knapp, R. D. (1980) Unusual dextrans: 13. Structural analysis of dextrans from strains of *Leuconostoc* and related genera, that contain 3-O-α-D-glucosylated α-D-glucopyranosyl residues at the branch points, or in consecutive linear position, *Carbohyd. Res.* **81**, 105–129.

Seymour, F. R., Knapp, R. D., Bishop, S. H. (1976) Determination of the structure of dextran by ^{13}C-nuclear magnetic resonance spectroscopy, *Carbohydr. Res.* **51**, 179–194.

Seymour, F. R., Slodki, M. E., Plattner, R. D., Jeanes, A. (1977) Six unusual dextrans: methylation structural analysis by combined G. L. C.-M. S. of per-O-acetylaldononitriles, *Carbohydr. Res.* **53**, 153–166.

Seymour, F. R., Chen, E. C. M., Bishop, S. H. (1979a) Methylation structural analysis of unusual dextrans by combined gas-liquid chromatography-mass spectrometry, *Carbohydr. Res.* **68**, 113–121.

Seymour, F. R., Knapp, R. D., Bishop, S. H. (1979b) Correlation of the structure of dextrans to their ^{1}H-NMR spectra, *Carbohydr. Res.* **74**, 77–92.

Sharpe, E. S., Stodola, F. H., Koepsell, H. J. (1960) Formation of isomaltulose in enzymatic dextran synthesis, *J. Org. Chem.* **25**, 1062–1063.

Shimamura, A., Tsumori, H., Mukasa, H. (1982) Purification and properties of *Streptococcus mutans* extracellular glucosyltransferase, *Biochim. Biophys. Acta* **702**, 72–80.

Shimizu, E., Unno, T., Ohba, M., Okada, G. (1998) Purification and characterization of an isomaltotriose-producing *endo*-dextranase from a *Fusarium* sp., *Biosci. Biotechnol. Biochem.* **62**, 117–122.

Shiroza, T., Ueda, S., Kuramitsu, H. K. (1987) Sequence analysis of the *gftB* gene from *Streptococcus mutans, J. Bacteriol.* **169**, 4263–4270.

Sidebotham, R. L. (1974) Dextrans, in: *Adv. Carbohydr. Chem. Biochem.* (Tipson, R. S., Horton, D., Eds.), New York: Academic Press, 371–444, Vol. 30.

Simonson, L. G., Liberta, A. E. (1975) New sources of fungal dextranase. *Mycologia* **4**, 845–851.

Sinha, J., Dey, P. K., Panda, T. (2000) Aqueous two-phase: the system of choice for extractive fermentations, *Appl. Microbiol. Biotechnol.* **54**, 476–486.

Slodki, M. E., England, R. E., Plattner, R. D., Dick Jr., W. E. (1986) Methylation analyses of NRRL dextrans by capillary gas-liquid chromatography, *Carbohyd. Res.* **156**,199–206.

Smith, M. R., Zahnley, J., Goodman, N. (1994) Glucosyltransferase mutants of *Leuconostoc mesenteroides* NRRL B-1355, *Appl. Environ. Microbiol.* **60**, 2723–2731.

Smith, M. R., Zahnley, J. C., Wong, R. Y., Lundin, R. E., Ahlgren, J. A. (1998) A mutant strain of *Leuconostoc mesenteroides* B-1355 producing a glucosyltransferase synthesizing α(1 → 2) glucosidic linkages, *J. Ind. Microbiol. Biotechnol.* **21**, 37–45.

Stiles, M. E., Holzapfel, W. H. (1997) Lactic acid bacteria of foods and their current taxonomy, *Int. J. Food Microbiol.* **36**, 1–29.

Stodola, F. H., Sharpe, E. S., Koepsell, H. J. (1956) The preparation, properties and structure of the disaccharide leucrose, *J. Amer. Chem. Soc.* **78**, 2514–2518.

Su, D., Robyt, J. F. (1994) Determination of the number of sucrose and acceptor binding sites for *Leuconostoc mesenteroides* B-512FM dextransucrase, and the confirmation of the two-site mechanism for dextran synthesis, *Arch. Biochem. Biophys.* **308**, 471–476.

Sun, J., Cheng, X., Zhang, Y., Yan, Z., Zhang, S. (1988) A strain of *Paecilomyces lilacinus* producing high quality dextranase, *Ann. N. Y. Acad. Sci.* **542**, 192–194.

Suzuki, M., Unno, T., Okada, G. (1999) Simple purification and characterization of an extracellular dextrin dextranase from *Acetobacter capsulatum* ATTC 11894, *J. Appl. Glycosci.* **46**, 469–473.

Swengers, D. (1991) Leucrose, a ketodisaccharide of industrial design, in: *Carbohydrates as Organic Raw Materials* (Lichtenthaler, F. W., Ed.), Weinheim, Germany: VCH, 183–195.

Takahashi, M., Okada, S., Uchimura, T., Kozaki, M. (1992) *Leuconostoc amelibiosum* Schillinger, Holzapfel, and Kandler 1989 is a later subjective synonym of *Leuconostoc citreum* Farrow, Facklam, and Collins 1989, *Int. J. Syst. Bacteriol.* **42**, 649–651.

Takazoe, I. (1989) Palatinose - an isomeric alternative to sucrose, in: *Progress in Sweeteners* (Grenby, T. H., Ed.), New York: Elsevier Science, 143–167.

Tarr, H. L. A., Hibbert, H. (1931) Studies on reactions relating to carbohydrates and polysaccharides. XXXVII. The formation of dextran by *Leuconostoc mesenteroides, Can. J. Res.* **5**, 414–427.

Taylor, C., Cheetham, N. W. H., Walker, G. J. (1985) Application of high-performance liquid chromatography to a study of branching dextrans, *Carbohydr. Res.* **137**, 1–12.

Tjerneld, F. (1992) Aqueous two-phase partitioning on an industrial scale, in: *Poly(Ethylene Glycol) Chemistry: Biotechnical and Biomedical Applications* (Harris, J. M., Ed.), New York: Plenum Press, 85–102.

Torii, M., Sakakibara, K., Misaki, A., Sawai, T. (1976) Degradation of alpha-linked D-gluco-oligosaccharides and dextrans by an isomalto-dextranase preparation from *Arthrobacter globiformis* T6, *Biochem. Biophys. Res. Comm.* **70**, 459–464.

Tsuchiya, H. M., Koepsell, H. J. (1954) Production of dextransucrase. U.S. Patent 2,686,147.

Tsuchiya, H. M., Hellman, N. N., Koepsell, H. J., Corman, J., Stringer, C. S., Rogovin, S. P., Bogard, M. O., Bryant, G., Feger, V. H., Hoffman, C. A., Senti, F. R., Jackson, R. W. (1955) Factors affecting molecular weight of enzymatically synthesized dextran, *J. Amer. Chem. Soc.* **77**, 2412–2419.

Tsuchiya, H. M., Jeanes, A., Bricker, H. M., Wilham, C. A. (1956) Production of dextranase. U.S. Patent 2,742,399.

Ueda, S., Shiroza, R., Kuramitsu, H. K. (1988) Sequence analysis of the *gtfC* gene from *Streptococcus mutans* GS-5, *Gene* **69**, 101–109.

Usher, T. C. (1989) Method for the manufacture of dextran sulfate and salts thereof. U.S. Patent 4,855,416.

Van Cleve, J. W., Schaefer, W. C., Rist, C. E. (1956) The structure of NRRL B-512 dextran. Methylation studies, *J. Amer. Chem. Soc.* **78**, 4435–4438.

Vandamme, E. J., Bruggeman, G., De Baets, S., Vanhooren, P. T. (1996) Useful polymers of microbial origin, *Agro-Food-Industry Hi-Tech* **Sept./Oct.**, 21–25.

Van Tieghem, P. (1878) On sugar-mill gum, *Ann. Sci. Nat. Bot. Biol. Veg.* **7**, 180–203.

Walker, G. J. (1978) Dextrans, in: *International Review of Biochemistry. Biochemistry of Carbohydrates II,* (Manners, D. J., Ed.), Baltimore, MD: University Park Press, 75–125, Vol. 16.

Webb, E., Spencer-Martins, I. (1983) Extracellular endodextranase from the yeast *Lipomyces starkeyi*, *Can. J. Microbiol.* **29**, 1092–1095.

Wilke-Douglas, M., Perchorowicz, J. T., Houck C. M., Thomas, B. R. (1989) Methods and compositions for altering physical characteristics of fruit and fruit products. PCT Patent WO89/12386.

Wolff, I. A., Mellies, R. L., Rist, C. E. (1955) Fractionation of dextran products. U.S. Patent 2,712,007.

Yalpani, M. (1986) Preparation and applications of dextran-derived products in biotechnology and related areas, *CRC Crit. Rev. Biotechnol.* **3**, 375–421.

Yamamoto, K., Yoshikawa, K., Kitahata, S., Okada, S. (1992) Purification and some properties of dextrin dextranase from *Acetobacter capsulatus* ATTC 11894, *Biosci. Biotechnol. Biochem.* **56**, 169–173.

Yamamoto, K., Yoshikawa, K., Okada, S. (1993a) Dextran synthesis from reduced maltooligosaccharides by dextrin dextranase from *Acetobacter capsulatus* ATTC 11894, *Biosci. Biotechnol. Biochem.* **57**, 136–137.

Yamamoto, K., Yoshikawa, K., Okada, S. (1993b) Effective dextran production from starch by dextrin dextranase with debranching enzyme, *J. Ferment. Bioeng.* **76**, 411–413.

21
Levan

Dr. Sang-Ki Rhee[1], Dr. Ki-Bang Song[2], Dr. Chul-Ho Kim[3], Dr. Buem-Seek Park[4], Ms. Eun-Kyung Jang[5], Dr. Ki-Hyo Jang[6]

[1] Biomolecular Engineering Laboratory, Korea Research Institute of Bioscience and Biotechnology (KRIBB), 52 Eoeun-dong, Yuseong, Daejeon 305-333, Korea; Tel.: +82-42-860-4450; Fax: +82-42-860-4594; E-mail: rheesk@mail.kribb.re.kr

[2] Biomolecular Engineering Laboratory, Korea Research Institute of Bioscience and Biotechnology (KRIBB), 52 Eoeun-dong, Yuseong, Daejeon 305-333, Korea; Tel.: +82-42-860-4457; Fax: +82-42-860-4594; E-mail: songkb@mail.kribb.re.kr

[3] Biomolecular Engineering Laboratory, Korea Research Institute of Bioscience and Biotechnology (KRIBB), RealBioTech Co., Ltd., #202 Bioventure Center, KRIBB, 52 Eoeun-dong, Yuseong, Daejeon 305-333, Korea; Tel.: +82-42-860-4452; Fax: +82-42-860-4594; E-mail: kim3641@mail.kribb.re.kr

[4] Biomolecular Engineering Laboratory, Korea Research Institute of Bioscience and Biotechnology (KRIBB), 52 Eoeun-dong, Yuseong, Daejeon 305-333, Korea; Tel.: +82-42-860-4454; Fax: +82-42-860-4594; E-mail: buemseekpk@mail.kribb.re.kr

[5] RealBioTech Co., Ltd., #202 Bioventure Center, KRIBB, 52 Eoeun-dong, Yuseong, Daejeon 305-333, Korea; Tel.: +82-42-863-4381; Fax: +82-42-863-4382; E-mail: levanis@realbio.com

[6] Department of Medical Nutrition, Graduate School of East-West Medical Science, Kyung Hee University, Suwon 449-701, Korea; Tel.: +82-2-961-0506; Fax: +82-2-961-9215; E-mail: kihyojang@hotmail.com

Biotechnology of Biopolymers. From Synthesis to Patents. Edited by A. Steinbüchel and Y. Doi

ISBN: 3-527-31110-6

1-kestose	*O*-β-D-fructofuranosyl-(2 → 1)-β-D-fructofuranosyl-(2 → 1)-β-D-glucopyranoside
1-SST	sucrose:sucrose 1-fructosyltransferase
6-SFT	sucrose:fructan 6-fructosyltransferase
DFA	di-β-D-fructofuranose dianhydride
DP	degree of polymerization
EPS	exopolysaccharides
FFT	fructan:fructan fructosyltransferase
FOS	fructo-oligosaccharides
HPr	histidine-containing phosphocarrier protein
LBT	levanbiosyl transfer
LFT	levanfructosyl transfer
LFTase	levan fructotransferase
PEG	polyethylene glycol
PTS	phosphoenolpyruvate-dependent carbohydrate
RBT	RealBioTech Co., Ltd.
TLC	thin-layer chromatography

1 Introduction

Fructan, one of the most highly distributed biopolymers in nature, is a homopolysaccharide composed of D-fructofuranosyl residues joined by β-(2,6) and β-(2,1) linkages. Two types of fructan, distinguishable by the type of linkage present, are inulin and levan. The term levan is used to describe the microbial polyfructan which consists of D-fructofuranosyl residues linked predominantly by β-(2,6) linkage as a main chain, but with some β-(2,1) branching points. The other polyfructan, inulin, is mainly isolated from natural vegetable sources and serves as a reserve carbohydrate in the Compositate and Gramineae (Vandamme and Derycke, 1983), although inulins from the microbial origin have also been reported in *Streptococcus mutans* and *Streptococcus sanguis*, the human pathogens involved in dental caries (Birkhed et al., 1979). The fructose homopolymer, levan, is found in plants and especially in bioproducts of microorganisms. Plant levans, graminans, and phleins have shorter residues (varying from 10 to ~200 fructose residues) than microbial levans, of which molecular weights are up to several million daltons, with multiple branches. Microbial levans are produced extracellularly from sucrose- and raffinose-based substrates by levansucrase (sucrose 6-fructosyltransferase, EC 2.4.1.10) from a wide range of taxa such as bacteria, yeasts, and fungi (Han, 1990; Hendry and Wallace, 1993). Microbial levans are produced mainly by bacteria such as *Bacillus subtilis, Zymomonas mobilis, Bacillus polymyxa, Aerobacter levanicum, Erwinia amylovora, Rhanella aquatilis* and *Pseudomonas.* The production and utilization of levan in the industrial field have been strictly limited until very recently, and very few reports have been made on the production of levan using fermentation techniques (Elisashvili, 1984; Beker et al., 1990; Han, 1990; Keith et al., 1991; Ohtsuka et al., 1992; Uchiyama, 1993). Recently, great interest in this fructan has been renewed to discover novel applications for levan as a new industrial gum in the fields of cosmetics, foods (e.g., as dietary fiber), and pharmaceuticals. In this chapter, we describe the production and degradation of levan by use of enzymatic reactions, the genetic regulation and control of such reactions, and outline the properties of levan and the current status of its industrial applications.

2
Historical Outline

Historically, levan was generally considered to be an undesirable byproduct of sugar and juice processing because it increases the viscosity of the processing liquor (Fuchs, 1959; Avigad, 1965). Levan was first described by Lippmann in Germany in 1881, when the name "laevulan" was proposed. Greig-Smith (1901) later showed that a strain of *Bacillus*, when grown on sucrose, produced fructans, and the name "levan" was then introduced as being analogous to dextran. The term laevulan now denotes partially degraded levan fractions. However, early reports on levan were confusing because the microbial nomenclature was unsystematic and the materials were inadequately described.

The biosynthesis of levan was elucidated some years later. The mechanism was shown to involve two enzymes, sucrose fructosyltransferase and fructan fructosyltransferase, and was proposed by Edelman and Jefford in 1968. The enzyme kinetics of the transfructosylation reaction was revealed by Chambert and Gonzy-Treboul in 1976. The enzyme which is now generally recognized as levansucrase was named by Hestrin et al. in 1943, and is responsible for the synthesis of levan from sucrose. The most extensive studies of levansucrase were performed in *B. subtilis*, and focused on the localization of the enzyme as well as its properties, expression regulation, genetic organization, and kinetics (Suzuki and Chatterton, 1993). As levan began to receive more attention based on its potential applications, many levan-producing microorganisms were identified (Han, 1990; Hendry and Wallace, 1993). The mass production of levan from *Z. mobilis*, and the secretion of levansucrase were reported relatively recently in detail by Song and coworkers (1996) and Ananthalakshmy and Gunasekaran (1999).

Although extensive research and searches for industrial applications have been conducted with dextran (which is also known as a bacterial biopolymer), much less attention has been focused on levan, mainly because of the very poor yields obtained in its industrial production. Nonetheless, great interest has been expressed in the diverse aspects of this fructose homopolymer over the past decade, despite the applications of levan having remained relatively few in number because of the limited supplies. Levans are now available commercially in reagent grade from microbial sources (e.g., from Sigma Chemical Co., IGI Biotechnology), but these are used only for research purposes. Since the time when levan was first produced on a large scale by using levansucrase from genetically engineered *Escherichia coli* (Song et al., 1996), attention has been renewed on the potential industrial application of levan and its derivatives in the fields of agriculture, cosmetics, food ingredients, animal feed and pharmaceuticals (Clarke et al., 1997; Kim et al., 1998; Vijn and Smeekens, 1999; Rhee et al., 2000d), as well as being a good source of pure fructose and di-β-D-fructofuranose dianhydride (DFA) (Saito and Tomita, 2000).

3
Chemical Structures of Levan

Fructans are chemically versatile molecules, and consist of a single glucose unit attached to two or more fructose units. Three fructan trisaccharides are known, each being produced through a glycosidic linkage of fructose to one of the three primary hydroxyl groups of sucrose. Fructose linked to the primary carbon of the fructose moiety of sucrose forms 1-kestose (also called isokestose), while fructose linked to the sixth carbon of the fructose moiety of sucrose forms 6-kestose (also called kestose). Both of these trisaccharides have a terminal glucose and a terminal fructose. Linkage of a fructose moiety to the sixth carbon of glucose moiety of sucrose forms neokestose, with both end groups being fructose (Nelson and Spollen, 1987).

Chemically, levan consists of β-D-fructofuranosyl residues linked predominantly through β-(2,6) as 6-kestose of the basic trisaccharide, with extensive branching through β-(2,1) linkages (Figure 1). In contrast, inulin is composed of β-D-fructofuranose attached by β-(2,1) linkages. The first monomer of the chain is either a β-D-glucopyranosyl or a β-D-fructofuranosyl residue. Although they are similar fructose homopolymers, it is evident that levan is different from inulin-type fructan since microbial inulin contains 5–7% of β-(2,6)-linked branches (Wolff et al., 2000).

The molecular shape of levan, as visualized by electron microscopy (Newbrun et al., 1971), is spheroidal, indicating that the constituent chains are extended radially at the same synthetic rate. The molecular

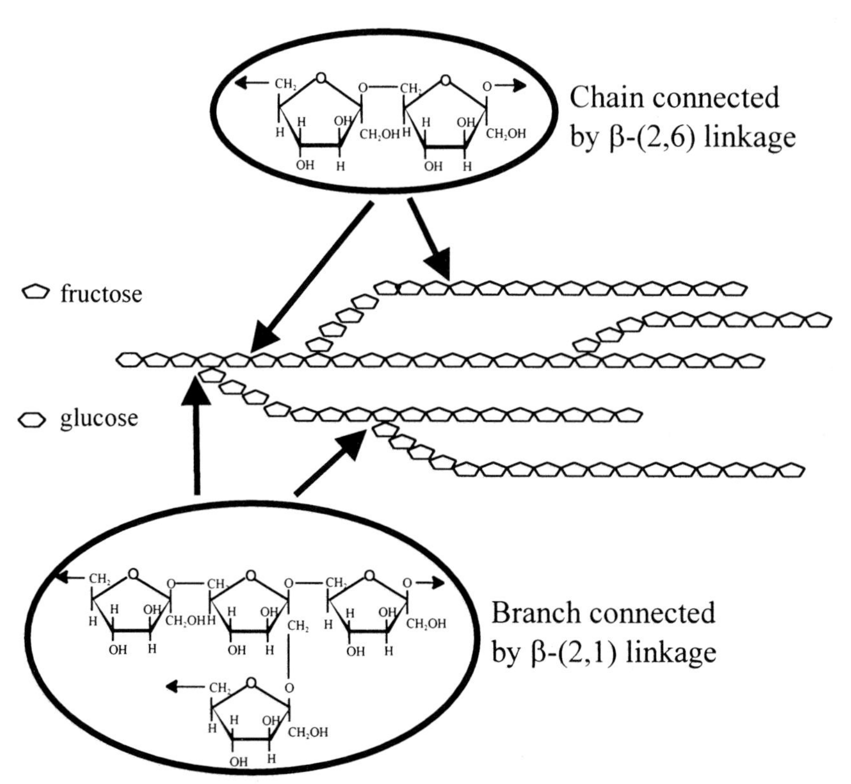

Fig. 1 Structure of levan. The main chain is connected by β-(2,6) linkages and the branch is connected to the main chain by a β-(2,1) linkage; the branch then continues with β-(2,6) linkages.

weight of bacterial levans is typically in the range of 2×10^6 to 10^8 (Keith et al., 1991), with the final molecular size being influenced by the synthesizing conditions such as ionic strength, temperature, and co-solutes. Although microbial levans have the similar structure, several types of (IX) levan are produced by different microorganisms, and this may be attributed to a varying degree of polymerization (DP) and branching of the repeating unit.

A cell-free enzyme system could be used to synthesize levans which have both β-(2,6) and β-(2,1) linked fructosyl units, and with similar structure to those of a whole-cell enzyme system. However, the structure of levans synthesized in the cell-free enzyme system is also known to differ in length compared with that synthesized by whole-cell systems (Han, 1990).

4 Occurrence

Various polysaccharides are produced as structural components in living organisms, and levan is one of the most diversely distributed components in plants, yeasts, fungi, and bacteria in particular. Levan is produced by grass (*Dactylis glomerata, Poa secunda* and *Agropyron cristatum*), wheat and barley (*Hordeum vulgare*), fungi (*Aspergillus sydawi* and *Aspergillus versicolor*) and yeasts (Han, 1990). Levans produced by microorganisms have been reported by Han (1990), and Hendry and Wallace (1993), and are listed in Table 1.

Previously, oral bacteria such as *Streptococcus, Rothis* and *Odontomyces* had received much attention due to their presence in human dental caries, together with soil microorganisms, especially *Bacillus*. Subsequently, focus was centered on the biological and functional aspects of levan rather than on its oral accumulation. The most extensive studies of levan were performed using *B. subtilis* (Suzuki and Chatterton, 1993). Furthermore, levans from *Bacillus polymyxa* (Aymerich, 1990) and *Pseudomonas* sp. (Hettwer et al., 1995, 1998) were identified as playing a role in the plant defense response. Levan from *B. subtilis* was shown to be tolerant against salt stress (Kunst and Rapoport, 1995), while that obtained from *Z. mobilis* exhibited antitumor activity (Calazans et al., 2000). The synthesis of levan using the genus *Lactobacillus* was also recently reported (Van Geel-Schutten et al., 1999).

Tab. 1 Levan-producing microorganisms

Microorganism	***Reference***
Acetobacter xylinum	Tajima et al. (1998)
Actinomyces naeslundii	Bergeron et al. (2000)
Bacillus circulans	Perez Oseguera et al. (1996)
Bacillus stearothermophilus	Li et al. (1997)
Gluconacetobacter (formerly *Acetobacter*) *diazotrophicus*	Arrieta et al. (1996)
Lactobacillus reuteri	Van Geel-Schutten et al. (1999)
Pseudomonas syringae pv. phaseolicola	Hettwer et al. (1995)
Pseudomonas syringae pv. glycinea	Hettwer et al. (1998)
Rahnella aquatilis	Ohtsuka et al. (1992); Song et al. (1998)
Serratia levanicum	Kojima et al. (1993)
Zymomonas mobilis	Song et al. (1993)

5 Physiological Functions of Levan

Bacterial polysaccharides are found either as a dense layer of more or less regularly arranged polymer structures attached to the bacterial cell walls (capsules) or as loosely associated exopolysaccharides (EPS) (Beveridge and Graham, 1991). Levans produced microbiologically have a number of interesting features. The levan which is synthesized extracellularly by bacteria may be visualized in a sucrose-containing medium, giving rise to a typical mucoid morphology. This type of mucoid feature provides a role in the symbiosis, phytopathogenesis, or participation in the defense mechanism against cold and dry conditions (Kunst and Rapoport, 1995). Extracellular levan produced by bacterial plant pathogens increases bacterial fitness and also acts as a detoxifying barrier against plant defence compounds (Hettwer et al., 1998). Among natural polysaccharides, glucans and fructans possess antitumor activity, and levan is included in this list. The antitumor activity of levan against sarcoma 180 depends on the molecular weight of the polysaccharides (Calazans et al., 2000), and this may indicate the polydiversity of levan.

Levans produced in plants are present as storage carbohydrates in the stem and leaf sheaths, and are degraded in a later stage of the growing season to provide plants with carbohydrates for grain filling (Pollock and Cairns, 1991). The biological role of polysaccharides in protection is less clearly understood, but a hypothesis has recently emerged. Levan penetrates into lipid membranes composed of monomolecular lipid layers, after which interactions occur which are orders of magnitude greater than the interaction between disaccharides and lipids. An extended layer of levan adheres to the lipids and partially protrudes into the aqueous phase. It is also possible that the membranes present are coated with levan; this coating of membranes imparts a reduction in accessibility of the membrane surface to proteins. In this way, in a biological system levan is able to protect membranes by interacting with the membrane lipid fraction (Vereyken et al., 2001).

6 Chemical Analysis and Detection

Several methods can be used in the qualitative and quantitative analysis of levan and the estimation of its concentration in solutions, with spectrophotometry and chromatography being the major techniques.

6.1 Spectrophotometry

Low concentrations of levan are measured by monitoring the optical density at 450–550 nm, as the presence of levan creates turbidity within the enzyme reaction mixture.

6.2 High-Performance Liquid Chromatography (HPLC)

The HPLC method is employed for both qualitative and quantitative determination of levan and other components (oligosaccharides, sucrose, fructose, and glucose). Details of the method are described here. The enzyme reaction mixture or levan solution is filtered using a 0.45 μm pore size membrane filter, and the filtrate is analyzed by HPLC equipped with a gel filtration column and refractive index detector (Shodex Ionpack KS-802, 300×8 mm; Showa Denko Co., Japan) (Jang et al., 2000). Deionized water is used as a mobile phase at a flow rate of

0.4 mL min^{-1}. The DP of levan is also determined by HPLC equipped with successive columns, GPC 4000–GPC 1000 (Polymer Laboratories, USA), and a refractive index detector (Jang et al., 2001). The analyses of sugar components and linkage type of levan are determined by using acid hydrolysis, methylation and nuclear magnetic resonance (NMR) shift experiments (Suzuki and Chatterton, 1993). In ^{13}C-NMR, signals of carbons of levan obtained from *Z. mobilis* and *Aerobacter levanicum* are identical, and show six main resonances at 104.9, 81.0, 76.9, 64.1, and 60.6 p.p.m. (Song and Rhee, 1994), these signals differing from those obtained with inulin.

6.3 Other Methods

Levan (nonmobiles) can be distinguished from oligosaccharides, sucrose, and other byproducts, either qualitatively or quantitatively, by the use of thin-layer chromatography (TLC). The sucrose-hydrolyzing activity of bacterial levansucrase is also used in the determination of levan concentration. The methods established are based on the fact that glucose is formed stoichiometrically in relation to the amount of fructose incorporated into levan (major product) and oligosaccharides (minor products). The amounts of glucose generated by the enzymatic reaction can be determined quantitatively by commercially available kits from the suppliers (Song et al., 1993).

7 Biosynthesis of Levan

The biosynthesis of levan requires the involvement of an extracellular enzyme levansucrase, which shows specificity for sucrose. Genetic characterization of the enzyme and the regulation of levan synthesis have been extensively studied, mostly using levansucrase genes from *B. subtilis* and *Z. mobilis*.

7.1 Enzymology of Levan Synthesis

Levansucrase (sucrose:2,6-β-D fructan:6-D-fructosyltransferase, sucrose 6-fructosyltransferase, EC 2.4.1.10.) was first named by Hestrin et al. (1943), and is responsible for the synthesis of levan from sucrose. Levansucrase exists as constituent intracellular and inducible extracellular forms in microorganisms (Han, 1990). The function of the levan-producing enzyme located intracellularly in some bacteria is not yet understood. The most abundant substrate for levansucrase in nature is sucrose, but raffinose also serves as a substrate.

Levansucrase is a type of transferase which catalyzes a fructosyl transfer from sucrose to various acceptor molecules. The enzyme catalyzes the following reactions:

1. Polymerization:

$(\text{Sucrose})_n \rightarrow (\text{Glucose})_n + \text{Levan} + \text{Oligosaccharides}$

2. Hydrolysis:

$\text{Sucrose} + H_2O \rightarrow \text{Fructose} + \text{Glucose}$

$(\text{Levan})_n + H_2O \rightarrow (\text{Levan})_{n-1} + \text{Fructose}$

3. Acceptor:

$\text{Sucrose} + \text{Acceptor molecules} \rightarrow \text{Fructosyl-acceptor} + \text{Glucose}$

4. Exchange:

$\text{Sucrose} + [^{14}\text{C}]\text{Glucose} \rightarrow \text{Fructose-}[^{14}\text{C}]\text{Glucose} + \text{Glucose}$

5. Disproportionation:

$$[\text{Levan}]_m + [\text{Levan}]_n \rightarrow [\text{Levan}]_{m-1} + [\text{Levan}]_{n+1}$$

The enzyme catalyzes hydrolysis and polymerization reactions concomitantly (Reaction 1), resulting in a fructose homopolymer (levan) and free glucose. This reaction occurs when sucrose exists as the sole fructosyl donor and acceptor, and involves three steps: initiation, propagation, and termination (Chambert et al., 1974). The chains of levan grow step-wise by repeated transfer of a hexosyl group from the donor to growing acceptor molecules. The enzyme primarily catalyzes a coupled reaction by a ping-pong mechanism, i.e., sucrose hydrolysis followed by transfructosylation involving a fructosyl-enzyme intermediate (Chambert et al., 1976).

When water acts as an acceptor, a free fructose is generated from both sucrose and levan (Reaction 2). This reaction occurs in all the levansucrase-catalyzed reactions mentioned above, but the rate is much slower when compared with a sugar acceptor. Reaction 3 occurs in the presence of an acceptor in the environment. The enzyme transfers the fructosyl residue of sucrose specifically to the C-1 hydroxyl group of aldose in the acceptor. Compounds containing hydroxyl groups, such as methanol, glycerol and oligosaccharides, can act as fructosyl acceptors.

The reaction mechanism yields a nonreducing sugar compound and a series of oligosaccharides, in which the sugar molecule with one more fructose moiety remains as a major reaction product. The reaction occurs predominantly in the presence of a high concentration of fructosyl donors, such as sucrose or raffinose. Reaction 4 might be considered analogous to Reactions 2 and 3, but differs in the regeneration of sucrose, which has a high-energy bond. The enzyme also catalyzes Reaction 5, a disproportionation reaction, in which the degree of polydispersity of levan or oligomers is modified. The above five reactions compete with one another, yielding a specific major product with some minor products but they are predominantly controlled by environmental factors.

At present, little is known of plant levans, and their biosynthesis is not fully understood (Heyer et al., 1999). One plant levan, known as "graminan", is synthesized by sucrose:fructan 6-fructosyltransferase (6-SFT) which catalyzes the formation and extension of β-(2,6)-linked fructans. The 6-SFT is closely related to vacuolar invertase and transfers the fructosyl residues from sucrose preferentially to 1-kestose or larger fructans (Sprenger et al., 1995). However, most fructan synthesis in plants occurs in two steps (Edelman and Jefford, 1968). Initially, sucrose:sucrose 1-fructosyltransferase (1-SST; EC 2.4.1.99.) catalyzes the formation of the trisaccharide 1-kestose and glucose from two molecules of sucrose. Later, fructan:fructan 1-fructosyltransferase (1-FFT; EC 2.4.1.100.) reversibly transfers fructosyl residues from one fructan with a DP of ≥ 3 to another DP of ≥ 2, producing a mixture of fructans with different chain lengths.

7.2 Genetic Basis of Levan Synthesis

As yet, levansucrase genes have been cloned and biochemically characterized in seven Gram-negative strains; namely, *Acetobacter diazotrophicus* (Arrieta et al., 1996), *Acetobacter xylinum* (Tajima et al., 2000), *Erwinia amylovora* (Geier and Geider, 1993), *P. syringae* pv. glycinea, *P. syringae* pv. phaseolicola (Hettwer et al., 1998), *Rahnella aquatilis* (Song et al., 1998) and *Z. mobilis* (Song et al., 1993). Several levansucrase genes have also been cloned in Gram-positive strains, such as *Bacillus* (Gay et al., 1983; Li et al., 1997) and *Streptococcus* species (Sato et al., 1984). All levansucrase genes share several conserved regions, which are thought to be important for the enzyme activity. Although conservation is observed, dissimilarity exists depending on the source of the enzyme. Levansucrase genes from a Gram-negative origin show relatively high similarity (>50%) when compared with the genes from Gram-positive bacterial enzymes. However, very little similarity (<30%) exists among the genes from two different sources (Song and Rhee, 1994). The deduced amino acid sequences are aligned in Figure 2.

Although the amino acid sequences of levansucrases do not show any considerable homology to those of sucrose-related enzymes, the third (-EWS/AGT/SP/A-) and the

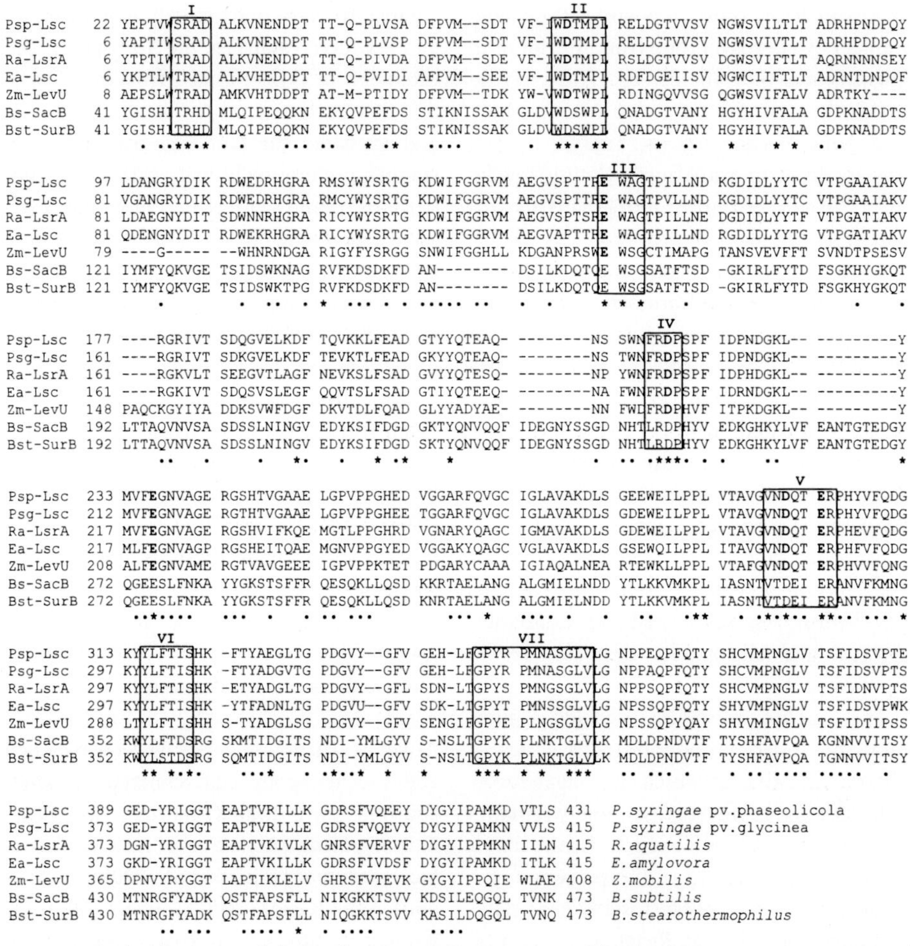

Fig. 2 Multiple alignment of deduced amino acid sequences of bacterial levansucrases. Origins of levansucrase are indicated in ends of the sequences. Asterisks indicate identical- and similar-residues in all levansucrases. Regions considered as important for activity are boxed (I–VII). Amino acid residues that are different between Gram-negative and Gram-positive origin are indicated by dots.

fourth (-FRDP-) conserved regions are found in all fructosyl- and glucosyltransferases, sucrase, sucrose-phosphate hydrolase and even in fructan-hydrolyzing enzymes. The fact that the regions are preserved in all of the sucrose-related enzymes implies that they may be catalytically important regions for the hydrolysis of sucrose. The serine residue in the sixth region (-YLFTI/DS-) has been proposed as the putative residue of the catalytic site (Chambert and Petit-Glatron, 1991).

7.3 Regulation of Levan Synthesis

The genes encoding sucrose-hydrolyzing enzymes may not be linked to each other on the chromosome, but linked only with accessory genes coding for proteins belonging to the phosphoenolpyruvate-dependent carbohydrate phosphotransferase (PTS) system. The expression of these genes is regulated by many regulatory protein systems such as the *glk* operon of *Z. mobilis* and the pleiotropic system of *B. subtilis*, etc.

7.3.1 Regulation at the Protein Level

In microorganisms, two types of sucrose utilization were found: intra- and extracellular. Commonly, these sucrose-uptake utilization systems exist within the cell; sucrose is transported by the PTS system. The sucrase system of this type is well known in *B. subtilis* (Klier and Rapoport, 1988). In contrast, some bacteria such as *Z. mobilis* that lack the PTS system first hydrolyze sucrose extracellularly to monomeric sugars, after which these sugars are transported inside the cell (Di Marco and Romano, 1985).

The co-contribution of both saccharolytic enzymes for the sucrose utilization of *Z. mobilis* has been well characterized. The glucose uptake and utilization system (*glk* operon), which is located very close to the *levU* operon and is also linked metabolically with the sucrose utilization system of *Z. mobilis*, is also regulated by the mechanism of tightly linked gene expression (Liu et al., 1992). In the intervening sequence of the *levU* and *glf* operon, two putative ORFs, encoding Lrp-like regulatory protein and aspartate racemase respectively, were found (Song et al., 1999).

At the molecular level, the genes encoding sucrose-hydrolyzing enzymes reported to date are not linked to each other on the chromosome, but are linked only with accessory genes coding for proteins belonging to the PTS system (Bruckner et al., 1993). The expression of these genes is modulated by regulatory mechanisms, such as anti-termination or repression, which is controlled by the complex regulatory network system including many regulatory proteins (Klier and Rapoport, 1988).

7.3.2 Regulation at Transcriptional and Translational Levels

The genes encoding the extracellular levansucrase and sucrase have been isolated and characterized. The nucleotide sequences of the DNA segment containing the genes encoding extracellular levansucrase and sucrase of *Z. mobilis* and *B. subtilis* were reported recently (Kyono et al., 1995). The two genes are located together in an operon on the chromosome, whereas almost all other genes coding for saccharolytic enzymes in other bacteria and yeasts are dispersed on the chromosomes (Carson and Botstein, 1983). The levansucrase gene of *B. subtilis* is activated in the presence of an inducer (sucrose or fructose), and is under a pleiotropic regulatory system controlling the expression of the sucrose operon (Lepesant et al., 1976; Shimotsu and Henner, 1986).

The pleiotropic system involving the *degS/degU*, *degQ* (formerly *sacU* and *sacQ*) and *degR* genes affects the expression of *sacB* (Débarbouillé et al., 1991). Levansucrase is encoded by the *sacB* gene and expressed from a constitutive promoter in the closely linked *sacR* locus. The *sacR* locus contains a palindromic structure acting as a transcription terminator. In the presence of sucrose, an anti-terminator, the *sacY* gene product that belongs to the *sacS* operon allows transcription of the *sacB* gene. The expression of this gene is also controlled by other regulatory genes such as two- component system *degS/degU* and also by *degQ* (Rapoport and Klier, 1990) (Figure 3).

8
Biodegradation of Levan

Although the biodegradation of levan involves several enzymes including levanase, levansucrase, and levan fructotransferase, the genetic characterization of these is limited to levanase.

8.1
Enzymology of Levan Degradation

Levan is degraded to D-fructose, levanbiose, sucrose, levan oligomers or low molecular-weight levan by the hydrolytic activity of levanase, levansucrase or levan fructotransferase from some plants and microorganisms. The mode and degree of hydrolysis depend on the enzyme sources and the reaction conditions.

8.1.1
Levanase

Many levan-forming microorganisms also produce hydrolytic enzymes–levanases–that degrade levan (Hestrin and Goldblum, 1953; Avigad, 1965). Certain strains of *Bacillus, Pseudomonas, Actinomyces, Aerobacter, Clostridium* and *Streptococcus* produce exocellular levanase (2,6-β-fructan 6-levanbiohydrolase, EC 3.2.1.64.) (Fuchs, 1959; Uchiyama, 1993). The enzyme hydrolyzes only levan, and the resulting product is usually levanbiose, indicating that a terminal fructosyl unit is removed. An exo-hydrolytic enzyme

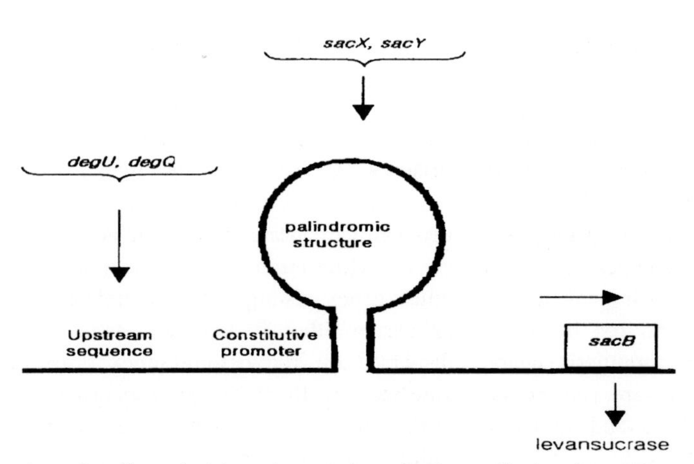

Fig. 3 Specific and pleiotropic control mechanisms affecting the *sacB* gene in *Bacillus subtilis*. *degU*: transcriptional regulator of degradation enzymes; *degQ*: pleiotropic regulatory gene; *sacX*: negative regulatory protein of *sacY*; *sacY*: positive levansucrase synthesis regulatory protein; *sacB*: gene encoding levansucrase. (Reprinted from Débarbouillé et al., 1991, p. 758, with permission from Elsevier Science.)

which has a 2,6-β-linkage-specific fructan-β-fructosidase activity (Marx et al., 1997) was reported from the grass *Lolium perenne*, and a 2,1-β-linkage-specific exohydrolase from Jerusalem artichoke. The other exo-levanase (fructan-β-fructosidase, EC 3.2.1.80; beta-D-fructofuranosidase, EC 3.2.1.26.) hydrolyzes levan to produce D-fructose. Endo-levanases (2,6-β-D-fructan fructanohydrolase, EC 3.2.1.65.) hydrolyze levan and levan oligomers consisting of more than three fructosyl units.

8.1.2
Levansucrase

Levan may be degraded not only by levanases, but also by levansucrase itself, which may catalyze the hydrolysis under certain conditions (Rapoport and Dedonder, 1963). The degree of levan hydrolysis depends on the enzyme sources and reaction conditions. For example, levansucrase from *R. aquatilis* showed a higher degradation activity than did that from *Z. mobilis* (Song et al., 1998), though for both enzymes a higher degradation activity of levan was seen as the reaction temperature was increased from 4 °C to 30 °C.

Although indirect evidence of the reversal of enzymatic synthesis of levan has been observed, little is known regarding the nature of such enzymatic degradation. Smith (1976) showed that beta-fructofuranosidase present in tall fescue degraded levan by removing one fructose residue at a time until a molecule of sucrose remained. Levansucrase of *B. subtilis* has a hydrolytic effect on small levans (Dedonder, 1966), the hydrolytic action stopping at branch points. Neither inulin, inulobiose, inulintriose, nor methyl D-fructofuranoside is hydrolyzed, despite these substrates being hydrolyzed by inulinase and yeast invertase. This hydrolytic activity may be responsible for the appearance of heterogeneous short-chain polysaccharides, rather than uniform high molecular-weight polymers, in the final product of many levan preparations.

8.1.3
Levan Fructotransferase

Microbial levan is an interesting starting material for the production of valuable oligosaccharides such as DFA IV (Yun, 1996; Saito and Tomita, 2000). DFA IV (di-D-fructose-2,6′:6,2′-dianhydride) is an oligosaccharide which is produced from levan by microbial enzymes, i.e., levan fructotransferase (LFTase) and a type of levanase (Tanaka et al., 1981, 1983; Saito et al., 1997). Currently, two LFTases have been isolated and cloned from *Arthrobacter nicotinovorans* GS-9 (Saito et al., 1997) and *A. ureafaciens* (Tanaka et al., 1981; Song et al., 2000). The enzymes have also been shown to degrade levan molecules from the nonreducing fructose end of the outer chains, and to catalyze intermolecular levanbiosyl and levanfructosyl transfer (LBT and LFT, respectively) reactions (Tanaka et al., 1983).

8.2
Genetic Basis of Levan Degradation

In *B. subtilis*, the expression of the levanase operon is inducible by fructose and is subjected to catabolite repression. A fructose-inducible promoter has been characterized 2.7 kb upstream from the gene *sacC*, which encodes levanase. The *sacC* gene is the distal gene of an operon containing five genes: *levD*, *levE*, *levF*, *levG*, and *sacC* (Martin et al., 1989; Débarbouillé et al., 1991) and is expressed under the regulated control of *sacR*, the inducible levansucrase leader region. The first four gene products are involved in a fructose-PTS system. In *Pseudomonas*, levanase is an exohydrolase of levan and produces levanbiose as a sole product; the limits of hydrolysis of levan from *Z. mobilis* and *Serratia* sp. were 65% and 80%, respectively (Jung et al., 1999).

8.3 Regulation of Levan Degradation

There are two levels on which the expression of the levanase operon in *B. subtilis* is controlled: (1) an induction by fructose, which involves a positive regulator, LevR, and the fructose phosphotransferase system encoded by this operon (lev-PTS); and (2) a global regulation of catabolite repression (Débarbouillé et al., 1991) (Figure 4).

The LevR protein is an activator for the expression of the levanase operon from *B. subtilis*. RNA polymerase containing the sigma 54-like factor sigma L recognizes the promoter of this operon. One domain of the LevR protein is homologous to activators of the NtrC family, and another resembles anti-terminator proteins of the BglG family (Débarbouillé et al., 1991). It has been proposed that the domain, which is similar to anti-terminators, is a target of phosphoenolpyruvate:sugar phosphotransferase system (PTS)-dependent regulation of LevR activity. The LevR protein is not only negatively regulated by the fructose-specific enzyme IIA/B of the phosphotransferase system encoded by the levanase operon (lev-PTS), but is also positively controlled by the histidine-containing phosphocarrier protein (HPr) of PTS (Martin et al., 1990; Stülke et al., 1995). This second type of control of LevR activity depends on phosphoenolpyruvate-dependent phosphorylation of HPr histidine 15, as demonstrated with point mutations in the ptsH gene encoding HPr. *In vitro* phosphorylation of partially purified LevR was obtained in the presence of phosphoenolpyruvate, enzyme I, and HPr. The dependence of truncated LevR polypeptides on stimulation by HPr indicates that the domain homologous to anti-terminators is the target of HPr-dependent regulation of LevR activity. This domain appears to be duplicated in the LevR protein. The first anti-terminator-like domain seems to be the target of enzyme I and HPr-dependent phosphorylation and the site of LevR activation, whereas the carboxy-terminal anti-terminator-like domain could be the target for negative regulation by lev-PTS (Débarbouillé et al., 1990).

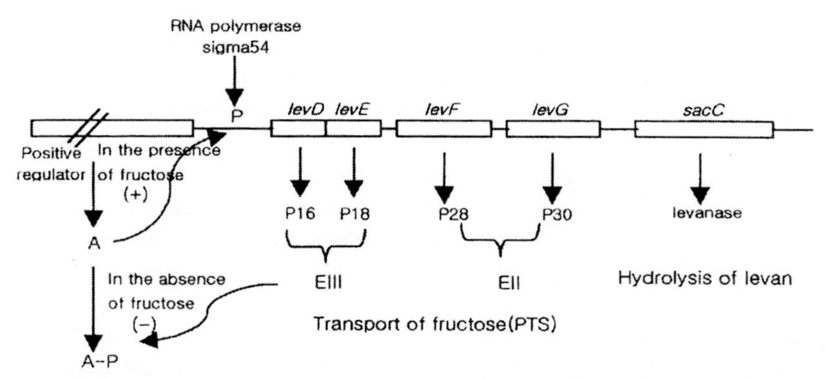

Fig. 4 Regulation model of levanase operon from *Bacillus subtilis*. P represents the fructose-inducible promoter. The *levD*, *levE*, *levF*, and *levG* gene products correspond to a fructose-specific phosphoenolpyruvate-dependent carbohydrate (PTS). *LevR* encodes a positive regulator. The activator may exist in two forms: (A-P), an inactive phosphorylated form; or (A), an active non-phosphorylated form. (Reprinted from Débarbouillé et al., 1991, p. 759, with permission from Elsevier Science.)

9 Biotechnological Production of Levan

Levan can be produced by either microbial fermentation or enzymatic synthesis, but the conversion yield of sucrose to levan is higher in the latter process than in the former.

9.1 Isolation and Screening for Levan-producing Strains

In the screening of levan-producing microbial strains, the levan formation activity can be determined using solid agar plates containing sucrose; the strains producing levan are then isolated by following the analytical procedures. The presence of levansucrase is positively selected by inducing mucoid morphology to the microorganisms. Subsequently, levan is collected from the agar plates by precipitation with alcohol (methanol, ethanol, or isopropanol). The identity of levan is then determined after acid hydrolysis of the polymers, followed by TLC analysis; fructose is identified as a single spot on the TLC plates.

9.2 Fermentative Production of Levan

The microbial production of levan requires fermentation and handling of highly viscous solutions. The conditions for producing levan by growing cultures of bacteria vary according to the microorganisms used, but yields of levan production are fairly low (Table 2); this is due to the utilization of sucrose as energy source, the formation of byproducts, the low level of levansucrase production, and the presence of levanase activity in bacteria. In addition, the recovery process of levan from the fermentation broth is often very difficult due to the high viscosity of levan. In theory, the yield of levan production by levansucrase is 50% (w/w) when sucrose is used as a substrate. Routinely, the yields of levan production based on the amount of sucrose consumed are no higher than 58% of the theoretical yield by fermentation (Table 2).

9.3 *In vitro* Biosynthesis of Levan

In the *in vitro* formation of levan by bacterial levansucrase, sucrose serves as fructosyl donor while the released glucose inhibits levan formation. The inhibitory action is influenced by competition with the glucose moiety of sucrose for the enzyme activity. The glucose moiety of sucrose can be replaced by D-xylose, L-arabinose, lactose, etc. Although the catalytic properties of levansucrase vary, the substrate specificity for acceptors is relatively broad where alco-

Tab. 2 Levan production by fermentation processes.

Strains	*Type of production*[a]	*Substrate conc. [%]*	*Levan Yield [%,w/w]*[b]	*Reference*
Bacillus spp.	BF	12	23.5	Elisashvili (1984)
Bacillus polymyxa	BF	15	26.6	Han (1990)
Erwinia herbicola	CF	5	19.2	Keith et al. (1991)
Gluconobacter oxydans	BF	6.2	23.3	Uchiyama (1993)
Rahnella aquatilis	BF	10	29	Ohtsuka et al. (1992)
Zymomonas mobilis	CF	12	23	Beker et al. (1990)

[a] BF, batch fermentation; CF, continuous fermentation. [b] Based on sucrose consumed.

hol, monosaccharides, disaccharides, sugar alcohols and levan are available, but not for levanbiose, levantriose, and levantetraose.

The optimal temperature range for the *in vitro* synthesis of levan is from 0 °C to 40 °C. Levansucrase prepared from *Z. mobilis* displayed an optimum levan formation activity at 0 °C (Song et al., 1996), from *B. subtilis* at >10 °C (Elisashvili, 1984), from *Pseudomonas* at 18 °C (Hettwer et al., 1995), and from *Rahnella* at 40 °C (Ohtsuka et al., 1992). Most levansucrase activities are inactivated at temperatures higher than 45 °C, with the exception of *R. aquatilis* ATCC 33071, which shows the maximum velocity of levan formation at 50 °C within 3 h, after which the rate declines slightly. Interestingly, at lower temperatures, transfructosylation rather than hydrolysis of sucrose is preferentially catalyzed; however, at higher temperatures hydrolysis is preferentially catalyzed, and this thermolabile feature may have advantages for large-scale levan production. In particular, levan production by *Z. mobilis* levansucrase was most active at the lowest temperature (0 °C), so that it could provide a stable operating condition with minimized contamination opportunities (Song et al., 1996). Plant fructosyltransferases lose 50% of their activity at 5 °C compared with that obtained at the optimum temperature of 20–25 °C (Koops and Jonker, 1996). The *Z. mobilis* levansucrase is stable at pH 4–7, and no activity is observed below pH 3 and above pH 9, similar to the enzymes from *Bacillus, Pseudomonas,* and *Rahnella,* the optimum pH of which was 6.0. However, the enzyme from *B. licheniformis* NRRL B-18962 retains 50% of its maximal activity at 55 °C and pH 4.

When *Z. mobilis* levansucrase was immobilized onto hydroxyapatite, the enzymatic and biochemical properties were similar to those of native enzyme towards salt and detergent effects (Jang et al., 2000). However, immobilization of the enzyme on the surface of a matrix shifts the optimum pH to acidic conditions (pH 4.0). The cell-free system synthesized two types of levan which differ in molecular weight. Levans produced by the immobilized system consisted of a higher proportion of low molecular-weight levan to total levan generated than those obtained by the native enzyme. Toluene-permeabilized whole-cell systems produced levan similarly to immobilized systems (Jang et al., 2001).

9.4 Recovery and Purification of Levan

In the microbial production of levan, the yield is low and, as a consequence, costly processes are required to extract the levan from the fermentation broth. The separation process of levan with high purity from the reaction mixture containing sucrose, glucose, fructose, and fructo-oligosaccharides is both laborious and inefficient. Likewise, the separation of levan by using solvents requires huge amounts of ethanol, methanol, isopropanol, or acetone to be used (Rhee et al., 1998). Subsequently, this solvent is lost as waste or recovered by distillation. Recently, membrane processes have been developed to separate polysaccharides from the fermentation broth or enzymatic reaction mixture, without organic waste. However, the resulting solution contains a low concentration (<5%) of levan, and this must be recovered using various types of drier.

9.5 Commercial Production of Levan

Currently, a Korean start-up company, RealBioTech Co., Ltd. (RBT), is the first and only company worldwide to produce levan on a commercial basis. RBT produces levan in large-scale quantities for supply to companies as a moisturizing ingredient in cosmetics, as a dietary fiber, as food and feed additives, and as a fertilizer (http://www.realbio.com). High-purity levan ($>99\%$) is used in cosmetics and functional foods, while low-grade levan ($<15\%$) containing glucose and oligosaccharides is used as a supplement for feeds and fertilizers.

9.6 Market Analysis and Cost of Levan Production

As levan became available commercially only recently, it is difficult to estimate the size of its current market. It is clear, however, that the expected world market is huge, since levan has a variety of functions and applications within the bioindustries. The production cost of levan depends mainly on the cost of the raw materials, including sucrose and levansucrase, for which the depreciation costs account for 30% and 18%, respectively (RealBioTech Co., unpublished data).

9.7 Levan Competitors

Many types of oligosaccharides and polysaccharides may represent potential competitors of levan in industrial applications. The strongest competitors in the food industry are the fructo-oligosaccharides (FOS), and especially inulin, which belongs to the same fructan category as levan. The use of inulin is limited as a dietary fiber by its insolubility in water at room temperature, whereas levan is a water-soluble and viscous fructan. Other potential competitors are dextran in the food and pharmaceutical industries, β-glucans in the feed industry, and hyaluronic acid in the cosmetic industry. In addition, levan could replace other potential competitors such as xanthan gum, pullulan, and mannan in the food industry. While most commercially available polysaccharides are produced either by microbial fermentation or direct extraction from natural sources, it is possible to produce levan from sucrose by a simple one-step enzyme reaction, which in turn makes it more competitive in terms of production costs.

10 Properties of Levan

While levan is highly soluble in water at room temperature, inulin is almost insoluble ($<0.5\%$), this difference being most likely due to the presence of β-(2,6) linkages in levan. Despite their highly branched molecular structures, microbial levans have several common interesting features in soluble form (Kasapis et al., 1994), including an exceptionally low intrinsic viscosity for a polymer of high molecular weight, an unusual sensitivity of viscosity to increasing concentration between the beginning and end of the intermediate zone, and an extreme concentration-dependence of viscosity at the intermediate zone (Figure 5). These properties may be derived from intermolecular interaction by physical entanglement rather than from any form of specific noncovalent association.

The viscosity of a solution of bacterial levan, originating from *P. syringae* pv. phaseolicola, exhibited Newtonian characteristics up to a concentration of 20%. The concentration-dependence of the 'zero-shear' specific viscosity for the levan solution was unusually high, as would be expected from the molecule's branched structure (Kasapis et al., 1994). In Figure 5, there are three linear regions with changes of slope at levan concentrations of about 4% and 20%. The linear region, including the intermediate region, was also observed in xanthan gum (Milas et al., 1990). However, the concentration-dependence of viscosity may

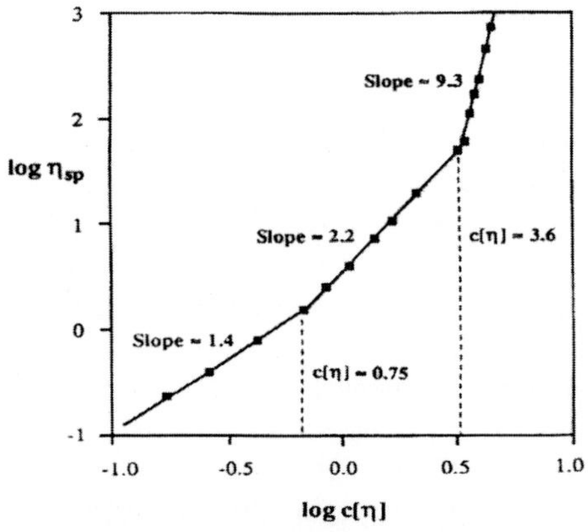

Fig. 5 Concentration dependence of 'zero-sheara-pos; specific viscosity for levan at 20 °C. The parameter η is used for solution viscosity and η_{sp} for specific viscosity which defines the fractional increase in viscosity due to the presence of the polymer ($\eta_{sp} = (\eta - \eta_s)/\eta_s = \eta_{rel} - 1$). η_{rel} indicates relative viscosity, which is the ratio of η to η_s (η_s for solvent viscosity). (Reprinted from Kasapis et al., 1994, p. 59, with permission from Elsevier Science.)

vary with DP, pH, temperature, and salt concentrations (Vina et al., 1998).

11 Applications of Levan

Levan has a wide range of industrial applications, for example in medicine, pharmacy, agriculture, and food. However, it is likely that the low production cost of levan will permit a much increased use of levan in the near future.

11.1 Medical Applications

Microbially produced levan has a direct effect on tumor cells that is related to a modification in the cell membrane, including changes in cell permeability (Leibovici and Stark, 1985; Calazans et al., 2000), as well as radioprotective and antibacterial activities (Vina et al., 1998). Levan derivatives have also been suggested as inhibitors of smooth muscle cell proliferation, as excipients in making tablets, and as agents to transit water into gels. Sulfated, phosphated and acetylated levans have also been suggested as anti-AIDS agents (Clarke et al., 1997).

11.2 Pharmaceutical Applications

Water-soluble polymers, including levan, can be used in a wide variety of applications in the pharmaceutical industry. Water-soluble polymers such as cellulose derivatives, pectin and carrageenan play key roles in the formulation of solid, liquid, semisolid and even controlled release dosage forms (Guo et al., 1998). The viscosity of levan varies with its DP and degree of branching, which relates to the number of side fructose chains attached to one fructose unit in the main fructose chain, and in this respect levan can be used in pharmaceutical formulations in various ways. Low molecular-weight, less branched levan usually provides a low viscosity, and can be used as a tablet binder in immediate-release dosage forms, while levans of medium- and high-viscosity grade are used in controlled-release matrix formulations. Levan has also been suggested as a possible substitute for blood expanders (Imam and Abd-Allah, 1974).

11.3 Agricultural Applications

Microbial levans were introduced in plants to promote their agronomic performance in temperature zones, as well as their natural storage capacities (Vijn and Smeekens, 1999). Transgenic tobacco plants expressing levansucrase genes from *B. subtilis* (Pilon-

Smits et al., 1995) or *Z. mobilis* (Park et al., 1999) showed an increased tolerance to drought and cold stresses. Transgenic plants accumulating fructan have been suggested as novel nutritional feed for ruminants (Biggs and Hancock, 1998, 2001). Recently, microbial levan produced enzymatically was developed as an animal feed (Rhee et al., 2000d) and also as a soil conditioner to improve the germination of various seeds (Imam and Abd-Allah, 1974).

11.4
Food Applications

A number of novel applications of levan have been suggested, particularly in food (Han, 1990; Suzuki and Chatterton, 1993). Levan may act as a prebiotic to change the intestinal microflora, thereby offering beneficial effects when present in the human diet. Levan and its partially hydrolyzed products are fermented by intestinal bacteria including bifidobacteria and *Lactobacillus* sp. (Müller and Seyfarth, 1997; Yamamoto et al., 1999; Marx et al., 2000). Levanheptaose was also suggested as a carbon source for selective intestinal microflora, including *Bifidobacterium adolescentis*, *Lactobacillus acidophilus*, and *Eubacterium limosum*, whereas *Clostridium perfringens*, *E. coli* and *Staphylococcus aureus* did not utilize levan (Kang et al., 2000). Cholesterol- and triacylglycerol-lowering effects of levan have also been reported (Yamamoto et al., 1999) and may be applied to develop levans as health foods or nutraceuticals.

Today, it also seems possible that levan might be used in the dairy industry, as *Lactobacillus reuteri* produces levan-type EPS. EPS-producing lactic acid bacteria, including the genera of *Streptococcus*, *Lactobacillus*, and *Lactococcus*, are used *in situ* to improve the texture of fermented dairy products such as yogurt and cheese. This group of food-grade bacteria produces a wide variety of structurally different polymers, including levan with potential use for new applications. A number of Japanese companies use microbial levans as additives in their milk products which contain *Lactobacillus* species. In addition, the replacement with levan of thickeners or stabilizers that are produced by nonfood-grade bacteria has recently emerged (Van Kranenburg et al., 1999).

11.5
Other Applications

One of the striking consequences of the densely branched structure of levan is its effectiveness in resisting interpenetration by other polymers, leading to macroscopic phase-separation (Kasapis et al., 1994; Chung et al., 1997). Dextran is often used to create two-phase liquid systems (e.g., polyethylene glycol (PEG)/dextran), and to purify biological materials of interest by selectively partitioning them into one phase (Albertsson and Tjerneld, 1994). Microbial levans also display phase-separation phenomena with pectin, locust bean gum, and PEG. Solutions of levan and locust bean gum showed a substantial reduction in viscosity, similar to the mixture levan/pectin. Levan/locust bean gum phases can be separated into discrete phases in mixed solutions in which the lower one consisted predominantly of the denser polymer, the levan phase (Kasapis et al., 1994). The PEG/levan two-phase system was prepared by combining PEG (60%, w/w) and levan (6.77%, w/w) (Chung et al., 1997). This aqueous two-phase system showed a good partitioning with six model proteins including horse heart cytochrome c, horse hemoglobin, horse heart myoglobin, hen egg albumin, bovine serum albumin, and hen egg lysozyme.

12 Patents

Many patents have been filed for the application of levan as functional food additives (Table 3). It was claimed that levan from *S. salivarius* can be used as a food additive with a hypocholesterolemic effect (Kazuoki, 1996). New applications of levan as food and feed additives (Rhee et al., 2000d) as well as a raw material for the production of difructose dianhydride IV (Rhee et al., 2000c) were also developed. Levan derivatives such as sulfated, phosphated or acetylated levans, were claimed to be anti-AIDS agents (Robert and Garegg, 1998), while a fructan *N*-alkylurethane which has excellent surface-active properties in combination with good biodegradability was patented as a surfactant for household use and industrial applications by means of replacing a hydroxyl group of fructose with an alkylaminocarbonyloxyl group (Stevens et al., 1999). A glycol/levan aqueous two-phase system which can substitute the glycerol/dextran system was developed for the partitioning of proteins (Rhee et al., 2000b). Besides levan, the fructosyl transferase activity of levansucrase has been used in the production of alkyl β-D-fructoside (Rhee et al., 2000a).

In spite of many studies on the production and application of levan, few of the levansucrase genes isolated from the microorganisms *Z. mobilis* (Rhee and Song, 1998), *R. aquatilis* (Rhee et al., 1999) and *A. diazotrophicus* (Juan et al., 1998) have been patented. A method for the production of levan using a recombinant levansucrase from *Z. mobilis* was claimed (Rhee et al., 1998), and a creative patent preparing transgenic plants harboring levansucrase gene was applied for by German researchers (Roeber et al., 1994). In the case of levansucrase, a process was claimed for the production of acid-stable

Tab. 3 Relevant patents for levan production and applications

Publication number	*Applicants*	*Inventors*	*Title of invention*	*Date*
US 4,769,254	IBI	Mays, T.D., Dally, E.L.	Microbial production of polyfructose.	September 6, 1988
US 4,927,757	IRFI	Hatcher, H.J. et al.	Production of substantially pure fructose.	May 22, 1990
WO 94/04692	IGF; Roeber, M.; Geier, G.; Geider, K.; Willmitzer, L.	Roeber, M. et al.	DNA sequences which lead to the formation of polyfructans (levans), plasmids containing these sequences as well as a process for preparing transgenic plants.	March 3, 1994
US 5,334,524	SOLVAY ENZYMES INC.	Robert, L.C.	Process for producing levansucrase using *Bacillus*.	August 2, 1994
US 5,527,784	–	Kazuoki, I.	Antihyperlipidemic and antiobesity agent comprising levan or hydrolysis products thereof obtained from *Streptococcus salivarius*.	June 17, 1996
US 5,547,863	USASA	Han, Y.W., Clarke, M.A.	Production of fructan (levan) polyfructose polymers using *Bacillus polymyxa*.	August 20, 1996

Tab. 3 (cont.)

Publication number	*Applicants*	*Inventors*	*Title of invention*	*Date*
WO 98/03184	Clarke, G; Margaret, A; SPRI -	Robert, E.J. et al.	Levan derivatives, their preparation, composition and applications including medical and food applications.	January 29, 1998
US 5,731,173		Juan, G.A.S. et al.	Fructosyltransferase enzyme, method for its production and DNA encoding the enzyme.	March 24, 1998
Korean patent 145946	KRIBB	Rhee, S. K. et al.	Method for production of levan using levansucrase.	May 6, 1998
Korean patent 176410	KRIBB	Rhee, S. K. et al.	A novel levansucrase.	November 13, 1998
Korean patent 0207960	KRIBB	Rhee, S. K. et al.	Base and amino acid sequence of levansucrase derived from *Rahnella aquatilis*.	April 14, 1999
EP 0964054 A1	TS N.V.	Stevens, C.V. et al.	Surface-active alkylurethanes of fructans.	December 15, 1999
Korean patent 0257118	KRIBB	Rhee, S. K. et al.	A process for preparation of alkyl β-D-fructoside using levansucrase.	Febraury 28, 2000
Korean patent 262769	KRIBB	Rhee, S. K. et al.	Novel polyethylene glycol/levan aqueous two-phase system and protein partitioning method using thereof.	May 6, 2000
WO 01/29185	KRIBB and RBT	Rhee, S. K. et al.	Enzymatic production of difructose dianhydride IV from sucrose and elevant enzymes and genes coding for them.	October 19, 2000
WO 01/49127	KRIBB and RBT	Rhee, S. K. et al.	Animal feed containing simple polysaccharides.	December 29, 2000

EP: European Patent; IBI: Igene Biotechnology Institute; IGF: INST GENBIOLOGISCHE FORSCHUNG (DE); IRFI: Idaho Research Foundation Institute; KRIBB: Korea Research Institute of Bioscience and Biotechnology; PCT: World Intellectual Property Organization; RBT: RealBioTech Co., Ltd.; SPRI: Sugar Processing Research Institute; TS N.V: Tiense Suikerraffinaderij N.V.; US: United States Patent; USASA: The United States of America as represented by the Secretary of the Agriculture; WO: World IPO(PCT).

levansucrase from *Bacillus* which is not induced by sucrose (Robert, 1994).

13 Current Problems and Limits

Although levan is a water-soluble and low-viscosity polysaccharide, its applications might be limited due to its turbidity and low fluidity. In order to expand the application areas of levan, the development of biosynthetic methods, including the involvement of novel enzymes, will be essential in order to control the DP (molecular weight) and the degree of branching. More important factors from a commercial aspect include process development for the large-scale production which is comparable with that used for other competitive polysaccharides, especially inulin from chicory. In order to produce levan more economically, a cost-effective purification process of levan from the reaction mixture containing sucrose, glucose, fructose and fructo-oligosaccharides must be developed. The membrane processes will likely serve as one of these solutions.

14 Outlook and Perspectives

Levan has a great potential as a functional biomaterial in the food, cosmetic, pharmaceutical and other industries. However, use of this biopolymer has yet not been practicable due to a paucity of information on its polymeric properties highlighting its industrial applicability, as well as a lack of feasible processes for large-scale production.

For technical applications, fructans with a high molecular mass and a low degree of branching would be desirable. However, microbial levans and their oligomers have been less well characterized with regard to carbohydrate structure analysis. In order to utilize the versatility of water-soluble levans to a maximum, a broader understanding of the behavior of levan is required. The characterization of polysaccharides has advanced considerably during the past two decades due to the introduction of powerful methods such as mass spectrometry, nuclear magnetic resonance, atomic force microscopy, scanning probe microscopy, small-angle X-ray scattering, small-angle neutron scattering, and molecular-mechanic-based carbohydrate modeling (Brant, 1999). As a consequence, the complete characterization of levan should be attained in the near future. Furthermore, the fundamental rheological properties of levan in solution, for example viscosity, thixotropy, dilatancy, elasticity, pseudoelasticity, and viscoelasticity will become increasingly important for new applications of this compound.

15
References

Albertsson, P. A., Tjerneld, F. (1994) Phase diagrams, *Methods Enzymol.* **228**, 3–13.

Ananthalakshmy, V. K., Gunasekaran, P. (1999) Overproduction of levan in *Zymomonas mobilis* by using cloned *sacB* gene, *Enzyme Microb. Technol.* **25**, 109–115.

Arrieta, J., Hernández, L., Coego, A., Suárez, V., Balmori, E., Menéndez, C., Petit-Glatron, M. F., Chambert, R., Selman-Housein, G. (1996) Molecular characterization of the levansucrase gene from the endophytic sugarcane bacterium *Acetobacter diazotrophicus* SRT4, *Microbiology* **142**, 1077–1085.

Avigad, G. (1965) In *Methods in Carbohydrate Chemistry*, New York, London: Academic Press, vol. V, 161–165.

Aymerich, S. (1990) What is the role of levansucrase in *Bacillus subtilis*? *Symbiosis* **9**, 179–184.

Beker, M. J., Shvinka, J. E., Pankova, L. M., Laivenienks, M. G., Mezhbarde, I. N. (1990) A simultaneous sucrose bioconversion into ethanol and levan by *Zymomonas mobilis*, *Appl. Biochem. Biotechnol.* **24/25**, 265–274.

Bergeron, L. J., Morou-Bermudez, E., Burne, R. A. (2000) Characterization of the fructosyltransferase gene of *Actinomyces naeslundii* WVU45, *J. Bacteriol.* **182**, 3649–3654.

Beveridge, T. J., Graham, L. L. (1991) Surface layers of bacteria, *Microbiol. Rev.* **55**, 684–705.

Biggs, D. R., Hancock, K. R. (1998) *In vitro* digestion of bacterial and plant fructans and effects on ammonia accumulation in cow and sheep rumen fluids, *J. Gen. Appl. Microbiol.* **44**, 167–171.

Biggs, D. R., Hancock, K. R. (2001) Fructan 2000, *Trends Plant Sci.* **6**, 8–9.

Birkhed, D., Rosell, K.-G, Granath, K. (1979) Structure of extracellular water-soluble polysaccharides synthesized from sucrose by oral strains of *Streptococcus mutans*, *Streptococcus salivarius*, *Streptococcus sanguis*, and *Actinomyces viscosus*, *Arch. Oral Biol.* **24**, 53–61.

Brant, D. A. (1999) Novel approaches to the analysis of polysaccharide structures, *Curr. Opin. Struct. Biol.* **9**, 556–562.

Bruckner, R., Wagner, E., Gotz, F. (1993) Cloning and characterization of the scrA gene encoding the sucrose-specific Enzyme II of the phosphotransferase system from *Staphylococcus xylosus*, *Mol. Gen. Genet.* **241**, 33–41.

Calazans, G. M. T., Lima, R. C., de França, F. P., Lopes, C. E. (2000) Molecular weight and antitumour activity of *Zymomonas mobilis* levans, *Int. J. Biol. Macromol.* **27**, 245–247.

Carson, M., Botstein, D. (1983) Organization of the SUC gene family in *Saccharomyces*, *Mol. Cell. Biol.* **3**, 351–359.

Chambert, R. G., Gonzy-Treboul, G. (1976) Levansucrase of *Bacillus subtilis*. Characterization of a stabilized fructosyl-enzyme complex and identification of an aspartyl residue as the binding site of the fructosyl group, *Eur. J. Biochem.* **71**, 493–508.

Chambert, R., Petit-Glatron, M. F. (1991) Polymerase and hydrolase activities of *Bacillus subtilis* levansucrase can be separately modulated by site-directed mutagenesis, *Biochem. J.* **279**, 35–41.

Chambert, R. G., Gonzy-Treboul, G., Dedonder, R. (1974) Kinetic studies of levansucrase of *Bacillus subtilis*, *Eur. J. Biochem.* **41**, 285

Chung, B. H., Kim, W. K., Song, K. B., Kim, C. H., Rhee, S. K. (1997) Novel polyethylene glycol/levan aqueous two-phase system for protein partitioning, *Biotech. Techn.* **11**, 327–329.

Clarke, M. A., Roberts, E. J., Garegg, P. J. (1997) New compounds from microbiological products of sucrose, *Carbohydr. Polym.* **34**, 425.

Débarbouillé, M., Arnaud, M., Fouet, A., Klier, A., Rapoport, G. (1990) The *sacT* genes regulating the *sacPA* operon in *Bacillus subtilis* shares strong homology with transcriptional antiterminators, *J. Bacteriol.* **172**, 3966–3973.

Débarbouillé, M., Martin. V, Arnaud, M., Klier, A., Rapoport, G. (1991) Positive and negative regulation controlling expression of the sac genes in *Bacillus subtilis, Res. Microbiol.* **142**, 757–764.

Dedonder, R. (1966) Levansucrase from *Bacillus subtilis, Methods Enzymol.* **8**, 500–505.

Di Marco, A. A., Romano, A. H. (1985) D-Glucose transport system of *Zymomonas mobilis, Appl. Environ. Microbiol.* **49**, 151–157.

Edelman, J., Jefford, T. G. (1968) The mechanism of fructosan metabolism in higher plants as exemplified in *Helianthus tuberosus, New Phytol.* **67**, 517–531.

Elisashvili, V. I. (1984) Levan synthesis by *Bacillus* sp., *Appl. Biochem. Microbiol.* **20**, 82–87.

Fuchs, A. (1959) *On the synthesis and breakdown of levan by bacteria*, Thesis, Uitgeverij Waltman, Delft.

Gay, P., Le Coq, D., Steinmetz, M., Ferrari, E., Hoch, J. A. (1983) Cloning structural gene *sacB*, which codes for exoenzyme levansucrase of *Bacillus subtilis*: expression of the gene in *Escherichia coli, J. Bacteriol.* **153**, 1424–1431.

Geier, G., Geider, K. (1993) Characterization and influence on virulence of levansucrase gene from the fireblight pathogen *Erwinia amylovora, Physiol. Mol. Plant Pathol.* **42**, 387–404.

Greig-Smith, R. (1901) The gum fermentation of sugar cane juice, *Proc. Linn. Soc. N.S.W.* **26**, 589.

Guo, J.-H, Skinner, G. W., Harcum, W. W., Barnum, P. E. (1998) Pharmaceutical applications of naturally occurring water-soluble polymers, *Pharmaceutical Science & Technology Today* **1**, 254–261.

Han, Y. W. (1989) Levan production by *Bacillus polymyxa, J. Ind. Microbiol.* **4**, 447–452.

Han, Y. W. (1990) Microbial levan, *Adv. Appl. Microbiol.* **35**, 171–194.

Han, Y. W., Clarke, M. A. (1996) Production of fructan(levan) polyfructose polymers using *Bacillus polymyxa*, U. S. Patent 5,547,863.

Hatcher, H. J., Gallian, J. J., Leeper, S. A. (1990) Production of substantially pure fructose, U. S. Patent 4,927,757.

Hendry, G. A. F., Wallace, R. K. (1993) The origin, distribution, and evolutionary significance of fructans, in: *Science and Technology of Fructans* (Suzuki, M., Chatterton, N. J., Eds.), Boca Raton: CRC Press, 119–139.

Hestrin, S., Goldblum, J. (1953) Levanpolyase, *Nature* **172**, 1047–1064.

Hestrin, S., Avineri-Shapiro, S., Aschner, M. (1943) The enzymatic production of levan, *Biochem. J.* **37**, 450.

Hettwer, U., Gross, M., Rudolph, K. (1995) Purification and characterization of an extracellular levansucrase from *Pseudomonas syringae* pv. phaseolicola, *J. Bacteriol.* **177**, 2834–2839.

Hettwer, U., Jaeckel, F. R., Boch, J., Meyer, M., Rudolph, K., Ullrich, M. S. (1998) Cloning, nucleotide sequence, and expression in *Escherichia coli* of levansucrase genes from the plant pathogens *Pseudomonas syringae* pv. glycinea and *P. syringae* pv. phaseolicola, *Appl. Environ. Microbiol.* **64**, 3180–3187.

Heyer, A. G., Lloyd, J. R., Kossmann, L. (1999) Production of polymeric carbohydrates, *Curr. Opin. Biotechnol.* **10**, 169–174.

Imam, G. M., Abd-Allah, N. M. (1974) Fructosan, a new soil conditioning polysaccharide isolated from the metabolites of *Bacillus polymyxa* AS-1 and its clinical applications, *Egypt. J. Bot.* **17**, 19–26.

Jang, K. H., Kim, J. S., Song, K. B., Kim, C. H., Chung, B. H., Rhee, S. K. (2000) Production of levan using recombinant levansucrase immobilized on hydroxyapatite, *Bioprocess Eng.* **23**, 89–93.

Jang, K. H, Song, K. B., Kim, C. H., Chung, B. H., Kang, S. A., Chun, U. H., Choue, R. W., Rhee, S. K. (2001) Comparison of characteristics of levan produced by different preparations of levansucrase from *Zymomonas mobilis, Biotechnol. Lett.* **23**, 339–344.

Juan, G. A. S., Lazaro, H. G., Alberto, C. G., Guillermo, S. H. S. (1998) Fructosyltransferase enzyme, method for its production and DNA encoding the enzyme, U. S. Patent 5,731,173.

Jung, K. E., Lee, S. O., Lee, J. D., Lee, T. H. (1999) Purification and characterization of a levanbiose-producing levanase from *Pseudomonas* sp. No. 43, *Biotechnol. Appl. Biochem.* **28**, 263–268.

Kang, S. K., Park, S. J., Lee, J. D., Lee, T. H. (2000) Physiological effects of levanoligosaccharide on growth of intestinal microflora, *J. Korean Soc. Food Sci. Nutr.* **29**, 35–40.

Kasapis, S., Morris, E. R., Gross, M., Rudolph, K. (1994) Solution properties of levan polysaccharide from *Pseudomonas syringae* pv. phaseolicola, and its possible primary role as a blocker of recognition during pathogenesis, *Carbohydr. Polym.* **23**, 55–64.

Kazuoki, I. (1996) Antihyperlipidemic and anti-obesity agent comprising levan or hydrolysis products thereof obtained from *Streptococcus salivarius*, U. S. Patent 5,527,784.

Keith, K., Wiley, B., Ball, D., Arcidiacono, S., Zorfass, D., Mayer, J., Kaplan, D. (1991) Continuous culture system for production of biopolymer levan using *Erwinia herbicola*, *Biotechnol. Bioeng.* **38**, 557–560.

Kim, M. G., Seo, J. W., Song, K.-B., Kim, C. H., Chung, B. H, Rhee, S. K. (1998) Levan and fructosyl derivatives formation by a recombinant levansucrase from *Rahnella aquatilis*, *Biotechnol. Lett.* **20**, 333–336.

Klier, A. F., Rapoport, G. (1988) Genetics and regulation of carbohydrate catabolism in *Bacillus*, *Annu. Rev. Microbiol.* **42**, 65–95.

Kojima, I., Saito, T., Iizuka, M., Minamiura, N., Ono, S. (1993) Characterization of levan produced by *Serratia* sp., *J. Ferment. Bioeng.* **75**, 9–12.

Koops, A. J., Jonker, H. H. (1996) Purification and characterization of the enzymes of fructan biosynthesis in tubers of *Helianthus tuberosus* Colombia. II. Purification of sucrose:sucrose 1-fructosyltransferase and reconstitution of fructan synthesis *in vitro* with purified sucrose:sucrose 1-fructosyltransferase and fructan:fructan 1-fructosyltransferase, *Plant Physiol.* **110**, 1167–1175.

Kunst, F., Rapoport, G. (1995) Salt stress is an environmental signal affecting degradative enzyme synthesis in *Bacillus subtilis*, *J. Bacteriol.* **177**, 2403–2407.

Kyono, K., Yanase, H., Tonomura, K., Kawasaki, H., Sakai, T. (1995) Cloning and characterization of *Zymomonas mobilis* genes encoding extracellular levansucrase and invertase, *Biosci. Biotechnol. Biochem.* **59**, 289–293.

Leibovici, J., Stark, Y. (1985) Increase in cell permeability to a cytotoxic agent by the polysaccharide levan, *Cell. Mol. Biol.* **31**, 337–341.

Lepesant, J. A., Kunst, F., Pascal, M., Steinmetz, M., Dedonder, R. (1976) Specific and pleiotropic regulatory mechanisms in the sucrose system of *Bacillus subtilis*, *Microbiology* **168**, 58–69.

Li, Y., Triccas, J. A., Ferenci, T. (1997) A novel levansucrase-levanase gene cluster in *Bacillus stearothermophilus* ATCC 12980, *Biochim. Biophys. Acta* **1353**, 203–208.

Liu, J., Barnell, W. O., Conway, T. (1992) The polycistronic mRNA of the *Zymomonas mobilis glf-zwf-edd-glk* operon is subject to complex transcript processing, *J. Bacteriol.* **174**, 2824–2833.

Martin, I., Débarbouillé, M., Klier, A., Rapoport, G. (1989) Induction and metabolic regulation of levanase synthesis in *Bacillus subtilis*, *J. Bacteriol.* **171**, 1885–1892.

Martin, I., Débarbouillé, M., Klier, A., Rapoport, G. (1990) Levanase operon of *Bacillus subtilis* includes a fructose-specific phosphotransferase system regulating the expression of the operon, *J. Mol. Biol.* **214**, 657–671.

Marx, S. P., Nosberger, J., Frehner, M. (1997) Hydrolysis of fructan in grasses: 2,6-β-linkage-specific fructan-β-fructosidase from stubble of *Lolium perenne*, *New Phytol.* **135**, 279–290.

Marx, S. P., Winkler, S., Hartmeier, W. (2000) Metabolization of β-(2,6)-linked fructose-oligosaccharides by different bifidobacteria, *FEMS Microbiol. Lett.* **182**, 163–169.

Mays, T. D., Dally, E. L. (1988) Microbial production of polyfructose, U.S. Patent 4,769,254.

Milas, M., Rinaudo, M., Knipper, M., Schuppise, J. L. (1990) Flow and viscoelastic properties of xanthan gum solutions, *Macromolecules* **23**, 2506–2511.

Müller, M., Seyfarth, W. (1997) Purification and substrate specificity of an extracellular fructan-hydrolase from *Lactobacillus paracasei* ssp. *paracasei* P 4134, *New Phytol.* **136**, 89–96.

Nelson, C. J., Spollen, W. G. (1987) Fructans, *Physiol. Plant.* **71**, 512–516.

Newbrun, E., Lacy, R., Christie, T. M. (1971) The morphology and size of the extracellular polysaccharide from oral streptococci, *Arch. Oral Biol.* **16**, 863–872.

Ohtsuka, K., Hino, S., Fukushima, T., Ozawa, O., Kanematsu, T., Uchida, T. (1992) Characterization of levansucrase from *Rahnella aquatilis* JCM-1638, *Biosci. Biotechnol. Biochem.* **56**, 1373–1377.

Park, J. M., Kwon, S. Y., Song, K. B., Kwak, J. W., Lee, S. B., Nam, Y. W., Shin, J. S., Park, Y. I., Rhee, S. K., Paek, K. H. (1999) Transgenic tobacco plants expressing the bacterial levansucrase gene show enhanced tolerance to osmotic stress, *J. Microbiol. Biotechnol.* **9**, 213–218.

Perez Oseguera, M. A., Guereca, L., Lopez-Munguia, A. (1996) Properties of levansucrase from *Bacillus circulans*, *Appl. Microbiol. Biotechnol.* **45**, 465–471.

Pilon-Smits, E. A. H., Ebskamp, M. J. M., Paul, M. J., Jeuken, M. J. W., Weisbeek, P. J., Smeekens, J. C. M. (1995) Improved performance of transgenic fructan-accumulating tobacco under drought stress, *Plant Physiol.* **107**, 125–130.

Pollock, C. J., Cairns, A. J. (1991) Fructan metabolism in grasses and cereals, *Annu. Rev. Plant Physiol. Plant Mol. Biol.* **42**, 77–101.

Rapoport, G., Dedonder, R. (1963) Le lévane-sucrase de *B. subtilis* III. Reaction d'hydrolyse, de transfert et d'échange avec des analogues du saccharose, *Bull. Soc. Chim. Biol.* (France) **45**, 515–535.

Rapoport, G., Klier, A. (1990) Gene expression using *Bacillus*, *Curr. Opin. Biotechnol.* **1**, 21–27.

Rhee, S. K., Song, K. B. (1998) A novel levansucrase, Korean Patent 176410.

Rhee, S. K., Song, K. B., Kim, C. H. (1998) Method for production of levan using levansucrase, Korean Patent 145946.

Rhee, S. K., Song, K. B., Seo, J. W., Kim, C. H., Chung, B. H. (1999) Base and amino acid sequence of levansucrase derived from *Rahnella aquatilis*, Korean Patent 0207960.

Rhee, S. K., Kim, C. H., Song, K. B., Kim, M. G., Seo, J. W., Chung, B. H. (2000a) A process for preparation of alkyl β-D-fructoside using levansucrase, Korean Patent 0257118.

Rhee, S. K, Chung, B. H., Kim, W. K., Song, K. B., Kim, C. H. (2000b) Novel polyethylene glycol/levan aqueous two-phase system and protein partitioning method using thereof, Korean Patent 262769.

Rhee, S. K., Song, K. B., Kim, C. H., Ryu, E. J., Lee, Y. B. (2000c) Enzymatic production of difructose dianhydride IV from sucrose and elevant enzymes and genes coding for them, PCT-KR00-01183.

Rhee, S. K., Song, K. B., Yoon, B. D., Kim, C. H. (2000d) Animal feed containing simple polysaccharides, PCT-KR00-01556.

Robert, E. J., Garegg, P. J. (1998) Levan derivatives, their preparation, composition and applications including medical and food applications, WO 98/03184.

Robert, L. C. (1994) Process for producing levan sucrase using *Bacillus*, U. S. Patent 5,334,524.

Roeber, M., Geier, G., Geider, K., Willmitzer, L. (1994) DNA sequences which lead to the formation of polyfructans (levans), plasmids containing these sequences as well as a process for preparing transgenic plants, WO-94/004692.

Saito, K., Tomita, F. (2000) Difructose anhydrides: their mass-production and physiological functions, *Biosci. Biotechnol. Biochem.* **64**, 1321–1327.

Saito, K., Goto, H., Yokoda, A., Tomita, F. (1997) Purification of levan fructotransferase from *Arthrobacter nicotinovorans* GS-9 and production of DFA IV from levan by the enzyme, *Biosci. Biotechnol. Biochem.* **61**, 1705–1709.

Sato, S., Koga, T., Inoue, M. (1984) Isolation and some properties of extracellular D-glucosyltransferases and D-fructosyltransferase from *Streptococcus mutans* serotypes c, e, and f, *Carbohydr. Res.* **134**, 293–304.

Shimotsu, H., Henner, D. J. (1986) Modulation of *Bacillus subtilis* levansucrase gene expression by sucrose and regulation of the steady-state mRNA level by *sacU* and *sacQ* genes, *J. Bacteriol.* **168**, 380–388.

Smith, A. E. (1976) Beta-fructofuranosidase and invertase activity in tall fescue culm bases, *J. Agric. Food Chem.* **24**, 476–478.

Song, K. B., Rhee, S. K. (1994) Enzymatic synthesis of levan by *Zymomonas mobilis* levansucrase overexpressed in *Escherichia coli*, *Biotechnol. Lett.* **16**, 1305–1310.

Song, K. B., Joo, H. K., Rhee, S. K. (1993) Nucleotide sequence of levansucrase gene (*levU*) of *Zymomonas mobilis* ZM1 (ATCC10988), *Biochim. Biophys. Acta* **1173**, 320–324.

Song, K. B., Belghith, H., Rhee, S. K. (1996) Production of levan, a fructose polymer, using an overexpressed recombinant levansucrase, *Ann. N. Y. Acad. Sci.* **799**, 601–607.

Song, K. B., Seo, J. W., Kim, M. K., Rhee, S. K. (1998) Levansucrase from *Rahnella aquatilis*: gene cloning, expression and levan formation, *Ann. N. Y. Acad. Sci.* **864**, 506–511.

Song, K. B., Seo, J. W., Rhee, S. K. (1999) Transcriptional analysis of *levU* operon encoding saccharolytic enzymes and two apparent genes involved in amino acid biosynthesis in *Zymomonas mobilis*, *Gene* **232**, 107–114.

Song, K. B., Bae, K. S., Lee, Y. B., Lee, K. Y., Rhee, S. K. (2000) Characteristics of levan fructotransferase from *Arthrobacter ureafaciens* K2032 and difructose anhydride IV formation from levan, *Enzyme Microb. Technol.* **27**, 212–218.

Sprenger, N., Bortlik, K., Brandt, A., Boller, T., Wiemken, A. (1995) Purification, cloning, and functional expression of sucrose:fructan 6-fructosyltransferase, a key enzyme of fructan synthesis in barley, *Proc. Natl. Acad. Sci. USA* **92**, 11652–11656.

Stevens, C. V., Karl, B., Isabelle, M.-A., Lucien, D. (1999) Surface-active alkylurethanes of fructans, EP 0964054 A1.

Stülke, J., Martin, V. I., Charrier, V., Klier, A., Deutscher, J., Rapoport, G. (1995) The HPr protein of the phosphotransferase system links induction and catabolite repression of the *Bacillus subtilis* levanase operon, *J. Bacteriol.* **177**, 6928–6936.

Suzuki, M., Chatterton, N. J. (1993) *Science and Technology of Fructans*. Boca Raton: CRC Press.

Tajima, K., Uenishi, N., Fujiwara, M., Erata, T., Munekata, M., Takai, M. (1998) The production of a new water-soluble polysaccharide by *Acetobacter xylinum* NCI 1005 and its structural analysis by NMR spectroscopy, *Carbohydr. Res.* **305**, 117–122.

Tajima, K., Tanio, T., Kobayashi, Y., Kohno, H., Fujiwara, M., Shiba, T., Erata, T., Munekata, M., Takai, M. (2000) Cloning and sequencing of the levansucrase gene from *Acetobacter xylinum* NCI 1005, *DNA Res.* **7**, 237–242.

Tanaka, K., Kawaguchi, H., Ohno, K., Shohji, K. (1981) Enzymatic formation of difructose IV from bacterial levan, *J. Biochem.* **90**, 1545–1548.

Tanaka, K., Karigane, T., Yamaguchi, F., Nishikawa, S., Yoshida, N. (1983) Action of levan fructotransferase of *Arthrobacter ureafaciens* on levanoligosaccharides, *J. Biochem.* **94**, 1569–1578.

Uchiyama, T. (1993) Metabolism in microorganisms. Part II. Biosynthesis and degradation of fructans by microbial enzymes other than levansucrase. in: *Science and Technology of Fructans* (Suzuki, M., Chatterton, N. J., Eds.), Boca Raton: CRC Press, 169–190.

Vandamme, E. J., Derycke, D. G. (1983) Microbial inulinase: fermentation process, properties, and applications, *Adv. Appl. Microbiol.* **29**, 139–176.

Van Geel-Schutten, G. H., Faber, E. J., Smit, E., Bonting, K., Smith, M. R., Ten Brink, B., Kamerling, J. P., Vliegenthart, J. F. G., Dijkhuizen, L. (1999) Biochemical and structural characterization of the glucan and fructan exopolysaccharides synthesized by the *Lactobacillus reuteri* wild-type strain and mutant strains, *Appl. Environ. Microbiol.* **65**, 3008–3014.

Van Kranenburg, R., Boels, I. C., Kleerebezem, M., de Vos, W. M. (1999) Genetics and engineering of microbial exopolysaccharides for food: approaches for the production of existing and novel polysaccharides, *Curr. Opin. Biotechnol.* **10**, 498–504.

Vereyken, I. J., Chupin, V., Demel, R. A., Smeekens, S. C. M., Kruijff, B. (2001) Fructans insert between the headgroups of phospholipids, *Biochim. Biophys. Acta* **1510**, 307–320.

Vijn, I., Smeekens, S. (1999) Fructan: more than reserve carbohydrate? *Plant Physiol.* **120**, 351–359.

Vina, I., Karsakevich, A., Gonta, S., Linde, R., Bekers, M. (1998) Influence of some physicochemical factors on the viscosity of aqueous levan solutions of *Zymomonas mobilis*, *Acta Biotechnol.* **18**, 167–174.

Wolff, D., Czapla, S., Heyer, A. G., Radosta, S., Mischnick, P., Springer, J. (2000) Globular shape of high molar mass inulin revealed by static light scattering and viscometry, *Polymer* **41**, 8009–8016.

Yamamoto, Y., Takahashi, Y., Kawano, M., Iizuka, M., Matsumoto, T., Saeki, S., Yamaguchi, H. (1999) *In vitro* digestibility and fermentability of levan and its hypocholesterolemic effects in rats, *J. Nutr. Biochem.* **10**, 13–18.

Yun, J. W. (1996) Fructooligosaccharides – occurrence, preparation, and application, *Enzyme Microb. Technol.* **19**, 107–117.

22 Hyaluronan

Prof. Dr. Peter Prehm
Institut für Physiologische Chemie und Pathobiochemie, Waldeyerstr. 15, D-48129 Münster, Germany; Tel.: +49-251-8355579; Fax: +49-251-8355596; E-mail: prehm@uni-muenster.de

Biotechnology of Biopolymers. From Synthesis to Patents. Edited by A. Steinbüchel and Y. Doi

ISBN: 3-527-31110-6

CHO Chinese hamster ovary
PMA phorbol-12-myristate-13-acetate
RHAMM receptor for hyaluronan-mediated motility

1 Introduction

Although hyaluronan has a very simple structure, almost everything else concerning the molecule is unusual. Sometimes its role is mechanical and structural (as in the synovial fluid, the vitreous humor, or the umbilical cord), whereas sometimes it interacts in tiny concentrations in cells to trigger important responses. Hyaluronan has an unusual mechanism of biosynthesis and exceptional physical properties; consequently, research on this compound was cumbersome, with progress often impeded by failures – often because established procedures from other fields were not applicable and new techniques needed to be developed before any progress could be made.

During the decades of hyaluronan research, several books and reviews have marked such progress including Balazs (1970), Laurent (1989, 1998), Laurent and Fraser (1992), Goa and Benfield (1994), Lapcik et al. (1998), and Abatangelo and Weigel (2000), whilst reviews are published continually on a new web-site: http://www.glycoforum.gr.jp/science/hyaluronan.

2 Historical Outline

In 1934, Karl Meyer described a procedure for isolating a novel glycosaminoglycan from the vitreous humor of bovine eyes, and named it hyaluronic acid (from the Greek, hyalos = glassy, vitreous) (Meyer and Palmer, 1934). These authors showed that this substance contained a uronic acid and an amino sugar, but no sulfoesters. At physiological pH all carboxyl groups are dissociated, and hence the polysaccharide should be called hyaluronate. Today, this macromolecule is most frequently referred to as hyaluronan, in order to emphasize its polysaccharide nature. During the 1930s and 1940s, hyaluronan was isolated from many sources such as the vitreous body, synovial fluids, umbilical cord, skin, and rooster comb (Meyer, 1947) and also from streptococci (Kendall et al., 1937).

The physico-chemical characterization of hyaluranon was carried out during the 1950s and 1960s. The molecular weight is in the order of several millions, whilst in solution the chain behaves as an extended random coil, with a diameter of ~500 nm. At concentrations as low as 0.1%, the chains are entangled, and this results in extremely high, shear-dependent viscosity (Laurent, 1970). These properties enable hyaluronan to regulate water balance, osmotic pressure and flow resistance, to interact with proteins, and also to act as a sieve, as a lubricant, and to stabilize structures by virtue of electrostatic interactions (Comper and Laurent, 1978).

In 1972, Hardingham and Muir discovered that hyaluronan interacts with cartilage proteoglycans and serves as the central structural backbone of cartilage. This was the first example of a specific interaction between hyaluronan and a protein, and many more such interactions were discovered during the 1990s.

After 1980, the research spread in many directions, mainly because until that time it had been assumed that hyaluronan belonged to the proteoglycans, and that its biosynthesis proceeded in a similar manner. In fact, many studies were conducted to detect the protein moiety, but this assumption was disproved when a plausible mechanism of biosynthesis was proposed (Prehm, 1983a,b). It had also been assumed that the synthesis of hyaluronan occurred in the Golgi body – as was the case for all other secretory eucaryotic polysaccharides – until it was shown that hyaluronan was in fact synthesized at plasma membranes and the chains were extruded directly into the extracellular matrix (Prehm, 1984). The catabolic pathways of hyaluranon were also elucidated at about this time (Fraser et al., 1981).

Subsequently, it became possible to measure hyaluronan specifically in body fluids with high sensitivity (Tengblad et al., 1980), and also to visualize it histochemically. These advances opened the way to assess the role of hyaluronan in many pathological disturbances.

Balazs pioneered the application of hyaluronan for medical purposes, and produced highly viscous and noninflammatory preparations on a commercial scale both as an aid for ophthalmic surgery and as viscosupplementation for synovial fluids in patients with osteoarthritis (Balazs, 1982; Balazs and Denlinger, 1989).

Although the importance of hyaluronan in cellular behavior had been recognized for decades, it was not until 1986 that its requirement for detachment in mitotic cell division was proven (Brecht et al., 1986). Underhill and Toole (1979, 1982) reported that hyaluronan was an adhesive cell surface component that formed large coats around untransformed fibroblasts, and smaller coats around transformed cells.

Cell surface hyaluronan-binding proteins were discovered during the late 1980s (Turley et al., 1987; Aruffo et al., 1990), and studied intensively during the 1990s. Although hyaluranon was already known to be involved in both metastasis (Toole et al., 1979) and cell differentiation (Toole et al., 1977), it was investigations into the molecular biology of the receptors which led to a fundamental understanding of these processes. In particular, the receptors CD44 and RHAMM (Receptor for Hyaluronan-Mediated Motility) have attracted much enthusiasm, mainly because they are believed to be involved in cancer metastasis (Arch et al., 1992; Hall et al., 1995). However, a sobering shock reached the scientific community, when CD44-deficient mice were bred that had only marginal physiological impairments (Schmits et al., 1997). In addition, the receptor RHAMM became a matter of bitter scientific debate when it was found to be located mainly intracellularly (Hofmann et al., 1998a; Turley et al., 1998). Subsequently, a number of other intracellular hyaluronan-binding proteins have been found (Huang et al., 2000), though their function remains somewhat of a mystery. During the 1990s, hyaluronan synthases were cloned from different sources (Weigel et al., 1997), each capable of producing hyaluronan of different chain lengths and quantities (Itano et al., 1999).

The actions of hyaluronan as an adhesive component and also as a detachment factor appeared paradoxical. This paradox has recently been solved however, when it was realized that the cellular functions are mediated through cell-surface receptors that are susceptible to proteases (Dube et al., 2001). It appears that hyaluronan acts as an amplifier for active proteases, but as an attachment factor when proteases are inactive.

It has long been known that hyaluronan is very sensitive to breakdown by oxygen radicals (Wong et al., 1981), and it has become clear that it is the breakdown products which mediate important biological functions. Oligosaccharides of hyaluronan induce angiogenesis (West et al., 1985) and also activate lymphocytes (McKee et al., 1997; Termeer et al., 2000). Radical degradation generates reactive aldehydes (Uchiyama et al., 1990) which modify proteins into the main antigenic structures of rheumatoid arthritis (Prehm, 2000). This discovery finally terminated a long and oppressive period of ignorance in a medically important problem, and may eventually lead to a curative treatment of these diseases that currently are treated only symptomatically.

3
Chemical Structure

The complete structure of hyaluranon was elucidated by the group of Karl Meyer, who characterized the oligosaccharides obtained by the action of testicular hyaluronidases (Weissman and Meyer, 1954). Hyaluronan consists of basic disaccharide units of D-glucuronic acid and D-*N*-acetylglucosamine, these being linked together through alternating β-1,4 and β-1,3 glycosidic bonds (Figure 1).

The number of repeat disaccharides in a completed hyaluronan molecule can reach 10,000 or more, with a molecular mass of ~4×10^6 daltons (each disaccharide is ~400 daltons). In a physiological solution, the backbone of a hyaluronan molecule is stiffened by a combination of the chemical structure of the disaccharide, internal hydrogen bonds, and interactions with solvent. In addition, the preferred shape in water features hydrophobic patches on alternating sides of the flat, tape-like secondary structure. The two sides are identical, so that hyaluronan molecules are ambidextrous, enabling them to aggregate via specific interactions in water to form meshworks, even at low concentrations (Scott et al., 1991). In physiological solutions a hyaluronan molecule assumes an expanded random coil structure which occupies a very large domain. The actual mass of hyaluronan within this domain is very low, and ~0.1% molecules would overlap each other at a hyaluronan concentration of 1 mg mL^{-1}, or higher. This domain structure of hyaluronan has interesting and important consequences. Small molecules such as water, electrolytes and nutrients can diffuse freely through the solvent, within the domain; however, large molecules such as proteins will be partially excluded from the domain because of their hydrodynamic sizes in solution. This leads both to slower diffusion of macromolecules through the network and to their lower concentration in the network compared with the surrounding hyaluronan-free compartments. At pH 7, the carboxyl groups are predominantly ionized, and the hyaluronan molecule is a polyanion that has associated exchangeable cation counterions to maintain charge neutrality.

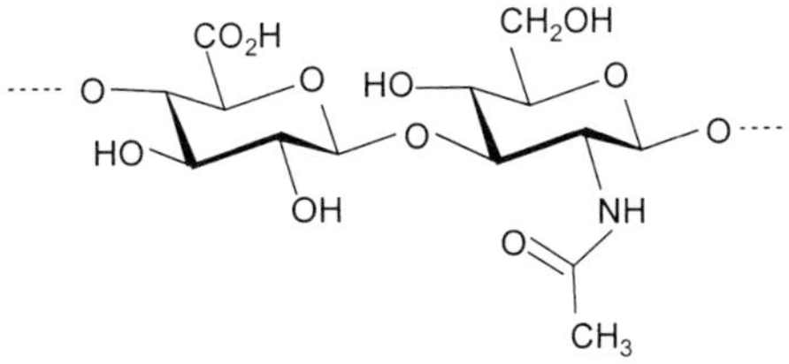

CO_2H, O, HO, CH_2OH, O, O, HO, O, OH, NH, O, CH_3

Fig. 1 Repeating unit of hyaluranon.

4
Occurrence of Hyaluronan

Hyaluronan is present in all vertebrates, and also in the capsule of some pathogenic bacteria such as *Streptococcus* sp. and *Pasteurella*. It is a component of extracellular matrices in most tissues, and in some tissues it is a major constituent. The concentration of hyaluronan is particularly high in rooster comb (7.5 mg mL^{-1}), in the synovial fluid (3–4 mg mL^{-1}), in umbilical cord (3 mg mL^{-1}), in the vitreous humor of the eye (0.2 mg mL^{-1}), and in skin (0.5 mg mL^{-1}). In other tissues that contain less hyaluronan, it forms an essential structural component of the matrix. In cartilage it forms the aggregation center for aggrecan, the large chondroitin sulfate proteoglycan, and re-

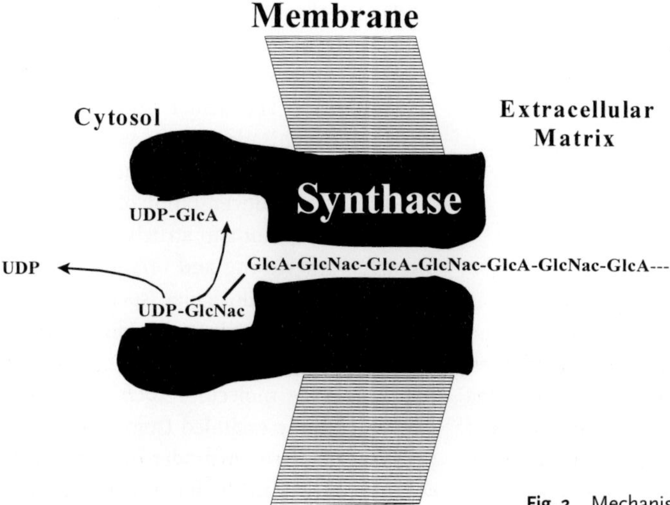

Fig. 2 Mechanism of hyaluronan synthesis.

tains this macromolecular assembly in the matrix by specific hyaluronan–protein interactions. It also forms a scaffold for binding of other matrix components around smooth muscle cells on the aorta, and on fibroblasts in the dermis of skin. The largest deposit of hyaluronan resides in the skin; in an adult human this totals ~8 g. Hyaluronan has also been detected intracellularly in proliferating cells (Evanko and Wight, 1999)

5 Mechanism of Hyaluronan Synthesis

The unusual mechanism of hyaluronan synthesis has impeded progress for a long time – a situation which has also occurred with other important polysaccharides such as cellulose and chitin. However, it now appears that two mechanisms have evolved independently – for mammalian cells and for streptococci on the one hand, and for *Pasteurella* on the other hand.

5.1 Chain Elongation

Hyaluronan synthesis in mammalian cells differs from that of other polysaccharides in many aspects. The molecule is elongated at the reducing end by alternate transfer of UDP-hyaluronan to the substrates UDP-GlcNac and UDP-GlcA, thereby liberating the UDP-moiety (Figure 2) (Prehm, 1983a,b).

Other glucosaminoglycans grow at the non-reducing end and hence require a protein backbone. Hyaluronan is synthesized at plasma membranes, the nascent chains being extruded directly into the extracellular matrix (Prehm, 1984). In contrast, other glucosaminoglycans are made in the Golgi body. Chain initiation does not require either a protein backbone (as for proteoglycans) or preformed oligosaccharides as starters; only the presence of the nucleotide sugar precursors is needed to initiate new chains. During elongation the chain is retained on the membrane-integrated synthase, this mechanism of synthesis being in operation for hyaluronan synthesis in both vertebrates and in Gram-positive streptococci. However, a different mechanism seems to exist for hyaluranon synthesis in Gram-negative *Pasteurella* in which the chains are elongated at the non-reducing end (DeAngelis, 1999).

5.2 Chain Size

One point for discussion is what determines the size of the synthesized hyaluronan, and this aspect of polymerization also applies to other macromolecular syntheses such as for proteins, DNA, or RNA. An answer was provided from experiments on isolated membranes from fibroblasts or streptococci, whereby the removal of nascent hyaluronan from the hyaluronan synthase enzyme stimulated its production. This was demonstrated in isolated streptococcal membranes (Nickel et al., 1998) and also in intact fibroblasts (Philipson et al., 1985). It thus appeared that high molecular-weight hyaluronan inhibited its own chain elongation, when it was retained on the synthase. This phenomenon may occur for solely thermodynamic reasons, because the decrease in entropy during the synthesis of a macromolecule must be compensated by free energy from cleavage of the nucleotide sugars and the subsequent formation of ordered structures. In fact, this explains why macromolecules such as proteins, RNA or DNA do not exceed a certain chain length (Peller, 1980).

5.3 Chain Export

The growing chain must be exported through a membrane pore, and consequently the proposal was made by Weigel that this pore is formed by the synthase itself, because the inactivation rate of the synthase by irradiation did not permit the participation of other proteins (Tlapak-Simmons et al., 1998). However, these authors did not show that transport of hyaluronan through the vesicle membrane was inactivated, and methods should be developed to confirm this finding.

5.4 Swelling

Hydration and swelling of nascent hyaluronan occurs at the site of synthesis on the cell surface. While swelling to enormous volumes (diameters up to 500 nm), one molecule of hyaluronan will displace many other cell-surface components by virtue of exclusion. Hence, it is conceivable that this swelling provides a mechanism whereby adhesive components are disrupted from the anchored cell.

5.5 Macromolecular Assembly

Macromolecular assembly with other matrix molecules such as proteoglycans also occurs at the cell surface. The compartmentation of hyaluronan and proteoglycan syntheses to the Golgi complex and the plasma membranes thus ensures that, during synthesis, very large aggregates are formed at the site of final deposition, and not intracellularly.

6 Hyaluronan Synthases

Hyaluronan is synthesized at the protoplast membrane of group A and group C streptococci (Markovitz and Dorfman, 1962), the enzymatic activity being solubilized by very mild detergents such as digitonin (Triscott and van de Rijn, 1986). Conventional purification procedures such as ion-exchange chromatography of detergent-solubilized membrane proteins yielded inhomogeneous protein mixtures that could not be separated into individual constituents without loss of enzymatic activity. Therefore, a new method, based on the phase separation of a detergent solution, was developed to allow purification of the synthase in its active form (Prehm et al., 1996). It was known that membrane proteins can be separated from soluble proteins by phase separation of a Triton X-114 extract. Phase separation can be induced in 1% Triton X-114 solutions by a temperature shift from 0 °C to 37 °C, with soluble proteins remaining in the aqueous phase and membrane proteins in the detergent phase. However, Triton X-114 was shown to inactivate the hyaluronan synthase. It was found that digitonin can undergo phase separation by the addition of polyethylene glycol 6000 at 0 °C, and that the synthase will remain in the aqueous phase, where it is associated with hyaluronan. Final purification of the hyaluronan synthase was achieved by ion-exchange chromatography and yielded an electrophoretically homogeneous protein of 42 kDa. This study proved that a single protein was sufficient to direct hyaluronan synthesis, and that the method may be generally applicable to other membrane proteins that are associated with polysaccharides, because it combines the advantages of the mild detergent digitonin with phase separation of all membrane proteins from polysaccharide-binding proteins.

Molecular cloning of the streptococcal hyaluronan synthase was reported independently by DeAngelis and van de Rijn (DeAngelis et al., 1993; Dougherty and van de Rijn, 1994). The gene was designated *HasA*. The *Streptococcus pyogenes* operon encodes two other proteins: HasB is a UDP-glucose-dehydrogenase, which is required to convert UDP-glucose to UDP-GlcA (Dougherty and van de Rijn, 1993), while HasC is a UDP-glucose-pyrophosphorylase, which is required to convert glucose-1-phosphate and UTP to UDP-glucose (Crater et al., 1995). Mammalian synthases were cloned simultaneously from a mutant mouse mammary carcinoma (Itano and Kimata, 1996) and *Xenopus laevis* (Meyer and Kreil, 1996). Now, three mammalian synthases are known and have been designated Has1, Has2, and Has3 (reviewed by Weigel et al., 1997). Because these proteins have 30% identity in terms of amino acid sequence with the streptococcal synthase, the genes may have a common ancestor. The synthase from *Pasteurella* has been cloned by DeAngelis et al. (1998), and is structurally unrelated to the other synthases.

7
Hyaluronan-binding Proteins and Receptors

Hyaluronan-binding proteins are constituents of the extracellular matrix, and stabilize its integrity. Hyaluronan receptors are involved in cellular signal transduction; one receptor family includes the binding proteins aggrecan, link protein, versican and neurocan and the receptors CD44, TSG6 (Lee et al., 1992), hyaluronectin (Delpech and Halavent, 1981), GHAP (Perides et al., 1990), and Lyve-1 (Banerji et al., 1999). The RHAMM receptor is an unrelated hyaluronan-binding protein, and the hyaluronan-binding sites contain a motif of a minimal site of interaction with hyaluronan. This is represented by B(X7)B, where B is any basic amino acid except histidine, and X is at least one basic amino acid and any other moiety except acidic residues. CD44 and RHAMM have attracted much attention, because they were believed to be involved in metastasis (Arch et al., 1992; Hall et al., 1995).

7.1 CD44

CD44 is a pleiomorphic extracellular matrix receptor that also binds to fibronectin and laminin, and also interacts with hyaluronan with relatively low affinity ($K_D = 10^{-8}$ M). CD44 exists as many isoforms, though some do not bind to hyaluronan. The hyaluronan-binding properties of CD44 have been implicated in promoting cell–cell aggregation and migration upon hyaluronan and collagen substrates. CD44 binds to oligosaccharides of up to 18 residues in a monovalent manner, and above 20 residues in a divalent manner, and this results in increased avidity (Lesley et al., 2000). When murine mammary carcinoma cells were transfected to produce high levels of soluble CD44, growth and metastasis *in vivo* was inhibited. In contrast, transfection with a CD44 mutant that does not bind CD44 has no effect (Peterson et al., 2000). The role of CD44 in hyaluronan synthesis shedding was analyzed in detail in metastatic and nonmetastatic melanoma cell lines that differed in degradation of CD44 and hyaluronan production (Lüke and Prehm, 1999). The nonmetastatic melanoma cell line IF6 did not significantly degrade CD44, while the metastatic cell line MV3 produced a soluble fragment. Increased hyaluronan synthesis and shedding correlated with proteolysis of CD44 on the melanoma cell lines. Intact cell surface CD44 retains hyaluronan to the vicinity of the synthase to inhibit shedding and the initiation of new chains.

The binding of hyaluronan to cells is also influenced by the level of CD44 expression (Takahashi et al., 1995; Miyake et al., 1998), by CD44 modifications such as expression of variants in different cell types (Van der Voort et al., 1995), by glycosylation (Skelton et al., 1998), by intracellular binding to ankyrin (Zhu and Bourguignon, 1998), or by phosphorylation (Puré et al., 1995). Treatment of cells with phorbol-12-myristate-13-acetate (PMA), calcium and forskolin stimulated phosphorylation of CD44, reduced the hyaluronan-binding activity (Liu et al., 1998), and stimulated hyaluronan synthesis (Klewes and Prehm, 1994). All these factors may be involved in the regulation of hyaluronan synthesis and exert profound effects on migration, growth and metastasis.

Reduced hyaluronan binding to CD44 might have two consequences:

- Loss of signal transduction from the extracellular hyaluronan to intracellular phosphorylation by the CD44-associated kinase.
- Overproduction of hyaluronan that swells to enormous volumes on the cell surface to displace cellular adhesion molecules.

Proteolysis of CD44 plays a key role in hyaluronan-mediated effects on cell growth, migration, metastasis, and adhesion (Dube et al., 2001) (Figure 3). Hyaluronan serves as an additional adhesive component that inhibits migration and proliferation, when it is retained on the cell surface by hyaluronan-binding proteins. In contrast, most transformed cells produce higher levels of surface proteases that also degrade CD44, resulting in the release of surface-bound hyaluronan and the stimulation of synthesis. In these cells, hyaluronan serves as an additional detachment factor, as the hyaluronan molecules which swell on the cell surface also displace undegraded adhesive components. Thus, hyaluronan acts as an amplifier of both active and inactive cell-surface proteases. It will therefore be difficult to identify those proteases which cleave CD44 at the surface of different cell lines. Recently, a chymotrypsin-like sheddase was thought responsible for shedding CD44 from a myoepithelial cell line (Lee et al., 2000).

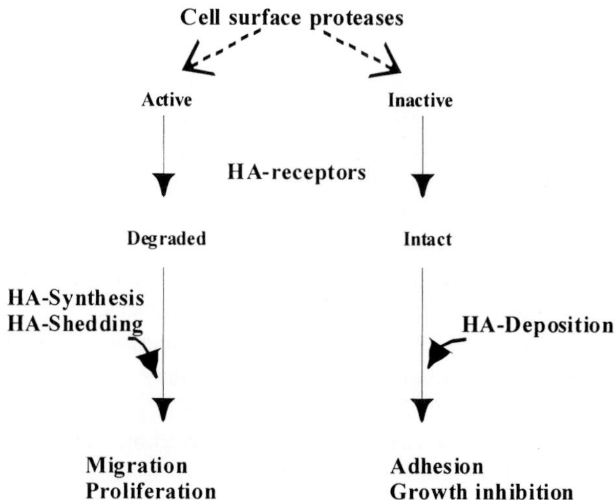

Fig. 3 Dual functions of hyaluronan.

The belief in a central role for CD44 in migration and metastasis met with severe disillusionment when CD44-deficient mice showed only marginal health disturbances (Schmits et al., 1997). This is an apparent contradiction to many studies which showed severe defects in the tissues of originally healthy animals when the function of CD44 was inhibited. A possible explanation might be that an unknown substitute of CD44 is expressed during development in CD44-defective mice (Ponta et al., 1998).

7.2 RHAMM

The hyaluronan-binding protein RHAMM was originally identified as a cell-surface protein that was also thought to be involved in hyaluronan-mediated migration and metastasis (Turley, 1989; Hall et al., 1995). Recently, a bitter scientific debate questioned its location and its function (Hofmann et al., 1998a; Turley et al., 1998), but it now appears that RHAMM is expressed either at the cell surface at low amounts under specific growth conditions, or is not expressed at all and is mainly localized intracellularly (Assmann et al., 1998; Hofmann et al., 1998b).

7.3 Other Hyaluronan-binding Proteins

Other intracellular hyaluronan-binding proteins have also been detected (Huang et al., 2000), but their functions remain elusive. Proliferating cells contain also detectable amounts of intracellular hyaluronan (Evanko and Wight, 1999) though again, the function of this is unknown.

8 Mechanisms of Hyaluronan Release from the Cell Surface

Hyaluronan could be released from cells by either enzymatic or radical degradation, or by dissociation as the intact macromolecule. Hyaluronidases were present in some transformed cells (Orkin et al., 1982), but their pH optima favor an intralysosomal function (Bernanke and Orkin, 1984); moreover, hyaluronidase-deficient cells also shed hyaluronan (Klein and von Figura, 1980). The mechanism of hyaluronan shedding from eucaryotic cells lines was analyzed by Prehm (1990). All cell lines shed identical sizes of hyaluronan as retained on the surface, but differed in the amount of hyaluronan synthesized and the ratio of released and retained hyaluronan. A method was developed which discriminated between intramolecular degradation and dissociation as the intact macromolecule. The cells were pulse-labeled to form hyaluronan chains with labeled and unlabeled segments, after which the sizes of labeled hyaluronan released into the media during the pulse extension period were determined by gel filtration. B6 cells released the same sizes as were retained on the cell surface, indicating that no intramolecular degradation occurred, and hyaluronan dissociated as the intact macromolecule. In contrast, SV3T3 cells released hyaluronan of varying molecular weight distribution during extension of the labeled segment. Shedding of smaller fragments could be prevented by radical scavengers such as superoxide dismutase and tocopherol. Therefore, SV3T3 cells released hyaluronan not only by dissociation, but also by radical degradation.

9 Regulation of Hyaluronan Synthesis

Almost any disturbance of cellular homeostasis results in a stimulation of hyaluronan production, though very few reports have investigated the inhibition of hyaluronan synthesis. Three levels of regulation have been recognized:

1) Expression of the synthase.
2) Stimulation and inhibition of the synthase by growth or differentiation factors acting on a specific target cell.
3) Disturbances in the integrity of the extracellular matrix, particularly the degradation of cell surface-bound hyaluronan.

9.1 Expression of the Synthase

The synthase has a high turnover rate and a half-life of 82 min, indicating a strict control of its activity by the transcription rate (Bansal and Mason, 1986). The three mammalian synthases are expressed in different tissues, thereby producing hyaluronan in different amounts and sizes (Itano et al., 1999).

9.2 Stimulation and Inhibition of the Synthase

Many growth or differentiation factors have been shown to stimulate hyaluronan synthesis (Tomida et al., 1975, 1977a,b; Lembach, 1976; Prehm, 1980; D'Arville and Mason, 1983; Hamerman and Wood, 1984; Hamerman et al., 1986; Pulkki, 1986; Tammi and Tammi, 1986; Wu and Wu, 1986; Heldin et al., 1989). During the early part of the 20th century, Kabat (1939) had already recognized enhanced hyaluronan production in tumors of sarcosis and leukosis. Viral transformation by Rous sarcoma virus (RSV) stimulates hyaluronan production in chondrocytes (Mikuni Takagaki and Toole, 1981), in chicken fibroblasts (Hamerman et al., 1965; Ishimoto et al., 1966), and in myoblasts (Yoshimura, 1985), whilst SV40 transformation induces a similar increase in 3T3 fibroblasts (Hopwood and Dorfman, 1977; Goldberg et al., 1984).

It is surprising that only two inhibitors of hyaluronan synthases have been discovered. Periodate-oxidized UDP-GlcA or UDP-GlcNac have been used as suicide inhibitors of the synthase (Prehm, 1985). These inhibitors do not surmount cell membranes, and must be imported into the cytoplasm by osmotic lysis of pinocytotic vesicles. Vesnarinone suppresses the synthase activity in intact myofibroblasts (Ueki et al., 2000). Hence, it may be worthwhile developing or isolating inhibitors of hyaluronan synthesis that might be used as curative drugs in pathological disturbances related to hyaluranon production, for example edema.

9.2.1
Signal Transduction at Membranes

The hyaluronan receptor CD44 is responsible for outside-in signaling (Perschl et al., 1995) and, intracellularly, this is associated with the tyrosine kinase $p56^{lck}$. Binding of hyaluronan stimulates intracellular phosphorylation (Taher et al., 1996). Similar phosphorylation was observed with bivalent antibodies to CD44, indicating that receptor cross-linking on the cell surface instigates the signal transduction cascade. Thus, the mechanism resembles growth factor-dependent signal transduction pathways. It is also possible that the angiogenic effects of hyaluronan oligosaccharides are transduced through CD44, though this mechanism does not apply to all cell types because CD44-deficient lymphocytes might also be activated by hyaluronan oligosaccharides (Termeer et al., 2000). Hence, other unknown receptors or signal transduction pathways must exist.

9.2.2
Intracellular Signal Transduction

Intracellular signal transduction pathways are dependent on protein synthesis and activation of protein kinase C (Heldin et al., 1992), and have been studied in detail in both B6 cells and 3T3 fibroblasts (Klewes and Prehm, 1994). Activation by fetal calf serum was inhibited by cycloheximide or by the protein kinase inhibitors H-7 or H-8, indicating that transcription as well as phosphorylation were required for activation. The activation by serum was markedly prolonged, when serum was added together with cholera toxin or theophylline. Without serum stimulation the hyaluronan synthase could also be activated by PMA, by dibutyryl-c-AMP, or forskolin. Increasing the intracellular Ca^{2+} concentration with a Ca-ionophore also led to an activation. In isolated plasma membranes the synthase activity could be decreased by phosphatase treatment, and enhanced by ATP in B6 cells and by ATP in the presence of PMA in 3T3 fibroblasts. In conclusion, hyaluronan synthase is induced by transcription and activated by phosphorylation by protein kinase C, c-AMP-dependent protein kinases or Ca^{2+}-dependent protein kinases.

9.3
Influence of Chain Length on Further Elongation

Another factor is the amount of cell surface hyaluronan which influenced its own production. Notably, hyaluronidase treatment of intact cells stimulated synthesis (Philipson et al., 1985). Interesting features on the regulation of hyaluronan synthesis by the growing hyaluronan chain itself have been obtained from studies with streptococci (Nickel et al., 1998). Group A and C streptococci differ in their capacity to retain hyaluronan as a coat on their cell surface. In group C streptococci, a 56-kDa hyaluronan receptor was closely associated with the synthase; this protein had an intrinsic kinase activity that performed autophosphorylation in response to extracellular ATP. Autophosphorylation of the 56-kDa protein led in turn to a reduction in hyaluronan binding and increased shedding of the hyaluronan capsule. Simultaneously, the synthase increased its activity to replace the lost hyaluronan chains. It thus appears that a large hyaluronan chain inhibited its own elongation, when it was retained in the vicinity of the synthase.

A similar mechanism for the regulation of hyaluronan synthesis operates on eucaryotic cells that express the CD44 receptor (Lüke and Prehm, 1999). Proteolytic degradation of CD44 in melanoma cells correlated with the metastatic potential and activation of the hyaluronan synthesis. This activation was the result of facilitated dissociation of growing hyaluronan chains from plasma membranes. These results add a new function to the CD44 receptor as a regulator of hyaluronan synthesis.

10 Turnover and Catabolism

Within the circulation of an adult human, a total of 34 mg of hyaluronan is turned over each day (Fraser and Laurent, 1989; Lebel et al., 1994). The major origins of hyaluronan are the joints, skin, eyes, and intestine. In skin and joints the half-life is about 12 h (Reed et al., 1990; Coleman et al., 2000), whilst in the anterior chamber of the eye it is 1–1.5 h (Laurent et al., 1993), and in the vitreous humor it is 70 days. The rapid turnover is somewhat surprising because hyaluronan has been regarded as a structural component of connective tissue.

Hyaluronan is drained through the lymph and reaches the circulating blood, where the serum concentration may reach ~31 ng L^{-1}. Hyaluranon is effectively endocytosed by the liver, the liver endothelial cells expressing a membrane receptor that endocytically clears hyaluronan from the circulation (Zhou et al., 1999). It is then transported into lysosomes that contain hyaluronidase, β-glucuronidase and β-*N*-acetyl-glucosaminidase (Roden et al., 1989).

11 Functions of Hyaluronan

Many different functions have been assigned to hyaluronan, but these can be grouped into cellular, physiological, and pathological functions. Most functions are determined either by the physical properties or by interactions with hyaluronan-binding proteins.

11.1 Cellular Functions

Fibroblasts are surrounded by a coat of hyaluronan that is lost upon transformation (Underhill and Toole, 1982) but which modifies cell–cell aggregation (Underhill, 1982) and cell–substratum adhesion (Erickson and Turley, 1983). Hyaluronan synthesis is coordinated with cell growth, with proliferating cells producing more hyaluronan than resting cells (Tomida et al., 1975; Mian, 1986). Synchronized fibroblasts show the highest hyaluronan production during mitosis, because hyaluronan synthesis is required for the detachment and mitosis of fibroblasts (Brecht et al., 1986).

The function of hyaluronan on cellular behavior appeared paradoxical, as hyaluronan synthesis was seen to be increased during cell migration (Toole, 1991), mitosis (Brecht et al., 1986), and tumor invasion (Toole et al., 1979). Overproduction of hyaluronan in the human tumor cell line HT1080 enhanced anchorage-independent growth and tumorigenicity (Kosaki et al., 1999). In contrast, hyaluronan promotes cell adhesion (Miyake et al., 1990) and cell–cell aggregation (Underhill, 1982), but inhibits cell proliferation (Goldberg and Toole, 1987; West and Kumar, 1989).

A similar paradoxical situation applies for hyaluronidases in tumor progression. In some tumors, hyaluronidase treatment blocks lymph node invasion by tumor cells for T-cell lymphoma (Zahalka et al., 1995), but overexpression of hyaluronidase correlates with disease progression in bladder, breast and prostate cancer (Lokeshwar et al., 1996, 1997; Bertrand et al., 1997).

A solution to this paradox was obtained from a comparison of the cellular behavior of hyaluronan-deficient Chinese hamster ovary (CHO) cells and CHO cells transfected with the hyaluronan synthase. Surprisingly, hyaluronan synthesis reduced initial cell adhesion, migration, growth and the density at contact inhibition. All these apparent contradictions were combined into a model for the cellular functions of hyaluronan (Dube et al., 2001). Thus, hyaluronan serves as an adhesive component when it is retained on the cell surface by intact CD44, and as a detachment factor when it is released. Migration- and growth- inhibitory effects of hyaluronan are mediated by the proteolytic cleavage of CD44. The cellular function of hyaluronan might also be an amplifier of cell-surface proteases: when proteases are inactive, hyaluranon amplifies the action of cellular adhesion factors by deposition and binding to receptors. However, when proteases are active, it amplifies cell detachment from the environment by activation of synthesis and shedding (see Figure 3).

11.2 Physiological Functions

A prominent physiological function of hyaluronan is the creation of hydrated pathways that allow the cells to penetrate cellular and fibrous barriers. Such hydrated pericellular matrices are required not only for cell rounding in mitosis, but also for cell migration during morphogenesis and wound healing.

11.2.1 Differentiation and Morphogenesis

Studies carried out on embryonic limb development have illustrated how hyaluronan modulates differentiation *in vivo* (Toole, 1991). Early limb mesodermal cells are surrounded and separated by an extensive, hyaluronan-enriched matrix. At this stage, pericellular hyaluronan appears to be tethered to the membrane-integrated synthase. This hydrated matrix facilitates the proliferation and migration of early mesenchymal precursors. Subsequently, the mesoderm condenses, i.e., the intercellular matrix decreases in volume at the sites of future cartilage and muscle differentiation. Mesodermal cells lose the ability to form hydrated matrices, the level of lysosomal hyaluronidases increases, and much of the pericellular hyaluronan is removed. The remaining hyaluronan is now retained at the cell surface by receptors. Similar events occur in the early mesodermal development of skin and teeth. Further differentiation of condensed limb mesoderm into cartilage is accompanied by recovery of matrix formation and by extensive production of proteoglycans that are tethered to the cell surface via hyaluronan and CD44.

Has2$^{-/-}$ mouse mutants contain virtually no hyaluranon at the E9.5 (embryo day 9.5) stage of development (Camenisch et al., 2000), and at this stage they exhibit multiple defects, including yolk sac, vasculature and heart abnormalities. The cardiac jelly and cardiac cushions fail to form valves and other structures.

11.2.2 Wound Healing

The following scenario has been proposed as a model for wound healing by (Weigel et al., 1986). After a wound has been sealed by the platelet plug and fibrin clot, platelet activators stimulate the inflammation and surrounding cells to synthesize hyaluronan into the matrix of the fibrin clot. The clot swells and becomes more porous in order to facilitate the migration of cells into the fibrin matrix. As more cells migrate into the wound, the hyaluronan–fibrin matrix is degraded and replaced by collagens and proteoglycans. The hyaluronan degradation products stimulate angiogenesis and the formation of new blood vessels.

Wound healing in fetal tissues occurs without scarring and is correlated with the prolonged presence of hyaluronan (Longaker et al., 1991). Hyaluronidases that are present in adult wounds were considered responsible for fibrotic healing (West et al., 1997).

11.2.3 Synovia

Hyaluronan is produced by the fibroblast-like type A cells in the upper synovial lining (Pitsillides et al., 1993), and not only serves as a lubricant but also enhances the resistance of the synovial linings to fluid outflow (Coleman et al., 1997). The concentration of hyaluronan in the synovial fluid ranges from 1.4 to 3.6 mg mL^{-1}. Its molecular mass is variable, and in normal synovial fluid has been estimated as 7.0×10^6 daltons, though this is decreased to $3-5 \times 10^6$ daltons in rheumatoid synovial fluids (Balazs et al., 1967; Dahl et al., 1985). The turnover rate depends on the size, indicating a partial reflection of hyaluronan by the lining (Coleman et al., 2000). Hyaluronan is drained through the lymphatic vessels and catabolized in lymph nodes (Fraser et al., 1996).

11.3 Pathological Functions

Aberrant hyaluronan synthesis will lead to disturbances of cell behavior and tissue integrity. For *Streptococcus* sp. and *Pasteurella*, this serves as a non-antigenic disguising capsule in the infected host.

11.3.1 Metastasis

Most malignant solid tumors contain elevated levels of hyaluronan; such enrichment of hyaluronan in tumors may be caused by increased production by the tumors themselves, or by induction in the surrounding stromal cells (Toole et al., 1979; Knudson et al., 1984). The mechanisms whereby hyaluronan–receptor interactions influence tumor cell behavior are not clearly understood, and this is currently a highly active area of investigation.

11.3.2 Edema

Inflammation of various organs is often accompanied with an accumulation of hyaluronan. This can cause edema due to the osmotic activity, and can in turn lead to dysfunction of the organs. For example, hyaluronan accumulation in the rheumatoid joint impedes flexibility of the joint (Engström-Laurent, 1997), whilst accumulation in rejected transplanted kidneys can cause edema and increased intracapsular pressure (Hällgren et al., 1990). Hyaluranon also accumulates in pulmonary edema (Nettelbladt et al., 1989) and during myocardial infarction (Waldenstrom et al., 1991).

11.3.3 Streptococci

Hyaluronan is produced by group A and C streptococci and deposited into a capsule which serves as a major virulence factor of pathogenic streptococci (Kass and Seastone, 1944; Wessels et al., 1991, 1994) because it protects the bacteria against phagocytosis (Whitnack et al., 1981) and oxygen damage (Cleary and Larkin, 1979). Another virulence factor is the hyaluronidase, which is a spreading factor and facilitates penetration of the bacteria through infected tissue (McClean, 1941; MacLennan, 1956). When streptococci enter the stationary phase, they lose the capsule (van de Rijn, 1983), but remain virulent. A constant rate of hyaluronan shedding from the capsule might be advantageous for bacteria, because they will eliminate host components such as attacking antibodies.

A 56-kDa protein was identified as the first example of a prokaryotic extracellular protein kinase. This is expressed on the surface of group C streptococci, and can bind and retain hyaluronan in the absence of ATP. The capsule is shed in the presence of extracellular ATP (Nickel et al., 1998). An equivalent hyaluronan-binding protein was not detected on group A streptococci that do not retain their capsule to the same extent.

Group A and C streptococci differ both in their hosts and their infection routes. Group A streptococci are virulent human pathogens and cause tonsilitis, scarlet fever and rheumatic fever, the infection route being via the throat. In contrast, group C streptococci are mainly animal pathogens, but can also infect humans; their infection route is mainly opportunistic through wounds in the skin.

Group C streptococci have clearly evolved a mechanism to retain their hyaluronan capsule that protects them against desiccation when they are localized on the skin surface. When they enter a wound, they encounter high concentrations of ATP from necrotic tissue and also need to defend themselves against attacking antibodies and macrophages. This they do by shedding the capsule, thereby preventing any attachment.

The cell surface-bound kinase on group C streptococci also elicits antibodies in the infected host. These antibodies show an immunological cross-reaction with cell surfaces proteins from fibroblasts, and are cytotoxic in the presence of complement (Prehm et al., 1995). This protein has recently been used as a vaccine to protect mice against fatal group C streptococci (Chanter et al., 1999).

12 Hyaluronan Degradation

As a very large molecule, hyaluronan is prone to mechanical degradation either by ultrasonic treatment or by thermal degradation at elevated temperatures. Physiologically, hyaluronan can be degraded by oxygen free radicals or by hyaluronidases.

12.1 Degradation by Free Radicals

Oxygen free radical degradation is mostly a side reaction of activated neutrophil granulocytes or monocytes in an inflammation. Radical degradation of hyaluronan results in a dramatic drop of the viscosity and function of the synovial fluid in the inflamed joints of patients with rheumatoid arthritis. This mechanism was hypothesized by Greenwald and Moak (1986). Although a hyaluronan structure modified by radical damage has been detected in the synovial fluid of the inflamed rheumatoid joint (Grootveld et al., 1991), direct biochemical proof for hyaluronan degradation was obtained from organ cultures of healthy and rheumatoid synovial tissue (Schenck et al., 1995). Healthy tissues and some arthritic tissues did not contain significant amounts of granulocytes and produced high molecular-weight hyaluronan. In contrast, arthritic tissue infiltrated with granulocytes released low molecular-weight hyaluronan. Hyaluronan degradation was accompanied by massive oxygen radical production. Radical scavengers protected hyaluronan from degradation in synovial tissue, in particular by the iron chelators DETAPAC or Desferal that block the formation of hydroxyl radicals. Hydroxyl radicals degrade hyaluronan with an efficiency of 100%, with 65% being cleaved within 15 min and the remaining radicals giving rise to thermally labile products that eventually lead to chain scission (Al Assaf et al., 2000). A variety of free aldehydes and oligosaccharides are produced in the rheumatoid synovial fluid (Chapman et al., 1989; Grootveld et al., 1991). The structures formed by oxygen radical damage of hyaluronan have been analyzed in detail by Uchiyama et al. (1990), the main degradation product being L-threotetradialdose.

12.2 Degradation by Hyaluronidases

Hyaluronidases are classified into three groups (Kreil, 1995).

1) Mammalian-type hyaluronidases (EC 3.2.1.35) are endo-β-*N*-acetylhexosaminidases and produce tetrasaccharides and hexasaccharides as the major end-products. They have both hydrolytic and transglycosidase activities, and can degrade hyaluronan and chondroitin sulfates.
2) Bacterial hyaluronidases (EC 4.2.99.1) are endo-β-*N*-acetylhexosaminidases and yield primarily disaccharides. They operate by a β-elimination reaction.
3) Hyaluronidases (EC 3.2.1.36) from leeches, other parasites, and crustaceans are endo-β-glucuronidases that generate tetrasaccharide and hexasaccharide end-products.

Hyaluronidase from testis has long been known as a spreading factor (Chain and Duthie, 1940). Other hyaluronidases from vertebrate tissues are more difficult to purify and analyze, because they occur in low concentrations and are often unstable (Csóka et al., 1997).

13 Production

Hyaluronan is produced either from rooster combs (Pharmacia AB, Uppsala, Sweden, have developed a special strain of roosters with highly luxuriant combs) or from streptococci. Commercially available hyaluranon is produced in molecular weights ranging from less than 10^6 daltons to as high as 8×10^6 daltons. There are many hyaluronan producers worldwide. For example, Genzyme Corp. (Framingham, MA, USA) merged with Biomatrix and operates a plant producing hyaluronan from mammalian sources in Canada. Anika (Woburn, MA, USA); Lifecore Biomedical (Chaska, MN, USA) are other suppliers. Pharmacia produces hyaluronan in Sweden, Fidia Advanced Biopolymers (Brindisi) in Italy, Bio-Technology General Corp. (Iselin, NJ, USA) in Israel, and a number of companies, including Kibun Food Chemifa Co. and Seikagaku Corp. (both Tokyo), in Japan.

Today there are still many products on the market that contain hyaluronan isolated from rooster combs, because this has set the standards for high molecular weight, purity and noninflammatory properties. Even if hyaluronan from streptococci meets these criteria, its market penetration is hampered by the reluctance of customers to change to cheaper medical or cosmetic products. In addition many of the streptococcal preparations are not characterized as thoroughly, or are less pure (Manna et al., 1999). The development of large-scale production from *Streptococcus zooepidemicus* cultures had to overcome many obstacles, including: growth in chemically defined media (Kjems and Lebech, 1976; van de Rijn and Kessler, 1980; Armstrong et al., 1997; Cooney et al., 1999); the production of high molecular-weight material (Kim et al., 1996); the elimination of toxic impurity such as streptolysin; and increasing the yield to about 7 g L^{-1} (Lowther and Rogers, 1956; Thonard et al., 1964; Gerlach and Kohler, 1970).

13.1 Patents

The patent literature concerning hyaluranon is extensive, the number of annual patent applications having increased from fewer than 10 before 1985 to about 200 in 2000. The major inventors in this field and their corporate affiliations are summarized in Table 1. Many different processing techniques and uses for hyaluronan have been developed and patented by Balazs and his coworkers. For example, the important Balazs patent (which was issued in 1979 but now has expired) on hyaluronan isolated from animal tissue that does not cause an inflammatory response when tested in the eye of the owl monkey (Balazs, 1979). This is marketed by Upjohn-Pharmacia as Healon®, a sterile, pyrogen-free, nonantigenic and noninflammatory, high molecular-weight fraction of hyaluronan.

13.2 Market

The world market of hyaluronan is difficult to estimate, because many companies produce it for medical and cosmetic applications. The current US market of $157 million for viscosupplementation in osteoarthritis is driven by two products: Synvisc® from Biomatrix (now Genzyme) and Hyalgan® from Fidia Pharmaceutical. In Europe, Fidia's Hyalgan® is the leading hyaluronan-based viscosupplement product. Details of the major pharmaceutical products are listed in Table 2.

Tab. 1 Some hyaluronan-related patents

Patent no.	*Patent holder*	*Inventors*	*Title*	*Date*
U.S. 4,141,973	Biomatrix	Balazs	Ultrapure hyaluronic acid and the use thereof	1979
U.S. 4,713,448	Biomatrix	Balazs	Chemically modified hyaluronic acid preparation and method of recovery thereof from animal tissues	1987
U.S. 4,957,744	Fidia	della Valle	Esters of hyaluronic acid	1990
U.S. 4,636,524	Biomatrix	Balazs	Cross-linked gels of hyaluronic acid and products containing such gels	1987
U.S. 4,937,270	Genzyme	Hamilton	Water-insoluble derivatives of hyaluronic acid	1990
U.S. 5,644,049	Murst Italian	Giusti	Biomaterial comprising hyaluronic acid and derivatives thereof in interpenetrating polymer networks	1997
U.S. 4,663,233	Biocaot	Beavers	Lens with hydrophilic coating	1987
U.S. 5,585,361	Burns	Genzyme	Method for the inhibition of platelet adherence and aggregation	1996

Tab. 2 Some commercially available pharmaceuticals containing hyaluronan

	Concentration [mg mL⁻¹]	*Size [mL]*	*Manufacturer*	*Application*	*Price (US $)*
AMO Vitrax Syringe	30	65	Allergen	Ophthalmology	138
Amvisc Plus Syringe	16	5	Chiron	Ophthalmology	145
Amvisc Plus Syringe	16	8	Chiron	Ophthalmology	190
Amvisc Syringe	12	5	Chiron	Ophthalmology	112
Healon GV Syringe	14	55	Kabi	Ophthalmology	101
Healon GV Syringe	14	85	Kabi	Ophthalmology	131
Healon Syringe	10	55	Kabi	Ophthalmology	94
Hyalgan SDV	10	2	Sanofi	Osteoarthritis	166
Hyalgan Syringe	10	2	Sanofi	Osteoarthritis	166
Provisc Syringe	10	4	Alcon	Ophthalmology	117
Provisc Syringe	10	55	Alcon	Ophthalmology	142
Provisc Syringe	10	85	Alcon	Ophthalmology	178
Synvisc Syringe	8	3×2.25	Wyeth	Osteoarthritis	705
Viscoat Syringe	40	5	Alcon	Ophthalmology	151

14 Medical Applications

Hyaluronan preparations or higher cross-linked products are increasingly used for many medical applications, such as ophthalmic viscosurgery, supplementation of the synovial fluid in patients with osteoarthritis (Balazs and Denlinger, 1989), as membranes for postsurgical separation of tissues (Burns et al., 1997), and as drug delivery systems (Vercruysse and Prestwich, 1998). Many cosmetics contain hyaluronan as an ingredient, because it is believed to keep skin young and fresh by preventing dryness as a result of its water-binding capacity, though scientific evidence on this subject is lacking.

14.1 Ophthalmics

Hyaluronan was first marketed in the early 1980s as Healon® in the ophthalmic field as a viscous gel which could be injected into the anterior chamber of the eye to protect tissues such as the corneal endothelium (Miller and Stegmann, 1983). A number of other hyaluronan products are now marketed by competing firms in the ophthalmic market.

14.2 Arthritis

Intra-articular administration of hyaluronan has been used in animals and man with reported clinical efficacy. In man, hyaluronan is being used to relieve pain and improve joint mobility in the treatment of osteoarthritis with intra-articular injections of Hyalgan® (Sanofi Pharmaceuticals), Orthovisc® (Anika Therapeutics), and SynVisc® (Biomatrix, now Genzyme). It has also been proposed for several degenerative joint diseases as an alternative to the traditional steroid therapy (Altman and Moskowitz, 1998; Wobig et al., 1999; Adams et al., 2000).

14.3 Wound Healing and Scarring

Hyaluronan products have been developed to foster the healing process. For burn and chronic ulcer patients, a line of modified hyaluronan products based on a HYAFF™

polymer is being marketed by Convatec in Europe, and is currently undergoing clinical trials in the US (Goa and Benfield, 1994).

14.4 Adhesion Prevention

Most surgical procedures are accompanied by undesired tissue damage caused by cutting, desiccation, lack of adequate blood supply and manipulative abrasion, and undesired connective tissue bridges (adhesions) are often formed on the damaged surfaces of organs. Hyaluronan preparations such as Seprafilm® from Genzyme can reduce such adhesions and improve the surgical outcome (Beck, 1997).

14.5 Drug Delivery

Hyaluronan is an ideal molecule for use as a carrier of drugs, particularly for local administration. As the polysaccharide is a ubiquitous component of tissues and fluids, it is immunologically inert. It can be metabolized in the lysosomes of certain cells, and the backbone of the molecule provides different chemical groups for drug attachment. Investigations are on-going for topical and intravenous drug delivery systems using modified hyaluronans (Vercruysse and Prestwich, 1998).

15 Effects of Hyaluronan Oligosaccharides

Hyaluronan oligosaccharides augment fibroblast proliferation, migration, stimulate the formation of new blood vessels, and also activate macrophages. These effects have been studied in detail in cell culture, where they stimulate proliferation of synovial fibroblasts *in vitro* (Goldberg and Toole, 1987), induce angiogenesis (Montesano et al., 1996), stimulate chemokine production on macrophages (McKee et al., 1996, 1997; Horton et al., 1998), and activate lymphocytes (Termeer et al., 2000).

16 Outlook and Perspectives

In times characterized by an enthusiasm for the achievements of molecular biology that are often announced with great fanfare for the welfare of mankind, it may be wise to issue a reminder that the understanding and management of some diseases require a horizon beyond the application of gene technology. Indeed, research on hyaluronan may be a typical illustration that pertinent yet silent progress contributes significantly to the solution of medical problems.

Nonetheless, many cellular and physiological functions of hyaluronan remain elusive. There is a notion amongst the scientific community that hyaluronan participates in the pathogenesis of metastasis and rheumatoid autoimmune diseases, and research into hyaluronan–cell interactions will clearly contribute to therapeutic interventions in these diseases. A second promising area is the development of hyaluronan-based biomaterials and hyaluronan-modified surfaces. Indeed, a number of important products have already reached the market, and the introduction of many hyaluronan-derived devices and drugs are eagerly anticipated during the next decade.

17 References

Abatangelo, G. and Weigel, P.H. (2000) *Redefining Hyaluronan*. Amsterdam: Wheley.

Adams, M. E., Lussier, A. J., Peyron, J. G. (2000) A risk-benefit assessment of injections of hyaluronan and its derivatives in the treatment of osteoarthritis of the knee. *Drug Safety* **23**, 115–130.

Al Assaf, S., Meadows, J., Phillips, G. O., Williams, P. A., Parsons, B. J. (2000) The effect of hydroxyl radicals on the rheological performance of hylan and hyaluronan. *Int. J. Biol. Macromol.* **27**, 337–348.

Altman, R. D., Moskowitz, R. (1998) Intraarticular sodium hyaluronate (Hyalgan®) in the treatment of patients with osteoarthritis of the knee: a randomized clinical trial. *J. Rheumatol.* **25**, 2203–2212.

Arch, R., Wirth, K., Hofmann, M., Ponta, H., Matzku, S., Herrlich, P., Zoller, M. (1992) Participation in normal immune responses of a metastasis-inducing splice variant of CD44. *Science* **257**, 682–685.

Armstrong, D. C., Cooney, M. J., Johns, M. R. (1997) Growth and amino acid requirements of hyaluronic-acid-producing *Streptococcus zooepidemicus*. *Appl. Microbiol. Biotechnol.* **47**, 309–312.

Aruffo, A., Stamenkovic, I., Melnick, M., Underhill, C. B., Seed, B. (1990) CD44 is the principal cell surface receptor for hyaluronate. *Cell* **61**, 1303–1313.

Assmann, V., Marshall, J. F., Fieber, C., Hofmann, M., Hart, I. R. (1998) The human hyaluronan receptor RHAMM is expressed as an intracellular protein in breast cancer cells. *J. Cell Sci.* **111**, 1685–1694.

Balazs, E. A. (1970) Chemistry and Molecular Biology of the Intercellular Matrix. London: Academic Press.

Balazs, E. A. (1979) Ultrapure hyaluronic acid and the use thereof. U.S. Patent 4,141,973.

Balazs, E. A. (1982) Use of hyaluronic acid in eye surgery. *Ann. Ther. Clin. Ophthalmol.* **33**, 95–110.

Balazs, E. A. (1987a) Chemically modified hyaluronic acid preparation and method of recovery thereof from animal tissues. U.S. Patent 4,713,448.

Balazs, E. A. (1987b) Cross-linked gels of hyaluronic acid and products containing such gels. U.S. Patent 4,636,524.

Balazs, E. A., Denlinger, J. L. (1989) Clinical uses of hyaluronan. *Ciba Found. Symp.* **143**, 265–275.

Balazs, E. A., Watson, D., Duff, I. F., Roseman, S. (1967) Hyaluronic acid in synovial fluid. I. Molecular parameters of hyaluronic acid in normal and arthritis human fluids. *Arthritis Rheum.* **10**, 357–376.

Banerji, S., Ni, J., Wang, S.X., Clasper, S., Su, J., Tammi, R., Jones, M., Jackson, D. G. (1999) LYVE-1, a new homologue of the CD44 glycoprotein, is a lymph-specific receptor for hyaluronan. *J. Cell Biol.* **144**, 789–801.

Bansal, M. K., Mason, R. M. (1986) Evidence for rapid metabolic turnover of hyaluronate synthetase in Swarm rat chondrosarcoma chondrocytes. *Biochem. J.* **236**, 515–519.

Beavers, E. M. (1987) Lens with hydrophilic coating. U.S. Patent 4,663,233.

Beck, D. E. (1997) The role of Seprafilm™ bioresorbable membrane in adhesion prevention. *Eur. J. Surg.*, **163** (Suppl. 577), 49–55.

Bernanke, D. H., Orkin, R. W. (1984) Hyaluronate binding and degradation by cultured embryonic chick cardiac cushion and myocardial cells. *Dev. Biol.*, **106**, 360–367.

Bertrand, P., Girard, N., Duval, C., D'Anjou, J., Chauzy, C., Ménard, J. F., Delpech, B. (1997)

Increased hyaluronidase levels in breast tumor metastases. *Int. J. Cancer* **73**, 327–331.

Brecht, M., Mayer, U., Schlosser, E., Prehm, P. (1986) Increased hyaluronate synthesis is required for fibroblast detachment and mitosis. *Biochem. J.* **239**, 445–450.

Burns, J. W., Valeri, C. R. (1996) Method for the inhibition of platelet adherence and aggregation. U.S. Patent 5,585,361.

Burns, J. W., Colt, M. J., Burgess, L. S., Skinner, K. C. (1997) Preclinical evaluation of Seprafilm™ bioresorbable membrane. *Eur. J. Surg.* **163** (Suppl. 577), 40–48.

Callegaro, L., Giusti, P. (1997) Biomaterial comprising hyaluronic acid and derivatives thereof in interpenetrating polymer networks. U.S. Patent 5,644,049.

Camenisch, T. D., Spicer, A. P., Brehm-Gibson, T., Biesterfeldt, J., Augustine, M. L., Calabro, A., Jr., Kubalak, S., Klewer, S. E., McDonald, J. A. (2000) Disruption of hyaluronan synthase-2 abrogates normal cardiac morphogenesis and hyaluronan-mediated transformation of epithelium to mesenchyme. *J. Clin. Invest.* **106**, 349–360.

Chain, E., Duthie, E. S. (1940) Identity of hyaluronidase and spreading factor. *Br. J. Exp. Pathol.* **21**, 324–338.

Chanter, N., Ward, C. L., Talbot, N. C., Flanagan, J. A., Binns, M., Houghton, S. B., Smith, K. C., Mumford, J. A. (1999) Recombinant hyaluronate associated protein as a protective immunogen against *Streptococcus equi* and *Streptococcus zooepidemicus* challenge in mice. *Microbiol. Pathog.* **27**, 133–143.

Chapman, M. L., Rubin, B. R., Gracy, R. W. (1989) Increased carbonyl content of proteins in synovial fluid from patients with rheumatoid arthritis. *J. Rheumatol.* **16**, 15–18.

Cleary, P. P., Larkin, A. (1979) Hyaluronic acid capsule: strategy for oxygen resistance in group A streptococci. *J. Bacteriol.* **140**, 1090–1097.

Coleman, P. J., Scott, D., Ray, J., Mason, R. M., Levick, J. R. (1997) Hyaluronan secretion into the synovial cavity of rabbit knees and comparison with albumin turnover. *J. Physiol.* **503** , 645–656.

Coleman, P. J., Scott, D., Mason, R. M., Levick, J. R. (2000) Role of hyaluronan chain length in buffering interstitial flow across synovium in rabbits. *J. Physiol.* **526**, 425–434.

Comper, W. D., Laurent, T. C. (1978) Physiological function of connective tissue polysaccharides. *Physiol. Rev.* **58**, 255–315.

Cooney, M. J., Goh, L. T., Lee, P. L., Johns, M. R. (1999) Structured model-based analysis and control of the hyaluronic acid fermentation by *Streptococcus zooepidemicus*: physiological implications of glucose and complex nitrogen-limited growth. *Biotechnol. Prog.* **15**, 898–910.

Crater, D. L., Dougherty, B. A., van de Rijn, I. (1995) Molecular characterization of *has*C from an operon required for hyaluronic acid synthesis in group A streptococci – Demonstration of UDP-glucose pyrophosphorylase activity. *J. Biol. Chem.* **270**, 28676–28680.

Csóka, T. B., Frost, G. I., Wong, T., Stern, R. (1997) Purification and microsequencing of hyaluronidase isozymes from human urine. *FEBS Lett.* **417**, 307–310.

D'Arville, C., Mason, R. M. (1983) Effects of serum and insulin on hyaluronate synthesis by cultures of chondrocytes from the Swarm rat chondrosarcoma. *Biochim. Biophys. Acta* **760**, 53–60.

Dahl, L. B., Dahl, I. M., Engstrom Laurent, A., Granath, K. (1985) Concentration and molecular weight of sodium hyaluronate in synovial fluid from patients with rheumatoid arthritis and other arthropathies. *Ann. Rheum. Dis.* **44**, 817–822.

DeAngelis, P. L. (1999) Molecular directionality of polysaccharide polymerization by the *Pasteurella multocida* hyaluronan synthase. *J. Biol. Chem.* **274**, 26557–26562.

DeAngelis, P. L., Papaconstantinou, J., Weigel, P. H. (1993) Molecular cloning, identification, and sequence of the hyaluronan synthase gene from group A *Streptococcus pyogenes*. *J. Biol. Chem.* **268**, 19181–19184.

DeAngelis, P. L., Jing, W., Drake, R. R., Achyuthan, A. M. (1998) Identification and molecular cloning of a unique hyaluronan synthase from *Pasteurella multocida*. *J. Biol. Chem.* **273**, 8454–8458.

della Valle, F. (2001) Esters of hyaluronic acid. U.S. Patent 4,957,744.

Delpech, B., Halavent, C. (1981) Characterization and purification from human brain of a hyaluronic acid-binding glycoprotein, hyaluronectin. *J. Neurochem.* **36**, 855–859.

Dougherty, B. A., van de Rijn, I. (1993) Molecular characterization of hasB from an operon required for hyaluronic acid synthesis in group A streptococci. Demonstration of UDP-glucose dehydrogenase activity. *J. Biol. Chem.* **268**, 7118–7124.

Dougherty, B. A., van de Rijn, I. (1994) Molecular characterization of hasA from an operon required for hyaluronic acid synthesis in group A streptococci. *J. Biol. Chem.* **269**, 169–175.

Dube, B., Luke, H. J., Aumailley, M., Prehm, P. (2001) Hyaluronan reduces migration and pro-

liferation in CHO cells. *Biochim. Biophys. Acta* **1538**, 283–289.
Engström-Laurent, A. (1997) Hyaluronan in joint disease. *J. Intern. Med.* **242**, 57–60.
Erickson, C. A., Turley, E. A. (1983) Substrata formed by combination of extracellular matrix components alter neural crest cell migration. *J. Cell Sci.* **61**, 299–323.
Evanko, S. P., Wight, T. N. (1999) Intracellular localization of hyaluronan in proliferating cells. *J. Histochem. Cytochem.* **47**, 1331–1341.
Fox, E. M., Walts, A. E., Acharya, R. A., Hamilton, R. (1990) Water insoluble derivatives of hyaluronic acid. U.S. Patent 4,937,270.
Fraser, J. R., Laurent, T. C. (1989) Turnover and metabolism of hyaluronan. *Ciba Found. Symp.* **143**, 41–53.
Fraser, J. R., Laurent, T. C., Pertoft, H., Baxter, E. (1981) Plasma clearance, tissue distribution and metabolism of hyaluronic acid injected intravenously in the rabbit. *Biochem. J.* **200**, 415–424.
Fraser, J. R. E., Cahill, R. N., Kimpton, W. G., Laurent, T. C. (1996) Lymphatic system, in: *Extracellular Matrix. 1. Tissue Function* (Comper, W. D., Ed.), Amsterdam: Harwood Academic Publications, 110–131.
Gerlach, D., Kohler, W. (1970) [Production and isolation of streptococcal hyaluronic acid]. *Zentralbl. Bakteriol. Orig.* **215**, 187–195.
Goa, K. L., Benfield, P. (1994) Hyaluronic acid. A review of its pharmacology and use as a surgical aid in ophthalmology, and its therapeutic potential in joint disease and wound healing. *Drugs* **47**, 536–566.
Goldberg, R. L., Toole, B. P. (1987) Hyaluronate inhibition of cell proliferation. *Arthritis Rheum.* **30**, 769–778.
Goldberg, R. L., Seidman, J. D., Chi Rosso, G., Toole, B. P. (1984) Endogenous hyaluronate-cell surface interactions in 3T3 and simian virus-transformed 3T3 cells. *J. Biol. Chem.* **259**, 9440–9446.
Greenwald, R. A., Moak, S. A. (1986) Degradation of hyaluronic acid by polymorphonuclear leukocytes. *Inflammation* **10**, 15–30.
Grootveld, M., Henderson, E. B., Farrell, A., Blake, D. R., Parkes, H. G. H.-P. (1991) Oxidative damage to hyaluronate and glucose in synovial fluid during exercise of the inflamed rheumatoid joint. Detection of abnormal low-molecular-mass metabolites by proton-n.m.r. spectroscopy. *Biochem. J.* **273**, 459–467.
Hall, C. L., Yang, B., Yang, X., Zhang, S., Turley, M., Samuel, S., Lange, L. A., Wang, C., Curpen, G. D., Savani, R. C., Greenberg, A. H., Turley, E. A. (1995) Overexpression of the hyaluronan receptor RHAMM is transforming and is also required for H-*ras* transformation. *Cell* **82**, 19–28.
Hamerman, D., Wood, D. D. (1984) Interleukin 1 enhances synovial cell hyaluronate synthesis. *Proc. Soc. Exp. Biol. Med.* **177**, 205–210.
Hamerman, D., Todaro, G. J., Green, H. (1965) The production of hyaluronate by spontaneously established cell lines and viral transformed lines of fibroblastic origin. *Biochim. Biophys. Acta* **101**, 343–351.
Hamerman, D., Sasse, J., Klagsbrun, M. (1986) A cartilage-derived growth factor enhances hyaluronate synthesis and diminishes sulfated glycosaminoglycan synthesis in chondrocytes. *J. Cell Physiol.* **127**, 317–322.
Hardingham, T. E., Muir, H. (1972) The specific interaction of hyaluronic acid with cartilage proteoglycans. *Biochim. Biophys. Acta* **279**, 401–405.
Hällgren, R., Gerdin, B., Tufveson, G. (1990) Hyaluronic acid accumulation and redistribution in rejecting rat kidney graft. Relationship to the transplantation edema. *J. Exp. Med.* **171**, 2063–2076.
Heldin, P., Laurent, T. C., Heldin, C. H. (1989) Effect of growth factors on hyaluronan synthesis in cultured human fibroblasts. *Biochem. J.* **258**, 919–922.
Heldin, P., Asplund, T., Ytterberg, D., Thelin, S., Laurent, T. C. (1992) Characterization of the molecular mechanism involved in the activation of hyaluronan synthetase by platelet-derived growth factor in human mesothelial cells. *Biochem. J.* **283**, 165–170.
Hofmann, M., Assmann, V., Fieber, C., Sleeman, J. P., Moll, J., Ponta, H., Hart, I. R., Herrlich, P. (1998a) Problems with RHAMM: a new link between surface adhesion and oncogenesis? *Cell* **95**, 591–592.
Hofmann, M., Fieber, C., Assmann, V., Göttlicher, M., Sleeman, J., Plug, R., Howells, N., Von Stein, O., Ponta, H., Herrlich, P. (1998b) Identification of IHABP, a 95kDa intracellular hyaluronate binding protein. *J. Cell Sci.* **111**, 1673–1684.
Hopwood, J. J., Dorfman, A. (1977) Glycosaminoglycan synthesis by cultured human skin fibroblasts after transformation with Simian virus 40. *J. Biol. Chem.* **252**, 4777–4785.
Horton, M. R., McKee, C. M., Bao, G., Liao, F., Farber, J. M., Hodge DuFour, J., Pure, E., Oliver, B. L., Wright, T. M., Noble, P. W. (1998) Hyaluronan fragments synergize with inter-

feron-gamma to induce the C-X-C chemokines Mig and interferon-inducible protein-10 in mouse macrophages. *J. Biol. Chem.* **273**, 35088–35094.

Huang, L., Grammatikakis, N., Yoneda, M., Banerjee, S. D., Toole, B. P. (2000) Molecular characterization of a novel intracellular hyaluronan-binding protein. *J. Biol. Chem.* 275, 29829–29839.

Ishimoto, N., Temin, H. M., Strominger, J. L. (1966) Studies of carcinogenesis by avian sarcoma viruses. II. Virus- induced increase in hyaluronic acid synthetase in chicken fibroblasts. *J. Biol. Chem.* **241**, 2052–2057.

Itano, N., Kimata, K. (1996) Expression cloning and molecular characterization of HAS protein, a eukaryotic hyaluronan synthase. *J. Biol. Chem.* **271**, 9875–9878.

Itano, N., Sawai, T., Yoshida, M., Lenas, P., Yamada, Y., Imagawa, M., Shinomura, T., Hamaguchi, M., Yoshida, Y., Ohnuki, Y., Miyauchi, S., Spicer, A. P., McDonald, J. A., Kimata, K. (1999) Three isoforms of mammalian hyaluronan synthases have distinct enzymatic properties. *J. Biol. Chem.* **274**, 25085–25092.

Kabat, E. A. (1939) A polysaccharide in tumors due to a virus of leukosis and sarcoma in fowls. *J. Biol. Chem.* **130**, 143–147.

Kass, E. H., Seastone, C. V. (1944) The role of the mucoid polysaccharide hyaluronic acid in the virulence of group A hemolytic streptococci. *J. Exp. Med.* **70**, 319–330.

Kendall, F. E., Heidelberger, M., Dawson, M. H. (1937) A serologically inactive polysaccharide elaborated by mucoid strains of group A hemolytic streptococci. *J. Biol. Chem.* **118**, 61–69.

Kim, J. H., Yoo, S. J., Oh, D. K., Kweon, Y. G., Park, D. W., Lee, C. H., Gil, G. H. (1996) Selection of a *Streptococcus equi* mutant and optimization of culture conditions for the production of high molecular weight hyaluronic acid. *Enzyme Microb. Technol.* **19**, 440–445.

Kjems, E., Lebech, K. (1976) Isolation of hyaluronic acid from cultures of streptococci in a chemically defined medium. *Acta Pathol. Microbiol. Scand. B.* **84**, 162–164.

Klein, U., von Figura, K. (1980) Characterization of dermatan sulfate in mucopolysaccharidosis VI. Evidence for the absence of hyaluronidase-like enzymes in human skin fibroblasts. *Biochim. Biophys. Acta* **630**, 10–14.

Klewes, L., Prehm, P. (1994) Intracellular signal transduction for serum activation of the hyaluronan synthase in eukaryotic cell lines. *J. Cell Physiol.* **160**, 539–544.

Knudson, W., Biswas, C., Toole, B. P. (1984) Interactions between human tumor cells and fibroblasts stimulate hyaluronate synthesis. *Proc. Natl. Acad. Sci. USA* **81**, 6767–6771.

Kosaki, R., Watanabe, K., Yamaguchi, Y. (1999) Overproduction of hyaluronan by expression of the hyaluronan synthase Has2 enhances anchorage-independent growth and tumorigenicity. *Cancer Res.* **59**, 1141–1145.

Kreil, G. (1995) Hyaluronidases – A group of neglected enzymes. *Protein Sci.* **4**, 1666–1669.

Lapcik, L., De Smedt, S., Demeester, J., Chabrecek, P. (1998) Hyaluronan: preparation, structure, properties, and applications. *Chem. Rev.* **98**, 2663–2684.

Laurent, T. C. (1970) Structure of hyaluronic acid, in: *Chemistry and Molecular Biology of the Intercellular Matrix* (Balazs, E. A., Ed.), New York: Academic Press, 703–732.

Laurent, T. C. (1989) *The Biology of Hyaluronan.* Ciba Foundation Symposium **143**. Chichester: Wiley.

Laurent, T. C. (1998) *The Chemistry, Biology and Medical Applications of Hyaluronan and its Derivatives.* Wenner-Gren International Series **72**. London: Portland Press.

Laurent, T. C., Fraser, J. R. (1992) Hyaluronan. *FASEB J.* **6**, 2397–2404.

Laurent, T. C., Dahl, L. B., Lilja, K. (1993) Hyaluronan injected in the anterior chamber of the eye is catabolized in the liver. *Exp. Eye Res.* **57**, 435–440.

Lebel, L., Gabrielsson, J., Laurent, T. C., Gerdin, B. (1994) Kinetics of circulating hyaluronan in humans. *Eur. J. Clin. Invest.* **24**, 621–626.

Lee, M. C., Alpaugh, M. L., Nguyen, M., Deato, M., Dishakjian, L., Barsky, S. H. (2000) Myoepithelial-specific CD44 shedding is mediated by a putative chymotrypsin-like sheddase. *Biochem. Biophys. Res. Commun.* **279**, 116–123.

Lee, T. H., Wisniewski, H. G., Vilcek, J. (1992) A novel secretory tumor necrosis factor-inducible protein(TSG-6) is a member of the family of hyaluronate binding proteins, closely related to the adhesion receptor CD44. *J. Cell Biol.* **116**, 545–557.

Lembach, K. J. (1976) Enhanced synthesis and extracellular accumulation of hyaluronic acid during stimulation of quiescent human fibroblasts by mouse epidermal growth factor. *J. Cell Physiol.* **89**, 277–288.

Lesley, J., Hascall, V. C., Tammi, M., Hyman, R. (2000) Hyaluronan binding by cell surface CD44. *J. Biol. Chem.* **275**, 26967–26975.

Liu, D. C., Liu, T., Sy, M. S. (1998) Identification of two regions in the cytoplasmic domain of CD44 through which PMA, calcium, and forskolin differentially regulate the binding of CD44 to hyaluronic acid. *Cell. Immunol.* **190**, 132–140.

Lokeshwar, V. B., Lokeshwar, B. L., Pham, H. T., Block, N. L. (1996) Association of elevated levels of hyaluronidase, a matrix-degrading enzyme, with prostate cancer progression. *Cancer Res.* **56**, 651–657.

Lokeshwar, V. B., Öbek, C., Soloway, M. S., Block, N. L. (1997) Tumor-associated hyaluronic acid: a new sensitive and specific urine marker for bladder cancer. *Cancer Res.* **57**, 773–777.

Longaker, M. T., Chiu, E. S., Adzick, N. S., Stern, M., Harrison, M. R., Stern, R. (1991) Studies in fetal wound healing. V. A prolonged presence of hyaluronic acid characterizes fetal wound fluid. *Ann. Surg.* **213**, 292–296.

Lowther, D. A., Rogers, H. J. (1956) The role of glutamine in the biosynthesis of hyaluronate by streptococcal suspensions. *Biochem. J.* **62**, 304–314.

Lüke, H. J., Prehm, P. (1999) Synthesis and shedding of hyaluronan from plasma membranes of human fibroblasts and metastatic and non-metastatic melanoma cells. *Biochem. J.* **343**, 71–75.

MacLennan, A. P. (1956) The production of capsules, hyaluronic acid and hyaluronidase by group A and group C streptococci. *J. Gen. Microbiol.* **14**, 134–142.

Manna, F., Dentini, M., Desideri, P., De Pita, O., Mortilla, E., Maras, B. (1999) Comparative chemical evaluation of two commercially available derivatives of hyaluronic acid (hylaform from rooster combs and restylane from *Streptococcus*) used for soft tissue augmentation. *J. Eur. Acad. Dermatol. Venereol.* **13**, 183–192.

Markovitz, A., Dorfman, A. (1962) Synthesis of capsular polysaccharide hyaluronic acid by protoplast membrane preparations of group A streptococci. *J. Biol. Chem.* **237**, 273–279.

McClean, D. (1941) The capsulation of streptococci and its relation to diffusion factor (hyaluronidase). *J. Pathol. Bacteriol.* **53**, 13.

McKee, C. M., Penno, M. B., Cowman, M., Burdick, M. D., Strieter, R. M., Bao, C., Noble, P. W. (1996) Hyaluronan (HA) fragments induce chemokine gene expression in alveolar macrophages – The role of HA size and CD44. *J. Clin. Invest.* **98**, 2403–2413.

McKee, C. M., Lowenstein, C. J., Horton, M. R., Wu, J., Bao, C., Chin, B. Y., Choi, A. M. K., Noble, P. W. (1997) Hyaluronan fragments induce nitric-oxide synthase in murine macrophages through a nuclear factor kappaB-dependent mechanism. *J. Biol. Chem.* **272**, 8013–8018.

Meyer, K. (1947) The biological significance of hyaluronic acid and hyaluronidase. *Physiol. Rev.* **27**, 335–359X.

Meyer, K., Palmer, J. W. (1934) The polysaccharide of the vitreous humor. *J. Biol. Chem.* **107**, 629–634.

Meyer, M. F., Kreil, G. (1996) Cells expressing the DG42 gene from early *Xenopus* embryos synthesize hyaluronan. *Proc. Natl. Acad. Sci. USA* **93**, 4543–4547.

Mian, N. (1986) Analysis of cell-growth-phase-related variations in hyaluronate synthase activity of isolated plasma-membrane fractions of cultured human skin fibroblasts. *Biochem. J.* **237**, 333–342.

Mikuni Takagaki, Y., Toole, B. P. (1981) Hyaluronate-protein complex of Rous sarcoma virus-transformed chick embryo fibroblasts. *J. Biol. Chem.* **256**, 8463–8469.

Miller, D., Stegmann, R. (1983) Healon. A Guide to its use in Ophthalmic Surgery. New York: Wiley.

Miyake, H., Hara, I., Okamoto, I., Gohji, K., Yamanaka, K., Arakawa, S., Saya, H., Kamidono, S. (1998) Interaction between CD44 and hyaluronic acid regulates human prostate cancer development. *J. Urol.* **160**, 1562–1566.

Miyake, K., Underhill, C. B., Lesley, J., Kincade, P. W. (1990) Hyaluronate can function as a cell adhesion molecule and CD44 participates in hyaluronate recognition. *J. Exp. Med.* **172**, 69–75.

Montesano, R., Kumar, S., Orci, L., Pepper, M. S. (1996) Synergistic effect of hyaluronan oligosaccharides and vascular endothelial growth factor on angiogenesis in vitro. *Lab. Invest.* **75**, 249–262.

Nettelbladt, O., Tengblad, A., Hällgren, R. (1989) Lung accumulation of hyaluronan parallels pulmonary edema in experimental alveolitis. *Am. J. Physiol.* **257**, L379–L384.

Nickel, V., Prehm, S., Lansing, M., Mausolf, A., Podbielski, A., Deutscher, J., Prehm, P. (1998) An ectoprotein kinase of group C streptococci binds hyaluronan and regulates capsule formation. *J. Biol. Chem.* **273**, 23668–23673.

Orkin, R. W., Underhill, C. B., Toole, B. P. (1982) Hyaluronate degradation in 3T3 and simian virus-transformed3T3 cells. *J. Biol. Chem.* **257**, 5821–5826.

Peller, L. (1980) Thermodynamic considerations in the synthesis and assembly of biological macromolecules. *Macromolecules* **13**, 609–615.

Perides, G., Asher, R., Dahl, D., Bignami, A. (1990) Glial hyaluronate-binding protein (GHAP) in optic nerve and retina. *Brain Res.* **512**, 309–316.

Perschl, A., Lesley, J., English, N., Trowbridge, I., Hyman, R. (1995) Role of CD44 cytoplasmic domain in hyaluronan binding. *Eur. J. Immunol.* **25**, 495–501.

Peterson, R. M., Yu, Q., Stamenkovic, I., Toole, B. P. (2000) Perturbation of hyaluronan interactions by soluble CD44 inhibits growth of murine mammary carcinoma cells in ascites. *Am. J. Pathol.* **156**, 2159–2167.

Philipson, L. H., Westley, J., Schwartz, N. B. (1985) Effect of hyaluronidase treatment of intact cells on hyaluronate synthetase activity. *Biochemistry* **24**, 7899–7906.

Pitsillides, A. A., Wilkinson, L. S., Mehdizadeh, S., Bayliss, M. T., Edwards, J. C. (1993) Uridine diphosphoglucose dehydrogenase activity in normal and rheumatoid synovium: the description of a specialized synovial lining cell. *Int. J. Exp. Pathol.* **74**, 27–34.

Ponta, H., Wainwright, D., Herrlich, P. (1998) The CD44 protein family. *Int. J. Biochem. Cell Biol.*, **30**, 299–305.

Prehm, P. (1980) Induction of hyaluronic acid synthesis in teratocarcinoma stem cells by retinoic acid. *FEBS Lett.* **111**, 295–298.

Prehm, P. (1983a) Synthesis of hyaluronate in differentiated teratocarcinoma cells. Characterization of the synthase. *Biochem. J.* **211**, 181–189.

Prehm, P. (1983b) Synthesis of hyaluronate in differentiated teratocarcinoma cells. Mechanism of chain growth. *Biochem. J.* **211**, 191–198.

Prehm, P. (1984) Hyaluronate is synthesized at plasma membranes. *Biochem. J.* **220**, 597–600.

Prehm, P. (1985) Inhibition of hyaluronate synthesis. *Biochem. J.* **225**, 699–705.

Prehm, P. (1990) Release of hyaluronate from eukaryotic cells. *Biochem. J.* **267**, 185–189.

Prehm, P. (2000) Antigene von rheumatischen Autoimmunerkrankungen. Patent PCT/EP00/05279.

Prehm, S., Herrington, C., Nickel, V., Völker, W., Briko, N. I., Blinnikova, E. A., Schmiedel, A., Prehm, P. (1995) Antibodies against proteins of streptococcal hyaluronate synthase bind to human fibroblasts and are present in patients with rheumatic fever. *J. Anat.* **187**, 271–277.

Prehm, S., Nickel, V., Prehm, P. (1996) A mild purification method for polysaccharide binding membrane proteins: phase separation of digitonin extracts to isolate the hyaluronate synthase from *Streptococcus* sp. in active form. *Protein Expression and Purification* **7**, 343–346.

Pulkki, K. (1986) The effects of synovial fluid macrophages and interleukin-1 on hyaluronic acid synthesis by normal synovial fibroblasts. *Rheumatol. Int.* **6**, 121–125.

Puré, E., Camp, R. L., Peritt, D., Panettieri, R. A., Jr., Lazaar, A. L., Nayak, S. (1995) Defective phosphorylation and hyaluronate binding of CD44 with point mutations in the cytoplasmic domain. *J. Exp. Med.* **181**, 55–62.

Reed, R. K., Laurent, T. C., Taylor, A. E. (1990) Hyaluronan in prenodal lymph from skin: changes with lymph flow. *Am. J. Physiol.* **259**, H1097–H1100.

Roden, L., Campbell, P., Fraser, J. R., Laurent, T. C., Pertoft, H. T.-J. (1989) Enzymic pathways of hyaluronan catabolism. *Ciba Foundation Symposium* **143**, 60–76.

Schenck, P., Schneider, S., Miehlke, R., Prehm, P. (1995) Synthesis and degradation of hyaluronate by synovia from patients with rheumatoid arthritis. *J. Rheumatol.* **22**, 400–405.

Schmits, R., Filmus, J., Gerwin, N., Senaldi, G., Kiefer, F., Kundig, T., Wakeham, A., Shahinian, A., Catzavelos, C., Rak, J., Furlonger, C., Zakarian, A., Simard, J. J., Ohashi, P. S., Paige, C. J., Gutierrez, R. J., Mak, T. W. (1997) CD44 regulates hematopoietic progenitor distribution, granuloma formation, and tumorigenicity. *Blood* **90**, 2217–2233.

Scott, J. E., Cummings, C., Brass, A., Chen, Y. (1991) Secondary and tertiary structures of hyaluronan in aqueous solution, investigated by rotary shadowing-electron microscopy and computer simulation. Hyaluronan is a very efficient network-forming polymer. *Biochem. J.* **274**, 699–705.

Skelton, T. P., Zeng, C. X., Nocks, A., Stamenkovic, I. (1998) Glycosylation provides both stimulatory and inhibitory effects on cell surface and soluble CD44 binding to hyaluronan. *J. Cell Biol.* **140**, 431–446.

Taher, T. E. I., Smit, L., Griffioen, A. W., Schilder-Tol, E. J. M., Borst, J., Pals, S. T. (1996) Signaling through CD44 is mediated by tyrosine kinases – Association with $p56^{lck}$ in T lymphocytes. *J. Biol. Chem.* **271**, 2863–2867.

Takahashi, K., Stamenkovic, I., Cutler, M., Saya, H., Tanabe, K. K. (1995) CD44 hyaluronate binding influences growth kinetics and tumorigenicity of human colon carcinomas. *Oncogene* **11**, 2223–2232.

Tammi, R., Tammi, M. (1986) Influence of retinoic acid on the ultrastructure and hyaluronic acid synthesis of adult human epidermis in whole skin organ culture. *J. Cell Physiol.* **126**, 389–398.

Tengblad, A., Caputo, C. B., Raisz, L. G. (1980) Quantitative analysis of hyaluronate in nanogram amounts. *Biochem. J.* **185**, 101–105.

Termeer, C. C., Hennies, J., Voith, U., Ahrens, T., Weiss, M., Prehm, P., Simon, J. C. (2000) Oligosaccharides of hyaluronan are potent activators of dendritic cells. *J. Immunol.* **165**, 1863–1870.

Thonard, J. C., Migliore, S. A., Blustein, R. (1964) Isolation of hyaluronic acid from broth cultures of streptococci. *J. Biol. Chem.* **239**, 726–728.

Tlapak-Simmons, V. L., Kempner, E. S., Baggenstoss, B. A., Weigel, P. H. (1998) The active streptococcal hyaluronan synthases (HASs) contain a single HAS monomer and multiple cardiolipin molecules. *J. Biol. Chem.* **273**, 26100–26109.

Tomida, M., Koyama, H., Ono, T. (1975) Induction of hyaluronic acid synthetase activity in rat fibroblasts by medium change of confluent cultures. *J. Cell Physiol.* **86**, 121–130.

Tomida, M., Koyama, H., Ono, T. (1977a) A serum factor capable of stimulating hyaluronic acid synthesis in cultured rat fibroblasts. *J. Cell Physiol.* **91**, 323–328.

Tomida, M., Koyama, H., Ono, T. (1977b) Effects of adenosine 3′:5′-cyclic monophosphate and serum on synthesis of hyaluronic acid in confluent rat fibroblasts. *Biochem. J.* **162**, 539–543.

Toole, B. P. (1991) Proteoglycans and hyaluronan in morphogenesis and differentiation, in: *Cell Biology of the Extracellular Matrix* (Hay, E. D., Ed.), New York: Plenum Press, 305–341.

Toole, B. P., Okayama, M., Orkin, R. W., Yoshimura, M., Muto, M., Kaji, A. (1977) Developmental roles of hyaluronate and chondroitin sulfate proteoglycans. *Soc. Gen. Physiol. Ser.* **32**, 139–154.

Toole, B. P., Biswas, C., Gross, J. (1979) Hyaluronate and invasiveness of the rabbit V2 carcinoma. *Proc. Natl. Acad. Sci. USA* **76**, 6299–6303.

Triscott, M. X., van de Rijn, I. (1986) Solubilization of hyaluronic acid synthetic activity from streptococci and its activation with phospholipids. *J. Biol. Chem.* **261**, 6004–6009.

Turley, E. A. (1989) The role of a cell-associated hyaluronan-binding protein in fibroblast behaviour. *Ciba Foundation Symposium* **143**, 121–133.

Turley, E. A., Moore, D., Hayden, L. J. (1987) Characterization of hyaluronate binding proteins isolated from 3T3 and murine sarcoma virus transformed 3T3 cells. *Biochemistry* **26**, 2997–3005.

Turley, E. A., Pilarski, L., Nagy, J. I. (1998) Problems with RHAMM: a new link between surface adhesion and oncogenesis? Response. *Cell* **95**, 592–593.

Uchiyama, H., Dobashi, Y., Ohkouchi, K., Nagasawa, K. (1990) Chemical change involved in the oxidative reductive depolymerization of hyaluronic acid. *J. Biol. Chem.* **265**, 7753–7759.

Ueki, N., Taguchi, T., Takahashi, M., Adachi, M., Ohkawa, T., Amuro, Y., Hada, T., Higashino, K. (2000) Inhibition of hyaluronan synthesis by vesnarinone in cultured human myofibroblasts. *Biochim. Biophys. Acta* **1495**, 160–167.

Underhill, C. B. (1982) Interaction of hyaluronate with the surface of simian virus40- transformed 3T3 cells: aggregation and binding studies. *J. Cell Sci.* **56**, 177–189.

Underhill, C. B., Toole, B. P. (1979) Binding of hyaluronate to the surface of cultured cells. *J. Cell Biol.* **82**, 475–484.

Underhill, C. B., Toole, B. P. (1982) Transformation-dependent loss of the hyaluronate-containing coats of cultured cells. *J. Cell Physiol.* **110**, 123–128.

van de Rijn, I. (1983) Streptococcal hyaluronic acid: proposed mechanisms of degradation and loss of synthesis during stationary phase. *J. Bacteriol.* **156**, 1059–1065.

van de Rijn, I., Kessler, R. E. (1980) Growth characteristics of group A streptococci in a new chemically defined medium. *Infect. Immun.* **27**, 444–448.

Van der Voort, R., Manten-Horst, E., Smit, L., Ostermann, E., Van den Berg, F., Pals, S. T. (1995) Binding of cell-surface expressed CD44 to hyaluronate is dependent on splicing and cell type. *Biochem. Biophys. Res. Commun.* **214**, 137–144.

Vercruysse, K. P., Prestwich, G. D. (1998) Hyaluronate derivatives in drug delivery. *Crit. Rev. Therap. Drug Carrier Systems* **15**, 513–555.

Waldenstrom, A., Martinussen, H. J., Gerdin, B., Hällgren, R. (1991) Accumulation of hyaluronan and tissue edema in experimental myocardial infarction. *J. Clin. Invest.* **88**, 1622–1628.

Weigel, P. H., Fuller, G. M., LeBoeuf, R. D. (1986) A model for the role of hyaluronic acid and fibrin in the early events during the inflammatory response and wound healing. *J. Theor. Biol.* **119**, 219–234.

Weigel, P. H., Hascall, V. C., Tammi, M. (1997) Hyaluronan synthases. *J. Biol. Chem.* **272**, 13997–14000.

Weissman, B., Meyer, K. (1954) The structure of hyalobiuronic acid and of hyaluronic acid from umbilical cord. *J. Am. Chem. Soc.* **76**, 1753–1757.

Wessels, M. R., Moses, A. E., Goldberg, J. B., DiCesare, T. J. (1991) Hyaluronic acid capsule is a virulence factor for mucoid group A streptococci. *Proc. Natl. Acad. Sci. USA* **88**, 8317–8321.

Wessels, M. R., Goldberg, J. B., Moses, A. E., DiCesare, T. J. (1994) Effects on virulence of mutations in a locus essential for hyaluronic acid capsule expression in group A streptococci. *Infect. Immun.* **62**, 433–441.

West, D. C., Kumar, S. (1989) The effect of hyaluronate and its oligosaccharides on endothelial cell proliferation and monolayer integrity. *Exp. Cell Res.* **183**, 179–196.

West, D. C., Hampson, I. N., Arnold, F., Kumar, S. (1985) Angiogenesis induced by degradation products of hyaluronic acid. *Science* **228**, 1324–1326.

West, D. C., Shaw, D. M., Lorenz, P., Adzick, N. S., Longaker, M. T. (1997) Fibrotic healing of adult and late gestation fetal wounds correlates with increased hyaluronidase activity and removal of hyaluronan. *Int. J. Biochem. Cell Biol.* **29**, 201–210.

Whitnack, E., Bisno, A. L., Beachey, E. H. (1981) Hyaluronate capsule prevents attachment of group A streptococci to mouse peritoneal macrophages. *Infect. Immun.* **31**, 985–991.

Wobig, M., Bach, G., Beks, P., Dickhut, A., Runzheimer, J., Schwieger, G., Vetter, G., Balazs, E. A. (1999) The role of elastoviscosity in the efficacy of viscosupplementation for osteoarthritis of the knee: a comparison of hylan G-F 20 and a lower-molecular-weight hyaluronan. *Clin. Ther.* **21**, 1549–1562.

Wong, S. F., Halliwell, B., Richmond, R., Skowroneck, W. R. (1981) The role of superoxide and hydroxyl radicals in the degradation of hyaluronic acid induced by metal ions and by ascorbic acid. *J. Inorg. Biochem.* **14**, 127–134.

Wu, R., Wu, M. M. (1986) Effects of retinoids on human bronchial epithelial cells: differential regulation of hyaluronate synthesis and keratin protein synthesis. *J. Cell Physiol.* **127**, 73–82.

Yoshimura, M. (1985) Change of hyaluronic acid synthesis during differentiation of myogenic cells and its relation to transformation of myoblasts by Rous sarcoma virus. *Cell Differ.* **16**, 175–185.

Zahalka, M. A., Okon, E., Gosslar, U., Holzmann, B., Naor, D. (1995) Lymph node (but not spleen) invasion by murine lymphoma is both CD44- and hyaluronate-dependent. *J. Immunol.* **154**, 5345–5355.

Zhou, B., Oka, J. A., Singh, A., Weigel, P. H. (1999) Purification and subunit characterization of the rat liver endocytic hyaluronan receptor. *J. Biol. Chem.* **274**, 33831–33834.

Zhu, D., Bourguignon, L. Y. W. (1998) The ankyrin-binding domain of CD44s is involved in regulating hyaluronic acid-mediated functions and prostate tumor cell transformation. *Cell Motil. Cytoskeleton* **39**, 209–222.

23
Exopolysaccharides of Lactic Acid Bacteria

Ir. Isabel Hallemeersch[1], Ir. Sophie De Baets[2], Prof. Dr. Ir. Erick J. Vandamme[3]

[1] Laboratory of Industrial Microbiology and Biocatalysis, Department of Biochemical and Microbial Technology, Faculty of Agricultural and Applied Biological Sciences, Ghent University, Coupure links 653, B-9000 Gent, Belgium; Tel.: +32-9-264-6029; Fax: +32-9-264-6231; E-mail: Isabel.Hallemeersch@rug.ac.be

[2] Laboratory of Industrial Microbiology and Biocatalysis, Department of Biochemical and Microbial Technology, Faculty of Agricultural and Applied Biological Sciences, Ghent University, Coupure links 653, B-9000 Gent, Belgium; Tel.: +32-9-264-6028; Fax: +32-9-264-6231; E-mail: Sophie.DeBaets@rug.ac.be

[3] Laboratory of Industrial Microbiology and Biocatalysis, Department of Biochemical and Microbial Technology, Faculty of Agricultural and Applied Biological Sciences, Ghent University, Coupure links 653, B-9000 Gent, Belgium; Tel.: +32-9-264-6027; Fax: +32-9-264-6231; E-mail: Erick.Vandamme@rug.ac.be

Biotechnology of Biopolymers. From Synthesis to Patents. Edited by A. Steinbüchel and Y. Doi

ISBN: 3-527-31110-6

EPS exopolysaccharides
GRAS generally recognized as safe
LAB lactic acid bacteria
NMR nuclear magnetic resonance

1 Introduction

The exopolysaccharides (EPS) produced by lactic acid bacteria (LAB) can be divided into three major groups, based on their composition: (1) glucans, namely dextrans, alternans, and mutans; (2) fructans such as levan; and (3) heteropolysaccharides produced by mesophilic and thermophilic LAB (Cerning, 1990). As details of the glucans and fructans are described in Chapters 12–14 of this Volume, only the latter group will be discussed here. These EPS are composed of linear and branched repeating units, and vary in size from disaccharides to heptasaccharides (Petry et al., 2000).

Over the past few decades, there has been a growing interest in ropy LAB, such as *Lactoccocus lactis* subsp. *cremoris, L. lactis* subsp. *lactis, Lactobacillus delbreuckii* subsp. *bulgaricus, Lb. helveticus, Streptococcus salivarius* subsp. *thermophilus,* etc. Based on the "generally recognized as safe" (GRAS) status of these bacteria, their EPS preparations are widely applied as thickening, gelling and stabilizing agents in the food industry (Sutherland, 1994). Their most important application area in this respect is undoubtedly in the dairy industry, where they are used in the production of various fermented milk products and contribute to both texture and mouthfeel (Cerning, 1995). Moreover, it was suggested that they also possess advantageous biological functions and are beneficial for human health (Nakajima et al., 1992; Kitazawa et al., 1993; Gibson and Roberfroid, 1995; Hosono et al., 1997).

2 Historical Outline

Although ropy LAB such as *S. thermophilus* and *Lb. delbreuckii* subsp. *bulgaricus* have been available commercially as dairy starter cultures since the early 1900s, the chemical composition and structures of their EPS were reported in detail only as recently as the 1980s (Marshall and Rawson, 1999). Little information existed regarding the level of polysaccharide produced, the culture conditions and the rheological properties, in contrast to the situation with glucans such as dextran, on which most attention was focused and which had already undergone intensive study (Forsén and Häivä, 1981).

Nonetheless, during the 1950s and 1970s, a number of articles were published on the composition of heteropolysaccharides produced by LAB. These early investigations did not refer to any precise culture conditions, nor analytical details, and should therefore be treated with some caution (Sundman, 1953; Nilsson and Nilsson, 1958; Groux, 1973; Tamime, 1978).

It was not until the 1980s that EPS from LAB received the attention of numerous investigators, particularly in France and the Netherlands, where the use of stabilizing agents from plant or animal origin was prohibited. Among the early publications on the chemical composition of EPS no clear picture emerged (Cerning et al., 1986), and EPS were seen either as proteinaceous material, a carbohydrate–protein complex, or simply as a complex carbohydrate. These differences were due to different isolation and purification methods, each with their variable efficiencies. This was especially so because EPS purification was hampered by the use of complex culture media, such as milk or whey; moreover, some bacterial strains were able to synthesize more than one type of EPS (Marshall et al., 1995).

Subsequently, a great deal of research was carried out on various aspects of EPS synthesis, such as the influence of nutritional and physico-chemical fermentation parameters, the isolation and characterization of the EPS, and the rheological properties. Detailed structural and rheological studies were performed in order to provide a better insight into the mechanisms by which the LAB and their EPS influence the consistency of dairy products (Staaf et al., 2000; Faber et al., 2001).

During the late 1980s, investigations into the instability of the mucoid character of LAB revealed the involvement of plasmids in EPS synthesis (Vedamuthu and Neville, 1986; Neve et al., 1988). This was the beginning of a series of studies on the molecular biology and genetics of EPS produced by various LAB.

During the past decade, a large number of genes involved in the synthesis of repeating units, polymerization, chain length determination and export have been characterized and functionally analyzed. In addition, the entire genome of *L. lactis* has been reported recently (Bolotin et al., 1998; Kleerebezem et al., 2000; Ricciardi and Clementi, 2000). Models were developed in order to predict the behavior and the effect of EPS addition to food products, and this has led to the identification of several important properties, though these models require further validation and fine-tuning (Kleerebezem et al., 1999).

Nowadays, the challenge is to increase production levels and to modify the structure and hence the properties of EPS, by using genetic approaches and enzyme technology. The purpose of this type of polysaccharide engineering is the synthesis of tailor-made oligo- and polysaccharides, for a variety of specific applications (De Vuyst and Degeest, 1999a; Kleerebezem et al., 2000).

3 Chemical Structure

Heteropolysaccharides produced by LAB have been investigated to a lesser extent when compared with other polymers. The overall level of EPS production is low and is often characteristically variable (Cerning, 1990). Most strains produce a limited quantity of polymer, perhaps up to 200 mg L^{-1}, although some strains have been found to produce up to 4 g L^{-1}.

Many LAB synthesize heteropolysaccharides with a molecular weight in excess of 10^6 daltons. In addition, their composition, structure and physico-chemical properties are highly variable, these are being influenced by the composition of the culture medium (Ricciardi and Clementi, 2000).

Today, it is generally recognized that EPS from LAB are composed of linear and branched repeating units that vary in size from disaccharides to heptasaccharides, and which contain α- and β- linkages. Most EPS contain D-glucose, D-galactose and/or L-rhamnose in different ratios, but other hexoses such as D-mannose, D-fructose and pentoses such as L-fucose, D-xylose and D-arabinose also appear. Hexosamines and uronic acids are also found in minor quantities (Andaloussi et al., 1995; Marshall et al., 1995; Petry et al., 2000). The chemical composition of a few EPS is summarized in Table 1, while Table 2 shows the structure of typical repeating units.

The EPS from mesophilic LAB have a more varied composition than those from thermophilic LAB, and contain sometimes acetylated and phosphorylated residues (Ricciardi and Clementi, 2000). Different authors have described the chemical composition of EPS produced by *Lactobacillus* strains (Toba et al., 1991; Kojic et al., 1992; Cerning, 1994; Van den Berg et al., 1995; Robijn et al., 1996). Several lactococci have also been described that produce EPS; these usually contain glucose and galactose, but rhamnose and charged residues are also found quite commonly (Cerning, 1994). Most *Pediococcus* strains, which occur most often as spoilers in beer and wine, produce β-glucans of high molecular weight (Ricciardi and Clementi, 2000).

Galactose is seen as the major monosaccharide in the EPS produced by thermophilic

Tab. 1 Chemical composition of several exopolysaccharides (EPS) synthesized by lactic acid bacteria (LAB)

Strain	***Monosaccharides***				***Reference***
	Gal	***Glc***	***Rha***	***Other***	
Lb. delbreukii subsp	+	+	+	–	
bulgaricus NCFB 2772	+	+	–	–	Grobben et al. (1995, 1996)
Lb. delbreukii subsp. *bulgaricus* CNRZ 1187	+	+	–	–	Petry et al. (2000)
Lb. delbreukii subsp. *bulgaricus* CNRZ 416	+	+	+	–	Petry et al. (2000)
Lb. casei CG11	–	+	+	–	Kojic et al. (1992)
Lb. casei CG11	+	+	+	–	Cerning (1994)
Lb. helveticus var. *jugurti*	+	+	–	–	Oda et al. (1983)
Lb. helveticus LB161	+	+	–	Ac, P	Staaf et al. (2000)
Lb. kefiranofaciens	+	+	–	–	Toba et al. (1991)
Lb. paracasei	+	+	+	–	Van Calsteren (2001)
Lb. paracasei 34-1	+	–	–	GP	Robijn et al. (1996)
Lb. rhamnosus	+	+	+	–	Van Calsteren (2001)
Lb. sake 0-1	–	+	+	Ac, GP	Robijn et al. (1996)
Lb. sake 0-1	–	+	+	–	Van den Berg et al. (1995)
L. lactis subsp. *cremoris* LC330	+	+	–	GlcNAc	Marshall et al. (1995)
L. lactis subsp. *cremoris* LC330	+	+	+	GlcNAc P	Marshall et al. (1995)
L. lactis subsp. *cremoris* SBT 0495	+	+	+	P	Nakajima and Toyoda (1990)
Pediococcus	–	+	–	–	Cerning (1994)
S. thermophilus EU20	+	+	+	–	Marshall et al. (2001)
S. thermophilus CNCMI 733	+	+	–	GalNAc	Doco et al. (1991)
S. thermophilus OR 901	+	–	+	–	Ariga et al. (1992)
S. thermophilus S3	+	–	+	Ac	Faber et al. (2001)

Gal, galactose ; Glc, glucose ; Rha, rhamnose ; Ac, acetate ; P, phosphate ; GP, glycerol-3-phosphate ; Nac, *N*-acetyl.

Tab. 2 Primary structure of several EPS synthesized by LAB

Strain	*Structure*	*Reference*
Lb. delbreukii subsp. *bulgaricus* NCFB 2772	β-D-Gal*p* β-D-Gal*p* α-L-Rha*p* ↓ ↓ ↓ →2)-α-D-Gal*p*-(1→3)-β-D-Glc*p*-(1→3)-β-D-Gal*p*-(1→4)-α-D-Gal*p*-(1→	Grobben et al. (1995, 1996)
Lb. helveticus LB161	β-D-Glc*p* β-D-Glc*p* ↓ ↓ →4)-α-D-Glc*p*-(1→4)-β-D-Gal*p*-(1→3)-α-D-Gal*p*-(1→2)-α-D-Glc*p*-(1→3)-β-D-Glc*p*-(1→	Staaf et al. (2000)
Lb. paracasei 34-1	*sn*-Glycerol-3-phosphate-3 ↓ →3)-β-D-Gal*p*NAc-(1→4)-β-D-Gal*p*-(1→6)-β-D-Gal*p*-(1→6)-β-D-Gal*p*-(1→	Robijn et al. (1996)
Lb. sake 0-1	β-D-Glc*p*-(1→6) (Ac) 0.85 ↓ ↓ →4)-β-D-Glc*p*-(1→4)-α-D-Glc*p*-(1→3)-β-L-Rha*p*-(1→ ↑ -*sn*-Glycerol-3-phosphate-3→4)-α-L-Rha*p*-(1→3)	Robijn et al. (1996)
L. lactis **subsp.** *cremoris* **SBT 0495**	α-L-Rha*p*-(1→2) ↓ →4)-β-D-Glc*p*-(1→4)-β-D-Gal*p*-(1→4)-β-D-Glc*p*-(1→ ↑ α-D-Gal*p*-1-phosphate	Nakajima and Toyoda, 1990

Tab. 2 (cont.)

Strain	*Structure*	*Reference*
S. thermophilus CNCMI 733	α-D-Gal*p*-(1→6) ↓ →3)-β-D-Gal*p*-(1→3)-β-D-Glc*p*-(1→3)-α-D-Gal*p*NAc-(1→	Doco et al., 1991
S. thermophilus OR 901	β-D-Gal*p*-(1→6)-β-D-Gal*p*-(1→4) ↓ →2)-α-D-Gal*p*-(1→3)-α-D-Gal*p*-(1→3)-α-L-Rha*p*-(1→2)-α-L-Rha*p*-(1→	Ariga et al., 1992
S. thermophilus S3	β-D-Gal*f*2Ac ↓ →3)-β-D-Gal*p*-(1→3)-α-D-Gal*p*-(1→3)-α-L-Rha*p*-(1→2)-α-L-Rha*p*-(1→2)-α-D-Gal*p*-(1→	Faber et al., 2001

Gal = Galactose, Glc = Glucose, Rha = Rhamnose, Ac = Acetate

LAB, most likely because it is metabolized less quickly than glucose and hence is available for polymer synthesis (Cerning, 1990).

Many EPS types from *S. thermophilus* strains have been investigated as to their composition and characteristics. Most such heteropolysaccharides contain D-glucose and D-galactose and, on occasion, also L-rhamnose.

Heteropolysaccharide production was also observed in other thermophilic LAB such as *Lactobacillus helveticus* and *Lactobacillus delbreuckii* subsp. *bulgaricus*. Grobben et al. (1995, 1996) demonstrated with a *Lb. delbreuckii* subsp. *bulgaricus* strain, that the EPS composition varied with different sugars present in the growth medium. When using glucose or lactose as a carbon source, the EPS consisted of glucose, galactose, and rhamnose in a ratio of 1:6.8:0.7, but in the presence of fructose, the EPS formed contained only glucose and galactose in a ratio of 1:2.4. However, when glucose and fructose are used together, the composition of the EPS is similar to that obtained when either glucose or lactose was the carbon source. Marshall et al. (1995) isolated two EPS types from the culture broth of the same strain–a high and a low molecular fraction. Finally, De Vuyst and Degeest (1999b) showed that both fractions were similar in their monomeric composition and that the growth medium was in fact influencing EPS composition.

Petry et al. (2000) also investigated the EPS of two *Lb. delbreuckii* subsp. *bulgaricus* strains and concluded that the monomeric sugar composition remained the same, but that the relative proportions of the individual monosaccharides varied under different fermentation conditions.

In order to obtain insight into the structure–function relationship, it is necessary to study the conformation of the polymer in solution as well as its dynamic behavior. The three-dimensional structure of a polysaccharide is dependent on the time-averaged ring conformation of the monosaccharides and the relative orientations of adjacent monosaccharides. Besides the conformations around the glycosidic linkages, the intramolecular hydrogen bonds also contribute to the conformation, and this can be important for a complete understanding of the solution behavior. NMR spectroscopy does not provide sufficient information to elucidate the complete conformation (Vliegenthart et al., 2001). As the secondary and tertiary conformation of a polysaccharide is mainly dependent on its primary (chemical) structure, relatively small changes in the chemical structure might have a major effect on the conformation and the physical and chemical properties of the polysaccharide. However, despite all information already currently available, prediction of the properties on basis of polymer structure is not yet possible.

4 Occurrence

The LAB form a diverse group of bacteria, including the genera *Lactobacillus, Lactococcus, Streptococcus, Enterococcus, Leuconostoc,* and *Pediococcus.* They are Gram positive, have a low DNA GC-content (32–53 mol.%), are catalase-, reductase-, and oxidase-negative, and are also nonmotile and nonsporulating. *Bifidobacteria* are often associated with these genera as they are also added to traditional yogurt cultures (Dellaglio, 1994).

EPS-producing LAB can be isolated from dairy products such as the well-known Scandinavian fermented milk products “viili” and “longfil”, yogurts, fermented milk drinks, and kefir (Kandler and Kunath, 1983; Toba et al., 1991; Ariga et al., 1992). Cheese and fermented meat and vegetables are also an important source of ropy LAB (Kojic et al., 1992; Van den Berg et al., 1993). Recently, Smitinont et al. (1999) isolated two EPS producing *Pediococcus* strains from traditional Thai fermented foods; although grown on sucrose, the EPS differed from dextran.

EPS-producing LAB can also be disadvantageous however, for example when causing spoilage of beer, wine, and vacuum-packed cooked meat products (Korkeala et al., 1988; Morin, 1998).

More severe problems arise when biofilm formation occurs on heat exchanger plates in cheese and milk factories, resulting in either excessive openness in cheese texture, or taste changes in milk. In particular, thermophilic LAB can cause significant problems by contaminating heat exchanger plates on the downstream side of the pasteurizers, whereby already pasteurized milk may become re-contaminated (Bouwman et al., 1982; Neu and Marshall, 1990).

5 Physiological Function

Although EPS are not essential to ensure the viability of bacterial cells, their exact function in nature has not yet been clearly defined and is most likely highly complex (Cerning, 1994).

In the past, it has generally been accepted that EPS do not serve as storage materials, since most EPS-producing strains are not able to catabolize EPS, nor use them as a carbon source. In the case of LAB this remains doubtful however, as the amount of EPS produced often decreases upon prolonged fermentation. One possible explanation is that the synthesized EPS are degraded enzymatically by the LAB themselves. Recently, Pham et al. (2000) reported on the synthesis of several glycohydrolases by a *Lb. rhamnosus* strain which was able to hydrolyze the EPS to a limited extent.

The role of EPS in pathogenesis has been investigated, since their synthesis–which results mainly in capsule formation–is widespread among pathogenic bacteria. The presence of EPS alone does not appear to be sufficient to turn bacterial cells virulent. It has been suggested that microorganisms producing EPS are more hydrophilic, and thus are less susceptible to phagocytosis (Jann and Jann, 1977). EPS also seem to play a role in the protection against phage attack, antibiotics and other toxic compounds.

Another proposed function is the protection against desiccation and other extreme physical conditions; this is due to the fact that EPS are present as a highly hydrated layer surrounding the bacterial cells. EPS also play an important role in the adhesion of microorganisms onto solid surfaces, although the adsorption process is complex and the exact impact of EPS in the process has not yet been completely elucidated (Cerning, 1994).

6 Chemical Analysis and Detection

6.1 Separation of EPS and Microbial Cells

The first step in the purification of EPS is the separation of microbial cells from the EPS-containing culture broth. This is normally done by centrifugation, and when working at laboratory scale, ultracentrifugation can be used. It is difficult to define an optimal g-value as this depends on the specific viscosity of the culture broth (Sutherland, 1972). Sometimes a too-high viscosity may prevent easy sedimentation of bacterial cells. The addition of electrolytes (NaCl) may facilitate the separation by neutralizing the charges on the polysaccharides (Cerning, 1994). Because EPS are thermostable, heat treatment can also be used to reduce the viscosity.

When present as a capsule, the EPS must first be dissociated from the cells and, depending on the nature of the association between the cells and the capsule, extreme conditions such as alkaline treatment, sonication or heating may be required (Morin, 1998).

6.2 Isolation and Purification

EPS are mainly recovered by precipitation with organic solvents, such as ethanol, acetone, or isopropanol. Organic solvents permit separation by lowering the solubility of the EPS. Although the EPS:solvent ratio is variable, about one up to three volumes of solvents are normally used. The precipitate is finally recovered by centrifugation, filtration, and pressing or settling, after which dialysis and freeze-drying of the final product is carried out (Garcia-Garibay and Marshall, 1991; Marshall et al., 1995).

EPS from LAB are often synthesized in complex media such as milk, milk ultrafiltrate and whey; the proteins present in these media, such as casein, seem to co-precipitate with the EPS and should be removed prior to the isolation step. This can be achieved enzymatically by the addition of proteases such as pronase, trypsin, or proteinase K (Cerning, 1994). Residual peptides can also be removed by repeated trichloroacetic acid precipitation, followed by centrifugation and precipitation of the EPS. Gel filtration is another possibility, although it is much slower and problems often arise due to the high viscosity of the media. Further purification of the EPS can be achieved by anion-exchange chromatography (Nakajima and Toyoda, 1990; Doco et al., 1991; Andaloussi et al., 1995).

Although EPS yields are higher in milk or whey-based media, the EPS isolation from such complex media is often tedious and time-consuming. More recently, several semi-defined and chemically defined media have been developed; these are more suitable to investigate the influence of nutrients on the growth and on the EPS production, the metabolic pathways involved, the composition of the EPS, and the rheological properties (Petry et al., 2000).

6.3 Structural Analysis

Several structures of repeating units of branched EPS from LAB have been elucidated. General methods used to study the composition and specific linkages present in polysaccharides are normally applied, including acid hydrolysis, periodate oxidation, methylation analysis, enzymatic degradation, Smith-degradation and NMR-spectroscopy. With these methods it is possible to determine not only the neutral sugars but also the anomeric configuration, the specific linkages present, and the sequence of the repeating units.

The molecular weights of polysaccharides are determined by gel filtration and high-performance liquid chromatography.

On occasion, a strain produces more than one type of EPS, and these can usually be separated using chromatography based on the differences in either molecular weight or charge (Toba et al., 1991; Marshall et al., 1995; Robijn et al., 1996; Grobben et al., 1997).

7 Biosynthesis

Unlike the biochemical pathways involved in the biosynthesis of exocellular homopolysaccharides, those of exocellular heteropolysaccharides are much more complex (Cerning, 1994).

Heteropolysaccharides are synthesized at the cytoplasmic membrane through polymerization of precursors, and are formed in the cytoplasm. These precursors are mainly UDP-nucleotide diphosphate sugars. As with the large number of enzymes involved, the sugar nucleotide precursors are not all unique to EPS synthesis; some are also involved in the synthesis of cell-wall polymers such as peptidoglycans, lipopolysaccharides, and teichoic acids. However, as they are freely soluble in the cytoplasm, they can be readily channeled to the appropriate biosynthetic process (Sutherland, 1990).

The sugar nucleotides serve different functions. First, they play an important role in sugar activation, supplying energy for the assembly of glycosyl units on appropriate carrier molecules, with the release of a diphosphate nucleotide. Second, they play a role in sugar interconversions, which involve several mechanisms such as epimerization (UDP-D-glucose → UDP-D-galactose), oxidation (UDP-D-glucose → UDP-D-glucuronic acid), decarboxylation (UDP-D-glucose → UDP-D-xylose), reduction and rearrangement (GDP-D-mannose → GDP-L-fucose) (Sutherland, 1972, 1990).

LAB can metabolize a variety of mono- and disaccharides, but as EPS are mainly produced in milk or whey-based media, the main carbon source present is lactose. Based on studies investigating the influence of sugars on EPS synthesis and the enzymes involved in their anabolism, it might be concluded that glucose or the glucose moiety of lactose is the most important sugar for

EPS production in LAB (De Vos and Vaughan, 1994; Grobben et al., 1996). Hence, glucose-1-phosphate serves as an important precursor for EPS synthesis, and phosphoglucomutase might be seen as a key enzyme linking energy generation and sugar nucleotide formation (Kleerebezem et al., 2000) (Figure 1).

De Vos (1996) suggested that if the flux via phosphoglucomutase was high enough, then EPS overproduction could be achieved. As such, the galactose moiety of lactose would be catabolized via glycolysis, while the glucose moiety would serve for EPS production. A major problem is the inability of *S. thermophilus* and *Lb. delbreuckii* subsp. *bulgaricus* to catabolize galactose, this being due to the absence of the enzyme galactokinase; hence galactose is excreted via the lactose/galactose antiport transport system. However, the gene coding for this enzyme is present but is not transcribed. Galactose-fermenting mutants were constructed and isolated by Kleerebezem et al. (1999) through repair of the promotor mutations, and as a result galactose can be fully metabolized.

The composition and amount of EPS produced are not only dependent on the type of carbon source and on the sugar nucleotide level, but also on the assembly process of the repeating units.

The involvement of an isoprenoid glycosyl lipid carrier was reported in 1971 by Troy et al. This lipid is a C55-isoprenyl phosphate (bactoprenyl phosphate, undecaprenyl phosphate), and is the same acceptor lipid that functions in the formation of several cell-wall polymers (Figure 2). As such, the availability of the isoprenoid carrier is one of the most important factors affecting EPS production. Because of competition for the same acceptor lipid, it was postulated that EPS production is stimulated under conditions which lead to a reduction in growth (e.g., a lower temperature), and so cell-wall polymer biosynthesis is reduced (Sutherland, 1972).

Few authors have reported the exact role of the lipid carrier, and details of the mechanism remain unknown. However, possibilities include facilitation of the formation of the repeating units, solubilization of hydrophilic oligosaccharides in a hydrophobic membrane environment, and transport across the membrane (Troy, 1979; Cerning, 1990).

Once transported through the cell membrane, the polysaccharide can be excreted into the environment or will remain attached to the cell as a capsule. The transport of polysaccharides is an energy-demanding process and has not been fully clarified (Van den Berg et al., 1995).

Recently, *in vitro* experiments have been conducted to elucidate the biosynthesis of the polysaccharide backbone of *L. lactis* NIZO B40. As shown in Figure 3, EPS synthesis occurs through sequential addition of sugar residues by specific glycosyl transferases from sugar nucleotides to a growing repeating unit anchored to the lipid acceptor, thereby yielding the EPS (Kleerebezem et al., 1999, 2000).

8
Genetics and Regulation

A well-known problem of LAB in the dairy industry is the instability of EPS production at the genetic level, as well as the instability of the ropy texture itself (Cerning 1990, 1994). Loss of the ropy character may occur after numerous transfers and prolonged incubation periods, even at optimum growth temperatures. Therefore, strains have to be reselected regularly from the master culture to conserve the ropy character in industrially applied strains.

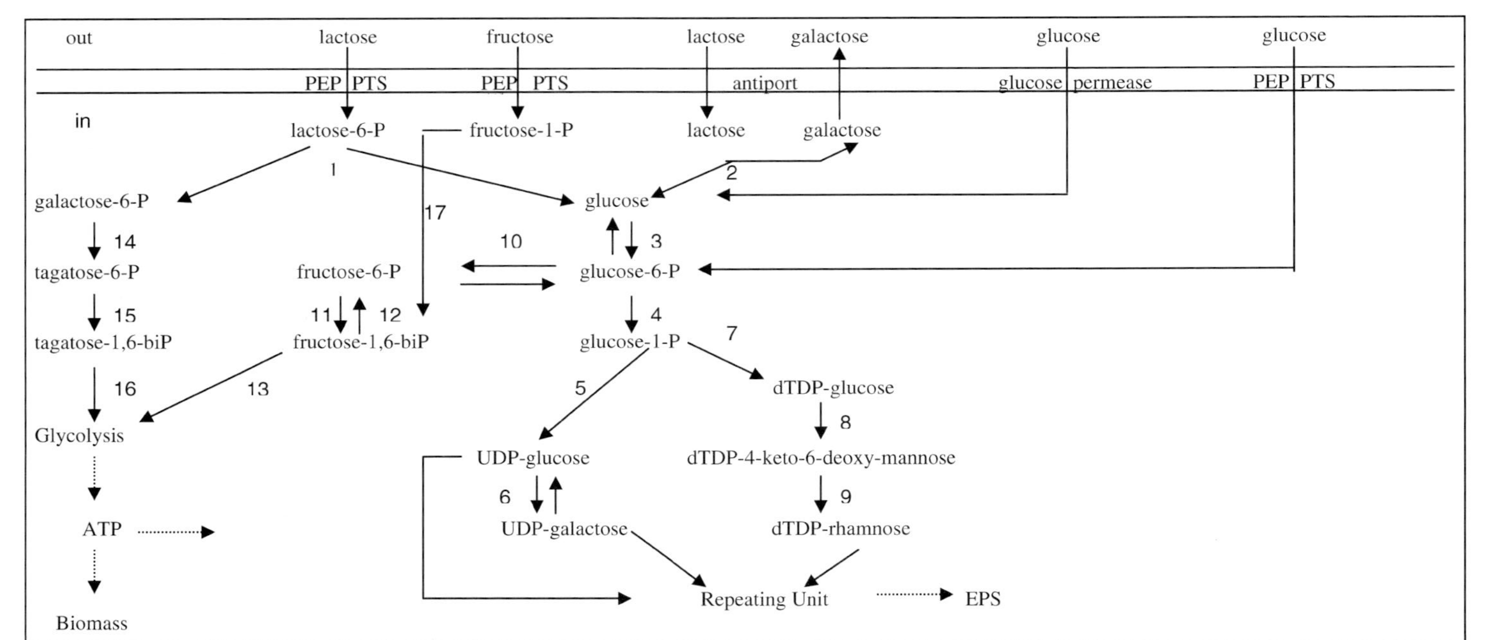

Fig. 1 Schematic representation of the lactose-, glucose-, and fructose-utilizing pathways and exopolysaccharide biosynthesis. *Lactococcus lactis* transports lactose via the lactose phosphoenolpyruvate (PEP) -dependent phosphotransferase system (PTS), while the galactose-negative *Streptococcus thermophilus* and *Lactobacillus delbreuckii* subsp. *bulgaricus* transport lactose via a lactose/galactose antiport transport system. Enzymes involved are: 1, phospho-β-galactosidase; 2, β-galactosidase; 3, glucokinase; 4, phosphoglucomutase; 5, UDP-glucose pyrophosphorylase; 6, UDP-galactose-4-epimerase; 7, dTDP-glucose pyrophosphorylase; 8, dehydratase; 9, epimerase reductase; 10, phosphoglucoisomerase; 11, 6-phosphofructokinase; 12, fructose-1,6-biphosphatase; 13, fructose-1,6-biphosphate aldolase; 14, galactose-6-phosphate isomerase; 15, tagatose-6-phosphate kinase; 16, tagatose-1,6-biphosphate aldolase; 17, fructokinase (Kleerebezem et al., 1999).

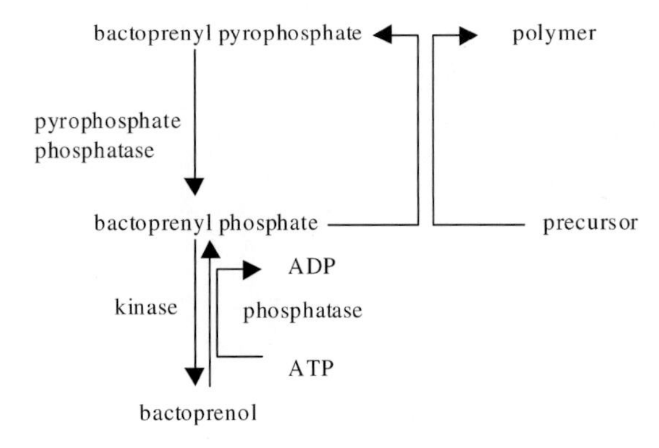

Fig. 2 Structure of the active form of the isoprenoid lipid carrier. (b) Activation and deactivation of the isoprenoid lipid; the polymer can be EPS, lipopolysaccharide, or peptidoglycan (Cerning, 1994).

Numerous plasmids encoding for EPS production have already been identified in mesophilic LAB. Moreover, it has been proven that the loss of the EPS-producing capacity in these bacteria is associated with the loss of plasmids (Vedamuthu and Neville, 1986; Neve et al., 1988). In thermophilic LAB, the EPS production is not encoded by plasmids, but the required genes are located on the chromosome. It is likely that the genetic instability is due to mobile genetic elements or to a generalized genomic instability, including deletions and rearrangements. Both phenomena were observed for *Lb. delbreuckii* subsp. *bulgaricus* and *S. thermophilus* (Stingele et al., 1996).

During the past few years, specific gene clusters have been characterized for some EPS-producing lactococci and streptococci; in addition, the entire genome sequence of *L. lactis* has recently been completed (Kleerebezem et al., 2000). These gene clusters contain *eps* genes, which are involved in the synthesis of the repeating units, export, polymerization and chain length determination. In addition to these *eps* genes, a number of housekeeping genes are also required for EPS synthesis. These genes are involved in the metabolic pathways leading to the EPS building blocks, namely the sugar nucleotides. Elucidation of the function of the glycosyltransferase genes might create opportunities for EPS engineering, while the identification and characterization of the housekeeping genes allows the design of metabolic engineering strategies, leading to increased EPS production levels (Kleerebezem et al., 1999).

The chromosomally located gene cluster of *S. thermophilus* Sfi6 consists of a 15.25 kb region coding for 15 genes; only 13 genes, *eps*A to *eps*M, were involved in EPS synthesis (Stingele et al., 1996). Low et al. (1998) partially identified and characterized the *eps* gene cluster of *S. thermophilus* MR-1C and found an organization similar to that of *S. thermophilus* Sfi6.

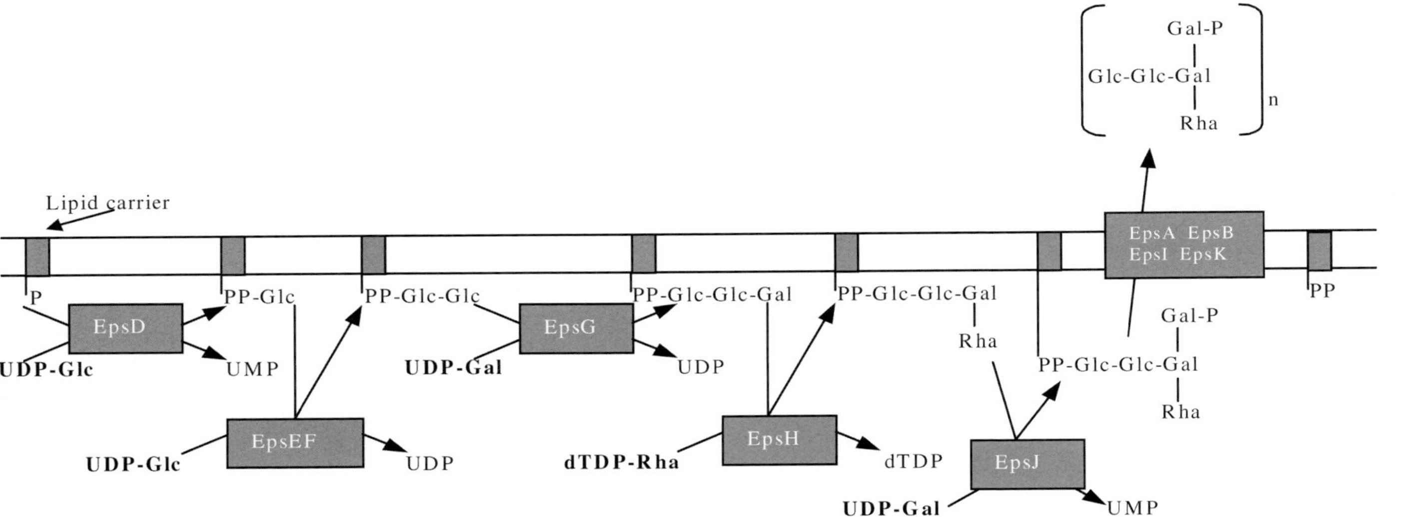

Fig. 3 Schematic representation of EPS biosynthesis by *L. lactis* NIZO B-40. Nucleotide sugars are indicated in bold; enzymes are indicated in the gray boxes by their encoding genes using established genetic nomenclature. Gal, galactose; Glc, glucose; Rha, rhamnose (Kleerebezem et al., 1999).

For *L. lactis*, the *eps* genes are encoded on large plasmids (>20 kb) and can be transferred from one strain to another by conjugation. The best-characterized lactococcal strain is *L. lactis* NIZO B40, which produces a phosphopolysaccharide. A 12-kb gene cluster specific for EPS biosynthesis, contains 14 coordinately expressed genes and is inserted between IS sequences on a 40-kb plasmid.

Major similarities exist between the organization of the genes in the different gene clusters involved in EPS biosynthesis; four functional regions can be distinguished: a region involved in regulation is located at the beginning of the gene cluster; a central region which is involved in the synthesis of the EPS repeating units; and two flanking regions that are involved in polymerization and export (Stingele et al., 1996).

Further detailed studies can offer perspectives for EPS engineering and evaluation of the possibilities and restrictions towards the production of tailor-made EPS. This concerns not only changing the chemical structure of the polymers but also other relevant features, such as its chain length.

It was shown recently that a *L. lactis* NIZO B40 EpsD deletion mutant could functionally be complemented by the homologous *Eps*D gene itself, but more important also by heterologous glucosyltransferase encoding genes, yielding an EPS with a similar molecular weight but with a different structure (Van Kranenburg et al., 1999). Other examples, which indicate possibilities for the synthesis of heterologous or new EPS structures, have also been reported (Gilbert et al., 1998; Stingele, 1998).

Another approach would be to engineer the metabolic pathways in order to increase EPS production levels, perhaps by increasing the flux towards sugar nucleotides. The so-called "housekeeping" genes are involved in the synthesis of sugar nucleotides; in this respect, the physiology of sugar nucleotide synthesis should be investigated in relation to EPS production levels (Kleerebezem et al., 1999).

At present, all genes of *L. lactis* MG1363 encoding for enzymes involved in the biosynthesis of sugar nucleotides from glucose-1-phosphate have been cloned, and specific enzyme activities, which form bottlenecks in EPS synthesis, can now be investigated. The potential of these metabolic engineering strategies to improve production levels can also be evaluated (Boels et al., 1998). The role of the genes which encode for important enzymes in these metabolic pathways is now also being investigated in other LAB.

It is expected that engineering of EPS and their biosynthetic pathways in LAB will be an interesting challenge in future metabolic research. However, the use of genetically modified microorganisms or their products will require not only legal approval, but also acceptance by the consumer (De Vuyst and Degeest, 1999b; Kleerebezem et al., 2000).

9 Factors Influencing Growth and EPS Production

9.1 Effect of Physico-chemical Parameters

Generally, EPS are produced at temperatures below the optimal growth temperature. For mesophilic LAB, EPS production increased by 50–60% if the strains are cultured at 25°C instead of at 30°C (Cerning et al., 1992). *L. lactis* subsp. *lactis* exhibited the highest EPS production at 22°C; at temperatures higher than 37°C, no further EPS production occurred. The highest EPS production for *L. lactis* subsp. *cremoris* was noted at 20°C (Marshall et al., 1995), while for *Lb. casei* EPS are produced at 25°C as well as at 30°C (Cerning, 1994; Mozzi et al., 1996).

Mozzi et al. (1995) investigated the influence of the temperature on EPS production for the thermophilic species *Lb. delbreuckii* subsp. *bulgaricus, Lb. acidophilus* and *S. salivarius* subsp. *thermophilus*. In milk-based media without pH control, the optimum temperature for EPS production was 42°C for *Lb. acidophilus* and 37°C for *Lb. delbreuckii* subsp. *bulgaricus*, which coincides with optimal growth temperatures. Garcia-Garibay and Marshall (1991) found a maximal EPS production for *Lb. delbreuckii* subsp. *bulgaricus* at 45°C. Grobben et al. (1995) noted an increased EPS production with an increased temperature during batch fermentation, using a defined medium with controlled pH; the EPS production was maximal at 40°C.

For *S. salivarius* subsp. *thermophilus*, the highest EPS production was noted at 30°C. These results were confirmed by Gancel and Novel (1994) and Gassem et al. (1995).

As mentioned previously, EPS synthesis is expected to be stimulated by reduced growth temperatures; as such, cell growth is reduced as well as the synthesis of cell-wall polymers. This would result in a better availability of the lipid isoprenoid carrier which is necessary for export of EPS (Sutherland, 1982).

For some LAB, the effect of the initial pH of the growth medium on EPS biosynthesis was investigated; during this type of experiment the pH was no longer controlled during fermentations and was allowed to vary freely. On occasion, no clear effect was noted, but for *S. thermophilus* S22 an optimal initial pH of 7.0 for lactose as a carbon source and an optimal initial pH of 5.5 for sucrose was determined (Gancel and Novel, 1994).

A maximal EPS yield was obtained for *Lb. casei* CRL 87 when the pH was controlled at 6.0 during the fermentation; moreover, no decrease of EPS level occurred upon prolonged incubation (Mozzi et al., 1994).

In a continuous fermentation with *Lb. delbreuckii* subsp. *bulgaricus*, and using a whey-based medium, maximal EPS production occurred at pH 6.5, while the optimal pH for EPS production in batch fermentations was 5.8 (Van den Berg et al., 1995).

With *Lb. delbreuckii* subsp. *bulgaricus* CNRZ 1187 and CNRZ 416, about three-fold and four-fold respectively more EPS was produced when the pH was controlled at 6.0 (Petry et al., 2000).

In general, higher EPS yields are obtained when fermentations are carried out under controlled pH conditions. The optimal pH control range for EPS production by LAB is often close to pH 6–6.5 (Mozzi et al., 1994), and this may be a limiting factor when considering industrial use of these bacteria in fermented milk drinks, as the pH is allowed to vary freely during such fermentations (De Vuyst and Degeest, 1999a).

With regard to temperature, EPS production appears to increase under less than optimal pH conditions for growth, though this might also be explained by a better availability of the lipid carriers (Ricciardi and Clementi, 2000).

According to Cerning (1990), EPS biosynthesis by LAB is highest under aerobic growth conditions, and this explains why more EPS is excreted during growth on solid media rather than in liquid cultures. However, another reason might be that adhesion of the bacterial cells onto a solid surface stimulates EPS formation.

Mäkelä and Korkeala (1992) investigated the effect of aeration, temperature and medium composition for different LAB. Aerobic or anaerobic conditions resulted in better EPS production, depending on the strain.

9.2
Effect of Nutritional Parameters

In general, milk-based media or other complex media containing peptones or yeast extract are used for culturing EPS-producing

LAB. Higher EPS yields are indeed obtained, but isolation and purification of the polymers are hampered. Recently, some chemically defined media were developed which greatly facilitate the study of the influence of nutrients on EPS production, growth, chemical composition, and rheological properties.

Cerning et al. (1986) found that the addition of casein enhanced growth, but not EPS synthesis, although contrasting results were subsequently reported which indicated a positive effect on EPS production by *Lb. delbrueckii* subsp. *bulgaricus* following the addition of hydrolyzed casein (Garcia-Garibay and Marshall, 1991; Cerning, 1994).

Another point identified was that the carbon source influences not only the composition of the EPS produced by LAB but also the total amount produced. *Lb. casei* CG11 produced more EPS when glucose was added to milk or milk ultrafiltrate; moreover, the monomeric composition changed when glucose was used as the major monosaccharide rather than galactose (Cerning et al., 1992). In a later study, Mozzi et al. (1995) obtained the highest EPS yield for *Lb. casei* CRL 87 when using galactose as the carbon source.

Considerably more EPS was synthesized in a chemically defined medium by *Lb. delbreuckii* subsp. *bulgaricus* NCFB 2772 when grown on glucose, lactose, or glucose and fructose rather than on mannose or fructose. A continuous culture was used to study the effect of glucose and/or fructose on EPS production and composition. When grown on glucose or glucose and fructose, the EPS contained rhamnose, while in the presence of fructose alone, no rhamnose was present in the EPS and the overall EPS yield remained low. It was found that in this latter case, the enzymes involved in the synthesis of rhamnose were absent; moreover, when glucose was present in the culture medium, higher activities were determined for various enzymes involved in EPS synthesis (Grobben et al., 1995, 1996).

Later, it was shown that this strain produces two types of EPS: a high-molecular fraction, which appears to be influenced by the carbon source; and a low-molecular fraction, which is produced more consistently (Grobben et al., 1997).

For another *Lb. delbreuckii* subsp. *bulgaricus* strain, the composition of the EPS was found to be independent of the nature of the carbon source (Manca de Nadra et al., 1985). No influence of the carbon source on the amount of EPS produced by *Lb. sake* 01 was noted by Van den Berg et al. (1995).

Petry et al. (2000) designed a chemically defined medium that allowed good growth and EPS production for several *Lb. delbreuckii* subsp. *bulgaricus* strains. The presence of excess sugar in the growth medium (10–20 g L^{-1}) did not stimulate EPS production, as was proven for *Lb. casei* and *Lb. rhamnosus*. Independently of the carbon source, the monomeric composition remained similar.

In the case of *S. thermophilus* LY03, a clear influence of the type of carbon source on the amount of EPS produced was noted, while the composition remained constant. This strain produces two polysaccharides with identical composition, but different molecular weights. The proportion in which these polysaccharides are synthesized is strongly dependent on the C/N ratio of the culture medium (De Vuyst and Degeest, 1999b).

Other nutrients such as vitamins, amino acids and minerals might also affect the composition and quantity of EPS produced. *Lb. delbreuckii* subsp. *bulgaricus* NCFB 2722 required only riboflavin, nicotinic acid and calcium pantothenate for EPS synthesis. The absence of glutamine, asparagine and threonine reduced growth considerably (Grobben et al., 1997).

Clearly, contrasting results have often been obtained, and so it is advisable to evaluate separately the influence of physicochemical and nutritional parameters on EPS synthesis for each strain. Moreover, additional studies on the effect of medium composition and of culture conditions on the amount and composition of EPS might result in a better understanding of the biosynthesis regulatory mechanisms.

9.3 Kinetics of EPS Biosynthesis

In general, mesophilic LAB produce greater amounts of EPS under conditions not optimal for growth, and so EPS production appears to be nongrowth-associated. In contrast, in thermophilic LAB, EPS production appears to be growth-associated (Andaloussi et al., 1995; Grobben et al., 1995; Manca de Nadra et al., 1985).

Growth-associated EPS production was noted for *S. thermophilus* LY03 (De Vuyst and Degeest, 1999a) and for *Lb. delbreuckii* subsp. *bulgaricus* NCFB 2772 (Grobben et al., 1995). However, Gancel and Novel (1994) reported that the EPS production of a *S. thermophilus* strain was not growth-associated, and Petry et al. (2000) noted that EPS production occurred mainly during the stationary growth phase for a *Lb. delbreuckii* subsp. *bulgaricus* strain.

For some mesophilic LAB, a growth-associated EPS production was observed, this being the case for *Lb. sake* 01 and a *Lb. rhamnosus* strain (Van den Berg et al., 1995). Other investigations observed a continued EPS synthesis during the stationary growth phase (Manca de Nadra et al., 1985; Gancel and Novel 1994).

Often, the amount of EPS produced declines upon prolonged fermentation. Several explanations were proposed, such as an enzymatic degradation, or a change in the physical parameters of the culture. Moreover, this trend is not rare with regard to some other microbial EPS types such as gellan (Kennedy and Sutherland, 1994) or sphingan (Hashimoto and Murata, 1998).

Pham et al. (2000) investigated this phenomenon for a *Lb. rhamnosus* strain and identified the presence of a large spectrum of glycohydrolases, three of them extracellular and the others either cell-wall bound or intracellular. When incubated with EPS, the enzymes were capable of liberating reducing sugars and of lowering the viscosity; indeed, a viscosity loss of 33% was noted after 27 h.

Gassem et al. (1995) observed a reduction in viscosity of the culture liquid after 18 h of fermentation with *S. salivarius* subsp. *thermophilus* ST3. Cerning (1994) noted similar results for *S. salivarius* subsp. *thermophilus*, although EPS degradation was also influenced by pH and temperature.

Various authors have reported this phenomenon, but few have investigated it further; as such, the complete elucidation of this degradation mechanism requires further study.

10 Applications

One way to improve the rheological properties and functional characteristics of dairy products, such as yogurt, is the addition of stabilizers. Normally, hydrocolloids of animal or plant origin are applied, but their use is prohibited in several countries, including France and the Netherlands. EPS from LAB, which are GRAS bacteria, may serve as alternative biothickeners (Grobben et al., 1995). In order to be suitable, the polymers should be compatible with other food components and with applied processing conditions, and should exhibit thixotropic or pseudoplastic behavior, notably when considering aspects such as processing costs, mouthfeel and texture (Van den Berg et al., 1995).

Since the amount of EPS produced by LAB is rather limited, these bacteria are mostly used to produce EPS *in situ*, to improve consistency and viscosity as well as to avoid syneresis (whey separation) in yogurts. Also, for the production of other fermented milk products and cheeses, EPS-producing LAB are used in starter cultures (Cerning, 1995; Marshall and Rawson, 1999; Ricciardi and Clementi, 2000).

Over the past 40 years, yogurt production has evolved substantially. One variation, which is now dominating the market, is stirred yogurt in various forms, obtained by further processing of set yogurt. Due to different mechanical processing steps, stirred yogurt tends to lose some of its initial viscosity. This might be solved by using ropy LAB as starter cultures, because yogurt containing viscosifying EPS is less susceptible to damage by mechanical pumping, blending and filling (Cerning, 1990; Rawson and Marshall, 1997).

Based on rheological studies, it was found that no clear correlation existed between the EPS concentration and apparent viscosity. The effect of EPS appeared to be much more complex, however. The contribution of EPS to specific food properties depends not only on the properties of the EPS itself, such as molar mass and structural composition, but also on the interaction of the EPS with various milk constituents, and in particular caseins (Kleerebezem et al., 1999; Marshall and Rawson, 1999).

Yogurt consists of a network of casein micelles, associated to form a gel-like structure within which small spaces are filled with liquid whey phase, while larger spaces contain the EPS and the LAB. The typical viscoelastic property of yogurt is obtained by retention of the whey. By using scanning electron microscopy, EPS strands were observed between the cells and the protein network; hence, besides the protein strands and protein–protein bonds, protein–polysaccharide bonds also occur (Wacher-Rodarte et al., 1993; Ricciardi and Clementi, 2000).

The use of ropy strains results in a yogurt which is more stable towards shear forces. Due to EPS synthesis, protein strand formation and protein–protein bond formation is partly prevented, resulting in less rigid gels with an increased viscosity, an ability to recover lost viscosity, and a certain adhesiveness. Although these characteristics are strain-dependent, it appears likely that EPS synthesis contributes to adhesiveness while the protein network has a greater influence on firmness and elasticity (Rawson and Marshall, 1997).

The thermophilic *S. thermophilus* and *Lb. delbreuckii* subsp. *bulgaricus* are traditionally used in yogurt and are both able to form EPS. Ropy, mesophilic LAB are also widely used in starter cultures, but they produce less EPS and their ropiness is often an unstable characteristic (Wacher-Rodarte et al., 1993).

In Scandinavian countries, slime-producing mesophilic LAB are used in the manufacture of different fermented milks. In Finnish "viili" and Nordic "longfil", a ropy strain of *L. lactis* subsp. *cremoris* is essential for correct consistency (Forsén and Häivä, 1981; Kontusaari and Forsén, 1988; Toba et al., 1990).

Kefir is an acidic, alcoholic beverage that is popular in Eastern Europe. The EPS produced by *Lb. kefiranofaciens* comprises the mass of the kefir grain, and was named kefiran. Other microorganisms are also associated with kefir, such as *Lb. acidophilus, Lb. kefir granum, Lb. kefir, Lb. parkefir* and the yeast *Candida kefir* (Yokoi et al., 1990).

The capacity of EPS to increase water retention was used to improve the functional properties of cheese. Perry et al. (1988) produced low-fat mozzarella with a significantly higher moisture content and a better stretching ability, using capsulated nonropy cultures of *S. thermophilus* and *Lb. delbreuckii* subsp. *bulgaricus*. Eventually, Low et al. (1998) showed that *S. thermophilus* was mainly responsible for the increased moisture content. Scanning electron microscopy was used to elucidate the positive effect on the microstructure of the cheese, and revealed that bridges were formed between the EPS filaments and the casein fibers, thus preventing collapse of the whey channels.

EPS produced by LAB may also contribute to human health because of their prebiotic (Gibson and Robertfroid, 1995), cholesterol-lowering (Nakajima et al., 1992), immunomodulating (Kitazawa et al., 1993; Hosono et al., 1997) or anti-tumor (Oda et al., 1983; Toba et al., 1991) activities.

Other health-promoting activities can be attributed to the LAB themselves, for example the removal of lactose in lactose-intolerant persons. While the pH permits metabolic activity of the LAB, lactose is utilized; however, by re-routing the sugar metabolism in *L. lactis* towards L-alanine rather than towards lactic acid, complete conversion of lactose was obtained. Moreover, the sweetness and general taste of the yogurt were enhanced (Kleerebezem et al., 2000).

Although LAB are known to be vitamin-auxotrophic bacteria, *S. thermophilus* is able to produce folic acid, an essential compound in human nutrition, and which is further metabolized by another yogurt bacterium, *Lb. delbreuckii* subsp. *bulgaricus*. Applying high folic acid-producing strains or relatively high numbers of *S. thermophilus* should lead to the production of yogurt with an increased folic acid content (Kleerebezem et al., 2000).

During the fermentation of dairy products, LAB are responsible for the synthesis of typical flavor compounds such as acetaldehyde in yogurt and diacetyl in buttermilk (Kleerebezem et al., 2000). Moreover, LAB also produce antimicrobial peptides ("bacteriocins") which have an antagonistic effect against genetically closely related strains; they are mostly thermostable, digestible and active in low concentrations. The antimicrobial spectrum of most bacteriocins is restricted to Gram-positive bacteria, but bacteria responsible for food spoilage, such as *Bacillus cereus*, *Clostridium perfringens*, *Listeria monocytogenes* and *Staphylococcus aureus* are often included. Consequently, bacteriocins may serve as a natural food preservative (De Vuyst and Vandamme, 1994).

It is clear that the metabolic activity of LAB results in "totally natural" products with a clear added value, and which have not only the desired nutritional properties but also contribute to human health in general. However, most bacterial activities may still be improved as they do not reach maximal functionality during *in situ* milk fermentation.

11
References

Andaloussi, S. A., Talbaoui, H., Marczak, R., Bonaly, R. (1995) Isolation and characterization of exocellular polysaccharides produced by *Bifidobacterium longum*, *Appl. Microbiol. Biotechnol.* **43**, 995.

Ariga, H., Urashima, T., Michihata, E., Ito, M., Morizono, N., Kimura, T., Takahashi, S. (1992) Extracellular polysaccharide from encapsulated *Streptococcus salivarius* subsp. *thermophilus* OR 901 isolated from commercial yogurt, *J. Food Sci.* **57**, 625–628.

Boels, I. C., Kleerebezem, M., Hugenholtz, J., de Vos, W. M. (1998) Metabolic engineering of exopolysaccharide production in *Lactococcus lactis*, in: Proceedings, 5th ASM on the genetics and molecular biology of streptococci, enterococci and lactococci, 66.

Bolotin, A., Mauger, S., Malarme, K., Sorokin, A., Ehrlich, D. S. (1998) *Lactococcus lactis* IL1403 diagnostic genomics, in: Proceedings, 5th ASM on the genetics and molecular biology of streptococci, enterococci and lactococci, 10–11.

Bouwman, S., Lund, D., Driessen, F. M., Schmidt, D. G. (1982) Growth of thermoresistant streptococci and deposition of milk constituents on plates of heat exchangers during long operating times, *Food Prot.* **45**, 806–812.

Cerning, J. (1990) Exocellular polysaccharides produced by lactic acid bacteria, *FEMS Microbiol. Rev.* **87**, 113–130.

Cerning, J. (1994) Polysaccharides exocellulaires produits par les bactéries lactiques, in: *Bactéries Lactiques: Aspects fondamentaux et technologiques* (de Roissart, H., Luquet, F. M., Eds), Uriage: Lorica, 309–329.

Cerning, J. (1995) Production of exopolysaccharides by lactic acid bacteria and dairy propionibacteria, *Le Lait* **75**, 463–472.

Cerning, J., Bouillanne, C., Desmazeaud, M. J. (1986) Isolation and characterization of exocellular polysaccharide produced by *Lactobacillus bulgaricus*, *Biotechnol. Lett.* **8**, 625–628.

Cerning, J., Bouillanne, C., Landon, M., Desmazeaud, M. J. (1990) Comparison of exocellular polysaccharide production by thermophilic lactic acid bacteria, *Sciences des Aliments* **10**, 443–451.

Dellaglio, F. (1994) Caractéristiques générales des bactéries lactiques, in: *Bactéries Lactiques: Aspects fondamentaux et technologiques* (de Roissart, H., Luquet, F. M., Eds), Uriage: Lorica, 10–58.

De Vos, W. M. (1996) Metabolic engineering of sugar catabolism in lactic acid bacteria, *Antonie van Leeuwenhoek* **70**, 223–242.

De Vos, W. M., Vaughan, E. E. (1994) Genetics of lactose utilization in lactic acid bacteria, *FEMS Microbiol. Rev.* **15**, 216–237.

De Vuyst, L., Degeest, B. (1999a) Exopolysaccharides from lactic acid bacteria: technological bottlenecks and practical solutions, *Macromol. Symp.* **140**, 31–41.

De Vuyst, L., Degeest, B. (1999b) Heteropolysaccharides from lactic acid bacteria, *FEMS Microbiol. Rev.* **23**, 153–177.

De Vuyst, L., Vandamme, E. J. (1994) Lactic acid bacteria and bacteriocins: their practical importance, in: *Bacteriocins of lactic acid bacteria, microbiology, genetics and applications* (De Vuyst, L., Vandamme, E. J., Eds), London: Blackie Academic and Professional, 1–11.

Doco, T., Carcano, D., Ramos, P., Loones, A., Fournet, B. (1991) Rapid isolation and estimation of polysaccharide from fermented skim milk with *Streptococcus salivarius* subsp. *thermophilus* by coupled anion exchange and gel permeation high-performance liquid chromatography, *J. Dairy Res.* **58**, 147–150.

Faber, E. J., van den Haak, M. J., Kamerling, J. P., Vliegenthart, J. F. G. (2001) Structure of the exopolysaccharide produced by *Streptococcus thermophilus* S3, *Carbohydr. Res.* **331**, 173–182.

Forsén, R., Häivä, V. (1981) Induction of stable slime-forming and mucoid states by p-fluorophenylalanine in lactic streptococci, *FEMS Microbiol. Lett.* **12**, 409–413.

Gancel, F., Novel, G. (1994) Exopolysaccharide production by *Streptococcus salivarius* subsp. *thermophilus* cultures: two distinct modes of polymer production and degradation among clonal variants, *J. Appl. Bacteriol.* **77**, 689–695.

Garcia-Garibay, M., Marshall, V. M. (1991) Polymer production by *Lactobacillus delbreuckii* ssp. *bulgaricus, J. Appl. Bacteriol.* **70**, 325–328.

Gassem, M. A., Schmidt, K. A., Frank, J. F. (1995) Exopolysaccharide production in different media by lactic acid bacteria, *Cult. Dairy Prod. J.* **30**, 18–21.

Gibson, G. R., Roberfroid, M. D. (1995) Dietary modulation of the human colonic microbiota: introducing the concept of prebiotics, *J. Nutr.* **124**, 1401–1412.

Gilbert, C., Robinson, K., LePage, R. W. F., Wells, J. M. (1998) Heterologous biosynthesis of pneumococcal type 3 capsule in *Lactococcus lactis* and immunogenicity studies, in: Proceedings, 5th ASM on the genetics and molecular biology of streptococci, enterococci and lactococci, 67.

Grobben, G. J., Sikkema, J., Smith, M. R., De Bont, J. A. M. (1995) Production of extracellular polysaccharides by *Lactobacillus delbreuckii* ssp. *bulgaricus* NCFB 2772 grown in a chemically defined medium, *J. Appl. Bacteriol.* **79**, 103–107.

Grobben, G. J., Smith, M. R., Sikkema, J., De Bont, J. A. M. (1996) Influence of fructose and glucose on the production of exopolysaccharides and the activities of enzymes involved in the sugar metabolism and the synthesis of sugar nucleotides in *Lactobacillus delbreuckii* subsp. *bulgaricus* NCFB 2772, *Appl. Microbiol. Biotechnol.* **46**, 279–284.

Grobben, G. J., Van Casteren, W. H. M., Schols, H. A., Oosterveld, A., Sala, G., Smith, M. R., Sikkema, J., De Bont, J. A. M. (1997) Analysis of the exopolysaccharides produced by *Lactobacillus delbreuckii* subsp. *bulgaricus* NCFB 2772 grown in continuous culture on glucose and fructose, *Appl. Microbiol. Biotechnol.* **48**, 516–521.

Groux, M. (1973) Kritische betrachtungen der heutigen joghurt-herstellung mit berücksichtigung des proteinabbaues, *Schweiz. Milchz.* **4**, 2–8.

Hashimoto, W., Murata, K. (1998) α-L-Rhamnosidase of *Sphingomonas* sp. R1 producing an unusual exopolysaccharide of sphingan, *Biosci. Biotechnol. Biochem.* **62**, 1068–1074.

Hosono, A., Lee, J., Ametani, A., Natsume, M., Adachi, T., Kaminogawa, S. (1997) Characterization of a water-soluble polysaccharide fraction with immunopotentiating activity from *Bifidobacterium adolescentis* M101-4, *Biosci. Biotechnol. Biochem.* **61**, 312–316.

Jann, K., Jann, B. (1977) Bacterial polysaccharide antigens, in: *Surface Carbohydrates of the Prokaryotic Cell* (Sutherland, I. W., Ed.), New York: Academic Press, 247–287.

Kandler, O., Kunath, P. (1983) *Lactobacillus kefir* sp. nov., a component of the microflora of kefir, *Syst. Appl. Microbiol.* **4**, 286–294.

Kennedy, L., Sutherland, I. W. (1994) Gellan lyases – novel polysaccharide lyase, *Microbiology* **140**, 3007–3013.

Kitazawa, H., Yamaguchi, T., Miura, M., Saito, T., Itoh, H. (1993) B-cell mitogen produced by slime-forming, encapsulated *Lactococcus lactis* subspecies *cremoris* isolated from ropy sour milk, viili, *J. Dairy Sci.* **76**, 1514–1519.

Kleerebezem, M., van Kranenburg, R., Tuinier, R., Boels, I. C., Zoon, P., Looijesteijn, E., Hugenholtz, J., de Vos, W. M. (1999) Exopolysaccharides produced by *Lactococcus lactis*: from genetic engineering to improved rheological properties, *Antonie van Leeuwenhoek* **76**, 657–665.

Kleerebezem, M., Hols, P., Hugenholtz, J. (2000) Lactic acid bacteria as a cell factory: rerouting of carbon metabolism in *Lactococcus lactis* by metabolic engineering, *Enzyme Microb. Technol.* **26**, 840–848.

Kojic, M., Vujcic, M., Banina, A., Cocconcelli, P., Cerning, J., Topisirovic, L. (1992) Analysis of exopolysaccharide production by *Lactobacillus casei* CG11, isolated from cheese, *Appl. Environ. Microbiol.* **58**, 4086–4088.

Kontusaari, S., Forsèn, R. (1988) Finnish fermented milk "Viili": involvement of two cell surface proteins in production of slime by *Streptococcus lactis* spp. *cremoris, J. Dairy Sci.* **71**, 3197–3202.

Korkeala, H., Suortti, T., Mäkelä, P. (1988) Ropy slime formation in vacuum-packed cooked meat products caused by homofermentative *Lactobacilli* and a *Leuconostoc* species. *Int. J. Food Microbiol.* **7**, 339–347.

Low, D., Ahlgren, J. A., Horne, D., McMahon, D. J., Oberg, C., Broadbent, J. R. (1998) Role of *Streptococcus thermophilus* MR-1C capsular exopolysaccharide in cheese moisture retention, *Appl. Environ. Microbiol.* **64**, 2140–2147.

Mäkelä, P. M., Korkeala, H. J. (1992) The ability of the ropy slime-producing lactic acid bacteria to form ropy colonies on different culture media and at different incubation temperatures and atmospheres, *Int. J. Food Microbiol.* **16**, 161–166.

Manca de Nadra, M. C., Strasser de Saad, A. M., Pesce de Ruiz Holgado, A. A., Oliver, G. (1985) Extracellular polysaccharide production by *Lb. bulgaricus* CRL 420, *Milchwissenschaft* **40**, 409–411.

Marshall, V. M., Rawson, H. L. (1999) Effects of exopolysaccharide-producing strains of thermophilic lactic acid bacteria on the texture of stirred yoghurt, *Int. J. Food Sci. Technol.* **34**, 137–143.

Marshall, V. M., Cowie, E. N., Moreton, R. S. (1995) Analysis and production of two exopolysaccharides from *Lactobacillus casei*, *J. Dairy Res.* **62**, 621–628.

Marshall, V. M., Dunn, H., Elvin, M., McLay, N., Gu, Y., Laws, A. P. (2001) Structural characterisation of the exopolysaccharide produced by *Streptococcus thermophilus* EU20, in: Proceedings, 1st International Symposium on Exopolysaccharides from Lactic acid bacteria. Brussels, Belgium: VUB, 12.

Morin, A. (1998) Screening for polysaccharide-producing microorganisms, factors influencing the production, and recovery of microbial polysaccharides, in: *Polysaccharides: Structural Diversity and Functional Versatility* (Dumitriu, S., Ed.), New York: Marcel Dekker, Inc., vol. 8, 275–296.

Mozzi, F., de Giori, G. S., Oliver, G., de Valdez, G. F. (1994) Effect of culture pH on the growth characteristics and polysaccharide production by *Lactobacillus casei*, *Milchwissenschaft* **49**, 667–670.

Mozzi, F., Oliver, G., de Giori, G. S., de Valdez, G. F. (1995) Influence of temperature on the production of exopolysaccharides by thermophilic lactic acid bacteria, *Milchwissenschaft* **50**, 80–82.

Mozzi, F., de Giori, G. S., Oliver, G., de Valdez, G. F. (1996) Exopolysaccharide production by *Lactobacillus casei* under controlled pH, *Biotechnol. Lett.* **18**, 435–439.

Nakajima, H., Toyoda, S. (1990) A novel phosphopolysaccharide from slime-forming *Lactococcus lactis* subspecies *cremoris* SBT 0495, *J. Dairy Sci.* **73**, 1472–1477.

Nakajima, H., Suzuki, Y., Kaizu, H., Hirota, T. (1992) Cholesterol-lowering activity of ropy fermented milk, *J. Food Sci.* **57**, 1327–1329.

Neu, T. R., Marshall, K. C. (1990) Bacterial polymers: physicochemical aspects of their interactions at interfaces, *J. Biomater. Appl.* **5**, 107–133.

Neve, H., Geis, A., Teuber, M. (1988) Plasmid-encoded functions of ropy lactic acid streptococcal strains from Scandinavian fermented milk, *Biochimie* **70**, 437–442.

Nilsson, R., Nilsson, G. (1958) Studies concerning Swedish ropy milk, the antibiotic qualities of ropy milk, *Arch. Microbiol.* **31**, 191–197.

Oda, M., Hasegawa, H., Komatsu, S., Kambe, M., Tsuchiya, F. (1983) Anti-tumor polysaccharide from *Lactobacillus* sp., *Agr. Biol. Chem.* **47**, 1623–1625.

Perry, D. B., McMahon, D. J., Oberg, C. J. (1998) Manufacture of low fat mozzarella cheese using exopolysaccharide-producing starter cultures, *J. Dairy Sci.* **81**, 561–563.

Petry, S., Furlan, S., Crepeau, M. J., Cerning, J., Desmazeaud, M. (2000) Factors affecting exocellular polysaccharide production by *Lactobacillus delbreuckii* subsp. *bulgaricus* grown in a chemically defined medium, *Appl. Environ. Microbiol.* **66**, 3427–3431.

Pham, P. L., Dupont, I., Roy, D., Lapointe, G., Cerning, J. (2000) Production of exopolysaccharide by *Lactobacillus rhamnosus* R and analysis of its enzymatic degradation during prolonged fermentation, *Appl. Environ. Microbiol.* **66**, 2302–2310.

Rawson, H. L., Marshall, V. M. (1997) Effect of ropy strains of *Lactobacillus delbreuckii* ssp. *bulgaricus* and *Streptococcus thermophilus* on rheology of stirred yogurt, *Int. J. Food Sci. Technol.* **32**, 213–220.

Ricciardi, A., Clementi, F. (2000) Exopolysaccharides from lactic acid bacteria: structure, production and technological applications, *Ital. J. Food Sci.* **12**, 23–45.

Robijn, G. W., Wienk, H. L. J., Van den Berg, D. J. C., Haas, H., Kamerling, J. P., Vliegenthart, J. F. G. (1996) Structural studies of the exopolysaccharide produced by *Lactobacillus paracasei* 34-1, *Carbohydr. Res.* **285**, 129–139.

Smitinont, T., Tansakul, C., Tanasupawat, S., Keeratipibul, S., Navarini, I., Bosco, M., Cescutti, P. (1999) Exopolysaccharide-producing lactic acid bacteria strains from traditional Thai fermented foods: isolation, identification and exopolysaccharide characterization, *Int. J. Food Microbiol.* **51**, 105–111.

Staaf, M., Yang, Z., Huttunen, E., Widmalm, G. (2000) Structural elucidation of the viscous exopolysaccharide produced by *Lactobacillus helveticus* Lb161, *Carbohydr. Res.* **326**, 113–119

Stingele, F. (1998) Exopolysaccharide production and engineering in dairy *Streptococcus* and *Lactococcus*, in: Proceedings, 5th ASM on the genetics and molecular biology of streptococci, enterococci and lactococci, 31–32.

Stingele, F., Neeser, J. R., Mollet, B. (1996) Identification and characterization of the eps

gene cluster from *Streptococcus thermophilus* Sfi6, *J. Bacteriol.* **178**, 1680–1690.
Sundman, V. (1953) On the protein character of a slime produced by *Streptococcus cremoris* in Finnish ropy sour milk, *Acta Chem. Scand.* 7, 558–560.
Sutherland, I. W. (1972) Bacterial exopolysaccharides, *Adv. Microbiol. Physiol.* **8**, 143.
Sutherland, I. W. (1982) Biosynthesis of microbial exopolysaccharides, in: *Advances in Microbial Physiology* (Rose, A.H., Ed.), London: Academic Press, Vol. 33, 78–150.
Sutherland, I. W. (1990) *Biotechnology of Microbial Exopolysaccharides.* Cambridge, Sydney: Cambridge University Press, 163.
Sutherland, I. W. (1994) Structure–function relationships in microbial exopolysaccharides, *Biotech. Adv.* **12**, 393–448.
Tamime, A. Y. (1978) Some aspects of the production of a concentrated yoghurt (labneh) popular in the Middle East, *Milchwissenschaft* **33**, 209–212.
Toba, T., Nakajima, H., Tobitani, A., Adachi, S. (1990) Scanning electron microscopic and texture studies on characteristic consistency of Nordic ropy sour milk, *Int. J. Food Microbiol.* **11**, 313–320.
Toba, T., Kotani, T., Adachi, S. (1991) Capsular polysaccharide of a slime-forming *Lactococcus lactis* ssp. *cremoris* LAPT 3001 isolated from Swedish fermented milk longfil, *Int. J. Food Microbiol.* **12**, 167–172.
Troy, F. A. (1979) The chemistry and biosynthesis of selected bacterial polymers, *Annu. Rev. Microbiol.* **33**, 519–560.
Troy, F. A., Frerman, F. A., Heath, E. C. (1971) Biosynthesis of capsular polysaccharides in *Aerobacter aerogenes*, *J. Biol. Chem.* **243**, 118–133.
Van Calsteren, M. (2001) Structural characterisation of the exopolysaccharide produced by different strains of *Lactobacillus rhamnosus* and *Lactobacillus paracasei*, in: Proceedings, 1st International Symposium on Exopolysaccharides from Lactic acid bacteria, 13.
Van den Berg, D. J. C., Smits, A., Pot, B., Ledeboer, A. M., Kersters, K., Verbrakel, J. M. A., Verrips, C. T. (1993) Isolation, screening and identification of lactic acid bacteria from traditional food fermentation processes and culture collections, *Food Biotechnol.* 7, 189–205.
Van den Berg, D. J. C., Robijn, G. W., Janssen, A. C., Giusepin, M. L. F., Vreeker, R., Kamerling, J. P., Vliegenthart, J. F. G., Ledeboer, A. M., Verrips, C. T. (1995) Production of a novel extracellular polysaccharide by *Lactobacillus sake* 0-1 and characterization of the polysaccharide, *Appl. Environ. Microbiol.* **61**, 2840–2844.
Van Kranenburg, R., Vos, H. R., Van Swam, I., Kleerebezem, M., De Vos, W. M. (1999) Functional analysis of glycosyltransferase genes from *Lactococcus* and other Gram-positive cocci: complementation, expression and diversity, *J. Bacteriol.* **181**, 6347–6353.
Vedamuthu, E. R., Neville, J. M. (1986) Involvement of a plasmid in production of ropiness in milk cultures by *Streptococcus cremoris, Appl. Environ. Microbiol.* **61**, 2840–2844.
Vliegenthart, J. F. G., Faber, E. J., Kamerling, J. P. (2001) Studies on structure of exopolysaccharides from lactic acid bacteria, in: Proceedings, 1st International Symposium on Exopolysaccharides from Lactic acid bacteria. Brussels, Belgium: VUB, 11.
Wacher-Rodarte, C., Galvan, M., Farres, A., Gallardo, F., Marshall, V. M. E., Garcia-Garibay, M. (1993) Yogurt production from reconstituted skim milk powders using different polymer and non-polymer forming starter cultures, *J. Dairy Res.* **60**, 247–254.
Yokoi, H., Watanabe, T., Fuji, Y., Toba, T., Adachi, S. (1990) Isolation and characterization of polysaccharide-producing bacteria from kefir grains, *J. Dairy Sci.* **73**, 1684–1689.

24
Scleroglucan

Ioannis Giavasis[1], Dr. Linda M. Harvey[2], Dr. Brian McNeil[3]
[1] Strathclyde Fermentation Centre, Department of Bioscience and Biotechnology, University of Strathclyde, 204 George Street, G1 1XW Glasgow, UK; Tel.: +44-141-548-3388; Fax: +44-141-553-4124; E-mail: ioannis.giavasis@strath.ac.uk
[2] Strathclyde Fermentation Centre, Department of Bioscience and Biotechnology, University of Strathclyde, 204 George Street, G1 1XW Glasgow, UK; Tel.: +44-141-548-2056; Fax: +44-141-553-4124; E-mail: l.m.harvey@strath.ac.uk
[3] Strathclyde Fermentation Centre, Department of Bioscience and Biotechnology, University of Strathclyde, 204 George Street, G1 1XW Glasgow, UK; Tel.: +44 141 548 4379; Fax: +44-141-553-4124; E-mail: b.mcneil@strath.ac.uk

Biotechnology of Biopolymers. From Synthesis to Patents. Edited by A. Steinbüchel and Y. Doi

ISBN: 3-527-31110-6

BSE	bovine spongiform encephalitis
CoA	coenzyme A
DMSO	dimethyl sulfoxide
DO	dissolved oxygen
LALS	low-angle light scattering
NMR	nuclear magnetic resonance
PGI	phosphoglucose isomerase
PGM	phosphoglucose mutase
r.p.m.	revolutions per minute
SEC	size-exclusion chromatography
UDP	uridine diphosphate
UGP	uridine glucose pyrophosphorylase
ZrC	zirconium citrate

1 Introduction

For many years, biopolymers have been key components in several industrial applications. Microbial biopolymers in particular, have been the core of extended research with a view to developing new products with novel properties in the food or pharmaceutical industries, or facilitating industrial processes such as oil drilling. Microbial polymers are produced by fermentation under controlled and reproducible conditions, thereby ensuring consistent product quality and constant availability. Thus, these polymers have significant advantages over traditional plant-derived gums which are not readily available throughout the year and also have changeable quality characteristics. In addition, animal-derived polymers such as gelatin, although widely used, are currently experiencing reduced consumer faith and acceptability following the recent outbreaks of bovine spongiform encephalitis (BSE) in Europe.

One main area of application for several microbial polymers is in foodstuffs, where they act as gelling and thickening agents, or as stabilizers and viscosifiers. They have also been used as encapsulating matrices for microorganisms or for controlled drug release. In addition, they are used as lubricants and emulsifiers for drilling fluids. Scleroglucan, in particular, has been proposed and used in the latter application, as well as for enhanced oil recovery (Gallino et al., 1996).

Scleroglucan is a polysaccharide produced by the filamentous fungus *Sclerotium glucanicum*, *Sclerotium rolfsii* and other closely related species. It is produced industrially in bioreactors, offering all the advantages of a controlled process, as explained above. However, at the same time the production of a fungal polymer involves some difficulties related to the synthesis of the product itself and the accumulation of filamentous biomass. More specifically, the rheological changes that occur in the fermentation fluid can cause mixing and mass transfer problems, which may in turn affect the physiology of the organism.

In this chapter we discuss the physiological and engineering aspects that influence the production of the polymer, current applications, and the biochemical characteristics and functions of scleroglucan that make it useful in various roles.

2 Historical Outline

The production of scleroglucan was first reported by Halleck (1967), who observed that this extracellular polysaccharide is secreted by *Sclerotium glucanicum*. Because of its physical properties, academia and industry soon realized that scleroglucan might have several potential applications. The ability of native scleroglucan to dissolve readily in water and act as a thickener or stabilizer, or the gelling properties of modified scleroglucan, combined with biopolymer thermal stability, were its main advantages.

Scleroglucan was introduced to the market by the Pillsbury Co. (Minneapolis) with the tradename Polytran®, and in 1976 it was commercialized by CECA S.A. (France) under the name Biopolymer CS® (Sandford, 1979). Subsequently, SATIA – a division of Mero Rousselot del produced scleroglucan under the name of Actigum CS6 (Morris, 1987), but nowadays Sanofie Bio Industries (a division of Elf Aquataine) has obtained the right to trade it from SATIA (Ouriaghli et al., 1992). Initially, the main use of scleroglucan was in chemically enhanced oil recovery, acting as a viscosifier in watered-out oil reservoirs, as well as in oil drilling where it contributed to lubricating and stabilizing oil mud and improving mud mobility. Scleroglucan had better performance than its rival polymer xanthan gum, as it resulted in reduced dilution rates and faster drilling, and also exhibited better thermal stability (Gallino et al., 1996). Scleroglucan can also be used in the food industry, where it functions as a bodying, gelling or stabilizing agent. Xanthan – which has broadly similar properties to scleroglucan – currently has a lower production cost due to higher yields and a shorter production process, and already dominates this market (Harvey and McNeil, 1998). Moreover, xanthan already possesses FDA approval for use in foodstuff.

The use of scleroglucan as an antitumor, antiviral and antimicrobial compound has also been investigated (Singh et al., 1974; Jong and Donovick, 1989; Pretus et al., 1991; Mastromarino et al., 1997). Scleroglucan has shown immune stimulatory effects compared with other biopolymers, and its potential contribution to the treatment of many diseases should be taken into account in therapeutic regimens.

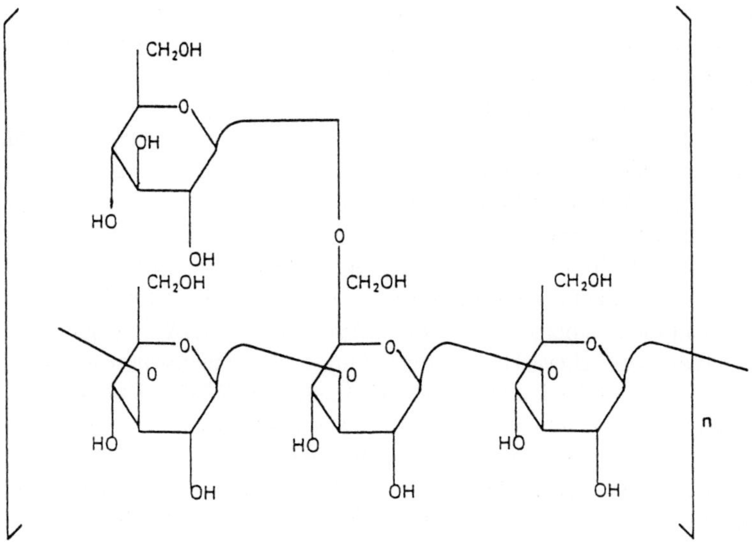

Fig. 1 Scleroglucan repeating unit (Brigand, 1993).

3 Chemical Structure

Scleroglucan is a neutral homopolysaccharide which consists of a linear chain of β-D-(1-3)-glucopyranosyl groups and β-D-(1-6)- glucopyranosyl groups. On average, about every third β-D-(1-3) glucose residue is linked with a β-D-(1-6) glucose residue (Nardin and Vincendon, 1989; Rinaudo et al., 1990). This structure was first elucidated by periodic oxidation analysis (Johnston et al., 1963), and later verified by methylated sugar analysis (Heyrad and Salemis, 1982) and ^{13}C nuclear magnetic resonance (NMR) analysis (Rinaudo and Vincendon, 1982; Brigand, 1993). The chemical structure of the tetrasaccharide repeating unit of scleroglucan, as established by NMR analysis, is depicted in Figure 1.

The length of the polymer chain – and hence the molecular weight – varies according to the microbial cultures used. The molecular weight as defined by low-angle light scattering (LALS) analysis ranges from $5-12\times10^6$ Daltons (Pretus et al., 1991; Stokke et al., 1992; Brigand, 1993). However, Lecacheux et al. (1986) estimated the molecular weight of scleroglucan by size-exclusion chromatography (SEC) coupled by LALS and found it to be in the region of 5.7×10^6 Daltons ($\pm5\%$). These authors also concluded that the polysaccharide has low polydispersity and stable molecular weight under different fermentation conditions. When dissolved in water, scleroglucan forms linear triple helices (Lecacheux et al., 1986) where the side glucose groups protrude and prevent helices from attracting each other and aggregating. This effect makes the polysaccharide more soluble, but reduces its gelling ability (Brigand, 1993). The triple helix structure is quite thermostable, but is dissolved into single random coils when dispersed in dimethyl sulfoxide (DMSO) or at a pH over 12.5 (Yanaki et al., 1981; Bluhm et al., 1982).

Scleroglucan has an identical structure to schizophyllan, which is produced by *Schizophyllum commune* (Steiner et al., 1987) and has a somewhat higher molecular weight than scleroglucan (Rau et al., 1989). Scleroglucan is also quite similar to curdlan (produced by *Agrobacterium* species) (Lee et al., 1999), which is composed of polymerized β-D-(1-3) glucose residues, but lacks the β-D-(1-6) glucose side groups. This reduces polymer solubility in water, but gives curdlan better gelling properties (Bluhm et al., 1982).

4 Occurrence

Scleroglucan is synthesized extracellularly by species of the genus *Sclerotium*, i.e., *S. glucanicum*, *S. rolfsii*, and *S. delphinii*. Other polysaccharides that are very similar structurally to scleroglucan are produced by *Corticum rolfsii* and *Schizophyllum commune* (schizophyllan) (Halleck, 1967). The two main species used for production are *Sclerotium glucanicum* (Wang and McNeil, 1995a, b, c, d) and *Sclerotium rolfsii* (Farina et al., 1998, 1999; Schilling et al., 1999, 2000).

S. glucanicum and S. rolfsii are heterotrophic, filamentous fungi which are characterized as plant pathogens and parasites. They possess (hydrolytic) enzymes, including cellulases, phosphatidase, arabinase, exogalactanase, polygalacturonase, galactosidase and exomannase, which enable them to attack plant cells. Oxalic acid is also produced by these organisms, which facilitates plant cell lysis (Wang and McNeil, 1996).

Sclerotium species have brown or black sclerotia (aggregated bodies of hyphae) or light-colored mycelia, and do not sporulate (Willets, 1978). The difference between sclerotia and mycelia is that the former are more resistant to biological or chemical degradation (Wang and McNeil, 1996). In liquid media the organism forms pellets with central capsules from which hyphal residues extend. On solid media, aerial hyphae are formed and organized in mycelia (Rodgers, 1973).

5 Functions

In this chapter we will discuss two aspects: first, the possible role of scleroglucan in the life cycle of the cell; and second, the functions and properties of the polysaccharide when used in several applications.

The role of scleroglucan in the life of the organism is mainly to assist in the attachment to (plant) surfaces and the protection of sclerotia against unfavorable environmental conditions, e.g., desiccation (Backhouse and Stewart, 1987). More specifically, scleroglucan probably forms an adhesive glycocalyx surrounding the cells, thus enabling the transfer of biodegradative enzymes and acids (mainly oxalate) out of the cells and onto the plant material under attack. In addition, this glycocalyx may act as a mechanical barrier that protects the fungus from potential plant defenses or antibiotic-like substances. Moreover, scleroglucan may be vital to the microorganism under conditions of environmental stress, such as desiccation or osmotic pressure. In these cases it can seal the cell and prevent the diffusion of water and equilibration of moisture inside and outside the cell. Similar functions to those described above have been proposed for other biopolymers (Kawahara et al.,1994; Takeuchi et al., 1995; Evans et al., 1998). In addition, scleroglucan may have a role as an energy reserve. The presence of hydrolytic enzymes synthesized by *Sclerotium* species (Rapp, 1989), which can degrade scleroglucan into glucose molecules, indicates that the biopolymer may be utilized by the microorganism when other carbon sources are depleted.

In a pure form, scleroglucan has many interesting properties. It can readily be dissolved in water at ambient temperature, due to the side glucose groups of the repeating unit, and it forms viscous, pseudoplastic solutions, which are very resistant to temperature and pH changes (Wang and McNeil, 1996). The viscosity of 0.5% and 2% solutions does not change over a range of 10 to 90°C (Brigand, 1993). Furthermore, scleroglucan was the single polymer, among 140 tested for polymer flooding in oil reservoirs, that had a practically stable viscosity for 500 days at 90°C (Brigand, 1993). With regard to the effect of pH, the viscosity of scleroglucan solutions remains the same over a range of pH from 1 to 11. Viscosity loss is only observed at pH higher than 12.5, or after addition of DMSO. In these cases a disruption of the triple helix conformation occurs and single strands of polymer coils are formed (Gamini et al., 1984; Nardin and Vincendon, 1989).

The behavior of scleroglucan in salt solutions depends on the concentration of the electrolyte. It remains soluble in 5% NaCl, 20% $CaCl_2$, 5% $NaSO_4$ and 10% Na_2HPO_4 solutions, but if the electrolyte concentration increases, then solubility is eventually lost and solutions may become gels (Brigand, 1993). This ability to maintain its functional properties in the presence of salts is of particular relevance in many drilling or oil recovery applications (where a mixture of scleroglucan with brine is essential).

Scleroglucan also exhibits good compatibility with other biopolymers, including xanthan, gelatin, carrageenans, cellulose derivatives, locust bean gum and guar gum. This is important in modifying the properties of scleroglucan and therefore extending its potential applications. The formulation with another polymer could produce a gel which is thermostable (due to scleroglucan), as well as having another characteristic deriving from the second polysaccharide, such as flavor release, elasticity, or brittleness. The cross-linking of scleroglucan with gellan gum, for example, has been used as a controlled drug delivery matrix, where the degree of (sustained) release is controlled by Ca^{2+} cations (Coviello et al., 1998). In this application, Ca^{2+} cations (added in the form of CaCl solution) interact with the ionic carboxylic groups of gellan, increasing its firmness. The increase in calcium concentration in this formulation is followed by a marked decrease in the delivery rate of a model molecule (theophylline). The formation of scleroglucan hydrogels for controlled drug release has received much attention, due to some attractive features of the biopolymer such as biocompatibility and bioadhesivity (Grassi et al., 1996). In these applications, scleroglucan has the ability to absorb water and form swollen hydrogels which have increased (but controlled) permeability.

In addition, the properties of the biopolymer can be altered by physical treatment. For instance, the hydration capacity and thickening properties of scleroglucan have been improved by a controlled pressure drop method established by Rezzoug et al. (2000). In this process, steam pressure of 1 to 6 bar was applied to scleroglucan for a short time, followed by a stage of instant pressure drop and incubation under a vacuum of 15 mbar. These conditions brought about conformational changes in the molecule (as revealed by mass spectroscopy and nuclear magnetic resonance analysis), which enhanced interactions with water. As a result, rheological measurements on rehydrated scleroglucan samples treated by a controlled pressure drop showed that viscometer torque (a measure of resistance to shear) was twice that produced by rehydrated control samples dried in a conventional rotary drier.

A very important attribute of scleroglucan is its function as a biological response modifier. Many β-glucans show antitumor and antimicrobial activity, but scleroglucan seems to have a greater effect than other polysaccharides (Singh et al., 1974; Pretus et al., 1991). The mode of action against herpes virus (Marchetti et al., 1996) and rubella virus (Mastromarino et al., 1997) has been investigated. The binding of the polysaccharide on the host cell membrane may prevent or reduce the attachment and entry of the virus into the cell. This inhibitory effect occurs only at an early stage of virus multiplication, that is, before infection of host cells with virus RNA. The key reaction seems to be the binding of scleroglucan with glycoproteins of the cell membrane, which impedes interactions between the virus and the host cell plasma (Marchetti et al., 1996; Mastromarino et al., 1997). A similar mechanism is proposed for the inhibition of HIV virus replication, which is reported for sulfated (modified) biopolymers (Parish et al., 1990; McClure et al., 1992). For example, curdlan – a biopolymer similar to scleroglucan – has been used *in vitro* in a sulfated form to inhibit partly HIV attachment to cells (Jagodzinski et al., 1994). Another possible explanation for antiviral activity can be that after the virus has entered the host cell, scleroglucan is also internalized in the cell and encapsulates the virus, thus inhibiting its activity. However, host cell penetration is unlikely for the case of a high molecular-weight polysaccharide such as scleroglucan (Mastromarino et al., 1997). As far as the antitumor activity of scleroglucan is concerned, Pretus et al. (1991) suggested that it occurs through the increase in number and activity of macrophages in the presence of water-soluble β-D-glucans. Studies on the immunostimulatory effect of the biopolymer (Pretus et al. 1991; Ohno et al., 1995) have revealed that the structural conformation of clinically used Sinofilan (the immunopharmacological form of scleroglucan), plays a vital role. In some cases, as in macrophage stimulation, the existence of a triple helix is important. In contrast, in the case of lipopolysaccharide-triggered tumor necrosis factor synthesis, or nitric oxide and hydrogen peroxide synthesis, a single helix scleroglucan (prepared after denaturation of the natural triple helix with alkali) was more effective (Ohno et al., 1995). The immunological effects of neutral glucans (including scleroglucan) were also studied by Jamas et al. (1996a, b, 1997, 1998a, b), who revealed that scleroglucan has high affinity for human monocytes and possesses two primary biological activities: (1) the stimulation of phagocytic cells; and (2) monocyte, neutrophil and platelet hematopoietic activity. These authors prepared neutral, water-soluble β-glucans for therapeutic or prophylactic treatment, which enhanced the human immune system without side effects such as fever or inflammation.

6 Physiology

The physiology of *Sclerotium glucanicum* has been studied by several researchers in order to understand how the microorganism functions and responds in a controlled environment, and to optimize process conditions for scleroglucan production. The polysaccharide is produced when one nutrient, for example nitrogen or phosphorous is depleted, and therefore becomes a growth-limiting factor (Sutherland, 1979). Environmental and other process conditions, such as temperature, pH, dissolved oxygen, aeration-agitation and composition of fermentation medium influence the physiology of the microorganism, and thus the synthesis of scleroglucan. The effects of these factors are discussed below.

6.1 Effect of Temperature

Temperature is a crucial parameter that affects both culture growth and polysaccharide production. As for every microorganism, there is an optimum temperature for growth of *S. glucanicum*, but optimal temperature for exopolysaccharide production is different to that for culture growth. Scleroglucan formation occurs between 20–37°C (Halleck, 1967), but the optimum temperature for biopolymer production was found to be 28°C (Wang and McNeil, 1995a). Optimal biomass production occurs at temperatures above 28°C, while below 28°C byproduct (oxalic acid) formation is enhanced, which has an adverse effect on scleroglucan production (Wang and McNeil, 1995a). Thus, one major obvious difference between *Sclerotium* species in a natural environment (at low ambient temperatures), and bioreactor cultures, may be the abundant synthesis of oxalate by the former. This seems reasonable if the proposed role of oxalate in plant pathogenicity is taken into account (Magro et al., 1984; Punja, 1985; Punja et al., 1985; Stone and Armentrout, 1985).

6.2 Effect of pH

pH is a significant factor that influences the physiology of a microorganism by affecting nutrient solubility and uptake, enzyme activity, cell membrane morphology, byproduct formation and oxidative/reductive reactions (Cooney, 1981; Forage et al., 1985). As described for temperature, the appropriate pH for maximum production of a polysaccharide can differ from that for optimum growth. According to Kang and Cotrell (1979) fungal biopolymer synthesis is optimal in the range of pH 4–5.5. Based on these observations, researchers have developed bi-stage processes, including one stage designed for optimal culture growth and a second stage for maximum polysaccharide production. Lacroix et al. (1985) conducted fermentations for the production of pullulan, another fungal glucan, where a first stage with a pH of 2 was maintained for best biomass and growth rate. Once high levels of biomass were achieved, the pH was adjusted to 5.5, which was found to lead to maximum pullulan production.

In a similar mode, Wang and McNeil (1995b) reported an improved scleroglucan production via a bi-stage process. In the first phase pH was controlled at 3.5 for optimal growth, after which a pH value of 4.5 was used to promote polysaccharide synthesis. The increased scleroglucan concentration achieved under these conditions was combined with a 10% reduction of byproduct (oxalic acid) formation. This probably indicates that at pH levels higher than those for optimal growth (4.5), carbon flux to biopolymer synthesis is increased. In some studies, in order to simplify the process and reduce costs, pH was not controlled after an initial adjustment, but scleroglucan production was comparatively low (Maxwell and Bateman, 1968; Wernau, 1985), showing that pH control is necessary during the fermentation.

6.3 Dissolved Oxygen

In aerobic microorganisms such as *S. glucanicum* and *S. rolfsii*, oxygen occupies a key role in the cell life cycle. It regulates – by inducing or repressing – several enzyme systems of primary or secondary metabolism and enables oxidative reactions for nutrient utilization and energy generation. Oxygen can also have negative functions in cell metabolism, however. Toxic peroxide and superoxide radicals may be formed in the presence of oxygen, which can lead to cell membrane destruction and potentially induce cell death (Forage et al., 1985).

In the production of some exopolysaccharides such as pullulan (Rho et al., 1988) and gellan (Dreveton et al., 1994; Giavasis et al., 2000), oxygen is both vital and stimulatory to polymer synthesis. However, when the effect of dissolved oxygen (DO) on biopolymer production by *S. glucanicum* and *S. commune* was studied, it was noted that while a high oxygen supply increased cell growth, it decreased glucan production. In contrast, the culture response to oxygen limitation was a limited growth and a specific stimulation of glucan formation (Rau et al., 1992; Schilling et al., 1999). This is somewhat unexpected for an aerobic microorganism, but it is possible that a reduction in DO affects fungal morphology or broth rheology, which in turn affects cell growth and metabolism (Olsvik, 1992). Sutherland (1979) offered another explanation by indicating that high DO enhances respiration; this means that more carbon is converted to carbon dioxide, leaving less available for scleroglucan production. Wang and McNeil (1995c) suggested that the stimulation of glucan synthesis at low DO was a result of limited cell growth (in a microaerobic environment). Under these conditions, more carbon of the substrate was utilized for scleroglucan production. In addition, as described in Section 6.7, a low DO causes a decrease in byproduct (oxalate) production, by repressing the synthetic enzyme glycolate oxidase (Schilling et al., 2000). Thus, the flow of carbon towards glucan production is again favored. The possible effects of oxygen limitation on scleroglucan production can be depicted diagrammatically (Figure 2).

6.4
Effect of Aeration and Agitation

Aeration and agitation each control mixing, mass and heat transfer in the bioreactor. The two factors are linked, and it is difficult to distinguish between them in the common stirred tank reactor. They determine the availability of nutrients and oxygen to the cultures, and control the rate of metabolite release from the cells, including biopolymers, byproducts and carbon dioxide (Fumitake, 1982; Forage et al., 1985; Parton and Willis, 1990). In polysaccharide production the fermentation medium becomes very viscous and exhibits nonNewtonian (pseudoplastic) behavior. With filamentous fungi, apart from the concentration of the biopolymer, the biomass may also contribute significantly to broth rheology. These phenomena restrict mixing in the bioreactor, and may in turn change culture morphology. At high stirring and aeration rates fungal hyphae may become fragmented, reducing broth viscosity (Nielsen, 1992). On the other hand, the formation of mycelial pellets may occur at low agitation rates (Forage et al., 1985; Brown et al., 1987). Consequently, heat and mass transfer into and out of the pelleted cells are restricted, and some me-

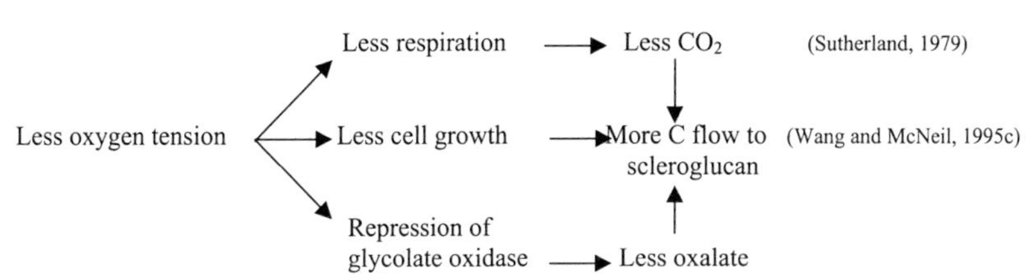

Fig. 2 The possible impact of low dissolved oxygen tension on scleroglucan synthesis (Schilling et al., 2000).

tabolites in the pellets may reach toxic levels.

Vigorous agitation and aeration are usually beneficial for polysaccharide production, although there are contradictory reports (Wecker and Onken, 1991). McNeil and Kristiansen (1987) indicated that increased agitation improved glucan synthesis by *A. pullulans* as well as polymer quality (high molecular weight), and influenced culture morphology by promoting the formation of yeast-type (instead of filamentous) cells, which seem to produce more glucan. For *S. glucanicum* a higher growth rate was achieved with increased aeration rate at 600 r.p.m. agitation, but more glucan was produced as aeration was reduced (Rau et al., 1992). The same authors also reported that at high stirring rates, glucan mechanical degradation and cell damage may occur, but if the shear applied in the bioreactor is not adequate, only low molecular-weight scleroglucan is released from the cells, while larger molecules of the biopolymer remain attached to the cell surface. Schilling et al. (1999) confirmed that when the stirring rate is high the obtained scleroglucan has a low molecular weight compared with that obtained after moderate agitation. When moderate agitation was combined with high aeration, a maximum molecular weight for scleroglucan was attained. This indicates that, although good agitation may be essential for mixing, it may adversely affect the culture or the quality of the product, whereas aeration contributes to good mixing in a milder manner that does not disturb the culture or the molecular size of the polysaccharide.

6.5 Effect of Medium Composition

The composition of a fermentation medium is always crucial for the outcome of the process, and the optimal design of a medium is a necessary step in order to achieve maximal product formation. Fungi require several macronutrients and micronutrients. Among the macronutrients, carbon, nitrogen, phosphorus and sulfur are the most important, while potassium and magnesium are the principal micronutrients (functioning mainly as enzyme co-factors) (Dube, 1983; Dunn, 1985).

Carbon is a major structural component of cells, as well as being essential for the production of polysaccharides and as an energy source (Dube, 1983; Dunn, 1985). Usually, glucose or sucrose are used as carbon sources for biopolymer production, although other carbohydrates can also be utilized, according to the microorganism involved. Most studies on scleroglucan report a glucose or sucrose concentration of either 30 g L^{-1} (Wang and McNeil, 1995a, b, c, d; Farina et al., 1999) or 35 g L^{-1} (Schilling, 2000). Under these conditions a maximum value of 8.5–10 g L^{-1} was obtained for scleroglucan concentration, which is much lower than the highest reported concentration of 27 g L^{-1} of xanthan, the main rival of scleroglucan (DeVuyst et al., 1987). Sucrose concentrations higher than 45 g L^{-1} have been found to inhibit growth of *S. glucanicum* and limit further glucan production (Taurhesia and McNeil, 1994). In contrast to this, Farina et al. (1998) studied the effect of high sucrose concentrations on scleroglucan formation by *S. rolfsii* and concluded that an increase in sucrose led to a clear improvement in glucan concentration and yield (glucan/sucrose). While only 7 g L^{-1} of scleroglucan was produced with 30 g L^{-1} initial sucrose, a three-fold increase (21 g L^{-1}) of product occurred when 150 g L^{-1} sucrose was supplied in the culture medium. Despite this improvement, residual sucrose at the end of that fermentation was as high as 100 g L^{-1}, thus questioning the economic benefits of this strategy (as opposed to the complete utilization of carbon observed with more modest glucose/sucrose concentrations; Schilling et al., 2000).

Nitrogen is a component of proteins and enzymes and is necessary in cell metabolism. It is supplied to the culture in the form of ammonium or nitrate salts, or more complex (organic) compounds such as yeast extract, casein hydrolyzate, soya hydrolyzate and corn steep liquor (Dreveton et al., 1996; Wang and McNeil, 1996). In the case of a glucan such as pullulan (Harvey, 1984) the production of the polysaccharide is stimulated by depletion of nitrogen. Generally, the addition of extra nitrogen source favors biomass concentration but diminishes glucan formation (Seviour and Kristiansen, 1983; Harvey, 1984). Similarly, high concentrations of nitrogen in the form of ammonium sulfate have been reported to reduce scleroglucan production (Farina et al., 1998). Also, it was found that the highest glucan level was reached when the culture was growing on nitrate ($NaNO_3$) rather than on ammonium [$(NH_4)_2SO_4$] (Farina et al., 1998, 1999). Ammonium, it was proposed, exhibited an inhibitory effect on glucan-synthesizing enzymes.

Phosphorus is an important element for secondary metabolism, and also regulates lipid and carbohydrate uptake by the cells. Phosphate salts, such as K_2HPO_4 or KH_2PO_4, also serve as a pH buffer in the fermentation medium (Dube, 1983; Dunn, 1985). Farina et al. (1998) indicated that an increase of total phosphorus (from 0.12 to 0.28 g L^{-1} resulted in improved scleroglucan concentration (from 4 to 5 g L^{-1}). Although a clear explanation for this was not given, it is possible that high phosphorus levels (to a certain extent) increase glucose uptake and metabolism.

6.6
Effect of Other Factors

It has been found that light, radiation and hydrostatic pressure may affect the growth and metabolism of the culture. Miller and Liberta (1976) found that white or blue light facilitated glucan formation in *S. rolfsii*. In addition, CO_2 is known potentially to influence fungal cell morphology (hyphal length, diameter and branching) (Smith and Ho, 1985), which in turn can affect the rheology and the mixing and mass transfer in the fermentor (McNeil and Harvey, 1993), especially at the base of large industrial bioreactors where CO_2 may reach relatively high levels, due to the combined effects of microbial growth and hydrostatic pressure.

6.7
Byproduct Formation

Along with scleroglucan, significant amounts of oxalate (as well as malate and fumarate in lesser amounts) are synthesized; this is an undesirable byproduct and reduces the amount of carbon available for glucan production (Wang and McNeil, 1995a). In a natural environment, this serves as a means of disrupting plant tissue (in synergy with pectinases and polygalacturonases), which facilitates the role of *Sclerotium* as a plant pathogen (Magro et al., 1984; Punja, 1985; Punja et al., 1985; Stone and Armentrout, 1985).

Several factors influence byproduct formation. Wang and McNeil (1995a) found that at low temperature, less scleroglucan and more oxalate is accumulated. At 20°C, acid concentration was 7.3 g L^{-1}, higher than both biomass and product concentration. High aeration, high pH, and increased carbohydrate concentration seem to enhance acid formation (Berry, 1975; Wang and McNeil, 1995b; Schilling, 2000).

Low pH has been reported to activate oxalate decarboxylase which breaks down oxalate to formate and CO_2 (Shimizano, 1955). Alternatively, Schilling et al. (2000) proposed that at low pH the activity of oxalic acid synthetic enzymes is reduced, and hence recommended a pH of 2 for this process, in order to minimize carbon losses to byproducts. These authors also found that oxygen limitation leads to a decrease in oxalate production, as the synthetic enzyme glycolate oxidase is repressed under anaerobic conditions, and suggested that an early low oxygen supply can be beneficial for glucan production.

7 Biochemistry

There is little information available on the biosynthetic pathway for scleroglucan formation in *S. glucanicum* and *S. rolfsii*, but generally this should resemble the biosynthetic steps encountered in the production of other glucans. First, glucose is transferred into the cells via a hexokinase and is then phosphorylated by the action of phosphoglucomutase (PGM) and phosphoglucose isomerase (PGI). A pyrophosphorylase (UGP) catalyzes the formation of uridine diphosphate glucose (UDP-glucose) which reacts with a lipid carrier and begins to polymerize. This proposed pathway is depicted schematically in Figure 3. As men-

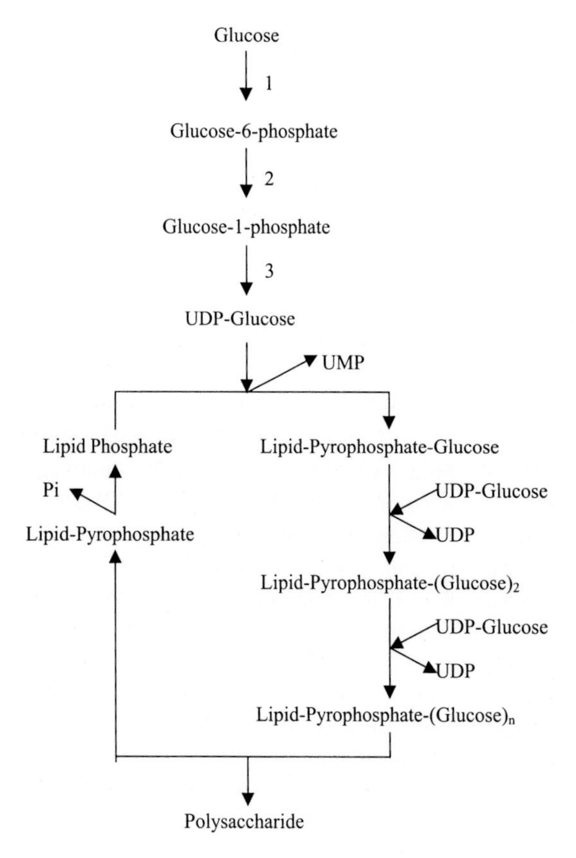

Fig. 3 Possible pathway for (sclero)-glucan synthesis (Rodgers, 1973). Key to enzymes: (1) hexokinase; (2) phosphoglucomutase; (3) phosphoglucose isomerase. UDP, uridine diphosphate; UMP, uridine monophosphate.

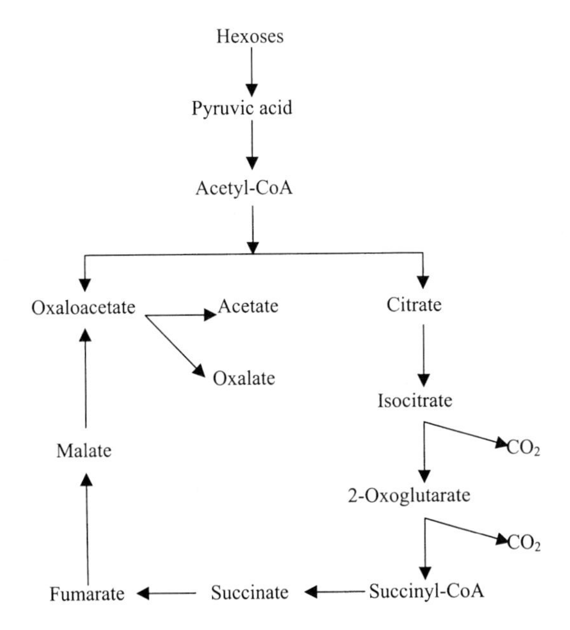

Fig. 4 Proposed route for glucose catabolism and oxalate synthesis in *S. rolfsii* (Hodgkinson, 1977).

tioned above, oxalate can – under certain circumstances – be a major byproduct and may be the dominant product of *Sclerotium* species in many natural environments. The biosynthetic steps of this byproduct are depicted in Figure 4.

8 Molecular Genetics

The genes encoding scleroglucan synthesis and polymerization have not yet been identified. As yet, insufficient studies have been performed in this particular field, possibly because there is no major drive to achieve higher yields as the rival polymer xanthan is already dominant in the market. However, if these genes could be overexpressed in *Sclerotium* cells, the disadvantage of lower yield and prolonged cultivation time (in comparison with xanthan production) might be overcome. A number of genes regulating xanthan synthesis have already been isolated and cloned, resulting in mutants which increased xanthan production by 10% (Harding et al., 1987; Thorne et al., 1987), or produced variants of xanthan with distinct structure and properties (Betlach et al., 1987). A similar approach for scleroglucan may be worth investigating.

9 Biodegradation

Under this topic we must consider two aspects, namely, the utilization of the biopolymer itself as a carbon source by the producing organism, and the likely fate of scleroglucan in the environment.

With regard to the first aspect, it would be highly surprising if enzymes capable of degrading β-1,3 and β-1,6 glucosidic bonds were not present at virtually all stages of growth of *Sclerotium* species, given the key role of β-glucans in cellular wall structure. The activities of such enzymes are likely to be most apparent during periods where either considerable wall degradation is occurring, and/or no readily available carbon source is present. In fact, it has been demonstrated that scleroglucan-degrading enzymes are present in *S. glucanicum*. 1,3-β-Glucanase and β-glucosidase activity was observed towards the end of the fermentation process (Rapp, 1989). It was noted that degradative activity can lead to complete hydrolysis of the polysaccharide to glucose molecules, and is stimulated by carbon source depletion. This is probably a response mechanism of the culture, as it is trying to liberate readily utilizable monosaccharides from scleroglucan under adverse conditions (Rapp, 1989). This also explains the reduced molecular weight levels and broth viscosity attained at that stage (Wernau, 1985; Lecacheux et al., 1986; Rau et al., 1989).

As far as the biodegradation of the biopolymer in a natural environment is concerned, it is possible that scleroglucan is also utilized as an energy source by other microorganisms found in soil or on plant surfaces. Curdlan, a biopolymer with a chemical structure similar to that of scleroglucan, can be degraded by enzymes of *Bacillus* species isolated from soil, namely, *B. kobensis* and *B. circulans* (Bertram et al., 1993; Kansawa et al., 1994, 1995) as well as by *Ruminococcus flavefaciens* (which can grow on plants). The glucanases of the latter microorganism were able to hydrolyze 1,3-β-glucosidic bonds, but not when a 1,6-β-glucose was present in the polymer. Therefore, they could not degrade scleroglucan or schizophyllan to glucose (Erfle and Teather, 1991).

10 Production

10.1 Producers

The first company to bring scleroglucan to the market was Pillsbury Co. (USA), after which it was traded by CECA S.E. (France) and SATIA (Division of Mero Rousselot). Nowadays, SANOFI Bio-Industries (Carentan, France), which obtained the rights from SATIA, and CECA are the main scleroglucan producers, trading scleroglucan under the commercial names Polytran® and Actigum®, respectively (Sandford, 1979; Morris, 1987; Ouriaghli et al., 1992; Vincendon, 1999; De Nooy et al., 2000).

10.2 World Market

The greatest potential market for scleroglucan today is the oil industry, where the biopolymer facilitates crude oil recovery, as described in Section 10.3. Xanthan is also used for the same purpose, although it is not considered to be as effective, or as stable. As mentioned earlier, scleroglucan is not at present as economical as xanthan, but the potential use of crude scleroglucan (biopolymer + cells) might be a cheap alternative for use in oil recovery and drilling. By extrapolation from the market of xanthan, if unit production costs could be significantly reduced, the oil industry would consume most of the polymer produced. However, this market is potentially subject to restrictions, such as exhaustion of energy reserves, or political instability in some oil-producing countries. Recently, the attractive properties of the polysaccharide in controlled drug release – and especially in immunopharmaceutical applications – have created a demand in the medical market, though in terms of quantity this market is comparatively limited. Currently, scleroglucan is not used in the food industry to any great extent because of its high cost, despite its thickening and stabilizing abilities that might be exploited in several foodstuffs. It is possible that, if genetic engineering were to be applied in order to improve biopolymer yields, then production costs might fall, allowing much wider use of the biopolymer.

10.3 Applications

The initial application of scleroglucan was in oil recovery where it proved to be more effective and stable than xanthan over a wide range of temperature and pH (Brigand, 1993). In oil recovery, scleroglucan has contributed to increasing the viscosity and thus the hydraulic pressure of (sea)water or brine used to extract oil. Especially in watered-out reservoirs where (sea)water pressure is no longer sufficient to recover the oil, the addition of scleroglucan can improve the process significantly. In addition, either in the crude form of dehydrated fermentation broth, or as clean, precipitated polymer, scleroglucan can lubricate drills and control the back-pressures created during drilling (Williams, 1968; Sutherland and Elwood, 1979). Scleroglucan is very useful as an oil mud thickener and stabilizer; that is, it serves to increase the viscosity of very thin oil muds which cannot otherwise be drilled. The presence of the biopolymer is also important in preventing the mixture of water with oil/gas, by interacting with the aqueous phase and in removing fluids from oil pipelines (Williams, 1968; Holzworth, 1985). When tested for polymer flooding in North Sea, scleroglucan retained over 90% of its viscosity for 500 d at 90°C in seawater, and was the most stable of 140 polymers tested.

In comparison with xanthan, scleroglucan has presented several advantages, for example low sensitivity to shear and temperature, which makes it suitable for drilling fluid applications. The rheology of scleroglucan-oil muds remained unchanged between 20 and 80°C, and the system where scleroglucan was applied had higher tolerance to field contaminants (Gallino et al., 1996).

Improved formulations of scleroglucan have also been developed for the above uses. Zirconium citrate (ZrC) has been used as a deflocculant and dispersing agent in scleroglucan-oil muds, and was found to improve field performance by its thinning properties. It resulted in reduced mud dilution requirements, and improved the total fluid costs by up to 44% (Burrafato et al., 1997). If needed, the scleroglucan-zirconium system can be controlled by the addition of oxyacids (e.g., citrate, malate, tartrate and lactate) which compete with the biopolymer for zirconium complex formation (Omari, 1995). In addition, scleroglucan (among other polymers) exhibited high thermostability in drilling fluids when it was associated with polyglycols through intermolecular H-bonds and hydrophobic interactions (VanOort et al., 1997).

Another important use of scleroglucan is in the pharmaceutical industry as a matrix for drug delivery in the form of tablets or films (Coviello et al., 1999). Here, it offers the advantages of controlled release as well as biocompatibility, biodegradability, bioadhesiveness and thermal and chemical stability (Sheth and Lachman, 1967; Lachman and Sheth, 1968; Grassi et al., 1996). The kinetics of drug release from scleroglucan hydrogels has been discussed earlier (see Section 5).

Like other glucans, scleroglucan has been proven to have immunestimulatory, antiviral, and antitumor properties (Pretus et al., 1991; Mastromarino et al., 1997). The mechanism of action has also been described in

Section 5. These properties of scleroglucan seem very promising in this field, and are worth further investigation.

Other recent uses of scleroglucan include personal care products, such as washing liquids (Dubief, 1996) and cement preparation for (road) building (Cartalos et al., 1994; Donche and Isambourg, 1994).

In addition, scleroglucan may be used in food, as a thickener, gelling or stabilizing agent. However, as mentioned previously, xanthan has similar properties and applications and at present dominates the food market. If the problem of high cost and low-scale production of scleroglucan be overcome, then it could replace xanthan in several foodstuffs such as jams and marmalades, soups, confectionery products and water-based gels, dairy products such yogurt or ice cream, "low-calorie" or nonfat products, or in fabricated/structured food (Halleck, 1967; Rodgers, 1973; Rodgers and Gofrett, 1976; San-Ei Chemical Industries, 1982; Nakao et al., 1991; Funami et al., 1998). Scleroglucan could be especially useful in food manufacturing where a heating process is involved, because of the thermal stability that it exhibits.

11 Outlook and Perspectives

It is clear from the earlier discussions on the properties and applications of scleroglucan, as well as from recent patents, that this biopolymer has received much attention from both the oil and pharmaceutical industries. Continued research has also led to the formulation of many variations of the original product, thereby altering its properties and extending its applications.

From a biotechnological and engineering perspective, the production process can be further improved, by optimizing the fermentation conditions, and by producing genetically engineered, highly productive mutants of *S. glucanicum* or *S. rolfsii*, in order to achieve better yields and reduce the cost of the process. With the possibility of creating a more economical product, the use of unrefined scleroglucan (polymer + cells) is worth considering, as has been demonstrated in oil (or generally fluid) drilling applications (Ladret and Donche, 1996). Unfortunately, the current high price and low-scale production of scleroglucan limits its competitiveness in the market, especially in relation to rival biopolymer xanthan, though a larger-scale production process may reduce the basic cost. Nonetheless, scleroglucan possesses some interesting properties in certain applications, such as immunostimulation, or the (thermal and physical) stability of drilling fluids. These properties should be further studied, in order for the industry to exploit the polysaccharide's full potential.

Whilst production levels remain very modest (at present), high cost may lead to scleroglucan being utilized mainly in areas which are "cost-insensitive", such as immunostimulation, or modulation and formulation into drug release systems. Despite its clear technical superiority to xanthan, the use of scleroglucan in industrial oil recovery will be limited until production levels are significantly increased.

12 Patents

A list of the available (recent) patent literature on scleroglucan is given in Table 1.

Tab. 1 Recent patent literature on scleroglucan

Patent no.	*Patent holder*	*Patent inventor*	*Patent title*	*Date of publication*
6,117,850	The Collaborative Group, Ltd. (New York, USA)	Patchen, L., Bleicher, P.	Mobilisation of peripheral blood precursor cells by beta (1,3)-glucan	12/9/2000
5,886,054	Not available (USA Patend Holder)	Amerongen, V. N.	Therapeutic method for enhancing saliva	23/3/1999
5,817,643	Alpha-Beta Technology, Inc. (Worcester, MA, USA)	Jamas, S., Easson, J., Davidson, D., Ostroff, G.	Underivatised, aqueous soluble beta (1,3)-glucan, composition and method of making same	6/10/1998
5,783,569	Alpha-Beta Technology, Inc. (Worcester, MA, USA)	Jamas, S., Easson, J., Davidson, D., Ostroff, G.	Uses for underivatised, aqueous soluble beta (1,3)-glucan and compositions comprising same	21/6/1998
5,622,940	Alpha-Beta Technology, Inc. (Worcester, MA, USA)	Ostroff, G.	Inhibition of infection-stimulated oral tissue destruction by beta (1,3)-glucan	22/4/1997
5,622,939	Alpha-Beta Technology, Inc. (Worcester, MA, USA)	Jamas, S., Easson, J., Davidson, D., Ostroff, G.	Glucan preparation	22/4/1997
5,612,294	Societe Nationale Elf Aquitaine (Courbevoie, France)	Vaussard, A., Ladret, A., Donche, A.	Scleroglucan based drilling mud	18/3/1997
5,555,936	Societe Nationale Elf Aquitaine (Courbevoie, France)	Pirri, R., Gadioux, J., Riveno, R.	Scleroglucan gel applied to the oil industry	17/9/1996
5,536,493	L'Oreal (Paris, France)	Dubief, C.	Composition for washing keratinous materials, in particular hair and/or skin	16/8/1996
5,532,223	Alpha-Beta Technology, Inc. (Worcester, MA, USA)	Jamas, S., Easson, J., Davidson,D.; Ostroff, G.	Use of aqueous soluble glucan preparations to stimulate platelet production	2/8/1996
5,525,587	Societe Nationale Elf Aquitaine (Courbevoie, France)	Ladret,A.; Donche, A	Application of muds containing scleroglucan to drilling large diameter wells	11/6/1996
5,504,062	Baker Hughes Inc. (Houston, TX, USA)	Johnson, M.	Fluid system for controlling fluid losses during hydrocarbon recovery operations	2/5/1996
5,488,040	Alpha-Beta Technology, Inc. (Worcester, MA, USA)	Jamas, S.; Easson, J; Davidson, D., Ostroff, G.	Use of neutral soluble glucan preparations to stimulate platelet production	30/1/1996

Tab. 1 (cont.)

Patent no.	*Patent holder*	*Patent inventor*	*Patent title*	*Date of publication*
5,330,015	Societe Nationale Elf Aquitaine (Courbevoie, France)	Donche, A.	Application of scleroglucan muds to drilling deviated wells	19/8/1994
5,323,857	Societe Nationale Elf Aquitaine (Courbevoie, France)	Pirri, R., Huet, Y., Donche, A.	Process for enhancing oil recovery using scleroglucan powders	28/6/1994
5,322,123	Institut Francais Du Petrole (Rueil Malmaison, France)	Kohler, N., Pirri, R.	Use of gel-based compositions for reducing the production of water in oil- or gas-producing wells	21/6/1994
5,306,340	Societe Nationale Elf Aquitaine (Courbevoie, France)	Donche, A., Isambourg, P.	Scleroglucan-based compositions and their use as a cementation spacer	26/4/1994
5,301,753	Institut Francais Du Petrole (Rueil Malmaison, France)	Cartalos,U., Lecourtier, J., Rivereau, A.	Use of scleroglucan as high temperature additive for cement slurries	12/4/1994
5,246,986	Societe Nationale Elf Aquitaine (Courbevoie, France)	Pierre, C., Demangeon, F., Yves, H.	Bituminous binder emulsion with a viscosity controlled by addition of scleroglucan	21/9/1993
5,224,988	Societe Nationale Elf Aquitaine (Courbevoie, France)	Pirri, R., Huet, Y., Donche, A.	Process for improving the dispersability and filterability of sclerglucan powders	6/7/1993
5,215,752	Vectorpharma International S.P.A. (Trieste, Italy)	Lovrecich, M., Riccioni, G.	Pharmaceutical tablets and capsule granulates of scleroglucan and active substance	1/6/1993
5,215,681	Societe Nationale Elf Aquitaine (Courbevoie, France)	Truong, D., Gadioux, J., Sarazin, D.	Concentrated liquid solutions of scleroglucan	1/6/1993
5,068,111	Vectorpharma International S.P.A. (Trieste, Italy)	Lovrecich, M., Riccioni, G.	Pharmaceutical tablets and capsule granulates of scleroglucan and active substance	26/11/1991
4,640,358	Mobil Oil Corp. (New York, USA)	Sampath, K.	Oil recovery process employing a complexed polysaccharide	2/3/1987
4,647,312	Mobil Oil Corp. (New York, USA)	Sampath, K.	Oil recovery process employing a complexed polysaccharide	3/3/1987
216,661	Not available (European patent holder)	Kohler, N., Tabary, R., Zaitoum, A.	Not available	1987
2,570,755	Not available (French patent holder)	Desbrieres, J.	Not available	1986

Tab. 1 (cont.)

Patent no.	*Patent holder*	*Patent inventor*	*Patent title*	*Date of publication*
2,570,756	Not available (French patent holder)	Desbrieres, J.	Not available	1986
2,570,754	Not available (French patent holder)	Desbrieres, J.	Not available	1986
2,570,753	Not available (French patent holder)	Desbrieres, J.	Not available	1986

13
References

Amerongen, V.N. (1999) Therapeutic method for enhancing saliva. U.S. Patent No. 5,886,054.

Backhouse, D., Stewart, A. (1987) Anatomy and histochemistry of resting and germinating sclerotia of *Sclerotium cepivorum*, *Trans. Br. Mycol. Soc.* **89**, 561–567.

Berry, D. (1975) The environmental control of the physiology of filamentous fungi, in: *The Filamentous Fungi* (Smith, J.E., Berry, D., Eds.), London: Edward Arnold, Vol. 1, 30–31.

Bertram, P., Buller, C., Stewart, G., Akagi, J. (1993) Isolation and characterisation of a *Bacillus* strain capable of degrading the extracellular glucan from *Cellulomonas flavigena* strain KU, *J. Appl. Bacteriol.* **74**, 460–466.

Betlach, M., Capage, M., Doherty, D., Hassler, R., Henderson, N., Vanderslice, R., Marelli, J., Ward, M. (1987) Genetically engineered polymers: manipulation of xanthan biosynthesis, in: *Industrial Polysaccharides* (Yalpani, M., Ed.), Amsterdam: Elsevier, 35.

Bluhm, C., Deslands, Y., Marchessault, R., Perz, S., Rinaudo, M. (1982) Solid-state and solution conformations of scleroglucan, *Carbohydr. Res.* **100**, 117–130.

Brigand, G. (1993) Scleroglucan, in: *Industrial Gums* (Whistler, R. L., BeMiller, J. N., Eds.), San Diego, CA: Academic Press, 3rd Edition, 461–472.

Brown, C., Campell, I., Priest, F. (1987) Growth and fermentation system, in: *Introduction to Biotechnology*, in *Basic Microbiology* (Brown, C., Campell, I., Priest, F., Eds.), Oxford: Blackwell Scientific Publications, Vol. 10, 49–66.

Burrafato, G., Guameri, A., Lockhart, T., Nicora, L. (1997) *Zirconium additive improves field performance and cost of biopolymer muds*, Society of Petroleum Engineers, USA.

Cartalos, U., Lecourtier, J., Rivereau, A. (1994) Use of scleroglucan as high temperature additive for cement slurries. U.S. Patent No. 5,301,753.

Cooney, C. (1981) Growth of microorganisms, in: *Biotechnology* (Rehmand, H., Reed, G., Eds.), London: Academic Press, Vol. 1, 101.

Coviello, T., Dentini, M., Rambone, G., Desideri, P., Carafa, M., Murtas, E., Riccieri, F., Alhaique, F. (1998) A novel co-crosslinked polysaccharide: studies for a controlled delivery matrix, *J. Control. Release* **55**, 57–66.

Coviello, T., Grassi, M., Rambone, G., Santucci, E., Carafa, M., Murtas, E., Riccieri, F., Alhaique, F. (1999) Novel hydrogel systems from scleroglucan: synthesis and characterisation, *J. Control. Release* **60**, 367–378.

Desbrieres, J. (1986a) French Patent No. 2,570,755.

Desbrieres, J. (1986b) French Patent No. 2,570,756.

Desbrieres, J. (1986c) French Patent No. 2,570,754.

Desbrieres, J. (1986d) French Patent No. 2,570,753.

De Nooy, A., Rori, V., Masci, G., Dentini, M., Crescenzi, V. (2000) Synthesis and preliminary characterisation of charged derivatives and hydrogels from scleroglucan, *Carbohydr. Res.* **324**, 116–126.

De Vuyst, L., Van Loo, J., Vandamme, E. (1987) Two-step fermentation process for improved xanthan production by *Xanthomonas campestris* NRRL-B-1459, *J. Chem. Technol. Biotechnol.* **39**, 263–273.

Donche, A. (1994) Application of scleroglucan muds to drilling deviated wells. U.S. Patent No. 5,330,015.

Donche, A., Isambourg, P. (1994) Scleroglucan-based compositions and their use as a cementation spacer. U.S Patent No. 5,306,340.

Dreveton, E., Monot, F., Ballerini, D., Lecourtier, J., Choplin, L. (1994) Effect of mixing and mass transfer conditions on gellan production by *Aeromonas elodea*, *J. Ferment. Bioeng.* **273**, 225–233.

Dreveton, E., Monot, F., Ballerini, D., Lecourtier, J., Choplin, L. (1996) Influence of fermentation hydrodynamics on gellan gum physico-chemical characteristics, *J. Ferment. Bioeng.* **82**, 272–276.

Dube, H. (1983) Nutrition of fungi, in: *An Introduction to Fungi* (Dube, H., Ed.), India: Vicks Publishing House PVT Ltd., 481–507.

Dubief, C. (1996) Composition for washing keratinous materials, in particular hair and/or skin. U.S. Patent No. 5,536,493.

Dunn, G. (1985) Nutritional requirements of microorganisms, in: *Comprehensive Biotechnology* (Moo-Young, M., Ed.), Oxford, New York: Pergamon Press, Vol. 1, 113–125.

Erfle, J., Teather, R. (1991) Isolation and properties of a (1,3)-β-D-glucanase from *Ruminococcus flavefaciens, Appl. Environ. Microbiol.* **57**, 122–129.

Evans, R., Nelson, C., Bowen, W., Kleve, M., Hickmon, S. (1998) Visualisation of bacterial glycocalyx with a scanning electron microscope, *Clin. Orthopaed. Rel. Res.* **347**, 243–249.

Farina, J., Sineriz, F., Molina, O., Perotti, N. (1998) High scleroglucan production by *Sclerotium rolfsii*: influence of medium composition, *Biotechnol. Lett.* **20**, 825–831.

Farina, J., Santos, V., Perotti, N., Casa, J., Molina, O., Garcia-Ochoa, F. (1999) Influence of the nitrogen source on the production and the rheological properties of scleroglucan produced by *Sclerotium rolfsii* ATCC 211126, *World J. Microbiol. Biotechnol.***15**, 269–275.

Forage, R., Harisson, D., Pitt, D. (1985) Effect of environment on microbial activity, in: *Comprehensive Biotechnology* (Moo-Young, M., Ed.), Oxford, New York: Pergamon Press, Vol. 1, 253–279.

Fumitake, Y. (1982) Aeration and mixing in fermentation, in: *Annual Reports on Fermentation Processes* (George, T. T., Ed.), New York, London: Academic Press, Vol. 5, 1–34.

Funami, T., Yada, H., Nakao, Y. (1998) Curdlan properties for application in fat mimetics for meat products, *J. Food Sci.* **63**, 283–287.

Gallino, G., Guarneri, A., Poli, G., Xiao, L. (1996) *Scleroglucan biopolymer enhances WBM performance,* Society of Petroleum Engineers, USA.

Gamini, A., Crescenzi, V., Abruzzese, R. (1984) Influence of the charge density on the solution behaviour of polycarboxylate derived from the polysaccharide scleroglucan, *Carbohydr. Polym.* **4**, 461–472.

Giavasis, I., Harvey, L., McNeil, B. (2000) Gellan gum. *Crit. Rev. Biotechnol.* **20**, 177–211.

Grassi, M., Lapasin, R., Pricl, S., Colombo, I. (1996) Apparent non-Fickian release from a scleroglucan gel matrix, *Chem. Eng. Commun.* **155**, 89–112.

Halleck, F. (1967) Polysaccharides and methods for production thereof. U. S. Patent No. 3,301,848.

Harding, N., Cleary, J., Cabanas, D., Rosen, I., Kang, K. (1987) Genetic and physical analysis of genes for xanthan gum biosynthesis in *X. campestris, J. Bacteriol.* **169**, 2854–2861.

Harvey, L. M. (1984) Production of microbial polysaccharides by the continuous culture of fungi, Ph.D. thesis, University of Strathclyde, UK.

Harvey, L. M., McNeil, B. (1998) Thickeners of microbial origin, in: *Microbiology of Fermented Foods,* 2nd edition (Wood, B. J. B., Ed.), London: Blackie Academic & Professional, Vol. 1, 150–171.

Heyraud, A., Salemis, P. (1982) Investigation of scleroglucan structure using sugar methylation analysis, *Carbohydr. Res.* **107**, 123.

Hodgkinson, A. (1977) *Oxalic Acid in Biology and Medicine,* New York: Academic Press.

Holzworth, G. (1985) Xanthan and scleroglucan: structure and use in enhanced oil recovery, *Dev. Ind. Microbiol.* **26**, 271–280.

Jagodzinski, P., Wiaderkiewicz, R., Kursawski, G., et al. (1994) Mechanism of the inhibitory effect of curdlan sulphate on HIV-1 infection *in vitro, Virology* **202**, 735–745.

Jamas, S., Easson, J., Davidson, D., Ostroff, G. (1996a) Use of aqueous soluble glucan preparations to stimulate platelet production. U. S. Patent No. 5,532,223.

Jamas, S., Easson, J., Davidson, D., Ostroff, G. (1996b) Use of neutral soluble glucan preparations to stimulate platelet production. U. S. Patent No. 5,488,040.

Jamas, S., Easson, J., Davidson, D., Ostroff, G. (1997) Glucan preparation. U. S. Patent No. 5,622,939.

Jamas, S., Easson, J., Davidson, D., Ostroff, G. (1998a) Underivatised, aqueous soluble beta (1,3)-glucan, composition and method of making same. U. S. Patent No. 5,817,643.

Jamas, S., Easson, J., Davidson, D., Ostroff, G. (1998b) Uses for underivatised, aqueous soluble beta (1,3)-glucan and compositions comprising same. U. S. Patent No. 5,783,569.

Johnson, M. (1996) Fluid system for controlling fluid losses during hydrocarbon recovery operations. U. S. Patent No. 5,504,062.

Johnston, J., Kirkwood, S., Misaki, A., Nelson, T., Scaletti, J., Smith, F. (1963) Structure of a new glucan, *Chem. Ind.* 820–822.

Jong, S., Donovick, R. (1989) Antitumour and antiviral substances from fungi, *Adv. Appl. Microbiol.* **34**, 183–262.

Kang, K., Cotrell, I. (1979) Polysaccharides, in: *Microbial Technology* (Peppler, H., Perlman, D., Eds.), Second edition. New York, London and San Francisco: Academic Press, 417–481.

Kansawa, Y., Harada, A., Takeuchi, M., Yokota, A., Harada, T. (1994) *Bacillus curdlanolyticus* sp. nov. and *Bacillus kobansis* sp. nov., which hydrolyse resistant curdlan, *Int. J. Syst. Bacteriol.* **45**, 515–521.

Kansawa, Y., Kurasawa, T., Kanegae, Y., Harada, A., Harada, T. (1995) Purification and properties of a new exo-(1-3)-β-D-glucanase from *Bacillus circulans* YK9 capable of hydrolysing resistant curdlan with formation of only laminaribiose, *Microbiology* **140**, 637–642.

Kawahara, K., Mizuta, I., Katabami, W., Koizumi, M., Wakayama, S. (1994) Isolation of *Sphingomonas* strains from rice and other plants of family Graminaea, *Biosci. Biotechnol. Biochem.* **58**, 600–601.

Kohler, N., Pirri, R. (1994) Use of gel-based compositions for reducing the production of water in oil- or gas- producing wells. U. S. Patent No. 5,322,123.

Kohler, N., Tabary, R., Zaitoum, A. (1987) European Patent No. 216,661.

Lachman, L., Sheth, P. (1968) U. S. Patent No. 3,415,929.

Lacroix, C., LeDuy, A., Noel, G., Choplin, L. (1985) Effect of pH on the batch fermentation of pullulan from sucrose medium, *Biotechnol. Bioeng.* **27**, 202–207.

Ladret, A., Donche, A. (1996) Application of muds containing scleroglucan to drilling large diameter wells. U. S. Patent No. 5,525,587.

Lecacheux, D., Mustiere, Y., Panaras, R. (1986) Molecular weight of scleroglucan and other extracellular microbial polysaccharides by size-exclusion chromatography and low angle laser scattering, *Carbohydr. Polym.* **6**, 477–492.

Lee, J., Lee, I., Kim, M., Park, Y. (1999) Optimal control of batch processes for production of curdlan by *Agrobacterium* species, *J. Ind. Microbiol. Biotechnol.* **23**, 143–148.

Lovrecich, M., Riccioni, G. (1991) Pharmaceutical tablets and capsule granulates of scleroglucan and active substance. U. S. Patent No. 5,068,111.

Lovrecich, M., Riccioni, G. (1993) Pharmaceutical tablets and capsule granulates of scleroglucan and active substance. U. S. Patent No. 5,215,752.

Marchetti, M., Pisani, S., Petropaolo, V., Seganti, L., Nicoletti, R., Degener, A., Orsi, N. (1996) Antiviral effect of a polysaccharide from *Sclerotium glucanicum* towards herpes simplex virus type 1 infection, *Planta Medica* **62**, 303–307.

Magro, P., Marciano, P., Di Lenna, P. (1984) Oxalic acid production and its role in pathogenesis of *Sclerotinia sclerotiorum, FEMS Microbiol. Lett.* **24**, 9–12.

Mastromarino, P., Petruzziello, R., Macchia, S., Rieti, S., Nicoletti, R., Orsi, N. (1997) Antiviral activity of natural and semisynthetic polysaccharides on early steps of rubella virus infection, *J. Antimicrob. Chemother.* **39**, 339–345.

Maxwell, D., Bateman, D. (1968) Oxalic acid biosynthesis by *Sclerotium rolfsii, Phytopathology* **58**, 1635–1642.

McClure, M., Moore, P., Blanc, D., Scotting, P., Cook, G., Keynes, R. (1992) Investigation into the mechanism by which sulfated polysaccharides inhibit HIV infection *in vitro, AIDS Res. Human Retrovir.* **8**, 19–26.

McNeil, B., Harvey, L. M. (1993) Viscous fermentation products, *Crit. Rev. Biotechnol.* **13**, 275–304.

McNeil, B., Kristiansen, B. (1987) Influence of impeller speed upon the pullulan fermentation, *Biotechnol. Lett.* **9**, 101–104.

Miller, R., Liberta, A. (1976) The effects of light on acid-soluble polysaccharide accumulating in *S. rolfsii, Sacc. Can. Microbiol.* **22**, 967.

Morris, V. (1987) New and modified polysaccharides, in: *Food Biotechnology* (King, R., Cheetham, P., Eds.), First edition. London, New York: Elsevier Applied Science, 233.

Nakao, Y., Konno, A., Taguchi, T., Tawada, T., Kasai, H., Toda, J., Terasaki, M. (1991) Curdlan: properties and application to foods, *J. Food Sci.* **56**, 3.

Nardin, P., Vincendon, M. (1989) Isotopic exchange study of the scleroglucan chain in solution, *Macromolecules* **22**, 3551–3554.

Nielsen, J. (1992) Modelling the growth of filamentous fungi, *Adv. Biochem. Eng. Biotechnol.* **46**, 188–221.

Ohno, N., Miura, N., Chiba, N., Adachi, Y., Yadomae, T. (1995) Comparison of the immunopharmacological activities of triple and single-helical schizophyllan in mice, *Biol. Pharmaceut. Bull.* **18**, 1242–1247.

Olsvik, E. (1992) *Rheological properties of a filamentous fermentation broth*, Ph.D. Thesis, University of Strathclyde, UK.

Omari, A. (1995) Gelation control of the scleroglucan zirconium system using oxyacids, *Polymer* **36**, 4263–4265.

Ostroff, G. (1997) Inhibition of infection-stimulated oral tissue destruction by beta (1,3)-glucan. U. S. Patent No. 5,622,940.

Ouriaghli, T., Francois, J., Sarazin, D., Dinh, N. (1992) Influence of nonionic surfactant on aggregation state of scleroglucan in aqueous solution, *Carbohydr. Polym.* **17**, 301–312.

Parish, C.R., Low, L., Warren, H.S., Cunnigham, A. L. (1990) A polyanion binding site on the CD4 molecule. Proximity to the HIV-gp120 binding region, *J. Immunol.* **145**, 1185–1195.

Parton, C., Willis, P. (1990) Strain preservation, inoculum preparation and inoculum development, in: *Fermentation: A Practical Approach* (McNeil, B., Harvey, L., Eds.), Oxford: IRL Press, 39–44.

Patchen, L., Bleicher, P. (2000) Mobilisation of peripheral blood precursor cells by beta (1,3)-glucan. U. S. Patent No. 6,117,850.

Pierre, C., Demangeon, F., Yves, H. (1993) Bituminous binder emulsion with a viscosity controlled by addition of scleroglucan. U. S. Patent No. 5,246,986.

Pirri, R., Huet, Y., Donche, A. (1993) Process for improving the dispersability and filterability of scleroglucan powders. U. S. Patent No. 5,224,988.

Pirri, R., Huet, Y., Donche, A. (1994) Process for enhancing oil recovery using scleroglucan powders. U. S. Patent No. 5,323,857.

Pirri, R., Gadioux, J., Riveno, R. (1996) Scleroglucan gel applied to the oil industry. U. S. Patent No. 5,555,936.

Pretus, H., Eusley, H., McNamee, R., Jones, E., Browder, I., Williams, D. (1991) Isolation, physicochemical characterisation and pre-clinical efficacy evaluation of a soluble scleroglucan, *J. Pharmacol. Exp. Ther.* **257**, 500–510.

Punja, Z. (1985) The biology, ecology and control of *Sclerotium rolfsii*, *Annu. Rev. Phytopathol.* **23**, 97–127.

Punja, Z., Huang, J., Jenkins, S. (1985) Relationship of mycelial growth and production of oxalic acid and cell wall degrading enzymes to virulence in *Sclerotium rolfsii*, *Can. J. Plant. Pathol.* **2**, 109–117.

Rapp, P. (1989) 1,3-β-glucanase, 1,6-β-glucanase, and β-glucosidase activities of *Sclerotium glucanicum*: synthesis and properties, *J. Gen. Microbiol.* **135**, 2847–2855.

Rau, U., Muller, R., Cordes, K., Klein, J. (1989) Process and molecular data of branched 1,3-β-D glucans in comparison with xanthan, *Bioprocess Eng.* **5**, 89–93.

Rau, U., Gura, E., Olszewski, E., Wagner, F. (1992) Enhanced glucan formation of filamentous fungi by effective mixing, oxygen limitation and fed-batch processing, *Ind. Microbiol.* **9**, 19–26.

Rezzoug, S., MaacheRezzoug, Z., Mazoyer, J., Jeannin, M., Allaf, K. (2000) Effect of instantaneous controlled pressure drop process on the hydration capacity of scleroglucan: optimisation of operating conditions by response surface methodology, *Carbohydr. Polym.* **42**, 73–84.

Rho, D., Mulchandani, A., Luong, J., LeDuy, A. (1988) Oxygen requirement in pullulan fermentation, *Appl. Microbiol. Biotechnol.* **28**, 361–366.

Rinaudo, M., Vincendon, M. (1982) ^{13}C-NMR structural investigation of scleroglucan, *Carbohydr. Polym.* **2**, 135–144.

Rinaudo, M., Milas, M., Tinland, A. (1990) Relation between molecular structure and physicochemical properties for some microbial polysaccharides, *NATO ASI Series, Ser. E.* **180**, 349–370.

Rodgers, N. (1973) Scleroglucan, in: *Industrial gums* (Whistler, R., BeMiller, J., Eds.), Second edition, New York: Academic Press, 499–511.

Rodgers, N., Gofrett, H. (1976) Polytran: trademark of scleroglucan. CECA S.A. Velizy-Villacowblay, France.

Sampath, K. (1987a) Oil recovery process employing a complexed polysaccharide. U. S. Patent No. 4,647,312.

Sampath, K. (1987b) Oil recovery process employing a complexed polysaccharide. U. S. Patent No. 4,640,358.

San-Ei Chemical Industries, Ltd. (1982) Japan Kokai Tokyo Koho, 57, 163, 450. *Chem. Abstr.* **98**, 15733.

Sandford, P. (1979) Exocellular, microbial polysaccharides. *Adv. Carbohydr. Chem. Biochem.* **36**, 265–312.

Sheth, P., Lachman, L. (1967) French Patent No. 1,480,874.

Shimizano, H. (1955) Oxalic acid decarboxylase, a new enzyme from the mycelium of wood destroying fungi, *J. Biochem.* **42**, 526–530.

Shu, C., Yang, S. (1990) Effects of temperature on cell growth and xanthan production in batch cultures of *Xanthomonas campestris*, *Biotechnol. Bioeng.* **35**, 454–468.

Schilling, B. (2000) *Sclerotium rolfsii* ATCC 15205 in continuous culture: economical aspects of scleroglucan production, *Bioproc. Eng.* **22**, 57–61.

Schilling, B., Rau, U., Maier, T., Fankhauser, P. (1999) Modeling and scale-up of the unsterile scleroglucan production process with *Sclerotium rolfsii* ATCC 15205, *Bioproc. Eng.* **20**, 195–201.

Schilling, B., Henning, A., Rau, U. (2000) Repression of oxalic acid biosynthesis in the unsterile scleroglucan production process with *Sclerotium rolfsii* ATCC 15205, *Bioproc. Eng.* **22**, 51–55.

Seviour, R., Kristiansen, B. (1983) Effect of ammonium ion concentration on polysaccharide production by *Aureobasidium pullulans* in batch culture, *Eur. J. Appl. Microbiol. Biotechnol.* **17**, 1983–1989.

Singh, P., Wisler, R., Tokuzen, R., Nakahara, W. (1974) Scleroglucan, an antitumor polysaccharide from *Sclerotium glucanicum*, *Carbohydr. Res.* **37**, 245–247.

Smith, M., Ho, C. (1985) The effect of dissolved CO_2 on penicillin production: mycelial morphology, *J. Biotechnol.* **2**, 347–363.

Steiner, W., Lafferty, R., Gomes, I., Esterbauer, H. (1987) Studies on a wild strain of *Schizophyllum commune*: cellulase and xylanase production and formation of the extracellular polysaccharide schizophyllan, *Biotechnol. Bioeng.* **30**, 169–178.

Stokke, B., Elsgaeter, A., Bjornestad, E., Lund, T. (1992) Rheology of xanthan gum and scleroglucan in synthetic sea water, *Carbohydr. Polym.* **17**, 209–220.

Stone, H., Armentrout, V. (1985) Production of oxalic acid by *Sclerotium cepivorum* during infection of onion, *Mycologia* **77**, 526–530.

Sutherland, I. (1979) Microbial polysaccharides: control of synthesis and acylation, in: *Microbial Polysaccharides and Polysaccharases* (Berkeley, R., Gooday, G., Ellwood, D., Eds.), London: Academic Press, 1–28.

Sutherland, I., Ellwood, D. (1979) Microbial polysaccharides, industrial polymers of current and future potential, in: *Microbial Technology: Current state, Future Prospects* (Bull, A., Ellwood, D., Ratledge, C., Eds.), Cambridge: Cambridge University Press, 107–150.

Takeuchi, M., Sakane, T., Yanagi, M., Yamasato, K., Hamana, K., Yokota, A. (1995) Taxonomic study of bacteria isolated from plants – proposal of *Sphingomonas rosa* sp. nov., *Sphingomonas pruni* sp. nov., *Sphingomonas asaccharolytika* sp. nov. and *Sphingomonas mali* sp. nov., *Int. J. Syst. Bacteriol.* **45**, 334–341.

Taurhesia, S., McNeil, B. (1994) Production of scleroglucan by *S. glucanicum* in batch and supplemented batch cultures, *Enzyme Microbiol. Technol.* **16**, 223–228.

Thorne, L., Tansey, L., Pollock, T. (1987) Clustering of mutations blocking synthesis of xanthan gum by *X. campestris*, *J. Bacteriol.* **169**, 3593–3600.

Truong, D., Gadioux, J., Sarazin, D. (1993) Concentrated liquid solutions of scleroglucan. U. S. Patent No. 5,215,681.

VanOort, E., Bland, R., Howard, S., Wiersma, R., Robertson, L. (1997) *Improving HPHT stability of water based drilling fluids*, Society of Petroleum Engineers, USA.

Vaussard, A., Ladret, A., Donche, A. (1997) Scleroglucan based drilling mud. U. S. Patent No. 5,612,294.

Vincendon, M. (1999) Scleroglucan derivatives: aromatic carbamates, *J. Polym. Chem.* **37**, 3187–3192.

Wang, Y., McNeil, B. (1995a) The effect of temperature on scleroglucan synthesis and organic acid production by *Sclerotium glucanicum*, *Enzyme Microb. Technol.* **17**, 893–899.

Wang, Y., McNeil, B. (1995b) pH effects on exopolysaccharide and oxalic acid production in cultures of *Sclerotium glucanicum*, *Enzyme Microb. Technol.* **17**, 124–130.

Wang, Y., McNeil, B. (1995c) Dissolved oxygen and the scleroglucan fermentation process, *Biotechnol. Lett.* **13**, 155–160.

Wang, Y., McNeil, B. (1995d) Production of the fungal exopolysaccharide scleroglucan by cultivation of *Sclerotium glucanicum* in an airlift reactor with an external loop, *J. Chem. Technol. Biotechnol.* **63**, 215–222.

Wang, Y., McNeil, B. (1996) Scleroglucan, *Crit. Rev. Biotechnol.* **16**, 185–215.

Wecker, A., Onken, U. (1991) Influence of dissolved oxygen concentration and shear rate on the production of pullulan by *A. pullulans*, *Biotechnol. Lett.* **13**, 155–160.

Wernau, W. (1985) Fermentation methods for the production of polysaccharides, *Dev. Ind. Microbiol.* **26**, 263–269.

Willets, H. (1978) Sclerotium formation, in: *The Filamentous Fungi* (Smith, J., Berry, D., Eds.), London: Edward Arnold, Vol. 3, 197–221.

Williams, S. (1968) U. S. Patent No. 3,373,810.

Yanaki, T., Kojima, T., Norisuge, T. (1981) Triple helix of scleroglucan in dilute aqueous sodium hydroxide, *Polymer J.* **13**, 1135–1140.

25
Schizophyllan

PD Dr. Udo Rau
Technical University Braunschweig, Institute of Biochemistry and Biotechnology, Spielmannstr. 7, D-38106 Braunschweig, Germany; Tel.: +49-531/391-5740; Fax: +49-531/391-5763; E-mail: U.Rau@tu-bs.de

Biotechnology of Biopolymers. From Synthesis to Patents. Edited by A. Steinbüchel and Y. Doi

ISBN: 3-527-31110-6

ADP	adenosine diphosphate
ATP	adenosine triphosphate
BDM	bio dry mass
cpm	counts per minute
DMSO	dimethylsulfoxide
EOR	enhanced oil recovery
GDP	guanosine diphosphate
Glc	glucose
GTP	guanosine triphosphate
L	liquid phase
Me	metal
MW	molecular weight
ORF	open reading frame
O_2	oxygen
P	phosphate
pO_2	oxygen partial pressure of the liquid phase
Pro-A	proteinic acceptor
Re	Reynold's number
r.p.m.	revolutions per minute
S	substrate
SDS-PAGE	sodium dodecyl sulfate-polyacrylamide gel electrophoresis
UDP	uridine diphosphate
UTP	uridine triphosphate
X	biomass
*	gas/liquid interface
C (g L^{-1})	concentration
k_La (h^{-1})	volume-related oxygen transfer coefficient
K_m	Michaelis-Menten constant
K_S (mol L^{-1})	substrate saturation constant
qO_2 (h^{-1})	specific oxygen uptake rate (oxygen uptake rate divided by actual biomass)
R	feedback rate
V (L)	volume
V (L h^{-1})	volume feed
$Y_{X/O2}$	yield coefficient (biomass formed/oxygen consumed)
δ (ppm)	chemical shift
$[\eta]$ (cm^3 g^{-1})	intrinsic viscosity

1 Introduction

Many fungi are able to secrete exopolysaccharides, the molecular structure of which varies between homopolysaccharides, which are composed of a single monomer, and heteropolysaccharides, which contain chemically different monomers. However, in contrast to bacteria where for example xanthan is synthesized by *Xanthomonas campestris* in large amounts, fungi secrete only relatively small amounts of heteropolysaccharides (Seviour et al., 1992).

The fungal homopolysaccharides frequently contain D-glucose as the monomer which is connected glycosidically by either α- or β-linkages. Branched β-glucans are secreted (among other fungi) by *Sclerotium rolfsii* (Pilz et al., 1991), *Sclerotium glucanicum* (Rau et al., 1992b), *Monilinia fructigena* (Cordes, 1990), and *Botrytis cinerea* (Gawronski et al., 1996). These β-glucans possess a uniform, primary molecular structure that is identical with that of schizophyllan, which forms the subject of this review.

Acknowledgement
This review is dedicated to Prof. Dr. Wolf-Dieter Deckwer on the occasion of his 60th birthday.

2 Historical Outline

According to Essig (1922), the wood-rotting and filamentously growing basidiomycete *Schizophyllum commune* was first described by Dillenius in 1719, and was found to secrete a neutral homoglucan called schizophyllan.

Later, Wang and Miles (1964) performed a physiological characterization of the dikaryotic mycelia of *S. commune*, whilst during the early 1970s two Japanese articles not only described the formation and properties of schizophyllan (Kikumoto et al., 1970) but also elucidated its structure by means of enzymatic degradation studies (Kikumoto et al., 1971). These were the first and, for a long time also the last, publications that dealt with the production of schizophyllan. It was during the late 1970s that the group of Wessels (Sietsma and Wessels, 1977, 1979) first began to analyze the cell wall of *S. commune* and to investigate how the organism carried out β-glucan synthesis. During the 1980s, a number of Japanese authors investigated the physico-chemical characteristics of schizophyllan, notably the triple helical arrangement in aqueous solution (Norisuye et al., 1980; Kashiwagi et al., 1981), conformation analysis in gels at raised pH (Saito et al., 1979) and conformation transition by the addition of dimethylsulfoxide (Kitamura, 1989), as well as its controlled degradation by hydrodynamic shear (Kojima et al., 1984) and ultrasonic treatment (Tabata et al., 1981). The non-Newtonian flow behavior of colloid disperse aqueous schizophyllan solutions was investigated by Rau and Wagner (1987).

Studies on the production (Rau et al., 1989) and additional downstream processing (Cordes et al., 1989) of schizophyllan were continued by the group of Rau. These authors established an oxygen controlled batch process (Brandt et al., 1993; Rau and Brandt, 1994) as well as an oxygen-limited continuous process (including cell feedback) for the enhanced production of schizophyllan (Brandt, 1995), as well as creating downstream processing (Rau et al., 1992a; Rau, 1999). At the same time, Steiner and co-workers were investigating the cultivation of *S. commune* (Steiner et al., 1988, 1993), albeit with their emphasis directed towards the differing enzymatic activities of this fungus (Haltrich and Steiner, 1994; Steiner et al., 1987).

The oil crisis in 1973 provided the first initiative to use schizophyllan as a polymer additive in enhanced oil recovery (EOR) (Lindoerfer et al., 1988; Rau et al., 1992c), but this role declined as the price of oil subsequently fell. Although, depending on the availability and price of crude oil, the use of schizophyllan in EOR may well return. It are the immune-stimulating properties of schizophyllan and subsequent use in the pharmaceutical industry that have recently proved exciting (Tsuzuki et al., 1999; Kidd, 2000; Miura et al., 2000; Mueller et al., 2000; Ooi and Liu, 2000).

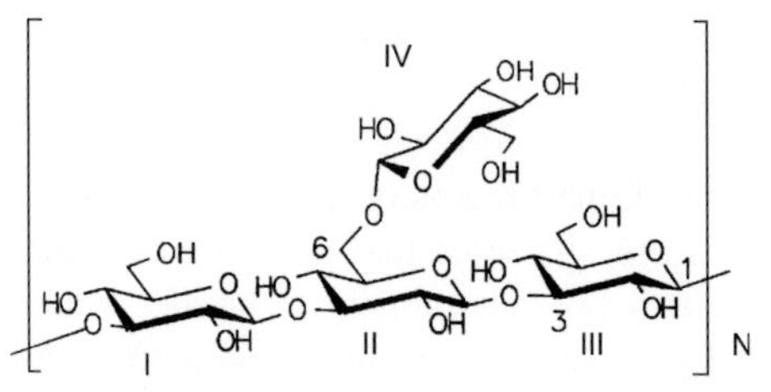

Fig. 1 Subunit of schizophyllan. N = 9000–18 000.

3 Chemical Structure

In contrast to xanthan, which is an anionic heteropolysaccharide, schizophyllan does not carry any charged groups and contains only β-D-glucose. This neutral homoglucan consists of a backbone chain of 1,3-β-D-glucopyranose units linked with single 1,6-bonded β-D-glucopyranoses at about every third glucose molecule in the basic chain (Figure 1). The molecular weight ranges from 6 to 12×10^6 g mol^{-1} (Rau et al., 1990).

The following data were obtained from the ^{13}C NMR-spectra (Figure 2) of schizophyllan: The signal at 102.7 and 102.8 ppm is due to C1 carbons of all sugars. The area of these signals was set to 100%. The signals in the 86 ppm region can be attributed to C3 carbons involved in the 1,3-β-glycosidic linkage. The chemical shifts of the free hydroxymethyl C6 carbons are found in the 60 ppm region (Rinaudo and Vincendon, 1982). The integral intensity of the signals from C1 and linked C3 gave a ratio of 1:0.74, and implies that a single glucose molecule is 1,6-β-linked

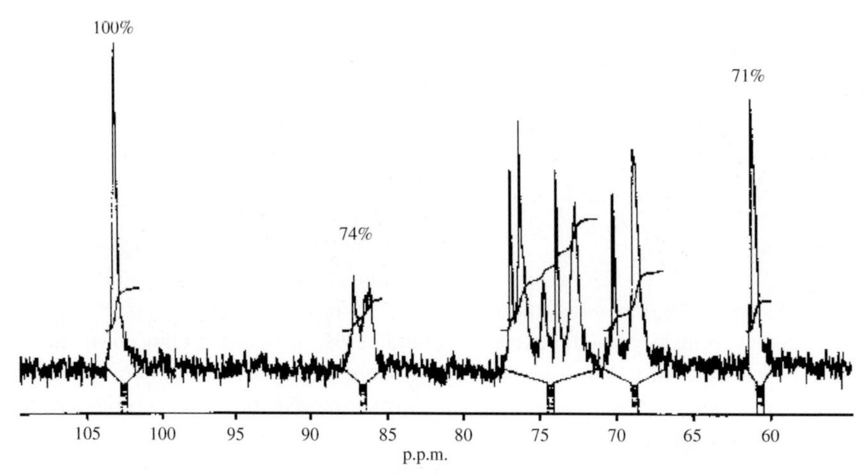

Fig. 2 ^{13}C-NMR spectra of schizophyllan in DMSO-d_6 at 80 °C. δ = 60.6 (C-6_{I}), 60.7 (C-6_{II}), 60.9 (C-6_{IV}), 68.2 (C-$4_{II,III}$), 68.4 (C-4_{I}), 70.0 (C-4_{IV}), 72.4 (C-$2_{II,III}$), 72.5 (C-2_{I}), 73.4 (C-2_{IV}), 74.5 (C-5_{III}), 75.9 (C-5_{II}), 76.1 (C-$5_{I,IV}$), 76.4 (C-3_{IV}), 85.8 (C-3_{III}), 86.1 (C-3_{II}), 86.4 (C-3_{I}), 102.7 (C-$1_{I\text{-}IV}$), 102.8 (C-$1_{I\text{-}IV}$). (Recorded by V. Wray, GBF, Braunschweig, Germany.)

to every third monomer of the main 1,3-β-chain.

Schizophyllan dissolves in water and dilute (≤0.01 M) sodium hydroxide solution to form highly viscous solutions. Norisuye et al. (1980) concluded from sedimentation equilibrium, light scattering and viscosity measurements that in aqueous

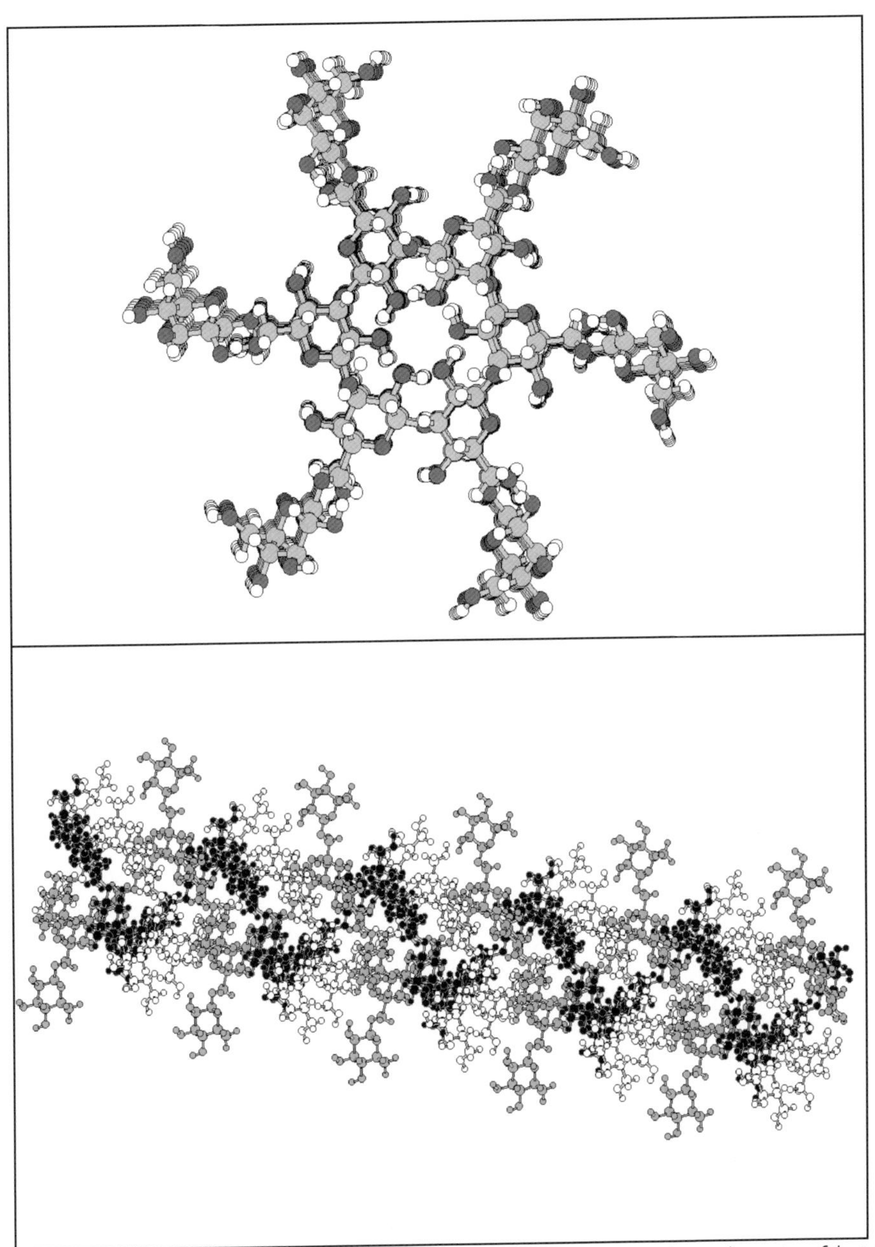

Fig. 3 Triple helical arrangement of schizophyllan in aqueous solution. Top: view to the center of the triplex. Bottom: side view.

solutions a trimer with triple helical structure is formed (Figure 3).

This structure (triple helix II) is stabilized by interchain hydrogen bonds at C-2 position with the β-1,6-glucose residues protruding outside the helix backbone. Using the data of Kashiwagi et al. (1981), it was calculated that there was a pitch of 0.3 ± 0.01 nm per glucose residue, a hydrodynamic diameter of 2.6 ± 0.4 nm, and a persistence length of 180 ± 30 nm for the triplex. Up to a molecular weight of 5×10^5 g mol^{-1}, the triple helix is almost perfectly rigid, underlined by a Mark-Houwink exponent of 1.49 (0.5 = random coil, 2 = rigid rod); however, the structure behaves like a semiflexible rod at increased molecular weights (Norisuye et al., 1980; Rau et al., 1992b).

At a midpoint temperature of 6 °C it is assumed that side-chain glucose residues slightly vary their positions, the result being a higher organized structure together with surrounding water molecules which form a helical chain at the outside of the triple helix. This ordered structure (triple helix I) surrounding the helical core increases the rigidity of the molecule and leads to gel formation (Asakawa et al., 1984; Van et al., 1984; Maeda et al., 1999; Sakurai and Shinkai, 2000).

High-sensitivity differential scanning calorimetry and optical rotatory dispersion (ORD) were used by Kitamura et al. (1996) to study the conformational transitions of schizophyllan in aqueous alkaline solution. They proposed a phase diagram for the conformation of schizophyllan as a function of temperature and pH. A more ordered structure (triple helix I) was attained at a temperature < 10 °C and pH < 11. This structure is stable up to 60 °C at pH 13. The triple helix II, attained at > 10 °C and pH < 11, converts to single coiled chains at > 135 °C or pH > 13 (Kitamura, 1989). The renatured samples at schizophyllan concentrations < 1 g L^{-1}, which were prepared by dissolution in 0.25 M KOH followed by neutralization with HCl, were observed as mixture of globular, linear and circular structures, and larger aggregates with less-defined morphology by electron microscopy. Subsequent annealing at 115–120 °C increases the proportion of circular species. This irreversibility of the triple helix reconversion is in sharp contrast to the investigations performed by Young and Jacobs (1998), who observed a slow (2–7 days) conversion of single helix to triple helix conformation. The renaturation took place between 39 and 84%, and was verified by fluorescence detection using aniline blue, which binds only to single-helix schizophyllan.

4 Chemical Analysis

Periodate cleaves oxidatively sugars with two vicinal hydroxyl groups. The β-1,3-backbone chain of schizophyllan does not carry adjacent hydroxyl groups and, therefore, will be not attacked. However, the side-chain glucose molecule is cleavable by 2 mol of periodate between C2–C3 and C3–C4 with simultaneous release of 1 mol of formic acid. Schizophyllan consumed between 0.52–0.55 mol periodate and released 0.22–0.26 mol formic acid per "anhydroglucose" unit, in accordance with the established structure (see Figure 1) of a 1,3-linked β-D-glucan with one β-D-glucopyranosyl group attached to position 6 of every third unit (Schulz and Rapp, 1991).

Partially oxidized schizophyllan is subsequently reduced by sodium borohydride (Smith degradation) and all nonacetalized aldehydes are transferred into hydroxy groups. Simultaneously, the acetals of the precleaved sugars are hydrolyzed and 0.23–0.25 mol glycerol is released per anhydroglucose unit. These procedures are described in detail elsewhere (Muenzer, 1989).

5 Occurrence and Physiological Function

The white rot basidiomycete *S. commune* belongs to the family *Aphyllophorales* (Donk, 1964) and forms "gilled" fruit-bodies (Figure 4). This fungus is a cosmopolitan organism that grows under extremely variable conditions throughout the temperature and tropic zones (Wessels, 1965). *S. commune* is frequently found on wood because it degrades lignin efficiently, but not cellulose, which is left as white fibers. Cooke (1961) reported that many native tribes used the fruit-bodies of *S. commune* as food or even as a type of chewing gum.

The water-insoluble portion of the hyphal wall consists of three types of polysaccharides: glucosoaminoglycans (chitin, chitosan and heteropolysaccharides containing both *N*-acetylglucosamine and glucosamine); the alkali-soluble *S*-glucan (α-1,3-glucan); and the alkali-insoluble or alkali-resistant *R*-glucan (β-1,3- and β-1,6-linked glucan), together comprising about 70% of the dry weight. The *R*-glucan is highly branched with β-1,3- and β-1,6-linkages; different structures may even be found in the same strain. The partial covalent linkage between the *R*-glucan and chitin is the reason for insolubility of the complex. Enzymatic or chemical degradation of the *R*-glucan chitin complex renders all of the β-glucan soluble in water or alkali (Sietsma and Wessels, 1979).

Schizophyllan is the water-soluble, extracellular glucan fraction excreted by the fungus that partially adheres to the walls as a jelly-like slime or mucilage (Figure 5). It is, therefore, a matter of opinion to consider schizophyllan a wall component, or not. The biological role of this gum is not yet well understood, though Wang and Miles (1964) assumed that it acts as a reserve carbon source because under glucose-limited conditions (Rau et al., 1990), β-glucan-degrading enzymes – the β-glucanases – are released (Rapp, 1992). The extracellular release of these enzymes underlies a glucose-induced repression. Wessels (1978) suggested that schizophyllan formation may be a result of unbalanced synthesis of *R*-glucan, or to a defect in the assembly of this component in the inner layer of the cell wall.

6 Biosynthesis

Selitrennikoff (1995) has provided an excellent review on the synthesis of β-glucans in

Fig. 4 Fruit-body of *Schizophyllum commune*. The characteristic lamellar structure is clearly recognizable.

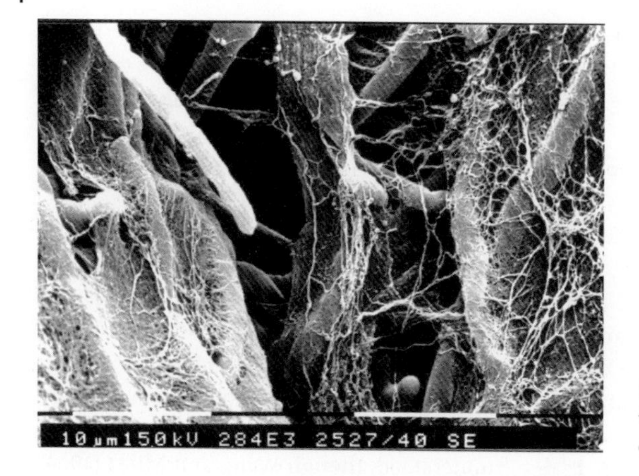

Fig. 5 Micrograph (×2800) of hyphae from *S. commune* covered with a mucilage of schizophyllan.

fungi. The 1,3-β-glucan synthase (E.C. 2.4.1.34; UDP-glucose: 1,3-β-D-glucan 3-β-D-glucosyltransferase) formed in the endoplasmic reticulum is transferred to the dictyosomes, which produce large apical vesicles (Ruiz-Herrera, 1991). The enzyme is transported by these vesicles to the hyphal tip, and is released at this location by fusing with the plasma membrane (Figure 6).

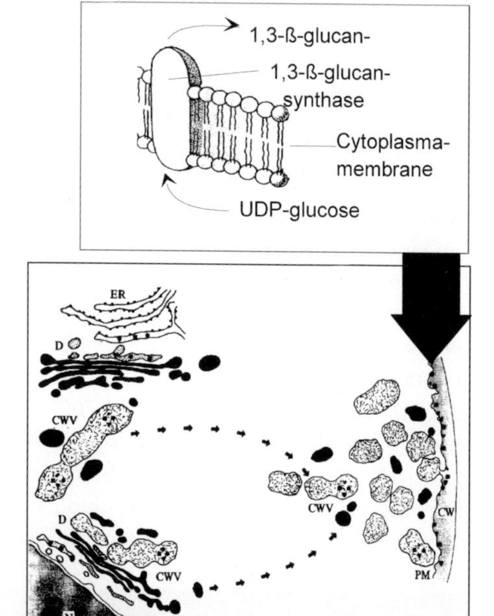

Fig. 6 Synthesis, transport and function of the β-glucan synthase. ER, endoplasmic reticulum; D, dictysome; N, nucleus; CWV, cell wall vesicle; PM, plasma membrane; CW, cell wall. (Adapted from Fèvre, 1979.)

The β-glucan synthase is not active during transport, and only becomes active after plasma membrane–vesicle fusion. Enzyme activity does not require a divalent metal ion, and does not use a lipid-linked intermediate, as is known for prokaryotes (Sutherland, 1982). However, a proteinic acceptor molecule (Andaluz et al., 1986, 1988) has been identified. It is possible that this is the same as the GTP-binding protein which interacts with the enzyme and stimulates its activity (Mol et al., 1994). Interestingly, a primer is not required to induce enzyme activity.

The assay for 1,3-β-glucan synthase activity of *S. commune* was performed as follows (Kottuz and Rapp, 1990; Rau, 1997). Intensive mixing disrupted the micelles, and the enriched membrane fractions were isolated by high-speed centrifugation (34,000 r.p.m.) at 4 °C. The crude enzyme was dissolved in buffer (0.05 M Tris-HCl, pH 7.2, 1 mM EDTA, 7 mM cellobiose), and UDP-[U-^{14}C]-glucose was used as precursor. After incubation at 26 °C for 45 min, the unincorporated radioactive [^{14}C]-glucose was separated by filtration. The amount of radioactivity bound to the enzyme and retained on the filter was determined by scintillation counting.

The apparent K_m values ranged from 0.1 to 5 mM, depending on the source of the enzyme (Selitrennikoff, 1995). The crude enzyme is highly unstable, and has a half-life of only a few minutes at 25 °C. However, a substantial increase in stability, with only 20% loss of activity, is achieved during 19-day storage in buffer solution at low temperatures of –80 or –196 °C (Rau, 1997). Stimulation of the enzyme activity is also possible by the addition of nucleotide di- and triphosphates (Figure 7).

This behavior was also found for the glucan synthase of *Pyricularia oryzae* (Kominato et al., 1987), *Saccharomyces cerevisiae* (Shematec and Cabib, 1980), *Saprolegnia monoica* (Girard and Fèvre, 1984) and *Neurospora crassa* (Lourdes et al., 1995). Kang and Cabib (1986) showed that 1,3-β-glucan synthase is composed of multiple protein subunits, which can be separated by treatment of the membranes with salt and detergents. The soluble fraction contains at least one GTP-binding protein, while the membrane fraction retains the core catalytic center.

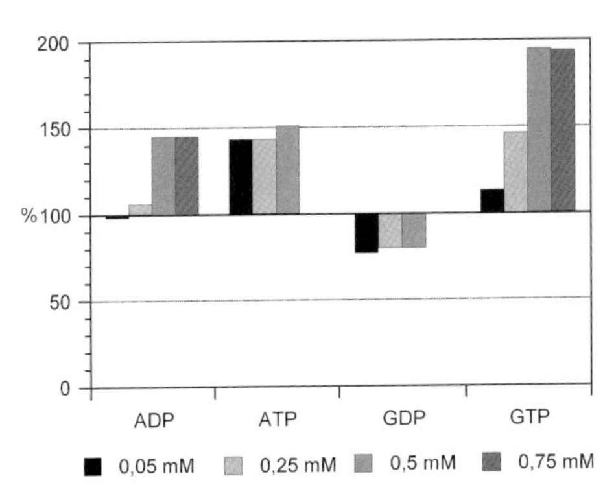

Fig. 7 Variation of 1,3-β-synthase activity by addition of nucleotide phosphates. Total (100%) activity is equal to a control assay without nucleotides (Rau, 1997).

These authors identified a 20-kDa GTP-binding protein that appeared to be the regulatory subunit, and a GTPase-activating protein that may regulate the GTP-binding protein (Mol et al., 1994).

β-Linked disaccharides such as cellobiose and laminaribiose also stimulate 1,3-β-glucan synthase activity (Wang and Bartnicki-Garcia, 1982). Quigley and Selitrennikoff (1984) found that disaccharides did not stimulate enzyme activity by acting as a primer, but rather interacted with the enzyme at a nonsubstrate site, thereby reducing the saturation constant for UDP-glucose.

The 1,3-β-glucan synthase operates as an integral membrane enzyme (Rau, 1997; Jabri et al., 1989) that uses UDP-Glc as substrate and vectorially synthesizes its β-1,3-glucan from the site of substrate hydrolysis (cytoplasmic facing) to the extracytoplasmic face of the plasma membrane (Jabri et al., 1989). Once external, the glucans are assembled into the cell wall by mechanisms that are not fully understood. The process that terminates the glucan polymerization process is also unknown. Chitin and 1,3-β-glucan become covalently cross-linked through lysine linkages (Sietsma and Wessels, 1979, 1981; Wessels and Sietsma, 1979, 1981), but the responsible enzyme(s) has (have) not been identified or assayed (Figure 8).

Different mechanisms have been proposed for the cell wall assembly, including: (1) a balance between synthesis and degradation of the polymers (Bartnicki-Garcia and Lippman, 1972); and (2) the steady-state model of Wessels (1988). These models were recently discussed by Bartnicki-Garcia (1999).

Sietsma et al. (1985) found, by using autoradiographic studies, that the alkali-insoluble β-glucan, which is linked with chitin and deposited at the apex in the extension zone, is mainly 1,3-β-linked and

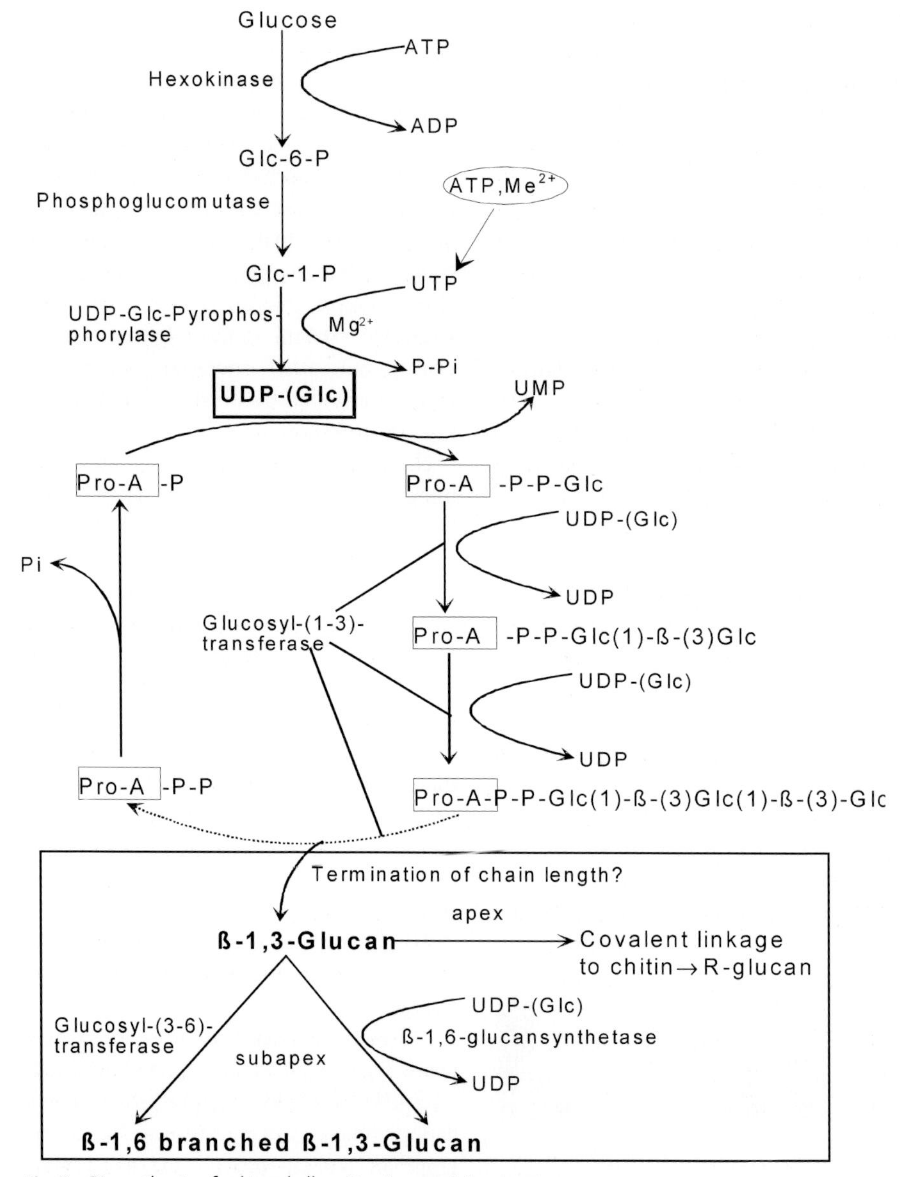

Fig. 8 Biosynthesis of schizophyllan. Pro-A, proteinic acceptor.

may be devoid of any 1,6-linkages at the extreme apex. This unbranched structure facilitates the formation of stable triple helices (Marchessault and Deslandes, 1980) from the chains that is required for cross-linking the chitin, and for subsequent microfibril formation. The mature wall contains very few pure 1,3-β-glucan chains. The 1,6-β-linked glucosyl residues occur in the extension zone, but form the majority in the subapically synthesized glucan. In the mature wall about half of the 1,3-β-glucan chains attached to chitin contain single glucose branches that are 1,6-β-linked to the main chain, while the other half also contains longer branches of 1,6-β-glucan chains. At least the presence of single glucose branches does not eliminate the tendency of the 1,3-β-glucan to form triple helices (Sato et al., 1981). These results do not permit conclusions to be made concerning the subapical synthesis of one or more glucans with variation in the 1,3-β- and 1,6-β-linkages that they contain. Thus, it seems probable that pure 1,3-β-glucans deposited in the cell wall are subsequently modified in subapical areas by attachment of 1,6-β-linked branches. Brown and Bussey (1993) proposed another concept, namely that the 1,6-β-glucans appear to be synthesized in the secretory pathway, transported to the plasma membrane, exocytosed, and then assembled into the cell wall. This is in sharp contrast to the procedure of 1,3-β-glucan synthesis mentioned earlier.

The characterization of 1,3-β-glucan synthase can be summarized as follows:

1) Enzyme activity is particulate, i.e., an integral membrane enzyme activity.
2) UDP-glucose is the preferred substrate.
3) The enzyme is not a zymogen, i.e., does not require proteolytic cleavage for *in vitro* activity.
4) The *in vitro* pH optimum is slightly basic (7.2).
5) The K_m value varies between 0.1 and 5 mM.
6) Enzyme activity can be stimulated by nucleoside phosphates and β-linked disaccharides.
7) A lipid-linked intermediate is not used.
8) The *in vitro* product is a linear 1,3-β-glucan with no detectable other linkages or branch points.

7 Molecular Genetics

S. commune has been used in many very different molecular genetics investigations, and it is not possible to cover all of these in this review. Hence, only a few examples will be described, including: the control of development (Marion et al., 1996; Shen et al., 1996; Lengeler and Kothe, 1999); the expression of heterologous genes (Schuren and Wessels, 1998; Scholtmeijer et al., 2001); the influence of the gene *FRT1* on homo- and dikaryotic fruiting (Horton et al., 1999); and the expression of the hydrophobin gene *SC3* (van Wetter et al., 2000; Wessels, 1999).

The 1,3-β-D-glucan synthase (EC 2.4.1.34) is a membrane enzyme which is activated by GTP and has been fractionated into soluble (GTP-binding) and membrane-bound (catalytic) components (Kang and Cabib, 1986; Mol et al., 1994). Genes encoding this enzyme have still not been isolated or sequenced in *S. commune*. However, this enzyme is essential for wall growth in yeasts and, therefore, related genetic investigations have been performed in other microorganisms, for example *S. cerevisiae* (Mazur and Baginski, 1996), *N. crassa* (Polizeli et al., 1995), and *Candida albicans* (Kondoh et al., 1997).

Two genes encode this multisubunit enzyme. *FKS1* (ORF: YLR342W) alias *GSC1*, *CND1*, *CWH53*, *ETG1* and *PBR1* encode a 215-kDa integral membrane protein (Fks1p) which mediates sensitivity to the echinocandin class of antifungal glucan synthase inhibitors and is a subunit of this enzyme. The residual glucan synthase activity present in *fks1* disruption mutants, the nonessential nature of the gene, and hybridization analysis of yeast chromosomal DNA pointed to the existence of a homologous gene encoding a functionally redundant product (Douglas et al., 1994). *FKS2* (ORF: YGR032W) alias *GSC2*, the homologue of *FKS1* encodes a 217-kDa also integral membrane protein (Fks2p) which at the amino acid level is 88% identical to Fks1p. The topological similarity of these proteins to many transporters suggests a possible role in transport of the growing glucan polymer across the membrane. Hydropathy profiles of Fks1p and Fks2p suggest that the genes encode integral membrane proteins which can be assumed to have approximately 16 transmembrane domains. The association of the *FKS* gene products with the catalytic activity suggests that these genes may encode catalytic subunits. The isolation of a neutralizing monoclonal antibody provides additional evidence that these subunits may be catalytic. However, the possibility of other glucan synthase subunits being present in the membrane fraction or the solubilized enzyme cannot be ruled out (Mazur et al., 1995). Disruption of each gene was not lethal, but disruption of both genes was lethal (Inoue et al., 1995).

An additional protein is necessary for activation of the 1,3-β-D-glucan synthase (Qadota et al., 1996). The gene *RHO1* (ORF: YPR165W) encodes a 22-kDa GTPase (Rho1p). The deduced amino acid sequence predicts that *C. albicans* Rho1p is 82.9% identical to *Saccharomyces* Rho1p and contains all the domains conserved among Rho-type GTPases from other organisms (Kondoh et al., 1997). *C. albicans* Rho1p was shown to interact directly with *C. albicans* Gsc1p in a ligand overlay assay and a cross-linking study. These results indicate that *C. albicans* Rho1p acts in the same manner as *S. cerevisiae* Rho1p to regulate as subunit the β-1,3-glucan synthesis (Arellano et al., 1996; Mazur and Baginsky, 1996).

8
Biodegradation

Schizophyllan is degradable by β-glucanases, which are secreted by *S. commune* when glucose is consumed as the carbon source. 1,3-β-glucanases are widely distributed among filamentous fungi (Fèvre, 1979; Prokop, 1990; Bielecki and Galas, 1991; Sutherland, 1999), and biodegradation of schizophyllan in soil, for example after use in enhanced oil recovery, is therefore possible.

Lo et al. (1990) described the kinetics and specificities of two closely related β-glucosidases secreted by *S. commune* that had similar molecular weights (102 and 96 kDa) and which were competitively inhibited by their glucose product. Both enzymes had binding sites for three glucose residues.

Chiu and Tzean (1995) investigated glucanolytic enzyme production by *S. commune* Fr. during mycoparasitism, and showed that the extracellular endo-1,3(4)-β-glucanase attacked 16 out of 50 fungi, representing oomycetes, zygomycetes and hyphomycetes, and which are either saprophytic, soilborne or foliar plant pathogens.

The production of extracellular β-1,3-glucanase activity by a monokaryotic *S. commune* strain was monitored by Prokop et al. (1994). The results showed that the β-glucanase activity consisted of an endo-β-1,3-glucanase activity, in addition to a negligible amount of β-1,6-glucanase and β-glucosidase activities. Unlike the β-1,3-glucanase production of the dikaryotic parent strain *S. commune* ATCC 38548, the β-1,3-glucanase formation of the monokaryon was not regulated by catabolite repression. The latter enzyme was purified from the culture filtrate by lyophilization, anion exchange chromatography on Mono Q, and gel filtration on Sephacryl S-100. It appeared homogeneous on SDS–PAGE, with a molecular mass of 35.5 kDa and an isoelectric point of

3.95. The enzyme was only active toward glucans containing β-1,3-linkages, including lichenan, a β-1,3/β-1,4-D-glucan. It attacked laminarin in an endo-like manner to form laminaribiose, laminaritriose, and high oligosaccharides. Whilst the extracellular β-glucanases from the dikaryotic *S. commune* ATCC 38548 degraded significant amounts of schizophyllan, the endo-β-1,3-glucanase from the monokaryon showed greatly reduced activity toward this high molecular mass β-1,3-/β-1,6-glucan, the K_m (using laminarin as substrate) being 0.28 mg mL^{-1}. The optimal pH was 5.5, and optimal temperature 50 °C; the enzyme was stable between pH 5.5 and 7.0.

9 Production

A wide range of carbon sources sustain both growth and schizophyllan formation. Simple mono- or disaccharides, such as glucose, xylose, mannose, sucrose, maltose and galactose, containing lactose, are utilized as well as polysaccharides such as starch or xylan (Steiner et al., 1993; Rau, 1997). In the following examples of cultivation only glucose was used as the carbon source, however.

9.1 Batch Cultivation

The production of schizophyllan is strongly coupled with growth and, as known for primary metabolites, secretion under nitrogen starvation reduces to zero when the stationary phase is reached. The medium for cultivation of *S. commune* is simple and requires only four components: glucose 30 g L^{-1}, yeast extract 3 g L^{-1}, KH_2PO_4 1 g L^{-1} and $MgSO_4.7H_2O$ 0.5 g L^{-1}.

If a filamentously growing fungus is used to produce highly viscous pseudoplastic polysaccharide suspensions, not only short mixing times and high mass transfer but also shear stress (which depends on the type of impeller used and stirrer speed influencing physiology of the fungus) must be considered. In other words, in proportion to the agitator used, the impeller must present a compromise between micro- and macromixing and schizophyllan release from the cell wall on the one hand, and low shear stress for the fungus and glucan on the other hand. Furthermore, the impeller should enable easy construction for a subsequent scale-up. Therefore, various types of impellers (Intermig™, Fundaspin™, helicon-ribbon, Rushton-turbine, fan) were tested in relation to their power requirement as well as yield, productivity and quality (expressed as specific shear viscosity) of the formed schizophyllan (Rau et al., 1992b; Gura and Rau, 1993). As a result of these investigations, the four-bladed fan impeller with a width ratio (impeller to vessel diameter) of 0.64 and a blade pitch of 45 ° gave the best product in relation to the highest specific shear viscosity (193 mPa s; shear rate 0.3 s^{-1}, 25 °C, 0.3 g L^{-1} aqueous schizophyllan solution) and highest yield at low formation of biomass (Figure 9).

The variation of intrinsic viscosity and molecular weight of schizophyllan during a cultivation run is shown in Figure 10.

The increase of these molecular data may be explained by premature release of mainly shorter schizophyllan molecules from the cell wall due to a low solution viscosity at the beginning of cultivation. The decrease at the end of cultivation is solely related to degradation by shear stress caused by the impeller and/or attack by released β-glucanases. Corresponding with intrinsic and shear viscosity, the flow behavior index shows a minimum of 0.18, i.e., a maximum of pseudoplasticity (data not shown).

As can be seen from Figure 9, most schizophyllan is produced during the oxy-

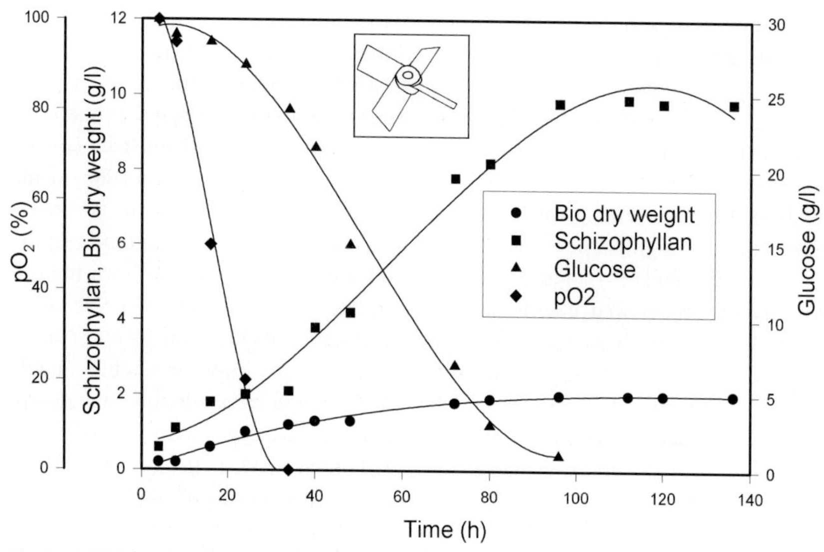

Fig. 9 A 30-L batch cultivation of *S. commune* equipped with three fan impellers at 100 r.p.m., 27 °C, an initial pH of 5.3 and an aeration rate of 150 L h^{-1}. pO_2, oxygen partial pressure of the liquid phase.

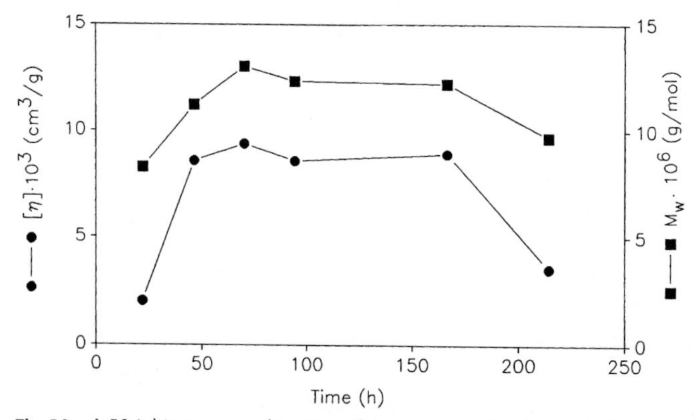

Fig. 10 A 50-L bioreactor cultivation of *S. commune*. Cell-free samples were used for the determination of molecular weight M_w (low angle light scattering) and intrinsic viscosity [η] (Rau et al., 1990).

gen-limited phase between 30 and 100 h. During this time interval, the partial pressure of oxygen in the liquid phase is almost zero and the additional formation of ethanol occurred (Figure 11). *S. commune* always produces ethanol as a result of a redundant fermentative pathway. Partly existing anaerobic domains inside the mycelia can never be fully avoided, as seen by the fact that ethanol is always formed in spite of a pO_2 > 5% throughout the whole cultivation run. However, the pO_2 is not a representative quantity in this connection as it reflects only one local point of the liquid phase and not the situation inside the micelles. Otherwise, local excess of oxygen is primarily used for biomass formation. This characteristic indicates that the fungus requires an optimum oxygen supply for a maximum of schizophyllan secretion. This fact is proved by the determination of an optimum constant specific oxygen uptake rate which depends

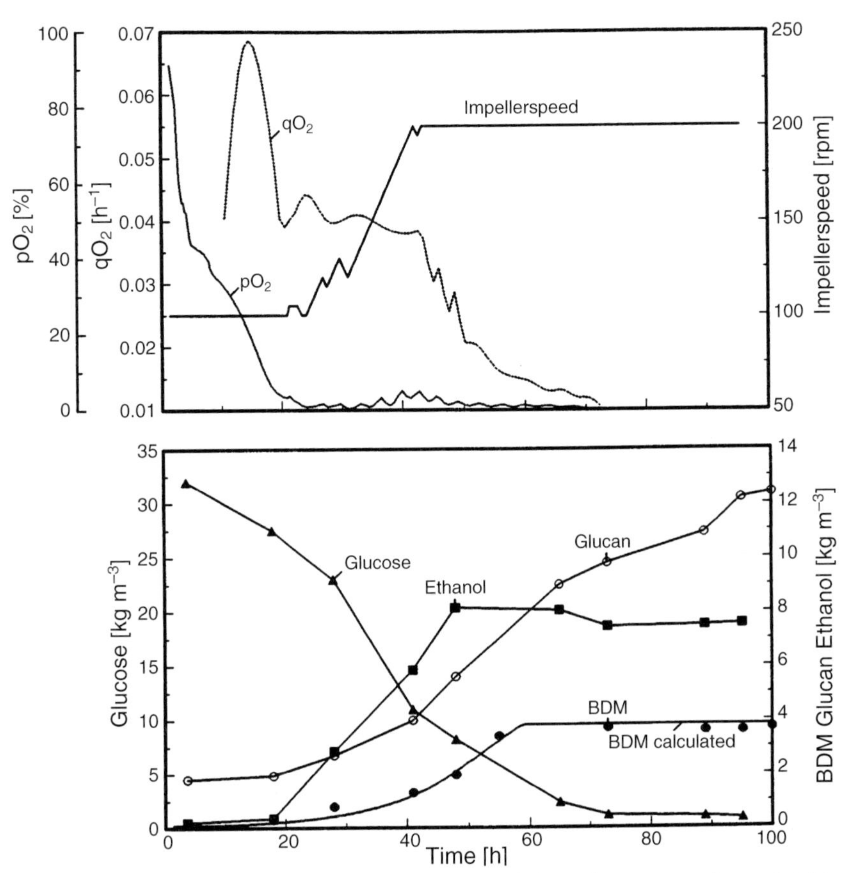

Fig. 11 A 30-L-batch cultivation of *S. commune* with controlled specific oxygen uptake rate (qO_2) by variation of impeller speed (fan impeller). For further cultivation conditions, see Figure 9.

on the process conditions (Rau et al., 1992d). Application of a constant specific oxygen uptake rate (oxygen consumed per time [$g\ L^{-1} h^{-1}$] divided by the actual biomass [$g\ L^{-1}$]) by controlling the stirrer speed resulted in an increased yield of schizophyllan (13 $g\ L^{-1}$) at equal biomass formed (see Figure 11). The on-line modeling of biomass for the calculation of the specific oxygen uptake rate was achieved by using the following equation (Rau and Brandt 1994):

$$\frac{dC_X}{dt} = k_L a \left(C^*_{O_2,L} - C_{O_2,L} \right) \cdot Y_{X/O_2} \cdot \frac{C_S}{K_S + C_S} \quad (1)$$

A further important result is the decreased amount of ethanol (from 12 to 7 $g\ L^{-1}$) that was channeled primarily into schizophyllan as a result of the improved oxygen supply (Rau and Brandt, 1994).

9.2
Continuous Cultivation

S. commune underlies a glucose-induced repression of the formation of β-glucanases (Rapp, 1992; Prokop et al., 1994), and therefore cultivations were terminated when glucose consumption was complete. Prolonged cultivation under carbon-limited conditions led to the release of β-glucan-degrading enzymes which caused a slight increase in glucose concentration, accompanied by a decrease in schizophyllan concentration as well as a sharp drop in the specific viscosity (mPa g^{-1}). Furthermore, Figure 11 shows that an increase of schizophyllan formation is possible by using a controlled oxygen supply. Independent of the process mode (batch or continuous), an optimum specific oxygen uptake rate exists for maximum yield and productivity of schizophyllan as well as minimum biomass formation because the biomass must be separated in further downstream processing. This characteristic was proven by variation of the oxygen supply during continuous cultivations in an oxygen-limited chemostat (Rau, 1997) (Table 1).

The addition of oxygen to the air feed enhanced growth and the associated schizophyllan formation until their maxima were reached. The specific oxygen uptake rate also increased with biomass, but specific schizophyllan productivity decreased. In spite of increased oxygen supply, all cultivations were oxygen-limited and therefore the various specific oxygen uptake rates reflected different states of oxygen limitation. Related to primary mycelial (homogeneous) growth of the fungus and using the same bioreactor set-up described in Figures 9 and 11, then the optimum specific oxygen uptake rate ranged between 0.04 and 0.06 h^{-1}. However, if another bioreactor or process configuration is used this specific value can change drastically and must be determined individually.

Compared with batch cultivation, the continuous mode revealed an increase in productivity (Rau et al., 1992d). The schematic set-up of the continuous process is shown in Figure 12.

In order to achieve a further increase of productivity combined with facilitated downstream processing, biomass feedback was used. A cross-flow filtration unit comprising a stainless steel membrane (200 μm) was employed for separation of biomass from the viscous culture suspension. For schizophyllan, a maximum productivity of 40 g L^{-1} per day was achieved at a feedback rate (permeate flow/medium input flow) of 0.92 and dilution rate of 0.2 h^{-1} (maximum specific growth rate 0.12 h^{-1}). Optimized process and filtration conditions resulted in a near cell-free and undiluted β-glucan solution at the outlet of the bioreactor (Figure 13).

10 Downstream Processing

An economic design for the downstream process of highly viscous and mycelia-con-

Tab. 1 Influence of oxygen supply on growth and formation of schizophyllan by addition of N_2 or O_2 to the air-feed during oxygen-limited continuous cultivations of *S. commune*. The 30-L bioreactor was equipped with three fan impellers (150 r.p.m.). All concentrations are related to stationary conditions. Dilution rate 0.04 h^{-1}, constant aeration rate 144 L h^{-1}. Note: all cultivations showed a $pO_2 \approx 0\%$.

$\%N_2$	40	10	–	–	–	–
%Air	60	90	100	90	70	30
$\%O_2$	–	–	–	10	30	70
Biomass [g L^{-1}]	0.65	0.9	1	1.2	1.7	2.4
Schizophyllan [g L^{-1}]	3.3	3.5	3.6	4.1	4.6	3.7
Productivity [g L^{-1} day^{-1}]	3.2	3.4	3.5	3.9	4.4	3.6
Specific productivity [g g^{-1} day^{-1}]	4.9	3.8	3.5	3.2	2.6	1.5
qO_2 [h^{-1}]	0.041	0.047	0.049	0.056	0.061	0.071

Specific producitivity: g schizophyllan formed per g biomass per day.qO_2: specific oxygen uptake rate.

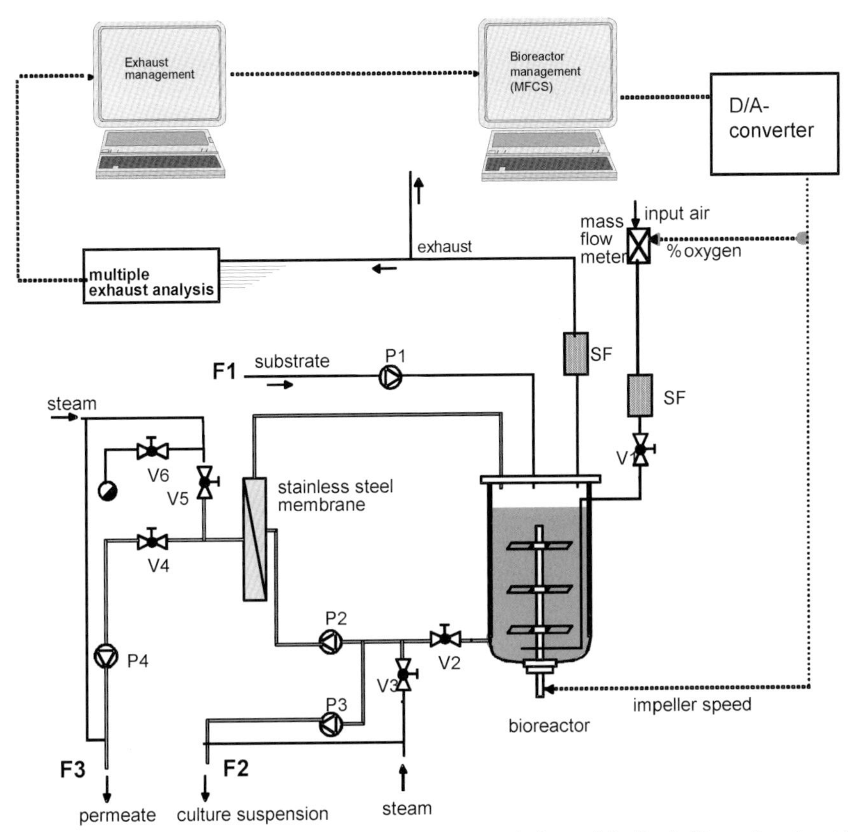

Fig. 12 Set-up of the continuous schizophyllan process including cell feedback. F1, medium input flow; F2, suspension output flow; F3, permeate; F1 = F2 + F3; feedback rate R = F3/F1; SF, sterile filter; V, valve; p, pump; MFCS, micro fermenter control system (B. Braun Biotech International; now Sartorius AG, Göttingen, Germany.)

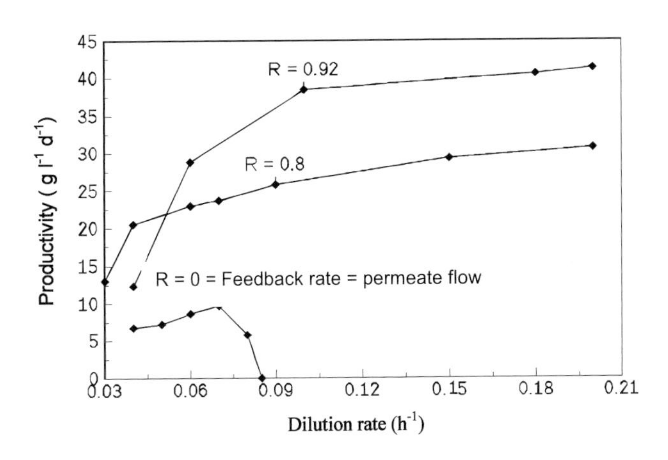

Fig. 13 Influence of feedback rate on the productivity of schizophyllan during 30-L oxygen-limited continuous cultivations with *S. commune*. Substrates ($g\ L^{-1}$) input feed: glucose 30; yeast extract 0.8; KH_2PO_4 0.2; $MgSO_4$ 0.1. For further cultivation conditions, see Figure 9.

taining culture suspensions is a challenge for the bioengineer. A three-stage, dead-end filtration (Jahn-Held et al., 1990) works in principle for cell separation, but this procedure is not recommended for scale-up. Therefore, centrifugation or high-speed cross-flow microfiltration were chosen which were also applicable for higher volumes. Again, microfiltration showed the best performance for purification and concentration of cell-free schizophyllan solutions.

10.1 Cell Separation

The suspensions harvested from batch cultivations contain cells that must be separated by either centrifugation or microfiltration. The best results with centrifugation are obtained when the diluted (≤ 1 g L^{-1} schizophyllan, ≤ 0.2 g L^{-1} biomass) and homogenized suspension is fed to a solid ejecting disc separator (e.g., CSA-1, 5700 r.p.m.; Westfalia, Oelde, Germany). The resulting supernatant contains only small amounts of hyphal fragments (concentration < 0.1 g L^{-1}), and this can easily be separated by dead-end filtration using glass-fiber filters.

A more effective alternative method for cell separation is cross-flow microfiltration. An undiluted suspension can be used if a sintered stainless steel membrane (10 µm; Krebsoege, Radevormwald, Germany) is used at high tangential feed velocity. Cell-free schizophyllan solutions without fragments, but with the same concentration as at the end of batch cultivation, are obtained as the permeate (Haarstrick et al., 1991; Haarstrick, 1992; Rau, 1997, 1999).

10.2 Purification and Concentration of Schizophyllan

The cell-free schizophyllan solution obtained either by high-speed microfiltration or by continuous cultivation with integrated biomass feedback must be purified (diafiltration mode) or eventually further concentrated (concentration mode) by using a cross-flow microfiltration technique (Figure 14).

Parallel investigations using different cross-flow systems (Haarstrick et al., 1991) led to the recommended use of low-shear PROSTAK™ (Millipore Corp., USA) flat membrane modules (0.1 µm). The best results related to a high permeation rate were achieved when the tangential feed velocity was at its individual maximum, avoiding a transmembrane pressure > 0.8 bar. Purification of the solution was attained by using the diafiltration mode when all schizophyllan molecules were fully rejected and low molecular-weight compounds (< 0.1 µm) such as proteins, glucose and salts, permeate the membrane. The permeate flow corresponds to the input solvent (water) flow, so that the volume of the retentate remains constant. The concentration mode was started by cutting the input solvent flow. During this process the negative influence of fouling at the membrane surface was increased, with the consequence of a continual decrease in permeation rate (Figure 15). This results in a highly viscous, colorless and transparent solution. Drying or lyophilization of the product solution must be avoided because only 50% (w/w) of the dried schizophyllan can be redissolved in water. Dimethylsulfoxide (DMSO) can be used to dissolve the dried schizophyllan totally, but this solvent degrades the triple helix to single coiled chains, with a drastic reduction in viscosity.

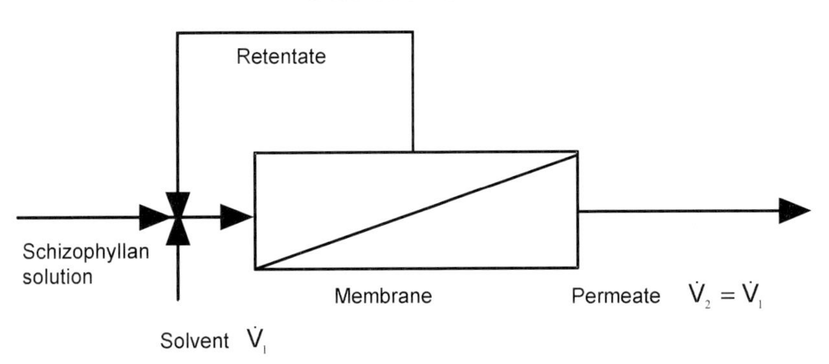

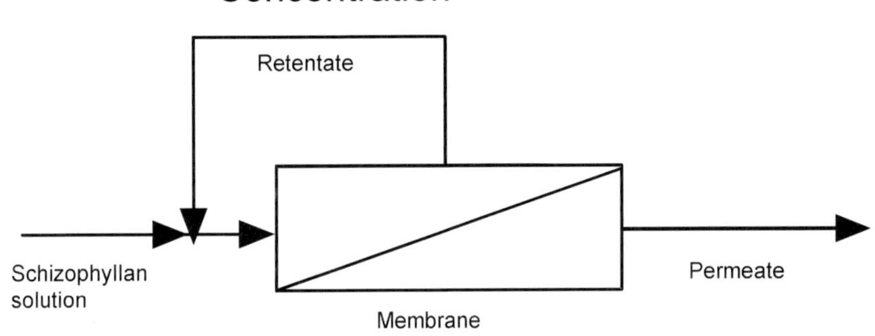

Fig. 14 Scheme of cross-flow microfiltration for the purification (diafiltration) and concentration of cell-free schizophyllan solutions. $\dot{V}$ = volume feed, L h^{-1}.

11
World Market

Schizophyllan is not the only trivial name for the exopolysaccharide of *S. commune*; elsewhere (primarily in Asiatic countries) it is also known as sizofilan and sizofiran. To the best of the author's knowledge, only two Japanese companies currently manufacture schizophyllan: Kaken Pharmaceutical Co., Ltd., Tokyo (Anonymous, 2001a); and Taito Co., Ltd., Kobe (Anonymous, 2001b).

Kaken offers schizophyllan/sizofiran under the brand name Sonifilan™ as an antimalignant tumor agent. Each ampoule (2 mL) contains 20 mg sizofiran with a molecular weight of approx. 450,000 daltons that should be injected intramuscularly in one to two divided doses each week. Its use is indicated for the enhancement of the direct effect of radiotherapy in the treatment of uterine cervical carcinoma. In 1998, Sonifilan was licensed to Kwang Dong in Korea. Information concerning production facilities and prices are not currently available.

12
Properties

In aqueous solution, schizophyllan is arranged as a triple helix with protruding

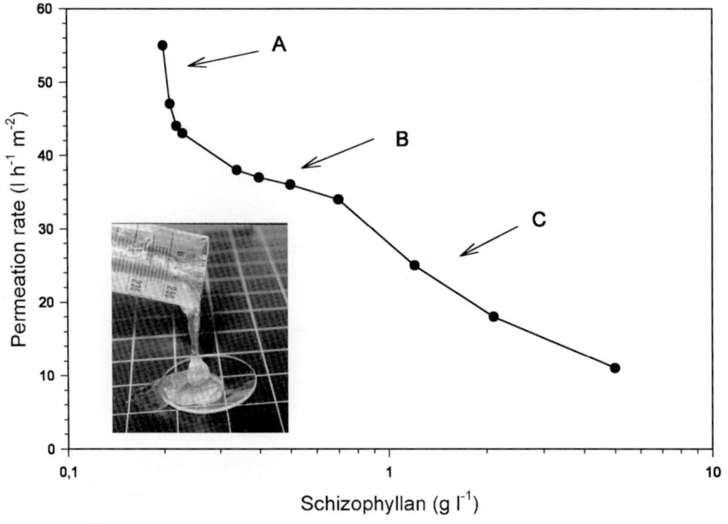

Fig. 15 Decrease of the permeation rate during the concentration run of a cell-free schizophyllan solution applying cross-flow microfiltration. Pore size 0.1 μm, transmembrane pressure 0.8 bar, tangential feed velocity 6.6 m s^{-1} (Rau, 1999). A = adsorption with increasing concentration polarization; B = steady-state of mass transport from and to membrane; C = gel-layer formation (fouling).

pendent β-1,6-linked D-glucose units originating from the outside of the triplex. In DMSO, at temperatures >135 °C and at a pH >12, the triple helix melts to single coiled strains, equivalent to reducing the average molecular weight by one-third (Norisuye et al., 1980). Thermal degradation is enhanced by the presence of oxygen (Zentz et al., 1992). Aqueous solutions show thixotropic, pseudoplastic (Figure 16) and viscoelastic behavior (Oertel and Kulicke, 1991). Native suspensions, additionally containing the producing fungus, reveal enhanced non-Newtonian characteristics due to the filamentous network of the internal woven hyphae.

The viscosity is decreased with increasing shear rate. Due to this flow behavior an individual viscosity is strongly connected to a single shear rate and characterizes the quality of schizophyllan because when used as a viscosifier, a high viscosity at low concentration is required. Furthermore, the shear viscosity depends on the concentration of the schizophyllan. For example, Figure 16 shows the flow behavior of a solution with 5 g L^{-1} schizophyllan. The comparison of different solutions >5 g L^{-1} yielded viscosities >10 Pa s at low shear rate (0.3 s^{-1}). The mean value of a 0.3 g L^{-1} solution (shear rate 0.3 s^{-1}) varies between 50 and 150 mPa s, depending on the quality of the schizophyllan.

Shear and intrinsic viscosity is slightly increased by the addition of NaCl or Mg^{2+} and Ca^{2+}ions. This is an unusual behavior, because synthetic polymers such as polyacrylamide show a slight decrease in intrinsic viscosity when salt concentration is increased, this being due to reduced solvent quality (coil contraction). An explanation for the increase in schizophyllan volume could be enhanced energetic interactions inside the triple helix in a poorer solvent. Additional intramolecular forces increase the stiffness of the triple helix and induce an expansion of the macromolecule (Rau et al.,

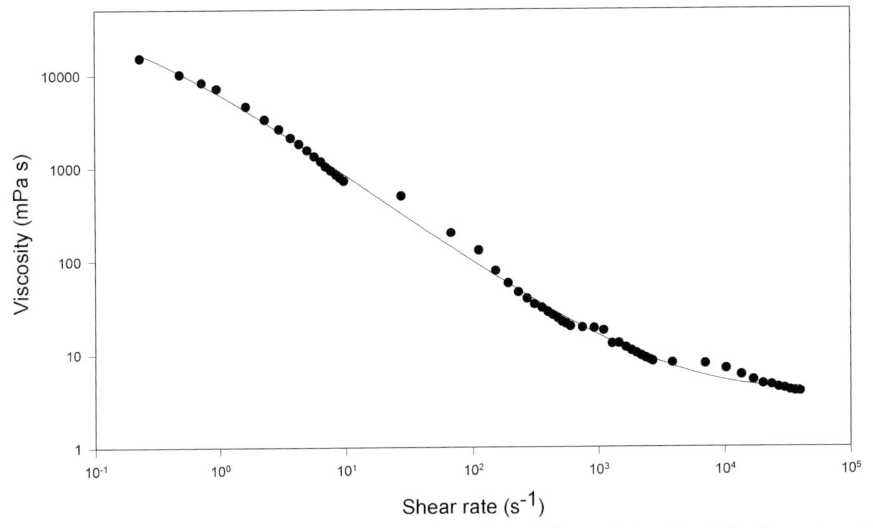

Fig. 16 Pseudoplastic flow behavior of an aqueous schizophyllan solution (5 g L^{-1}). Shear viscosity was measured by a rotary viscometer (Haake, Karlsruhe, Germany) at 25 °C and at different constant shear rates until a constant shear stress resulted (Rau, 1999).

1990). Although schizophyllan undergoes no gelation by itself in aqueous solution, the addition of sorbitol results in the formation of a transparent gel at lower temperatures. Based on the results of small-angle X-ray scattering, sorbitol is found to disentangle a part of triple-stranded helices and bridge the disentangled parts, which serve as a cross-linking domain (Maeda et al., 1999). Schizophyllan also forms gels in the presence of borate ions (Rau et al., 1992c; Grisel and Muller, 1997). The gelation occurs by both hydrogen bonding and chelation of borate ions through the hydroxyl groups of the biopolymer as far as they are in favorable position, e.g. in the side chain.

Sakurai and Shinkai (2000) found that a mixture of schizophyllan and poly(ethylene oxide) in aqueous solution underwent phase separation at around 3–4 °C, and this temperature was independent of both polymer concentration and the difference in poly(ethylene oxide) molecular weight (6000 and 70,000).

Oertel and Kulicke (1991) described the formation of aqueous, lyotropic phases of ultrasonically degraded (MW 335,000 g mol^{-1}) schizophyllan investigated by stationary shear flow, as well as with the aid of polarization microscopy. In oscillatory measurements, schizophyllan exhibited a maximum for the storage modulus. The ORD of liquid crystal solutions (MW 400,000 g mol^{-1}) was also investigated by Van et al. (1984). An abrupt change in ORD behavior occurred when an isotropic solution was cooled to a temperature close to the isotropic–biphasic boundary temperature, indicating the occurrence of a pretransition from isotropic to cholesteric phases. Yanaki (1982) claimed the cholesteric liquid crystal formation of schizophyllan. Furthermore, the rotational dynamics of schizophyllan was determined by transient electric birefringence (Fuglestad et al., 1996).

Schizophyllan possesses no thermoplastic characteristics; it has no specific melting point, but it decomposes at temperatures

>180 °C. The moisture content of the dried product was ~10% (w/w), this being bound as water of crystallization (Schulz, 1992), while the density was 1429 kg m^{-3} (Creszenzi and Gamani, 1988).

Acetylation of the polymer yielding schizophyllan acetate can be carried out under strong acidic catalysis; the resultant product is only slightly soluble in water or DMSO (Albrecht and Rau, 1994).

Muenzberg et al. (1995) found that hydrolysis of aqueous schizophyllan solutions was possible by incubation in DURAN™ borosilicate glass 3.3 (DIN ISO 3585) at 121 °C and 1 bar. A slight decrease in pH, a rapid loss of viscosity, and a constant increase in reducing end-groups indicated that the degradation of schizophyllan proceeded regioselectively by cleaving only the main chain, although normally the protruding 1,6-linkages would be expected to be less stable against hydrolytic influences. Maintenance of the side chains was additionally verified by ^{13}C-NMR spectra. Stepwise ultrafiltration of degraded solutions yielded fractions with varying molar masses, with the mass ratio of fractions depending on the total incubation time. The regioselectivity of this degradation method was explained by a pore theory.

13 Applications and Patents

Claims exist for the production of schizophyllan using different strains of *S. commune*, and for its application as an additive for polymer flooding in the field of enhanced oil recovery (Lindoerfer et al., 1988, 1991). Due to the viscosity-related stability of schizophyllan against high shear rates, increased temperatures and high salinities, this material is very useful in this type of application (Rau et al., 1992c).

Aqueous schizophyllan solutions were shown effectively to reduce drag in pipe flow, the drag reduction depending on schizophyllan concentration and being minimal at 0.5 g L^{-1}. The transition from laminar to turbulent flow was delayed at a higher Reynold's (*Re*) value (Haarstrick and Rau, 1993).

Schulz et al. (1992) described the preparation of films with native schizophyllan and with a polyalcohol derived from schizophyllan by chemical treatment. The films could only be prepared by casting from aqueous solutions because the polymers were not thermoplastic. They possess a low permeability to oxygen (schizophyllan: <2 mL m^{-2} day^{-1} bar^{-1}), but present a high permeability to water vapor and hence can be used to protect foods against oxygen-mediated spoilage. The tensile strength of the films was 45–58 N mm^{-2} for schizophyllan, and 12–18 N mm^{-2} for the polyalcohol.

Besides schizophyllan, *S. commune* also secretes a 24-kDa hydrophobin. Hydrophobins were initially discovered in the mid-1980s and represent a unique family of small, amphipathic proteins (about 100 amino acids) that play an important role in forming stable coatings on various surfaces. Enzymatic digestion of the hydrophobin eliminates the ability of the remaining schizophyllan to assemble as a stable entity on a hydrophobic surface. By using water contact angle measurements and atomic force microscopy, Martin et al. (1999) showed that schizophyllan and hydrophobin form a synergistic complex that allows facile surface modification of both hydrophilic and hydrophobic surfaces.

The tetrasaccharide subunit (see Figure 1) can be prepared by chemical degradation of schizophyllan, and is useful as oligomerization building block (Kunz et al., 1991). Schizophyllan, in combination with colloidal silica particles and water-soluble salts, can be used for the fine polishing of wafers (Sasaki, 1990) and as a membranous material in terms of a polysaccharide associate after first dissolving it in DMSO (random coil) with subsequent addition of water or methanol to return the polysaccharide to a state capable of assuming the helical structure (Yanaki, 1981). The production of a monoclonal antibody that reacts specifically with a β-1,3-glycosidic bond and could be used to determine small amounts of schizophyllan has been described (Hirata, 1992; Hirata et al., 1993). A polyclonal antibody was also used in a sandwich-type enzyme immunoassay (Adachi et al., 1999) for quantifying schizophyllan, as well as for estimating its ultrastructure (triple/single helix).

Due to its shear-thinning characteristics combined with a high specific viscosity, schizophyllan can be generally used as viscosifier, e.g., in cremes and lotions. Within the scope of special cosmetic applications, schizophyllan is also useful as a skin anti-aging, depigmenting and healing agent – generally termed 'skin care' (Kim et al., 1999). Schizophyllan is an active ingredient that can increase skin cell proliferation, collagen biosynthesis and also aid in recovery after sunburn. In addition, schizophyllan effectively reduces skin irritation and is therefore suitable for cosmetic and dermatological applications (Kim et al., 1999, 2000).

To date, a vast amount of data has been generated in relation to the use of schizophyllan as a pharmaceutical compound (for reviews, see Bohn and BeMiller, 1995; Kraus, 1990). Schizophyllan acts as biological response modifier and has been known to be a nonspecific stimulator of the immune system, generally as a result of macrophage activation. In 1991, Czop and Kay identified a macrophage cell surface receptor that is specific for a small oligosaccharide with β-1,3-D-glucose linkages. It is generally known that a glucan with β-1,3-linkages has greater macrophage stimulatory activity than any other linkage type. The receptor binding in a human monocyte-like cell line was confirmed specifically for schizophyllan by Mueller et al. (2000).

Schizophyllan has been used as an immunotherapeutic agent for cancer treatment in Japan since 1986. It is used in conjunction with chemotherapy or radiotherapy. The additional application of schizophyllan to radiation increased both macrophage and T-lymphocyte infiltration in local lung tumor cells in mice (Inomata et al., 1996). An overdose of schizophyllan must be avoided however as this reduces the antitumor activity (Miura et al., 2000). Clinical studies have shown that the administration of schizophyllan, along with antineoplastic drugs, prolongs the lives of patients with lung, gastric or uterine cancers (Furue, 1987; Yoshio et al., 1992; Kimura et al., 1994). Schizophyllan has no antigenicity or mitogenic effect on the T cell; its antitumor activity is exerted only through the activation of macrophages, which subsequently augment the T-cell cascade (Kidd, 2000; Ooi and Lui, 2000). However, an evaluation of schizophyllan in patients with advanced head and neck squamous cell carcinoma did not reveal any significant improvements in immunological parameters (Mantovani et al., 1997).

With regard to its immune-stimulating activities, schizophyllan was claimed as anti-(AIDS) virus agent (Shigero et al., 1989; Hagiwara and Kikuchi, 1992), as prevention against fish diseases (Yano, 1990), and as an immune effect enhancer for vaccines (Honma, 1994). An ultrasonicated schizophyllan, neoschizophyllan, with novel pharmacological activity was described by Kikumoto et al. (1978).

It has been shown that the molar mass, the degree of branching, conformation and chemical modification significantly affect the pharmaceutical activity of schizophyllan.

However, it is difficult to establish a uniform structure–functional activity relationship because reported results have differed widely. Structural features such as β-1,3-linkages in the main chain and β-1,6-branch points (branching degree 0.2–0.33) are needed because loss of the side-chain glucose reduced both antitumor action (Kishida, 1992; Suda et al., 1994) and water solubility.

Ooi and Lui (2000) stated that glucans with high molecular weight appear to be more effective than those with low molecular weight. They regarded glucans with a molecular mass of 800,000 g mol^{-1} to be maximally effective, though very small schizophyllan molecules also showed some efficacy. Investigations with native schizophyllan ($6-10\times10^6$ g mol^{-1}) showed that it is likely to be effective against attack by tobacco mosaic virus (Michiko, 1989; Stuebler and Buchenauer, 1996; Rau, 1997) in tobacco leaves (*Nicotiana tabacum*) as a regioselectively degraded (Muenzberg et al., 1995) and purified molar mass fraction (1000–5000 g mol^{-1}) of schizophyllan.

The conformation of schizophyllan also appears to influence the immunological activity, though this point is controversial. Native schizophyllan is arranged as triple helix, but molecules of molecular mass $<50,000$ g mol^{-1} (Kojima et al., 1984) or schizophyllan treated with NaOH show single-helix conformation. Kulicke et al. (1997) presented results, based on a Congo red assay, that helical structures are not essential – nor even advantageous – for immunological activity. Saito et al. (1991) also found that single-helix conformation stimulates antitumor activity. In contrast, Ooi and Lui (2000) stated that single-helix schizophyllan showed a reduced ability to inhibit tumor growth as compared with the native material. The truth will lie somewhere between these situations because, depending on the immunological assay used, either the triple- or single-helical conformer showed increased activities (Miura et al., 1995; Ohno et al., 1995; Tsuzuki et al., 1999).

Table 2 provides a summary of patents relating to the numerous applications of branched β-glucans.

14 Outlook and Perspectives

Using batch or continuous cultivations, schizophyllan can be produced from glucose in high yields (13 g L^{-1}) and high productivities (40 g L^{-1} per day). The downstream processing has been improved and can easily be carried out using high-speed, cross-flow filtration techniques without producing solvent waste that is harmful to the environment. In relation to these data, and depending on the scale of production, an economically realistic selling price for schizophyllan of ≤ 50 DM per kg should be possible.

In addition, depending on the political and economical situations, a prospective intensified use in enhanced oil recovery is conceivable. Should the application of schizophyllan as a bulk product prove unattractive however, its pharmaceutical properties might provide sufficient stimulus for its production, and hopefully industries outside of Japan will add schizophyllan to their product lists.

Tab. 2 Summary of patents related to schizophyllan, in order of date of publication

No. of patent	*Patent holder*	*Inventors*	*Title of patent*	*Date of publication*
US 4098661	Taito Co., Ltd. (JP)	Kikumoto et al.	Method of producing neoschizophyllan having novel pharmalogical activity	1978-07-04
JP 56127603	Taito Co., Ltd. (JP)	Yanaki	Preparation of polysaccharide associate	1981-10-06
JP 57147576	Taito Co., Ltd. (JP)	Yanaki	Novel liquid crystal of polysaccharide	1982-09-11
EP 0271907	Wintershall AG (GER)	Lindoerfer et al.	Homopolysaccharides with a high molecular weight, process for their extracellular preparation and their use, as well as the corresponding fungus strains	1988-06-22
JP 1272509	JAPAN TOBACCO INC.	Michiko	Production of controlling agent of plant virus by microorganism	1989-10-31
JP 1287031	Taito Co., Ltd. (JP)	Shigero et al.	Anti-AIDS virus agent	1989-11-23
US 4921615	Wintershall AG (GER)	Jahn-Held et al.	Separation of solid particles of various sizes from viscous liquids	1990-05-01
EP 0373501	Mitsubishi Monsanto Chem (JP)	Sasaki	Fine polishing composition for wafers	1990-06-20
JP 2218615	Taito Co., Ltd. (JP)	Yano	Preventive for fish desease comprising water-soluble glucan	1990-08-31
DE 4012 238	Wintershall AG (GER)	Lindoerfer et al.	Verfahren zur Erhöhung der volumenbezogenen Produktivität (g/ld) nichtionischer Biopolymere	1991-01-03
EP 0416396	Merck Patent GmbH (GER)	Kunz et al.	Tetrasaccharide and process for their preparation	1991-03-13
JP 4054124	Taito Co., Ltd. (JP)	Hagiwara et al.	Anti-virus agent	1992-01-02
JP 4346791	Taito Co., Ltd. (JP)	Hirata	Monoclonal antibody	1992-12-02
JP 6172217	Taito Co., Ltd. (JP)	Honma	Immune effect enhancer for vaccine	1994-06-21
JP 11313667	Pacific Co., Ltd. (KR)	Kim et al.	Production of beta-1,6-branched beta-1,3-glucan useful as skin anti-aging, depigmenting and healing agent	1999-10-07

15
References

Adachi, Y., Miura, N. N., Ohno, N., Tamura, H., Tanaka, S., Yadomae, T. (1999) Enzyme immunoassay system for estimating the ultrastructure of (1,6)-branched (1,3)-β-glucans, *Carbohydr. Polym.* **39**, 225–229.

Albrecht, A., Rau, U. (1994) Acetylation of β-1,6-branched β-1,3-glucan yielding schizophyllan-acetate, *Carbohydr. Polym.* **24**, 193–197.

Andaluz, E., Guillen, A., Larriba, G. (1986) Preliminary evidence for a glucan acceptor in the yeast *Candida albicans, Biochem. J.* **240**, 495–502.

Andaluz, E., Ridruejo, J. C., Ramirez, M., Ruiz-Herrera, J., Larriba, G. (1988) Initiation of glucan synthesis in yeast, *FEMS Microbiol. Lett.* **49**, 251–255.

Anonymous (2001a).

Anonymous (2001b).

Arellano, M., Duran, A., Perez, P. (1996) Rho1 GTPase activates 1–3-β-D-glucan synthase and is involved in *Schizosaccharomyces pombe* morphogenesis, *EMBO J.* **15**, 4584–4591.

Asakawa, T., Van, K., Teramoto, A. (1984) A thermal transition in a cholesteric liquid crystal of aqueous schizophyllan, *Mol. Cryst. Liq. Cryst.* **116**, 129–139.

Bartnicki-Garcia, S. (1999) Glucans, walls, and morphogenesis: on the contributions of J. G. H. Wessels to the golden decades of fungal physiology and beyond. *Fungal Genet. Biol.* **27**, 119–127.

Bartnicki-Garcia, S., Lippman, E. (1972) The bursting tendency of hyphal tips of fungi: presumptive evidence for a delicate balance between wall synthesis and wall lysis in apical growth, *J. Gen. Microbiol.* **73**, 487–500.

Bielecki, S., Galas, E. (1991) Microbial β-glucanases different from cellulases, *Crit. Rev. Biotechnol.* **10**, 275–304.

Bohn, J. A., BeMiller, J. A. (1995) 1,3-β-D-glucans as biological response modifiers: a review of structure-functional activity relationships, *Carbohydr. Polym.* **28**, 3–14.

Brandt, C. (1995). *O_2-geregelte β-Glucanproduktion mit Schizophyllum commune* ATCC 38548 *im Batch- und Chemostatbetrieb*, PhD Thesis, Technical University Braunschweig, GER.

Brandt, C., Schilling, B., Gura, E., Rau, U., Wagner, F. (1993) Definierte Sauerstoffversorgung im Batch und im O_2-limitierten Chemostaten mit Zellrückführung zur Produktion von β-1,3-Glucanen. *DECHEMA Biotechnology Conferences* **2**, 465–466.

Brown, J., Bussey, H. (1993) The yeast KRE9 gene encodes an O-glycoprotein involved in cell surface β-glucan assembly, *Mol. Cell. Biol.* **13**, 6346–6356.

Chiu, S. C., Tzean, S. S. (1995) Glucanolytic enzyme production by *Schizophyllum commune* Fr. during mycoparasitism, *Physiol. Mol. Plant Path.* **46**, 83–94.

Cooke, W. B. (1961). The genus *Schizophyllum, Mycologia* **53**, 575–599.

Cordes, K. (1990) Produktionsoptimierung und Charakterisierung der von *Monilinia fructigena* ATCC 24976 und ATCC 26106 gebildeten extrazellulären Glucane, PhD Thesis, Technical University of Braunschweig, GER.

Cordes, K., Rau, U., Wagner, F. (1989) Influence of processing parameters and downstream processing on viscometric data of aqueous glucan solutions, *DECHEMA Biotechnology Conference* **3**, 1067–1070.

Crescenzi, V., Gamini, A. (1988) On the solid state and solution conformations of a polycarboxylate derived from the polysaccharide scleroglucan, *Carbohydr. Polym.* **9**, 169–184.

Czop, J. K., Kay, J. (1991) Isolation and characterization of β-glucan receptors on human mononuclear phagocytes, *J. Exp. Med.***173**, 1511–1520.

Donk, M. A. (1964) A conspectus of the families of Aphyllophorales, *Persoonia* **3**, 199–324.

Douglas, C. M., Foor, F., Marrinan, J. A., Morin, N., Nielsen, J. B., Dahl, A. M., Mazur, P., Baginsky, W., Li, W., El-Sherbeini, M., Clemas, J. A., Mandala, S. M., Frommer, E. R., Kurtz, M. B. (1994) The *Saccharomyces cerevisiae FKS1* (*ETG1*) gene encodes an integral membrane protein which is a subunit of 1,3-β-D-glucan synthase, *Proc. Natl. Acad. Sci. USA* **91**, 12907–12911.

Essig, F. M. (1922) The morphology, development and economic aspects of *Schizophyllum commune* Fries, *Am. J. Bot.* **36**, 360–363.

Fèvre, M. (1979). Glucanases, glucan synthases and wall growth in *Saprolegnia monoica*, in: *Fungal Walls and Hyphal Growth* (Burnett, J. H., Trinci, A. P. J., Eds.), Cambridge: Cambridge University Press, 225–263.

Fuglestad, G. A., Mikkelsen, A., Elgsaeter, A., Stokke, B. T. (1996) Transient electric birefringence study of rod-like triple-helical polysaccharide schizophyllan, *Carbohydr. Polym.* **29**, 277–283.

Furue, H. (1987) Biological characteristics and clinical effect of schizophyllan, *Drugs Today* **23**, 335–346.

Gawronski, M., Conrad, H., Springer, T., Stahmann, K.-P. (1996) Conformational changes of the polysaccharide cinerean in different solvents from scattering methods, *Macromolecules* **24**, 7820–7825.

Girard, V., Fèvre, M. (1984) Distribution of (1-3)-β- and (1-4)-β-glucan synthases along the hyphae of *Saprolegnia monoica*, *J. Gen. Microbiol.* **130**, 1557–1562.

Grisel, M., Muller, G. (1997) The salt effect over the physical interactions occurring for schizophyllan in the presence of borate ions, *Macromol. Symp.* **114**, 127–132.

Gura, E., Rau U. (1993) Comparison of agitators for the production of branched β-1,3-D-glucans by *Schizophyllum commune*, *J. Biotechnol.* **27**, 193–201.

Haarstrick, A. (1992). Mechanische Trennverfahren zur Gewinnung zellfreier, hochviskoser Polysaccharidlösungen von *Schizophyllum commune ATCC 38548*, PhD Thesis, Technical University Braunschweig.

Haarstrick, A., Rau, U. (1993) Strömungscharakteristik pseudoplastischer Polysaccharidlösungen von *Schizophyllum commune* ATCC 38548, *Chem. Ing. Tech.* **65**, 556–559.

Haarstrick, A., Rau, U., Wagner, F. (1991) Crossflow filtration as a method of separating fungal cells and purifying the polysaccharide produced, *Bioprocess Eng.* **6**, 179–186.

Hagiwara, K., Kikuchi, M. (1992) Anti-virus agent, JP 4054124.

Haltrich, D., Steiner, W. (1994) Formation of xylanase by *Schizophyllum commune*: effect of medium components, *Enzyme Microb. Technol.* **16**, 229–235.

Hirata, A. (1992) Monoclonal antibody, JP 4346791.

Hirata, A., Itoh, W., Tabata, K., Kojima, T., Itoyama, S., Sugawara, I. (1993) Preparation and characterization of murine anti-schizophyllan monoclonal antibody, *Biosci. Biotech. Biochem.* **57**, 125–126.

Honma, M. (1994) Immune effect enhancer for vaccine, JP 6172217.

Horton, J. S., Palmer, G. E., Smith, W. J. (1999) Regulation of dikaryon-expressed genes by *FRT1* in the basidiomycete *Schizophyllum commune*, *Fungal Genet. Biol.* **26**, 33–47.

Inomata, T., Goodman, G. B., Fryer, C. J., Chaplin, D. J., Palcic, B., Lam, G. K., Nishioka, A., Ogawa, Y. (1996) Immune reaction induced by X-rays and ions and its stimulation by schizophyllan (SPG), *Br. J. Cancer Suppl.* **27**, 122–125.

Inoue, S. B., Takewaki, N., Takasuka, T., Mio, T., Adachi, M., Fujii, Y., Miyamoto, C., Arisawa, M., Furuichi, Y., Watanabe, T. (1995) Characterization and gene cloning of 1,3-beta-D-glucan synthase from *Saccharomyces cerevisiae*, *Eur. J. Biochem.* **231**, 845–854.

Jabri, E., Quigley, D. R., Alders, M., Hrmova, M., Taft, C. S., Phelps, P., Selitrennikoff, C. P. (1989) (1-3)-β-Glucan synthesis of *Neurospora crassa*, *Curr. Microbiol.* **19**, 153–161.

Jahn-Held, W., Lindoerfer, W., Sewe, K.-U., Wagner, F., Ziebolz, B. (1990) Separation of solid particles of various sizes from viscous liquids, US Patent No. 4921615.

Kang, M., Cabib, R. (1986) Regulation of fungal cell wall growth: a guanine nucleotide-binding, proteinaceous component required for activity of 1,3-β-D-glucan synthase, *Proc. Natl. Acad. Sci. USA* **83**, 5808–5812.

Kashiwagi, Y., Norisuye, T., Fujita, H. (1981) Triple helix of *Schizophyllum commune* polysaccharide in dilute solution: light scattering and viscosity in dilute aqueous sodium hydroxide, *Macromolecules* **14**, 1220–1225.

Kidd, P. M. (2000) The use of mushroom glucans and proteoglycans in cancer treatment, *Altern. Med. Rev.* **5**, 4–27.

Kikumoto, S., Miyajima, T., Yoshizumi, S., Fujimoto, S., Kimura, K. (1970) Polysaccharide produced by *Schizophyllum commune*. I. Formation and some properties of an extracellular polysaccharide. *Nippon Nogei Kagaku Kaishi* **44**, 337–342.

Kikumoto, S., Miyajima, T., Kimura, K., Okubo, S., Komatsu, N. (1971) Polysaccharide produced by *Schizophyllum commune.* II. Chemical structure of an extracellular polysaccharide, *Nippon Nogei Kagaku Kaishi* **45**, 162–168.

Kikumoto, S., Yamamoto, O., Komatsu, N., Kobayashi, H., Kamasuka, T. (1978) Method of producing neoschizophyllan having novel pharmacological activity, US Patent No. 4098661.

Kim, J. S., Kim, M. S., Lee, D. C., Lee, S. G., So, S., Kim, Y. T., Park, B. H., Park, K. M. (1999) Production of beta-1,6-branched beta-1,3-glucan useful as skin anti-aging, depigmenting and healing agent, JP 11313667.

Kim, M.-S., Park, K. M., Chang, I.-S., Kang, H.-H., Sim, Y.-C. (2000) β-1,6-branched β-1,3-glucans in skin care, *Allured's Cosmetic & Toiletries Magazine* **115**, 79–86.

Kimura, Y., Tojima, H., Fukase, S., Takeda, K. (1994) Clinical evaluation of sizofilan as assistant immunotherapy, *Otolaryngology* (Stockholm) **511**, 192–195.

Kishida, E., Yoshiaki, S., Misaki, A. (1992) Effects of branch distribution and chemicals modifications of antitumor (1-3)-β-D-glucans, *Carbohydr. Polym.* **17**, 89–95.

Kitamura, S. (1989) A differential scanning calorimetric study of the conformational transitions of schizophyllan in mixtures of water and dimethylsulfoxide, *Biopolymers* **28**, 639–654.

Kitamura, S., Hirano, T., Takeo, K., Fukada, H., Takahashi, K., Falch, B. H., Stokke, B. T. (1996) Conformational transitions of schizophyllan in aqueous alkaline solutions, *Biopolymers* **39**, 407–416.

Kojima, T., Tabaka, K., Ikumoto, T., Yanaki, T. (1984) Depolymerization of schizophyllan by controlled hydrodynamic shear, *Agric. Biol. Chem.* **48**, 915–921.

Kominato, M., Kamimiy, S., Tanake, H. (1987) Preparation and properties of β-glucan synthase of *Pyricularia oryzae* P_2, *Agric. Biol. Chem.* **51**, 755–761.

Kondoh, O., Tachibana, Y., Ohya, Y., Arisawa, M., Watanabe, T. (1997) Cloning of the *RHO1* gene from *Candida albicans* and its regulation of beta-1,3-glucan synthesis, *J. Bacteriol.* **179**, 7734–7741.

Kottutz, E., Rapp, P. (1990) 1,3-β-glucan synthase in cell-free extracts from mycelium and protoplasts of *Sclerotium glucanicum*, *J. Gen. Microbiol.* **136**, 1517–1523.

Kraus, J. (1990) Biopolymere mit antitumoraler und immunmodulierender Wirkung, *Pharmazie in unserer Zeit* **19**, 157–164.

Kulicke, W. M., Lettau, A. I., Thielking, H. (1997) Correlation between immunological activity, molar mass, and molecular structure of different 1,3-β-D-glucans, *Carbohydr. Res.* **297**, 135–143.

Kunz, H., Klinkhammer, U., Kinzy, W., Neumann, S., Radunz, H.-E. (1991) Tetrasaccharide and process for their preparation, EP 0416396.

Lengeler, K. B., Kothe, E. (1999) Identification and characterization of *brt1*, a gene down-regulated during B-regulated development in *Schizophyllum commune*, *Curr. Genet.* **35**, 551–556.

Lindoerfer, W., Sewe, K.-U., Wagner, F., Münzer, S., Nachtwey, S., Rapp, P., Rau, U., Stephan D. (1988) Homopolysaccharides with a high molecular weight, process for their extracellular preparation and their use, as well as the corresponding fungus strains, EP 0 271 907.

Lindoerfer, W., Sewe, K.-D., Wagner, F., Münzer, S., Rau, U., Veuskens, J. (1991) Verfahren zur Erhöhung der volumenbezogenen Produktivität (g/ld) nichtionischer Biopolymere, DE- 40 12 238.

Lo, A. C., Barbier, J.-R., Willick, G. E. (1990). Kinetics and specificities of two closely related β-glucosidases secreted by *Schizophyllum commune*, *Eur. J. Biochem.* **192**, 175–181.

Lourdes, M., Polizeli, T. M., Noventa-Jordao, M. A., DaSilva, M. M., Jorge, J. A., Terenzi, H. F. (1995) 1,3-β-D-glucan synthase activity in mycelial and cell wall-less phenotypes of the fz, sg, os-1 ("slime") mutant strain of *Neurospora crassa*, *Exp. Mycol.* **19**, 35–47.

Maeda, H., Yuguchi, Y, Kitamura, S., Urakawa, H., Kajiwara, K., Richtering, W., Fuchs, T., Burchard, W. (1999) Structural aspects of gelation in schizophyllan/sorbitol aqueous solution, *Polymer J.* **31**, 530–534.

Mantovani, G., Bianchi, A., Curreli, L., Ghiani, M., Astara, G., Lampis, B., Santona, M. C., Dessi, D., Esu, S., Lai, P., Massa, E., Maccio, A., Proto, E. (1997) Clinical and immunological evaluation of schizophyllan (SPG) in combination with standard chemotherapy in patients with head and neck squamous cell carcinoma, *Int. J. Oncol.* **10**, 213–221.

Marchessault, R. H., Deslandes, Y. (1980). Texture and crystal structures of fungal polysaccharides, in: *Fungal Polysaccharides. A.C.S. Symposium series* (Sandford, P. A., Matsuda, K., Eds.), Washington, DC: American Chemical Society, 221–250, vol. 126.

Marion, A. L., Bartholomew, K. A., Wu, J., Yang, H., Novotny, C. P., Ullrich, R. C. (1996) The Aα mating type locus of *Schizophyllum commune*: structure and function of gene *X*, *Curr. Genet.* **29**, 143–149.

Martin, G. G., Cannon, G. C., McCormick, C. L. (1999) Adsorption of a fungal hydrophobin onto surfaces as mediated by the associated polysaccharide schizophyllan, *Biopolymers* **49**, 621–633.

Mazur, P., Baginsky, W. (1996) In vitro activity of 1,3-β-glucan synthase requires the GTP-binding protein *RHO1*, *J. Biol. Chem.* **271**, 14604–14609.

Mazur, P., Moring, N., Baginsky, W., El-Sherbeini, M., Clemas, J. A., Nielsen, J. B., Foor, F. (1995) Differential expression and function of two homologous subunits of yeast 1,3-β-D-glucan synthase, *Mol. Cell. Biol.* **15**, 5671–5681.

Michiko, A. (1989) Production of controlling agent of plant virus by microorganism, JP 1272509.

Miura, N. N., Ohno, N., Adachi, Y., Aketagawa, J., Tamura, H., Tanaka, S., Yadomae, T. (1995) Comparison of the blood clearance of triple- and single-helical schizophyllan in mice, *Biol. Pharm. Bull.* **18**, 185–189.

Miura, T., Miura, N. N., Ohno, N., Adachi, Y., Shimada, S., Yadomae, T. (2000) Failure in antitumor activity by overdose of an immunomodulating beta-glucan preparation, sonifilan, *Biol. Pharm. Bull.* **23**, 249–253.

Mol, P. C., Park, H. M., Mullins, J. T., Cabib, E. (1994) A GTP-binding protein regulates the activity of 1,3-beta-glucan synthase, an enzyme directly involved in yeast cell wall morphogenesis, *J. Biol. Chem.* **269**, 31267–31274.

Mueller, A., Raptis, J., Rice, P. J., Kalbfleisch, J. H., Stout, R. D., Ensley, H. E., Browder, W., Williams, D. L. (2000) The influence of glucan polymer structure and solution conformation on binding to (1 → 3)-beta-D-glucan receptors in a human monocyte-like cell line, *Glycobiology* **10**, 339–346.

Muenzberg, J., Rau, U., Wagner, F. (1995) Investigations to the regioselective hydrolysis of a branched β-1,3-glucan, *Carbohydr. Polym.* **27**, 271–276.

Muenzer, S. (1989) Produktion und Charakterisierung eines von *S. commune ATCC 38548* gebildeten extrazellulären β-1,3-Glucans, PhD Thesis, Technical University Braunschweig.

Norisuye, T., Yanaki, T., Fujita, H. (1980) Triple helix of a *Schizophyllum commune* polysaccharide in aqueous solution, *J. Polym. Sci. Polym. Phys.* **18**, 547–558.

Oertel, R., Kulicke, W.-M. (1991) Viscoelastic properties of liquid crystals of aqueous biopolymer solutions, *Rheol. Acta* **30**, 140–150.

Ohno, N., Miura, N. N., Chiba, N., Adachi, Y., Yadomae, T. (1995) Comparison of the immunopharmacological activities of triple and single-helical schizophyllan in mice, *Biol. Pharm. Bull.* **18**, 1242–1247.

Ooi, V. E., Liu, F. (2000) Immunomodulation and anti-cancer activity of polysaccharide-protein complexes, *Curr. Med. Chem.* **7**, 715–729

Pilz, F., Auling, G., Rau, U., Stephan, D., Wagner, F. (1991) A high affinity Zn^{2+} uptake system and oxygen supply control growth and biosyntheses of an extracellular, branched β-1,3-glucan in *Sclerotium rolfsii* ATCC 15205, *Exp. Mycol.* **15**, 181–192.

Polizeli, M. T. M., Noventa, J. M. A., Silva, M. M., Jorge, J. A., Terenzi, H. F. (1995) (1,3)-beta-D-glucan synthase activity in mycelial and cell wall-less phenotypes of the *fz, sg, os-1* ("Slime") mutant strain of *Neurospora crassa*, *Exp. Mycol.* **19**, 35–47.

Prokop, A. (1990) Protoplastenmutagenese und Protoplastenfusion von Schizophyllum commune: Einfluß auf die Synthese von β-1,3-Glucanen sowie Reinigung und Charakterisierung einer Endo-β-1,3-Glucanase, PhD Thesis, Technical University Braunschweig.

Prokop, A. Rapp, P. Wagner, F. (1994) Production, purification, and characterization of an extracellular endo-beta-1,3-glucanase from a monokaryon of *Schizophyllum commune* ATCC 38548 defective in exo-beta-1,3-glucanase formation, *Can. J. Microbiol.* **40**, 18–23.

Qadota, H., Python, C. P., Inoue, S. B., Arisawa, M., Anraku, Y., Zheng, Y., Watanabe, T., Levin, D. E., Ohya, Y. (1996) Identification of yeast Rho1p GTPase as a regulatory subunit of 1,3-beta-glucan synthase, *Science* **272**, 279–281.

Quigley, D. R., Selitrennikoff, C. P. (1984) β(1-3)Glucan synthase activity of *Neurospora crassa*: stabilization and partial characterization, *Exp. Mycol.* **8**, 202–214.

Rapp, P. (1992) Formation, separation and characterization of three β-1,3-glucanases from *Sclerotium glucanicum*, *Biochim. Biophys. Acta* **1117**, 7–14.

Rau, U. (1997) Biosynthese, Produktion und Eigenschaften von extrazellulären Pilz-Glucanen. Aachen, Germany: Shaker Verlag.

Rau, U. (1999) Production of schizophyllan, in: *Methods in Biotechnology – Carbohydrate Biotechnology Protocols* (Bucke, C., Ed.), Totowa, NJ, USA: Humana Press, Inc., 43–57, vol. 10.

Rau, U., Brandt, C. (1994) Oxygen controlled batch cultivation of *Schizophyllum commune* for enhanced production of branched β-1,3-glucans, *Bioprocess Eng.* **11**, 161–165.

Rau, U., Wagner, F. (1987) Non-Newtonian flow behaviour of colloid-disperse glucan solutions, *Biotechnol. Lett.* **9**, 95–100.

Rau, U., Gura, E., Schliephaake, A., Wagner, F. (1989) Influence of processing on the formation of exopolysaccharides by filamentously growing fungi, *DECHEMA Biotechnology Conferences* **3**, 571–574.

Rau, U., Müller, R.-J., Cordes, K., Klein, J. (1990) Process and molecular data of branched 1,3-β-D-glucans in comparison with xanthan, *Bioprocess Eng.* **5**, 89–93.

Rau, U., Gura, E., Haarstrick, A. (1992a) Prozessintegrierte Aufarbeitung verzweigter β-1,3-Glucane (schizophyllan), *GIT* **12**, 1233–1238.

Rau, U., Gura, E., Olszewski, E., Wagner, F. (1992b) Enhanced glucan formation of filamentous fungi by effective mixing, oxygen limitation and fed-batch processing, *J. Ind. Microbiol.* **9**, 19–26.

Rau, U., Haarstrick, A., Wagner, F. (1992c) Eignung von Schizophyllanlösungen zum Polymerfluten von Lagerstätten mit hoher Temperatur und Salinität, *Chem.-Ing.-Tech.* **64**, 576–577.

Rau, U., Olszewski, E., Wagner, F. (1992d) Gesteigerte Produktion von verzweigten β-1,3-Glucanen mit *Schizophyllum commune* durch Sauerstofflimitierung, *GIT* **4**, 331–337.

Rinaudo, M., Vincendon, M. (1982) ^{13}C-NMR structural investigation of scleroglucan, *Carbohyd. Polym.* **2**, 135–144.

Ruiz-Herrera, J. (1991) Biosynthesis of β-glucans in fungi, *Antonie van Leeuwenhoek* **60**, 73–81.

Saito, H., Ohki, T., Sasaki, T. (1979) A 13C-Nuclear magnetic resonance study of polysaccharide gels, *Carbohydr. Res.* **74**, 227–240.

Saito, H., Yoshioka, Y., Uehara, N. (1991) Relationship between conformation and biological response for (1-3)-β-D-glucans in the activation of coagulation Factor G from limulus amebocyte lysate and host-mediated antitumor activity. Demonstration of single-helix conformation as a stimulant, *Carbohydr. Res.* **217**, 181–190.

Sakurai, K., Shinkai, S. (2000) Phase separation in the mixture of schizophyllan and poly(ethylene oxide) in aqueous solution driven by a specific interaction between the glucose side chain and poly(ethylene oxide), *Carbohydr. Res.* **324**, 136–140.

Sasaki, S. K. J. (1990) Fine polishing composition for wafers, EP 0373501.

Sato, T., Norisuye, T., Fukita, H. (1981) Melting behaviour of *Schizophyllum commune* polysaccharides in mixtures of water and dimethylsulfoxide, *Carbohydr. Res.* **95**, 195–204.

Scholtmeijer, K., Wösten, H. A. B., Springer, J., Wessels, J. G. H. (2001) Effect of introns and AT-rich sequences on expression of the bacterial hygromycin B resistance gene in the basidiomycete *Schizophyllum commune*, *Appl. Environ. Microbiol.* **67**, 481–483.

Schulz, D. (1992) Untersuchungen zur Folien- und Gelbildung von natürlichen und modifizierten β-1,3-Glucanen, PhD Thesis, Technical University Braunschweig, Germany.

Schulz, D., Rapp, P. (1991) Properties of the polyalcohol prepared from the β-D-glucan schizophyllan per periodate oxidation and borohydrate reduction, *Carbohydr. Res.* **222**, 223–231.

Schulz, D., Rau, U., Wagner, F. (1992) Characteristics of films prepared by native and modified branched β-1,3-D-glucans. *Carbohydr. Polym.* **18**, 295–299.

Schuren, F. H. J., Wessels, J. G. H. (1998) Expression of heterologous genes in *Schizophyllum commune* is often hampered by the formation of truncated transcripts, *Curr. Genet.* **33**, 151–156.

Selitrennikoff, C. P. (1995) Antifungal drugs: (1,3)β-glucan synthase inhibitors, in: *Molecular Biology Intelligence Unit* (Molsberry, D. M., Ed.), Austin, Texas, USA: L. R. G. Landes Company, 45–89.

Seviour, R. J., Stasinopoulos, S. J., Auer, D. P. F., Gibbs, P. A. (1992) Production of pullulan and other exopolysaccharides by filamentous fungi, *Crit. Rev. Biotechnol.* **12**, 279–298.

Shematec, E. M., Cabib, E. (1980) Biosynthesis of the yeast cell wall. II. Regulation of β-(1-3)glucan synthetase by ATP And GTP, *J. Biol. Chem.* **255**, 895–902.

Shen, G.-P., Park, D.-C., Ullrich, R. C., Novotny, C. P. (1996) Cloning and characterization of a *Schizophyllum* gene with *Aβ6* mating-type activity, *Curr. Genet.* **29**, 136–142.

Shigero, M., Fisamu, S., Wataru, I. (1989) Anti-aids virus agent. JP 1287031.

Sietsma, J. H., Wessels, J. G. H. (1977) Chemical analysis of the hyphal wall of *Schizophyllum commune*, *Biochim. Biophys. Acta* **496**, 225–239.

Sietsma, J. H., Wessels, J. G. H. (1979) Evidence for covalent linkages between chitin and β-glucan in a fungal wall, *J. Gen. Microbiol.* **114**, 99–108.

Sietsma, J. H., Wessels, J. G. H. (1981) Solubility of (1-3)-β-D-(1-6)-β-D-glucan in fungal walls: importance of presumed linkage between glucan and chitin, *J. Gen. Microbiol.* **125**, 209–212.

Sietsma, J. H., Sonnenberg, A. M. S., Wessels, J. G. H. (1985) Localization by autoradiography of synthesis of (1-3)-β- and (1-6)-β linkages in a wall glucan during hyphal growth of *Schizophyllum commune*, *J. Gen. Microbiol.* **131**, 1331–1337.

Steiner, W., Lafferty, R. M., Gomes, I., Esterbauer, H. (1987) Studies on a wild strain of *Schizophyllum commune*: cellulase and xylanase production and formation of the extracellular polysaccharide schizophyllan, *Biotechnol. Bioeng.* **30**, 169–178.

Steiner, W., Divjak, H., Lafferty, R. M., Steiner, E., Esterbauer, H., Gomes, I. (1988) Production and properties of schizophyllan and scleroglucan, *DECHEMA Biotechnology Conferences* **1**, 149–154.

Steiner, W., Haltrich, D., Lafferty, R. M. (1993) Production, properties and practical applications of fungal polysaccharides, in: *Biosurfactants* (Kosaric, N., Ed.), New York, USA: Marcel Dekker, 175–204, vol. 48.

Stuebler, D., Buchenauer, H. (1996) Antiviral activity of the glucan lichenan (poly-β-(1,3,-1,4)D-anhydroglucose): 1. Biological activity in tobacco plants, *J. Phytopath.* **144**, 37–43.

Suda, M., Ohno, N. Adachi, Y. Yadomae, T. (1994) Preparation and properties of metabolically 3H- or 13C-labeled 1,3-β-D-glucan SSG from *Sclerotinia sclerotiorum* IFO 9395, *Carbohydr. Res.* **254**, 213–220.

Sutherland, I. W. (1982). Biosynthesis of microbial exopolysaccharides, in: *Advances in Microbial Physiology* (Rose, A. H., Gareth-Morris, J. G., Eds.), New York, USA: Academic Press, 79–150, vol. 23.

Sutherland, I. W. (1999) Polysaccharases for microbial exopolysaccharides. *Carbohydr. Polym.* **38**, 319–328.

Tabata, K., Ito, W., Kojima, T. (1981) Ultrasonic degradation of schizophyllan, an antitumor polysaccharide produced by *Schizophyllum commune* Fries, *Carbohydr. Res.* **89**, 121–135.

Tsuzuki, A., Tateishi, T., Ohno, N., Adachi, Y., Yadomae, T. (1999) Increase of hematopoietic responses by triple or single helical conformer of an antitumor 1,3-β-D-glucan preparation, Sonifilan, in cyclophosphamide-induced leukopenic mice, *Biosci. Biotechnol. Biochem.* **63**, 104–110.

Van, K., Asakawa, T., Teramota, A. (1984) Optical rotatory dispersion of liquid crystal solutions of a triple-helical polysaccharide schizophyllan, *Polymer J.* **16**, 61–69.

van Wetter, M.-A., Wösten, H. A. B., Sietsma, J. H., Wessels, J. G. H. (2000) Hydrophobin gene expression affects hyphal wall composition in *Schizophyllum commune*, *Fungal Genet. Biol.* **31**, 99–104.

Wang, C., Miles, P. G. (1964) The physiological characterization of dikaryotic mycelia of *Schizophyllum commune*, *Physiol. Plant.* **17**, 573–588.

Wang, M. C., Bartnicki-Garcia, S. (1982) Synthesis of noncellulose cell-wall β-glucan by cell-free extracts from zoospores and cysts of *Phytophtora palmivora*, *Exp. Mycol.* **6**, 125–135.

Wessels, J. G. H. (1965) Morphogenesis and biochemical processes in *S. commune*, *Wentia* **13**, 1–113.

Wessels, J. G. H. (1978) Incompatibility factors and the control of biochemical processes, in: *Genetics and Morphogenesis in Basidiomycetes* (Schwalb, M. N., Miles, P. G., Eds.), New York, USA: Academic Press, 81–104.

Wessels, J. G. H. (1988) A steady-state model for apical wall growth in fungi, *Acta Bot. Neerl.* **37**, 3–16.

Wessels, J. G. H. (1999) Fungi in their own right, *Fungal Genet. Biol.* **27**, 134–145.

Wessels, J. G. H., Sietsma, J. H. (1979) Wall structure and growth in *Schizophyllum commune*, in: *Fungal Walls and Hyphal Growth* (Burnett, J. H., Trinci, A. P. J., Eds.), Cambridge: Cambridge University Press, 27–48.

Wessels, J. G. H., Sietsma, J. H. (1981) Significance of linkages between chitin and β-glucan in fungal walls, *Microbiology*, **127**, 232–234.

Yanaki, T. (1981) Preparation of polysaccharide associate, JP 56127603.

Yanaki, T. (1982) Novel liquid crystal of polysaccharide, JP 57147576.

Yano, T. (1990) Preventive for fish disease comprising water-soluble glucan, JP 2218615.

Yoshio, S., Katsuhiko, H., Kazumasa, M. (1992) Augmenting the effect of sizofiran on the immunofunction of regional lymph nodes in cervical cancer, *Cancer* **69**, 1188–1194.

Young, S.-H., Jacobs, R. R. (1998) Sodium hydroxide-induced conformational change in schizophyllan detected by the fluorescence dye, aniline blue, *Carbohydr. Res.* **310**, 91–99.

Zentz, F., Verchere, J.-F., Muller, G. (1992) Thermal denaturation and degradation of schizophyllan, *Carbohydr. Polym.* **17**, 289–297.

26
Alginates from Algae

Dr. Kurt Ingar Draget[1], Prof. Dr. Olav Smidsrød[2], Prof. Dr. Gudmund Skjåk-Bræk[3]
[1] Norwegian Biopolymer Laboratory, Department of Biotechnology, Norwegian University of Science and Technology, Sem Saelands vei 6-8, N-7491 Trondheim, Norway; Tel.: +47-73598260; Fax: +47-73591283;
E-mail: Kurt.I.Draget@chembio.ntnu.no
[2] Norwegian Biopolymer Laboratory, Department of Biotechnology, Norwegian University of Science and Technology, Sem Saelands vei 6-8, N-7491 Trondheim, Norway; Tel.: +47-735-98260; Fax: +47-735-93337;
E-mail: Olav.Smidsroed@chembio.ntnu.no
[3] Norwegian Biopolymer Laboratory, Department of Biotechnology, Norwegian University of Science and Technology, Sem Saelands vei 6-8, N-7491 Trondheim, Norway. Tel.: +47-735-98260; Fax: +47-735-93340;
E-mail: Gudmund.Skjaak-Braek@chembio.ntnu.no

Biotechnology of Biopolymers. From Synthesis to Patents. Edited by A. Steinbüchel and Y. Doi

ISBN: 3-527-31110-6

DP degree of polymerization
EDTA etylenediamine tetraacetic acid
G α-L-guluronic acid
GDL D-glucono-δ-lactone
M β-D-mannuronic acid (M)
$N_{G>1}$ average G-block length larger than 1
NMR nuclear magnetic resonance spectroscopy
PGA propylene glycol alginate
pK_a dissociation constants for the uronic acid monomers

1 Introduction

Alginates are quite abundant in nature since they occur both as a structural component in marine brown algae (*Phaeophyceae*), comprising up to 40% of the dry matter, and as capsular polysaccharides in soil bacteria (see Chapter 8 on bacterial alginates in Volume 5 of this series). Although present research and results point toward a possible production by microbial fermentation and also by post-polymerization modification of the alginate molecule, all commercial alginates are at present still extracted from algal sources. The industrial applications of alginates are linked to its ability to retain water, and its gelling, viscosifying, and stabilizing properties. Upcoming biotechnological applications, on the other hand, are based either on specific biological effects of the alginate molecule itself or on its unique, gentle, and almost temperature-independent sol/gel transition in the presence of multivalent cations (e.g., Ca^{2+}), which makes alginate highly suitable as an immobilization matrix for living cells.

Traditional exploitation of alginates in technical applications has been based to a large extent on empirical knowledge. However, since alginates now enter into more knowledge-demanding areas such as pharmacy and biotechnology, new research functions as a locomotive for a detailed further investigation of structure–function relationships. New scientific breakthroughs are made, which in turn may benefit the traditional technical applications.

2 Historical Outline

The British chemist E. C. C. Stanford first described alginate (the preparation of "algic acid" from brown algae) with a patent dated 12 January 1881 (Stanford, 1881). After the patent, his discovery was further discussed in papers from 1883 (Standford, 1883a,b). Stanford believed that alginic acid contained nitrogen and contributed much to the elucidation of its chemical structure.

In 1926, some groups working independently (Atsuki and Tomoda, 1926; Schmidt and Vocke, 1926) discovered that uronic acid was a constituent of alginic acid. The nature of the uronic acids present was investigated by three different groups shortly afterwards (Nelson and Cretcher, 1929, 1930, 1932; Bird and Haas, 1931; Miwa, 1930), which all found D-mannuronic acid in the hydrolysate of alginate. The nature of the bonds between the uronic acid residues in the alginate molecule was determined to be β1,4, as in cellulose (Hirst et al., 1939)

This very simple and satisfactory picture of the constitution of alginic acid was, however, destroyed by the work of Fischer and Dörfel (1955). In a paper chromatographic study of uronic acids and polyuronides, they discovered the presence of a uronic acid different from mannuronic acid in the hydrolysates of alginic acid. This new uronic acid was identified as L-guluronic acid. The quantity of L-guluronic acid was considerable, and a method for quantitative determination of mannuronic and guluronic acid was developed.

Alginate then had to be regarded as a binary copolymer composed of α-L-guluronic and β-D-mannuronic residues. As long as alginic acid was regarded as a polymer containing only D-mannuronic acid linked together with β-1,4 links, it was reasonable to assume that alginates from different raw materials were chemically identical and that any given sample of alginic acid was chemically homogeneous. From a practical and a scientific point of view, the uronic acid composition of alginate from different sources had to be examined, and methods for chemical fractionation of alginates had to be developed. These tasks were undertaken mainly by Haug and coworkers (Haug, 1964), as described in Section 3 below. The discovery of alginate as a block-copolymer, the correlation between physical properties and block structure, and the discovery of a set of epimerases converting mannuronic to guluronic acid in a sequence-dependent manner also are discussed further in later sections.

3
Chemical Structure

Being a family of unbranched binary copolymers, alginates consist of (1 → 4) linked β-D-mannuronic acid (M) and α-L-guluronic acid (G) residues (see Figure 1a and b) of widely varying composition and sequence. By partial acid hydrolysis (Haug, 1964; Haug et al., 1966; Haug and Larsen, 1966; Haug et al., 1967a; Haug and Smidsrød, 1965), alginate was separated into three fractions. Two of these contained almost homopolymeric molecules of G and M, respectively, while a third fraction consisted of nearly equal proportions of both monomers and was shown to contain a large number of MG dimer residues. It was concluded that alginate could be regarded as a true block copolymer composed of homopolymeric regions of M and G, termed M- and G-blocks, respectively, interspersed with regions of alternating structure (MG-blocks; see Figure 1c). It was further shown (Painter et al., 1968; Larsen et al., 1970; Smidsrød and Whittington, 1969) that alginates have no regular repeating unit and that the distribution of the monomers along the polymer chain could not be described by Bernoullian statistics. Knowledge of the monomeric composition is hence not sufficient to determine the sequential structure of alginates. It was suggested (Larsen et al., 1970) that a second-order Markov model would be required for a general approximate description of the monomer sequence in alginates. The main difference at the molecular level between algal and bacterial alginates is the presence of *O*-acetyl groups at C2 and/or C3 in the bacterial alginates (Skjåk-Bræk et al., 1986).

4
Conformation

Knowledge of the monomer ring conformations is necessary to understand the polymer properties of alginates. X-ray diffraction studies of mannuronate-rich and guluronate-rich alginates showed that the guluronate residues in homopoly-meric blocks were in the $^1C_4$ conformation (Atkins et al., 1970), while the mannuronate residues had

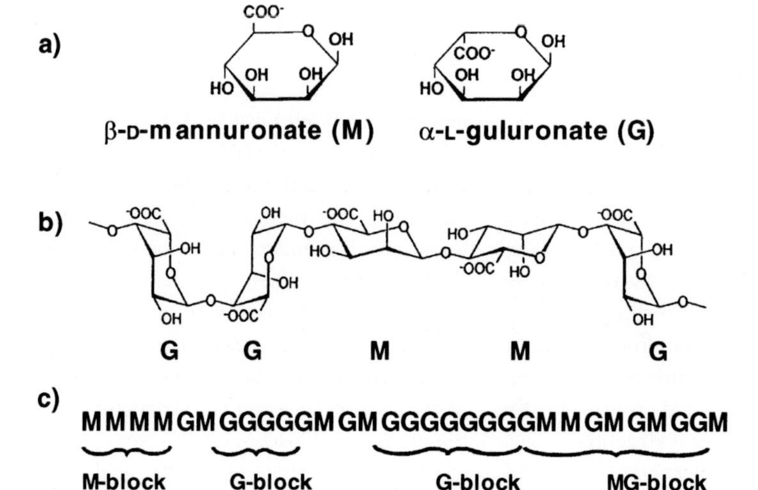

Fig. 1 Structural characteristics of alginates: (a) alginate monomers, (b) chain conformation, (c) block distribution.

the $^4C_1$ conformation (see Figure 1a). Viscosity data of alginate solutions indicated that the stiffness of the chain blocks increased in the order MG < MM < GG. This series could be reproduced only by statistical mechanical calculations when the guluronate residues were set in the $^1C_4$ conformation (Smidsrød et al., 1973) and was later confirmed by ^{13}C-NMR (Grasdalen et al., 1977). Hence, alginate contains all four possible glycosidic linkages: diequatorial (MM), diaxial (GG), equatorial-axial (MG), and axial-equatorial (GM) (see Figure 1b).

The diaxial linkage in G-blocks results in a large, hindered rotation around the glycosidic linkage, which may account for the stiff and extended nature of the alginate chain (Smidsrød et al., 1973). Additionally, taking the polyelectrolyte nature of alginate into consideration, the electrostatic repulsion between the charged groups on the polymer chain also will increase the chain extension and hence the intrinsic viscosity. Extrapolation of dimensions both to infinite ionic strength and to θ-conditions (Smidsrød, 1970) yielded relative dimensions for the neutral, unperturbed alginate chain being much higher than for amylose derivatives and even slightly higher than for some cellulose derivatives.

Another parameter reflecting chain stiffness and extension is the exponent in the Mark-Houwink-Sakurada equation,

$$[\eta] = K \cdot M^a$$

where M is the molecular weight of the polymer, $[\eta]$ is the intrinsic viscosity, and the exponent a generally increases with increasing chain extension. Some measurements on alginates (Martinsen et al., 1991; Smidsrød and Haug, 1968a, Mackie et al., 1980), yielded a-values ranging from 0.73 to 1.31, depending on ionic strength and alginate composition. Low and high a-values are related to large fractions of the flexible MG-blocks and the stiff and extended G-blocks, respectively (Moe et al., 1995).

5
Occurrence and Source Dependence

Commercial alginates are produced mainly from *Laminaria hyperborea, Macrocystis pyrifera, Laminaria digitata, Ascophyllum nodosum, Laminaria japonica, Eclonia maxima, Lessonia nigrescens, Durvillea antarctica,* and *Sargassum* spp. Table 1 gives some sequential parameters (determined by high-field NMR-spectroscopy) for samples of these

Tab. 1 Composition and sequence parameters of algal alginates (Smidsrød and Draget 1996)

Source	F_G	F_M	F_{GG}	F_{MM}	$F_{GM,MG}$
Laminaria japonica	0.35	0.65	0.18	0.48	0.17
Laminaria digitata	0.41	0.59	0.25	0.43	0.16
Laminaria hyperborea, blade	0.55	0.45	0.38	0.28	0.17
Laminaria hyperborea, stipe	0.68	0.32	0.56	0.20	0.12
Laminaria hyperborea, outer cortex	0.75	0.25	0.66	0.16	0.09
Lessonia nigrescens[a]	0.38	0.62	0.19	0.43	0.19
Ecklonia maxima	0.45	0.55	0.22	0.32	0.32
Macrocystis pyrifera	0.39	0.61	0.16	0.38	0.23
Durvillea antarctica	0.29	0.71	0.15	0.57	0.14
Ascophyllum nodosum, fruiting body	0.10	0.90	0.04	0.84	0.06
Ascophyllum nodosum, old tissue	0.36	0.64	0.16	0.44	0.20

[a] Data provided by Bjørn Larsen

alginates. The composition and sequential structure may, however, vary according to seasonal and growth conditions (Haug, 1964; Indergaard and Skjåk-Bræk, 1987). High contents of G generally are found in alginates prepared from stipes of old *Laminaria hyperborea* plants, whereas alginates from *A. nodosum*, *L. japonica*, and *Macrocystis pyrifera* are characterized by low content of G-blocks and low gel strength.

Alginates with more extreme compositions containing up to 100% mannuronate can be isolated from bacteria (Valla et al., 1996). Alginates with a very high content of guluronic acid can be prepared from special algal tissues such as the outer cortex of old stipes of *L. hyperborea* (see Table 1), by chemical fractionation (Haug and Smidsrød, 1965; Rivera-Carro, 1984) or by enzymatic modification *in vitro* using mannuronan C-5 epimerases from *A. vinelandii* (Valla et al., 1996; see Section 9.2). This family of enzymes is able to epimerize M-units into G-units in different patterns from almost strictly alternating to very long G-blocks. The epimerases from *A. vinelandii* have been cloned and expressed, and they represent at present a powerful new tool for the tailoring of alginates. It is also obvious that commercial alginates with less molecular heterogeneity, with respect to chemical composition and sequence, can be obtained by a treatment with one of the C-5 epimerases (Valla et al., 1996).

6 Physiological Function

The biological function of alginate in brown algae generally is believed to be as a structure-forming component. The intercellular alginate gel matrix gives the plants both mechanical strength and flexibility (Andresen et al., 1977). Simply speaking, alginates in marine brown algae may be regarded as having physiological properties similar to those of cellulose in terrestrial plants. This relation between structure and function is reflected in the compositional difference of alginates in different algae or even between different tissues from the same plant (see Table 1). In *L. hyperborea*, an alga that grows in very exposed coastal areas, the stipe and holdfast have a very high content of guluronic acid, giving high mechanical rigidity. The leaves of the same algae, which float in the streaming water, have an alginate characterized by a lower G-content, giving it a more flexible texture. The physiological function of alginates in bacteria will be covered elsewhere in this series.

7 Chemical Analysis and Detection

Since alginates are block copolymers, and because of the fact that their physical properties rely heavily on the sequence of these blocks, it obvious that the development of techniques enabling a sequence quantification is of the utmost importance. Additionally, molecular mass and its distribution (polydispersity) is a significant parameter in some applications.

7.1 Chemical Composition and Sequence

Detailed information about the structure of alginates became available by introduction of high-resolution ^{1}H and ^{13}C NMR-spectroscopy (Grasdalen et al., 1977, 1979; Penman and Sanderson, 1972; Grasdalen, 1983) in the sequential analysis of alginate. These powerful techniques make it possible to determine the monad frequencies F_M and F_G; the four nearest neighboring (diad) frequencies F_{GG}, F_{MG}, F_{GM}, and F_{MM}; and the eight next nearest neighboring (triad) frequencies. Knowledge of these frequencies enables, for example, the calculation of the average G-block length larger than 1:

$$N_{G>1} = (F_G - F_{MGM}) \,/\, F_{GGM}.$$

This value has been shown to correlate well with gelling properties. It is important to realize that in an alginate chain population, neither the composition nor the sequence of each chain will be alike. This results in a composition distribution of a certain width.

7.2 Molecular Mass

Alginates, like polysaccharides in general, are polydisperse with respect to molecular weight. In this aspect they resemble synthetic polymers rather than other biopolymers such as proteins and nucleic acids. Because of this polydispersity, the "molecular weight" of an alginate is an average over the whole distribution of molecular weights.

In a population of molecules where N_i is the number of molecules and w_i is the weight of molecules having a specific molecular weight M_i, the number and the weight average are defined respectively as:

$$\overline{M_n} = \frac{\Sigma_i N_i M_i}{\Sigma_i N_i}$$

$$\overline{M_w} = \frac{\Sigma_i w_i M_i}{\Sigma_i w_i} = \frac{\Sigma_i N w_i M_i^2}{\Sigma_i N_i M_i}$$

For a randomly degraded polymer, we have $\overline{M_w} \approx 2\overline{M_n}$ (Tanford, 1961). The fraction $\overline{M_w}/\overline{M_n}$ is called the polydispersity index. Polydispersity index values between 1.4 and 6.0 have been reported for alginates and have been related to different types of preparation and purification processes (Martinsen et al., 1991; Smidsrød and Haug, 1968a; Mackie et al., 1980; Moe et al., 1995).

The molecular-weight distribution can have implications for the uses of alginates, as low-molecular-weight fragments containing only short G-blocks may not take part in gel-network formation and consequently do not contribute to the gel strength. Furthermore, in some high-tech applications, the leakage of mannuronate-rich fragments from alginate gels may cause problems (Stokke et al., 1991; Otterlei et al., 1991) and a narrow molecular-weight distribution therefore is recommended.

7.3 Detection and Quantification

Detection and quantification of alginates in the presence of other biopolymers, such as proteins, are not straightforward mainly because of interference. Once isolated, a number of colorimetric methods can be applied to quantify alginate. The oldest and most common is the general procedure for carbohydrates, the phenol/sulfuric acid method (Dubois et al., 1956), but there are also two slightly refined formulas specially designed for uronic acids (Blumenkrantz and Asboe-Hansen, 1973; Filisetti-Cozzi and Carpita, 1991).

8 Biosynthesis and Biodegradation

Our knowledge of the alginate biosynthesis mainly comes from studying alginate-producing bacteria. Figure 2 shows the principal enzymes involved in alginate biosynthesis, and the activity of all enzymes (1–7) has been identified in brown algae. During the last decade, the genes responsible for alginate synthesis in *Pseudomonas* and *Azotobacter* have been identified, sequenced, and cloned. For further information on alginate biosynthesis, please see Chapter 8 on bacte-

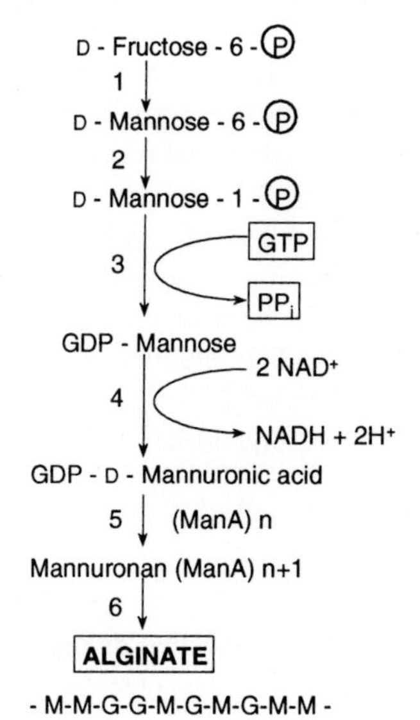

Fig. 2 Biosynthetic pathway of alginates.

rial alginates in Volume 5 of this series. Because of their potential use in alginate modification, the only enzymes we will comment on here are the alginate lyases and the mannuronan C-5 epimerases.

Alginates are not degraded in the human gastric-intestinal tract, and hence do not give metabolic energy. Some lower organisms have, however, developed lyases that degrade alginates down to single components, resulting in alginates that function as a carbon source. Alginate lyases catalyze the depolymerization of alginate by splitting the 1–4 glycosidic linkage in a β-elimination reaction, leaving an unsaturated uronic acid on the non-reducing end of the molecules. Alginate lyases are widely distributed in nature, including in organisms growing on alginate as a carbon source such as marine gastropods, prokaryotic and eukaryotic microorganisms, and bacteriophages. They also are found in the bacterial species producing alginate such as *Azotobacter vinelandii* and *Pseudomonas aeruginosa*. All of them are endolyases and may exhibit specificity to either M or G. Since the aglycon residue will be identical for both M and G, the use of lyases for structural work is limited. Table 2 lists a range of lyases and their specificities.

9
Production: Biotechnological and Traditional

There has been significant progress in the understanding of alginate biosynthesis over the last 10 years. The fact that the alginate molecule enzymatically undergoes a post-polymerization modification with respect to chemical composition and sequence opens up the possibility for *in vitro* modification and tailoring of commercially available alginates.

9.1
Isolation from Natural Sources / Fermentative Production

As already described, all commercial alginates today are produced from marine brown algae (Table 1). Alginates with more extreme compositions can be isolated from the bacterium *Azotobacter vinelandii*, which, in contrast to *Pseudomonas* species, produces polymers containing G-blocks. Production by fermentation therefore is technically possible but is not economically feasible at the moment.

9.2
Molecular Genetics and *in vitro* Modification

Alginate with a high content of guluronic acid can be prepared from special algal tissues by chemical fractionation or by *in vitro* enzymatic modification of the alginate *in vitro* using mannuronan C-5 epimerases from *A. vinelandii* (Ertesvåg et al., 1994, 1995, 1998b; Høydal et al., 1999). These epimerases, which convert M to G in the

Tab. 2 Substrate specificity and biochemical properties of some alginate lyases (Gacesa, 1992; Wong et al., 2000)

Source	Localization	Sequence specifty	Major end-product	pHopt	Mw (kDa)	pI	Reference
K.aerogenes	Extracellular	G↓X	Trimer	7.0	31.4	8.9	Boyd and Turvey, 1978
Enterobacter cloacae	Extracellular	G	–	7.8	32–38	8.9	Nibu et al., 1995
	Intracellular	G↓G	Dimer/pentamer	7.5	31–39	8.9	Shimokawa et al., 1997
P. aeruginosa (AlgL)	Periplasmic	M-X	Trimer	7.0	39	9.0	Boyd and Turvey, 1977
A. vinelandii	Periplasmic (AlgL)	M↓X M_{Ac}↓X	Trimer/tetramer	8.1–8.4	39	5.1	Ertesvåg et al., 1998a
	Extracellular (AlgE7)	G↓X	Tetramer- septamer	6.3–7.3	90.4	–	Ertesvåg et al., 1998a
P. alginovora	Extracellular	G↓G	–	7.5	28	5.5	Boyen et al., 1990a
	Intracellular	M↓M	–	–	24	5.8	Boyen et al., 1990a
Haliotis tuberculata	Hepato-pancreas	M↓X, G↓M	Trimer/ dimer	8	34	–	Boyen et al., 1990b
Sphingomonas sp. ALYI-III	Cytoplasmic	M↓X M_{Ac}↓X	–	5.6–7.8	38	10.16	Murata et al., 1993
Littorina sp.	Hepato-pancreas	M↓M	Trimer	5.6	~40	–	Elyakova and Favorov, 1974
A. vinelandii phage	Extracellular	M↓X M_{Ac}↓X	Trimer	7.7	30–35	–	Davidson et al., 1977

polymer chain, recently have allowed for the production of highly programmed alginates with respect to chemical composition and sequence. *A. vinelandii* encodes a family of 7 exocellular isoenzymes with the capacity to epimerize all sorts of alginates and other mannuronate-containing polymers, as shown in Figure 3, where the mode of action of AlgE4 (giving alternating introduction of G) is presented. Although the genes have a high degree of homology, the enzymes they encode exhibit different specificities. Different epimerases may give alginates with different distribution of M and G, and thus alginates with tailored physical and chemical properties can be made as illustrated in Figure 4. None of the enzymatically modified polymers, however, are commercially available at present. Table 3 lists the modular structure of the mannuronan C-5 epimerase family and its specific action.

9.3 Current and Expected World Market and Costs

Industrial production of alginate is roughly 30,000 metric tons annually, which is probably less than 10% of the annually biosynthesized material in the standing macroalgae crops. Because macroalgae also may be cultivated (e.g., in mainland China where 5 to 7 million metric tons of wet *Laminaria japonica* are produced annually) and because production by fermentation is technically possible, the sources for industrial production of alginate may be regarded as unlimited even for a steadily growing industry.

It is expected that future growth in the alginate market most likely will be of a qualitative rather than a quantitative nature. Predictions suggest that manufacturers will move away from commodity alginate production toward more refined products, e.g., for the pharmaceutical industry.

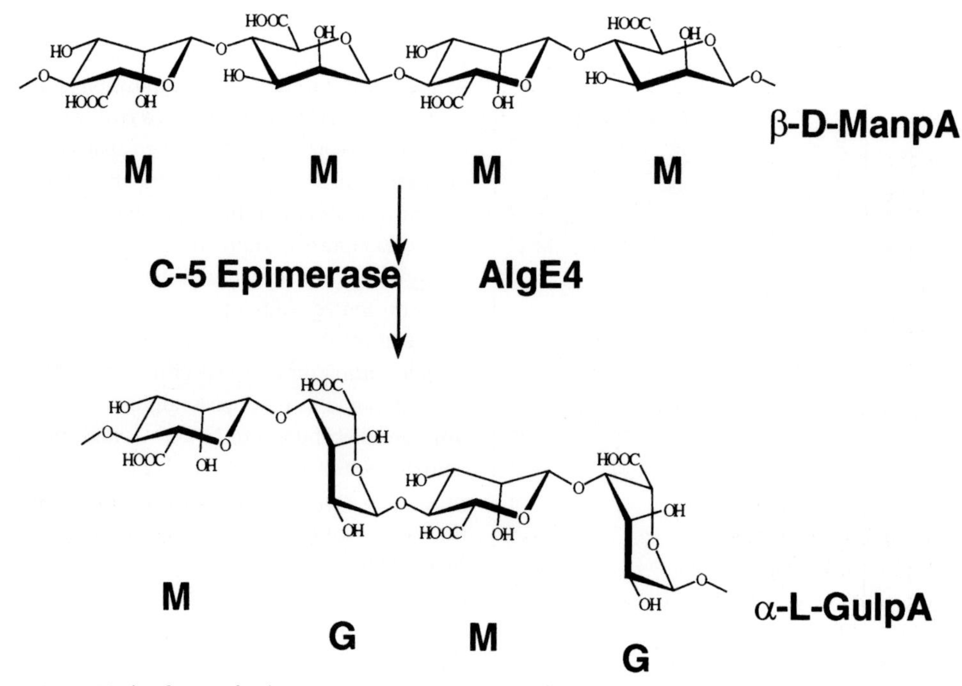

Fig. 3 Mode of action for the mannuronan C5-epimarase AlgE4.

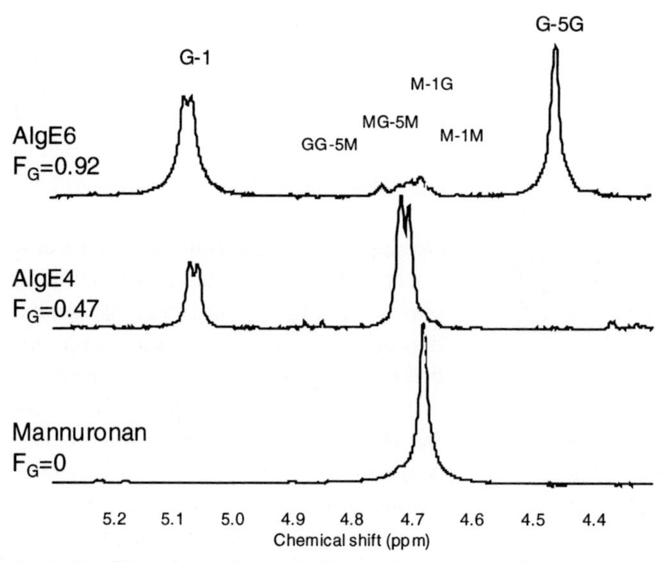

Fig. 4 Resulting chemical composition and sequence after treating mannuronan with different C5-epimerases.

Tab. 3 The seven AlgE epimerases from *A. vinelandi*[a]

Type	*[kDa]*	*Modular structure*	*Products*
AlgE1	147.2	A1 R1 R2 R3 A2 R4	Bi-functional G-blocks + MG-blocks
AlgE2	103.1	A1 R1 R2 R3 R4	G-blocks (short)
AlgE3	191	A1 R1 R2 R3 A2 R4 R5 R6 R7	Bi-functional G-block + MG-blocks
AlgE4	57.7	A1 R1	MG-blocks
AlgE5	103.7	A1 R1 R2 R3 R4	G-blocks (medium)
AlgE6	90.2	A1 R1 R2 R3	G-blocks (long)
AlgE7	90.4	A1 R1 R2 R3	Lyase activity + G-blocks + MG-blocks

[a] A – 385 amino acids; R – 155 amino acids

The cost of alginates can differ extremely depending on the degree of purity. Technical grade, low-purity alginate (containing a substantial amount of algae debris) can be obtained from around 1 USD per kilogram, and ordinary purified-grade alginate can be obtained from approximately 10 USD per kilo, whereas ultra-pure (low in endotoxins) alginate specially designed for immobilization purposes typically costs around 5 USD per gram.

9.4 Alginate Manufacturers

The alginate producers members list of the Marinalg hydrocolloid association includes six different companies. These are China Seaweed Industrial Association, Danisco Cultor (Denmark), Degussa Texturant Systems (Germany), FMC BioPolymer (USA), ISP Alginates Ltd. (UK), and Kimitsu Chemical Industries Co., Ltd. (Japan). In addition to these, Pronova Biomedical A/S (Norway) now commercially manufactures ultra-pure alginates that are highly compatible with mammalian biological systems following the increased popularity of alginate as an immobilization matrix. These qualities are low in pyrogens and facilitate sterilization of the alginate solution by filtration due to low content of aggregates.

10 Properties

The physical properties of the alginate molecule were revealed mainly in the 1960s and 1970s. The last couple of decades have exposed some new knowledge on alginate gel formation

10.1 Physical Properties

Compared with other gelling polysaccharides, the most striking features of alginate's physical properties are the selective binding of multivalent cations, being the basis for gel formation, and the fact that the sol/gel transition of alginates is not particularly influenced by temperature.

10.1.1 Solubility

There are three essential parameters determining and limiting the solubility of alginates in water. The pH of the solvent is important because it will determine the presence of electrostatic charges on the uronic acid residues. Total ionic strength of the solute also plays an important role (salting-out effects of non-gelling cations), and, obviously, the content of gelling ions in the solvent limits the solubility. In the latter case, the "hardness" of the water (i.e., the content of Ca^{2+} ions) is most likely to be the main problem.

Potentiometric titration (Haug, 1964) revealed that the dissociation constants for mannuronic and guluronic acid monomers were 3.38 and 3.65, respectively. The pK_a value of the alginate polymer differs only slightly from those of the monomeric residues. An abrupt decrease in pH below the pK_a value causes a precipitation of alginic acid molecules, whereas a slow and controlled release of protons may result in the formation of an "alginic acid gel". Precipitation of alginic acid has been studied extensively (Haug, 1964; Haug and Larsen, 1963; Myklestad and Haug, 1966; Haug et al., 1967c), and addition of acid to an alginate solution leads to a precipitation within a relatively narrow pH range. This range depends not only on the molecular weight of the alginate but also on the chemical composition and sequence. Alginates containing more of the "alternating" structure (MG-blocks) will precipitate at lower pH values compared with the alginates containing more homogeneous block structures (poly-M and poly-G). The presence of homopolymeric blocks seems to favor precipitation by the formation of crystalline regions stabilized by hydrogen bonds. By increasing the degree of alternating "disorder" in the alginate chain, as in alginates isolated from *Ascophyllum nodosum* (see Table 1), the formation of these crystalline regions is not formed as easily. A certain alginate fraction from *A. nodosum* is soluble at a pH as low as 1.4 (Myklestad and Haug, 1966). Because of this relatively limited solubility of alginates at low pH, the esterified propylene glycol alginate (PGA) is applied as a food stabilizer under acidic conditions (see Section 10.3).

Any change of ionic strength in an alginate solution generally will have a profound effect, especially on polymer chain extension and solution viscosity. At high ionic strengths, the solubility also will be affected. Alginate may be precipitated and fractionated to give a precipitate enriched with mannuronate residues by high concentrations of inorganic salts like potassium chloride (Haug and Smidsrød, 1967; Haug, 1959a). Salting-out effects like this exhibit large hysteresis in the sense that less than 0.1 M salt is necessary to slow down the kinetics of the dissolution process and limit the solubility (Haug, 1959b). The gradient in the chemical potential of water between the bulk solvent and the solvent in the alginate particle, resulting from a very high counterion concentration in the particle, is most probably the drive of the dissolution process of alginate in water. This drive becomes severely reduced when attempts are made to dissolve alginate in an aqueous solvent already containing ions. If alginates are to be applied at high salt concentrations, the polymer should first be fully hydrated in pure water followed by addition of salt under shear.

For the swelling behavior of dry alginate powder in aqueous media with different concentrations of Ca^{2+}, there seems to be a limit at approximately 3 mM free calcium ions (unpublished results). Alginate can be solubilized at $[Ca^{2+}]$ above 3 mM by the addition of complexing agents, such as polyphosphates or citrate, before addition of the alginate powder.

10.1.2 Selective Ion Binding

The basis for the gelling properties of alginates is their specific ion-binding characteristics (Haug, 1964; Smidsrød and Haug, 1968b; Haug and Smidsrød, 1970; Smidsrød, 1973, 1974). Experiments involving equilibrium dialysis of alginate have shown that the selective binding of certain alkaline earth metals ions (e.g., strong and cooperative binding of Ca^{2+} relative to Mg^{2+}) increased markedly with increasing content of α-L-guluronate residues in the chains. Poly-mannuronate blocks and alternating blocks were almost without selectivity. This is illustrated in Figures 5 and 6, where a marked hysteresis in the binding of Ca^{2+} ions to G-blocks also is seen.

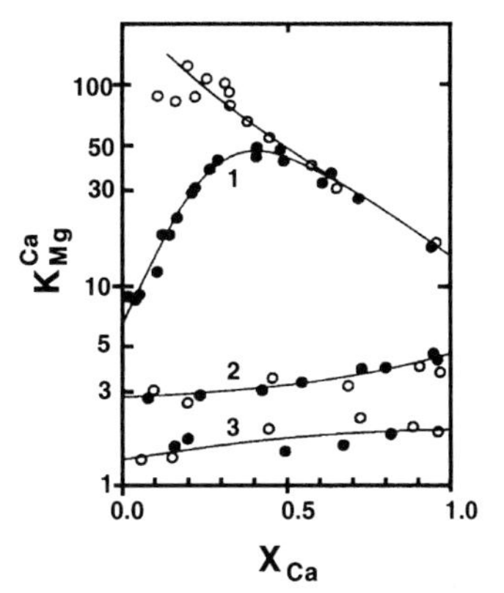

Fig. 6 Selectivity coefficients, K_{mg}^{Ca} as a function of ionic composition (X_{Ca}) for different alginate fragments. Curve 1: Fragments with 90% guluronate residues. Curve 2: Alternating fragment with 38% guluronate residues. Curve 3: Fragment with 90% mannuronate residues. •: Dialysis of the fragments in their Na^+ form. ○: Dialysis first against 0.2 M $CaCl_2$, then against mixtures of $CaCl_2$ and $MgCl_2$.

The high selectivity between similar ions such as those from the alkaline earth metals indicates that some chelation caused by structural features in the G-blocks takes place. Attempts were made to explain this phenomenon by the so-called "egg-box" model (Grant et al., 1973), based upon the linkage conformations of the guluronate residues (see Figure 1b). NMR studies (Kvam et al., 1986) of lanthanide complexes of related compounds suggested a possible binding site for Ca^{2+} ions in a single alginate chain, as given in Figure 7 (Kvam 1987).

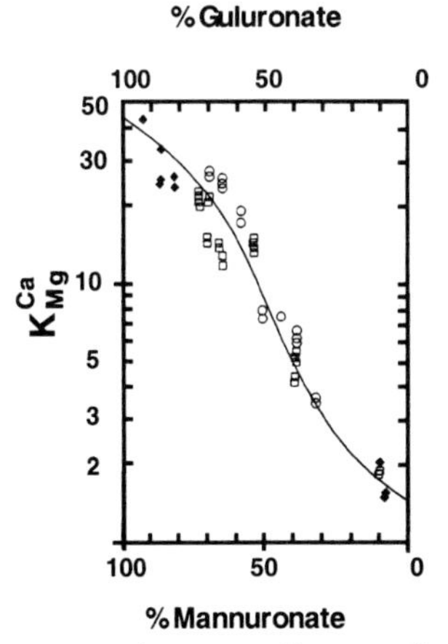

Fig. 5 Selectivity coefficients, K_{mg}^{Ca}, for alginates and alginate fragments as a function of monomer composition. The experimental points are obtained at $X_{Ca}=X_{Mg}=0.5$. The curve is calculated using $K_{mg\ guluronate}^{Ca}=40$ and $K_{mg\ mannuronate}^{Ca}=1.8$

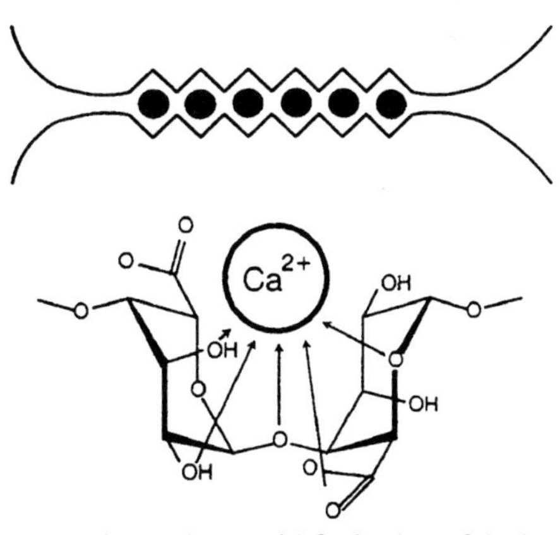

Fig. 7 The egg-box model for binding of divalent cations to homopolymeric blocks of α-L-guluronate residues, and a probably binding site in a GG-sequence.

Although more accurate steric arrangements have been suggested, as supported by x-ray diffraction (Mackie et al., 1983) and NMR spectroscopy (Steginsky et al., 1992), the simple "egg-box" model still persists, as it is principally correct and gives an intuitive understanding of the characteristic chelate-type of ion-binding properties of alginates. The simple dimerization in the "egg-box" model is at present questionable, as data from small-angle x-ray scattering on alginate gels suggest lateral association far beyond a pure dimerization with increasing [Ca^{2+}] and G-content of the alginate (Stokke et al., 2000). In addition, the fact that isolated and purified G-blocks (totally lacking elastic segments; typically DP = 20) are able to act as gelling modulators when mixed with a gelling alginate suggests higher-order junction zones (Draget et al., 1997).

The selectivity of alginates for multivalent cations is also dependent on the ionic composition of the alginate gel, as the affinity toward a specific ion increases with increasing content of the ion in the gel (Skjåk-Bræk et al., 1989b) (see Figure 6). Thus, a Ca-alginate gel has a markedly higher affinity toward Ca^{2+} ions than has the Na-alginate solution. This has been explained theoretically (Smidsrød, 1970; Skjåk-Bræk et al., 1989b) by a near-neighbor, auto-cooperative process (Ising model) and can be explained physically by the entropically unfavorable binding of the first divalent ion between two G-blocks and the more favorable binding of the next ions in the one-dimensional "egg-box" (zipper mechanism).

10.1.3
Gel Formation and Ionic Cross-linking

A very rapid and irreversible binding reaction of multivalent cations is typical for alginates; a direct mixing of these two components therefore rarely produces homogeneous gels. The result of such mixing is likely to be a dispersion of gel lumps ("fish-eyes"). The only possible exception is the mixing of a low-molecular-weight alginate with low amounts of cross-linking ion at high shear. The ability to control the introduction of the cross-linking ions hence becomes essential.

A controlled introduction of cross-linking ions is made possible by the two fundamental methods for preparing an alginate gel: the diffusion method and the internal setting method. The diffusion method is characterized by allowing a cross-linking ion (e.g., Ca^{2+}) to diffuse from a large outer reservoir into an alginate solution (Figure 8a). Diffusion setting is characterized by rapid gelling kinetics and is utilized for immobilization purposes where each droplet of alginate solution makes one single gel bead with entrapped (bio-) active agent (Smidsrød and Skjåk-Bræk, 1990). High-speed setting is also beneficial, e.g., in restructuring of foods when a given size and shape of the final product is desirable. The molecular-weight dependence in this system is negligible as long as the weight average molecular weight of the alginate is above 100 kDa (Smidsrød, 1974).

The internal setting method differs from the diffusion method in that the Ca^{2+} ions are released in a controlled fashion from an inert calcium source within the alginate solution (Figure 8b). Controlled release usually is obtained by a change in pH, by a limited solubility of the calcium salt source, and/or the by presence of chelating agents. The main difference between internal and diffusion setting is the gelling kinetics, which is not diffusion-controlled in the former case. With internal setting, the tailor-making of an alginate gelling system toward a given manufacturing process is possible because of the controlled, internal release of cross-linking ions (Draget et al., 1991). Internally set gels generally show a

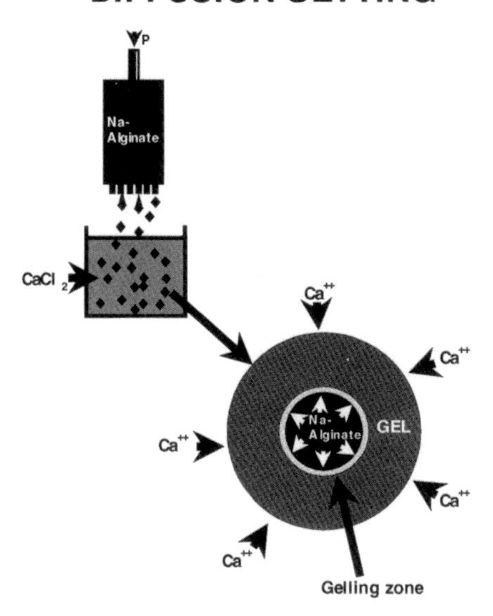

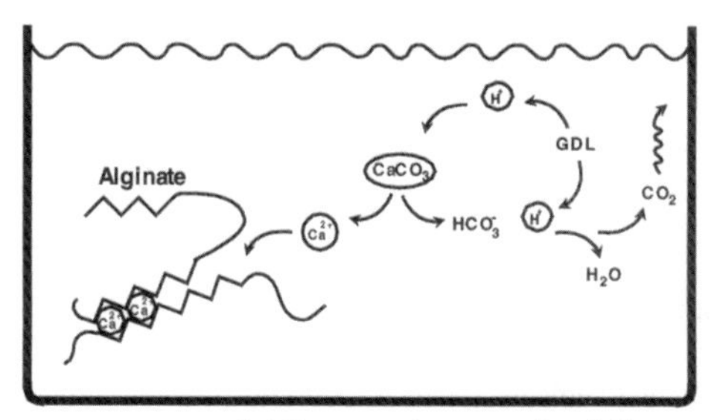

Fig. 8 Principal differences between the diffusion method exemplified by the immobilization technique and the internal setting method exemplified by the $CaCO_3$/GDL technique.

more pronounced molecular weight dependence compared with diffusion set gels. It has been reported that the internally set gels depend on molecular weight even at 300 kDa (Draget et al., 1993). This could be due to the fact that internally set gels are more calcium-limited compared with the gels made by diffusion, implying that the non-elastic fractions (sol and loose ends) at a given molecular weight will be higher in the internally set gels.

10.1.4
Gel Formation and Alginic Acid Gels

It is well known that alginates may form acid gels at pH values below the pK_a values of the uronic residues, but these alginic acid gels traditionally have not been as extensively studied as their ionically cross-linked counterparts. With the exception of some pharmaceutical uses, the number of applications so far is also rather limited. The preparation of an alginic acid gel has to be performed with care. Direct addition of acid to, e.g., a Na-alginate solution leads to an instantaneous precipitation rather than a gel. The pH must therefore be lowered in a controlled fashion, and this is most conveniently carried out by the addition of slowly hydrolyzing lactones like D-glucono-δ-lactone (GDL).

10.2
Material Properties

Since alginates are traditionally used for their gelling, viscosifying, and stabilizing properties, the features of alginate based materials are of utmost importance for a given application. Recently some quite unique biological effects of the alginate molecule itself have been revealed.

10.2.1
Stability

Alginate, being a single-stranded polymer, is susceptible to a variety of depolymerization processes. The glycosidic linkages are cleaved by both acid and alkaline degradation mechanisms and by oxidation with free radicals. As a function of pH, degradation of alginate is at its minimum nearly neutral and increases in both directions (Haug and Larsen, 1963) (Figure 9). The increased instability at pH values less than 5 is explained by a proton-catalyzed hydrolysis, whereas the reaction responsible for the

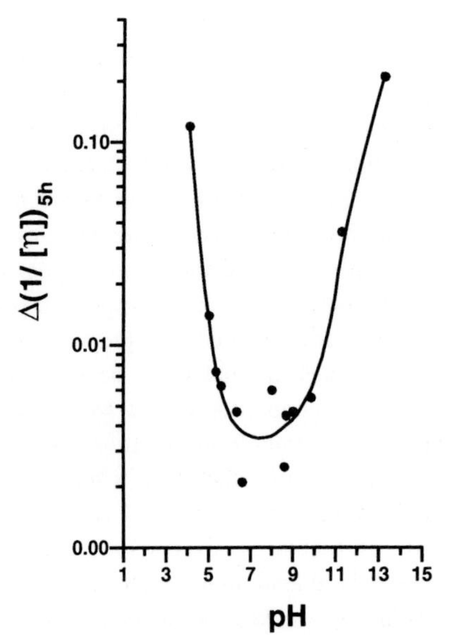

Fig. 9 Degradation of alginate isolated from *Laminaria digitata* measured as the change (Δ) in intrinsic viscosity ([η]) after 5 h at different pH and at 68°C.

degradation at pH 10 and above is the β-alkoxy elimination (Haug et al., 1963, 1967b). Free radicals degrade alginate mainly by oxidative-reductive depolymerization reactions (Smidsrød et al., 1967; Smidsrød et al., 1963a,b) caused by contamination of reducing agents like polyphenols from the brown algae. Since all of these depolymerization reactions increase with temperature, autoclaving generally is not recommended for the sterilization of an alginate solution. Since alginate is soluble in water at room temperature, sterile filtering rather than autoclaving has been recommended as a sterilization method for immobilization purposes in order to reduce polymer breakdown and to maintain the mechanical properties of the final gel (Draget et al., 1988).

Sterilization of dry alginate powder is also troublesome. The effect of γ-irradiation is often disastrous and leads to irreversible damage. It is generally believed that, under these conditions, O_2 is depleted rapidly with formation of the very reactive $OH^{\bullet}$ free radical. A short-term exposure in an electron accelerator could be an alternative to long-term exposure from a traditional ^{60}Co source. It has been shown that sterilization doses applied by ^{60}Co irradiation reduce the molecular weight to the extent that the gelling capacity is almost completely lost (Leo et al., 1990).

10.2.2
Ionically Cross-linked Gels

In contrast to most gelling polysaccharides, alginate gels are cold-setting, implying that alginate gels set more or less independent of temperature. The kinetics of the gelling process, however, can be strongly modified by a change in temperature, but a sol/gel transition will always occur if gelling is favored (e.g., by the presence of cross-linking ions). It is also important to realize that the properties of the final gel most likely will change if gelling occurs at different temperatures. This is due to alginates being non-equilibrium gels and thus being dependent upon the history of formation (Smidsrød, 1973).

Alginate gels can be heated without melting. This is the reason that alginates are used in baking creams. It should be kept in mind that alginates, as described earlier, are subjected to chemical degrading processes. A prolonged heat treatment at low or high pH might thus destabilize the gel because of an increased reaction rate of depolymerizing processes such as proton-catalyzed hydrolysis and the β-elimination reaction (Moe et al., 1995).

Because the selective binding of ions is a prerequisite for alginate gel formation, the alginate monomer composition and sequence also have a profound impact on the final properties of calcium alginate gels. Figure 10 shows gel strength as a function of

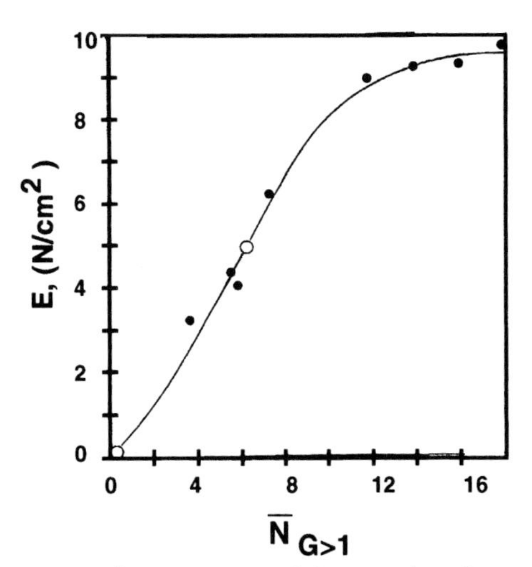

Fig. 10 Elastic properties of alginate gels as function of average G-block length.

the average length of G-blocks larger than one unit ($N_{G>1}$). This empirical correlation shows that there is a profound effect on gel strength when $N_{G>1}$ changes from 5 to 15. This coincides with the range of G-block lengths found in commercial alginates.

The polyelectrolyte nature of the alginate molecule is also important for its function, especially in mixed systems where, under favorable conditions, alginates may interact electrostatically with other charged polymers (e.g., proteins), resulting in a phase transition and altering the rheological behavior. Generally, it can be stated that if the purpose is to avoid such electrostatic interactions, the mixing of alginate and protein should take place at a relatively high pH, where most proteins have a net negative charge (Figure 11). These types of interactions also can be utilized to stabilize mixtures and to increase the gel strength of some restructured foods. In studies involving gelling of bovine serum albumin and alginate in both the sodium and the calcium form, a consid-

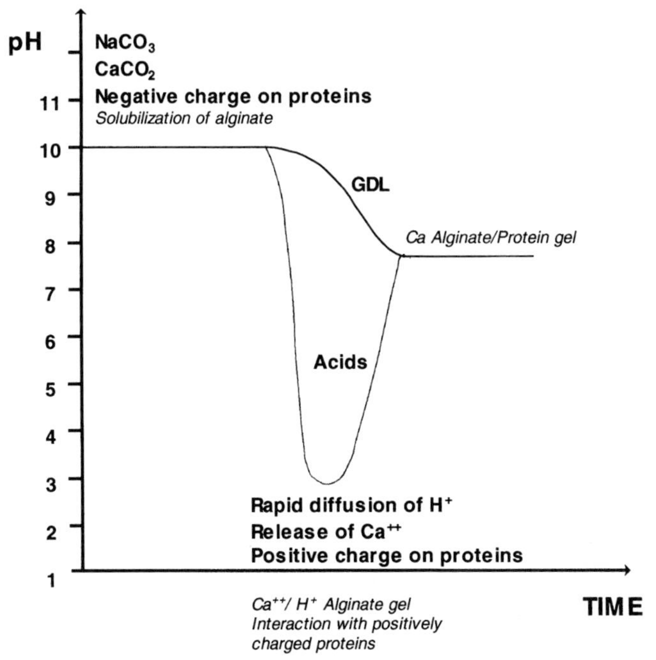

Fig. 11 Alginate/protein mixed gels exemplified by the internal gelation with $CaCO_3$. Release of Ca^{2+} is achieved by either a slow pH-lowering agent (GDL) or by a fast lowering with acids.

erable increase in Young's modulus was found within some range of pH and ionic strength (Neiser et al., 1998, 1999). These results suggest that electrostatic interactions are the main driving force for the observed strengthening effects.

An important feature of gels made by the diffusion-setting method is that the final gel often exhibits an inhomogeneous distribution of alginate, the highest concentration being at the surface and gradually decreasing towards the center of the gel. Extreme alginate distributions have been reported (Skjåk-Bræk et al., 1989a), with a five-fold increase at the surface (as calculated from the concentration in the original alginate solution) and virtually zero concentration in the center (Figure 12). This result has been explained by the fact that the diffusion of gelling ions will create a sharp gelling zone that moves from the surface toward the center of the gel. The activity of alginate (and of the gelling ion) will equal zero in this zone, and alginate molecules will diffuse from the internal, non-gelled part of the gelling body toward the zero-activity region

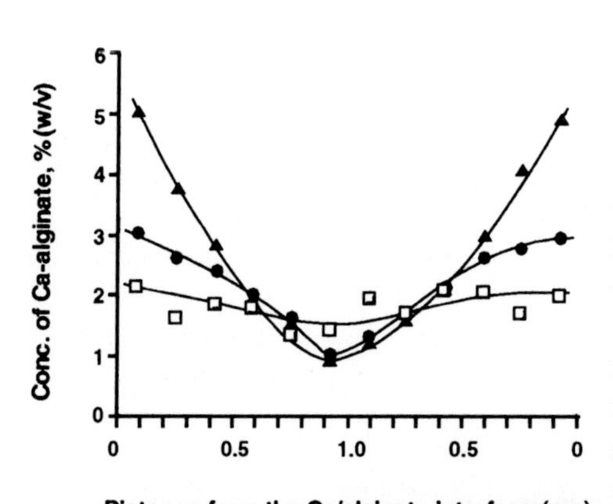

Fig. 12 Polymer concentration profiles of alginate gel cylinders formed by dialyzing a 2% (w/v) solution of Na-alginate from *Laminaria hyperborea* against 0.05 M $CaCl_2$ in the presence of NaCl. ❑: 0.2 M; •: 0.05 M; ▲: no NaCl.

(Skjåk-Bræk et al., 1989a, Mikkelsen and Elgsæter, 1995). Inhomogeneous alginate distribution may or may not be beneficial in the final product. It is therefore important to know that the degree of homogeneity can be controlled and to know which parameters govern the final alginate distribution. Maximum inhomogeneity is reached by placing a high-G, low-molecular-weight alginate gel in a solution containing a low concentration of the gelling ion and an absence of non-gelling ions. Maximum homogeneity is reached by gelling a high-molecular-weight alginate with high concentrations of both gelling and non-gelling ions (Skjåk-Bræk et al., 1989b).

The presence of non-gelling ions in alginate-gelling systems also affects the stability of the gels. It has been shown that alginate gels start to swell markedly when the ratio between non-gelling and gelling ions becomes too high and that the observed destabilization increases with decreasing F_G (Martinsen et al., 1989).

Swelling of alginate gels can be increased dramatically by a covalent cross-linking of preformed Ca-alginate gels with epichlorohydrin, followed by subsequent removal of Ca^{2+} ions by etylenediamine tetraacetic acid (EDTA) (Skjåk-Bræk and Moe, 1992). These Na-alginate gels can be dried, and they exhibit unique swelling properties when re-hydrated. The forces affecting the swelling of a polymer network can be split into three terms. Two of these terms favor swelling and can be said to constitute what might be called "swelling pressure": (1) the mixing term (Π_{mix} = the osmotic pressure generated by polymer/solvent mixing) and (2) the ionic term (Π_{ion} = the osmotic effect of an unequal distribution of the polymer counter-ions between the inside and the outside of the gel; the Donnan equilibrium). Of these two terms, the ionic part has been shown to contribute approximately 90% of the swelling pressure, even at 1 molar ionic strength, for highly ionic gels like Na-alginate (Moe et al., 1993). The third term (Π_{el} = the reduction in osmotic pressure due to the elastic response of the polymer network) balances the "swelling pressure" so that the total of these three terms equals zero at equilibrium.

These Na-alginate gels would function well as water absorbents in hygiene and pharmaceutical applications. However, Π_{ion} depends upon the ionic strength of the solute; with increasing ionic strength, the difference in chemical potential is reduced because of a more even distribution of the mobile ions between the inside and the outside of the gel. Therefore, reduced swelling will be observed at physiological ionic conditions compared with deionized water, but this reduction will be less pronounced than that for other water-absorbing materials, such as cross-linked acrylates, as a result of the inherent stiffness of the alginate molecule itself (Skjåk-Bræk and Moe, 1992) (Figure 13).

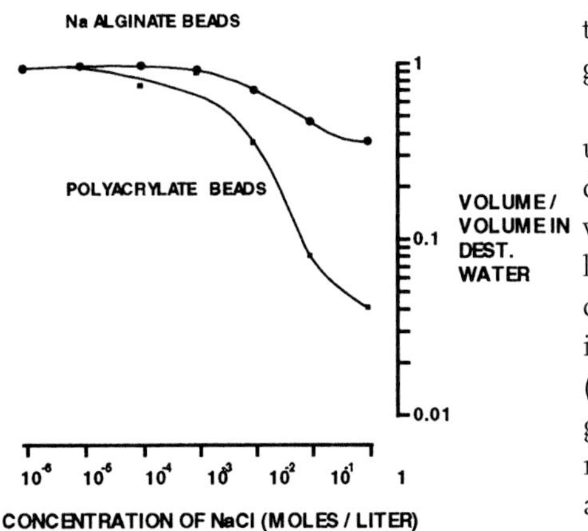

Fig. 13 Salt tolerance of covalently cross-linked Na-alginate and polyacrylate gel beads measured as swelling at different ionic strengths.

10.2.3
Alginic Acid Gels

It has been shown (Draget et al., 1994) that the gel strength of acid gels becomes independent of pH below a pH of 2.5, which equals 0.8 M GDL in a 1.0% (w/v) solution. Table 4 shows the Young's moduli of acid gels prepared (1) by a direct addition of GDL and (2) by converting an ionic cross-linked gel to the acid form by mineral acid. The modulus seems to be rather independent of the history of formation. Therefore, a most important feature of the acid gels compared with the ionic cross-linked gels seems to be that the former reaches equilibrium in the gel state.

Figure 14 shows the observed elastic moduli of acid gels made from alginates with different chemical composition, together with expected values for ionically cross-linked gels. From these data, it can be concluded that acid gels resemble ionic gels in the sense that high contents of guluronate (long stretches of G-blocks) give the strongest gels. However, it is also seen that poly-mannuronate sequences support alginic acid gel formation, whereas poly-alternating sequences seem to perturb this transition. The obvious demand for homopolymeric sequences in acid gel formation suggests

Tab. 4 E_{app} (kPa) for gels made from three different high-G alginates at 2% (w/v) concentration

Ca-alginate gel	*Ca-gel to alginic acid gel*	*Syneresis correction*	*Direct addition of GDL*
105 ± 4.6	52 ± 4.3	15.5 ± 0.3	15 ± 1.1
116 ± 11	64 ± 8.1	17.1 ± 1.8	17.8 ± 1.4
127 ± 6.4	79 ± 5.8	19.8 ± 1.3	20.4 ± 0.7

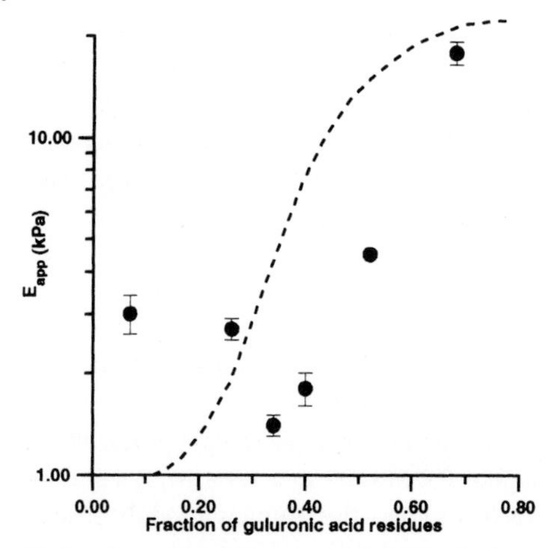

Fig. 14 Young's modulus E_{app} of alginic acid gels at apparent equilibrium as function of guluronic acid content. Dashed line refers to expected results for Ca^{2+} cross-linked alginate gels.

that cooperative processes are involved just as in the case of ionic gels (Draget et al., 1994). High molecular weight dependence has been observed, and this dependence becomes more pronounced with increasing content of guluronic acid residues.

A study of the swelling and partial solubilization of alginic acid gels at pH 4 has confirmed the equilibrium properties of the gels (Draget et al., 1996). By comparing the chemical composition and molecular weight of the alginate material leaching out from the acid gels with the same data for the whole alginate, an enrichment in mannuronic acid residues and a reduction in the average length of G-blocks were found together with a lowering of the molecular weight.

10.3 "Biological" Properties

Through a series of papers, it has been established that the alginate molecule itself has different effects on biological systems. This is more or less to be expected because of the large variety of possible chemical compositions and molecular weights of alginate preparations. A biological effect of alginate initially was hinted at in the first animal transplantation trials of encapsulated Langerhans islets for diabetes control. Overgrowth of alginate capsules by phagocytes and fibroblasts, resembling a foreign body/inflammatory reaction, was reported (Soon-Shiong et al., 1991). In bioassays, induction of tumor necrosis factor and interleukin 1 showed that the inducibility depended upon the content of mannuronate in the alginate sample (Soon-Shiong et al., 1993). This result directly explains the observed capsule overgrowth; mannuronate-rich fragments, which do not take part in the gel network, will leach out of the capsules and directly trigger an immune response (Stokke et al., 1993). This observed immunologic response can be linked in part to $(1 \rightarrow 4)$ glycosidic linkages, as other homopolymeric di-equatorial polyuronates, like D-glucuronic acid (C6-oxidised cellulose), also exhibit this feature (Espevik and Skjåk-Bræk, 1996). The immunologic potential of polymannuronates have now been observed in *in vivo* animal models in such diverse areas as for protection against lethal bacterial infections and irradiation and for increasing non-specific immunity (Espevik and Skjåk-Bræk, 1996).

11 Applications

Given the large number of different applications, alginates must be regarded as one of the most versatile polysaccharides. These applications span from traditional technical utilization, to foods, to biomedicine.

11.1
Technical Utilization

The quantitatively most important technical application of alginates is as a shear-thinning viscosifyer in textile printing, in which alginate has gained a high popularity because of the resulting color yield, brightness, and print levelness. Alginates also are used for paper coating to obtain surface uniformity and as binding agents in the production of welding rods. In the latter case, alginate gives stability in the wet stage and functions as a plasticizer during the extrusion process. As a last example of technical applications, ammonium alginate frequently is used for can sealing. The ammonium form is used because of its very low ash content (Onsøyen, 1996).

11.2
Medicine and Pharmacy

Alginates have been used for decades as helping agents in various human-health applications. Some examples include use in traditional wound dressings, in dental impression material, and in some formulations preventing gastric reflux. Alginate's increasing popularity as an immobilization matrix in various biotechnological processes, however, demonstrates that alginate will move into other and more advanced technical domains in addition to its traditional applications. Entrapment of cells within Ca-alginate spheres has become the most widely used technique for the immobilization of living cells (Smidsrød and Skjåk-Bræk, 1990). This immobilization procedure can be carried out in a single-step process under very mild conditions and is therefore compatible with most cells. The cell suspension is mixed with a sodium alginate solution, and the mixture is dripped into a solution containing multivalent cations (usually Ca^{2+}). The droplets then instantaneously form gel-spheres entrapping the cells in a three-dimensional lattice of ionically cross-linked alginate. The possible uses for such systems in industry, medicine, and agriculture are numerous, ranging from production of ethanol by yeast, to production of monoclonal antibodies by hybridoma cells, to mass production of artificial seed by entrapment of plant embryos (Smidsrød and Skjåk-Bræk, 1990).

Perhaps the most exciting prospect for alginate gel immobilized cells is their potential use in cell transplantation. Here, the main purpose of the gel is to act as a barrier between the transplant and the immune system of the host. Different cells have been suggested for gel immobilization, including parathyroid cells for treatment of hypocalcemia and dopamine-producing adrenal chromaffin cells for treatment of Parkinson's disease (Aebisher et al., 1993). However, major interest has been focused on insulin-producing cells for the treatment of Type I diabetes. Alginate/poly-L-lysine capsules containing pancreatic Langerhans islets have been shown to reverse diabetes in large animals and currently are being clinically tested in humans (Soon-Shiong et al., 1993, 1994). Table 5 lists some biomedical applications of alginate-encapsulated cells.

Tab. 5 Some potential biomedical application of alginate-encapsulated cells

Cell type	*Treatment of*
Adrenal chromaffin cells	Parkinson's disease[a]
Hepatocytes	Liver failure[a]
Paratyroid cells	Hypocalcemia[a]
Langerhans islets (β-cells)	Diabetes[b]
Genetically altered cells	Cancer[c]

[a] Aebischer et al., 1993 [b] Soon-Shiong et al., 1993, 1994 [c] Read et al., 2000

11.3 Foods

Alginates are used as food additives to improve, modify, and stabilize the texture of foods. This is valid for such properties as viscosity enhancement; gel-forming ability; and stabilization of aqueous mixtures, dispersions, and emulsions. Some of these properties stem from the inherent physical properties of alginates themselves, as outlined above, but they also may result from interactions with other components of the food product, e.g., proteins, fat, or fibers. As an example, alginates interact readily with positively charged amino acid residues of denatured proteins, which are utilized in pet foods and reformed meat. Cottrell and Kovacs (1980), Sime (1990), and Littlecott (1982) have given numerous descriptions and formulations on alginates in food applications. A general review on this topic is given by McHugh (1987).

Special focus perhaps should be placed on restructured food based on Ca-alginate gels because of its simplicity (gelling being independent upon temperature) and because it is a steadily growing alginate application. Restructuring of foods is based on binding together a flaked, sectioned, chunked, or milled foodstuff to make it resemble the original. Many alginate-based restructured products are already on the market (see Figure 15), as is exemplified by meat products (both for human consumption and as pet food), onion rings, pimento olive fillings, crabsticks, and cocktail berries.

Fig. 15 An example of alginate used in the restructuring of foods: the pimiento fillings of olives. (picture kindly provided by FMC BioPolymer).

For applications in jams, jellies, fruit fillings, etc., the synergetic gelling between alginates high in guluronate and highly esterified pectins may be utilized (Toft et al., 1986). The alginate/pectin system can give thermoreversible gels in contrast to the purely ionically cross-linked alginate gels. This gel structure is almost independent of sugar content, in contrast to pectin gels, and therefore may be used in low calorie products.

The only alginate derivative used in food is propylene glycol alginate (PGA). Steiner (1947) first prepared PGA, and Steiner and McNeely improved the process (1950). PGA is produced by a partial esterification of the carboxylic groups on the uronic acid residues by reaction with propylene oxide. The main product gives stable solutions under acidic conditions where the unmodified alginate would precipitate. It is now used to stabilize acid emulsions (such as in French dressings), acid fruit drinks, and juices. PGA also is used to stabilize beer foam.

As for the regulatory status, the safety of alginic acid and its ammonium, calcium, potassium, and sodium salts was last evaluated by the Joint FAO/WHO Expert Committee on Food Additives (JECFA) at its 39th meeting in 1992. An ADI "not specified" was allocated. JECFA allocated an ADI of 0 to 25 mg/kg bw to propylene glycol alginate at its 17th meeting.

In the U.S, ammonium, calcium, potassium, and sodium alginate are included in a list of stabilizers that are generally recognized as safe (GRAS). Propylene glycol alginate is approved as a food additive (used as an emulsifier, stabilizer, or thickener) and in several industrial applications (used as a coating for fresh citrus fruit, as an inert pesticide adjuvant, and as a component of paper and paperboard in contact with aqueous and fatty foods). In Europe, alginic acid and its salts and propylene glycol are all listed as EC-approved additives other than colors and sweeteners.

Alginates are inscribed in Annex I of the Directive 95/2 of 1995, and as such can be used in all foodstuffs (except those cited in Annex II and those described in Part II of the Directive) under the Quantum Satis principle of the EU.

12 Relevant Patents

A search in one of the international databases for patents and patent applications yielded well above 2000 hits on alginate. It is outside the scope of this chapter, not to mention beyond the capabilities of its authors, to systematically discuss all this literature, which covers inventions for improved utilization of alginates in the technical, pharmaceutical, food, and agricultural areas. We have therefore limited this section to only a handful of *prior art* inventions (Table 6), with the present authors as co-inventors, that point toward a production of alginate and alginate fractions with novel structures and some new biomedical applications based on the physical and biological properties of certain types of alginates with specified chemical structures.

US Patent 5,459,054 may represent a large number of patents dealing with the immobilization of living cells in immuno-protective alginate capsules for implantation purposes as discussed in Section 10.2. US Patents 5,169,840 and 6,087,342 cover the use of alginates enriched with mannuronate for the stimulation of cytokine production in monocytes, which could be of future importance in the treatment of microbial infections, cancer, and immune deficiency and autoimmune diseases. This stimulating effect has been discussed and connected to the use of alginate fibers in wound-healing dressings. A closer look at these specific effects is presented in Section 9.3.

When the calcium ions in alginate gels are exchanged by covalent cross links, the resulting gel with monovalent cations as counter-ions has the ability to swell several hundred times its own weight in water or salt solutions at low ionic strength, as shown in US Patent 5,144,016. This super-absorbent system, further elaborated in Section 9.2.2, still has not found any commercial uses, mainly because of competition from similar materials based on starch and cellulose derivatives, but certain biomedical applications may be foreseen.

A patent on the genes encoding the different C5-epimerases (US Patent 5,939,289) points to the possibility of producing alginates with a large number of different predetermined compositions and sequences and opens up the possibility for the tailor-making of different alginates, as discussed in Sections 8.2 and 12.

An alternative way of manufacturing alginate fractions with extreme compositions is by using selective extraction techniques, as disclosed in Patent WO 98/51710. One possible use of such fractions as gelling modifiers is revealed in Patent WO 98/02488, where it is suggested that these purified low-molecular-weight guluronate blocks give a gel enforcement at high concentrations of calcium ions by connecting and shortening elastic segments that otherwise would be topologically restricted. In conclusion, it may be argued that this relative high rate of patent filing suggests

Tab. 6 Summary of prior art patents on alginates pointing toward specialty applications

Patent number	***Holder***	***Inventors***	***Patent title***	***Public date***
U.S. 5,459,054	Neocrine	G. Skjåk-Bræk O. Smidsrød T. Espevik M. Otterlei P. Soon-Shiong	Cells encapsulated in alginate containing a high content of α-L-guluropnic acid	2 July 1993
U.S. 5,169,840	Nobipol, Protan Biopolymer	M. Otterlei T. Espevik O. Smidsrød G. Skjåk-Bræk	Diequatorially bound β-1,4-polyuronates and use of same for cytokine stimulation	27 March 1991
U.S. 6,087,342	FMC BioPolymer	T. Espevik G. Skjåk-Bræk	Substrates having bound polysaccharides and bacterial nucleic acids	15 May 1998
U.S. 5,144,016	Protan Biopolymer	G. Skjåk-Bræk S. Moe	Alginate gels	29 May 1991
U.S. 5,939,289	Pronova Biopolymer, Nobipol	H. Ertesvåg S. Valla G. Skjåk-Bræk B. Larsen	DNA compounds comprising sequences encoding mannuroran C-5-epimerase	9 May 1995
WO 98/51710	FMC BioPolymer	M. K. Simensen O. Smidsrød K. I. Draget F. Hjelland	Procedure for producing uronic acid blocks from alginate	11 November 1998
WO 98/02488	FMC BioPolymer	M. K. Simensen K. I. Draget E. Onsøyen O. Smidsrød T. Fjæreide	Use of G-block polysaccharides	22 January 1998

that new alginate-based products are being developed and that there is continuous stable demand for alginates and their products.

13 Outlook and Perspectives

From a chemical point of view, the alginate molecule may look very simple, as it contains only the two monomer units M and G linked by the same 1,4 linkages. This simplification of its chemical structure may lead potential commercial users of alginate to treat it as a commodity like many of the cellulose derivatives. In this chapter, we have shown that alginate represents a very high diversity with respect to chemical composition and monomer sequence, giving the alginate family of molecules a large variety of physical and biological properties. This may represent a challenge to the unskilled users of alginate, but it may be an advantage for the producers and new users of alginate who are interested in developing research-based, high-value applications. When microbial alginate and epimerase-modified alginate enter into the marked in the future, the possibility of alginate being tailor-made to diverse applications will be increased even further.

We therefore see a future trend, which has already started, in which the exploitation of alginate gradually shifts from low-tech applications with increasing competition from cheap alternatives to more advanced, knowledge-based applications in the food, pharmaceutical, and biomedical areas. We then foresee continuous, high research activity in industry and academia to describe, understand, and utilize alginate-containing products to the benefit of society.

Acknowledgements
The authors would like to thank Anne Bremnes and Hanne Devle for most skillful assistance in preparing graphic illustrations and Nadra J. Nilsen for collecting the data on alginate lyases.

14 References

Aebischer, P., Goddard, M., Tresco, P. A. (1993) Cell encapsulation for the nervous system, in: *Fundamentals of Animal Cell Encapsulation and Immobilization* (Goosen, M. F. A., Ed.), Boca Raton, FL: CRC Press, 197–224.

Andresen, I.-L., Skipnes, O., Smidsrød, O., Østgaard, K., Hemmer, P. C. (1977) Some biological functions of matrix components in benthic algae in relation to their chemistry and the composition of seawater, *ACS Symp. Ser.* **48**, 361–381.

Atkins, E. D. T., Mackie, W., Smolko, E. E. (1970) Crystalline structures of alginic acids, *Nature* **225**, 626–628.

Atsuki, K., Tomoda, Y. (1926) Studies on seaweeds of Japan I. The chemical constituents of Laminaria, *J. Soc. Chem. Ind. Japan* **29**, 509–517.

Bird, G. M., Haas, P. (1931) XLVII. On the constituent nature of the cell wall constituents of *Laminaria* spp. Mannuronic acid, *Biochem. J.* **25**, 26–30.

Boyd. J., Turvey, J. R. (1977) Isolation of a poly-"-L-guluronate lyase from *Klebsiella aerogenes*, *Carbohydr. Res.* **57**, 163–171.

Boyd, J., Turvey, J. R. (1978) Structural studies of alginic acid, using a bacterial poly-L-guluronate lyase, *Carbohydr. Res.* **66**, 187–194.

Boyen, C., Bertheau, Y., Barbeyron, T., Kloareg, B. (1990a) Preparation of guluronate lyase from *Pseudomonas alginovora* for protoplast isolation in Laminaria, *Enzyme Microb. Technol.* **12**, 885–890.

Boyen, C., Kloareg, B., Polne-Fuller, M., Gibor, A. (1990b) Preparation of alginate lyases from marine molluscs for protoplast isolation in brown algae, *Phycologia* **29**, 173–181.

Blumenkrantz, N., Asboe-Hansen, G. (1973) New method for quantitative determination of uronic acids, *Anal. Biochem.* **54**, 484–489.

Cottrell, I.W., Kovacs, P. (1980). Alginates, in: *Handbook of Water-Soluble Gums and Resins* (Crawford, H.B., Williams, J., Eds.), Auckland, New Zealand: McGraw-Hill, 21–43.

Davidson I.W., Lawson, C.J., Sutherland, I.W. (1977) An alginate lyase from *Azotobacter vinelandii* phage, *J. Gen. Microbiol.* **98**, 223–229.

Draget, K. I., Myhre, S., Skjåk-Bræk, G., Østgaard, K. (1988) Regeneration, cultivation and differentiation of plant protoplasts immobilized in Ca-alginate beads, *J. Plant Physiol.* **132**, 552–556.

Draget, K.I., Østgaard, K., Smidsrød, O. (1991) Homogeneous alginate gels; a technical approach, *Carbohydr. Polym.* **14**, 159–178.

Draget, K.I., Simensen, M.K., Onsøyen, E., Smidsrød, O. (1993) Gel strength of Ca-limited gels made *in situ*, *Hydrobiologia*, **260/261**, 563–565.

Draget, K.I., Skjåk-Bræk, G., Smidsrød, O. (1994) Alginic acid gels: the effect of alginate chemical composition and molecular weight, *Carbohydr. Polym.* **25**, 31–38.

Draget, K. I., Skjåk-Bræk, G., Christensen, B. E., Gåserød, O., Smidsrød, O. (1996) Swelling and partial solubilization of alginic acid gel beads in acidic buffer, *Carbohydr. Polym.* **29**, 209–215.

Draget, K.I., Onsøyen, E., Fjæreide, T., Simensen M. K., Smidsrød O. (1997) Use of G-block polysaccharides', *Intl. Pat. Appl.#* PCT/NO97/00176.

Dubois, M., Gilles, K.A., Hamilton, J.K., Rebers, P.A., Smith, F. (1956) Colorimetric method for determination of sugars and related substances, *Anal. Chem.* **28**, 350–356.

Elyakova, L. A., Favorov, V. V. (1974) Isolation and certain properties of alginate lyase VI from the mollusk *Littorina* sp., *Biochim. Biophys. Acta* **358**, 341–354.

Ertesvåg, H., Larsen, B., Skjåk-Bræk, G., Valla, S. (1994) Cloning and expression of an *Azotobacter vinelandii* mannuronan-C-5 epimerase gene, *J. Bacteriol.* **176**, 2846–2853.

Ertesvåg, H., Høidal, H.K., Hals, I.K., Rian, A., Doseth, B., Valla, S. (1995) A family of moddular type mannuronana C-5 epimerase genes controls the alginate structure in *Azotobacter vinelandii*, *Mol. Microbiol* **16**, 719–731.

Ertesvåg, H., Erlien, F., Skjåk-Bræk, G., Rehm, B.H.A., Valla, S. (1998a) Biochemical properties and substrate specificities of a recombinantly produced *Azotobacter vinelandii* alginate lyase, *J. Bacteriol.* **180**, 3779–3784.

ErtesvågH., Høydal, H., Skjåk-Bræk, G., Valla, S. (1998b) The *Azotobacter vinelandii* mannuronan C-5 epimerase AlgE1 consists of two separate catalytic domains, *J. Biol. Chem.* **273**, 30927–30938.

Espevik, T., Skjåk-Bræk, G. (1996) Application of alginate gels in biotechnology and biomedicine, *Carbohydr. Eur.* **14**, 19–25.

Filisetti-Cozzi, T. M. C. C., Carpita, N. C. (1991) Measurement of uronic acids without interference from neutral sugars, *Anal. Biochem.* **197**, 157–162.

Fisher, F. G., Dörfel, H. (1955) Die Polyuronsäuren der Braunalgen (Kohlenhydrate der Algen), *Z. Physiol. Che.*, **302**, 186–203.

Gacesa P., (1992) Enzymic degradation of alginates, *Int. J. Biochem.* **24**, 545–552.

Grant, G. T., Morris, E. R., Rees, D. A., Smith, P. J. C., Thom, D. (1973) Biological interactions between polysaccharides and divalent cations: The egg-box model, *FEBS Lett.* **32**, 195–198.

Grasdalen, H., Larsen, B., Smidsrød O. (1977) ^{13}C-NMR studies of alginate, *Carbohydr. Res.* **56**, C11–C15.

Grasdalen, H., Larsen, B., Smidsrød, O. (1979) A PMR study of the composition and sequence of uronate residues in alginate, *Carbohydr. Res.* **68**, 23–31.

Grasdalen, H. (1983) High-field ^{1}H-nmr spectroscopy of alginate: Sequential structure and linkage conformations, *Carbohydr. Res.* **118**, 255–260.

Haug, A. (1959a) Fractionation of alginic acid, *Acta Chem. Scand.* **13**, 601–603.

Haug, A. (1959b) Ion exchange properties of alginate fractions, *Acta Chem. Scand.* **13**, 1250–1251.

Haug, A., Larsen B. (1963) The solubility of alginate at low pH, *Acta Chem. Scand.* **17**, 1653–1662.

Haug, A., Larsen, B., Smidsrød, O. (1963) The degradation of alginates at different pH values, *Acta Chem. Scand.* **17**, 1466–1468.

Haug, A. (1964) Composition and properties of alginates, Thesis, Norwegian Institute of Technology, Trondheim.

Haug, A., Smidsrød O. (1965) Fractionation of alginates by precipitation with calcium and magnesium ions, *Acta Chem. Scand.* **19**, 1221–1226.

Haug, A., Larsen B. (1966) A study on the constitution of alginic acid by partial acid hydrolysis, *Proc. Int. Seaweed Symp.* **5**, 271–277.

Haug, A., Larsen B., Smidsrød O. (1966) A study of the constitution of alginic acid by partial hydrolysis, *Acta Chem. Scand.* **20**, 183–190.

Haug, A., Smidsrød O. (1967) Precipitation of acidic polysaccharides by salts in ethanol-water mixtures, *J. Polym. Sci.* **16**, 1587–1598.

Haug, A., Larsen B., Smidsrød O. (1967a) Studies on the sequence of uronic acid residues in alginic acid, *Acta Chem. Scand.* **21**, 691–704.

Haug, A., Larsen, B., Smidsrød, O. (1967b) Alkaline degradation of alginate, *Acta Chem. Scand.* **21**, 2859–2870.

Haug, A., Myklestad, S., Larsen, B., Smidsrød, O. (1967c) Correlation between chemical structure and physical properties of alginate, *Acta Chem. Scand.* **21**, 768–778.

Haug, A., Smidsrød, O. (1970) Selectivity of some anionic polymers for divalent metal ions, *Acta Chem. Scand.* **24**, 843–854.

Hirst, E. L., Jones, J. K. N., Jones, W. O., (1939) The structure of alginic acid. Part I, *J. Chem. Soc.* 1880–1885.

Høydal, H., Ertesvåg, H., Stokke, B. T., Skjåk-Bræk, G., Valla, S. (1999) Biochemical properties and mechanism of action of the recombinant *Azotobacter vinelandii* mannuronan C-5 epimerase, *J. Biol. Chem.* **274**, 12316–12322.

Indergaard, M., Skjåk-Bræk, G. (1987) Characteristics of alginate from *Laminaria digitata* cultivated in a high phosphate environment, *Hydrobiologia* **151/152**, 541–549.

Kvam, B. J., Grasdalen, H., Smidsrød, O., Anthonsen, T. (1986) NMR studies of the interaction of metal ions with poly-(1,4-hexuronates). VI. Lanthanide(III) complexes of sodium (methyl ∀-D-galactopyranosid)uronate and sodium (phenylmethyl ∀-D-galactopyranosid)uronate, *Acta Chem. Scand.* **B40**, 735–739.

Kvam, B. (1987) Conformational conditions and ionic interactions of charged polysaccharides. Application of NMR techniques and the Poisson-Boltzmann equation, Thesis, Norwegian Institute of Technology, Trondheim.

Larsen, B., Smidsrød O., Painter T. J., Haug A. (1970) Calculation of the nearest-neighbour frequencies in fragments of alginate from the yields of free monomers after partial hydrolysis, *Acta Chem. Scand.* **24**, 726–728.

Leo, W. J., McLoughlin, A. J., Malone, D. M. (1990) Effects of terilization treatments on some properties of alginate solutions and gels, *Biotechnol. Prog.* **6**, 51–53.

Littlecott, G. W. (1982). Food gels–the role of alginates, *Food Technol. Aust.* **34**, 412–418.

Mackie, W., Noy, R., Sellen, D. B. (1980) Solution properties of sodium alginate, *Biopolymers* **19**, 1839–1860.

Mackie, W., Perez, S., Rizzo, R., Taravel, F., Vignon, M. (1983) Aspects of the conformation of polyguluronate in the solid state and in solution, *Int. J. Biol. Macromol.* **5**, 329–341.

Martinsen, A., Skjåk-Bræk, G., Smidsrød, O. (1989) Alginate as immobilization material: I. Correlation between chemical and physical properties of alginate gel beads, *Biotechnol. Bioeng.* **33**, 79–89.

Martinsen, A., Skjåk-Bræk, G., Smidsrød, O., Zanetti, F., Paoletti, S. (1991) Comparison of different methods for determination of molecular weight and molecular weight distribution of alginates, *Carbohydr. Polym.* **15**, 171–193.

McHugh, D.J. (1987). Production, properties and uses of alginates, in: Production and Utilization of Products from Commercial Seaweeds, FAO Fisheries Technical Paper No. 288 (McHugh, D. J., Ed.), Rome: FAO, 58–115.

Miawa, T. (1930) Alginic acid, *J. Chem. Soc. Japan* **51**, 738–745.

Mikkelsen, A., Elgsæter, A. (1995) Density distribution of calcium-induced alginate gels, *Biopolymers* **36**, 17–41.

Moe, S. T., Skjåk-Bræk, G., Elgsæter, A., Smidsrød, O. (1993) Swelling of covalently cross-linked ionic polysaccharide gels: Influence of ionic solutes and nonpolar solvents, *Macromolecules* **26**, 3589–3597.

Moe, S., Draget, K., Skjåk-Bræk, G., Smidsrød, O. (1995) Alginates, in: *Food Polysaccharides and Their Applications* (Stephen, A. M., Ed.), New York: Marcel Dekker, 245–286.

Murata, K., Inose, T., Hisano, T., Abe, S., Yonemoto, Y. (1993) Bacterial alginate lyase: enzymology, genetics and application, *J. Ferment. Bioeng.* **76**, 427–437.

Myklestad. S., Haug A. (1966) Studies on the solubility of alginic acid *from Ascophyllum nodosum* at low pH, *Proc. Int. Seaweed Symp.* **5**, 297–303.

Neiser, S., Draget, K. I., Smidsrød O. (1998) Gel formation in heat-treated bovine serum albumin–sodium alginate systems, *Food Hydrocolloids* **12**, 127–132.

Neiser, S., Draget, K. I., Smidsrød O. (1999) Interactions in bovine serum albumin–calcium alginate gel systems, *Food Hydrocolloids* **13**, 445–458.

Nelson, W. L., Cretcher, L. H. (1929) The alginic acid from *Macrocystis pyrifera, J. Am. Chem. Soc.* **51**, 1914–1918.

Nelson, W. L., Cretcher, L. H. (1930) The isolation and identification of *d*-mannuronic acid lactone from the *Macrocystis pyrifera, J. Am. Chem. Soc.* **52**, 2130–2134.

Nelson, W. L., Cretcher, L. H. (1932) The properties of *d*-mannuronic acid lactone, *J. Am. Chem. Soc.* **54**, 3409–3406.

Nibu, Y., Satoh, T., Nishi, Y., Takeuchi, T., Murata, K., Kusakabe, I. (1995) Purification and characterization of extracellular alginate lyase from Enterobacter cloacae M-1, *Biosci. Biotechnol. Biochem.* **59**, 632–637.

Onsøyen, E. (1996) Commercial applications of alginates, *Carbohydr. Eur.* **14**, 26–31.

Otterlei, M., Østgaard, K., Skjåk-Bræk, G., Smidsrød, O., Soon-Shiong, P., Espevik, T. (1991). Induction of cytokine production from human monocytes stimulated with alginate, *J. Immunother.* **10**, 286–291.

Painter, T. J., Smidsrød O., Haug A. (1968) A computer study of the changes in composition-distribution occurring during random depolymerisation of a binary linear heteropolysaccharide, *Acta Chem. Scand.* **22**, 1637–1648.

Penman, A., Sanderson G. R. (1972) A method for the determination of uronic acid sequence in alginates, *Carbohydr. Res.* **25**, 273–282.

Read, T.-A., Sorensen, D.R., Mahesparan, R., Enger, P.Ø., Timpl, R., Olsen, B.R., Hjelstuen, M.H.B., Haraldseth, O., Bjerkevig, R. (2000) Local endostatin treatment of gliomas administered by microencapsulated producer cells, *Nature Biotechnol.* **19**, 29–34.

Rivera-Carro H. D. (1984) Block structure and uronic acid sequence in alginates, Thesis, Norwegian Institute of Technology, Trondheim.

Schmidt, E., Vocke, F. (1926) Zur Kenntnis der Polyglykuronsäuren, *Chem. Ber.* **59**, 1585–1588.

Shimokawa, T., Yoshida, S., Kusakabe, I., Takeuchi, T., Murata, K., Kobayashi, H. (1997) Some properties and action mode of "-L-guluronan lyase from *Enterobacter cloacae* M-1. *Carbohydr. Res.* **304**, 125–132.

Sime, W. (1990). Alginates, in: *Food Gels* (Harris, P., Ed.), London: Elsevier, 53–78.

Skjåk-Bræk, G., Larsen B., Grasdalen H. (1986) Monomer sequence and acetylation pattern in some bacterial alginates, *Carbohydr. Res.* **154**, 239–250.

Skjåk-Bræk, G., Grasdalen, H., Smidsrød, O. (1989a) Inhomogeneous polysaccharide ionic gels, *Carbohydr. Polym.* **10**, 31–54.

Skjåk-Bræk, G., Grasdalen, H., Draget, K. I., Smidsrød, O. (1989b). Inhomogeneous calcium alginate beads, in: Biomedical and Biotechnological Advances in Industrial Polysaccharides (Crescenzi, V., Dea, I. C. M., Paoletti, S., Stivala, S. S., Sutherland, I. W., Eds.), New York: Gordon and Breach, 345–363.

Skjåk-Bræk, G., Moe, S. T. (1992) Alginate gels, US Patent 5,144,016.

Smidsrød, O., Haug, A., Larsen, B. (1963a) The influence of reducing substances on the rate of degradation of alginates, *Acta Chem. Scand.* **17**, 1473–1474.

Smidsrød, O., Haug, A., Larsen, B. (1963b) Degradation of alginate in the presence of reducing compounds, *Acta Chem. Scand.* **17**, 2628–2637.

Smidsrød, O., Haug, A., Larsen, B. (1967) Oxidative-reductive depolymerization: a note on the comparison of degradation rates of different polymers by viscosity measurements, *Carbohydr. Res.* **5**, 482–485.

Smidsrød, O., Haug, A. (1968a) A light scattering study of alginate, *Acta Chem. Scand.* **22**, 797–810.

Smidsrød, O., Haug, A. (1968b) Dependence upon uronic acid composition of some ion-exchange properties of alginates, *Acta Chem. Scand.* **22**, 1989–1997.

Smidsrød, O., Whittington S. G. (1969) Monte Carlo investigation of chemical inhomogeneity in copolymers, *Macromolecules* **2**, 42–44.

Smidsrød, O. (1970) Solution properties of alginate, *Carbohydr. Res.* **13**, 359–372.

Smidsrød, O. (1973). Some physical properties of alginates in solution and in the gel state, Thesis, Norwegian Institute of Technology, Trondheim.

Smidsrød, O., Glover, R. M., Whittington, S. G. (1973) The relative extension of alginates having different chemical composition, *Carbohydr. Res.* **27**, 107–118.

Smidsrød, O. (1974) Molecular basis for some physical properties of alginates in the gel state, *J. Chem. Soc. Farad. Trans* **57**, 263–274.

Smidsrød, O., Skjåk-Bræk, G. (1990) Alginate as immobilization matrix for cells, *Trends Biotechnol.* **8**, 71–78.

Smidsrød, O., Draget, K. I. (1996) Alginates: chemistry and physical properties, *Carbohydr. Eur.* **14**, 6–13.

Soon-Shiong, P., Otterlei, M., Skjåk-Bræk, G., Smidsrød, O., Heintz, R., Lanza, R. P., Espevik, T. (1991) An immunologic basis for the fibrotic reaction to implanted microcapsules, *Transplant Proc.* **23**, 758–759.

Soon-Shiong, P., Feldman, E., Nelson, R., Heints, R., Yao, Q., Yao, T., Zheng, N., Merideth, G., Skjåk-Bræk, G., Espevik, T., Smidsrød, O., Sandford P. (1993) Long-term reversal of diabetes by the injection of immunoprotected islets, *Proc. Natl. Acad. Sci. USA* **90**, 5843–5847.

Soon-Shiong, P., Heintz, R. E., Merideth, N., Yao, Q. X., Yao, Z. W., Zheng, T. L., Murphy, M., Moloney, M. K., Schmehl, M., Harris, M., Mendez, R., Mendez, R., Sandford, P. A. (1994) Insulin independence in a type 1 diabetic patient after encapsulated islet transplantation, *Lancet* **343**, 950–951.

Stanford, E. C. C. (1881). British Patent 142.

Stanford, E. C. C. (1883a) New substance obtained from some of the commoner species of marine algæ; Algin, *Chem. News* **47**, 254–257.

Stanford, E. C. C. (1883b) New substance obtained from some of the commoner species of marine algæ; Algin, *Chem. News* **47**, 267–269.

Steginsky, C. A., Beale, J. M., Floss, H. G., Mayer, R. M. (1992) Structural determination of alginic acid and the effects of calcium binding as determined by high-field n.m.r., *Carbohydr. Res.* **225**, 11–26.

Steiner, A. B. (1947). Manufacture of glycol alginates, US Patent 2,426,215.

Steiner, A. B., McNeilly, W. H. (1950). High-stability glycol alginates and their manufacture, US Patent 2,494,911.

Stokke, B. T., Smidsrød, O., Bruheim, P., Skjåk-Bræk, G. (1991). Distribution of uronate residues in alginate chains in relation to alginate gelling properties, *Macromolecules* **24**, 4637–4645.

Stokke, B. T., Smidsrød, O., Zanetti, F., Strand, W., Skjåk-Bræk G. (1993) Distribution of uronate residues in alginate chains in relation to gelling properties 2:Enrichment of -D-mannuronic acid and depletion of "-L-guluronic acid in the sol fraction, *Carbohydr. Polym.* **21**, 39–46.

Stokke B. T., Draget K. I., Yuguchi Y., Urakawa H., Kajiwara K. (2000) Small angle X-ray scattering and rheological characterization of alginate gels. 1 Ca-alginate gels, *Macromolecules* **33**, 1853–1863.

Tanford, C. (1961). *Physical Chemistry of Macromolecules.* New York: John Wiley & Sons, Inc.

Toft, K., Grasdalen, H., Smidsrød, O. (1986). Synergistic gelation of alginates and pectins, *ACS Symp. Ser.* **310**, 117–132.

Valla, S., Ertesvåg, H., Skjåk-Bræk, G. (1996) Genetics and biosynthesis of alginates, *Carbohydr. Eur.* **14**, 14–18.
Wong, T. Y., Preston, L. A., Schiller, N. L. (2000) Alginate lyase: review of major sources and enzyme characteristics, structure-function analysis, biological roles, and applications, *Ann. Rev. Microbiol.* **54**, 289–340.